Separations Division 2017

Core Programming Area at the 2017 AIChE Annual Meeting

Minneapolis, Minnesota, USA
29 October - 3 November 2017

Volume 1 of 2

ISBN: 978-1-5108-5791-9

Printed from e-media with permission by:

Curran Associates, Inc.
57 Morehouse Lane
Red Hook, NY 12571

CURRAN ASSOCIATES INC. proceedings.com

Some format issues inherent in the e-media version may also appear in this print version.

Printed by Curran Associates, Inc. (2018)

AIChE
120 Wall Street, FL 23
New York, NY 10005-4020

Phone: (800) 242-4363
Fax: (203) 775-5177

www.aiche.org

Additional copies of this publication are available from:

Curran Associates, Inc.
57 Morehouse Lane
Red Hook, NY 12571 USA
Phone: 845-758-0400
Fax: 845-758-2633
Email: curran@proceedings.com
Web: www.proceedings.com

TABLE OF CONTENTS

VOLUME 1

VOLUME 2

Some abstracts in this book refer to images that were not made available by the publisher

498654 Novel Membranes, Membrane Processes and Membrane Devices

Abstract:

Membranes have created an alternate and expanding universe in the field of separations. I will talk first about membranes and membrane processes leading to membrane extractors, membrane contactors and membrane reactors. I will then talk about our research into various novel membranes, membrane processes and membrane devices. The novel membranes will include organic solvent nanofiltration membranes, interfacially polymerized membranes, and liquid membranes. Systems studied include biofuel separation and recovery from clarified fermentation broths, CO_2 separation, and various separations encountered in organic solvent-based systems. Next I will focus on membrane distillation for desalination, specifically our studies on direct contact membrane distillation dealing with a novel membrane structure with hydrophobic hollow fiber membranes, membrane-cross flow configuration in concentrating sea water up to 19-20% salt level with scaling salt precipitates flowing all around and recently developed larger-scale devices. I will describe in addition a novel but simple configuration for highly productive air gap membrane distillation. I will then touch on hollow fiber membrane crystallizers based on anti-solvent crystallization and cooling crystallization and their subsequent adoption to develop continuous processes for producing polymer-coated drug crystals/nanoparticles. In membrane separations we always look for highly-selective membranes to avoid multi-staging. Yet membranes in practice often have limited selectivity. The concept of internal staging in gas separation and ultrafiltration in a membrane device containing multiple membranes led to much higher selectivity than achievable otherwise with a single membrane. The conception of such structures for hollow fiber membranes was stimulated by the notion of hollow fiber contained liquid membranes developed earlier by us. I hope to provide you with a brief introduction to these developments from our laboratory in membranes and membrane separations.

486727 Ion Sorption, Diffusion and Transport in Charged Polymer Membranes

Abstract:

Charged polymer membranes are widely used for water purification applications involving control of water and ion transport, such as reverse osmosis and electrodialysis. Additional applications, such as energy recovery ventilation and capacitive deionization, rely on polymer membranes to control transport rates of water, ions, or both. Improving membranes for such processes would benefit from more complete fundamental understanding of the relation between membrane structure and ion sorption, diffusion and transport properties in both cation and anion exchange membrane materials. Ion-exchange membranes often contain strongly acidic or basic functional groups that render the materials hydrophilic, but the presence of such charged groups also has a substantial impact on ion (and water) transport properties through the polymer.

We are exploring the influence of polymer backbone structure, charge density, and water content on ion transport properties. Results from some of these studies will be presented, focusing on transport of salt, primarily NaCl, through various neutral, positively charged and negatively charged membranes via concentration gradient driven transport (i.e., ion permeability) and electric field driven transport (i.e., ionic conductivity).Â One long-term goal is to develop and validate a common framework to interpret data from both electrically driven and concentration gradient driven mass transport in such polymers and to use it to establish structure/property relations leading to rational design of membranes with improved performance.

Ion sorption and permeability data were used to extract salt diffusion coefficients in charged membranes. Concentrations of both counter-ions and co-ions in the polymers were measured via desorption followed by ion chromatography or flame atomic absorption spectroscopy.Â Salt permeability, sorption and electrical conductivity data were combined to determine individual ion diffusion coefficients in neutral, cation exchange and anion exchange materials. Manningâ€™s counter-ion condensation models and the Mackie/Meares model were used to correlate and, in some cases, predict the experimental data.

488457 Bioinspired to Multifunctional Nanostructured Membranes

Abstract:

The development of responsive, multifunctional materials and membranes for environmental and energy-related applications requires a high level of control of both the characteristics of the base polymeric or inorganic support layer, as well as, its corresponding surface properties. The special features for membrane processes that make them attractive for industrial applications are their compactness, ease of fabrication, operation, and modular design. Although membrane processes such as, Reverse Osmosis, Nanofiltration, Ultrafiltration, and Microfiltration have provided many successful applications ranging from high quality water production to material recovery, but incorporation of nanostructured aspects with tunable properties (learning from life sciences) in membranes has added immense value in the area of separations, reactions, water applications, and green synthesis. The use of macromolecules, such as, poly-acrylic acid (PAA), poly-glutamic acid (PLGA) provide pH responsive behavior through helix-coil transitions, whereas poly-N-isopropylacrylamide (pNIPAAm) provides temperature responsive behavior. Further, one can synthesize catalytic metal nanoparticles in the functionalized membrane for water remediation applications. The presentation will include bio-inspired functionalized membranes with tunable and responsive properties, layer-by layer assembly, synthesis of metal nanoparticles and metal films in membranes, and applications (including joint work with industries) in water/energy area. Special acknowledgement goes to his current and recent graduate/undergraduate students (Cassandra Porter, Minghui Gui, Li Xiao, Doug Davenport, Scott Lewis, S. Islam, Mike Detisch,). In addition to various industrial collaborators, NIH-NIEHS-SRC and NSF EPSCoR program provided high level financial support.

498606 FRI/John G. Kunesh Award Lecture:Â in Search of New Paradigms for Crystal Engineering: Old Tricks, Novel Discoveries, and Future Challenges

Abstract:

Crystal engineering is a broad area of research that focuses on methods of designing and/or optimizing materials for diverse applications in fields spanning from energy to medicine. The ability to selectively control crystallization to achieve desired material properties requires detailed understandings of the thermodynamic and kinetic factors regulating crystal nucleation and growth. Combining this fundamental knowledge with innovative approaches to tailor crystal size, structure, and morphology can lead to the production of materials with superior properties beyond what is achievable by conventional routes. In this talk, I will discuss two mechanisms of crystal growth: (1) classical pathways involving 2-dimensional layer nucleation and advancement on crystal surfaces through monomer addition; and (2) nonclassical pathways, termed *crystallization by particle attachment* (CPA), involving the formation of metastable precursors that play a direct role in crystal nucleation and growth. Our group and collaborators have employed a combination of colloidal assays, surface science techniques, applied studies, and ab initio modeling to investigate crystallization *in situ* and elucidate the mechanisms of growth modification. To this end, we investigate modifiers that control crystallization in two distinctly different, yet fundamentally similar, areas of research. Here, I will discuss the progress we have made to develop modifiers as new drugs to prevent the formation of crystals in pathological diseases, such as kidney stones (calcium oxalate) and malaria (hematin). Our findings have spurred additional research thrusts, including the use of bio-inspired approaches to tailor the properties of commercial materials, such as zeolites, which are microporous aluminosilicates utilized in many separations processes. The adaptation of crystal growth modification to these materials poses many challenges, central among them being the complex mechanism(s) by which they grow. In this talk, I will discuss our ability to elucidate and control the growth of these materials; and summarize our recent efforts to develop commercially-viable methods of tailoring the physicochemical properties of crystals relevant to the separations industry.

498330 Clarence G. Gerhold Award Lecture:Â Flexibility-Based Separations â€“ a New Paradigm for Membranes

Abstract:

Membrane systems have been designed for separations based on both size and solubility (e.g., partitioning). The size-based separations dominate in ultrafiltration processes used for the concentration and purification of proteins. However, there is also considerable interest in using membrane systems for the separation of large flexible biomolecules like DNA and PEGylated proteins (proteins with attached polyethylene glycol chains). The behavior of these biomolecules is considerably more complex, with the elongational flexibility playing a critical role in their ultrafiltration characteristics. This talk examines experimental data for DNA transmission demonstrating the importance of flow-induced elongation on the ultrafiltration behavior. Results show how differences in elongational flexibility can be exploited for the separation of topological isoforms of the DNA, with a linear isoform having much greater transmission than a supercoiled isoform with the same base-pair sequence. It is also possible to design novel membranes to enhance the separation by pre-stretching the DNA. These results provide the first quantitative demonstration of a flexibility-based separation using membrane ultrafiltration.

505008 Atomistic Simulations of Ordered and Disordered Carbons: "Mimicking" Versus "Targeting"

Abstract:

Atomistic simulation techniques have become extremely valuable tools for the elucidation of the structure of disordered or (partially) ordered materials. These techniques not only allow for the production of large-scale three-dimensional representations of the materials from the atomic scale, but also to predict most of their properties, and in some cases, to assess their formation mechanisms. In this talk, I will focus on carbon materials that can be ordered as well as disordered. There two routes to obtain realistic molecular-scale reconstruction of porous carbons: mimicking their formation process and targeting a â€œfinalâ€ state by constraining/guiding the simulation using experimental texture information. Obviously the first is a priori more predictive than the second but requires dealing upfront with chemistry (through reactive force fields) than can be rather complex with kinetically hindered processes.

To illustrate the mimicking route, I will present (i) the Grand-Canonical numerical synthesis of ordered carbon phases by carbon vapor deposition on zeolites and MCM-type mesoporous oxides, (ii) the Replica Exchange MD simulation of the geological degradation of the main components of wood (lignin and cellulose that are both organic molecules) into disordered nanoporous carbons and kerogen phases with the production of a complex liquid phase. The so-called targeting route will be illustrated with the example of the kerogen phases using the Hybrid-Reverse Monte-Carlo approach with an extension towards their mechanical properties from elasticity to failure (fracture).

505007 Graphene Oxide Membranes: A Molecular Simulation Approach

Abstract:

Graphene oxide membranes have attracted a lot of attention from the scientific community due to their potential application in challenging separations. Nevertheless, to optimise a material for an engineering application, it is necessary to understand their structures and the features that can improve their performance. Following on Keith Gubbins philosophy that realistic models of materials at a molecular level can be obtained from mimicking some aspects of their synthesis using molecular simulations, we have developed a protocol to create model graphene oxide sheets and mimic the membrane formation procedure to obtain model materials, with properties comparable to those measured experimentally.

495702 Polyethylene Oxide (PEO) in a Polyethylene (PE) Framework: A Simple Model for Simulation Studies of Scaling and Solvent Effects on Polymers in an Open Framework

Abstract:

We report molecular dynamics simulations studies of a polyethylene oxide polymer in a cubic polyethylene framework. Radius of gyration, shape, and other structural parameters are computed. Self-diffusion and displacement relative to the framework are also computed to investigate mobility and transport behavior. To investigate scaling effects, polyethylene oxide polymers of three different lengths and framework of three different sizes are simulated. Explicit water molecules are introduced to investigate solvation effects in different dimensions of polymer and framework.

Confinement of the polymer in smaller framework lead to closer interactin of polymer and framework and a more elongated configuration. The elongated configuration is favorable for diffusion through small windows of the framework. A larger mean square displacement is observed for the smaller framework, particularly in the presence of water. Adjusting the framework interaction potential sees a critical threshold dividing normal diffusion behavivoir and a regime dominated by hopping between cavities of the framework.

Although the model framework contains only simple van der Waalsl groups, confinement and solvent effects on polymer observed agree qualitatively with those in experiments reported recently for polyeletrolyte threaded in metal organic frameworks [1,2]

[1] L. Gao, C.Y.V. Li, and K.Y. Chan, *Chem. Mater.* 27(10) (2015) 3601-3608.

[2] L. Gao, C.Y. V. Li, K.Y. Chan, and Z.N. Chen, *J. Am. Chem. Soc.* 136 (2014) 7209-7212.

505006 Molecular Understanding and Design of Zwitterionic Materials

Abstract:

An important challenge in many applications, ranging from medical devices to ship hulls, is the prevention of nonspecific protein adsorption on surfaces. To address this challenge, our goals are twofold. First, we strive to provide a fundamental understanding of nonfouling mechanisms at the molecular level using an integrated experimental and simulation approach. Both simulation and experimental results show that the strong hydration of zwitterionic materials is responsible for their excellent nonfouling properties (JACS, **127**, 14473, 2005). Further studies examine zwitterionic structures and reveal the role of these structures in nonfouling properties beyond hydration (Advanced Materials, **27**, 15, 2015). Second, we aim to develop biocompatible and environmentally benign ultra low fouling materials based on the molecular principles learned. We have demonstrated that zwitterionic and mixed charge materials and surfaces are highly resistant to nonspecific protein adsorption, cell adhesion and bacteria adhesion/biofilm formation from complex media. Recent results show that zwitterionic materials induce no capsule formation upon implantation (Nature Biotechnology, **31**, 553, 2013) and no immunological response in blood circulation (PNAS, **112**, 12046, 2015). Furthermore, these materials are shown to preserve protein (Nature Chemistry, 4, 59, 2012) and cell (Angewandte Chemie, **53**, 12729, 2014) bioactivity. Because of these unique properties, zwitterionic materials are excellent alternatives to widely used poly(ethylene glycol) (PEG)-based materials.

504992 Interfacial Transport of Protons on a 2-Dimensional Functionalized Graphane Surface

Abstract:

Proton transport is crucial for many industrial and biological processes. The movement of protons is most often facilitated by the presence of water because water forms an extended hydrogen bonded network allowing for fast hopping of protons from one molecule to another. We have developed a novel 2-dimensional material that is capable of facilitating fast proton transport in the complete absence of water. Anhydrous proton transport is important for applications such as high- and intermediate-temperature fuel cell membranes. Our material was designed and tested using first principles density functional theory calculations. The material is based on graphane, which is an sp3 analogue of graphene. We have functionalized the graphane surface with hydroxyl groups and have used ab initio molecular dynamics and transition state calculations to study the mechanism and barriers of proton transport. We examine the limit of 1-dimensional proton hopping along a line of OH groups and the limit of fully hydroxylated graphane, forming 2-dimensional hydrogen bonding networks. We report diffusivities and transition state theory calculations. We find that anhydrous functionalized graphane has higher transport rates than hydrated Nafion.

504993 Gas Adsorption Behavior in Ionic Polyimide Composite Membranes

Abstract:

Industrial gas separation and storage processes present many technical and economic challenges to the natural gas industry and to power plants. In particular, palliating CO_2 from emission sources is a critical need in industry, and it is necessary in order to meet current and future environmental regulations. Polyimides are at the forefront of advanced membrane materials for CO_2 capture and gas purification processes. Recently, â€œionic polyimidesâ€ (i-PIs) have been reported as a new class of condensation polymers which combine structural components of both ionic liquids (ILs) and polyimides through covalent linkages. In this study, we report CO_2 and CH_4 adsorption and structural analyses of an i-PI and an i-PI + IL composite containing [C_4mim][Tf_2N]. The combination of molecular dynamics (MD) and grand canonical Monte Carlo (GCMC) simulations are used to compute the gas solubility and the adsorption performance with respect to the density, fractional free volume (FFV), and surface area of the materials.

Our results highlight the polymer relaxation process, and its correlation to the gas solubility. In particular, the surface area can provide meaningful guidance with respect to the gas solubility, and it tends to be a more sensitive indicator of the adsorption behavior versus only considering the system density and FFV. For instance, as the polymer continues to relax, the density, FFV, and pore-size distribution remain relatively constant, while the surface area can continue to increase, enabling more adsorption. Structural analyses are also conducted to identify the nature of the gas adsorption once the ionic liquid is added to the polymer. The presence of the IL significantly displaces the CO_2 molecules from the ligand nitrogen sites in the neat i-PI to the imidazolium rings in the i-PI + IL composite. Whereas, the CH_4 molecules move from the imidazolium ring sites in the neat i-PI to the ligand nitrogen atoms in the i-PI + IL composite. These molecular details can provide critical information for the experimental design of highly selective i-PI materials, as well as provide additional guidance for the interpretation of the simulated adsorption systems.

493614 Computational Simulation of Supported Nanocatalysts Under Realistic Conditions

Abstract:

Supported noble metal catalysts have played an important role in the conversion of energy and resources. Computational simulations ranging from quantum chemistry density functional calculations, molecular dynamic have been widely used to investigate the reaction mechanisms and the relationship between structure and properties of nanocatalysts. However, there are still huge gaps between computational simulations and realistic experiments. During the recent years, we try to computational simulate the nanocatalysts by consideration of realistic conditions in order to reduce the pressure and materials gap.

503118 Diffusion in Nanopores Under the Microscope

Abstract:

In his reflections on "Past Progress and Future Challenges in Adsorption Research" at the turn to the new millennium [1], Professor Douglas Ruthven characterizes the relevance of the first technique of microscopic diffusion measurement by saying "The measurement of micropore diffusion in zeolite crystals and other microporous solids has proved a far more challenging task than might have been anticipated. The development during the 1970s of the pulsed field gradient nuclear magnetic resonance (PFG NMR) technique for measuring self-diffusion under equilibrium conditions proved a major milestone. Remarkably, the key developments were made by Professor Pfeifer and his students at the University of Leipzig in East Germany." We consider it equally important to emphasize that all this progress did, to a large extent, benefit from the support and advice which, in turn, we had the privilege and pleasure to receive from Professor Ruthven. Some highlights in this cooperation shall be presented in this talk, which is co-authored by the fellow-combatants of his stays in Leipzig.

Deviating from, e.g., the gravimetric method commonly applied in former time for diffusion measurement in zeolites, PFG NMR permits following the diffusion path of guest molecules over distances much shorter than the crystal diameter, enabling the truly "microscopic" measurement. The introduction of microimaging by interference and IR microscopy gave rise to a dramatic enhancement in measurement performance since now, with the option of recording transient concentration profiles, diffusion phenomena could as well be studied under non-equilibrium conditions – as reviewed in, e.g., ref. [2]. Owing to the small size of the samples (namely single crystals rather than beds of crystals, corresponding with maximum surface-to-volume ratios), microimaging offers the additional advantage that, notably during fast ad- and desorption, corrupting influences due to the final rate of heat release may, in general, be ruled out [3].

As the main outcome of microscopic diffusion measurements, zeolite morphology was found to notably deviate, as a rule, from the regular pattern as generally implied on analyzing the outcome of gravimetric – or, more generally, of "macroscopic" – diffusion measurement. These deviations were, in addition, found to dramatically depend on the way of synthesis and sample pretreatment. As a main constituent of these deviations, transport resistances ("barriers") acting in addition to the diffusion resistance of the genuine pore network were found to exist both on the external surface of the individual crystals/particles [4] and in their interior [5]. It is worthwhile mentioning that, after PFG NMR has over decades been exclusively applied to studying the dynamics of guest molecules, now such measurements have become also possible with exchangeable cations [5].

In our joint search for a model system which, upon investigation by both microscopic and macrsocopic techniques, gave rise to a compatible set of diffusion data, we were finally successful on considering cyclo- and n-hexane in nanoporous glass [6]. This material proved an excellent host in also subsequent applications. Thus, using dispersed nickel as a catalyst, nanoporous glass appeared as the system of choice in demonstrating the ability of IR microimaging for the one-shot determination of effectiveness factors during catalytic reaction, in the given case for the hydrogenation of benzene to cyclohexane [7]. Nanoporous glass did, further on, serve as an ideal host system for the direct observation of "uphill diffusion", namely of ethane towards increasing ethane concentration within zeolite DDR, "driven" by the concentration gradient of the second component, propene, [8] and for demonstrating the validity of Fick's law by direct experimental evidence, notably in also such cases where, in the literature, this validity is put under question [9].

The potentials of microimaging will, most likely, notably transgress the limits so far encountered during its application. This might, among others, in particular concern the investigation of the dynamics of phase transitions between the gaseous and liquid and the liquid and frozen states in mesopores. First investigations along these lines with mesoporous silicon [10] revealed substantial deviations between the common view on their structure and reality [11].

1. D.M. Ruthven, Past Progress and Future Challenges in Adsorption Research, *Ind. Eng. Chem. Res.* **39** (2000) 2127–2131.
2. J. Kärger, and D.M. Ruthven, Diffusion in nanoporous materials: Fundamental principles, insights and challenges, *New J. Chem.* **40** (2016) 4027–4048.
3. L. Heinke, C. Chmelik, P. Kortunov, D.B. Shah, S. Brandani, D.M. Ruthven, and J. Kärger, Analysis of thermal effects in infrared and interference microscopy: n-Butane-5A and methanol-ferrierite systems, *Microporous Mesoporous Mater.* **104** (2007) 18–25.
4. F. Hibbe, J. Caro, C. Chmelik, A. Huang, T. Kirchner, D. Ruthven, R. Valiullin, and J. Kärger, Monitoring molecular mass transfer in cation-free nanoporous host-crystals of type AlPO-LTA, *J. Am. Chem. Soc.* **134** (2012) 7725–7732.
5. S. Beckert, F. Stallmach, H. Toufar, D. Freude, J. Kärger, and J. Haase, Tracing Water and Cation Diffusion in Hydrated Zeolites of Type Li-LSX by Pulsed Field Gradient NMR, *J. Phys. Chem. C* **117** (2013) 24866–24872.
6. C. Chmelik, D. Enke, P. Galvosas, O.C. Gobin, A. Jentys, H. Jobic, J. Kärger, C. Krause, J. Kullmann, J.A. Lercher, S. Naumov, D.M. Ruthven, and T. Titze, Nanoporous glass as a model system for a consistency check of the different techniques of diffusion measurement, *ChemPhysChem* **12** (2011) 1130–1134.
7. T. Titze, C. Chmelik, J. Kullmann, L. Prager, E. Miersemann, R. Gläser, D. Enke, J. Weitkamp, and J. Kärger, Microimaging of transient concentration profiles of reactant and product molecules during catalytic conversion in nanoporous materials, *Angew. Chem. Int. Ed.* **54** (2015) 5060–5064.
8. A. Lauerer, T. Binder, C. Chmelik, E. Miersemann, J. Haase, D.M. Ruthven, and J. Kärger, Uphill diffusion and overshooting in the adsorption of binary mixtures in nanoporous solids, *Nat. Comms.* **6** (2015) 7697.
9. T. Titze, A. Lauerer, L. Heinke, C. Chmelik, N.E.R. Zimmermann, F.J. Keil, D.M. Ruthven, and J. Kärger, Transport in nanoporous materials including MOFs: The applicability of Fick's laws, *Angew. Chem. Int. Ed.* **54** (2015) 14580–14583.
10. A. Lauerer, P. Zeigermann, J. Lenzner, C. Chmelik, M. Thommes, R. Valiullin, and J. Kärger, Micro-imaging of liquid–vapor phase transition in nano-channels, *Microporous Mesoporous Mater.* **214** (2015) 143–148.
11. D. Kondrashova, A. Lauerer, D. Mehlhorn, H. Jobic, A. Feldhoff, M. Thommes, D. Chakraborty, C. Gommes, J. Zecevic, P. de Jongh, A. Bunde, J. Kärger, and R. Valiullin, Scale-dependent diffusion anisotropy in nanoporous silicon, *Scientific reports* **7** (2017) 40207.

502012 Transport in Small Pore Zeolites

Abstract:

Professor Doug Ruthven has made numerous profound contributions to the field of adsorption and adsorption technology. One of these contributions is chromatographic techniques for diffusion measurements, namely the Zero Length Chromatography (ZLC). In this talk we will highlight how the ZLC technique is used in an industrial setting, provide comparisons with other microscopic and macroscopic methods, and demonstrate how it enables the development of advanced materials for gas separations.

494808 Process-Based Adsorbent Screening and Design for Post-Combustion CO_2 Capture By Vacuum Swing Adsorption

Abstract:

Over the past 20 years, adsorption based post-combustion CO2 capture has attracted keen interests from both process engineers and material scientists. The traditional approach to process development involves synthesis of adsorbents; screening using metrics such as selectivity etc.; perform lab-scale breakthrough experiments; computer-based optimization and scale-up the separation. Very often, materials that are touted to be promising do not show success at a process-scale. This is primarily because the complexities of the process are not considered in screening materials.

This talk will summarize the outcome of activities in our research groups that have focused on understanding the importance of material properties that lead to the desired outcomes of post-combustion CO2 capture, i.e., achieving product purity and recovery, while minimizing parasitic energy and maximizing productivity. These studies provide insight into what properties should be targeted in material synthesis to make them successful at a process level.

496137 Using a Volumetric Apparatus to Distinguish between Diffusion and Surface Resistance Mass Transfer Kinetics in Commercial Adsorbents

Abstract:

When designing fast cycle or kinetically controlled pressure swing adsorption processes an understanding of the prevailing mechanism of mass transfer and the knowledge of the kinetic constants is an essential requirement. Professor Ruthven has made significant contributions in this area and in this paper we present a methodology that allows to distinguish clearly between internal diffusion and surface resistances when studying the kinetics of adsorption on commercial pellets.

For fast adsorption kinetics, the traditional method of converting the pressure signal of the dosing cell into an uptake curve and then plotting this vs the square root of time is not easily applicable because of the initial effect of the flow through the valve between the dosing and uptake cells. The new approach we propose is based on a semi-log plot of the reduced pressure in the dosing cell vs time, and provided that the volumetric experiment is carried out under isothermal and linear conditions, a simple inspection of this plot reveals clearly which process is prevailing. A detailed analysis of the models indicates also that the long-time asymptotic decay is very sensitive to the mass transport mechanism, because the equilibrium properties are fully defined by the initial and final state of the system.

The methodology will be demonstrated on commercial adsorbents used in air separation. To discuss both limiting behaviours experimental results will be presented on LiLSX beads which are macropore diffusion controlled and on CMS beads which exhibit primarily a surface resistance. The talk will conclude with a discussion of how to determine the kinetic constants when both processes contribute to the overall mass transport.

497167 Development of Sour PSA Technology for Gasification Markets

Abstract:

Doug Ruthven is one of the most prominent adsorption technologists of our time. He has achieved that status in part by his own personal dedication and hard work in the area, and also by partnering and collaborating with numerous colleagues in academia and in industry. His example of collegial engagement has been noted and often replicated in the subsequent careers of the large group of past graduate students and post docs.

Â Air Productsâ€™ Sour PSA technology is an example of an industrial collaborative effort. Sour PSA is a competitive technology to solvent-based AGR systems currently employed to separate sulfur containing species, along with CO_2 and other impurities, from gasifier syngas streams. It utilizes pressure swing adsorption concepts to generate a sweet H_2 or syngas product stream along with an enriched CO_2/H_2S stream. The latter can be transformed into 1) an elemental sulfur or sulfuric acid stream and 2) a sulfur-free CO_2 stream applicable to CO_2â€“capture or non-CO_2-capture opportunities. Development efforts have ranged from laboratory tests of adsorbent feasibility to demonstration of cyclic PSA operation in the field. In this talk, I will highlight these collaborative efforts and the technical insight they have provided.

501322 Separations Inspired from Adsorption: From Adsorbents to Adsorbent Membranes

Abstract:

Everybody who worked with Doug has been inspired by adsorption and adsorbents. The extension of this knowledge sometimes leads to other research areas such as transport phenomena including adsorbents. Some examples of these processes include inorganic membranes containing adsorbents such as zeolites, as well as mixed matrix membranes containing any adsorbent materials. These membranes can be used in different gas and liquid separation applications.

In this presentation, examples of adsorption and adsorption inspired membrane separations will be discussed. These include gas separations with adsorbents; gas separations with inorganic membranes made out of adsorbents; pervaporation separation and recovery of alcohols from fermentation of agricultural by-products using mixed matrix membranes for them to be blended into gasoline.

Incorporation of adsorption and diffusion theories into the characterization of membranes will be discussed with separation examples mentioned above.

511765 Zeolite and Metal-Organic-Framework Nanosheet Crystallization for Membrane Separations and Catalysis

Abstract:

The ability to control crystal size and morphology is essential for enabling optimal use of micro- and meso-porous materials in separation and catalysis applications. In the last 15 years, our group is developing methods to synthesize zeolite and, more recently, metal-organic-framework nanosheets. They hold promise for ultra-selective highly-permeable gas separation membranes because their molecularly-sized pores can ensure the desirable separation selectivity and increased permeance along their thin dimension. Moreover, pillared or self-pillared (intergrown) nanosheets provide reduced diffusion lengths that are desirable in certain adsorption and catalysis applications. I will describe a crystal growth methodology based on controlling the frequency of rotational intergrowths (twining) that enabled direct-crystal growth of the zeolite framework MFI as self-pillared [1] and free-standing nanosheets [2]. I will also describe recent results on direct synthesis of metal-organic-framework nanosheets and their use in thin film formation.

Zhang, X.; Liu, D.; Xu, D.; Asahina, S.; Cychosz, K.; Varoon, K.; Al Wahedi, Y.; Bhan, A.; Al Hashimi, S.; Terasaki, O.; Thommes, M.; Tsapatsis M., *Direct Synthesis of Self-Pillared Zeolite Nanosheets by Repetitive Branching. Science*, 336, 1684-1687, **2012**.

Jeon, M.Y.; Kim, D., Kumar; P., Lee, P.S.; Rangnekar, N.; Bai, P.; Shete, M.; Elyassi, B.; Lee, H.S.; Narashimharao, K.; Basahel, S.N.; Al-Thabaiti, S.; Xu, W.Q.; Cho, H.J.; Fetisov, E.O.; Thyagarajan, R.; DeJaco, R.F.; Fan, W.; Mkhoyan, K.A.; Siepmann, J.I.; Tsapatsis, M., *Ultra-Selective High-Flux Membranes from Directly Synthesized Zeolite Nanosheets, Nature*, 543(7647), 690-694, **2017**.

PSE

On Advances and Challenges in crystallization Process Modelling: Systems Approaches, Product Performance and Measurement Interpretation

Dr Sean Bermingham, Head of PSE Formulated Products
AIChE Annual meeting, Minneapolis, 30 October 2017

gPROMS

Outline

- A proven engineering capability? Practical tips (lessons learnt)
  - GSK: characterising kinetics
  - Eli Lilly: designing a continuous process from batch experiments
  - Syngenta: scale down process to address filterability issues
  - Solvay: changing the feed location to reduce fines
  - Pfizer: linking crystallization performance to downstream processing and end-user performance
- Making this capability more widely accessible
- No room for complacency (where next?)
- Concluding remarks

Characterisation of kinetics

GSK – batch cooling crystallization of an API

Presented at AIChE 2011

Model-based decision support for design and operation of crystallisation processes

Efficient workflows for validation against experiments and scale-up

Sean Bermingham, Mark Pinto
Process Systems Enterprise Ltd (UK)
s.bermingham@psenterprise.com

Ugo Cocchini, Alex Caillet
GlaxoSmithKline (UK)

AIChE Annual Meeting
Minneapolis
17 October 2011

Case study
Recipe and experimental data

- Batch seeded cooling crystallisation of an API from an organic solvent
- Sampling strategy
  - solute concentration and PSD throughout the crystallisation
  - not standard sampling regime
- Dataset of 8 experiments varying
  - agitation
  - supersaturation
  - seed PSD
  - T seeding
  - cooling rate

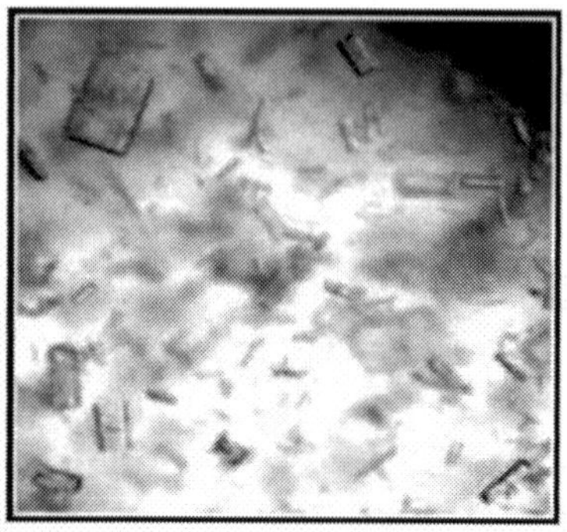

Exp		Run 1	Run 2	Run 3	Run 4	Run 5	Run 6	Run 7	Run 8
API	g/l	69.3	66.8	67.0	47.0	52.4	45.7	64.4	63.8
Seed Load	%	0.7	0.5	0.4	0.5	0.5	0.7	0.5	0.4
Stirrer speed RPM		216	407	217	220	217	379	407	217
Seeding Temp	degC	70	70	70	70	70	70	70	70
Holding time	hrs	3.0	2.0	2.0	6.0	3.0	6.0	2.0	2.0
Final Temp	degC	5	5	5	5	5	5	5	5
Cooling time	hrs	1	3	3	no cooling	1	1	3	3
Rate of cooling	deg/min	1.08	0.36	0.36	na	1.08	1.08	0.36	0.36
Holding at final temp	hrs	14	14	14	0	14	14	14	14
Total exp time	hrs	18.0	19.0	19.0	6.0	18.0	21.0	19.0	19.0

gCRYSTAL

Case study
Parameter estimation

- Use data from 3 experiments (out of 8 available)
  - concentration data: 10 to 15 time points per experiment
  - quantiles of PSD (d_{10}, d_{16}, d_{50}, d_{84}, d_{90}, d_{95}): 3 time points per experiment
- Parameters to estimate
  - 2/4/2 for nucleation/growth/breakage mechanisms
  - 2 for PSD of seeds – approximated by a Roslin Rammler distribution
- Maximum likelihood objective function

$$\Phi = \frac{N}{2}\ln(2\pi) + \frac{1}{2}\min_{\theta}\left\{\sum_{i=1}^{NE}\sum_{j=1}^{NV_i}\sum_{k=1}^{NM_{ij}}\left[\ln\left(\sigma_{ijk}^2\right) + \frac{(\tilde{z}_{ijk}-z_{ijk})^2}{\sigma_{ijk}^2}\right]\right\}$$

  - measurement accuracy (variance) determines weighting!

gCRYSTAL

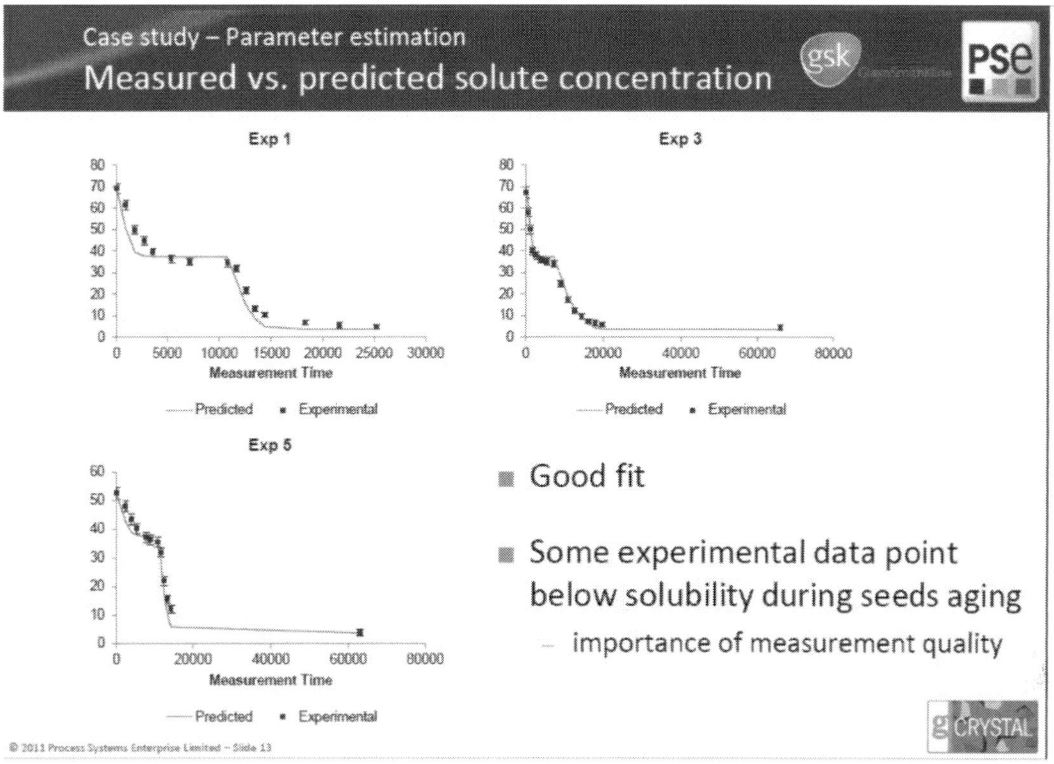
Case study – Parameter estimation
Measured vs. predicted solute concentration
gsk
PSE
Exp 1
Measurement Time
Predicted
Experimental
Exp 3
Measurement Time
Predicted
Experimental
Exp 5
Measurement Time
Predicted
Experimental
Good fit
Some experimental data point below solubility during seeds aging
importance of measurement quality
gCRYSTAL
© 2011 Process Systems Enterprise Limited – Slide 13

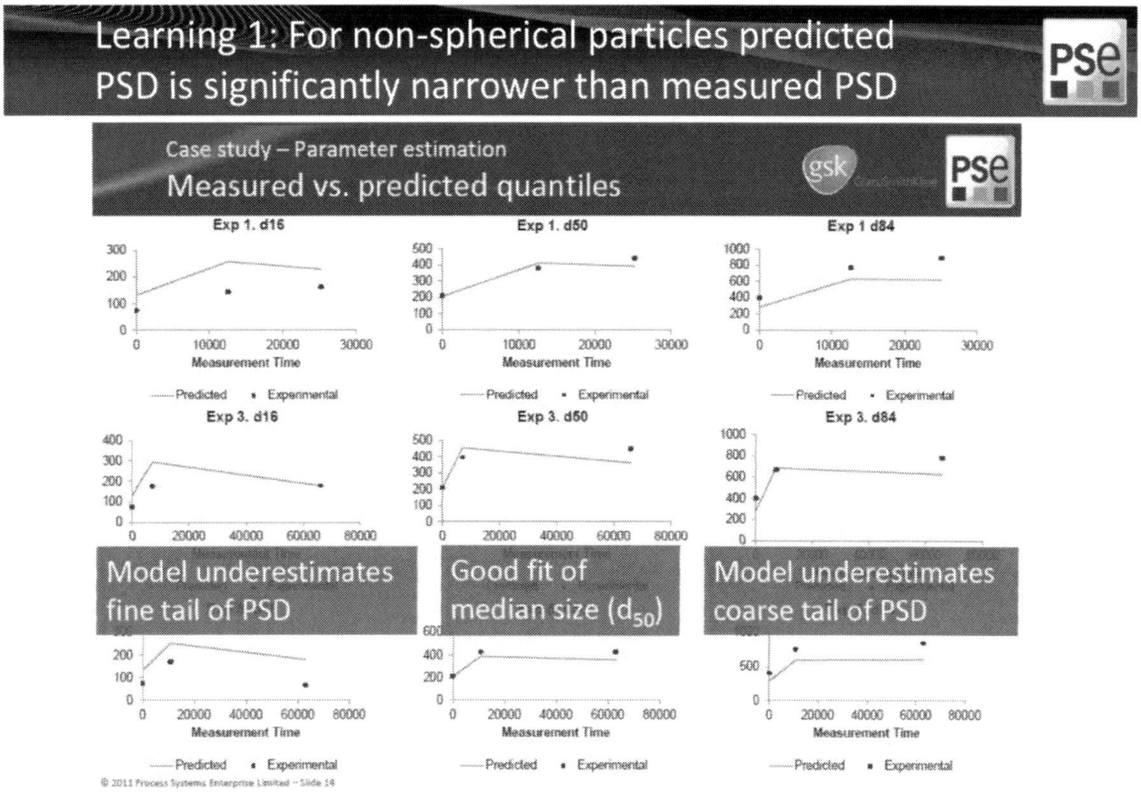
Learning 1: For non-spherical particles predicted PSD is significantly narrower than measured PSD
PSE
Case study – Parameter estimation
Measured vs. predicted quantiles
gsk
Exp 1. d16
Exp 1. d50
Exp 1 d84
Exp 3. d16
Exp 3. d50
Exp 3. d84
Measurement Time
Predicted
Experimental
Model underestimates fine tail of PSD
Good fit of median size (d50)
Model underestimates coarse tail of PSD
© 2011 Process Systems Enterprise Limited – Slide 14

Learning 1: For non-spherical particles predicted PSD is significantly narrower than measured PSD

DE GRUYTER OPEN

INTERNATIONAL Agrophysics
www.international-agrophysics.org

Int. Agrophys., 2014, 28, 195-200
doi: 10.2478/intag-20014-0008

Influence of the sand particle shape on particle size distribution measured by laser diffraction method**

Cezary Polakowski[1], *Agata Sochan*[1]*, *Andrzej Bieganowski*[1], *Magdalena Ryżak*[1], *Rita Földényi*[2], *and Judit Tóth*[3]

[1]Institute of Agrophysics, Polish Academy of Sciences, Doświadczalna 4, 20-290 Lublin, Poland
[2]Department of Earth and Environmental Sciences, University of Pannonia, Egyetem 10, H-8200 Veszprém, Hungary
[3]Institute of Materials and Environmental Chemistry, Research Centre for Natural Sciences, Hungarian Academy of Sciences, Pusztaszeri 59-67, H-1025 Budapest, Hungary

Received November 5, 2013; accepted January 3, 2014

Learning 2: Initial PSD needs to be estimated

Case study
Why estimate PSD of seeds (initial condition)?

- Because like all PSD measurements, the measured PSD is not 100% accurate
  - measurement variance
  - variance from sampling
  - instrument bias

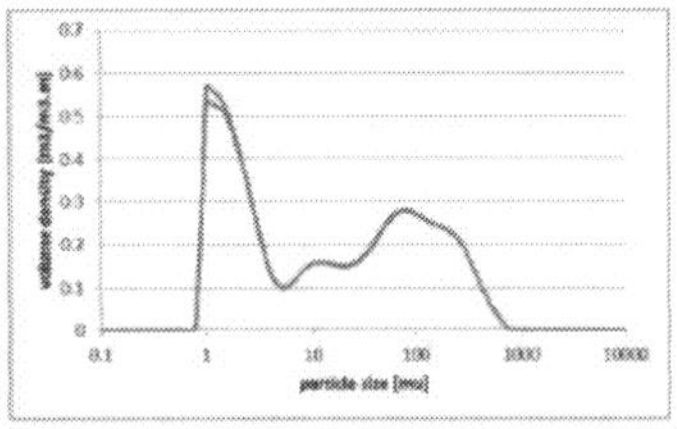

- Simulation using bimodal (left) and unimodal (right) seed PSD

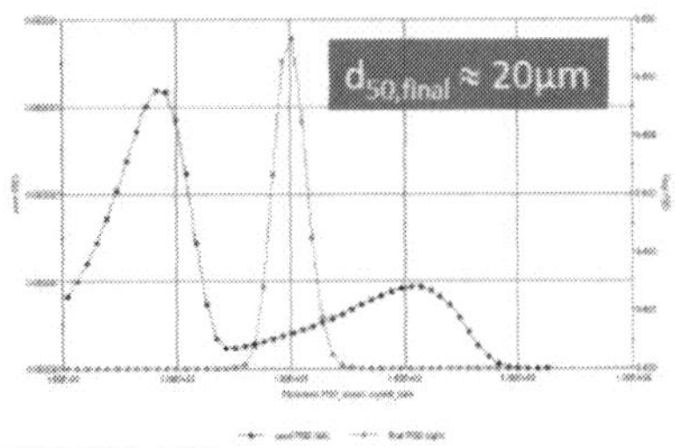

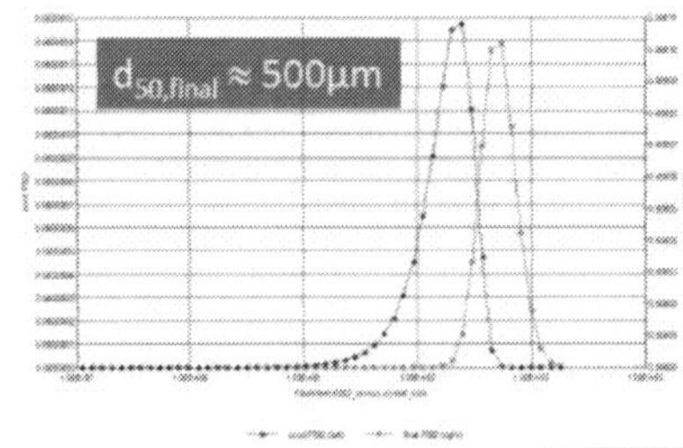

Batch-to-continuous

Eli Lilly – utilization of population balance models to develop a continuous crystallization process

Presented at APMF 2013 in London

Utilization of Population Balance Models to Develop a Continuous Crystallization Process: Process Design and Optimization

Christopher L. Burcham
Christopher S. Polster
Michael A. Lovette
Eli Lilly and Company

Sean K. Bermingham
Hassan S. Mumtaz
Process Systems Enterprise

Acknowledgements
Marty Johnson
Derek Griffin
Dan Jarmer
Bret Huff
Paul Collins
Eoin McManus

Answers That Matter.

Motivation

Pharma has recently "rediscovered" continuous processing.

Crystallization:

- Workhorse for separations.
  - Undesired impurities are preferentially removed (potency and purity)
  - Acceptable yield is achieved
  - Desired crystal form is maintained
  - Target particle size (or specific surface area)
  - Morphology (crystal shape)

Moving from Batch to Continuous Crystallization Process Development.

- 2012 APM Forum "Tailoring API Physical Properties through Crystal Engineering: Moving from Batch to Continuous Crystallization Development"
- How to integrate continuous process development into a traditional batch process development environment?
  - Desire to deliver same/better potency, impurity profile, particle size and morphology
- *How to achieve this in the same development timeframe?*
- *How to achieve the same (or better) physical properties?*

Continuous Anti-Solvent Crystallization Development Example – Two MSMPRs

- 3 variables – τ_1, τ_2, X_1
- 3 levels, 27 total experiments, 4 experiments/wk
- ~7 weeks

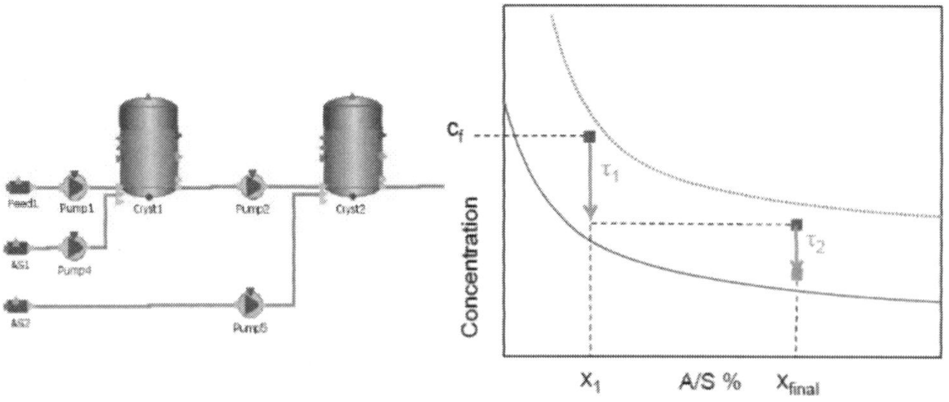

Continuous Crystallization Development Workflow

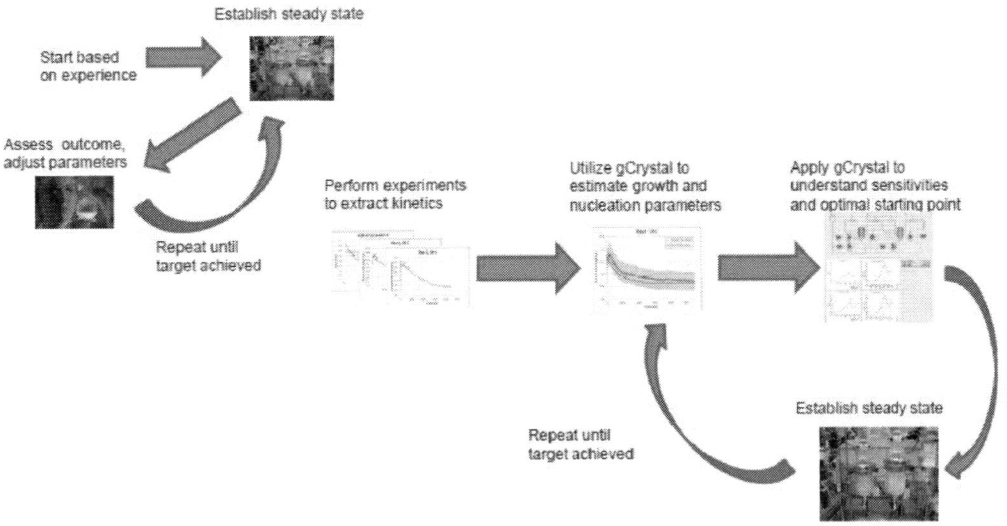

10

Solute Concentration vs. Time Fits

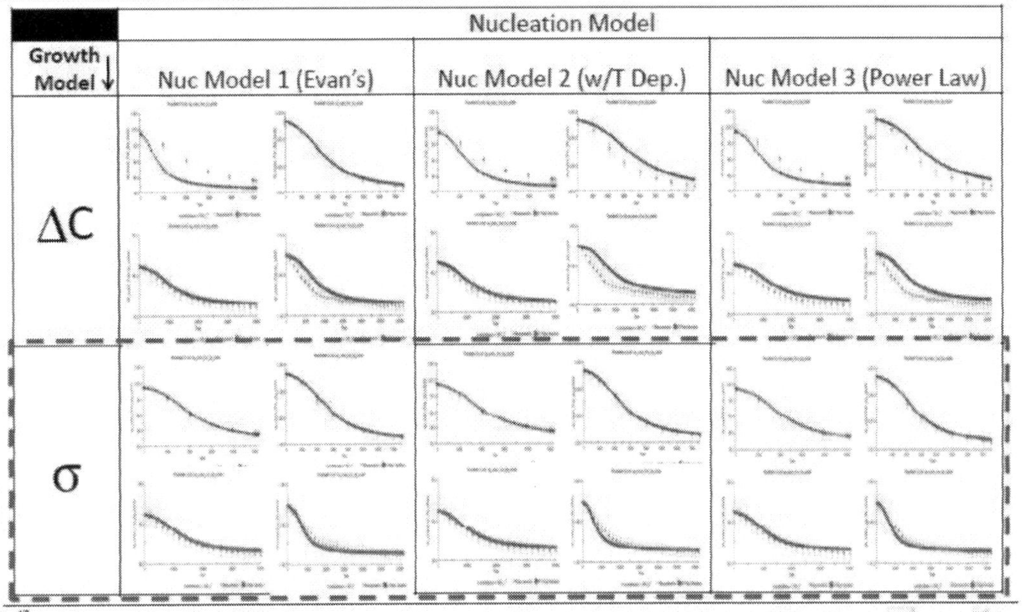

17

Learning 1 (again!): For non-spherical particles predicted PSD is significantly narrower than measured PSD

PSE

Particle Size Quantile Fits

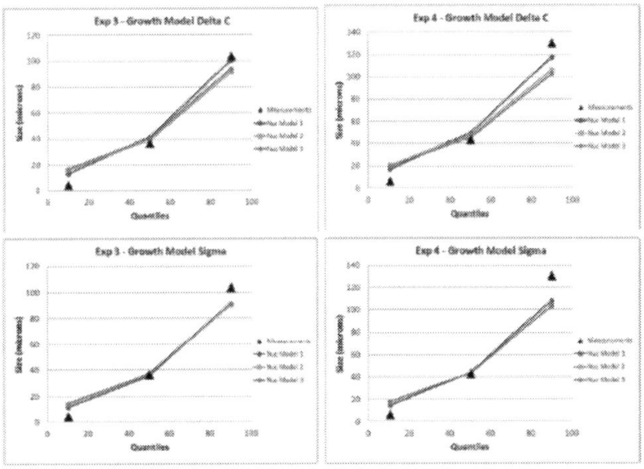

Observations

- Nucleation Model 1 predicts broader PSD (lower d10, higher d90) across the board. This is more representative of the measured values.
- Predictions using relative supersaturation growth model less dependent on nucleation model choice.
- Nucleation Models 2 and 3 predict very similar values.

18

Learning 3: You can use existing investment in batch lab equipment to help design a continuous process

PSE

gCRYSTAL Model Verification – Steady State Conditions, Iteration 1

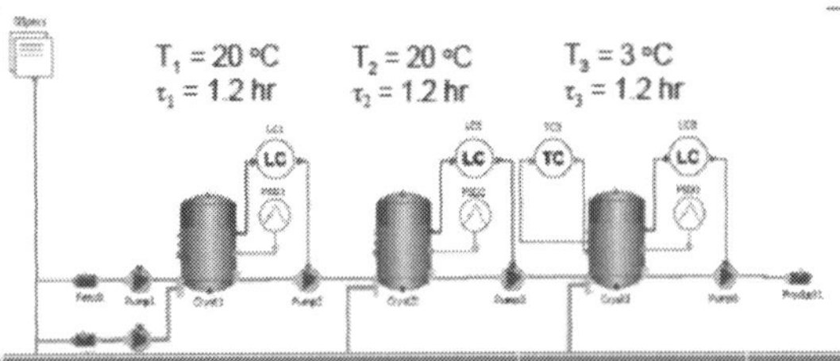

A/S_1 = 5 L EtOH/kg Solute

A/S_2 = 0.9 L H2O/kg Solute

A/S_3 = None

Value	Actual	Model 1	Model 2
Solubility MSMPR 1 (wt%)	5.7	5.7	5.7
Solubility MSMPR 2 (wt%)	1.2	1.2	1.2
Solubility MSMPR 3 (wt%)	0.76	0.73	0.73
Solute in Solution MSMPR 1 (wt%)	**7.9**	**8.2**	10
Solute in Solution MSMPR 2 (wt%)	**1.7**	**1.6**	2.0
Solute in Solution MSMPR 3 (wt%)	**0.83**	**0.86**	1.0
Rel. Supersaturation MSMPR 1 (%)	40	44	78
Rel. Supersaturation MSMPR 2 (%)	35	29	63
Rel. Supersaturation MSMPR 3 (%)	9.2	18	41

22

Scaling down (and up)

Syngenta – commercial scale batch cooling crystallization

Presented at APMF 2013 in London

syngenta

Applying Crystallization Modelling to Improve the Understanding of a Batch Cooling Process of an Agrochemical Active Ingredient

Rhea Brent, Manish Parmar, Pauline Sillers, Michael Bryce, Gillian Clelland and Neil George (SYNGENTA)
Hassan Mumtaz and Niall Mitchell (PSE)

Classification: EXTERNAL USE

Syngenta AI-X Current Process

- AI-X currently manufactured and isolated by a cooling crystallisation process
- Currently isolation has batch to batch variability– in extreme cases rate decreases four-fold.
- Occasionally, subsequent batches must be held in a slurry hold vessel until the centrifuge is freed up.
- The AI is not likely to be polymorphic
- Poor filtration has been attributed to the width of the particle size distribution.

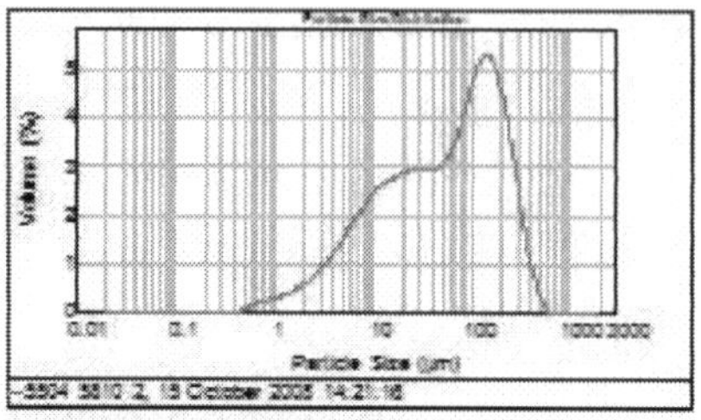

Typical particles from plant batches

Crystallisation Scale-down of AI-X

- Experiments performed at 10 L based on a scaled-down version of the plant crystalliser.
- Replicate plant cooling profile
- Two different cooling profiles:

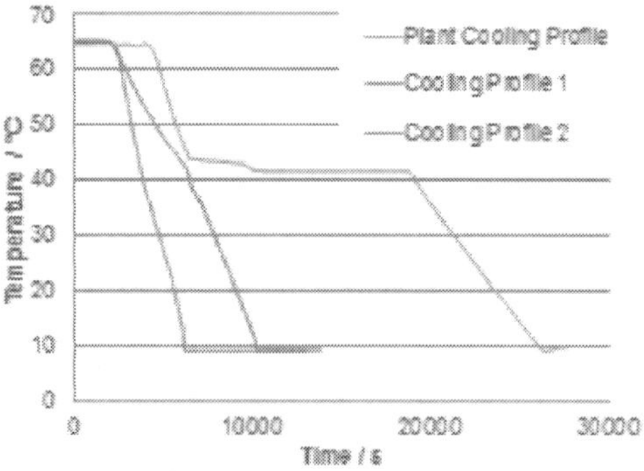

- Seeding with 2 different seeding quantities (1 and 2 % w/w of small PSD)- slightly modified cooling profile and agitation

Experimental Results: Seeded Crystallisations

Conditions	Mean volume D4,3	Span (D90-D10)/D50
unseeded	118.5	2.1
Seeded with 1 % w/w	61.2	1.8
Seeded with 2 % w/w	63.6	1.7

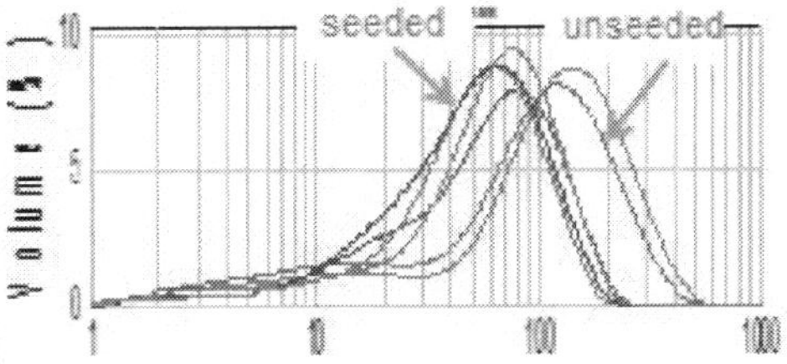

- Effect of seeding: mean size and overall distribution of the particles was reduced
- Apparently no difference between the effects of using different amounts of seed
- The viscosity of the resulting crystal slurry was high – stirring was difficult

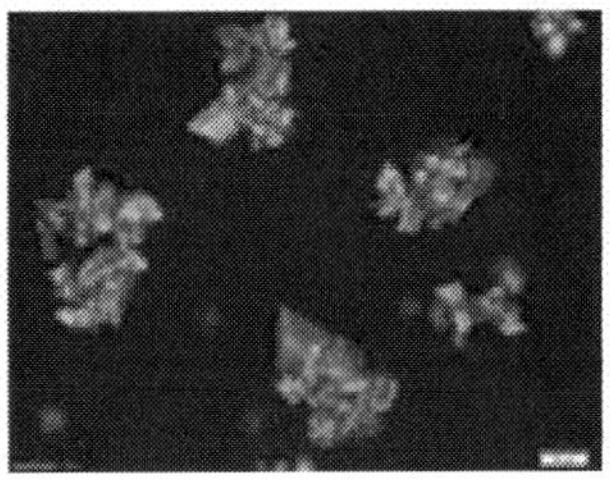

Results from the Model: Experiments at 10 L Scale

Phenomena	Estimated Parameter	Value Obtained
Secondary Nucleation	Rate constant (ln k_n)	17.8
	Order (dependence on supersaturation) (σ^n)	0
	Particle size at which crystals undergo attrition (L_{min})	73.2 µm
Growth	Order (surface integration dependence on supersaturation) (g)	1.4

- Secondary nucleation and growth good statistical fit by model.
- Dominant phenomena in process are the favourable crystal growth kinetics and attrition of crystals growing larger than 73 µm.
- Primary nucleation poorer statistical fit by model: likely to be due to dominance of secondary nucleation in the system

Learning 4: Mechanistic models reduce amount of plant data required to characterise kinetics

PSE

Building the Plant Scale Model and Optimisation

- Scale up to plant vessel, modified agitation and added experimental data on end concentration and particle size
- Used estimated growth parameters from 10 L and estimated:
  - Primary nucleation rate constant
  - Secondary nucleation rate constant
  - Minimum size for attrition

Phenomena	Estimated Parameter	Value Obtained (10 L model)	Value Obtained (plant model)
Secondary Nucleation	Rate constant (ln k_n)	17.8	15.5
	Particle size at which crystals undergo attrition (L_{min})	73.2 µm	76.8 µm

19 Classification: EXTERNAL USE syngenta

Learning 5: Sometimes it is a good idea to heat during cooling crystallization

PSE

Modelling Results: Effect of Modifying Cooling Profile with Low Loading of Seeds

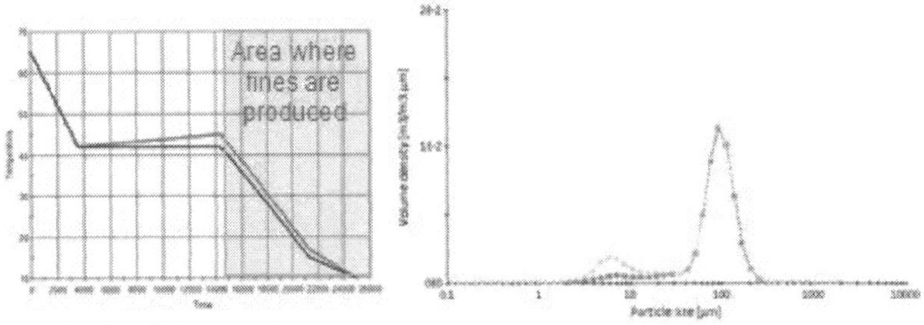

Original plant cooling profile
Modified cooling profile

Seeded low loading plant cooling profile
Seeded low loading modified cooling profile

- Reduction of fines apparent when a temperature ramp is implemented during the hold period

21 Classification: EXTERNAL USE syngenta

Changing the feed location to reduce fines

Solvay – commercial scale crystallization

Presented at APMF 2016 in London

Limitation of the plant production capacity

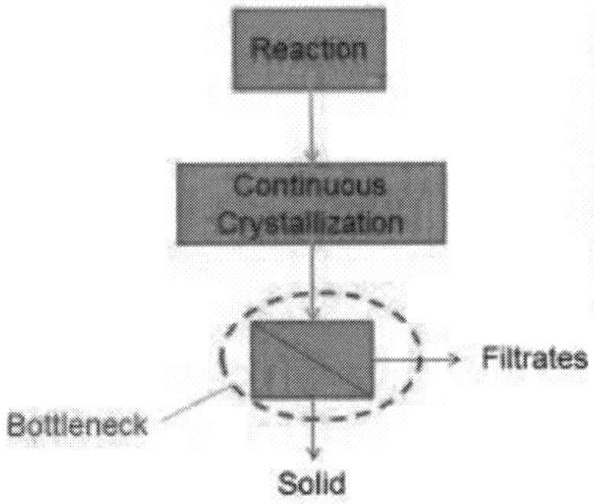

SOLVAY

Objective: Production increase

with limited Investments

- Modification of the current separation system

➔ tests of different operating conditions and filtration medium: No success ☹

➔ A new liquid / solid separation technology has been found, but it is expensive and there are still some technical risks

Improvement of the feed quality

➔ Work on the crystallization operation !

SOLVAY

Controling Crystal's filter-ability

Multiscale approach

1. Experiments at lab scale
2. Cystallization modelling within gCRYSTAL – Lab scale
3. Simulation of the industrial crystallizer
4. Input of the CFD / gCRYSTAL Multizonal
5. Optimization of the crystallizer design and operating parameters

6 | APMF - Mathieu Oullion
20th of April 2016

SOLVAY

Using CFD simulation in g-CRYSTAL

g-CRYSTAL Multizonal

CFD simulation

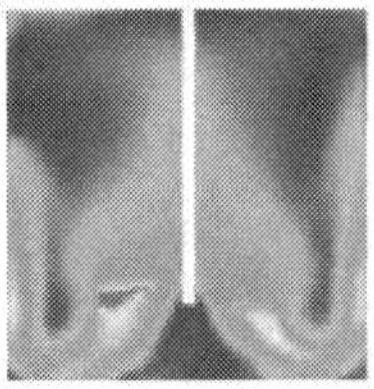

1 crystallization stage

Choosing the « Zones »

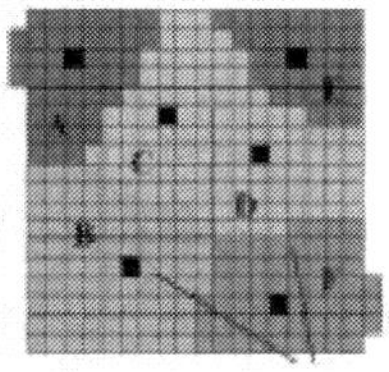

Extraction of CFD data

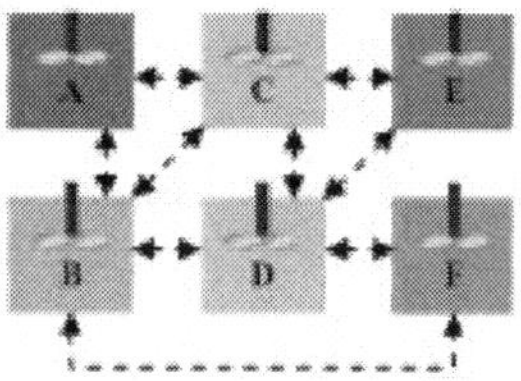

1. Experiments | 2. Lab. model | 3. Indus. model | 4. CFD Input | 5. Optimization

15 | APMF - Mathieu Oullion
20th of April 2016

SOLVAY

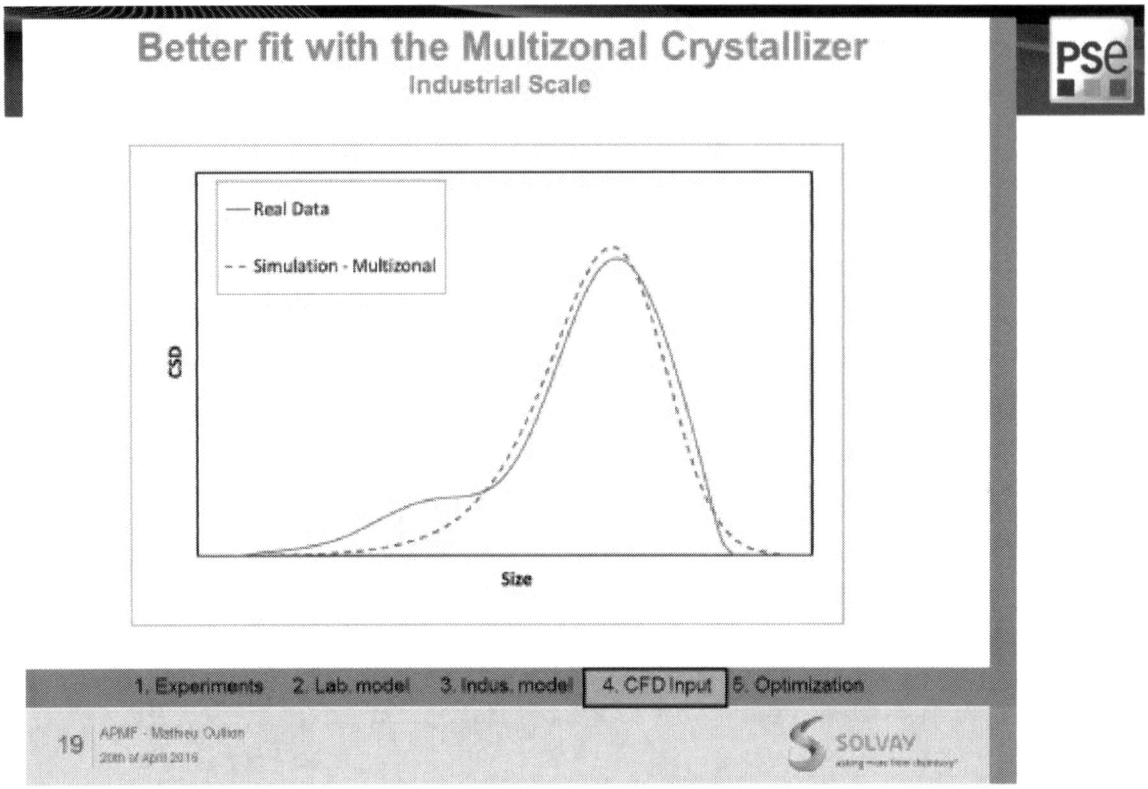
Better fit with the Multizonal Crystallizer
Industrial Scale
PSE
Real Data
Simulation - Multizonal
CSD
Size
1. Experiments
2. Lab. model
3. Indus. model
4. CFD Input
5. Optimization
19
SOLVAY

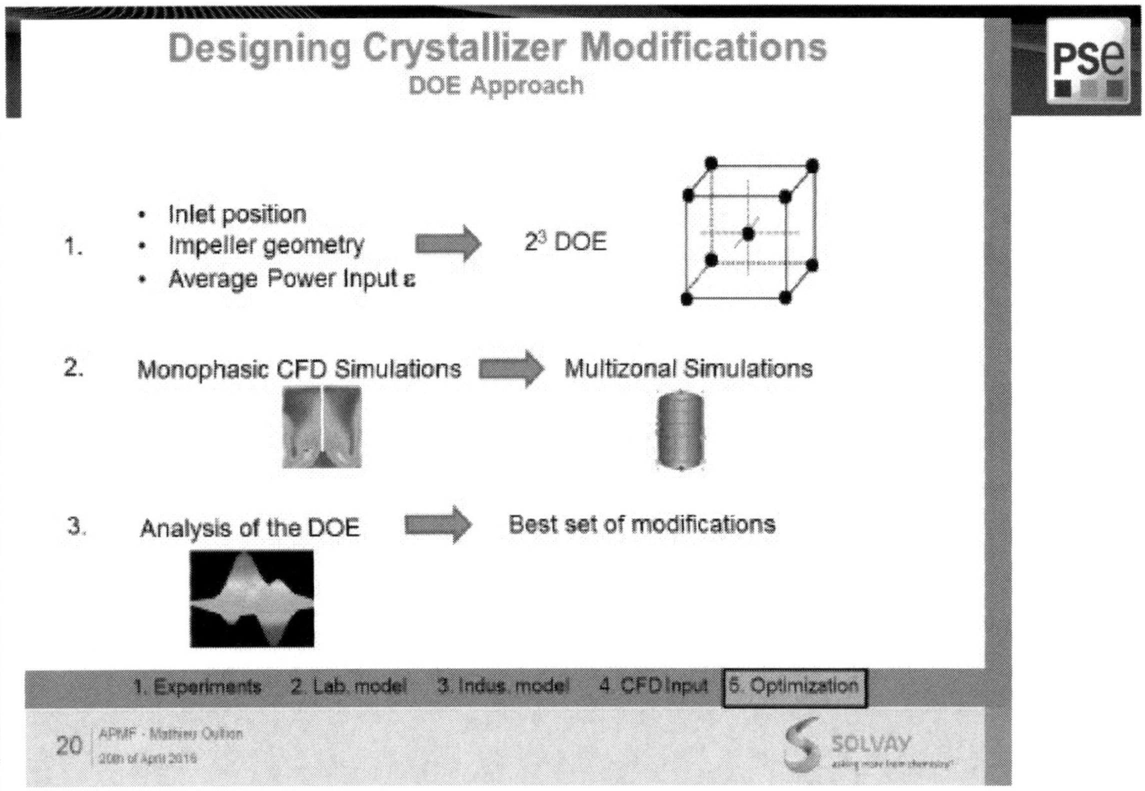
Designing Crystallizer Modifications
DOE Approach
PSE
1.
• Inlet position
• Impeller geometry
• Average Power Input ε
2³ DOE
2.
Monophasic CFD Simulations
Multizonal Simulations
3.
Analysis of the DOE
Best set of modifications
1. Experiments
2. Lab. model
3. Indus. model
4. CFD Input
5. Optimization
20
SOLVAY

Capacity increase after modifications

At low investment

\+

ε

||

Narrower CSD

||

Higher filter-ability

We achieved a 25% increase in filtration capacity

1. Experiments 2. Lab. model 3. Indus. model 4. CFD Input 5. Optimization

21 APMF

SOLVAY

Conclusions

- CFD + Population Balance Modeling is a very efficient tool for
  - Optimizing existing crystallizer ➔ reduce experimental costs (time and money !)
  - Designing New equipments ➔ Allow the implementation of breaktrough technologies

&

Future Work

- A purification model can be extracted from the population balance

$$C_i = f(r_{N2}, G, Agg)$$

⬇

optimization with the same model

22 APMF

SOLVAY

Linking crystallization performance to downstream processing and end product performance

Pfizer – Integrated flowsheets and sensitivity analysis

Submitted to AAPS 2017 in San Diego

Integrated flowsheet – Sensitivity analysis per unit op

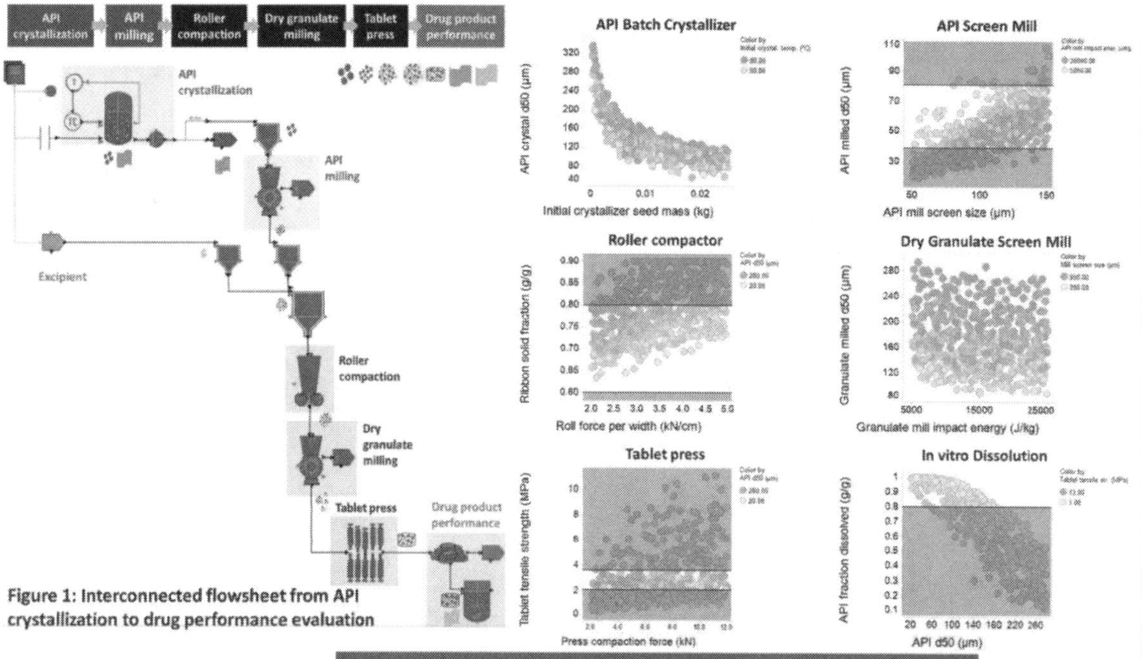

Figure 1: Interconnected flowsheet from API crystallization to drug performance evaluation

Pfizer work: Moreno-Benito et al., AAPS Annual Meeting, November 2017

Integrated flowsheet – Sensitivity analysis on flowsheet

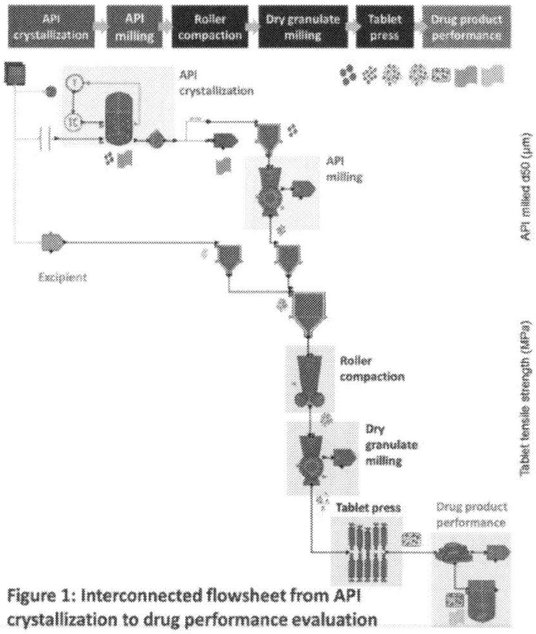

Figure 1: Interconnected flowsheet from API crystallization to drug performance evaluation

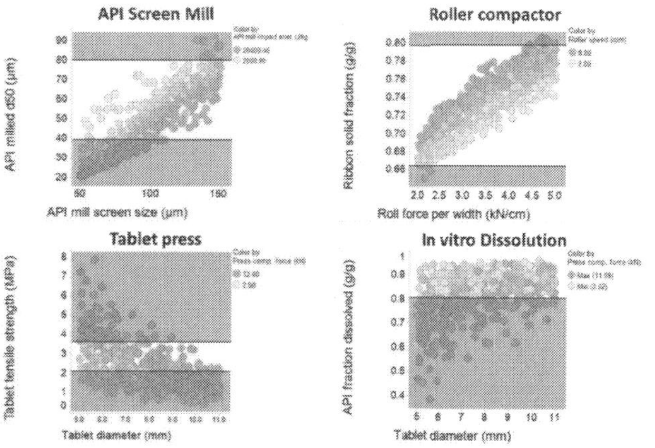

Pfizer work: Moreno-Benito et al., AAPS Annual Meeting, November 2017

Making this capability more widely accessible

Other users of mechanistic modelling for crystallization

Mechanistic modelling of crystallization processes has become a proven engineering capability, but ...

Making this capability more widely accessible

More focus on users and workflows

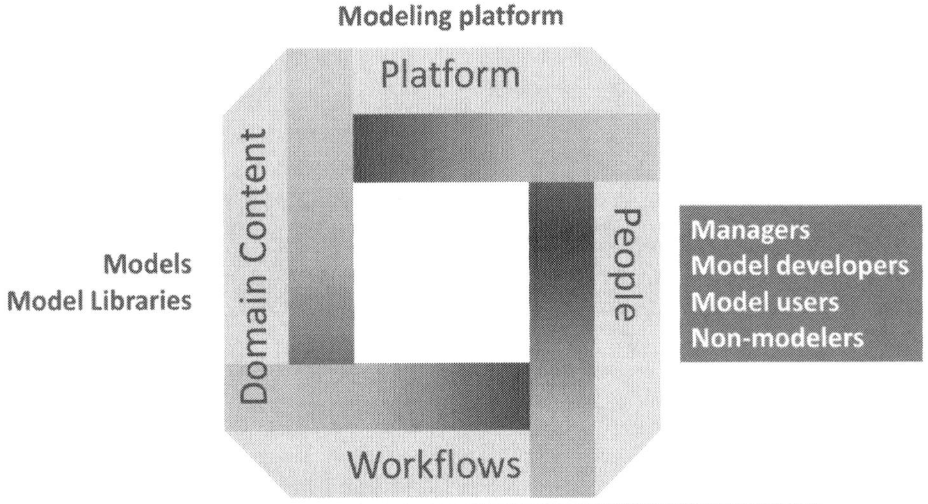

Integration within corporate workflows
Improved workflows taking advantage of new capabilities

Making this capability more widely accessible
Modeling technology deployment

PSE

Tier I	Tier II	Tier-III	Tier-IV
First-principles modellers ("custom modeling")	Drag-and-drop flowsheeting using model libraries	"Non-modellers" requiring access to model-based calculations	Model-based applications embedded in operations decision-support systems
Primarily R&D Subject-Matter Experts	R&D Engineering	R&D Engineering Operations Commercial	Operations
		E.g. Web-Based Applications	Embedded Applications

All using the same model (basis).
Capturing and consistently deploying organisation's knowledge

Making this capability more widely accessible

PSE

- Automating data import
  - Better integration with vendors of analytical techniques
- Streamlining parameter estimation and external model validation
  - Learning set vs (blind) test set
- Automatic discrimination between many models

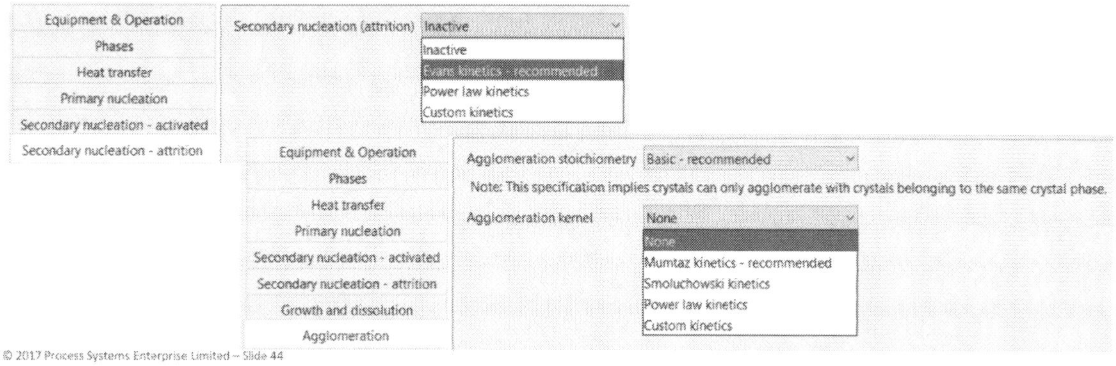

Making this capability more widely accessible

- Automating data import
  - Better integration with vendors of analytical techniques
- Streamlining parameter estimation and external model validation
  - Learning set vs (blind) test set
- Automating model discrimination
- Etc.

- It is widely recognised that improved usability will have a significant impact on value that mechanistic modelling brings
  - See for instance the 2016 ETC request for proposals http://www.etconsortium.org/population-balance-model

No room for complacency (where next?)

Activity based driving forces

- With the improvement in thermodynamic models (e.g. SAFT-γ Mie), it is time to move from concentration based to activity based driving forces in process models
  - This will allow us to predict how changes in liquid phase composition may not only affect solubility, but supersaturation → ability to use single set of kinetic parameters across range of liquid phase compositions

- Same is true for reaction kinetics

Modelling of analytical techniques

- Converting PSD to CLD (not the ill-posed inverse problem) is a well known example
- Widening of PSD in laser diffraction of non-spherical particles
- ...

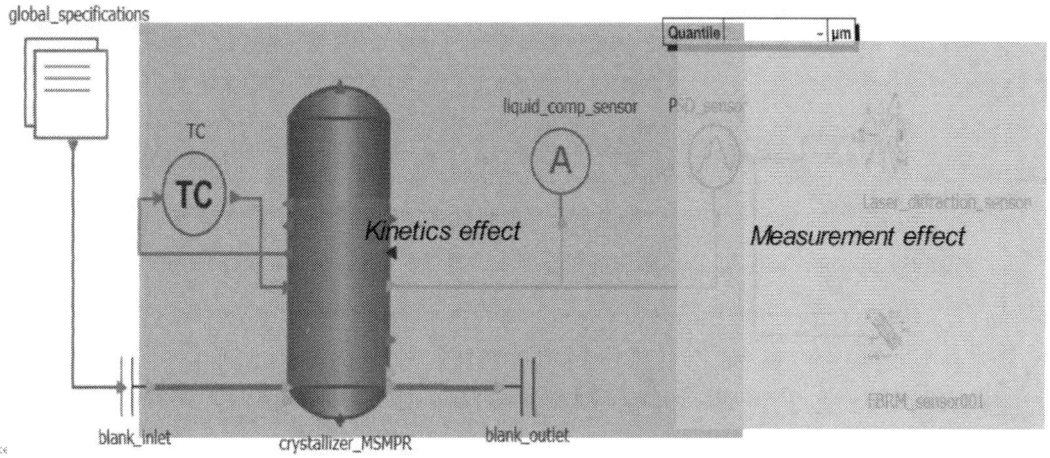

Use of multi-dimensional population balances

- Predict evolution of size and shape distribution
- Good progress made in academia
- So far, limited uptake in industry
  - Not because of computational challenges
  - But data requirements for 1D PBE are already often considered challenging
- Progress expected by
  - Linking to theories for relative growth rate of crystal faces (e.g. Doherty group) should be able to address data concerns; and/or
  - Development of new analytical techniques
    e.g. ETC have today published a RFI for "a multidimensional particle size and shape measurement tool"

Taking mechanistic models into operations

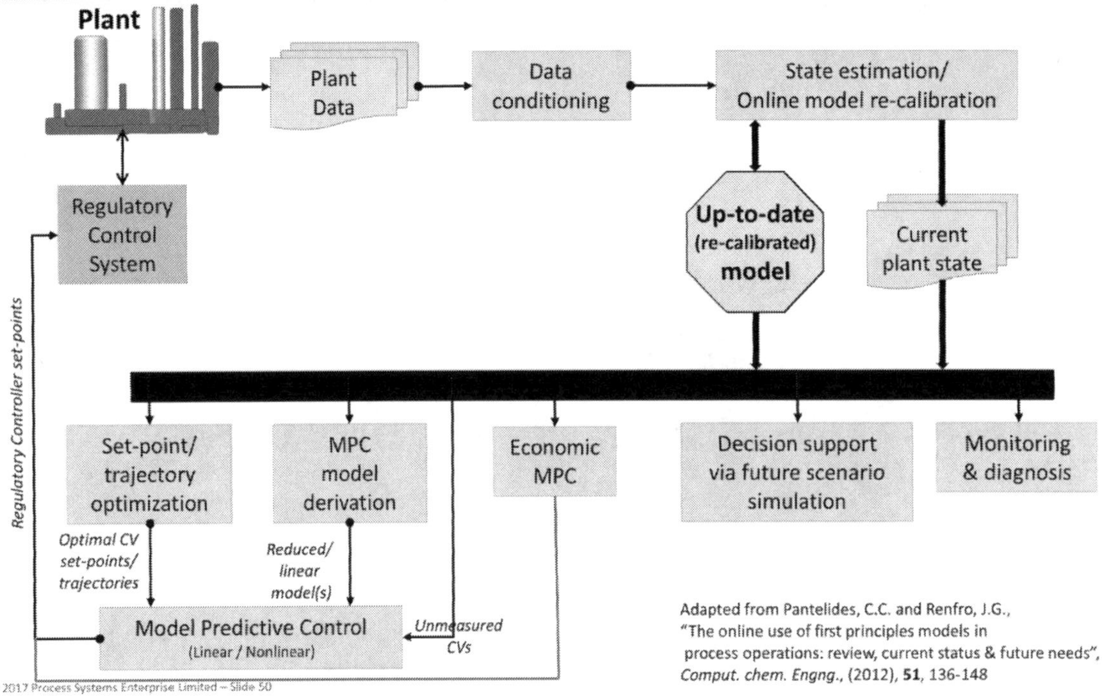

Adapted from Pantelides, C.C. and Renfro, J.G., "The online use of first principles models in process operations: review, current status & future needs", *Comput. chem. Engng.*, (2012), **51**, 136-148

Taking mechanistic models into operations

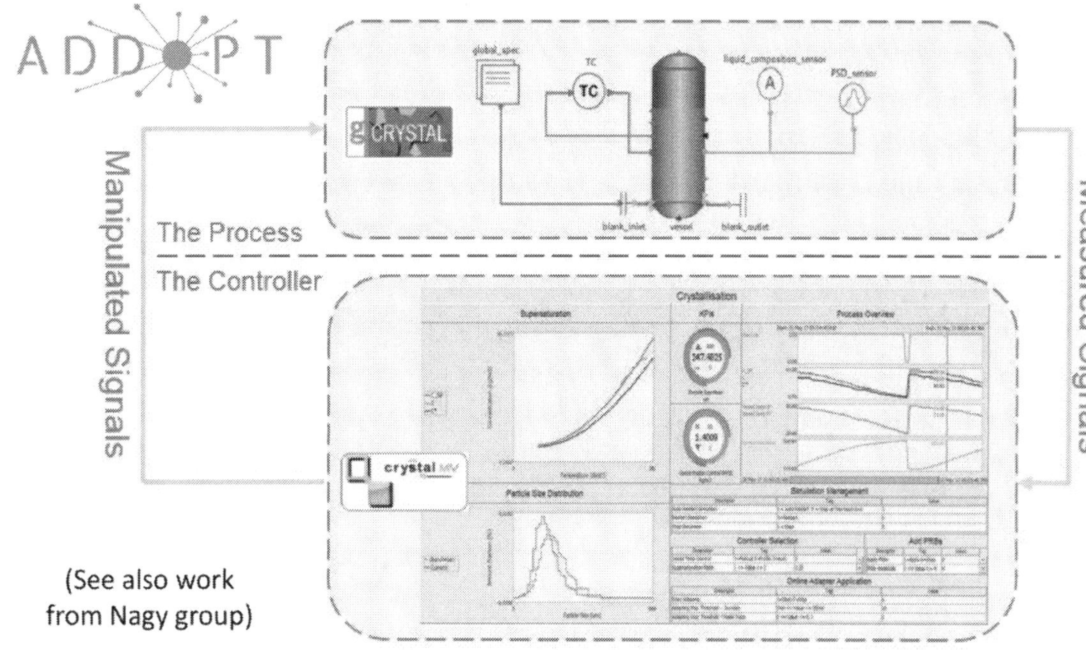

(See also work from Nagy group)

Linking models from multiple length scales

- Aim: prediction of material behaviour from molecular structure
- Examples:
  - Molecular dynamics → Maxwell-Stefan diffusivities
  - *Ab initio* crystal structure prediction → solubility, crystal growth kinetics

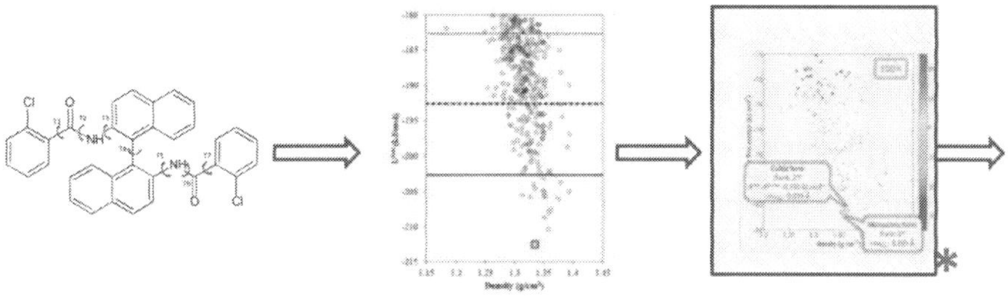

$$g^S_{XXVI}(T) = \mu^L_{XXVI}(T, x)$$

Molecule XXVI
6th Blind Test (2015)

Lattice energy landscape
Sugden, I.J., Adjiman, C.S., and Pantelides, C.C., *Acta Cryst.* (2016). B72, 864-874.

Free energy landscape
Vasileiadis, M.,. PhD thesis, Imperial College London, (2013)

Solubility prediction

* *Not* molecule XXVI (unfortunately)

  - Predicting growth kinetics of different crystal faces (e.g. Doherty group)

Concluding remarks

Concluding remarks

- Mechanistic modelling of crystallization processes has become a proven engineering capability, but ...
- Still a lot of room for increased adoption in industry, by focussing on better usability and support for integrated experimentation-modelling workflows
- And also plenty of room and need for incorporating emerging science into this engineering capability
  - Further reduction in experiments required → increased R&D efficiency
  - Better predictive capability → risk reduction in scale-up / tech transfer

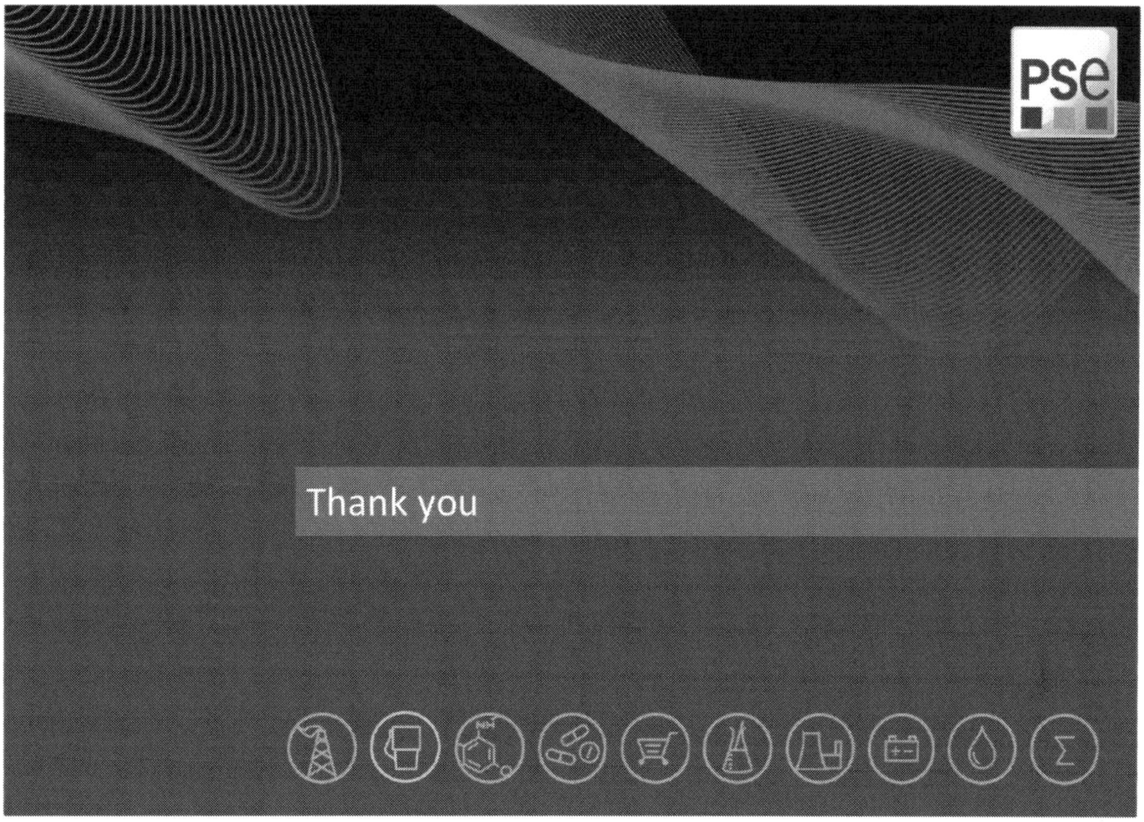
PSE
Thank you

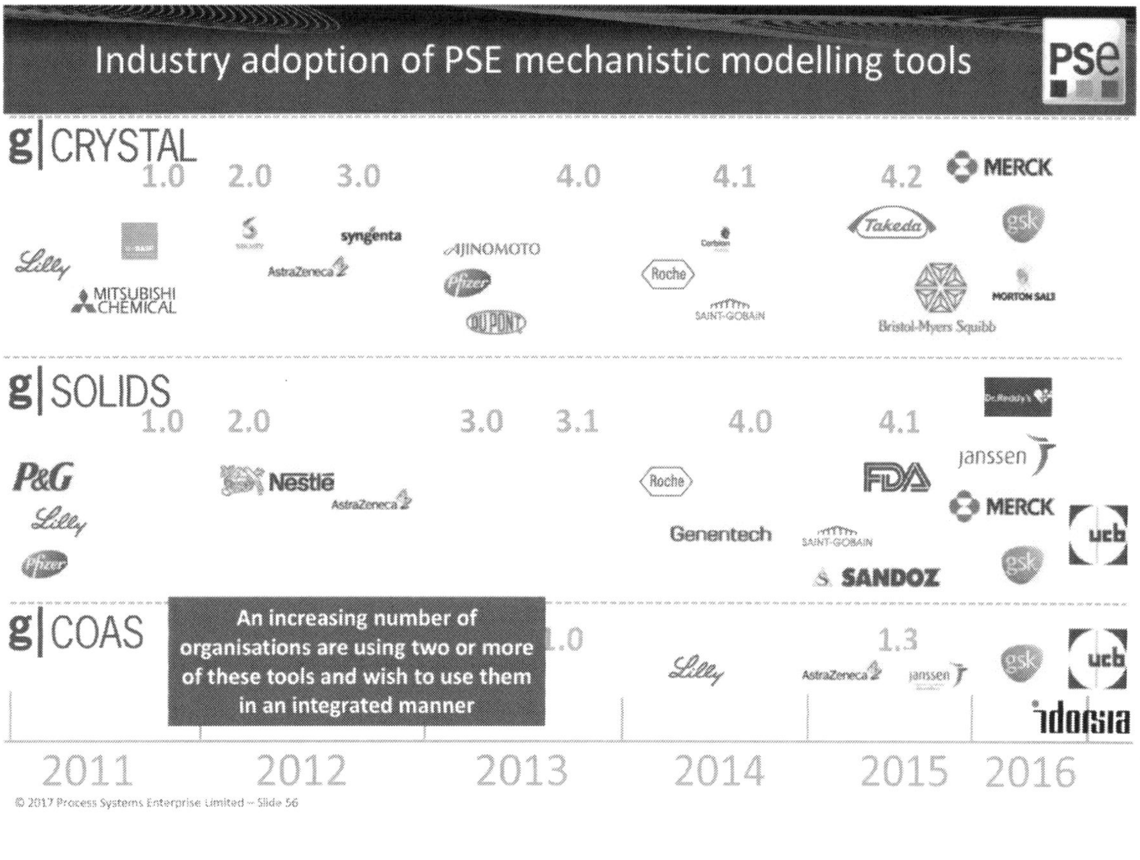
Industry adoption of PSE mechanistic modelling tools
PSE
g|CRYSTAL
1.0 2.0 3.0 4.0 4.1 4.2
MERCK
Lilly
MITSUBISHI CHEMICAL
syngenta
AstraZeneca
AJINOMOTO
Pfizer
DUPONT
Roche
SAINT-GOBAIN
Takeda
gsk
MORTON SALT
Bristol-Myers Squibb
g|SOLIDS
1.0 2.0 3.0 3.1 4.0 4.1
P&G
Lilly
Pfizer
Nestlé
AstraZeneca
Roche
Genentech
SAINT-GOBAIN
FDA
SANDOZ
Dr.Reddy's
janssen
MERCK
ucb
gsk
g|COAS
1.0 1.3
An increasing number of organisations are using two or more of these tools and wish to use them in an integrated manner
Lilly
AstraZeneca
janssen
gsk
ucb
idorsia
2011 2012 2013 2014 2015 2016
© 2017 Process Systems Enterprise Limited – Slide 56

Development approach and pre-competitive ecosystem for gPROMS FormulatedProducts

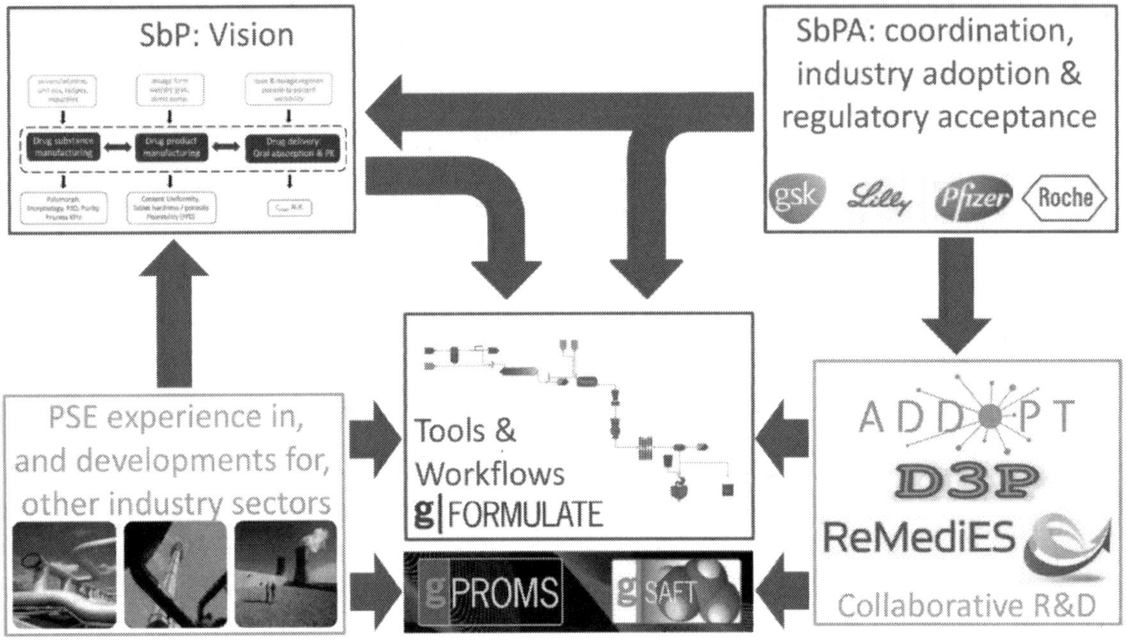

Customer applications

- SOLVAY – from optimization of industrial crystallizers to design of new equipment using advanced modelling tools
- Eli Lilly – utilization of population balance models to develop a continuous crystallization process
  - Process design and optimisation
- Syngenta – commercial scale batch cooling crystallization
  - Characterisation of kinetics and scale-down
- FrieslandCampina – batch cooling crystallization of pharma grade lactose
  - identification of kinetics
  - recipe optimisation
  - operator training
- GSK – batch cooling crystallization of an API
  - identification of kinetics
- BASF – A step towards rational crystallization modelling using gCRYSTAL
  - Characterisation of kinetics
- PURAC – batch crystallization of lactitol monohydrate
  - process optimisation / advanced control
- ROQUETTE – continuous cooling crystallization of dextrose
  - process optimisation / advanced control
- Anonymous PTA manufacturer – multiple stage, continuous, flash cooling
  - product quality control

498780 Metastable Phase Transitions

Abstract:

Several model systems exhibit a low-temperature first-order transition between two disordered phases that are metastable with respect to an ordered phase. The preponderance of experimental evidence is consistent with the possibility that water may behave in this way. The confluence of structural, kinetic and thermodynamic factors that make this interesting phase behavior possible is an active area of inquiry. I will present a theoretical and computational overview of metastable phase transitions, including discussions of finite size effects, the interplay of nucleation and relaxation times, the stability of interfaces, and metastable criticality.

498824 Computational Investigation of Multipolar Colloidal Particle

Abstract:

Colloids with anisotropic charge distributions hold promise for creating a number of useful new materials including optic materials with novel symmetries, materials for information storage, and dampers for controlling vibrations in structures. Experimentally manipulating properties of anisotropic particles in a controlled fashion can sometimes be difficult and is time consuming. The search for novel anisotropic colloidal materials can be enhanced and even guided by simulations of colloidal system assembly. We report the results of two projects. In the first we use discontinuous molecular dynamics (DMD) simulation to investigate how the internal charge separation of anisotropic, dipolar colloidal rods with 4:1 aspect ratio affects their phase behavior. The differences between the head-tail and side-side ordering displayed in these simulations could be exploited to design gels whose properties (especially those related to percolation: viscosity, conductivity, etc.) are responsive to external stimuli. In the second project we perform quasi-2D Monte Carlo simulations of systems of colloidal spheres with off-center, extended dipoles. These colloidal spheres are found to aggregate into cyclical structures whose interior angle is equal to the angle between the charges on a single dipolar sphere.

498803 Interfacial Tensions from SAFT: Connecting Equations of State to Molecular Simulations

Abstract:

A description of fluid systems with molecular-based algebraic equations of state (EoSs) and by direct molecular simulation is common practice in chemical engineering and the physical sciences, but the two approaches are rarely closely coupled. The key for an integrated representation is through a well-defined force field and Hamiltonian at the molecular level. We discuss the latest developments in a top-down representation of fluids, where we use an accurate EoS to link the macroscopic properties of the fluid and the force-field parameters, with a particular focus on a coarse-grained formulation of the statistical associating fluid theory (SAFT-Î³ Mie). The EoS is used to estimate the parameters of the Mie force field, which can then be used with confidence in direct molecular simulations to obtain structural and dynamical properties that are otherwise inaccessible from the EoS, including the interfacial tension, which is a property not included in the original parametrization.

A further theoretical framework is developed based on scaling arguments applied to the influence parameter of the van der Waals theory of inhomogeneous fluids. The molecular model stems from the application of the square gradient theory to the SAFT-Î³ Mie equation of state. The theory is validated against computer simulation results for homonuclear pearl-necklace linear chains made up to six Mie (Î»-6) beads with repulsive exponents spanning Î» = 8 to 44 by combining the theory with a corresponding states correlation to determine the intermolecular potential parameters. We show how this can be employed to provide a predictive tool to determine interfacial tensions for a wide range of molecules including hydrocarbons, fluorocarbons, polar molecules, among others. The proposed methodology is tested against comparable existing correlations in the literature, proving to be vastly superior, exhibiting an average absolute deviation of 2.2%

493586 Generalized Gibbs Free Energy of Confined Nanoparticles

Abstract:

The nanoparticles generally show abnormal properties compared to those in the bulk phase, and they exhibit significant potential in various applications such as catalysis and energy conversion. However, the theoretical work for describing the properties of nanoparticles is limited with poor prediction capacity. In this work, the Gibbs free energy was studied, from both marcoscope and microscope, predictive models were proposed to study the thermodynamic properties of nanoparticles with a generalized description of the Gibbs free energy considering the effects of surface-energy and the substrate contacted. The proposed model from the microscope was based on the corresponding states theory to describe the effect of the substrate on the Gibbs free energy of nanoparticles, in which the molecular parameter with a generalized parameter was obtained from the melting point of metals due to sufficient information. The comparison with the new measured experimental results proves the reliability of the model prediction.

498818 Molecular Modeling of Polymeric Systems

Abstract:

In this talk, I will present the results of three exciting opportunities to comprehensively investigate nano- materials through the use of molecular simulations. Tailoring and optimization of these materials in- clude: 1) shape-directed packing of rigid 2D-like polymer chains (PIM-TMN-Trip), that provides ul- trapermable polymers with highly interconnected micropores facilitating rapid and selective gas transport of polymers, 2) ultra-thin molecular layer-by-layer polyamide membranes which can play a crucial role in the design and understanding of the next generation of reverse osmosis membranes, and 3) bio-conjugates, protein-polymer systems, which will benefit from a predictive knowledge of how an attached polymer will interact with a protein and its environment.
In all cases, it is anticipated that experimentalists will avoid the peril of performing a large number of difficult synthetic chemistry that would result in no enhancement of the target property. Concurrent to the above goals, is the generation of large-data sets and the extraction of critical information from that data (e.g., structure factors to understand the intrinsic correlations between structure and properties/ behavior) to catalyze materials breakthroughs.

492444 Effect and Regulation of Surface Wettability on Molecular Transport and Reaction

Abstract:

Whereas massive experimental works have been contributed into the investigation of surface wettability effect on fluid transport and reaction, theoretical study on molecular scale is still few. The obstacle mainly originates from the lack of both microscopic description for the surface wettability and accurate theoretical method for quantifying the effect. Here, we demonstrate that the surface wettability effect on molecular transport and reaction can be quantitatively accessed from molecular perspective. The surface wettability is characterized with the ratio of fluid-fluid interaction to fluid-substrate interaction, whose relation to contact angle is given by an empirical yet analytical equation. This mesoscale relation, connecting macroscopic measurement and microscopic modeling, allows for seeking regulation method by identifying the thermodynamic and/or dynamic properties of fluid nearby substrate, and these properties can be well predicted with the help of modern statistical theories including classical density functional theories and molecular simulation. By discussing two specific examples associated with the flux through nanopores and conversion rate of a surface catalytic reaction, we show that this route is generic and has directive and practical importance to the design of efficient processes.

496030 Mixed-Matrix Membranes Based on Polymers of Intrinsic Microporosity and Metal-Organic Frameworks for Gas Separation Applications

Abstract:

The concept of potentially enhancing both permeability and selectivity of polymeric membranes by embedding microporous carbon, zeolites or metal-organic frameworks (MOFs) in a mechanically strong polymer matrix by forming mixed-matrix membranes (MMMs) has been one of the many research areas pioneered by Bill Koros and co-workers. Inspired by his groundbreaking work on 6FDA-DAM polyimide/ZIF-8 MMMs for propylene/propane separation1 our group developed MMMs using hydroxyl-functionalized polyimides of intrinsic microporosity that served as ideal polymer matrices due to their excellent compatibility with ZIF-8. Here, we report the fabrication and exceptional pure- and mixed-gas gas permeation properties of PIM-PI-based MMMs containing ZIF-8 loadings ranging from 20% to 65% by weight. The MMM membranes were tested with a 50:50 C3H6/C3H8 feed mixture at pressures up to 7 bar (p/psat ~0.8). Exceptional C3H6/C3H8 mixed-gas selectivities ranging from 20 at 30% loading to 30 at 65% loading were achieved. Furthermore, minimal onset of plasticization was observed, highlighting the strong potential of PIM-6FDA-OH/ZIF-8 mixed-matrix membranes for C3H6/C3H8 separations.

References

1. Chen Zhang, Ying Dai, Justin R. Johnson, Oguz Karvan, William J. Koros, â€œHigh performance ZIF-8/6FDA-DAM mixed matrix membrane for propylene/propane separationsâ€, J. Membr. Sci. 389 (2012) 34-42.

488950 Revolutionary Ultrathin Carbon Molecular Sieve Hollow Fiber Membranes

Abstract:

Revolutionary ultrathin carbon molecular sieve hollow fiber membranes

Chen Zhang, William J. Koros

Georgia Institute of Technology, Atlanta, Georgia U.S.A.

Carbon molecular sieves (CMSs) are membrane materials with tailored microporosity offering attractive mechanical properties and extraordinary permselectivities for many gas pairs. The extraordinary permselectivities well exceed polymer upper bounds, which rely upon entropically-enabled diffusion selectivities and synergistic sorption selectivities.[1] CMSs can be transformed into the preferred scalable hollow fiber geometry by controlled pyrolysis of asymmetric polymer precursor hollow fibers. The porous substrate of CMS fibers tends to densify during pyrolysis. In that case, the CMS hollow fiber membrane completely or partially loses its asymmetry, thereby producing thick separation layer (30-50 μm) with unattractive permeance.[2]

In this work, we present formation of CMS hollow fiber membranes with ultrathin skin (separation) layers for gas and vapor separations. By tuning the membrane structure and formation process, we have created advanced CMS hollow fiber membranes with excellently-preserved asymmetry and defect-free sub-micron skin layers. The ultrathin CMS hollow fiber membranes show permeances that are an order of magnitude higher than state-of-the-art CMS hollow fiber membranes for a given precursor material[3], while maintaining highly attractive separation factors. Combining high scalability, high permeance and high selectivity, the ultrathin CMS hollow fiber membranes offer an attractive and uniquely-flexible platform for next-generation membrane-enabled fluid separations.

References

1 Koros, W. J. & Zhang, C. *Nat Mater* **16**, 289-297, (2017).

2 Xu, L., Rungta, M. & Koros, W. J. *J. Membr. Sci.* **380**, 138-147, (2011).

3 Bhuwania, N. *et al. Carbon* **76**, 417-434, (2014).

496624 Engineering the Performance of Carbon Molecular Sieve Membranes through the Use of Oxygen Doping

Abstract:

Carbon molecular sieve (CMS) membranes are formed from the high temperature pyrolysis of polymeric membranes. Due to the molecular sieving nanostructure of CMS membranes, the separation performance has been proven to exceed that of polymer membranes, even those produced from pre-commercial, high performance polymers. The separation performance of CMS membranes is influenced by a variety of processing parameters including the precursor polymer composition, the pyrolysis temperature profile, and various pre-and post-treatments This study will review the discovery and development of another tool for engineering the nano-structure of CMS membranes through the use of oxygen-doping. This method has been developed and utilized to tune the microstructure of CMS membranes for a variety of industrially important gas pairs including CO2/CH4, O2/N2, and ethane/ethylene.

495834 Enhanced Transport Properties of Metal-Organic Framework Mixed-Matrix Membranes for Gas Separation

Abstract:

Membrane gas separation technology offers energy-saving and smaller footprints compared to traditional thermally driven gas amine absorption processes. Development of high-performance membrane materials is a key component of membrane technology implementation. However, polymeric membranes, the dominant commercial membrane products for gas separation, are limited by a performance due to trade-off between permeability and selectivity. Although inorganic membranes exhibit much higher gas separation performance, their scalable fabrication remains a challenge. On the other hand, mixed-matrix membranes (MMMs) offer an excellent balance between separation performance and scalability, driven by highly selective and permeable molecular sieve particles dispersed in an appropriate polymer matrix. Such MMMs offer both high separation performance assisted by high performing sieves, and economical fabrication due to reliance on simple extension of processing approaches used for pure polymeric membranes. Purely inorganic zeolite generally shows poor compatibility with organic polymer and often requires additional surface modification to eliminate non-selective interfacial voids. Alternatively, metal-organic frameworks (MOFs) consisting of metal ions/clusters connected by organic linkers are another important class of crystalline molecular sieve porous materials, can be a suitable candidate for MMMs, owing to the diverse functionalities and tunable pore structures.

By matching the transport properties and interphase compatibility between sieve and polymer, we developed a diverse array of MOF MMMs by incorporating different kinds of MOF crystals (e.g., UiO-66, RE-**fcu**-MOF) in various polymers (e.g., 6FDA-polyimides, PEBAX) for natural gas purification, butane isomers separation and CO_2 capture. Transport properties were systematically studied via interpreting the sorption and diffusion behaviors. Theoretical framework was employed to predict permeation properties of pure MOF membrane and MOF MMMs using other polymer matrices. Pure-gas and mixed-gas permeations showed that both permeability and selectivity of the polymeric membranes were highly improved by rationally incorporating MOF crystals. The performance of the resulting MOF MMMs were beyond the upper-bounds for traditional polymeric membranes. We demonstrated that MMM is a promising platform for providing high-performance and scalable gas separation membranes, as well as a powerful tool for studying the transport properties of pure MOF membrane.

503403 Carbon Molecular Sieve Membranes As Enablers for Organic Solvent Reverse Osmosis

Abstract:

We created free-standing carbon molecular sieve membranes that translate the advantages of reverse osmosis for aqueous separations to the separation of organic liquids. High-performance membranes derived from carbon materials have shown excellent chemical resistance, high molecular selectivity and fast mass transport across the membrane. Carbon molecular sieve (CMS) membranes with tailored ultramicropore and micropore dimensions show both high processability of polymeric membrane and the high selectivity of inorganic membranes. CMS membranes have been proven to be effective in a variety of gas separation processes such as olefin/paraffin separation, natural gas separation and air separations. However, the low fluxes observed in CMS hollow fibers (due to porous substructure collapse during pyrolysis) hinder scale-up of CMS membranes for industrial separation applications. The performance of carbon molecular sieve separation membranes, which exploit the effect of mass transport across a selective diffusion barrier to separate molecules, can be improved by reducing the thickness of the membrane. Here, a novel post-spinning cross-linking treatment is introduced to maintain the asymmetric porous morphology in CMS hollow fiber membranes even after high temperature pyrolysis processes. A spinning process was used to prepare asymmetric poly(vinylidene fluoride) (PVDF) hollow fibers with thin selective skin layers. Cross-linking of as-spun hollow fibers "locked-in" the porous substructure and the fiber morphology remained unchanged after the cross-linking process. Moreover, this technique effectively prevented substructure collapse in asymmetric CMS hollow fiber membranes. We show that these membranes operate in "organic solvent reverse osmosis" (OSRO) separation modalities and purify p-xylene at room temperature without requiring any phase change.

488459 Carbon Molecular Sieve Structure Development and Membrane Performance Relationships

Abstract:

Characterization techniques beyond microscopy, scattering and spectroscopy approaches are needed to understand and improve sub-angstrom discrimination between penetrants in carbon molecular sieve (CMS) membranes. Here we use a method based on molecular scale gas diffusion probes to understand relevant membrane properties at the required level of detail. We further use this method to consider hypotheses about the evolution of structure responsible for fundamental properties of CMS materials derived from a high performance CMS precursor polymer, 6FDA:BPDA-DAM. While 6FDA:BPDA-DAM derived CMS membranes display a ~230% improvement in CO_2 permeability when compared to Matrimid® derived CMS formed under the same conditions, the CO_2/CH_4 diffusional selectivity for these two materials are very similar at 35 and 38.5, respectively. Furthermore, the performance can be improved by applying techniques such as oxygen doping. These results indicate a non-trivial connection between CMS precursor material structure and resulting performance. Linking hypotheses about structural changes likely to occur during pyrolysis with the probe data provides insights regarding transformation of the random coil polyimide into CMS, with exquisite size and shape diffusion selectivity. The results provide a framework for understanding and tuning properties of this special class of materials with important technological advantages in energy-intensive gas separations.

References

1. Rungta, M.; Wenz, G.B.; Zhang, C.; Xu, L.; Qiu, W., Adams, J.S.; Koros, W.J.; Carbon 115 (2017) 237-248
2. Rungta, M.; Xu, L..; Koros, W.J.; Carbon 85(2015) 429-442

503670 Digital Alchemy for Assembly Engineeringc

Abstract:

Back in the day, Stuart Churchill was a pioneer in the use of computers for modeling chemical and thermal processes. Since that time, the role of computing in chemical engineering has become indispensable. In this talk, we present new computational methods we call "digital alchemy" that allow for predictive modeling of thermodynamic assembly, and permit the design of building blocks optimized for assembly into target structures with desired properties. In this approach, standard thermodynamic ensembles are generalized to include alchemical variables and their thermodynamic conjugates, allowing for rapid, real-time computational investigation of large, complex design spaces.

503671 Improved Algebraic, Numerical, and Graphical Representations in Fluid Mechanics

Abstract:

The concept of equivalents is applied to fully developed flows through straight channels, and two applications are introduced: the equivalents that produce equal rates of flow through channels of different cross-section, and the diminishment that occurs between the regimes of laminar and turbulent flow. The validity of the eddy diffusivity and the invalidity of the mixing-length as measures of the shear stress due to turbulence have perhaps never before been demonstrated so clearly. The possible choices of primary variables and the consequences are also examined. The objective of these explorations is an improved understanding of the basic fluid mechanics by teachers, students, and industrial practitioners.

503672 The Scaling of Turbulence Near the Wall and the Churchill Turbulent Flux Correlation: Insights with Lagrangian Simulations

Abstract:

The evaluation of mean velocity and mean temperature in wall-bounded turbulent flows has been based on the "classical" theory of turbulence and the law of the wall. The viscous wall parameters, namely the friction velocity, the friction temperature, and the kinematic viscosity of the fluid have been used to scale not only the mean values of velocity, but also other mean turbulent quantities close to the wall, including the Reynolds stresses and the turbulent velocity and temperature fluctuations. Stuart Churchill, as he questioned the validity of the Reynolds analogy between momentum and heat or mass transfer in his works [1], he realized that extending the scaling of momentum to the scaling of scalar quantities in turbulence was not a sound approach. He proposed a scaling of turbulence that depends on the turbulent momentum flux and the turbulent heat flux. In his AIChE Institute Lecture (in Miami, 1998), Churchill presented a summary of his work on this topic, and closed form solutions for the mean velocity and temperature [2]. The turbulent Prandtl number, the most significant parameter for the modeling of turbulent transport not only in theory but also in modern CFD commercial software packages, was also interpreted physically based on these ideas [3]. We have used direct numerical simulation results and results from Lagrangian simulations [4] for the transport of passive scalars in turbulent channel flows to explore Churchill's ideas on turbulent flow scaling and to develop an updated theory for the scaling of heat transport in turbulent flows [5-7]. His passion to remove empiricisms and approximations from the development of predictive correlations in turbulent transport has been an inspiration. In this talk, we will revisit the Churchill scaling in view of our simulation results, and will discuss the counterintuitive finding that wall turbulence can be used to separate rather than mix substances under special circumstances [8].

1. Churchill, S. W., 1996, "A Critique of Predictive and Correlative Models for Turbulent Flow and Convection," Ind. Eng. Chem. Res., 35, pp. 3122–3140.
2. Churchill, S.W., 2000, "Progress in the Thermal Sciences: AICHE Institute Lecture," AIChE Journal, 46(9), pp. 1704-1722.
3. Churchill, S. W., 2002, "A reinterpretation of the Turbulent Prandtl Number," Ind. Eng. Chem. Res., 41, pp. 6393–6401.
4. Papavassiliou, D.V, 2002, "Turbulent transport from continuous sources at the wall of a channel," Int. J. Heat Mass Transfer, 45(17), pp. 3571-3583.
5. Le, P.M., and Papavassiliou, D.V., 2008, "On the scaling of heat transfer using thermal flux gradients for fully developed turbulent channel and Couette flows," Int. Commun. Heat Mass Transfer, 35(4), pp. 404-412.
6. Srinivasan, C., and Papavassiliou, D. V., 2011, "Prediction of Turbulent Prandtl Number in Wall Flows With Lagrangian Simulations," Ind. Eng. Chem. Res., 50(15), pp. 8881–8891.
7. Srinivasan, C., and Papavassiliou, D.V., 2013, "Heat transfer scaling for wall bounded turbulent flows," Applied Mechanics Review 65(3), Art. 031002.
8. Nguyen, Q., Srinivasan, C., and Papavassiliou, D.V., 2015, "Flow induced separation in wall turbulence", Phys Rev E, 91, Art. 033019.

503674 Flow Boiling Using a Piranha Pin Fin Heat Sink

Abstract:

Flow Boiling using a Piranha Pin Fin Heat Sink

Corey Woodcock, Xiangfei Yu, Yoav Peles[1] and Joel Plawsky

Rensselaer Polytechnic Institute; [1]University of Central Florida

The *Piranha Pin Fin* (*PPF*) microdevice is a heat exchanger engineered to dissipate extreme thermal loads using either dielectric fluids such as HFE7000 or water. Fabrication of the device is CMOS compatible and the first generations of the device have been fashioned from silicon substrates. The PPF operates on the principle that heat transfer would be more efficient if heated fluids could be removed nearly as fast as they are created. This process would keep the driving force as high as possible. Thus the heat exchanger is built as two layers; a primary substrate that dissipates the bulk of the heat and shunts the heated fluids to a booster section that continues to heat the fluid and extends the total surface area of the device. Three generations of devices were fabricated. The first two generations were single-layer systems designed to help understand the operation of the primary layer and optimize the shape and location of the pin fins. The third generation was the first, two-layer system to be fabricated and tested.

Figure 1 Representative PPF array

Correlations for the heat transfer coefficient, h_{tp}, based on the Boiling Number and the Weber Number in the PPF design was developed and validated with Gen II. The discrepancy of heat transfer coefficient between predicted and measured values was within 20%. The ratio of two-phase to single-phase pressure drop was correlated with the Boiling Number (Bo) and the Jakob Number (Ja). The discrepancy of the two-phase multiplier between prediction and experimental measurements were within 12%.

The third generation *PPF* heat sinks were the first series to provide both a primary substrate populated with *PPF*s, as well as a booster substrate populated with solid micro pin fins. The substrates are micro-aligned and fusion-bonded to provide a full three dimensional (3D) architecture with bulk silicon properties. Experimental results from the Gen III *PPF* microdevices provide a proof of concept of the true *PPF* heat sink concept with fluid venting into multiple substrates. At low flowrates (G = 670 [kg/s/m^2]) the Gen III *PPF* heat sink has been shown to dissipate heat loads approaching 1.1 [kW/cm^2] at an exit quality of 95%, while at higher flowrates (G = 3,300 [kg/s/m^2]) the thermal limit of the heat sink remains unknown due to exceeding the available power supply at >1.4 [kW/cm^2].

503673 Transport Problems in the Spirit of Stuart Churchill for Teaching and Research at the University of Michigan

Abstract:

Stuart Churchillâ€™s solutions to important heat transport problems have greatly influence both pedagogy and research. The correlation of Churchill and Chu for the natural convection heat transfer to or from a vertical plate, over an astonishing range of Rayleigh numbers is based partly on a similarity solution of a type common in heat, mass, and momentum transfer for problems in chemical engineering. One such problem is asphaltene or wax transport and deposition onto the wall of an oil pipeline, which can be solved using similar methods, as will be discussed. In addition, such problems, and the wealth of their applications make for excellent motivations and examples in the teaching of graduate or undergraduate transport, as will be discussed in the talk.

500861 Screening, Purification and Concentration of Membrane Proteins

Abstract:

Screening, purification and concentration of membrane proteins

Membrane proteins (MPs) are important targets for pharmaceutical products and for the design of novel sensors, synthetic membranes and bioelectronics devices. The hydrophobic exterior of MPs necessitates their extraction and stabilization in detergent micelles. Low-yield MPs are typically concentrated by ultrafiltration (UF) in commercially available centrifugal filers which is accompanied by concomitant increase in concentration of detergent micelles. Excessive detergent can interfere with stability and characterization of MPs. Here, we discuss two methods of obtaining high concentration of MPs without increasing detergent concentration using light-driven sodium and chloride transporters, KR2 and halorhodopsin (pHR), as model proteins. From an earlier study, we know that concentration polarization (CP) during UF promotes detergent passage while retaining MPs. Thus to promote CP, we designed a flat-bottomed centrifugal filter unit that retrofits into commercially available falcon tubes and tested its performance using four commonly employed non-ionic MP detergents, octyl-beta-D maltoside (OM), decyl-beta-D maltoside (DM), dodecyl-beta-D maltoside (DDM) and octylÐbeta-D glucoside (OG). Detergent transmission was significantly greater for our new flat-bottomed filter compared to currently available modules due to the higher degree of CP. Alternately, increasing the extraction of MPs from lipid bilayers using longer alkyl chain detergents and higher salt concentration is another means of improving yield. We attribute this improvement in extraction to the increase in micellar hydrophobicity with chain length and salt concentrations; this was verified using StokeÕs shift of the solvatochromic chromophore, DSSN+. However, while longer chain detergents show better extraction, they also form larger micelles that are difficult to remove by UF. Thus, there exists a trade-off between choosing a detergent that will enable efficient extraction of MP while showing easier removal during subsequent ultrafiltration. To screen and optimize aqueous detergent and salt/ buffer conditions for histidine-tagged MP purification, we used nickel-functionalized membranes in a 96 well plate format. In addition to MP downstream processing, our findings can be extended to other detergentÐmediated separation/extraction processes.

Figure. A. Detergents play an important role in membrane protein (MP) solubilization, purification and post-purification processing, i.e., concentration for subsequent characterization assays. Detergents with increasing chain length promotes MP extraction but also lead to simultaneous concentration of detergent during ultrafiltration. **B.** Ultrafiltration of different detergent and membrane protein-detergent systems show higher transmission of detergent in flat (designed) versus commercial centrifugal filters.

501411 The Effects of Filtration Condition on Virus Clearance

Abstract:

Validation of adequate virus clearance is critical in order to obtain FDA approval for a manufacturing process. Clearance can be due to inactivation or removal from the process stream. During the purification, manufacturers of monoclonal antibodies must demonstrate reduction of 10^3 to 10^5 or more virus particles than is estimated in a single dose equivalent of the unprocessed bulk. Depending on the antibody titer, estimates of the number of virus particles in a single dose equivalent could be as high as 10^{10} to 10^{15} retrovirus-like particles per mL. In addition to the removal of adventitious virus, using model parvovirus is required. Today unit operations such as virus filtration are frequently used to validate virus clearance.

Virus filtration is conducted industrially in dead end mode. Membrane performance is shown to be affected by feed pH, ionic strength and feed buffer components. In this work, one of the FDA required model viruses, minute virus of mice (MVM) was spiked and used to investigate the effects of feed condition including pH, buffer and ionic strength on membrane performance and virus clearance using commercial virus filters in the presence and absence of model proteins. The interactions and possible associations among virus particles, model proteins and membrane materials modulated by the feed conditions were investigated. Their effects on membrane filtration performance and MVM clearance were elucidated.

492433 Modification of Polycarbonate Membranes with EDC/NHS Coupling: The Impact of Electrostatic Particle-Pore Interactions on Rejection

Abstract:

It has long been recognized that electrostatic and other long range particle-pore interactions can play a dominant role in MF and UF separations involving biological systems. In this work, the impact of physicochemical interactions between particles and porous membranes has been examined from both experimental and theoretical perspectives. To facilitate interpretation of results in terms of fundamental membrane and pore characteristics, experimental work involved model systems comprised of polycarbonate track etch membranes with well-defined pore geometry and monodisperse polystyrene particles.

Commercially available polycarbonate track etched membranes and polystyrene particles both carry a strong negative surface charge. To create an experimental system with different particle-pore wall interactions, membranes were modified using an EDC/NHS coupling reaction with tetraethylpentamine, creating a set of membranes with reduced surface charges and no measurable change in pore size. Streaming potential measurements were used to determine pore surface characteristics. The rejection of negatively charged red dyed spherical polystyrene particles was measured using a 10 ml dead-end filtration cell with both modified and non-modified membranes. The impact of flow rate, rotational stir speed and feed concentration on measured rejections was examined for both modified and non-modified membranes.

When compared to particle rejections observed with non-modified membranes with the same pore size, larger rejections (up to 36% increase) were observed with the modified membranes. The pore surface modification reduces electrostatic repulsion between particle and membrane, resulting in an increase in particle deposition on the membrane surface. This explanation is supported by visual examination of the membranes which shows a stronger red color on the modified membranes at the end of filtration. Examination of the impact of flow rate and stir speed on rejections illustrate the complexity of the particle-pore wall interactions for these systems.

A macroscale model of the filtration cell (using the k-epsilon turbulent flow model) was developed and coupled with a Brinkman model of the membrane and a microscale model of a single pore. Model results were used to track particle trajectories as they flowed from the reservoir outside the membrane into the pore. In this novel modelling approach, the kinetic energy of a particle at the pore entrance was calculated and coupled with steric and electrostatic particle-pore wall repulsive forces to examine particle concentration profiles within the pore and to estimate the probability of particle deposition to the membrane surface. Model predictions were found to be consistent with experimentally measured rejection values.

The results from this study provide insight into the impact of electrostatic and hydrodynamic particle-pore wall interactions on particle trajectories and resulting rejections from porous membranes. The results as well as the impact

of these interactions on membrane design considerations will be discussed in this presentation.

496591 Smart Nanogel-Containing Membranes in Microchip for Temperature- and Ethanol-Responsive Permeability Regulation

Abstract:

Recently, integration of membranes in microchips has shown great potential for biomedical fields such as analysis, separation, microreaction and cell-based study, due to their portability and lower consumption of sample and energy. In particular, microchips containing smart membranes that can change their permeability in response to external chemical/physical stimuli are important and necessary, as they can be used as sensors, separators and controlled release systems. Here we report on novel types of smart in-chip membranes containing stimuli-responsive nanogels as nanovalves for self-regulation of permeability. Such smart in-chip membranes can be integrated into microchips by assembly of a pre-fabricated nanogel-containing membrane into microchannels, or *in situ* fabrication of a nanogel-containing membrane in the microchannel. To demonstrate the first strategy for fabricating smart in-chip membranes, a polyethersulfone membrane, prepared by blending poly(*N*-isopropylacrylamide) (PNIPAM) nanogels into the membrane-forming solution via vapor-induced phase separation process, is sandwiched between two polydimethylsiloxane modules of a microchip. To demonstrate the second strategy, a chitosan membrane containing PNIPAM nanogels, is *in situ* fabricated in microchannels, based on crosslinking reaction at the interface between two parallel laminar flows. The PNIPAM nanogels, that allow temperature-responsive and ethanol-responsive swelling/shrinking volume transitions, serve as smart nanovalves for controlling the diffusional permeability of solutes across the membrane. Based on this, both membranes can show excellent temperature- and ethanol-dependent permeability, rapid responsive rate, and good repeatability and stability. These studies provide promising strategies for creation of versatile nanogel-containing smart membranes within microchips via simply changing the functional nanogels for developing micro-scale detectors, sensors, separators and controlled release systems.

504861 Membrane Water Treatment - Process Design Considerations and Economics

Abstract:

Although two-thirds of the planet's surface is covered by water, most of this water is saline and requires some treatment before use. The growing global population, and economic growth, both drive up the demand for water. After use, there is additional need for water treatment. This is growing in importance due to both stringent discharge requirements and increasing competition for water. Water treatment processes are at the nexus of the factors described above. There is a lot of variety in the design of these processes but there are some common considerations. This presentation would focus on design considerations for membrane-based processes (e.g. ultrafiltration, reverse osmosis). One of the design considerations is economics in the form of capital and operating costs. This presentation also describes cost associated with procurement and disposal of water.

500159 Molecular Simulations to Inform Polymeric Membrane Science and Technology

Abstract:

Molecular dynamics simulations offer the opportunity to address key knowledge gaps in Angstrom-scale transport in polymeric membranes. These simulations can illuminate atomic level details for membrane/solvent/solute interactions governing membrane performance and provide new ways to model how the tortuous transport pathways, associated energy landscapes, and membrane charge structure influence solvent flux and solute rejection. However, there are many challenges. Models of polymeric membranes that accurately reflect the membrane molecular structure and still provide length and time scales that can represent real membrane processes are difficult. Appropriately representing the interactions between the membrane, solvent, and solutes sometimes requires unique modeling methods. Accurately representing the non-equilibrium physical processes that govern transport through a membrane requires special approaches. The vast number of calculations necessary to simulate even a few nanoseconds of transport through a membrane is computationally daunting. And finally, extracting meaningful information from the simulations that can lead to a better understanding of solute and solute transport through a membrane is nontrivial. While the focus of this talk will be on molecular dynamics simulations of polymeric nanofiltration and reverse osmosis membranes, the general principles can be applied to membranes more broadly. The ultimate goal is to use molecular-level simulations of polymeric membranes to not only broaden the understanding of membrane transport processes, but also to extend the approach to achieve "membrane-by-design" molecular structures for specific applications of membrane technology.
Funded by the Institute for Sustainability and Energy at Northwestern with computing resources from XSEDE, which is supported by NSF grant ACI-1053575.

504863 Tutorial on Membrane Separations Technology for the Biopharmaceutical Industry

Abstract:

Membrane processes play a critical role in the purification of biopharmaceuticals. This presentation will review current applications of membrane technology in downstream processing, including clarification of cell culture fluid, bioburden reduction, virus removal filtration, sterile filtration, and ultrafiltration for final product concentration and formulation. Specific emphasis will be placed on the special characteristics of the membranes and modules used for commercial production of biotherapeutics. In addition, increasing demand for biopharmaceutical products and competition from biosimilars has generated interest in transitioning to continuous biopharmaceutical processes, creating exciting new opportunities for development of membrane technology for bioprocessing.

504856 A Methodology for the Evaluation of Membrane Robustness and Lifetime

Abstract:

After determining whether a membrane is able to conduct a needed separation, economic factors are normally key drivers in the selection of a UF or MF membrane solution. Typically this would include the number of modules required, required pressures and the associated energy costs, as well as chemical costs for any required chemical dosing and cleaning. As a result much attention is paid to a membraneâ€™s surface chemistry, charge, and roughness (cleanability), the feed side fluid dynamics (sustainable flux/number of elements), and the structure and morphology of the separating layer (permeance and operating pressure). As a result of this, introductions of new membrane products often feature deep dives into these well-travelled aspects of membrane science. Missing in this analysis is a discussion of membrane lifetime. This may be partly attributable to the fact that each system ends up seeing a unique combination of conditions and stresses that can lead to a multitude of potential failure modes.

This paper will present a process that can be used to validate membrane lifetime for a new membrane product without known end of life examples. Application of this method to a new ceramic ultrafiltration module will be discussed including the identification of specific criteria used to quantify the end of life point, determination of potential stresses that could result in failure, and development and execution of accelerated tests to probe each potential stress as well interactions amongst the stresses. Also to be discussed are models relating the accelerated conditions with conditions to be experienced in normal operation. Stresses to be discussed include temperature, pressure, vibration, and chemical exposure.

502799 Membrane Processes Utilizing Coupled Heat and Moisture Transfer

Abstract:

This presentation will review fundamentals and recent research on thermally-driven membrane processes, which involve coupled heat and mass transfer. These processes often include a vapor/liquid phase change, where this fluid is removed from (or added to) a liquid or gas stream. Understanding the coupled heat and mass transfer is important for modeling, designing, and/or selecting devices and systems for thermally driven processes. Applications that use coupled heat and mass transfer include traditional separation processes, like liquid purification (e.g., membrane/osmotic distillation, pervaporation), but also processes relevant to the heating, ventilation, and air conditioning industry, such as dehumidification of air streams (e.g., liquid-desiccant air drying), and processes where heat is the desired output (e.g., absorption heat pumps).

Membrane and osmotic distillation use a vapor-pressure differences to drive a vapor (often water) from one liquid to another. Pervaporation is a similar process, but relies on a vapor-pressure gradient between a liquid and gas stream, often set up by reducing the pressure of the gas.

Liquid desiccant air drying has received considerable attention in recent years as an efficient method for dehumidifying air. It relies on a liquid-to-air membrane contactor, where the liquid desiccant is a low-vapor-pressure solution that absorbs moisture from the air. Heating the desiccant in a liquid-to-air membrane contactor drives vapor out of the desiccant, regenerating it so it can be used for drying again.

Absorption heat pumps traditionally use a closed system with ammonia-water or LiBr-water working fluids. These closed systems operate under pressures either much higher or lower than ambient. Alternative systems using membranes allow for semi-open or fully-open heat pump processes, which can be advantageous in certain cases.

The presentation will review these processes, their key modeling aspects, recent research, and future research needs.

504865 Transport Processes in Batteries: A Synergistic Research Direction for Membrane Scientists

Abstract:

From electric and plug-in hybrid vehicles gaining a foothold in the automotive market to the 787 airline and Galaxy Note 7 battery fires, battery technology has, for better or worse, found itself in the popular spotlight in recent years. This spotlight is likely to shine increasingly brighter, as improvements in lifetime and size of portable electronic devices will make batteries and battery research more prevalent in mobile and stationary (e.g., distributed storage of wind and solar energy) applications in the coming decade. Nevertheless, the 787 and Galaxy Note 7 incidents highlight an interesting dichotomy in battery research: although decades of development have allowed rechargeable batteries to be used in advanced applications, our understanding of how to design a safe, high energy density, low cost rechargeable battery still needs to be improved. Designing materials that allow improved mass and charge transport processes, while maintaining low cost and high safety, is critical to this cause. This talk will provide a basic introduction to battery technology, focusing on the most common battery chemistry, Li-ion. The second part of the talk will outline the importance of ion transport properties through the batteryâ€™s primary components, namely its porous electrodes and separator, particularly highlighting the challenges and opportunities that exist for polymer and ceramic-based ion conductors.

497394 Role of Electrokinetics in Glomerular Capillary Filtration: Toward an Artificial-Kidney

Abstract:

According to the U.S. Department of Health and Human Services, 1 in 10 people have some form of chronic kidney disease (CKD) worldwide. CKD has no cure, kidney transplants are accompanied by long wait times and the possibility of tissue rejection, and dialysis fails to provide a good quality of life. An artificial kidney is an ideal solution to combat suffering due to kidney disease and failure. As a path to an artificial kidney, the micro-rheological properties of blood in the glomerular capillary are analyzed at the microscopic scale based on continuum mechanic approaches. Once the proper understanding of the microscopic behavior is attained (with the possible impact on the macroscopic scale), the system will be downscaled to a size ideal for potentially designing an artificial kidney. As the system is downscaled the importance of electrokinetics comes into question in a more detailed manner. In fact, the role of electrokinetics in kidney function (as well as its malfunction) has come into consideration recently. A glomerular capillary has three main layers of which have various characteristics that determine their filtering specificity due to size and charge. As a preliminary step toward a more detailed understanding, in this presentation, a single capillary of cylindrical geometry will be used as a potential domain for a â€œnephronâ€- the most crucial element of the kidney filtration. Principles of EKHD (see Pascal et al., 2016) will be used in modeling the glomerular capillary domain to understand the role of electrokinetics in the filtration efficiency.Â The role of electrokinetics in influencing the separation performance of the capillary-nephron domain under a variety of operating conditions will be considered in the development of an asymptotic solution for separation (filtration) efficiency.

496314 Patterning Various Commercial Nanofiltration and Reverse Osmosis Membranes

Abstract:

Progressive decline in permeability due to fouling is one of the largest costs associated with membrane processes in water treatment. Innovations in membrane surface modification that limit fouling would have significant economic and societal impacts. Prior studies in this area largely have focused on chemical modifications to the membrane surface. Based on considerable literature and experience, we know that chemical modification strategies alone can be effective but not sufficient for controlling fouling. The discovery that patterning of a membrane surface can improve its fouling resistance provides an orthogonal membrane design parameter. Our hypothesis is that combining chemical modification and patterning will yield membrane surfaces that are more effective at fouling control than either method alone. This strategy is unique in its use of both chemical coating and patterning to combat fouling.

This contribution describes our method to apply a wide variety of patterns to commercial nanofiltration and reverse osmosis membranes produced using various chemistries and coatings. The goal is to better understand what factor(s) contribute to differences in patterning of thin-film nanocomposite membranes, leading to a set of heuristics that can be used by the membrane community for patterning, which is becoming a hot topic in the field. Direct patterning of 15 commercial thin-film nanocomposite membranes was done by thermal embossing using nano- and micropatterned stamps. Patterning studies began by embossing all membranes at one set of fixed conditions followed by AFM characterization. These initial studies uncovered the strong membrane chemistry dependence on pattern replication that sparked a more comprehensive study using controlled conditions of force and time with varying temperatures. AFM, SEM, and laser scanning microscopy (LEXT) were used to determine the extents of patterning. AFM and LEXT were used to determine the local strain as a function of embossing conditions. Membrane performance was evaluated by measuring flux and salt rejection using a magnesium sulfate solution. Flux and salt rejection properties were compared to local strain values and Youngâ€™s Modulus values of the patterned membranes to establish correlations.

493270 Preparation of ZIF-8 Membranes Supported on Polymer Hollow Fibers Using Microwave-Assisted Seeding and Secondary Growth Method

Abstract:

Preparation of ZIF-8 Membranes supported on Polymer Hollow Fibers using Microwave-assisted Seeding and Secondary Growth Method

Moon Joo Lee[1], Mohamad Rezi Abdul Hamid[1], Jongmyeong Lee[3], Ju Sung Kim[3], Young Moo Lee[3] and Hae-Kwon Jeong*[1,2]

[1]Artie McFerrin Department of Chemical Engineering and [2]Department of Materials Science and Engineering, Texas A&M University, College Station, TX 77843-3122

[3]Department of Energy Engineering, College of Engineering, Hanyang University, Seoul 133-791, Republic of Korea

* Corresponding author: hjeong7@tamu.edu

Separation of olefin/paraffin mixtures, e.g. propylene/propane, is one of the most challenging separations due to the similarity of their physicochemical properties. Currently, an energy intensive cryo-distillation process is employed. Though energy-efficient membrane-based separation processes are drawing significant interest, no membranes are commercially available for olefin/paraffin separations mainly due to the limitations of current membrane materials and processing methods to fabricate their membranes with large membrane areas.[1-3]

Zeolitic-imidazolate frameworks (ZIFs), a sub-class of metal-organic frameworks (MOFs) with zeolite topology, are highly crystalline nanoporous materials with aperture sizes of typically less than 5 Å. ZIFs composed of metal ions (typically Zn^{2+} and Co^{2+}) and imidazole-derived organic linkers,[4] have been extensively studied for gas separation applications. Specifically, ZIF-8 is shown very promising as a new membrane material for propylene/propane separation because the effective pore size is in between the size of propylene and propane. While several ZIF-8 membranes showed high propylene/propane separation factors, however, there are critical issues for their commercial applications: 1) high cost of membranes and 2) low productivities. To increase the productivity and to reduce the cost of ZIF-8 membranes, polymeric hollow fibers have been utilized as substrates, but only a few of the resulting ZIF-8 membranes showed promising propylene/propane separation performances.[6-9]

Herein, we would like to present a microwave-based method to prepare well inter-grown ZIF-8 membranes on polymeric hollow fibers readily. The method is modified based on our previously reported microwave-assisted seeding and secondary growth technique.[10] Densely-packed nano-sized ZIF-8 seed crystals were rapidly prepared on either inner or outer surface of hollow fibers under microwave heating. Subsequently, seed crystals were secondarily grown into continuous ZIF-8 films on polyimide hollow fibers. The thickness of the resulting ZIF-8 membranes was much smaller than any ZIF-8 membranes grown on polymer hollow fibers reported, which are several micrometers in thickness. The effects of synthesis parameters and fiber morphologies on the microstructures of ZIF-8 membranes will be discussed. Finally, the gas separation performances of ZIF-8 membranes will be presented.

References and Notes

1 Koros, William J. and Fleming, G.K. (1993) "Membrane-based gas separation", *Journal of Membrane Science*, 83, pp 1-39

2 Baker RW., (2002) "Future Directions of Membrane Gas Separation Technology" *Industrial & Engineering Chemistry Research.*, 41 (6), pp 1393-1411

3 Tanaka, K., Taguchi, A., Hao, JQ, Kita, H., and Okamoto, K., (1996) "Permeation and separation properties of polyimide membranes to olefins and paraffins", *Journal of Membrane Science*, 121, pp 197-207

4 Park KS, Ni Z, Côté AP, et al., (2006) "Exceptional chemical and thermal stability of zeolitic imidazolate frameworks", *Proceedings of the National Academy of Sciences*, 103 (27), pp 10186-10191

5 Zhang C, Lively RP, Zhang K, Johnson JR, Karvan O, Koros WJ., (2012) "Unexpected Molecular Sieving Properties of Zeolitic Imidazolate Framework-8", *The Journal of Physical Chemistry Letters*., 3 (16), pp 2130-2134.

6 Andrew J. Brown, Nicholas A. Brunelli, Kiwon Eum, Fereshteh Rashidi, J. R. Johnson, William J. Koros, Christopher W. Jones, Sankar Nair, (2014) "Interfacial microfluidic processing of metal-organic framework hollow fiber membranes", *Science*, 345, 6192, pp 72-75

7 FernandoCacho-Bailo, SilviaCatalán-Aguirre, MirenEtxeberría-Benavides, Oguz Karvan, VictorSebastian, CarlosTéllez, JoaquínCoronas (2015) "Metal-organic framework membranes on the inner-side of a polymeric hollow fiber by microfluidic synthesis", *Journal of membrane science*, 476, pp 277-285

8 Kiwon Eum, Chen Ma, Ali Rownaghi, Christopher W. Jones, and Sankar Nair (2016) "ZIF-8 Membranes via Interfacial Microfluidic Processing in Polymeric Hollow Fibers: Efficient Propylene Separation at Elevated Pressures", *ACS Applied materials and interfaces*, 8, pp 25337 – 25342

9 Anne M. Marti, Wasala Wickramanayake, Ganpat Dahe, Ali Sekizkardes, Tracy L. Bank, David P. Hopkinson, and Surendar R. Venna (2017) "Continuous Flow Processing of ZIF-8 Membranes on Polymeric Porous Hollow Fiber Supports for CO2 Capture", *ACS Applied materials and interfaces*, 9, pp 5678-5682

10 Kwon HT, Jeong H-K., (2013), "Highly propylene-selective supported zeolite-imidazolate framework (ZIF-8) membranes synthesized by rapid microwave-assisted seeding and secondary growth", *Chemical Communications*, 49, pp 3854-3856

495664 A Zeolitic Imidazolate Framework (ZIF-8) Film for H_2/CO_2 Separation

Abstract:

A Zeolitic Imidazolate Framework-8 (ZIF-8) Film for H_2/CO_2 separation

Eunhee Jang[1], Jungkyu Choi[1*]

[1] *Department of Chemical & Biological Engineering, Korea University, Seoul 02841, Republic of Korea*

ehphoenix@korea.ac.kr, jungkyu_choi@korea.ac.kr

Zeolitic imidazolate framework-8 (ZIF-8) membranes are highly suitable as H_2 selective walls in membranes reactors used for water gas shift reactions because of their efficient molecular sieving properties that favor H_2 transport, coupled with their high thermal stability and processability. In this study, we adopted an *in-situ* counter diffusion method (CD method) to fabricate a ZIF-8 membrane; Zn sources, already placed inside a porous support, were allowed to diffuse out and react with the 2-methylimidazole (mim) ligands in the bulk phase. Because the reaction rates between the Zn source and the mim ligand were very high, their diffusion rates played a key role in determining the final properties of the membranes. To exploit the effect of diffusion rate on preparation of ZIF-8 membrane via the CD method, a hierarchically structured support, i.e., a gamma alumina layer-coated alpha alumina disc (gamma/alpha alumina disc), was used in addition to an intact alpha alumina disc. Existence of ZIF-8 membrane was demonstrated with SEM, EDX of Zn atom, XRD, and contact angle. The position of ZIF-8 grains in the supports was estimated based on the direct measurement of N_2 physisorption. ZIF-8 membranes in the alpha alumina disc (membrane ZIF-8_A) were primarily formed on top similar to a conventional supported-membrane, whereas those in the gamma/alpha alumina disc (membrane ZIF-8_GA) were embedded inside the modified support. From the N_2 physisorption analysis, it was found that the ZIF-8 grains in membrane ZIF-8_GA was pre-dominantly formed inside the alpha alumina disc, while filling ~70% of the mesopores in the gamma alumina layer. As desired, the embedded membrane ZIF-8_GA showed marked H_2 separation performance with a maximum H_2/CO_2 separation factor (SF) of ~9.9 plus-minus 1.2 at 250 deg C (vs. a maximum H_2/CO_2 SF of ~7.5 plus-minus 0.2 for membrane ZIF-8_A). Although both type membranes persisted at 200 and 250 deg C for up to 72 h, at a higher temperature of 300 deg C, the membrane performance started deteriorating after ~2 h and ~10 h for membranes ZIF-8_A and ZIF-8_GA, respectively. This indicates that the gamma alumina layer served as a protective layer for preserving the performance of the embedded ZIF-8 membrane. Their performance was completely degraded due to the eventual conversion of ZIF-8 into ZnO phases.

Figure. 1. H_2/CO_2 separation performance test of ZIF-8-A and ZIF-8-GA at 300 ¡ÆC for 72 h. Here, performance test result obtained for 48 h was expressed. Schematic images for ZIF-8-A and ZIF-8-GA was depicted at the bottom of the performance test.

489268 Scale-up of Electrochemical Carbon Dioxide Separation Using Membrane Electrode Assemblies

Abstract:

Scale-up of multi-cell membrane electrode assembly (MEA) systems require reproducible fabrication of reliable electrodes and membrane materials. Optimization of electrode support and catalyst microstructures ensures a high rate of CO_2 separation through the anion exchange membrane. Separation of CO_2 from a 5–7 mol% CO_2, high-humidity gas mixture, representative of an exhaled breath, was demonstrated by a sub-scale system. Screen-printed Pt and IrO_2 electrodes were fabricated and evaluated as single MEAs to measure baseline CO_2 removal rates and electrochemical behavior. Catalyst microstructures were characterized using scanning electron microscope (SEM) imaging. Electrode mass and charge transfer resistances were quantified using electrochemical impedance spectroscopy (EIS). The electrode efficiency and CO_2 flux were measured over a range of equivalent full-scale breathing flow rates (40 L/min–120 L/min) for 3, 7, 11, and 22 cell stacks at variable current densities. The multi-cell system displayed performance linearity and uniformity with increasing number of cells. For a full-scale flow rate of 80 L/min, the multi-cell system achieved a CO_2 flux of 1.57 $L/m^2 \cdot min$ with an electrode efficiency of 75%. Flow field design and cell hydrodynamics were addressed for scale-up of the electrochemical system.

494146 Iron/Palladium Nanoparticle Functionalized Membrane for Chlorinated Contaminates Treatment

Abstract:

Functionalized polyvinylidene fluoride (PVDF) membrane platforms were developed for environmentally benign in-situ nanostructured Fe/Pd synthesis and remediation of chlorinated contaminates (PCB and TCE). To prevent leaching and aggregation, nanoparticle catalysts were integrated into membrane domains functionalized with poly (acrylic acid). Furthermore, nanoparticles of 16-19 nm were observed inside the membrane pores. The quantification of nanoparticles properties (size, distribution and composition) versus depth underneath the membrane surface was also investigated by cross-sectioning the membrane samples with focused ion beam (FIB). The Fe/Pd nanoparticles immobilized membrane showed excellent performance in the degradation of chlorinated organics and also the repeated degradation experiments were conducted to demonstrate the reusability of Fe/Pd immobilized membranes.

To increase the accuracy of kinetic analysis and further simulation of PCB dechlorination, the correlation between nanoparticle properties and depth inside membrane pores was studied by using FIB. Particle size was uniform inside membrane pores at different depths (particle size: 16.7±0.7 nm) but slightly smaller than those nanoparticles located on the surface (19.4±3.2 nm). Furthermore, the fairly similar surface coverage values (0.169±0.011) demonstrated the evenly distributed nanoparticles inside the membrane pores. Besides, the element composition of particles as well as the functionalized membrane domain were analyzed in line-scan mode by using energy dispersive spectroscopy (EDS).

Over 96% degradation of 3,3',4,4',5-pentachlorobiphenyl was achieved at a residence time of 14.7 seconds in the membrane pores. The ksa was calculated (using LFR model) to be 0.171 L/(m^2h) in convective flow mode, which is 2.5 times the rate obtained in batch mode. The effects of temperature and pH values were also investigated.

In XRD analysis, the Fe/Pd particle samples (which were deliberately oxidized and then reduced) exhibited the same crystalline patterns as the original samples. The membrane was tested for reactivity after four degradation cycles with regeneration between each cycle. The increase of surface particle size of 22% resulted in a decrease of 9.7% PCB conversion for the 4 hr reaction time. [1]

This research is supported by the NIEHS-SRP grant P42ES007380.

Reference:

[1] Wan, Hongyi, L. Ormsbee, and D. Bhattacharyya. "Pore functionalized PVDF membranes with in-situ synthesized metal nanoparticles: Material characterization, and toxic organic degradation." Journal of Membrane Science 530 (2017): 147-157.

494584 *In situ* Growth of MOF Membranes Assisted By Electro-Deposition

Abstract:

Metal–organic frameworks (MOFs) have great potential for use a membranes in separation processes because their pore size can be tuned and they have highly designable structures and adsorption properties.[1] The preparation of supported MOF membranes has attracted increasing attention in separation technology.[2] We develop a facile and efficient electro-deposition method to modify inexpensive porous stainless-steel nets for use as substrates in the in situ growth of metal–organic framework membranes, such as
ZIF-8, ZIF-67 and HKUST-1. Using this method, different metal precursors can be electro-deposited depending on the central metals required in the target metal–organic frameworks. The inorganic modifiers prepared by this approach are sufficiently reactive for the one-step growth of continuous metal–organic framework membrane.

We chose stainless-steel nets (SSNs) as the substrate as they are inexpensive
and easily welded. The preparation concept is shown schematically in Fig. 1. A metal oxide/hydroxide with a morphology exactly corresponding to the metal centres of the targeted MOFs is introduced onto the SSN via electro-deposition. For example, ZnO nanorods, $Co(OH)_2$ nanosheets and Cu_2O nanocubes can be electrodeposited onto substrates quickly and provide active sites for the crystallization of Zn–ZIF-8, Co–ZIF-67 and Cu–HKUST-1,
followed by in situ growth to form continues MOF membranes.

The metal precursors of ZnO nanorods, $Co(OH)_2$ nanosheets and Cu_2O nanocubes are electro-deposited onto SSN respectively as inorganic modifiers. As a representative, the microstructure images of ZnO nanorods are shown in Fig. 2a. These reactivity of these modifiers is excellent because electro-deposition is operated at low temperature and also some intrinsic defect exists. The following growth of three different MOF membranes can be finished in one step without any pre-treatment. The microstructure images of ZIF-8 membranes are shown in Fig. 2b, 2c. The pure phase is confirmed by XRD analysis (Fig 2d). This strategy is proved to be general and all the membranes are tested for gas separation performance (shown in Table 1).

An efficient electro-deposition strategy is proposed for the introduction of diverse metal clusters into inexpensive SSN substrates. These metal clusters universally act as inorganic modifiers for the subsequent facile in situ growth of the MOF
membranes. The fast and mild process of modification assisted by electro-deposition is promising for scaled-up production as well as low-cost manufacturing. The excellent reactivity of these metal oxide/hydroxide precursors arising from electro-deposition plays a vital part in the one-step in situ synthesis of various MOF membranes. The proposed electro-deposition-assisted strategy simplifies the procedure, exhibits excellent generality and can be applied to the preparation of different kinds of metal-centered MOF membranes.

495767 Pd/Ta Composite Metallic Membranes for High Purity Hydrogen Separation: Permeability and Durability

Abstract:

Hydrogen is considered as one of the clean energy carrier, but still, more than 90% of hydrogen is produced from hydrocarbon reforming processes involving a hydrogen purification step. Dense metallic membranes are drawing attention as a high purity hydrogen separation technique either as a stand-alone purification unit or to be used with catalyst as a membrane reactor to replace a series of end processes. Palladium based thin metallic membranes have been intensively developed due to their stable nature, but they are costly and permeability is significantly lower than that of body centered cubic (BCC) metal based membranes . In this study, as an effort to overcome this problem, Tantalum/Palladium (Ta/Pd) composite membrane has been fabricated using electroless deposition of Pd on a chemically treated Ta tube. The amount of palladium is significantly reduced while maintaining high permeability of Tantalum. The fabricated membrane had permeability value higher than that of the best Pd based metallic membrane reported, with hydrogen purity of 99.999% under mixed gas operations including N2, CO2, CO, H2O. It was possible to operate the membrane in a desired temperature range of 400 °C ~500 °C with no failure due to hydrogen embrittlement. Durability behavior of the membrane has been evaluated as a function of operating temperature and hydrogen flux, and optimal operating conditions are suggested as a function of Pd thickness.

501077 Fabrication and Characterization of Silicalite Membranes Subject to Knudsen and Surface Diffusion Transportation Regimes

Abstract:

Inorganic membranes composed of zeolitic crystals inside ceramic membrane supports show particular promise for gas separation applications due their high permeability, chemical stability, and low operating costs. The fabrication of membranes of this type is however very challenging due to the significance of even a small number of defects on membrane performance. An elegant characterization model has therefore been developed which can be used to quantify the physical characteristics of defects in novel membranes based on a limited number of pure gas permeation tests, and evaluate novel membranes. Additional refinements to standard fabrication and testing conditions have also been developed which allow greater control of experimental variables, and are presented in this work.

To demonstrate this characterization model, several batches of silicalite membranes have been hydrothermally synthesized using ceramic membrane supports with active layers whose pore sizes are 0.45 and 0.8 μm. Pure gas permeation experiments were then conducted using these membranes with He, N_2, and CH_4 gases in turn. The transport of these gases through membranes of this type proceed according to the regimes of Knudsen diffusion through defects, and surface diffusion through silicalite crystals (excepting He, which is sufficiently small that it is predominantly transported by Knudsen diffusion through silicalite crystals at standard conditions). The results of these permeation experiments, in combination with previously reported diffusion data and adsorption isotherms have then been used to determine the effective silicalite channel path length, and defect size to length ratio that are unique to each membrane. Through the application of the proposed model, effective silicalite channel lengths and defect ratios for membranes that were fabricated in different batches at similar conditions have been found to vary up to 20% and 700% respectively.

Given these results, characteristic values such as permselectivity can be determined that are normalized to exclude the defect transport of gases, and are therefore expected to be more comparable. Furthermore, this model can be applied to many different inorganic type membranes, and enable more reliable simulations of scaled up membrane processes.

Effects of Cyanuric Chloride and Its Derivatives on Gas Separation Properties of Polyurethane Membranes

A. A. Shamsabadi[1], M. Sadeghi[2], M. Dinari[3], A. Ronasi[2], M. Soroush[1]

[1] Chemical and Biological Engineering, Drexel University, Philadelphia, USA
[2] Department of Chemical Engineering, Isfahan University of Technology, Iran
[3] Department of Chemistry, Isfahan University of Technology, Iran

CO_2>400 ppm | Membrane Technology | Mixed Matrix Membranes

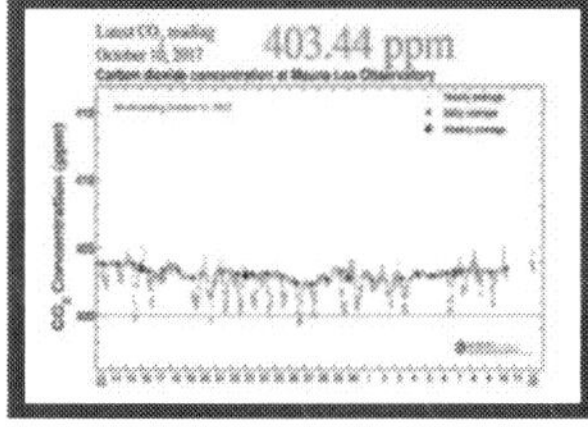

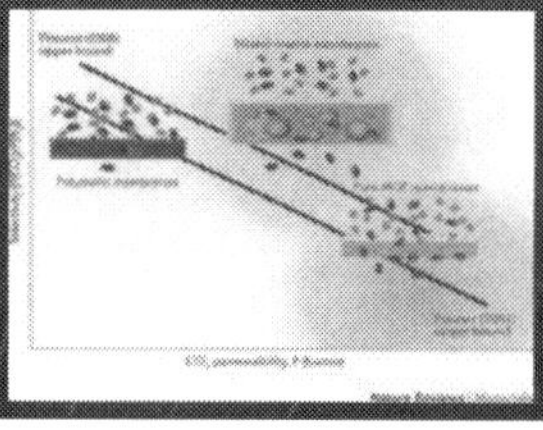

Membranes Fabrication and Characterization

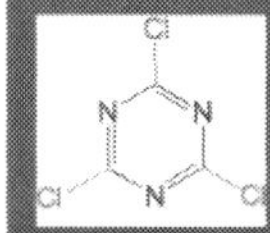

Cyanuric chloride (CC)

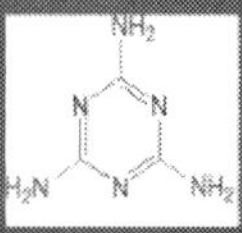

Melamine (M)

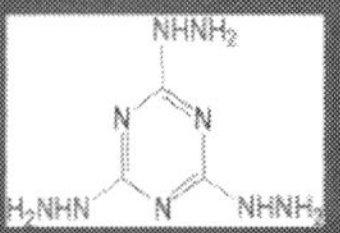

2,4,6-Trihydazino-1,3,5-triazine (THDT)

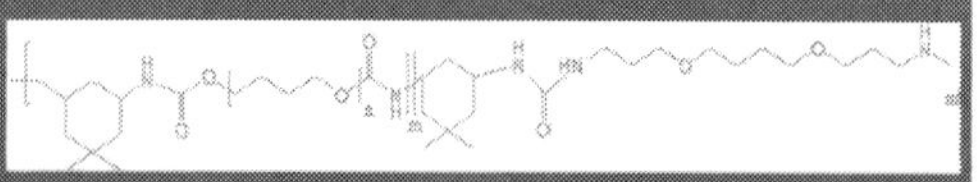

Polyurethane (PU)
Polytetramethylene glycol (PTMG)
+
Isophorone diisocyanate (IPDI)

$$\text{Hydrogen bonding index (HBI)} = \frac{A_{C=O,bonded}}{A_{C=O,free}}$$

FTIR: (a) PU-CC ,(b) PU-M , and (c) PU-THDT

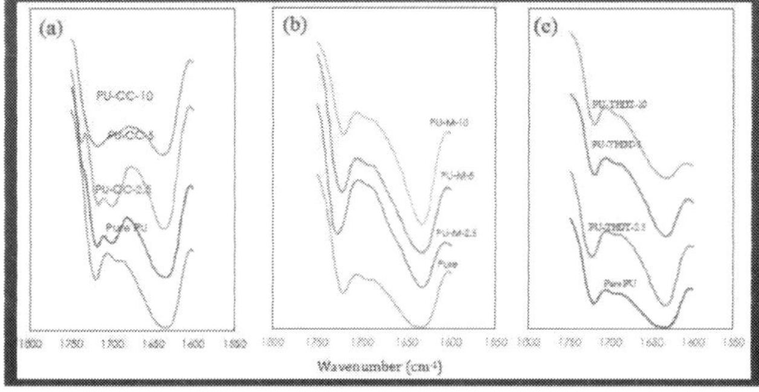

Sample	HBI
PU	1.23
PU-CC-2.5	1.12
PU-CC-5	1.08
PU-CC-10	1.03
PU-M-2.5	1.35
PU-M-5	1.38
PU-M-10	1.86
PU-THDT-2.5	1.44
PU-THDT-5	1.54
PU-THDT-10	1.62

2

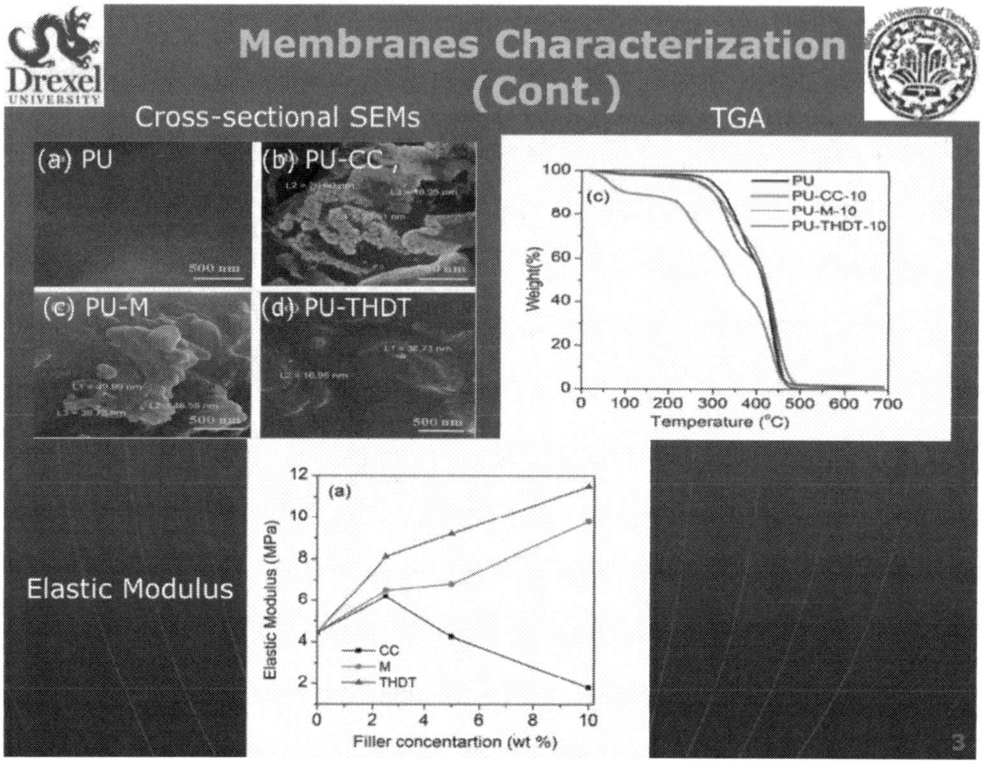
Drexel
UNIVERSITY
Membranes Characterization (Cont.)
Cross-sectional SEMs
(a) PU
(b) PU-CC
(c) PU-M
(d) PU-THDT
500 nm
TGA
PU
PU-CC-10
PU-M-10
PU-THDT-10
(c)
Weight(%)
Temperature (°C)
Elastic Modulus
(a)
Elastic Modulus (MPa)
Filler concentartion (wt %)
CC
M
THDT
3

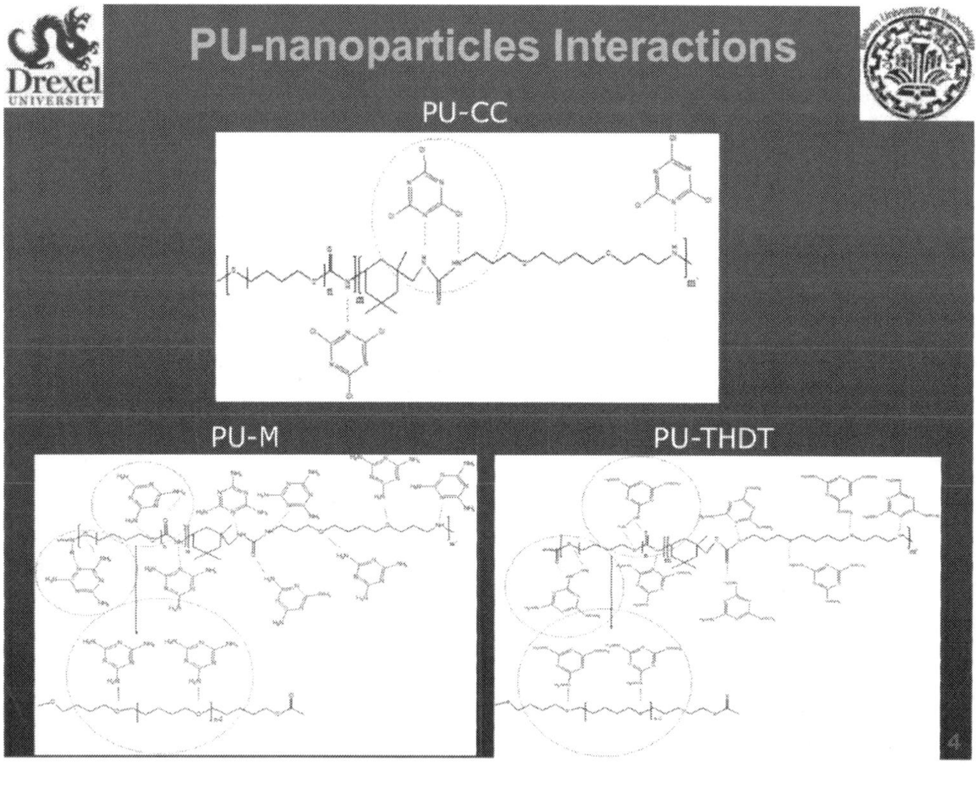
Drexel
UNIVERSITY
PU-nanoparticles Interactions
PU-CC
PU-M
PU-THDT
4

Membranes Performance

Sample	Permeability (barrer)				Selectivity		
	CO_2	CH_4	O_2	N_2	CO_2/CH_4	CO_2/N_2	O_2/N_2
PU	83.44	11.71	7.26	3.48	7.12	24.01	2.09
PU-CC-2.5	97.21	12.15	8.66	3.6	8	27.02	2.41
PU-CC-5	107.06	12.36	9.13	3.74	8.67	28.66	2.44
PU-CC-10	132.68	14.24	10.73	4.19	9.31	31.65	2.56
PU-M-2.5	90.5	10.55	6.64	3.07	8.58	29.5	2.17
PU-M-5	85.33	8.75	5.82	2.55	9.75	33.49	2.28
PU-M-10	73	7.31	4.92	2.11	9.98	34.57	2.33
PU-THDT-2.5	97.22	10.56	6.68	3.17	9.21	30.7	2.11
PU-THDT-5	104.27	9.58	5.73	2.72	10.89	38.35	2.11
PU-THDT-10	87.9	7.05	4.58	1.91	12.46	46.14	2.4

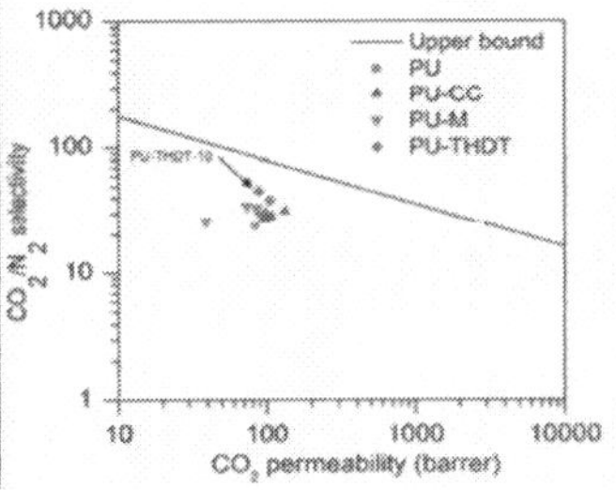

5

Concluding Remarks

- Cyanuric chloride nanoparticles are dispersed in the hard segments, but melamine and THDT particles in the soft segments
- Considerable enhancement of selectivity (75% for CO_2/CH_4 and 92% for CO_2/N_2) with higher CO_2 permeability via embedding 10% THDT into PU matrix
- THDT nanoparticles provide strong H-bonding and fixed facilitated transport for CO_2 removal
- Cyanuric chloride nanoparticles improve the gas permeability and selectivity simultaneously

6

Acknowledgment

Dr. Ali Pournaghshband

7

498772 The Growth of Glycidyl Methacrylate on Ultrafiltration Membrane: Spatial Control on Surface Initiated Aget-ATRP with Chain End Potential Functionalities

Abstract:

The growth of glycidyl methacrylate on Ultrafiltration Membrane: Spatial control on surface initiated AGET-ATRP with chain end potential functionalities

Arijit Sengupta[1], Blaine M. Carter[1], Xianghong Qian[2], S. Ranil Wickramasinghe[1]*

[1]Ralph E Martin Department of Chemical Engineering, [2]Department of Biomedical Engineering, University of Arkansas, Fayettteville, AR 72701

Corresponding author: swickram@uark.edu

Abstract

Surface modification is frequently used to tune membrane properties [1]. Atom Transfer Radical Polymerization (ATRP) was found to be one of the most convincing methods for controlled polymer growth on the outer surface and inside the pore of the membrane depending upon the associated chemistry and pore size [2]. ATRP has been shown to provide great control over the grafted nano structure [3]. Though the ATRP showed very attractive features, the most challenging step of the ATRP is to keep air free or oxidant free environment to sustain the modification [4]. The less stable Cu in +1 oxidation states is mainly responsible for the sustaining of ATRP reaction. The surface initiated Activator generated electron transfer (AGET) ATRP provides the advantages of ATRP removing the stringent condition of traces of oxidant free environment [5]. Ultrafiltration membranes are having large pore size (1-100 nm) frequently being used for size exclusion, pressure driven membrane based separation for protein concentration and buffer exchange. Though the controlled surface modification is the remedy protein fouling of UF membrane; but no spatial control on the modification was so far reported involving AGET-ATRP on UF membrane which is very important because of the large pore size provides possibilities of outside surface or inside pore modification [6].

An attempt was made to have spatial control of the modification on external versus internal pore surface of regenerated cellulose (RC) ultra-filtration (UF) membrane by glycidil methachrylate (GMA) polymers using activator generated electron transfer (AGET) atom transfer radical polymerization (ATRP). This controlled polymerization was optimized for GMA on RC UF resulting epoxy functionalized polymer brush end which on subsequent ring opening generates variety of chain end functionalities. The optimization of the experimental parameters of the surface initiated AGET ATRP revealed that the increase in relative concentration of the reductant i.e. ascorbic acid to the Cu makes a constant ratio of the two oxidation states of Cu for sustaining the AGET-ATRP, while the increased ATRP time can lead to the formation of longer polymer chain on the membrane. With increase in the initiator concentration during initiator immobilization step, its density either on the outside surface of the membrane or on the inside surface of the pore enhanced resulting decrease in the water flux. 100 mM initiator concentration with 1 min of immobilization followed by AGET ATRP for 2 hours with 2M monomer concentration with ascorbic acid to copper ratio 0.4 was found to be the optimized experimental condition. The appearance of carbonyl peak in Infra-Red spectroscopy confirmed the initiator immobilization step on the RC UF membrane while the peak due to epoxy ring confirmed the AGET-ATRP. The opening of epoxy ring by the treatment of sodium sulphonete and monoethanol amine was signature by X-ray Photoelectron Spectroscopic analysis. The performance of the membrane in terms of change in water flux and rejection of specified protein also confirmed the tuning of modification on the outer surface of the membrane vis a vis internal pore surface.

The interfacial modification using different pore filling solvents were found to bring the surface selective modification of the RC-UF membrane compared to the non-selective solution phase modification. The nature of the pore filling solvents, their viscosity coefficient, the interaction of initiator molecules with the pore filling solvents and the solvent of reaction played significant role in tuning the spatial selectivity. Depending upon the interaction of the initiator with these solvents the spatial selectivity of the plugging of initiator and subsequent polymer growth will be determined. The initiator needs to be partitioned between the pore filling solvent and the solvent of reaction and followed by diffusion to the pore to be immobilized. More the partitioning of the initiator into the pore filling solvent more will be the chance of immobilization inside the pore. Lesser the viscosity of the pore filling solvent lesser will be the probability for the initiator to diffused through it and plugged on the surface of the pore. The ratio of flux of ATRP to the flux of initiator immobilization was found to follow the trend: glycerol > L 64 > PEG 400 > ethanol > acetonitrile ~ solution. Since acetonitrile is the solvent used for initiator immobilization, using it as pore filling solvent showed the similar scenario of solution phase immobilization and hence no spatial selectivity was observed. For glycerol, the ratio was found to be 0.9 indicated the selective plugging of the initiator on the external surface of the membrane not on the surface of the pore. High immiscibility of glycerol and acetonitrile makes the partitioning of BIB into the glycerol difficult and again the higher viscosity coefficient of the glycerol makes the diffusion of the BIB even more difficult to reach to the surface of the pore. Therefore, probably, only external surface got modified in case of glycerol as pore filling solvent. In case of other solvent, depending upon their miscibility with acetonitrile and viscosity coefficient, there will be selective initiator immobilization on the external surface over internal surface of the pore as indicated by the flux ratio measurement. The maximum selectivity was found to be achieved by using glycerol as pore filling solvent due to its highly viscous nature as well as immiscibility with reaction medium.

References

1. M. Ulbricht, Advanced functional polymer membranes, Polymer, 47 (7) (2006) 2217
2. Z. Ma, X. Lu, C. Wu, Q. Gao, L. Zhao, H. Zhang, Z. Liu, Functional surface modification of PVDF membrane for chemical pulse cleaning, J. Memb. Sci, 524 (2017) 389
3. K. Matyjaszewski, P. J. Miller, N. Shukla, B. Immaraporn, A. Gelman, B. B. Luokala, T. M. Siclovan, G. Kickelbick, T. Vallant, H. Hoffmann, T. Pakula, Polymers at Interfaces: Using Atom Transfer Radical Polymerization in the Controlled Growth of Homopolymers and Block Copolymers from Silicon Surfaces in the Absence of Untethered Sacrificial Initiator, *Macromolecules*, *32* (26) (1999) 8716
4. J.-Q. Meng, C.-L. Chen, L.-P. Huang, Q.-Y. Du, Y.-F. Zhang, Surface modification of PVDF membrane via AGET ATRP directly from the membrane surface, Applied Surface Science 257 (2011) 6282
5. W. Jakubowski and K. Matyjaszewski, Activator generated by electron transfer for atom transfer radical polymerization, Macromolecules, 2005, 38, 4139.
6. N. Singh, Z. Chen, N. Tomar, S. R. Wickramasinghe, Modification of regenerated cellulose ultrafiltration membranes by surface â€“initiated atom transfer radical polymerization, J Memb Sci, 2008, 311, 225

503250 Carbon Molecular Sieves for Binary Permeation of N_2/CH_4 and CO_2/CH_4Â Gas Pairs

Abstract:

Carbon Molecular Sieve Membrane (CMSM) is derived from the pyrolysis of polymeric precursors. It possesses good mechanical strength, inert surface chemistry, and above all, high permeability and selectivity. It is one of the ideal candidates for the purification of natural gas (NG). The research aims at the removal of N_2 out of NG stream by using membrane technology. It compares favorably to the traditional method of cryogenic distillation at about -180 ^{0}C, which is energy intensive and costly. For N_2/CH_4 gas pair, CMSM 700 at 20 ^{0}C showed a good perm-selectivity of ~ 16.35 with N_2 permeability of 2.370 Barrer and CH_4 permeability of 0.145 Barrer at 250 KPa. The membrane is working more on sieving mechanism than surface diffusion mechanism for N_2/CH_4 gas pair. For CO_2/CH_4 gas pairs, CMSM 600 showed very good selectivity of 129.30 and CO_2 and CH_4 permeability of 342.64 and 2.65 Barrer, respectively at the permeation temperature of 20 ^{0}C.

499961 A Framework for Optimization-Based Design of Heat-Integrated Crude Oil Distillation Units Using a Surrogate Model and Support Vector Machine

Abstract:

Global demand for petroleum-based fuels for industrial, transportation and domestic use has been increasing since inception. These fuels (gasoline, kerosene, diesel, liquefied petroleum gas, fuel oil, etc.) are produced in refineries via crude oil distillation. Crude oil distillation is a complex and capital-intensive process; the crude oil distillation system comprises distillation columns, pump-around loops, side-strippers and a heat recovery system, known as the preheat train. The large number of degrees of freedom (column structure and operating conditions) and complex interactions between individual units makes the design of crude oil distillation unit a highly challenging task. Rigorous tray-by-tray column models are useful for designing crude oil distillation columns, but the approach requires trial and error, or, if optimization is applied, good initial estimates to obtain convergence and significant computational time and poses numerical difficulties.

Surrogate models, including those based on artificial neural networks (ANNs), have enabled significant capabilities in modelling and optimization of various chemical processes, including crude oil distillation. Researchers have shown that artificial neural networks can provide simple, accurate and robust simulation models. Artificial neural network-based distillation models have been implemented in optimization frameworks to perform operational optimization of existing crude oil distillation unit. To date, however, surrogate models for crude oil distillation reported in the literature have not considered design of the distillation column, i.e. optimization-based design, especially considering discrete variables representing the number of trays in each column section.

This paper presents a novel systematic approach for the design of crude oil distillation units that combines artificial neural network-based distillation models and a support vector machine (SVM). The support vector machine aims to ensure that only feasible design alternatives â€" i.e. those that meet constraints related to product specifications and hydraulic performance â€" are considered during the search process. This filtering helps to produce an optimal solution that is feasible; it also reduces computational effort by limiting the search to points within the demarcated feasible region.

The proposed framework comprises five main steps. First, an initial feasible design of a crude oil distillation unit is built using a rigorous tray-by-tray model in Aspen HYSYS. In step 2, the most important degrees of freedom are identified and represented using probability distribution functions. Latin hypercube sampling is then applied on these distributions to generate samples (design inputs) that will be used in the calculations. In step 3, all the samples are simulated and the corresponding responses (design outputs) are collected and automatically separated into feasible and infeasible solutions. In step 4, feasible samples are used to build a surrogate (ANN) distillation column model, while the entire set of samples is used to build a support vector machine. In the final step, the resulting models are implemented in an optimization framework, in which a genetic algorithm is used to search for design alternatives that minimize the total annualized cost (the sum of annualized capital cost and annual operating cost). Pinch analysis is used within the design procedure to determine minimum utility targets and estimate the utility costs (i.e. operating costs). The capabilities of the approach are illustrated through its application to an industrially-relevant case study.

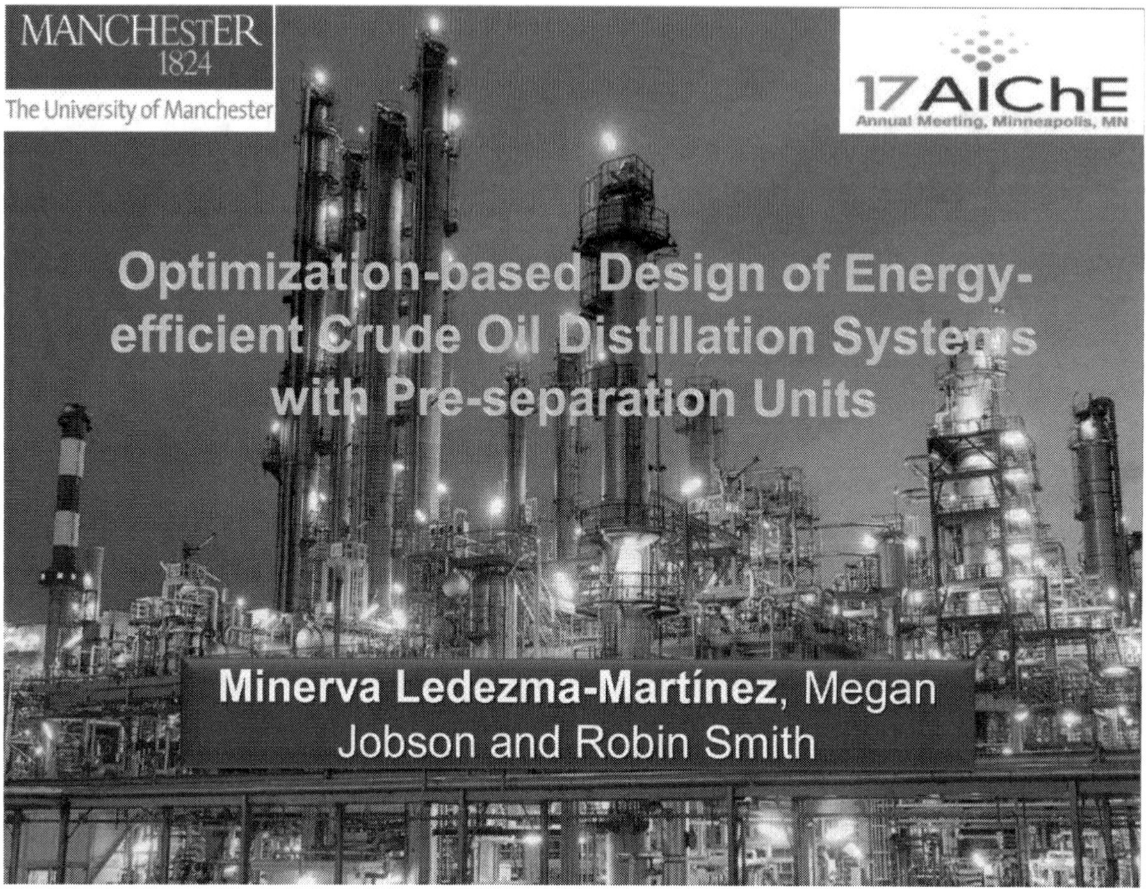

MANCHESTER 1824

The University of Manchester

Contents

1. Introduction
2. Research Objective
3. Optimization-based Design Methodology
4. Case Study
5. Conclusions
6. Future Work

MANCHESTER 1824
The University of Manchester

1. Introduction

MANCHESTER 1824
The University of Manchester

Crude Oil Distillation

- Crude oil distillation systems play a central role in petroleum refining.
- Energy-intensive systems: 2–7% of the crude oil input is consumed for crude oil distillation[1].
- Energy savings can be achieved by exploiting interactions between separation units and the heat recovery system.

Catlettsburg Oil Refinery, Kentucky USA

1. A. Szklo, and R. Schaeffer, 2007, Fuel Specification, Energy Consumption and CO_2 Emission in Oil Refineries, Energy, 32, 7, 1075–1092.

Crude Oil Distillation System

- Crude oil distillation system:

  1) A preheat train.
  2) Crude oil distillation units with side strippers and pump-arounds.
  3) Pre-separation units.

- **Preflash units** help to enhance energy efficiency (i.e. to reduce fired heat demand).

- Energy efficiency benefits to reduce CO_2 emissions and operating costs.

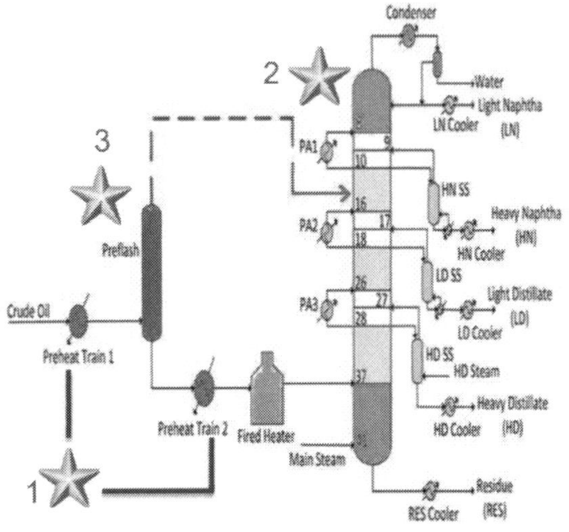

High operating and equipment costs and the complexity of the system motivate the development of systematic approaches for optimal system design.

Crude Oil Distillation System Design

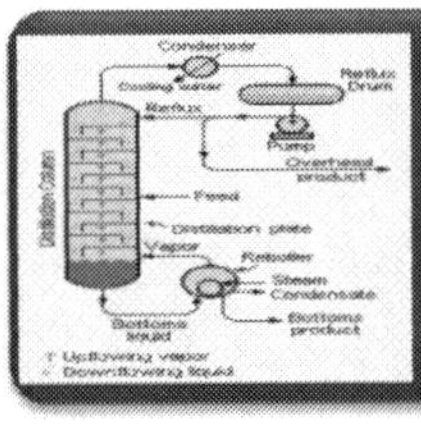

Traditional methods

- Heuristic rules
- Involves trial and error

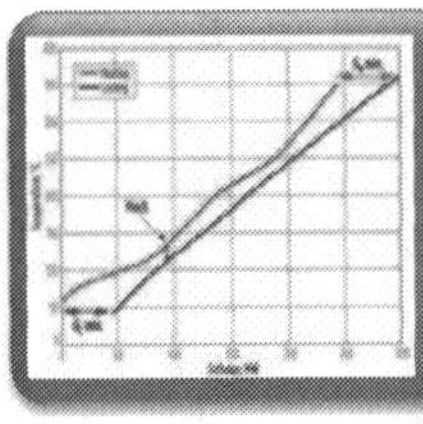

Integrated methods

- Simulations and pinch analysis
- Capital–energy trade offs not considered

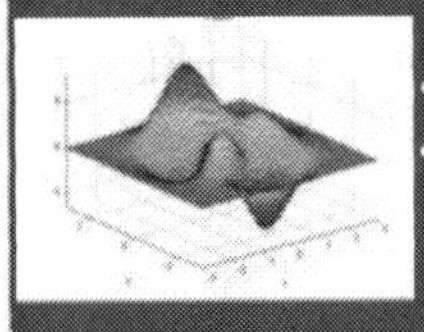

Optimization-based design methods

- Design supported by optimization
- Lack of design methodologies for crude oil distillation systems with pre-separation units

The University of Manchester

Pinch Analysis

- Grand Composite Curve shows net heating and cooling needs by temperature interval (after heat recovery).
  - Useful for conceptual design of column.
  - No heat exchanger network details.
- Stream data (supply and return temperatures and duties) are simulation results.
  - Pump-arounds.
  - Product coolers.
  - Condenser.
  - Reboilers.
- High temperature hot utility (**fired heating** = furnace duty) dominates operating costs.

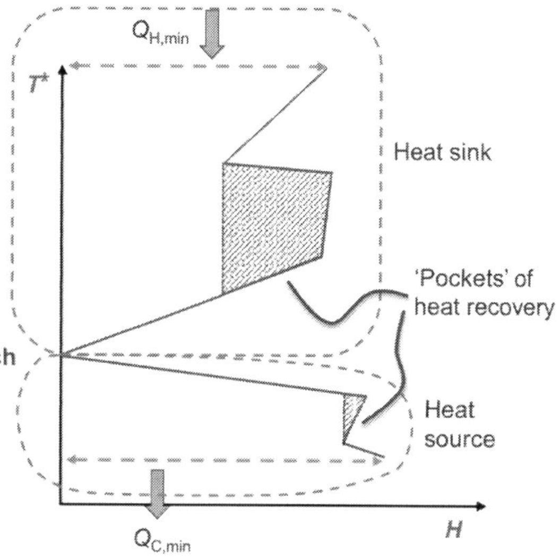

The University of Manchester

Optimization (Genetic Algorithm)

- Genetic Algorithms (GA) have found wide acceptance in Chemical Engineering because of their simplicity, and ability to determine optimum solutions for both single and multi-objective, non linear optimization problems[1].

Advantages

- Less prone to getting trapped in local optima than deterministic searches.
- It is possible to handle a variety of decision variables (binary, integer or continuous).

Structure of Genetic Algorithm

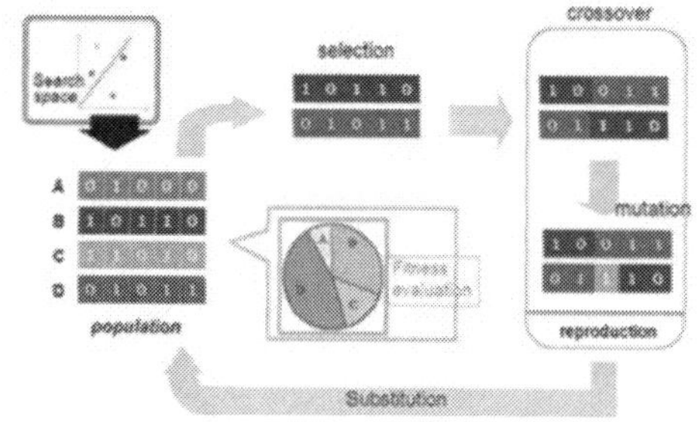

1. Prakash R. Kotecha, Mani Bhushan and Ravindra D. Gudi, 2010, Constraint Programming and Genetic Algorithm in *Stochastic Global Optimization: Techniques and Applications in Chemical Engineering* ,Vol. 2, (ed. G.P. Rangaiah), World Scientific Co., Singapore.

The University of Manchester

2. Research Objective

The University of Manchester

Research Objective

To develop a systematic approach to design cost-effective, energy-efficient crude oil distillation systems with pre-separation units, accounting for product quality and heat integration.

Preflash/ Prefractionator
- Key opportunity for energy saving.
- Light components bypass fired heater.
- Vapour feed location affects separation within column.

Crude Oil Distillation Unit
- Many degrees of freedom interacting with each other.
- Interactions with Heat Exchanger Network (HEN).
- Product quality specifications must be within allowed range.

Heat Integration
- Reduces energy cost.
- Grand Composite Curve (GCC) sets energy targets.

MANCHESTER 1824
The University of Manchester

3. Optimization-based Design Methodology

MANCHESTER 1824
The University of Manchester

Heat-integrated crude oil distillation unit with a preflash – Degrees of freedom

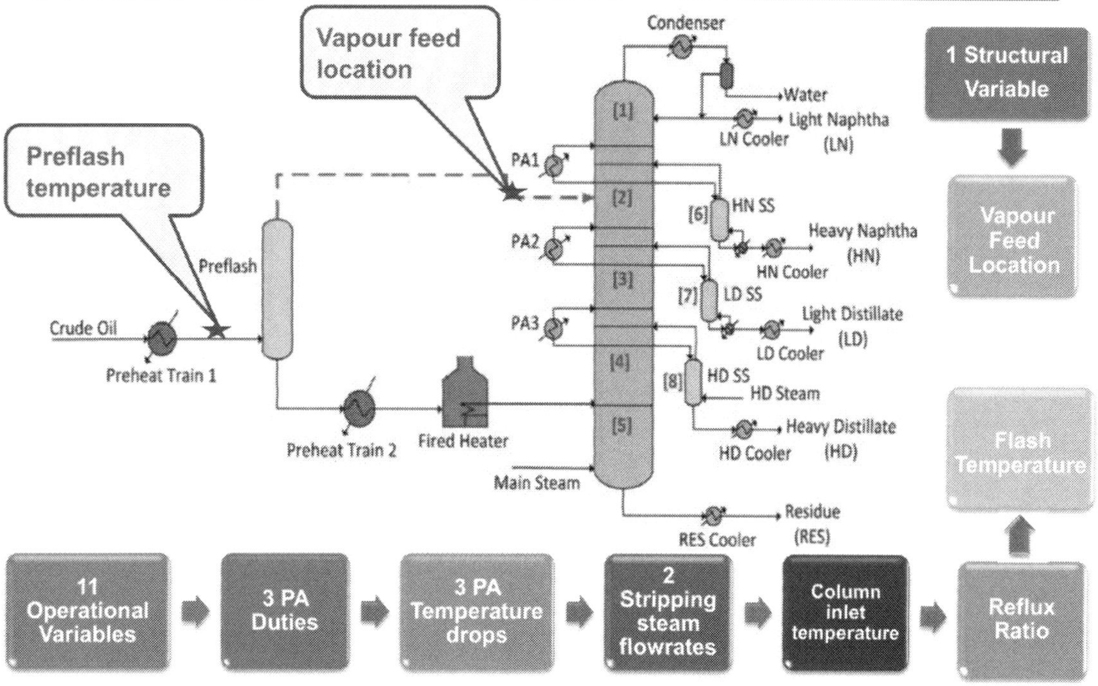

MANCHESTER
1824
The University of Manchester

Simulation-optimization-based design methodology

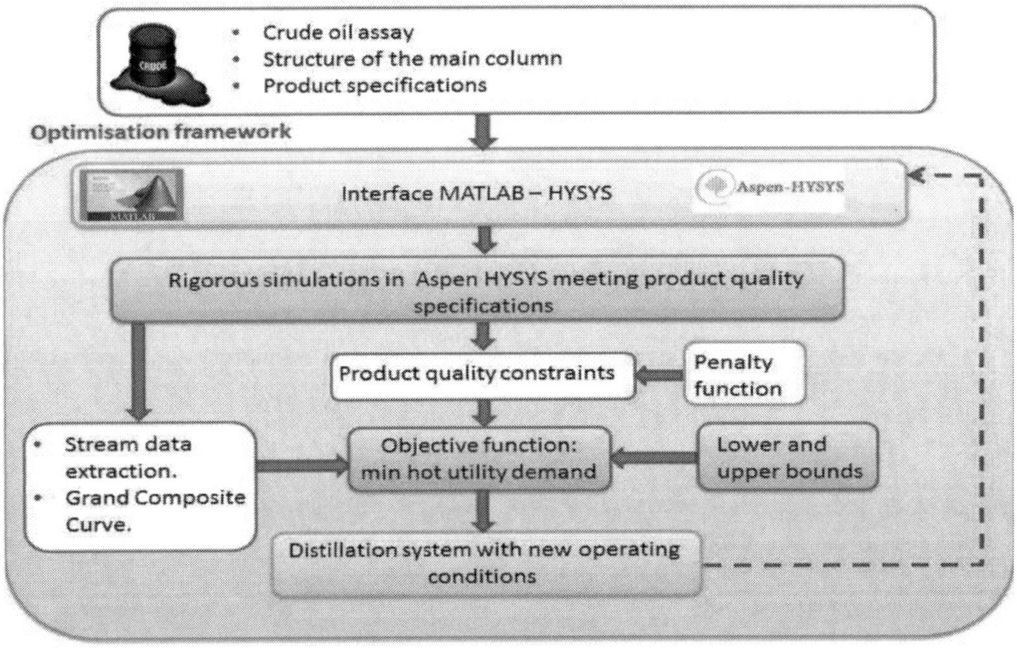

MANCHESTER
1824
The University of Manchester

Direct Optimization with GA

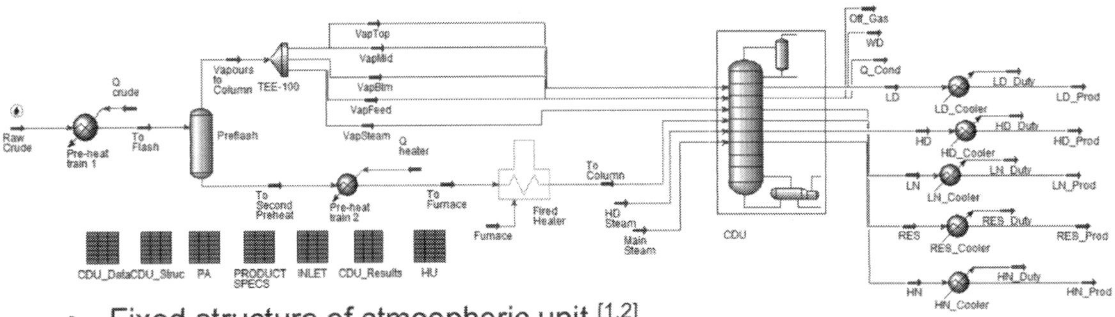

- Fixed structure of atmospheric unit [1,2].
- Optimize preflash temperature and flash vapour feed stage.
  Flash vapour connected to 5 different stages in the main column (1 per section).
- Minimise hot utility demand.
- Constraints: product quality specifications (T5%, T95% ASTM).
- Optimization: Genetic Algorithm.
  Population size = 100; 500 generations.

[1] Watkins R.N., 1979, Petroleum Refinery Distillation, Gulf Pub. Co., Houston, USA.
[2] Chen L., 2008, Heat-integrated Crude Oil Distillation System Design, PhD Thesis, University of Manchester, Manchester, UK.

MANCHESTER
1824
The University of Manchester

4. Case Study

MANCHESTER
1824
The University of Manchester

Crude Oil Distillation Unit with Preflash

Venezuela Tia Juana Light crude oil
100,000 bbl/day (0.184 m^3/s)
25°C, 2.5 bar

Main Steam:
260°C, 4.5 bar
1200 kmol/h

HD Steam:
260°C, 4.5 bar
250 kmol/h

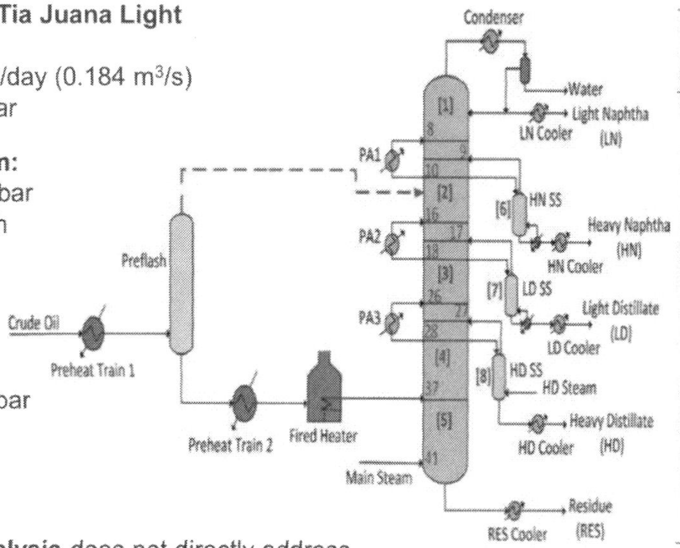

Column section	Stage numbering
Section 1	1–9
Section 2	10–17
Section 3	18–27
Section 4	28–36
Section 5	**37–41**
Section 6 (HN SS)	42–47
Section 7 (LD SS)	48–54
Section 8 (HD SS)	55–59

- **Pinch analysis** does not directly address HEN design.
- Assumption: ΔT_{min} = 30°C.

MANCHESTER 1824
The University of Manchester

Product specifications and flowrates

Product	T5% ASTM D86 (°C)	T95 % ASTM D86 (°C)	Flow rate (m^3/h)
LN	26	110*	102
HN	140	196*	87
LD	217*	300*	128
HD	308	354*	54
RES	362	754	292

* Specified in HYSYS.

- For all products, the allowed range was ±10°C.

MANCHESTER 1824
The University of Manchester

Objective function and constraints

- **12 Degrees of freedom**
  - Pump-around duties (3)
  - Pump-around temperature drops (3)
  - Main steam
  - HD steam
  - Column inlet temperature
  - Reflux ratio
  - Flash temperature
  - Vapour feed location (main column)
- **10 Constraints**
  - ✓ Product quality specifications (ASTM D86 T5% and T95% temperatures) for five products

Objective function

$$\min F(x) = f(x) + \gamma_1 |h(x)| + \sum_{j=1}^{n} \gamma_{1+j} \max\left(0, g_j(x)\right)$$

$F(x)$ = Unconstrained objective function

$f(x)$ = Constrained objective function

$g_j(x) \leq 0$ $h(x) = 0$

γ = Penalty factor

$$T5_i^{lb} \leq T5_i \leq T5_i^{ub}$$

$$T95_i^{lb} \leq T95_i \leq T5_i^{ub}$$

i=1, 2, …..$N_{products}$

Ochoa-Estopier, L. M., Jobson, M., 2015, Optimization of heat-integrated crude oil distillation systems. Part I: The distillation model, *Industrial & Engineering Chemistry Research*, 54 (18) 4988–5000 doi: 10.1021/ie503802j.

MANCHESTER
1824
The University of Manchester

Optimization variables and bounds

Variable	Units	Base Case	Lower bound	Upper bound
1. Main steam flow	kmol/h	**1200**	900	1300
2. HD steam flow	kmol/h	**250**	200	375
3. PA1 duty	MW	**12.8**	14	6
4. PA2 duty	MW	**17.8**	18	6
5. PA3 duty	MW	**11.2**	12	6
6. PA1 ΔT	°C	**30**	20	50
7. PA2 ΔT	°C	**50**	15	60
8. PA3 ΔT	°C	**20**	10	40
9. Column inlet temperature	°C	**365**	350	375
10. Reflux ratio	-----	**4.17**	3.0	4.5
11. Flash temperature	°C	**115**	110	230
12. Vapour feed location (column section)	-----	**5**	1	5

✓ Bounds defined based on sensitivity studies.

MANCHESTER
1824
The University of Manchester

Optimization Results

Variable	Units	Base Case	Optimization Results	
			No Preflash	With Preflash
1. Main steam flow	kmol/h	1200	900	1190
2. HD steam flow	kmol/h	250	200	211
3. PA1 duty	MW	12.8	6.0	6.8
4. PA2 duty	MW	17.8	9.8	8.3
5. PA3 duty	MW	11.2	11.9	8.8
6. PA1 ΔT	°C	30	48.6	25.5
7. PA2 ΔT	°C	50	33.8	23.0
8. PA3 ΔT	°C	20	39.9	39.3
9. Column inlet temperature	°C	365	350	354
10. Reflux ratio	°C	4.17	3.7	3.0
11. Flash temperature		115	----	230
12. Vapour feed location (column section)		**5**	---	3
Minimum hot utility demand	**MW**	**58.3**	**43.4**	**35.9**

MANCHESTER 1824
The University of Manchester

Grand Composite Curves

Optimized base case (no preflash)

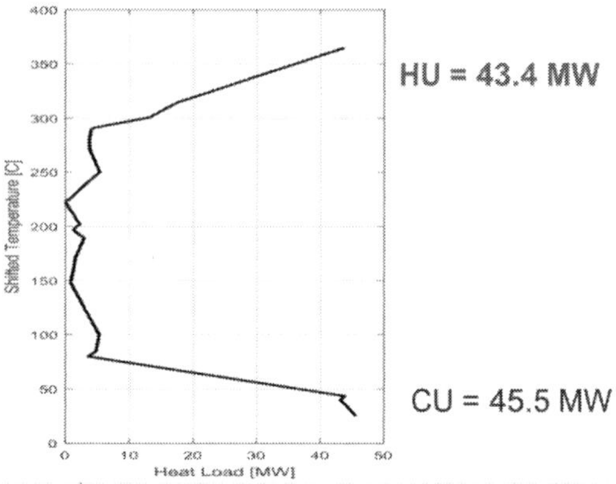

Optimized design (with preflash)

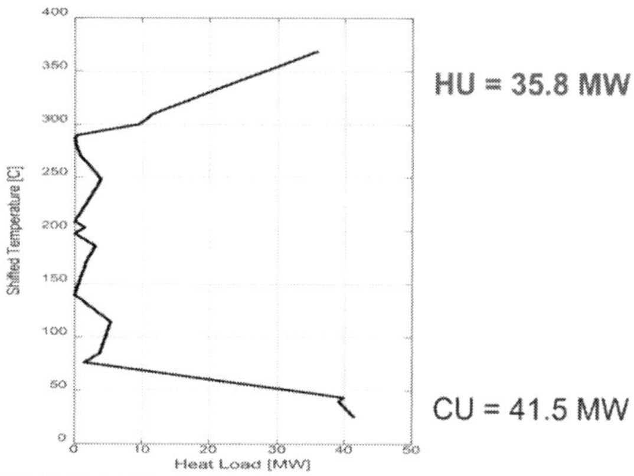

✓ **Preflash reduced hot utility demand by 21%, compared to optimised base case.**

✓ Energy targets are reduced when implementing a preflash unit within the crude oil distillation system.

Significant potential to reduce operating costs

➢ $\Delta T_{min} = 30°C$

MANCHESTER 1824
The University of Manchester

Product specifications and flowrates

Product	Base Case ASTM (°C)		Optimization Results No Preflash ASTM (°C)		Optimization Results With Preflash ASTM (°C)	
	T5%	T95%	T5%	T95%	T5%	T95%
RES	363	755	354	753	353	753
LN	27	110*	25	110	25	110
HN	143	196*	133	196	133	196
LD	218*	300*	218	300	218	300
HD	308	354*	305	354	302	354

* Specified in HYSYS.

✓ All product specifications are within the allowed tolerance of ± 10°C.

Product Flowrates in m³/h

Product	Base Case	Optimized No Preflash	Optimized With Preflash
RES	292.1	299.4	301.0
LN	102.4	101.1	100.9
HN	86.8	88.3	88.7
LD	127.6	125.5	122.9
HD	53.7	48.4	49.2

Summary ΔT_{min} analysis

Variable	Units	Base Case	Optimization Results No Preflash			With Preflash		
	°C		ΔT 30	ΔT 20	ΔT 10	ΔT 30	ΔT 20	ΔT 10
1. Main steam flow	kmol/h	1200	900	901	1054	1190	949	1103
2. HD steam flow	kmol/h	250	200	206	284	211	241	365
3. PA1 duty	MW	12.8	6.0	6.0	6.0	6.8	6.0	6.0
4. PA2 duty	MW	17.8	9.8	12.5	11.8	8.3	8.4	8.3
5. PA3 duty	MW	11.2	11.9	11.9	11.7	8.8	11.0	12.0
6. PA1 ΔT	°C	30	48.6	28.6	50.0	25.5	40.1	23.1
7. PA2 ΔT	°C	50	33.8	46.8	48.3	23.0	36.7	41.5
8. PA3 ΔT	°C	20	39.9	40.0	30.0	39.3	22.5	20.2
9. Column inlet temperature	°C	365	350	350	351	354	362	358
10. Reflux ratio		4.17	3.7	3.5	3.7	3.0	3.0	3.0
11. Flash temperature	°C	115	---	----	----	230	230	230
12. Vapour feed location		----	---	----	----	3	3	3
Minimum hot utility demand	**MW**	**58.3**	**43.4**	**38.8**	**34.7**	**35.9**	**31.1**	**26.3**
Energy Savings						**21%**	**24%**	**32%**

Case Study: Summary

- A systematic design optimization approach is demonstrated for a crude oil distillation system with a preflash unit.
- Genetic Algorithm was successfully implemented to optimize a crude oil distillation system with a preflash unit.
  - CPU time: 5.8h using an HP desktop Intel(R) Core(TM) i5-4570 CPU @ 3.20GHz with 16.0 GB in memory RAM.
  - Product quality specifications are all met.
- Minimum hot utility demand was reduced by:
  - 32% for ΔTmin = 10°C.
  - 24% for ΔTmin = 20°C.
  - 21% for ΔTmin = 30°C.

  by introducing and optimizing a preflash unit through optimization.

5. Conclusions

Conclusions

1. Methodology has been developed for simulation-optimization-based design of a crude oil distillation system with a preflash unit.
   - Implemented using Aspen HYSYS–MatLab interface with genetic algorithm.
2. An optimization framework accounts for operational and structural variables.
   - Constraints applied to product quality.
   - Heat integration considered simultaneously using pinch analysis.
3. Case study results indicate energy demand can be reduced by using a preflash unit in the crude oil distillation system.
4. The methodology has the merits of using rigorous process models, of being systematic, of using robust optimization techniques and of accounting for configuration as well as operating conditions.

MANCHESTER
1824
The University of Manchester

6. Future Work

MANCHESTER
1824
The University of Manchester

Prefractionation option

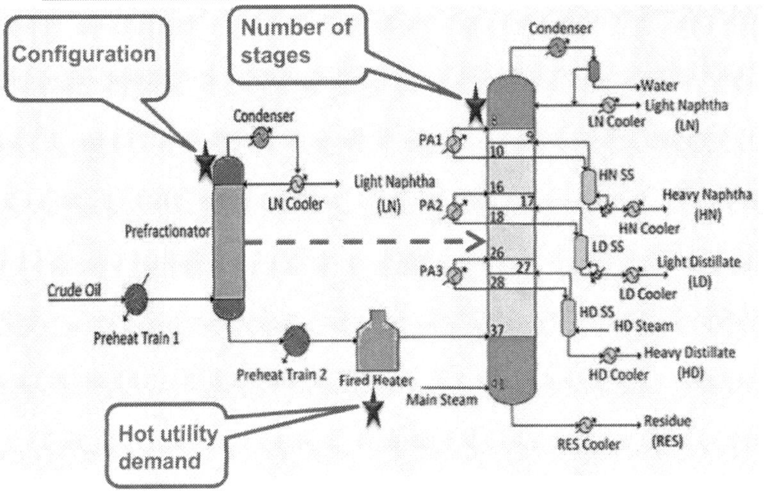

System design

- Main column structure.
- Main column operating conditions.
- Prefractionator design.

Key design parameters

- Number of stages.
- Feed locations.
- Pump-around duties.
- Pump-around temperature drops.
- Stripping steam flow rates

Methodology:

- Rigorous simulation of columns.
- Fuel consumption of the fired heater.

Optimization:

- Continuous and discrete variables.
- Stochastic optimization (GA).

MANCHESTER 1824
The University of Manchester

Thank you

National Council of Science and Technology (Mexico) for the financial support granted for the development of this work.

CONACYT

Contact Details:

Minerva Ledezma-Martínez

minerva.ledezmamartinez@manchester.ac.uk

Paper 175c
(Proceedings 483295)

Vacuum Distillation:

Slop Wax Disposition

Andrew W. Sloley
Advisian (WorleyParsons Group)
Energy Center II
575 N Dairy Ashford St
Houston, TX 77079

Prepared for Presentation at the
American Institute of Chemical Engineers
2017 National Meeting
Minneapolis, MN October 29-November 3, 2017

UNPUBLISHED

1. Summary

Refinery vacuum units include a deentrainment, or wash zone, for removal of heavy material from desired products. A continuous wash oil stream prevents the wash zone (deentrainment zone) from coking. Underneath the wash zone a collector tray gathers the liquid from the wash zone plus any entrainment removed. The total liquid collected on this tray is called slop wax.

Traditionally, slop wax has been drawn from the vacuum tower as a stream and recycled to the feed upstream of the vacuum feed heater. Less common alternate dispositions include sending the slop wax to a stripping section inside the vacuum tower or making the slop wax a product stream.

This review covers:

- Survey of operating units to identify the most common slop wax dispositions
- Evaluation of the major options for towers without stripping sections
- Evaluation of the major options for towers with stripping sections
- Review of the effect of entrainment on slop wax dispositions

2. Refinery Slop Wax

Refinery slop wax is a stream created inside the refinery crude distillation unit. It normally refers to a mixed distillate-entrainment stream from the vacuum distillation tower. However, some plants may use the term slop wax to refer to mixed distillate-entrainment streams from other towers in the crude distillation unit that include de-entrainment sections.

Figure 1 shows the configuration of the de-entrainment section of a refinery vacuum tower wash zone. The two-phase feed enters the vacuum tower below the wash zone. Due to high velocity in the feed line, some liquid droplets entrain up the column with the vapor. These liquid droplets contaminate the distillate products. A wash bed helps de-entrain the droplets. Wash oil irrigates the top of the wash bed to prevent plugging and improve unit reliability. Distillate liquid leaving the bottom of the wash bed (overflash) combines with entrainment. The mixed distillate-entrainment stream is called slop wax. Previous work covered required rates for the wash oil and overflash [1].

The term slop wax comes from lubricants processing. Figure 2 shows an example vacuum tower for manufacturing lubricant base stocks. The lowest lubricant product is "lube base 3". Among other requirements, specifications on this product normally include some combination of viscosity and viscosity index [2] ranges. To make lube base 3 on specification, the rate of production is controlled to reject material with too high a viscosity. The rejected material is left in the liquid inside the vacuum tower.

In lubricant units the vacuum tower bottoms (VTB) often goes to downstream processing to produce "bright stock" lubricants. The bright stock specifications also include some combination of viscosity and viscosity index ranges.

Commonly, material that has too high a viscosity for the lowest lubricant base oil (lube base 3 in this example) has too low a viscosity for bright stock. Without a slop wax stream, any material not drawn in lube base 3 ends up in the vacuum tower bottoms. The unit cannot produce both the highest viscosity lubricant product (lube base 3) and the bright stock on specification at the same time.

The slop wax stream is a drag stream that removes intermediate viscosity material from the process. With the slop wax stream as a product, both the lowest lubricant product and the bright stock meet specifications. Units with bright stock production usually do not recycle the slop wax. Recycled slop wax in these units builds up an inventory of off-specification material inside the unit. The slop wax rate is set by the amount of undesirable material in the crude oil feed that falls into the gap between the product specifications. The slop wax is a product from the crude distillation unit. Conventional dispositions include blending into asphalt, sending it to a fuels crude unit, or delayed coking.

Fuels vacuum units produce feeds for downstream catalytic processing to produce transportation fuels. Fuels vacuum units do not have a gap between product specifications that prevents recycling the slop wax. While the recycle problem does not exist, the term slop wax has been retained for the stream name.

Since the slop wax can be recycled in fuels vacuum units, two immediate question are raised:

- When is slop wax recycle desirable?
- If desirable, what is the best way to recycle slop wax for distillate recovery?

3. Survey

Basic configurations for slop wax recycle in the refining industry include

- no recycle, slop wax is a product (Figure 3);
- no recycle, external piping with gravity flow (Figure 4);
- no recycle, external piping with a pump (Figure 4);
- no recycle, internal flow (Figure 5, Figure 8);
- recycle to upstream of fired heater (Figure 6); or
- recycle to an upstream tower boot or other upstream location (Figure 7).

Table 1 summarizes a recent survey of vacuum towers and shows the percent use of major options.

Table 1 Summary of Recycle Dispositions and Configurations

Slop Wax Disposition	Occurrence, %	
Product	14	
Recycled, total	36	
Recycled, vacuum tower has no stripping section		28
Recycled, vacuum tower has stripping section		8
Not recycled, total	50	
Not recycled, vacuum tower has no stripping section		20
Not recycled, vacuum tower has stripping section		30
Total	100	
Recycled, Destination (Always pumped external configuration)		
To vacuum heater feed	67	
To upstream tower boot	33	
Total	100	
Not Recycled, Destination		
To boot or tower bottoms	40	
To stripping section	60	
Total	100	
Not Recycled, Configuration		
Gravity flow, external	32	
Gravity flow, internal	52	
Pumped, external	16	
Total	100	

Figure 3 to Figure 7 show the different configurations in use. All the figures show vacuum towers without stripping sections. However, all configurations shown can be used with or without a stripping section. Figure 8 shows an example with internal flow and a stripping section in the tower. When a stripping section is present, systems without recycle return the slop wax to the top of the stripping tray. The slop wax then mixes and becomes part of the feed to the stripping section.

4. Slop Wax Recycle

Slop wax is the liquid collected on the slop wax collector. Slop wax is the combination of two components. The first component is entrainment carried by the vapor rising from the feed zone. The second is overflash, distillate liquid leaving the bottom of the wash bed. The overflash is material that was originally flashed as vapor in the feed. All vapor in the feed is potential distillate product from the vacuum tower. Overflash is vapor in the feed, but ends up in the slop wax instead of the distillate product.

Since the slop wax contains entrained liquid (residue), it is not suitable for downstream use as a gas oil product in most plants. The concept behind recycling the slop wax is that reprocessing the slop wax allows for recovery of the distillate contained in the slop wax. Based on a recent survey of slop wax disposition, 36% of the surveyed units recycled slop wax (Table 1).

Figure 9 illustrates the possible benefits of slop wax recycle in an example unit. The example is for a realistic commercial unit [3] with a feed very close to the average currently processed in U.S. refineries (Tia Juana Light, 31.6°API, SG=0.8676 [4]).

At a best-practice liquid irrigation rate of 0.15 gpm/ft^2 (6.11 lpm/m^2) in a tower without a stripping section, the slop wax recycle increases yield from a cutpoint[a] of 1,026°F to 1,038°F (552°C to 559°C). This is equivalent to a liquid distillate yield increase of 0.62 volume percent. The relative benefit is nearly the same for a tower with a stripping section. Slop wax recycle improves unit yields at fixed tower inlet conditions and fixed tower operating pressure.

Given the possible benefits, the survey results surprise. They show that only 36% of units have slop wax recycle compared to 50% of units that do not (the 14% of units that have slop wax as a product lie outside the scope of the comparison). For every 2 units with recycle, 3 units do not have recycle.

5. Alternatives to Slop Wax Recycle

To understand why slop wax recycle is not more widely used, alternatives to slop wax recycle must be examined. Refiners have three straightforward alternatives to maintain yield.

- Operate the vacuum tower at lower pressure.
- Operate the vacuum tower at higher temperature.
- Add a steam stripping section (or make it bigger).

5.1. Operate The Vacuum Tower at Lower Pressure

Reducing the vacuum tower pressure increases feed vaporization. Table 2 outlines the main effects of lower pressure operation.

a Cutpoint is a measure of distillate yield. It is the atmospheric equivalent temperature on an ASTM D2892 test, or equivalent, that gives the same liquid distillate yield as seen on the operating unit.

Table 2 Effects of Lower Pressure Operation (For Same Yield)

Effect	Cost or Benefit
Larger tower diameter for same capacity	Cost
Larger vacuum system	Cost
Higher vacuum system utility demand	Cost
Lower temperature streams for heat integration	Cost

Compared to slop wax recycle, the costs are higher. Both more capital is required and operating costs rise for the same capacity. Economically, lower operating pressure instead of slop wax recycle is not attractive.

5.2. Operate the Vacuum Tower at Higher Temperature

Many units operate at as high a temperature as possible within the constraint of a target operating duration between shutdowns to decoke the furnace. Crude oil thermal stability can limit operating temperatures. However, some units can raise temperature or heaters can be modified to allow operation at higher temperatures.

Vacuum tower effects of higher operating temperatures versus slop wax recycle are more complex. The higher temperature increases the vapor load by heating up the entire vapor stream but allows for reducing the vapor load by eliminating the slop wax recycle. Figure 10 shows the relative vacuum tower cross-section required in the wash section for a slop wax recycle system versus a higher heater outlet temperature. In Figure 10 the vapor load is defined by a maximum C-factor of 0.45 ft/sec at the top of the wash section [5]. Overall, the vacuum tower is slightly larger with higher heater outlet temperatures.

Table 3 summarizes the effects of using higher heater temperature to get the same yield as slop wax recycle. Compared to slop wax recycle, the costs are higher. Both more capital is required and operating costs rise slightly for the same capacity. Economically, higher heater outlet temperature instead of slop wax recycle is not attractive.

Table 3 Effects of Higher Temperature Operations (For Same Yield)

Effect	Cost or Benefit
Larger tower diameter for same capacity	Cost
Larger vacuum system (more cracked gas from heater)	Cost
Higher vacuum system utility demand	Cost
Higher temperature streams for heat integration	Benefit
Higher coking rates in the fired heater	Cost

5.3. Add a Steam Stripping Section

Steam stripping the residue recovers light material from the residue [6]. Stripping adds steam and removes recycle vapor from the tower but adds stripping steam load. Figure 10 includes a comparison of using stripping steam to get the same yield as slop wax recycle. The basis used is a one-stage stripping section with enough steam to match the slop wax recycle yield at the same tower inlet conditions. The net effect of steam stripping is that a larger tower diameter is required for the same capacity and yield.

Table 4 Effects of Stripping Steam (For Same Yield)

Effect	Cost or Benefit
Larger tower diameter for same capacity, Entire Tower	Cost
Taller tower to add stripping trays	Cost
Larger vacuum system (steam load)	Cost
Higher vacuum system utility demand	Cost

Table 4 summarizes the effects of using a stripping section to get the same yield as slop wax recycle. Compared to slop wax recycle, both the capital costs and the operating costs are significantly higher. Economically, stripping is less attractive than slop wax recycle.

6. Problems with Slop Wax Recycle

Evidently, if most plants to not use slop wax recycle, it must have other problems. Most of the problems with slop wax recycle come from two areas, thermal cracking and system reliability.

6.1. Thermal Cracking

Slop wax recycle takes hot liquid from the slop wax collector and recycles it to upstream of the heater feed. This liquid gets reprocessed back to heater outlet temperatures. The initial feed material may be subject to some thermal cracking at high heater outlet temperatures. The thermal cracking results in products that are less thermally stable. Heating them a second time can result in aggressive coking.

The coking may take place in either the fired heater or the vacuum tower. Several plants have eliminated slop wax recycle to extend plant run length between shutdowns. Native crudes with high paraffin and asphaltic content tend to have more coking problems with slop wax recycle. Syncrude blends also have more problems as they usually contain reactive material. Highly aromatic crudes tend to be more thermally stable and have fewer coking problems.

Coking is the result of a time-temperature-reactivity profile. The same crude may have significantly different behavior depending on the unit layout and operating point. Coking problems will often be found only by experience with specific crudes on a given unit. Several plants suspend slop wax recycle when running specific crudes.

6.2. Pump and Mechanical Issues

The major problem with slop wax recycle comes from the recycle pump, its control, and associated equipment. The slop wax recycle pump has the following characteristics:

- High operating temperatures
- Low flow rate in all except very large units
- High discharge head
- Unstable feed conditions
- Fluid with a high freezing temperature

- Difficult level control

These combine to make pump reliability in this service difficult. Many refiners find the cost and potential safety problems of maintaining the pump outweigh the benefits.

Operating temperatures routinely exceed 700°F (371°C). High temperatures create potential problems with seals, piping stress, and nozzle loads. Proper pump selection and installation may be extremely expensive.

Nearly all pumps in this service are built to API-610 [7] standards to maximize reliability and safety. API-610 pump specifications set criteria that make building low-flow pumps challenging. Additionally, if the recycle is returned immediately upstream of the fired heater (Figure 6), the pump will have a high discharge head requirement. The combination of low flow plus high discharge head can make selection of a reliable pump difficult [8].

To reduce discharge head requirements, the recycle may return to the boot of an upstream tower (Figure 7). This reduces the head requirement and makes the pump more reliable. However, the recycle now has to go through two pumps and the bottoms pump on the upstream tower needs to be larger.

The pump feed conditions tend to be unstable in both state and flow rate. Solutions to these problems tend to create other problems.

The entrainment and the overflash mix on the collector tray. The entrainment is hotter than the overflash. The mixing can cause vaporization. If the outlet nozzle of the tower is too small or the liquid level on the collector is too low, vapor can be drawn into the pump suction. Vapor in the feed line to the slop wax pump creates cavitation, impeller damage, and seal problems.

Entrainment rates may vary dramatically from the average value. If liquid hold-up on the slop wax collector is low, the level may vary dramatically. Large level changes can result in needing to change the recycle pump rate quickly.

One solution might be to hold a substantial liquid level on the slop wax collector. This allows for degassing the mixture and for control to avoid flow rate changes. However, large liquid inventory in the column increases thermal cracking. The liquid is at a high temperature. Excessive residence time results in coke formation on the tray, eventually plugging it. Thermal cracking also produces light gas that increases the load on the vacuum system. High vacuum system load increases tower operating pressure.

Economics strongly favor keeping the tower pressure low. In nearly all cases, the refinery attempts to minimize slop wax inventory. This makes control and reliability problems difficult to solve.

Inevitably, pumps in this service require more-than-normal amounts of maintenance. With a pump out of service, slop wax in the system can cool off. Slop wax has a high freezing temperature. Even a little bit of excessive cooling can lead to a cascade of colder slop wax-higher viscosity-lower flow velocity-colder slop wax until the system solidifies. Returning systems to service after repairs or maintenance can be difficult.

Reliable level control is also difficult in this service. Many instruments fail. Poor level control can lead to pump problems. The most reliable equipment uses gamma ray level detection on the pump suction pipe or on the slop wax collector.

All these problems can be solved. The solutions cost money and may require considerable effort and attention to stay in service. For these reasons, the majority of plants do not use slop wax recycle.

7. Alternatives

If the slop wax is not recycled and cannot be used as a product, it must still go somewhere. The alternative is to put the slop wax into the bottoms product. Figure 4 shows two options for external flow from the slop wax collector. Figure 5 shows internal flow inside the vacuum tower. Both the external flow and internal flow configurations can be used in units with or without a stripping section. In vacuum towers without stripping sections, the external flow may go to the bottoms product line from the vacuum tower.

7.1. External Flow to Stripping Section, Boot, or Bottoms

Table 1 shows that approximately ½ of the units with the slop wax going to the tower stripping section, boot, or bottoms, have external flow. Of the units with slop wax with external flow, 2/3 use gravity flow and 1/3 use pumped flow. Gravity flow is shown on the right side of Figure 4 and pumped flow is shown on the left side of Figure 4.

The major purpose of external flow is to measure the flow rate of the slop wax. An accurate flow measurement makes optimization to minimize wash oil more straightforward and accurate. Based on field review of flow measurement installations, most flow meters in this service are incorrectly installed. In addition, if special measures are not taken, the slop wax may contain vapors that make the measurement inaccurate. With properly installed equipment [9], flow measurements can be made in this system.

Full optimization of wash oil requires analyzing the slop wax stream as well as measuring its flow rate. With an analysis, the amount of entrainment versus the amount of overflash can be estimated. The analysis method uses the amount of a non-volatile material, such as sodium, in the sample to estimate the entrainment rate. The purpose of the pumped systems is to have enough pressure to take a sample for analysis. With both the slop wax rate and the slop wax entrainment known, the wash oil rate can be fully optimized.

However, only about half the towers with no slop wax recycle have either a flow meter or a sampling point. The same problems that plague operability and maintenance of slop wax recycle systems also create problems with external flow meters and sampling stations.

If present, the pump required in this service is a low-head pump so has fewer problems. Nevertheless, all the problems with pump suction level control persist, as do problems with slop wax freezing.

In addition, taking and cooling the sample is difficult. The sample stream is very hot and handling requires extensive safety precautions. The sample systems tend to be unreliable (they plug up) and expensive. Overall, many plants have decided the problems and costs of the sample outweigh the potential benefit.

7.2. Internal Flow to Stripping Section or Boot

Internal flow is the simplest method of running the system. It has few maintenance or safety issues. Special piping and collector tray arrangements can make it highly reliable. The cost of internal flow is that neither the flow rate nor the composition are known. The most effective control method with internal flow is to use temperature profiles in the wash bed to estimate the wash oil rate [1]. Internal flow is recommended for most units.

8. Anecdotes

Based on interviews with plant operating personnel, two points merit further discussion.

- Based on operating experience, units with functioning stripping trays show dramatic performance changes when the stripping trays are damaged. This is evidence that the stripping trays function as expected.
- Based on operating experience, units with slop wax recycle often show little to no performance changes when the slop wax pumps are out of service. This is evidence that slop wax recycle systems often under-perform expectations.

The major reason that slop wax systems might underperform is that the slop wax contains little overflash. In some units slop wax is nearly 100% entrainment. In these units the slop wax is material that did not vaporize at feed conditions. That material will not vaporize the second time it is heated and sent to the tower either.

9. Conclusions

Slop wax is a combined distillate-entrainment stream from the de-entrainment section of a refinery vacuum tower. Slop wax recycle should be an effective way to increase distillate yields in vacuum units. Nevertheless, just over a third of vacuum units recycle the slop wax. A small number take the slop wax as a product. However, half of the surveyed vacuum units send the slop wax to a stripping section, the vacuum tower boot, or directly to the tower bottoms stream. The reason for this is mechanical difficulty and expense in keeping slop wax systems running. Maintenance cost and effort often exceed benefits.

Additionally, many units show little to no performance change when the slop wax system is out of service. Under these conditions, the survey results showing that many plants use systems without slop wax recycle are unsurprising.

More expensive alternatives to increase unit yield include higher heater outlet temperatures and enhanced stripping sections. However, these have limits and are more expensive in most cases. They are more reliable within their appropriate limits.

External flow systems to improve control of the vacuum tower also suffer from many of the problems associated with slop wax recycle. The systems are expensive and difficult to keep in service. The alternative of sending slop wax to the tower bottoms with internal piping has the highest reliability and lowest capital cost, even though it may have slightly lower yields. Many plants find the reliability improvement and lower capital cost worthwhile.

10. References

[1] A. W. Sloley, "Vacuum Distillation: Wash Zone Configuration," presented at the AIChE Spring Meeting, San Antonio, TX, 2017, p. Paper 34a, Proceedings 479206.

[2] ASTM, "D2270-93, Standard Practice for Calculating Viscosity Index from Kinematic Viscosity at 40 and 100 C (Reapproved 1998)," ASTM International, West Conshohocken, PA, 1998.

[3] A. W. Sloley, "Crude Distillation Sequences," presented at the AIChE National Meeting, San Francisco, CA, 2013.

[4] R. N. Watkins, "Crude Assay: Tia Juana Light," in *Petroleum Refinery Distillation*, Houston, TX: Gulf Publishing Company, 1973, pp. 251–283.

[5] A. W. Sloley, "Refinery Vacuum Towers: State of the Art," presented at the AIChE National Meeting, San Francisco, CA, 2013.

[6] A. W. Sloley, "Crude Unit Stripping Sections," presented at the AIChE Southwest Process Technology Conference, Galveston, TX, 2017.
[7] API, "STD 610, Centrifugal Pumps for the Petroleum, Petrochemical and Natural Gas Industries," American Petroleum Institute, Washington, D. C.
[8] F. Korkowski and L. Glassburn, "Low Flow / High Head Centrifugal Pump Options," presented at the Calgary Pump Symposium, Calgary, ALB, Canada, 2005.
[9] A. W. Sloley, "Get Some Inside Information," *Chem Proc*, vol. 76, no. 8, pp. 43–44, Aug. 2014.

11. Figures

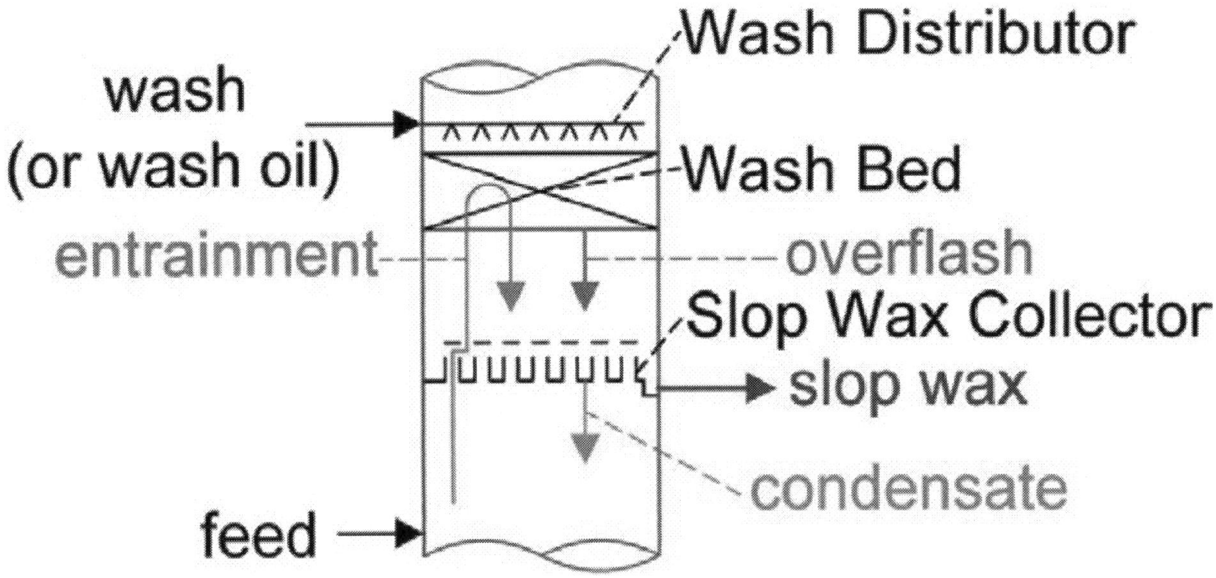

Figure 1 Wash Zone Terminology

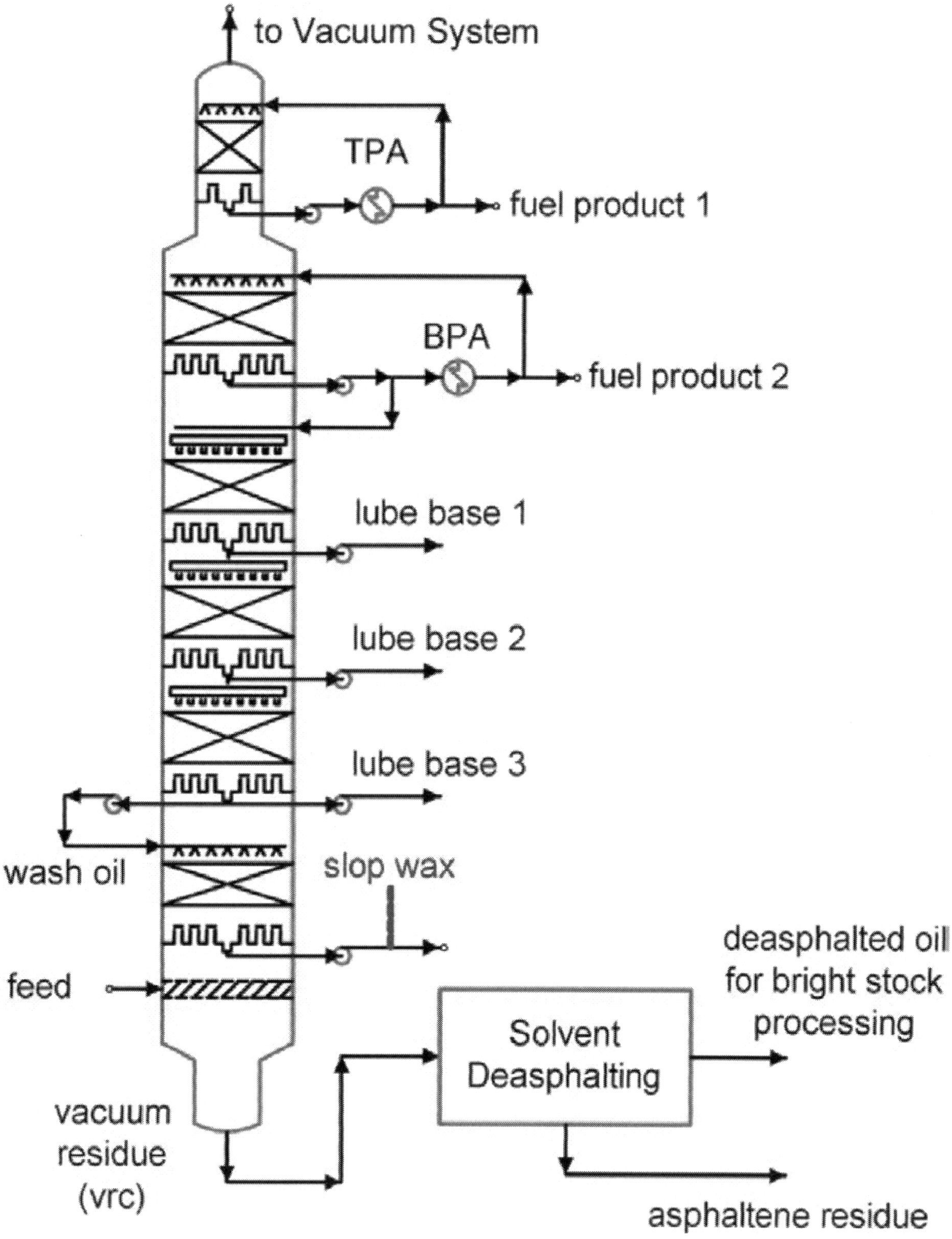

Figure 2 Vacuum Tower and Downstream Lubricant Processing

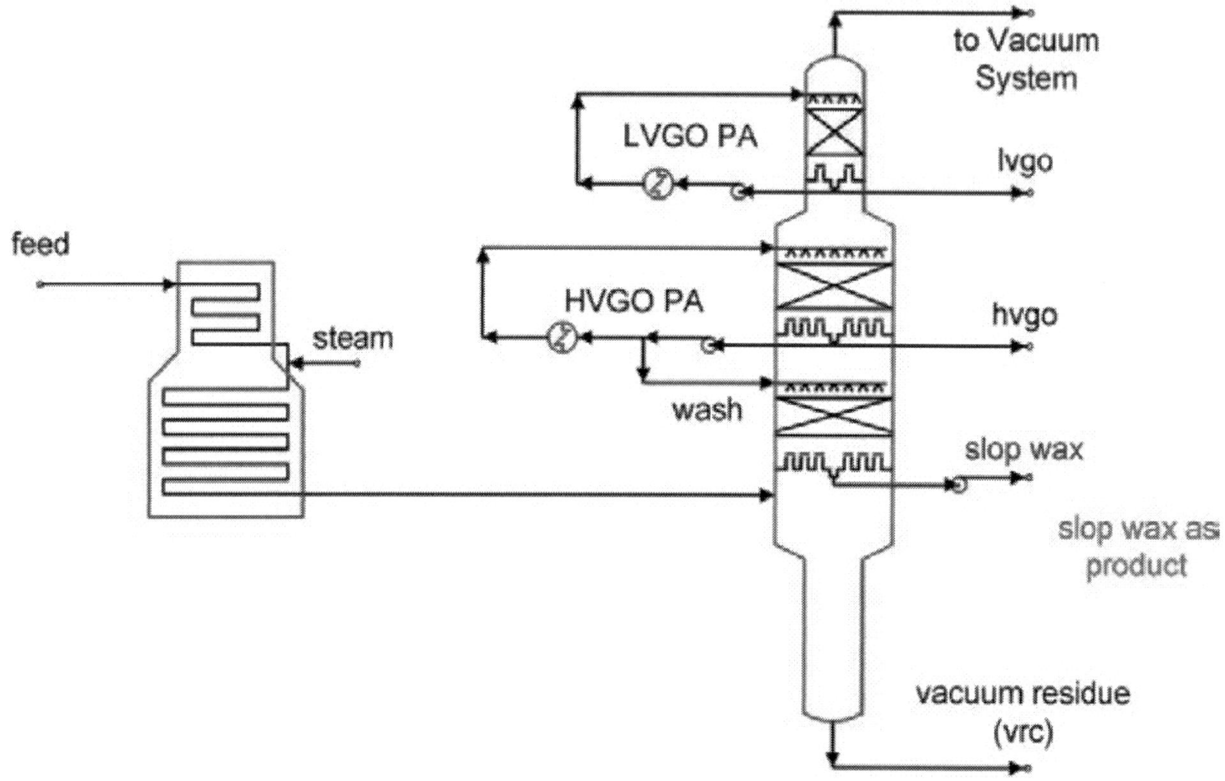

Figure 3 Slop Wax as Product

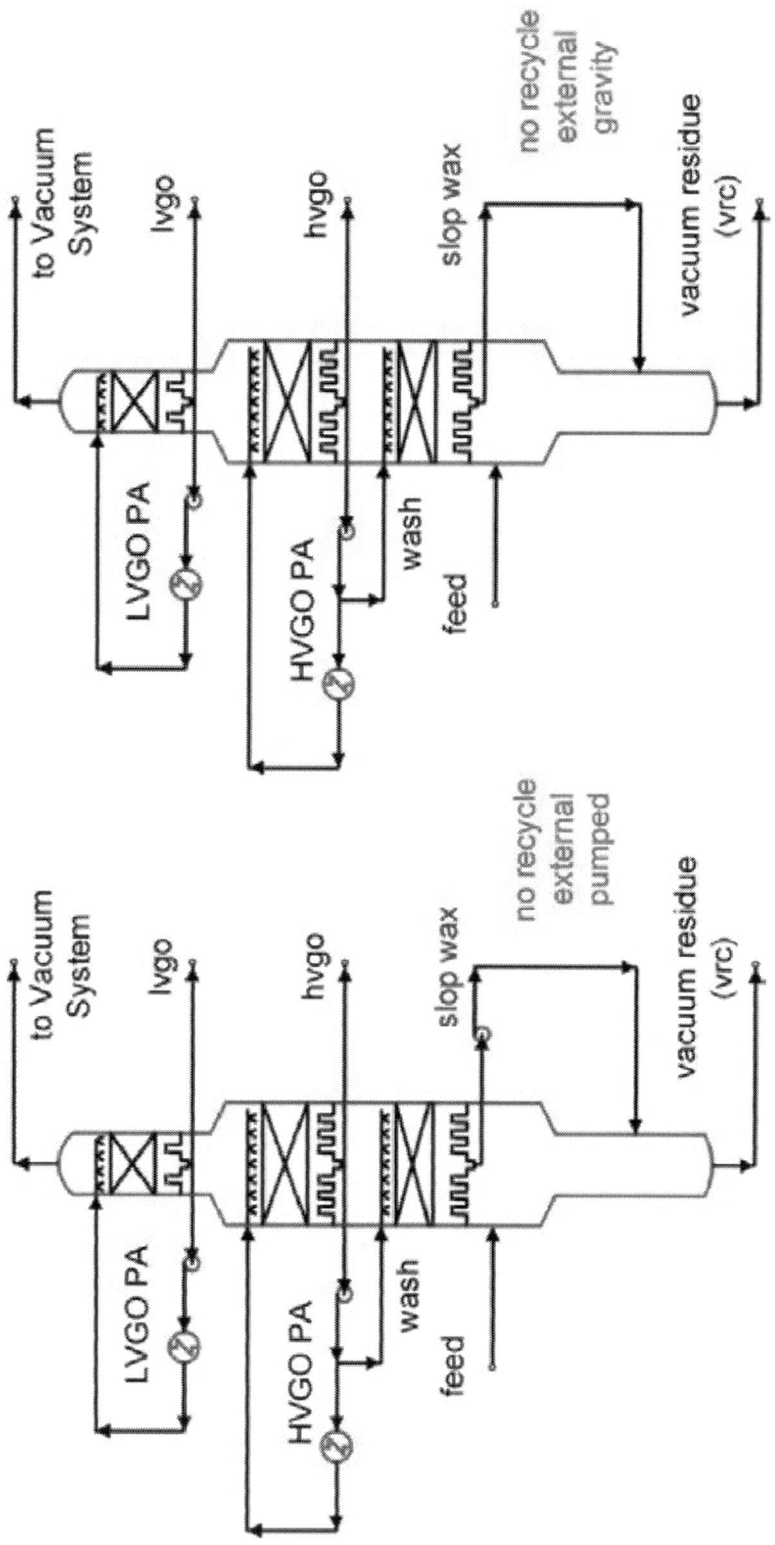

Figure 4 External Slop Wax Recycle (Pumped and Gravity Flow)

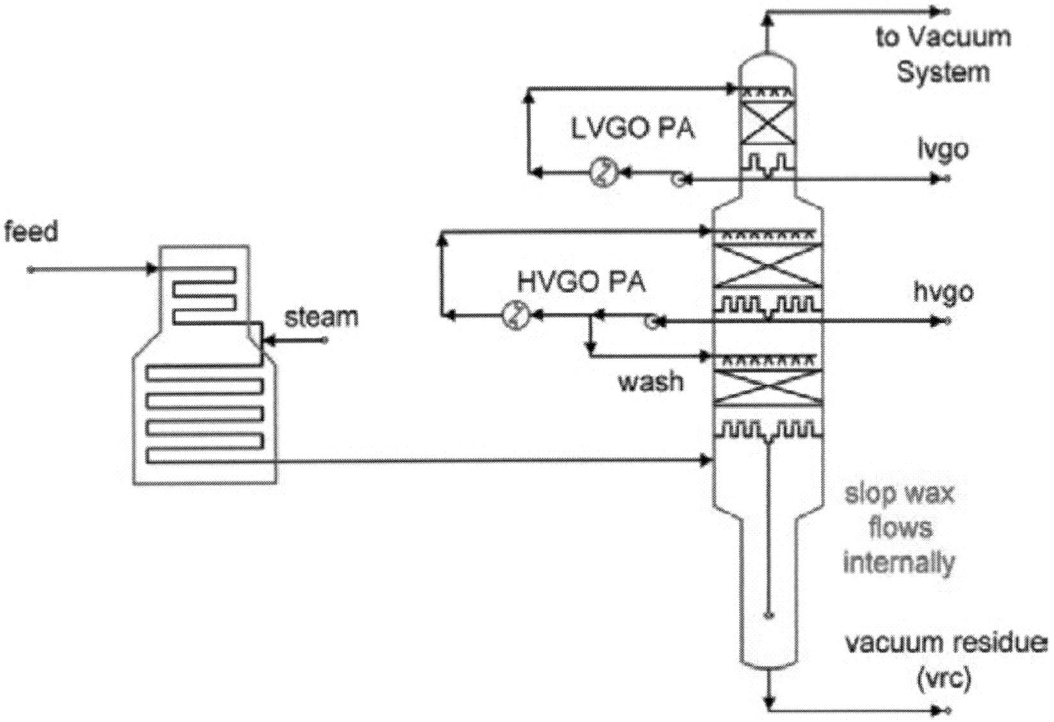

Figure 5 Slop Wax Internal Flow

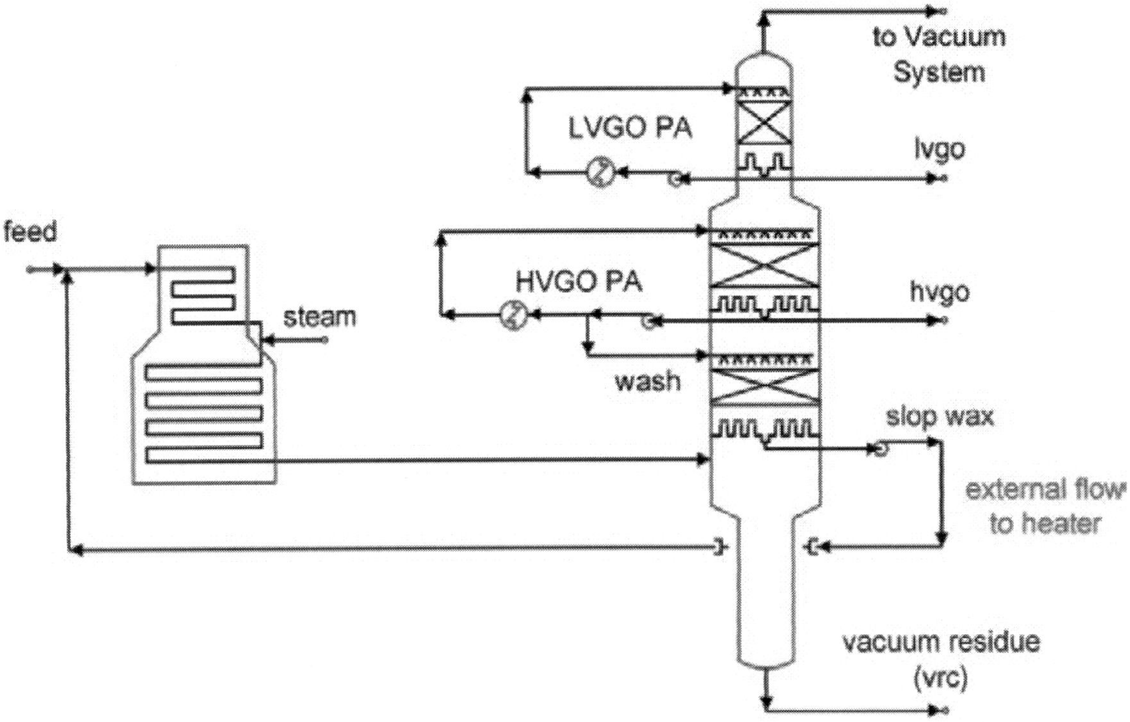

Figure 6 Slop Wax Recycle to Heater Feed

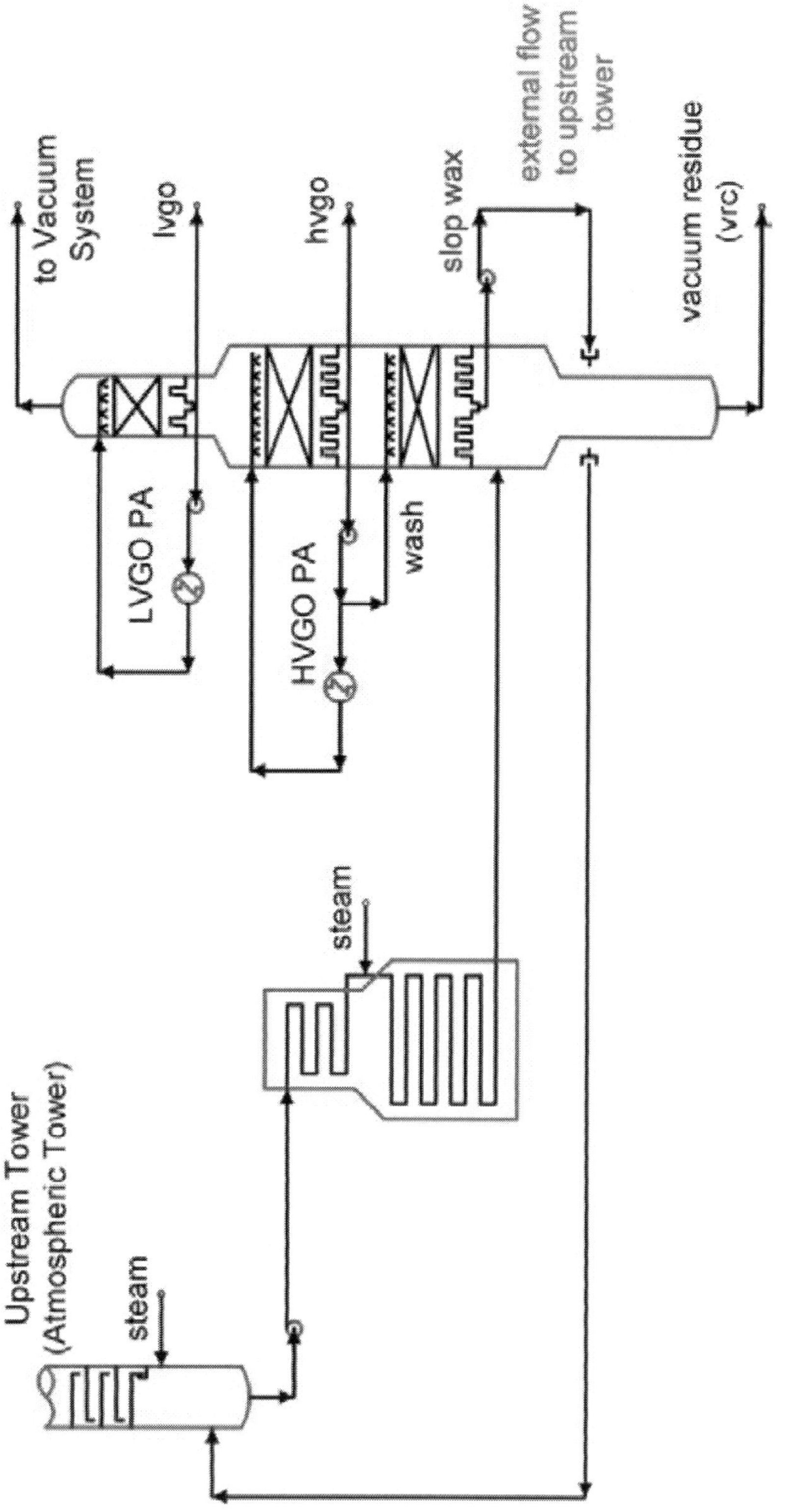

Figure 7 Slop Wax Recycle to Upstream Tower

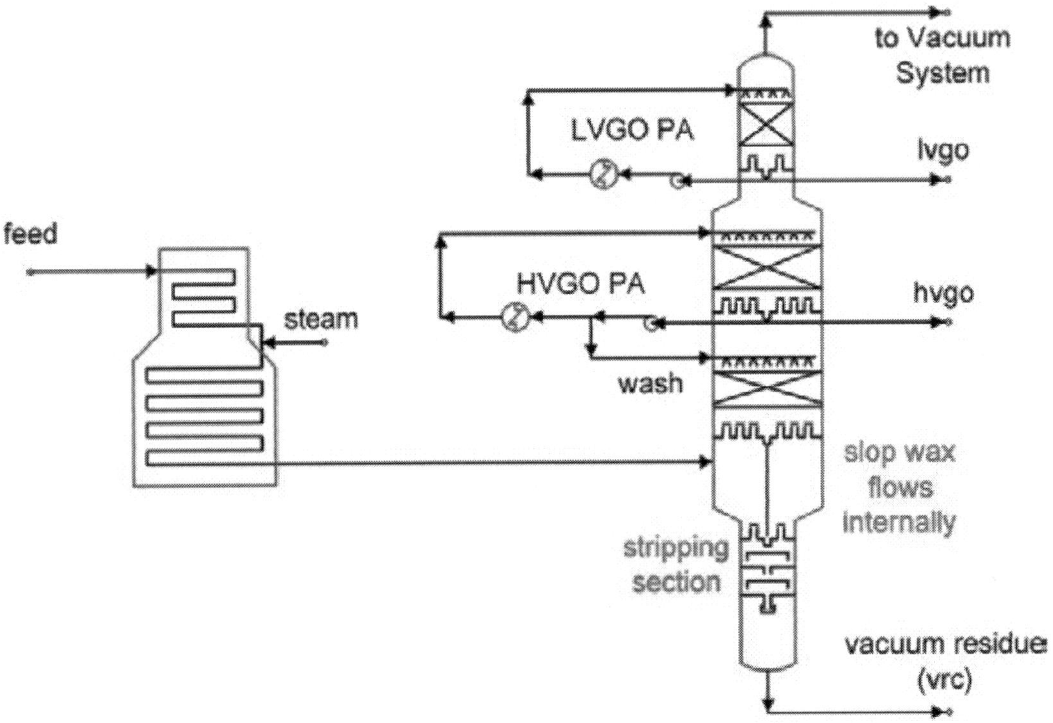

Figure 8 Slop Wax, No Recycle, Internal Flow to Stripping

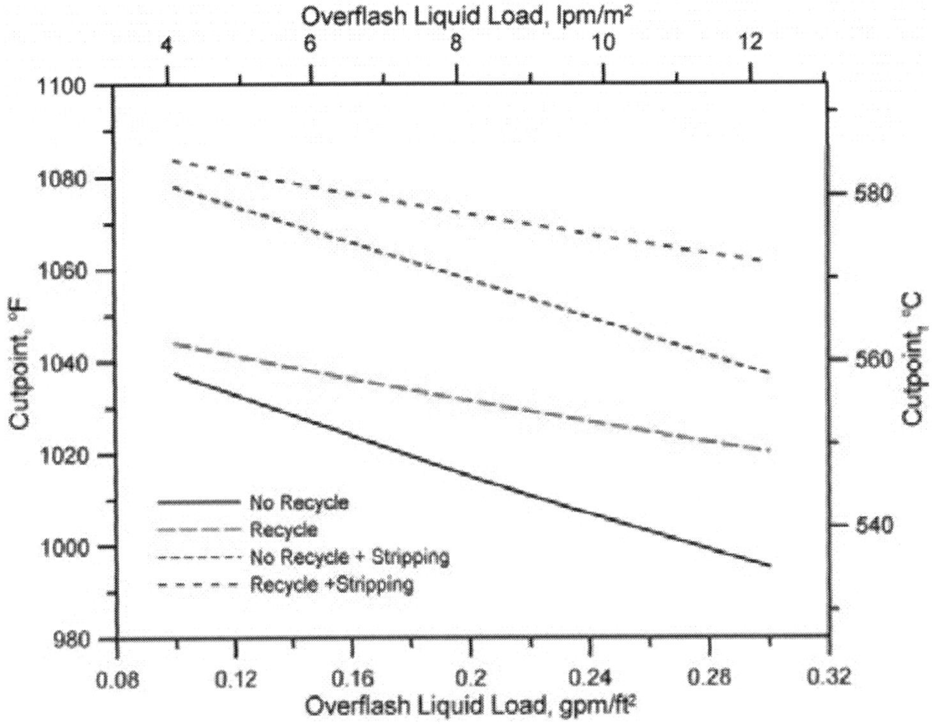

Figure 9 Slop Wax Recycle Yield Benefits

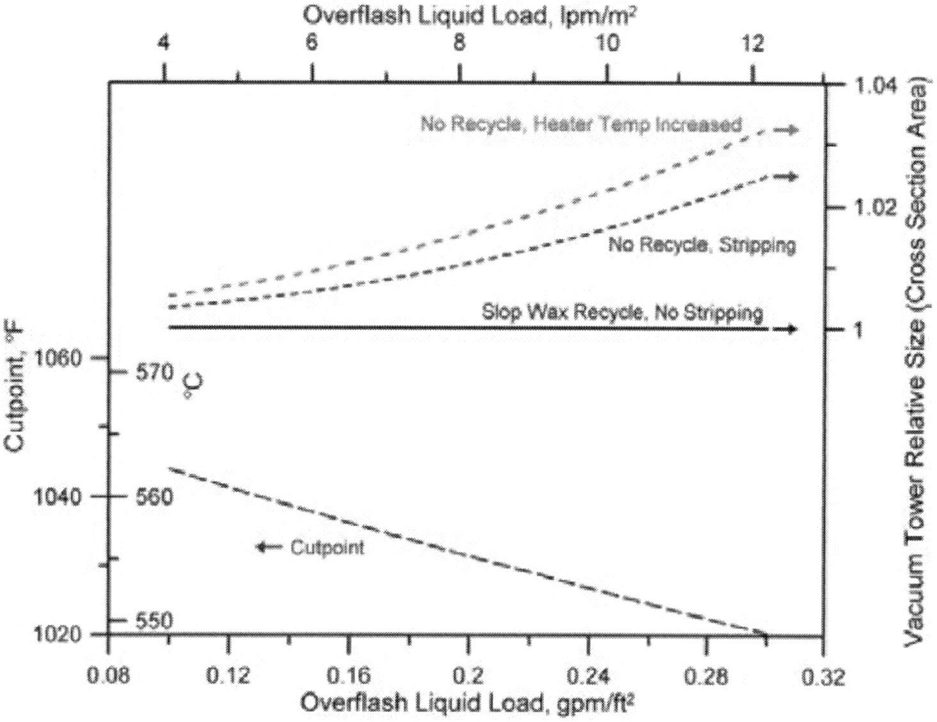

Figure 10 Slop Wax Recycle versus Higher Heater Outlet Temperatures or Stripping, Fixed Yields

489909 Acid Regenerator Optimization for Increased Unit Throughput

Abstract:

This discussion presents steps taken to increase unit throughput by pushing existing equipment beyond historical maximum rates using an acid regenerator tower case study in an HF Alkylation unit. Process changes at the acid regenerator contributed to record breaking alkylate production in March of 2017.

Acid regenerator optimization included process step tests, tower scans, and evaluation of adjacent equipment capacity. Regenerator optimization allowed for over 15% tower feed rate increase. This supports demonstrated acid strength recovery at increased unit throughputs for record breaking monthly average alkylate production.

Vacuum Distillation: Fractionating Wash Zones

Paper 175e

Proceedings 483296

Andrew W. Sloley, Principal Consultant, Advisian

Prepared for Presentation at the
American Institute of Chemical Engineers
2017 National Meeting
Minneapolis, MN, October 29 – November 3, 2017

www.**advisian**.com

Advisian
WorleyParsons Group

Overview

Wash Section

- Mainly considered a de-entrainment device
- Limited literature discussion on wash section as fractionating bed for controlling content of volatile metal-containing compounds:
  - — Importance of vacuum system type
  - — Importance and costs vs. benefits of reflux vs. product quality
- Nickel and vanadium porphyrins most important commercially
- Nickel and vanadium affect downstream fluid catalytic cracking and hydrotreating catalysts

Advisian / 3

Wash Section

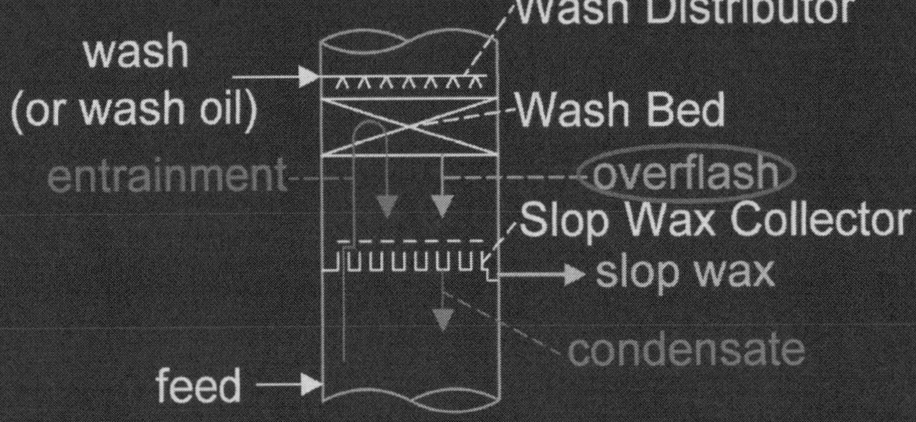

Advisian / 4

Without Stripping

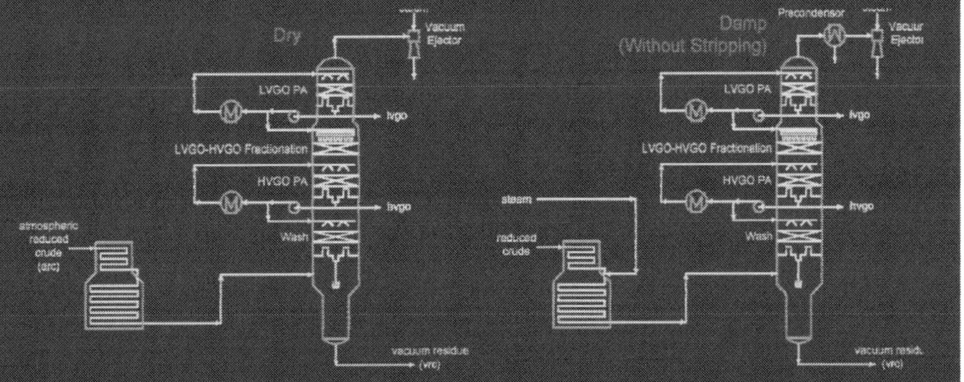

Advisian / 5

With Stripping

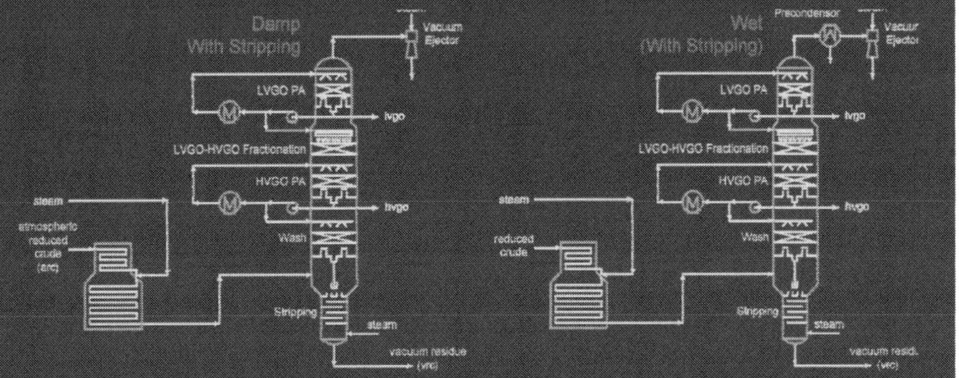

Advisian / 6

Basis

- Volatile metal data extremely difficult to obtain
- Comparison against previous work
- Reverse-calculate metals distributions
- Duplicated 1991 publication results to within round-off error
- Looking at:
  - Nickel (Ni)
  - Vanadium (Va)
  - Conradson Carbon Residue (CCR)

Advisian / 7

Feed TBP (Arab Atmospheric Residue)

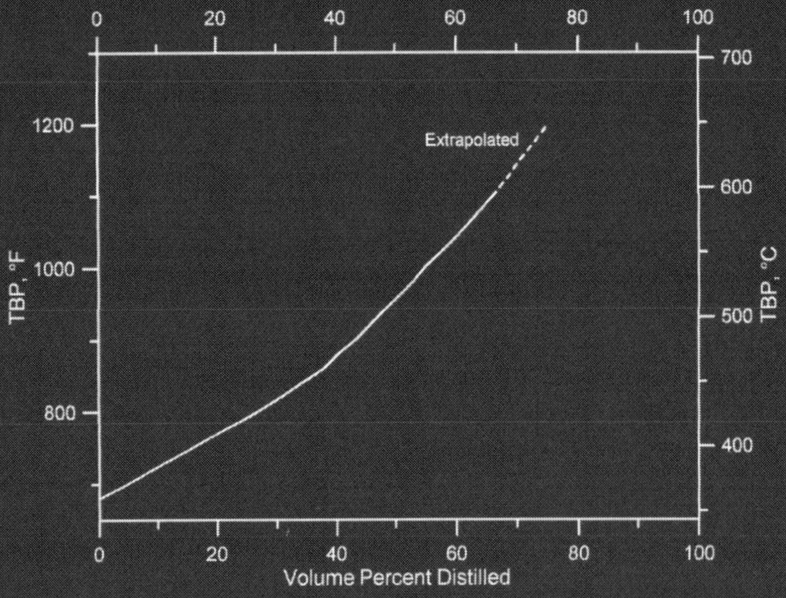

Advisian / 8

Feed Properties

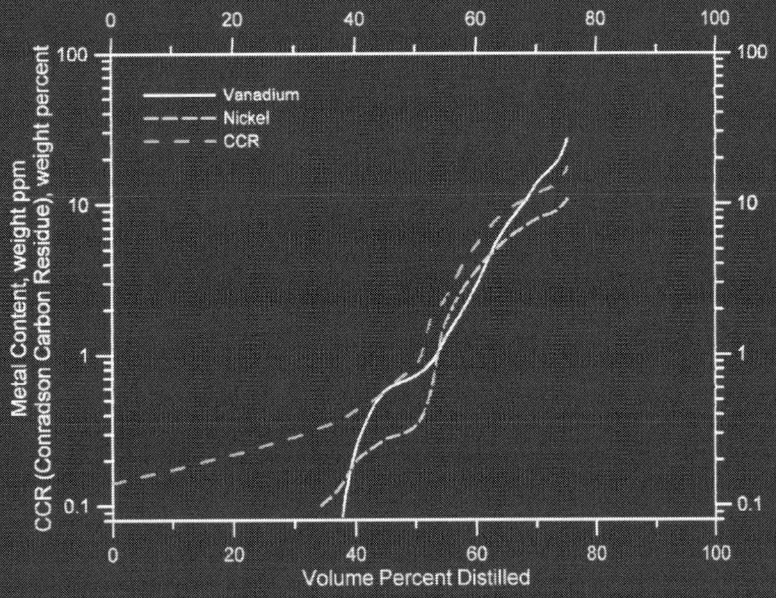

Advisian / 9

Type Comparison (Duplicates 1991)

Operation	Dry	Damp 1	Damp 2	Wet
Heater steam, lb/hr		5918	2948	2948
Stripping steam, lb/hr			4004	4004
Upstream cutpoint, °F	680	680	680	680
Vacuum tower cutpoint, °F	1,063	1,063	1,063	1,063
Overflash rate, vol % on feed	1	1	1	1
Flash zone pressure, mmHg	18	35	35	66
Wash stages		1	1	1
Stripping stages			1	1
Ni, wppm	1.93	2.10	1.56	1.79
Va, wppm	3.10	3.40	2.43	2.84
CCR, wt percent	2.09	2.16	1.84	1.94

Advisian / 10

Old Conclusion and Follow-Up

- Stripping sections inherently improve product quality (lower Va, Ni, CCR)
- No real mechanism proposed
- Always thought partial pressure and relative volatility was the answer
- Time to test conclusion
- Keep operating types constant and vary partial pressures
- Vary damp partial pressures by varying both system pressure and steam rates

Partial Pressure Determines Separation

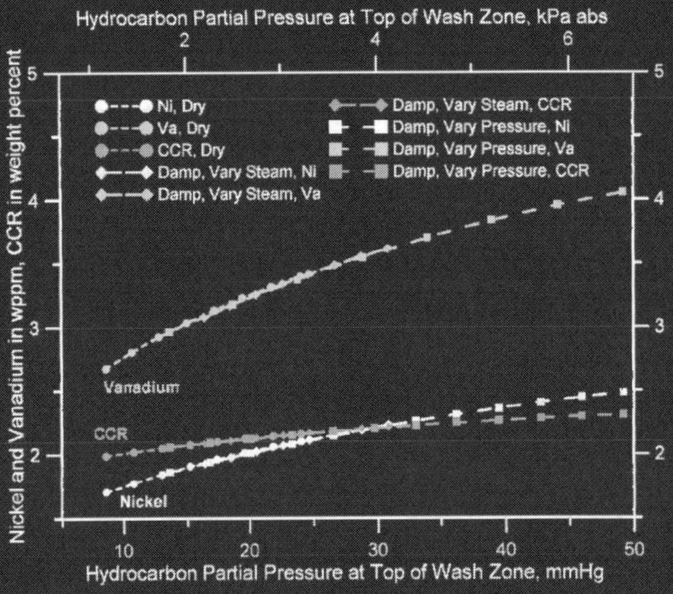

1063°F (573°C) Cutpoint

New Conclusion

- Partial pressure determines product quality improvement
- At lower partial pressures hydrocarbon compounds have higher relatively volatilities
- Low partial pressure comes from either:
  — Steam injection
  — Low operating pressures

Advisian / 13

CCR vs. Overflash Rate

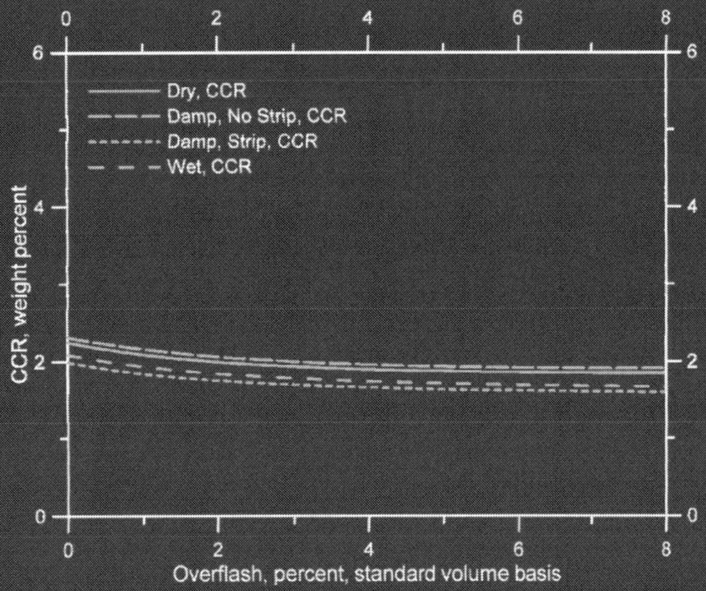

1063°F (573°C) Cutpoint

Advisian / 14

Nickel Concentration vs. Overflash Rate

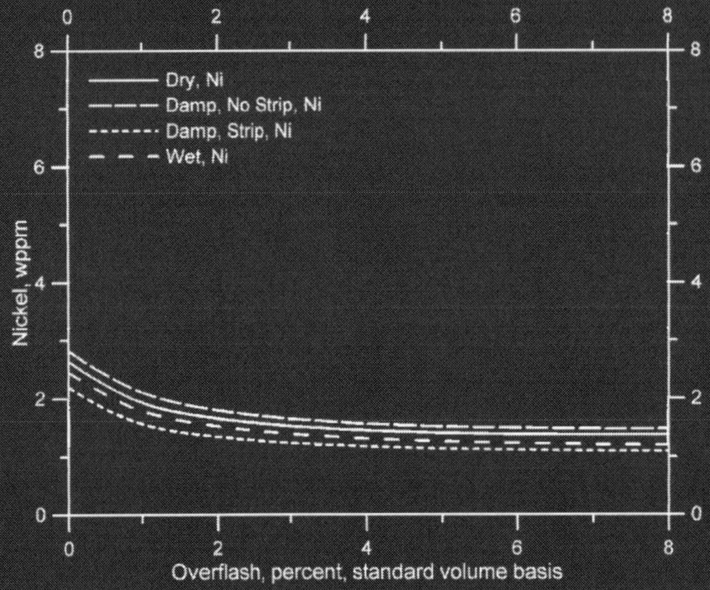

1063°F (573°C) Cutpoint

Advisian / 15

Vanadium Concentration vs. Overflash Rate

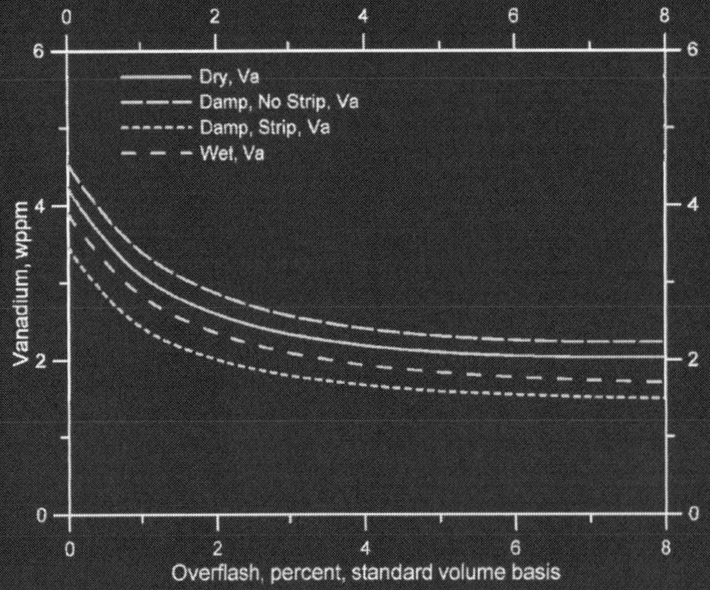

1063°F (573°C) Cutpoint

Advisian / 16

Other Factors Limiting Overflash

- Irrigation required for reliable operations (reduces coking rate):
  - ~3 to 4% overflash required for reliable commercial operation
- This pushes operation into flat part of performance curve
- Little benefit for overflash vs. fractionation on most commercial units: agrees with plant observation
- Impurity concentration becomes highly correlated with yield (cutpoint)
- Other observations:
  - Vanadium has largest response for this crude
  - Vanadium is "best case" for justifying process changes
  - Only show vanadium results for other options

Advisian / 17

Alternatives

Advisian / 18

Alternatives to Reduce Volatile Metals

- More wash stages
- More stripping stages
- Slop wax recycle

- Lower pressure operation – large equipment, high capital

Wash Stages

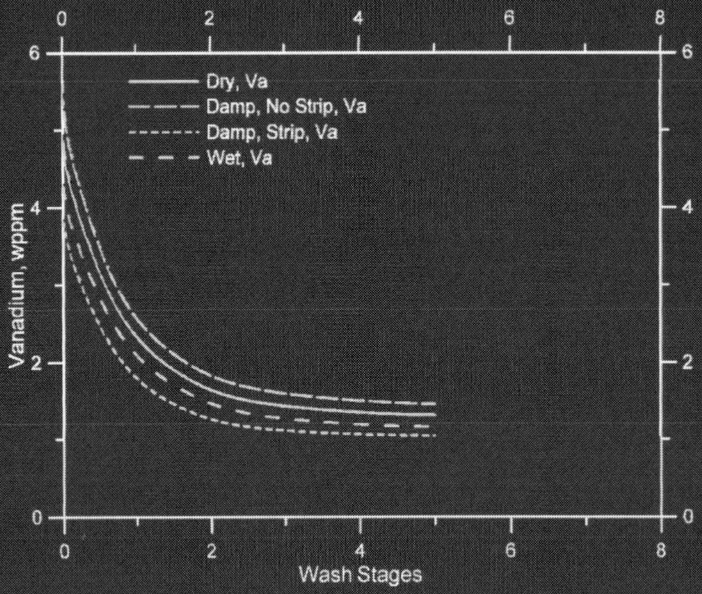

1063°F (573°C) Cutpoint, 3% Overflash

Stripping Stages

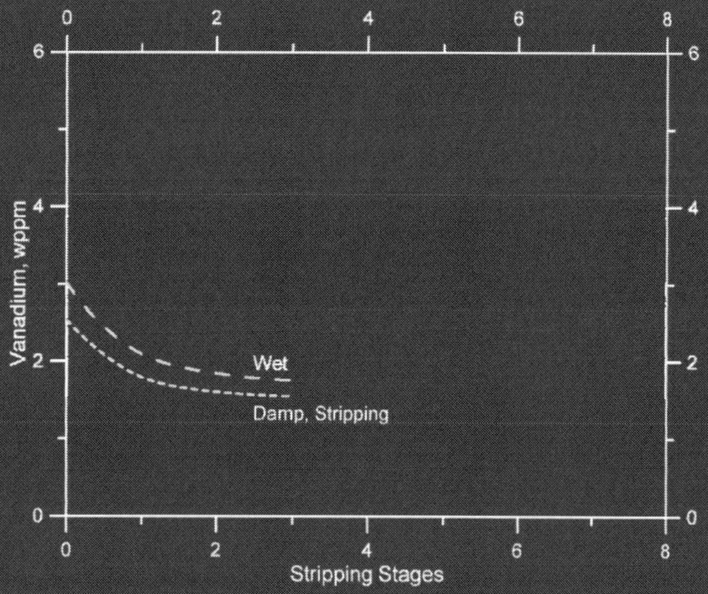

1063°F (573°C) Cutpoint, 3% Overflash

Advisian / 21

Slop Wax Recycle

Operation, without recycle	Dry	Damp 1	Damp 2	Wet
Ni, wppm	1.52	1.65	1.24	1.39
Va, wppm	2.33	2.57	1.79	2.09
CCR, wt percent	1.94	2.00	1.70	1.78

Operation, with recycle	Dry	Damp 1	Damp 2	Wet
Ni, wppm	1.44	1.56	1.19	1.34
Va, wppm	2.17	2.40	1.70	1.99
CCR, wt percent	1.89	1.96	1.68	1.76

1063°F (573°C) Cutpoint, 3% Overflash

Advisian / 22

Reducing Volatile Metals

- More wash stages: modest improvement
  - — Very little benefit past 2 stages
- More stripping stages: small improvement
  - — Very difficult to do
- Slop wax recycle: very small improvement
  - — Not worth the cost

Advisian / 23

Conclusions

Advisian / 24

Reducing Volatile Metals

- Commercial units have little benefit for increasing overflash rate:
  — Reliability sets a minimum rate high enough that extra overflash has little value
- More wash stages produce a modest, but possibly useful benefit
- More stripping stages and slop wax recycle produce small benefits – usually not worth the cost

Advisian / 25

Disclaimers

DISCLAIMER

This presentation has been prepared by a representative of Advisian.

The presentation contains the professional and personal opinions of the presenter, which are given in good faith. As such, opinions presented herein may not always necessarily reflect the position of Advisian as a whole, its officers or executive.

Any forward-looking statements included in this presentation will involve subjective judgment and analysis and are subject to uncertainties, risks and contingencies—many of which are outside the control of, and may be unknown to, Advisian.

Advisian and all associated entities and representatives make no representation or warranty as to the accuracy, reliability or completeness of information in this document and do not take responsibility for updating any information or correcting any error or omission that may become apparent after this document has been issued.

To the extent permitted by law, Advisian and its officers, employees, related bodies and agents disclaim all liability—direct, indirect or consequential (and whether or not arising out of the negligence, default or lack of care of Advisian and/or any of its agents)—for any loss or damage suffered by a recipient or other persons arising out of, or in connection with, any use or reliance on this presentation or information.

Advisian / 27

About Us

Advisian
WorleyParsons Group

Consulting and Engineering in

Refinery and Petrochemicals

- Molecule Management
- Economics and Planning
- Technology Selection
- Refinery-Petrochemical Integration

Logistics

- Hydrocarbon Logistics
- Bulk liquid terminals
- Railway corridors
- Market analysis
- Risk Analysis
- Safety and Environment

Sulfur and Gas Treating

- Oxygen enrichment
- Gas treating solvents

Infrastructure and Environmental

- Water
- Materials handling

EcoNomics

Advisian / 28

Advisian
WorleyParsons Group

Backup Information
Basis Crude
Performance
(Tia Juana Light)

CCR vs. Cutpoint

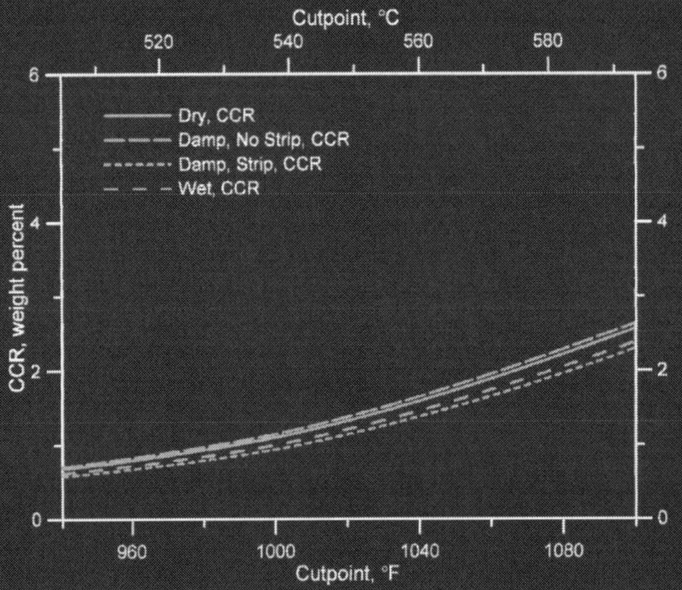

Advisian / 31

Nickel vs. Cutpoint

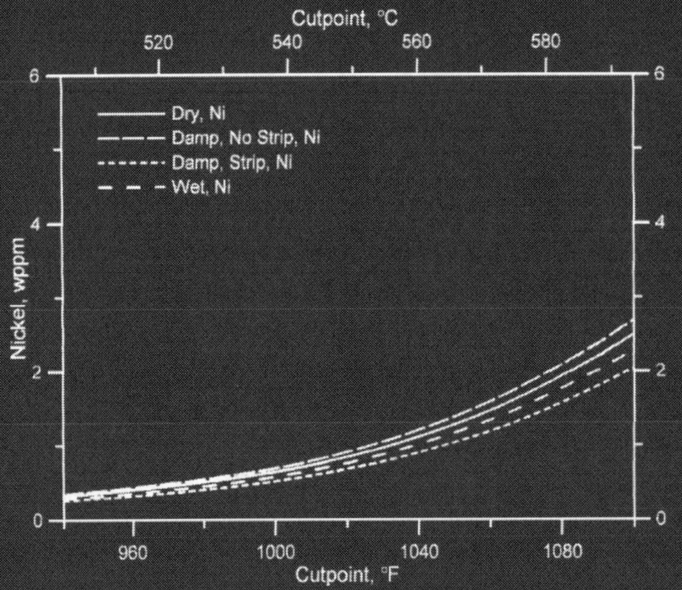

Advisian / 32

Vanadium vs. Cutpoint

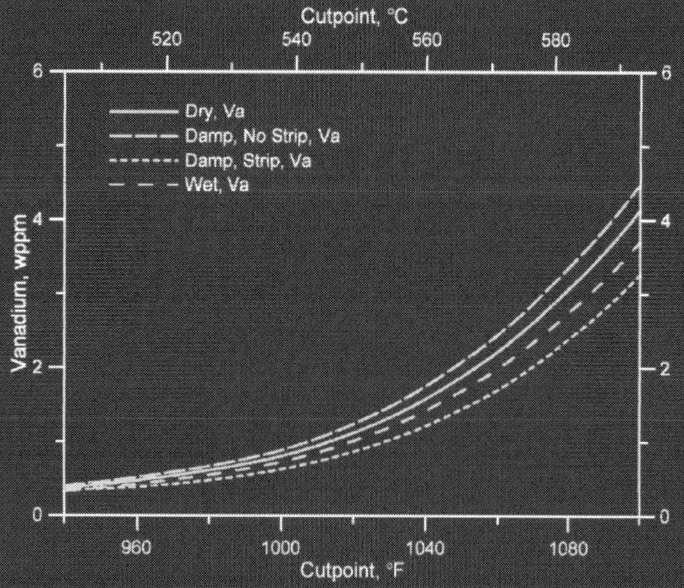

Backup Information
Crude Metals Contents

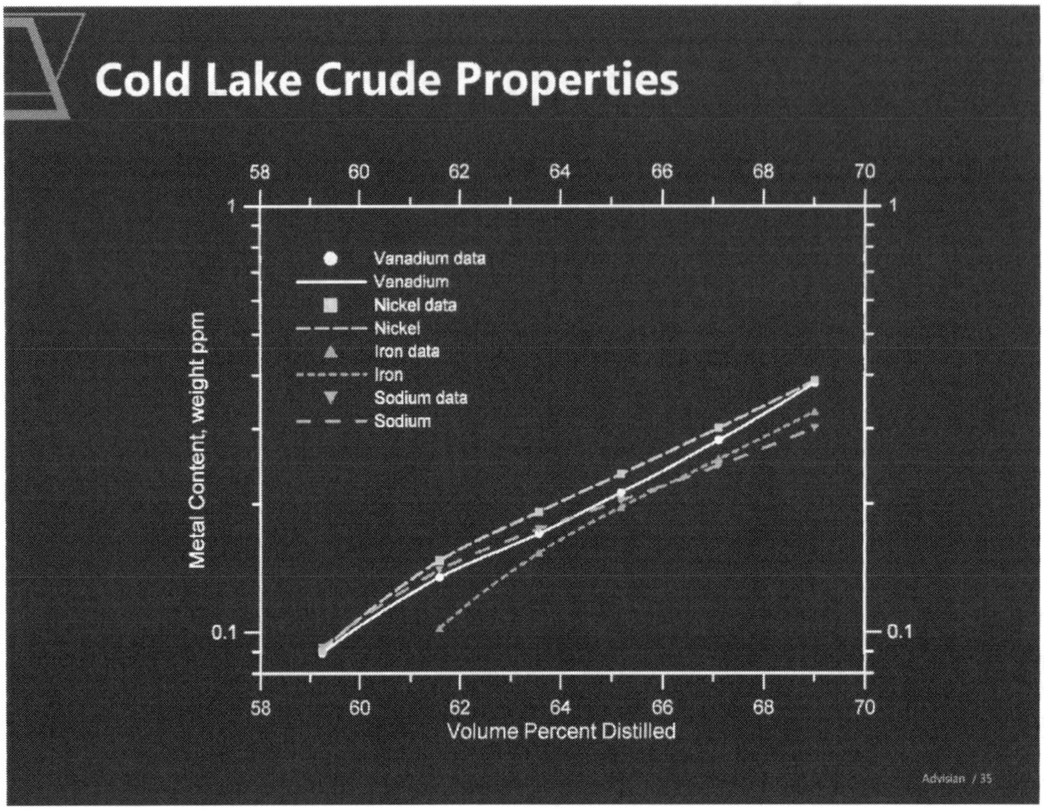
Cold Lake Crude Properties
Vanadium data
Vanadium
Nickel data
Nickel
Iron data
Iron
Sodium data
Sodium
Metal Content, weight ppm
Volume Percent Distilled
58
60
62
64
66
68
70
1
0.1
Advisian / 35

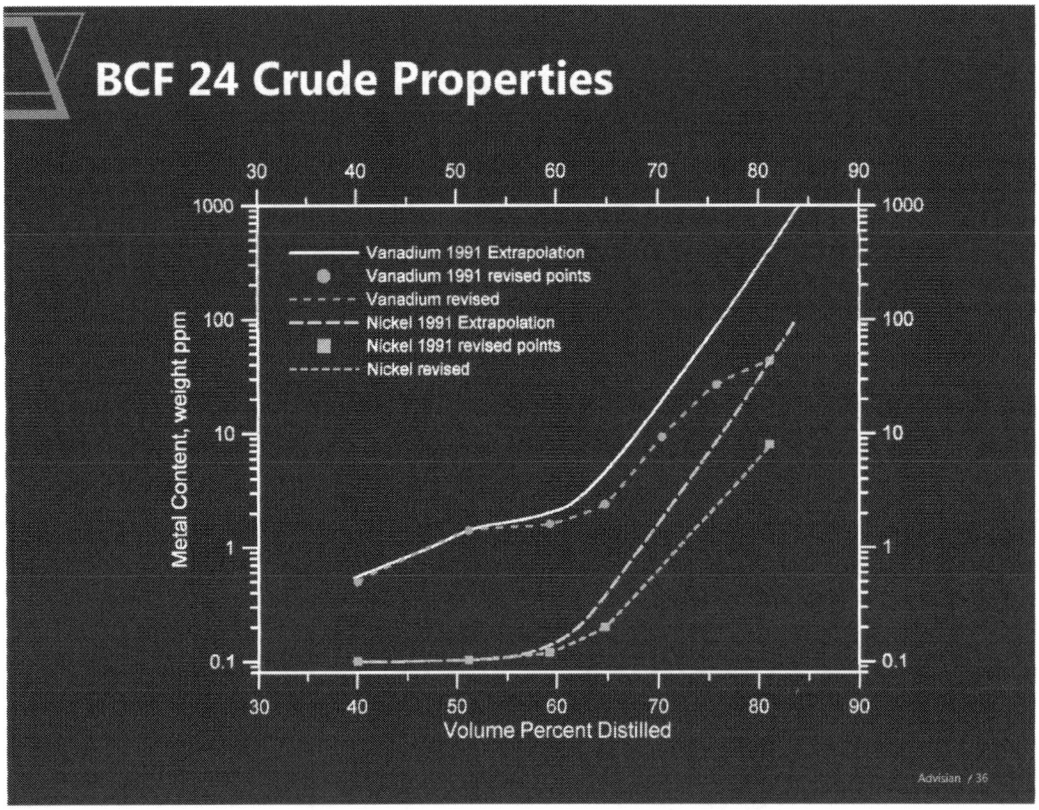
BCF 24 Crude Properties
Vanadium 1991 Extrapolation
Vanadium 1991 revised points
Vanadium revised
Nickel 1991 Extrapolation
Nickel 1991 revised points
Nickel revised
Metal Content, weight ppm
Volume Percent Distilled
30
40
50
60
70
80
90
1000
100
10
1
0.1
Advisian / 36

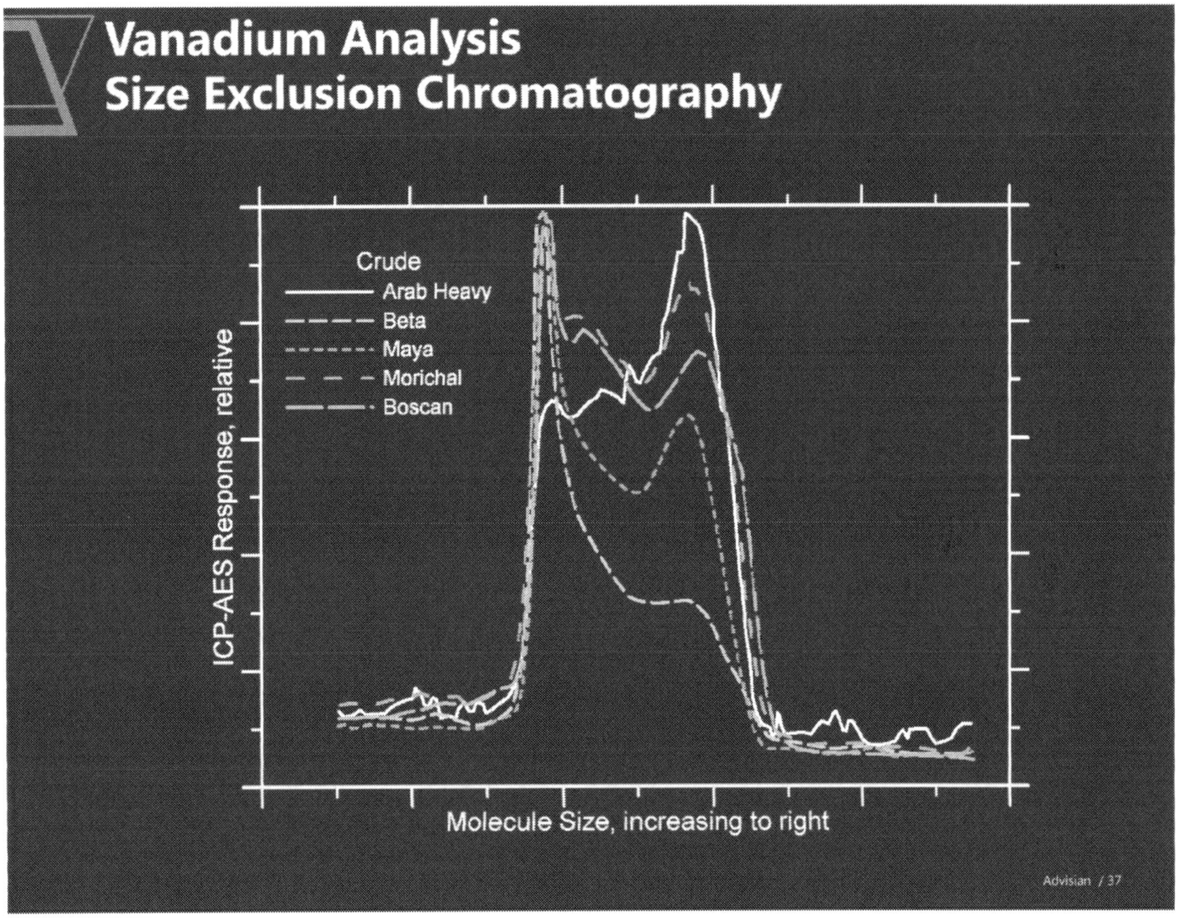
Vanadium Analysis
Size Exclusion Chromatography
Crude
Arab Heavy
Beta
Maya
Morichal
Boscan
ICP-AES Response, relative
Molecule Size, increasing to right
Advisian / 37

Advances in Salt Point Management Practices: Mitigating Overhead Corrosion with Big Data

Collin W. Cross, Ph.D.; Wan Zhaoyang, Ph.D.; Keyur Patel, Ph.D.
9-27-2017

Abstract

In the world of overhead corrosion, it has long been recognized colloquially that ninety percent of metal loss occurs during ten percent of the time in operation. In many types of fractionating tower overheads, a confluence of fluctuating dynamical factors lead to the formation of amine hydrochloride salts and ongoing associated corrosion. Because of the constantly changing tendency to form corrosive salts in an operating unit, it has remained difficult using traditional approaches to systematically identify the proverbial ten percent of time where damage-causing episodes occur so that effective intervention and mitigation are possible.

The challenges above have caused the fundamental principles of overhead corrosion mitigation to remain largely unchanged in the last decades. While advances in metallurgy, inhibitors, injection practices, and monitoring techniques have occurred, overhead corrosion measurement and predictive practices have not. The result is that crude unit overhead systems continue to experience corrosion that leads to unexpected failures, fouling, outages, all contributing significantly to lost profit opportunities.

Salt point management practices continue to rely largely on series of unconnected salt point calculations over time using sparse data sets. Here we will present a new computational framework that allows continuous and quantitative sensitivity computations for amine hydrochloride salt formation in an operating crude unit. This capability then leads to the rapid identification of seemingly random corrosion events and their precise causes. During an emerging corrosion event, the onset can be rapidly flagged and alerted while the driving forces of salt formation and their sensitivities are quantitatively revealed. The demonstrated method takes near real-time data from operating crude units and uses it to compute how changing factors, such as crude diet, processing objectives, operations, control schema, physical constraints, set points, and randomly varying amine and chloride levels, cause corrosion events. The practical objective of the methodology is to drive timely, precise, and proactive mitigation options in a granular and systematic fashion across a wide variety of event types. Real-world examples taken from operating refineries will be used to demonstrate the technique.

Introduction

In the world of overhead corrosion, it has long been recognized colloquially that ninety percent of metal loss occurs during ten percent of the time in operation. In many types of fractionating tower overheads, a confluence of fluctuating dynamical factors leads to the formation of amine hydrochloride salts and ongoing associated corrosion. Because of the constantly changing tendency to form corrosive salts in an operating unit, it has remained difficult using traditional approaches to systematically identify the proverbial ten percent of time where damage-causing episodes occur so that effective intervention and mitigation are possible.

The challenges above have caused the fundamental principles of overhead corrosion mitigation to remain largely unchanged in the last decades. While advances in metallurgy, inhibitors, injection practices, and monitoring techniques have occurred, overhead corrosion measurement and predictive practices have not. The result is that crude unit overhead systems continue to experience corrosion that leads to unexpected failures, fouling, outages, all contributing significantly to lost profit opportunities. Bundle washes continue to yield unequivocal evidence of amine hydrochloride salt-induced corrosion that went un-detected and, therefore, un-mitigated.

Salt point management practices surrounding overhead systems continue to largely rely on series of unconnected salt point calculations over time using sparse data sets around key thermodynamic drivers. Salt point deposition temperature computations continue to be largely performed manually, meaning users must provide and manage over time all data necessary for calculating results. The result of this situation is that overhead corrosion mitigation practices fall into a spectrum of sophistication surrounding mitigation options. At the most basic level of the spectrum, corrosion control depends primarily upon injecting corrosion inhibitors and maintaining traditional service and monitoring metrics, such as pH, chlorides, iron throw, ER data, coupons, inspection, etc.

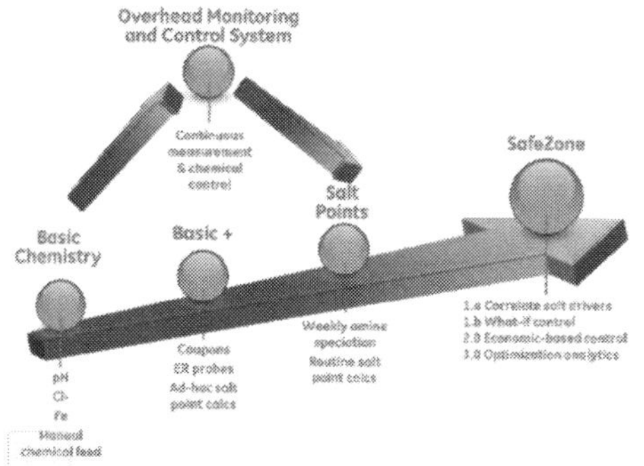

Figure 1: Continuum of corrosion mitigation and monitoring practices, shown as moving from more simple to more complex.

Figure 1 shows a way to conceptualize the progression of corrosion management practices from more basic to more advanced. Currently, the bulk of refineries globally do not use high frequency, wide spectrum amine speciation as a regular part of their corrosion control program. Salt point deposition temperature computations are often used, but are not granular enough, predictive enough, reliable enough, nor do they provide enough operational or economical information to be practically useful towards optimizing a crude unit in a quantitative way. In cases where amine speciation analyses are routinely performed at sufficient frequency, often the number of amines sampled are insufficient. Alternatively, if enough amine species are sampled, either routinely or for specific events, the labor-intensive traditional method of performing calculations with such large data sets largely prevent effective salt point analysis from being carried out daily as a routine procedure.

Here, we will present a new computational framework and supporting set of infrastructure and processes that allows rapid, continuous, and quantitative computations for amine hydrochloride salt formation in an operating crude unit to be easily performed. From this basis, the dynamic sensitivities, or gradient, of the salt point to its input variables is available and can be used towards the realization of many practical capabilities. Furthermore, in addition to rendering the salt point and its gradient in close to real-time, the dynamic variabilities of each input parameter, as well as their impact on the output, are additionally calculated. This capability then leads to automatic detection of emerging corrosion events, enabling their rapid and dynamic identification, as well as their precise causes and sensitivities. During an emerging corrosion event, the onset of the conditions making salt deposition likely are flagged and alerts are sent digitally to concerned parties. At the same time, the driving forces of salt formation and their sensitivities are quantitatively revealed and deconstructed graphically and tabularly into a group of indicators and actionable options towards mitigation. The demonstrated method automatically takes multiple sources of asynchronous, high-frequency data from operating crude units and uses it to compute how changing factors, such as crude diet, processing objectives, operations, control schema, physical constraints, set points, randomly varying amine and chloride levels, cause corrosion events, and other factors affect salt formation. Sophisticated signal processing and analytics algorithms are then used to make a wide array of valuable capabilities become practical, seamless, and potentially fully automated. All of these computations are done subject to the actual physical and operational constraints on the unit during its operation, which are handled quantitatively. The practical objective of the methodology and its supporting infrastructure is to drive timely, precise, and proactive mitigation options in a granular and systematic fashion across a wide variety of event types to improve reliability and productivity of the asset being monitored.

Computational Framework for Automated Salt Point Calculation

Salt point calculations require operational data and chemistry data from the crude distillation tower overhead system. Operational data required for salt deposition temperature calculations are temperature, pressure, reflux flow rate, product flowrate, and feed flow rate, etc., whereas required chemistry data are a regularly repeating wide list of individual amine concentrations and the chloride concentrations. Usually, high frequency operational and process data are readily available from controllers installed in a crude distillation column. Chloride levels in accumulator boot water can be cost effectively and reliably obtained at high frequency by a variety of methods. Additionally, accumulator boot water pH data are also easily measured and tabulated as a time series. These data are all then retrieved into a central data storage platform, which is a key part of the computational pipeline. After being received, data is then trended and analyzed to identify and monitor specific corrosion events as they emerge. This enables proactive actions to be continuously evaluated towards optimal mitigation by means which are in line with processing objectives.

Accumulator boot water amine concentrations are normally the rate limiting factor in the deployment of high frequency salt deposition computations. The lack of a suitable processing schema has also made the routine manual calculation for salt points of all amines widely known to affect corrosion seem cost prohibitive. The value and usage of such a data set to provide tangible operational predictive power has remained elusive due to both the expense and complexity of amine speciation as well as the inability to effectively convert a series of broad-based, high-frequency amine data into reliable and actionable information. Here, we will show that using such data is very worthwhile. The combination of broad-based, high-frequency amine data and a high speed automated processing pipeline can allow previously large, unseen, and/or randomly occurring corrosion events to be systematically identified and revealed for mitigation in a proactive rather than reactive way. The ability to identify, understand, and prioritize a practical list of weighted options towards the mitigation of large and previously unseen corrosion events on a regular daily basis can have a dramatic impact on unit reliability over the length of a production run. This is especially true during the processing of challenging crudes.

To complete the computational process, all the high-frequency telemetry discussed above is extracted hourly, daily, or semi-weekly, in an automatic fashion, and is used in a new platform called **SafeZone*** – a proprietary, automated, salt-point analytics framework, offered exclusively by **SUEZ - Water & Solutions Technologies**. This is done via a suite of integrated hardware, data warehousing, high performance computing, analytics, expert analysis, and reporting. The platform uses the output of the computations to provide a set of tools which enable the ability for a team of people to graphically optimize salting risk against a refinery's production objectives on an ongoing basis with little training or effort.

The suite of tools and the data analysis pipeline for the SafeZone tool suite is designed with digital, team-oriented communication and decision-making support as key objectives. Because of this system, the tools are rigorous, systematic, and automatic, but distill the data into a simple set of streamlined information widgets that are capable of being displayed using a diverse set of reports built around them. This system is then able to provide a large array of powerful and standardized information to a dispersed group of people operating in a busy environment. The information is provided in a very flexible, yet structured and quantitative way. This provides a platform for rapid, streamlined, and quantitative decision making by the team over a wide variety of dynamic situations as they emerge during a production run.

At the heart of the **SafeZone*** process is a sophisticated set of proprietary algorithms developed by **SUEZ** scientists and engineers that use a state-of-the-art high-performance computing infrastructure to implement a stochastic, feed-forward control schema in a crude fractionators multi-dimensional state space and subject to all its constraints, which can also be modeled for evaluation. The results of the complex mathematical analysis are used to provide graphical output as a set of indicators in a variety of expert tools, visualization dashboards, and structured reports. This platform is then used to iteratively understand the drivers of both systematic and episodic corrosion and relate them to the objectives of the crude unit as well as its constraints and control scheme. These elements are then used to iteratively optimize an operating crude unit on a day-to-day basis (or faster) using a team-oriented stochastic feed-forward decision loop facilitated by the **SafeZone*** framework.

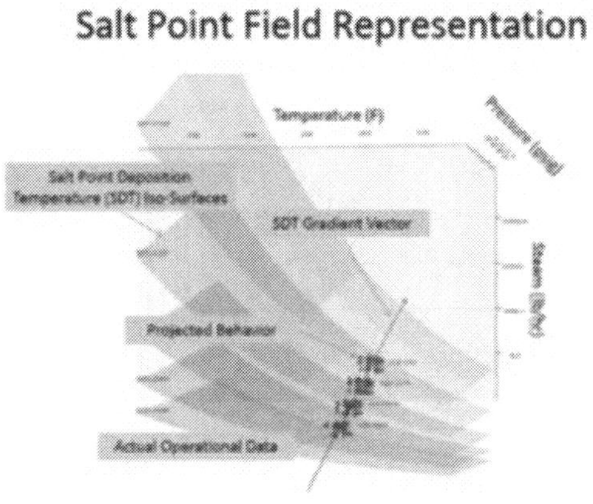

Figure 2**: Cross section of multidimensional state space containing the scalar and vector fields computed via the **SafeZone high performance computing platform.*

Figure 2 shows a cross section of the multidimensional state space containing the scalar and vector fields for the margin differential between salt point deposition temperature (SDT) and tower top temperature and the SDT's gradient. **Figure 3** shows a graphical schematic of the telemetry gathering process and computational framework as well as a snapshot of an example dashboard.

For ensuring reliable mitigation of corrosion for arbitrary emerging events, it must be ensured that the key telemetry has sufficient granularity to enable a sufficient numerical estimate of salt deposition temperatures as well as their sensitivities and variabilities to both the controlled and stochastic aspects of the multidimensional input space. The algorithms are then used to predict the probability of a salting event occurring daily given the state of the system, its immediate history, and its path-dependent trajectory through the local input space. The probability of salting computed at the core of the process is called the "SPI," or Salting Probability Index. This index represents the probability that salting will occur during a current or historical time, given the location of the system in state space and the variability of the input data and confidence intervals around the output data. The dynamic rendering of the SPI is what drives the core of the **SafeZone*** process. **Figure 4** shows a two-dimensional, graphical representation of the SPI within a manifold of its input space.

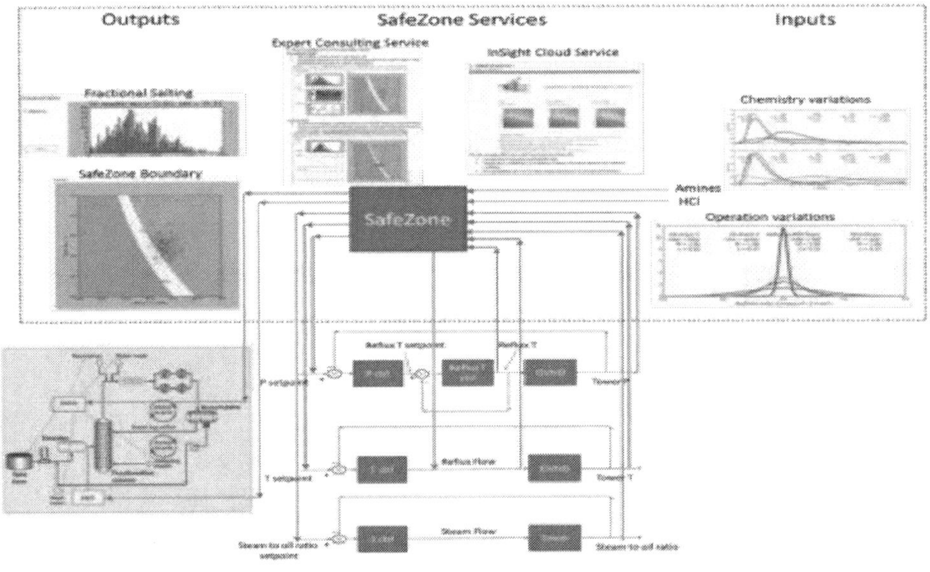

Figure 3: *Computational Framework for Automatic Salt Point Calculations*

Amine Testing Requirements

Traditionally, to eliminate corrosion risk from amine hydrochloride salt deposition, a typical margin, or differential, of 15-25 degrees is assumed and then maintained between either the mixed exit temperature (MET) or tower top temperature (TT), respectively, and the salt deposition temperature (SDT). However, because amine and HCl measurement frequencies are often limited by sparse data for a variety of reasons, using these various margins as indicators has not been found to be a reliable means to predict corrosion risk. To address this challenge, the salt point calculation algorithms used here consider also the random variations of operational and chemistry variables over a given timeframe. These are then used to compute a more representative indicator of salt formation potential, called the salting probability index (SPI). Over a specified timeframe, and for several random variations of operational and chemistry variables, the SPI indicates the number of times salt formation occurs, divided by the total number of operational points being sampled. The SPI represents

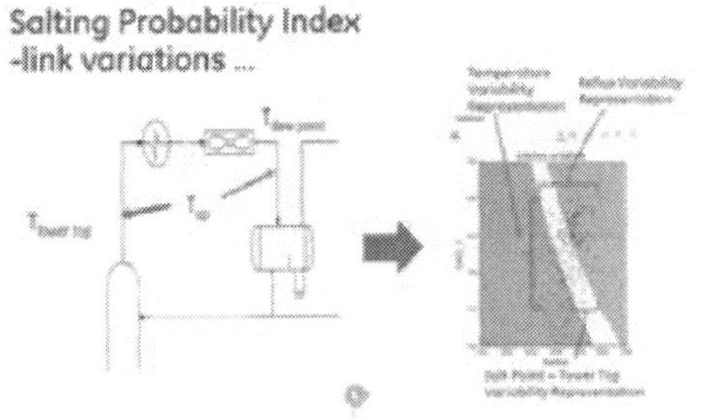

Figure 4: *Graphical representation of the integral SPI and its operational ensemble within a two-dimensional cross section of the fields input space.*

the dynamical probability of salting based on neutralizer, tramp amine, chloride concentrations, and the current operational situation.

After automatically gathering the data and performing the calculations above, the platform then reduces the information to graphical and numerical output and delivers high-frequency, advanced reporting via a variety of possible channels. The executive summary of the daily report shown below gives visual indications of salting in both the overhead condensing system and the tower top by showing both respective SPIs independently. **Figure 5** shows a snapshot of the executive summary section of the daily report. While the report itself is beyond scope, this is the first of many pages designed to offer unique insight into dynamic aspects of corrosion and their relationships to both operations and chemistry so emerging events can be detected, identified, categorized, and acted upon. Additional examples of routine weekly, monthly, and strategic reports are not shown here. Within the daily report, SPI values are plotted over a specified period and color coded to indicate whether salting occurs in the tower top or overhead system, respectively. Not only are SPIs presented, but sensitivities are also output for both operational and chemistry variables. This allows prioritization and ranking of multiple potential mitigation options to best bring the tower top or overhead system out of the salting region if desired. A large range of competing mitigation options can easily and systemically be evaluated and all stakeholders can have a similar viewpoint and a reference system by which to make quantitative comparisons over time.

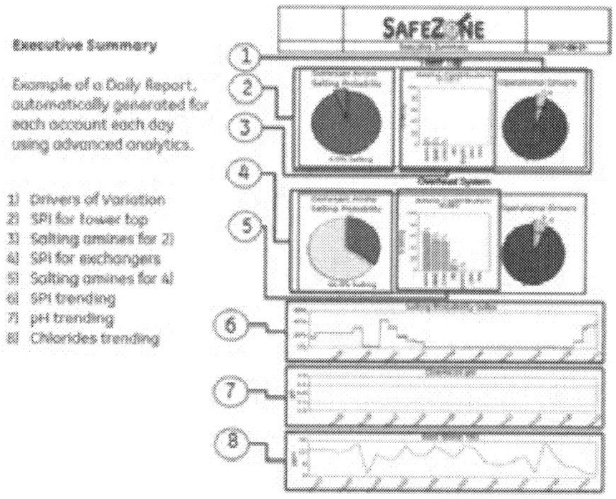

Figure 5: SafeZone daily report executive summary page. Detailed explanation is beyond scope. Executive summary answers 3 questions each morning. Q1) Is there a problem? Q2) If so, what is causing it? Q3) How can it best be addressed? The answer to Q3) extends onto the reports other 10 pages.*

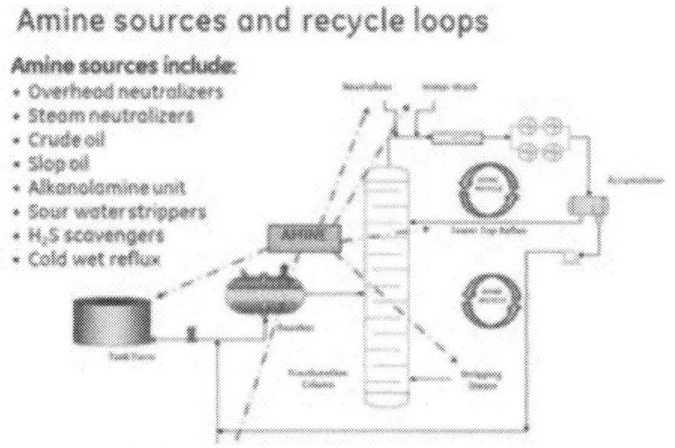

Figure 6: Multiple varying sources of entry to the system, along with multiple complex recycle mechanisms having different initiation criteria and time scales, along with the different physical properties of amines with respect to pH, boiling point, and relative oil/water solubility, cause the overhead of a typical refinery to experience a randomly fluctuating tendency (stochastic) for all the various amines to impact the amine hydrochloride deposition process and dependent ongoing corrosion.

The system also calculates many other useful parameters such as HCL margins. This is the dynamical amount of HCl reduction needed to move the system from a salting region to a non-salting region, or vice-versa. The HCl margins are one of many parameters automatically calculated and trended daily and used towards weighting and evaluating multiple potential mitigation options when needed. This parameter will become important for the case study discussed below. Additionally, along with salt deposition temperatures, mixed exit temperatures, and dew point temperatures, many other metrics are reported for easy comparison with

SPI charts to help unravel the driving causes of ongoing events and their relative sensitivities towards mitigating corrosion upon manipulation.

Amine Dynamics

As discussed above, the lack of an adequate computational platform as well as the expense and complexity of broad-based high-frequency amine speciation, have made it difficult to observe, document, and understand the complex, random, dynamical aspects of amine hydrochloride salt deposition that have been shown to underpin ongoing corrosion. While detailed discussion is beyond scope of this document, **Figures 6, 7, and 8** show various data representations that detail how different amines participate in the corrosion process over time and that MEA, while a predominate actor, is only one of many problematical amines that affect the corrosion process over time. Often, common suspect amines are completely misidentified or focused on as a primary driver of corrosion while other hidden causes remain unknown. This lack of granular visibility during operations often leads to the misidentification of more significant drivers, and thus often prevent ongoing problems from being effectively revealed, identified, and eliminated, eventually leading to subsequent unplanned impacts on reliability and production.

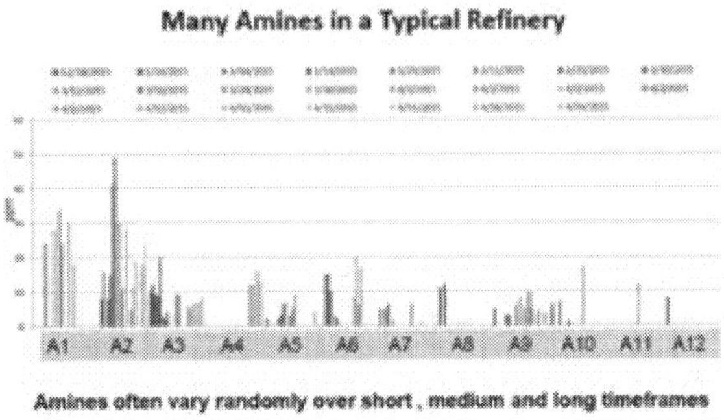

Figure 7: *Overhead receiver water amine concentrations sampled over a 4-month timeframe. Continuous daily sampling will reveal large random variations over time.*

Chloride	DMAE	Morph	MDEA	MA	DMA	TMA	MEA	NH3
7	171	249		4	11		10	16
10	129	178		3		4		17
10	82	122		8	4	10	22	10
12	92	138		4		12	14	25
14	40	54	34	3	3	10	24	11
15	76	124		3			13	14
16	132	196		4		6		15
20	198	270		4	4	14		15
60	145	241		2		10	23	12

Crude Unit Neutralizer
Steam Neutralizer
Plant Amines
Scavenger, Byproducts or Other

amine:chl ratio above SPBP @ 260°F

Figure 8: *Multiple samples of Salt Point Deposition (SDT) temperatures for a series of receiver samples taken over a 2-week timeframe. Color coded boxes are examples where the SDT is problematical from a corrosion perspective. Some of the randomly varying tendencies of the corrosion potential for multiple amines can be seen here.*

To effectively identify, understand, de-construct, categorize, analyze, down select, and prioritize effective and dynamic corrosion mitigation actions that have a meaningful affect upon the trajectory of overhead corrosion during an ongoing run, it is necessary to use high-frequency, broad-based amine speciation. Along with a comprehensive feed of amine speciation data, as well as the other telemetry, a high-performance processing and signal analysis platform makes it possible to gain visibility, understanding, and higher orders of control over the fluctuating and stochastic tendencies of amine hydrochloride deposition potential leading to active corrosion in the overhead of refinery fractionators. This, in-turn, has an ongoing and direct impact on operational reliability.

Example Study

To see how the **SafeZone*** technique differs from traditional methods of overhead corrosion control an example case is shown that illustrates how the **SafeZone*** technique was used to smoothly, reliably, and profitably, handle a very difficult opportunity crude with a bad reputation and a deep discount. The opportunity crude referenced here is corrosive to the overhead because it contains very high levels of volatile organic acids and the refinery had no previous experience running the crude. It is important to realize that there are three physical realities largely defining the data granularity requirements needed to achieve the given objectives. These three physicalities are:

1) Salt points cannot be computed unless there is a valid set of amine partial pressure estimates available.

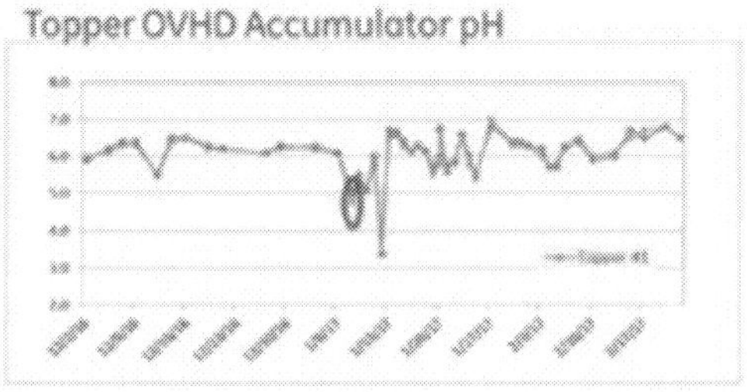

Figure 9: pH trace of overhead receiver during the introduction of a corrosive opportunity crude. pH quickly drops to near 3 until it is stabilized with an increase of the neutralizing amine DMAE.

2) The set of amine partial pressures driving overhead corrosion in most refineries is extremely variable and highly random.

3) The specific individual natures and patterns of the amine variations driving corrosion are often specific to a given situation, but fluctuate widely due to operations and other factors.

Because of these physical constraints, any salt point calculation methodologies attempting to predict salt points by use of recycle models of partial pressure and neutralizer feed rates are not sufficient. Furthermore, techniques seeking to base calculations on sets of assumptions wherein past amine values continue to be used longer than relevant (a few days usually) cannot reliably detect episodic corrosion events. This is true for both short and medium timeframes, which are normally the most important timeframes of interest from a proactive refinery overhead corrosion mitigation perspective.

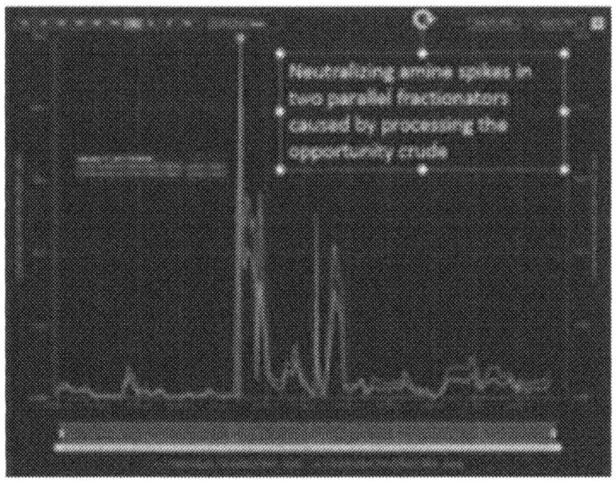

*Figure 10: **Insight*** data system which feeds the **SafeZone*** engine. The first DMAE spike on the left is the one covered by this case study.*

In this example, as the opportunity crude was introduced, the pH of the overhead receiver rapidly dropped to near 3, as shown in **Figure 9**. As a response to this observed change in the corrosion landscape, the neutralizer was quickly and manually increased to bring pH back into the proper control range of 5.5-6.5. Continuing throughout the run, which lasted several months, chlorides remained well controlled and always less than 15 ppm, with less than 10 ppm being the norm. Furthermore, no other indicators of corrosion had manifested until after the run was well in progress. Caustic was used initially at less than 0.5 pounds per thousand barrels to control chlorides and keep then less than 10 ppm and there was also an upper sodium spec on the residua streams of 10 ppm.

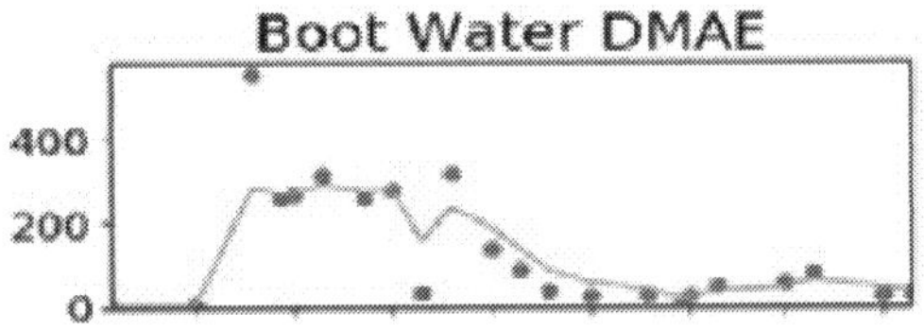

*Figure 11: Chart showing a trace of the high frequency amine data used to detect the DMAE spike, whose corrosion impact is being mitigated using **SafeZone***. The timescale of the trace is 2 months and begins just before the introduction of the increased neutralizer injection used to combat the drop in overhead pH.*

Normally, after the depressed pH was brought into range upon introducing the opportunity crude, no further actions would have appeared necessary using traditional corrosion mitigation techniques. This is because pH was in range, chlorides and caustic were low and controlled, and no iron throw or ER probe readings yet indicated a problem. Traditionally corrective actions would have ceased at this point because there was no reliable indication of active corrosion. Since organic acids and amines are not routinely speciated at high frequencies as part of normal service, the DMAE increase and other drivers of this problem, as well as many other potential dynamical contributions, would not likely have been detected or triggered any further corrective actions.

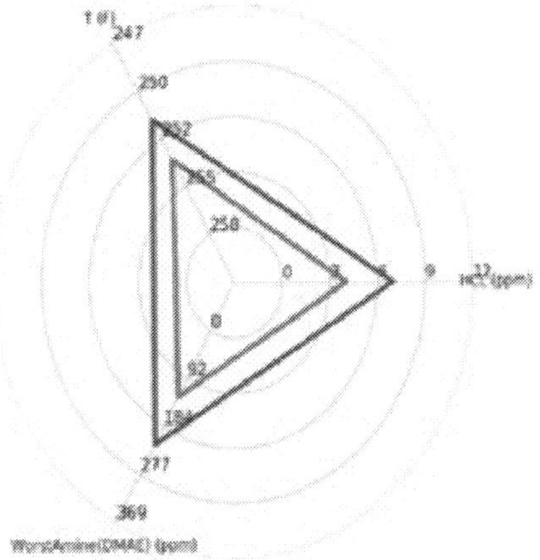

*Figure 12: Constrained optimization envelope chart. If the blue triangle is outside the red triangle the system is in the salting zone. The diagram also shows three geometrically equal ways to bring the system back into the **SafeZone***. Since the system is now in the salting zone, it would theoretically require approximately 3 degrees of temperature, vs. 3 ppm of chloride reduction, vs. nearly 100 ppm of DMAE reduction to achieve the necessary lowering of salt point. Combinations of the above changes can also be utilized if desired.*

In this case, however, when the neutralizer was increased to bring pH into range, the **SafeZone*** SPI indicator immediately alerted a high probability of induced salting and forced a response from the **SafeZone*** team. The salting event was quickly found to be due to the neutralizing amine itself, and its salt point information was dynamically computed based on the actual low levels of amines and chlorides in the system and the actual operational state of the unit at that particular time.

Figure 10 shows a data trace of two DMAE spikes occurring during the opportunity run, while **Figure 11** shows only the first spike in expanded scale. This is the spike of interest and is the one being primarily covered in this case. The first spike occurred between January 13th and February 14th and caused significant corrosion. The second event was better controlled and corrosion was mitigated effectively. Because the **SafeZone*** program detected a high likelihood of salting directly at the event's onset, the chloride margins defined above were evaluated. The team determined from the daily report and offline "what-if" analysis tools that if chlorides could be lowered to 4 ppm then salting would no longer be favorable at the current conditions. **Figure 12** shows how a constrained envelope chart from the **SafeZone*** report can be used to quantitatively evaluate the impact of several options. Many such charts are available for a wide variety of situations as the complete gradient vector is always available.

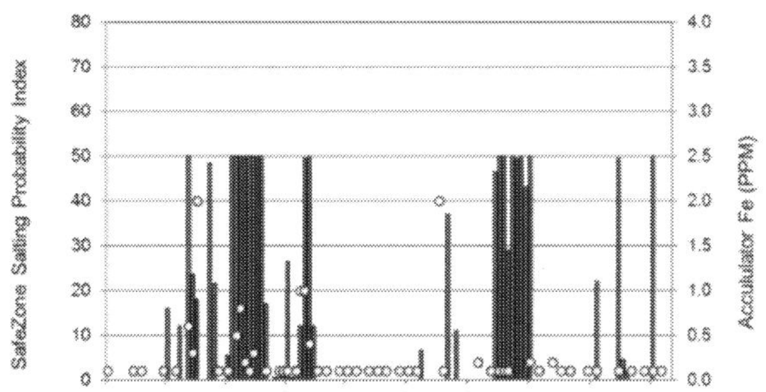

Figure 13: Chart of iron throw vs SPI. Only the initiation of positive SPI and the initiation of iron throw should correlate. Not the relative magnitudes of the two responses. On the far right, chlorides could now be driven lower than 2 ppm using allowed caustic. Thus, even though the SPI was sometimes positive and roughly equal to the responses on the left, there were now many fewer chloride ions available to deposit as a salt. Therefore, actual corrosion was dramatically attenuated relative to the SPI signal. This is like a forecast for 100% probability of rain. Such a forecast equally describes both a light mist and a torrential downpour. Both are given a 100% probability of precipitation. The outcomes are very different however, on one case a person gets very wet, in the other they will not. Further information must be invoked in both cases to detail the degree of impact after its initiation.

Around this point in the chronology of the case study, Iron throw and ER probe signals were also positively detected as a delayed response to the corrosive conditions, thus validating the **SafeZone*** pipelines early warning SPI predictions. **Figure 13** shows the iron throw and the SPI on the same chart. The initiation of the SPI correlates very well with the initiation of corrosion for events on the left-hand side of the chart. The magnitude of the SPI and the magnitude of the iron throw are not expected to correlate, instead on only the initiation. Other indicators then predict the severity of the expected corrosion. Towards the left side of the chart, chlorides were systematically higher and could not be effectively reduced below the needed 4 ppm threshold as described above, thus the salt deposition process remained active. On the right-hand side of the chart, however, more aggressive desalting led to improved ability to lower chlorides with caustic. This allowed the nearly complete elimination of detected corrosion, well before the run was completed.

When active corrosion potential was first detected by the **SafeZone*** engine, refinery operations gave approval to increase caustic to the upper limit to drive chlorides low enough to stop induced salting and subsequent corrosion. Unfortunately, this action alone was not enough to mitigate the event. This is because as chlorides were decreased and approached 8 ppm, the sodium spec in the residua was reached and no further caustic could be used. At this point, several other options were evaluated, leading the team to decide upon using a 2-degree temperature increase as the next best move. Several options were theoretically computed and ranked by the **SafeZone*** platform, as the minimum move needed to push the system from the salting zone to the non-salting zone. Because of the small change in tower temperature needed after that other actions were taken, the SPI's upwards pressure was largely reduced. By simultaneously lowering and balancing neutralizer feed as the event passed and the corrosive species left the system, good corrosion control was maintained throughout the event.

The weighted list of factors and information quantitatively displayed by the **SafeZone*** platform allowed the best balance of options to be chosen each day by the team to move the unit as minimally needed to mitigate corrosion to the greatest degree possible. Later, as mentioned above, further manipulations at the desalter allowed chlorides to be driven even lower and corrosion was nearly eliminated, making the remainder of the opportunity run a large and well controlled success. **Figure 14** shows the margin report during the event. It can be seen how the various indicators interplay during the case. It is important to note that keeping traditional KPIs inside traditional ranges were not sufficient to effectively mitigate this transient event. This was especially true because in this case, the local teams had no a-priory knowledge or experience running this challenging crude.

Conclusion

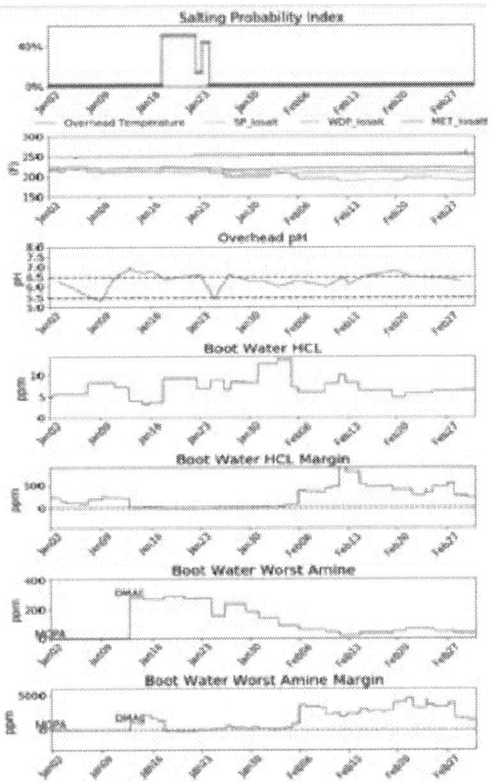

Figure 14: SafeZone Margin report. Stacked plots showing how the chlorides and pH were well controlled during the run. When the neutralizer (DMAE) was introduced to stabilize pH, volatile acid attack was transferred to amine hydrochloride attack. The technique allowed all the most key factors to be identified, evaluated, and balanced to achieve a smooth and successful run.*

The new techniques and infrastructure described here have been shown to proactively detect a wide range of previously un-observed corrosion events and enable their effective and practical mitigation as they are occurring. Furthermore, the signal processing techniques used to derive the SPI enable the specific drivers and response curves of key variables affecting corrosion and production to be quantitatively evaluated and balanced in a proactive and team oriented way. Advanced analysis of the information output by the platform is proving to enable a wide range of novel and key actions to be quickly and effectively deployed towards a range of mitigation solutions well beyond the traditional. Using traditional low frequency manual techniques to derive salt points or corrosion rates would not have been able to detect the event in question, detail its drivers, or affect any actions in the required time scales. In such cases previously, events have often remained hidden until a surprise shutdown and inspection are utilized due to unplanned shutdowns. In many cases, by the time such a unit is opened, the damage cannot be recovered and most of the evidence showing the stream of dynamic corrosion episodes giving rise to the overall problem are no longer understandable.

MANCHESTER 1824
The University of Manchester

Contents

- Introduction
- HEN and Fouling Modelling
- Case Study
- Conclusions and Future work

17 AIChE Annual Meeting

MANCHESTER 1824
The University of Manchester

Introduction

MANCHESTER 1824
The University of Manchester

Fouling in Crude Oil Refineries

- ❑ Undesired accumulation of solid material
- ❑ In crude oil refineries, mostly located within preheat trains
- ❑ Associated costs include capital, energy, maintenance, among others
- ❑ Almost 50% of fouling related costs attributed to distillation unit (US$ 13 per year for 100,000 bbl/day production)

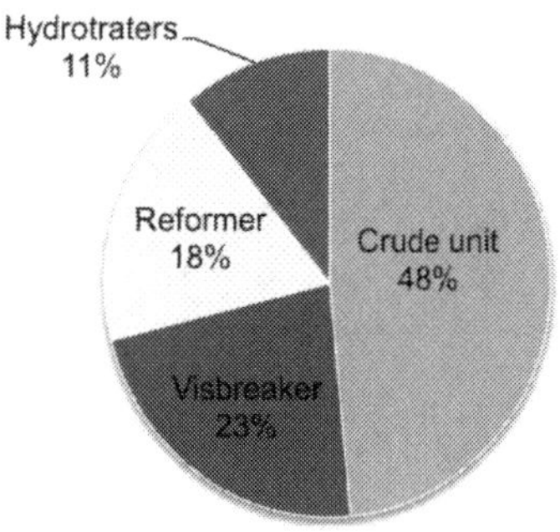

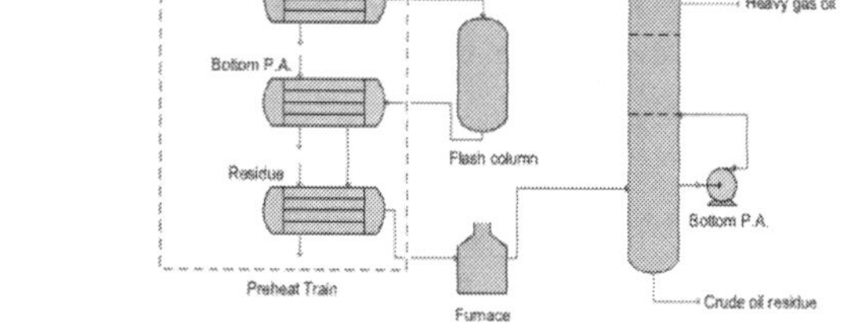

Coletti, F., Joshi, H. M., Macchietto, S. and Hewitt, G. F. (2015). Chapter One: Introduction, in *Crude Oil Fouling*. Boston: Gulf Professional Publishing, 1-22

MANCHESTER 1824
The University of Manchester

Fouling Rate – Definition

- Fouling rate: Dynamic change in fouling resistance

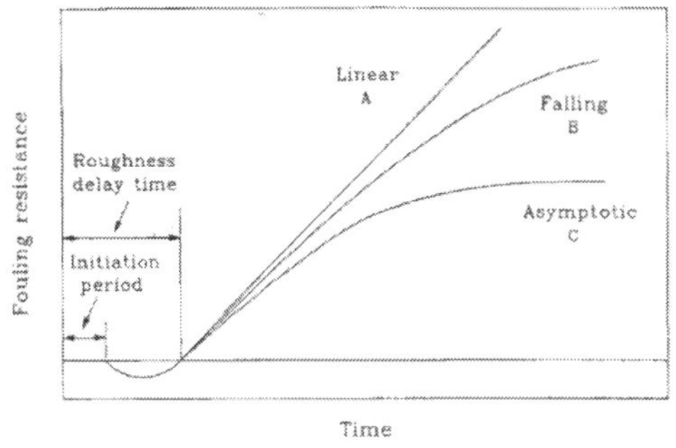

Fouling Rate = Formation - Removal

- Operational conditions set formation and removal rates
- Certain combinations can mitigate fouling occurrence
- Different effects for different fouling mechanisms
- Occurrence limited to existence of fouling threshold

Ebert, W. A. and Panchal, C. B. (1995). Analysis of Exxon crude-oil-slip stream coking data, in Panchal, C. B., Bott, T. R., Somerscales, E. F. C. and Toyama, S., (eds.) *Fouling Mitigation of Industrial Heat-Exchange Equipment*, 18th-23th June 1995. San Luis Obispo, California. USA: Begell House

MANCHESTER 1824
The University of Manchester

Fouling Rate – Modelling

- Linear rate

$$\frac{dR_f}{dt} = \alpha_2'$$

Model parameters need to be determined

- dR_f/dt: Fouling rate
- Re: Reynolds number
- Pr: Prandtl number
- E_A: Activation energy
- R: Gas constant
- T_{film}: Film temperature
- T_{wall}: Wall temperature
- τ: Wall shear stress
- α, β, γ: Fouling model parameters

- Chemical reaction rate
  - Ebert and Panchal

$$\frac{dR_f}{dt} = \alpha_1' Re^{\beta'} exp\left(\frac{-E_A'}{RT_{film}}\right) - \gamma' \tau$$

  - Panchal et. al

$$\frac{dR_f}{dt} = \alpha_1' Re^{\beta'} Pr^{-0.33} exp\left(\frac{-E_A'}{RT_{film}}\right) - \gamma' \tau$$

  - Polley et. al

$$\frac{dR_f}{dt} = \alpha_1' Re^{-0.8} Pr^{-0.33} exp\left(\frac{-E_A'}{RT_{wall}}\right) - \gamma' Re^{0.8}$$

Ebert, W. A. and Panchal, C. B. (1995). Analysis of Exxon crude-oil-slip stream coking data, in Panchal, C. B., Bott, T. R., Somerscales, E. F. C. and Toyama, S., (eds.) *Fouling Mitigation of Industrial Heat-Exchange Equipment*, 18th-23th June 1995. San Luis Obispo, California. USA: Begell House.
Panchal, C.B., Kuru, W.C., Liao, C.F., Ebert, W.A. and Palen, J.W., 1999. Threshold conditions for crude oil fouling. Understanding heat exchanger fouling and its mitigation, 273, p.282.
Polley, G.T., Wilson, D.I., Yeap, B.L. and Pugh, S.J., 2002. Evaluation of laboratory crude oil threshold fouling data for application to refinery pre-heat trains. Applied Thermal Engineering, 22(7), pp.777-788.

MANCHESTER 1824
The University of Manchester

Fouling Rate – Modelling

- Fouling considerations in heat exchanger networks
  - Assessment for operational conditions for fouling mitigation
  - Optimisation of cleaning schedules
  - Design and retrofit for fouling mitigation
- However
  - How to determine each fouling model parameter?
  - How to account for different fouling mechanisms and different crudes?
  - How to identify the dominant fouling mechanism on each side of heat exchangers?
  - How to get a representative and free of error sample for reliability of output results?

↓ Use of operational data

↓ Accounting for shell and tube side contributions

17 AIChE Annual Meeting

MANCHESTER 1824
The University of Manchester

Fouling Rate – Split Contributions

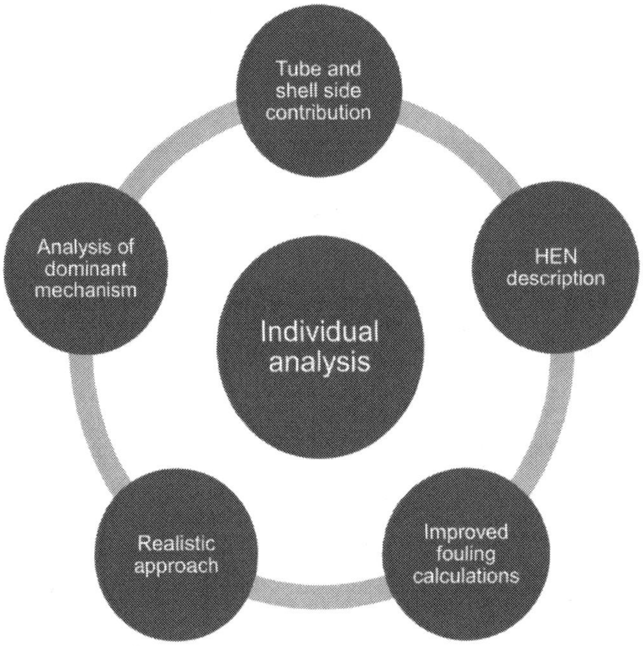

17 AIChE Annual Meeting

MANCHESTER 1824
The University of Manchester

Research Objectives

To determine fouling rate models in crude oil preheating systems from online reconciled data using parametric regression

1) Parametric fitting of fouling models for shell and tube sides on heat exchanger networks
2) Application of fitted models for predicting fouling behaviour
3) Identification of faulty measurements by means of data reconciliation and gross error detection
4) Optimisation of cleaning schedules and operational conditions for fouling mitigation

MANCHESTER 1824
The University of Manchester

HEN and Fouling Modelling

General Modelling Scheme

The University of Manchester

HEN modelling
- ➢ Matrix representation based on graph theory
- ➢ Accounts for fouling dynamics and contributions from shell and tube side

Data reconciliation
- ➢ Mitigation of random and gross errors
- ➢ Identification of faulty instruments
- ➢ Optimisation based on non linear programming (NLP)

Parametric regression
- ➢ Calculation of fouling model parameters for HEN on each side
- ➢ Optimisation approach based on Genetic Algorithm and deterministic solvers

HEN Simulation and Fouling Modelling

The University of Manchester

❑ HEN simulation: Matrix representation (de Oliveira Filho et.al, 2007)

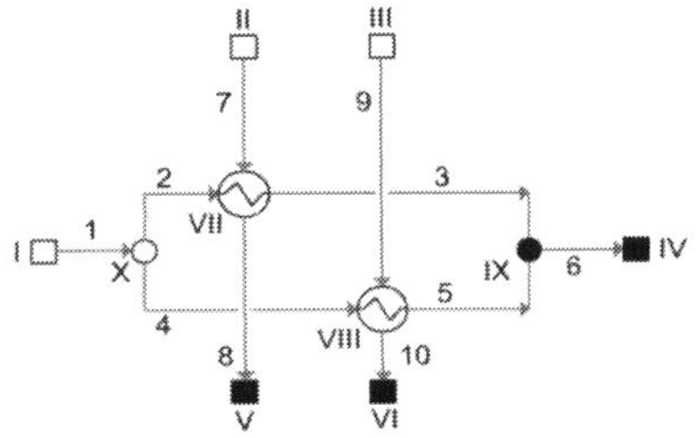

	1	2	3	4	5	6	7	8	9	10
I	-1									
II							-1			
III									-1	
IV						1				
V								1		
VI										1
VII		1	-1				1	-1		
VIII				1	-1				1	-1
IX			1		1	-1				
X	1	-1		-1						

❑ 3 supply and demand units

❑ 2 heat exchanger

❑ 1 mixer, 1 splitter

$$M = \begin{bmatrix} M^{PS} \\ M^{PD} \\ M^{HE} \\ M^{MX} \\ M^{SP} \\ M^{PR} \end{bmatrix} = \begin{bmatrix} M_c^{PS} & M_h^{PS} \\ M_c^{PD} & M_h^{PD} \\ M_c^{HE} & M_h^{HE} \\ M_c^{MX} & M_h^{MX} \\ M_c^{SP} & M_h^{SP} \\ M_c^{PR} & M_h^{PR} \end{bmatrix}$$

Mass and energy balance are solved using matrix *M*

MANCHESTER 1824
The University of Manchester

HEN Simulation and Fouling Modelling

- Modelling for fouling contributions

$$R_f = \frac{1}{U_D} - \frac{1}{U_C} = R_{fi} \cdot \left(\frac{d_o}{d_i} \right) + R_{fo}$$

- Different fouling models can be used based on heat exchanger location and operating conditions

$$R_f \Big|_{n+1} = R_f \Big|_n + \Delta t \cdot \frac{dR_f}{dt} \Big|_n$$

- U_C: Clean overall heat transfer coefficient
- U_D: Fouled overall heat transfer coefficient
- R_{fi}: Tube side fouling resistance
- R_{fo}: Shell side fouling resistance
- h_i: Tube side heat transfer coefficient
- h_o: Shell side heat transfer coefficient

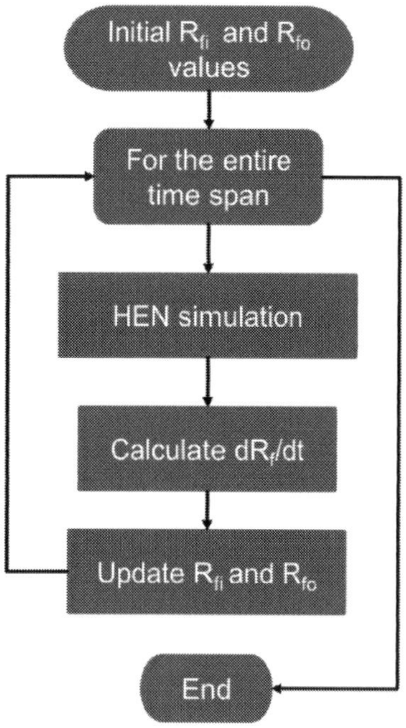

MANCHESTER 1824
The University of Manchester

HEN Simulation and Fouling Modelling

- Data reconciliation

$$\begin{aligned} \underset{y_R}{\text{minimise}} \quad & (y_M - y_R)^T \psi^{-1} (y_M - y_R) \\ \text{subject to} \quad & f(y_R) = 0 \\ & g(y_R) \geq 0 \\ & y_R^L \leq y_R \leq y_R^U \end{aligned}$$

- Gross error detection

$$\begin{aligned} \underset{y_R}{\text{minimise}} \quad & (y_M - y_R)^T \psi^{-1} (y_M - y_R) \\ \text{subject to} \quad & f(y_R - B_\xi \omega_\xi) = 0 \\ & g(y_R - B_\xi \omega_\xi) \geq 0 \\ & y_R^L \leq y_R \leq y_R^U \end{aligned}$$

- Where:
  - y_M: Measured value
  - y_R: Real value
  - B_ξ: Position matrix for single and multiple gross errors
  - ω_ξ: Gross error magnitude
  - ψ: Covariance matrix

- Mass and energy balance are used as process constraints
- Data reconciliation solved as a NLP problem
- Gross error solution based on combinatorial approach
- Maximum number of gross errors limited by number of independent constraints

The University of Manchester

HEN Simulation and Fouling Modelling

- Parametric fitting
  - Needs to account for parameters of different magnitudes
  - Needs to account for sensitivity on initial estimation
  - Needs to avoid local optima
- Minimisation of error between simulated and predicted values

$$\text{minimise} \quad \sum_{i=1}^{n} \frac{\left(y_i^{fit} - y_i^{msr}\right)^2}{y_i^{msr}}$$

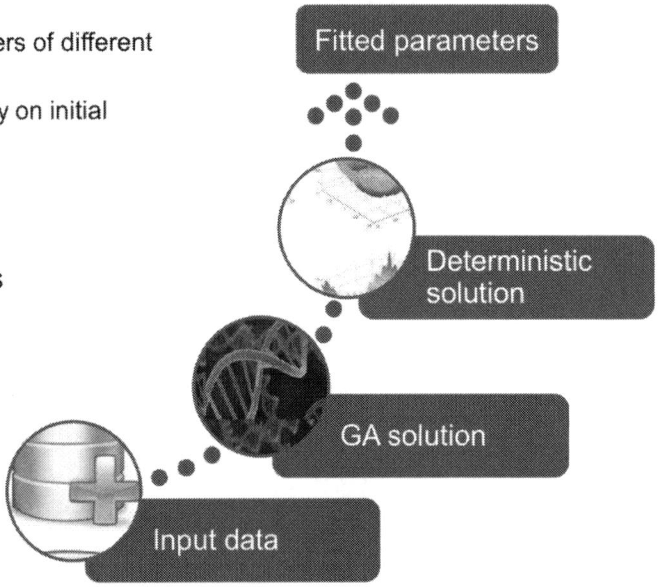

The University of Manchester

Case Study

Case study – Crude Oil Preheat Train

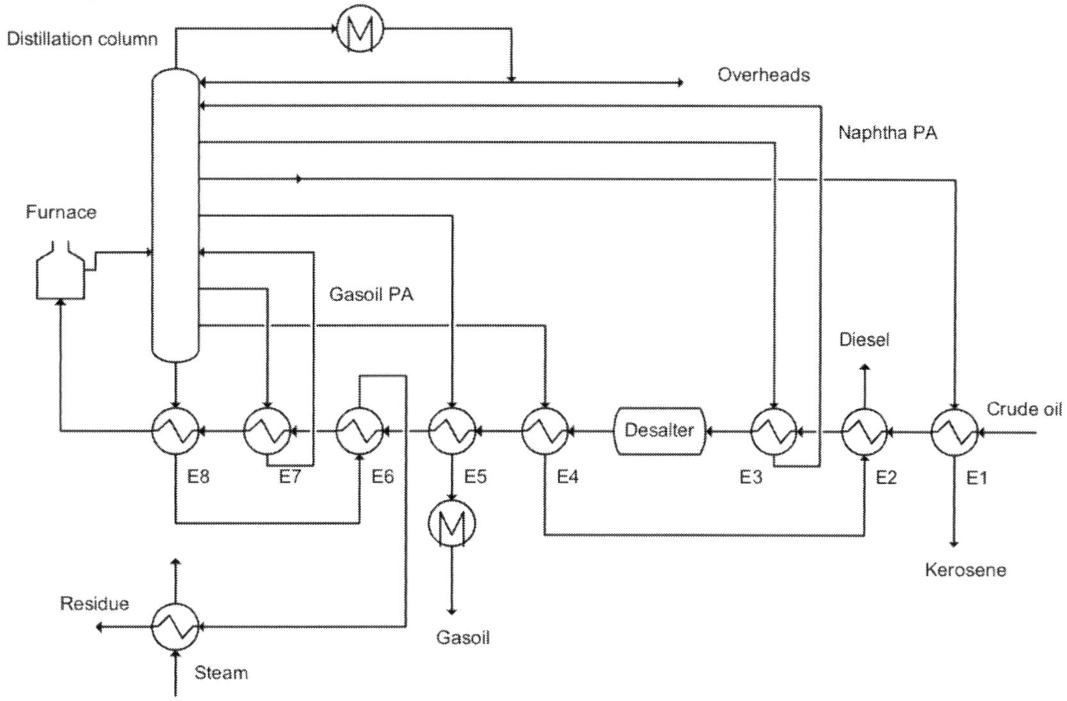

Case study – Crude Oil Preheat Train

Problem formulation

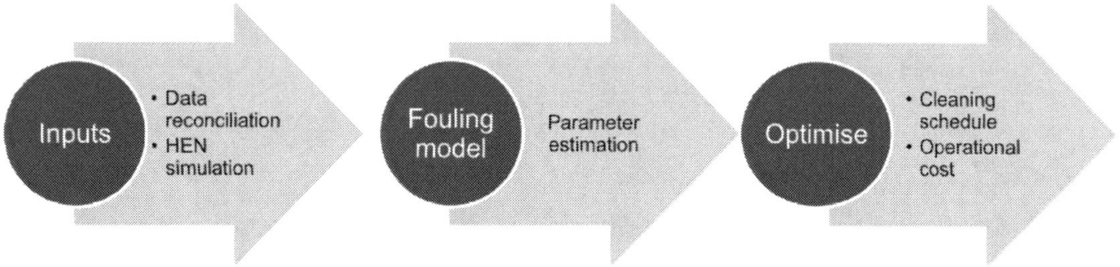

- Measured flow rates and temperatures
- Linear fouling rate at cold end of preheat train
- Polley's model at hot end for tube side
- Heat exchangers specified by area
- Hot and cold utilities specified by heat duty

MANCHESTER 1824
The University of Manchester

Case study – Crude Oil Preheat Train

Base case: No cleaning actions on Preheat Train

Total operating cost (MM$/y)	Time period (y)	Temperature drop (°C)
17.34	1	55.91

- ❑ Furnace inlet temperature decreases
- ❑ Furnace input duty increases
- ❑ Decreasing in energy recovery
- ❑ Fouling deposits accumulate over time

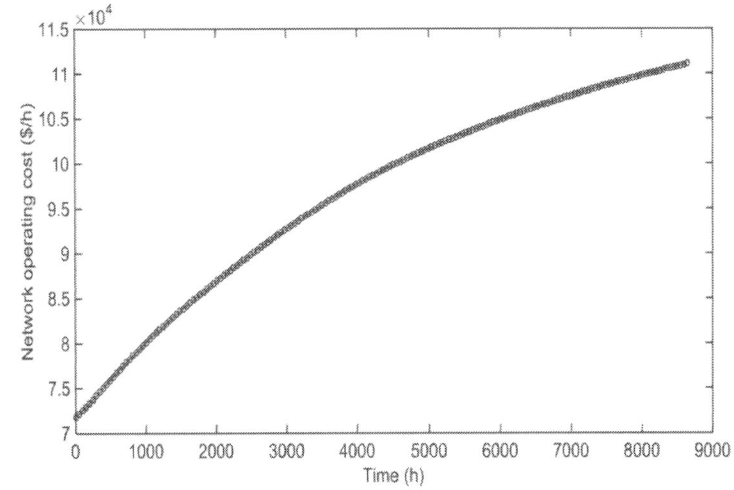

17 AIChE Annual Meeting

MANCHESTER 1824
The University of Manchester

Case study – Crude Oil Preheat Train

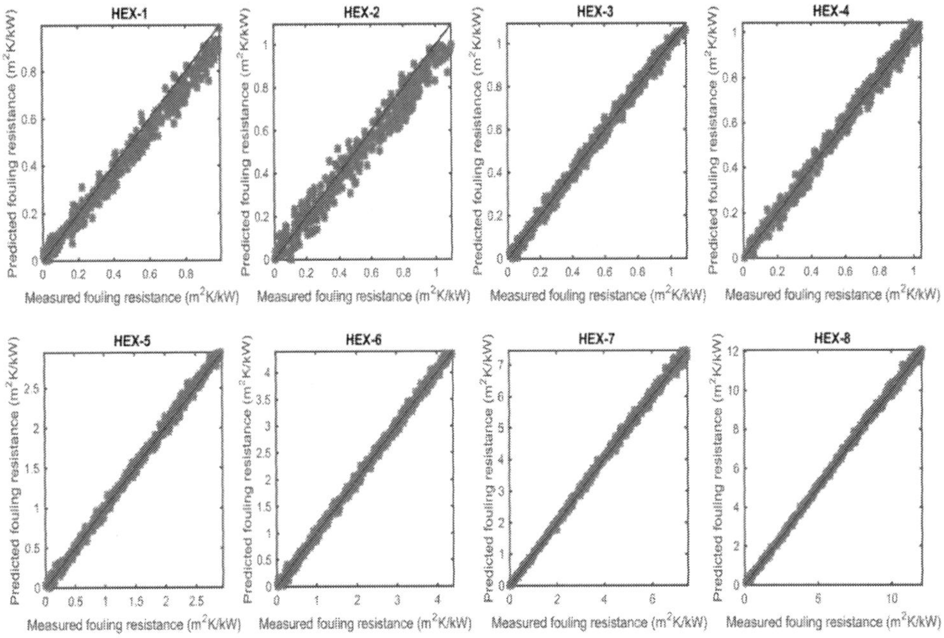

17 AIChE Annual Meeting

Case study – Crude Oil Preheat Train

- ❑ Parameters for simulations (base):

α'_1 ($m^2K\ kW^{-1}h^{-1}$)	E'_A ($kJ\ mol^{-1}$)	γ' ($m^2K\ kW^{-1}h^{-1}$)	α'_2 ($m^2K\ kW^{-1}h^{-1}$)
1.00e+06	48.00	1.50e-09	5.42e-05

- ❑ Fitted parameters for optimisation

	α'_1 ($m^2K\ kW^{-1}h^{-1}$)	E'_A ($kJ\ mol^{-1}$)	γ' ($m^2K\ kW^{-1}h^{-1}$)	α'_2 ($m^2K\ kW^{-1}h^{-1}$)
E1	-	-	-	5.04e-05
E2	-	-	-	5.09e-05
E3	-	-	-	5.41e-05
E4	-	-	-	5.40e-05
E5	5.74e+03	16.38	1.50e-07	4.09e-07
E6	9.98e+07	67.09	1.62e-08	2.58e-04
E7	3.83e+04	32.80	1.20e-07	1.97e-04
E8	3.49e+04	31.25	1.42e-07	6.04e-05

Case study – Crude Oil Preheat Train

Optimised case: Optimal cleaning schedule

Total operating cost (MM$/y)	Cleaning actions	Cost savings (%)
16.13	11	6.98

- ❑ Use of fitted parameters allows for optimisation
- ❑ Cleaning events reduce decrease in FIT
- ❑ Fouling impact is alleviated
- ❑ Different constraints can be applied for different scenarios

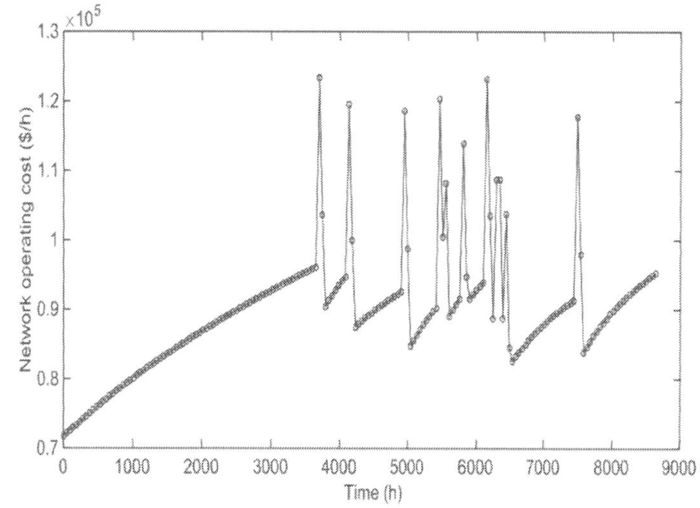

MANCHESTER 1824
The University of Manchester

Conclusions and Future work

MANCHESTER 1824
The University of Manchester

Conclusions and future work

- ❑ Fouling can lead to significant impact in thermal and hydraulic performance
- ❑ Fouling can be modelled by using fouling rate models and determining their specific parameters
- ❑ Including fouling into design and optimisation of heat exchanger networks allows for mitigation strategies
- ❑ A methodology has been developed for stablishing fouling models using reconciled data
- ❑ Fouling from shell and tube side are determined by means of different fouling model parameters
- ❑ Fitted parameters are used for optimisation of cleaning schedules, leading to operational savings of 7%

- ➢ Use of industrial data for testing rigor of methodology
- ➢ Identification of minimum bias magnitude for isolation of gross errors
- ➢ Application of methodology for operational optimisation and retrofit of heat exchanger networks

MANCHESTER
1824
The University of Manchester

Acknowledgements

CONICYT
Ministerio de Educación
Gobierno de Chile

To CONICYT (Chile) for the financial support provided

MANCHESTER
1824
The University of Manchester

Thank you

17 AIChE Annual Meeting

509054 Solve this! Fundamental Approach to Problem Solving in Industrial Processes I (Invited Talks)

Abstract:

Solving complex plant problems is an everyday exercise that affects all projects and budgets, and it is very rarely introduced to students at any level. In fact, by the time young engineers are few years into working in industry, most have lost the engineering skills that they developed in school. We hope to encourage future engineers to retain those skills and learn the importance of plant problem-solving.

490649 Scale-up and Optimization of Filtration Processes: Small Scale Pressure Filtration to Centrifugal Filtration

Abstract:

The separation of solid particles from liquids via filtration is a key unit operation in the pharmaceutical sector. This processing step commonly occurs after the isolation of an Active Pharmaceutical Ingredient (API) via crystallization and may involve a number of wash cycles in order to deliqour the cake and remove any residual solvents below specified limits. For the purposes of filtration, pressure or vacuum filters and centrifuges are most commonly employed.

Figure 1: Filtration scale-up workflow - pressure filter to centrifuge

In this work we detail a model based workflow for the scale-up of filtration unit operations, as summarized in Figure 1 above. This workflow begins with the estimation of the following cake properties from small scale pressure filter experiments:

- Specific cake resistance
- Medium resistance
- Compressibility index of the cake
- Reference pressure drop for compressible cakes

The above parameters are then regressed from experimental data from the lab scale, namely the change in filtrate volume as a function of time for a range of conditions, as follows:

- Pressure drop
- Particle size distribution of particles in cake

These cake properties were subsequently employed within a centrifuge model to predict the filtration behavior of the system at plant scale. Additionally, the schedule of operation of the centrifuge was optimized with an objective of achieving the maximum product throughput through the centrifuge. The various trade-offs observed in the optimization objective will also be outlined. This filtration workflow for scale-up outlined above was applied in this work for a number of industrial cases, involving compressible cakes. This model-based approach allows the filtration unit step to be scaled-up and optimized in a fast and efficient manner, with minimal experimental effort and consumption of

valuable materials.

489004 Steady State Elutriation of Fines from Binary Particle Mixtures in Bubbling Fluidized Cold Flow Model

Abstract:

Abstract

Chemical looping combustion (CLC) technology is being perused as a promising clean and high efficiency method of power generation [1]. This technology has already been researched extensively for gaseous fuels, but has more recently drawn interest in solids fuel. The basis of CLC is to supply oxygen to the fuel reaction process via solid "oxygen carriers" preventing air from the reaction. The "oxygen carrier" is oxidized in an air reactor which is separate from the fuel reactor and then looped through the fuel reactor where it is reduced, thus providing pure oxygen for combustion of the solid fuel. As a benign benefit of the CLC process is the potential for significant reduction to carbon capture penalties due to the flue gas consisting of mainly CO_2 and H_2O. Therefore, the H_2O in the flue gas can be condensed leading to efficient separation and capture of the CO_2.

One issue with solid fuel CLC is that typical combustion beds are estimated to produce 1% of unburnt carbon in the flow stream. In a CLC reactor, the ash/char has the potential to be carried to the air reactor along with the oxidizer material resulting in an efficiency penalty. Therefore, the US DOE has a set a goal to demonstrate that over 90% of the unburnt carbon can be separated from the oxygen carrier being transported back into the air reactor. To meet this objective, multiple fluidized bed technologies are being pursued. Among them is a bubbling bed to measure steady state elutriation rates. Elutriation rates of binary materials in bubbling beds have been studied in the past [2], but most consist of a batch process methodology that only investigate transient elutriation rates. Steady state elutriation rates are need to assess the separation performance of a continuous process.

To investigate the steady state elutriation process of a binary mixture a cold flow bubbling bed riser model was constructed. A rendering of the bed is provided in Figure 1 which consists of a 180 cm riser with a 10 cm diameter and a distributor plate with 32 holes with 0.159 cm diameters. A cyclone is in series with exit port of the riser to drive elutriated particles into a collection bin. To achieve quasi steady state conditions three overflow ports of about 15, 25, and 36 cm heights are built into lower portion of the riser to maintain consistent fluidized bed heights as binary solids mixtures are continuously screw fed into the system. Digital scales are used in both the cyclone collection and the overflow bins to measure the elutriation rate of particles and the bed overfill. The particles considered for this effort consist of a plastic ZQ-17 "Day-glow" to simulate unburnt carbon, and two different sizes of glass beads (P-0230 and P-0337) with the particle properties given in Table 1 to simulate potential size ranges of oxygen carriers. Parameters currently considered for investigation include varying the gas velocity within the region of bubbling for the glass beads and varying the concentration of the day-glow particles between 1% and 5% of the total mass.

Current test methodology consists of prefilling the bed to a desired static bed height with the particle mixture. A constant fed rate was set to feed the same mixture into the bed while a constant gas flow was supplied. The system is operated for a set period of time to allow steady state conditions to be reached with the goal of measuring the time dependent weight of the elutriated particles. The gas velocities considered for these tests should be within the bubbling regime of the glass bead particles.

A series of preliminary runs have been performed with a mixture of P-0337 with 5% ZQ-17 particles by mass at gas velocities of 45 and 62 cm/sec. Both of these velocities are within the bubbling regime of the P-0337 particles. The mixture feed rate the system was 1.5 g/sec to maintain a desired fluidized bed height of 25 cm for both runs. Initial results show promising separation performance. Figure 2 shows the system prefilled with the mixture prior to running the system post run. The ZQ-17 is yellow so the post image run clearly demonstrates that this material was entrained and achieved some elutriation based on material that adhered to the surfaces and the a nominal 0.635 cm sedimentation on top of the bed.

Steady state elutriation rates are needed to assess the performance of continuous particle separation techniques. A cold flow bubbling bed was designed and built to achieve this in order to help achieve the NETL DOE goal of 90% carbon removal from oxygen carriers in a CLC. Initial results show promise in separation of fines analogous to carbon from larger particles.

References

[1] H.E.J. Andrus, J.H. Chiu, P.R. Thibeault, A. Brutsch, Akstom's calcium oxide chemical looping combustion coal power technology development, Proceedings of the 34th International Technical Conference on Clean Coal and Fuel Systems, May31 – June 4, 2009 (Clearwater FL).

[2] E.R. Monazam, R.W. Breault, J.M. Weber, K. Layfield, Elutriation of fines from binary particle mixtures in bubbling fluidized bed cold model, Powder Technology 305 (2017) 340-346.

Table 1 – Particle physical properties where SMD is sauter mean diameter).

Material	Size Range (μm)			Density (kg/m³)	Sphericity (-)	$U_{t\_estimated}$ (m/s)	
	Max	Min	SMD			Largest	Smallest
Glass Beads (P-0230)	600	430	550	2500	0.94	4.25	0.13
Glass Beads (P-0337)	850	600	707	2500	0.94	6.31	0.24
Day-glow (ZQ-17)	60	20	39	1200	0.74	0.18	1.4E-4

Figure 1 – Schematic of 10 cm diameter continuous bubbling bed

Figure 2 – Continuous bubbling bed, Left: pre-run, Right: Post run

497380 Sustainable Fractionation of Plant-Derived Proteins with Pneumatic Tribo-Electrostatic Separation

Abstract:

The potential of a dry fractionation approach for the production of native plant protein concentrates has been validated for navy bean (*Phaseolus vulgaris*) flour as a model system utilizing a lab-scale pneumatic tribo-electrostatic separator. This methodology employed a tribo-electrostatic approach to selectively charge proteins, carbohydrates, and fibers in the bean flour and separate them based on the magnitude and type of their charge. For this approach, a wide variety of tribo-charging tubes with different sizes, shapes, and materials were screened to determine the most effective for use in tribocharging of the bean flour. The optimized process consisted of a fluidized bed flour reservoir, polytetrafluoroethylene (PTFE) tribo-charging tube, and a plate-type separation chamber. The location and distribution of the charged protein-rich particles along the surface of the electrode plate was investigated as a function of air flow rate (laminar *vs*. turbulent) and electric field strength. Protein-rich fractions were collected from bottom, middle and top sections of the electrode plate and analyzed for protein content and particle size distribution. Particles were imaged using scanning electron microscopy (SEM). A turbulent air flow rate at a variety of electric field strengths resulted in the formation of protein-rich particles and starch granule agglomerates affecting the production of high-purity protein concentrates. Charging the flour particles at a laminar air flow rate followed by separation under a low electric field strength enabled the production of fine protein-rich fractions with considerably higher protein contents. To further improve the separation efficiency of the navy bean protein concentrate without compromising its protein content, a two-stage tribo-electrostatic separation approach was evaluated at a laminar air flow but different plate voltages. The combined protein-rich fraction produced by the two-stage approach had a protein content of ~38% accounting for 60% of the total protein, facilitating the scale-up of this dry fractionation technique for pilot-plant applications. Protein recovery by this approach was higher than other solvent-free separation approaches and could be improved by further optimization of the flow rate and design configuration of the separator. This novel solvent-free tribo-electrostatic separation approach is of considerable commercial significant as it preserves the bio-functionality of the protein and averts the likelihood of toxic microbial contamination common in currently used wet processes.

499105 Particle Image Velocimetry (PIV): An Important Tool for Understanding the Fluid Dynamics of Magnetically Responsive Membranes

Abstract:

Particle image velocimetry (PIV) is an approach for measuring two dimensional fluid velocity field in the region of interest at a very high spatial resolution. Visualizing the flow field at the surface of the membrane can help in designing magnetically responsive membranes that actively break up the concentration polarization boundary layer at the surface of the membrane. The present investigation deals with understanding the effect of the movement of magnetically responsive polymer brushes on the concentration polarization boundary layer. The interaction of external magnetic field with the superparamagnetic nano particle attached via thermos-responsive Poly(N-isopropylacrylamide), PNIPAM polymer on regenerated cellulose ultra-filtration (RC-UF) membrane leads to local perturbation and is proposed break up concentration polarization. Laser illuminated fluorescent particles were used as probes to get two consecutive images at a definite time interval. The small displacement of these particles was used to define the velocity vectors. The influence of the experimental conditions on the velocity vectors of the fluid at the surface of the membranes was investigated. The results revealed that the nanoparticle attachment was more homogeneous for 100 kDa membrane surface compared to 300 kDa. The PIV images of the modified membrane by longer atom transfer radical polymerization (ATRP) time, were found to show more perturbation in the fluid field and can be attributed to the longer polymer chains. The outcome of the PIV image analysis was found to be in good agreement with the data obtained Bovine serum albumin (BSA) rejection studies.

Overview of Real Time Monitoring Techniques for Drop Size Distributions in Research and Industrial Applications

Sebastian Maaß[1], Jörn Emmerich[1,2], Matthias Kraume[2]
[1]SOPAT GmbH; Berlin, Germany
[2]Technische Universität Berlin, Department of Chemical Engineering, Berlin, Germany

SOPAT
Smart Online Particle Analysis Technology

Agenda

1. Introduction
2. Experimental methods
   1. Measurement technique overview
   2. Experimental set-up
   3. Comparitative results
3. Test in a three phase system
   1. Imaging hardware
   2. Gas-solid-liquid (G/S/L) results
4. Summary and outlook

Extraction in the pharma industry

TU SOPAT

Extraction in the pharmaceutical industry has been used primarly for:

- isolation of antibiotics from fermenter broth
- preparation of naturally and synthetic vitamins
- preparation from drugs from naturally occuring materials

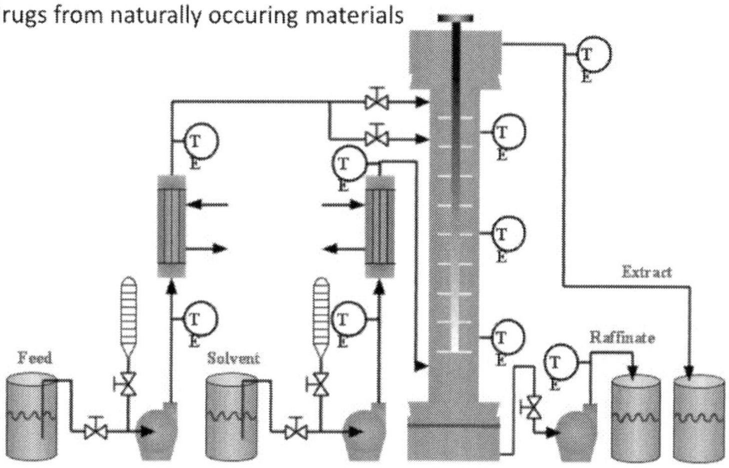

Agenda

TU SOPAT

1. Introduction
2. Experimental methods
   1. Measurement technique overview
   2. Experimental set-up
   3. Comparitative results
3. Results in a three phase system
   1. Imaging hardware
   2. Gas-solid-liquid (G/S/L) results
4. Summary and outlook

Measurement technique overview

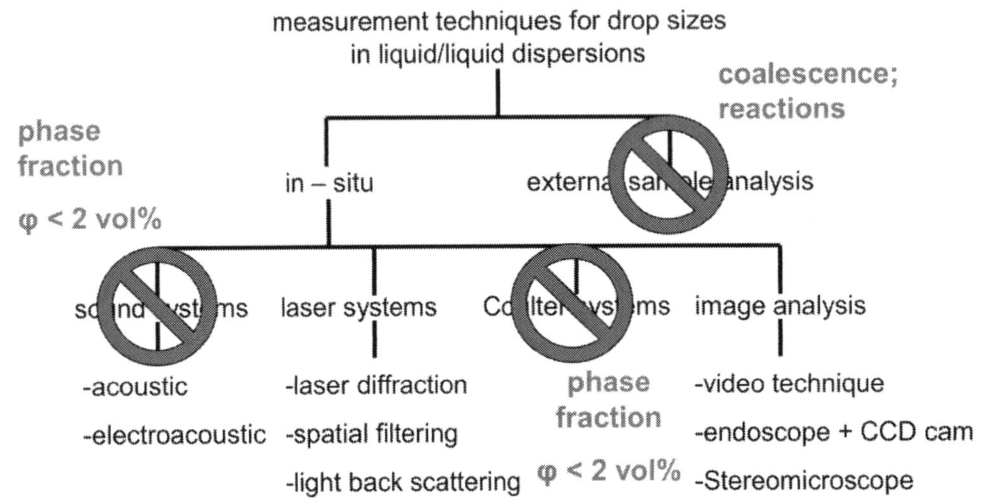

Measurement technique overview

probe	supplier	method
FBRM®	Mettler Toledo	laser
IPP 30	Parsum	laser
2D-ORM®	Messtechnik Schwartz	laser
FBR	Sachweh Mannheim	laser
PVM®	Mettler Toledo	photo-optical
SOPAT-VI	SOPAT GmbH	photo-optical
IVM	S&E Sequip	photo-optical
ISPV	Perdix	photo-optical
EnviroCam	EnviroCam	photo-optical
PetroCam	Fluid Imaging	photo-optical
Pixact Probe	Pixact	photo-optical
InFlow	Canty	photo-optical

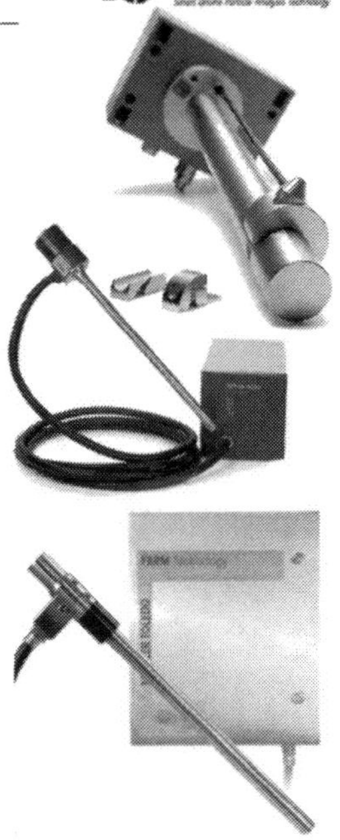

2D ORM®

TU SOPAT

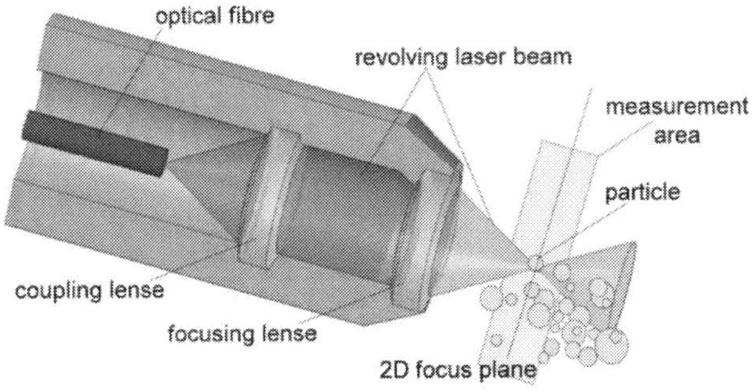

Maaß et al. **2011**. Experimental comparison of measurement techniques for drop size distributions in liquid/liquid dispersions, Experiments in Fluids, 50(2): 259-269.

requirements	2D-ORM®
applicable in various ways	+
in-situ	+
on-line	+
local analysis	+
probe diameter	22 mm
measurement range [µm]	1-1000
output quantity	particle / chord l.
results	?

FBRM®

TU SOPAT

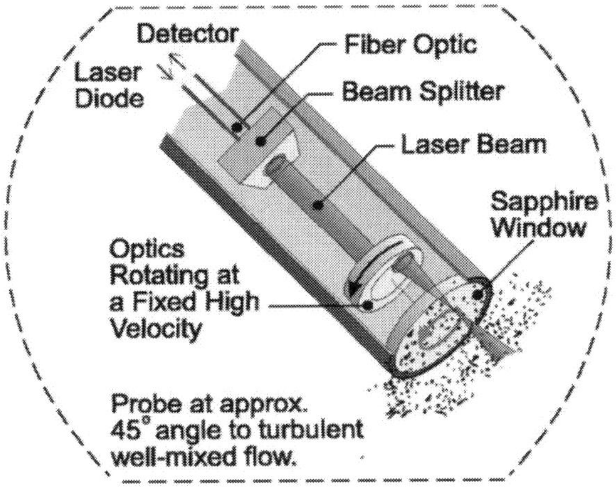

Ruf et al. 2000

requirements	FBRM®
applicable in various ways	+
in-situ	+
on-line	+
local analysis	+
probe diameter	25 mm
measurement range [µm]	1-1000
output quantity	chord lengths
results	?

FBR (forward-backward ratio) sensor

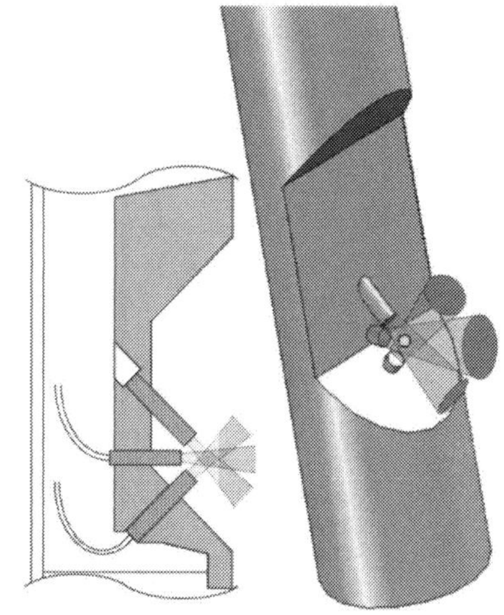

Sachweh et al. 1998

requirements	FBR-sensor
applicable in various ways	+
in-situ	+
on-line	+
local analysis	+
probe diameter	4 mm
measurement range [µm]	1-300
output quantity	mean diameter
quantitative results	?

IPP 30 (Inline particle probe)

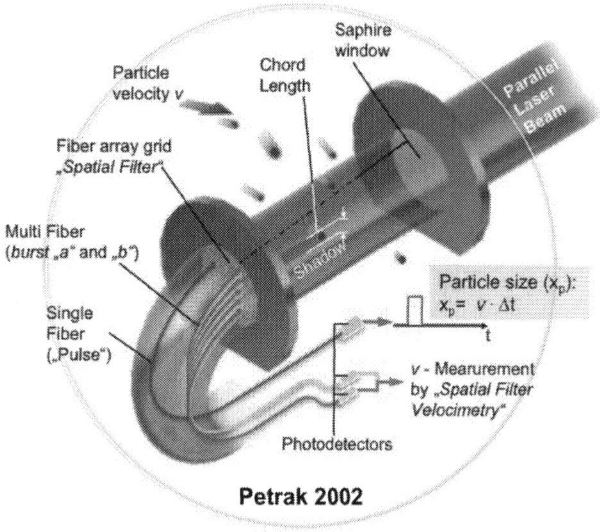

Petrak 2002

requirements	IPP 30 Parsum
applicable in various ways	+
in-situ	+
on-line	+
local analysis	+
probe diameter	10-35 mm
measurement range [µm]	30-5000
output quantity	chord lengths
quantitative results	?

Agenda

1. Introduction
2. Experimental methods
   1. Measurement technique overview
   2. Experimental set-up
   3. Comparitative results
3. Results in a three phase system
   1. Imaging hardware
   2. Gas-solid-liquid (G/S/L) results
4. Summary and outlook

Experimental set-up

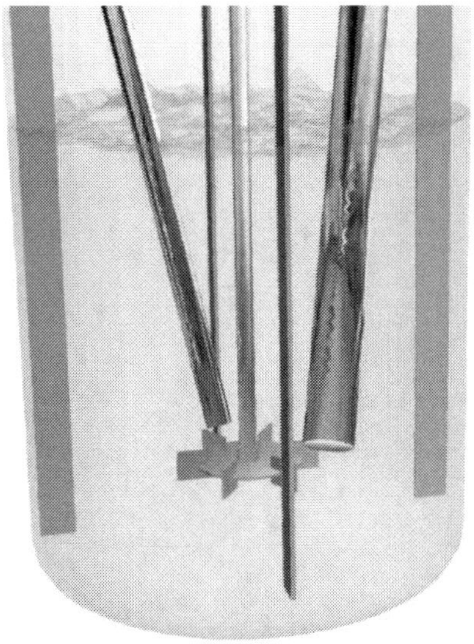

Maaß et al. **2011**. Experimental comparison of measurement techniques for drop size distributions in liquid/liquid dispersions, Experiments in Fluids, 50(2): 259-269.

reactor: DN 150

- H/D = 1.0
- Rushton turbine
- four baffels
- torospherical head
- full turbulent (Re > 20,000)

species:

- pH 7, deionized water
- air (g/l), flow rate 1 to 300 L/h
- toluene φ = 20 vol.% (l/l)

Results DSD

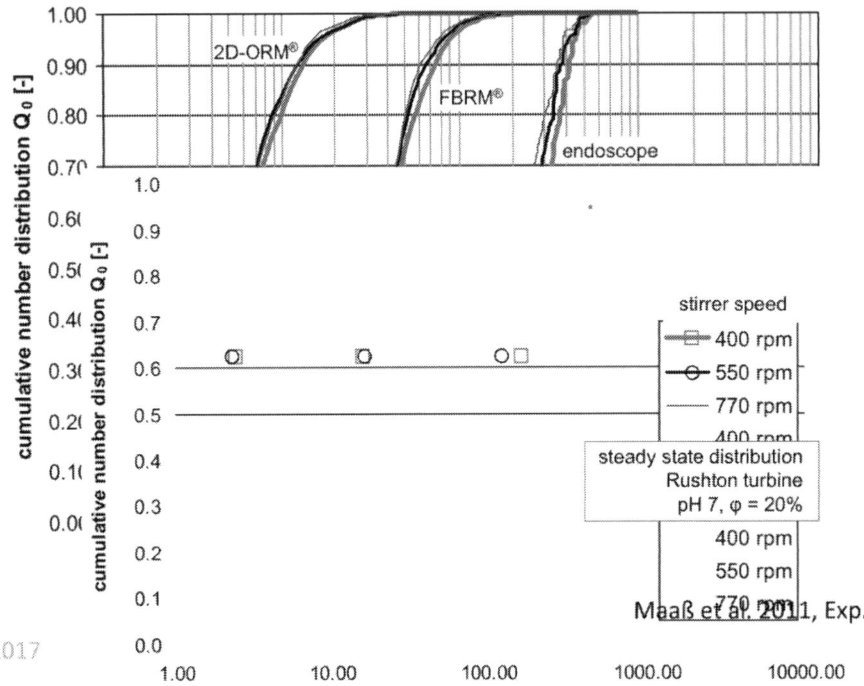

Maaß et al. 2011, Exp. Fluids

drop size distribution

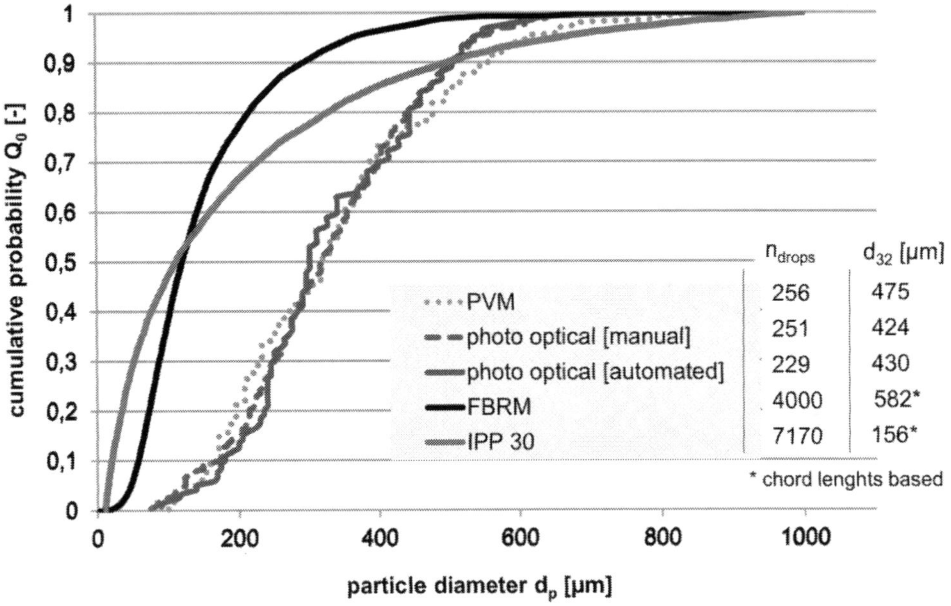

Transient drop size results

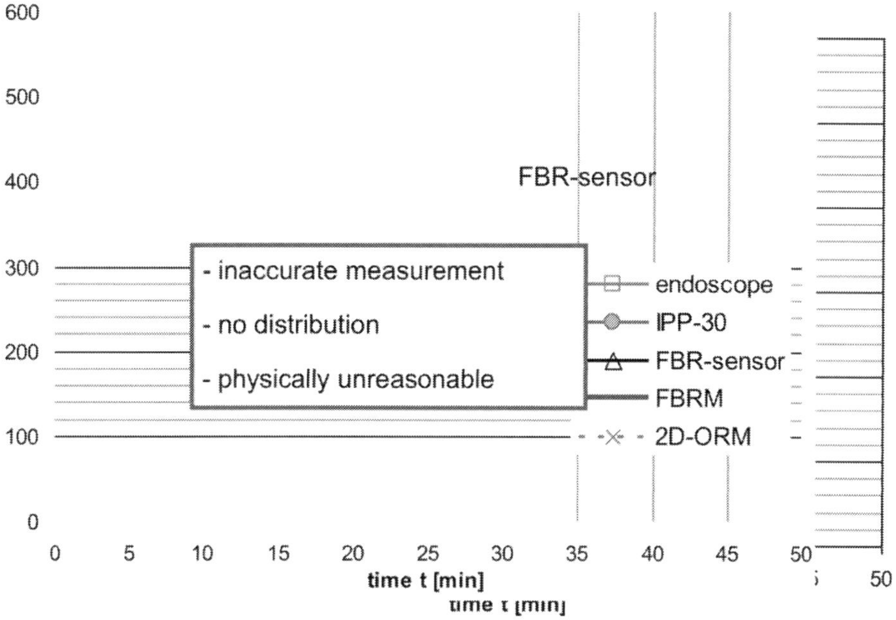

Agenda

1. Introduction
2. Experimental methods
    1. Measurement technique overview
    2. Experimental set-up
    3. Comparitative results
3. Tests in a three phase system
    1. Imaging hardware
    2. Gas-solid-liquid (G/S/L) results
4. Summary and outlook

SOPAT Probes

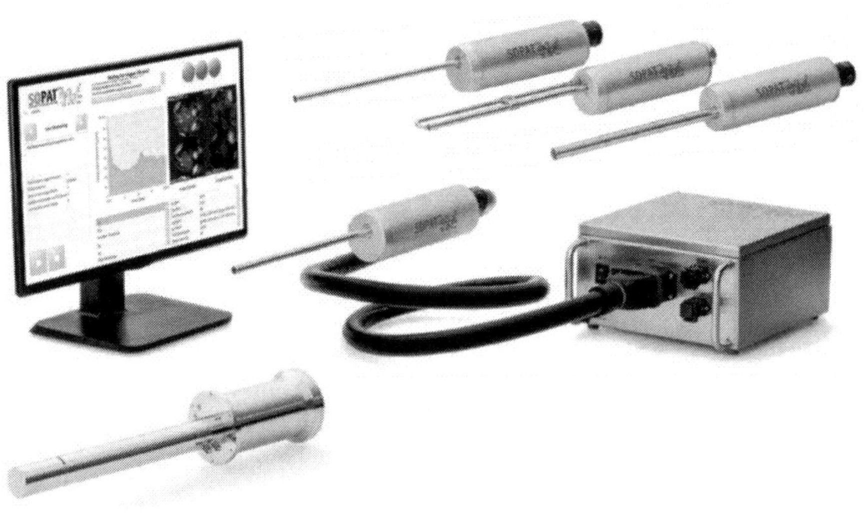

Illumination – image quality

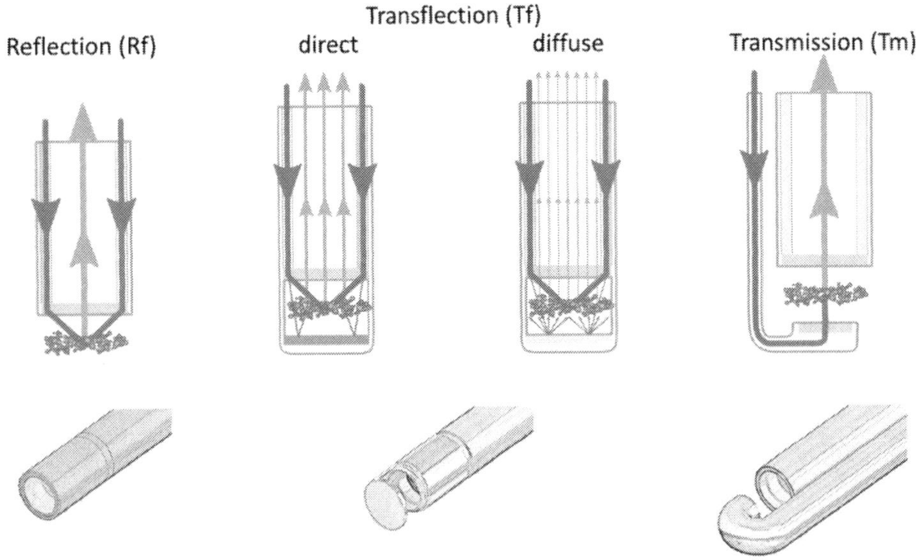

Panckow et al. **2017**. Photo-Optical In-Situ Measurement of Drop Size Distributions: Applications in Research and Industry, Oil & Gas Science and Technology – Rev. IFP Energies nouvelles, 72, 14: 2-19.

Illumination – image quality

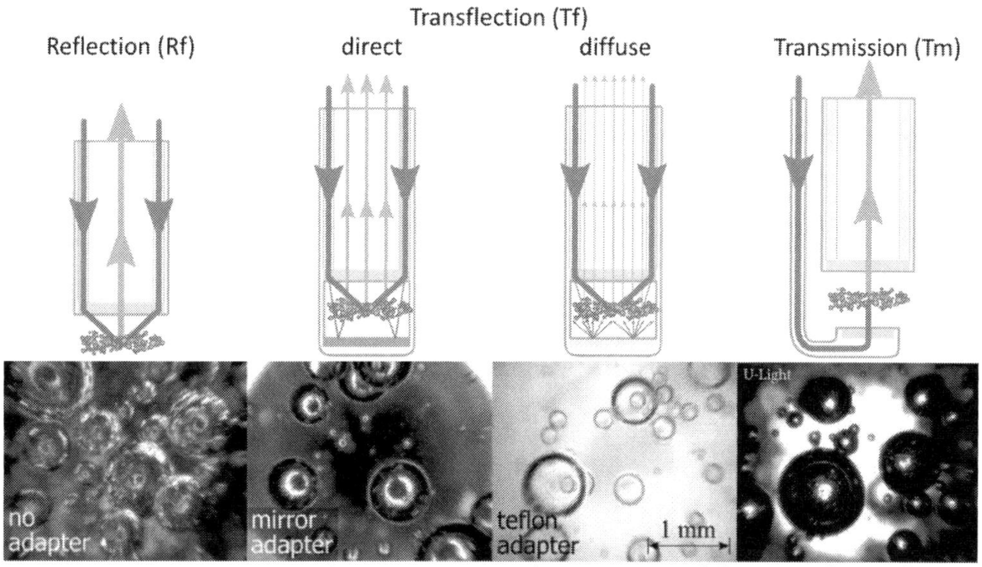

Air-water system with 120 ppm PVA (Alcotex B72) as surfactant

Agenda

1. Introduction
2. Experimental methods
    1. Measurement technique overview
    2. Experimental set-up
    3. Comparitative results
3. Tests in a three phase system
    1. Imaging hardware
    2. Gas-solid-liquid (G/S/L) results
4. Summary and outlook

Example of 3-Phase image analysis

- System: solid crystals (dispersed phase 1) and air bubbles (dispersed phase 2) in water (continuous phase)

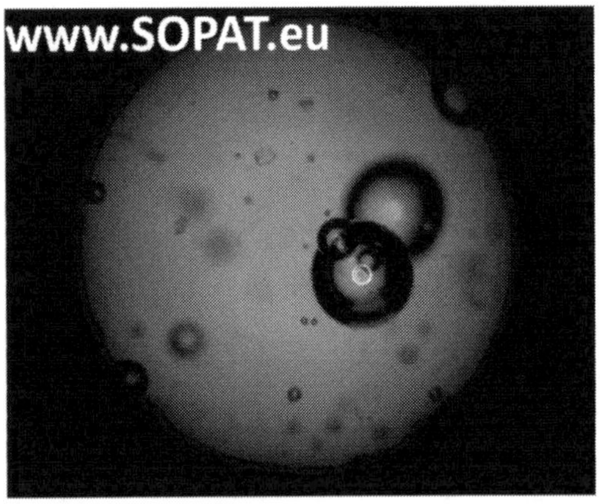

Original Image.

Example of 3-Phase image analysis

- System: solid crystals (dispersed phase 1) and air bubbles (dispersed phase 2) in water (continuous phase)

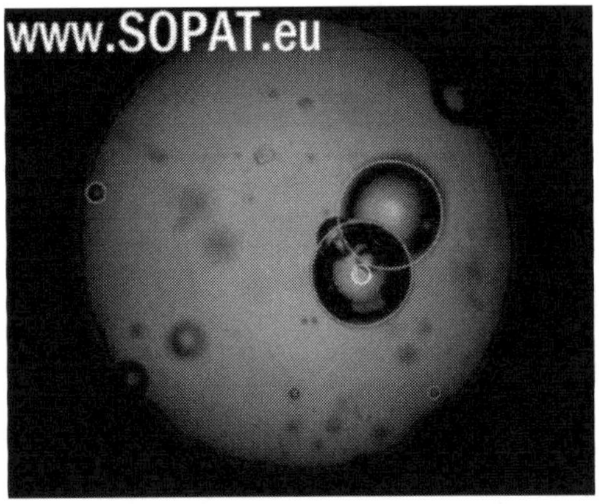

In a first step, the software detects the bubbles.

Example of 3-Phase image analysis

- System: solid crystals (dispersed phase 1) and air bubbles (dispersed phase 2) in water (continuous phase)

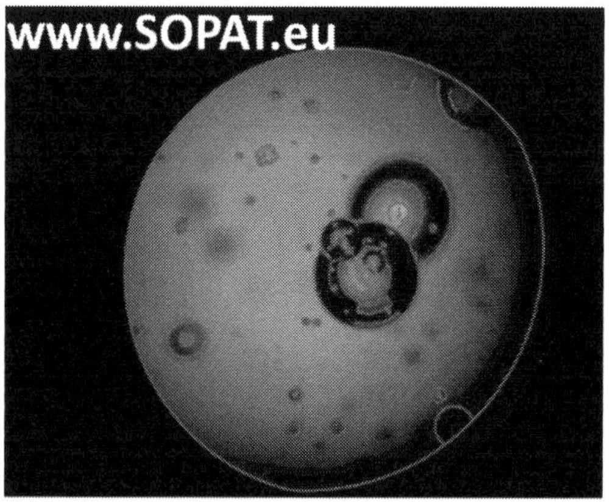

In an intermediate step, the software detects the boundaries of both bubbles and crystals.

3-Phase example

- System: solid crystals (dispersed phase 1) and air bubbles (dispersed phase 2) in water (continuous phase)

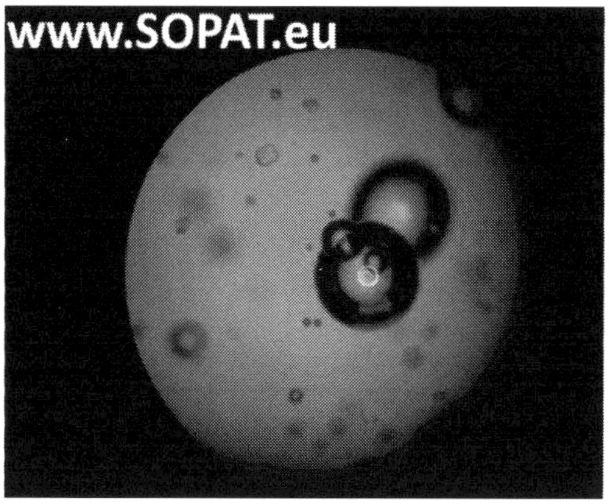

In the next step, the boundaries are closed.

3-Phase example

- System: solid crystals (dispersed phase 1) and air bubbles (dispersed phase 2) in water (continuous phase)

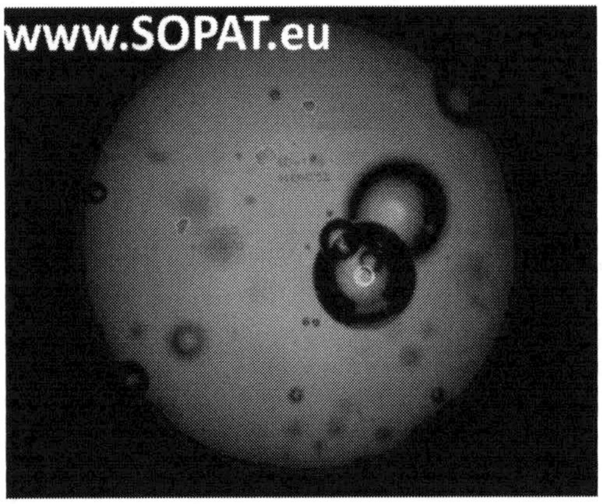

The last step uses subtraction to isolate the crystals.

Agenda

1. Introduction
2. Experimental methods
   1. Measurement technique overview
   2. Experimental set-up
   3. Comparitative results
3. Tests in the multiphase system
   1. Gas-liquid (G/L) results
   2. Gas-solid-liquid (G/S/L) results
4. Summary and outlook

Summary and Outlook

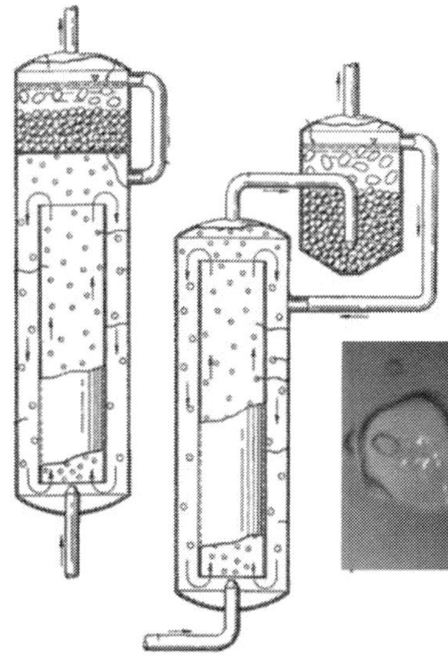

- Photo-optical analysis prefered over laser back scattering
- Size, shape, concentration even from multiple phase is possible via image analysis
- The control of the drop size possible → leads to the control of the reaction

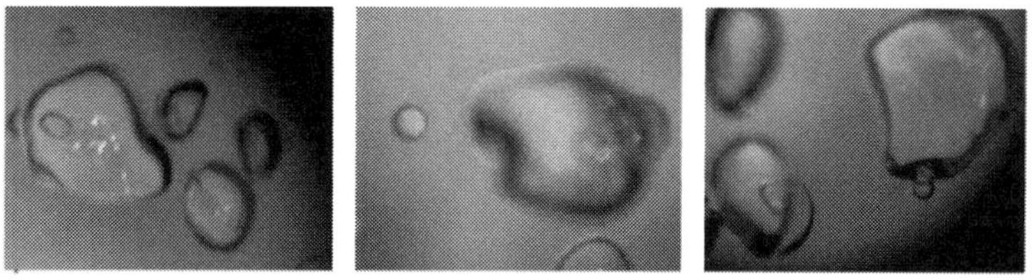

Extraction column

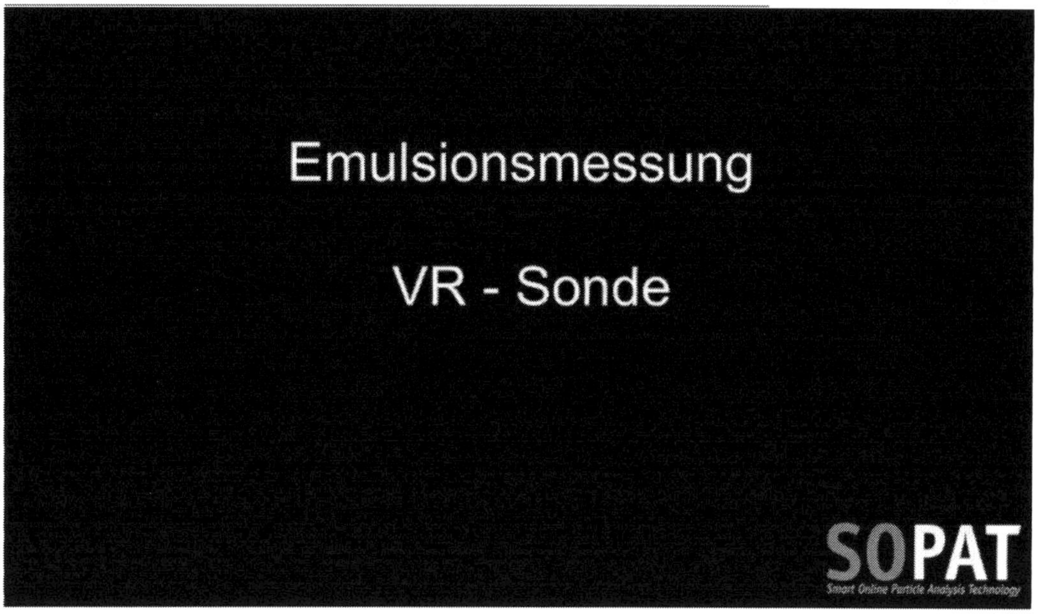

...未完待续

Thank you!

joern.emmerich@sopat.de

SOPAT-Hardware Specifications

Measurement Range	Min 1μm
Probe Diameter	12 mm (standard)
Probe Length	320 mm (standard)
Image Resolution	6 MegaPixel
Image Frequency	20 s^{-1}
Pressure Range	0.5 - 300 bar
Temperature Range	-10 - 125°C
Temperature Range (incl. Heat exchanger)	-50 - 450°C
Tubus Material	Resistant to all known mediums, pH 0 - 14; (1.4401, 1.4571, Hastelloy C22), Titanium

光学照片 vs. 激光后向散射

SOPAT

- 各种出版物认为光学照片分析更加可靠 (例如：SOPAT, PVM 和其他显微镜和内窥镜探头）而且，这种方法生成的数据相比激光后向散射和超声波检测法生成的数据更加精确. [Boxall et al. 2010, Hu et al. 2006, Maaß et al. 2011, Pacek et al. 1994].
- 弦长转化成真正的粒子大小是十分困难的.
- 光学照片分析能够分析不同颗粒相 (各种固体，固体 vs. 液体 vs. 气体...). 探针总能找到不同相的平均值并将它们一起显示出来.
- SOPAT 系统是唯一能投提供高度集成的，充分自动化的图像分析系统。

来源:

Hu, B., et al., Evaluation of drop size distribution from chord length measurements. AIChE Journal, 2006. 52(3): p. 931-939.

Boxall, J.A., et al., Measurement and calibration of droplet size distributions in water-in-oil emulsions by particle video microscope and a focused beam reflectance method. Industrial & Engineering Chemistry Research, 2010. 49(3): p. 1412-1418.

Maaß, S., et al., Experimental comparison of measurement techniques for drop size distributions in liquid/liquid dispersions. Experiments in Fluids, 2011. 50(2): p. 259-269.

Pacek, A.W., et al., Video technique for measuring dynamics of liquid-liquid dispersion during phase inversion. AIChE Journal, 1994. 40(12): p. 1940-1949.

497781 Sedimentation of Agglomerates Consisting of Polydisperse Nanoparticles

Abstract:

Sedimentation of agglomerates consisting of polydisperse nanoparticles

Anastasia Spyrogianni,[1] Katerina S. Karadima,[2,3] Eirini Goudeli,[1] Vlasis G. Mavrantzas[1-3] and Sotiris E. Pratsinis[1]

[1] *Department of Mechanical and Process Engineering, ETH Zurich, Zurich, 8092, Switzerland*

[2] *Department of Chemical Engineering, University of Patras, Patras, 26504, Greece*

[3] *Institute of Chemical Engineering Sciences, FORTH/ICE-HT, Patras, 26504, Greece*

The precipitation of agglomerates is of interest in water cleaning, pharmaceutical processing, nanotoxicology[1] and nanomedicine as well as for the stability of engineered nanofluids. Here, the mobility of nanosized fractal-like SiO_2 agglomerates in water is investigated and their settling rate in infinitely dilute suspensions is calculated by thoroughly validated Brownian dynamics (BD) tracking the agglomerate translational and rotational dynamics by a Cartesian vector formalism. Agglomerates are generated by an event-driven method[2] and have constant mass fractal dimension of 1.80 ± 0.01 but varying primary particle (PP) size distribution, mass and relative shape anisotropy. The calculated diffusion coefficient tensor is used to obtain their mobility diameter d_m, which is in excellent agreement with the d_m from scaling laws for fractal-like agglomerates. The agglomerate mobility to gyration diameter ratio, d_m/d_g, decreases with increasing relative shape anisotropy κ^2. For constant d_m, the agglomerate settling rate, u_s, increases with increasing PP diameter and geometric standard deviation $\sigma_{p,g}$ (polydispersity). A linear relationship between u_s and agglomerate mass to mobility diameter ratio, m/d_m, is revealed leading to an analytical expression for calculation of u_s for agglomerates consisting of polydisperse PPs. The latter shows that the commonly-made assumption of monodisperse PPs underestimates u_s by a fraction depending on $\sigma_{p,g}$ and the agglomerate mass mobility exponent. The simulations are in excellent agreement with precipitation rate measurements of commercial fumed SiO_2 agglomerates in water[1].

1. Spyrogianni, A.; Sotiriou, G. A.; Brambilla, D.; Leroux, J.-C.; Pratsinis, S. E. The effect of settling on cytotoxicity evaluation of SiO_2 nanoparticles. *J. Aerosol Sci.* **108**, 56–66 (2017).

2. Goudeli, E.; Eggersdorfer, M.L.; Pratsinis, S.E. Coagulation – Agglomeration of Fractal-like Particles: Structure and Self-Preserving Size Distribution, *Langmuir*, **31**, 1320-1327 (2015).

486786 Numerical Study on the Separation Characteristics and Kinematic Behavior of Particles in a Hydrocyclone

Abstract:

During the production process of titanium dioxide by sulfuric acid method, the incomplete reaction produces huge amount of acid hydrolysis residue which contain unreactive ilmenite. It was reported that more than 10 million tons of acid hydrolysis residue have been accumulated without any processing, which resulted in the waste of resources and serious environmental pollution problems. Hydrocyclone were regarded as a method to separate and recover them efficiently. In this work, to improve the classification sharpness, the separation characteristic and kinematic behavior of particles in a 75 mm hydrocyclone was studied by numerical method. The results showed that the particle motion was jointly controlled by the drag force, the pressure gradient force and the centrifugal force. For particles of different sizes, the drag force played a decisive role in the radial direction, the centrifugal force and the pressure gradient force remained about the same. While with the increasing of particle size, the drag force decreased exponentially. As a result, the influence of the centrifugal force and the pressure gradient force was strengthened, manifested as the increasing of the separation efficiency. Among the overflow particles, most particles moved rapidly to the bottom of the vortex finder, entering the inner vortex. Among the underflow particles, with the increasing of particle size, particles entered the outer vortex in a faster rotation speed with a clearer motion trajectory. Besides, it also investigated the influence mechanism of particle density, fluid viscosity, inlet flow rate, particulate arrangement and split ratio on the separation characteristic and the motion behavior of the particle phase.

493646 Evaluation of the Efficacy and Mechanism of Cactus (*Opuntia ficus-indica*) As a Natural Coagulant for Pre-Treatment of Oil Sand Process Affected Water

Abstract:

Abstract

Third-largest demonstrated crude oil reserve in the world, next to Saudi Arabia and Venezuela occurs in the oil sands of western province of Canada. Oil sands production in Canada is expected to upsurge from 4.1 million barrels per day in 2015 to 5 million barrels per day in 2024. About three to five barrels of fresh water per barrel of crude produced is used by mining operations and 0.5 barrels for in situ operations. Due to high concentrations of organic acids leached from the bitumen during extraction, the oil sand process affected water (OSPW) is brackish, alkaline in nature and contains dissolved organic and inorganic pollutants. The existence of these substances in processed water makes it highly toxic to aquatic life [1]. To address this critical issue, Environment and Climate Change Canada (ECCC) introduced zero discharge policy. As per the mandate of this policy, oil sand industries are required to store all OSPW and to recycle it after treatment. The treatment of OSPW has become a serious issue for Canada's oil sand industry, because the fine particles in tailing pond wastewater take decades to consolidate and dry out, which can delay reclamation. In currently available technologies, such as composite tailings (CT), whereby gypsum is added to mature fine tailings, further increases the salinity and hardness of process water [2]. In conventional pre-treatment techniques, chemical coagulants are used, namely, aluminum sulphate ($Al_2(SO_4)_3$), ferric chloride ($FeCl_3$) etc. Though the effectiveness of these chemicals as coagulants is well-recognized, still there are disadvantages associated with their usage, such as reduced efficiency in low-temperature water, relatively high procurement costs, detrimental effects on human health and production of large sludge volumes. Moreover, they significantly affect pH of the treated water. There is also strong evidence linking aluminium-based coagulants to the development of Alzheimer's disease and detrimental impact on the environment and health [3].

Maintaining good recyclable water quality and ensuring more efficient and limited use of fresh water are the most important challenges that must be met in the oil sands production sector. Therefore, it is critical to develop an effective and appropriate strategy to treat OSPW to sustain production. The combinations of the concerns with chemical coagulants and limited available technologies have led the researchers to focus on development of plant based natural coagulants. Due to their inherently renewable characteristics, abundance, low toxicity, and biodegradability, these naturally occurring and environmentally benign agents are rapidly gaining interest in water treatment. The mucilage extracted from *Opuntia ficus-indica* (OFI), a species of cactus, was evaluated for turbidity removal from synthetic turbid water and oil sand process affected water (OSPW), and an effort was made towards elucidating the underlying coagulation mechanism. For the first time, a natural coagulant derived from OFI was used to treat OSPW. Fourier transform infrared spectroscopy (FTIR) analysis was performed to understand chemical properties of the extracted mucilage. The functional groups observed in FTIR analysis signify the presence of l-arabinose, D-galactose, l-rhamnose, and D-xylose, as well as galacturonic acid. Preliminary jar test of synthetic turbid water achieved maximum turbidity removal of 98% under alkaline condition with coagulant dosing of 60 mg/l. The highest turbidity removal from OSPW was obtained up to 76% for 1200 mg/l coagulant dosage at neutral and alkaline pH (as shown in Figure 1 below). The effect of different ions commonly present in OSPW, on the process performance was also studied. The comparison of zeta potential measurements and microscopic images of thread like large flocs reveals the possibility that mechanism of coagulation by OFI is inter-particle bridging (as shown in Figure 2 below). Measurement of total organic carbon (TOC) was done and slight increment in TOC was observed after treatment of OSPW. A comparative study between natural coagulant and alum was also done and it was found that less quantity of compact sludge was produced by natural coagulant compared to alum. Once optimized, application of this readily available plant may offer a practical, inexpensive, and appropriate solution for pre-treatment of OSPW.

Keywords: OSPW; Natural coagulant; Cactus.

Figure 1. Effect of natural coagulant dosages and pH on turbidity removal from OSPW.

Figure 2. (a) Flocs produced while using alum (b) Flocs produced while using bio coagulant

References

[1] Lari, E., et al. "Determining the effect of oil sands process-affected water on grazing behaviour of Daphnia magna, long-term consequences, and mechanism." *Chemosphere* 146 (2016): 362-370.

[2] Renault, S., et al. "Effect of high salinity tailings waters produced from gypsum treatment of oil sands tailings on plants of the boreal forest." *Environmental Pollution* 102.2 (1998): 177-184.

[3] Martyn, C. N., et al. "Geographical relation between Alzheimer's disease and aluminium in drinking water." *The Lancet* 333.8629 (1989): 61-62.

500471 Thermogravimetric and Breakthrough Studies on the Adsorption Reversibility of SO_2, NO_2 and NO on Type a and X Zeolites

Abstract:

The adsorption reversibility of trace level flue-gas contaminants such as SO_2, NO_2 and NO gas was investigated on type A and X zeolites. The zeolites considered in this work included 13X, calcium exchanged 13X, 5A and 4A. This study was carried out to evaluate the influence of these species on the CO_2 removal from flue gas by PSA using zeolites. Six individual gas mixtures containing these species were utilized that included 74 ppm NO, 74 ppm NO_2, 42 ppm SO_2, 1 vol% NO, 1 vol% NO_2, and 500 ppm SO_2, all using either nitrogen or helium as the balance gas. A 15.03 mol% CO_2 mixture (balance nitrogen) was used to evaluate the effect of SO_2, NO and NO_2 on the adsorption behavior of CO_2 on the selected zeolites. This study included single cycle thermogravimetric analyses (TGA) individually with each gas mixture and each zeolite. It also included consecutive CO_2 breakthrough studies through an adsorption bed containing 5A or 13X before and after saturating the bed with the contaminant gas of interest. The roles of purging with nitrogen and temperature on the contaminated samples were also investigated. All TGA and breakthrough runs were carried at 70 ^{o}C and 1 atm absolute pressure. The results highlight the importance of the cation (Ca^{2+} versus Na^{+}) in the stability of the zeolite against the negative effect exerted by the presence of the contaminant.

496356 The Correct Use of the Ideal Adsorbed Solution Theory for High Pressure Systems

Abstract:

The Ideal Adsorbed Solution Theory (IAST) of Myers and Prausnitz is used in adsorption engineering to predict multicomponent mixtures from pure component adsorption isotherms. It is the equivalent of Rault's law in vapour liquid equilibrium, but the reference state is calculated from the Gibbs isotherm and cannot be measured directly and very accurately as is the case for pure component vapour pressures. Because the reference state is calculated, any error in this calculation leads to the incorrect evaluation of the adsorbed phase excess Gibbs energy and thus makes it not possible to develop robust correlations also for the more general case where non-ideality in the adsorbed phase is introduced.

In theory many gas-solid systems should be represented well by the IAST because in microporous solids the main interaction is between the solid and the adsorbed molecule. The only main exception would be due to entropic effects, where a molecule can access regions of the micropore volume that a second molecule cannot. A recent review by Wu and Sircar (2016) shows that data reported in the literature can deviate substantially and what is more worrying is that even for the same binary system different authors report different deviations from the IAST, amongst these, oxygen and nitrogen on LiLSX is an example of roughly equally sized molecules which should follow closely the IAST but in some cases literature results point to poor predictions.

Clearly one possible explanation for the observed discrepancy may be that the binary experimental data are inconsistent, but this is not the only reason for such differences. This contribution is aimed at addressing two main points:

1) The correct representation of pure component isotherms when the gas phase is not ideal;

2) The discussion of the assumptions that are implied by the actual procedure with which the ideal adsorbed phase is calculated;

Examples from the literature will be used to highlight several inconsistencies which point to the incorrect use of the IAST. As a more in depth case study, the binary mixtures of nitrogen and carbon dioxide on NaX zeolites at pressures up to 10 bar will be considered. This system is chosen for its industrial relevance, especially in carbon capture applications. It provides also a stringent test for the IAST since the ideal selectivity is > 100 and therefore allows to highlight issues that may not be as evident when selectivities are relatively low.

Mapping the Limitations of Breakthrough Analysis in Fixed-Bed Adsorption

James Clinton Knox

NASA Marshall Space Flight Center, Huntsville, AL, USA

October 30, 2017

Introduction: Field of Study

- Gas separation processes using fixed beds of adsorbents are important in many existing and emerging industries
- Adsorbent processes are typically cyclic, with adsorption and desorption steps that swing between high and low pressures, temperatures, or concentrations (PSA, TSA, CSA)
- Computer simulations are typically used to select hardware and process parameters (adsorbent type, layer sizes, pressure and temperature set points, cycle time, flow rate, etc.)
- A cyclic steady-state condition is only reached after many cycles
- Simulation execution speed is critical due to the large trade space and extensive number of cycles required for a single data point

2

Skarstrom Patent

- Pressure swing adsorption process
- Air drying using silica gel beds
- Cyclic steady state only achieved after 6 days of operation

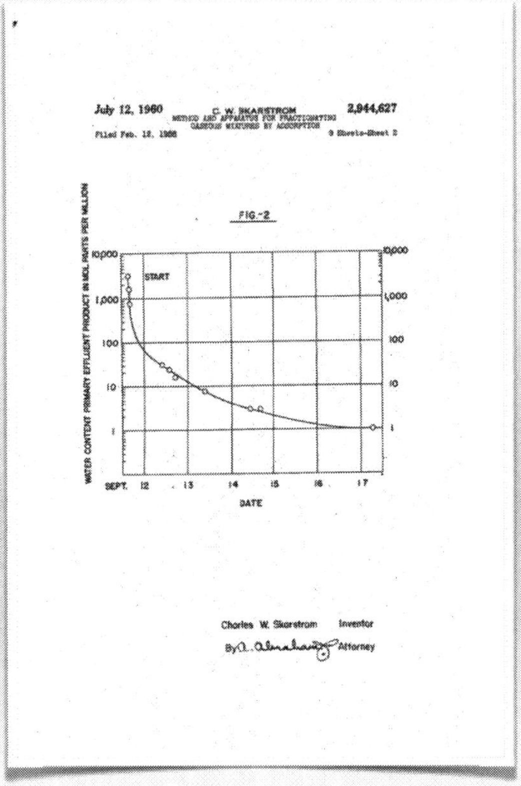

3

Introduction: Area of Concern

- Execution speed requirements dictate use of simplified one-dimensional model for cyclic process simulations
- 1-D axially dispersed plug flow equation predominantly used in process simulations based on current literature
- Simplifying assumptions include two lumped mass transfer terms requiring empirical determination
- Mass transfer term determination is generally via breakthrough analysis in sub-scale fixed-beds with a low tube diameter to particle diameter ratio

$$\frac{\partial c}{\partial t}+\left(\frac{1-\varepsilon}{\varepsilon}\right)\frac{\partial \bar{q}}{\partial t}-\boxed{D_L}\frac{\partial^2 c}{\partial x^2}=-\frac{\partial \upsilon_i c}{\partial x}$$

$$\frac{\partial \bar{q}}{\partial t}=\boxed{k_n}(q^*-\bar{q})$$

4

Introduction: Research Findings (1)

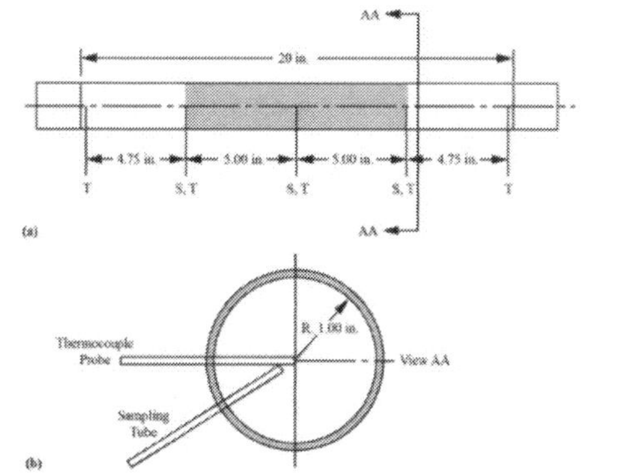

(a) Breakthrough test apparatus and (b) cross-sectional view of a typical temperature measurement and gas sampling location. "T" indicates thermocouple probe location, and "S" indicates sampling tube location. Shading in (a) indicates location of sorbent packing.

Knox, J. C. Finite Difference Modeling and Experimental Investigation of Carbon Dioxide Adsorption on a Molecular Sieve Sorbent Material Used in Spacecraft Carbon Dioxide Removal Systems : A Thesis. University of Alabama, Huntsville, 1992.

- Standard breakthrough analysis is based on measurement taken after mixing of two flow regimes: in the bed core, and channeled flow along the walls
  - Resultant inaccuracies in mass transfer term cause errors during simulation-based design of full-scale separation process
  - *An improved approach was developed to use a centerline measurement in addition to the mixed measurement and determine terms individually*

5

Introduction: Research Findings (2)

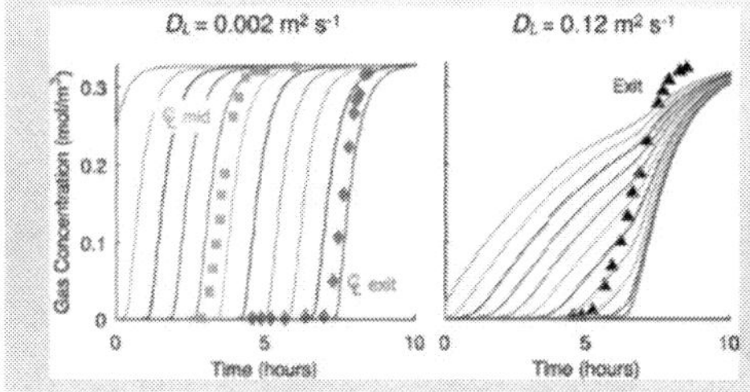

Knox, J. C.; Ebner, A. D.; LeVan, M. D.; Coker, R. F.; Ritter, J. A., Limitations of Breakthrough Curve Analysis in Fixed-Bed Adsorption. *Ind Eng Chem Res* **2016**.

- Axial dispersion term derived based on Fickian (molecular) diffusion, but is used to model dispersion arising from flow around pellets and wall effects
  - *For strongly adsorbed species, interaction of a large dispersion term with the ill-posed Danckwerts boundary condition causes hidden nonphysical simulation result*
  - *To prevent nonphysical behavior, limiting expressions for the mass transfer terms were derived for specific sorbent/sorbate pairs and inlet conditions*
  - *A generalized expression was derived to limit the mass transfer terms for any sorbent/sorbate pair based on the strength of adsorption*

6

Principle Equations in 1-D Model

$$\frac{\partial c}{\partial t}+\left(\frac{1-\varepsilon}{\varepsilon}\right)\frac{\partial \bar{q}}{\partial t}-D_L\frac{\partial^2 c}{\partial x^2}=-\frac{\partial \upsilon_f c}{\partial x}$$

$$\frac{\partial \bar{q}}{\partial t}=k_n(q^*-\bar{q})$$

$$\varepsilon a_f \rho_f c_{pf}\frac{\partial T_f}{\partial t}-\varepsilon a_f k_{eff}\frac{\partial^2 T_f}{\partial x^2}=-\varepsilon a_f \rho_f \upsilon_f c_{pf}\frac{\partial T_f}{\partial x}+a_f a_s h_s\left(T_s-T_f\right)+P_i h_i\left(T_w-T_f\right)$$

$$(1-\varepsilon)\rho_s c_{ps}\frac{\partial T_s}{\partial t}=a_f a_s h_s\left(T_f-T_s\right)-(1-\varepsilon)a_f \lambda\frac{\partial q}{\partial t}$$

$$a_w \rho_w c_{pw}\frac{\partial T_w}{\partial t}-a_w k_w\frac{\partial^2 T_w}{\partial x^2}=p_i h_i\left(T_f-T_w\right)+P_o h_o\left(T_o-T_w\right)$$

$$n=\frac{ap}{\left[1+(bp)^t\right]^{1/t}};\quad b=b_0\exp(E/T);\quad a=a_0\exp(E/T);\quad t=t_0+c/T$$

$$\frac{1}{Pe}=\frac{20}{\varepsilon}\left(\frac{D}{2\upsilon R_p}\right)+\frac{1}{2}=\frac{20}{ReSc}+\frac{1}{2}$$

$$\frac{1}{Pe}=\frac{0.73\varepsilon}{ReSc}+\frac{1}{2\left(1+\frac{13\cdot 0.73\varepsilon}{ReSc}\right)}\quad 0.0377<2R_p<0.607\text{ cm}$$

$$h_i=\frac{k_f}{2R_i}Nu\quad\text{with}\quad Nu=2.03Re^{0.8}\exp\left(-6\frac{R_p}{R_i}\right)$$

$$c_p=a_o+a_1T_f+a_2T_f^2+a_3T_f^3$$

$$Sh=2+1.1Sc^{\frac{1}{3}}Re^{0.6}\qquad h_s=\frac{ShD}{2R_p}$$

$$k_e=k_f\left(\frac{k_s}{k_f}\right)^n\quad\text{with}\quad n=0.280-0.757\log_{10}\varepsilon-0.057\log_{10}\left(\frac{k_s}{k_f}\right)$$

$$k_{eff}=k_f\left(\frac{k_e}{k_f}+0.75PrRe\right)\quad\text{where}\quad Pr=\frac{c_p\mu}{\rho_f k_f}$$

All variables in Mass and Heat Balance Equations are determined except D_L, k_n, *and* h_o

7

Experimental Results

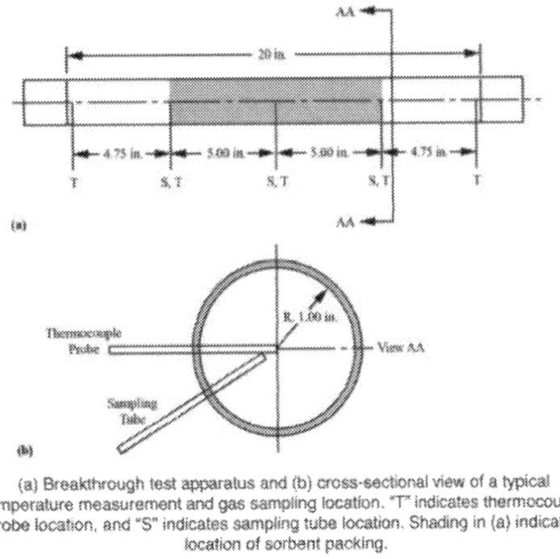

(a) Breakthrough test apparatus and (b) cross-sectional view of a typical temperature measurement and gas sampling location. "T" indicates thermocouple probe location, and "S" indicates sampling tube location. Shading in (a) indicates location of sorbent packing.

Knox, J. C. Finite Difference Modeling and Experimental Investigation of Carbon Dioxide Adsorption on a Molecular Sieve Sorbent Material Used in Spacecraft Carbon Dioxide Removal Systems : A Thesis. University of Alabama, Huntsville, 1992.

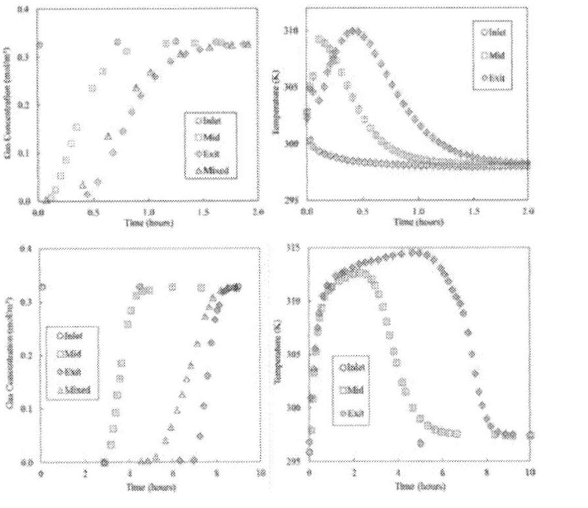

Left panels: Experimental gas-phase concentration profile history breakthrough curves for CO2 (top) and H2O vapor (bottom) on zeolite 5A at 3 centerline locations in the bed (circles: 2.5%, squares: 50%, and diamonds: 97.5%) and just outside the bed (triangles). Right panels: Corresponding experimental temperature profile histories for CO2 (top) and H2O vapor (bottom) on zeolite 5A at 3 centerline locations in the bed (circles: 2%, squares: 50%, and diamonds: 98%). Error bars show experimental uncertainty.

Knox, J. C.; Ebner, A. D.; LeVan, M. D.; Coker, R. F.; Ritter, J. A., Limitations of Breakthrough Curve Analysis in Fixed-Bed Adsorption. *Ind Eng Chem Res* **2016**.

8

Step 1: Wall to Ambient Heat Transfer Coefficient

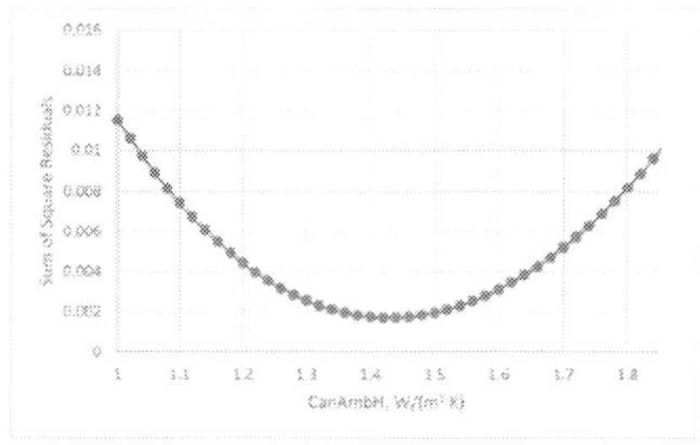

Sum of Square Residuals for Experimental and Simulation Data

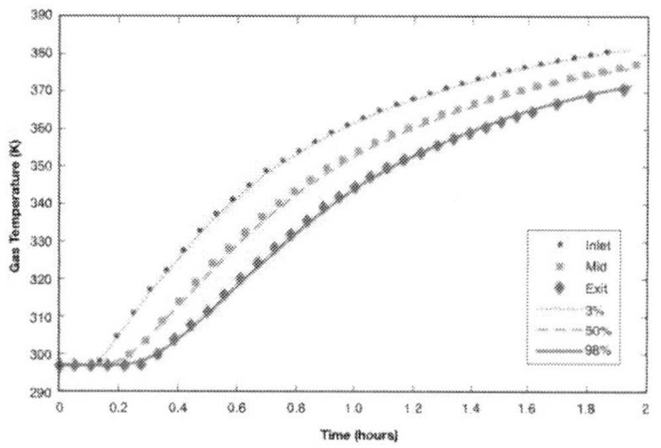

Temperature history data for the thermal characterization test with N_2

h_o is empirically derived via a Thermal Characterization Test

9

Step 2: Linear Driving Force Mass Transfer Coefficient

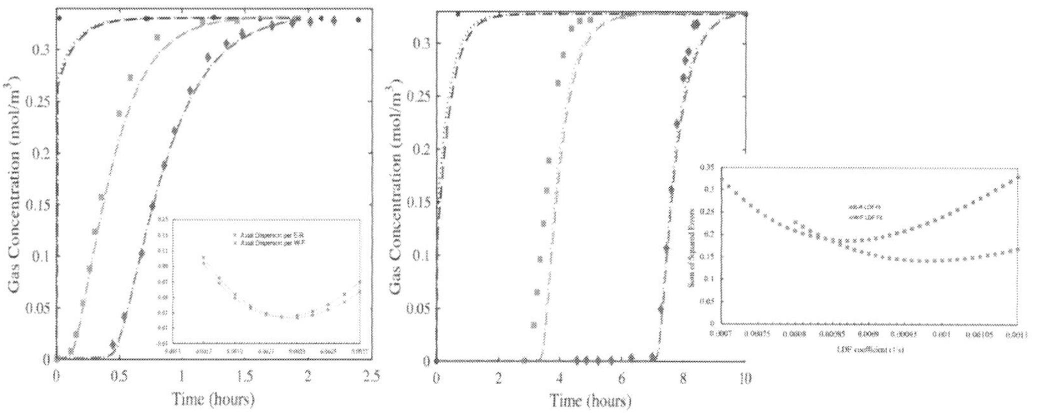

Fits of the 1-D axial dispersed plug flow model to the 97.5% location (diamonds) experimental centerline gas-phase concentration breakthrough curves for CO_2 (left) and H_2O vapor (right) on zeolite 5A, and corresponding predictions from the model of the 2.5% (circles) and 50% (squares) locations. The saturation term in the CO_2-zeolite 5A isotherm was increased by 15%. The saturation term in the H_2O vapor-zeolite 5A isotherm was decreased by 3%. **The void fraction was reduced to 0.33 based on the Cheng distribution (Cheng *et al.*, 1991) with C = 1.4 and N = 5, as recommended by Nield and Bejan (1992)**

k_n is empirically derived via fitting to centerline concentration breakthrough curve. For this step, dispersion is taken to result from pellet effects only (no wall effects). Choice of dispersion correlation has a small impact on k_n

10

Step 3: Axial Dispersion Coefficient (CO_2 Case)

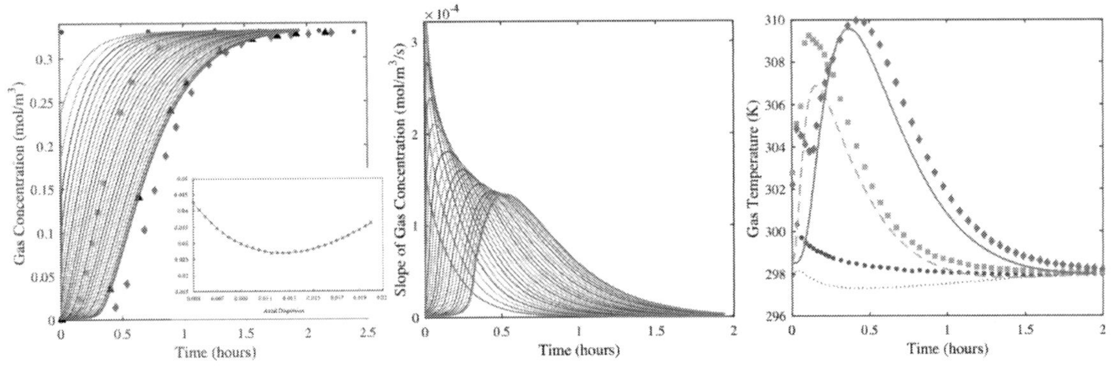

CO_2 on zeolite 5A: Fit of the 1-D axial dispersed plug flow model to the outside bed (triangles) experimental breakthrough curve using a value of D_L 7 times greater than that from the Wakao and Funazkri correlation and the fitted LDF k_a = 0.0023 s^{-1} (left panel). The reported saturation term for the CO_2-zeolite 5A isotherm was used, along with **the reported void fraction of 0.35**. Predictions from the model (lines) of the gas-phase concentration breakthrough curves at 0, 4, 8, 12, ..., 92, 96 and 100% locations in the bed are also shown in the left panel, along with the 2.5% (circles), 50% (squares) and 97.5% location (diamonds) experimental center line gas-phase concentration breakthrough curves (left panel). The corresponding derivative (or slope) of the predicted gas-phase concentration breakthrough curves in the bed are shown in the middle panel. Predictions from the model (lines) of the 2.5% (circles), 50% (squares) and 97.5% location (diamonds) experimental center line temperature profile histories are shown in the right panel.

D_L term is fit to mixed gas concentration (far downstream), but requires value 7 times the correlation value to compensate for wall channeling. Fit is specific to the size of the column; for a much larger column wall channeling may be neglected and correlated values of D_L used (but not for fixed beds with a tube to pellet ratio of 20 as in this case, or less)

11

Step 3: Axial Dispersion Coefficient (H_2O Case)

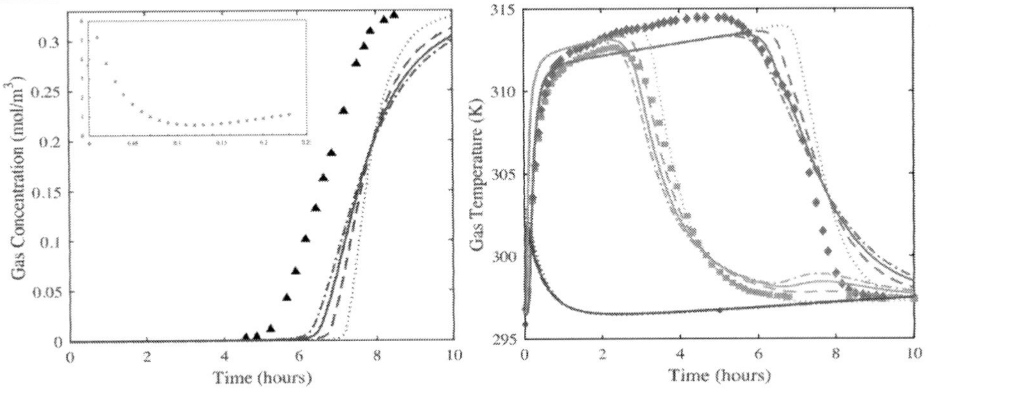

H_2O vapor on zeolite 5A: Predictions from the 1-D axial dispersed plug flow model of the outside the bed (triangles) experimental breakthrough curve when varying the value of D_L. **D_L = 10 (dotted lines), 30 (dashed lines), 50 (solid lines) and 70 (dash-dot lines) times greater than Wakao and Funazkri correlation** with the LDF k_a = 0.00083 s^{-1} (left panel). The reported saturation term for the H_2O-zeolite 5A isotherm was used, along with the reported void fraction of 0.35. The corresponding predictions from the model (lines) of the 2.5% (circles), 50% (squares) and 97.5% location (diamonds) experimental center line temperature profile histories are shown in the right panel.

D_L term is fit to mixed gas concentration (far downstream), but requires value ***50(!)*** times the correlation value to compensate for wall channeling. However the temperature profiles deviate increasingly from the test data with increasing D_L indicating a breakdown of the axial dispersed plug flow model.

12

Step 3: Axial Dispersion Coefficient (H_2O Case)

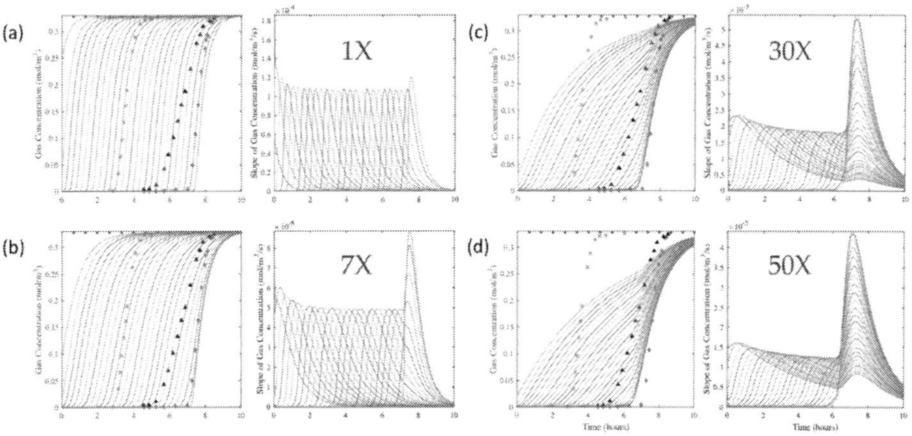

H_2O vapor on zeolite 5A: Predictions from the model (lines) shown in Figure 9 of the gas-phase concentration breakthrough curves at 0, 4, 8, 12, . . ., 92, 96 and 100% locations in the bed (left panels). The 2.5% (circles), 50% (squares) and 97.5% location (diamonds) experimental centerline gas-phase concentration breakthrough curves are also shown for comparison in the left panels. The corresponding derivatives (or slopes) of the gas-phase concentration breakthrough curves in the bed are shown in the right panels. (a) D_L = Wakao-Funazkri correlation, and (b) D_L = 7, (c) 30 and (d) 50 times greater than Wakao and Funazkri correlation.

At 7X, internal concentration history slope matches mixed concentration just as for CO_2 case. This indicates that same dispersive mechanism occurs regardless of sorbate. To overcome non-physical breakthrough sharpening, D_L must be increased by 50X to decrease breakthrough slope. Expected CPB is lost entirely for this condition.

13

Modeling Conclusions

- Breakthrough tests with tube diameter to pellet diameter ratios of around 20 (or less), are subject to wall channeling, an mechanism not captured in standard dispersive correlations. Breakthrough tests are generally sub-scale to conserve sorbent materials and gas flow equipment costs and thus frequently in this range.
- The typical breakthrough measurement is taken far downstream, after mixing. Fitting the mass transfer coefficient to this measurement will provide erroneous results for a larger (or smaller) diameter column due to the influence of channeling.
- A method has been demonstrated where a centerline measurement is used to derive a mass transfer coefficient that captures physics free of wall effects and thus appropriate for scale-up to large diameter columns.
- Using the mass transfer coefficient derived above, this method uses the mixed concentration data for fitting of a dispersion coefficient D_L specific to the tube diameter, as needed for processes that utilize small diameter tubes.
- However fitting D_L blindly to the breakthrough curve (as apparent in many published breakthrough analyses) can, in specific cases, result in a complete breakdown of the axially dispersed plug flow model, and result in fitted coefficients that are incorrect.
- Thus it is important to map the set of conditions where significant breakthrough sharpening occurs in order to avoid nonphysical and non-predictive simulation behavior.

14

Mapping the Sensitivity of Sorbate/Sorbent Systems to D_L and k_n

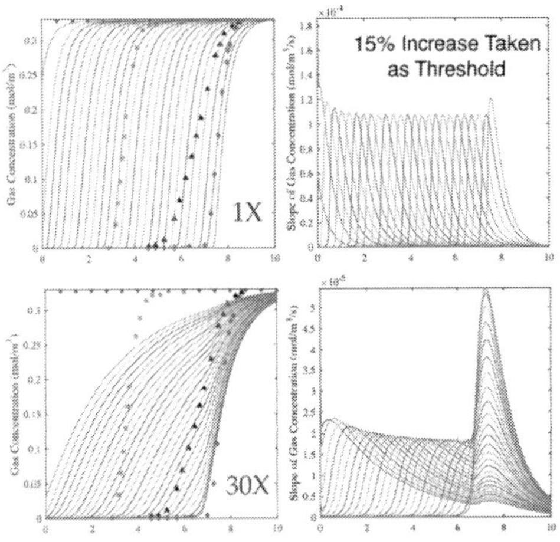

Slope ratio provides metric for breakthrough sharpening and departure from constant pattern behavior

Mass Balance Equations

$$\frac{\partial c}{\partial t}+\left(\frac{1-\varepsilon}{\varepsilon}\right)\frac{\partial \bar{q}}{\partial t}-D_L\frac{\partial^2 c}{\partial x^2}=-\frac{\partial v_i c}{\partial x}$$

$$\frac{\partial \bar{q}}{\partial t}=k_n(q^*-\bar{q})$$

CO₂ and H₂O Capacity Isotherms

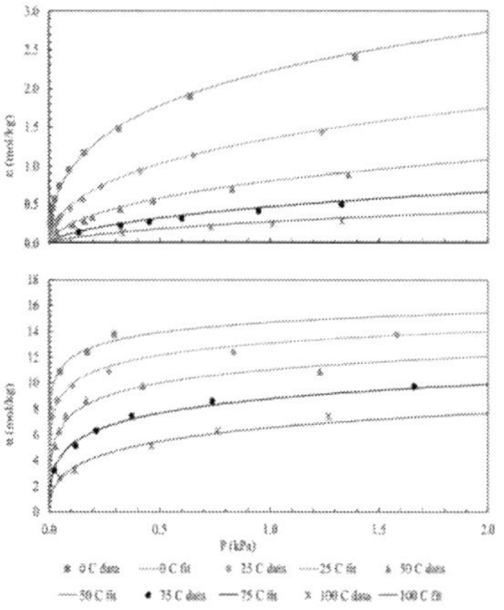

Equilibrium adsorption isotherms for CO2 (top) and H2O vapor (bottom) on zeolite 5A at temperatures from 0°C to 100°C as indicated. Symbols represent experimental data; Toth isotherm fits are shown as lines (Wang and LeVan, 2009)

$$n=\frac{ap}{\left[1+(bp)^t\right]^{1/t}};\quad b=b_0\exp(E/T);\quad a=a_0\exp(E/T);\quad t=t_0+c/T$$

15

Threshold parameter determination for H_2O on 5A

(similar analysis for CO_2 on 5A, H_2O on 13X, and CO_2 on 13X)

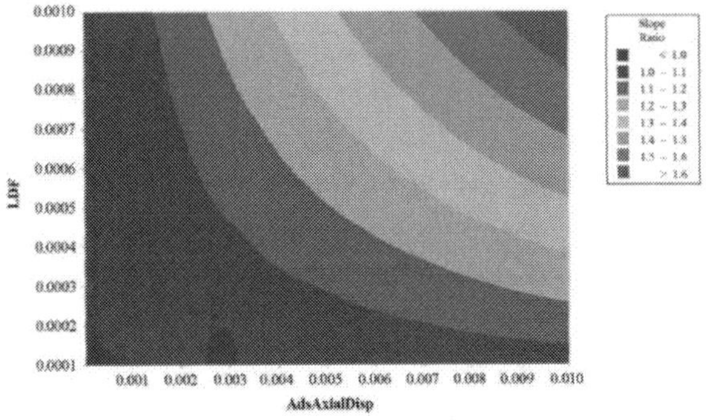

Contour plot of slope ratio for H_2O/5A system based on 100 breakthrough simulations

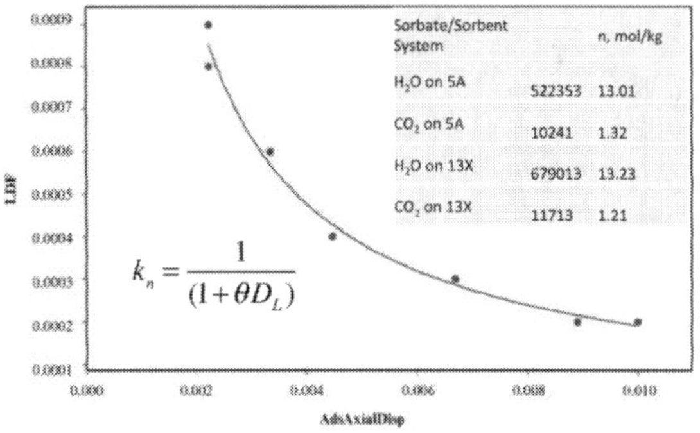

Sorbate/Sorbent System	θ	n, mol/kg
H_2O on 5A	522353	13.01
CO_2 on 5A	10241	1.32
H_2O on 13X	679013	13.23
CO_2 on 13X	11713	1.21

$$k_n=\frac{1}{(1+\theta D_L)}$$

Curve fit of $k_n = f(D_L)$ for simulation runs with slope ratio values between 1.13 and 1.16 for H_2O on 5A. Coefficient of determination (R^2) is 0.998.

For this system, the simulation will result in a slope ratio limit below the threshold if $k_n(1+\theta D_L)-1<0$

16

Generalization to any sorbent/sorbate system

$$K_d = \frac{q^*}{q_0^* - q^*}\frac{c_0 - c}{c}$$ Distribution Factor

$$\frac{q^*}{q_m} = \frac{bc}{1+bc}$$ Langmuir Isotherm

$$K_d = 1 + bc_0$$ Distribution Factor for Langmuir Isotherm

Threshold parameter values and distribution factor values

Sorbate/Sorbent System	θ	K_d (fitted)
CO_2 on zeolite 5A	10241	5.213
CO_2 on zeolite 13X	11713	4.862
H_2O on zeolite 5A	522353	33.106
H_2O on zeolite 13X	679013	75.309

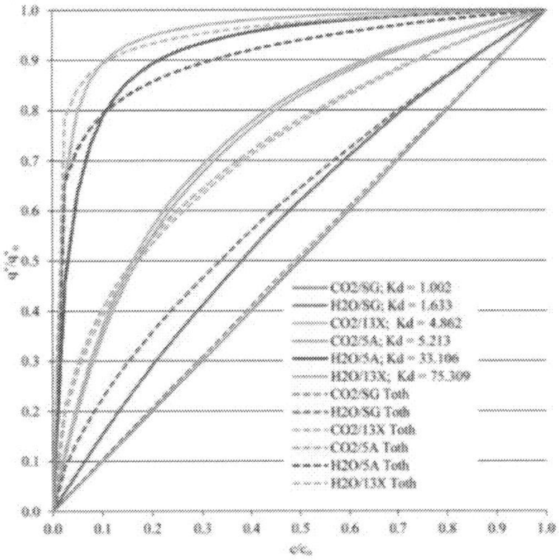

Normalized concentration vs. normalized bed loading for six sorbate/sorbent systems for conditions of 10°C and 1.0 kPa. Solid lines: Langmuir isotherms; Dashed lines: Toth isotherms.

17

Correlation between threshold parameter and distribution factor

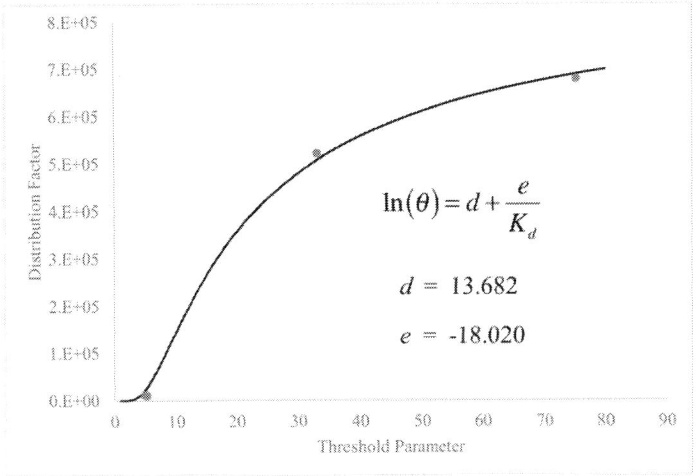

$$\ln(\theta) = d + \frac{e}{K_d}$$

d = 13.682

e = -18.020

Estimated threshold value q vs. distribution factor Kd for four sorbate/sorbent systems (filled circles) and fitted relationship shown in Equation 5.6 (line). Coefficient of determination (R^2) is 0.997.

Threshold parameter values and distribution factor values

Sorbate/Sorbent System	θ	K_d (fitted)
CO_2 on zeolite 5A	10241	5.213
CO_2 on zeolite 13X	11713	4.862
H_2O on zeolite 5A	522353	33.106
H_2O on zeolite 13X	679013	75.309

18

Three-step process to prevent excessive breakthrough sharpening

1. Determine distribution factor (K_d) for sorbent/sorbate system of interest by fitting to Langmuir isotherm
2. Calculate threshold parameter: $$\ln(\theta) = d + \frac{e}{K_d}$$
3. Map limits on D_L vs. k_n plot based on: $$k_n = \frac{1}{(1+\theta D_L)}$$

For values where $k_n(1+\theta D_L)-1<0$ excessive breakthrough sharpening and breakdown of the constant pattern behavior will be avoided

19

Conclusions for Parameter Mapping

- The axially dispersed plug flow equation and the Danckwerts boundary condition works well for values of dispersion within bounds of accepted correlations
- However, for specific combinations of K_d, D_L and k_n this model breaks down due to the elimination of dispersion at the outlet boundary. In these cases, significant breakthrough sharpening occurs as well as distortion of the internal concentration, deviating from the accepted CPB for these systems.
- This work presents a methodology where a threshold parameter may be calculated based on K_d, D_L and k_n, and applied to avoid non-physical model distortion

20

Overall Conclusions

- The separation of gases through adsorption plays an important role in the chemical processing industry, where the separation step is often the costliest part of a chemical process and thus worthy of careful study and optimization.
- This work developed a number of new, archival aspects on the computer simulations used for the refinement and design of these gas adsorption processes:
    1. Presented a new approach to fit the undetermined heat and mass transfer coefficients in the axially dispersed plug flow equation and associated balance equations
    2. Examined and described the conditions where non-physical simulation results can arise
    3. Presented an approach to determine the limits of the axial dispersion and LDF mass transfer terms above which non-physical simulation results occur

21

Backup

Gas Separation Processes

- Separation processes are defined as those that transform a mixture of substances into two or more product streams (King, 1980)
- The study of separations is of critical importance as they are the costliest step in many chemical processes, as they reverse the mixing of substances and thus require a decrease in entropy (Yang 2003)

Table 2.1 Common Commercial and Industrial Uses for Sorbents (Keller, 1983; Yang, 2003)

Component to be adsorbed	*Other components*	*Adsorbent(s)*
Gas purification		
H_2O	Olefin-containing cracked gas, natural gas, air, synthesis gas, etc.	Silica, alumina, zeolite (3A)
CO_2	C_2H_4, natural gas, etc.	Zeolite, carbon molecular sieve
Hydrocarbons, halogenated organics, solvents	Vent streams	Activated carbon, silicalite, others
Sulfur compounds	natural gas, hydrogen, liquefied petroleum gas (LPG)	Zeolite, activated alumina
SO_2	Vent streams	Zeolite, activated carbon
Gas bulk separations		
Normal paraffins	iso-paraffins, aromatics	Zeolite
N_2	O_2	Zeolite
O_2	N_2	Carbon molecular sieve
CO	CH_4, CO_2, N_2, Ar, NH_3/H_2	Zeolite, activated carbon
Acetone	Vent streams	Activated Carbon
C_2H_4	Vent streams	Activated Carbon

Table 2.2 Gas Separation and Purification Applications Enabled by New Sorbents (Yang, 2003)

Application	Sorbent and Notes
N_2/CH_4 separation for natural gas upgrading	Clinoptilolite, tinanosilicates by kinetic separation, single-wall carbon nanotubes
CO removal from H2 to < 1 ppm for fuel cell applications	π-complexation sorbents such as CuCl/γ-Al_2O_3, CuY, and, AgY
NO_x removal	Fe-Mn-Ti oxides, Fe-Mn-Zr oxides, Cu-Mn oxides, multi-wall carbon nanotubes
C_3H_6/C_3H_8 (+hydrocarbons) separation	π-complexation sorbents such as CuCl/γ-Al_2O_3, AgNO3/SiO2, AgNO3/clays, aluminophosphate
C_2H_4/C_2H_6 (+hydrocarbons) separation	π-complexation sorbents such as CuCl/γ-Al_2O_3, AgNO3/SiO2, AgNO3/clays

Table 2.3 CO_2 Capture Technologies Funded Under DOE (Vora, 2013).

Project Name	**Project Focus**	**Sorbent Materials**	**Process Approach**	**Technology Maturity**	**Ref.**
Bench-Scale Development & Testing of a Novel Adsorption Process for Post-Combustion CO2 Capture	Novel Adsorption Process	Micro-porous carbon	TSA, fixed-beds	Bench-Scale, Actual Flue Gas	Jain, 2012
Low-Cost Sorbent for Capturing CO2 Emission Generated by Existing Coal-Fired Power Plants	Low-Cost Solid Sorbent	Alkalized Alumina sorbent	Simulated Moving Beds	Bench-Scale Using Actual Flue Gas	Elliot, 2012
CO2 Removal from Flue Gas Using Microporous Metal Organic Frameworks	Micro-porous MOFs	Alumina and Mg/ DOBDC	VPSA	Laboratory-Scale, Simulated Flue Gas	Benin, 2012

23

Adsorbents and Fixed Beds

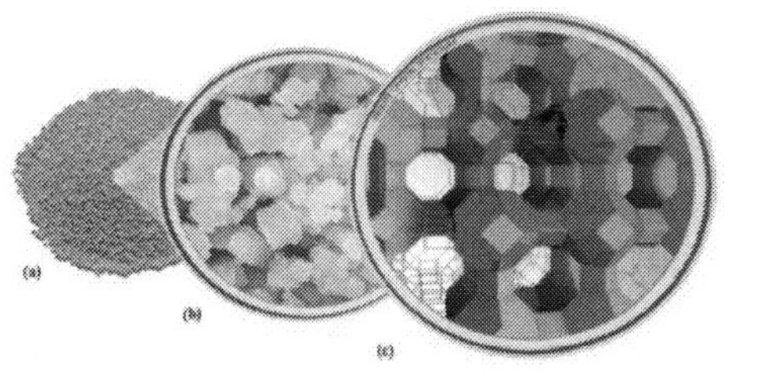

Figure 2.3 (a) Pelletized zeolite pellets, (b) crystals, and (c) framework structure http://www.grace.com/engineeredmaterials/productsandapplications/InsulatingGlass/SieveBeads/Grades.aspx)

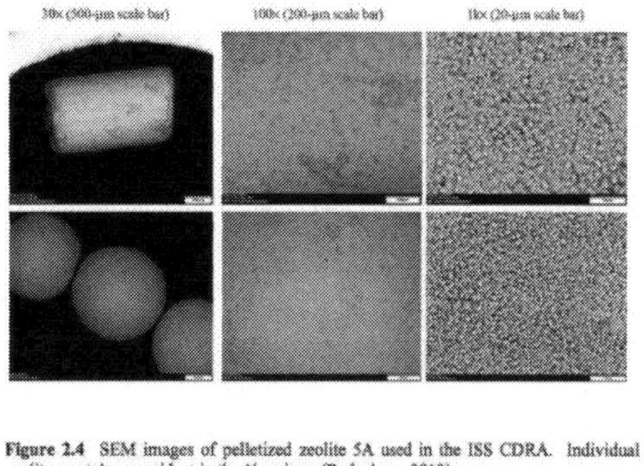

Figure 2.4 SEM images of pelletized zeolite 5A used in the ISS CDRA. Individual zeolite crystals are evident in the 1kx views (Radenburg, 2013).

24

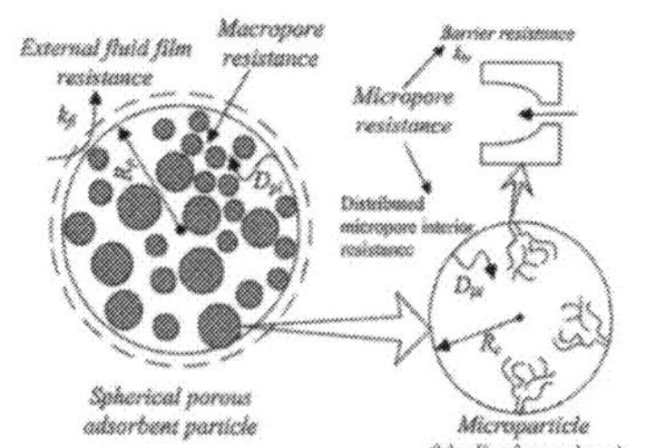

Figure 2.8 Depiction of fixed bed and zeolite mass transfer mechanisms (Shareeyan et al, 2014)

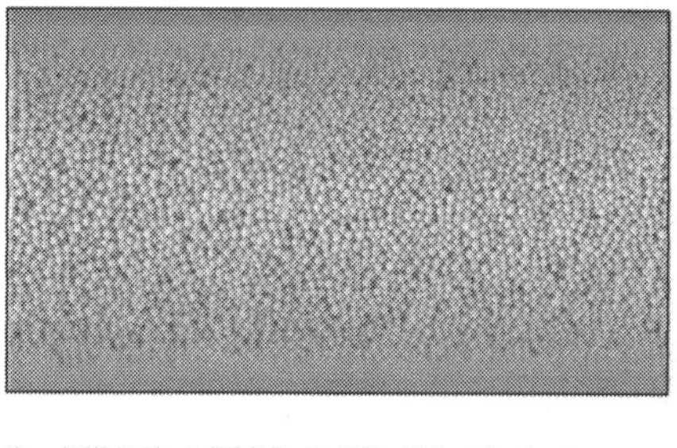

Figure 2.7 Packed (or fixed) bed of zeolite 13X beads. Photo taken by author.

Literature Review of Fixed Bed Gas Adsorption Models

Table A.1 Literature Review of Fixed Bed Gas Adsorption Models

#	Application	Experimental System	Spatial Dimensions	Tube ID/Particle Diameter	Mass Transfer				Reference
					Gas to Particle Rate Expression	Method to Determine Gas to Particle Rate	Axial Dispersion	Internal Profile Shown?	
1	Post-combustion CO_2 capture, steam methane reforming	CO_2 on hydrotalcite (dry and wet)	1	14	Modified LDF: function of loading and isotherm (pore diffusion model)	Breakthrough analysis (adsorb and desorb)	Estimated per Edwards and Richardson 1968	no	Ding and Alpay 2000
2	Hydrogen purification	H2, CO_2, CH4, CO on 5A, Activated Carbon	1	8 and 15	LDF	Breakthrough analysis (Park et al. 1998)	none	no	Park et al., 2000
3	CO_2 scrubbing	H_2O and CO_2 on 5A	1	20	LDF	Breakthrough analysis	Estimated per Edwards and Richardson 1968	yes	Mohamadinejad et al., 2000
4	Dehumidification of organic solvents	H_2O and Acetone on 4A	2	40	Bi-disperse pellet modeled	Not specified	Breakthrough Analysis	no	Pentchev and Seikova, 2002
5	Dehumidification of organic solvents	H2O on Silica gel, 13X	1	10	LDF	Uptake curve analysis (Yang and Lee, 1998)	Estimated per Wakao and Funazkri, 1978	no	Ahn and Lee, 2003
6	CO_2 scrubbing	CO_2 and H_2O on 5A	2	20	LDF	Breakthrough analysis	Estimated per Edwards and Richardson 1968	Yes (CO_2 only)	Mohamadinejad et al., 2003
7	Drying of instrument air	H2O on 5A	1	10	Modified LDF as a function of pressure, temperature and molar fraction	Breakthrough analysis	none	no	Gorbach et al., 2004
8	Dehumidification of organic solvents	H2O on Silica gel, Alumina, 13X	1	10	LDF	Breakthrough analysis (Ahn and Lee, 2003)	Estimated per Wakao and Funazkri, 1978	no	Ahn and Lee, 2004
9	Oxygen purity	O2, Ar, and N2 on CMS	1	n/a	Modified LDF (concentration-dependent diffusivity combined with	Estimated per Rao and Lee (2005)	yes	no	Jee et al., 2005
10	Post-combustion CO_2 capture	CO_2 and N2 on silicalite	1	11	LDF	Estimated per Farooq and Ruthven 1990	Estimated per Wakao et al., 1978	no	Delgado et al., 2006
11	Gas Separation	CH4 and CO_2 on 5A, butane and pentane on silicalite	1	5	Double LDF	Estimated per method reported in this paper, adjusted via breakthrough analysis	Included but not defined	no	Leinekugel-le-Cocq et al., 2007
12	Hydrogen purification	CO on 5A (4 samples) and AC in H2	1	7 and 11	LDF	Breakthrough analysis	Estimated per Lopes et al., 2009	no	Bastos-Neto et al., 2011
13	Post-combustion CO_2 capture	CO_2 and N2 on 13X	1	6	Combined LDF and QDF	Breakthrough analysis	None	No	Won et al., 2012
14	Capture of hydrocarbon emissions from gasoline engines during the cold start period of the engine	Propane on Na-ZSM-5	1	296	LDF	Estimated from film and micropore calculations per Brandani et al. 2003	Yes (not specified)	Yes	Puértolas et al., 2012
15	Post-combustion CO_2 capture	CO_2 on 13X	1	21	Macropore and micropore solved simultaneously	Estimated + breakthrough analysis	Estimated per Wakao and Funazkri, 1978	Yes	Mulgundmath et al., 2012
16	Post-combustion CO_2 capture	CO_2 and N2 on carbon	1	9, 3	LDF	Breakthrough analysis and estimated via Ruthven 1984	None	No	Gonzalez et al., 2013
17	Development process of a reactor design for open thermochemical energy storage	H_2O on 13X	2	25	LDF	Breakthrough analysis	Yes	Temperature only	Mette et al., 2014
18	Drying of ethanol for fuel production	Ethanol and H_2O on 3A	1	6	LDF	Breakthrough analysis	None	Yes	Kupiec et al., 2014
19	Dehumidification of gases containing organic component	H_2O and benzene on Silica gel, 13X, activated carbon	1	n/a	LDF	Breakthrough analysis (Ko et al., 2002, Park and Knaebel 1992, and Ahn and Lee 2003)	Estimated per Edwards and Richardson 1968	No	Nastaj and Ambrozek, 2015

- Criteria for inclusion:
  - Published in 2000 or later
  - Includes description of experimental data used for validation
- 1-D model used in 16 publications
- Tube to particle diameter ratio ≤ 20 for 16 (most much lower)
- LDF used in 12 publications
- Axial dispersion used in 14 publications
- Breakthrough curve only shown in 14 publications

25

498832 Calculation of the Isosteric Enthalpy of Adsorption in Monte Carlo Molecular Simulation: New Equations Addressing Bulk Phase Nonideality and Isosteres of Total Adsorption

Abstract:

In engineering applications of adsorption, it is important to consider the heat released by the adsorption process, as the various choices made in managing that heat will have different effects on the overall process. It may represent any or all of 1) the energy cost of controlling column temperature, 2) a change in adsorption equilibrium to lower loading capacity, or 3) the energy cost of desorption during adsorbent regeneration. A differential heat of adsorption, that is the change in molar enthalpy of during transfer of gas to the adsorbed phase, is the usual thermodynamic metric for describing the heat evolved by adsorption. The preferred version of the differential heat is now usually called the isosteric enthalpy of adsorption (IEA), *i.e.*, the heat released by the transfer of gas at fixed adsorbate loading [1]. Whether from experiment or simulation, this quantity may be obtained by isostere analysis of a sequence of isotherms at closely spaced temperatures, via the Clausius-Clapeyron equation. For molecular simulations, this IEA may alternatively be computed from fluctuations of the internal energy and molecule count. Yet, there are competing fluctuation definitions of the IEA, which yield markedly different results [2-4], and controversies regarding which measure of adsorbate loading (absolute, total, or excess) should be held fixed [5-8]. Furthermore, equations for the IEA for multi-component adsorption are underdeveloped compared to their single-component counterparts.

In the present work, we reexamine equations for the IEA in light of recent work [5-8] that stresses the importance of analyzing *absolute adsorption* isotherms during isostere analysis, and accordingly present new fluctuation-based expressions for the IEA for both single- and multi-component adsorption. These expressions account for non-ideality of both the adsorbed and bulk gas phases, regardless of the gas composition. Using flat-histogram sampling in Monte Carlo molecular simulation [9], we apply these equations to both single- and multi-component gas adsorption to show how TMMC can yield the IEA for an entire isotherm from statistics collected in a single simulation. We also demonstrate that these calculations compare favorably with isostere-based calculations from either the simple Clapeyron equation or the Clausius-Clapeyron, depending departure from ideality in the bulk gas phase. Additionally, we present sample calculations that highlight the differences between the various definitions of the IEA and discuss when these differences may prove critically important. Overall, we aim to place simulation-based IEA calculations on a firmer foundation so that this important thermodynamic quantity may be used more confidently for the design and analysis of adsorption processes.

[1] F. Rouquerol, J. Rouquerol, and K. Singh, Adsorption by Powders and Porous Solids, Academic Press: London, 1999.
[2] D. Nicholson and N. G. Parsonage, Computer Simulation and the Statistical Mechanics of Adsorption, Academic Press: New York, 1982.
[3] D.D. Do, H.D. Do, and D. Nicholson, *J Phys Chem B*, 113:1030 (2009).
[4] D.D. Do, D. Nicholson, and C. Fan, *Langmuir*, 27:14290 (2013).

[5] A. L. Myers, *AIChE J*, 48:145 (2002).
[6] A. L. Myers and P. A. Monson, *Adsorption*, 20:591 (2014).
[7] S. Brandani, E. Mangano, and L. Sarkisov, *Adsorption*, 22:261 (2016).
[8] Y. Tian and J. Wu, *Langmuir*, 33:996 (2017).
[9] V. K. Shen and D. W. Siderius, *J Chem Phys*, 140:244106 (2014).

500550 Compressibility of Argon Confined in Nanopores: Effect of the Pore Geometry

Abstract:

Compressibility is a fundamental thermodynamic property of a fluid. When a fluid is confined in a nanopore, many of its thermodynamic properties differ from that of a fluid in the bulk. Experimental measurements of sound velocity on fluid-saturated nanoporous materials can give information on the compressibility of the fluids confined in the pores [1,2]. Those experiments showed in particular that the compressibility of confined argon is different from the bulk [2].

The experimental observations have been confirmed using Monte Carlo simulations in the grand canonical ensemble (GCMC), where the compressibility was calculated based on the fluctuation of the number of particles in the simulated pores [3,4]. These results, performed for a spherical pore model, showed that the isothermal elastic modulus, which is the reciprocal of compressibility, has a linear relationship with respect to pore size. This method has been shown to be quite accurate for spherical pores larger than 2.5 nm, but has complications for smaller pores. A spherical pore ca. 2.5 nm can accommodate only about 125 molecules. Below this number, the fluctuations become non-Gaussian, making compressibility calculations unsuitable. Cylindrical pores can surpass this difficulty by increasing the length of the pore and thus the number of molecules in the system, while keeping the fluid nanoconfined. Here we extend this method for cylindrical pores.

The model system for our simulations was Lennard-Jones argon in cylindrical silica pores of diameters ranging from 1 to 6 nm and lengths 3.4, 6.8, and 13.6 nm. We found that the isothermal modulus displays a similar linear relationship with the pore size as for the fluid confined in the spherical pore model for pores of size 3 to 6 nm. Interestingly, the calculations for the micropore size range (~2 nm or less) showed that this regularity breaks even when the system is large enough that the fluctuations are Gaussian. Therefore, the behavior in the 2 nm pore did not coincide with the linear relationship found for the larger pores, showing a qualitative difference between the confinement in meso- and micro- pores.

[1] J.H. Page, et al., Phys. Rev. E 52, 2763 (1995).
[2] K. Schappert & R. Pelster, Europhys. Lett.105, 56001 (2014).
[3] G.Y. Gor, et al., J. Chem. Phys. 143, 194506 (2015).
[4] G.Y. Gor, et al., J. Chem. Phys. 145, 164505 (2016).

495186 Ligand-Assisted Displacement Chromatography for Rare Earth Elements Separations

Abstract:

Rare earth elements (REE), which include scandium, yttrium, and 15 elements of the lanthanides, are critical components of many high-technology products. Since the ions of REE have the same valence and little differences in size and other physical and chemical properties, they cannot be separated using conventional ion exchange processes. Spedding and Powell demonstrated the feasibility of ligand-assisted displacement chromatography (LAD) for REE purification using polymeric ion exchangers and ligand elution on the preparative chromatography scale in the 1950s and 1960s. Three lanthanides were recovered with relatively high purity (>99%) and with yields from 83 to 93%. However, each separation run took more than three weeks, resulting in very low adsorbent productivity. More importantly, the complex mechanisms in this system were not well understood, and no systematic design/optimization method or scaling rules have been reported for the systems with mass transfer effects.

In this study, general LAD design and optimization methods were developed to recover high-purity REE with high yield and high productivity. The methods were tested with the purification of three REE. The results showed that the average yield of high-purity (>99%) products was more than 96%, and sorbent productivity was two orders of magnitude higher than that of Spedding and Powell. Rate model simulations were developed and verified using the experimental data and literature data. The verified model and model parameters were used to elucidate the dynamic separation phenomena in LAD. The simulation results showed that the mechanism of displacement is different from that of conventional displacement chromatography. The reaction of ligand with presaturant, which has the highest affinity for the ligand, drives the isotachic train in LAD. After the presaturant, solute with the highest affinity to the ligand elutes next. The constant pattern of shock waves forms due to the reactions of ligand with solutes. The shock layer thickness depends on the ligand selectivity relative to sorbent selectivity and mass transfer resistance. The maximum yield and productivity were greatly affected by the mass transfer resistances, selectivity, feed compositions, feed loading volume, and effective ligand concentration. These important parameters were incorporated into key dimensionless groups, which can be used in the efficient design and optimization method to optimize product purity, yield, or productivity for LAD systems at different scales.

498009 Adsorption Studies of Tryptophan in a BSA Coated Surface in Semi-Batch Fractionation and Microchannel

Abstract:

In bio processing industries, foam fractionation is extensively used for the separation and purification of solutes from a mixture at low concentrations at room temperature. In this work we discuss the absorption of tryptophan using a semi-batch foam fractionation method and a microfluidic channel. The concentrations of tryptophan is analyzed using a spectrophotometer. First the adsorption of BSA is studied at different ionic concentration to determine the optimal conditions i.e pH and initial concentration. For these optimized condition in a foam fractionation column, adsorption studies are further carried out in a microchannel.

Foam fractionation in a micro channel can be operated with both phases flowing continuously. This eliminates the challenge of vertical drainage. In this work we have designed a microchannel where the two phases i.e. foamate and the residual aqueous solution are separated at the outlet. The hydrodynamics of the system is first analysed[2,3]. Nitrogen is used as the gas phase and the liquid phase is an aqueous solution of CTAB, BSA and tryptophan in Milli Q. The flow patterns observed are analysed and the parameter space in which they occur are determined. The absorption of tryptophan on the bubble coated with BSA is analysed in this system for varying enrichment ratio of BSA. A mathematical model is used to predict the performance of the absorption of tryptophan on the bubble coated with BSA. The model predictions are compared with experimental results.

499144 Modeling Continuous Enzymatic Reactive Crystallization of β-Lactam Antibiotics

Abstract:

Modeling continuous enzymatic reactive crystallization of β-lactam antibiotics

Matthew A. McDonald*, Andreas S. Bommarius*, Martha A. Grover*, and Ronald W. Rousseau*

School of Chemical & Biomolecular Engineering, Georgia Institute of Technology, Atlanta, Georgia

Keywords: reactive crystallization, β-lactam antibiotics, continuous manufacturing

Introduction

β-lactam antibiotics, despite a long and storied history, find continued widespread use in the fight against infections. In 2010 the two most common classes of β-lactam antibiotics, the penicillins and cephalosporins, accounted for nearly 60% of world antibiotic consumption, with delivery of over 20 billion doses of penicillin-derived antibiotics.[1] Traditionally, semi-synthetic β-lactam antibiotics have been made batch-wise which lends itself to sharing equipment, such as crystallizers, between different active pharmaceutical ingredients (APIs).[2] β-lactam antibiotics can also cause a severe allergic reaction; a 2012 study found that 92 out of 11,761 patients at Mount Sinai Hospital in New York City suffered anaphylaxis from penicillin, a prevalence of 0.7%.[3] This rate is significant enough for the U.S. Food and Drug Administration (FDA) recently to require that all penicillin-derived antibiotics be manufactured, processed, and packaged in facilities separate from those used for any other pharmaceuticals, removing the ability to share batch equipment between APIs.[4] The need to establish new β-lactam-specific facilities presents an opportunity to implement novel, more efficient, processes such as enzymatic synthesis. However, a new design concept is required to overcome some drawbacks of the enzymatic route. The key to this process is enzymatic reaction enhanced by simultaneous crystallization to manufacture semi-synthetic β-lactam antibiotics while decreasing the risk of cross-contamination and without creating large amounts of waste.

We use ampicillin and cephalexin as a model penicillin and cephalosporin, respectively. The enzyme, penicillin G acylase (PGA), catalyzes the condensation of 6-aminopenicillanic acid (6-APA) and phenylglycine methyl ester (PGME) and 7-aminodesacetoxy-cephalosporanic acid (7-ADCA) and PGME into ampicillin and cephalexin, respectively. PGA also hydrolyzes both antibiotics and PGME into phenylglycine, a sparingly soluble contaminant.

Batch Processing

Catalytic hydrolysis of the β-lactam product by the same enzyme that catalyzes its synthesis renders the desired product an intermediate in a kinetically-controlled system. By selectively removing this intermediate from the solution via a reactive crystallization process we have shown that the process selectivity can be enhanced approximately 1.5-fold compared to an equivalent non-crystallizing batch process, resulting in an equivalent increase in yield. The increase in yield and selectivity stems from the concentration of product in solution, and hence the concentration available to hydrolysis by the enzyme, remaining at the solubility limit while the actual amount of product formed (dissolved and crystallized) far exceeds this level.[5]

Continuous Processing

More recently we have explored continuous reactive crystallization as a way of increasing the productivity and efficiency of semi-synthetic β-lactam synthesis. Continuous manufacturing may enable greater sustainability, product consistency, and access to different chemistries. The accumulation of phenylglycine by-product—a potential contaminant—poses the greatest challenge to increasing conversion. Continuous operation could increase conversion over batch by tailoring the conditions to optimize reactive crystallization; for example, a plug flow reactor/crystallizer can operate with excess 6-APA or 7-ADCA relative to PGME to hinder hydrolysis of the PGME while a recycle stream could act to seed the plug flow apparatus to slow antibiotic hydrolysis via crystallization.

The reaction kinetics are predicted via a previously described model[5] and the crystallization kinetics have been determined via online chord-length distribution and concentration measurements, by focused beam reflectance measurement (FBRM) and attenuated total-reflectance Fourier-transform infrared (ATR-FTIR) spectroscopy, respectively.[6] Mapping the operation space requires reconciliation of the kinetically controlled enzymatic synthesis and crystallization of the antibiotic. Many of the parameters that are typically used to control crystallization are restricted by the simultaneous reaction; the enzyme prevents use of anti-solvents or temperature changes and the pH changes both the enzyme activity and ampicillin solubility. At times, the reaction and crystallization processes can be at odds with each other; high concentrations of 6-APA have been shown to inhibit nucleation and growth of ampicillin trihydrate crystals while increasing enzyme selectivity.[7] Our model examines these interactions and allows for prediction of a wide range of operating conditions for continuous reactive crystallization synthesis of β-lactam antibiotics.

References

[1] Gelbrand, H., Miller-Petrie, M., Pant, S., Gandra, S., Levinson, J., Barter, D., White, A., Laxminarayan, R., Ganguly, N. K., and Kariuki, S. (2015) The State of the World's Antibiotics 2015, *Wound Healing Southern Africa 8*, 30-34.

[2] Elander, R. P. (2003) Industrial production of β-lactam antibiotics, *Applied microbiology and biotechnology 61*, 385-392.

[3] Albin, S., and Agarwal, S. Prevalence and characteristics of reported penicillin allergy in an urban outpatient adult population, 6 ed., pp 489-494, OceanSide Publications, Inc.

[4] (2015) 21 C.F.R § 211.42 (The Code of Federal Regulations of the United States), (Administration, F. a. D., Ed.).

[5] McDonald, M. A., Bommarius, A. S., and Rousseau, R. W. (2017) Enzymatic reactive crystallization for improving ampicillin synthesis, *Chemical Engineering Science 165*, 81-88.

Monitoring Tools and Robust Parameter Estimation, *Industrial & Engineering Chemistry Research 55*, 2153-2162.

[7] Ottens, M., Lebreton, B., Zomerdijk, M., Rijkers, M., Bruinsma, O. S. L., and van der Wielen, L. M. (2004) Impurity effects on the crystallization kinetics of ampicillin, *Industrial & engineering chemistry research 43*, 7932-7938.

499608 Continuous Crystallization-Milling Processes: Guaranteeing the Manufacture of Stable Polymorphs

Abstract:

Manufacturing the thermodynamically stable polymorph of a given solute is typically desired to prevent changes of crystal form upon storage (which impacts the performance of the final product). While the stable polymorph is readily accessible in batch processes through seeding, efficient polymorph control strategies for continuous crystallization are still absent. However, continuous processes have considerable advantages over batch processes, such as higher productivity and steady state operation. In this contribution, we therefore present a strategy to achieve the production of the stable polymorph from such continuous processes.

Figure 1. Process flowsheet.

In this work, we investigate the polymorphs, yields and productivities obtained from a continuous process using models and experiments. The flowsheet is shown in Figure 1 where a crystallizer is fed from a surge tank with clear supersaturated solution and a milling unit is connected to the crystallizer in a loop. Using the mill in this way allows to increase the surface area of the crystals in the crystallizer which is expected to lead to an increased yield. We show that it also ensures the production of the stable polymorph in a much wider range of operating conditions than without the mill.

To study this process configuration, we use L-glutamic acid (LGA) as model compound. The two known polymorphs of LGA exist in a monotropic system (a metastable; b stable) and its thermodynamics and kinetics have been well characterized previously [2-6]. We introduce a model consisting of two population balance equations (PBEs) that represent the two polymorphs in the crystallizer (Eq. 1):

(1)

Figure 2. Solute concentration during the start-up of the processes (left) and the polymorph ratios of the b polymorph (right).

To conceptually show the effect of the mill on the process outcome, we report results with and without the mill under otherwise identical operating conditions (residence time: 1 h, feed concentration: 40 g/kg, temperature: 45ûC, equal starting masses of both polymorphs) in Figure 2. The model results show that the process including milling achieves a higher yield (Figure 2 left) and achieves production of the stable b polymorph (Figure 2 right). Conversely, the process without milling delivers a lower yield and produces the metastable a polymorph. The process concentration trajectory drawn in black shows that the steady state concentration drops below the a solubility when milling is used, hence all a crystals dissolve, are washed out, and no new crystals can be formed. Hence, the process only yields b crystals. In addition to that more solute is crystallized increasing both, the yield and the productivity of the process.

Concluding, the suspension mill effectively maintains a high surface area of the crystals in a continuous crystallization process, which allows achieving higher productivities, but can also be used to steer the polymorphic outcome towards the stable polymorph. For instance, using this setup, it becomes possible to obtain the stable polymorph at high productivities under processing conditions that would lead to the metastable polymorph without the mill.

In an attempt to confirm our modeling results experimentally, we performed experiments using a 500 mL crystallizer and an IKA MagicLab suspension mill. The experiments were conducted at 25ûC, 40 g/kg feed concentration and 2 h residence time. The feed is prepared in a 10 L reactor kept at 80ûC. The suspension/solution is transported between the vessels using peristaltic pumps. Qualitatively confirming our modelling results, we observed that operating the crystallizer with milling yields the b polymorph. Operation without the mill yields the metastable a polymorph.

We highlight here that the stable b polymorph was previously only obtained from continuous crystallizers that were either operated at higher temperature (i.e., lower potential yield) or with very long residence times exceeding 17.4 h at 25ûC [1]. In contrast, we show that using the combined milling-crystallization process, the same result can be achieved with a residence time of 2 h at 25ûC. Therefore, the present combination achieves both higher yield and higher productivity.

Through a combination of further experiments and modeling, we obtain a map of polymorphic outcomes in dependence of operating conditions (residence time, feed concentration, milling intensity), similar to what has been presented for systems without milling by Farmer et al. [6]. We show that our combined milling/crystallization process shows larger regions of operating conditions where the stable polymorph is reliably obtained in comparison to the crystallizer-only configuration.

References:

[1] T. Lai, S. Ferguson, L. Palmer, B.L. Trout, A.S. Myerson Org. Process Res. Dev. **18**, 1382-1390 (2014).

[2] J. Cornel, C. Lindenberg, M. Mazzotti Cryst. Growth Des. **9**, 243-252 (2008).

[3] M. Kitamura Cryst. Growth Des. **96**, 541-546 (1989).

[4] J. Schšll, D. Bonulami, L. Vicum, M. Mazzotti Cryst. Growth Des. **6**, 881-891 (2006).

[5] C. Lindenberg, J. Schšll, L. Vicum, M. Mazzotti Cryst. Growth Des. **8**, 224-237 (2008).

[6] T.C. Farmer, C.L. Carpenter, M.F. Doherty, AIChE J. **62**, 3505-3514 (2016).

CMAC
FUTURE MANUFACTURING RESEARCH HUB

PSE

Micro-scale process development and optimization for crystallization processes

AIChE Conference, Minneapolis Convention Center, 30 October 2017

N.A. Mitchell[1], *C. J. Brown*[2]

[1] *Process Systems Enterprise (PSE) Ltd., London, UK*

[2] *EPSRC CMAC Future Manufacturing Research Hub, University of Strathclyde, Glasgow, UK*

CMAC FUTURE MANUFACTURING RESEARCH HUB

Continuous Manufacturing and Advanced Crystallisation

PSE

Co-created with industry to address key manufacturing challenges and skills needs

- World leading manufacturing research platform
- A partnership approach for world-class:
  - Research
  - Training & Skills
  - Translation to Industry & Impact
  - Facilities & Infrastructure

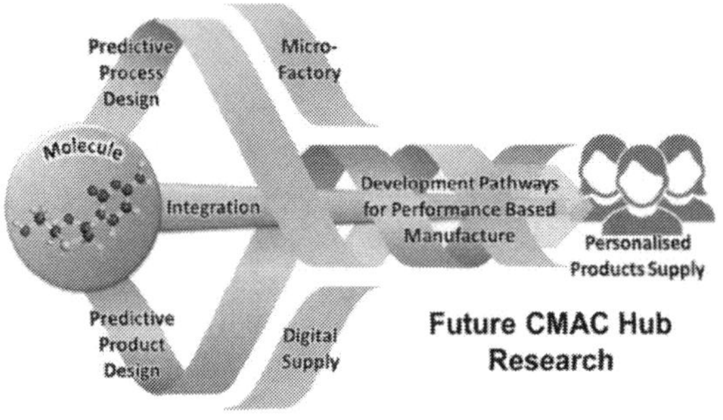

University of Strathclyde Glasgow · University of BATH · UNIVERSITY OF CAMBRIDGE · The University of Sheffield · Imperial College London · UNIVERSITY OF LEEDS · Loughborough University

EPSRC · gsk · AstraZeneca · NOVARTIS · BAYER · Lilly · Takeda · Roche

CMAC FUTURE MANUFACTURING RESEARCH HUB — Contents

PSE

1. Introduction and motivation
2. Typical collected data
3. Data integration
4. Case study
5. Conclusions

Intro > Data > Integration > Case study > Conclusions

CMAC FUTURE MANUFACTURING RESEARCH HUB — Introduction and motivation

PSE

- Initial experimental phases of crystallization process development;
    - Carried out at small scales (1-5mL)
    - Aim to select the solvent based on solubility
    - Aim to select the crystal solid state
    - Conducted in high throughput reactor systems
        - (Crystalline – Technobis Crystallization Systems)

- This data is currently typically not utilized for the development and calibration of crystallization process models
    - Probed experimentally and more quantitatively at much larger scales (typically between 100 – 1000 mL)

- More quantitative usage of the data generated at small scale for the development of process models may;
    - Significantly reduce the number of larger scale experiments required
    - Aid in addressing the increasing constraints on time and materials in pharmaceutical development

Intro > Data > Integration > Case study > Conclusions

CMAC FUTURE MANUFACTURING RESEARCH HUB — Typical raw Crystalline data

PSE

- Sample turbidity and block temperature are recorded throughout
- Addition of *in situ* camera module allows for the imaging of suspended particles and the determination of particle size distribution (PSD) through image analysis

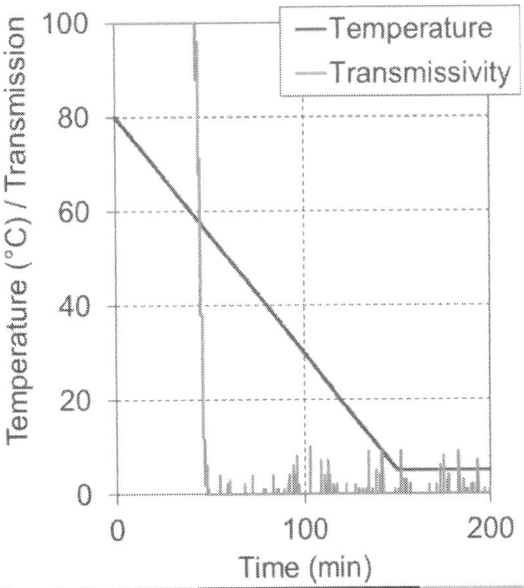

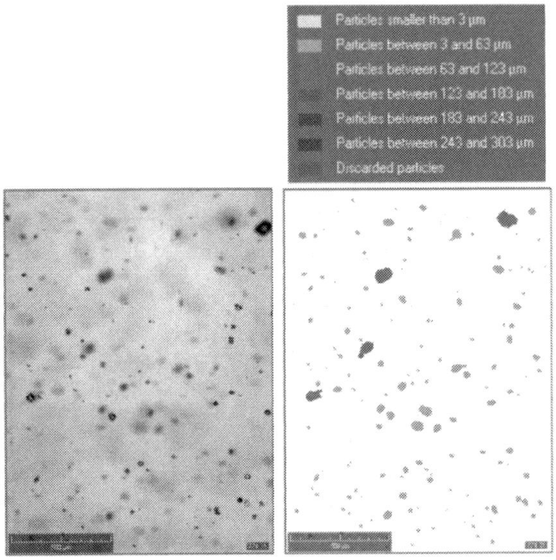

Intro | **Data** | Integration | Case study | Conclusions

CMAC FUTURE MANUFACTURING RESEARCH HUB — Particle count thresholds

PSE

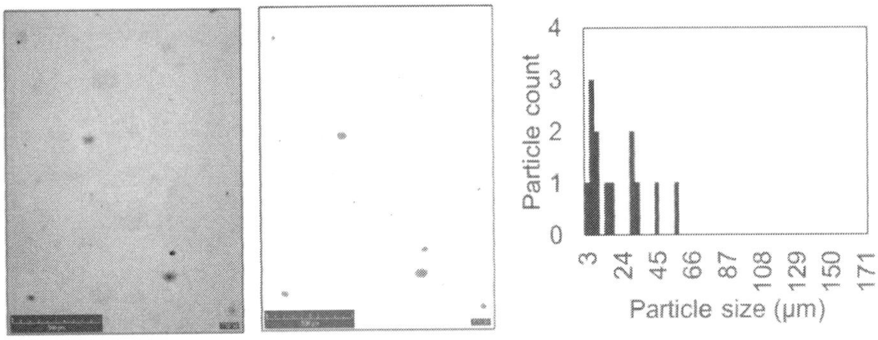

< 10 particles
Clear segregation of particles. Insufficient particles for statistical PSD

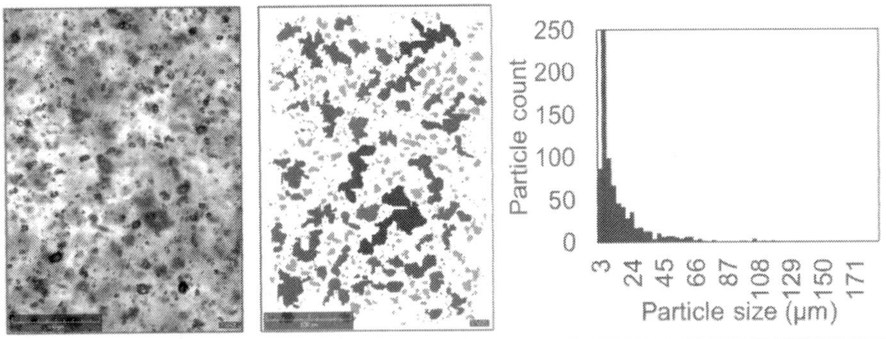

> 200 particles. Significant probability of particle overlap. Sufficient particles for statistical PSD.

Intro | Data | **Integration** | Case study | Conclusions

CMAC FUTURE MANUFACTURING RESEARCH HUB Data integration (1)

PSE

- Step 1: Convert number based PSD to cumulative volume based PSD

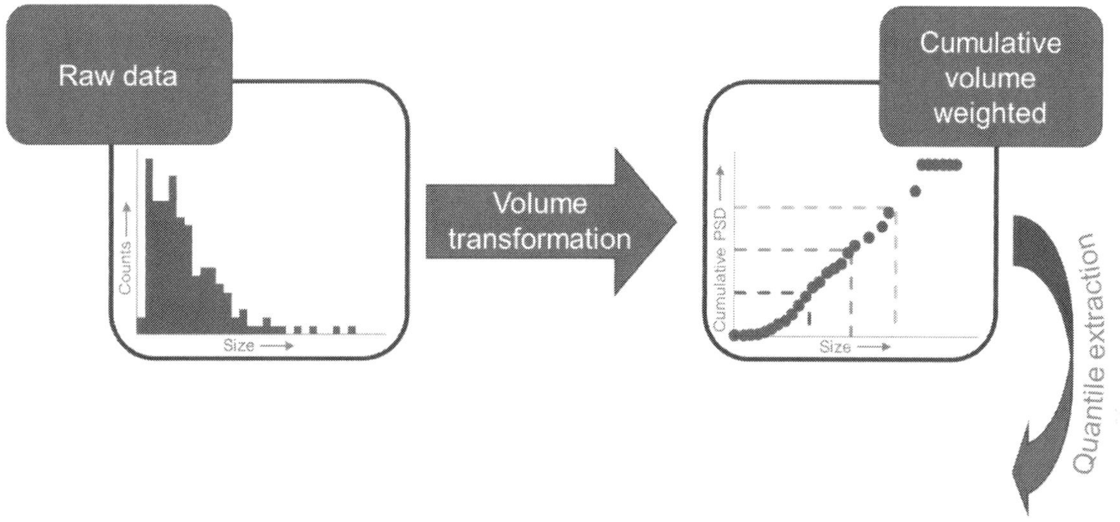

Intro > Data > Integration > Case study > Conclusions

CMAC FUTURE MANUFACTURING RESEARCH HUB Data integration (2)

PSE

- Step 2: Evaluate the PSD quantiles from the cumulative volume PSD
- Step 3: Apply count thresholds (10 < particle count < 200)

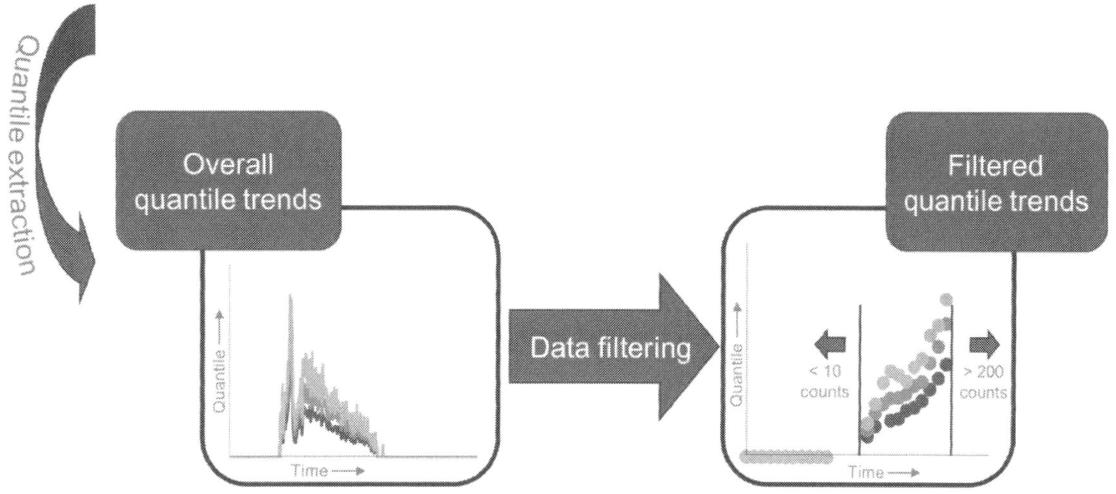

Intro > Data > Integration > Case study > Conclusions

CMAC Case study

Data collected as part of a larger project developing a continuous crystallisation for paracetamol in 3-methyl-1-butanol

Microscale estimation data:

- Collected as part of routine solubility measurements:
    - 6 mL aliquots of concentrations from 25 to 200 g/L
        - Cooled to 5 °C, held for 1 hr
        - Heated to 10 °C below solvent boiling point at 0.1 °C/min, held for 1 hr
        - Cool to 5 °C at 0.1 °C
    - Turbidity and *in-situ* images recorded throughout

1 L scale estimation data:

- Specifically designed experiments for estimation of growth parameters
- Seeded crystallisations using a combination of constant supersaturation (using FTIR feedback control) and isothermal experiments
- Seed mass (1 – 5 g), seed size (40 - 100 μm) and seeding supersaturation (S = 1.08 to 1.20) were varied
    - 1 L of solvent and approximately 171 g solute per experiment

Intro > Data > Integration > Case study > Conclusions

CMAC Parameter estimation – Small Scale (1)

- Small scale data available
- Solid side measurement
    - Evolution of the **volume based d25, d50 and d75 of PSD**
        - In the range **10 > particle count < 200**
    - Detect onset of primary nucleation (**particle count > 10**)

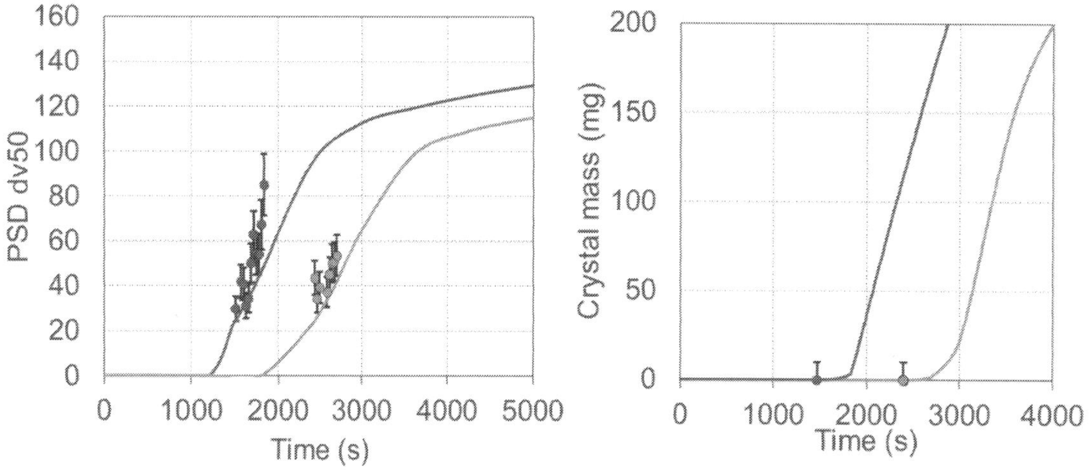

Intro > Data > Integration > Case study > Conclusions

CMAC Parameter estimation – Small Scale (2)

- Liquid side information
  - An estimate of final solute concentration is required to help complete the material balance
  - A large variance is placed on this as it is unknown if the solution reached equilibrium.

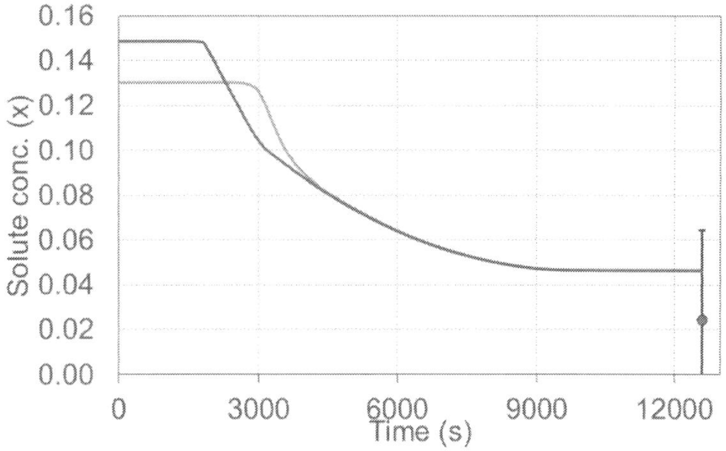

Intro › Data › Integration › Case study › Conclusions

CMAC Mechanism discrimination

- Using small scale measurements
  - Enabling mechanism discrimination through sensitivity analysis at an earlier stage of process development

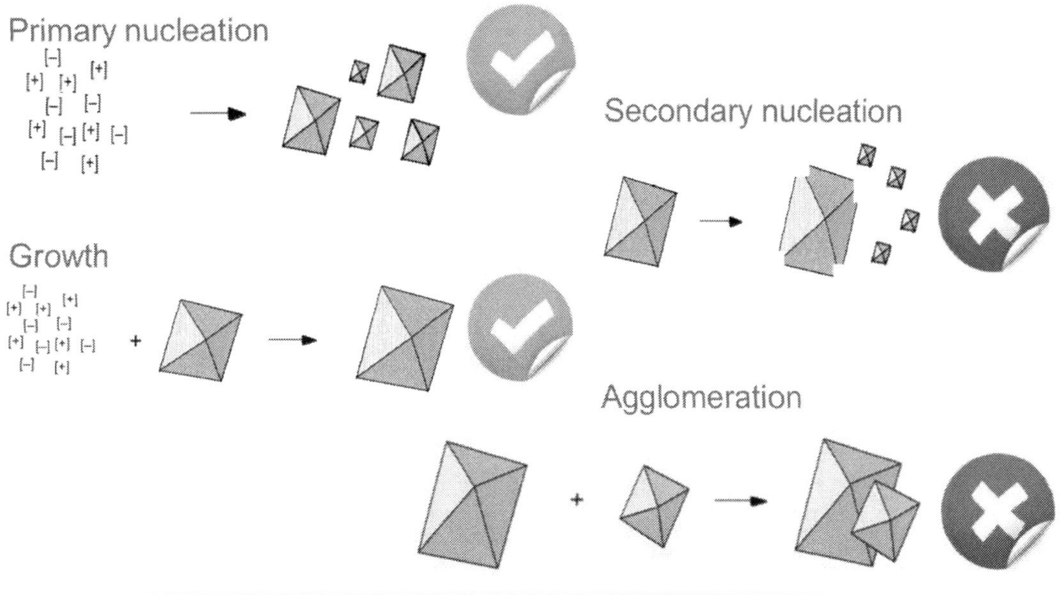

Intro › Data › Integration › Case study › Conclusions

CMAC FUTURE MANUFACTURING RESEARCH HUB — Model predictions – Large scale — PSE

- Parameters estimated at microscale and at 1 L scale used to predict outcome of an experiment performed at 1 L scale
- Close agreement in model predictions at 1L scale

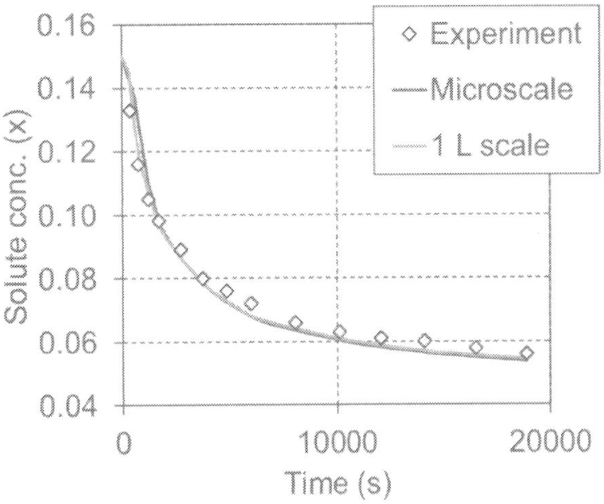

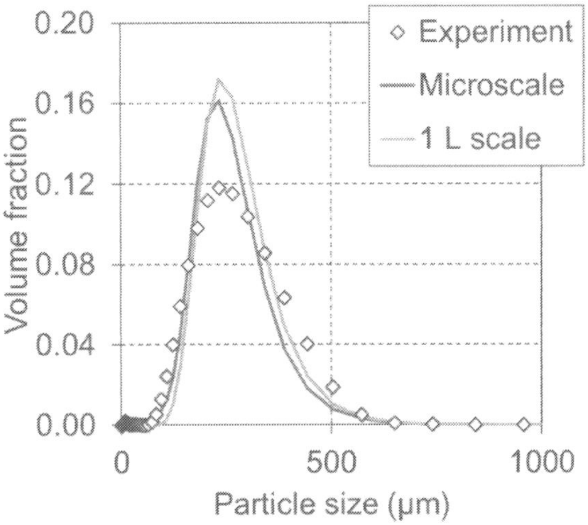

Intro › Data › Integration › **Case study** › Conclusions

CMAC FUTURE MANUFACTURING RESEARCH HUB — Comparison of material usage — PSE

	Crystalline 5mL	Optimax 1L	Combination
Experimental scales & systems			

	Estimation at microscale	Estimation at 1 L scale	Estimation at microscale with refinement at 1 L scale
Microscale experiments (5 mL each)	8 (as part of solubility measurement)	-	8 (as part of solubility measurement)
1 L scale experiments	-	7	3
Total API usage	**4 g**	**1200 g**	**520 g**

Intro › Data › Integration › **Case study** › Conclusions

CMAC FUTURE MANUFACTURING RESEARCH HUB Conclusions

PSE

- Initial **estimates of crystallisation kinetics parameters** can be obtained at the microscale as part of typical solubility measurement experiments.
- **Crystallization mechanism and model discrimination** can be performed to identify the most dominant crystallisation phenomena.
- Using these initial estimates a **57 % reduction in API usage** can be achieved when compared with estimating parameters solely at lab scale (1 L).

Intro > Data > Integration > Case study > **Conclusions**

CMAC FUTURE MANUFACTURING RESEARCH HUB Acknowledgements

PSE

CMAC Future Manufacturing Research Hub
- Bilal Ahmed
- Stephanie Yerdelen

This work was supported by:

University of Strathclyde Glasgow, University of Bath, University of Cambridge, The University of Sheffield, Imperial College London, University of Leeds, Loughborough University

EPSRC, gsk, AstraZeneca, Novartis, Bayer, Lilly, Takeda, Roche

491320 Continuous Crystallization Process for Resolution of Diastereomeric Salts: Ibuprofen Lysine Case Study

Abstract:

Continuous crystallization process for resolution of diastereomeric salts: Ibuprofen lysine case study

M. Simon[1], R.C. Jones[1], P. Donnellan[1] and B. Glennon[1]

[1] *Synthesis and Solid State Pharmaceutical Centre (SSPC), University College Dublin, School of Chemical and BioProcess Engineering, University College Dublin, Belfield, Dublin 4, Ireland*

Keywords: diastereomeric salt, continuous crystallization, phase diagram

Introduction

More than half of the active pharmaceutical ingredients are found to be chiral[1]; hence, enantioseparation is of significant interest to the pharmaceutical industry.[2] In the case of a racemic mixture, typically only one enantiomer possesses the active therapeutic activity while the other is regarded as an impurity constituting 50% of the mixture.[3] This undesired enantiomer can cause undesired side effects[4] e.g., the tragic incident of the drug thalidomide in the 1960s, which resulted in physical birth defects (the R-enantiomer was proven to be a teratogen whereas the S-enantiomer was used to treat morning sickness for pregnant women).[5]

Crystallization is broadly applied at small and large scales to eliminate impurities, including enantiomeric impurities.[2] An in-depth understanding of the solid-liquid equilibrium (SLE) data in terms of phase diagrams is vital for devising and optimizing crystallization processes.[6]

One method to isolate high purity enantiomer from a racemic mixture is by diastereomeric salt formation. Racemic basic (or acidic) target compounds are reacted with optically active compounds such as enantiopure acidic (or basic) resolving agents to form a pair of diastereomeric salts that can [7] be separated by fractional crystallization.[8] The resulting salt is subjected to neutralization process to give a pure enantiomer. The crystallization process can be either kinetically (separation of salts depending on the rate of crystals formed) or thermodynamically (separation of salts depending on their solubility differences) controlled.[7, 9] Unlike enantiomers, diastereomeric salts do not possess symmetry.[10] The properties of diastereomeric salts are as follows: high solubility difference, distinct crystal morphology, good thermal stability, rapid dissolution rate, and varied inclination to crystallize.[11] The high solubility difference between the salt pairs is accomplished by the appropriate choice of resolving agents and/or a solvent system during the resolution.[12]

A systematic study of the diastereomeric salt formation of R-ibuprofen-S-lysinate (R-salt) and S-ibuprofen-S-lysinate (S-salt) were explored in the binary and ternary phase diagram using 94 wt% aqueous ethanol as a solvent. Based on the phase diagram, a continuous crystallizion process with crystal-free mother liquor exchange streams was devised.

Section 1: Solubility Studies

Thermodynamic data (binary and ternary phase diagram of the two diastereomeric salts in the solvent system at different temperatures) and kinetic properties (metastable zone widths in an appropriate solvent) are required to design appropriate crystallization process.[13] The ternary phase diagram of the salt pairs in 94 wt% aqueous ethanol were determined by classical isothermal and polythermal methods. The metastable zone widths of the pure salts were measured. The ternary phase diagram depicted in Fig. 1, confirms that the eutectic point exists at a salt composition of ~34:66 wt% R-salt:S-salt. The phase diagram and X-ray Powder diffraction patterns of the salt mixtures confirmed the formation of a simple eutectic behaviour in the system of the diastereomeric salts.

Section 2: Design of a Continuous Coupled Crystallization Process

The thermodynamic data depicted in *Fig. 1* was used to design the resolution method to resolve both salts with high purity. The aim was to investigate the applicability of a continuous coupled MSMPR crystallizers for simultaneous separation of the diastereomeric salts by crystal growth with pure seed crystals. Crystal-free mother liquor exchange streams were operated continuously between the two MSMPR.

Figure 1: Ternary phase diagram of diastereomeric salts of ibuprofen lysine: R-salt & S-salt (axes in weight fraction (W); lines are guides to the eye; tie lines are drawn for 15 °C and 40 °C)

Acknowledgements:

The financial support of the Synthesis and Solid State Pharmaceutical centre (SSPC) and Science Foundation Ireland (SFI, 12/RC/2275) are gratefully acknowledged.

References:

[1] J. Chen and A. S. Myerson, *CrystEngComm* **2012**, *14*, 8326-8329.

[2] Y. Wang and A. M. Chen, *Organic Process Research & Development* **2008**, *12*, 282-290.

[3] B. S. Sekhon, *Int J Pharm Technol Res* **2010**, *2*, 1584-1594.

[4] A. Singh and A. S. Myerson, *Journal of pharmaceutical sciences* **2010**, *99*, 3931-3940.

[5] J. E. Rekoske, *AIChE Journal* **2001**, *47*, 2-5.

[6] H. Lorenz, D. Sapoundjiev and A. Seidel-Morgenstern, *Engineering in life sciences* **2003**, *3*, 132-136.

[7] F. C. Ferreira, N. F. Ghazali, U. Cocchini and A. G. Livingston, *Tetrahedron: Asymmetry* **2006**, *17*, 1337-1348.

[8] P. R. Anandamanoharan, P. W. Cains and A. G. Jones, *Tetrahedron: Asymmetry* **2006**, *17*, 1867-1874.

[9] G. Springuel, K. Robeyns, B. Norberg, J. Wouters and T. Leyssens, *Crystal Growth and Design* **2014**, *14*, 3996-4004.

[10] J. Jacques, A. Collet and S. H. Wilen, *Enantiomers, racemates, and resolutions*, Wiley, **1981**, p. 289.

[11] H. Wu and A. G. Jones, *Chemical Engineering and Technology* **2012**, *35*, 1031-1038.

[12] N. Kolla, C. R. Elati, M. Arunagiri, S. Gangula and P. J. Vankawala, *Organic process research & development* **2007**, *11*, 455-457.

[13] V. S. Sistla, J. V. Langermann, H. Lorenz and A. Seidel-Morgenstern, *Crystal Growth & Design* **2011**, *11*, 3761-3768.

489234 Ultrasound-Assisted Crystallization in a Two-Stage Continuous Msmpr Crystallizer System

Abstract:

Ultrasound-assisted crystallization in a two-stage continuous MSMPR crystallizer system

Zhenguo Gao [1,2] Dan Zhu [1,2] Yuanyi Wu [1] Sohrab Rohani [1]* Junbo Gong [2], Jingkang Wang [2]

1. The University of Western Ontario, Department of Chemical and Biochemical Engineering, London, Ontario N6A 5B9, Canada

2. Tianjin University, School of Chemical Engineering and Technology, Tianjin, 300072, P. R. China

Authors have contributed the same on this paper.

* Corresponding author. Email: srohani@uwo.ca

Keywords: Ultrasonic irradiation, Continuous crystallization, Crystal size distribution

Introduction

Continuous pharmaceuticals manufacturing is an attractive method to improve product quality, consistency and also process efficiency and robustness. Since the promotion proposed by U.S. Food and Drug Administration (FDA) to integrate continuous manufacturing into pharmaceutical production and the practices implemented by leading pharmaceutical companies e.g. Vertex and Novartis, there has been a growing interest in transforming a batch production plant into a continuous process in pharmaceutical industry (Adamo et al., 2016; Mascia et al., 2013). Compared with a batch process, which contains many intermittent steps, a continuous process shows its advantages in better product quality and consistency, greater process robustness and safety, lower cost and higher efficiency. Continuous crystallization as a key unit operation to optimize solid product quality e.g. purity, polymorphism, size/size distribution and stability, has been attempted in several geometries. Sigle stage or multistage mixed suspension mixed product removal (MSMPR) crystallizers have been studied to optimize the process efficiency and product quality (Alvarez et al., 2011; Wong et al., 2012). A continuous tubular crystallizer has shown potential application because of optimizing mixing performance, supersaturation profile and diminishing shear force. Oscillatory baffled crystallizer (OBC) has been successfully commercialized and various strategies such as seeded, self-seeded, laminar, plug flow etc., have been studied (Besenhard et al., 2015; McGlone et al., 2015).

Studies have proven that ultrasonic irradiation affects nucleation and growth during crystallization process (Castro and Priego-Capote, 2007; Bhangu et al., 2016). The shock wave caused by cavitation of ultrasonic irradiation can promote nucleation at a lower supersaturation level and a shortened induction time. This promotes nucleation, which would result in a smaller size crystal. The ultrasonic irradiation would also help to separate crystals at high surface tension to prevent agglomeration (Narducci et al., 2011). Thus, with the help of ultrasound, a smaller crystal size and a narrower crystal size distribution (CSD) would result in a short time. In general, bigger crystals with narrower CSD are preferred considering purity and downstream processing in pharmaceutical industry. Therefore, in this study, a two-stage continuous MSMPR crystallizer series is used in which uniform small crystals are produced in the first stage and further grow in the second stage. In the present study, large a-form L-glutamic acid crystals around 150 μm were obtained with a narrow CSD and a higher process efficiency.

Experiments and Discussions

The two-stage continuous MSMPR crystallizer series was built to optimized the CSD of a-form L-glutamic acid (Alfa Aesar, USA) in this study. The schematic of experimental setup is shown in Figure 1. The first step was to prepare the saturated solution (3.0 g/L) in the feed tank, and then flowed through (50 mL/min) the two-stage MSMPR crystallizers with the help of two peristaltic pumps. The first stage MSMPR crystallizer was coupled with an ultrasonic processor (20KHz, 14-30W, VCX 500, Sonic & Materials Inc., USA) to promote nucleation and also to optimize the size and population of crystals by tuning the residence time and the intensity of ultrasonic irradiation (with and without pause). A magnetic stirrer (200 rpm, Cimarec+^{TM}, Thermo Fisher Scientific, USA) was used to provide moderate mixing to keep the crystals suspended. *In-situ* focused beam reflectance measurement (FBRM, S400, Mettler Toledo, USA) was used to measure the size/chord length and population/FBRM counts per sec to characterize the crystal slurry in the two-stage MSMPR crystallizers. There was a moderate mechanical stirring (200 rpm, a four-bladed stirrer, 45° blade angle) in the second stage MMSPR crystallizer. Further characterizations like scanning electron microscope, laser scattering crystal size measurement etc., were conducted to characterize the crystalline product and optimize the performance of the system.

Figure 1. Schematic of experimental setup of two-stage ultrasound assisted MSMPR crystallizer series.

With the help of ultrasonic irradiation, the nucleation occurs easily (short induction time and lower supersaturation level required). As shown in Figure 2, there are four lines that indicate the free energy required by primary nucleation, secondary nucleation, nucleation under ultrasonic irradiation and crystal growth, respectively. The primary nucleation requires the highest free energy in contrast to crystal growth which requires the least free energy. This shows that primary nuclei are difficult to generate as well as hard to control. Crystals grow in any supersaturated solution. In this system, the ultrasound-assisted nucleation occurred in the first stage MSMPR crystallizer, which required lower free energy compared with primary and secondary nucleation (Jiang, 2012). In the second stage MSMPR crystallizer, secondary nucleation will not occur in the crystal slurry since the free energy required was not sufficient. In the presence of moderate mechanical stirring in the second stage, there is a difference of required free energy for nucleation between those two stages, as shown in Figure 2. Thus, the transferred crystal slurry from first stage to second would not generate secondary nucleation appreciably because of higher free energy required with the moderate mechanical mixing. The crystals from the first stage with narrow distribution would further grow in the second stage without significant breakage. The inserted images (a) and (b) are micrographs of the crystals at the outlet of the two stages. Uniform crystals around 35μm were obtained under paused ultrasonic irradiation (15s on, 10s off, 15W, residence time one minute) in the first stage, and the crystals grew to 150 μm without obvious breakage and agglomeration.

Figure 2. Schematic analysis of free energy required by different kinds of nucleation processes and corresponding operation point in the two-stage MSMPR crystallizer system. Image (a) and (b) are crystals at the outlet of the first stage and the second stage MSMPR crystallizer. The scale bar is 200 microns.

Conclusions

In this study, a two-stage continuous MSMPR crystallizer system was used to optimize the CSD of a-form L-glutamic acid and to improve the process efficiency with the help of ultrasonic irradiation. The final product crystals around 150 μm were obtained with a narrow crystal size distribution. The total residence time in this system was 5 to 15 minutes, which is an obvious improvement of process efficiency with the help of ultrasonic irradiation. With the improved crystal qualities and process efficiency, challenges still exist in continuous crystallization process: i) a long time may be required to reach steady state; ii) any unexpected process malfunction may cause the shut down of the whole production process; iii) there is a limited application in small scale, low output systems, especially for newly released drugs. The properties of model compound are highly dependent on the design of continuous crystallization process and operation recipe.

Acknowledgments

The authors acknowledge the financial support provided by the Natural Science and Engineering Research Council (NSERC) of Canada and the China Scholarship Council.

References

Adamo, A., Beingessner, R.L., Behnam, M., Chen, J., Jamison, T.F., Jensen, K.F., Monbaliu, J.-C.M.C., Myerson, A.S., Revalor, E.M., Snead, D.R., Stelzer, T., Weeranoppanant, N., Wong, S.Y., Zhang, P., 2016. On-demand continuous-flow production of pharmaceuticals in a compact, reconfigurable system. Science 352, 61–7.

Alvarez, A., Singh, A., Myerson, A., 2011. Crystallization of Cyclosporine in a Multistage Continuous MSMPR Crystallizer. Cryst Growth Des 11, 4392–4400.

Besenhard, M., Neugebauer, P., Ho, C.-D., Khinast, J., 2015. Crystal Size Control in a Continuous Tubular Crystallizer. Crystal Growth & Design 15, 1683–1691.

Bhangu, S., Ashokkumar, M., Lee, J., 2016. Ultrasound Assisted Crystallization of Paracetamol: Crystal Size Distribution and Polymorph Control. Cryst Growth Des 16, 1934–1941.

Castro, M.D., Priego-Capote, F., 2007. Ultrasound-assisted crystallization (sonocrystallization). Ultrason Sonochem 14, 717–724.

Jiang, S., 2012. An Examination of Sonocrystallization Kinetics of L-Glutamic Acid. University of Leeds.

Mascia, S., Heider, P.L., Zhang, H., Lakerveld, R., Benyahia, B., Barton, P.I., Braatz, R.D., Cooney, C.L., Evans, J.M., Jamison, T.F., Jensen, K.F., Myerson, A.S., Trout, B.L., 2013. End-to-end continuous manufacturing of pharmaceuticals: integrated synthesis, purification, and final dosage formation. Angew. Chem. Int. Ed. Engl. 52, 12359–63.

McGlone, T., Briggs, N., Clark, C., Brown, C., Sefcik, J., Florence, A., 2015. Oscillatory Flow Reactors (OFRs) for Continuous Manufacturing and Crystallization. Organic Process Research & Development 19, 1186–1202.

Narducci, O., Jones, A.G., Kougoulos, E., 2011. Continuous crystallization of adipic acid with ultrasound. Chem Eng Sci 66, 1069–1076.

Wong, S., Tatusko, A., Trout, B., Myerson, A., 2012. Development of Continuous Crystallization Processes Using a Single-Stage Mixed-Suspension, Mixed-Product Removal Crystallizer with Recycle. Cryst Growth Des 12, 5701–5707.

491224 Membrane Assisted Antisolvent Crystallization for the Continuous and Accurate Control of Pharmaceuticals Manufacture

Abstract:

Membrane Assisted Antisolvent Crystallization for the Continuous and Accurate Control of Pharmaceuticals Manufacture

Xiaobin Jiang, Linghan Tuo, Gaohong He*

State Key Laboratory of Fine Chemicals, School of Chemical Engineering, Research and Development Center of Membrane Science and Technology, Petrochemical Energy-efficient Separation Technology Engineering Lab of Liaoning Province, Dalian University of Technology, Dalian, P. R. China

*Corresponding author: Email: xbjiang@dlut.edu.cn, hgaohong@dlut.edu.cn

Abstract

Different from batch process, membrane assisted antisolvent crystallization (MAAC) is proposed as a novel process to optimize pharmaceuticals manufacture significantly in recent year [1, 2]. MAAC could achieve continuous production and obtain the high quality of crystal product[3, 4].

In this work, without introducing vacuum and temperature gradient, the proposed MAAC process is operated under atmospheric pressure and isothermal conditions in the hollow fiber membrane module to achieve the continuous and accurate pharmaceuticals manufacture. Due to the high specific surface area of hollow fiber membrane module, the accuracy of the interfacial mass transfer rate is greatly higher than conventional antisolvent crystallization. The accurate mass transfer provides homogenous supersaturation through the membrane module, intensified micro mixing and more stable supersaturation gradient for nucleation and growth.

The driving force of MAAC is determined by the concentration difference, pressure difference and turbulence of membrane outside (shell side). Thus, the antisolvent permeation rate can be adjusted flexibly. As reported in the literature[5], an antisolvent liquid layer could be established on the membrane outer surface by regulating the process operation parameters. According to Bernoulli equation and liquid membrane renewal theory [6], high flow rate accelerates liquid membrane renewal and decreases pressure simultaneity. Thus, the interaction relationship of membrane inner and outer flow rates have obvious impact on solution mixing effect. Suitable flow rates keep nucleation occurring on the boundary layer of the liquid membrane, which would avoid membrane pore blockage and membrane fouling efficiently. Based on the advantages above, the continuous MAAC process could be adapted to the various requirements of pharmaceuticals during a production cycle, timely discharge and deal with the crystal produce could prevent tube blockage, which is one of the important problems of high viscosity pharmaceuticals in the convention batch manufacture process [7]. The proposed MAAC system is utilized in erythritol production to acquire crystals with desire morphology and size distribution (showed in Fig. 1). Crystals obtained by MAAC are obvious subuliform polyhedron shape, as a comparison, crystals obtained by conventional antisolvent crystallization are commonly spheroid or irregular shape. Moreover, the length-width ratio of MAAC crystals are closer to the simulated ideal morphology (MAAC: 1.49, conventional: 1.30, ideal: 1.52)[8]. The coefficient of variation (CV) of crystals decreases from 25.2 % (conventional) to 19.3 % (MAAC), which indicates narrower crystal size distribution.

With the stable and accurate mass transfer process controlled by the introduced membrane system, MAAC can be expected to expand the application and improve the performance of antisolvent crystallization in not only pharmaceuticals, but also food, biomacromolecule, fine chemicals fields.

Fig. 1 Erythritol crystal size distribution, morphology, interfacial mass transfer rate and CV comparison between MAAC and conventional antisolvent crystallization.

Refrences

[1] Di Profio, G., Stabile, C., Caridi, A., Curcio, E., and Drioli, E., Antisolvent membrane crystallization of pharmaceutical compounds. J Pharm Sci, **98**(12). 4902-13(2009).

[2] Lu, D., Li, P., Xiao, W., He, G., and Jiang, X., Simultaneous recovery and crystallization control of saline organic wastewater by membrane distillation crystallization. AIChE Journal, (2016).

[3] Drioli, E., Di Profio, G., and Curcio, E., Progress in membrane crystallization. Current Opinion in Chemical Engineering, **1**(2). 178-182(2012).

[4] Zarkadas, D.M. and Sirkar, K.K., Antisolvent crystallization in porous hollow fiber devices. Chemical Engineering Science, **61**(15). 5030-5048(2006).

[5] Ren, Z., Yang, Y., Zhang, W., Liu, J., and Wang, H., Modeling study on the mass transfer of hollow fiber renewal liquid membrane: Effect of the hollow fiber module scale. Journal of Membrane Science, **439**. 28-35(2013).

[6] Chaturabul, S., Srirachat, W., Wannachod, T., Ramakul, P., Pancharoen, U., and Kheawhom, S., Separation of mercury(II) from petroleum produced water via hollow fiber supported liquid membrane and mass transfer modeling. Chemical Engineering Journal, **265**. 34-46(2015).

[7] Blagden, N., de Matas, M., Gavan, P.T., and York, P., Crystal engineering of active pharmaceutical ingredients to improve solubility and dissolution rates. Adv Drug Deliv Rev, **59**(7). 617-30(2007).

[8] Traini, D., Young, P.M., Jones, M., Edge, S., and Price, R., Comparative study of erythritol and lactose monohydrate as carriers for inhalation: atomic force microscopy and in vitro correlation. Eur J Pharm Sci, **27**(2-3). 243-51(2006).

498277 Rapid Crystallization Development through Automation of Solubility Screening Coupled with XRPD

Abstract:

Rapid crystallization development was enabled through a series of automated screenings using robotics to enable automated solid and liquid dispensing, solubility screening, and crystal form testing. Screenings including an initial form and rough solubility screen, followed by a 10 g screen focused on 96 solvent and antisolvent combinations at 20°C and 50°C coupled with HPLC and XRPD. Solvent and antisolvents were then selected based upon the form produced, solvent classification, and potential impurities.

487669 Performance Evaluation of a Hybrid Hydrogen-Vanadium Reversible Fuel Cell

Abstract:

Hybrid hydrogen-vanadium reversible fuel cells offer a feasible solution for storing electrical energy from the grid or directly from renewable energy sources such as wind and solar. In this presentation, hydrogen-vanadium reversible fuel cells are tested using a Pt/C hydrogen electrode, carbon vanadium electrode and interdigitated flow fields at both electrodes. Vanadium electrolyte flow rate is varied to study its effect on mass transport performance. Two types of vanadium electrodes are explored, a single layer of high surface area carbon nanotube (CNT) and three layers of nitric acid-treated carbon paper. Finally, four types of Nafion and two types of electrospun Nafion/PVDF single-fiber blend membranes are examined to determine the effect of membrane type and thickness on fuel cell charge and discharge performance.

Acknowledgement

The work presented was partially funded by the National Science Foundation under Grant No. EFRI-1038234.

494299 Fuel Cell Membranes with Enhanced Durability and Performance Based on Fluoroelastomers Functionalized with Heteropoly Acids

Abstract:

There is still a need for membranes that operate in proton exchange membrane (PEM) fuel cells at hotter and drier conditions than can be achieved with current materials such as Nafion, i.e. >100°C and <50%RH. One approach pursued by us and independently others is the use of inorganic super acids, such as the heteropoly acids (HPAs). HPAs are a sub group of the large class of metal oxygen clusters known as polyoxometalates in which a central heteroatom is surrounded by a number of W or Mo oxygen octahedra. For proton conductivity it is desirable that a strong negative charge be delocalized across the whole anion so that the proton will be as dissociated as possible. This limits the choice of HPA to the spherical tungsten based Keggin anion with as light as possible a heteroatom. This limit is reached with Si as the P based HPA is known to decompose in the presence of peroxide and the electron deficient nature of B renders the spherical Keggin anion unstable. Many fundamental studies have been undertaken on solid state HPA systems. These studies indicated that despite the original report in the early 80's that HPA had the highest proton conductivity reported at that time, that when dry there proton conductivity was disappointingly low at moderate temperatures in crystaline HPA. The key to making these systems work, thefore is make them amorphous.

Our approach is to make monomers from HPA and immobilize the HPA by polymerization into hybrid systems. In order to functionalize the Keggin anion one W oxygen octahedra is removed and a Si or P based organic functionality introduced that may be a monomer or a tether to a functionalized polymer backbone. Our first generation materials based on divinyl functionalized HPA and acrylate chemistry produced films with impressive conductivities, >100 mS cm^{-1} at T >80°C and 50% RH. This model system contained ester linkages that we think would be hydrolysed under the harsh conditions of fuel cell operation and so we attached HPA via di-phosphonate linkages to perfluorinated polymers. This is acheived by dehydrofluorinating the based perflourinated polymer and functioalizing it with phosphonate groups taht serve as covalnent attcahmnet points to the HPA. Very recently we have fully perfected this chemistry and can now produce large area thin, 10 μm, high loaded HPA films for fuel cell cell operation. Not only do the materials have very low ASRs, <0.02 Ω cm^2, under all operating conditions (except freeze), but there is very little cross-over of H_2 and O_2. The materials also survive the DOE mechanical stability test and exceed the chemical stability test. The chemical stability is proof a theory; published by us using PFSA/HPA composite films, the silicotungstic HPA moieties catalytically decompose peroxy radicals. The measured water flux is also superior to PFSA materials allowing back diffusion of water during fuel cell operation. Unfortunately, we do not yet have a suitable ionomer for the electrodes to operate in an hot and dry environment and so fuel cell data using PFSA ionomers will be presented under more standard operating conditions.

This work is sponsored by DOE EERE and ARPA-E.

492002 Hydroxylated Graphane : An Anhydrous Proton Exchange Membrane

Abstract:

Operating proton exchange membrane (PEM) fuel cells at higher temperatures is desirable for many reasons, namely increased tolerance of catalyst to CO poisoning and improved waste heat rejection. Traditionally, water facilitates the transport of protons in the PEM material and therefore imposes a limitation on the operating temperature of the fuel cell. We have designed a novel material based on graphane functionalized with hydroxyl groups. We show that this material can rapidly conduct protons under anhydrous conditions through a contiguous network of hydroxyl group hydrogen bonds. Density functional theory calculations predict remarkably low barriers to diffusion of protons along a 1D chain of surface hydroxyls. Diffusion is controlled by the local rotation of hydroxyl groups, a mechanism that is very different from that found in 1D water wires in confined nanopores or in bulk water. Completely hydroxylated graphane provides a 2D hydrogen bonded network, having more pathways for the diffusion of protons. A charge analysis reveals that the charge on the proton is essentially equally shared by all hydrogens bound to oxygens, effectively delocalizing the proton; this charge delocalization reduces the barrier to proton hopping. We have also investigated the influence of morphological and functional group defects to proton transport.

492397 Interfacial Transport Resistances and Coupled Transport

Abstract:

Thermodynamic fluxes in membranes are proportional to thermodynamic forces (gradients of chemical potentials, temperature, etc.) via a matrix of phenomenological coefficients. Onsager's relations imply that the matrix is symmetric, which is a crucial result as the number of unknown coefficients is reduced. We show that for certain classes of non-equilibrium thermodynamic models, in addition to Onsager's relations the transport coefficients have to share the same functional dependence on the local thermodynamic state of the system. We prove this statement for thermodynamic models of coupled water and proton transport in ionomer membranes. Thence classical irreversible thermodynamic models and experimental data should be checked for consistency. Additionally, these newly identified constraints further reduce the number of experiments needed to describe the phenomenological coefficients.

These functional constraints have been confirmed by detailed studies of water and proton fluxes across Nafion based membrane electrode assemblies. For small electric potential differences ($\Delta\Phi$) the water and proton fluxes are correlated; both increase linearly with applied potential. The electro-osmotic drag coefficient is constant, functionally independent of the applied electrical potential, water activity and temperature. For larger applied electric potential differences the water and proton fluxes are no linearly dependent on applied potential; the electro-osmotic drag coefficient increases with increasing applied electrical potential and increasing water activity. The transition between intramembrane transport limited flux and interfacial transport limited flux corresponds to the limits of applicability of the standard Onsager reciprocal relationships for coupled transport.

494363 Exploring Ionomer-Related Transport Phenomena in Polymer-Electrolyte Fuel Cells

Abstract:

Ion-conducting membranes serve critical functions within polymer-electrolyte fuel cells (PEFCs), both as a standalone separator as well as an ionic network within the catalyst layers. They are also critical components in enabling one to realize high power density and thus reduce stack count and cost. For proton-conducting PEFCs, the local ionomer within the cathode catalyst layer dominates the mass transport for such low loadings. For hydroxide-conducting PEFCs, such membranes can allow for less expensive membranes and catalysts, but have possible issues with water management and carbon dioxide interactions. In this talk, we will explore the limiting transport mechanisms and effects related to fuel-cell membranes through mathematical modeling and advanced diagnostics.

Thus, we will explore the gas transport limitations within PEFC catalyst layers, with a focus on the observed local transport resistance as measured using hydrogen and oxygen limiting current measurements. We will examine test cases for diagnosis of polarization curves in terms of different limiting phenomena. Different oxygen concentrations, catalyst-layer properties, and operating humidity and temperature will be introduced.

Water management is a more serious concern in hydroxide PEFCs, because OH^- conductivity is more highly dependent on water content and the oxygen-reduction reaction consumes water. Compared to proton PEFCs, the lower performance of hydroxide PEFCs is mostly caused by a extremely nonuniform distribution of water in the ionomer phase between the anode and cathode as well as the increased overpotential for the hydrogen oxidation reaction. In this presentation, we will discuss the performance-limiting mechanisms specific to different operating conditions (e.g. varying inlet relative humidity (RH)) based on a cell-level mathematical model. In addition, the issue of carbonate formation with the membrane will be discussed.

Acknowledgements

We would like to thank helpful discussions and samples from Toyota Motor Company. This work was mainly funded under the Fuel Cell Performance and Durability Consortium (FC PAD) funded by the Energy Efficiency and Renewable Energy, Fuel Cell Technologies Office, of the U. S. Department of Energy, Program Development Manager Dimitrios Papageorgopoulos

494508 Structured Ion-Exchange Membrane-Electrode Interfaces Fabricated Via Conventional and Advanced Block Copolymer Lithography

Abstract:

Intimate electrode-electrolyte interfaces for fuel cell and water electrolysis applications are central for extracting excellent thermodynamic device efficiencies. This talk will report on our efforts to prepare microstructured and nanostructured ion-exchange membrane-electrode interfaces via micropatterning and advanced block copolymer lithography (i.e., nanopatterning platform). The structured interfaces enable greater interfacial area that minimizes charge-transfer resistances. However, increasing the interfacial area by patterning deeper into the membrane (rather than shrinking the lateral feature sizes) gives rise to large mass transfer related resistances that eventually offset gains from the reduced charge-transfer resistances. The talk will highlight appropriate design heuristics for fabricating structured electrochemical interfaces from patterned membranes that optimize fuel cell and electrolyzer performance. The talk will also show promising new results to make periodic, nanostructured electrocatalyst on ion-exchange membranes with sub-15 nm feature sizes over large areas using block copolymer lithography.

Note: The term ion-exchange membrane encompasses both proton exchange and hydroxide ion exchange membranes

492202 Highly Durable Aromatic Anion Exchange Membranes for Solid Alkaline Fuel Cells

Abstract:

Introduction Solid alkaline fuel cell using anion exchange membrane (**AEM**) attracts increasing attention since non-noble metal electrode catalyst and high density liquid fuel can be used. However, development of chemically durable membrane is critical for long-term operation. **AEM** consisting of only aromatic component and anion exchange group such as polyphenylene **AEM** is potential candidate for highly durable membranes, but is suffered from low solubility due to the rigid linear backbone.

To overcome this problem, we have synthesized a spirobifluorene- based aromatic **AEM** as a highly soluble and durable **AEM** and used this **AEM** as an ionomer of methanol fuel cell.

Method Target **AEM** was synthesized by polymerization of the spirobifluorene and the phenyl monomers and successive quaternization. Alkaline and radical stability of this membrane were evaluated by comparing ^{1}H-NMR spectra before and after soaking into 1M NaOH aq, and fentone solution. Finally, MEA was fabricated by using this **AEM** ionomer and direct Methanol fuel cell performance and durability were investigated at 80 ^{o}C.

Results and Discussion The synthesized membrane can be dissolved well into chloroform, tetrahydrofran and mixed solvent of THF and 1-propanol but not dissolved into fuel methanol.

^{1}H-NMR spectra of this membrane did not change at all even after soaking into 1M NaOH aq. for one week indicating high alkaline stability. After soaking into fentone solution, the aromatic region of ^{1}H-NMR did not change indicating high radical stability of the polymer backbone.

Direct methanol fuel cell using this membrane shows high and durable performances at 80 ^{o}C. These performances did not change even after one week operation indicating that highly durable solid alkaline fuel cell was realized by using the membrane (Fig 1).

499732 Sustainion™ Membranes for Carbon Dioxide and Alkaline Water Electrolyzers

Abstract:

This paper will describe the performance of Sustainion™ membranes in CO_2 and alkaline water electrolyzers. The membranes are imidazolium functionalized styrene. The membranes have a conductivity of 90 mS/cm in 1 M KOH at 60 °C . No changes in conductivity are detected after 1500 hours in 1 M KOH at 60 °C. NMR indicates that there is interesting cross linking in the polymers. The original polymers were brittle, but improvements have made them much less brittle.

We have tested the membranes for use in CO_2 electrolyzers and alkaline water electrolyzers. Tests in an alkaline water electrolyzer with iron/nickel/cobalt catalysts that show a current density of 1 A/cm^2 at 1.9 V in 60 °C 1 M KOH. The performance is stable for 2000 hours, showing only 3-5 μV/hr degradation. Tests in a CO_2 electrolyzer show 600 mA/cm^2 at 3 V at 50 °C in 0.01 M $KHCO_3$. These and other results will be discussed.

486728 Membrane Materials for Gas Separations

Abstract:

Large scale industrial gas separations with polymer membranes began in earnest in about 1980, around the time that Bill Koros began his academic career. As the industry grew and new applications and challenges arose, Bill was a key contributor, providing crucial theoretical concepts that now underpin the science in this area, developing key structure/property insights that provide the foundation of modern materials design, exploring and developing novel concepts in materials design that many groups around the world have adopted, and supplying the field with the current generation of leaders both industrially and academically. No one has had such a profound effect on this field as Bill Koros. In recent times, he continues to push the envelope on development of novel materials and leads efforts to reduce them to practice. While being keenly connected with the fundamental science in the area, Bill is always driving towards real applications and meaningful engineering contributions. This presentation will highlight a few of many outstanding contributions he has made to the field of gas separations over 40+ years of outstanding practice in the field.

497649 Molecular Sieve Nanosheets for Membrane Applications

Abstract:

Single-unit-cell thick zeolite nanosheets (2-dimensional zeolites; AIChE Journal 60(7), 2374-2381 (2014)) with intact crystal and micropore structure can be made by exfoliation (Science 334, 72-75 (2011)). Recently, few-unit-cell thick zeolite nanosheets were prepared by direct synthesis (Nature 543, 690-694 (2017)). The nm-thickness and high aspect-ratio of these microporous nanosheets open exciting possibilities for technological breakthroughs in molecular sieve membrane fabrication (Johnson and Koros, Journal of the Taiwan Institute of Chemical Engineers 40(3), 268-275 (2009)). Recent progress on synthesis of high aspect ratio zeolite and other crystalline nanoporous nanosheets, methods to characterize their structure and properties, along with their processing and assembly to create membranes and catalysts will be the focus of this talk.

487396 Carbon Molecular Sieve Membranes for Applications in Olefins Units

Abstract:

The separation of olefins from paraffins is important in the petrochemical industry. The current dominant technology, cryogenic distillation, is highly energy intensive. A hybrid membrane-distillation process is proposed to retrofit the current platform.

Carbon molecular sieve (CMS) membranes are able to overcome the polymer upper bound of the productivity-efficiency tradeoff. Making the novel carbon membrane into the asymmetric hollow fiber form is of great interest as well as tailoring the microscopic molecular sieving microstructure.

Dow has been collaborating with Prof. Bill Koros on the development of CMS membranes for cracked gas processing. Major advancements made in the past years in both membrane materials and new processes using these materials have led to new discoveries that will be reported during this presentation.

493325 High Flux Thermally Rearranged (TR) Hollow Fiber Membranes for Gas Separation

Abstract:

Thermally rearranged (TR) polymer membranes have been developed from ortho-functionalized polyimides by a post thermal conversion process. They demonstrate bimodal cavity size of 0.3-0.4 nm and 0.7-0.9 nm and narrow cavity size distributions characterized by PALS measurement which is an efficient transport path for small molecules. As a result, TR polymer membranes exhibit both high permeability and high selectivity based on high diffusion of gas molecules.

In gas separation processes using membranes, for example CO2/N2 or O2/N2 separation, membrane performance itself as represented by permeability (or permeance) and selectivity is important in deposal capacity and quality of resultant gas (purity). Particularly, process optimization achieved by controlling operation variations such as feed flow rate, stage-cut, pressure ratio, operation temperature, etc. plays a critical role influencing separation factor and overall operation cost in mixed gas separation system. In addition, the understanding of gas transport mechanism in presence of water vapor is essential in a process where water cannot be removed before it is in contact with membranes.

In this study, we report highly permeable thermally rearranged polybenzoxazole (TR-PBO) hollow fiber membranes and their modules to examine gas permeation performance in mixed gas system similar to real flue gas exhausted from post-combustion process [1, 2]. The optimal operation condition for CO2 capture application was identified from simulation data [3]. Additionally, we proposed a conceptual transport mechanism in TR-PBO hollow fiber membranes under humidified condition. Separation factor showed an increase with higher water vapor content without severe permeation flux drop. Moreover, the TR-PBO membrane modules under permeate vacuum condition showed improved gas separation performance with no permeation flux drop.

References

1. Woo, K. T., Lee, J., Dong, G., Kim, J. S., Do, Y. S., Jo, H. J., & Lee, Y. M., Journal of Membrane Science, 498 (2015) 125-134.

2. Woo, K. T., Lee, J., Dong, G., Kim, J. S., Do, Y. S., Hung, W. S., ... & Lee, Y. M., Journal of Membrane Science, 490 (2015) 129-138.

3. Dong, G., Woo, K. T., Kim, J., Kim, J. S., & Lee, Y. M., Journal of Membrane Science, 496 (2015) 229-241

4. Woo, K.T., Dong, G., Lee, J., Yu, S.D., Lee, W.H., Lee, H.S., Lee, Y.M., Journal of Membrane Science, 510 (2016) 472-480

499075 Polymeric Membranes for High Temperature Gas Separations

Abstract:

Bill Koros has unquestionably been one of the key contributors to the discovery of new polymeric materials and advanced membrane structures for the effective separation of gases, vapors, and liquids. In his 40+ years conducting and directing academic research in this area, he established the intellectual underpinnings of the structure/property relationships that have guided advances in material design and facilitated polymeric membrane adoption in an ever-increasing range of applications. Perhaps equally important are the researchers trained in the Korosâ€™ Group and now contributing to the scientific and engineering communities around the world.

Based upon the foundations laid by the Korosâ€™ Group, we have developed polymers and membranes for use in the separation of gases at temperatures above 100 Â°C. This presentation will examine the findings from this work. Additionally, the career paths and impacts of the Koros Group alumni will be reviewed.

502490 Mixed Matrix Membranes for Organic Solvent Nanofiltration

Abstract:

Organic liquids are ubiquitous in chemical science based industries, which range in scale from refining to pharmaceutical production. It is generally accepted that 40-70% of capital and operating costs in these industries are dedicated to separations; and a substantial fraction of this cost is related to processing of organic liquids. Membrane technology has the potential to provide game changing alternatives to conventional concentration and purification technologies such as evaporation, liquid extraction, adsorption and chromatography, through Organic Solvent Nanofiltration (OSN). The membranes must offer resistance to organic environments, attractive selectivies and permeabilities. Ideally they should also be resistant to physical aging under use.

This presentation will focus on research into mixed matrix membranes for OSN. Metal-organic frameworks (MOFs) have been used to create mixed matrix membranes in two distinct approaches.

In the first approach, we have used In-Situ-Growth (ISG) to prepare MOFs inside cross-linked polyimide ultrafiltration support membranes. The MOF, in this case HKUST-1, is prepared from the MOF precursors by submerging the pre-formed polyimide membrane in HKUST-1 precursor solutions. ISG is presented as an alternative to mixed matrix membranes (MMMs) in which MOF crystals are fabricated ex-situ of the membrane and added to the polymer dope solution. The presence of HKUST-1 in the ISG membranes was shown to increase the rejection of styrene oligomers on average by 30%, and also reduced pressure-induced flux decline from above 40% flux decline over 24 hours for pure polymer membranes, to less than 2% for the ISG membranes.

In the second approach, thin film nanocomposite (TFN) membranes containing 50-150 nm MOF nanoparticles [ZIF-8, MIL-53(Al), NH_2-MIL-53(Al) and MIL-101(Cr) were employed] have been synthesized via interfacial polymerisation on top of a cross-linked P84 support. MOF nanoparticles were homogeneously distributed in the polyamide layer, and their presence was inferred by a combination of contact angle measurements, FT-IR spectroscopy, SEM, EDX, XPS and TEM. TFN membranes showed dramatically increased MeOH and THF permeances when compared to the same membranes with no MOFs added, without sacrificing rejection. A direct trend was observed between the MeOH permeance and the porosity of the MOFs, suggesting preferential flow paths provided by the MOF pores. The best performance was obtained with the incorporation of nanosized mesoporous MIL-101(Cr), with the largest pore size and surface area. This TFN membrane led to an exceptional increase in permeance, from 1.5 to 3.9 and from 1.7 to 11.1 $L \cdot m^{-2} \cdot h^{-1} \cdot bar^{-1}$ for MeOH and THF respectively, whereas the rejection remained higher than 90% (molecular weight cut off of 232 $g \cdot mol^{-1}$ for MeOH and 295 $g \cdot mol^{-1}$ for THF).

510033 Olefin/Paraffin Separations Enabled by New High Capacity/Flux CMS Materials

Abstract:

Olefin/paraffin separations by fractional distillation are very energy intensive due to the small differences in boiling points; e.g., 5 °C for C_3H_8 vs. C_3H_6. Pressure swing adsorption (PSA) and membrane systems have the potential to significantly reduce the energy consumption. The heart of future successful PSA or membrane systems will be advanced materials. The development of novel Carbon Molecular Sieve (CMS) materials for olefin/paraffin separations derived from two families: (1) cation exchange resins and (2) polyvinylidene chloride copolymers will be discussed. Both the C2 and C3 olefin/paraffin separations can be addressed using these types of carbon molecular sieves because their micropore sizes can be fine-tuned by precursor copolymerization and thermal treatment. The micropore volumes in these new CMS materials are 30-75% higher than ones produced with Matrimid® polyimide resins, which translates into higher capacity for sorbent and flux for membrane applications.

There is significant barrier to scale up (> 10, 000 m^2) thin film CMS membranes. PSA using CMS adsorbent pellets can potentially be commercialized sooner than membrane technology to provide an interim solution. However, PSA has much more complex set of transport processes and is an inherently dynamic process. The engineering of the PSA process using benchtop PSA experiments coupled with modelling (Aspen based) will also be discussed. PSA processes with additional steps from the basic Skarstrom cycle, i.e. co-current blowdown and effluent rinse, were investigated. Propylene purity of 89% at 89% propylene recovery was demonstrated using a 25% propylene-75% propane feed.

497623 Nanostructured Redox Interfaces for Electrochemically-Mediated Bioseparations

Abstract:

The purification and selective separation of biomolecules is one of the most important steps in modern biochemical industry. Processes based on electrosorption have emerged as a promising candidate in many fields of separation, especially in liquid-phase deionization â€“ however, its application for protein purification has been limited, both due to low adsorption capacity and to selectivity challenges. Redox-active materials are an attractive platform for engineering specific interactions with charged species by electrochemical control.[1]Â Here, we present the advantage of nanostructured, poly(vinyl)ferrocene-based redox-electrodes for the electrosorption and recovery of a range of proteins, with high adsorption capacity and modulated solely by electrochemical potential, thus with no need for chemical regenerants or pH changes.[2]Â Our system was selective for various proteins based on size and charge distribution, exhibiting fast kinetics (<120 s for a charge-discharge cycle) and high uptake capacities (>200 mg/g) under moderate overpotentials (+0.4 V vs Ag/AgCl). Furthermore, the preservation of bioactivity at the surface of electrodes indicates the potential for our systems to to be used as heterogeneous supports for enzyme catalysis. This work draws on the molecular selectivity of ferrocene-functionalized materials towards organic anion groups, and suggests that these smart redox-active materials may possibly be used at-scale for practical bioseparations.

References. (1) Su, X.; Kulik, H. J.; Jamison, T. F.; Hatton, T. A. *Advanced Functional Materials* **2016**, *26*, 3394. (2) Su, X.; Hubner, J.; Kauke, M. J.; Dalbosco, D.; Thomas, J.; Gonsalez, C.; Zhu, E.; Franzreb, M.; Jamison, T. F.; Hatton, T. A. *Submitted* **2017**.

485516 Breaking through the Limits of Chromatography Operations: Integration of a Novel Modular Chromatography Scaffold with Bed Design Features to Achieve High Speed and High Productivity

Abstract:

Much attention has been paid to the recent proliferation of new resin types such as protein A and mixed mode resins. Likewise there has been an increased focus on operating modes including SMB and multicolumn operations. All possess a variety of attributes that can affect purification operations and costs. However very little innovation has been made on the chromatography bed itself and how that can enhance productivity, affect resin usage and enable new operating domains. Additionally, very little innovation has been made in the design of the adsorptive device to enable simple and reliable scale-up from discovery to production scale. Chromassette® technology is a novel modular chromatography platform with a prepacked bed supported by an integral structural scaffold. Chromassette devices are linearly scalable chromatography columns in the form of a cassette. The supported bed enables operation at velocities exceeding 1000 cm/hr and high pressure pressure drops, further enabling smaller bead sizes and high speed operation even with compressible beads. Mechanistic modeling was used to identify the optimal design point of bed and bead resulting in 5-10 X productivity gains over current capture & release processes. The novel Chromassette device design will be reviewed along with results for protein A Mab capture and polishing. Future direction and possibilities of this new technology will also be illustrated.

493094 Multi-Column Continuous Chromatography for Process Intensified Capture and Polishing of Monoclonal Antibodies

Abstract:

Multi-column continuous chromatography (MCC) offers significant economic advantages over traditional batch methods for purification of monoclonal antibodies (mAbs), including increased resin capacity utilization, smaller columns, reduced buffer consumption, and faster process time. The Protein A capture step is a primary target to apply MCC due to its high cost, which is driven even higher as improvements in upstream processing have produced a steady increase in mAb titers. In this study we consider the key factors in designing an MCC process that optimizes productivity. Process design begins with determination of the feed duration, which is governed by the mAb titer, column dimensions, residence time, binding capacity, and flow properties of the resin. Use of 3 columns in the capture zone enables efficient utilization of the resin even at short residence times, thereby increasing productivity. Scheduling constraints experienced at high titer with MCC approaches using only 2â€“4 columns are relieved by including additional columns to perform all process operations simultaneously. Productivities exceeding 100 g mAb/L resin/h were achieved with 10.8 g/L titers using a Semba ProPDâ„¢ System and 8-column MCC Protein A affinity process. Downstream flow-through and bind-elute polishing steps including ion exchange and mixed mode have been converted from batch to MCC toward realization of a completely continuous biomanufacturing platform.

499035 Improved Chromatographic Distributors through CFD Modeling

Abstract:

Chromassette® cassettes are modular chromatographic devices under development for biopharm applications. These novel devices have a rectangular geometry, multi-level distributors and a bed-supporting scaffold. CFD modeling was used to explore alternative distributor designs as well as alternative scaffolds. The results of that exploratory work will be presented.

488978 Methods for Optimizing Bio-Chromatography Processes for Repeated Cyclic Operations

Abstract:

Simple methods for calculating the productivity of chromatography processes are proposed, which consider both the productivity P defined by P = (mass recovered)/[(bed volume)(cycle time)] and the buffer consumption per cycle, V_C. The method was applied to model separation systems (polyphenol separations by polystyrene divinylbenzene resins with the ethanol-water mixture mobile phase, and protein separations by ion-exchange chromatography). The distribution coefficient K was determined as a function of mobile phase concentration I by linear gradient elution experiments. The HETP-mobile phase velocity u curves were determined as a function of I. By using K and HETP, the iso-resolution curve was calculated, from which the productivity was determined as a function of I. It was found that there is an optimum I, where the highest productivity with the minimum amount of mobile phase consumption is obtained.

506551 Advective Flow Membrane Chromatography: Utilization of Flow-through Mode for Scalable, Single-Use Downstream mAb Processing

Abstract:

Membrane chromatography can be an effective downstream tool, overcoming many of the traditional problems associated with packed bed chromatography such as high pressure drops, low flow rates, dependence on intra-particle diffusion, and potential complications due to radial/axial dispersion limitations and other transport phenomenon. However, many membrane devices are limited by both dynamic binding capacity and resolution, thus limiting their effectiveness beyond pilot scales. This presentation demonstrates how these limitations are effectively overcome through the utilization of a train consisting of the Natrix HD-Q (quaternary amine) followed by the HD-Sb (sulfonic acid/t-butyl) single-use, advective flow membrane devices operated entirely in a flow through mode for mAb downstream polishing.

These devices allow effective removal of impurities such as host cell protein, DNA, and HMW species at high capacities (5 kg/L) while maintaining rapid flow rates (10 MV/min). Furthermore, both devices were optimized to run at the same pH and conductivity, eliminating the need for inter-step eluate adjustments. Together, the high capacities and flow rates, along with single-use functionality and the development of adjustment-free downstream processing, make these membranes a good option for large scale flexible and disposable downstream processing.

496842 Development of Tie Lines for Optimization of Porcine Parvovirus Recovery in Aqueous Two-Phase System

Abstract:

Improvements are needed in the downstream processing of viral vaccines, due to the increased complexity of the upstream processing, elevated demand, and the urge to reduce production costs. To escalate the purification process throughput, there is a demand for a shift from batch processes to continuous. Aqueous two—phase systems (ATPS) provides an inexpensive, biocompatible, and environmentally friendly method to purify viruses that can easily be adapted to continuous processing. Recent progress in increasing ATPS recovery has created a space for ATPS to become a potential substitute for more expensive unit operations, like density gradient centrifugation and chromatography. The main drawback that is holding back commercial production from using ATPS is the poor understanding of the partition mechanism of biomolecules. We are taking a thermodynamic approach to ATPS by studying the separation of virus particles along the tie-lines that connect the binodal two-phase region. Moving along the tie-lines changes the volume of each phase while keeping the concentration of the PEG and citrate in each phase constant. This shift in volume of the phase is hypothesized to change the purification and concentration of the viral product. The common system of aqueous polymer—salt was used to optimize the virus recovery. Porcine parvovirus (PPV), a 20 nm diameter, non-enveloped, single-stranded-DNA virus, was used as a model virus. We created the binodal curves for the PEG 12000-citrate salt system at pH 7 using the turbidimetric method and the tie lines were determined by measuring the concentration of citrate in each phase. Infectious virus titer was used to determine the partitioning of the virus to the different phases. To this date we have checked different PEG-citrate configurations on the tie line length (TLL) of 36%w/w. We obtained a similar infectivity of ~log 5 MTT_{50}/ml of the PEG-rich phase by shifting the system on the selected TLL but the increasing trend for the yield was seen with the increasing volume of the PEG-rich phase as we move down the tie line. The citrate-rich phase contained minimum infectivity and the main shift was the amount of virus that was sequestered at the interface. We are looking for the same trend with different TLL and obtain a trend to optimize the recovery process. This approach to optimize the recovery/purification of PPV will aid in faster process development and will help to develop the key process parameters that effect ATPS separation of viral particles.

489605 Direct Synthesis and Tuning of Highly Porous Boron Nitride: Towards a New Class of Adsorbent

Abstract:

Porous boron nitride (BN) is being established as a new class of high performance adsorbent materials, primarily due to its high thermal and chemical stability along with its bond polarity. This permits a wide range of possible applications for BN, owing to its tolerance to extreme conditions. However, control over the porosity of BN has thus far proven difficult to achieve, limiting BNâ€™s potential in the field of adsorption.

Herein, we present how a series of porous BN samples with exceptional and tunable surface area and pore volume was produced via a simple and cost effective template-free method. We examine their chemical and structural features from the nano- to the macro-scale using a range of analytical techniques. In particular, we show how state-of-the art 3D tomography techniques could directly visualize the pore network. The highest surface area samples showed promising performances for high pressure gas sorption applications. A number of analyses were conducted to identify the mechanisms behind the pore formation and the porosity enhancement involved in this new synthetic approach. Overall, our study provides fundamental knowledge on the formation of porous BN as well as a facile and reproducible method for its synthesis.

502680 Exploiting Unique Properties of Porous Polymers for Air Pollution Control

Abstract:

Microporous activated carbon (AC) and zeolites are industrially applied for air pollution control, but pore blockage, irreversible adsorption, and co-adsorption of moisture reduce their contaminant removal efficiencies and/or prevent reuse. Porous polymers are gaining traction in environmental engineering because they can have tunable physical and chemical properties that allow tailoring for specific applications. Our previous studies show that hydrophobic styrene-divinylbenzene (St-DVB) polymers maintain their catalytic NO oxidation performance when moisture is present, while AC fails due to competitive water adsorption caused by increased hydrophilicity and surface oxidation during use. It was also observed that polymers provide faster adsorption/desorption kinetics and higher desorption efficiency than AC at low temperatures. However, commercial St-DVB polymers have low micropore (d_p < 2 nm) volume, reducing adsorption capacity for low concentration contaminants. To overcome, the Friedel-Crafts reaction is applied to increase micropore volume by hyper-cross-linking polymer segments and reducing the network mesh size. The Friedel-Crafts reaction can also be directly used to synthesize microporous polymers from aromatic compounds such as benzene, phenol, or methylbenzene using external cross-linkers. This method not only produces cost-effective microporous polymers, but also allows user-defined surface functional groups by changing the aromatic building blocks or by post-synthesis modifications.

In this work, commercial St-DVB polymers were first hyper-cross-linked with dichloroalkanes to increase micropore volume by up to 333%, without sacrificing the inherent hydrophobicity of the material. Consequently, hexane adsorption capacity increased by up to 218% at relative pressure (p/p_0) = 0.05 and, in a separate series of experiments, catalytic NO oxidation rate increased by 100% ([NO] = 500 ppm_v). Next, microporous polymers were directly prepared using the Friedel-Crafts reaction with benzene or phenol monomers, formaldehyde dimethyl acetal or 4,4â€²-Bis(chloromethyl)-1,1â€²-biphenyl (BCMBP) cross-linkers, and $FeCl_3$ as the reaction catalyst. Maximum surface area and micropore volume were 1600 m^2/g and 0.43 cm^3/g, respectively, when BCMBP self-condensed (reacted as monomer and cross-linker). The surface area and micropore volume decreased by 50% and 95% when 0.2 ml and 2 ml benzene was used as a co-monomer, respectively. Using co-solvents of dichloroethane and hexane reduced surface area, micropore volume, and total pore volume by 50%, 40%, and 70%, respectively. Overall, this work describes novel polymer modification and production processes that highlight opportunities to tailor the productâ€™s physical and chemical properties. The overarching goal is to exploit this flexibility, in an effort to present microporous polymers as competitive materials for use in air pollution control applications.

494821 Developing Nanometal Oxide Composite from Solid Waste to Make an Efficient Adsorbent for Heavy Metal Removal

Abstract:

Over the last few decades, there is a generous rise in use of heavy metals in different human and industrial activities. Unlike organic pollutants heavy metals are not biodegradable and they easily get accumulated in living tissues and become concentrated in food chain. Among those heavy metals, Ni(II) and its compounds are being widely used in many industries such as, electroplating, surface treatment and metal finishing facilities. Nickel is suspected to cause lung, nose and bone cancer moreover, it leads to extreme weakness, dermatitis, headache, dizziness, respiratory distress, chest pain, shortage of brain and dry cough. The toxic nature of crops, algae and fishes are also reported. Keeping the harmful effects of Ni(II) in mind, more research attention needs to be paid for finding an efficient and cost effective method for removing the said compound from waste water.

In this aspect, a number of methods including ion exchange, chemical precipitation, coagulation, flocculation, membrane separation, electro dialysis, electro chemical treatment and biological treatment are there. However, they have many disadvantages (like high energy and cost consumption, complex in operation, incomplete removal and production of huge solid sludge). Due to all these factors adsorption has drawn attention for the separation of heavy metals from aqueous solution. Selection of a suitable adsorbent is a critical issue for the adsorption process.

This work explores the feasibility of alumina based nanocomposite in removing heavy metal ions (Ni(II)) from waste water. The nanocomposite is made from the alumina extracted from scrap aluminium with a suitable nanometal oxide (i.e., ZnO) introduced in it, to enhance its adsorption capacity. The prepared adsorbent is characterised by Scanning Electron Microscopy (SEM), Brunauerâ€"Emmettâ€"Teller (BET) analysis and X-ray Diffraction (XRD) analysis in order to observe the surface morphology and crystalline nature. The adsorption quality is studied for Ni(II) adsorption. The effect of pH, initial concentration of Ni(II), contact time, adsorbent dosage and temperature on adsorption are investigated and the adsorption kinetics and equilibrium adsorption isotherm is also observed. It is noted that up to 96% of Ni(II) concentration can be reduced with this alumina based nanocomposite incorporated with ZnO.

500068 Ionic Liquids and Ionic Liquid-Functionalized Carbons for Sorption of Gaseous Toxins

Abstract:

Highly efficient and selective adsorption systems are needed for gas masks and security-sensitive buildings to protect against exposure to warfare toxins. Ionic liquids (ILs), molten salts with melting points below 100°C composed only of anions and cations, offer an opportunity to enhance adsorption of low-concentration warfare toxins and improve selectivity compared to traditionally-used activated carbons. By tuning the IL, the properties of sorption can be controlled. ILs can be functionalized on the surface of carbons to utilize the unique sorption properties, while creating a filter-stable system. Additionally, the minimal vapor pressure of ILs will not create air quality concerns post-filtration, which is an added benefit compared to other possible chemical modifiers.

To test the feasibility of ionic liquids to capture warfare toxins, dimethyl methylphosphonate (DMMP) was used as a model simulant. Equilibrium absorption capacities for DMMP were measured in neat ILs to identify basic structure-property relationships between absorption and IL anions and cations. Imidazolium, pyridinium and cholinium-based ILs with dicyanamide and bis(trifluoromethylsulfonyl) imide anions were investigated. It was found that both the anion and cation impacted the equilibrium absorption capacity. Promising ILs were tethered to activated carbons (AC) to investigate DMMP adsorption on to AC-ILs. AC-ILs were synthesized by physical adsorption of the IL to the AC from a solvent, followed by drying under vacuum. Characterization of the AC-ILs was performed using X-ray Photoelectron Spectroscopy (XPS) and thermal analysis techniques. Adsorption of DMMP on the AC-ILs was performed to determine breakthrough and equilibrium capacities.

489617 Preparation and Its Excellent $CO_2/CH_4/N_2$ Adsorption Selectivity of Novel Glucose-Based Adsorbents (C-GLC) with High BET Surface Area

Abstract:

Preparation and Its Excellent $CO_2/CH_4/N_2$ Adsorption Selectivity of Novel Glucose-based Adsorbents (C-GLC) with High BET Surface Area

Xingjie Wang, Binqin Yuan, Xin Zhou**, Qibin Xia, Zhong Li*

School of Chemistry and Chemical Engineering, South China University of Technology, Guangzhou 510640, PR China

*Corresponding author e-mail address: cezhli@scut.edu.cn

**Corresponding author e-mail address: xinzhou@scut.edu.cn

Abstract

In this work, we used starch sugar£¨e.g. glucose£©, a relatively inexpensive and renewable source, as carbon source to prepare glucose-based carbons (C-GLCs) materials with high specific surface area and pore volume. The resultant C-GLCs were characterized, and the $CO_2/CH_4/N_2$ adsorption isotherms of C-GLCs were measured at different temperatures separately. The isosteric heats of $CO_2/CH_4/N_2$ adsorption on the samples were estimated. IAST theory was also applied to predict the selectivity of the samples toward binary gas mixture. Results showed that BET area and pore volume of C-GLC-4 were up to 3153 m^2/g and 2.056 cm^3/g, respectively. C-GLCs possessed not only excellent stability, but also super-high CO_2 adsorption capacity up to 22.4 mmol/g at 30 bar, much higher than many metal-organic frameworks (MOFs). More importantly, C-GLC-4 possessed excellent adsorption selectivity toward CO_2/N_2 and CO_2/CH_4 binary mixtures, which increased with pressure. For CO_2/N_2 binary mixture, its IAST-predicted selectivity at 30 bar reached as high as 20 for CO_2/N_2 (0.5:0.5) mixture and 66 for CO_2/N_2 (0.15:0.85) mixture separately. For CO_2/CH_4 binary mixture, its IAST-predicted selectivity at 30 bar reached as high as 4.5 for CO_2/CH_4 (0.5:0.5) mixture and 27 for CO_2/CH_4 (0.1:0.9) mixture separately. The multiple cycles of CO_2 adsorption-desorption experiment showed that the C-GLC-4 has excellent regenerability. C-GLCs is a promising adsorbent for the separation of CO_2/N_2 and CO_2/CH_4 mixtures

Keywords: Glucose-based carbon, CO_2 adsorption, Isotherms, $CO_2/CH_4/N_2$ selectivity, Isosteric heats

Fig. 1 SEM images of C- GLCs

Fig. 2 $CO_2/CH_4/N_2$ adsorption isotherms of C-GLC-4 at 298 K

487919 Hybrid of Metal-Organic Framework and Ionic Liquid As Adsorbent for Adsorptive Separation of Acetylene and Ethylene

Abstract:

During the process of petroleum and natural gas cracking to produce ethylene, one of the most industrially important raw chemicals, a small amount of acetylene (C_2H_2) as a by-product of about 1% is unavoidably generated, and will take serious detriment during the process of producing polyethylene. The commercial methods of removing C_2H_2 from C_2H_4 include partial hydrogenation and solvent extraction are costly, energy-intensive, and have several drawbacks. As an outstanding alternative for gas separation, MOFs, as well as functionalized MOFs by post-synthetic modification, have been tentatively utilized for removal of C_2H_2 from C_2H_4, but the adjustability of adsorption capacity and selectivity is not flexible enough to satisfy different demands from chemical industries. Therefore, Ionic liquids (ILs) catch our attention due to their unique properties, such as high thermal and chemical stability, negligible vapor pressures and high designability as a result of nearly unlimited number of possible combinations of different anion-cation pairs.

Herein, we first reported an IL@MOF composite by incorporating ILs into a mesoporous MOF material for efficient separation of C_2H_2 and C_2H_4. It was well established that the selectivity of C_2H_2/C_2H_4 was greatly improved from 3.0 to 9.6 through the encapsulation of IL, at the meantime it maintained an adequate adsorption capacity for C_2H_2. Furthermore, the potential of industrial practice was examined by breakthrough and regeneration experiments. It not only satisfies the industrial production of removal of low level of acetylene from ethylene, but also is notably stable during the adsorption-desorption process. The high designability of ILs combined with richness of MOFsâ€™ structures exploits a novel blueprint for gas separation.

498236 Binary Gas Adsorption Equilibria on MIL-53(Al)

Abstract:

Flexible MOFs are interesting class of materials which alter their structures based on external stimulus. MIL-53(Al) is a well-studied flexible MOF that exhibits structural transformation between narrow pore (np) and large pore (lp) forms depending on suitable external stimulus/history. The phase diagrams for transformation between the two phases at various temperature/pressure/adsorbate concentrations are well documented in literature. The two pore forms have widely different adsorption characteristics for adsorption of pure gases; the np form of the framework exhibits higher capacity and selectivity to CO_2 compared to gases such as N_2 and CH_4. Literature for extensive binary data on this MOF is scare in literature. In this work, we present the measurement and analysis of binary gas adsorption equilibria of CO_2/N_2 and CO_2/CH_4mixtures on np and lp pore forms of the MIL-53 (Al) framework.

499654 Recent Advancements in Reactive Electrochemical Membrane Development for Water Treatment

Abstract:

Membrane filtration and advanced oxidation processes (AOPs) are two technologies that are being increasingly used in water treatment applications. These methods are often used in tandem to treat impaired water supplies or to facilitate water recycling. However, both of these technologies are often characterized with high capital and operating costs and a large reactor footprint. The development of more compact, less expensive hybrid technologies could significantly contribute to water treatment applications on many levels. In this talk I will discuss our ongoing research efforts aimed at the synthesis and characterization of novel, conductive reactive electrochemical membranes (REMs) that both filter and electrochemically destroy water contaminants. The REMs consist of conductive ceramic substoichiometric titanium dioxide (Ti_nO_{2n-1}) membranes and have shown exceptional promise for the removal of a broad range of contaminants, including nitrate, ammonia, and various organic contaminants and pathogens. A combination of electrochemical oxidation/reduction experiments, electrochemical impedance spectroscopy measurements, and density functional theory modeling was used to develop a mechanistic understanding of the interaction of water contaminants and membrane foulants at the REM. Results from this work are being used to develop more efficient and compact treatment technologies, including â€˜self cleaningâ€™ membranes and multi-functional point-of-use water treatment devices.

502453 Rationale Design of Ion-Exchange Membranes for Low Energy Brackish Water Desalination Via Membrane Capacitive Deionization

Abstract:

The energy-water nexus, a 21st century challenge, represents a symbiotic relationship as it takes a lot of energy to purify water and a significant amount of water is needed to produce energy. Prudent manage of these precious resources is paramount for a sustainable planet. Today, numerous industrial processes, such as thermal electric power plants, consume 40% of fresh water withdrawals within the United States.[1] Industry demand is often met by treating brackish water streams, which are about 5,000 ppm in total dissolved salts (TDS), and the current annual energy consumption in the U.S. amounts to 2.7×10^{11} kWh annually. Based on thermodynamic analysis constrained to a 50% minimal water recovery, todayâ€™s state-of-the-art for water desalination of brackish water streams is 20% efficient. Improving the efficiency from 20% to 30% could amount to a savings of $4 billion dollars worth of electricity annually in the U.S. and a reduction of energy needs by 20 GW. One technology capable of providing low energy water desalination of brackish water streams with over 30% thermodynamic efficiency is membrane capacitive deionization (MCDI).

MCDI, an electrochemical technology, offers the enticing prospect of low energy water desalination as it can recover energy (demonstrated up to 83%)[2] when discharging the saturated electrodes as a supercapacitor. Most research today has focused on porous carbon electrodes to increase MCDI capacity. However, material innovation with regard to the ion-exchange membranes used in MCDI has been neglected. Todayâ€™s MCDI currently uses electrodialysis (ED) membranes â€“ not originally designed for MCDI. The ion-exchange membranes for ED tend to be thick as they serve as a separator in addition to an ion-conducting electrolyte in electrodialysis. In MCDI, the ion-exchange membranes do not serve as separator and they are employed to prevent co-ion adsorption to the porous carbon electrodes.

This talk will demonstrate how thinner ion-exchange membranes with higher ionic conductivity substantially reduces the energy footprint for MCDI - approximately 35 kT per ion removed with ED membranes to 15 kT per ion removed with new ion-exchange membranes. Additionally, the new ion-exchange membranes displayed a significantly higher Coulombic efficiency. The talk will showcase how ex-situ properties of ion-exchange membranes, such as ionic conductivity, thickness, and permselectivity, are instrumental for lowering the energy footprint of MCDI.

References:

[1] D. Bauer, M. Philbrick, B. Vallario in The Water-Energy Nexus: Challenges and Opportunities, **2014**.

https://energy.gov/sites/prod/files/2014/07/f17/Water%20Energy%20Nexus%20Full%20Report%20July%202014.pdf

[2] P. Dlugolecki, A. van der Wal *Environmental Science & Technology*. **2013**, *47*, 4904-4910.

500393 Inkjet Printing of Charge Mosaic Membranes As a Method for Tailored Aqueous Separation

Abstract:

Charge mosaic membranes are capable of separating dissolved ions from neutral molecules of comparable size by facilitating the transport of charged species through oppositely charged domains. We have developed a new method for producing charge mosaic membranes in a scalable manner by ink-jet printing polymeric materials onto a structural template. Using a striped pattern, positively charged and negatively charged polyelectrolytes are alternatively printed to form the mosaic. Printing of nanomaterials allows for facile design and control over microstructure. This microstructure, in turn affects overall membrane performance. To demonstrate the versatility of this method, we printed membranes with different polyelectrolytes and printing conditions. Enrichment of small ionic salts (i.e. KCl) was first observed using this robust, reliable, and rapid method. Printing enabled us to extend this application to both acetic acid extraction and functionalization for pH-responsive salt separation. The contribution of interactions beyond electrostatics was investigated by printing with polyelectrolytes of varying solubility parameters and challenging the membrane with acetic acid at different concentrations and feed solution pH. To observe pH-responsive salt permeation, membranes were printed with a weak polyelectrolyte while the other domain was printed with a strong polyelectrolyte whose ionization does not vary with pH. The ability to easily tailor membranes for very different separations is what makes inkjet printing such a unique and attractive method for fabricating membranes.

499350 Ion Specific Effects in Charged Polymer Membranes for Water Purification

Abstract:

Global need for sustainable water and energy sources presents an opportunity for polymer science as many technologies for water purification and some emerging power generation technologies rely on polymeric membranes to control rates of water and ion transport. Ion exchange membranes are used in many of these technologies, and the presence of charged groups on the polymer backbone of such materials influences ion sorption and, subsequently, both rates of ion transport and the selective nature of ion transport. Membranes are often characterized using sodium chloride, which is widely prevalent in nature, but several applications expose membranes to a variety of different ions. For electric field-driven processes, ion transport properties, such as ionic resistance and electrochemical permselectivity, are sensitive to the specific nature of the ions present in the system. This presentation will discuss the influence of ion properties and ion-polymer interactions on the ion transport properties of polymer membrane materials. The observed ion specific effects can be understood in terms of both the thermodynamic and kinetic (mobility) contributors to the overall ion transport process.

511518 Polymeric Ion Pumps: Using an Oscillating Stimulus to Drive Solute Transport in Reactive Membranes

Abstract:

The development of membranes that separate molecules based on chemical factors, rather than physical factors, is one promising approach to meeting the demand for highly-permeable membranes that are more selective. In this study, the design of multifunctional, pH-responsive membranes that selectively pump a target solute is detailed. The membranes consist of a gate layer made from an amine-functionalized copolymer and a reactive matrix lined by iminodiacetic acid groups, which bind divalent cations reversibly. These chemistries exhibit concurrent changes in the cation binding affinity and gate permeability in response to the pH value of the surrounding solution such that when they are exposed to an oscillating pH, the combination drives a facilitated transport mechanism that mimics ion pumps. In mixed solute systems, calcium permeated through the membrane four times faster than sucrose in the presence of an oscillating pH even though the solutes possess similar hydrodynamic sizes and permeated through the membrane at the same rate when the pH value was constant. The development of the polymeric ion pumps was guided by a model that provided several critical insights. First, the solute binding capacity and thickness of the membrane define the asymptotic limit for enhanced selectivity. Second, the maximum enhancement in selectivity is realized in the limit of infinitely rapid oscillations. The multifunctional membranes discussed here provide a platform for the development of processes that can fractionate molecules of similar size but varying chemistry.

499790 Thermodynamic Modeling of Ion Exchange Membranes with Enrtl Model

Abstract:

Ion Exchange Membranes are of great importance in the fields of separations, energy storage, electrochemical reaction et al. [1, 2] They are a special type of polyelectrolytes that have charged functional groups distributed on the backbone polymers. The charged functional groups are electrostatically attached with counterions that carry opposite charges. The counterions are able to dissociate in solutions. The interactions between counterions and other particles in the system might cause high nonidealities of mobile ions' behavior. Such nonidealities have often been ignored in membrane studies. [3] And there is no successful modeling work dealing with such nonidealities in membrane systems yet. Kamcev tried to calculate the activity coefficients of mobile ions in the highly charged membranes with Manning's limiting law, however, it does not work for less charged polymers and has not been examined beyond certain concentrations of the electrolyte solutions. [3, 4]

This work is able to model mobile ions in membranes in equilibrium with salt solutions. Manning's limiting law is used to capture the "point-to-line" electrostatic interaction between mobile ions and charged functional group, as well as counterion condensation. Counterion condensation is a phenomenon happens in highly charged polymer solutions where counterions attach to the polyion to reduce its charge density to a critical value. [4] The Pitzer-Debye-Hückel formulation is applied to account for the "point-to-point" electrostatic interactions between each charged particles in the solution. And the eNRTL theory is used to deal with the local short-range interactions between all molecular and ionic species. [6]

Sulfonated crosslinked hydrogel membranes, with Ion Exchange Capacity (IEC) values ranging from 0.01 to 1.93 meq/g(dry polymer) and water uptake levels ranging from 0.2 to 1.8 g(water)/g(dry polymer) are studied in this work. They were synthesized by UV-crosslinking. 2-acrylamido-2-methylpropane sulfonic acid (AMPS) was copolymerized into a hydrogel matrix based on poly (ethylene glycol) diacrylate (PEGDA). Anion sorption was measured using a desorption technique and ion chromatography and cation sorption was measured using a polymer ashing technique and atomic absorption spectrophotometry [7].

Dimensionless charge density of polymers and eNRTL short-range parameters are required to describe the membrane systems. The charge density of polymers can be either determined by the structure of the polymers or fitted from experimental data, while eNRTL short-range parameters have to be identified from data for species of significant concentrations. For membranes that have only one type of monomer with charged functional groups, the short-range parameter between water and the counterion-polyion ionic pair is necessary. In addition, the pair parameter between simple salts and the polyelectrolyte is needed when salt concentration is high. In a more interesting case where there are non-charged segments in the polymer, the neutral segments are treated as second solvent and its short range parameters with water have to be well identified.

This model is able to correlate and extrapolate the concentration of mobile ions in membranes equilibrated

in external brine solutions. With limited number of parameters, this model represents the experimental data of different salt concentrations and IECs very well, with much higher accuracy compared to using Manning's limiting law alone.

1. Jeon, Sung-il, et al. "Desalination via a new membrane capacitive deionization process utilizing flow-electrodes." Energy & Environmental Science 6.5 (2013): 1471-1475.
2. Cai, Zhijun, et al. "High performance of lithium-ion polymer battery based on non-aqueous lithiated perfluorinated sulfonic ion-exchange membranes." Energy & Environmental Science 5.2 (2012): 5690-5693.
3. Kamcev, Jovan, Donald R. Paul, and Benny D. Freeman. "Ion activity coefficients in ion exchange polymers: applicability of Manning's counterion condensation theory." Macromolecules 48.21 (2015): 8011-8024.
4. Kamcev, Jovan, et al. "Partitioning of Mobile Ions Between Ion Exchange Polymers and Aqueous Salt Solutions: Importance of Counter-ion Condensation." Physical Chemistry Chemical Physics (2016).
5. Manning, Gerald S. "Limiting laws and counterion condensation in polyelectrolyte solutions I. Colligative properties." The Journal of Chemical Physics 51.3 (1969): 924-933.
6. Song, Yuhua, and Chau-Chyun Chen. "Symmetric electrolyte nonrandom two-liquid activity coefficient model." Industrial & Engineering Chemistry Research 48.16 (2009): 7788-7797.
7. G.M. Geise, L.P. Falcon, B.D. Freeman, D.R. Paul, Sodium chloride sorption in sulfonated polymers for membrane applications, Journal of Membrane Science, 423-424 (2012) 195-208.

484243 Ion Partitioning Between Brines and Ion Exchange Polymers

Abstract:

Ion partitioning between brines and ion exchange polymers

Michele Galizia[â€¡], Donald R. Paul, Benny D. Freeman

J. McKetta Department of Chemical Engineering, 200 E. Dean Keeton Street, 78712 Austin,

and Center for Energy and Environmental Resources, 10100 Burnet Road, 78758 Austin,

The University of Texas at Austin, TX (USA)

â€¡ *Permanent address: Department of Chemical, Biological and Materials Engineering, 100 E. Boyd St., 73019 Norman, University of Oklahoma, OK (USA)*

For long time, the validity of the Donnan model [1] to describe ion partitioning between electrolyte solutions and ion exchangers was questioned [2]. As recently reported in the literature [3], quantitative failure of Donnan model in predicting ion partitioning between diluted electrolyte solutions and charged polymers arises from neglecting non-idealities in the polymer phase. When a charged polymer is equilibrated with diluted salt solutions, ion activity coefficients in the membrane do not match those in the contiguous, external solution, so corrections for membrane non ideal behavior are needed. However, when salt concentration in the external solution exceeds 1 mol/L for NaCl and 0.1 mol/L for $CaCl_2$, ion activity coefficients in the membrane and in solution rapidly collapse on one another. In these conditions, the environment in the membrane becomes thermodynamically similar to that in the external solution. Indeed, since most of fixed charges are screened by sorbed counter-ions, the membrane environment is essentially made from water and ions, while polymer chains just slow down ion diffusion due to tortuosity/area effects.

In this study, a combined experimental and theoretical approach was used to test this hypothesis. Specifically, a fundamental, quantitative study of NaCl and $CaCl_2$ partitioning between a cation exchange membrane based on a cross-linked poly(p-styrene sulfonate-co-divinylbenzene) and concentrated brines is reported. In these conditions, the experimentally determined ion activity coefficients in membrane match those in the contiguous, external solution, and the Donnan model provides an accurate prediction of co-ion and counter-ion concentration in the polymer.

Experimental findings were interpreted using the recently developed Donnan-Manning model [3], which permits to track the change in polymer structure as a function of the external salt concentration, and build structure-property correlations for ion sorption in commercial membranes for water desalination and power generation applications.

[1] F.G. Donnan, Chem. Rev. 1924 **1** 73

[2] E. Glueckauf et al., Nature 1961 **191** 904

[3] J. Kamcev et al., PCCP 2016 **18** 6021

THE UNIVERSITY *of* EDINBURGH

SCCS

Towards multiscale screening of porous materials for carbon capture

<u>S. Krishnamurthy</u>, A. H. Farmahini, D. Friedrich, S. Brandani and L. Sarkisov

School of Engineering, University of Edinburgh, UK

s.krishnamurthy@ed.ac.uk

Presentation for the
2017 AIChE Annual Meeting, Minneapolis, USA
31/10/17

Introduction

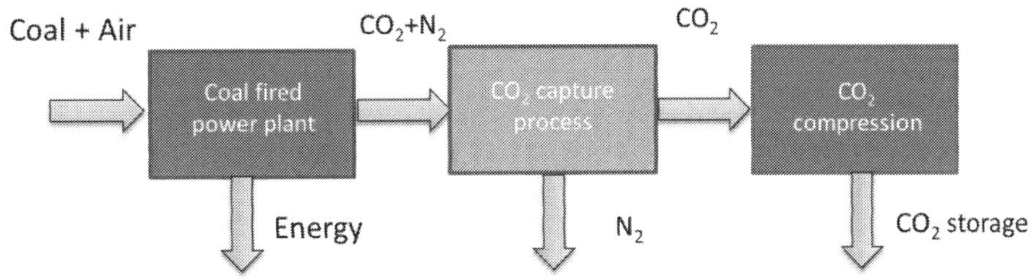

Typical flue gas composition 15% CO_2, 85% N_2 assuming dry flue gas

Challenge: Capturing 90% CO_2 at 95% purity with minimum energy penalty

SCCS

Adsorption processes for CO_2 capture

- Two modes
  - Pressure/Vacuum swing adsorption (PSA/VSA)
  - Temperature swing adsorption (TSA)
- TSA
  - Long cycle times affect throughput
- PSA
  - Pressurizing flue gas with large N_2 content not attractive
- Vacuum swing adsorption (VSA)
  - Adsorption at atmospheric pressure and desorption at vacuum
  - Current benchmark adsorbent: Zeolite 13X

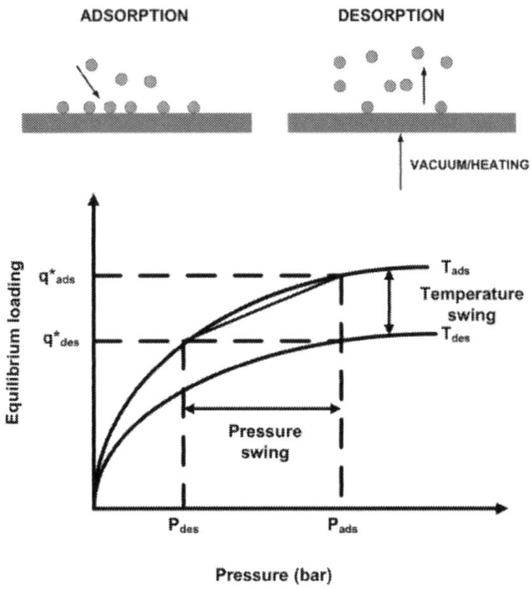

SCCS

Adsorbents for CO_2 capture

Adsorbents

- Zeolites [1,2]
- Carbons [3,4]
- Metal organic frameworks (MOFs)[5,6]
- Zeolite Imidazolate frameworks (ZIFs)[7]
- Amine impregnated Silica[8]
- Porous polymer networks (PPNs)[9]
- Coordinated organic frameworks (COFs)[10]

How do we select the right material for carbon capture processes ?

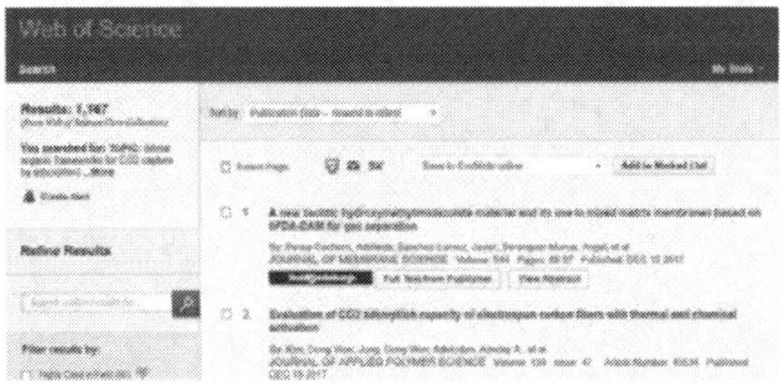

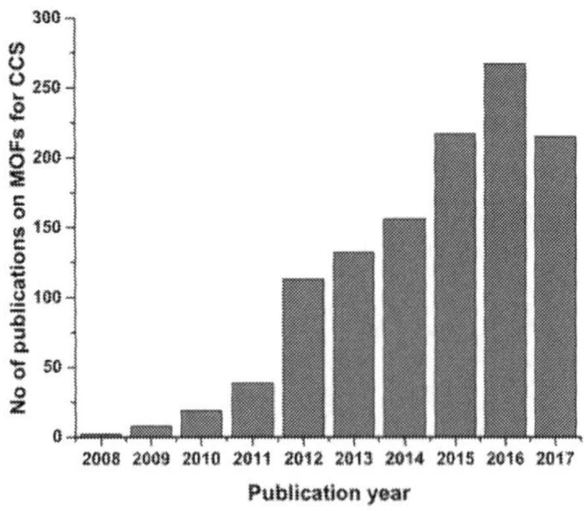

Source : Web of science database

1. Lee et al., J. Chem. Eng. Data 2002,47, 1237-1242
2. Pham et al., Langmuir, 2013, 29 (2), 832–839
3. Ribero et al., J. Chem. Eng. Data, 2008, 53 (10), 2311–2317
4. Shirani and Eic, Can. J. Chem. Eng. 94, 2016 2023–2034
5. B. Pato-Doldán, et al., ChemSusChem 2017, 10, 1710-1719.
6. Queen et al., Chem Sci 2014, 5, 4569-4581
7. Zhang et al., J. Phys. Chem. C, 2014, 118 (17), 8788–8794
8. Belmabkhout et al. Chem Eng Sci 65 (2010) 3695–3698
9. Lu et al., J.Phys.Chem.C 2013, 117, 4057–4061
10. Zheng et al., Adv. Mater. **2016**, 28, 2855–2873

SCCS

Material ranking : Adsorption isotherms

Experiments

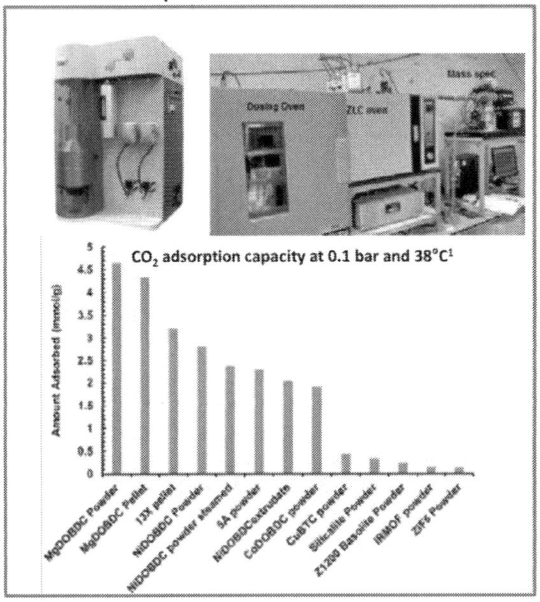

Molecular simulations

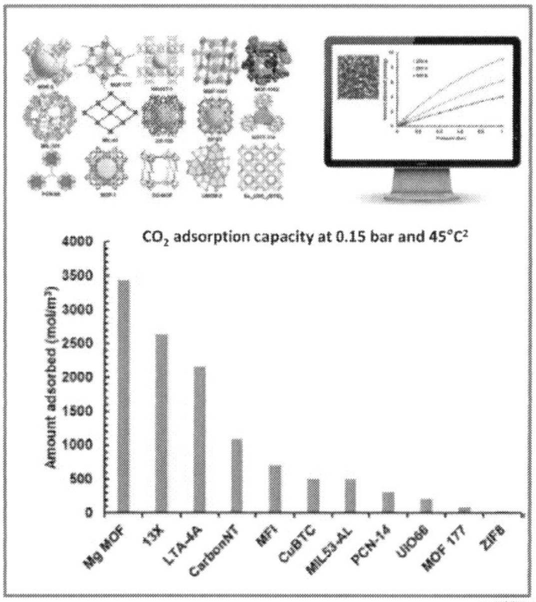

1. Hu et al., Ind. Eng. Chem. Res. 2015 54 6772–6780
2. Bahamon, D. Vega, L.F. Chemical Eng J 2016 284 438–447

SCCS

Process based material ranking

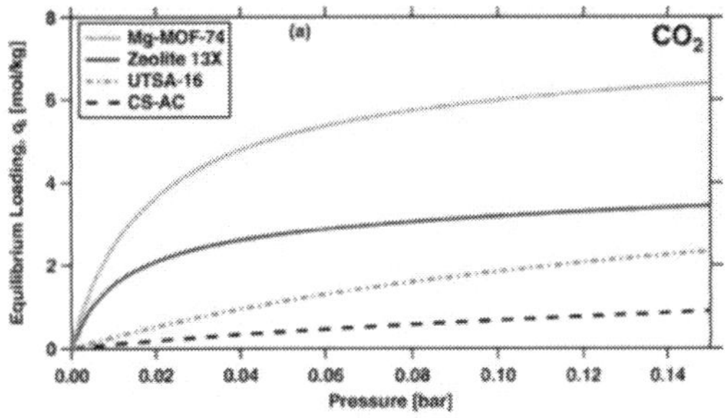

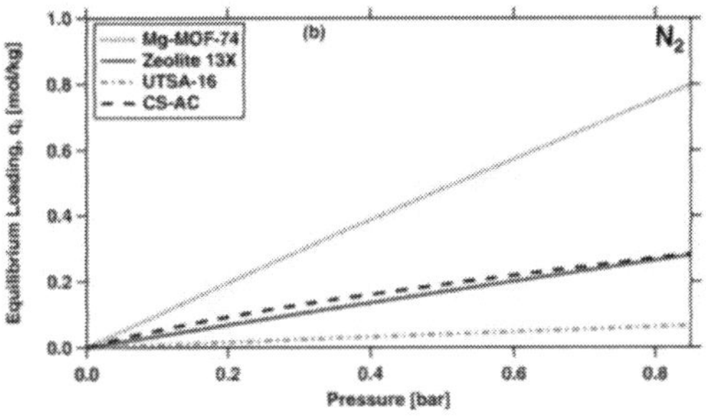

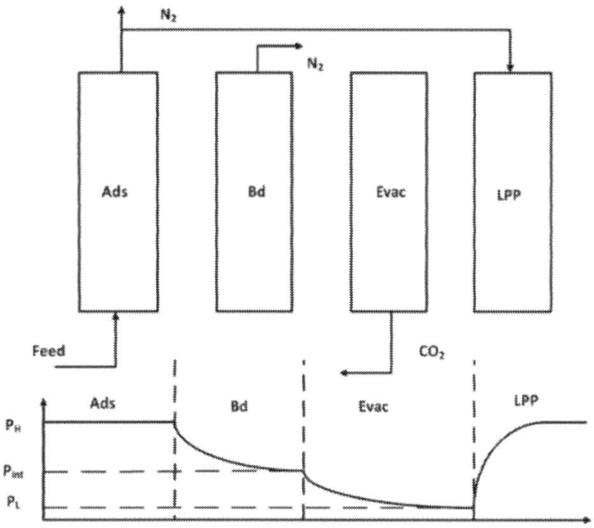

SCCS

Rajagopalan, A.K, Avila, A.M, Rajendran, A.. *Int J Greenhouse Gas Control.* 2016;46:76-85.

Process based material ranking

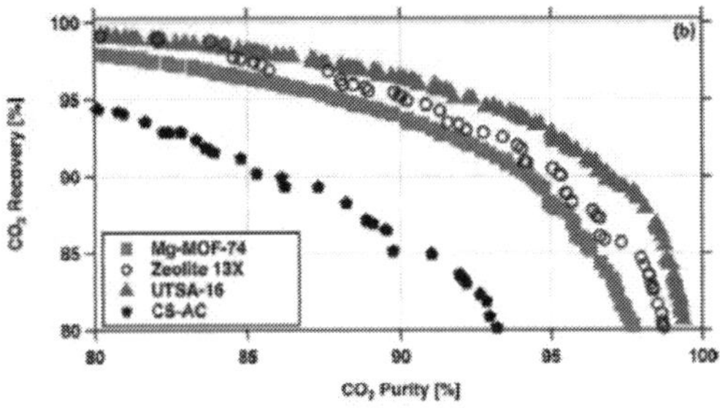

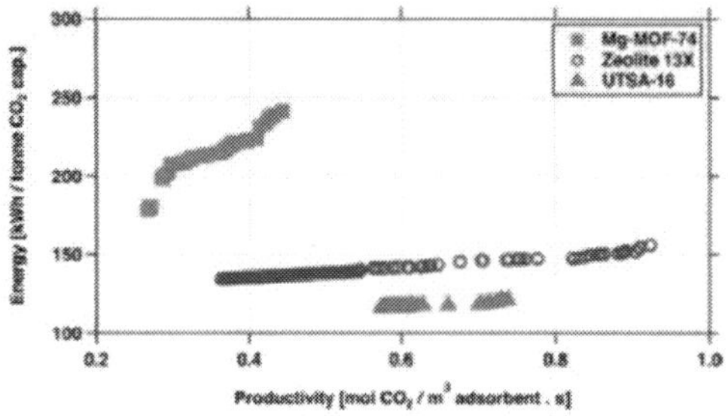

Maximise Purity & Recovery

Maximise Productivity
Minimize Energy

Main conclusions from this work

1. Detailed process optimization provides a more reliable picture than ranking materials based on adsorption capacity
2. Nitrogen adsorption plays an important role in determining the process performance

SCCS

Rajagopalan, A.K, Avila, A.M, Rajendran, A.. *Int J Greenhouse Gas Control.* 2016;46:76-85.

Multiscale screening of adsorbents

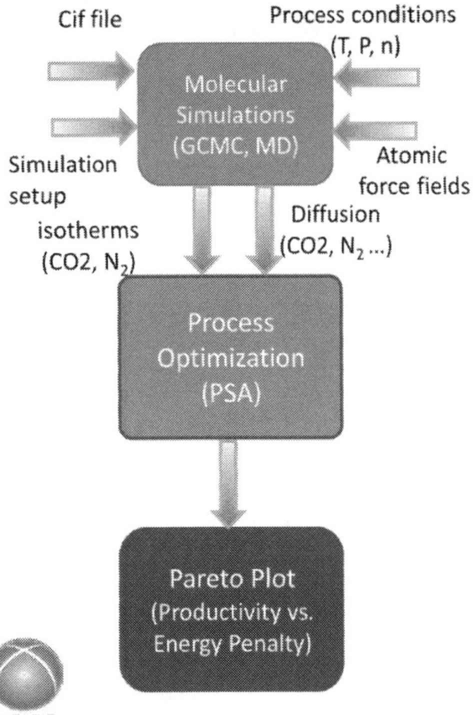

Scope of this work

- Predict isotherms on well known materials and compare with experimental data
- Process optimization with experimental and simulation isotherms
- Materials : Zeolite 13X and CPO-27-Ni

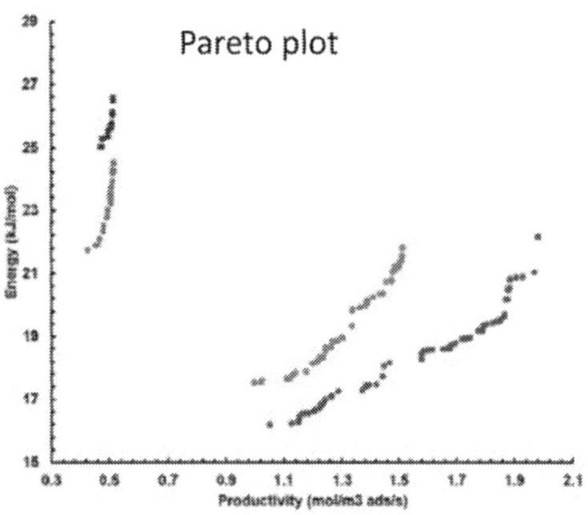

SCCS

Experimental isotherms: 13X

Zeolite 13X

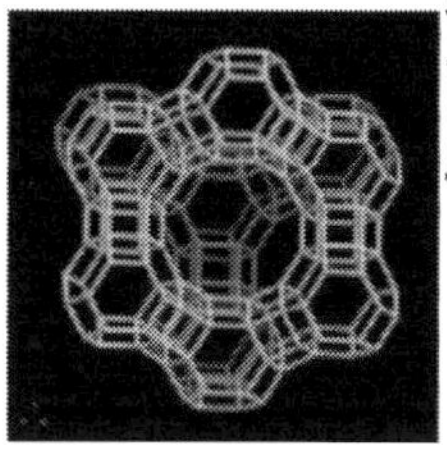

Source : IZA database

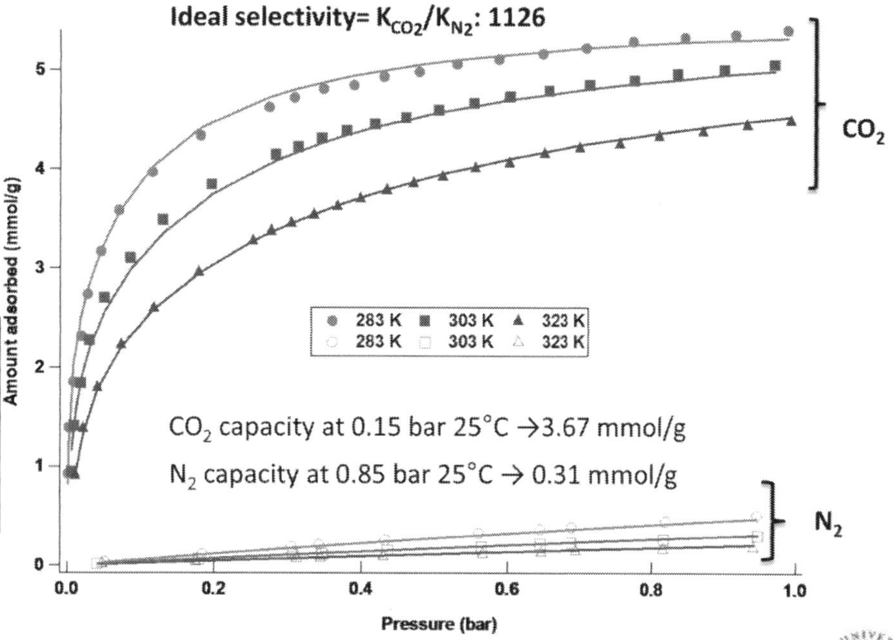

Experimental isotherms: CPO-27-Ni

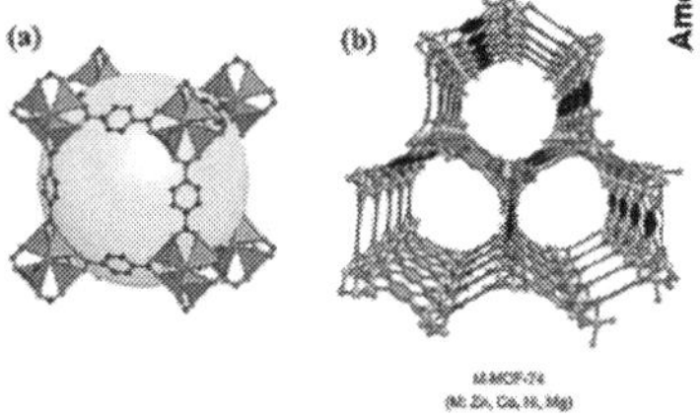

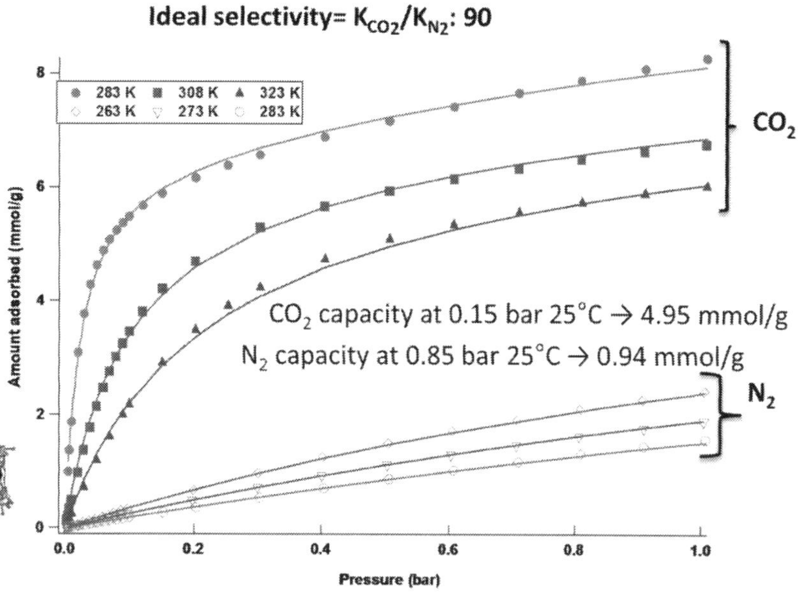

Liu et al., ***Chem. Soc. Rev.***, 2012,41, 2308-2322

Molecular simulations: Zeolite 13X

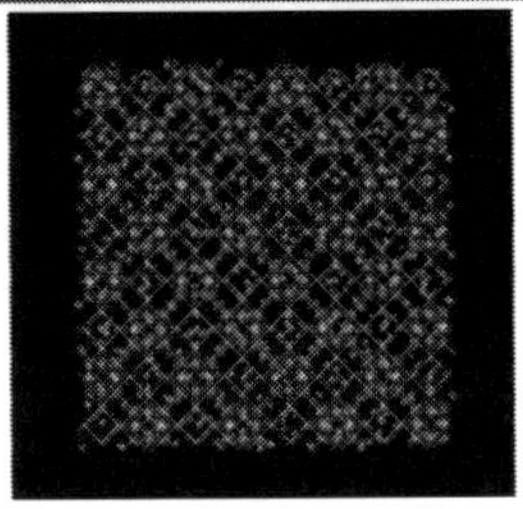

Zeolite 13X

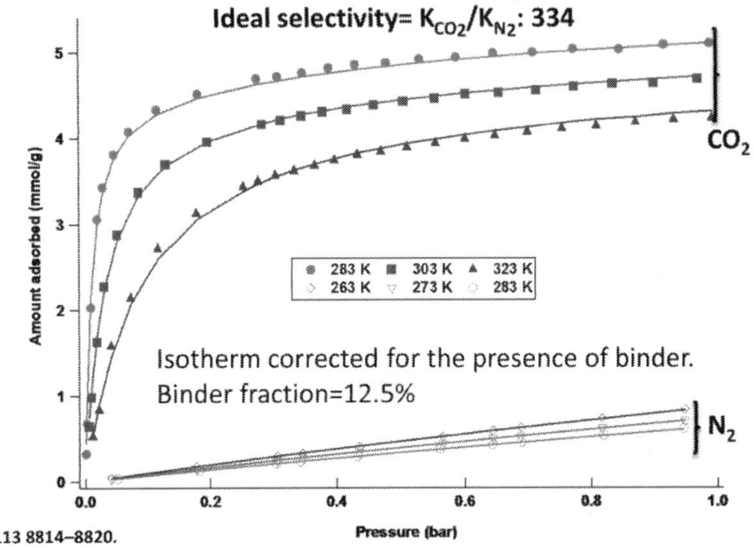

Crystal structure
D. H. Olson, Zeolites 15 1995 439-443.
Force fields
Garcia-Sanchez et al. J., Phys.Chem.C 2009 113 8814–8820.
Vujić, B., Lyubartsev, A. Modelling Simul. Mater. Sci. Eng. 2016
Jiang. J., Sandler, S. I., J. Am. Chem. Soc. 2005 127 11989-11997

SCCS

Experiment vs molecular simulation: 13X

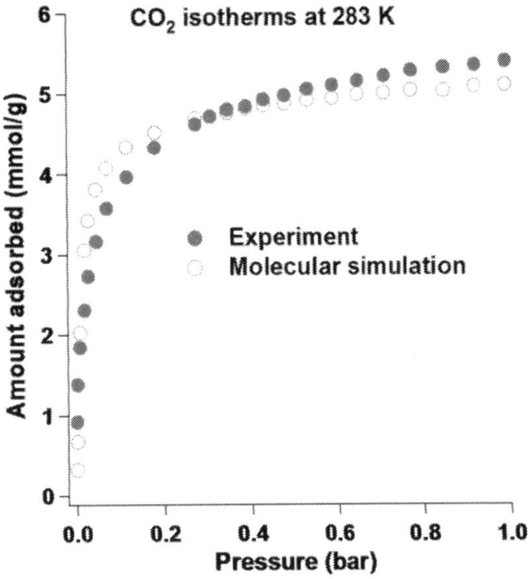

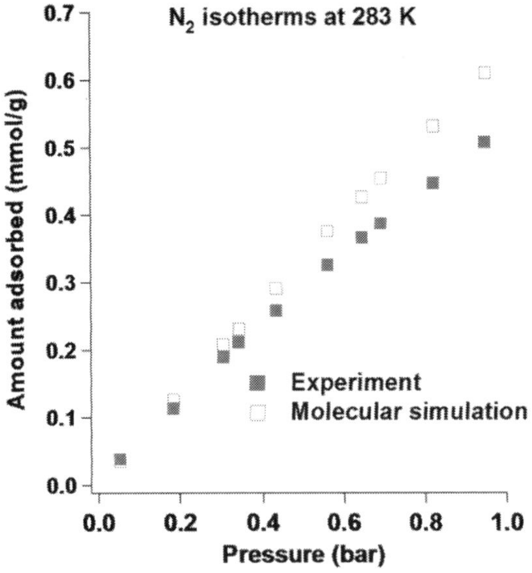

CO_2 capacity at 0.15 bar 25°C → 3.97 mmol/g

N_2 capacity at 0.85 bar 25°C→ 0.44 mmol/g

SCCS

Molecular simulation: CPO-27-Ni

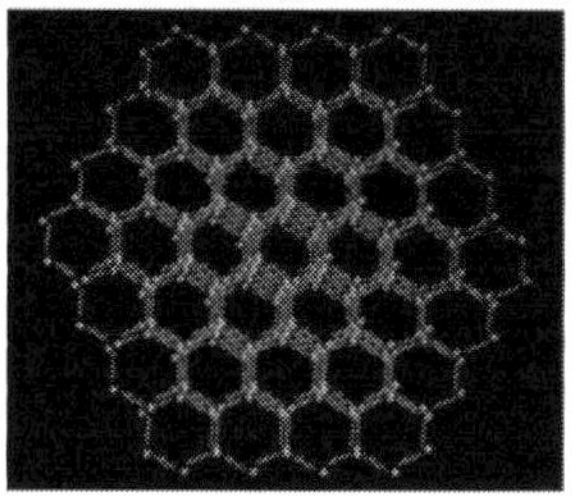

CPO-27-Ni

Ideal selectivity= K_{CO_2}/K_{N_2}: 145

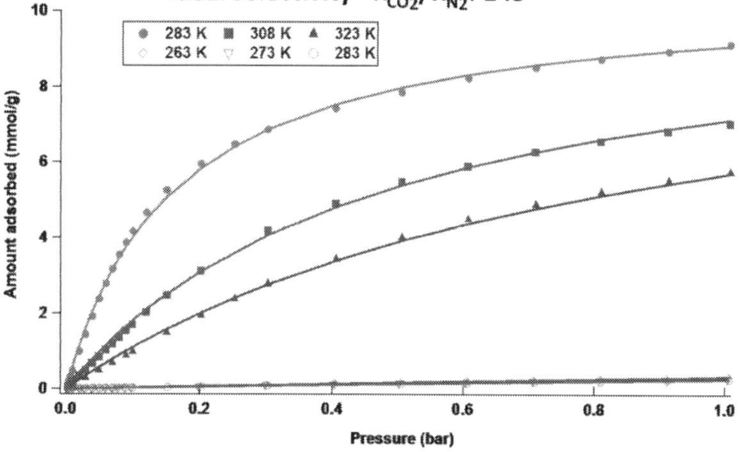

Crystal structure
Lee et al., *Chem. Mater.*, 2015 27 668–678.
Force fields
Mercado et al., *J. Phys. Chem C* 2016 120 12590–12604.
Bai et al., *J.Phys.Chem C* 2013 117 24375–24387
Rappe et al., J. Am. Chem. Soc. 1992 114 10024–10035

Isotherm corrected for the presence of binder.
Binder fraction=12.5%

SCCS

Molecular simulation: CPO-27-Ni

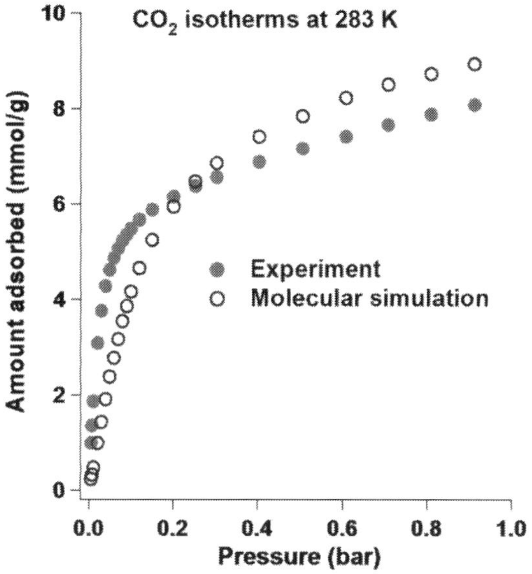

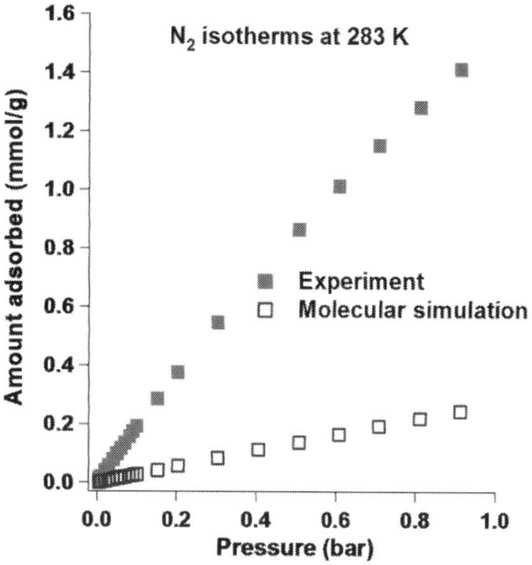

CO_2 capacity at 0.15 bar 25°C → 3.39 mmol/g

N_2 capacity at 0.85 bar 25°C → 0.19 mmol/g

SCCS

Process simulation

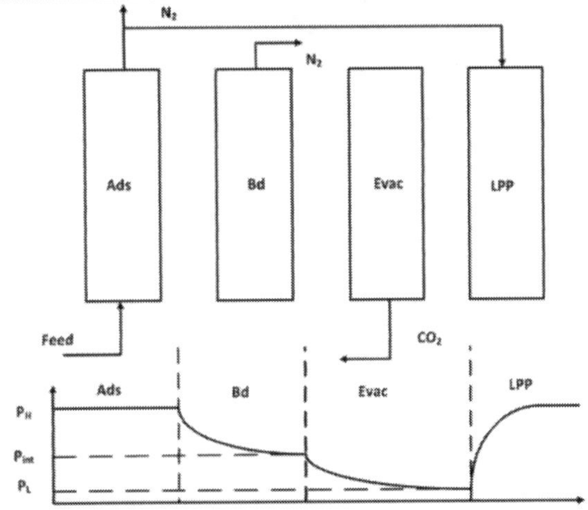

4-step VSA cycle with LPP [1,2]

- CySim adsorption simulator[3]
- Feed 15% CO_2 , 85% N_2 at 1atm and 25°C
- Adsorption isotherms
  - Pellet isotherms from experiment
  - Crystal isotherms from molecular simulations corrected for binder
- Same pellet size for the two materials
- Competitive dual-site Langmuir model for binary adsorption equilibrium

$$\mathrm{CO_2\ purity} = \frac{\mathrm{moles\ CO_2\ in\ evac}}{\mathrm{total\ moles\ evac}}$$

$$\mathrm{CO_2\ recovery} = \frac{\mathrm{moles\ CO_2\ in\ evac}}{\mathrm{moles\ CO_2\ in\ feed}}$$

$$\mathrm{Specific\ energy} = \frac{\mathrm{Compression+Evacuation\ energy}}{\mathrm{moles\ CO_2\ in\ evacuation}}$$

$$\mathrm{Productivity} = \frac{\mathrm{moles\ CO_2\ in\ evacuation}}{\mathrm{volume\ of\ adsorbent\ X\ cycle\ time}}$$

1. Haghpanah et al AIChE J (2013)
2. Krishnamurthy et al AIChE J (2014)
3. Dang et al., Energy Procedia (2013)

SCCS

Process optimization

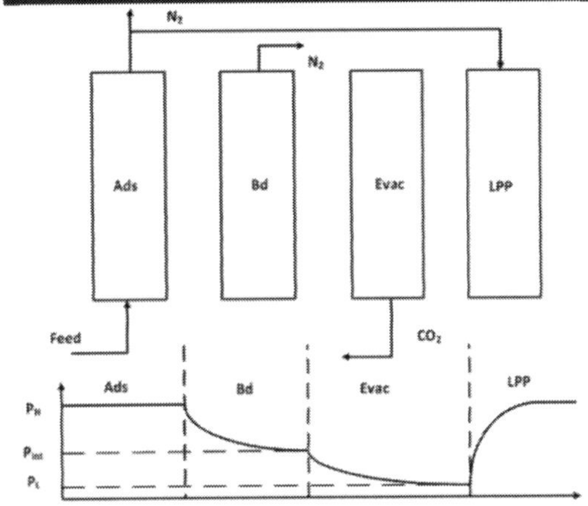

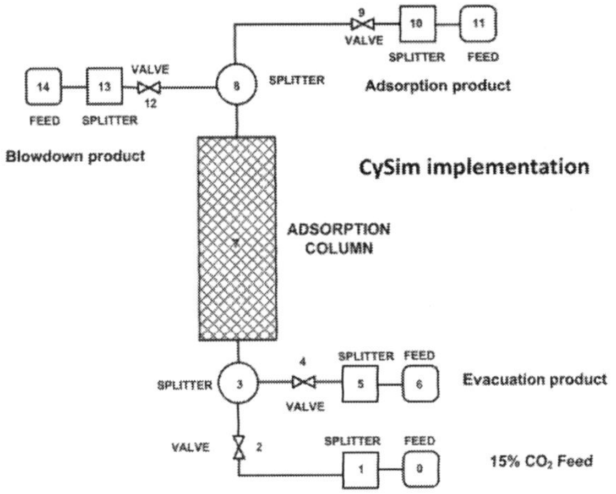

CySim implementation

Decision variables

Adsorption, blowdown, evacuation step times

Blowdown & evacuation pressures

Feed flow rate

valve coefficients of blowdown and evacuation steps

GA based optimization (NSGA-III)

Minimize energy and maximize productivity

Constraints : Purity > 95% Recovery >90%

SCCS

Performance analysis : 13X

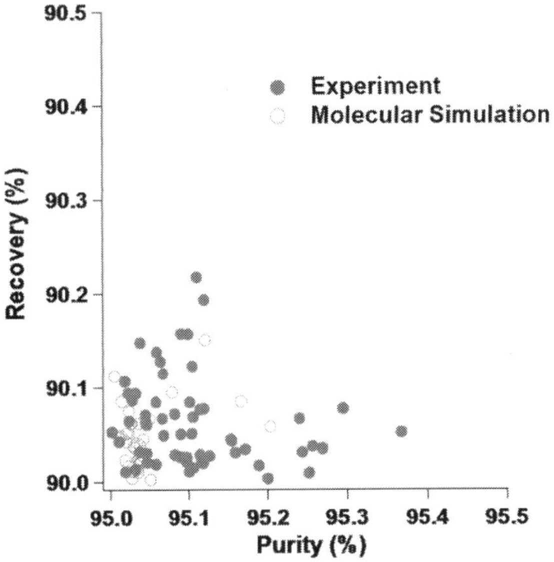

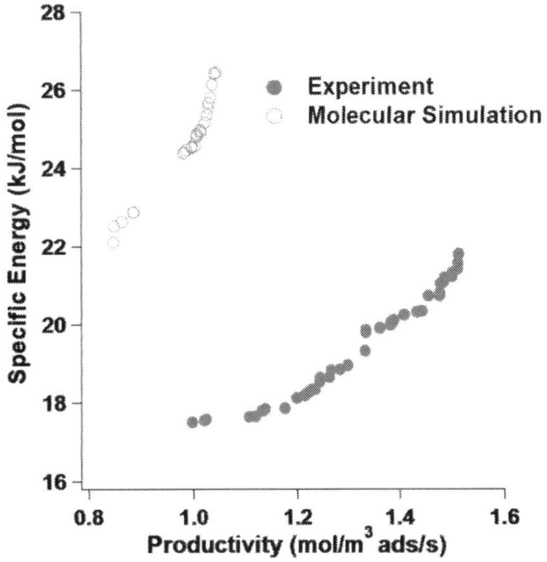

Minimum energy 17.51 kJ/mol (experiment isotherm) < 22.6 kJ/mol (simulated isotherm)

SCCS

Experiment vs Simulation: 13X

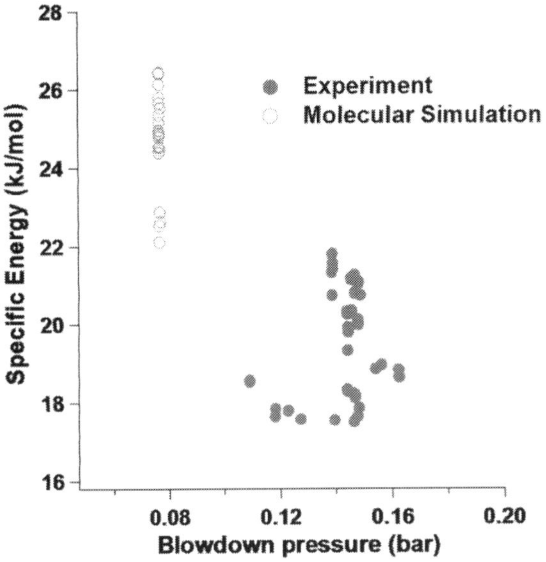

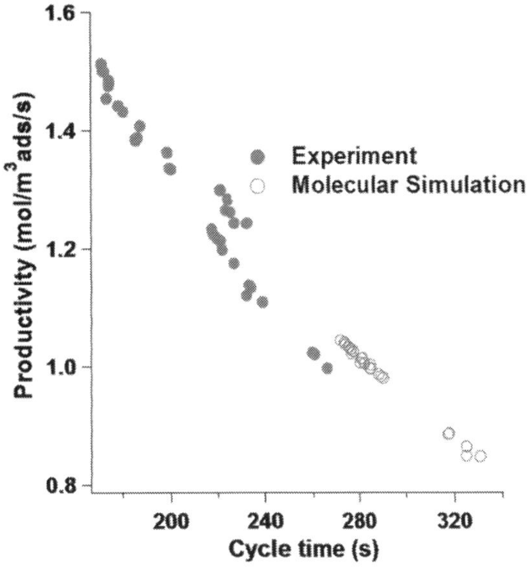

SCCS

Performance analysis: CPO-27-Ni

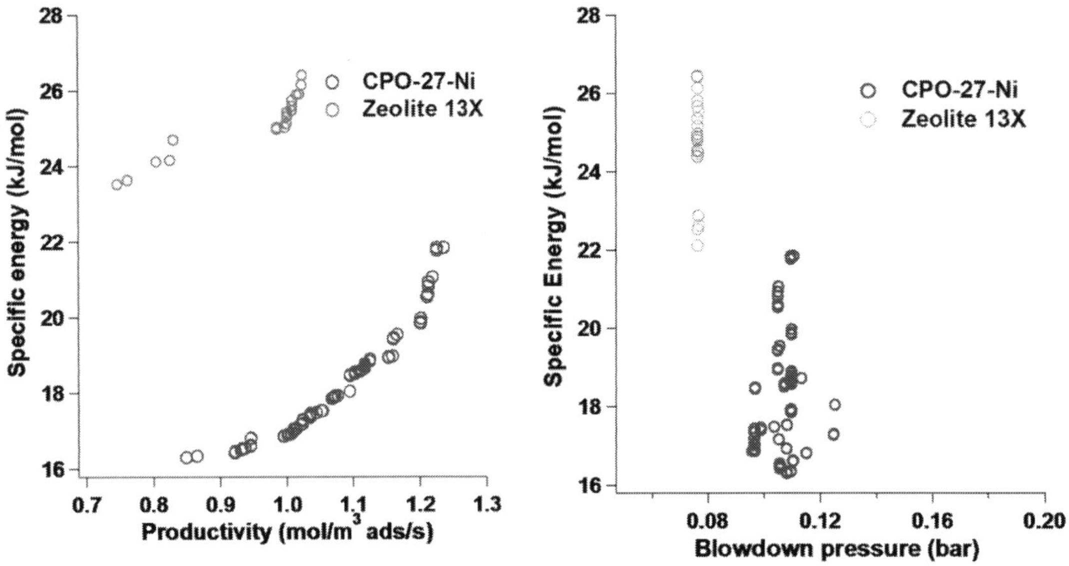

SCCS

95% Purity and 90% Recovery constraints not satisfied with experimental isotherm

Minimum specific energy 16.37 kJ/mol < 13X (22.6 kJ/mol)

Conclusions

- Multiscale screening of adsorbents from molecular scale to process optimization: first step in a toolbox for screening materials for CCS.
- Simulated isotherms compared with experimental isotherms:
  - Shape of CO_2 isotherm varied in both the samples
  - CO_2 capacity at 0.15 bar: over-predicted in 13X and underestimated in CPO-27-Ni
  - N_2 isotherms overestimated in 13X and underestimated in CPO-27-Ni
- Detailed process optimization carried out for two materials
- Prediction of N_2 isotherm has a profound impact on process performance
- Preliminary conclusion: accurate prediction of equilibrium properties is crucial for process optimization
- Yet, it still remains a substantial challenge for molecular simulations. Although predicted adsorption isotherms may look reasonable, they differ in essential features such as capacity in low pressure region
- Nitrogen adsorption largely ignored in simulations: needs to be addressed.

SCCS

Acknowledgements

EPSRC
Engineering and Physical Sciences
Research Council

Grant Ref : EP/N007859/1

SINTEF
Materials and Chemistry

17 AIChE
Annual Meeting, Minneapolis, MN

SCCS

501445 Evaluation of Diamine Appended Metal-Organic Frameworks for Post-Combustion CO_2 Capture Using a Vpsa

Abstract:

The recent invention of diamine appended metal organic frameworks [1] that show unusual step-shaped isotherms have opened up the possibility to tailor-make adsorbents for post-combustion CO2 capture. By varying the metal, it was possible to position the step at various pressure levelsl. The characteristic CO2 uptake of these materials indicate a possibility to obtain larger theoretical working capacities when compared to typical classical materials such as Zeolite 13X that show classical Langmuirian-type isotherms. The performance of five different materials with step-shaped isotherms was evaluated for Vacuum swing adsorption(VSA)-based post-combustion CO2 capture from a dry flue gas. In order to evaluate the true potential of these materials, a rigorous optimization based on detailed models that take into account mass+heat transfer, cycle configuration was performed. Using Genetic algorithms the operating space was throougly scanned to obtain the Pareto curves for Purity and Recovery for five diamine-appended MOFs. The optimization results revealed that some of these materials show promise as they are able to outperform the current benchmark material, i.e., Zeolite 13X. The optimization results also demonstrated that the flue gas temperature plays a critical role in deciding which of the MOFs should be used and provide intersting heuristics for choosing operating conditions depending on the type of the MOF and the flue gas temperature. Finally, detailed optimization to reduce parastic energy showed that some of the MOF s are capable of providing lower energy consumption compared to Zeolite 13X for CO2 purity> 95% and Recovery > 90%, making them attractive materials for further study.

1.MMcDonald, Thomas M., Jarad A. Mason, Xueqian Kong, Eric D. Bloch, David Gygi, Alessandro Dani, Valentina Crocella et al. "Cooperative insertion of CO2 in diamine-appended metal-organic frameworks." *Nature* 519, no. 7543 (2015): 303-308.

495090 Sub-Ambient Pressure-Swing Adsorption for Enhanced Post-Combustion Carbon Dioxide Capture Via Phase-Change Materials Application

Abstract:

Sub-ambient pressure-swing adsorption for enhanced post-combustion carbon dioxide capture via phase-change materials application

Héctor Octavio Rubiera Landa, Stephen J. A. DeWitt, Matthew J. Realff, Ryan P. Lively, and Yoshiaki Kawajiri.

School of Chemical & Biomolecular Engineering, Georgia Institute of Technology,

311 Ferst Drive N. W., Atlanta, GA, 30332-0100, USA.

As a transition step towards a carbon-free society, a number of process technologies are currently being investigated and developed. Amongst these, adsorption-based separations for carbon capture and storage (CCS) offer an attractive alternative for this global challenge **[1]**. Critical aspects to successful implementation of adsorption-based CCS technologies for post-combustion carbon capture include the application of high-performance adsorbents, suitable for high carbon dioxide recovery from flue gas streams.

As demonstrated experimentally in **[2]**, and analyzed via numerical simulations of breakthrough curves at sub-ambient conditions **[3]**, the application of a phase-change material (PCM) embedded in fiber adsorbents has the potential to enhance the performance of adsorption cycles by providing suitable heat management, therefore aiming at lower plant footprint and enabling a quasi-isothermal operation with optimal adsorptive capacity usage. In combination with tailored metal-organic frameworks operating at sub-ambient conditions, wherein significant adsorption of carbon dioxide occurs, 'PCM-assisted' cyclic adsorption processes promise to enable higher productivity adsorption systems.

The present work discusses firstly simple 'PCM-assisted' modeling and numerical simulation for sub-ambient pressure-swing adsorption cycles (PSA/VPSA), whereby a smooth-interface phase-change model is applied to represent melting/freezing transitions of the PCM, occurring by the intrinsic non-isothermal, periodic process operation. Next, as preliminary step for further process optimization work, we conduct a parametric study, useful in obtaining insight into the process dynamics governing 'PCM-assisted' PSA/VPSA cycles.

References

[1] J. Wilcox, R. Haghpanah, J. He, K. Lee, E. Rupp. Advancing Adsorption and Membrane-Based Separation Processes for the Gigaton Carbon Capture Challenge. *Annual Review of Chemical and Biomolecular Engineering* **2014**, *5(1)*, 479-505. http://dx.doi.org/10.1146/annurev-chembioeng-060713-

040100.

[2] R. P. Lively, N. Bessho, D. A. Bhandari, Y. Kawajiri, W. J. Koros. Thermally moderated hollow fiber sorbent modules in rapidly cycled pressure swing adsorption mode for hydrogen purification. *International Journal of Hydrogen Energy* **2012**, *37(20)*, 15227-15240. http://dx.doi.org/10.1016/j.ijhydene.2012.07.110.

[3] H. O. Rubiera Landa, M. J. Realff, R. P. Lively, Y. Kawajiri. *2016 AIChE Annual Meeting* **2016**, San Francisco, CA, USA.

497486 Pressure Swing Adsorption Cycle Synthesis Utilizing Artificial Neural Networks As Surrogate Models

Abstract:

Pressure Swing Adsorption Cycle Synthesis Utilizing Artificial Neural Networks as Surrogate Models

K.T. Leperi,[1] F. You,[2] R.Q. Snurr[1]

[1]Northwestern University, Evanston, IL, USA

[2] Cornell University, Ithaca, NY, 14850

With fossil fuels expected to be a significant portion of the world's energy mix for the near future, it is important to minimize CO_2 emissions from current power plants through carbon capture and sequestration (CCS).[1] In post-combustion CCS, CO_2 is separated from the power plant flue gas emissions, containing mainly N_2 and CO_2. Of the technologies currently available for CCS, pressure swing adsorption (PSA) is one of the more promising due to low energy requirements and short cycle times compared to other adsorption based technologies. However, one challenge that exists in using PSA for CCS is designing the cycle to match newly developed materials for this application. Although the steps in all PSA cycles can be classified into six different possibilities (pressurization, feed, depressurization, light reflux, heavy reflux and pressure equilibration), the arrangement of the steps and interactions between steps lead to hundreds of potential different combinations. The objective of this work is to develop a new approach that is capable of synthesizing PSA cycles to capture the CO_2 from flue gas at the required purities and recoveries while minimizing energy requirements and maximizing adsorbent productivities.

In this work, we present a new framework for synthesizing the PSA cycle with the lowest predicted CO_2 capture costs. In this framework, we train artificial neural networks (ANNs) using Bayesian regularization methods as surrogate models for the various PSA steps. The ANNs are trained on simulation data collected from our PSA model consisting of a system of partial differential algebraic equations incorporating mass and energy balances, pressure drop across the column, competitive multi-site Langmuir isotherms and the linear driving force model.[2] With the ANN surrogate models, we propose a mixed integer nonlinear programming (MINLP) model to determine the ideal ordering and duration of steps in order to minimize the energy requirements and maximize the adsorbent productivity. We evaluate this model with several adsorbents, including Ni-MOF-74, UTSA-16 and zeolite 13X in order to compare the adsorbents under optimized cycle conditions.

1. MIT. *The Future of Coal*; 2007. http://web.mit.edu/coal/.

2. Leperi KT, Snurr RQ, You F. Optimization of Two-Stage Pressure/Vacuum Swing Adsorption with Variable Dehydration Level for Postcombustion Carbon Capture. *Ind. Eng. Chem. Res.* 2016;55:3338-3350.

501398 Process Flow Sheet Optimization and Techno-Economic Assessment of Post-Combustion CO_2 Capture Using Heat Integrated, Sub-Ambient PSA

Abstract:

Solid sorbents are a promising low-energy approach to post-combustion CO2 capture. However, the relatively low swing capacities (i.e., ~ 1 mmol CO2/g) of most solid sorbent systems drives up both capital and operating costs. These challenges can be overcome by an unorthodox approach: flue gas cooling and compression matched with sorbent materials that have large isotherm steps. These two factors enable high swing capacity (>10 mmol CO2/g) sorbent that can drive down capital costs relative to other state-of-the-art sorbent systems by many folds. Further, the sorbents are embedded within hollow fiber modules that have been shown to be 3-5x more capital and energy efficient than traditional pellet-packed fixed beds. Critically, the energy associated with flue gas cooling and compression system is largely recouped by employing sub-ambient heat exchange and power recovery from the clean N2 exhaust. The goal of this study is to show the viability of this heat exchange/power recovery approach for our sub ambient pressure swing adsorption (PSA) system.

Flue gas compression and cooling is major cost elements and tradeoffs between the capital cost of the heat exchangers and compressor train and utility use are explored. In the optimization model, the major components are considered including the gas-gas heat exchanger to recover the cold utility from the CO2 stripped flue gas and the work recovery from the stream for gas compression. The optimization problem is solved by considering a superstructure that embeds the different options of multistage compression and cooling. Some optimal designs are considered as case studies to perform high-level technoeconomic analysis. After the technoeconomic analysis, the optimization models discussed earlier are refined and updated, thus making the entire effort iterative. To estimate the capital cost, only major components are considered such as fiber modules, compressors, turbines, and heat exchangers utilizing the existing models. The specifications given by NETL for a 550MW (net) power plant is used and our initial studies show the CO_2 capture cost to be below \$25.0/ton.

494609 Fast Sorbent Mediated Water-Gas Shift (C-SHIFT) Process for Pre-Combustion CO_2 Capture

Abstract:

Fast Sorbent mediated water-gas shift (C-SHIFT) process for pre-combustion CO_2 capture

M.E. Boot-Handford, P.S. Fennell, J.A.Z. Pieterse, P.D. Cobden, J. Boon, M.G. Sceats, B.N.C. Sweeney, B. Arstad, R. Blom

The carbonated shift (C-SHIFT) process is part of the ASCENT project aiming at proof of concept of three different advanced CO_2 capture process. C-SHIFT, a sorption enhanced water-gas shift process, uses a highly innovative fluidized bed reactor system to match the heat requirements of reaction and regeneration of medium temperature carbon dioxide acceptor materials (300-600°C) with pressure swing technology. The aim is to produce a fuel stream consisting of high temperature H_2 and H_2O, at a pressure above the compressor temperature of a gas turbine, with a CO_2 capture efficiency greater that 90%. Alkali carbonate promoted Mg/Al-hydrotalcites (HTCs) and magnesium oxides were considered the most promising of the bifunctional CO_2 sorption and water-gas shift reactive materials reviewed. In particular, high MgO containing Mg/Al-hydrotalcites impregnated with K_2CO_3 and high magnesium contents were identified as being strong candidates for high pressure, mid-temperature CO_2 capture in the presence of large quantities of steam. However, in the context of C-SHIFT, the cyclic performance information provided thus far is limited. There is a significant discrepancy between the long carbonation periods under which these materials have been investigated and the short residence time that the CO_2 sorbent material is expected to reside within the proposed C-SHIFT carbonator (< 60 s). At present there are no published studies reporting either the uptake capacity or kinetics for the reaction between CO_2 and HTC-derived materials over time scales relevant to C-SHIFT. The work performed within ASCENT project aims to address this gap in the knowledge and provide information relating to the kinetics and capacities for CO_2 uptake by K_2CO_3-promoted HTC-derived materials at conditions relevant to a C-SHIFT process. This has been achieved using a pressurized spouted fluidized bed and fixed bed reactor and for pressure-swing operation at pressures up to 10-20 bar with large concentrations of steam. Fluidized beds are particularly suited for kinetic studies based on their superior heat and mass transfer characteristics when compared with other reactor configurations such as fixed beds and thermo-gravimetric analysis (TGA). The reactor has now been commissioned for use at pressures up to 10 bar with steam concentrations up to 40 % v/v. Modifications to the water trap design and drainage protocol that enables continuous removal of water from the system without disruption to the system pressure stability. Results from initial testing of the cyclic performance of the selected material over carbonation/ calcination cycles are also provided serving as a proof-of-concept for the pressurized fluidized bed reactor testing. The aim of this initial cyclic test was to provide empirical measurements of the carbonation and calcination kinetics and cyclic capacities of K_2CO_3-promoted HTC-derived mixed metal oxide materials at conditions and within the time period relevant to an industrial C-SHIFT process. The capabilities of the pressurized fluidized bed reactor to test the cyclic performance of candidate mid-temperature CO_2 sorbents with pressure-swing regeneration was demonstrated with carbonation at 10 bar 400 °C, 20 % v/v CO_2 and 40 % v/v steam and regeneration at 1 bar, 400 °C under mild condition (100% N_2). The presence of steam was also found to dramatically enhance both the carbonation rate and short-term carbonation capacity. Furthermore, the presence of H_2S or operation with syngas did not affect the cyclic performance of either of the sorbent materials tested. The materials reacted reversibly with H_2S presenting the possibly of a consolidating CO_2 and sulfur gas removal into a single step in a sour sorption enhanced water gas shift reaction.

Acknowledgements

The presented work is funded within the ASCENT project as part of the European Union's Seventh Framework Programme (FP7/2007-2013) under grant agreement n° 608512. Note: "The present publication reflects only the authors' views and the European Union is not liable for any use that may be made of the information contained therein".

500673 Chiral Control of Enantiomers through Crystallization: Moving from Ternary Phase Diagrams to Design Spaces

Abstract:

Although chiral purity is a critical quality attribute for the majority of synthetic drugs being developed, the understanding of process design for chiral purity in terms of conventional batch crystallization processes is limited. The industrial implementation within this area has relied heavily on the construction of ternary phase diagrams, system classification, and the use of subject matter experts to determine thermodynamic operating policies. These operating policies have been developed to yield resulting solids with 100% ee; regardless of the product specification/target and the potential upgrades to ee remaining between the molecule being isolated and the drug substance. In the present work (which for simplicity is limited to the case of a single chiral center in the absence of salts, polymorphs, and solvates/hydrates), a new policy is proposed for obtaining resulting solids with ee < 100% but passing specifications for arbitrary systems. This operating policy can provide significant isolation yield improvements when compared to the conventional policy. Additionally, this work demonstrates an approach for the creation of design spaces for arbitrary ternary systems based on the collection of limited small (<10 g total) scale solubility data. This approach has been extended to processes with sequential re-crystallizations and can be used to determine â€œglobalâ€ optimal policies for an entire synthetic route.

499039 Leveraging Mechanistic Understanding of Impurity Rejection during Crystallization for Rational Process Design

Abstract:

The control of impurities in the chemical synthesis of a new drug substance is a requirement in registering pharmaceuticals with worldwide regulatory agencies. Crystallization is an essential tool in controlling the purity profile of a drug substance and its intermediates. An overview of the various mechanisms of impurity incorporation during crystallization will be discussed along with remediation strategies. A case study will be provided on the replacement of chromatography with crystallization during the development of a second generation commercial manufacturing route for Ceftolozane, the cephalosporin antibiotic found in Zerbaxa. The implementation of a crystallization based purification process for Ceftolozane, replacing a chromatography/nanofiltration/diafiltration sequence, resulted in a more productive, greener, and more portable manufacturing process. Another case study will be presented on limiting substrate concentration on product crystals during a reactive crystallization via engineering design.

488309 Development of Combined Continuous Reaction to Crystallization Process Using PAT Tools

Abstract:

Introduction

Figure 1. Synthesis of (*Z*)-3-chloro-2-(phenylthio)-*N*-(p-tolyl)acrylamide

Traditionally the majority of active pharmaceutical compounds have been manufactured at large plant scales using batch methodologies. Continuous manufacturing is becoming an increasing attractive alternative due to apparent benefits such as smaller manufacturing footprints, increased process safety, lower manufacturing costs and increased supply-chain flexibility. In this work, a new approach to the design of continuous manufacturing platforms is being presented which is not restricted to using such conventional equipment but which is based around bespoke design and manufacture. As a model compound, 2-chloro-*N*-(4-methylphenyl)propanamide (**S1**), a precursor in the multi-step synthesis of (*Z*)-3-chloro-2-(phenylthio)-*N*-(p-tolyl)acrylamide (**P1**, Figure 1) is used. The aim of this work is to design a combined continuous reaction to crystallization process for **S1** and overcome the existing challenges between the two process.

Solubility Studies for S1

Figure 2. Solubility data of **S1** in toluene + hexane mixtures: *, $w_2 = 1$; +, $w_2 = 0.8$; ◊, $w_2 = 0.6$; □, $w_2 = 0.4$; -, Phase Equilibria with NRTL equation calculated. w_2 is the mass fraction of toluene in binary toluene + hexane mixtures.

The solubility behaviour of **S1** in ethyl acetate + hexane, toluene + hexane, acetone + hexane and butanone + hexane solutions were measured as the solubility characteristics have considerable influence upon the design and optimisation of any crystallisation process, affecting both the resulting crystal structure and crystal size distribution (CSD)[1]. The solubility study is done using the polythermal method and conducted in EasyMax (Mettler Toledo) reactor with in situ Focused Beam Reflectance Measurement (FBRM G400, Mettler Toledo) to characterize the saturation temperature of **S1**. The experimental data are correlated with Apelbat, λh and the Phase Equilibria with NRTL (non-random two liquid) model equations using the function *lsqcurvefit* in Matlab's optimization toolbox to enable both interpolation and extrapolation of the measured data, as shown in Figure 2.

Design of Continuous Reaction and Crystallization Process

Reaction: Using a modified procedure developed by Li *et al*[2], a batch process is employed where a-chloropropionyl chloride (12 mmol) was added dropwise to a suspension of *p*-toluidine (10 mmol) in toluene (20 mL) at 0°C and the resulting solution was heated to reflux for 1 h with vigorous stirring. After cooling, the solvent was removed under vacuum and the resulting off white solid was collected by filtration and washed thoroughly with cold cyclohexane. Spectral data of the obtained product matched that reported in the literature[3]. Therefore, it has been decided to use toluene as the reaction solvent as it can also be used as a crystallization solvent based on the solubility studies. However, this developed procedure produces a small amount of HCl as a by-product, which can be problematic for reaction product purity and its subsequent crystalisation steps. Therefore, it is decided to run the reaction in a novel 3D printed continuous biphasic reaction system in the presence of aqueous sodium hydroxide to neutralize HCl salt. The design of the biphasic reactor is based on the specific needs of the reaction which is capable of achieving extremely high dispersion rates between the two phases under the very low Reynolds numbers typically encountered in the pharmaceutical industry ($Re < 5$). Process steps are quantified in detail using methods such as computational fluid dynamics (CFD) leading to tailored equipment designs which can be manufactured with high precision using metallic 3D printing. The biphasic reactor is coupled with liquid-liquid separation such that both unit operations are designed using fundamental engineering principles combined with numerical analyses (based upon the requirements of the specific process), prior to being 3D printed and combined to create a bespoke, sealed, hard-piped manufacturing system.

Reaction to Crystallization: From the liquid-liquid separator, the organic phase is transferred in the feed/dissolution tank while aqueous phase goes to waste. In situ FTIR (React IR 15, Mettler Toledo) is used on the feed/dissolution tank to determine the concentration of **S1** prior to crystallization and the crystallization kinetic data obtained (explained below) enables the adjustment of crystallization parameters for better control and set up of the process.

Crystallization: Using the solubility curve obtained from solubility studies for toluene, a continuously operated single stage mixed suspension, mixed product removal (MSMPR) crystallizer with intermittent withdrawal via dip pipe using an automated pressure supply[4-6] is set up for the cooling crystallization of **S1**. Acquisition of reliable kinetic correlations that describe how rates of nucleation and growth vary over the design space for the crystallization process is necessary for enabling robust prediction of crystallizer performance. Therefore, simple extraction of the growth and nucleation parameters from the single stage MSMPR operating at steady state is performed using the population balance model. Comprehensive experimental data for model fitting and parameter estimation is obtained by varying crystallizer operating conditions (temperature, supersaturation and suspension density) to induce changes in the rate-affecting variables for growth and nucleation. FBRM is used for real-time monitoring of size distribution and identification of steady state and hence enabling on-demand adjustment of operating conditions to transition multiple steady-states for rapid acquisition of kinetic data while PVM (ParticleView V819, Mettler Toledo) provides microscopy quality images in real time to monitor the crystal habit. Malvern Mastersizer is used as an offline analysis to determine the crystal size distribution (CSD) at steady-state while the concentration of mother liquor at steady-state is determined gravimetrically. By estimating the six kinetic parameters for crystal growth and nucleation, accurate prediction of CSD and concentration results at steady-state can be achieved.

References:

[1] Mullin, J. W. *Crystallization*; Butterworth-Heinemann, 1997

[2] Li, X.; Wang, D.; Wu, J.; Xu, W. Novel Approach to 3-Methyl-1H-quinoxalin-2-ones. *Synthetic communications* **2005,** *35* (19), 2553.

[3] Murphy, M.; Lynch, D.; Schaeffer, M.; Kissane, M.; Chopra, J.; O'Brien, E.; Ford, A.; Ferguson, G.; Maguire, A. R. Investigation of the synthetic and mechanistic aspects of the highly stereoselective transformation of [small alpha]-thioamides to [small alpha]-thio-[small beta]-chloroacrylamides. *Organic & Biomolecular Chemistry* **2007,** *5* (8), 1228.

[4] Hou, G.; Power, G.; Barrett, M.; Glennon, B.; Morris, G.; Zhao, Y. Development and Characterization of a Single Stage Mixed-Suspension, Mixed-Product-Removal Crystallization Process with a Novel Transfer Unit. *Crystal Growth & Design* **2014,** *14* (4), 1782.

[5] Power, G.; Hou, G.; Kamaraju, V. K.; Morris, G.; Zhao, Y.; Glennon, B. Design and optimization of a multistage continuous cooling mixed suspension, mixed product removal crystallizer. *Chemical Engineering Science* **2015,** *133,* 125.

[6] Morris, G.; Power, G.; Ferguson, S.; Barrett, M.; Hou, G.; Glennon, B. Estimation of Nucleation and Growth Kinetics of Benzoic Acid by Population Balance Modeling of a Continuous Cooling Mixed Suspension, Mixed Product Removal Crystallizer. *Organic Process Research & Development* **2015,** *19* (12), 1891.

485308 Process Design to Preserve Particle Properties during Crystallization of Energetic Material

Abstract:

Obtaining desired particle size and morphology of isolated material from a quenched crystallization can be challenging during scale up. This work focuses on the scale-up of a quenched crystallization of an energetic compound while maintaining a large blocky morphology of the isolated material. The compound has low solubility in most solvents, therefore recrystallization options were limited. In this case, the key to obtaining the desired particle size and morphology was quenching the reaction mixture into a nitrate salt solution. This talk will discuss the process optimization and the effects of mixing, temperature, and salt concentration on this quenched crystallization. Optimization and development efforts were performed in the 1-L Mettler Toledo OptiMax reactor before being scaled up to the pilot plant.

491816 Multi-Scale and Multi-Purpose Modelling for *in silico* Designing and Optimization of Pharmaceutical Crystallization Process

Abstract:

Crystallization often serves as a key final and intermediate separation/purification unit operation in pharmaceutical production. More than 90% of small molecule drugs are produced in crystalline form[1]. Operation conditions of crystallization process determine process metrics such as crystal size distribution, yield, space-time yield (STY) and crystal quality attributes like crystal shape, morphology as well as solvates. The crystal product properties affect the efficiency of downstream unit operations such as filtration, drying and formulation. Moreover, in pharmaceutical products the crystal properties affect the stability and efficacy of drugs. Therefore, a comprehensive multi-scale and multi-purpose thermodynamic and kinetic modeling is the main objective of this work in order to understand crystallization mechanism and define optimum operation conditions. Optimization of crystallization operation conditions includes selection of the best solvent or mixture of solvents, cooling rate, anti-solvent flowrate, pH profile, mixing and seeding parameters as well as selection of mode of operation such as batch or continuous. Figure 1 shows a comprehensive multi-scale and multi-purpose modeling framework for *in silico* designing and optimization of pharmaceutical crystallization process.

Figure 1: A schematic diagram illustrating comprehensive multi-scale and multi-purpose modeling framework for *in silico* designing and optimization of pharmaceutical crystallization process

The framework will be implemented in Matlab-Simulink environment as a tool to support different engineering tasks from solvent selection to crystallization process design, optimization and control in life science and biopharmaceutical industries.

Acknowledgment: We would like to thank the Danish Council for Independent Research (DFF) for financing the project with grant ID: DFF-6111600077B.

[1] Narayan Variankaval, Aaron S. Cote, *AIChE* 54 (2008) 1682 -1688

492269 Sensitivity Analysis and Optimization Study of Pharmaceutical Crystallization Process

Abstract:

Crystallization is a prevalently applied technique in many industries including food, bulk and fine chemicals as well as manufacturing of active pharmaceutical ingredients (APIs). In the pharmaceutical industry, crystallization processes are employed mainly for separation,, purification as well as for formulation of APIs. Moreover, the process should guarantee the performance of the delivered APIs suitable for downstream processing, handling and drug efficacy.

The performance of an API crystal is strongly affected by the product quality attributes such as crystal size distribution, morphology, shape and purity, which are the main targets of the process output to be achieved within specific range. However, in a crystallization process, complex feedback interactions take place between the formed crystals and the growth and nucleation kinetics. These are at the same time dependent on the supersaturation and hydrodynamic conditions to which the crystals are subjected. Consequently, the quality characteristics of the crystal product have a complex dependency on multiple operational variables and input parameters.

In order to evaluate and quantify which process factors or variables have the most impact on the final quality attributes of the crystal product, a comprehensive sensitivity analysis of the system model is required. This study therefore focuses on a global sensitivity analysis of a predictive pharmaceutical crystallization model through assessing the sources of uncertainty and their influence on crystal size distribution as the measured process outcome/performance criteria in order to guide design and suggest the optimized operation conditions.

This process model of pharmaceutical batch cooling crystallization is developed based on compartmentalization technique which takes into account mixing phenomena as well. Several global sensitivity analysis methods including Morris screening, global derivative based sensitivity analysis (DGSM), variance decomposition methods such as Sobol's index and linear regression of Monte Carlo outputs are used to perform the sensitivity analysis. The developed crystallization process model is analyzed for a range of values for parameters like cooling rate, growth and nucleation kinetics etc. to rank their contributions on the measured outcome of the system "crystal size distribution". Comparison of two sensitivity analysis methods and evaluation of the results are used as a guide for process optimization and robust system design.

Acknowledgment

We would like to thank the Danish Council for Independent Research (DFF) for financing the project with grant ID: DFF-6111600077B.

495572 Segmented Continuous Crystallization Process Optimization

Abstract:

The continuous crystallization behavior of glycine in segmented tubular crystallizer has been investigated. The factors of liquid slug length and shape, Reynolds number, suspension density and arrangement were investigated respectively on the crystal growth rate, size distribution and morphology. The results found that the slug being closer to a sphere both in length and shape is the best compared to the other slug geometries, in which mixing within each stable self-circulating slug is maximized, thus the crystal growth rate is faster and the particle size distribution is more uniform. Besides, Higher Reynolds number improves the growth rate and uniform size distribution at the same residence time. The mass fraction of seed crystals at 2% was also determined to be the optimal condition. The arrangement of the tube in vertical and horizontal direction has little influence on the final yield and crystal characteristics. These results provide an insight into the segmented tubular continuous crystallization.

497853 Membranes with Chemical Structure-Based Selectivity from the Assembly of Functionalizable Random Copolymer Micelles

Abstract:

Regulating permeation through materials is crucial for many applications including selective membranes, controlled drug delivery, and packaging. An ideal membrane material would pass the desired molecules with little resistance (high permeability) while preventing the passage of all else (high selectivity). However, synthetic materials today cannot meet this challenge. There are no commercially available membranes that can separate small molecules of similar size in the liquid phase based on their chemical properties. In this study, we aim to prepare synthetic polymer membranes that mimic two key features of biological pores such as ion channels and porins: Constricted pores only 1-5 nm in diameter, lined with functional groups lining the pore that interact with the target during passage. The nanostructure constricts flow and confines all components passing through, forcing them to interact with the chemically functional walls. As an initial system, we focused on membranes capable of charge-based separation through electrostatic interactions. To build these nanostructured layers using scalable techniques, we deposited packed arrays of polymer micelles whose coronas exhibit carboxylic acid groups onto a porous support membrane. Random copolymers that combine highly hydrophobic fluorinated repeat units of trifluoroethyl methacrylate (TFEMA) with ionizable repeat units of methacrylic acid (MAA) form micelles and vesicles in methanol. When these micelles are coated onto the surface of a porous support membrane whose pores are smaller than the micelles and then immersed into water, a selective layer of micelles packed together is formed. The gaps between the micelles act as nanochannels functionalized with carboxylic acid groups. The membrane showed charge-based selectivity between organic molecules, efficiently rejecting negatively charged solutes while allowing the passage of neutral solutes in two-solute separation tests. Furthermore, the carboxyl groups can be post-functionalized to alter the selectivity of the membrane for desired separations. Such functionalization has enabled us to impart significant diffusion selectivity of about 10 between two solutes of identical size and charge, differentiated by minor chemical features. This demonstrates the potential of using polymer self-assembly and functionality to design membranes that mimic biological pores while maintaining scalable manufacturing methods. We believe these approaches will eventually lead to novel membranes that can separate molecules of similar size but different chemical structure.

498866 Fractionation of Ionic Liquid and Monomeric Sugars during Biomass Pretreatment By Nanofiltration

Abstract:

Ionic liquids could be an emerging solvent for pretreatment of lignocellulosic biomass. They act by dissolving the biomass which allows for much better access of the enzymes used for hydrolysis of cellulose. Further, ionic liquids reduce the crystallinity of the cellulose leading to higher glucose yields during enzymatic saccharification. The high cost of ionic liquids means that efficient recovery and recycle of the ionic liquid is essential for an economically viable process. Here, we investigate the feasibility of developing nanofiltration membranes for ionic liquid recovery.

We have created nanofiltration membranes with different barrier properties. Our aim is to maximize the ionic liquid recovered in the permeate while obtaining as high a rejection of saccharides as possible. Using base ultrafiltration membranes as a support structure, we have used interfacial polymerization (IP) to create a dense polyamide layer. Common reactive monomers are diamines such as piperazine (PIP), m-phenylenediamine (MDP) and p-phenylemediamine (PPD) and acid chloride monomers such as trimesoyl chloride (TMC), isophthaloyl chloride (IPC) and 5-isocyanato-isophthaloyl chloride (ICIC). Here we focus on the reaction of TMC with PIP. Recent publications highlight the ability to tune the performance of these nanofiltration membranes by inclusion of novel monomers as well as inorganic additives. Here we have incorporated 3-aminophenylboronic acid (BA) in the IP layer to control the sugar selectivity and permeance of the membrane. We have also developed unique nanofiltration membranes using layer-by-layer deposition on base ultrafiltration membranes. Membrane performance has been determined using feed streams consisting of aqueous saccharide solutions, water, ionic liquid and saccharide solutions as well as real biomass hydrolysates.

500993 Uranium-Selective Membrane Adsorbers for Use in Nuclear Forensics Applications

Abstract:

Clandestine activities involving the manipulation of special nuclear material (SNM) with the intent of developing a weapon of mass destruction is likely to contaminate environmental waters. The capability to conduct isotopic analyses for waterborne SNM would be a powerful nuclear forensics tool. Despite widespread interest, there currently is no fieldable, separation technique capable of selectively concentrating uranium directly from groundwater for isotopic analysis.

This contribution describes our efforts to develop a high throughput analytical technique for waterborne isotopic analysis using reactive, functional membranes. In this work, we develop functional ultrafiltration membranes that (1) selectively concentrate uranium directly from neutral pH ground water and (2) serve as the substrate for isotopic analysis with alpha spectroscopy. Ultrafiltration alpha spectroscopy substrates were prepared by grafting ethylene glycol methacrylate phosphate from the surface of polyethersulfone membranes using (UV)-initiated polymerization. Membranes were characterized by Fourier-transform infrared spectroscopy before and after modification to support polymer grafting.

The effect of membrane pore size on peak resolution in the alpha spectrum was investigated for pure uranium containing solutions at pH 6. To mimic more realistic conditions, the selectivity of the membrane was tested using uranium-233 in simulated ground water. Membranes exhibited good resolution and fast preparation time. Membranes prepared from both distilled water and simulated ground water produced resolutions of 80-100 keV in the alpha spectrum. Furthermore, baseline resolution was observed for uranium-238 and uranium-233 using a membrane substrate loaded directly from neutral pH ground water. The permeability coefficient of 100 kDa UV-grafted ultrafiltration membranes was determined to be 566 ± 80 LMH/bar which means 3.6 L of contaminated ground water can be processed per hour at 414 kPa through a 45 mm ultrafiltration cell. Uranium recovery from DI and ground water were both 85-90%, supporting the selectivity of the membranes for uranium over common competitor ions. Since the membranes are not thermodynamically limited, recoveries less than 100% are indicative of a low Damköhler number and potential kinetic limitations in the system; consequently, kinetic studies were performed to support this hypothesis. Our porous, uranium-selective substrate design allows for the filtration of uranium-contaminated ground water through the uranium-selective membrane, leading to a high throughput separation and sample mounting process.

492079 High-Performance Separation Membranes Fabricated from Directly-Synthesize MFI Nanosheets

Abstract:

Separation membranes based on porous crystalline materials are promising energy-efficient alternatives to conventional separation processes.[1] They can find industrial use provided their high fabrication cost can be reduced and/or performance can be further improved. One approach is membrane fabrication based on 2-dimensional nanosheet MFI materials, which allow to fabricate thin and defect-free membranes, resulting in high throughput per unit cost.[2] We have developed a direct synthesis method of MFI nanosheets, which can be used to fabricate thin and defect-free high-performance separation membranes.[3] Use of seed crystal enabled the preparation of single MFI nanosheets without orthogonal twinning, yielding ~2 um lateral dimension and predominantly 5-nm thickness. These high-aspect-ratio MFI nanosheets effectively cover the surface of the porous supports and form high-density seed coating layers, which are further intergrown into continuous membranes. Resultant MFI membranes exhibit unprecedented combination of high permeance and high selectivity for xylene isomer separation (*p*-/*o*-xylene binary mixture), which are consistent with our theoretical calculations. Microstructure analyses established the MFI membranes fabricated from the MFI nanosheets possess favorable properties, such as thin, wide and well intergrown *b*-axis-oriented grains. Permeation properties of the membranes assessed under various temperature and feed pressure conditions will be presented.

Reference

1. Gin, D. L.; Noble, R. D. *Science* **2011**, *332*, 674-676.

2. Tsapatsis, M. Toward High-Throughput Zeolite Membranes. *Science* **2011**, 334, 767â€“768.

3. Jeon, M. Y.; Kim, D.; Kumar, P.; Lee, P. S.; Rangnekar, N.; Bai, P.; Shete, M.; Elyassi, B.; Lee, H. S.; Narasimharao, K.; Basahel, S. N.; Al-Thabaiti, S.; Xu, W.; Cho, H. J.; Fetisov, E. O.; Thyagarajan, R.; DeJaco, R. F.; Fan, W.; Mkhoyan, K. A.; Siepmann, J. I.; Tsapatsis, M. *Nature* **2017**, *543*, 690â€“694.

501670 Preparation and Characterization of Fluoropolymeic Microcapsules for Gas Separation Applications

Abstract:

Fabrication of microcapsules with permeable shells has been an active research topic for the past two decades. A simple one-step approach for encapsulation of ionic liquid absorbents within a polymeric shell of fluoropolymers was developed. Using this approach, we can benefit from the advantages of liquid solvent sorbent, such as high selectivity and high capacity, and the enhanced specific mass transfer area, provided by the microcapsules. The effect of different parameters including choice of solvent, non-solvent pair, the molecular weight of polymers, dope solution concentration and different solvents such as acetone, triethyl phosphate (TEP), and dimethylacetamide (DMAc) were systemically investigated. Microcapsules with desired sizes and porosity were fabricated. By adjusting the experimental parameters, the mean diameters of these microcapsules were tuned in between 700 to 2000 μm. The cross-sectional scanning electron microscopy images of the shells, showed that the thickness was in between 60 to 160 μm, translating into the void density values of more than 75%. Moreover, using different non-solvent systems, roughness and surface structure of the capsules were adjusted. The physical stability and chemical resistance of these polymeric shells allow for these capsules to be used in a variety of processes such as gas sorption. Microcapsules, loaded with ionic liquid, were fabricated and tested for carbon dioxide adsorption from gaseous mixture.

501754 Mixed Matrix Membranes for Ammonium Removal from Wastewaters

Abstract:

Novel electrospun mixed matrix membranes were fabricated to investigate ammonium removal from low total ammonium nitrogen (TAN) wastewaters. Natural zeolite 13X particles as well as several other zeolites such as 3A, 4A and Y were effectively incorporated into polymer matrices by electrospinning method. These composite membranes are found to be efficient for ammonium removal with high capacity and high stability. No leaching of zeolite materials was detected. These mixed-matrix membranes demonstrate high water flux and are easy to regenerate. Over 90% of total ammonia nitrogen (TAN) can be removed from relatively low TAN concentration aquaculture wastewaters. The effects of pore size and Si to Al ratio on ion-exchange kinetics and ammonium removal capacity were elucidated. These novel adsorptive membranes are reusable and exhibit high viability for large scale industrial applications.

501841 Reliable Fabrication of Oriented Fe-MFI Membranes for Efficient Ethanol Recovery from its Dilute Aqueous Solution

Abstract:

The increased dependence on oil supplies from unstable countries and sensitive regions in the world have created a strong desire for the domestic production of alternative energies on a renewable stand [1]. Alternative energies including light-voltage, wind energy, biodiesel and bioethanol are being developed to replace fossil fuels worldwide. As a promising energy alternative for transportation, bioethanol is convenient to carry in place. Although there has been a furious debate on the production of bioethanol with corn-based feedstocks, the non corn-based bioethanol, predominantly produced from sugarcane or biomass such as corn kernels, switchgrass and perennial plants, can be converted from lignocellulose for global sustainability [2,3]. In recent years, major economies with a high demand of fossil fuels have set their own progressive goals to achieve an increasing proportion of biofuels in the energy system and reduce net greenhouse gas (GHG) emissions [4]. The required enzyme and energy to convert lignocellulose into fermentable sugars and the energy consumption to separate and purify ethanol from its dilute aqueous solution construct the main cost of the whole production process [5]. The cost for separating ethanol from its dilute aqueous solution can be substantially reduced by using membrane-based pervaporation processes either combined with distillation or not [6]. Nevertheless, the increase of separation factor with a relatively high permeation flux is desired to reduce the cost of ethanol recovery.

The Fe-MFI membranes were fabricated via the seeded growth method. The 1-3 micrometer-thick and preferentially *b*-oriented Fe-MFI membranes could be fabricated on alpha-alumina supported *b*-oriented seed layers. The preferentially *b*-oriented seed layer was synthesized using either the manual assembly method or the Langmuir-Blodgett technique. The preferentially *b*-oriented Fe-MFI membranes were formed under the induction of the seed layer without using a mesoporous intermediate layer and trimer-TPAOH. This could dramatically decrease the difficulties and costs for preparation of *b*-oriented MFI membranes. Furthermore, the 6-8 micrometer-thick and *h0h*-oriented Fe-MFI membranes were fabricated on randomly oriented seed layers. The fabrication of the two preferentially oriented Fe-MFI membranes with specific microstructures is highly reproducible.

A novel calcination method was proposed and developed to remove templates in the synthesized Fe-MFI membranes. The template removal was carried out at low-temperature conditions helped with catalytic decomposition of templates to keep the Fe-MFI membranes defect-free. Both UV-Raman and BET measurements were performed to demonstrate the efficiency of this low-temperature calcination method. At last, the pervaporation performance of the two oriented Fe-MFI membranes were checked with respect to ca. 5 wt.% ethanol aqueous solutions at feed temperatures from 30 to 60 °C. With respect to the fabricated Fe-MFI membranes of two microstructures, the separation factors ranged from 90 to over 500 could be achieved while keeping the permeation flux of ethanol relatively high.

Acknowledgements

This work was supported by the National Natural Science Foundation of China under three consecutive General-Program Grants (20476076, 20776100 and 21076154) and a Key-Program Grant (21136008).

References

[1] Koonin S. E. Getting serious about biofuels, Science 311 (2006) 435.

[2] Farrell A. E., Plevin R. J., Turner B. T., Jones A. D., O'Hare M., Kammen M. D. Ethanol can contribute to energy and environmental goals, Science 311 (2006) 506-508.

[3] Fairley P. Next generation biofuels, Nature 474 (2011) S2-S5.

[4] IEA Technology roadmap: biofuels for transport (2011), http://www.iea.org/publications/freepublications/publication/Biofuels_Roadmap.pdf

[5] Zhu J. Y., Zhuang X. S. Conceptual net energy output for biofuel production from lignocellulosic biomass through biorefining, Prog. Energ. Combust. 38 (2012) 583-598.

[6] Vane L. M. Separation technologies for the recovery and dehydration of alcohols from fermentation broths, Biofuels, Bioprod. Bioref. 2 (2008) 553-588.

[7] Nai S. F., Liu X. F., Liu W., Zhang B. Q. Ethanol recovery from its dilute aqueous solution using Fe-ZSM-5 membranes: Effect of defect size and surface hydrophobicity, Microporous Mesoporous Mater. 215 (2015) 46-50.

499815 Advanced Functionalized Polymeric Membranes for Molecularly Selective Gas Separations

Abstract:

Advanced functionalized polymeric membranes for molecularly selective gas separations

Shouliang Yi[a], Ingo Pinnau[b], William J. Koros[a]

[a] School of Chemical and Biomolecular Engineering, Georgia Institute of Technology, 311Ferst Drive, Atlanta, GA 30332, United States

[b] Advanced Membranes and Porous Materials Center, Physical Sciences and Engineering Division,

King Abdullah University of Science and Technology, Thuwal 23955-6900, Saudi Arabia

Membrane-based gas separations have attracted significantly increasing attention in many environmental and energy processes such as natural gas purification, CO_2 capture, biogas upgrading, and hydrogen production during the last two decades, because of its inherent advantages including small footprint, low capital and operating costs, and lower energy requirements. The acid gases carbon dioxide (CO_2) and hydrogen sulfide (H_2S) are among the most commonly found impurities in natural gas. These impurities can cause corrosion of natural gas pipelines and processing equipment, as such, must be removed from raw natural gas before its distribution to consumers or delivery to pipelines.

While many membrane materials have been developed for the separation of CO_2/CH_4 over the past several decades, only a few studies have focused on the development of materials for removal of overall sour gas components due to the high toxicity of H_2S. In this presentation, a number of advanced functionalized polymers, including silane-grafted industrial standard glassy polymer cellulose esters, crosslinkable polymers based on the glassy 6FDA-based copolyimide 6FDA-DAM:DABA (3:2), and hydroxyl-functionalized polyimides with intrinsic microporosity, have been developed for molecularly selective separations from sour natural gas feeds. Specifically, the permeation and separation properties of these functionalized membranes was tested with high concentrations of H_2S and CO_2 feeds as well as high feed pressures. Ternary mixed gas (containing up to 20%H_2S) permeation results showed that these advanced membranes maintained excellent separation performance even under exceedingly challenging feed conditions. Our results indicate that these advanced materials have great potential in reducing the energy requirements and capital costs for environmentally friendly energy processes in industrial applications, which makes membrane-based gas separation technology an attractive option for clean energy production and reducing greenhouse gas emissions.

492077 Moving Towards Practical Application of Mixed Matrix Membranes for High PerformanceÂ C_3H_6/C_3H_8 Separations

Abstract:

Propylene/propane (C_3H_6/C_3H_8) separation is one of the most important processes in the petrochemical industry due to the highly utile applications of both gases in our everyday life. Currently, C_3H_6/C_3H_8 separation is implemented by utilizing highly energy-intensive cryogenic distillation due to similar physical properties between them (i.e., T_b of C_3H_6 and C_3H_8 is -47.6 and -42.1 °C, respectively). If the drawbacks of polymeric membranes are resolved, membrane-based separation technique can be a good auxiliary candidate in reducing energy use over the conventional cryogenic distillation process.

Mixed matrix membranes (MMMs) are essentially hybrid membranes consisting of molecular sieves as a dispersed phase for exceptional separation performance and polymeric membranes as a continuous phase for good processability. Prof. Bill Koros made a significant contribution to the field of MMMs revolutionizing a new generation in gas separation membranes. Especially, their recent work [1] in his group promoted the development of MMMs for C_3H_6/C_3H_8 separations. In this talk, we will present the current progress in developing MMMs for C_3H_6/C_3H_8 separations [2].

[1] Zhang, C., Dai, Y., Johnson, J. R., Karvan, O., Koros, W.J., Journal of Membrane Science, 2012, 389, 34-42

[2] An, H., Park, S., Kwon, H.T., Jeong, H-K., Lee, J.S., Journal of Membrane Science, 2017, 526, 367-376

500118 Functional Nanofiber Network Membranes for Solubility Based Separation of Carbon Dioxide

Abstract:

Membranes have garnered interest as alternatives to amine absorption for selective recovery of CO_a from gas streams because of their small modular size, low energy usage, and capital cost. However, novel membrane system designs and materials are needed to address the energy costs associated with separations of CO2 from high temperature, low pressure flue gas streams. While many polymeric membranes focus on diffusion based separations, membrane material designed to target solubility selectivity through incorporation of groups that have a high affinity for CO2 have the potential to provide the high selectivities required for this application. The presentation focuses on development of a platform membrane material based on a functionalized nanofiber network within a polymer matrix for solubility based separations of CO_2 from N_2. Functional nanofiber network membranes (FNNM) consist of a network of amine functionalized carbon nanofibers that provide a highly selective CO_2 pathway within a continuous polymer matrix. FNNM produced using a scaffold of amine functionalized carbon nanofibers exhibited high permeabilities and reasonable selectivities for CO_2/N_2 at elevated temperatures and relative humidity relative to the based polymer, poly(dimethyl siloxane) (PDMS). The effect of carbon nanofiber functionalization and loading on the gas transport properties of the PDMS with emphasis on the CO_2/CH_4 and CO_2/H_2 systems will be addressed. Additionally, the effect of temperature and humidity on transport characteristics of the membranes will be discussed

503478 Engineering Polymeric Hollow Fiber Contactors for Separation and Reaction Applications

Abstract:

There are three objectives regarding the advancement of composite polymer hollow fibers for separation and reaction: (i) fabrication of polyamide-imide hollow fiber (PAI-HF) support with engineered surface porosity. A new approach for formation of a thin selective layer of phosphazene (MEEP) and polydimethylsiloxane (PDMS) intermediate layers on PAI-HF support was developed on the as-spun fibers without the need to undergo the final conventional solvent exchange and drying steps, thereby saving post-spinning processing steps. The effect of coating solution, re-wetting and crosslinking temperature, physical aging, moisture and fouling on the MEEP/PDMS/PAI-HFs features e.g. gas transport, skin thickness, skin integrity, and substructure resistance were investigated. The produced MEEP/PDMS/PAI-HF composites were used for CO_2/N_2 separation. (ii) creation of polymeric hollow fiber sorbents consisting of PAI polymer and silica particles by an *in-situ* process. This approach resulted in silica/polymer composite fibers with good amine distribution and accessibility after infusion with aminosilane, as well as adequate porosity. Moreover, the *in-situ* formed particles appear to be firmly retained within the fibers and enable rapid mass transfer and adsorption kinetics. Therefore, this process offers a low-cost method of fabricating and engineering the amine/silica/polymer hollow fiber sorbents with high equilibrium capacity for postcombustion CO_2 capture. (iii) development of a proof-of-concept study for a new type of microfluidic reactors based on porous PAI composite hollow fibers. These novel hollow fiber microfluidic reactors consist of bifunctional groups in the fiber wall that make them bifunctional catalysts for cooperative interactions (i.e., acid-base catalysis). Based on the optimized conditions for aldol and nitroaldol condensations, Heck coupling and CO_2 cycloaddition reactions, we identified a robust process window to produce composite hollow fiber microfluidic reactors for sustainable C-H functionalization and CO_2 utilization system in flow.

498961 Real Time Phase Imaging AFM of Palladium-Decorated Membranes for Hydrogenation

Abstract:

Heterogeneous catalysis for hydrogenation in the liquid phase is important in many industrial processes, and in emerging areas in bio-based renewable energy. Hydrogen mass transfer through the liquid phase to the catalytic sites is hampered by the low solubility of hydrogen in liquids. Hydrogen starvation of the catalyst can cause unwanted reactions, poor selectivity, and coking. Hydrogen supply directly to the catalyst palladium deposited on an integral-asymmetric membrane can alleviate mass transfer issues and decouple the liquid-phase hydrostatic pressure from the hydrogen supply to the catalyst. Phase imaging atomic force microscopy (AFM) can detect the appearance of hydrogen on the palladium in the liquid phase in real time, and is applied to track a hydrogenation reaction. This application of AFM opens new avenues to study liquid-phase catalysis over metals in real time, with nanometer-scale resolution.

495182 Polycrystalline MOF Membranes on Hollow Fibers: Processing, High-Performance Separations, and Tunable Molecular Sieving

Abstract:

This talk will describe our recent progress[1-6] in the fabrication and properties of ZIF-type MOF molecular sieving membranes on polymeric and other hollow fiber platforms.

Membrane separation is a key component of strategies targeting step changes (rather than incremental ones) in cost and energy-efficiency during production of fuels and chemicals from conventional and renewable feedstocks. Two important barriers to the development of MOF membrane separation technology are (i) the difficulty in obtaining tunable, high-performance, and robust molecular sieving materials; and (ii) the lack of low-cost, scalable membrane fabrication processes.

A 'multiscale' approach to nanoporous membrane fabrication processing is required to successfully exploit their fundamentally attractive characteristics. We will discuss how ZIF-type MOF materials can be processed into polycrystalline membranes on hollow fiber platforms, with control over nanometer-scale defects as well as micron-scale transport and film crystallization phenomena within the hollow fibers. These membranes allow high-performance separations of molecules such as hydrogen, olefins, and isomers. The talk will conclude with examples of generalizing the above strategy towards a versatile MOF membrane technology that addresses a wider range of molecular separation targets.

Overall, this talk highlights the fact that crystalline nanoporous materials and membranes with precise molecular sieving characteristics can overcome limitations associated with 'conventional' polymeric, carbon molecular sieve (CMS), or hybrid/mixed matrix membranes. At the same time, the above 'conventional' materials and concepts can still be useful in the development of advanced nanoporous membranes. Fruitful insights of this nature – obtained during collaborations of the Nair group and the Koros group – are highlighted at various junctures of this talk.

1. K. Eum, C. Ma et al, *Adv. Mater. Interfaces*, 2017 (*in press*).

2. K. Eum et al, *ACS Appl. Mater. & Interfaces*, 8 (38), 25337-25342 (2016).

3. K. Eum et al, *Adv. Funct. Mater.*, 26 (28), 5011-5018 (2016).

4. K. Eum, K. C. Jayachandrababu, F. Rashidi, K. Zhang et al, *J. Am. Chem. Soc.*, 137, p. 4191-4197 (2015).

5. A. J. Brown et al, *Science*, 345 (6192), 72-75 (2014).

6. A. J. Brown et al, *Angew. Chem. Intl. Ed.*, 124, p. 10767-10770 (2012).

487193 Membranes for Separation of Complex Refinery Streams

Abstract:

Large-scale hydrocarbon separation technologies are largely thermally driven phase-change based processes and account for roughly 45% of production energy in downstream petrochemical and manufacturing processes. Pressure-driven membranes based separation systems offer an energy-efficient, non-phase change based alternative strategy. Chemically, thermally and mechanically robust membranes can potentially lay the foundation for a â€œmembrane revolutionâ€ in the area of hydrocarbon purification similar to the revolution that occurred in water desalination with traditional reverse osmosis membranes.

We discuss class and boiling point separation of complex liquid hydrocarbon streams using commercial ceramic membranes currently in use for water treatment applications. We have evaluated whole crudes ranging from light (Arabian Light) to heavy (Maya), studying the impact of membrane pore size, operating temperature and pressure. The membrane performance was found to be stable in batch and long-term crossflow testing under harsh chemical, thermal (up to 200°C) and mechanical (~600 psi) conditions. Fluxes were comparable to ultrafiltration (~10 GFD) with molecular weight based rejection of multi-ring aromatics, >90% rejection of microcarbon residue (MCR) and metals. To better understand the composition of permeate and retentate streams; simulated distillation (SIMDIST), elemental, MCR and total acid number (TAN) analyses were conducted.

Next, we describe the development of a separations model for the transport of thousands of molecular species in the whole crude by combining fundamental transport equations with a compositional platform. Finally, we discuss a case study from the refining and petrochemicals space providing a perspective on the potential application of these membranes.

493933 Bibliography of Tony Cai

Abstract:

Dr. Tony Cai holds a B.S. and M.S. in Chemical Engineering from Tianjin University and a PhD in Mechanical Engineering from Oklahoma State University. Dr. Tony Cai has been FRIâ€™s senior engineer and Assistant director for over 20 years. During this period, Tony played a significant role in leading and shaping the FRI research program.

Tony is a renowned worldwide distillation expert, and provides expert distillation guidance to FRI members all over the world. He also has experience in the areas of fluid dynamic analysis, instrumentation, structural analysis and hydraulic modeling. Tony is an exceptional leader and mentor to FRI engineers, FRI operators, and Oklahoma State University students who work closely with FRI.

Tonyâ€™s contributions to distillation are in numerous areas, including but not limited to operational planning of the FRI experimental test program, data collection and analysis, safety, leading joint-technical collaborations, and correlation development.

This paper highlights many of the distillation contributions of Dr. Tony Cai in his tenure at Fractionation Research, Inc. (FRI).

493938 Dr. Tony Cai's Non-Distillation Contributions to Chemical Engineering

Abstract:

Dr. Tony Cai's contributions to distillation engineering have been appreciable and globally appreciated. His contributions to non-distillation areas have been of equal magnitudes. This presentation and paper summarize his work in the fields of kettle reboiler heat transfer, kettle reboiler hydraulics, absorption, PDI spray characterizations, spray absorption, mist elimination, process control software, Steady State Identification (SSID) software, flow metering, corrosion under insulation (CUI), gamma scanning, pressure relief valves and liquid orifice discharge coefficients. On Sunday afternoons, he rested - sometimes.

High Energy Saving by Commercially Operated Heat Integrated Distillation Column with New Heat Integration Arrangement

Toshihiro WAKABAYASHI, Toyo Engineering Corporation, Japan*

INTRODUCTION

Distillation is the most-used separation unit operation in chemical process industries. On the other hand, it is widely recognized as large thermal energy consumer due to its inherent characteristics. Accordingly, many energy conservation technologies have been developed in this area.

In view of the energy conservation, the reversible distillation provides the ultimate figure theoretically. Reversible distillation is the operation in which number of theoretical stage is infinite and each stage is equipped with side-condenser in the rectifying section and side-reboiler in the stripping section. The heat duty at each side heat exchangers is infinitesimal in this operation. With such scheme, the operation is implemented in reversible manner and subsequent exergy loss is ultimately minimized.

Heat integrated distillation column (hereinafter called HIDiC) draws high attention since it attempts to realize the reversible distillation in reality by incorporating the heat pump effect into the reversible distillation. Many researchers have worked on this research and development all over the world in attempting to commercialize the HIDiC in industries, however, its commercialization has not yet been achieved. [1),2),3),4)]

Authors focused on the relationship between the composition-change and the enthalpy-change inside the distillation column and came up with new concept of internal heat exchanges[5),6)]. Moreover, the new structure to realize new concept has been raised, which is called as *SUPERHIDIC®*. In 2014 Japanese petrochemical company, Maruzen Petrochemical Co., decided to adopt *SUPERHIDIC®* in commercial distillation column, which was started up in 2016. *SUPERHIDIC®* has been operated successfully in stable operation with 55%+ energy conservation to the conventional distillation column.

In this paper, the concept of *SUPERHIDIC®* is presented and actual operating data is presented to demonstrate its excellent performance.

CONCEPT of CONVENTIONAL HIDiC

First, the concept of the conventional HIDiC is provided so as to understand the concept of *SUPERHIDIC®* easily. In the conventional HIDiC, as shown in **Figure 1**, the rectifying section and the stripping section in the conventional distillation column is divided and the compressor is installed in between two sections. Subsequently, the temperature in the rectifying section can be higher than that in the stripping section. This makes it possible to transfer the heat rejected from the side condenser in the rectifying section to the side reboiler in the stripping section, which is called as internal heat exchanges. Accordingly, the enthalpy changes in both the rectifying section and the stripping section can be accomplished without supplying various ranges of utilities. In case that the distillation column has quite large number of theoretical stages and sufficient temperature difference can be derived, this scheme can bring the similar enthalpy profile inside the column as reversible distillation.

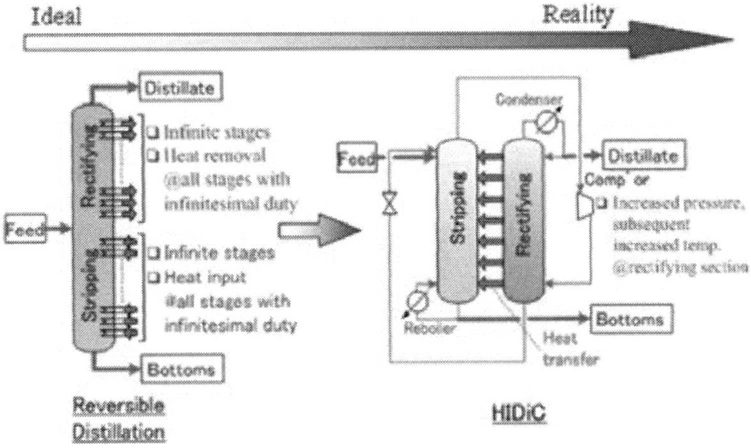

Fig. 1 Concept of reversible distillation and the conventional HIDiC

DRAWBACKS of CONVENTIONAL HIDiC

Although it appears that the conventional HIDiC can derive similar enthalpy profile as the reversible distillation, this is not true in most columns. This is because the actual column has finite stages whereas the reversible distillation assumes infinite stages. Once the column has finite stage, the ideal heat duty at each stage, *i.e.* at each composition, becomes finite whereas the reversible distillation operation

*: toshihiro.wakabayashi@toyo-eng.com

assumes infinitesimal heat duty. Intensive thermodynamics analysis made by author demonstrated that ideal heat duty is different at each stage as illustrated in **Figure 2** where such heat duty is shown as the length of the arrow in example[5),6)]. In the conventional HIDiC, the rectifying section and the stripping section are set in parallel and the internal heat exchanges are implemented between stages physically at the same level and those duties are simply dependent on the temperature difference between the involved stages. In such a scheme, the ideal internal heat exchange cannot be achieved. Therefore, the conventional HIDiC has inherent drawback to be solved.

Most of the conventional HIDiC has the concentric structure in which the high pressure rectifying section corresponds to the inner pipe and the low pressure stripping section corresponds to the spacing between the outer pipe and the inner pipe. Typically, the unit module is developed and the capacity is adjusted by changing the number of such unit module[4)]. Due to such concentric structure, the conventional HIDiC has the following drawbacks which are obstacles for the commercialization.

(1) Draw-off product from intermediate stage is very difficult

(2) Multiple-feed cannot be accommodated.

(3) Maintenance is impractical.

CONCEPT of SUPERHIDIC

As has been shown in Figure 2, in order to make the composition–enthalpy profile closer to that in the reversible distillation, the following points should be achieved.

(1) Heat should be supplied or removed only at the appropriate composition. Some of the small amount of heat may be summarized at the neighbor composition having large amount of heat demand. This makes the heat exchange manner simpler and the subsequent enthalpy changes become discretely. Typically around 4 enthalpy changing points are sufficient to have similar enthalpy profile as the reversible distillation operation.

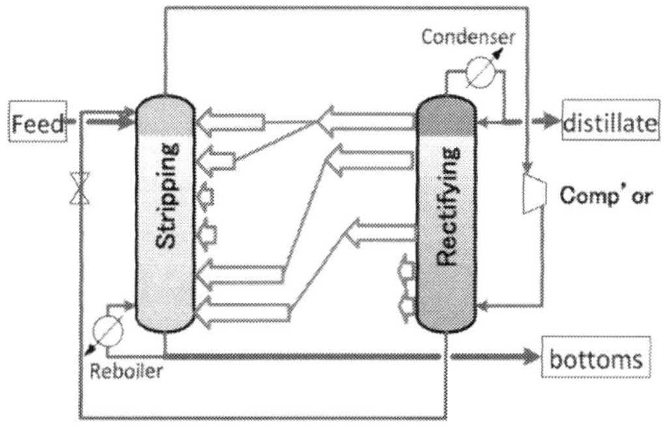

Fig. 2 Drawbacks of the conventional HIDiC and basic concept of *SUPERHIDIC*®

(2) Stages requiring same heat duty in the rectifying section and the stripping section should be heat-integrated. In case that the heat more than the required amount is supplied to the side reboiler, excess heat does not help reducing the reboiler duty but causes unnecessary evaporation, which results in unnecessary power at the compressor. In case that the heat less than the required amount only is supplied to the side reboiler, the composition misses the opportunity to reduce the reboiler duty. Here, one stage can be heat-integrated with plural stages if necessary. Also, the condenser stage in the rectifying section and the reboiler stage in the stripping section can be involved in the heat integration. Such heat integration is called side heat exchange hereinafter, distinguishing from the internal heat exchange in the conventional HIDiC.

(3) The selected side heat exchange should have sufficient temperature difference so as to achieve the required heat duty with economical heat transfer area.

To achieve such three points, *SUPERHIDIC*® illustrated in **Figure 3** has been developed. In this structure, the most outstanding features is that side heat exchanges are accomplished by employing the normal stabbed-in type heat exchangers (if necessary the heat exchanger can be installed externally to the column). Side heat exchanger is installed either of the stage in the rectifying section or the stripping section. The other stage is connected by external piping. With such a way, the side heat exchange can be implemented between the stages having the target compositions. At same time, by adjusting the heat transfer area, the required heat duty can be achieved by not simply depending on the temperature difference like the conventional HIDiC. Other outstanding feature in the *SUPERHIDIC*® is that the low

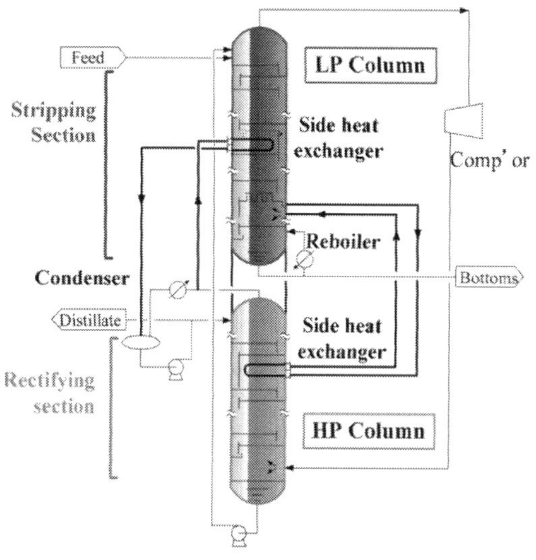

Fig. 3 Schematic process flow of *SUPERHIDIC*®

pressure stripping section is elevated above (not limited as stacked) the pressurized rectifying section. With such arrangement, the heat exchange can be accomplished without utilizing additional equipment like pump since the circulation can be implemented by thermos-siphon effect or the gravity. Although the structure is simple and the system comprises of well-proven equipment and component, it can achieve closer composition-enthalpy profile to reversible distillation than the conventional HIDiC. Accordingly, the resulted energy conservation is quite large.

At same time, all drawbacks highlighted in last section can be solved in *SUPERHIDIC*®.

SUCCESSFUL OPERATION of SUPERHIDIC in COMMERCIAL PLANT

Maruzen Petrochemical Co. Ltd., Japanese petrochemical producer, decided to apply *SUPERHIDIC*® technology to MEK(methyl ethyl ketone) fractionator that is large energy consumer in his Alcohol Ketone Process Unit in 2014. The facility has been in operation successfully since August, 2016.

Table 1 shows typical feedstock in the commercial distillation column. MEK is purified from mixture mainly consisting of MEK and SBA (*sec*-butanol). The column separates the feedstock to the distillate having MEK of 99.918 wt% and bottoms having MEK of 200 (wt)ppm. The conventional distillation, having theoretical stages corresponding to 1.11 times minimum reflux ratio, consumes the reboiler duty of 6055 kW and the pump power of 64 kW

Table 1 Feedstock of *SUPERHIDIC*® in commercial plant in Maruzen Petrochemical Co., Chiba, Japan

Flowrate	[kg/h]	15994
	[kmol/h]	219.9
Temperature	[K]	366.0
Phase	[-]	liquid
Composition	[weight frac.]	
light non-keys		0.000158
methyl ethyl ketone		0.720506
intermediate		0.001723
sec-butanol		0.270165
heavy non-keys		0.007448

for the required separation. The process simulation for the conventional distillation is based on the various conditions in existing column implementing same separation.

Figure 4 shows the process flow diagram of *SUPERHIDIC*® in commercial plant. Upper part of the stacked column, 5V-1B, is the low pressure column which consists of the stripping section and lower part of the rectifying section while the lower part of the stacked column, 5V-1A, is the high pressure column corresponding to the upper part of the rectifying section. Intensive thermodynamics study on side heat exchanger arrangement suggests that four side heat exchanges are suitable to have similar composition-enthalpy profile as the reversible distillation. In this particular application, three side heat exchangers, 5E-1A, 1B and 1C, involves condenser stage in the rectifying section and the intermediate stages in the stripping section while fourth side heat exchanger, 5E-1D, involves the reboiler stage in the stripping section and the

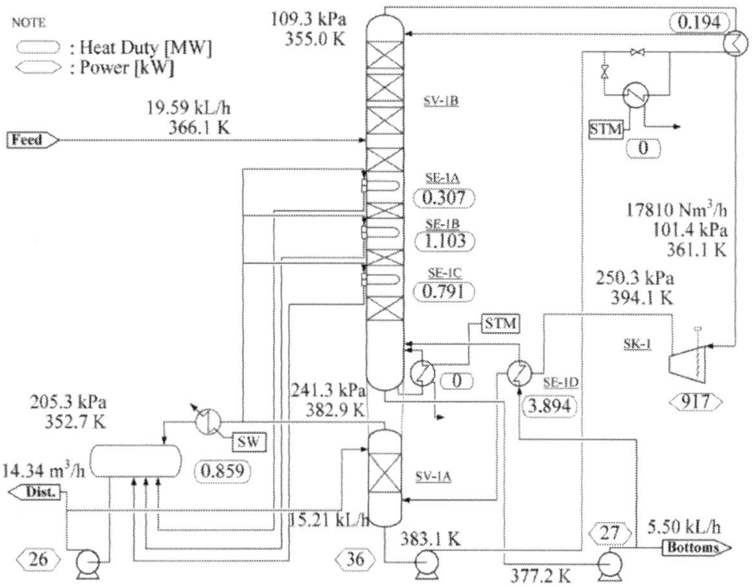

Fig. 4 Process flow diagram with operating conditions in 1st commercial application of thermally optimized HIDiC in Maruzen Petrochemical Co., Chiba, Japan

Fig. 5 Overview photograph of high performance HIDiC in commercial plant in Maruzen Petrochemical Co., Chiba, Japan

intermediate stage in the rectifying section. The overview of the facility can be seen in **Figure 5**.

The operating data in *SUPERHIDIC*® under stable commercial operation are presented in Figure 4 and **Table 2**. The reboiler duty becomes 0 kW while the compressor consumes the power of 917 kW. This power already includes mechanical loss and motor efficiency of 93.4 %. In addition, system consumes total 89 kW at three pumps.

In this paper, the power is converted into the primary energy with power generation efficiency of 36.6%. In accordance with the said procedure, the energy consumption in the conventional distillation is 6230 kW. On the other hand, *SUPERHIDIC*® consumes 2751 kW only, which results in 55.9 % energy conservation to the conventional distillation.

APPLICABLE SERVICES of SUPERHIDIC

SUPERHIDIC® is a sort of mechanical heat pump technology, thus, it still has some limitation to be applied. Good guideline is the temperature difference between the overhead and bottom when the service is operated in conventional distillation. In case that the said temperature difference is less than 80 degree Celsius, *SUPERHIDIC*® has potential to provide high energy conservation such as 50%+ to the conventional distillation. The process units which have the distillation column having high energy conservation are summarized in **Figure 6**. These process units are just examples and have been just confirmed so far.

Table 2 Operating data of *SUPERHIDIC*® in commercial plant in Maruzen Petrochemical Co., Chiba, Japan

Overall material balance			
Feed	[kL/h]	19.59	(= 219.9 kmol/h)
Distillate	[m3/h]	14.34	(= 160.0 kmol/h)
Bottoms	[kL/h]	5.50	(= 59.95 kmol/h)
Energy consumption			
Condenser duty	[MW]	0.859	
Reboiler duty	[MW]	0	
Compressor power	[kW]	917	
Pump power			
Reflux & distillate pump	[kW]	26	
Bottms pump	[kW]	27	
Recycle pump	[kW]	36	
Operating pressure			
High pressure column @top	[kPa]	241.3	
Low pressure column @top	[kPa]	109.3	
Compressor inlet	[kPa]	101.4	
Compressor outlet	[kPa]	250.3	

Side heat exchangers performance	Combination (*2)	Heat duty	Temp. difference
	[-]	[MW]	[K]
Side heat exchnager No.1	R1-S3	0.307	22.9
Side heat exchnager No.2	R1-S5	1.103	21.1
Side heat exchnager No.3	R1-S7	0.791	17.3
Side heat exchnager No.4(*1)	R8-S21	3.894	5.8

*1: Heat duty of side heat exchanger No.4 was determined based on heat balance around high pressure column while others were based on the enthalpy change in hot stream.

*2: R_i-S_j corresponds to the heat exchanges between stage i in the rectifying section and stage j in the stripping section.

CONCLUSION

The concept of HIDiC was re-considered and new heat integration arrangement has been developed to improve the energy conservation. Such new concept has been realized as *SUPERHIDIC*® technology, which has been applied to the commercial distillation in Japan. The commercial operation demonstrates excellent energy conservation of 55.9% to the conventional distillation under stable operation.

SUPERHIDIC® has large potential to be applied to wide range of application in chemical process industry.

Literature Cited

1) Mah RSH, Nicholas JJ, Wodnik RB. Distillation with secondary reflux and vaporization: a comparative evaluation. *AIChE J.* 1977; **23**: 651-658.
2) Nakaiwa M, Huang MK, Endo A, Ohmori T, Akiya T, Takamatsu T. Internally heat-integrated distillation columns: a review. *Chem Eng Res Des.* 2003; **81**: 162-177.
3) Kiss AA, Olujic Z. A review on process intensification in internally heat-integrated distillation columns. *Chem Eng Process.* 2014; **86**: 125-144.
4) Horiuchi K., Nakaiwa M, Iwakabe K, Matsuda K and Toda M. Intensification of the process flow in the pilot plant of an internally heat-integrated distillation column (HIDiC). *Kagaku Kogaku Ronbunshu*, 2008; **34**: 70-75.
5) Wakabayashi T, Hasebe S. Design of heat integrated distillation column by using *H-xy* and *T-xy* diagrams. *Comput Chem Eng.* 2003; **56**: 174-183.
6) Wakabayashi T, Hasebe S. Higher energy saving with new heat integration arrangement in heat-integrated distillation column. *AIChE J.* 2015; **61**: 3479-3488

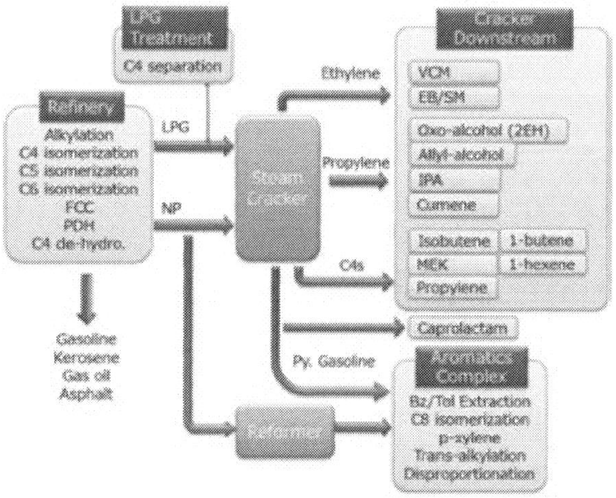

Fig. 6 Overview photograph of high performance HIDiC in commercial plant in Maruzen Petrochemical Co., Chiba, Japan

Designing Edible Oil Stripping Columns

Ž. Olujić
R. Schulz,
M. Fuhrmeister
H. Jansen
T. Rietfort

Your partner in chemical engineering

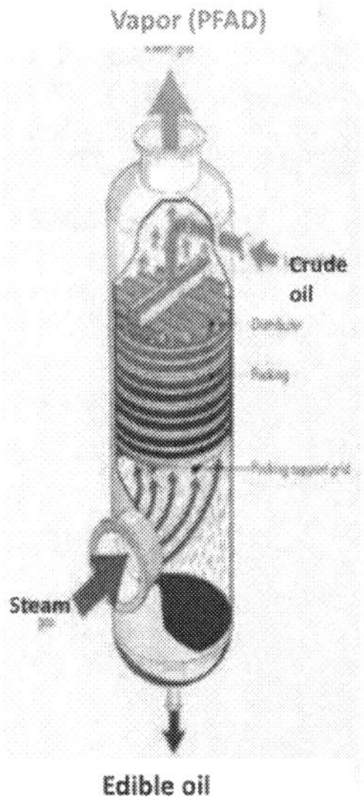

Separations Division - DAA Honors Session (**Honoree Dr. Tony Cai**)
2017 AIChE Annual Meeting
Minneapolis, MN, USA, October 29 – November 3, 2017

Edible Oil Refining

Vegetable oils processed on largest scale: **Palm oil, Soybean oil, Sunflower oil, ...**

Final processing step, from crude to edible oil, includes **removal of all volatile impurities** to get edible oil complying with national standards.

Impurities, mainly **FFAs** (free fatty acids: C12-C20) and a number of other useful volatile components as well as contaminants (depending on oil origin) are efficiently and cost-effectively removed from crude edible oil by **overheated steam stripping at high** (as high as tolerable) **temperatures and lowest** industrially viable **operating pressure**.

Extremely low operating pressure imposes "**exotic**" **fluid-dynamic conditions** that differ from those encountered in common stripping and distillation applications. A thorough understanding of peculiarities of this particular stripping process is needed to be able to act adequately and arrive at an optimal design.

Design practice: It is empirical, but sound and certainly adequate.

Challenge: Development of a simple, robust, and reliable engineering tool for the purposes of fast screening of potential (re)design options.

Edible Oil Deacidification & Deodorization

***State of art** - Three operations in one shell*

(1) Stripping of FFA & other valuables & contaminants!
(2) Recovery of FFA & other valuables (Scrubbing)!
(3) Deep deodorization, to remove odoriferous components
(lower pressure & temperature, longer residence time)

Operating conditions

T_{oil} = **230 - 270** °C ($T_{TOP} \geq T_{BOTTOM}$)

p_{top} = **2-5** mbar ($p_{BOTTOM} = p_{TOP} + \Delta p$)

$F_{G,top}/F_{G,bot}$ = 1.5 - 10

Crude edible oil entering stripping section usually contains ≤1% of FFA and other volatiles. Exceptional are palm oil strippers, which have much larger FFA loads (up to 5% or even more!)

Main part of FFAs contained in crude edible oil is usually separated in a **pre-flash vessel**.

Steam ejectors based vacuum generation (≥ **2 mbar** at the top), with **stripping steam** being **10 - 15% of total motive steam requirement.**

Operating pressure at the top of stripping section depends on the complexity of the scrubbing (condensation) section above it.

Δp relatively large and affects adversely stripping performance!!!

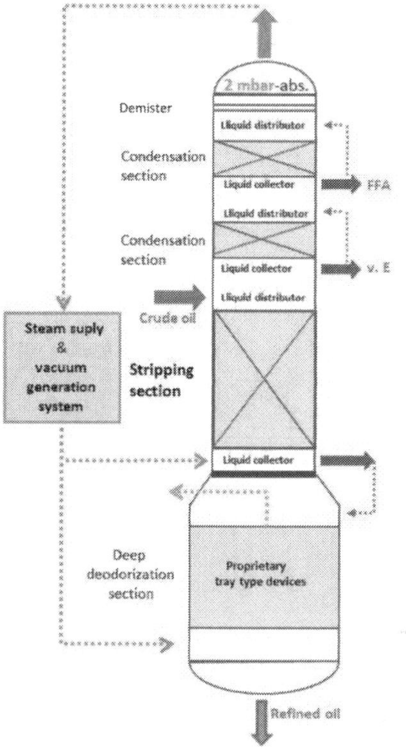

zarko.olujic@hotmail.com

Stripping and Scrubbing

Structured packings a natural choice!

1st generation: common corrugated gauze and sheet metal (45° and 60°) structured packings (250 – 500 m²/m³)

2nd generation: Optiflow and **Rombopak**
(preferred choice in 1990s, but **not available anymore**!)

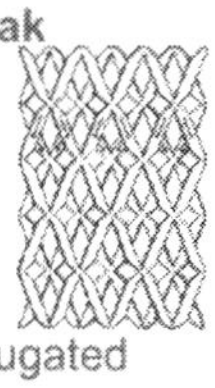

3rd generation: High performance corrugated sheet structured packings (BXPlus, Flexipac-HC, MellapakPlus, Montz-Pak M&MN, Raschig Super Pak)

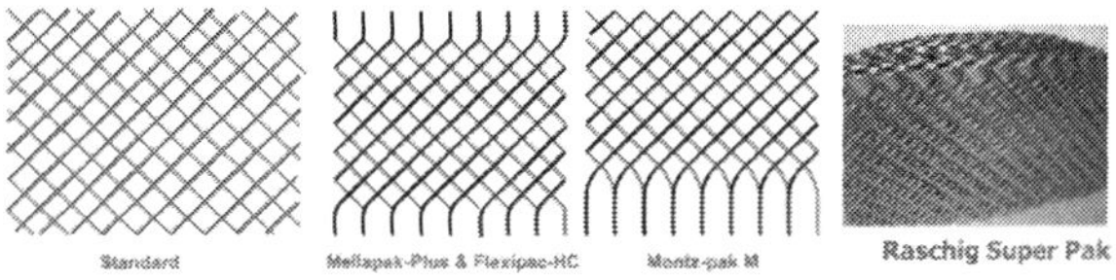

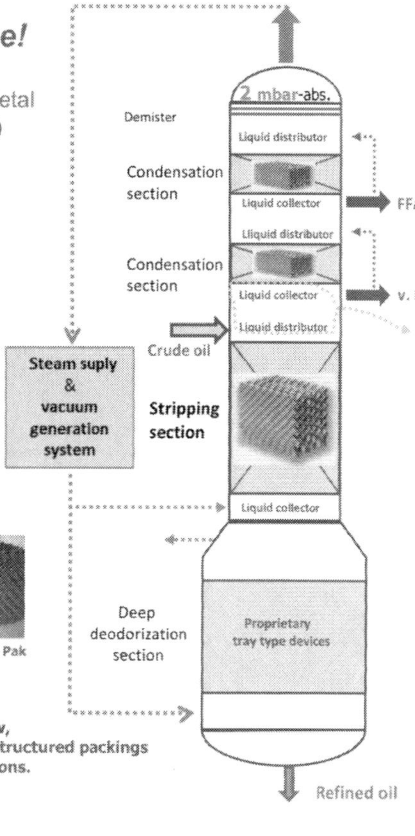

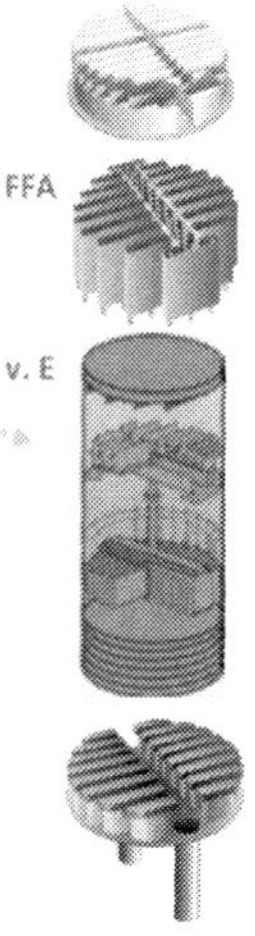

In deep vacuum applications, with prevailing laminar gas/vapor flow, advanced geometrical features of 3rd generation corrugated sheet structured packings do not bring benefits to the extent experienced in common applications.

zarko.olujic@hotmail.com

Stripping Process Model

Predicting stage requirement and bed height

h_{bed} = N HETP = $NTU_{oL}\ HTU_{oL}$

$$NTU_{oL} = \frac{S}{S-1}\ln\left[\frac{x_{in}}{x_{out}}\left(1-\frac{1}{S}\right)+\frac{1}{S}\right]$$

$$HTU_{oL} = \frac{L_t}{K_x a_p A_c} = HETP\frac{1}{S}\frac{S-1}{lnS}$$

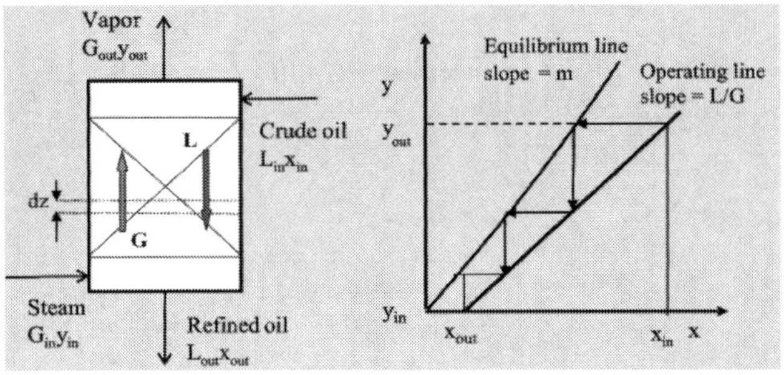

Design model of edible oil stripper:

$$h_{bed} = HETP\frac{ln\left[\frac{x_t}{x_b}\left(1-\frac{1}{S}\right)+\frac{1}{S}\right]}{\ln S}$$

Key design variable: *Stripping factor*

$$S = \frac{m}{\left(\frac{L_t}{G_b}\right)} = \frac{2p^*}{(2p_t+\Delta p)}\left(\frac{G_b}{L_t}\right)$$

Qualitatively, the effectiveness of edible oil stripping increases with increasing **(*G/L*)**, increasing ***T***, decreasing $\boldsymbol{p_{top}}$, and decreasing $\boldsymbol{\Delta p}$.

$p^* = f(T)$, and $p_{des} = (p_t + p_b)/2$

If T = const.: $m = p^*/p_{des} = 2p^*/(2p_t + \Delta p)$

If $T_t > T_b$: $m = (p_t^*/p_t + p_b^*/p_b) = (m_t + m_b)/2$

Note: (p_b, m, S, NTU_{oL})& $h_{bed} = f(\Delta p)$

$\Delta p_{DESIGN} = (\Delta p_t + \Delta p_b)/2$

*(p_b = **unknown**; iterative procedure required!)*

zarko.olujic@hotmail.com

5

ALGORITHM

Given: M_L, x_{in}, x_{out}; *packing dimensions and HETP*
μ_G & μ_L & ρ_L & $p_v = f(T)$, and $\rho_G = f(p,T)$

Fixed: ***T***, $\boldsymbol{p_{top}}$, *and* $\boldsymbol{M_G}$, Initial guesses: $\boldsymbol{\Delta p}$, $\boldsymbol{d_c}$, $\boldsymbol{h_{bed}}$

Calculated iteratively until no change in $\boldsymbol{\Delta p}$

m, S, and $\boldsymbol{h_{bed}}$ *calculated*

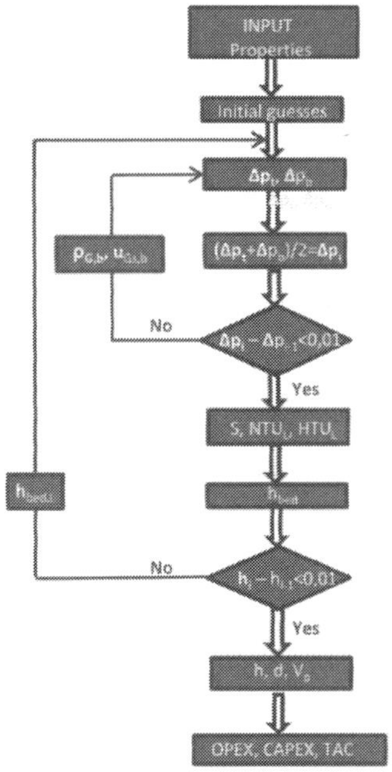

Calculated $\boldsymbol{h_{bed}}$ *used to start new iteration!*
($\boldsymbol{d_c}$ can also be adjusted)

Upon convergence,
*total annualized costs (**TAC**) estimated.*

As a guidance, m, S, and h_{bed} can be calculated separately as a function of $\boldsymbol{\Delta p}$
(in small Δp increments) to see the amount of pressure drop that can be tolerated in given case.

zarko.olujic@hotmail.com

7

Pressure Drop Estimation Method

A theoretically founded approach
(*adaptation of Hagen-Poisseuille laminar tube flow equation*)

$$\Delta p = 26.2\ \mu_G \left(\frac{h_{bed}/sin\alpha}{d_{hG}^2}\right) \frac{u_{Gs}}{(\varepsilon - h_L) sin\alpha}$$

Hydraulic diameter for corrugated sheet packing with parallel V-shaped gas flow channels:

$$d_{hG} = 4A/P = 4/a_p = s$$

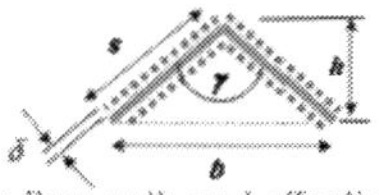

Effective gas flow angle α determines the length of the gas flow path and effective gas velocity (in first approximation taken equal to corrugation inclination angle!!)

Liquid holdup and film thickness:

$$h_L = \delta \cdot a_p \qquad \delta = \left(\frac{3\mu_L u_{Ls}}{\rho_L g\, a_p \sin\alpha_L}\right)^{\frac{1}{3}}$$

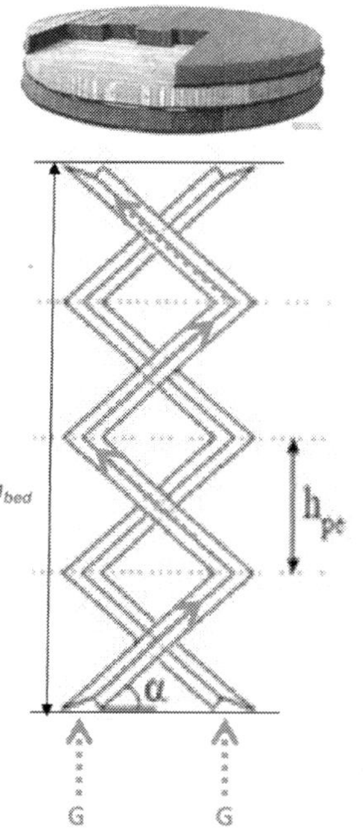

SULCOL empirical model

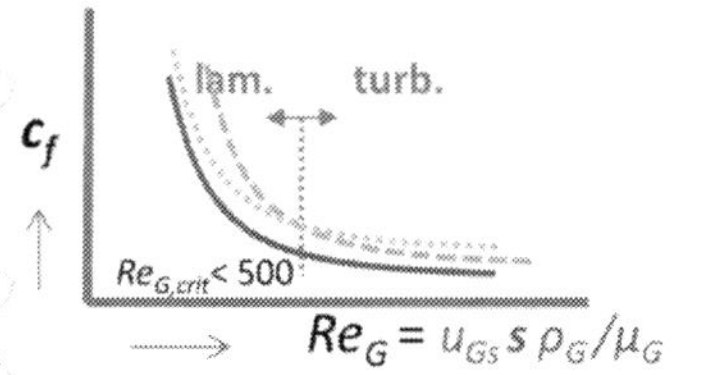

$$\frac{\Delta p}{\Delta z} = \frac{c_f}{d_{hG}} \frac{\rho_G u_{Gs}^2}{2} = \frac{c_f}{d_{hG}} \frac{F_{Gs}^2}{2}$$

BASE CASE - *Edible oil stripper design and rating*

(a typical, lowest top gas density case!)

Capacity:	M_L = 20 t/h
Impurities removed:	70 kg/h, including key component
Key component specs:	x_{in} = 1500 ppm, x_{out} = 500 ppm
Operating conditions:	p_{top} = 2,5 mbar, T = 250 °C ($p_v \approx$ 4 mbar)
Initial guesses:	d_c = 2.5 m, h_{bed} = 4 m, Δp = 1 mbar M_G = 250 kg/h (1.25 % of M_L)

Design alternatives considered

Packings:	A3-500,	B1-250.60, and	**B1-170.60**
Assumed packing HETP:	(0.5 m)	(1 m)	(1.5 m)

Operating top pressure:	p = 2, 3, and 3.5 mbar
Operating temperature:	T = 240, 260, and 270 °C
Stripping steam flow rates:	M_G = 1.1, 1.5, 1.75, and 2 % of M_L !
Existing plant rating:	increased capacity/purity.

Results - *"Optimum" designs based on same pressure drop*

$M_G = 1.25\ \%\ M_L$

Packing	Δp (mbar)	d_c (m)	h_{bed} (m)	V_p (m^3)	$F_{G,top}$ ($Pa^{0.5}$)	TAC (US $)
A3-500	**1.77**	2.87	1.88	12.1	0.395	704000
B1-250.60	**1.77**	1.96	3.76	11.3	0.847*	699000
B1-170.60	**1.77**	1.62	5.64	11.6	1.239*	695000

*Very high gas velocity (>20 m/s)!

Height to diameter ratio (h/d) and TAC as a function of stripping gas load for A3-500

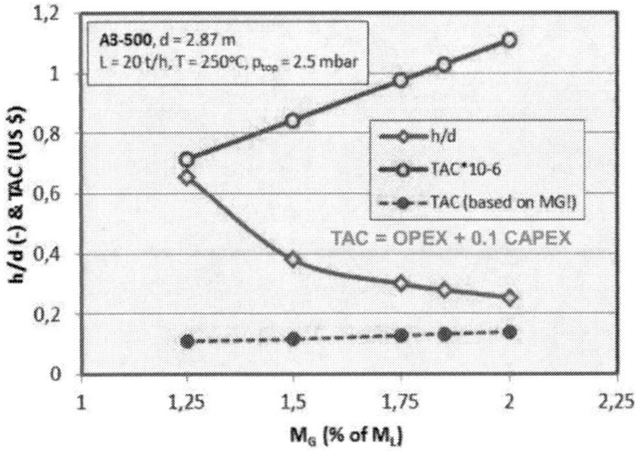

zarko.olujic@hotmail.com

Results

Comparison based on same column diameter and fixed stripping gas load

$M_G = 1.25\ \%\ M_L$

Packing	Δp (mbar)	d_c (m)	h_{bed} (m)	V_p (m^3)	$F_{G,top}$ ($Pa^{0.5}$)	TAC (US $)	CAPEX (US $)
A3-500	1.77	**2.87**	1.88	12.1	0.395	714000	431000
B1-250.60	0.65	**2.87**	2.66	17.2	0.395	711000	386000
B1-170.60	0.43	**2.87**	3.73	24.1	0.395	713000	410000

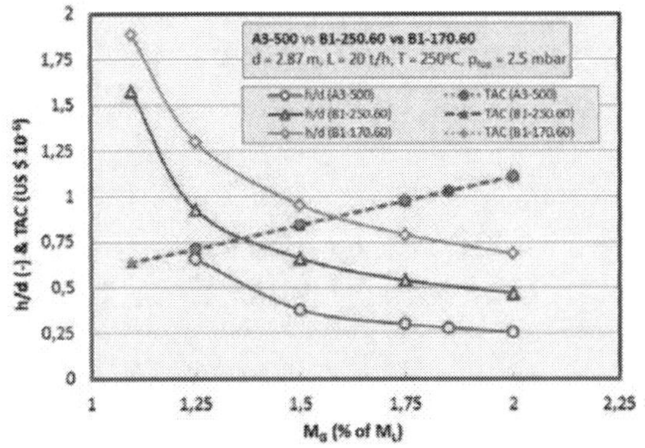

All structured packings suitable, but preference should go to ***coarser packings with a steep corrugation inclination angle.***

zarko.olujic@hotmail.com

Results

Column diameter effects

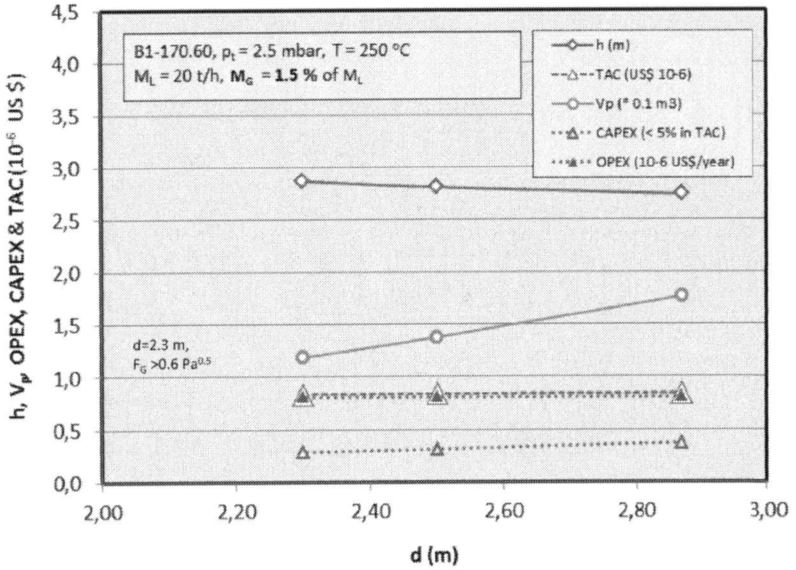

TAC not suitable to serve as design optimization criterion, ... !

Results

Operating temperature and pressure effects

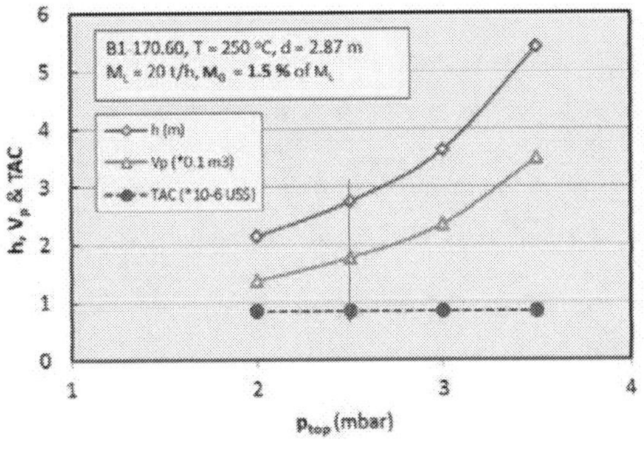

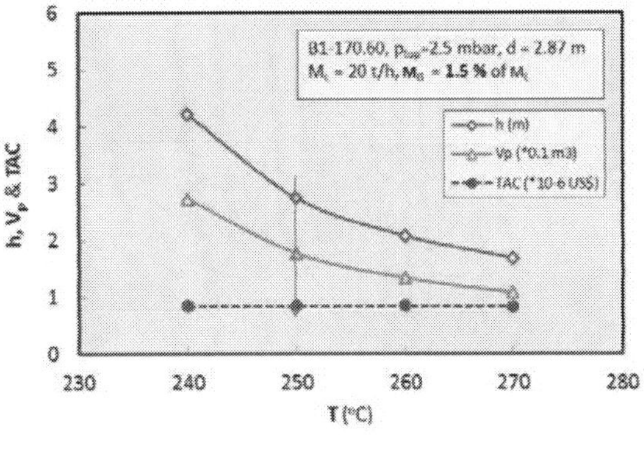

Packing	Δp (mbar)	d_c (m)	h_{bed} (m)	V_p (m^3)	$F_{G,top}$ ($Pa^{0.5}$)	TAC (US $)
p_t = 2 mbar	0.36	2.87	**2.19**	10.7	0.511	842000
p_t = 2.5 mbar	0.37	2.87	**2.81**	13.8	0.457	844000
p_t = 3 mbar	0.41	2.87	**3.77**	18.5	0.417	847000
p_t = 3.5 mbar	0.52	2.87	**5.42**	34.9	0.386	854000

⇩

Packing	Δp (mbar)	d_c (m)	h_{bed} (m)	V_p (m^3)	$F_{G,top}$ ($Pa^{0.5}$)	TAC (US $)
T = 240 °C	0.54	2.87	**4.22**	7.3	0.453	850000
T = 250 °C	0.37	2.87	**2.74**	7.7	0.457	844000
T = 260 °C	0.30	2.87	**2.08**	3.4	0.461	842000
T = 270 °C	0.25	2.87	**1.69**	0.9	0.466	840000

⇧

Results

Rating of an existing column

Packing	Δp (mbar)	T (°C)	M_G (%)	S (-)	h_{bed} (m)	$F_{G,\,top}$ ($Pa^{0.5}$)	OPEX (-)
B1-170.60, p_{top} = 2.5 mbar, d_c = 2.87 m, h_{bed} = 2.74 m, x_{in}/x_{out} = 3							
Increasing production capacity by 20 %							
M_L = 20 t/h	0.37	**250**	**1.50**	1.067	**2.74**	0.457	1
M_L = **24** t/h	0.44	**251**	1.50	1.079	2.69	0.550	1.20
M_L = **24** t/h	0.45	250	**1.52**	1.068	2.73	0.556	1.22
Increasing product purity by 200 ppm (x_{in}/x_{out} = **5**)							
M_L = 20 t/h	0.48	**260**	**1.95**	1.714	2.73	0.574	1.30
M_L = 20 t/h	0.45	**265**	**1.75**	1.729	2.71	0.526	1.17

zarko.olujic@hotmail.com

14

Concluding Remarks

Deep vacuum - high temperature operation related peculiarities *need to be recognized and considered properly when dimensioning hot steam stripping units for edible oil deacidification and deodorization.*

Pressure drop affects hydraulics and separation strongly*. Established predictive models tend to underpredict pressure drop, and this tendency grows stronger with decreasing gas phase Reynolds number, which may result in adopting a design with inadequate bed height.*

Fortunately, insufficiencies in design/operation can be repaired at affordable costs, simply by increasing ***operating temperature*** *accordingly.*

Proposed preliminary design method, utilizing ***a theoretically founded pressure drop expression*** *in conjunction with an adequate design equation for bed height, a suitable and* ***effective engineering tool*** *for fast screening of the edible oil stripping (re)design options.*

OPEX is dominating factor, *and lowest (****reasonable****) stripping gas flow rate should be chosen.*

Overdesign on bed height not, but on diameter side well a reasonable consideration.

Deep vacuum applications remain a domain of designer's creativity and preferences.

zarko.olujic@hotmail.com

15

All the best to you, Tony!

zarko.olujic@hotmail.com

494719 On the Selection of Optimal Configuration for Ternary Distillation

Abstract:

Ternary distillation has recently drawn much attention, owing to the recent extensive investigations on dividing-wall column (DWC), or its equivalence the Petlyuk column, and thermally coupled distillation configurations with their significant potential in energy saving. However, except the DWC and thermally coupled distillation, a number of alternatives for tenary distillation may be atractive in terms of Total Annual Cost (TAC). The cost of a ternary distillation configuration depends on the composition, relative voatilities, and mass fractions of the components in the feed, allocation of the components among the product streams and so on. In addition, a ternary separation can also be used for initial splits of multicomponent mixture into three streams of more than one component for further processing. In this case, the specifications for the key components may be rather flexible, or the effective ease of separetion can be different from those for three-component mixture separation. In the precent work, rigorous simulation and optimization were performed to analyze and compare, in terms of TAC, different promising distillation configurations, including simple columns sequences with or without heat-integration, complex column with side-rectifier or side-stripper, and fully thermally coupled, or DWC for ternary separation. The triangle diagrams of the feed composition were employed to show the partition of the diagram by the different distillation configurations that have the lowest TAC in the corresponding composition areas by the partition on the triangle diagram. We demonstrated that the feed composition, the relative volatilities between adjacent components have strong influnce on the partition and, thus, the choice of the distillation configuration of the lowest TAC for a specific feed composition. We also found in the present work that the purity specifications of the products also affect significantly the partition. We showed that, even with the same Ease of Separation Idex (ESI, the ratio the relative volatility between the light and middle components to that between the moddle and heavy components, traditionally used to indicate the suitibility for a DWC), the partition is quite different if the specified productsâ€™ purities are different. Based on this finding, a new index was introduced to represent the influence of the product purities on the selection of the optimal distillation configuration for a given feed composition. Rules were proposed, and systematic approach, based on the rules, for selecting optimal distillation configuration of ternary distillation was also proposed for some simplified cases.

497704 A Novel Microfluidic Platform for Screening of Pharmaceutical Polymorphs Under Hydrodynamically-Controlled Crystallization Conditions

Abstract:

The majority of oral drugs are delivered in solid forms such as capsules, tablets, powders or suspensions. The physicochemical properties and quality of such drugs strongly depend on their crystalline structure or polymorph, which is controlled during the crystallization process. Various factors such as solvent, temperature, co-solvent, supersaturation, and additives can affect the formation of polymorph. Screening of polymorphs with respect to these factors can be combinatorially tedious and time-consuming. High-density microtiter plates with as many as 1586 wells are commonly employed to screen ~10,000 factors per day. However, the application of microtiter plates for polymorph screening has some practical limitations as the environmental conditions in stagnant wells of the microtiter plate is much different than those in the batch crystallizers. Alternatively, the microfluidic platforms have been used to screen polymorphs using very small quantities (~100 μg) of the drugs in the early stages of developments. Such microfluidic devices locally produce a mixture of crystalline material which is not suitable for the study of nucleation and growth of polymorphs. Here we propose a novel microfluidic platform which mitigates the issues of the previous designs by facilitating nucleation and growth of single crystals. This microfluidic device can effectively screen the aforementioned factors to generate single crystals of polymorphs at the stagnation point while implementing the hydrodynamic conditions which are not considered in the microtiter plate method. In this work, we have screened polymorphs of Carbamazepine using various mole fractions of chloroform:ethanol mixtures and temperatures in the microfluidic device. Further analysis for the measurement of growth rate and morphology are also discussed. The high throughput screening system is also proposed to evaluate multiple crystallization conditions within one single device.

POLYMORPHISM IN DRUG COMPOUNDS BY CONTROLLING FLUID DYNAMICS DURING CRYSTALLIZATION

Stephanie Guthrie, Gaurav Giri

AICHE 2017: Solid Form Selection: Cocrystals, Salts, Solvates, Polymorphs, and Beyond

UNIVERSITY of VIRGINIA
Department of Chemical Engineering

Current drug development

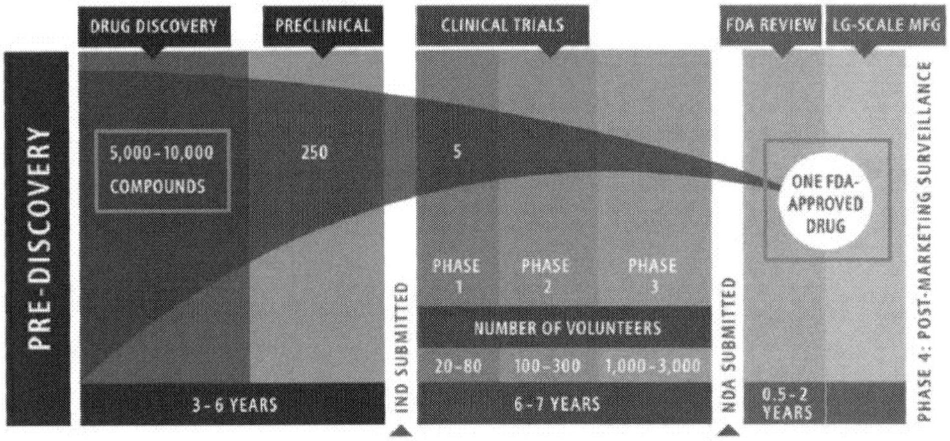

Source: Pharmaceutical Research and Manufacturers of America

Need to optimize polymorph detection

- Need ability to detect and isolate polymorphs
  - Polymorphs have unique physical and chemical properties
  - Color, solubility, bioavailability

ROY: 5-Methyl-2-[(2-nitrophenyl)amino]-3-thiophenecarbonitrile

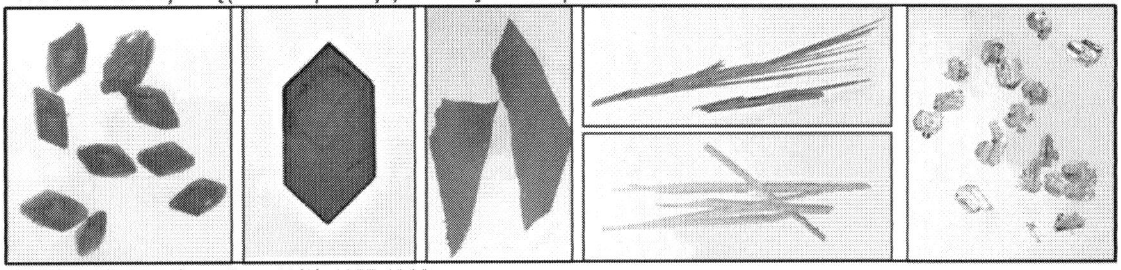

Yu, L. (2010). *Acc. Chem. Res.*, 43(9), 1257-1266.

Solid state changes solubility

Increased solubility

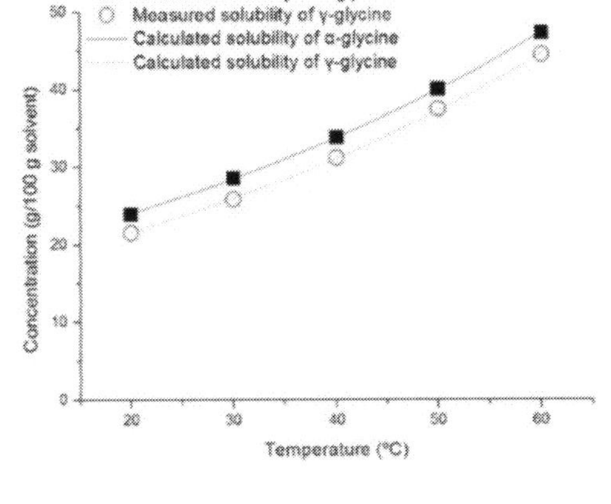

Kim, J. W., (2012). *Crystal Growth & Design*, 12(10), 4739-4744.

Enhanced dissolution rate

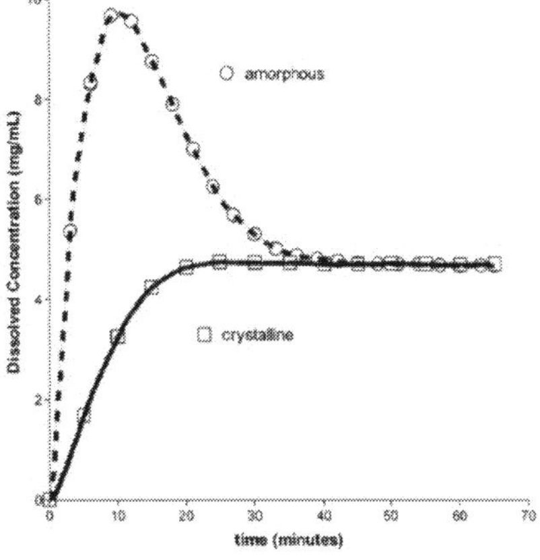

Brittain, H. (2014). *Am Pharmaceut Rev.*

5

Fluid dynamics can impact polymorphism

- Increased charge carrier mobility based on polymorphism in organic semiconductors

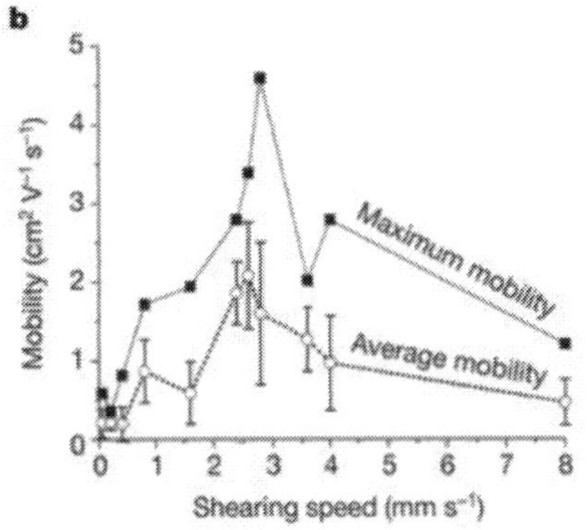

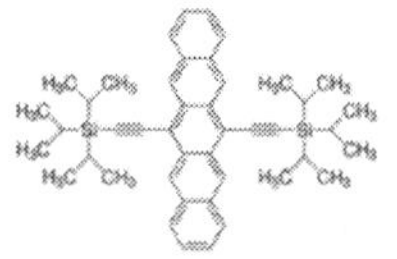

TIPS-pentacene

Giri, G. et al. (2011). *Nature* 480(7378), 504–508.

6

Solution shearing technique for pharmaceuticals

- Two pharmaceutically relevant molecules
  - Glycine, acetaminophen
- 10 mg/mL drug in deionized water
- Change processing parameters
  - Shearing speed: 0.01 – 3 mm s^{-1}
  - Temperature: 70-100 °C

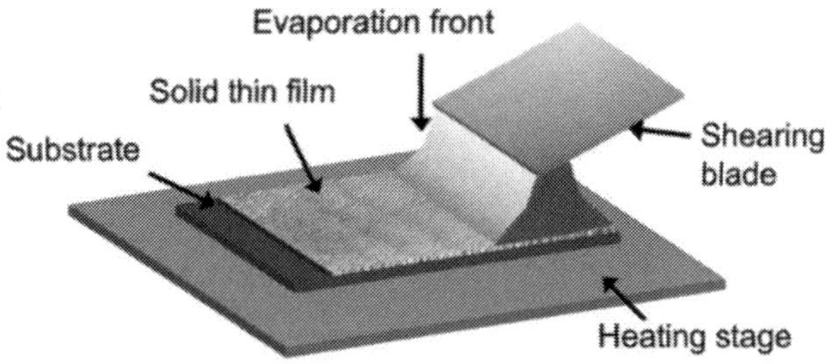

Shearing regimes

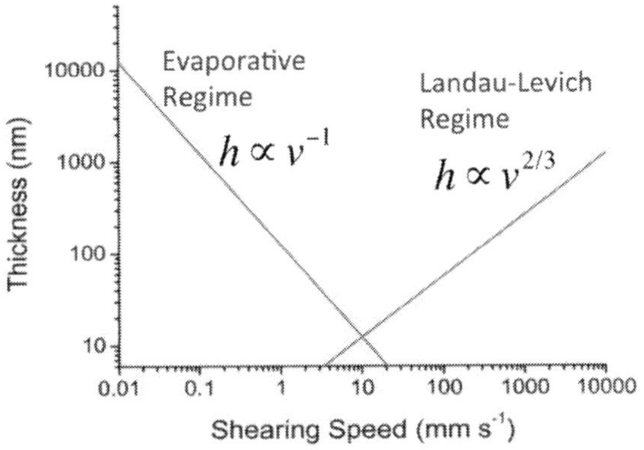

Le Berre, et al. (2009). *Langmuir*, *25*(5), 2554-2557.

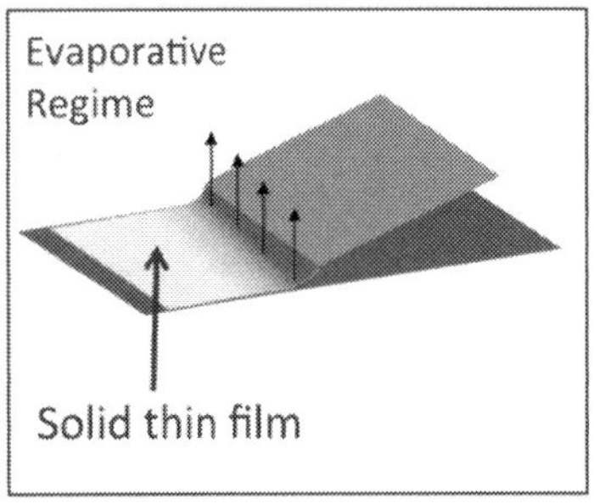

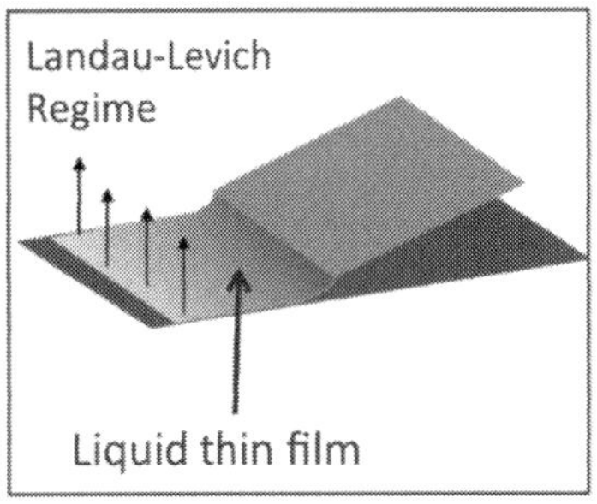

Film thickness characterization

70˚C

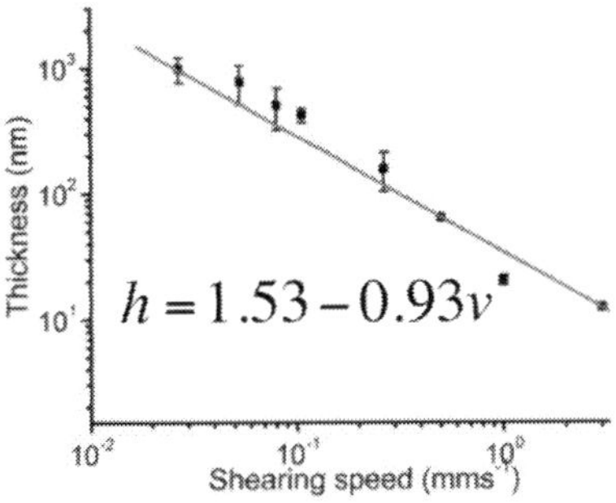

100˚C

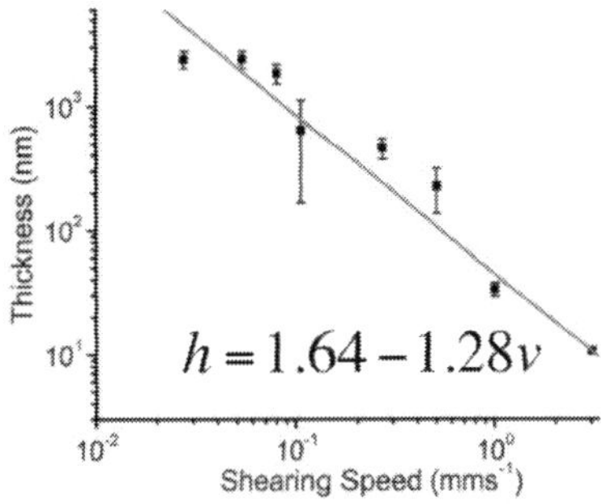

Glycine morphology

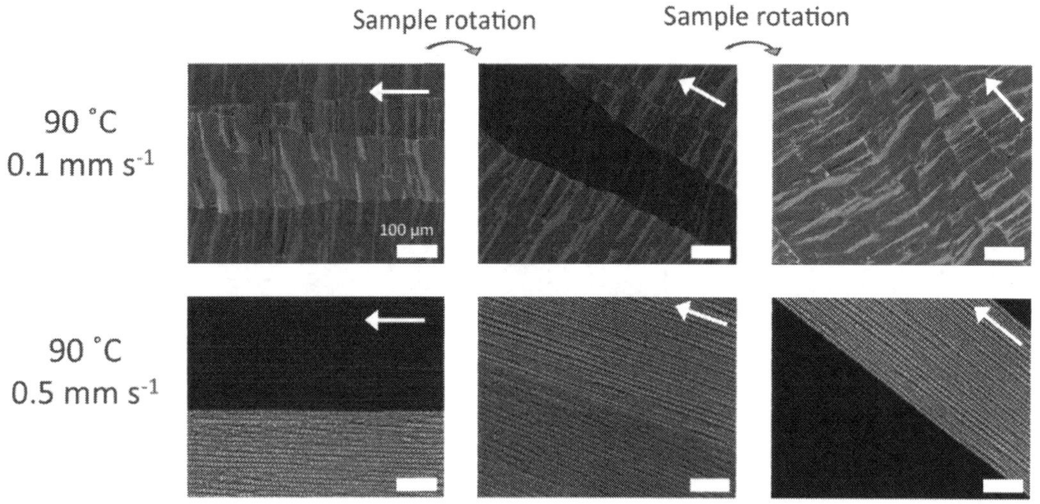

Acetaminophen morphology

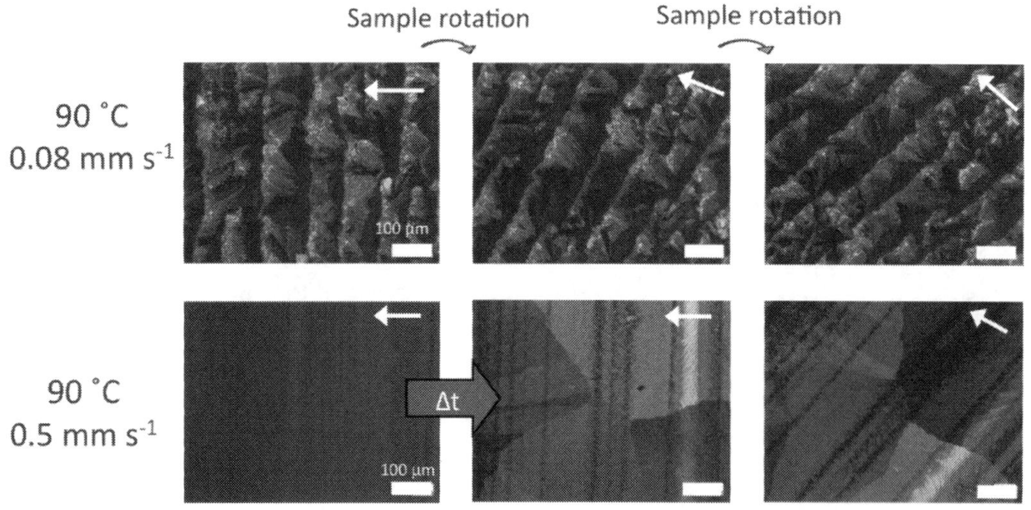

Grazing incidence X-ray diffraction (GIXD)

1. Incident X-ray beam is directed towards sample

2. X-rays are diffracted on film crystallites

3. Diffracted beam lands on area detector

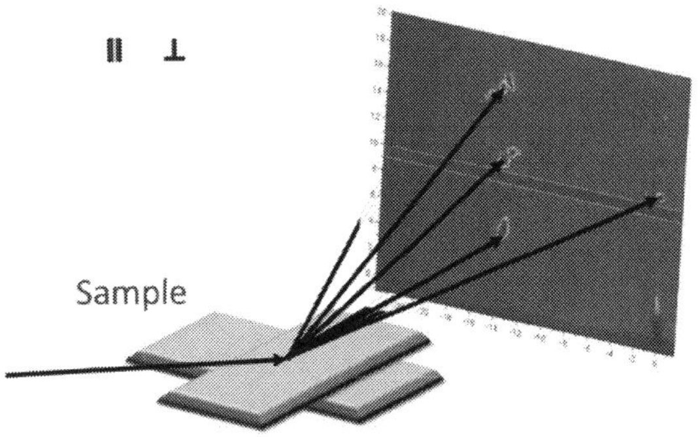

Samples are rotated parallel and perpendicular to shearing direction
- Orients new planes in the line of X-rays

Increasing order ↓

Sample type	Film Appearance	Characteristic GIXD	Description
Isotropic powder		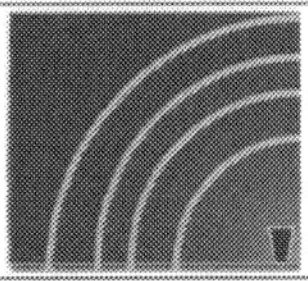	Annular rings
Anisotropic textured film			Peak smearing
Oriented film	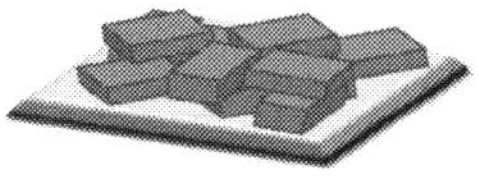	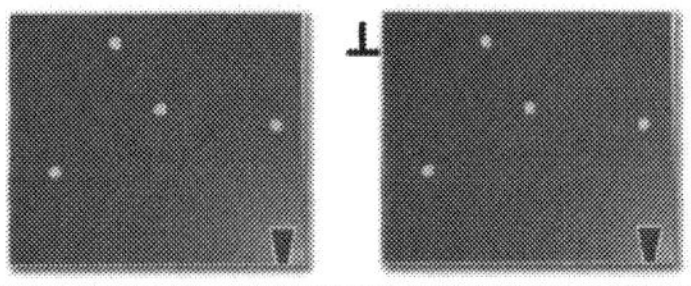 ∥ ⊥	Identical peaks upon rotation
Aligned film	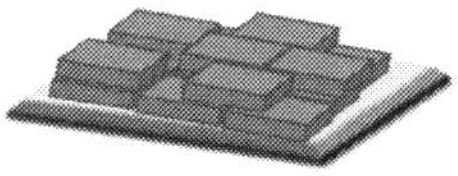	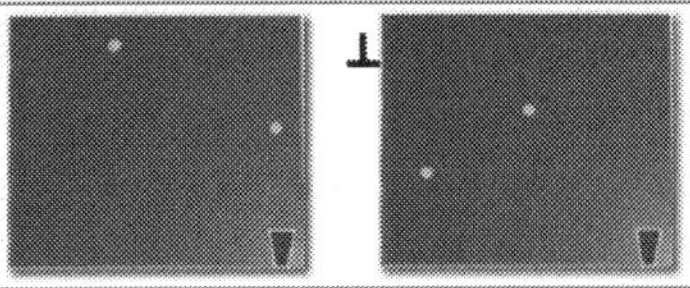 ∥ ⊥	Unique peaks upon rotation

13

Morphological characterization for glycine

Sample type 1

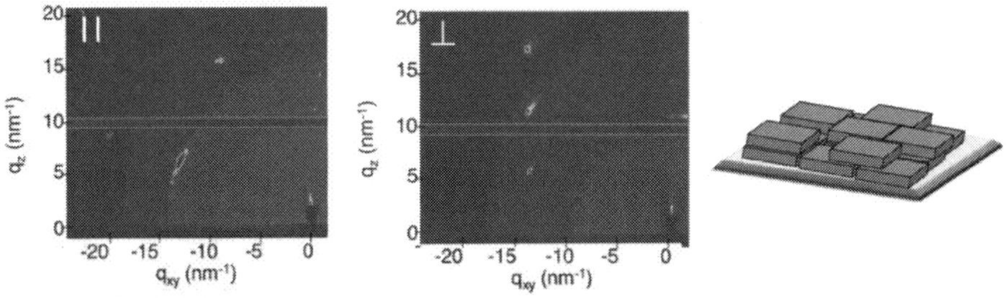

Sample type 2

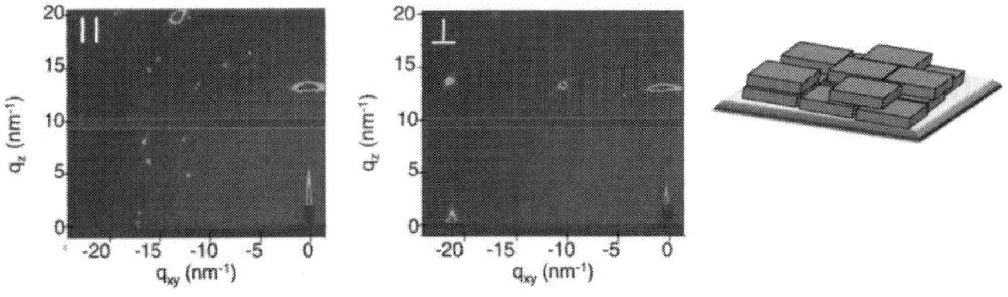

14

Morphology characterization for acetaminophen

Sample type 1

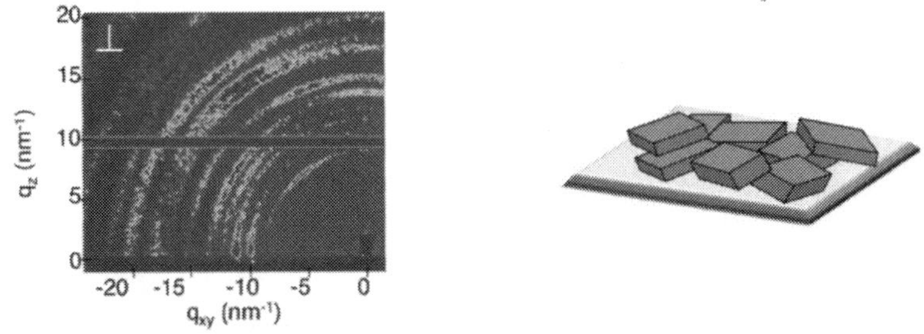

Sample type 2

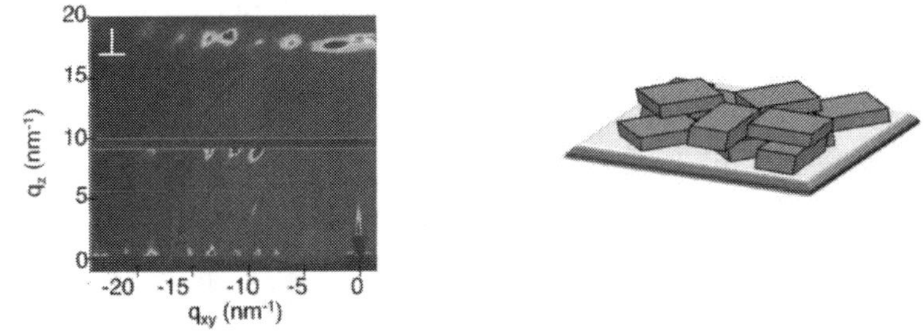

Integrated GIXD for glycine

- Confirmation of polymorphism

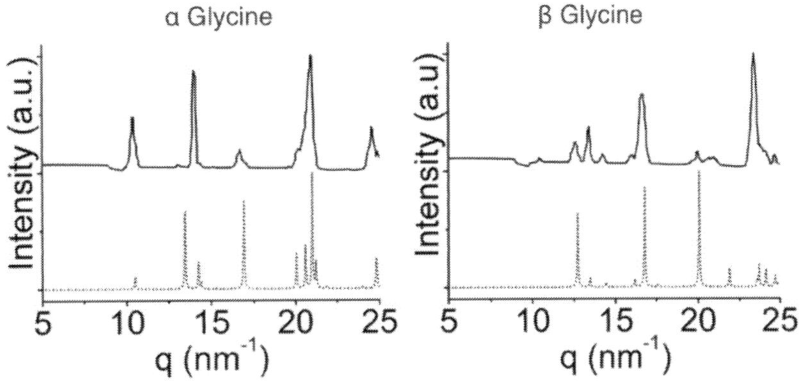

(—): Experimental (- -): Integrated from crystal structure

Integrated GIXD for acetaminophen

- Confirmation of polymorphism

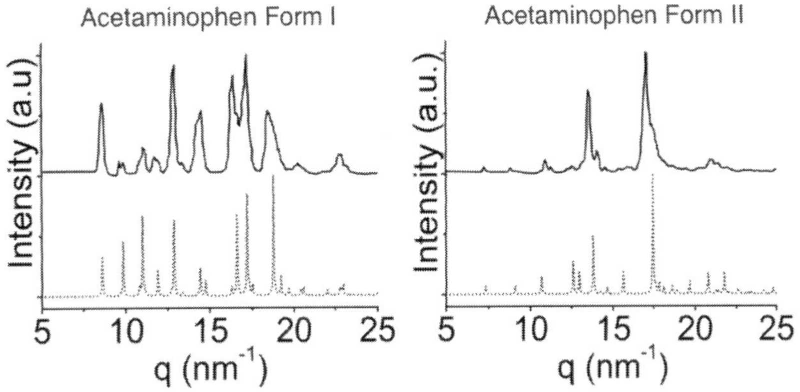

(—): Experimental (- -): Integrated from crystal structure

Data reduction

Acetaminophen, 0.27 mm s^{-1}, 90°C

Sample	Polymorph assignment
1	Form II
2	Form II
3	Form I
4	Both
5	Form I

3/5 samples contain Form I
3/5 samples contain Form II

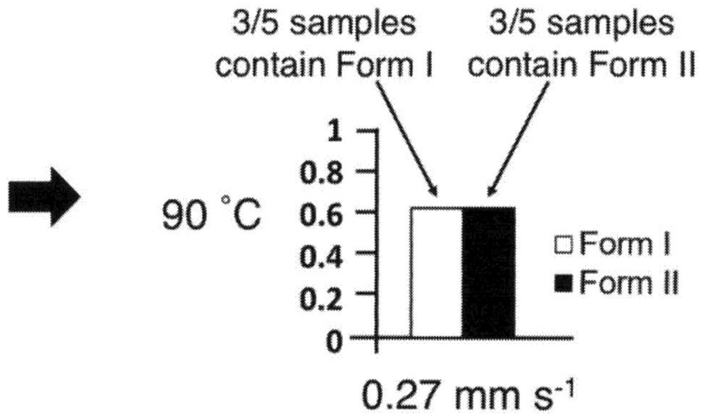

Tracking glycine polymorph formation

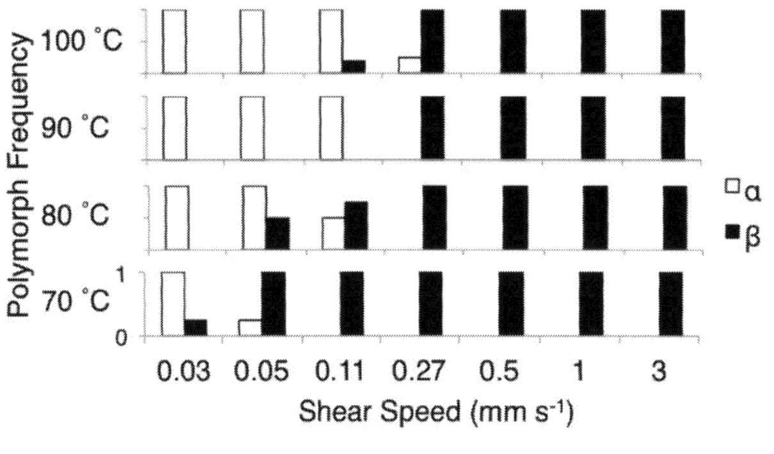

19

Tracking glycine polymorph formation

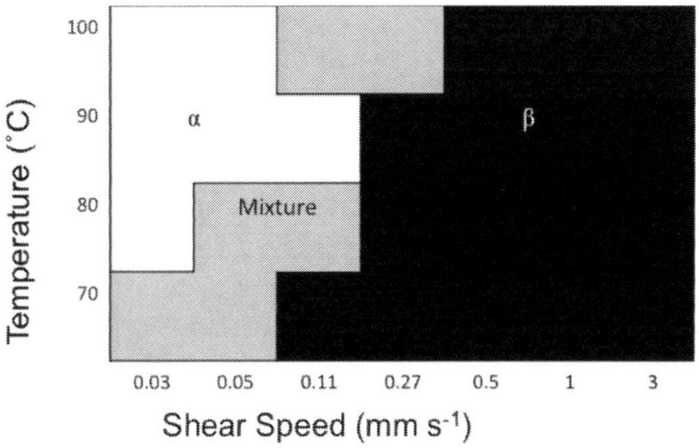

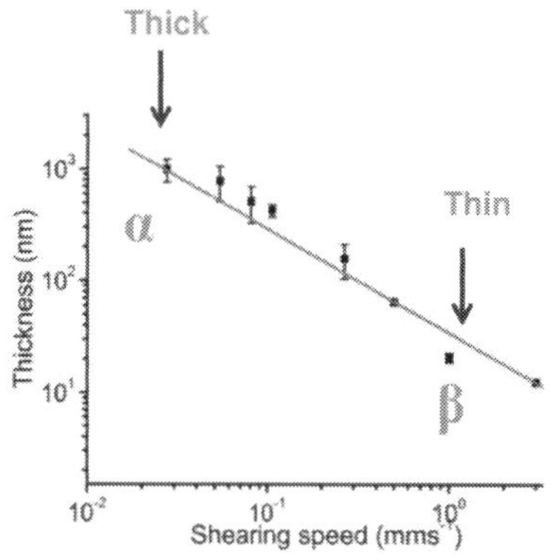

20

Tracking acetaminophen polymorph formation:

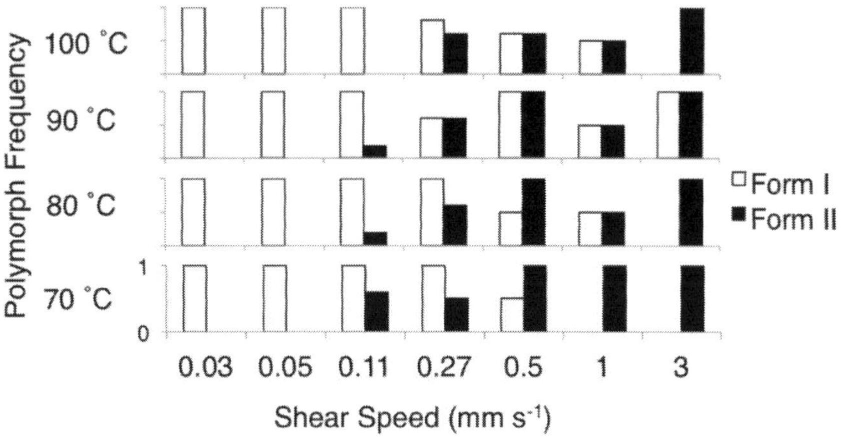

Tracking acetaminophen polymorph formation

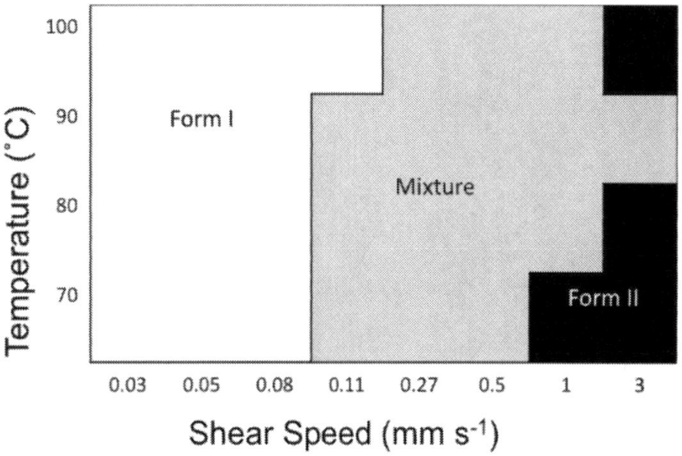

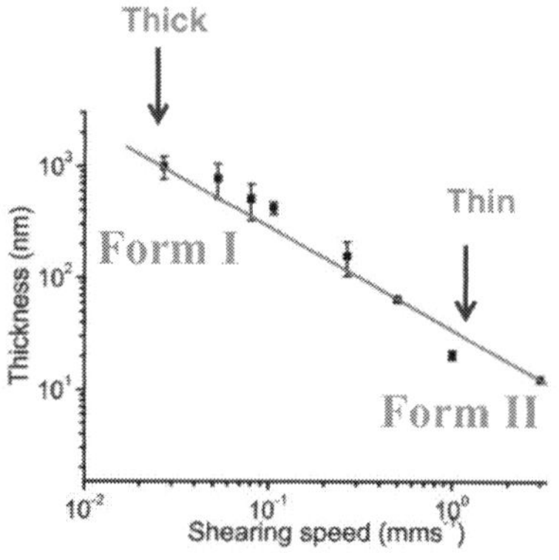

Rationale for polymorphism

- Thick films and thin films display polymorphic changes
- More stable form in thicker films
- Thin films provide a confined environment for isolation of metastable polymorphs

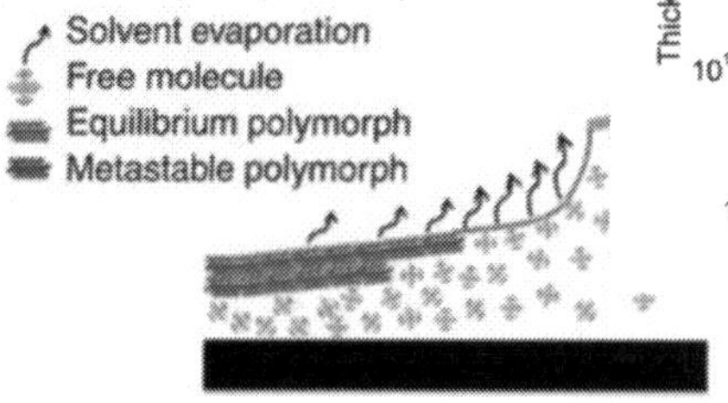

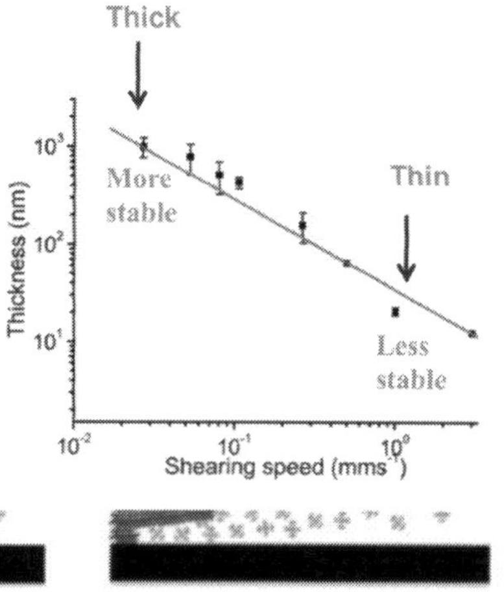

Giri, G. (2014). *Nature communications.* ***5**, 3573*

From II from amorphous transformation

- Amorphous phase isolated first
- Crystallizes over time

Acetaminophen

t =30 seconds t = 90 seconds t = 300 seconds

Carbamazepine

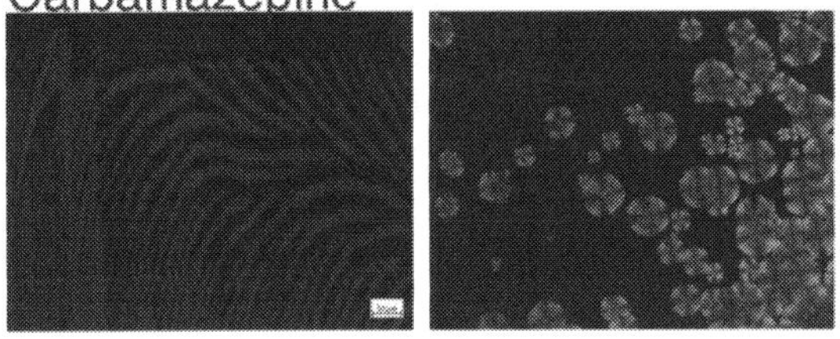

t = 0 seconds t = 10 minutes

Time temperature transformation

- Processing temperature impacts final phase formation
- Different microstructures
- Balance between nucleation and growth creates 'C' shape

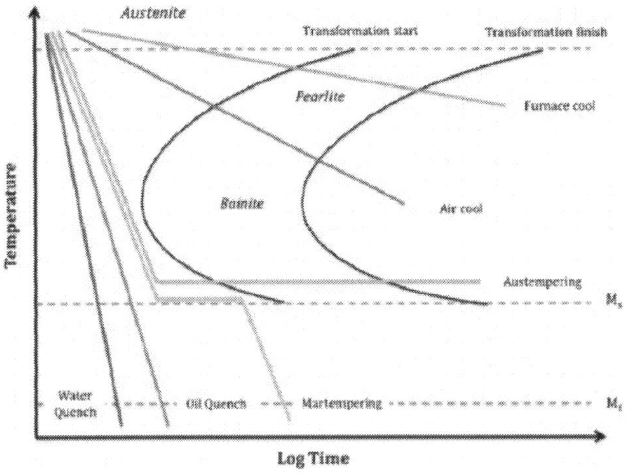

Brog, J. P (2013). *Rsc Advances*, 3(38), 16905-16931.

Crystallization kinetic diagram

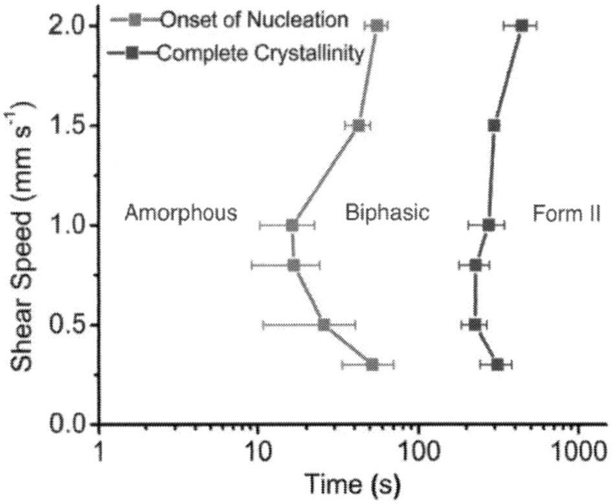

Summary

Processing techniques

Polymorphism

Processing diagram

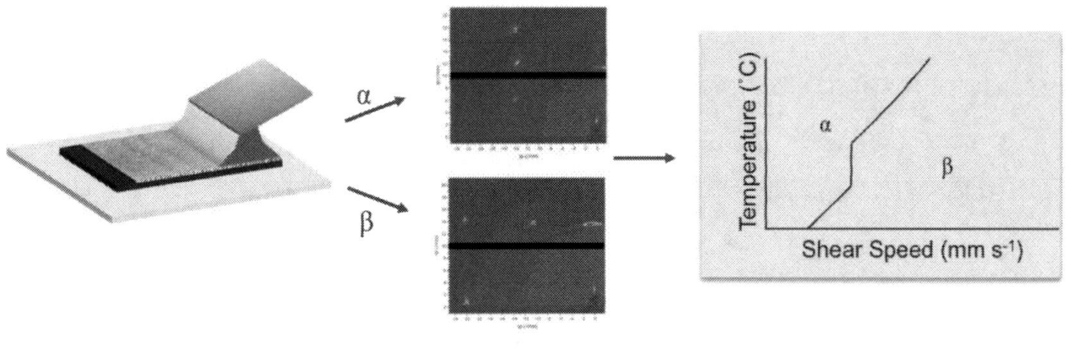

Acknowledgement

- Detlef Smilgies (CHESS)
- Luke Huelsenbeck
- Arian Ghorbanpour
- Giri group members

Virginia Space Grant Consortium
Aerospace Partnerships in Education, Research and Industry

Health Resources in Action®
Advancing Public Health and Medical Research

Cornell University
Cornell High Energy Synchrotron Source

Questions

488669 Mechanism of Selective Co-Crystallization of Cresol Isomers with Urea

Abstract:

The separation of *m*-cresol from mixtures of *m*-cresol and *p*-cresol has been a great challenge in industry since the physicochemical properties of *m*-cresol and *p*-cresol are pretty similar. In this work, co-crystallization method was used to separate *m*-cresol and *p*-cresol and the co-crystal preparation was attempted by cooling crystallization. The selective co-crystallization of *m*-cresol and *p*-cresol with urea was evaluated, based on the principle of supramolecular chemistry and crystal engineering strategy. It was found that the co-crystal former, urea, can only form co-crystal with m-cresol. One supramolecular synthon was formed between *m*-cresol and urea with the help of intermolecular mutual recognition and self-assembly process among them. Consequently, *m*-cresol can be effectively separated from *p*-cresol by the way of selective co-crystallization. The *m*-cresol _urea co-crystal (MC_U) was prepared on 1:1 molar ratio, which exhibits a regular hexagonal shape and belongs to orthorhombic system with space group P b c a and cell parameters a = 11.2502(5)Å, b = 7.1880(4) Å, c = 21.7868(15) Å. The hydroxyl group of *m*-cresol forms a H-bond with the carbonyl group in amide moiety of urea. The MC_U molecules adopt a chained layer structure. The solid-state properties of MC_U were characterized using X-ray Power Diffractometry (PXRD), Differential Scanning Calorimetry (DSC), ^{1}H nuclear magnetic resonance (^{1}H NMR) and Fourier Transformed Infrared Spectrometer (FTIR). Meantime, the selective co-crystallization process was on-line monitored by process analytical technologies/tools (PAT), such as Raman spectroscopy (Raman), Attenuated Total Reflection Fourier Transformed Infrared Spectroscopy (ATR-FTIR) and Focused Beam Reflectance Measurement (FBRM). Combined with the on-line monitoring data, ^{1}H NMR and the single crystal structure data, the mechanism of molecular recognition and self-assembly between m-cresol and urea is revealed. It was found that the whole co-crystal formation process can be divided into three steps: (1) *m*-cresol-urea dimers formation process, (2) MC_U nucleation process, and (3) MC_U growth process. And the reason for the selective co-crystallization is elucidated from the molecular level by using the corresponding calculation method. The mechanisms of selective co-crystallization as well as molecular recongnition and self-assembly between *m*-cresol and urea were proposed by analyzing the structure of *m*- and *p*-cresol, the single crystal structure data of cocrystal and the on-line monitoring data of Raman and ATR-FTIR. It was found that a chain supramolecular synthon was formed by the carbonyl oxygen (C=O) of the urea and the hydroxyl hydrogen (O-H) of the *m*-cresol relying on stronge hydrogen bonds. However, no obvious non-covalent bonds between *p*-cresol and urea molecules were observed. This failure is probably due to the position of the phenolic hydroxyl in toluene, which is affected by the methyl group. The electron cloud density of hydroxyl groups on *p*-cresol is reduced because of the presence of methyl groups, which could lead to the decrease of the electronegativity of oxygen, and thus will reduce the degree of exposure to the hydrogen atom attached to it. Obviously, *m*-cresol _urea co-crystal could be formed because of the contribution of the more exposed hydrogen atoms.

Continuous co-crystallization of benzoic acid and isonicotinamide by mixing-induced supersaturation: Exploring opportunities between reactive and antisolvent crystallization concepts

Vaclav Svoboda[1], Pól MacFhionnghaile[2], John McGinty[1], Lauren E. Connor[3], Iain D.H. Oswald[4], Jan Sefcik[2,*]

1. EPSRC Doctoral Training Centre in Continuous Manufacturing and Crystallisation, Department of Chemical and Process Engineering, University of Strathclyde, James Weir Building, 75 Montrose Street, Glasgow, G1 1XJ, United Kingdom

2. EPSRC Centre for Innovative Manufacturing in Continuous Manufacturing and Crystallisation, Department of Chemical and Process Engineering, University of Strathclyde, James Weir Building, 75 Montrose Street, Glasgow, G1 1XJ, United Kingdom

3. Collaborative International Research Programme, Strathclyde Institute of Pharmacy and Biomedical Sciences, University of Strathclyde, Technology and Innovation Centre, 99 George Street, Glasgow, G1 1RD, United Kingdom

4. Strathclyde Institute of Pharmacy and Biomedical Sciences, University of Strathclyde, 161 Cathedral Street, Glasgow, G4 0RE, United Kingdom

Corresponding Author:
Prof. Jan Sefcik
Department of Chemical and Process Engineering,
University of Strathclyde,
James Weir Building,
75 Montrose Street,
Glasgow,
G1 1XJ,
United Kingdom
jan.sefcik@strath.ac.uk

ABSTRACT

This study combines reactive and antisolvent crystallization concepts via mixing-induced supersaturation to demonstrate a wider range of options for solvent system selection in multicomponent crystallization. This approach was applied to investigate continuous crystallization of 1:1 and 2:1 co-crystals of benzoic acid and isonicotinamide. Design of Experiments was used to identify conditions where pure co-crystal phases are obtained and a continuous mixing-induced co-crystallization process was implemented to selectively produce either 1:1 or 2:1 co-crystals.

INTRODUCTION

Multicomponent crystalline materials are composed of more than one molecule in its crystal lattice. This includes co-crystals, solid solutions, solvates, and salts. The interest in multicomponent systems is a growing trend within the pharmaceutical industry[1]. Co-crystals are of benefit due to their ability to tailor physical and pharmaceutical properties of active pharmaceutical ingredients (APIs) such as solubility, bioavailability, stability, and the processability of the solid powder within industrial manufacturing processes[2]. They can also be utilized to expand IP portfolios[3]. However, co-crystallization is inherently more complex than single component crystallization as it involves an additional component with new potential solid phases. Having additional components and related process variables makes navigating the phase diagram and crystallization process more challenging.

Several methods for manufacturing co-crystals have been previously reported including both dry powder processes and solvent based processes, typically in batch processing mode. Dry powder processes are mechanochemistry based and utilize grinding at room temperature[4] or cryo-temperatures[5], polymer assisted grinding[6], high shear granulation[7] or hot melt extrusion[8]. Co-crystals from solution-based processes have been reported from evaporative and cooling[9], antisolvent[10] or reactive crystallization[11]. Cooling co-crystallization has previously been implemented in continuous platforms[12,13] and with co-crystal stoichiometry control[14]. Co-crystallization using antisolvent has been previously studied in batch processes[10], but not using continuous methods. Antisolvent crystallization of single component materials has been shown adaptable to continuous manufacturing and scale up in a number of processes[15–17]. Reactive co-crystallization mentioned in the work of Rodríguez-Hornedo et al.[11] does not involve a chemical reaction as such but a solution-based formation of a multicomponent solid phase from a mixture of two solutions undersaturated with respect to individual components using the same solvent.

Processes which rely on mixing to induce supersaturation such as reactive and antisolvent crystallizations, are well amenable to scale-up under continuous conditions[18]. Continuous manufacturing of pharmaceuticals can provide benefits such as decreased plant footprint, easier scale-up, shorter lead times and better unit performance through process intensification[19]. Since mixing can have strong impact on generation of supersaturation profile and subsequent nucleation, especially under kinetically controlled conditions[20], well-controlled mixing is essential for a control over final crystal properties, such as solid form and particle size distribution.

Antisolvent and reactive crystallization approaches can be combined in order to design a process where supersaturation is induced by mixing and is determined by the position in a quaternary phase diagram involving two crystal co-formers and two solvents. Depending on the shape of the phase diagram, which often shows a highly unsymmetrical nature[21], the co-crystal would be the thermodynamically most stable phase under some conditions. However, sometimes it may not be readily crystallized due to kinetic limitations (slow nucleation or growth) even if thermodynamically favored. Munshi and co-workers pointed out opportunities to use mixed solvents in controlling solid phase outcomes in cooling co-crystallization[22]. Solvent selection, a key design choice in crystallization, becomes more challenging in multicomponent systems. The key parameter of co-crystallization process design is the supersaturation with respect to the co-crystal phase, rather than the supersaturations of the individual co-formers. While having four components increases the complexity of mapping the phase diagram, it also allows for more options how to access solid phase regions which might not be easily accessible at a fixed solvent composition. For example, it may be possible to start with a solution of both coformers undersaturated in one solvent and add a second solvent to induce supersaturation, as in antisolvent crystallization. Alternatively, one can start with one co-former undersaturated in a given solvent mixture, and the other co-former undersaturated in the same solvent mixture and generate supersaturation by mixing these two solutions together, as in reactive crystallization. These decisions will be driven by the shape of the phase diagram and the nature of the target solid phase.

This study combines antisolvent and reactive crystallization concepts to develop a continuous co-crystallization process to produce benzoic acid (BZA) – isonicotinamide (INA) co-crystals. These molecules form two different co-crystals in 2:1 and 1:1 stoichiometric ratios which have been previously isolated and characterized from small scale cooling crystallization[23,24]. In this work, benchtop screening crystallization experiments were scaled to run in a continuous process. A Design of Experiment was carried out for the benchtop screening to better understand the effects of different process conditions on the crystallization of the two co-crystal phases. DoE screening was done in order to cut down the time and number of experiments to find suitable continuous crystallization conditions to selectively produce either 1:1 or 2:1 co-crystals by continuous crystallization.

EXPERIMENTAL SECTION

Materials. Benzoic acid (≥99.5%), isonicotinamide (99%), and ethanol (≥99.8%) were supplied by Sigma-Aldrich (Gillingham, UK). Deionised water was produced using the in-house Millipore Milli-Q system.

Batch Screening. Two sets of screening experiments were carried out. The first was an initial screening and the second was a systematic Design of Experiments (DoE) approach used for mapping of a limited region of the quaternary phase diagram (BZA, INA, Water, Ethanol) in order to identify suitable operating conditions for continuous crystallization. For both sets of screening experiments, an aqueous solution of isonicotinamide was added to an ethanolic solution of benzoic acid in a 20 mL vial, illustrated in Figure 1. Solutions were mixed at various ratios to obtain the total of 10g of solution and agitated using magnetic stirrer bar for 10 minutes past observed nucleation. All experiments were carried out at 25°C. For the initial screen, solutions of isonicotinamide in water and benzoic acid in ethanol were prepared at concentrations of 68.1 g/kg water and 213 g/kg ethanol, respectively, and mixed in ratios of 2:1, 1:1, 1:2, 1:4 and 1:9 by mass. The solid product obtained was filtered using 0.2 μm PTFE filter without washing, dried at 40°C for 24 hours and analyzed by X-ray powder diffraction (see solid characterization section). The design space for DoE was selected based on information from the initial screen. Experimental plan was created using MODDE in a 2 level full factorial design. The experimental worksheet, model fitting and results analysis was carried out in MODDE by Umetrics. Data were fitted using Partial Least Squares (PLS). The three variable parameters chosen for the DoE were: BZA-Ethanol solution concentration, INA-Water solution concentration, and the mass ratio of the two solutions mixed. Values of these three parameters combined determine the final composition of the mixture and thus the position on the phase diagram. The responses measured were induction time, solid yield, solid phases present and resulting slurry flow properties. Induction time was estimated as time after mixing until first particles were visually observed. Solid yield was taken as a percentage of total solute that was recovered as solids by filtration 10 minutes after the estimated induction time while agitating. Slurry flow was assessed by a rating system where a number was assigned on a scale from 1 to 5: 1 for an easily flowing mixture and 5 for a slurry too thick to flow with gravity. Solid phase was determined by XRPD (see solid characterization section). Based on this DoE, conditions for the production of 2:1 co-crystal were selected. Although the initial DoE was aimed to find suitable conditions for crystallization of both co-crystal forms, conditions leading to crystallization of 1:1 co-crystal resulted in solid loadings which would be too high for a continuous operation. Based on results from the initial DoE, a set of further experiments was carried out in an expanded design space to find suitable conditions for crystallization of 2:1 co-crystal. These experiments were carried out at lower solution concentrations and resulting solid phases, induction times and slurry flow ratings were determined in the same way as in the initial DoE runs.

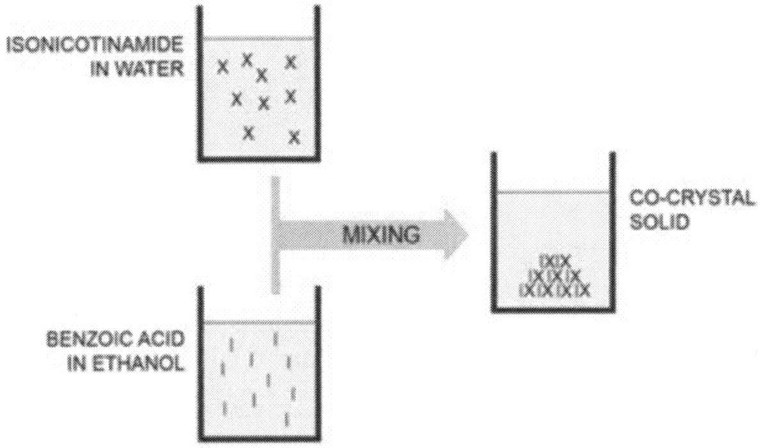

Figure 1. Solvent system used for mixing-induced co-crystallization

Solubility measurement. Solubility of the 2:1 co-crystal has been determined by a solvent addition method in the Crystalline Reactor System (Technobis), similar to a previously published procedure[25]. As compared to temperature variation or gravimetric methods, solvent addition relies on slow dilution of a suspension with a solvent (mixture) of a given composition under isothermal conditions until complete dissolution occurs, when a clear point can be detected. Solvent was added at a constant rate using PHD Ultra syringe pumps (Harvard Apparatus) to the Crystalline vials. The solubility measurement setup is shown in Figure 2. Multiple addition rates were tested (0.5 ml/hr and 0.75 ml/hr) to verify that dissolution kinetics have a negligible effect on the result at the addition rates used. Clear point was determined using both transmissivity measurement and images from the Crystalline camera. Image analysis produced more consistent results, which is in line with previous findings from Reus et al[25]. A study was also carried out to monitor the solid state transformation of the co-crystal in order to check phase stability in the same setup. Suspensions with concentrations as used for the start of solubility measurements were held for varying amount of time from 1 to 5 hours, after which the remaining solid was filtered, dried and analyzed by XRPD.

Figure 2. Technobis Crystalline and syringe pumps for solubility determination using the solvent addition method

Continuous crystallization. Continuous crystallization runs were performed using a concentric capillary mixer (Figure 3) as well as the Ehrfeld modular micro-reactor system equipped with Valve Mixer 30 module (Figure 4). The BZA-Ethanol solution was supplied to the concentric capillary mixer through the inner PEEK capillary, while the INA-Water solution was fed to the outer glass tube as shown in Figure 3. The capillary has internal and outer diameters of 0.51mm and 1.59mm, respectively. The outer glass tube has an internal diameter of 3mm. The capillary stream entered into the outer tube stream 6 cm from the T-junction. The feed solutions were pumped using Bronkhorst Mini CORI-FLOW system coupled with gear pumps allowing for accurate control of mass flow rates. The resulting slurry from the mixer was sampled 0.7m from the mixing point at the start and end of experiments. Collected slurry was filtered after a holding period in a Buchner funnel with a 0.45μm filter paper and dried overnight at 40°C. The slurry holding time without agitation was 10 minutes for 2:1 co-crystal runs, and 18 minutes for 1:1 co-crystal run. A camera was used for visual analysis of any fouling in the glass tube. Temperature of both streams was measured throughout experiments and it varied less than 2°C from 25°C. The total mass flow rates through the capillary mixer were 20, 40 and 60 g/min in 50:50 (w/w) ratio for producing the 2:1 co-crystal and 115 g/min at a 15:100 (w/w) ratio of benzoic acid solution flow rate to isonicotinamide solution flow rate for the 1:1 co-crystal. In order to demonstrate the transferability of the process to a commercial platform, the Ehrfeld modular micro-reaction system, shown in Figure 4, was used at the total mass flow rate of 20 g/min for producing the 2:1 co-crystal.

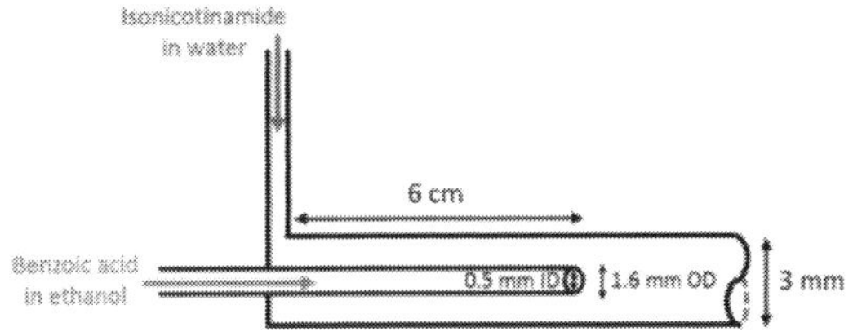

Figure 3. Concentric capillary mixer diagram with dimensions indicated

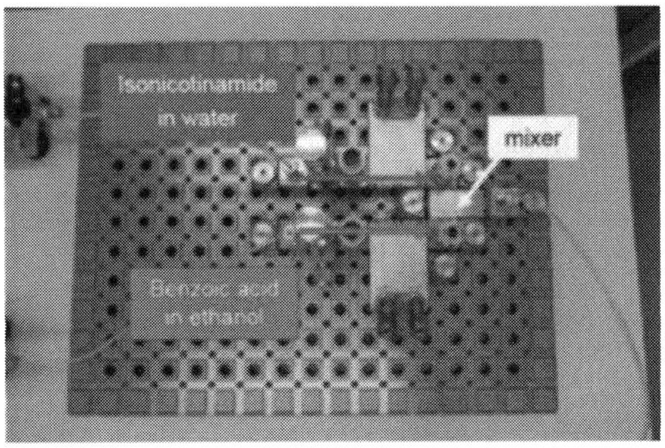

Figure 4. Ehrfeld modular micro-reaction system fitted with valve assited mixer 30

Solid characterization. Filtered and dried crystalline powder from the screening and continuous experiments was analyzed by X-ray Powder Diffraction (XRPD) and IR spectroscopy. Samples from continuous runs were further analyzed by Differential Scanning Calorimetry (DSC) paired with Thermogravimetric Analysis (TGA), and microscope image analysis for particle size measurement. Solids from continuous runs targeted to produce the 2:1 co-crystal were also analyzed by NMR. XRPD fingerprinting was performed on a sample placed in a 28 well plate, supported by Kapton film (7.5 μm thickness). Data were collected on a Bruker AXS D8 Advance transmission diffractometer equipped with θ/θ geometry, primary monochromatic radiation (Cu $K\alpha_1$ $\lambda = 1.54056$ Å), a Braun 1D position sensitive detector, and an automated multiposition x–y sample stage. Data were collected from 4 to 35° 2θ with a 0.015° 2θ step size and 1 s $step^{-1}$ count time. FT-IR measurements were taken using Bruker Tensor II, using 32 scans with 4 cm-1 resolution from 450 to 4000 cm^{-1} with diamond tip ATR sampling plate. DSC/TGA data was obtained from Netszch STA 449 F1 Jupiter. Measurements for DSC/TGA were taken from 20 to 180 °C with a ramp of 10°C/min. For Particle Size Distribution (PSD), image analysis was carried out on dry powder using Malvern Morphologi G3 using low pressure dispersion and 2.5x optics. ^{1}H and ^{13}C NMR analysis was carried out using Bruker Advance 3 at 400 MHz by dissolving the solid product in deuterated DMSO in 5mm NMR vials.

RESULTS and DISCUSSION

Mixing-induced supersaturation. A case study on benzoic acid – isonicotinamide co-crystallization presented here explores the space between reactive and antisolvent crystallization concepts in generating mixing-induced supersaturation to target specific solid phases of co-crystals. The illustrative phase diagram in Figure 5 is similar to the phase diagrams presented previously for the system investigated here[23]. We use it to demonstrate different modes of inducing supersaturation through mixing. The blue lines illustrate examples of reactive and antisolvent crystallization. In reactive crystallization, two undersaturated solutions, each containing a single coformer in the same solvent mixture, are mixed to supersaturate with respect to a desired co-crystal solid. In antisolvent crystallization, an undersaturated solution containing both coformers in one solvent is mixed with antisolvent to generate supersaturation. The red line illustrates a new approach used in this study, where antisolvent and reactive crystallization concepts were combined to target a specific region of the phase diagram, in this case 2:1 and 1:1 co-crystal regions. In this combined approach, two undersaturated solutions, each containing a single co-former in a single (pure) solvent, are mixed to generate supersaturated solution with respect to desired co-crystal phase in the mixed solvent.

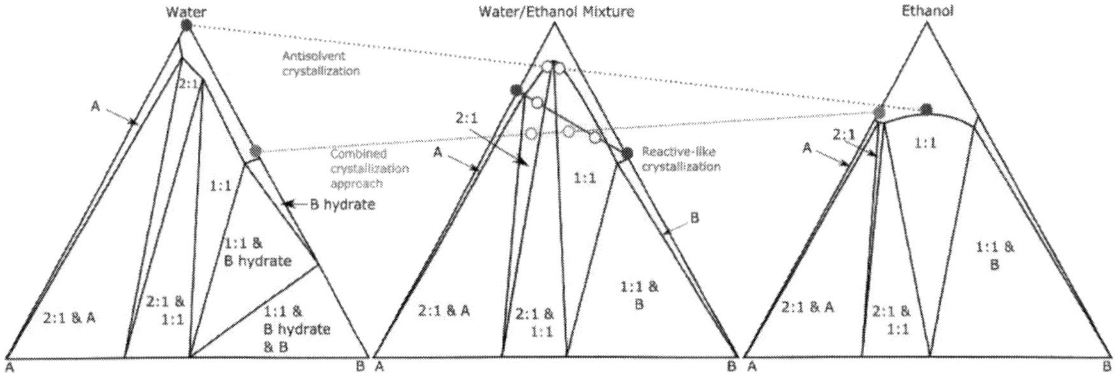

Figure 5. Illustration of three modes of inducing supersaturation through mixing: Reactive, antisolvent and combined approach. Targeted solid phases are 1:1 and 2:1 co-crystals of A and B and solvent is a mixture of water and ethanol. Dotted lines do not correspond to lever rule.

Screening. An initial screen was used to determine design space boundaries and was followed by a detailed DoE driven screen to map a design space corresponding to a region of interest in the phase diagram. XRPD from the initial screen has shown that the method of mixing solutions of isonicotinamide in water and benzoic acid in ethanol can produce both 2:1 and 1:1 co-crystals. DoE responses were chosen as key outcomes relevant for development of a continuous crystallization process. High solid loadings at the end of experiments limited solution concentrations from above, as too high solid loadings could lead to downstream blockages. On the other hand, long induction times limited solution concentrations from below, as relatively rapid nucleation is desirable. Due to the probabilistic nature of primary nucleation, induction time estimates from batch experiments are only intended to give an approximate guidance, nevertheless there was a good consistency among repeated experiments.

Despite not necessarily providing direct information on phase stability under thermodynamic equilibrium conditions, the DoE mapping provided a rapid way to find potentially suitable crystallization conditions while also giving indication about kinetics. Contour plots displayed in Figure 6 show responses at a fixed solution ratio where at higher concentrations of starting solutions the yield is higher, slurry flow rating diminishes due to higher solid loading and induction time decreases due to higher supersaturation. The co-crystal phase formed is primarily dependent on the solution flow ratio rather than the solution concentrations within the design space investigated. Contour plots for other solution ratios are located in supplementary information (ESI Figures 3-5). After targeting the desired phase by adjusting the solution flow ratio, slurry flow ratings of 2-3 were deemed as acceptable for a continuous crystallizer. The induction time model prediction is the least reliable as can be expected for a probabilistic process. The design space selected was large enough to find suitable conditions for the crystallization of the 2:1 co-crystal, while the conditions to crystallize the 1:1 co-crystal phase had to be extrapolated and subsequently validated. The extrapolation was done by choosing various concentrations of initial solutions outside of the original design space and analysing the same response parameters. The conditions from the DoE chosen as most suitable to produce the 2:1 co-crystal are 50/50 w/w solution flow ratio, with the solution concentrations of 109.4 g BZA/kg ethanol and 48.5 g INA/kg water. For the 1:1 co-crystal the optimal conditions were the solution flow ratio of 13/84 w/w (benzoic acid solution to isonicotinamide solution) with the solution concentrations of 30g BZA/kg ethanol and 40g INA/kg Water.

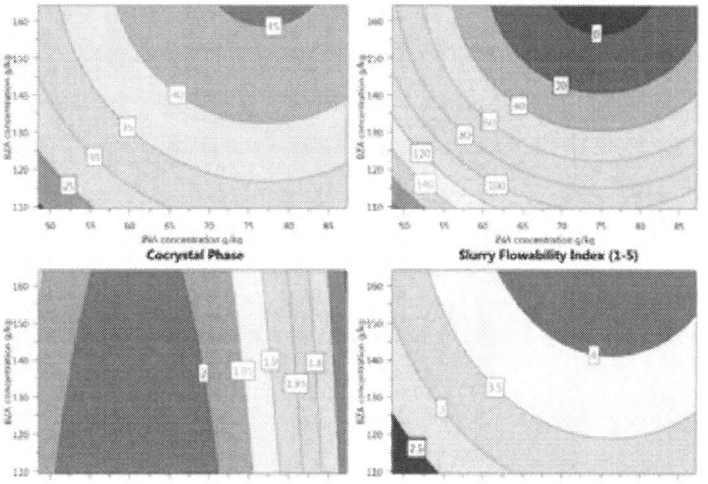

Figure 6. Contour plot showing responses from the Design of Experiments at a fixed solution ratio of 50:50 (w/w)

Solubility. During the phase stability test, the 2:1 co-crystal began to transform to the 1:1 co-crystal over time at the crystallizer outlet conditions. Transformation commenced between 158 to 264 minutes in an isothermal stirred environment as detected by XRPD, see Figure 7. This makes solubility determination more difficult as the 2:1 solid form is not thermodynamically stable at the chosen process conditions. However, the solid-liquid equilibrium of the metastable solid phase was determined for the purpose of thermodynamic yield calculations. The solubility test was timed to avoid any significant transformation. Solubility measurements were carried out in 48.8% (w/w) ethanol/water mixture, which is the same as the solvent mixture at the continuous mixer outlet during crystallization of 2:1 co-crystal. The results from six experiments suggest the solubility to be 0.0446 g/ml solvent with a high level of reproducibility based on clear points from image analysis. The spread of various measurements is illustrated in Figure 8. The high spread of clear point measurements based on transmission may be attributed to the limited ability to detect solids by transmission measurements at low concentrations, especially for thin needles. The different solvent addition rates show comparable results, thus the result is not dependent on the kinetics of crystal dissolution. The findings about the 2:1 co-crystal metastability point out that the outlet composition of the crystallizer slurry is either in the 1:1 co-crystal or a mixture of the 1:1 and 2:1 co-crystal regions on the phase diagram. The solubility of the 2:1 co-crystal has been previously reported between 0.0367 g/ml and 0.0482 g/ml for 50% water/ethanol mixture by[26] using two methods. However, the measurements reported therein have a high variability and the gravimetry method implemented did not seem to account for a possibility of a phase transformation. Therefore, it is possible that in some cases the solubility of the 1:1 co-crystal or a mixture of both co-crystal phases could have been measured. A recent study by Munshi et al.[22] shows the ternary phase diagram at 50% ethanol/water, although it was not reported whether this was 50% by mass or by volume[22]. As we have shown, solvent mediated transformation may occur in this system under conditions studied in this and previous works, and this may perhaps also help to explain the unusual shape of the phase diagram reported in the work of Munshi et al.

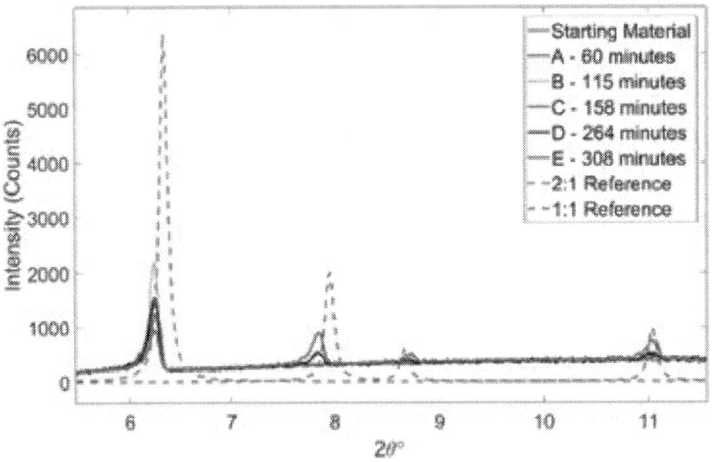

Figure 7. XRPD from suspension stability study of 2:1 co-crystal in 48.8% ethanol – water (w/w)

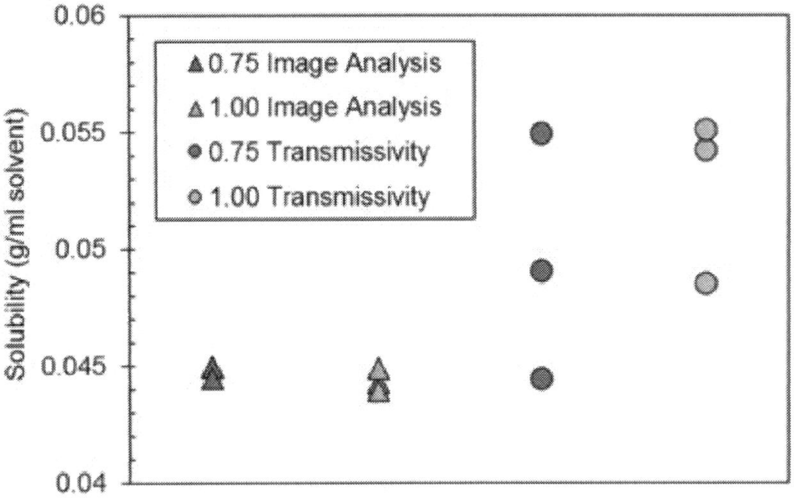

Figure 8. Solubility of 2:1 co-crystal in 48.8% ethanol – water (w/w) at 25 °C

Continuous crystallization. Moving from benchtop vial-based screening to a continuous platform introduces new process parameters and challenges. The overall flow rate was found to have a strong impact on the mixing regime and thus final particle properties. Three different overall flow rates were examined with the aim of producing the 2:1 co-crystal: 60 g/min, 40 g/min and 20 g/min at 50:50 ratio (w/w) of solution streams. The continuous runs revealed fouling and subsequent blockages to be a challenge, however blockage issues decreased with lower flowrates. Antifouling measures such as low power sonication could be used to mitigate this issue[27]. The solid output was 0.507 g/min, 0.342 g/min and 0.0259 g/min for overall flowrates of 60 g/min, 40 g/min and 20 g/min, respectively. The L50 of particles obtained at overall flow rates of 60 g/min, 40 g/min and 20 g/min was 361μm, 317μm and 238μm, respectively. Despite producing the desired solid form (Figure 10) from all conditions, mixing performance drastically reduced at the overall flow rate of 20 g/min resulting in a lower solid yield. The low velocity difference between the capillary jet and the outer flow did not provide sufficient mixing to rapidly achieve the desired supersaturation. The same process at the overall flow rate of 20 g/min resulted in 0.0484 g/min solid output in the Ehrfeld platform. Clearly, mixing conditions influence nucleation kinetics of co-crystal as the solid yield and resulting particle size distribution (see ESI Figure 6) varies with overall flowrate although the overall solution composition is the same in all cases. In order to achieve a higher yield in the crystallization process, increasing of residence time downstream from the mixing point combined with additional cooling stages should be considered. Further process optimization would benefit from more detailed knowledge of the relevant phase diagram and additional experiments in an expanded design space.

A continuous crystallization was also carried out to produce the 1:1 co-crystal, using flow rates of 15 g/min of benzoic acid solution through the capillary and 100 g/min of isonicotinamide solution through the outer tube. For this experiment the solutions were less concentrated to avoid blocking due to potentially high solid loading of the resulting slurry. The induction time at these concentrations was relatively long compared to the 2:1 co-crystal experiments and particles were not immediately visible in the mixer outlet. The solid output after holding the slurry for 18 minutes was 0.0726 g/min.

Solid characterization. Solid state characterization techniques were selected in order to establish solid phase identity and purity. Crystal unit cell parameters were previously documented for both co-crystal phases which makes powder X-ray diffraction suitable for phase verification. The co-crystal phases also have unique melting points verifiable by DSC. NMR was also used to independently verify the stoichiometry of co-formers in the crystalline sample. The morphological differences in crystals for the 2:1 and 1:1 forms were minor. The needle shaped morphology of both co-crystal forms can be seen in Figure 9. Results from Malvern Morphologi G3 are shown in supplementary information and similar aspect ratio distributions were found for both co-crystal forms. Analysis by XRPD on the dry samples from the capillary mixer is shown in Figure 10 (XRPD on samples from the Ehrfeld platform are in supplementary information ESI Figure 9). Lines A, B and C are samples from continuous runs designed to produce the 2:1 co-crystal. Simulated patterns are generated from unit cell data available from the Cambridge Structural Database under structure labels: BUDWEC[28] and MOVTOH[23] for the 1:1 and 2:1 co-crystals respectively. Comparison with D – the simulated pattern suggests that a pure 2:1 co-crystal phase has been produced by all runs. Similarly, line E shows pure 1:1 co-crystal by comparison to the 1:1 co-crystal simulated pattern – line F. This is further supported by results from DSC shown in Figure 11. DSC for the 2:1 co-crystal samples shows a sharp thermal event previously reported at 142 – 143°C for the 2:1 co-crystal[29]. The TGA curves also show a decrease in mass from 148-150°C onwards indicating sample decomposition. Similarly, DSC for 1:1 sample shows a unique endotherm above 160 °C, at the expected melting point of the 1:1 co-crystal[29] with subsequent decomposition.

For additional confirmation, the samples were analyzed using ATR-IR. The absorbance spectra are in Figure 12. The figures show a spectrum for each starting material as well as samples from continuous experiments. It is clear that the co-crystal phases have a well-defined unique spectra and it is possible to distinguish the 2:1 and 1:1 co-crystals using IR spectrometry.

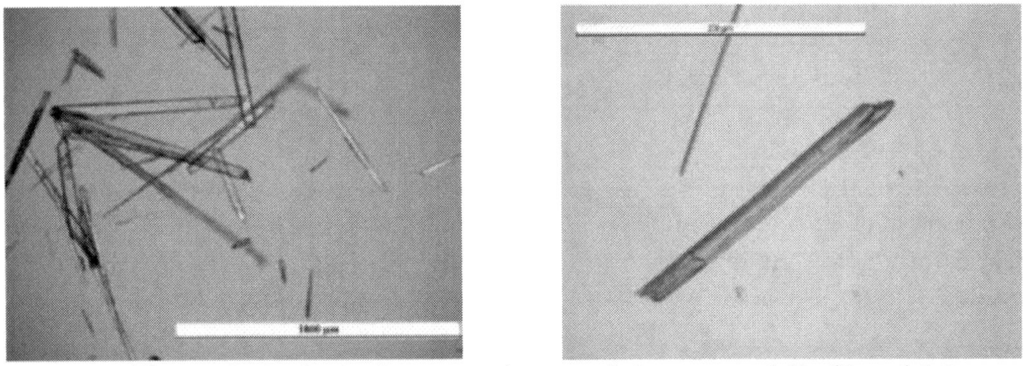

Figure 9. Optical microscope images: 2:1 co-crystal (Left) and 1:1 co-crystal (Right)

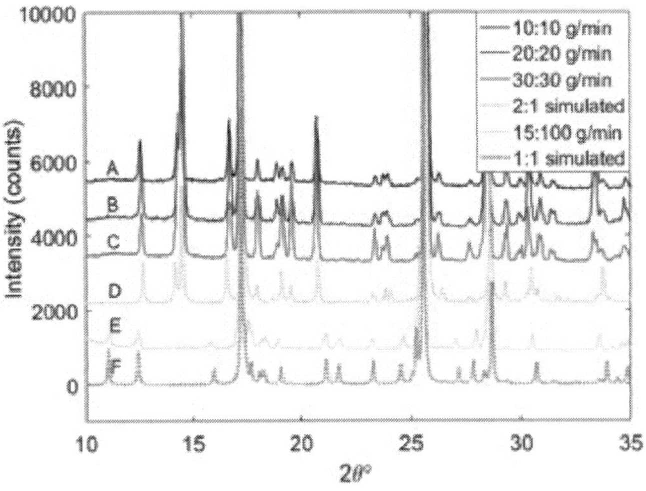

Figure 10. XRPD for samples from all continuous runs together with calculated patterns

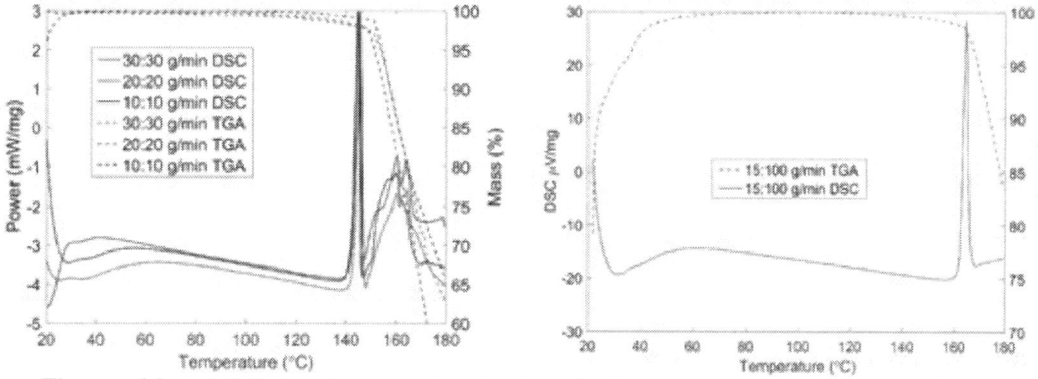

Figure 11. DSC/TGA for samples for 2:1 (left) and 1:1 (right) co-crystal from continuous runs

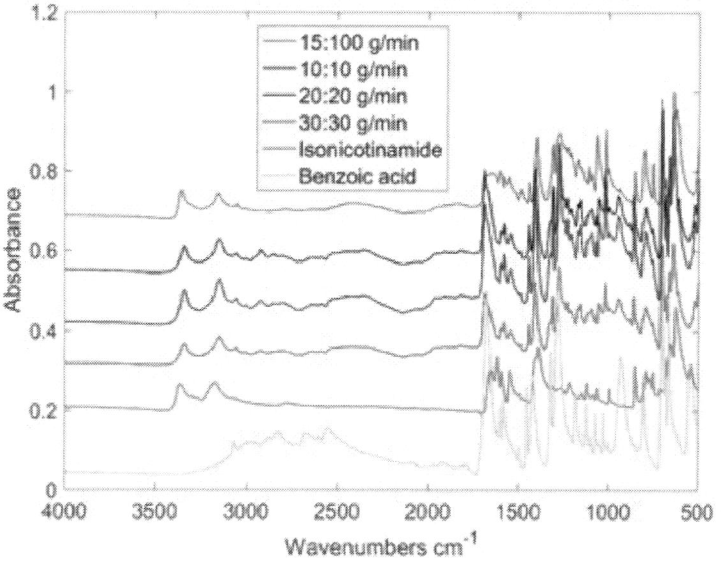

Figure 12. ATR-IR spectra for starting material and samples from continuous runs

NMR analyses on solid samples dissolved in deuterated DMSO from each continuous run confirmed (within 2% error) a 2:1 ratio for each sample from the continuous runs aimed to produce 2:1 co-crystal (ESI Figure 8). Continuous experiments using 40 g/min and 60 g/min flow rates produced particles of very similar particle size and shape characteristics. Results from 20 g/min continuous experiment show significantly smaller particles at much lower solid yield (ESI Figure 6). The aspect ratios for the particles from 20 g/min run are more needle-like with a narrower distribution.

We showed that both 2:1 and 1:1 pure forms of the co-crystal can be produced using the combined approach method based on mixing-induced supersaturation in a continuous crystallization process. The combined antisolvent and reactive crystallization approach allows for additional flexibility in reaching specific phase regions in multicomponent systems using multiple solvents.

CONCLUSIONS

We have used Design of Experiments to find conditions suitable for formation of 2:1 and 1:1 co-crystals of benzoic acid and isonicotinamide by mixing ethanolic solutions of benzoic acid and aqueous solutions of isonicotinamide in isothermal conditions. The input parameters were the two solution concentrations and the mass ratio of the two solutions, while the output variables were the solid form produced, solid yield, induction time and slurry flow rating. Optimal conditions were then identified for both 2:1 and 1:1 co-crystal formation and continuous crystallization experiments were performed at several total flow rates. Both solid forms were selectively produced in continuous crystallization as confirmed by XPRD, IR, DSC and NMR. Mixing conditions were found to influence the solid yield and particle size distribution of crystal produced although not the solid form.

The combined antisolvent and reactive crystallization approach can be a useful tool for reaching solid phases that are otherwise difficult to access through mixing-induced supersaturation. Under conditions of thermodynamic control, it may be possible to reach certain regions of phase diagrams that would be not accessible by other approaches due to non-ideal shapes of some phase diagrams. Under conditions of kinetic control, it may be possible to influence nucleation kinetics through the initial mixing of two solutions which would depend on compositions of both solutions as well as mixing conditions applied.

ACKNOWLEDGMENT

We would like to thank Maria Briuglia for assistance with Technobis Crystalline. The authors would like to thank EPSRC and the Doctoral Training Centre in Continuous Manufacturing and Crystallisation (Grant Ref: EP/K503289/1) for funding Vaclav Svoboda and John McGinty. The authors would like to thank EPSRC and Centre for Innovative Manufacturing in Continuous Manufacturing and Crystallisation (Grant Ref EP/I033459/1) for funding Pól MacFhionnghaile. The authors would like to thank the International Strategic Partnership between University of Strathclyde and Nanyang Technological University Singapore for funding Lauren E. Connor. The authors would like to thank EPSRC for Fellowship (Grant Ref EP/N015401/1) for funding Iain D.H. Oswald. The authors would like to acknowledge that this work was carried out in the CMAC National Facility supported by UKRPIF (UK Research Partnership Fund) award from the Higher Education Funding Council for England (HEFCE) (Grant ref HH13054).

REFERENCES

(1) Domingos, S.; André, V.; Quaresma, S.; Martins, I. C. B.; Minas da Piedade, M. F.; Duarte, M. T. J. Pharm. Pharmacol. 2015, 67, 830–846.
(2) Elder, D. P.; Holm, R.; Diego, H. L. de. Int. J. Pharm. 2013, 453, 88–100.
(3) Trask, A. V. Mol. Pharm. 2007, 4, 301–309.
(4) Friščić, T.; Jones, W. Cryst. Growth Des. 2009, 9, 1621–1637.
(5) Macfhionnghaile, P.; Hu, Y.; Gniado, K.; Curran, S.; Mcardle, P.; Erxleben, A. J. Pharm. Sci. 2014, 103, 1766–1778.
(6) Hasa, D.; Carlino, E.; Jones, W. Cryst. Growth Des. 2016, 16, 1772–1779.
(7) Rehder, S.; Christensen, N. P. A.; Rantanen, J.; Rades, T.; Leopold, C. S. Eur. J. Pharm. Biopharm. 2013, 85, 1019–1030.
(8) Moradiya, H. G.; Islam, M. T.; Halsey, S.; Maniruzzaman, M.; Chowdhry, B. Z.; Snowden, M. J.; Douroumis, D. CrystEngComm 2014, 16, 3573.
(9) Wittering, K.; King, J.; Thomas, L.; Wilson, C. Crystals 2014, 4, 123–140.
(10) Lee, M.-J.; Chun, N.-H.; Wang, I.-C.; Liu, J. J.; Jeong, M.-Y.; Choi, G. J. Cryst. Growth Des. 2013, 13, 2067–2074.
(11) Rodríguez-Hornedo, N.; Nehm, S. J.; Seefeldt, K. F.; Pagán-Torres, Y.; Falkiewicz, C. J. Mol. Pharm. 2006, 3, 362–367.
(12) Zhao, L.; Raval, V.; Briggs, N. E. B.; Bhardwaj, R. M.; McGlone, T.; Oswald, I. D. H.; Florence, A. J. CrystEngComm 2014, 16, 5769.
(13) Powell, K. A.; Bartolini, G.; Wittering, K. E.; Saleemi, A. N.; Wilson, C. C.; Rielly, C. D.; Nagy, Z. K. Cryst. Growth Des. 2015, 15, 4821–4836.
(14) Powell, K. A.; Croker, D. M.; Rielly, C. D.; Nagy, Z. K. Chem. Eng. Sci. 2016, 152, 95–108.
(15) Alvarez, A. J.; Myerson, A. S. Cryst. Growth Des. 2010, 10, 2219–2228.
(16) Brown, C. J.; Adelakun, J. A.; Ni, X. Chem. Eng. Process. Process Intensif. 2015, 97, 180–186.
(17) Ferguson, S.; Morris, G.; Hao, H.; Barrett, M.; Glennon, B. Chem. Eng. Sci. 2013, 104, 44–54.
(18) Lindenberg, C.; Schöll, J.; Vicum, L.; Mazzotti, M.; Brozio, J. Chem. Eng. Sci. 2008, 63, 4135–4149.
(19) Baxendale, I. R.; Braatz, R. D.; Hodnett, B. K.; Jensen, K. F.; Johnson, M. D.; Sharratt, P.; Sherlock, J.-P.; Florence, A. J. J. Pharm. Sci. 2015, 104, 781–791.
(20) Bałdyga, J.; Makowski, Ł.; Orciuch, W. Chem. Eng. Res. Des. 2007, 85, 745–752.
(21) Leyssens, T.; Springuel, G.; Montis, R.; Candoni, N.; Veesler, S. Cryst. Growth Des. 2012, 12, 1520–1530.
(22) Munshi, T.; Redha, B.; Feeder, N.; Meenan, P.; Blagden, N. Cryst. Growth Des. 2016, 16, 1817–1823.
(23) Seaton, C. C.; Parkin, A.; Wilson, C. C.; Blagden, N. Cryst. Growth Des. 2008, 9 (1), 47–56.
(24) Boyd, S.; Back, K.; Chadwick, K.; Davey, R. J.; Seaton, C. C. J. Pharm. Sci. 2010, 99, 3779–3786.
(25) Reus, M. A.; van der Heijden, A. E. D. M.; ter Horst, J. H. Org. Process Res. Dev. 2015, 19, 1004–1011.
(26) Redha, B. H. Impact of mixed solvent on co-crystal solubility, ternary diagrams and crystallisation scale-up. Crystallisations of Isonicotinamide- Benzoic Acid Co-crystals from Ethanol-Water Co-solvent System., University of Bradford, 2013.
(27) Al Nasser, W. N.; Pitt, K.; Hounslow, M. J.; Salman, A. D. Powder Technol. 2013, 238, 151–160.
(28) Aakeröy, C. B.; Beatty, A. M.; Helfrich, B. A. Angew. Chem. Int. Ed. 2001, 40, 3240–3242.
(29) Buanz, A.; Prior, T. J.; Burley, J. C.; Raimi-Abraham, B. T.; Telford, R.; Hart, M.; Seaton, C. C.; Davies, P. J.; Scowen, I. J.; Gaisford, S.; Williams, G. R. Cryst. Growth Des. 2015, 15, 3249–3256.

497864 A Multiscale Computational Method for Prediction of Polymorphs

Abstract:

Physical properties of solid pharmaceutical drugs stem from their crystalline structure or polymorph. Pharmaceutical molecules can undergo polymorphism to attain more than one crystalline structure depending on the crystallization conditions. Experimental screening of polymorphs is expensive and time-consuming. Therefore, the accurate methods to predict polymorphs have become essential to build in quality into the pharmaceutical products. The most prominent method to predict crystalline polymorph is the lattice energy minimization technique, which uses appropriate sampling and optimization algorithms to identify space groups and lattice energy. This approach has limitations in treating the solute-solvent interactions and capturing the effect of additives, solvents, co-solvents and hydrodynamic conditions of the crystallizer. Therefore, there is a need to develop a model which considers these effects involved in multiple length and time scales during the crystallization process. Here we propose a consistent multiscale approach to predict dynamics of polymorph formation, which includes hierarchical simulation of sequential events â€“ i) lattice formation via self-assembly of molecules, ii) growth of lattice and formation of molecular clusters, and iii) development of faces and evolution of faceted crystal.Â The effect of the solute-solvent interactions, additives, solvents, co-solvents and hydrodynamic conditions are embedded into the calculation of free energy barriers for molecular attachment/detachment in these three events. An example of polymorphism in Glutamic acid crystals using the multiscale method will be shown.

499540 Polymorph Selection By Continuous Precipitation

Abstract:

Due to polymorphic variation in mechanical, electrical, stability, bioavailability and physicochemical properties, there always exists an optimal polymorph for a given application. Sometimes, a thermodynamically metastable form is preferred. In these cases, intelligent process design must be utilized to produce the desired polymorph. By judiciously selecting a solvent, temperature, inlet supersaturation, and residence time, one can typically produce a pure, preferred polymorph in a continuous crystallization process regardless of the polymorphâ€™s thermodynamic stability.Â Design rules exist to guide the process engineer for systems in which heterogeneous nucleation and size independent linear crystal growth are the dominant mechanisms of solid deposition from solution. [1] Here, we hypothesize that these findings remain qualitatively true in precipitative systems in which homogeneous nucleation and/or agglomeration are also important crystal growth mechanisms. This hypothesis is supported by our recent experiments in which we continuously produce metastable forms of agglomerative $CaCO_3$. We have also extended the population balance analysis to include systems in which agglomeration is significant. Here we report these experiments and calculations and explain the implications for general polymorphic systems that exhibit significant agglomeration.

[1] Farmer, T. C., Carpenter, C. L., & Doherty, M. F. (2016). Polymorph selection by continuous crystallization. *AIChE Journal*, *62*(9), 3505-3514.

499259 Tray Pressure Drop Model Development

Abstract:

Fractionation Research, Inc., (FRI) first developed correlations for tray pressure drop in 1959 and continues to improve the models based on ongoing research. Pressure drop influences the design and selection of the distillation column and internals. The literature has documented extensive research made in modeling clear liquid height and dry pressure drop over the years. Previous FRI correlations and most correlations in the literature work well only for a narrow range of operating conditions and tray designs on which these correlations are based. This work discusses the parameters affecting the tray pressure drop, and limitations to the pressure drop correlations for sieve trays and valve trays. Different models are compared with each other and with experimental data, which provides a pathway for future model development.

494374 Tomographic Investigation of a Three-Phase System in Packed Columns

Abstract:

The fluid dynamic behavior of two poorly miscible liquids and a gas phase in packed columns, as it occurs in three-phase distillation, is very complex and poorly understood. Although structured packings provide significant separation improvements and less pressure losses at two-phase applications, trays are in industry the preferred internals for three-phase applications. The reason, among other aspects, is the unpredictable risk of a reduction of the separation performance by the second liquid in packed columns.

Therefore the objective of this work is to determine the fluid dynamics of a three-phase system in packed columns as the basis for understanding the phenomena on internals. For this purpose X-ray computed tomography (CT) is a suitable non-invasive measuring technique, which allows the determination of local parameters in packed columns under process conditions. The high resolution of the CT-images (80 µm) and the phase contrast of the system water/1,2-Dichlorobenzene/air enable, in combination with an image processing routine, a precise determination of the individual phases.

All experiments were performed in a 100 mm column with a packing height of 1 m. Several structured and random packings were investigated. The ratio and the total amount of the liquid loads were varied. Besides the investigations over the entire column, detailed images in 1 mm and 150 µm intervals enable a closer look at local flow patterns. Moreover long-term experiments over several hours were conducted to investigate temporal changes of the flow. As described in literature, it is important for the performance, which liquid phase is fed first into the column. Therefore and due to the influence of the initial distribution of the liquids on the flow morphology, different experiments were performed on this issue. Furthermore several fluid dynamic assumptions from literature and single sheet experiments were checked.

Whereas the focus was on the feasibility and qualitative results in the previous year, this year the focus is on quantitative results and validation. Segmented 3D visualizations of packing sections allow an insight from all perspectives and simplify the flow analysis for the first time. After identifying films, rivulets and bridges as the main flow pattern qualitatively, their shares on hold-ups and interfacial areas were quantified. The results confirm the qualitative analysis. A combination of this flow morphology identification with residence time results should allow a better understanding of the influence of the single flow pattern on mass transfer. A determination of the variance in long-term measurements gives information about the dynamics of three-phase systems in packed columns. Furthermore the hold-ups and the interfacial areas of both liquids were quantified and a dependency on the position in the column and as a function of the liquid loads was proven.

487779 Multiphase Flow Investigations in a Structured Packings Unit

Abstract:

Multiphase flow investigations in the structured packings unit

Rajesh K. Singh[1], Janine E. Galvin[2] and Xin Sun[1]

[1]Pacific Northwest National Laboratory, Richland, WA 99352, USA

[2]National Energy Technology Laboratory, Albany, Oregon 97321, USA

Post combustion carbon capture by solvent absorption has gained a significant commercial interest. The solvent absorption is carried out through counter-current gas liquid flow in a structured packed column. The local hydrodynamics in the packing, in particular wetting of sheets is a key factor dictating overall efficiency of the column. The CFD modelling of such system is a complex problem due to a wide range of length scales. In this context, multiphase flow simulation for wetting of sheets in the representative elementary unit (REU) of the structured packing can explain fundamental insights into meso-scale flow phenomenon. Specifically, Volume of Fluid (VOF) flow simulations are carried out over a wide range of parameters influencing the wetting dynamics such as solvent properties, contact angle, liquid load, etc. A scaling analysis for interfacial areas was performed based on simulation results since interfacial area actively participates in the transport phenomenon. Accordingly, a phenomenological correlation for interfacial area in terms of Kapitza number (Ka) is also proposed. The Kapitza number only depends on fluid properties and it decreases with increased solvent viscosity. The interfacial area decreases with increasing Ka number for a given contact angle and liquid load. Effects of contact angle are also extensively investigated at different Kapitza numbers and liquid loads. As expected, interfacial area decreases with increasing contact angle. Further, influence of liquid load at different contact angles and Kapitza number has also been explored.

McMaster University

IMPROVED CONTROL STRATEGIES FOR SEMICONTINUOUS DISTILLATION

Pranav Bhaswanth Madabhushi
Dr. Thomas A. Adams II

CATEGORIZATION OF OPERATIONS*

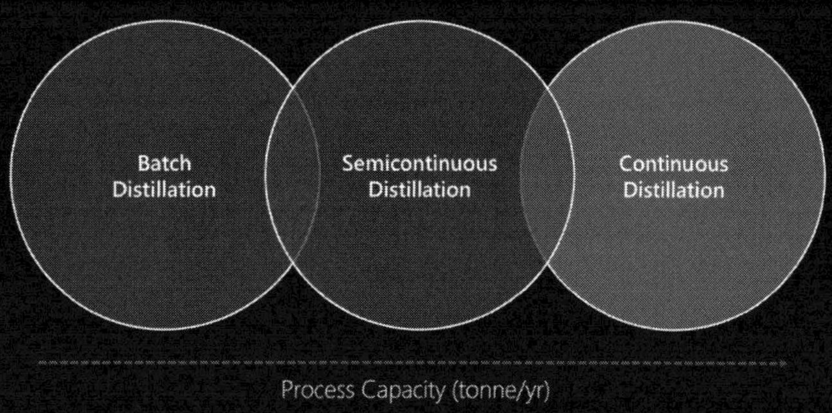

* Knopf, F.C., Okos, M.R. and Reklaitis, G.V., 1982. Optimal design of batch/semicontinuous processes. *Industrial & Engineering Chemistry Process Design and Development*, 21(1), pp.79-86.

CONTINUOUS DISTILLATION

- A steady state operation: operated at an economic optimal fixed point.
- Flexible designs.
- Economical for a large production rate.

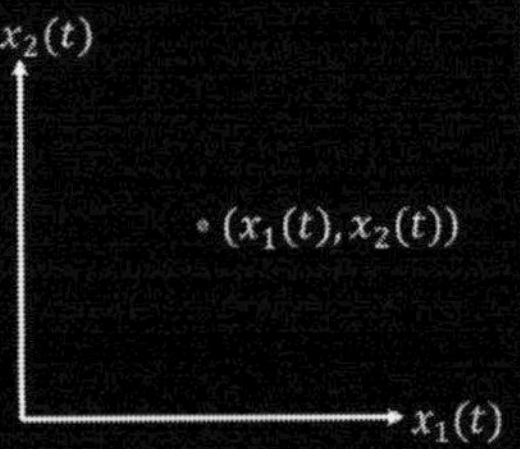

$\boldsymbol{x}(t) := \begin{bmatrix} x_1(t) \\ x_2(t) \end{bmatrix}$ is a fixed point.

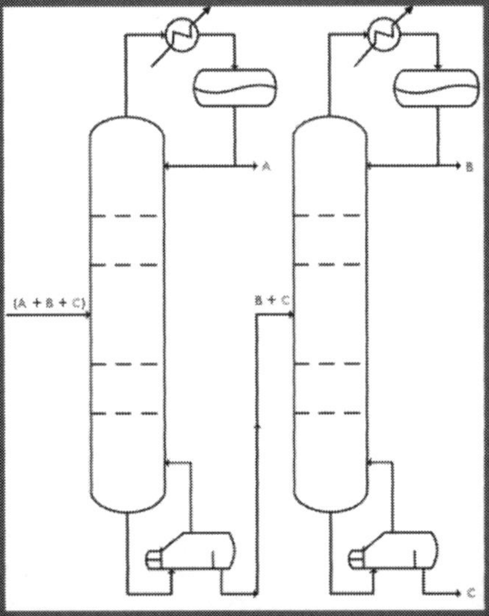

MCMASTER UNIVERSITY PRANAV BHASWANTH MADABHUSHI DR. THOMAS A. ADAMS II

SEMICONTINUOUS DISTILLATION

- A periodic operation (with three operating modes): operated at the economic optimal periodic orbit.
- Inherently flexible
- Economic for intermediate production rates.

$\boldsymbol{x}(t) := \begin{bmatrix} x_1(t) \\ x_2(t) \end{bmatrix}$ is a periodic orbit.

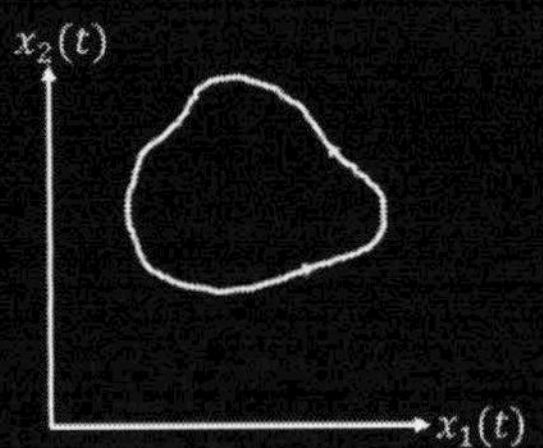

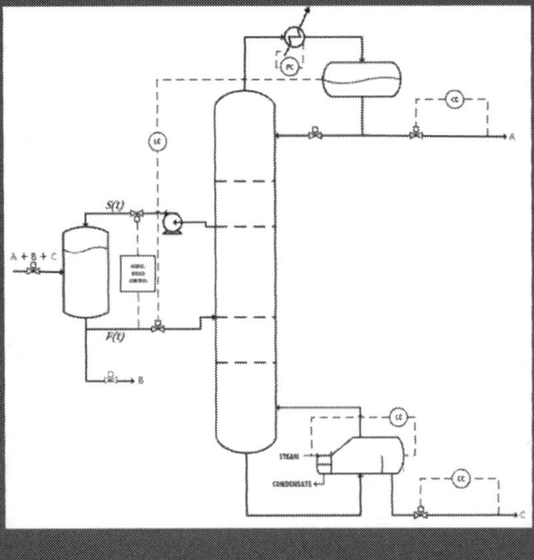

MCMASTER UNIVERSITY PRANAV BHASWANTH MADABHUSHI DR. THOMAS A. ADAMS II

MODES OF OPERATION (SEPARATING MODE)*

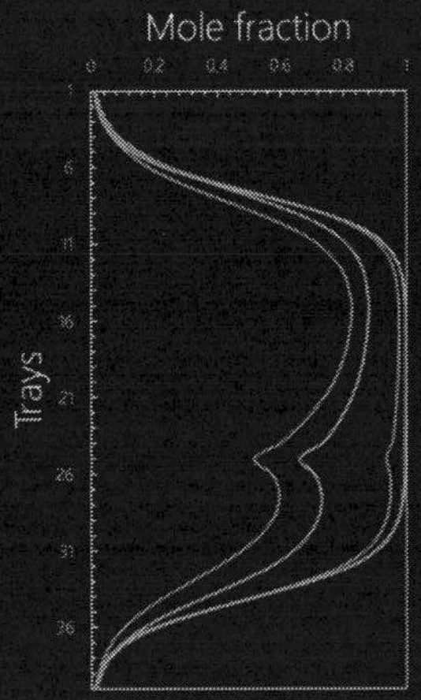

Concentration profile of component B on column trays at different times during the separating mode.

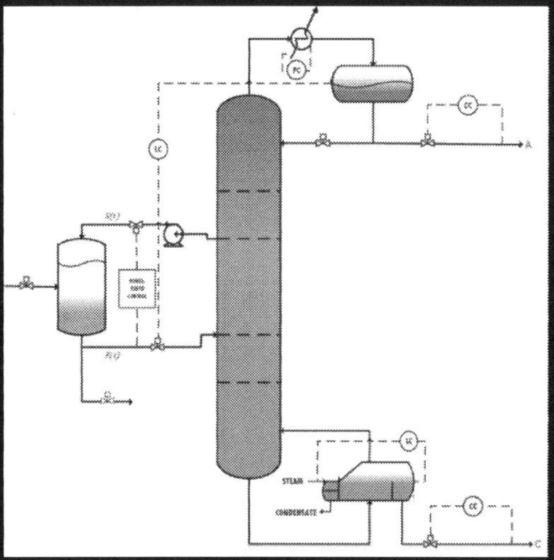

* Adams, T.A. and Pascall, A., 2012. Semicontinuous thermal separation systems. *Chemical Engineering & Technology*, 35(7), pp.1153-1170.

MODES OF OPERATION (DISCHARGING MODE)*

- Triggered by a state dependent, discrete input action on the discharge valve of the middle vessel.
- B at the desired purity ($x_{B,MV}^{desired}$) is collected periodically i.e., the valve opens at every $t + n\,T,\ n = 0, 1, 2, \dots, n$ for product collection. T is the period of the periodic orbit (process cycle time).

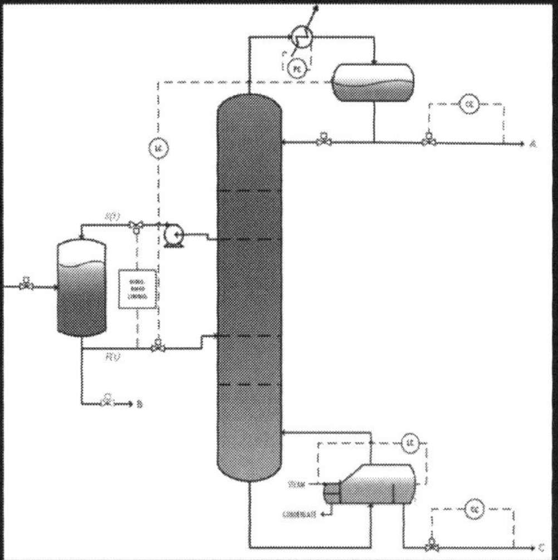

* Adams, T.A. and Pascall, A., 2012. Semicontinuous thermal separation systems. *Chemical Engineering & Technology*, 35(7), pp.1153-1170.

MODES OF OPERATION (CHARGING MODE)*

- Triggered by a state dependent, discrete input action on the feed valve of the middle vessel.
- Feed is periodically fed to the middle vessel i.e., the valve opens at every $t + nT,\ n = 0, 1, 2, \ldots, n$.

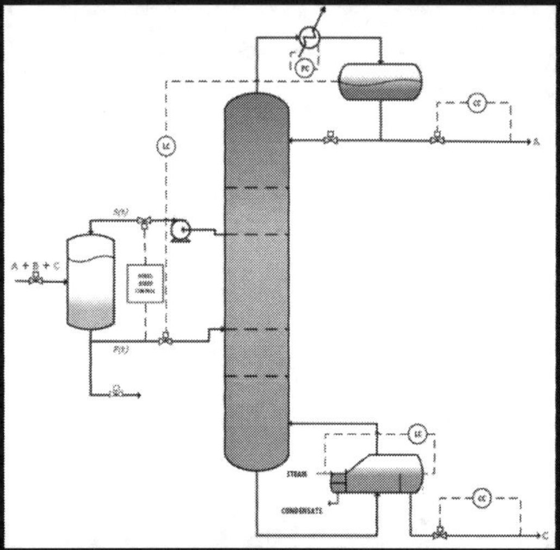

* Adams, T.A. and Pascall, A., 2012. Semicontinuous thermal separation systems. *Chemical Engineering & Technology*, 35(7), pp.1153-1170.

EXAMPLE TRAJECTORIES

This is for:

A: carbon dioxide

B: dimethyl ether

C: methanol/water

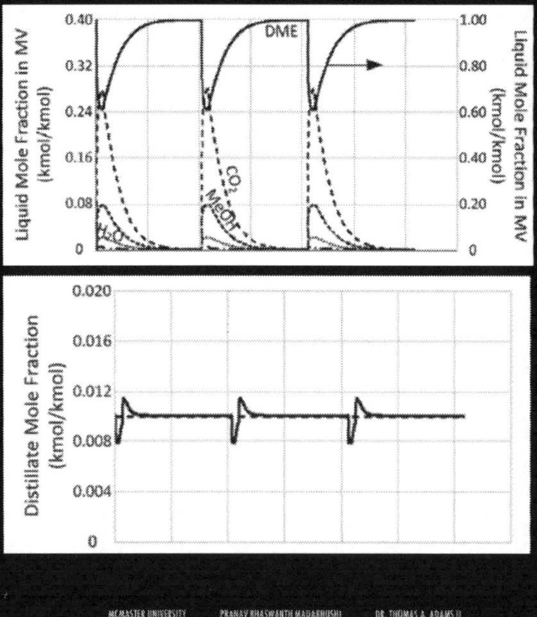

FLOODING AND WEEPING

- Flooding and weeping avoided at all times and on all stages

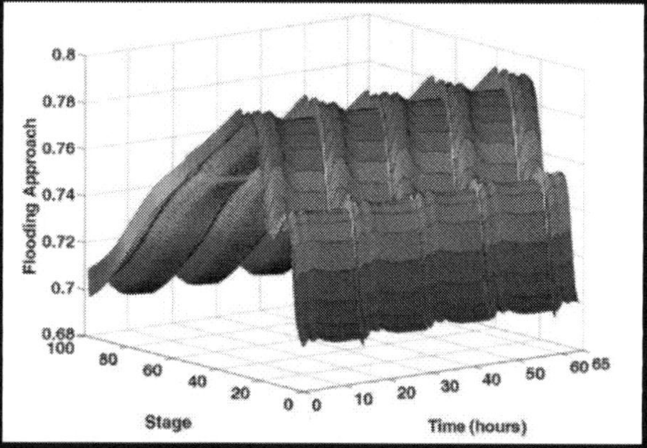

THE CONTROL CONFIGURATION*

- Side stream flowrate is controlled using a feed forward control model.

Inputs		Outputs
Distillate molar flowrate	→	Mole fraction of A in the distillate
Bottoms molar flowrate	→	Mole fraction of C in the bottoms
Feed molar flowrate to the column	→	Reflux Drum Level
Reboiler duty	→	Sump Level
Condenser duty	→	Column Pressure

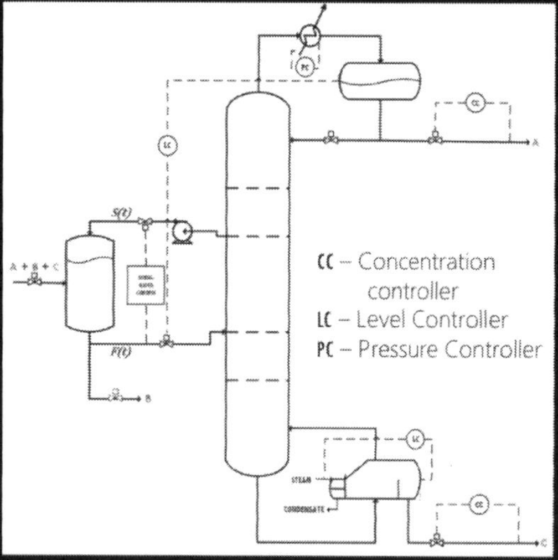

*Pascall, A. and Adams, T.A., 2013. Semicontinuous separation of dimethyl ether (DME) produced from biomass. *The Canadian Journal of Chemical Engineering*, 91(6), pp.1001-1021.

MODEL-BASED CONTROL OF SIDE STREAM MOLAR FLOWRATE

IDEAL SIDE-DRAW RECOVERY (ISR) CONTROL MODEL*

$$S(t) = F(t) \times x_{B,F}(t)$$

$F(t)$ is the molar flowrate of feed to the column

$x_{B,F}(t)$ is the mole fraction of B in feed to the column

- ISR control model was implemented using a P controller where the setpoint is changed according to $F(t) \times x_{B,F}(t)$ and $S(t)$ is allowed to track this setpoint.

*Adams, T.A. and Seider, W.D., 2008. Semicontinuous distillation for ethyl lactate production. AIChE journal, 54(10), pp.2539-2552.

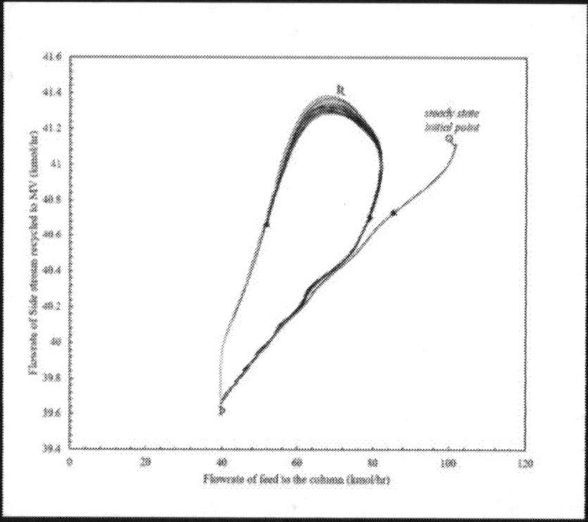

A phase plane plot which illustrates the evolution of $F(t)$ and $S(t)$ in a Benzene (B), Toluene (T), o-Xylene (X) ternary mixture semicontinuous separation simulation.

MCMASTER UNIVERSITY PRANAV BHASWANTH MADABHUSHI DR. THOMAS A. ADAMS II

EFFECT OF USING A PI CONTROLLER FOR BETTER SETPOINT TRACKING

- The steady state becomes a periodic orbit when τ_s is large (or PI is converging to be a P controller).

Integral Time of side stream controller (τ_S) (min)	Total time required to close the side stream valve below 5% (hr)
5	4314
50	> 8400
5000	>>8400

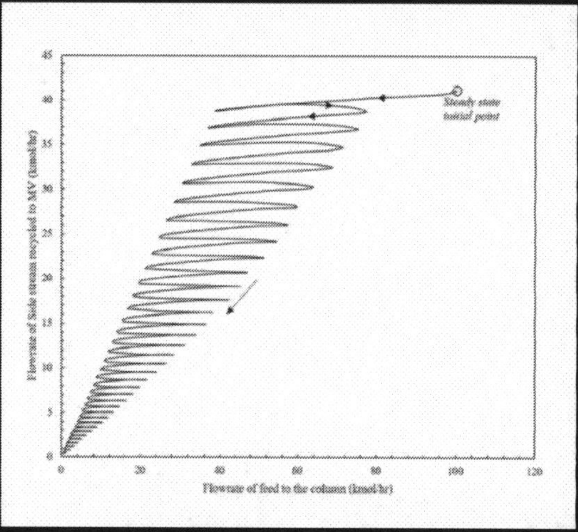

A phase plane plot which illustrates the evolution of $F(t)$ and $S(t)$ in a Benzene (B), Toluene (T), o-Xylene (X) ternary mixture semicontinuous separation simulation.

MCMASTER UNIVERSITY PRANAV BHASWANTH MADABHUSHI DR. THOMAS A. ADAMS II

CONTROL MODEL MODIFICATION

MODIFIED IDEAL SIDE-DRAW RECOVERY (MISR) CONTROL MODEL

$$S(t) = \frac{F(t) \times x_{B,F}(t)}{x_{B,S}(t)}$$

$F(t)$ is the molar flowrate of feed to the column.

$x_{B,F}(t)$ is the mole fraction of B in feed to the column.

$x_{B,S}(t)$ is the mole fraction of B in the side stream.

Previous Best TAC: $2.444 \left(\frac{\$}{\text{Kmol}}\right)$

New TAC: $2.063 \left(\frac{\$}{\text{Kmol}}\right)$

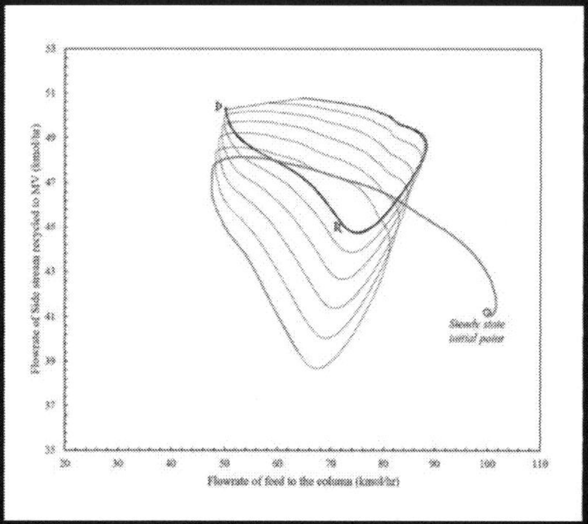

A phase plane plot which illustrates the evolution of $F(t)$ and $S(t)$ in a Benzene (B), Toluene (T), o-Xylene (X) ternary mixture semicontinuous separation simulation.

EXAMPLE VALVE TRAJECTORIES

- The new policy has much larger flow rates.
- So, what if we just maximize the flow rate?

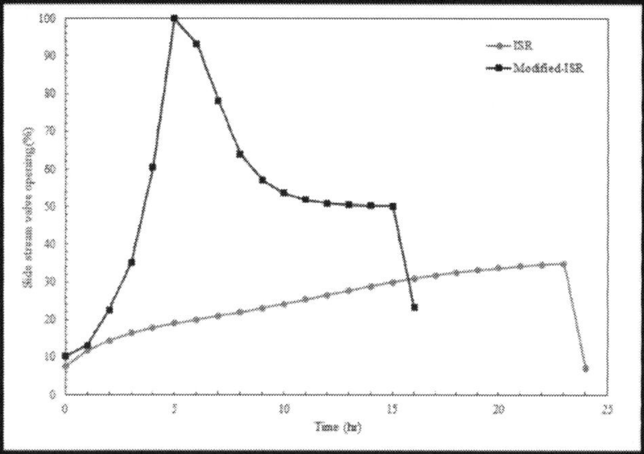

100% OPENING OF SIDE STREAM VALVE

$$T_{ISR} > T_{MISR} \approx T_{100\%}$$

16.14 hr > 13.58 hr ≈ 13.47 hr

- Total Annualized cost per Kmol of feed processed (TAC)

$$TAC_{ISR} > TAC_{MISR} \approx TAC_{100\%}$$

$$2.444\left(\frac{\$}{\text{Kmol}}\right) > 2.063\left(\frac{\$}{\text{Kmol}}\right) \approx 2.064\left(\frac{\$}{\text{Kmol}}\right)$$

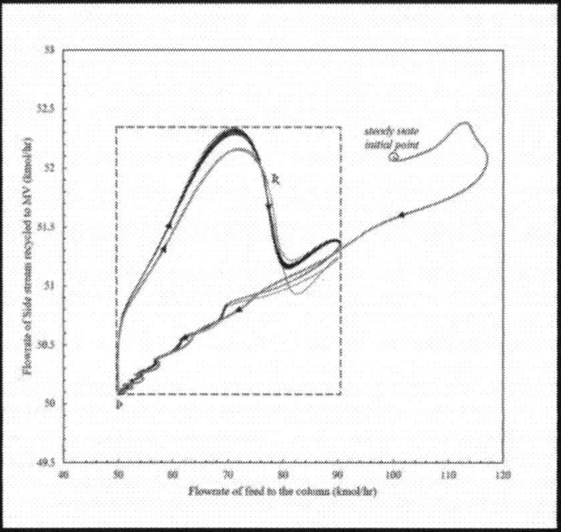

A phase plane plot which illustrates the evolution of $F(t)$ and $S(t)$ in a Benzene (B), Toluene (T), o-Xylene (X) ternary mixture semicontinuous separation simulation.

DYNAMIC OPTIMIZATION

INITIAL GUESS 1: Used side stream valve opening at 20 discrete time points between $0 \le t \le 13.58$ (hr) when MISR control model is applied.

- Length of each element (Δt) = $\frac{13.58}{20} \approx 0.68$

INITIAL GUESS 2: 100% opening of side stream valve

- Optimized result : 100% valve opening at each discretized time point.
- This is a local optimum.

$$\emptyset(t_f) := \frac{Total\ Annualized\ Cost}{Annual\ molar\ amount\ of\ feed\ processed}$$

$$\underset{K_j, \tau_j, t_f}{minimize} \quad \emptyset(t_f)$$

subject to

System Differential Algebraic Equations

Periodic Constraints

Variable bounds

MCMASTER UNIVERSITY PRANAV BHASWANTH MADABHUSHI DR. THOMAS A. ADAMS II

OPTIMIZATION RESULTS – INITIAL GUESS 1

$T_{MISR} > T_{optimum}$

13.58 hr > 13.34 hr

$TAC_{MISR} > TAC_{optimum}$

$2.063 \left(\frac{\$}{\text{Kmol}}\right) > 2.034 \left(\frac{\$}{\text{Kmol}}\right)$

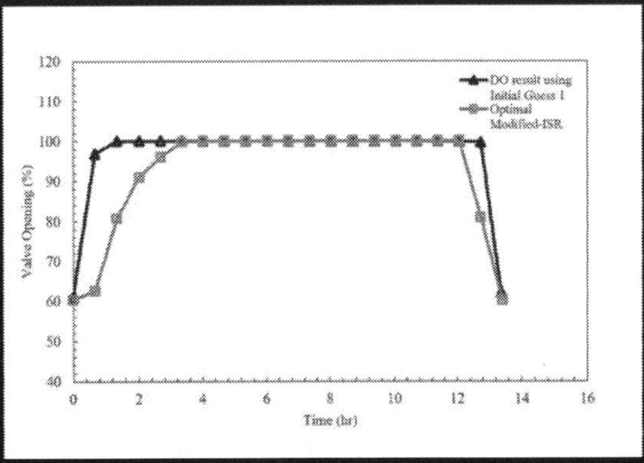

Plot comparing the optimal valve opening percentages as a piecewise linear function of time $v(t)$ with the initial guess – 1: optimal MISR control model side stream valve opening percentages (piecewise linear approximation) during the cycle.

APPLYING THE OPTIMAL RESULT

- "Almost Periodic" orbit. The period is within a small neighborhood and does not converge to a specific value asymptotically.
- Emphasizes the need to optimize each cycle independently, implying that **MPC based control** of side stream flowrate could be advantageous.

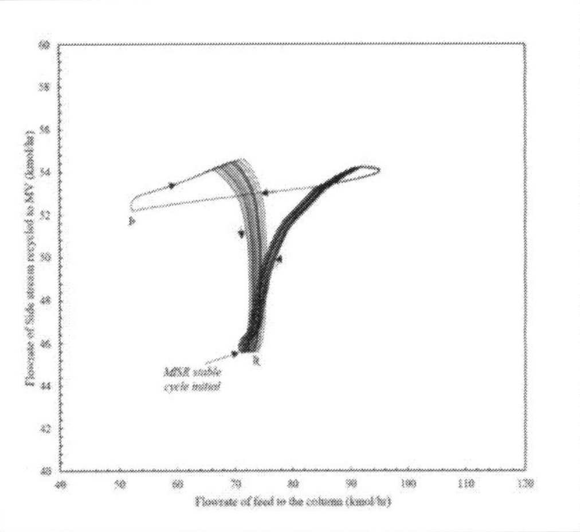

MCMASTER UNIVERSITY PRANAV BHASWANTH MADABHUSHI DR. THOMAS A. ADAMS II

CONCLUSION & FUTURE WORK

- A better control model (MISR) was presented, which led to a 15% reduction in TAC when compared to the optimized case where ISR control model (implemented by a P controller) is used.
- Showed that 100% opening of side stream valve can perhaps be used as an alternative to MISR control model in the case of equimolar separation of components.
- Concluded that MPC based control of side stream flowrate to converge to economic optimal periodic orbits starting from a given initial steady state could be beneficial.
- Understand the dynamics of the system to develop better design procedures.

MCMASTER UNIVERSITY PRANAV BHASWANTH MADABHUSHI DR. THOMAS A. ADAMS II

ACKNOWLEDGMENTS

Ontario Research Fund

MACC
McMaster Advanced Control Consortium

BTX EXAMPLE DESIGN DETAILS

Steady state conventional continuous side stream column design data	
Number of stages (Meidanshahi and Adams, 2014)	40 (stage 1: Total condenser and stage 40: Kettle reboiler)
Feed condition (Ling and Luyben, 2009)	Liquid, 358 K and 0.533 atm
Feed flowrate	100 kmol/hr
Feed stream location (Meidanshahi and Adams, 2014)	Above Stage 25
Side stream location (Meidanshahi and Adams, 2014)	On Stage 14
Condenser pressure (Ling and Luyben, 2009)	0.37 atm
Tray type	Sieve
Tray spacing	0.6096 m
Pressure drop across stages (Ling and Luyben, 2009)	0.0068 atm
Reflux ratio	4.066 (mole basis)
Boilup ratio	3.437 (mole basis)
Column Diameter	1.273 m
Murphee tray efficiency (Meidanshahi and Adams, 2014)	0.75
$\frac{F(t)}{S(t)}$	0.396

MCMASTER UNIVERSITY PRANAV BHASWANTH MADABHUSHI DR. THOMAS A. ADAMS II

National Taiwan University

PSE LABORATORY

Energy-Saving Heterogeneous Extractive Distillation System for the Separation of Close-boiling Cyclohexane/Cyclohexene Mixture

Chun-Cheng Yi, Wen-Chi Huang and I-Lung Chien*

2017 AIChE Annual Meeting
Minneapolis, MN

Speaker : Chun-Cheng, Yi

Department of Chemical Engineering, National Taiwan University

Outline

- Introduction
  - ✓ Background
  - ✓ Single Column System
  - ✓ Relevant Studies
- Heterogeneous Extractive Distillation System
  - ✓ Conceptual Design
  - ✓ Optimal Design
  - ✓ Comparisons
- Energy-Saving Designs
  - ✓ Design with entrainer recovery column operated at vacuum pressure
  - ✓ Design with a process-to-process heat exchanger
  - ✓ Design with a process-to-process heat exchanger combining of operating entrainer recovery column at vacuum pressure
- Conclusions

Introduction

➢ Cyclohexene can be used to synthesize many kinds of chemicals of high commercial value, such as adipic acid, an important reagent in the manufacture of nylon.

➢ However, processes used to produce cyclohexene essentially let a certain amount of cyclohexane present in the cyclohexene product stream.[1] [2]

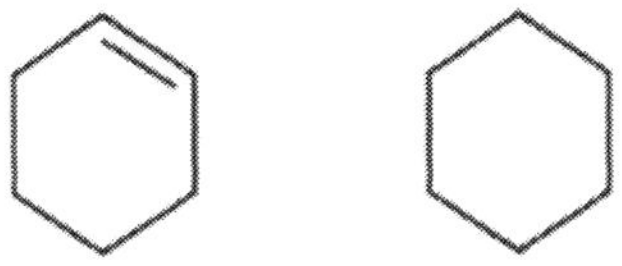

Cyclohexene (ENE) Cyclohexane (ANE)

T-xy diagram for ENE/ANE (1 atm)

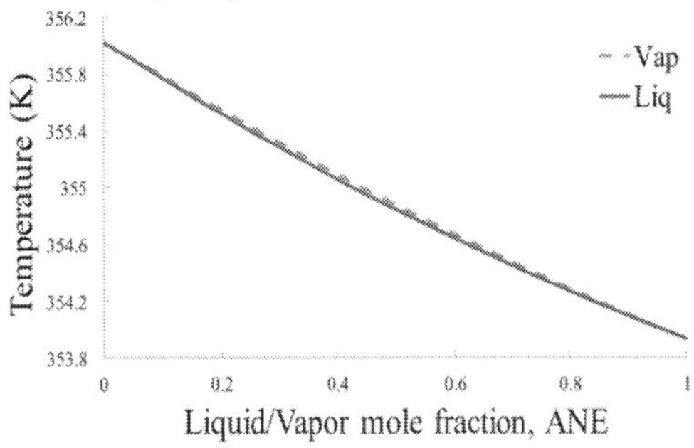

[1] Ralph C. Tallman, Conversion of cyclohexane to cyclohexene. US Patent 3,108,144, 1963.

[2] M. Larry Campbell, *Cyclohexane*. Wiley-VCH, Weinheim, Germany, 2012.

Introduction

➢ Theoretically, the separation of close boiling compounds can be performed by using just one regular distillation column.

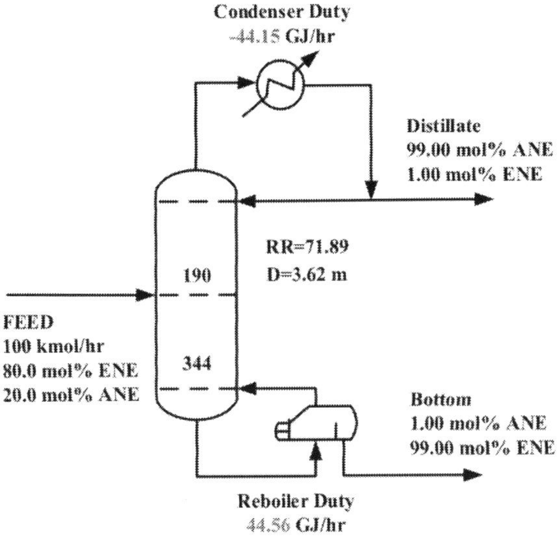

Top Pressure: 1 atm
Number of Stages : 345
Feed Stage: 190th
Reboiler Duty: 44.56 GJ/hr

TAC : 5129.77 (1000$/yr)

3

Introduction

- Steyer et al. (2002) proposed a conceptual design to reactively separate ANE and ENE using water as reactive entrainer.[3]

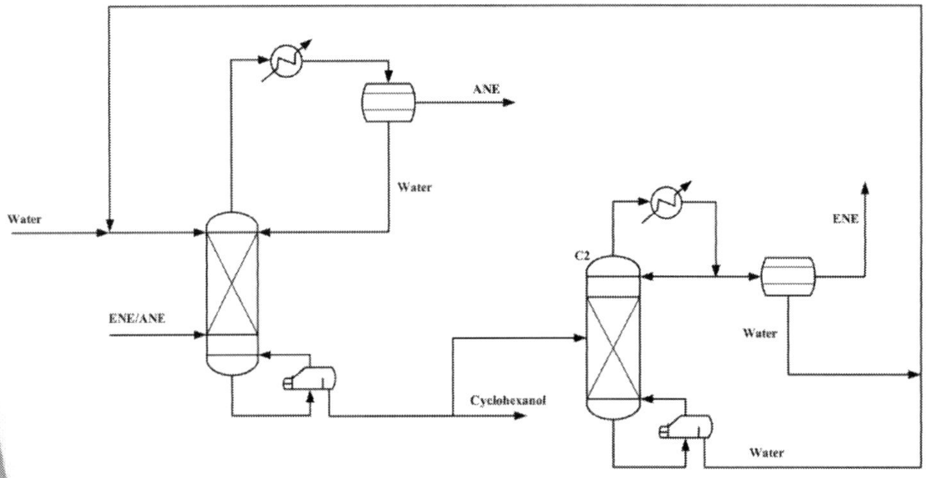

[3] Steyer, F.; Qi, Z.; Sundmacher, K., Synthesis of cylohexanol by three-phase reactive distillation: influence of kinetics on phase equilibria. *Chem. Eng. Sci.* **2002**, 57, 1511-1520.

4

Introduction

- Yu et al. (2016) proposed a thermally coupled reactive distillation system (TCRD) to eliminate the remixing effect in the bottom of the hydration column.[4]

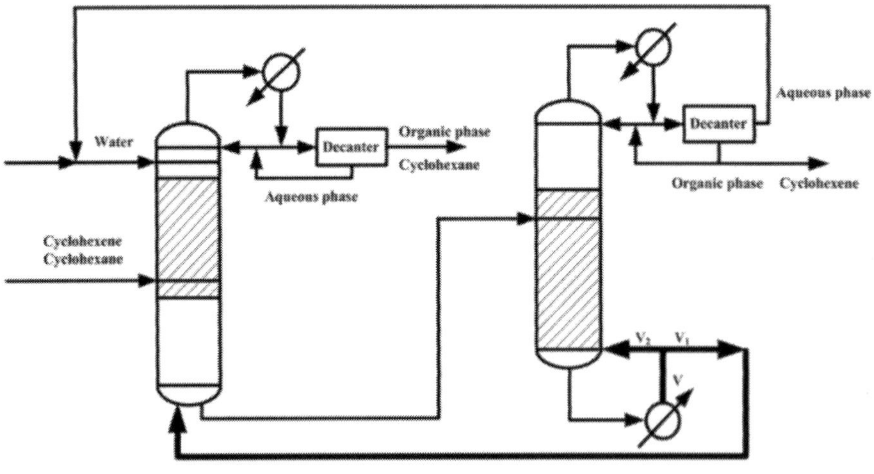

[4] Yu, J.; Shi, L.; Yuan, Y.; Chen, H.; Wang, S.; Huang, K.; Thermally Coupled Reactive Distillation System for the Separations of Cyclohexane/Cyclohexene Mixtures. *Ind. Eng. Chem. Res.* **2016**, 55, 311-322.

5

Introduction

- Yu et al. (2016) proposed a thermally coupled reactive distillation system (TCRD) to eliminate the remixing effect in the bottom of the hydration column.[4]

Number of Stages of Hydration Column : 107
Number of Stages of Dehydration Column : 42
Reboiler Duty: 28.21 GJ/hr

- It can be observed that the whole process is still energy intensive and the number of stages required in this complex design structure is quite large.

[4] Yu, J.; Shi, L.; Yuan, Y.; Chen, H.; Wang, S.; Huang, K.; Thermally Coupled Reactive Distillation System for the Separations of Cyclohexane/Cyclohexene Mixtures. *Ind. Eng. Chem. Res.* **2016**, 55, 311-322.

6

Heterogeneous Extractive Distillation System

- There are studies relevant to the effectiveness of various compounds on the separation of ANE and ENE.[5]-[7]
- Entrainers that can be applied in the extractive distillation process in industry were listed in Gmehling and Möllmann.[11]

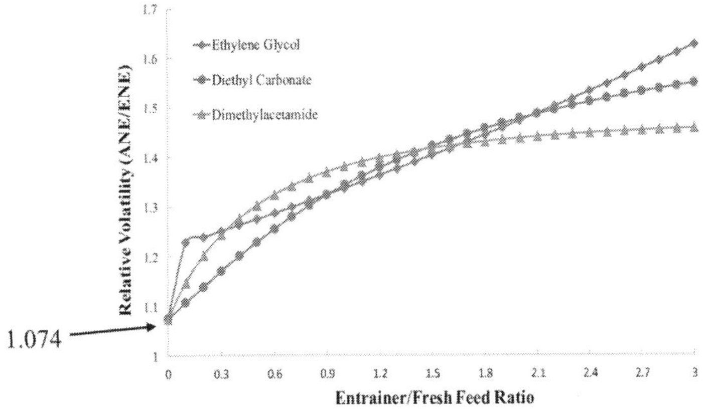

[11] Gmehling, J.; Möllmann, C., Synthesis of distillation processes using thermodynamic models and the Dortmund data bank. *Ind. Eng. Chem. Res.* **1998**, 37, 3112-3123.

7 Heterogeneous Extractive Distillation System

- NRTL model parameters of the studied systems

Comp. i	ENE	ANE	ENE
Comp. j	ANE	EG	EG
Source	Steyer and Sundmacher [8]	ASPEN LLE	UNIFAC estimated
a_{ij}	0	-10.0469	0
a_{ji}	0	-10.1035	0
b_{ij} (K)	5.10961	5152.694	671.7161
b_{ji} (K)	7331.87	5171.496	602.0344
c_{ij}	0.831053	0.2	0.3

Aspen Plus NRTL:

$$\ln \gamma_i = \frac{\sum_j x_j \tau_{ji} G_{ji}}{\sum_k x_k G_{ki}} + \sum_j \frac{x_j G_{ij}}{\sum_k x_k G_{kj}} \left[\tau_{ij} - \frac{\sum_m x_m \tau_{mj} G_{mj}}{\sum_k x_k G_{kj}} \right] \qquad Where: \; G_{ij} = \exp(-\alpha_{ij} \tau_{ij})$$

$$\tau_{ij} = a_{ij} + \frac{b_{ij}}{T}$$

$$\alpha_{ij} = c_{ij} \;,\; \tau_{ii} = 0 \;,\; G_{ii} = 1$$

Cyclohexane (ANE)
Cyclohexene (ENE)
Ethylene Glycol (EG)

[8] Steyer, F.; Sundmacher, K., VLE and LLE data for the system cyclohexane+ cyclohexene+ water+ cyclohexanol. *J. Chem. Eng. Data* **2004**, 49, 1675-1681.

8 Heterogeneous Extractive Distillation System

- Ternary Diagram at 1 atm
- With Vapor-Liquid-Liquid envelope

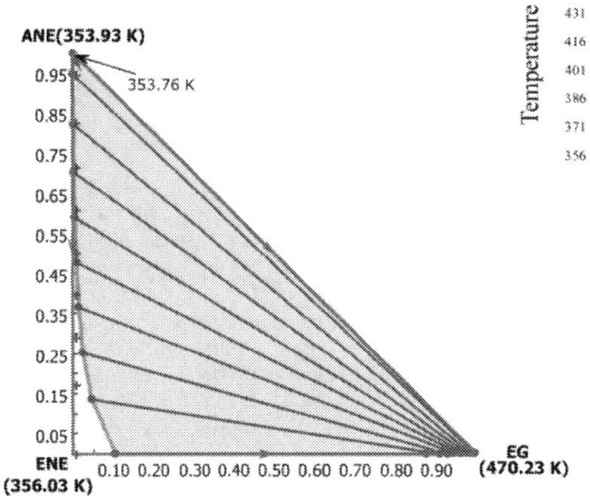

Heterogeneous Azeotrope
ANE : 0.9931
EG : 0.0069

T-xy diagram for ENE/EG (1 atm)

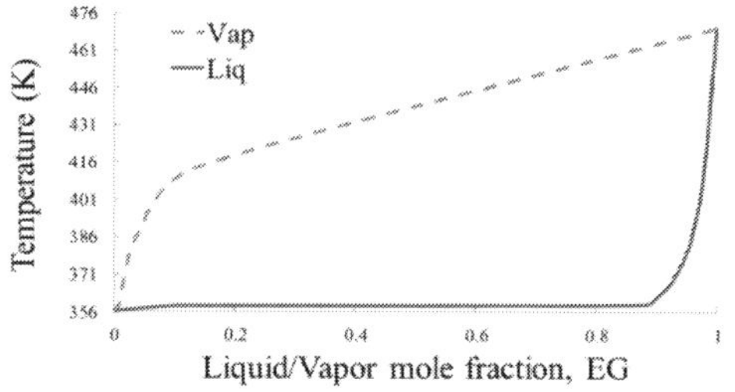

ANE
b.p. : 353.93 K (80.78 °C)

ENE
b.p. : 356.03 K (82.88 °C)

EG
b.p. : 470.23 K (197.08 °C)

- The entrainer is considered as heterogeneous if it causes liquid–liquid phase split over a range of compositions.[10]

9

Heterogeneous Extractive Distillation System

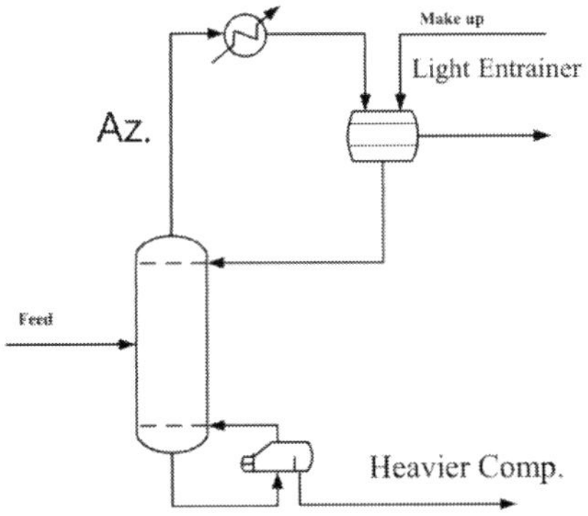

- To form a heterogeneous minimum-boiling azeotrope with at least one component in the mixture.
- The component with higher boiling point mainly exits the column from the bottom.
- Azeotrope drawn from the top of the column will split into two immiscible liquid phases .

- Heavy entrainer would enhance the relative volatility of two original components.
- Need rectifying section.

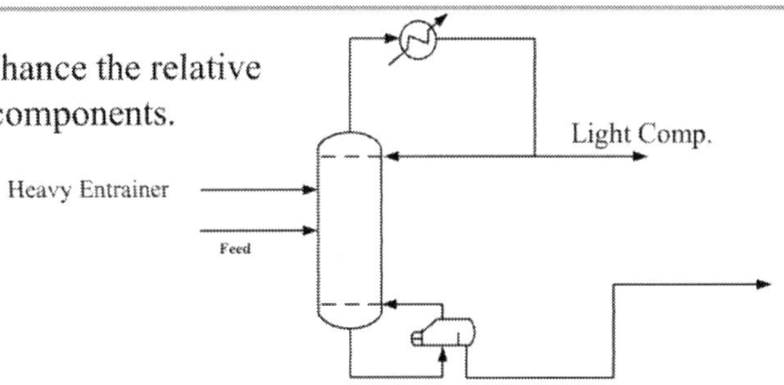

10

Heterogeneous Extractive Distillation System

- Design of the heterogeneous extractive distillation system

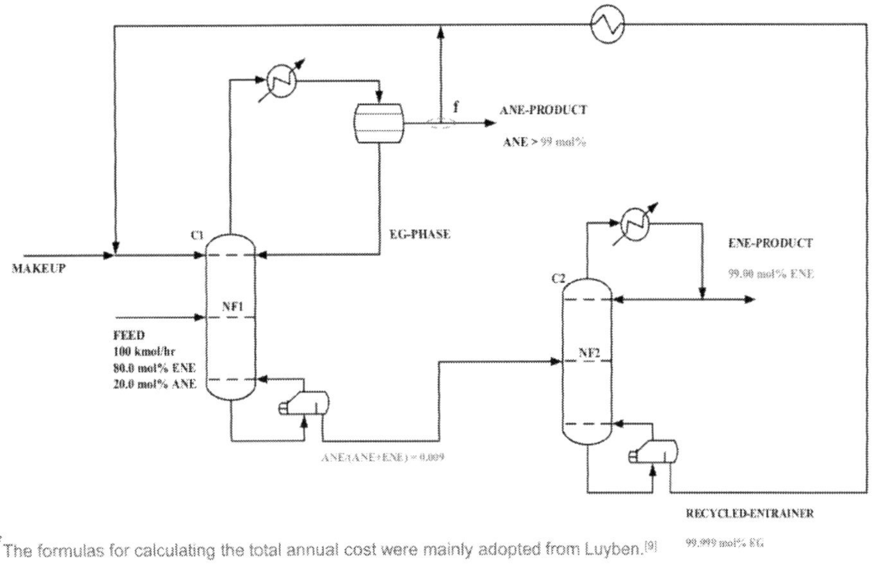

*The formulas for calculating the total annual cost were mainly adopted from Luyben.[9]

TAC1 TAC2

11

Heterogeneous Extractive Distillation System

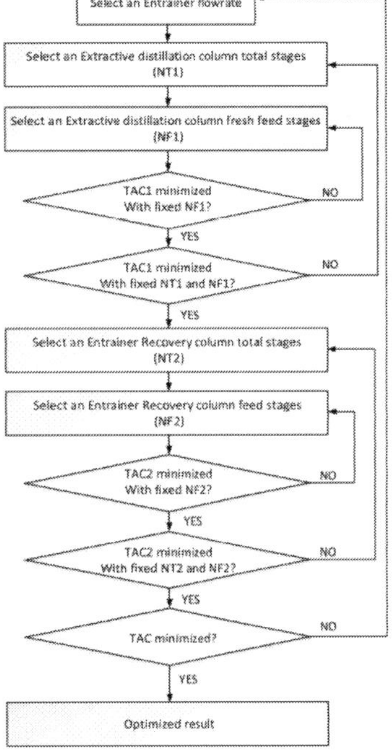

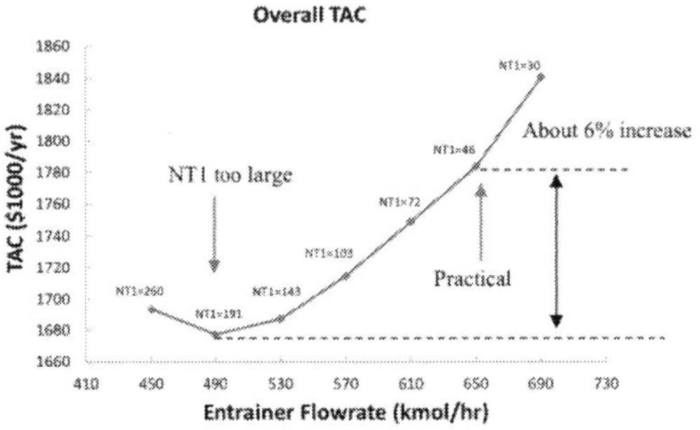

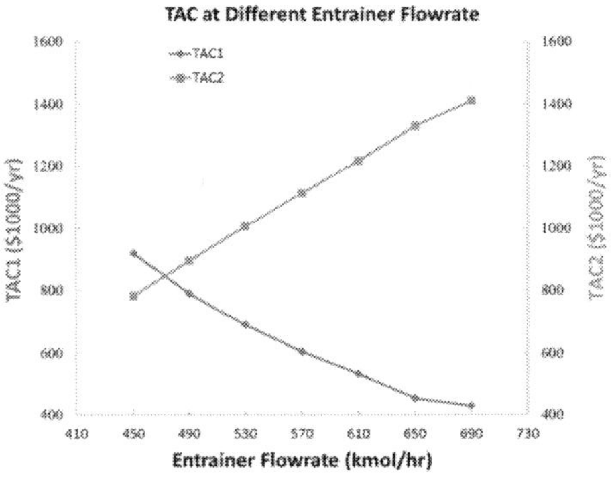

12

Heterogeneous Extractive Distillation System

650 kmol/hr

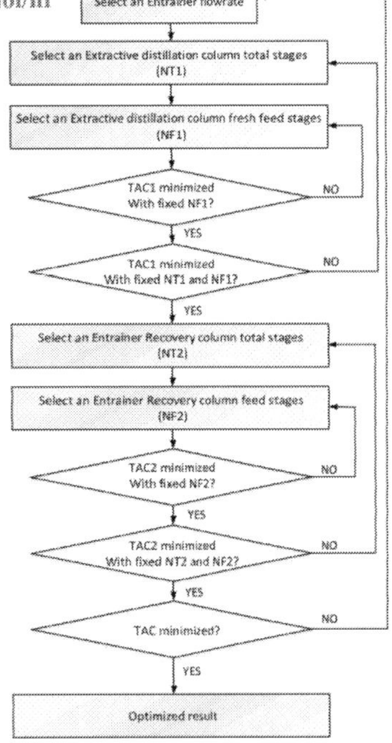

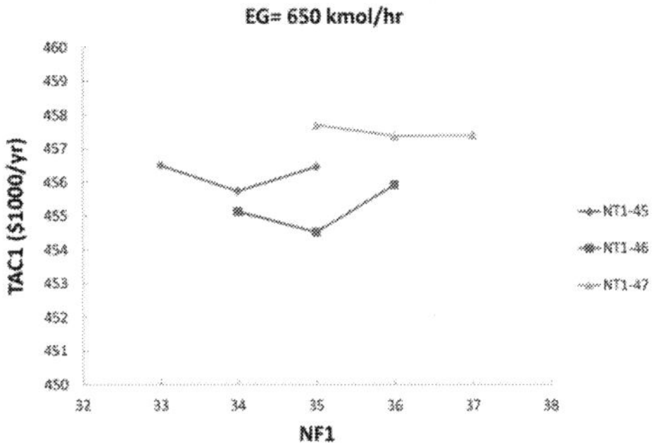

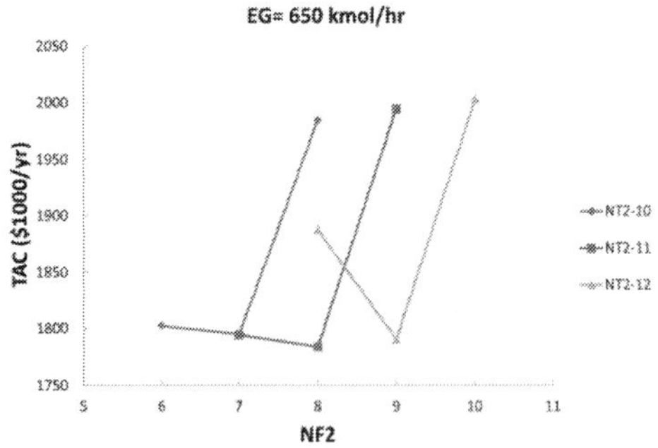

13

Heterogeneous Extractive Distillation System

- Optimized design flowsheet at 650 kmol/hr EG flowrate

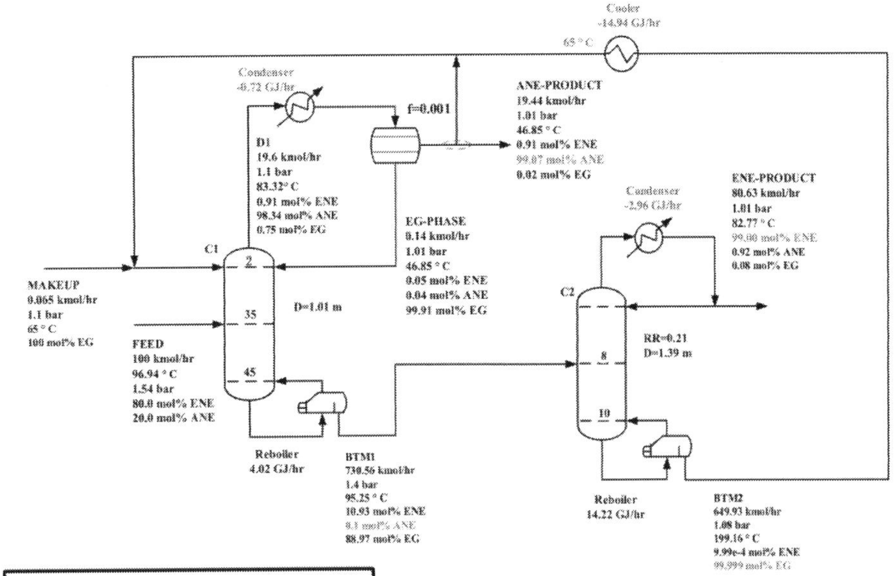

TAC : 1783.79 (1000$/yr)

14

Heterogeneous Extractive Distillation System

	Regular column	TCRD [4]	Heterogeneous extractive distillation system		
			C1	C2	Cooler
Column (1000 $)	5851.88	1444.92	288.81	123.68	
Reboiler (1000 $)	501.80	305.97	67.04	169.86	
Condenser (1000 $)	340.46	400.71	88.71	56.93	
Cooler (1000 $)					115.01
Decanter (1000 $)			45.27		
Steam cost (1000 $/yr)	2773.37	1755.77	249.95	1123.79	
Cooling water cost (1000 $/yr)	125.02	81.39	2.03	8.38	42.3
Makeup cost (1000 $/yr)			38.91		
Catalyst cost (1000 $/yr)		246.34			
Total operating cost (1000 $/yr)	2898.39	2083.5	1465.36		
TAC (1000 $/yr)	5129.77 (0%)	2800.70 (-45.4%)	1783.79 (-65.2%)		

Heterogeneous Extractive Distillation System

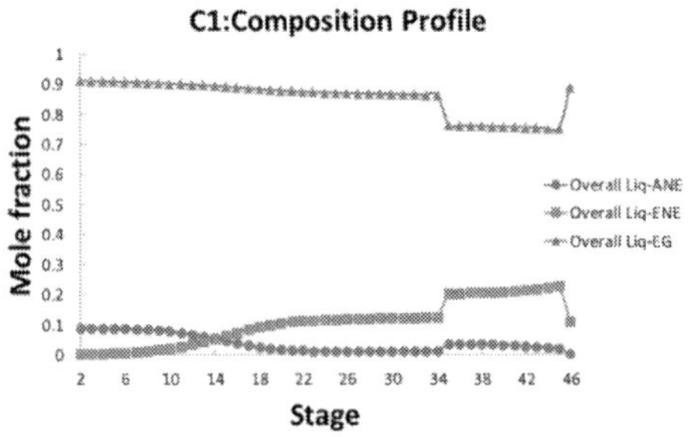

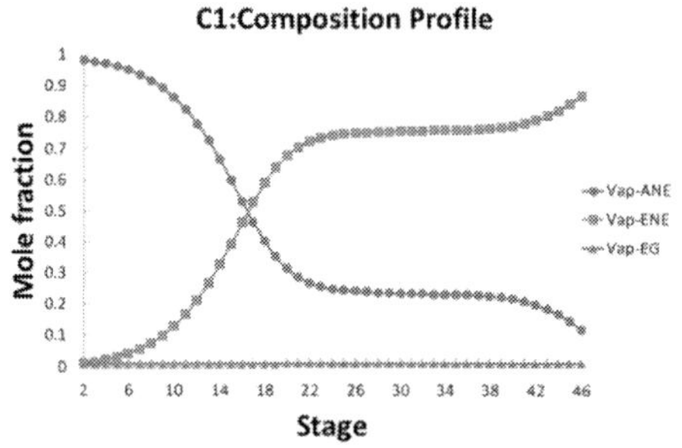

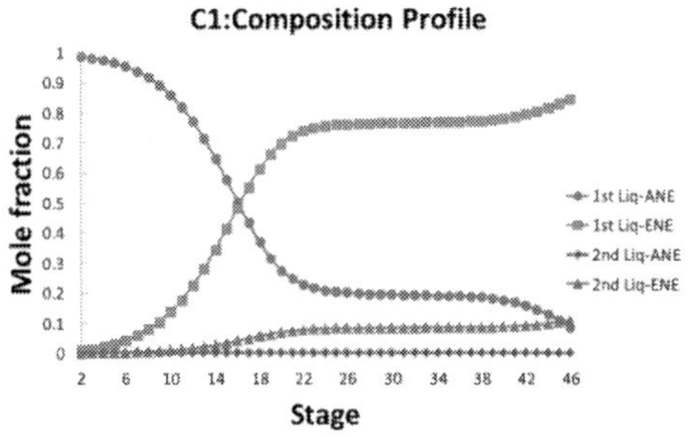

- Vap. : ANE/ENE
- 1st Liq. : ANE/ENE
- 2nd Liq. : EG/ENE

Energy-Saving Designs

■ Two Reasons that made the reboiler duty of C2 so high:

1) Large entrainer flowrate

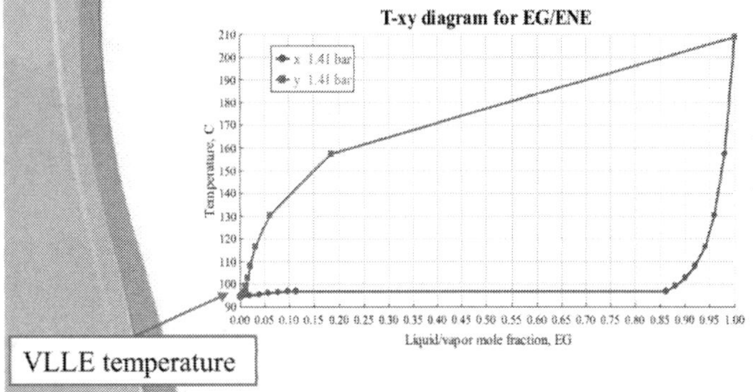

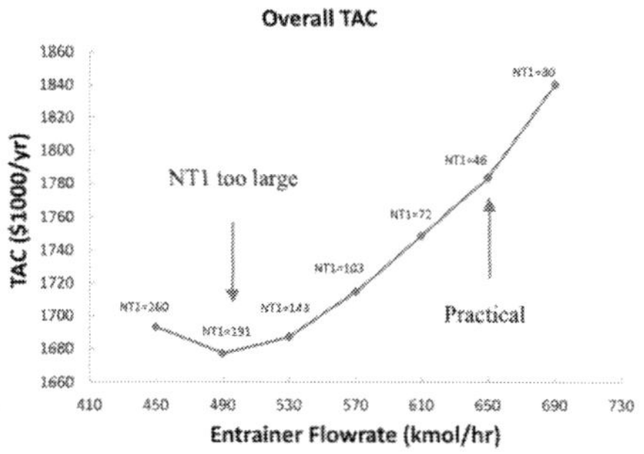

2) Low inlet temperature (C1 bottom temperature)

◆ C1 bottom temperature will be quite low even at a relative high EG composition.

Energy-Saving Designs

- **Design with entrainer recovery column operated at vacuum pressure**

  *** The calculation of liquid-ring pump investment cost is taken from Seider [13]**

- The concept of operating distillation column under vacuum pressure in extractive distillation process can be seen in open literature.[12]

1) Relative volatility
2) Reboiler temperature (Column temperature)

Condensers
  Differential temperature
  = condenser temperature – 315 K
Reboilers
  Differential temperature
  = steam temperature – base temperature (ΔT > 20 K)
Energy cost
  HP steam = \$9.88/GJ (41 barg, 254 °C)
  MP steam = \$8.22/GJ (10 barg, 184 °C)
  LP steam = \$7.78/GJ (5 barg, 160 °C)
  Cooling water = \$0.354/GJ
  Electricity = \$16.9/GJ

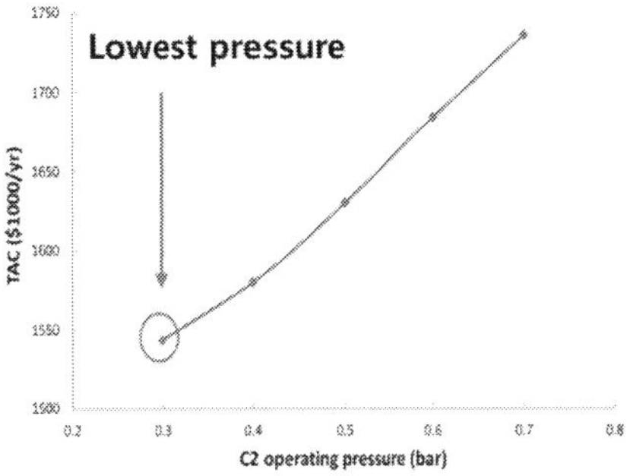

[12] Luyben, W. L., Distillation column pressure selection. *Separation and Purification Technology* **2016**, 168, 62-67.

[13] Seider, W. D.; Seader, J. D.; Lewin, D. R.; Widagdo, S. *Product and Process Design Principles Synthesis, Analysis, and Evaluation.* John Wiley & Sons, Inc: Hoboken, New Jersey, 2010.

Energy-Saving Designs

- **Design with entrainer recovery column operated at vacuum pressure**

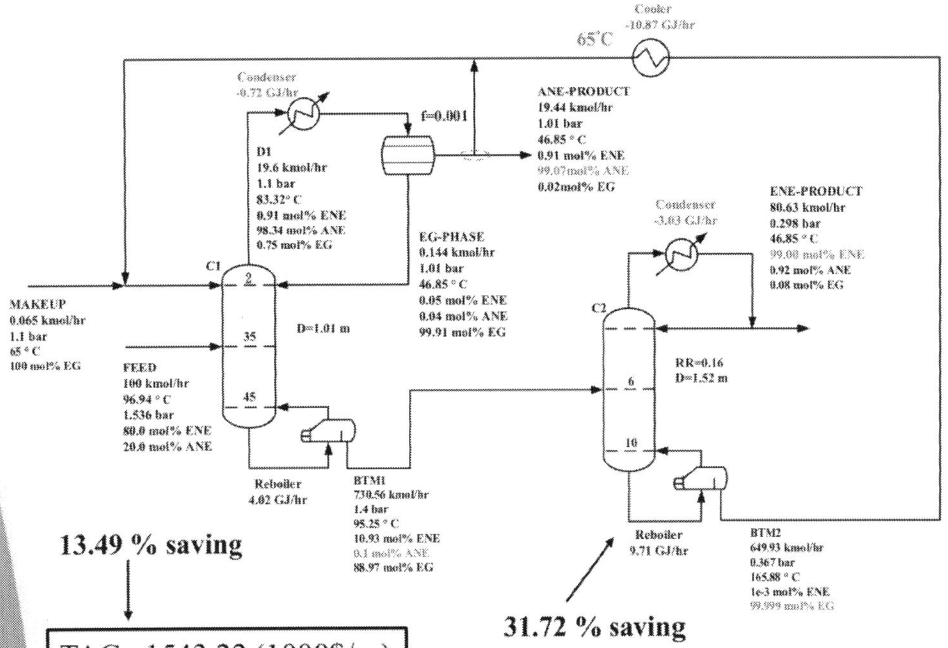

TAC : 1543.22 (1000\$/yr)

19

Energy-Saving Designs

- Design with a process-to-process heat exchanger

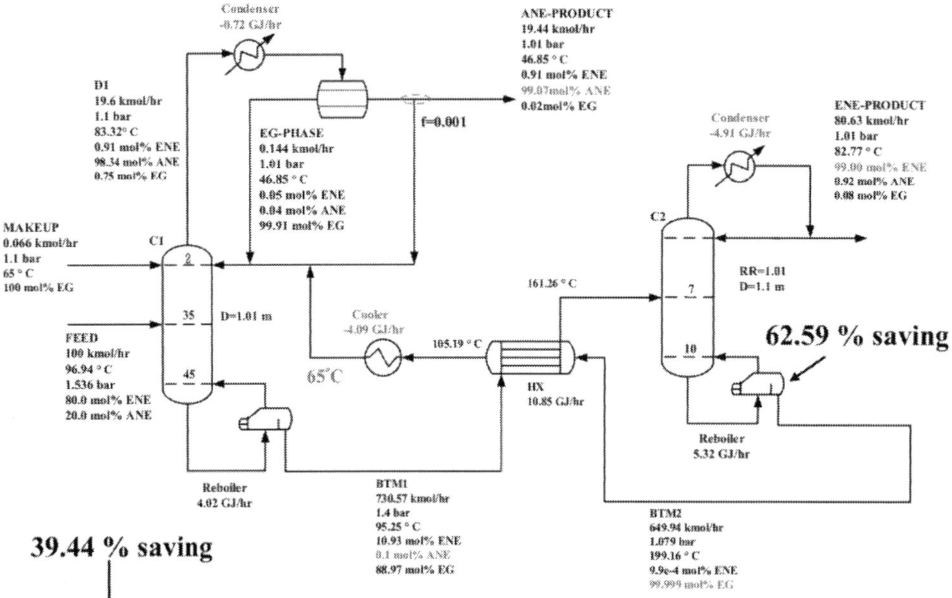

39.44 % saving

TAC : 1080.32 (1000$/yr)

20

Energy-Saving Designs

- Design with a process-to-process heat exchanger combining of operating entrainer recovery column at vacuum pressure

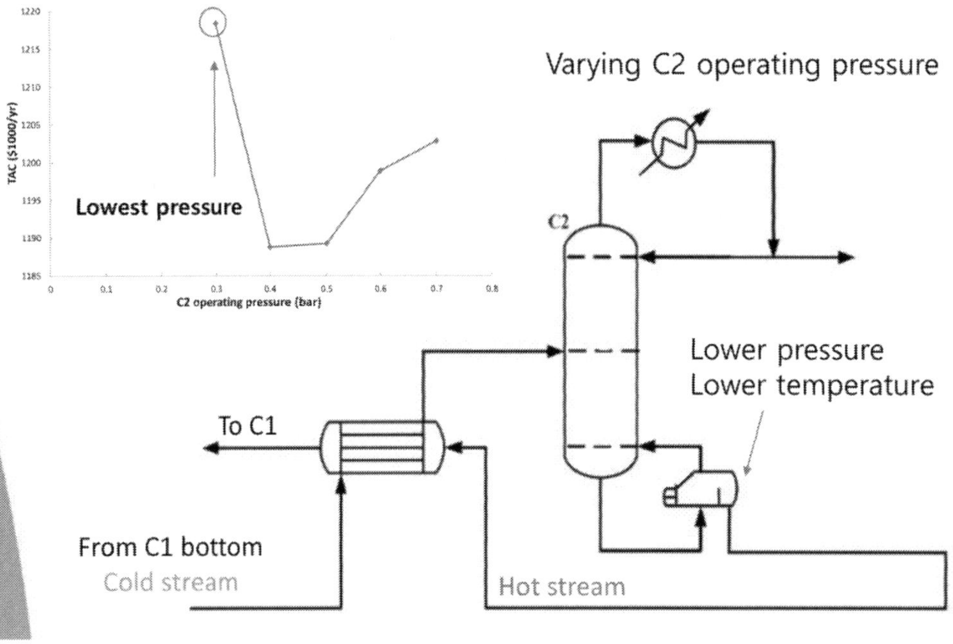

21

Energy-Saving Designs

- **Design with a process-to-process heat exchanger combining of operating entrainer recovery column at vacuum pressure**

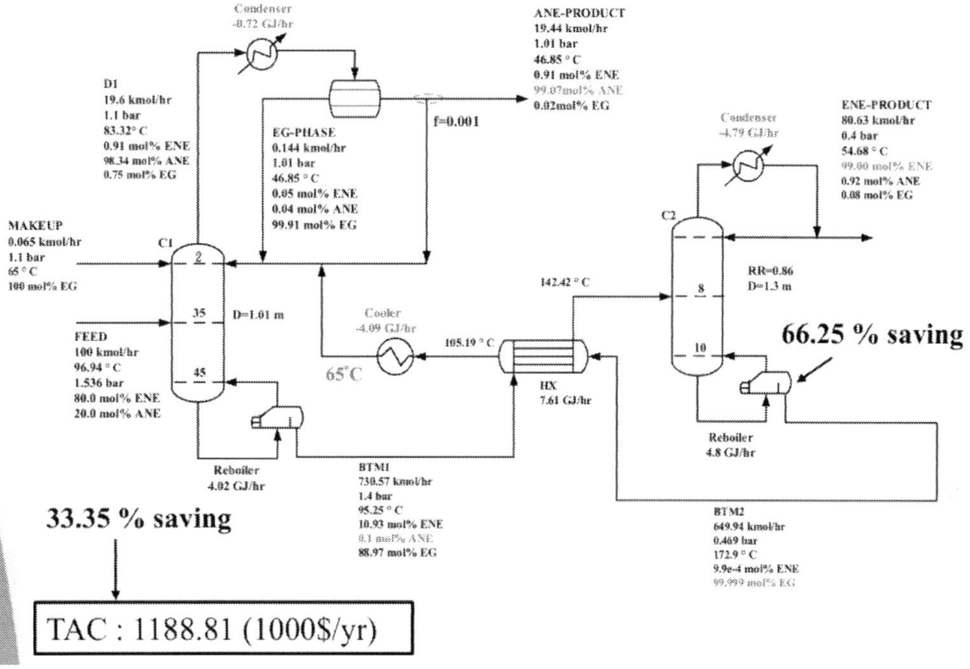

22

Energy-Saving Designs

	TCRD [4]	Heterogeneous extractive distillation system			Design with only process-to-process heat exchanger		
		C1	C2	Cooler	C1	C2	Cooler
Column	1444.92	288.81	123.68		288.81	95.08	
Reboiler	305.97	67.04	169.86		67.04	89.66	
Condenser	400.71	88.71	56.93		88.71	79.12	
Cooler				115.01			71.46
Decanter		45.27			45.27		
Heat exchanger					204.63		
Steam	1755.77	249.95	1123.79		249.95	420.52	
Cooling water	81.39	2.03	8.38	42.3	2.03	13.91	11.61
Makeup		38.91			39.04		
Catalyst cost	246.34						
Total operating cost	2083.50 (0.0%)	1465.36 (-29.67%)			737.06 (-64.62%)		
TAC	2800.70 (0.0%)	1783.79 (-36.31%)			1080.32 (-61.43%)		

Capital Cost : 1000$ Operating Cost : 1000$/yr TAC: 1000$/yr

23

Conclusions

- An industrial relevant cyclohexane/cyclohexene separation process is investigated.
- A novel heterogeneous extractive distillation system is proposed, and the corresponding TAC is considerably reduced compared to the single column system.
- Three energy-saving schemes are suggested and analyzed, and the results show that scheme merely adding a process-to-process heat exchanger is the most desirable one.
- The optimal result of this design flowsheet is much more economical with 61.4% savings in TAC and 64.6% savings in annual operating cost, compared to the result in Yu et al.[4]

Thanks for your attention!

臺灣大學

National Taiwan University

PSE

LABORATORY

References

[1] Ralph C. Tallman, Conversion of cyclohexane to cyclohexene. US Patent 3,108,144, 1963.

[2] M. Larry Campbell, *Cyclohexane*. Wiley-VCH, Weinheim, Germany, 2012.

[3] Steyer, F.; Qi, Z.; Sundmacher, K., Synthesis of cylohexanol by three-phase reactive distillation: influence of kinetics on phase equilibria. *Chem. Eng. Sci.* **2002**, 57, 1511-1520.

[4] Yu, J.; Shi, L.; Yuan, Y.; Chen, H.; Wang, S.; Huang, K.; Thermally Coupled Reactive Distillation System for the Separations of Cyclohexane/Cyclohexene Mixtures. *Ind. Eng. Chem. Res.* **2016**, 55, 311-322.

[5] Marrufo, B.; Loras, S.; Lladosa, E., Phase equilibria involved in the extractive distillation of cyclohexane+ cyclohexene using diethyl carbonate as an entrainer. *J. Chem. Eng. Data* **2011**, 56, 4790-4796.

[6] Mi, W.; Tong, R.; Hua, C.; Yue, K.; Jia, D.; Lu, P.; Bai, F., Vapor–Liquid Equilibrium Data for Binary Systems of *N,N*-Dimethylacetamide with Cyclohexene, Cyclohexane, and Benzene Separately at Atmospheric Pressure. *J. Chem. Eng. Data* **2015**, 60, 3063-3068.

[7] Vega, A.; Díez, F.; Esteban, R.; Coca, J., Solvent selection for cyclohexane-cyclohexene-benzene separation by extractive distillation using non-steady-state gas chromatography. *Ind. Eng. Chem. Res.* **1997**, 36, 803-807.

[8] Steyer, F.; Sundmacher, K., VLE and LLE data for the system cyclohexane+ cyclohexene+ water+ cyclohexanol. *J. Chem. Eng. Data* **2004**, 49, 1675-1681.

[9] Luyben, W. L., *Distillation design and control using Aspen simulation*. John Wiley & Sons, Inc: Hoboken, New Jersey, 2013.

[10] Ivonne Rodriguez-Donis; Katalin Papp; Endre Rev; Zoltan Lelkes; Vincent Gerbaud; Xavier Joulia, Column configurations of continuous heterogeneous extractive distillation. *AIChE Journal*. **2007**, 53, 1982-1993.

[11] Gmehling, J.; Möllmann, C., Synthesis of distillation processes using thermodynamic models and the Dortmund data bank. *Ind. Eng. Chem. Res.* **1998**, 37, 3112-3123.

[12] Luyben, W. L., Distillation column pressure selection. *Separation and Purification Technology* **2016**, 168, 62-67.

[13] Seider, W. D.; Seader, J. D.; Lewin, D. R.; Widagdo, S., *Product and Process Design Principles Synthesis, Analysis, and Evaluation*. John Wiley & Sons, Inc: Hoboken, New Jersey, 2010.

487873 An Experimental Validation of the Concurrent Approach for Simultaneous Isotherm Determination and Process Design for Simulated Moving Bed (SMB) System

Abstract:

Determining adsorption isotherms for multi-component systems requires substantial experimental efforts, especially when the number of components is large. In our previous work, a rigorous dynamic model for the simulated moving bed (SMB) process was implemented and optimized deterministically. The internal concentration profiles were used to uncover the compositions where experiments for isotherm determination should be carried out. By this iterative approach, experimental efforts were reduced to minimal, and model reliability was increased.

In this report, we present a SMB process for the separation of toluene and dodecane using cyclohexane as desorbent. Bench-scale SMB mini-plant is constructed and used to validate the results obtained from the concurrent approach. This investigation is a natural extension of our previous work. Dead volume models are added to the SMB model. A refined/improved composition-selection algorithm that include sensitivity analysis for uncertain parameters is reported. Moreover, modeling results and experimental results are compared to demonstrate the advantage of the concurrent approach.

Reference:

Guo, S., et al.. *A Concurrent Approach for Isotherm Determination and Adsorption Process Design for Multicomponent Systems.* Chemical Engineering Science. Currently being reviewed.

499552 Process Synthesis for Improved Bioseparations Using Multi-Resin Simulated Moving Bed Chromatography

Abstract:

Simulated moving bed (SMB) chromatography uses the principle of counter-current extraction in a set of chromatography columns by applying a counter-current flow of a solid phase and a mobile gas/liquid phase. SMB chromatography leads to an increase in productivity and lower eluent consumption over batch chromatography by using a four zone system divided by four steams: 1) the feed mixture, 2) the purified fast component, 3) the purified slow component, and 4) the incoming eluent.[1] To further improve the productivity, researchers have looked at optimizing the amount of columns in each zone and the time each stream switches, but have not considered using multiple types of resins in each zone simultaneously to increase the separation efficiency for bioprocess applications.

In this work, we optimize the profitability of the SMB chromatography by considering the process design and the resin material selection by improving upon an MINLP model developed by Hasan and Floudas.[2] We apply this improved model to the case study of the industrially-relevant fructose-glucose separation and beta-carotene extracted from *Saccharomyces cerevisiae* to improve the productivity and profitability. Separation parameters are determined experimentally for each resin and used in the modified MINLP model. The multi-resin SMB chromatography will be compared to the single-resin to SMB chromatography process to show the increase profitability.

1. S. V. Vignesh, K. Hariprasad, P. Athawale, and S. Bhartiya, "An optimization-driven novel operation of simulated moving bed chromatographic separation," IFAC-PapersOnLine, vol. 49, no. 7, pp. 165–170, Jan. 2016.
2. M. M. Faruque Hasan, E. L. First, and C. A. Floudas, "Discovery of novel zeolites and multi-zeolite processes for p-xylene separation using simulated moving bed (SMB) chromatography," Chemical Engineering Science, vol. 159, pp. 3–17, Feb. 2017.

Industrial Experience with Model-Based SMB System Design and Optimization: Discretization Scheme and Case Study

Wade Martinson, Cargill Inc.
wade_martinson@cargill.com

Cargill

Motivation and Background

- Chromatographic separation particularly well-suited to aqueous-phase separation of mono- and disaccharide molecules
- Cargill operates several different simulated moving bed separations
- Modeling capability supports efficient SMB process design and optimization
- Model solution speed and accuracy dependent on discretization method

Cargill Optimizing Services

Cargill

Kurganov-Tadmor Discretization

- The KT discretization scheme (Kurganov and Tadmor, 2000) is designed for nonlinear conservation laws and convection-diffusion equations
  - Low numerical viscosity, especially with smaller time steps
  - Does not require a Riemann solver
  - Semi-discrete form is non-oscillatory and can be generalized to multiple dimensions
- Presented for infinite or periodic domains
- SMB applications require modification for use on finite domains

Semidiscrete Scheme

For a general one-dimensional hyperbolic conservation law

$$\frac{\partial u}{\partial t} + \frac{\partial f(u)}{\partial x} = 0$$

the semidiscrete form of the KT scheme is

$$\frac{du_j(t)}{dt} = -\frac{H_{j+1/2}(t) - H_{j-1/2}(t)}{\Delta x}$$

where the flux terms

$$H_{j+1/2}(t) = \frac{f\left(u^+_{j+1/2}(t)\right) + f\left(u^-_{j+1/2}(t)\right)}{2} - \frac{a_{j+1/2}(t)}{2}\left[u^+_{j+1/2}(t) - u^-_{j+1/2}(t)\right]$$

are written in terms of maximum local speed a and intermediate values

$$u^+_{j+1/2}(t) = u_{j+1}(t) - \frac{\Delta x}{2}(u_x)_{j+1}(t) \qquad u^-_{j+1/2}(t) = u_j(t) + \frac{\Delta x}{2}(u_x)_j(t)$$

Application to Liquid SMB Models

Neglecting the forcing term for interphase transfer, the transport portion of a 1-D species conservation balance (constant density and volume) is

$$\frac{\partial C}{\partial t} - \frac{\partial (vC)}{\partial x} = \frac{\partial C}{\partial t} - v\frac{\partial C}{\partial x} = 0$$

Applying the semidiscrete KT scheme gives

$$H_i = \frac{vC_i^+ + vC_i^-}{2} - \frac{a_i}{2}(C_i^+ - C_i^-)$$

$$C_i^+ = C_{i+1} - \frac{\Delta x}{2}(C_x)_{i+1} \qquad C_i^- = C_i - \frac{\Delta x}{2}(C_x)_i$$

$$(C_x)_i = minmod\left(\frac{C_{i+1} - C_i}{\Delta x}, \frac{C_i - C_{i-1}}{\Delta x}\right) \qquad a_i = max(|v_{i-1/2}|, |v_{i+1/2}|) = v$$

Domain of Dependence

Domain of dependence for C_i is $\{C_{i-2}, C_{i-1}, C_i, C_{i+1}, C_{i+1}\}$, so adjustments to scheme are required for two intervals at each end of the domain

Assume specified inlet condition on node at upstream end of domain

For computer implementation, also require clarity of array indices

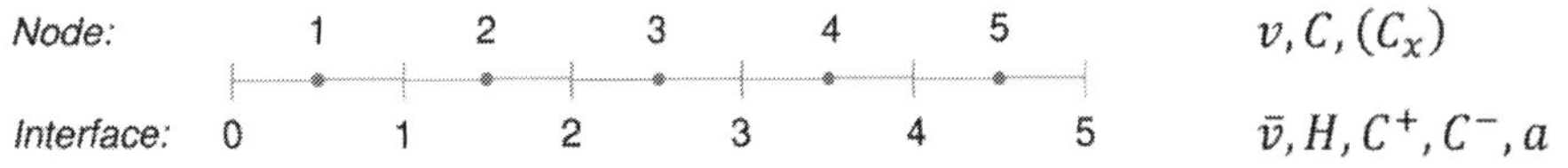

Modifications for Finite Domains

	0	1	j	4	5
C_x		$\frac{C_2 - C_1}{\Delta x}$	$minmod\left(\frac{C_{j+1} - C_j}{\Delta x}, \frac{C_j - C_{j-1}}{\Delta x}\right)$	$(same\ as\ j)$	$\frac{C_5 - C_4}{\Delta x}$
C^+	C_1	$C_2 - \frac{\Delta x}{2}(C_x)_2$	$C_{j+1} - \frac{\Delta x}{2}(C_x)_{j+1}$	$C_5 - \frac{\Delta x}{2}(C_x)_5$ $C_5\ for\ \frac{\partial vC}{\partial x}\ at\ 4\ if\ v < 0$	C_5
C^-	C_1	$C_1 + \frac{\Delta x}{2}(C_x)_1$ $C_1\ for\ \frac{\partial vC}{\partial x}\ at\ 2\ if\ v > 0$	$C_{j-1} + \frac{\Delta x}{2}(C_x)_{j-1}$	$C_4 + \frac{\Delta x}{2}(C_x)_4$	C_5
$\bar{v}$	v_1	${}^1/_2(v_1 + v_2)$ $v_1\ if\ v_1 > 0$	${}^1/_2(v_j + v_{j+1})$	${}^1/_2(v_4 + v_5)$ $v_5\ if\ v_5 < 0$	v_5
a	$\lvert\bar{v}_0\rvert$	$\lvert\bar{v}_1\rvert$	$\lvert\bar{v}_j\rvert$	$\lvert\bar{v}_4\rvert$	$\lvert\bar{v}_5\rvert$
$f(C^+)$	$\bar{v}_0 C_0^+$	$\bar{v}_1 C_1^+$	$\bar{v}_j C_j^+$	$\bar{v}_4 C_4^+$	
$f(C^-)$		$\bar{v}_1 C_1^-$	$\bar{v}_1 C_j^-$	$\bar{v}_4 C_4^-$	$\bar{v}_5 C_5^+$

Example: Square Pulse

- Modified method is conservative, retains low dispersion characteristics

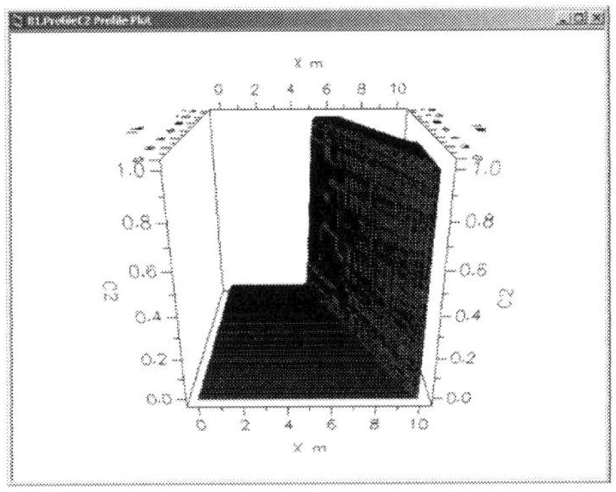

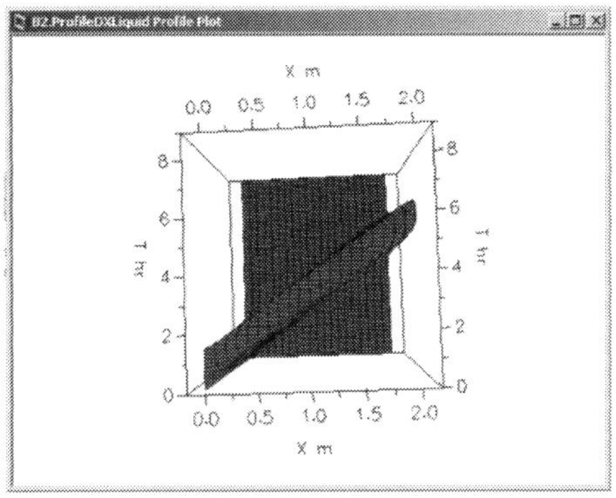

Numerical Dispersion

- First-order schemes like upwind finite differences exhibit greater numerical dispersion than centered / higher-order schemes

- What are implications for model-based parameter fitting?

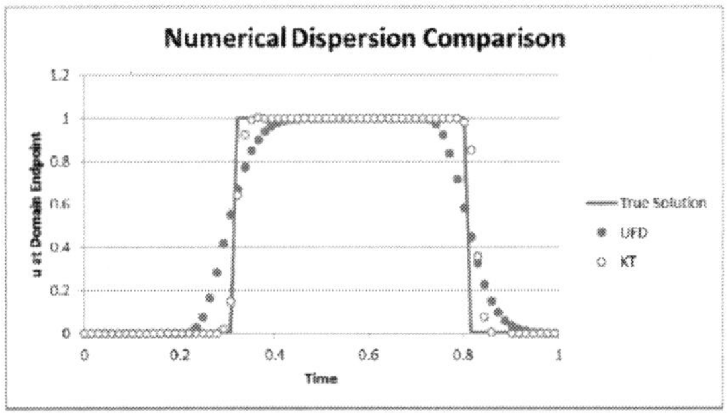

Cargill Optimizing Services

Cargill

Fitting Data

- Sample fits for reported fructose elution profile data (Avezedo 2001)
- Reasonable fits can be obtained with both UFD and KT schemes

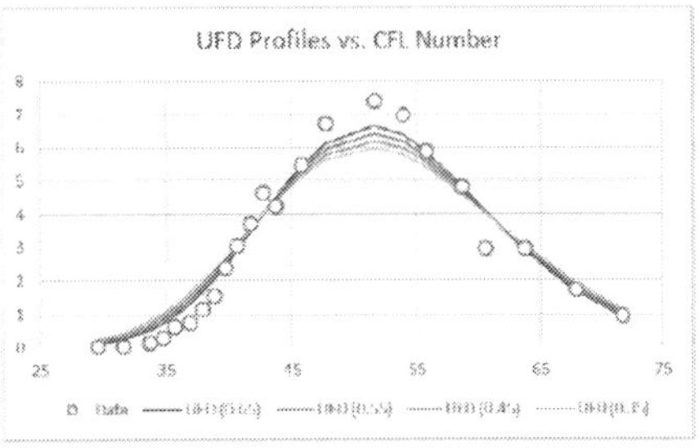

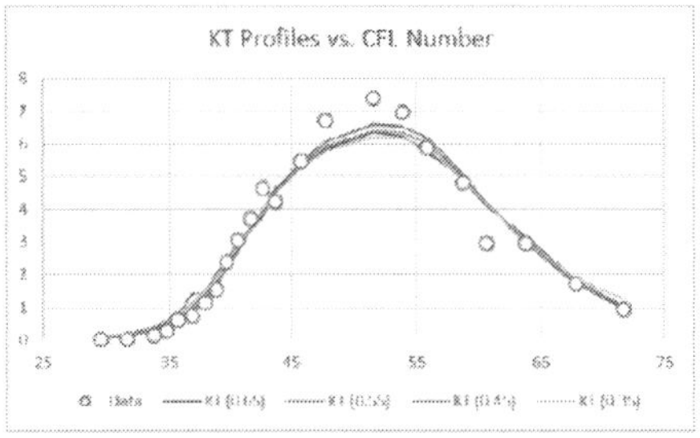

Cargill Optimizing Services

Cargill

Significance for Parameter Fitting

- Best-fit values from model-based parameter fitting for dispersion-like phenomena can be *strongly influenced by choice of discretization*
- Consider overall mass transfer coefficient values for fructose profile

Discretization Method	Fructose Ka (m/s)
UFD	0.25
KT	0.0035

- UFD estimate must "correct" for much higher numerical dispersion

Parameter estimates are more physically meaningful when regressed using higher-order discretization

Background – CFL Number

- The CFL (Courant-Friedrich-Lewy) number compares advection velocity with time step on a given spatial grid

$$N_{CFL} = v\frac{dt}{dx}$$

- CFL number gives the fraction of a grid interval traveled in one time step if moving at the system flow velocity

- Discretization scheme performance depends on CFL number
  - Poor / unstable above a critical value
  - Can recover analytical solution for some systems at the proper value

Flow Rate Dependence in SMB Models

- Flows in SMB system differ in each zone
- Spatial grids must be identical
- CFL numbers will vary across zones
- Lower numerical dispersion offered by higher-order schemes reduces impact of varying CFL numbers on calculated peak sharpness & spread

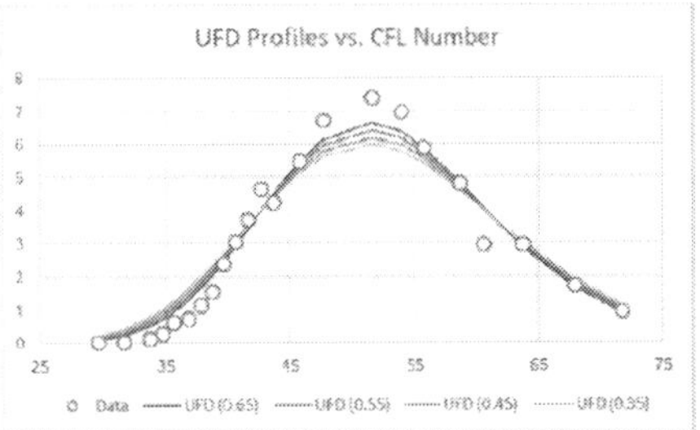

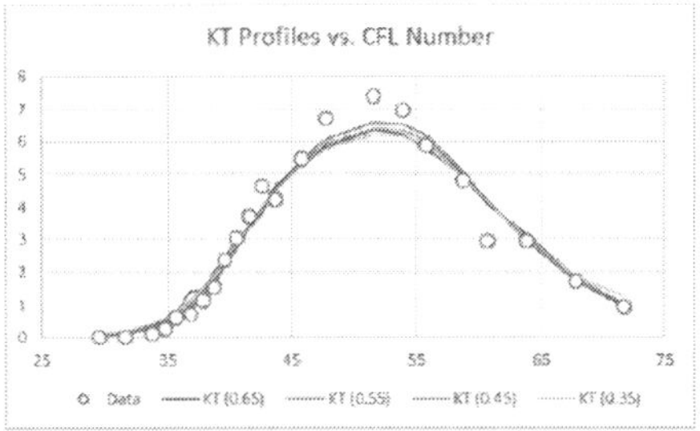

Case Study

- New process development effort – byproduct purification
- Working on crystallization- and SMB-based processes

- Pulse test data and single-bed model used to determine
  - Resin bed voidage
  - Isotherm parameters
  - Lumped mass transfer coefficient

Pulse Test Fits

- Pulse tests performed at different feed concentrations and flow rates
- No competitive behavior observed – linear isotherm captured equilibrium behavior
- Lumped mass transfer resistance able to capture observed peak spreads

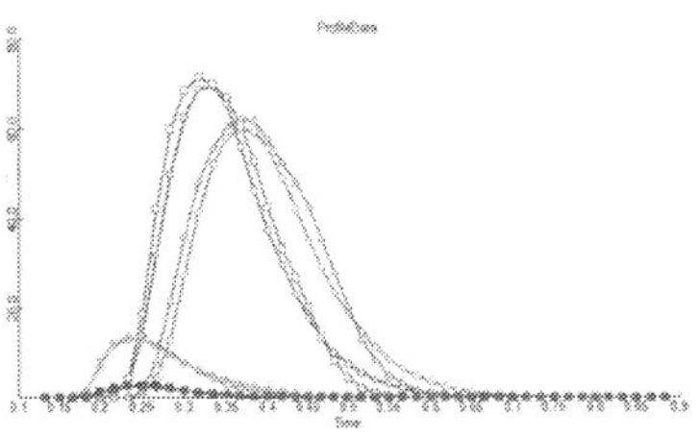

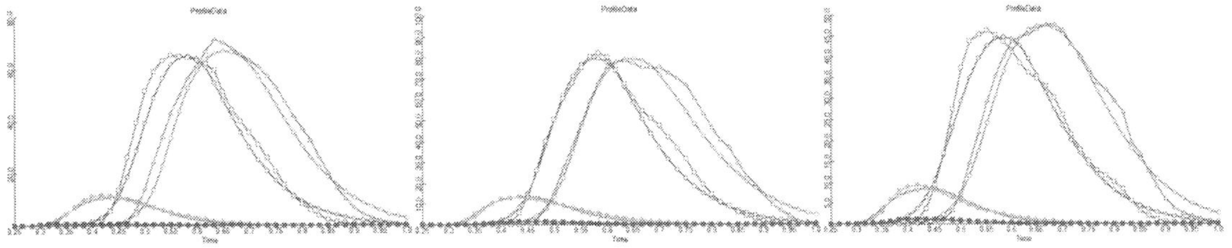

Final Parameter Values

- Tight confidence intervals achieved for all isotherm parameters
- Acceptable confidence intervals achieved for mass transfer coefficients

	Mean	*Std. Dev.*	*StdDev%*	*95% Limit*	*95% Limit*
Ki("A")	0.408	0.018	4%	0.373	0.444
Ki("B")	0.804	0.035	4%	0.735	0.872
Ki("C")	0.415	0.020	5%	0.377	0.453
Ki("D")	1.010	0.031	3%	0.949	1.071

	Mean	*Std. Dev.*	*StdDev%*	*95% Limit*	*95% Limit*
Ka("A")	1.88E-06	3.30E-07	18%	1.23E-06	2.53E-06
Ka("B")	5.27E-06	1.15E-06	22%	3.02E-06	7.52E-06
Ka("C")	1.99E-06	7.08E-07	36%	6.07E-07	3.38E-06
Ka("D")	8.39E-06	1.71E-06	20%	5.04E-06	1.18E-05

SMB System Design

- Tested SMB performance on four-column pilot system
  - Initial operating parameters determined by tuning heuristic-based values
  - Initial results did not meet all purity targets
  - Used initial operating parameters as starting point for model-based optimization
  - Performed optimization using fully discretized SMB model
  - Validated optimized conditions back on pilot system

- Outcome:
  - Key impurity concentration reduced by 50%, with moderate *increase* in throughput
  - Optimized operation met all required performance targets

Summary

- Kurganov-Tadmor discretization scheme can be modified for use on finite domains while preserving conservative and non-oscillatory properties

- Higher-order schemes like KT provide
  - More physically meaningful estimates of true parameter values
  - More consistent simulation results than simpler upwind schemes in SMB system simulations where flow rates / CFL numbers vary

- A model-based approach to SMB system design allowed Cargill to reduce impurity levels by 50% over a heuristic-only approach for a new process

Acknowledgments and References

- CC Chiang, Tiziano Furlan, David Gasperino, Matthew Kuure-Kinsey, Riccardo Socrato, and Chris Tervo all contributed to model development and the process design project

A. Kurganov and E. Tadmor, "New High-Resolution Central Schemes for Nonlinear Conservation Laws and Convection-Diffusion Equations", *Journal of Computational Physics* v.160 pp. 241-282, 2000.

D. Avezedo, *Separation/Reaction in Simulated Moving Bed: Application to the Production of Industrial Sugars*, PhD thesis, University of Porto, Portugal, 2001.

492389 Effects of Operating Parameters, Equipment Parameters, and Material Properties in Ternary Separations in SMB

Abstract:

Simulated moving bed (SMB) systems have been used in a variety of separation applications. Compared to batch systems, SMB systems require less solvent and are more productive per unit sorbent. While there are many advantages to SMB systems, there are several key barriers still in place that prevent their use in many industrial applications. One of the key barriers is the large number of design parameters for SMB processes. A four-zone, ternary, linear SMB has 27 design parameters for operation. Additionally, it is unclear how the design parameters, which include operating parameters, equipment parameters, and material properties, are related to solvent consumption and sorbent productivity in ternary systems.

The Standing Wave Design equations for ternary linear systems are solved in terms of dimensionless groups to obtain analytical expressions for solvent consumption and sorbent productivity. The Speedy Standing Wave Design (SSWD) equations that result provide a framework for exploring the relationships between the design parameters and the resulting separations.

This study utilizes the SSWD equations and rate-model simulations to elucidate the relationship between operating parameters such as cycle time and zone velocities to productivity and solvent efficiency. Longer cycle times result in lower productivities, but higher solvent efficiencies. By quantifying these effects on unit sorbent cost, solvent cost, and equipment cost, an optimal overall separation cost can be achieved. Because the equations are dimensionless, the resulting design principles are easily scalable. The equations for the column configuration that gives the maximum productivity in four zone SMB systems are derived. The lengths of zones I and IV have no effect on the productivity of the SMB and thus by minimizing these zone lengths, higher productivities are achieved. The optimal lengths of zones II and III can be determined based on material properties. It is also shown that changing the column configuration to maximize productivity has very little effect on the solvent efficiency of the system.

The effects of material properties on productivity and solvent efficiency are revealed through the SSWD equations. Larger differences in heavy key and light key retention factors are shown to increase both productivity and yield. However, for a set difference in heavy key and light key retention factors, minimizing the overall range of retention factors can significantly improve the solvent efficiency of a separation without sacrificing productivity. Larger diffusivities in all the components of the separation lead to higher productivity and higher solvent efficiency by minimizing mass transfer effects. These general design rules can be applied when selecting a solvent-sorbent system or when determining splitting strategies in ternary systems.

496262 Simulated Moving Bed Reactor for Enhancing the Productivities of Equilibrium Limited Reactions

Abstract:

Equilibrium-limited reactions are ubiquitous in the chemical industry. Those with highly unfavorable equilibrium constants face very energy intensive operations for conversion and recovery of the product, often needing extensive recycle. Chances are also that such severely equilibrium-limited reactions using conventional reactors are written off at an early stage and are never scaled up to production.

In this work, we demonstrate the potential of simulated moving bed reactor (SMBR) in increasing the productivity of strongly equilibrium-limited reactions and possibly rendering them economically feasible. In an SMBR, in-situ chromatographic separation of the products aids in shifting the equilibrium, thereby improving the conversion.

Liquid phase transesterification of propylene glycol methyl ether (DOWANOL™ PM) to propylene glycol methyl ether acetate (DOWANOL™ PMA) using a homogeneous catalyst [1] is considered as a case study to compare the performance of SMBR with that of a conventional batch reactor. Here, DOWANOL™ PM is used both as a reactant and desorbent for the chromatographic separation inside the SMBR. In the SMBR process, an anion exchange resin, AMBERLITE™ IRA904 in chloride form is used as a non-catalytic adsorbent, while sodium alkoxide of PM is supplied as a homogeneous catalyst in the desorbent. Using model based optimization, we demonstrate the advantage of employing a non-catalytic adsorbent in the SMBR for the homogeneously catalyzed reaction. With the ability to independently tune the separation of the product in the SMBR, it is demonstrated that the productivity of the process increases with increasing separation factor. Presence of an optimal product separation factor is observed with respect to the amount of DOWANOL™ PM used for reaction and separation. At a fixed conversion of 95%, SMBR outperforms a batch reactor by increasing the productivity by 85% and reducing DOWANOL™ PM usage to 50% of what is required by the batch reactor.

[1] J. Oh, B. Sreedhar, M. E. Donaldson, A. K. Schultz, T. C. Frank, A. S. Bommarius, Y. Kawajiri, "Transesterification of Propylene Glycol Methyl Ether in Chromatographic Reactors Using Anion Exchange Resin as a Catalyst," *J. Chromatography A* **2016**, *1466*, 84-95; DOI:10.1016/j.chroma.2016.08.072

497954 Engineering Studies of the Effect of pH and Temperature on the Adsorbent Surface Interaction for the Anion-Exchange Chromatographic Separation

Abstract:

This presentation explores a fundamental case of separation involving amino acids on a strong anion-exchange resin. The pH and temperature were increased and the retention time amino acids on a Sepharose Q Fast Flow column increased. When analyzing the data obtained from the chromatograms, the moments method by Schneider and Smith was employed. From the moments analysis, the adsorption and desorption rates were calculated. The adsorption and desorption rate were fitted to an Arrhenius expression to quantify the activation energy. The main outcomes of this investigation are that the retention increases with increasing pH and temperature. Along with that the adsorption rate was determined to be a diffusive phenomenon. Another important takeaway from this paper is that the adsorption rate has a positive activation energy and the desorption has a negative activation energy. This information can then be used to simulate the optimum pH and temperature conditions for the separation.

494812 Local Equilibrium Theory Analysis of Chromatographic Peak Shapes in the Presence of Adsorbing Modifiers

Abstract:

Modifiers(or co-solvents) are often added to the mobile phase in many chromatographic separations.Very often these modifiers compete for adsorption sites along with the solute. The nature of modifier adsorption (stronger or weaker than the solute) can have a distinct effect on the shape of the solute elution profiles. Although the adsorption isotherms of both the modifier and solute can be Langmurian, the elution profiles may display either a Langmuirian (sharp front followed by a dispersed rear) or an anti-Langmuirian shape (dispersed front followed by a sharp rear). In addition, the formation of rounded peaks (that are neither strictly Langmuirian nor anti-Langmuirian) has been observed in the literature [1,2] and an explanation for this has not been thusfar available. In this study, the local-equilibrium theory of chromatography has been used to arrive at explicit conditions under which Langmuirian and anti-Langmuirian peaks are obtained and for the first time the origin of the "rounded" peaks has been fundamentally explained. The extension of this study to bi-Langmuir isotherms will also be discussed.

1. T. Fornstedt, G. Guiochon, Theoretical study of high-concentration elution profiles and large system peaks in nonlinear chromatography, *Anal. Chem.* 66 (1994) 2116â€“2128.

2. T. Fornstedt, G. Guiochon, Comparison between experimental and theoretical profiles of high concentration elution bands and large system peaks in nonlinear chromatography, *Anal. Chem.* 66 (1994) 2686â€“2693.

492145 CO_2 capture in Nitrogen-Doped Porous Carbons Synthesized from Biomass

Abstract:

In this work, nitrogen-dope and highly porous carbons were synthesized from lignin biomass along with successive activation with potassium hydroxide and ammonia. The resultant carbons had a BET surface area of 1631-2922 m^2/g and nitrogen content 5.6-7.1 at.%. Different nitrogen functionalities include pyridinic, amino, pyrrolic/pyridonic, nitro and nitroso groups. High surface area and nitrogen functionalities resulted in high carbon dioxide adsorption capacity of 5.48 and 8.6 mmol/g in 298 K and 273 K, respectively, and pressure upto 1 bar. The selectivity for CO_2/N_2 was calculated using the Ideal Adsorbed Solution Theory (IAST). The results indicated that selectivity increases with the increase in nitrogen content. IAST calculations were also employed to calculate the binary adsorption isotherms to simulate mixed gas adsorption for CO_2 and N_2 as a function of total pressure and mole fraction of CO_2 in the feed gas mixture. A stability study with 10 continuous cycles of adsorption and desorption indicated no loss of adsorption capacity. A breakthrough experiment with (10/90) CO_2/N_2 mixture in a dynamic column experiment confirmed the physical separation of the two components. The overall results suggest that these carbons can be used as potential adsorbents for CO_2 capture purposes.

486819 3D-Printed SAPO-34 Structured Adsorbent for CO_2 Separation

Abstract:

Most adsorbents are produced as porous, micron-sized powder materials and are thus not suitable to be directly used in adsorption processes. In order to avoid excessive pressure drop during flow of gas or liquid streams through a packed bed of adsorbent, these powders are shaped into larger particles or structured adsorbents. Extrudates, beads or pellets with a size of several millimeters are widely used in industrial processes. The most important disadvantages of such particles include the relatively large pressure drop they generate at high flow rates and the presence of mass transfer limitations as a result of slow diffusion of molecules to the core of the particles. A trade-off between these two effects limits the possibilities to optimize packed bed adsorptive separation processes; e.g. decreasing pellet size allows to reduce mass transfer limitations but this in turn leads to larger pressure drops. In practice, bed geometry (length/width of the packed bed) is adapted to limit pressure drop. Nevertheless, classical packed beds are not ideal for processes in which very short cycle times or very high gas or liquid velocities are required.

Other types of adsorbent formulation that allow eliminating the limitations mentioned above are thus of large interest. Monolithic adsorbents are superior to classical packed bed adsorbents in terms of pressure drop and mass transfer kinetics. The honeycomb structure, mostly known from catalytic exhaust treatment in the automotive industry, is a well-known example, but monolithic structures are also used in liquid chromatography and heterogeneous catalysis. Nevertheless, the production of monoliths is complicated; classical extrusion processes only offer a very limited flexibility in the geometric properties of the monolith while polymerization processes are not suited for the production of materials for high temperature applications.

Recently, 3D-printing methods have tremendously expanded the possibilities in material synthesis, with much more degrees of freedom. In the present work, a new method to develop monolithic structures is tested for its use in gas and liquid adsorptive separation. Layers of adsorbent fibers are printed on top of each other, where each layer can have a different orientation. Fiber thickness and interdistance can be varied easily. This allows to generate structured adsorbents with a very high porosity, interconnected channels and high adsorption capacity. SAPO-34 based monoliths were produced in different geometries by this 3D-printing method. The obtained materials were characterized via Hg and Ar porosimetry to determine micro-, meso- and macroporosity. Adsorption capacities for CO_2, N_2 and CH_4 were obtained via pure component isotherms measurements to allow comparison with the adsorbents in their initial powder form. It was demonstrated that the method developed at VITO allows to obtain materials with very large adsorption capacity and very good accessibility. Subsequently, the monoliths were subjected to breakthrough separation experiments. The separation of CO_2 from flue gas was studied. Promising separation properties were obtained, with large selectivities, large capacities and low pressure drop. Depending on the chemical composition of the adsorbents, fast regeneration could be obtained under isothermal and thermal swing conditions.

494250 Double layer structured amine-impregnated adsorbent using modified polyethyleneimine for CO_2 capture

Abstract:

CO_2 emissions from fossil fuel combustion have been considered as a serious problem causing greenhouse effect in the international community. In order to reduce CO_2 emissions from large stationary sources including fossil fuel power plant, extensive researches have been carried out. Among various technologies for CO_2 capture, adsorption technologies using silica-supported amine sorbents have been shown high CO_2 adsorption capacity and amine efficiency. Also, for practical CO_2 capture, temperature swing adsorption (TSA) processes including high temperature regeneration with CO_2 or steam stripping have been proposed. Lots of researches have been conducted to improve CO_2 adsorption capacity but could not sufficient confirm physical and chemical stability of silica-supported amine sorbents for TSA cycle. Impregnated amines have physical stability problems (leaching out from the support at steam stripping) and chemical stability problems (irreversible formation of urea at high temperature and CO_2 condition in regeneration step). The lack of stability can cause the decrease of working capacity for long-term operation, so the physical and chemical stabilities of silica-supported amine sorbents are very important. This work investigated the stabilities of double layer structured adsorbent using modified polyethyleneimine (PEI). Epoxy modified PEI and diepoxide crosslinked PEI were impregnated in the first and second layers, respectively. The double layer structured adsorbent showed excellent physical and chemical stability by suppressing amine leaching and urea formation because of crosslinked PEI in the second layer.

494959 Identifying Best Core MOFs with Open Mg Sites for CO_2/N_2 Separation Using Computational Tools

Abstract:

Metal-organic frameworks (MOFs) are nanoporous, crystalline structures with favorable properties such as high pore volume, surface area, chemical diversity, flexibility that can be quite beneficial for gas storage, separation/filtering, catalysis applications. Since they are assembled from metal nodes and organic linkers as the building blocks, hypothetically, a limitless number of MOFs can be formed. MOFs with open metal sites are very promising candidates for separations because certain sorbates can have highly favorable interactions with these nodes. Recently, the CoRE (computation-ready experimental) MOF database[1] has been created to enable large-scale computational screening studies. To understand their potential for CO_2/N_2 separation, performing full quantum chemical calculations on each of these MOFs would be computationally prohibitive, but molecular mechanics force fields fail to capture chemisorption interactions.

In this study, the CoRE MOF database is screened for MOFs with open Mg sites that are optimal for CO_2/N_2 separation at flue gas conditions. To this extent, a hierarchical screening approach is employed. First, a geometric approach is used to identify MOFs with open Mg sites. Second, unary adsorption isotherms are determined from a multi-site Langmuir model in which the guest-host interaction energy for the volume element near the open Mg site is computed by periodic density functional theory (DFT) and the remainder of the interactions is calculated by a combination of the TraPPE force field for the sorbate molecules and of the UFF force field for the framework. Third, for MOFs with sufficient CO_2 capacity, the binary CO_2/N_2 adsorption isotherms and selectivities are estimated using the ideal adsorbed solution theory (IAST)[2]. Fourth, the self diffusivities and transport selectivities are estimated. Finally, first principles simulations are performed for the most promising materials.

References

1. Chung, Y. G. *et al.* Computation-Ready, Experimental Metalâ€"Organic Frameworks: A Tool To Enable High-Throughput Screening of Nanoporous Crystals. *Chem. Mater.* **26,** 6185â€"6192 (2014).

2. Myers, A. L. & Prausnitz, J. M. Thermodynamics of mixed-gas adsorption. *AIChE J.* **11,** 121â€"127 (1965).

495445 Oriented Growth of Mg_2(dobpdc) on Honey-Comb Monolith for CO_2 Capture

Abstract:

Metal organic framework (MOF) supported amine materials are a class of materials that have shown promising performance for CO2 capture. In the present talk, we will consider Mg_2(dobpdc) & Mn_2(dobpdc) MOFs functionalized with N-Nâ€™ dimethyl ethylene diamine (MMEN) for CO_2 capture. These adsorbents have previously shown exceptional equilibrium capacities for CO_2 capture. They will be discussed here in the context of their stability under humid conditions. Also, different approaches for practical applications of Mg_2(dobpdc) using honeycomb monolith supports will be presented. The uniformity of the MOF film on the monolith channels, as well as the mechanical and thermal stability of the film are some of the key parameters in the development of MOFs on monolith supports. Using our optimized approach, we demonstrate a highly effective, oriented and robust film for CO_2 capture. The adsorbent film shows higher capacities for CO_2 capture than most other MOF based adsorbent films on monoliths for simulated flue gas capture. Finally, a practical process using MOF films coated on monolith contactors, as well as the kinetics of CO_2 capture are briefly discussed.

497474 Spectroscopic Characterization of Humid CO_2 Adsorption on Solid Supported Tertiary Amines

Abstract:

Aminosilica materials are promising candidates for CO_2 capture from dilute streams such as ambient air and flue gas. While ammonium bicarbonate formation is known to occur in aqueous amines in the presence of CO_2 there has been little evidence of its formation on solid supported analogues. To probe if the ammonium bicarbonate species can exist on solid supported amines, tertiary amines were grafted on to SBA15 and the materials were further characterized using in-situ FTIR spectroscopy in the presence of humid CO_2. Dry and humid CO_2 capacities for these adsorbents were also evaluated using fixed bed experiments and thermogravimetric analysis. This work shows that ammonium bicarbonates can exist on solid supported amines.

503122 Perspectives on Development of Membranes for Vapor Separation Applications

Abstract:

Over its 35 year history, MTR has commercialized a number of new membrane applications and materials. In particular, MTR has been a pioneer in the use of vapor or solubility-selective membranes for industrial separations. This experience has provided many lessons learned on the challenges of bringing a new membrane technology to a market. This presentation will discuss some of these issues through the lens of MTRâ€™s history in the petrochemical and natural gas industries. Topics to be covered include:

- The development history of MTRâ€™s VaporSep and FuelSep membranes for olefin recovery and fuel gas conditioning, and insights on what makes a good membrane process design.
- A discussion on the qualities of a useful industrial membrane. Academic literature focuses heavily on new materials with better permeabilities and selectivities in an ongoing quest to exceed the upper bound. On the other hand, industry is less concerned with spectacular pure gas performance, but rather concentrates on processable materials with adequate mixture transport properties that can be reproducibly fabricated at large scale into robust, stable, low-cost membranes.
- Comments on membrane device versus film data; pure gas film measurements are a useful starting place for fundamental material understanding. However, this is the very beginning of the membrane development process. Testing of asymmetric/composite membranes in module form at expected operating conditions is crucial to understand how device and process issues dictate achievable performance. This information is useful to guide materials development.

Ideally discussion of these issues can help accelerate the introduction of next-generation membranes and the use of such membranes in new applications.

Chunqing Liu
Honeywell UOP

Advanced Membranes for Gas Separations
- From Dense Films to Asymmetric Membranes

October 31, 2017 | 2017 AIChE Meeting | Minneapolis, Minnesota

Honeywell UOP

UOP 6055-0

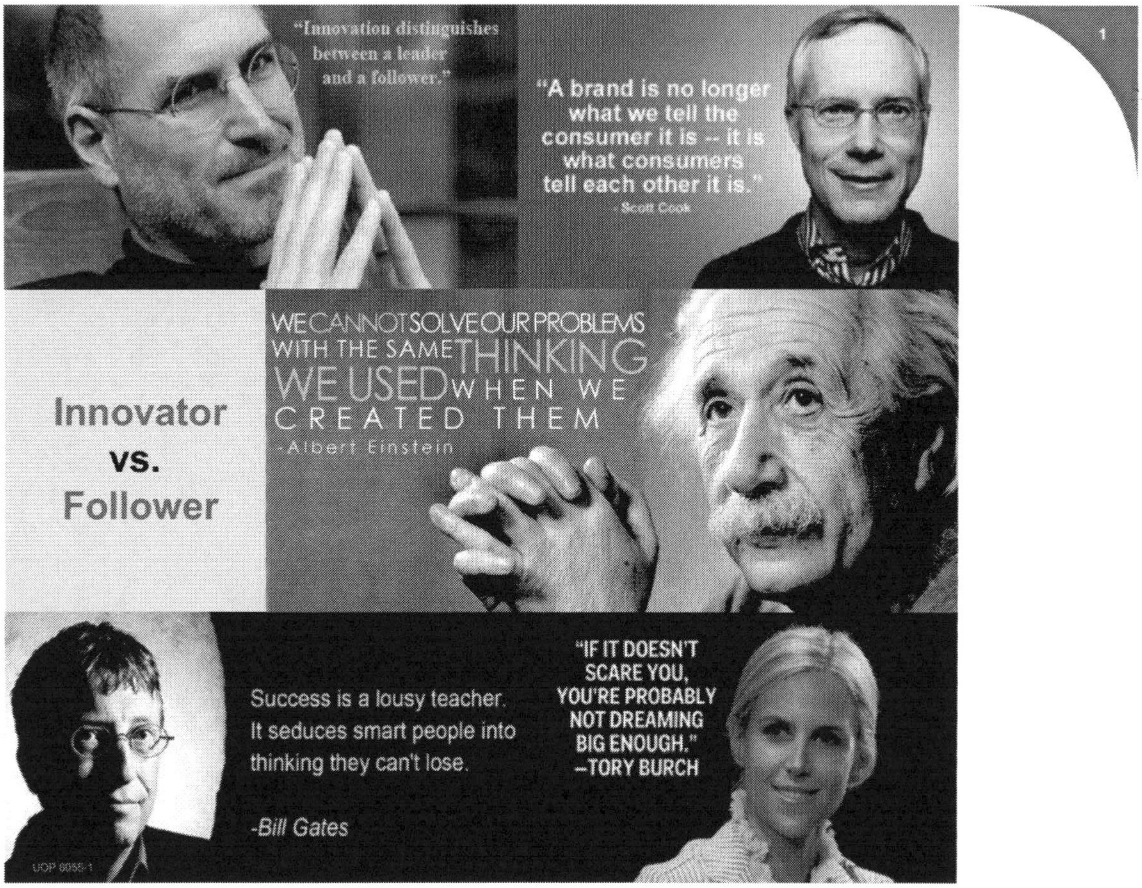

Membrane Technologies

2

- **Membranes for gas and vapor separations**
  - Polymeric gas separation membranes (Almost all current applications)
  - Metallic or microporous inorganic/polymer mixed matrix membranes
  - Metal membranes (Pd membranes for H_2 separation)
  - Ceramic/zeolite/MOF/CMS membranes
- **Membranes for pervaporation (organic or aqueous)**
  - Polymeric and mixed matrix membranes (org. sol./H_2O)
  - Ceramic/zeolite/MOF membranes (alcohol/H_2O, xylene isomers)
- **Membranes for liquid separations**
  - Reverse osmosis/forward osmosis (RO/FO/PRO/OSRO)
  - Nanofiltration (NF)
  - Ultrafiltration (UF)
  - Microfiltration (MF)
- **Membrane contactors/membrane distillation/membrane chromatography/membrane bioreactors/catalytic membrane reactors/fuel cell and battery membranes, etc.**

UOP 8055-2

Development of New GSMs

4

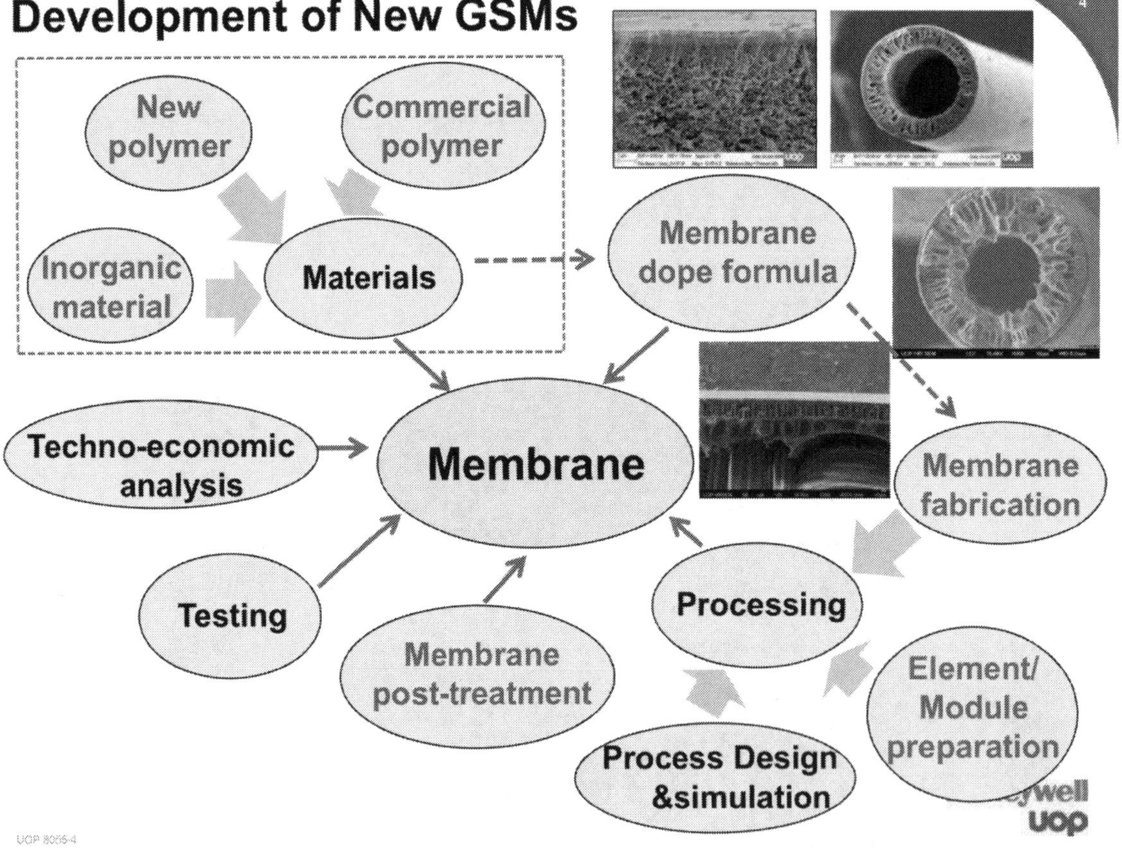

UOP 8055-4

Development of Polymeric Membranes for GSs

- **Polymer Synthesis and Characterization**
  - Polymer synthesis from monomers
  - Polymer modification/post-synthesis from polymers
  - In-situ polymerization on membrane support
  - Polymer modification from membranes
  - Polymer characterization
  - New polymer manufacturing
- **Polymeric Dense Film Preparation and Permeation testing**
  - Symmetric non-porous polymeric dense film preparation
  - Time-lag method to measure material intrinsic **permeability** and **selectivity**
- **Asymmetric Integrally-Skinned or Thin-Film Composite (TFC) Flat Sheet or Hollow fiber Membrane Fabrication**
  - AIS: No delamination, very thin skin layer; maybe high cost, difficult to process
  - TFC: Low cost, high performance new material; delamination, thicker skin layer
- **Hollow Fiber module, Flat Sheet Circle, and Spiral Wound Element Permeation Testing**
  - CO_2/CH_4, H_2/CH_4, He/CH_4, O_2/N_2, olefin/paraffin, NGL recovery, etc.

UOP 8065-5

Polymers for Commercial Gas Separation Membranes

- Glassy
  - Cellulose acetate/triacetate (CO_2/CH_4)
  - Polysulfone (N_2/air, H_2O/CH_4, H_2 separations)
  - Polyimides (CO_2/CH_4, H_2 separations, N_2/air, VOCs/gas)
  - Polyaramides (N_2/air, H_2 separations)
  - Polycarbonates (N_2/air, H_2 separations)
  - Polyphenylene oxide (N_2/air, VOCs/gas)

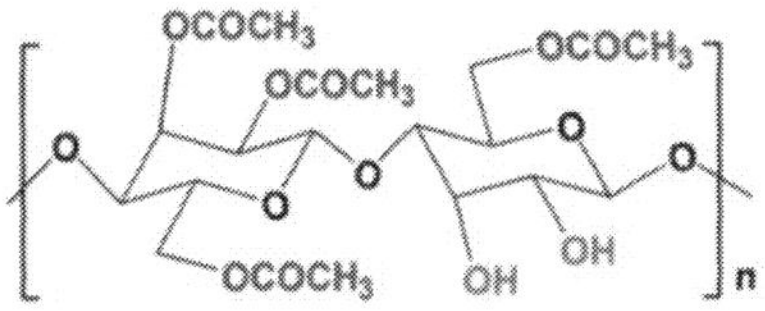

CA

* Rubbery
  - Silicone rubber
    (HCs/gas, solvents/H_2O, N_2/CH_4)

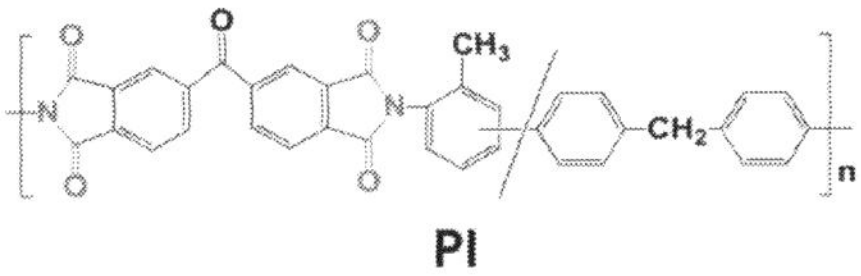

PI

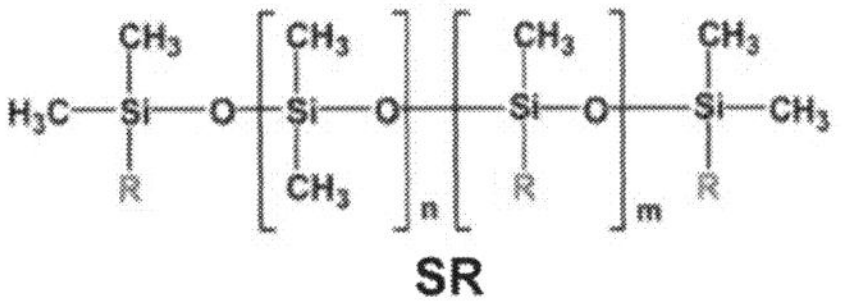

SR

UOP 8065-5

Polymeric Dense Films vs. Asymmetric Membranes

Symmetric Dense Film | *Asymmetric Integrally-skinned or Thin Film Composite Membrane*

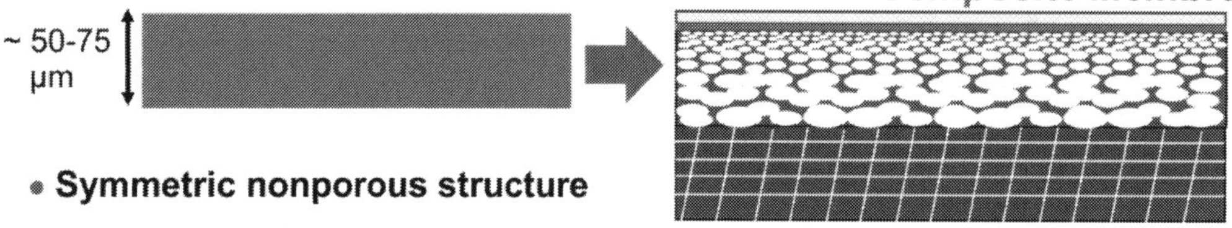

Symmetric Dense Film

- **Symmetric nonporous structure**
- **Screening polymeric membrane materials**
- **Measuring intrinsic properties of polymeric materials for gas separations using a time-lag method (permeability & selectivity)**

Asymmetric Integrally-skinned or Thin Film Composite Membrane

- **Asymmetric structure**
- **A thin nonporous selective layer (<100 nm) supported on a highly porous support layer**
- **Commercially viable structure (Permeance & selectivity)**

$$\text{Permeance (P/L)} = \frac{\text{Permeability (P)}}{\text{Thickness (L)}}$$

Permeance (P/L) ← (from asymmetric membrane)

Permeability (P) ← (from dense film)

Honeywell UOP

UOP 6055-6

The Challenge of Acid Gas in Natural Gas

- Acid Gas (CO_2, H_2S), in combination with water, are highly corrosive and can rapidly destroy pipelines and equipment.
- H_2S is toxic and at relatively modest levels can be life threatening.
- CO_2 reduces the heating value of a natural gas stream and wastes pipeline capacity.
- In LNG, CO_2 must be removed to prevent freezing in the low-temperature chillers.

Natural Gas Feed → Gas Processing Requirements → Treated Products

Gas Processing Requirements → Acid Gas & Waste Disposition

Acid Gas Removal Solution is Required to Achieve Product Specifications

8

UOP Separex™ Membrane Technology

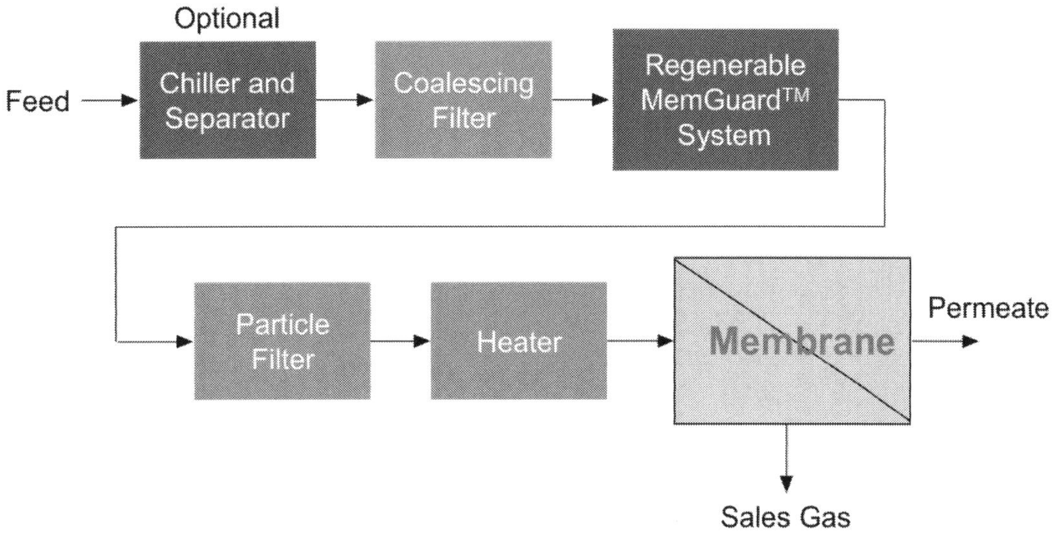

UOP Separex™ technology includes membrane and pretreatment unit

Honeywell UOP

UOP 8056-8

9

UOP Separex™ Membrane for CO_2/CH_4 Separation

*Membrane Element Performance**

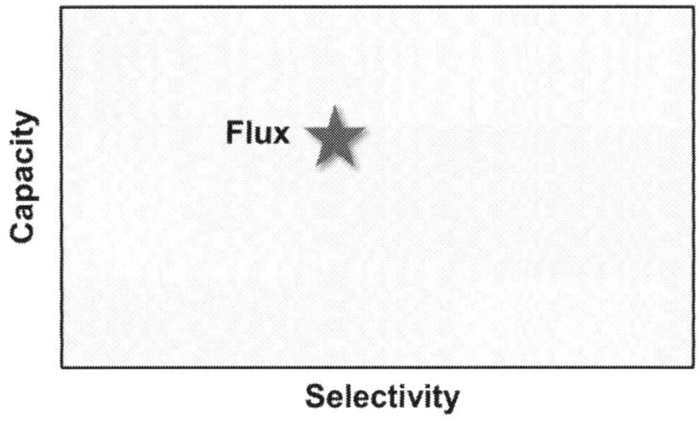

- UOP's commercial Separex™ Flux membrane for CO_2/CH_4 separation has been proven commercially since the 1990s
- Innovation is desired to enhance membrane permeance (higher gas treating capacity) or/and higher selectivity (higher HC recovery)

Honeywell UOP

UOP 8056-9

From Membrane Materials to Membrane Systems: Key Membrane Development Steps

10

Membrane Materials → Dense Films → Flat Sheet or Hollow Fiber Membranes → Spiral Wound Element or Hollow Fiber Module → Separation System

- Phase inversion is a key to asymmetric structure
- Thin, defect-free selective skin layer is a key to high permeance and high selectivity

UOP 8055-10

UOP New Membrane Materials for CO_2/CH_4 Separation (Dense Films)

11

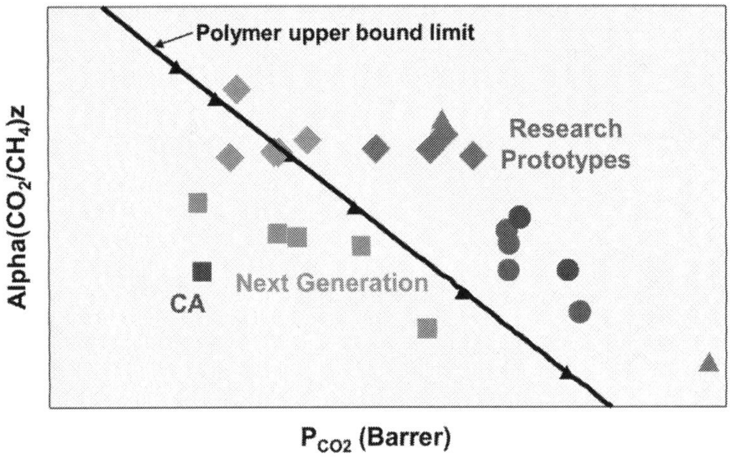

UOP proprietary membrane materials showed much better intrinsic performance than commercially available polymers for CO_2/CH_4 separation

Honeywell UOP

12

From Polymers (Dense Films) to Asymmetric GSMs

- **Cost**
- **Purity**
- **MW**
- **Commercial availability/manufacturing possibility**
- Intrinsic gas permeation property
  - High permeability
  - High selectivity
- Solubility (asymmetric membrane dope forming possibility)
  - Solvents for phase-inversion process
  - Solvents for coating process
- **Film/membrane forming property/**processability
- **Mechanical property**
- **Post-treatment possibility**
- **Chemical and thermal stabilities**
- **Hydrophilicity/hydrophobicity**
  - Controlled surface properties

Honeywell UOP

UOP 8055-12

13

From Polymers (Dense Films) to Asymmetric GSMs

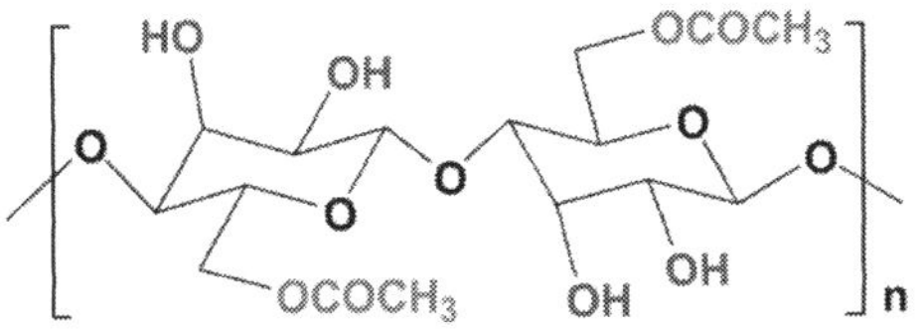

cellulose acetate (CA)

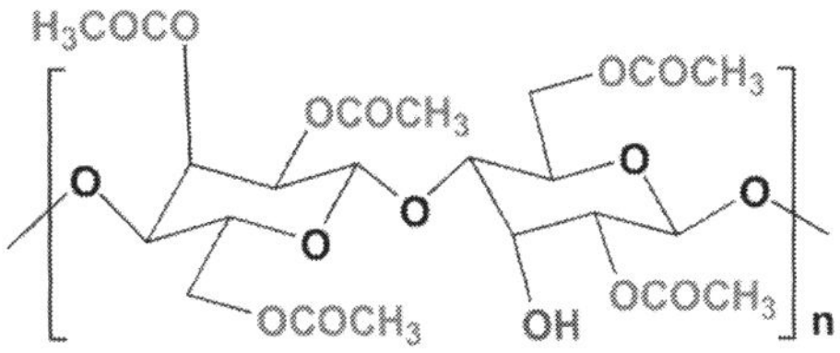

cellulose triacetate (CTA)

Dense film	P_{CO2} (Barrer)	$\alpha_{CO2/CH4}$
CA/CTA blend	7-8	21-22
Material A	**70**	**20-21**
Material B	6-7	40-50

Membrane casting/spinning dopes:

Polymer ratio	wt% Polymers	Solvents	Non-solvents	Sol./Non-sov. ratio	Viscosity	Stability (cloudy point)

Honeywell UOP

UOP 8055-13

Making Asymmetric Gas Separation Membranes

- **Advanced asymmetric membrane fabrication and testing capabilities**
- **Optimization of membrane dope formula and processing conditions is critical to achieve high performance asymmetric membranes**

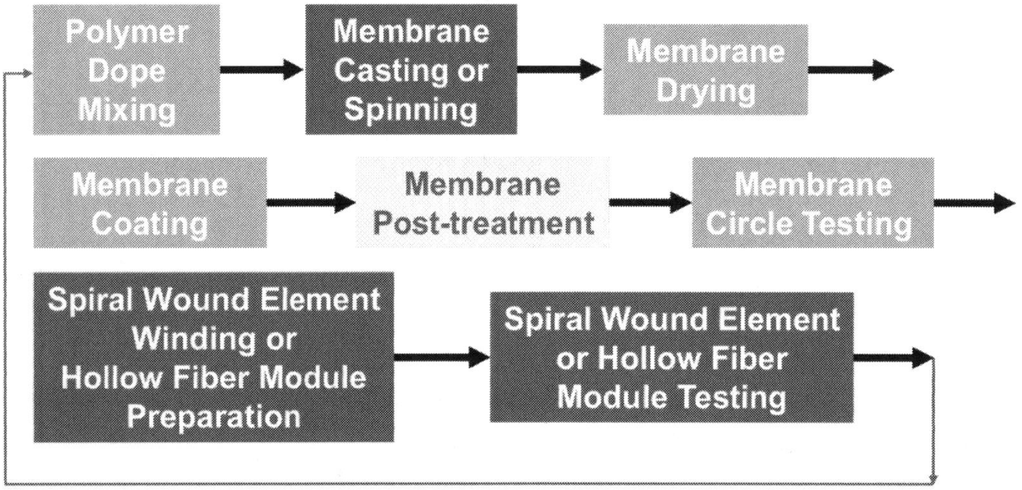

Honeywell UOP

UOP 8055-14

New UOP Separex™ Membranes for CO_2/CH_4 Separation

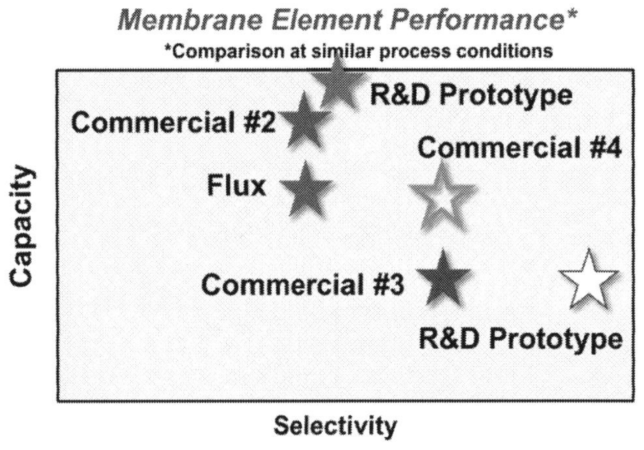

Continue to develop high performance membrane through introduction of new membrane materials and advanced membrane processing technology

- **UOP Separex™ commercial membrane #2 provides up to 25% higher processing capacity than Flux (CA) membrane.**
- **UOP Separex™ commercial membrane #3 provides up to 50% higher selectivity resulting in significantly higher hydrocarbon recovery compared to Separex™ Flux membrane.**
- **UOP Separex™ commercial membrane #4 provide up to 50% higher selectivity and same processing capacity compared to Separex™ Flux membrane.**

Honeywell UOP

UOP 8055-15

Current UOP Membrane Commercial Applications

UOP Membrane Commercial Experience

Installation Type	*No. of Units*
• Natural Gas	52
• Enhanced Oil Recovery (EOR)	5
• Hybrids	5
• Offshore	20
• Landfill Gas	7
• Digester Gas	1
• H_2 Purification	94
Total	>170

- **UOP supplied Separex™ spiral wound membranes for natural gas upgrading (CO_2 & H_2S from CH_4)**
- **Single-lift Separex™ membrane module offers smaller footprint & lower weight for offshore FPSO applications**

Honeywell
UOP

UOP 8055-16

UOP Separex™ Membrane Systems

Honeywell
UOP

UOP 8055-17

New UOP Separex™ Membranes for H_2 Purification and Recovery

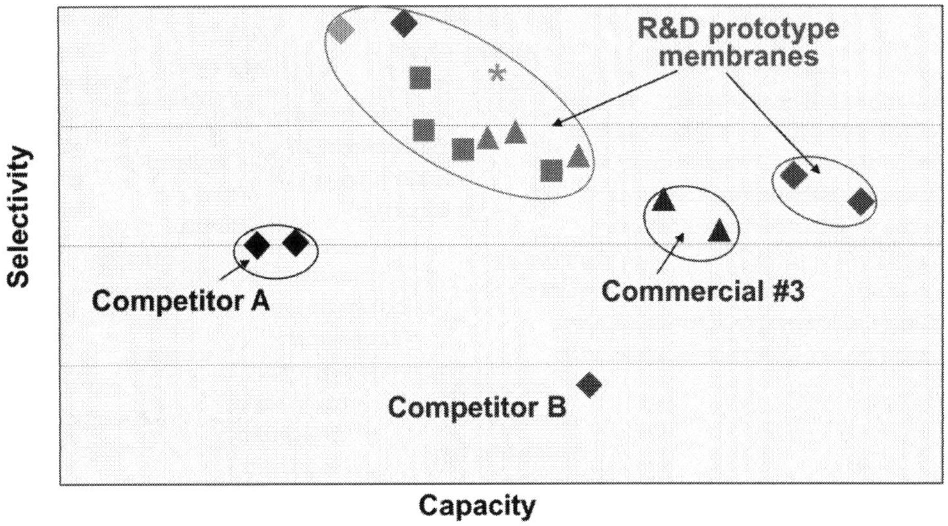

UOP Separex™ commercial membrane #3 and R&D prototype membranes showed high performance for H_2 purification and recovery applications

Honeywell UOP

New UOP Separex™ Membranes for He Recovery

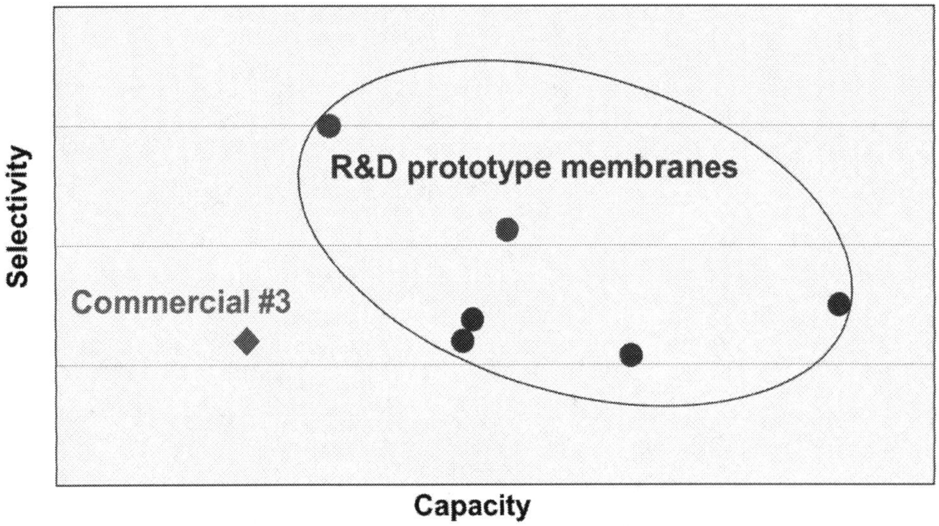

UOP Separex™ commercial membrane #3 and R&D prototype membranes showed high performance for He recovery from natural gas

Honeywell UOP

Hybrid Membrane-Distillation for Olefin/Paraffin Separations

- Extending the membrane platform to large volume organic vapors needs new membranes with high selectivity, high permeance, and high stability!

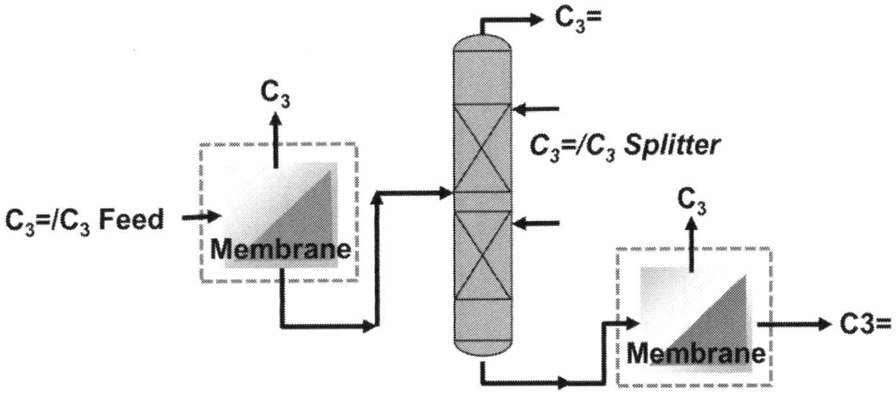

- Properly designed integrated distillation and membrane hybrid processes may offer the opportunity for lower energy intensity olefin/paraffin separations

Honeywell UOP

UOP 8055-20

UOP Separex^{TM} Membranes for C_3=/C_3 Separation

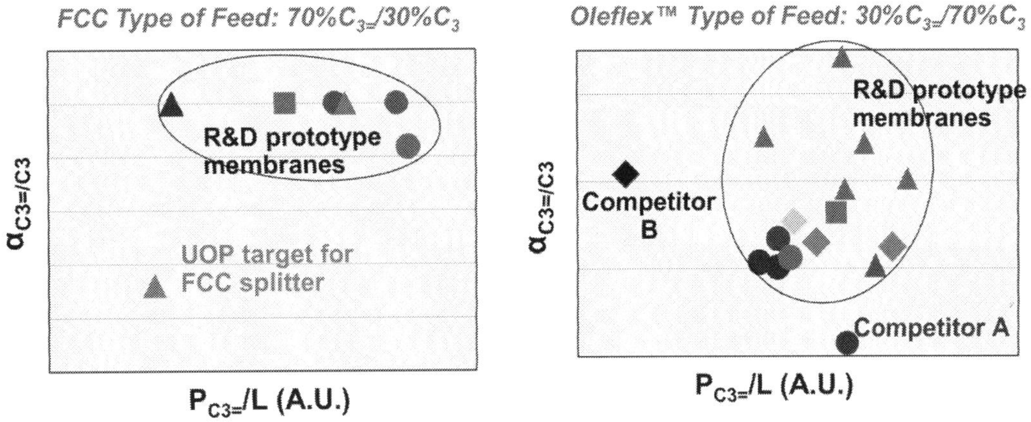

New UOP Separex^{TM} R&D prototype membranes showed high stable performance for C_3=/C_3 separation

Honeywell UOP

UOP 8055-21

22

Summary

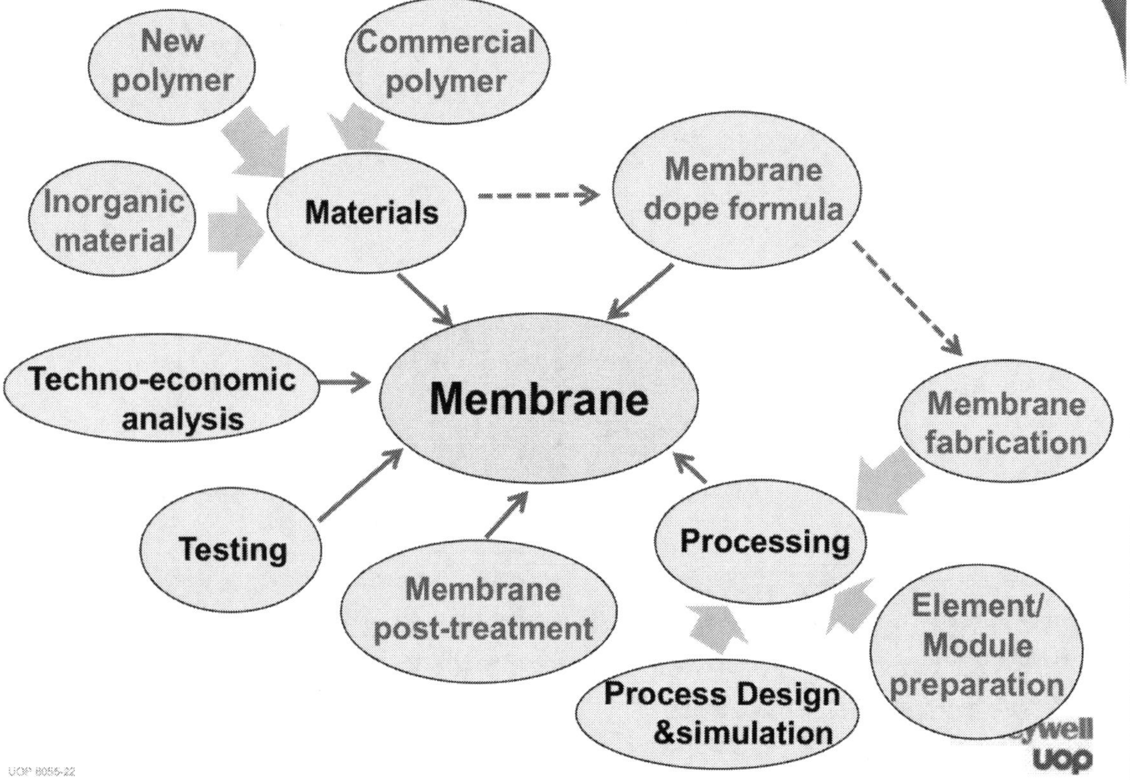

Honeywell UOP

UOP 8055-22

23

Honeywell UOP

UOP 8055-23

502107 Inorganic Membrane Research and Development at Exxonmobil – the Last 25 Years to Today

Abstract:

Highlights

1. Highlights from ExxonMobil's 25+ years or research in inorganic membranes technology

2. Focus on current research in NF and RO separations of complex hydrocarbon feeds

3. Provide perspectives for future membrane use in refining and petrochemical industry

ExxonMobil has been interested in the use of inorganic membranes for both gas and liquid separations for at least the last 25 years. In that time we have investigated numerous applications with and worked with inorganic membranes of all types. That work resulted in over 100 granted US patents and over 30 publications in the peer reviewed literature. This presentation will provide an overview of the work from the past 25 years – including key findings in our continued progression toward understanding molecular transport and materials engineering aspects of inorganic membrane separations. Specific examples will be presented on zeolite membranes for the separations of CO_2/CH_4 and xylenes as well as hybrid systems for aromatics/non-aromatics pervaporation.

The second half of the presentation will focus on our on-going work in the separation of complex liquid hydrocarbon mixtures using inorganic membranes. We will cover our efforts in novel materials development, compositional modeling for separation products in complex mixtures, and process concept development for nano-filtration and hydrocarbon reverse osmosis applications.

490623 Radially Aligned Carbon Nanotube Hollow Fiber Membranes

Abstract:

We report the first radially aligned carbon nanotube (CNT) hollow fiber (HF) membrane in which the pores of the membrane consist of open nanotubes traversing a non-porous polymer skin. The nanotubes are located on the outer surface of the fiber, allowing for outside-in flow with precise size selectivity. Two types of CNT HF membranes will be discussed â€“ membranes using Arc Discharge, and CoMoCat nanotubes. Performance data for both water and gas separations will be presented, and this data will be compared to flat sheet CNT membrane data shared previously. Perspectives on the development of the membrane technology in a start up company context will also be discussed.

491111 Membranes for Water Purification and Industrial Separations

Abstract:

Over the course of the last 5 decades, separations applications with membranes have significantly evolved, and membranes are becoming a preferred technology for various separations at industrial scale. Primary reasons for the evolution of membranes and their adoption are 1) the inherent energy efficiency of separations via membranes, as compared to other separations methods 2) new materials discovery and development, and 3) large scale manufacturing. In addition, economics presented by membrane processes due to efficiency in technologies like pumps, energy recovery devices, novel module designs and better fundamental understanding of membrane structure and property relationships due to advanced analytical methods, has also propelled their wide scale adoption.

In this talk some key industrial applications of membranes will be presented and analyzed from an efficiency and economics standpoint. Insight into the future of membrane technology development will also be provided.

487401 Virus Filtration: The Most Challlenging Filtration in Biotech Process Lessons from a Successful Product Development

Abstract:

Membrane separations are critical in biotechnology drug production. Membranes enable drug safety by removal of bacteria and viruses and provide a convenient tool for achieving the final drug formulation. One of the most challenging filtration applications in a biotech process is virus filtration. This application requires a 4-log high retention (R>0.9999) of a 20 nm sized parvovirus and a nearly complete passage (R<0.01) of a protein product that is only slightly less than a half of that size. In addition, a high filtration economy is expected, requiring a high filter throughput capacity despite a common presence of small amount of protein aggregates similar in size to the retained virus in the filtered streams.

Development of a successful product based on a new type of membrane will be described. Membrane morphology optimization and device design options will be discussed and trade-offs between performance and cost, robustness and ease of use will be described.

499625 Chemical Engineering Mobile Apps

Abstract:

Although we have seen a nearly complete shift among chemical engineering students from desktop to laptop computers, it has merely created an incremental change in their education. However, the benefits of mobile devices in engineering education still remain largely untapped. There is a great need and opportunity to put chemical engineering data and problem-solving tools in studentsâ€™ (and faculty membersâ€™) pockets. This presentation will discuss 5+ years of effort and motivation behind the development of several chemical engineering-oriented mobile apps, including *Chemical Engineering AppSuite*. Through the use of embedded analytics, measuring user demographics and interaction with the apps reveals interesting details about what kinds of calculations and what types of data chemical engineers are most interested in and use most frequently. While using to these mobile apps has a number of benefits for expediting calculations and facilitating access to data, the associated analytics can help shape teaching chemical engineering in the 21st Century.

495506 Process Design: Learning in Chemical Engineering

Abstract:

As chemical engineers, we are trained to identify and model the fundamental mechanisms of chemical and biological processes towards innovation and reliable production. This disposition is the cornerstone of a discipline that continues to make key contributions in traditional (e.g., petrochemical, pharmaceutical) and emerging (e.g., microelectronics, biotechnology, alternative energy) fields. This *chemical engineering approach* can also be applied to the complex processes of learning. Thus, it is not surprising that chemical engineers, like Phil Wankat, have been pioneers in the emerging field of engineering education research, where scholars seek systematic and generalizable understanding of the learning process and learning systems. This talk will focus on examples from the authorâ€™s research of learning processes and ways that he has explored foundational mechanisms and connected them to innovations and reliable delivery of course and unit programs.

Letters from the Editor: Reflections on Phil Wankat

Lynn Heasley, Richard Felder, and Lisa Bullard

"I am interested in your comments on my e-mail that describes the situation."

2

"Make any suggestions on diplomacy."

3

"Sooner or later we will have to deal with this. Since you are more diplomatic than I am, I suggest you be the lead and respond."

4

"Prof. [Blank]: Unfortunately, you appear to have misinterpreted our letter. It was meant to be a polite way of rejecting your paper. We wish you luck directing it or future revisions of it elsewhere."

5

"The author needs to look at the journal to see what types of papers are published."

6

"Please send the authors the following. They will be unhappy."

7

"In a nutshell, the current paper is not tightly focused, is too long, has unnecessary detail, does not follow CEE style guidelines, and needs to be rewritten."

8

"I must have sinned, and Dr. [Blank] is the penalty for my sins."

9

"Lynn, You have no idea how delighted I am to receive e-mails from you that start, *'This one has a lot of hallmarks of past problem papers.'* "

10

"There were a couple of places where your silence was golden."

11

"I do not think we need to review this book. It does not appear to be teaching oriented (do not say this to the author, as he will argue it is)."

12

"Dr. [Blank] is a problem. He ignores the advice on the most important point and changes minor points. I am not willing to inflict this revision on the reviewers. I am saying no, but with the carrot that if he follows the original advice, we will send it out for re-review."

13

"This is not the first time we have had rather far-out materials. I am willing to give him one trial to develop an acceptable paper, but I am not willing to play extended games trying to save the world."

14

INDIANA
MIGRATORY
WATER FOWL
STAMP

“Enjoy!”

15

“Sorry for the late response, but I was in Scotland for vacation and refused to be sucked in by my e-mail.”
INDIANA TROUT STAMP
1951
$1
DEPARTMENT OF CONSERVATION
DIVISION OF FISH AND GAME

16

"They have been good sports about fixing things."

17

"Unfortunately, Dr. [Blank]'s Teaching Tip did not receive as much attention from me as it deserved. Upon rereading it, I think it will benefit from more work. Going over items too quickly when I have too much to do is a recurring problem. I will have to try and slow down."

Followed shortly after by:

"On second thought, let's go with Dr. [Blank]'s current version. I do not want start requesting late changes of authors because I do not have my act together. Better to get my act together."

18

“Some people are just more fun to work with than others.”

19

Image Citations

1. www.istockphoto.com
2. https://about.usps.com/news/national-releases/2016/pr16 048.htm
3. http://www.stockunlimited.com/vector/white-house-stamp 1525799.html
4. https://www.pinterest.com/bluememaw/usa-state-birds-postage-stamps/
5. https://arago.si.edu/category 2028905.html
6. http://www.theus50.com/indiana/stamp.php
7. http://www.indianastampclub.org/
8. http://www.brookmanstamps.com/Netcat/DUCK/in.htm
9. https://postalmuseum.si.edu/stampstakeflight/firstusairmail.html
10. http://www.brookmanstamps.com/Netcat/DUCK/in.htm
11. https://www.mysticstamp.com/Products/United-States/4290/USA/
12. https://www.123rf.com/photo 34876810 usa-and-indiana-state-flag--old-postage-stamp.html
13. https://www.flickr.com/photos/shookphotos/5794984932
14. http://www.brookmanstamps.com/Netcat/DUCK/in.htm
15. https://about.usps.com/news/national-releases/2016/pr16 048.htm
16. https://postalmuseum.si.edu/stampstakeflight/firstusairmail.html
17. https://waterfowlstampsandmore.com/gallery-3/indiana-trout-stamps/
18. http://www.indianastampclub.org/
19. http://www.brookmanstamps.com/Netcat/DUCK/in.htm

20

487550 Blending Teaching, Research and Writing Textbooks

Abstract:

In the early years of teaching engineering as an academic discipline instead of an apprenticeship, teaching was the main function of professors. Research was one of several methods to stay abreast of changes in engineering and to stay intellectually active. This view remained common even after the Second World War had illustrated the power of research. For example, the opening of NSF brought forth the comment, "it would in fact be surprising if more than 500 or 1000 engineering professors would be interested in a federal research program" [1, p. 60]. A combination of war experiences, the availability of research money, the ASEE Grinter report, and the shock of Sputnik in 1957 led to the engineering science revolution in engineering education [2]. This change took time but was in place when I attended Purdue. Currently, about 2/3 of professors at research institutions are more interested in research than teaching [3, p. 421]. Regardless of one's interest, how do professors blend teaching, research and service, and in the process balance their lives? I will present my story of teaching, research, and writing textbooks in both separations and education. I am well aware that my story may be "a tale told by an idiot, full of sound and fury, signifying nothing." [4]

1. Stewart, *IRE Trans. Educ.*, **1** (2) 54–61, 1958.

2. Froyd, Wankat and Smith, *Proceedings of the IEEE, 100,* 1344-1360 (2012)

3. Wankat and Oreovicz, *Teaching Engineering,* 2nd ed., Purdue University Press: West Lafayette, IN (2015).

4. Shakespeare, *The Tragedy of Macbeth,* speech by Macbeth, probably first performed in 1606.

500241 Direct Numerical Simulations of Polarization Phenomena in Direct Contact Membrane Distillation

Abstract:

Direct contact membrane distillation (DCMD) is a relatively new thermal-driven membrane separation technology. In a DCMD system, warm feed and cool permeate flow on opposite sides of a hydrophobic membrane. The temperature difference causes water vapor to evaporate from the feed and condense in the permeate. To date, computational fluid dynamics (CFD) studies of DCMD focus primarily on the challenge of temperature polarization. For treating high concentration brines, however, concentration polarization is another major challenge that reduces system efficiency and leads to mineral scaling. Consequently, we perform a CFD study of DCMD systems, with a focus on applications to treating high concentration brines rejected by reverse osmosis systems.

We develop a dedicated in-house CFD method to simulate 3-D, unsteady, heat and mass transport within plate-and-frame DCMD systems. The coupled momentum, energy, and mass transport equations are discretized spatially using finite-volume methods. The equations are integrated in time using an efficient, non-iterative, semi-implicit projection method to solve for the pressure field in the Navier-Stokes equations. We present the results of a parametric study of 3-D temperature and concentration polarization under a variety of operating conditions, including feed and permeate flow rates and temperatures, feed concentration, and membrane properties. The numerical results are benchmarked with a parallel experimental study of permeate flux and thermal efficiency.

500098 Modeling Separations in Mesoporous Membranes Using Lattice Based and Molecular Simulation Techniques

Abstract:

Mesoporous membranes provide an energy efficient alternative for processes where enrichment or separation of condensable vapor is required given its mixture with a light gas. With continuous progress in synthesis methods for these membranes, it is now possible to fine tune both geometry and surface chemistry of mesopores. However, a rational design requires rules to establish link between transport mechanisms and mesopore properties. Absence of such correlations is one of the obstacles to the wide-spread usage of mesoporous membranes. Determination of these design rules would require a thorough knowledge of various transport phenomenon present in the membrane process and how they cooperate to give rise to macroscopic quantities such as selectivity and permeance. Dynamic mean field theory (DMFT), a lattice-based density functional theory which is computationally efficient and sufficiently descriptive, was applied to these systems to understand the underlying nature of transport processes. We initially applied DMFT to permporometry, an experimental technique where light gas permeates through condensable vapor under small pressure gradients. We then investigated a separation process where condensable vapor is separated from its mixture with light gas under significant pressure gradient. During our studies, we found a nonequilibrium steady state with capillary condensation confined to the high pressure (feed side) of the pore. We investigated the effect of this state on separation efficiency and permeation using different pore geometries and surface chemistries. Based on the understanding that we gained through our model, we attempted design better pore structure incorporating effects of partial capillary condensation which would provide higher efficiency and yield in separation.

501027 Mass Separation By Metamaterial Membranes

Abstract:

In this talk, we introduce a novel approach to perform mass separation based on mass-diffusion metamaterial devices that can shield one compound from reaching a certain spatial area while concentrating other compound towards that area. Such metamaterials provide the basis for a new method to manipulate mass diffusion and achieve separations in chemical and biomolecular engineering. We coordinate transform Fickâ€™s law diffusion equation to simultaneously manipulate the diffusion paths of different species in both transient and steady state regimes. The design of a novel mass-separator metamaterial device is first introduced. We also present a fabrication procedure that allows for experimental realization of such metamaterial using commercially available homogeneous and isotropic polymeric materials. A practical proof-of-concept is introduced by separating a binary mixture of oxygen and nitrogen diffusing through a polymeric matrix. This work paves the way to achieve mass separations using metamaterials devices in chemical and biomolecular industries, science, and technology. Mass flow manipulation is of utmost importance in many processes in chemical engineering and, in particular, separations are critically important in different areas ranging from chemical manufacturing, recovery of biological solutes from wastes, to semiconductor crystal growth, and generation of artificial kidneys. The systematic method presented in this talk to achieve precise manipulation of molecular diffusion paths by coordinate transformations and metamaterials provides a novel tool to engineer separation processes in chemical engineering.

498602 Modeling Bioethanol Enrichment Using Hydrophobic and Hydrophilic Zeolite Membranes

Abstract:

Zeolite membrane separation is considered to be a promising alternative to the traditional energy-intensive industrial separation techniques such as distillation. We describe the development of a detailed mathematical model based on the real adsorption solution theory and the Maxwell-Stefan formulation for transport to describe permeation through zeolite membranes. We also discuss the application of the model for bioethanol (obtained at ~ 5 wt. % ethanol) enrichment using hydrophobic all-silica MFI and hydrophilic NaA zeolite membranes.

Although significant improvements have been realized in the separation performance of zeolite membranes for water-ethanol separation, rigorous models, which can accurately predict the membrane performance are lacking. MFI zeolite membranes separate ethanol based on its greater adsorption affinity over water. However, unavoidable water co-adsorption due to hydrogen-bonding effects (i.e., water is brought in by adsorbed ethanol molecules) and due to favorable interaction of water molecules with silanol defects adversely affects the separation performance. As a result of these non-ideal intermolecular interactions, using ideal theories based on single-component adsorption and diffusion leads to over-estimation of their separation capability.

In this work, mixture adsorption is modeled using real adsorption solution theory. The co-adsorption of water molecules due to hydrogen bonding is accounted by using activity coefficients based on the two-constant Margules equation. The multi-component diffusion is modeled using the Maxwell-Stefan formulation with non-zero correlation coefficients. Furthermore, the effects due to support layer and hydrophilic defects are also incorporated. Ethanol permeance predicted by the model is found to be lower while the selectivity is found to be higher as compared to those determined in experiments. These results suggest that the zeolite layer in the simulated membrane is more permeable than the real membranes, and thus the overall permeation in the model is governed predominantly by the support layer. This discrepancy has been attributed to the reasoning that the diffusion coefficients used in the model are obtained using molecular simulations and reflect the behavior in crystals with ideal pore structure, while the presence of structural non-idealities, compressive stresses, thermodynamic and physical surface barriers in the real membranes results in reduced permeation. Hence, the diffusion coefficients in the model were reduced by a factor of 120 to obtain an experimentally validated model.

This experimentally-validated model is further used in flowsheeting thermally-integrated conceptual process designs to assess the viability of zeolite membranes over the current distillation technology for bioethanol enrichment. It is found that while hydrophobic zeolite membranes have potential for significant energy savings but lack in separation performance and hydrophilic zeolite membranes can achieve the separation target but result in no energy savings, a process with a combination of hydrophobic and hydrophilic membranes can achieve the separation target at the current performance of zeolite membranes with 10 % energy savings over distillation. However, techno-economic analysis suggests that 5- to 10-fold improvements in permeation or equivalent cost reductions are required for economic feasibility.

499605 CH_4 and CO_2 Transport Properties through Nanoporous Graphene and Graphene Oxide Membranes: A Molecular Dynamics Simulation Study

Abstract:

Molecular dynamics simulations were performed to investigate the gas transport properties of a CH_4/CO_2 gaseous mixture in single- and double-layered nanoporous graphene (NPG) and graphene oxide (NPGO) membranes. An increase in the CH_4 concentration was found to improve the through-the-pore diffusion coefficients of both CO_2 and CH_4 for both single- and double-layered membranes. The CO_2 permeance is smaller than that of CH_4 for all membrane configurations, in agreement with the diffusion coefficients of both gases. Therefore, a CH_4/CO_2 separation is realized in the NPG and NPGO membranes due to a faster permeance of CH_4 than CO_2. The highest and lowest permeances for CO_2 are for the single-layered NPG and the double-layered NPGO membrane, respectively. The perm-selectivities of both NPG and NPGO membranes were determined based on the number of gas molecules passing through the membranes and the gas permeances. The single- and double-layered NPG and NPGO membranes demonstrate a satisfactory performance for the CH_4/CO_2 separation. Furthermore, the adsorption- and perm-selectivity of the double-layered NPGO membranes show improvement when compared to the single-layered NPGO membranes.

490827 Understanding Water Sorption and Transport in Graphene-Based Membranes from First-Principles Based Atomistic Modeling

Abstract:

Graphene-like materials (GLMs) have gained significant attention for many applications because of their interesting features. One of interesting feature of graphene sheets is their layered structure. The planar (2D) structure makes GLMs possible to form an ultra-thin laminated structure which is suitable for use as a separation membrane. Many researchers have attempted to apply ultra-thin graphene-based membranes to various separation processes including water purification and gas separation. However, graphene-based membranes have a large drawback since the interlayer space of graphene sheets is too small for even small molecules to pass. To overcome this problem, researchers have used graphene oxide (GO), the oxidized form of graphene, as hydroxyl groups on the basal plane may increase the interlayer space while also attracting polar molecules such as water. Finally, as water molecules can permeate into the GO membranes, they can be utilized for desalination and CO_2 separation. However, the permeability of GO membranes tends to be too low for their practical application. There has been a substantial experimental effort to improve their permeability, but only limited fundamental studies have been undertaken to understand the factors influencing the sorption and diffusion of gaseous species in GO-based membranes.

In this talk, we will present our recent findings from intensive molecular simulations regarding the molecular mechanisms of water sorption and transport in GO sheets with layered structures, especially the influences of relative humidity, OH group content and distribution, and GO geometry. We will focus on discussing (i) how the amount and dynamics of intercalated water varies under different conditions of humidity, temperature, and GO structure and (ii) the effects of intercalated water on the separation of a target gas from a mixture of gases. The improved understanding helps the design of more efficient GO-based membranes for gas separation.

496008 Role of Solvent in Structural Reorganization of a Polymer Membrane: An Atomistic Simulation Study

Abstract:

The swelling of a polymer membrane in a solvent plays a significant role in the reorganization of the polymer chains, as well as the membrane performance for organic solvent nanofiltration (OSN). In this work, an atomistic simulation study is reported on the structural characteristics (accessible pores, pore size distribution and polymer mobility) for a polybenzimidazole (PBI) membrane swollen in three different solvents, including water, methanol and ethanol. The swelling degrees of the PBI membrane in water (7.85%), methanol (15.50%) and ethanol (17.20%) are obtained from reported experimental measurements. The simulation results indicate that the reorganization of polymer chains is mainly determined by the polarity of solvents, the hydrophobicity of polymer, and the competition of solvent-polymer and polymer-polymer interactions. Due to the higher capability of ethanol for PBI swelling, the pore size of the membrane swollen in ethanol is larger than swollen in other two solvents (methanol and water). Furthermore, acetone flux and solute rejection in three swollen membranes are estimated from simulation and it is found that the membrane swollen in ethanol shows a better performance than in the other two solvents. This simulation study provides microscopic insight into the reorganization of polymer chains in the PBI membrane, and suggests that the membrane structure and performance can be subtly tuned by varying the pre-treated solvents.

493770 Surface Dynamics of Calcium Oxalate Monohydrate Crystallization: Elucidating Mechanisms of Growth Inhibition

Abstract:

Calcium oxalate monohydrate (COM) crystals are the primary constituent of human kidney stones. The growth of these crystals *in vivo* is affected by different urinary constituents (e.g., ions and molecules), which act as modifiers by adsorbing onto different sites of the crystal surface, thereby influencing the rate(s) of anisotropic crystal growth with concomitant alterations in crystal size and habit.[1-2] At the molecular level, COM crystals grow classically (i.e. monomer-by-monomer addition) where single layers nucleate and advance across the surface by solute addition to surface sites (e.g. kinks, steps, or terraces).[3] Modifiers can alter crystal growth by adsorbing onto surfaces and blocking solute attachment and/or by complexing with free monomers in solution to reduce supersaturation (i.e., the primary driving force for crystal growth).[4] In order to design drugs that inhibit crystallization, it is critical to understand their modes of action. To this end, we conducted *in situ* atomic force microscopy to probe the effect of naturally occurring and synthetic modifiers on the dynamics of surface growth as a function of supersaturation and modifier concentration. Our studies reveal unique modes of action that vary significantly based on subtle differences in the physicochemical properties of modifiers[5]. Here, we will discuss the results of our studies and demonstrate how these techniques serve as a useful platform for the design of new generation therapeutics to treat diseases such as kidney stones.

1. Farmanesh, S.; Alamani, B. G.; Rimer, J. D., Identifying alkali metal inhibitors of crystal growth: a selection criterion based on ion pair hydration energy. *Chemical Communications* **2015,** *51* (73), 13964-13967.

2. Farmanesh, S.; Ramamoorthy, S.; Chung, J. H.; Asplin, J. R.; Karande, P.; Rimer, J. D., Specificity of Growth Inhibitors and their Cooperative Effects in Calcium Oxalate Monohydrate Crystallization. *Journal of the American Chemical Society* **2014,** *136* (1), 367-376.

3. Chung, J.; Granja, I.; Taylor, M. G.; Mpourmpakis, G.; Asplin, J. R.; Rimer, J. D., Molecular modifiers reveal a mechanism of pathological crystal growth inhibition. *Nature* **2016,** *536* (7617), 446-450.

4. Olafson, K. N.; Li, R.; Alamani, B. G.; Rimer, J. D., Engineering Crystal Modifiers: Bridging Classical and Nonclassical Crystallization. *Chem Mater* **2016**.

5. Farmanesh, S.; Chung, J. H.; Sosa, R. D.; Kwak, J. H.; Karande, P.; Rimer, J. D., Natural Promoters of Calcium Oxalate Monohydrate Crystallization. *Journal of the American Chemical Society* **2014,** *136* (36), 12648-12657.

500837 Use of the Constant Composition Method to Evaluate the Nucleation and Growth Kinetics of Calcium Biominerals

Abstract:

In most biomineralization processes many soluble biopolymers exist in minute quantities; these control the nucleation, growth, orientation, and polymorph selection, leading to the formation of minerals with diverse biological functions and distinct properties but also to pathological mineralization, e.g., bone disease, dental calculus, and renal stone formation. In this work, we study the crystallization kinetics of calcium oxalate (the major constituent of kidney stones) by employing a constant composition potentiostatic technique. We evaluate the effects of supersaturation, temperature, and additives on the nucleation and growth of calcium oxalate in aqueous solutions. We have demonstrated that certain anionic polyelectrolytes can be very effective toward kidney stone inhibition by reducing the growth rate, altering the crystal morphology, and favoring the formation of the metastable dihydrate (COD) over the thermodynamically stable monohydrate (COM) which is mainly responsible for kidney, gall, and bladder stones. We compare our results to citrate, a naturally occurring inhibitor in the urinary system and currently the standard pharmacological treatment for stones. Our findings offer a better understanding of the calcium oxalate crystallization kinetics and provide information on the viability of synthetic polymers as potential calcium oxalate stone therapies.

488682 Solution Mediated Polymorphic Transformation of Cefaclor: Insights into the Interaction Between Polymorph Surfaces

Abstract:

The solution mediated polymorphic transformation (SMPT) of the cefaclor monohydrate metastable Form I to the stable Form II was investigated in ethanolâ€"water mixture. During the transformation for a variety of initial conditions, bulk transformation experiments were performed by Raman to monitor the polymorphic composition combined with UV to analysis of the solution concentration over time. The main factors studied were solvent composition, temperature, the scale of operation, mass of solids loading as well as pretreatment of the loaded solids. The results revealed that the transformation time was governed by the nucleating and growth behavior of the stable Form II across the range of experiments. The stable Form II is consistent nucleated epitaxially on the metastable Form I during the transformation. Microscopic techniques, single-crystal X-ray diffraction and computational simulation methods were used to analyze the transformation and epitaxial behavior. However, no specific lattice matching or epitaxy was conclusively identified. Therefore lattice matching may not the dominant reason for the nucleation of Form II on the surface of Form I. Molecular simulation is carried out to understand the different probabilities of the Form II nucleation on the surfaces of Form II at the molecular level.

490645 Desorption of Gases from Ionic Liquids Using an Applied Electric Field

Abstract:

Ionic liquids (ILs) are defined as salts which melt at temperatures below 100 ^{o}C. ILs possess unique properties including negligible vapor pressures and molecular tunability achieved through the selection of the cation and anion with a broad range of applications in reaction engineering and separations processes. Ionic liquids have been extensively developed for selective gas separations using both physical and chemical absorption. However, alternative desorption processes to remove gas from ionic liquids remain largely uninvestigated. Desorption processes require external energy inputs to separate the gas from the ionic liquid. Conventionally this is done by applying thermal energy in the form of heat which in many cases reduces the overall efficiency and economic viability of the process. To lower the required thermal energy, we have studied the effect of applying an electric field to achieve gas desorption from ionic liquids. Preliminary results demonstrate that using electric fields to manipulate the orientation of ions within ionic liquids could be a new method for desorbing gases with industrial relevance to many gas separation processes.

511741 Green Solvent Use and Opportunities in Biomass Processing

Abstract:

As solvents comprise a large portion of processing waste and energy use for separations, the selection of alternatives with improved safety, environmental, and recyclability would significantly improve process sustainability. As new applications expand in the field of biomass refining and valorization, appropriate solvent development and usage will aid in viable implementation. This work presents an overview of green solvent applications with specific application geared towards biomass production. Particular solvent needs for biomass fractions to produce fuels and value-added products are presented. Green solvent metrics with respect to functionality, and environmental, safety, and health impacts from a process and life cycle view are also addressed and applied to conventional and neoteric solvents. Current and potential applications of various solvents for extractions and conversions are provided for consideration in a biorefinery setting with focus on research in supercritical CO_2 applications.

498338 Tempo Oxidized Cellulosic Membranes for Low Fouling Applications

Abstract:

Cellulose for membrane materials is a well developed technology. However, cellulose is still a difficult material to work with and the properties are seldom as consistent as you woudl like them to be. A variety of novel barrier membranes were cast using different concentrations of the proposed products of the tempo oxidation reaction. Film properties linked to multiple categories ranging from tensile strength to hydrophilicity were investigated. Additionally, data of physical and chemical alterations to the casting of cellulosic membranes have been examined and summarized. We found that using new methods of making tempo modified cellulose has decreased the hydrophilicity of the membrane below what has previously been reported in literature. Further, the fouling characteristics of this membranes have been improved based on the increased hydrophilicity. This paper will examine the reasoning for these improved properties and hyphotisize the possibilities of nanocellulosic membrane materials.

494908 Techno-Economic Analysis of Polymeric Membrane Systems for Post-Combustion Carbon Capture

Abstract:

The increasing consumption of fossil fuels has resulted in immense CO_2 emissions after the Industrial Revolution. Membrane processes have been suggested as an alternative to amine absorption for post-combustion carbon capture. Despite of the tremendous research efforts, CO_2-selective polymeric membranes typically exhibit a trade-off between CO_2 permeance and CO_2/N_2 selectivity. This presentation will show how to overcome the limitations as a result of this trade-off by noval membrane process designs. In the first membrane process, called the hybrid air sweep process, the flue gas flows to a vacuum membrane stage using a membrane with high CO_2/N_2 selectivity and moderate CO_2 permeance. The retentate gas then flows to an air-sweep membrane stage employing another membrane with high CO_2 permeance but low CO_2/N_2 selectivity. The CO_2-laden sweep air is fed to a power plant boiler as combustion air. A highly selective amine-containing membrane with a CO_2 permeance of 1100 GPU (1 GPU = 10^{-6} cm^3 (STP)·cm^{-2}·s^{-1}·$cmHg^{-1}$) and a CO_2/N_2 selectivity >140 is used for the vacuum stage to enrich the CO_2 to >95% purity whereas a highly permeable poly(ethylene oxide)-based membrane with a CO_2 permeance of 2000 GPU and a CO_2/N_2 selectivity of 20 is employed for the air sweep stage. This hybrid membrane configuration yields a capture cost of \$40.2/tonne CO_2 in the 2007 dollar basis for 90% CO_2 recovery, which nearly meets the target of the Department of Energy (DOE) set for 2025. In the second membrane process, i.e., retentate recycle process, the flue gas is passed to two stages employing the membrane with high CO_2/N_2 selectivity; these two stages are placed in an enriching cascade. A portion of the CO_2-depleted retentate of the first membrane stage is recycled as its own sweep gas. It is identified that the highly selective amine-containing membrane is needed for both stages and a 20% retentate recycle renders a capture cost of \$43.1/tonne CO_2. These cost calculations in the 2007 dollar basis have included both membrane module installation cost and 20% process contingency.

492718 An Outlook on Hydrogen Production from Bio-Ethanol Reforming Reactions By Membrane Reactor Technology: Opportunities and Challenges

Abstract:

Our dependence on fossil fuels as the primary energy source, has led to an energy crisis due to the depletion of fossil fuels and the subsequent warming of the globe [1]. In response to these two concerns, continuous efforts are ongoing to search for clean and renewable alternatives. In this context, hydrogen is recognized as one of the most promising energy carriers of the future. Unfortunately, hydrogen is not available in the Earth in elementary form and is currently produced primarily from steam reforming of natural gas, an energy-intensive process.

However, different methods, such as alcohol-reforming reactions, and alternative devices, such as membrane reactors, have being receiving great attention for the production and separation of high-purity hydrogen. Specifically, alcohols may be largely available from renewable sources and they could be used as â€˜bio-feedstockâ€™ for hydrogen generation via reforming reactions. In particular, bio-ethanol constitutes an important bio-source that, compared to other renewable liquid fuels such as methanol, glycerol, acetic acid, diethylether, etc., appears to be more suitable due to its very low toxicity and volatility in addition to its high hydrogen content [2]. Membrane reactor technology seems to represent a valid choice for substituting the conventional process due to its ability of combining reaction and separation in a single stage [3].

Hence, the current study reviews the most relevant scientific results on hydrogen production from bio-ethanol reforming processes performed through membrane reactor technology. The benefits and the main challenges of inorganic membrane reactors, when applied to ethanol-reforming processes, are examined and the performances in terms of hydrogen yield, hydrogen recovery, and conversion are compared to those of conventional systems.

References

[1] M. Ni, D.Y.C. Leung, M.K.H Leung, K. Sumathy, An overview of hydrogen production from biomass, *Int. J. Hydrogen En.*, 87 (2006) 461â€"472.

[2] G. Maggio, S. Freni, S. Cavallaro, Light alcohols/methane fuelled molten carbonate fuel cells: a comparative study, *J. Power Sou.*, 74 (1998) 17-23.

[3] A. Iulianelli, S. Liguori, A. Vita, C. Italiano, C. Fabiano, Y. Huang, A. Basile, The oncoming energy vector: hydrogen produced in Pd-composite membrane reactor via bioethanol reforming over Ni/CeO_2 catalyst, *Catal. Today*, 259 (2015) 368-375

495942 FO Separations with High-Performance TFC Membranes and Novel Draw Solutes

Abstract:

Forward osmosis (FO), an emerging membrane separation process utilizing the difference of osmotic pressure between the feed and draw solution as the driving force, has drawn growing attention ascribed to its unique advantages of relative low energy consumption, high water recovery, and low fouling propensity. Nevertheless, the development of a high-efficient FO system is still hindered by the lack of sustainable high-performance FO membranes and suitable draw solutes. Thin-film composite (TFC) membranes with excellent separation performance and convenient fabrication technology, have gained much attention in FO applications recently. Despite this, traditional TFC membranes formed by TMC and MPD still face problems of insufficient hydrophilicity and the permeability-rejection trade-off phenomenon. In our work, the surface modification by various hydrophilic amine monomers has been performed to improve the membrane water permeability and anti-fouling property. In-situ modification with various tertiary amines as the aqueous phase additives is also investigated, which can act as the catalyst to accelerate the interfacial polymerization. In comparison with the control TFC membrane, these modified membranes have the higher water flux, lower reverse salt flux and enhanced fouling resistance because of the smoother surface, as well as the thinner and denser PA layer with the higher crosslinking degree. In addition, the selection of proper draw solution is also crucial to guarantee a high-performance FO process with minimal reverse solute flux and easy draw solution recovery. Here, series of organic salt draw solutes with adjustable chemical structure and optimized molecular size are synthesized and applied for FO process. The results show that the introduction of carboxyl and phosphate groups to the molecular structure of organic draw solutes is a simple and effective way to enhance the osmotic pressure and FO performance of the draw solutions.

502860 Adsorption of PAHs on Al-MCM-41: Batch Equilibrium and Column Breakthrough Behavior

Abstract:

Polycyclic aromatic hydrocarbons (PAHs) are ubiquitous contaminants and are well known for their toxic, carcinogenic, and mutagenic effects. Several industrial applications use streams with PAHs as inputs or solvents, depending on the type and aromatics concentration. This has led to an upsurge of interest in developing and implementing methods for their removal from commercial products using adsorption. Extensive research has been carried out on the adsorption of organic compounds from aqueous solutions. However, there are few reports for the separation of PAHs from a hydrocarbon mixture in liquid phase. In this study, naphthalene and pyrene, in iso-octane solution, were used in batch and column experiments to assess the PAHs adsorption on Al-MCM-41. Batch experiments were performed and equilibrium isotherms were measured at different temperatures. Effective surface diffusivities were estimated using column breakthrough experiments for both PAHs in different feed concentrations.

495337 Effect of Different Epoxy Modifiers Applied to Amine â€“ Silica Adsorbents for CO_2 Capture

Abstract:

Recently, carbon dioxide (CO_2), one of the greenhouse gases that absorb radiant heat from the earth's surface and increase the atmospheric temperature, is produced from various industrial sectors. CO_2 is the major cause of global warming due to its large emitted amount and the carbon capture and storage (CCS) technology has attracted attention to curb the release of CO_2. The CCS technology can be broadly divided into three categories: capture, transport, and storage. From an economic point of view, it is important to develop effective capture technology because the capture occupies more than 80% of the total cost. CO_2 capture technology can be largely classified into absorption, membrane separation, and adsorption. Among them, adsorption has advantages of a relatively low operating cost and continuous CO_2 separation without byproducts. However, adsorption has main drawback of low CO_2 selectivity over N_2. In this study, we developed adsorbents having improved CO_2/N_2 selectivity and CO_2 adsorption performance by functionalizing amines on the silica support. Furthermore, the instability at high temperature during regeneration, a disadvantage of existing amine-based adsorbents, was compensated by transforming the amine structure using different epoxy-modifiers. The CO_2 adsorption performance of the adsorbents was evaluated by thermogravimetric analysis (TGA), and the surface characteristics and structure of the adsorbents were examined by N_2 adsorption analysis, Fourier transform infrared spectroscopy (FT-IR), and ^{13}C-nuclear magnetic resonance (NMR) spectroscopy.

495576 Simulation Study of Capturing CO_2 from Syngas after Water Gas Shift Reaction By Pressure Swing Adsorption

Abstract:

Abstract:

Since global warming is getting worse, the United Nations signed the Paris Agreement in April 2016 to control the rate of temperature rise. The agreement including restricting the increase of temperature within 1.5~2 C, setting a fund for climate change and balancing CO
agreement, countries show their determination to solve global warming.

There are four major technologies to capture carbon dioxide: chemical absorption, cryogenic separation, membrane separation and adsorption. Among these technologies, PSA has advantages of low cost and easy to operate. PSA can separate different gas mixture by various adsorbents, which makes it very competitive.

This simulation research studies the separation and concentration of carbon dioxide from syngas after water gas shift reaction (41.4% carbon dioxide, 1.3% carbon monoxide and 57.3% hydrogen after water removal) by pressure swing adsorption (PSA) process. The UOP 13X zeolite is used as adsorbent. Langmuir-Freundlich isotherm equation is used to regress the isotherm data to obtain the parameters of CO_2, N_2, CO and H_2. The mass transfer coefficient of linear driving force (LDF) model was calculated by theory. In order to test the reliability of the simulation program, we compare the results of simulation program with breakthrough curve experimental data (Figure 1), desorption curve experimental data (Figure 2) and the results of single-bed four-step PSA process experiments. The verification shows that the simulation program can be trusted.

Finally, a dual-bed six-step PSA process (Figure 3) with beds of inside diameter 2.32 cm is employed for syngas feed to find the optimal operating conditions and for temperature from 373 K to 413 K because syngas is usually at high temperature. The operating variables such as feed pressure, adsorption time/vacuum time, adsorption time/cocurrent depressurization time, pressure equalization time, bed length, vacuum pressure and feed/surrounding temperature are discussed.

The results show that when feed pressure increases, total amount of feed gas increases and so does the amount of gas adsorbed, which makes CO_2 purity increase, but makes recovery decrease because of more gas exhaust from top stream. As adsorption time/vacuum time increase, adsorber adsorbs more CO_2 and is regenerated better which make CO_2 purity increase and slight influence on CO_2 recovery. While adsorption time/cocurrent depressurization time increase, more weak adsorptive gas exits from top stream, which makes CO_2 purity increase but recovery decrease. There is only slight influence on CO_2 purity and recovery while pressure equalization time increases. As bed length increases, more gas is adsorbed and less gas exits at cocurrent depressurization step, which makes CO_2 purity decrease

but recovery increase. When vacuum pressure increases, CO_2 purity increases because more CO exits at cocurrent depressurization step, but recovery decreases because CO_2 also exits at cocurrent depressurization step. The last variable is feed and surrounding temperature. While temperature increases, less CO_2 adsorb, which makes CO_2 purity decrease but CO_2 recovery increase slightly.

After discussing the operation variables, the optimal operating conditions are feed pressure 3.95 atm, bed length 110 cm, vacuum pressure 0.05 atm, adsorption time 160 s, cocurrent depressurization time 10 s, vacuum time 130 s, pressure equalization time 5 s and feed temperature 393 K. The results of final conditions are 95.3% purity and 97.6% recovery of CO_2 in bottom product (Figure 4) and, at the same time, hydrogen with 97.6% purity and 98.3% recovery is obtained.

Figure 1. Breakthrough curve of CO_2 of gas mixture.

Figure 2. Desorption curve of CO_2.

Figure 3. Schematic diagram of dual-bed six-step PSA process.

Figure 4. Schematic diagram of optimal results for CO_2 capture from syngas feed.

500677 Effect of an Equalization Step on the Minimum Bed Size Factor of a Rapid Pressure Swing Adsorption Process

Abstract:

Rapid pressure swing adsorption (RPSA) is an advancement of PSA technology for the purpose of process size reduction. This is advantageous for small scale processes such as portable oxygen concentrators where the main objective is a lightweight device. The productivity of RPSA processes is typically represented by the bed size factor (BSF), which is inversely related to productivity. Significant BSF reduction is possible with small diameter particles and fast cycling. However, at some point, it is thought that there is a limit to the possible BSF reduction. It is still relatively unknown what causes a minimum BSF and if it the same for all types of cycles.

The main objective of this study was to determine what effect operating with a pressure equalization step has on the minimum BSF. An equalization step is often employed to reduce the power requirement of a RPSA process. The process in this study used a commercial LiLSX zeolite (~0.5 mm in diameter) to compare results of traditional 4-step feed pressurization cycles with cycles operating with a product end pressure equalization step. The results from the 4-step cycles demonstrate a definite minimum in BSF at a cycle time ~ 4 seconds for a product oxygen purity ~ 96% (argon free air feed). However, the equalization step cycles were able to go to even lower cycle times without a measurable minimum BSF. Possible reasons for why this is occurring will be discussed.

497759 Reactive Fibrous Materials for the Sorption and Self-Decontamination of Chemical Threats

Abstract:

Modification of surfaces with nucleophilic polymers enables the creation of self-decontaminating materials (SDM), granting them powerful reactive and catalytic properties. We present a series of heterogeneously immobilized copolymers for targeting a variety of chemical threats, especially chemical warfare agents (CWA), and show both chemical characterization and dynamic vapor sorption results.[1] Polymers with nucleophilic pyrrolidinopyridine (Pyr) and primary amino (vinylamine, VAm) groups were prepared by free-radical copolymerization of *N*,*N*-diallylpyridin-4-amine (DAAP) and *N*-vinylformamide (NVF), resulting in copolymers of poly(DAAP-*co*-VAm-*co*-NVF) which were then immobilized fibers by spray-coating and covalent treatment. Rayon fibers modified by poly(DAAP-*co*-VAm-*co*-NVF) were used for the sorption of dimethyl methylphosphonate (DMMP), in the presence of moisture using dynamic vapor sorption technique. These fibers modified with poly(DAAP-*co*-VAm-*co*-NVF) were found to facilitate hydrolysis of the sorbed DMMP in the presence of moisture. These were further tested using a CWA simulant diisopropylfluorophosphate (DFP) in aqueous media at pH 8.7, and were shown to facilitate hydrolysis of DFP with an order-of-magnitude faster kinetics. The deposition method and catalytic properties are expected to be of great potential for implementation in various industries, and to provide a novel route for the creation of reactive-sorption materials.

References

(1) Bromberg, L.; Su, X.; Martis, V.; Zhang, Y. F.; Hatton, T. A. *ACS Applied Materials & Interfaces* **2016**, *8*, 17555.

Low-pressure Performance Evaluation of CO_2, H_2O and CH_4 on Li-LSX as a Superior Sorbent for Air-Prepurification

Franklin E. Epiepang & Ralph T. Yang
Department of Chemical Engineering, University of Michigan, Ann Arbor, MI 48109

Air Pre-purification & Separation

High-volume production of O_2 and N_2 from atmospheric air is generally accomplished by cryogenic air distillation and prior to feeding the air into the separation unit, the ambient air must be pre-purified.

Pre-purification involves the removal of trace impurities such as CO_2, H_2O vapor, and light hydrocarbons (CH_4, C_2-hydrocarbons) to their tolerable limits of 1.0 ppm, 0.1 ppm and a few ppb respectively.

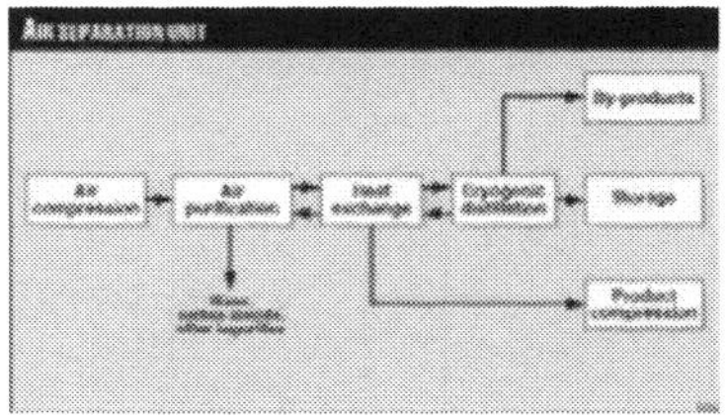

Fig. 1. Schematics of an Air Separation process.

Failure to remove these impurities will lead to not just operating issues but also serious safety hazards.

Research Theory

Adsorptive processes (Pressure Swing Adsorption & Temperature Swing Adsorption) are ideally suited for air pre-purification and can be achieved via fixed bed adsorption.

The most widely used sorbents includes activated alumina, silica gel, and 13X (Na-X) zeolite.

In this study, the adsorption characteristics of the synthetic zeolite Li-LSX (where LSX denotes Low-Silica type-X zeolite with Si/Al = 1.0) is evaluated for the removal of the aforementioned trace impurities and compared with other ion exchanged synthetic zeolites (Ca-LSX and K-LSX) and 13X. Besides the low pressure performance evaluation, other highlights of this work includes:

- Heats of adsorption.
- Rates of adsorption.
- Theoretical adsorbate-adsorbent potentials of studied gas/sorbent pairs.
- Fitting models.

Theoretical sorbate-sorbent potentials

$$\phi_T = \phi_D + \phi_R + \phi_{Ind} + \phi_{F\mu} + \phi_{FQ}$$

$$\phi_D = -\frac{A}{r^6} \qquad A = \frac{6mc^2\alpha_i\alpha_j}{(\alpha_i/\chi_i) + (\alpha_j/\chi_j)}$$

$$\phi_R = \frac{B}{r^{12}} \qquad B = \frac{Ar_0^6}{2}$$

$$\phi_{Ind} = -\frac{\alpha q^2}{2r^4(4\pi\varepsilon_0)^2} \qquad \phi_{F\mu} = -\frac{q\mu\cos\theta}{r^2(4\pi\varepsilon_0)} \qquad \phi_{FQ} = -\frac{Qq(3\cos^2-1)}{4r^3(4\pi\varepsilon_0)}$$

The total sorbate-sorbent interaction energy (ϕ_T) for physisorption was estimated by calculating the potential energies between each gas molecule and a free or isolated cation, e.g., Li^+, Na^+, etc.

Fitting Models

$$Q = \frac{Q_{sat}^* KC^n}{1 + KC^n} \qquad Q = \frac{aP}{(d + P^k)^{\frac{1}{k}}} \qquad n = n_o \exp\left[-\left(C \ln\frac{P_s}{P}\right)^m\right]$$

Langmuir-Freundlich (L-F) — Toth Model — Dubinin-Astakhov (D-A)

Cation Site Distribution

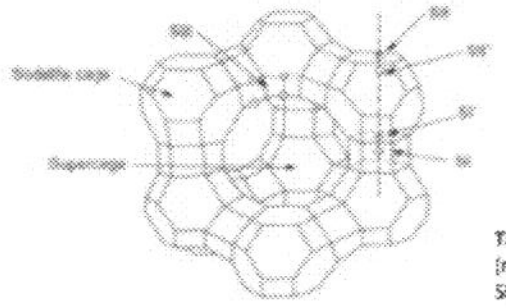

Fig. 2. Unit cell of faujasite-type (X and Y) zeolites, including cation sites.

Table 1. Cation site occupancies in dehydrated X zeolites (maximum 96 monovalent cations per unit cell for Si/Al=1).

Zeolite	Al/Unit Cell	Sites I	I'	II	II'	III
Li-LSX[a]	95.8		27.2	33.9		32.4
Li-LSX[b]	96		33	34		29
Na-X (13X)[c]	81	3.8	32.3	30.8		7.9
K-X[d]	87	9	13	26		38
Ca-X[e]	86	7.5	17.3	17.3	9.0	

a Hutson and Yang (2000). b Feuerstein et al.(1998). c Mortier (1982). d Hseu (1972). e Mortier et al.(1972).

Pure Component Isotherms

The degas/dehydration conditions: 375 °C for Li-LSX, 375 °C for 13X, and 350 °C for K-LSX and Ca-LSX. All samples were treated in-situ the Micromeritics ASAP 2020 overnight to a final pressure of 25 μmHg before each analysis.

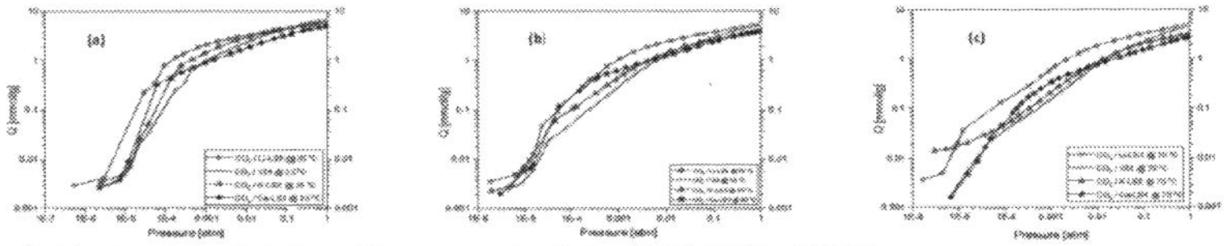

Fig. 3. Experimental adsorption isotherms of CO_2 on representative sorbents at (a) 25 °C, (b) 50 °C, and (c) 70 °C.

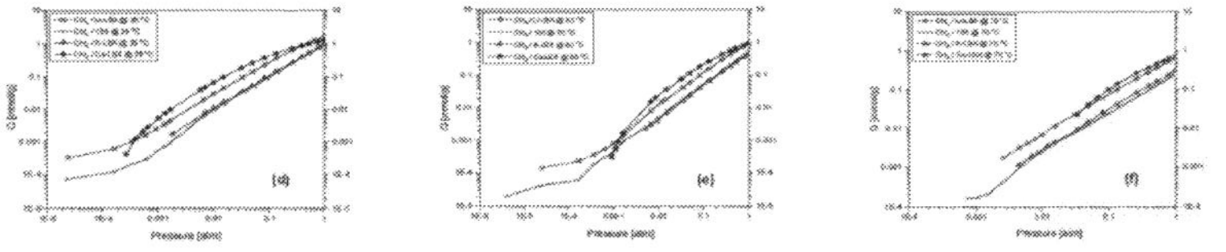

Fig. 4. Experimental adsorption isotherms of CH_4 on representative sorbents at (d) 25 °C, (e) 50 °C, and (f) 70 °C.

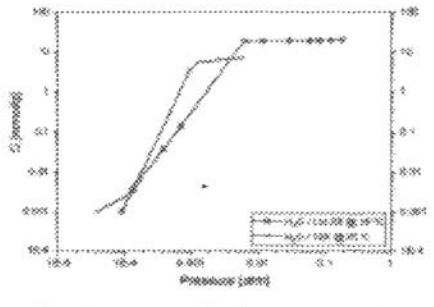

Fig. 5. Experimental adsorption isotherms of H_2O at 25 °C on Li-LSX and 13X.

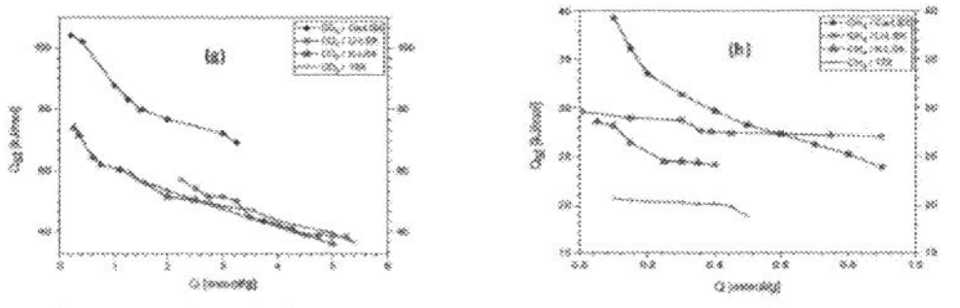

Fig. 6. Isosteric heats of adsorption on representative adsorbents. (g) CO_2 and (h) CH_4.

Acknowledgement: Jianlong Chemical – China

Prepared and Presented By:
Franklin E. Epiepang – epiepang@umich.edu

492385 Dmof-1 As a Representative MOF for SO_2 Adsorption in Both Humid and Dry Conditions

Abstract:

Acid gases are pervasive in energy applications ranging from CO_2 capture from flue gas to solid oxide fuel cells powered by syngas. MOFs are potentially useful adsorbents in many of these applications, but the behavior of MOFs in the presence of acid gases is relatively unstudied. DMOF is known for exhibiting partial water stability, and functionalization of its BDC ligand with methyl groups has resulted in increased stability towards water vapor. In addition, it has been shown that varying the metal node of the parent DMOF material results in different mechanisms of structural degradation in the presence of water. The variety of both metal substitution and ligand substitutions available to DMOF provides an interesting set of isostructural materials for assessing acid gas interactions and degradation in MOFs. In this work, DMOF has been successfully synthesized via ligand: 2,3,5,6-tetramethylterephthalic acid (TM), 9,10-anthracenedicarboxylic acid (ADC), 1,4-naphthalenedicarboxylic acid (NDC), 2,5-dimethyl terephthalic acid (DM); and metal (Zn, Cu, Ni, Co) substitutions to develop structure-property relationships for adsorption of water and SO_2, an acid gas commonly found in flue gas streams at ppm-level concentrations. The substitution of bulky ligands, TM and ADC, resulted in not only water stability, but also stability towards dry SO_2, whereas substitution with NDC and DM did not provide water or SO_2 stability. ADC-functionalized Zn-DMOF exhibited the greatest stability under humid SO_2 exposure of the ligand-functionalized materials that were tested. Metal substitution of Cu into the DMOF-TM structure resulted in the greatest stability of the M-DMOF-TM samples upon exposure to humid SO_2, and Ni provided the second most stable material. These results follow the prediction of the Irving-Williams series of metal node / ligand bond strengths.

489618 Preparation and Its Selective Adsorption Property of Asphalt-Based Carbon Materials for Effictive Separation of Light Hydrocarbons Methane/Ethane/Propane

Abstract:

Preparation and its Selective Adsorption property of Asphalt-based Carbon Materials for Effictive Separation of Light hydrocarbons Methane/Ethane/Propane

Wanwen Liang[1], Huiyu Xiao[1], Daofei Lv[1], Jing Xiao[2**], Qibin Xia[1], Zhong Li[1*]

1 School of Chemistry and Chemical Engineering, South China University of Technology, Guangzhou 510640, PR China

2 The Key Laboratory of Enhanced Heat Transfer and Energy Conversation Ministry of Education, South China University of Technology, Guangzhou 510640, PR China

ABSTRACT

We used asphalt as carbon source to prepare a series of novel asphalt"Cbased activated carbons (A-ACs) for efficient separation of light hydrocarbons in pectrochemistry industries, especially recovery C_2H_6 and C_3H_8 from Natural gas. A-ACs were prepared by an one-step preparation method and then characterized. The adsorption isotherms of C_3H_8, C_2H_6 and CH_4 on A-ACs were measured at different temperatures, and the dynamic separation performance of the samples were investigated by fixed bed experiments. The results showed that (a) the BET surface area of the resulting A-ACs reached as high as 2928-3131 m^2/g with pore volumes of 1.60-2.10 cm^3/g, and meanwhile the surface O and N contents of A-ACs gradually decreased with the KOH/asphalt ratio increasing at which A-ACs were prepared; (b) the sample prepared at 1073 K and the KOH/asphalt ratio = 4 (A-AC-4) exhibited a superior C_3H_8 and C_2H_6 adsorption capacity of 11.76 and 6.59 mmol/g at 298 K and 100 kPa, respectively; (c) the IAST-predicted C_2H_6/CH_4 selectivity of the samples A-AC-3 and A-AC-4 reached 16.9 and 15.1, respectively, at 298 K and 100 kPa, and meanwhile the IAST-predicted C_3H_8/CH_4 selectivity of the samples A-AC-3 and A-AC-4 reached as high as 76.6 and 88.8, respectively at 298 K and 100 kPa; (d) $C_3H_8/C_2H_6/CH_4$ (0.05/0.1/0.85) mixture can be well separated in the fixed bed packed with the sample A-AC-3 or the sample A-AC-4.

Keywords: Asphalt, Porous carbons, light hydrocarons, Adsorption separations, Isotherms, Selectivity

* Corresponding author. Tel: +86 20 87110608 Fax: +86 20 87110608 E-mail: cezhli@scut.edu.cn (Z. Li).

** Corresponding author Tel: +86 20 87113513 Fax: +86 20 87113513 E-mail: cejxiao@scut.edu.cn(J. Xiao);

Study on the adsorption behavior of activated red mud for Cr (VI) ions from aqueous solution

*Lei Wang, Pengjie Gao, Shuqin Liang, Dan Zhang, Suyu Jiang, Huiping Li**

School of Chemical Engineering and Energy, Zhengzhou University, Zhengzhou, Henan 450001, China

**Corresponding author phone: +86 0371 67781807; Fax: +86 0371 67781807;*

E-mail: huipingli@zzu.edu.cn (H.Li.)

Abstract

The red mud (RM), a kind of residue from alumina industry, was activated with different methods, followed by adsorption experiments for optimal conditions. The results indicated that the adsorption effect of RM modified by $FeCl_3$ solution was most efficient. The removal efficiency of Cr (VI) was 99.6 % under the optimal conditions with 20 mg/L initial concentration of Cr (VI), Fe-RM concentration 8.0 g/L, pH = 5, 180 min and 25 ℃. The adsorption capacity decreased with the temperature increasing from 25 to 45 ℃. The sorption behavior of Fe-RM-Cr (VI) system was consistent with the Freundlich model, which indicated that adsorption process existed in multi-molecular physical adsorption. The thermodynamic parameters showed that the sorption process of Cr (VI) onto Fe-RM was feasible, spontaneous and exothermic in nature. The pseudo-second-order kinetic model provided the best correlation with the experimental data compared to other models. It was concluded that the Fe-RM can be used as effective sorbent for treating the Cr (VI) ions from industrial wastewaters.

Keywords: Modified red mud; Chromium; Adsorption; Thermodynamic property; Kinetic

1. Introduction

Industrial effluents containing chromium compounds, from metal plating, textile dyeing, leather tanning, pigments and mining industries, pose a very severe threat to eco-environment and human health due to its high toxicity, carcinogenicity and oxidation[1]. Some treatment technologies such as adsorption[2], biological treatment, ion exchange[3], membrane filtration, and reduction-precipitation etc., have been applied for the removal of chromium ions from wastewaters.

Red mud (RM) is a fine-textured solid waste generated during alumina production by the Bayer process, sintering process and Bayer-sintering process from bauxite. Several studies have reported that the activated red mud has also demonstrated strong adsorption capability of different pollutants from industrial effluents[4].

The effect of different parameters such as adsorbent concentration, initial pH of solution and contact time on the removal efficiency of Cr (VI) were investigated by batch equilibrium experiments in this work. Several models of isotherms and sorption kinetics were chose to fit the experimental equilibrium data.

2. Experimental sections

(1) Instruments and materials

723 visible light spectrophotometer (Shanghai Analytical Instrument Co., Ltd.), AB204-N electronic balance (Mettler Toledo instruments Ltd.), PHS-3C pH meter (Hangzhou Orion Instrument Co. Ltd.), DF-101S collector type magnetic stirrer (Henan Yuhua Instrument Co. Ltd.), XL-1 muffle roaster (Xinghai Instruments Co. Ltd), ZKXFB-1 electrothermal vacuum drying box (Shanghai Establish Instrument and Meter Co. Ltd.), 80-1 centrifuge (Jiangsu Guohua Instrument Plant), Shaking incubator (Changzhou Guohua Electrical Appliance Co. Ltd.)

RM samples, from the Henan Branch of Aluminum Corporation (China), have the average particle diameter between 18.75 μm and 40.72 μm (Fig.1). Other chemicals used are all commercially available.

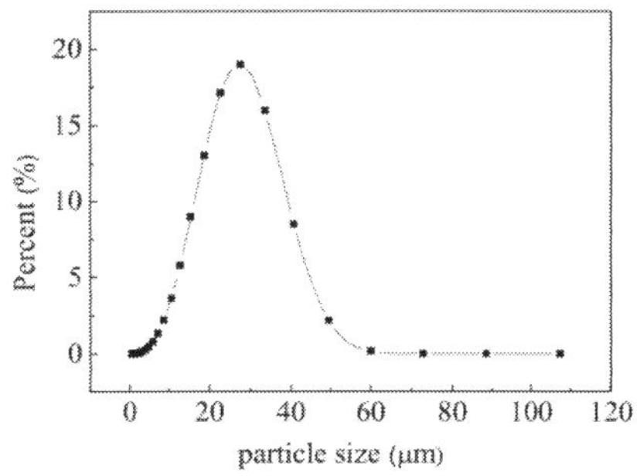

Fig.1 Particle size distribution of red mud

(2) Adsorption experiments

RM was modified by three kind of methods, thermal, HCl soaking and $FeCl_3$ solution modification and its sorption properties in the removal of Cr (VI) from solution were examined. Then the most effective method would be selected for the single factor experiments examination such as adsorbent concentration, initial pH value of solution, contact time and sorption temperature.

3. Results and discussion

3.1 Adsorption behavior of red mud

The removal efficiency η and uptake capacity q_e of Cr (VI) were calculated as follows:

$$\eta = \frac{(c_o - c_e)}{c_o} \times 100\% \qquad q_e = \frac{(c_o - c_e)\ V}{m}$$

where c_o and c_e (mg/L) are the initial and equilibrium concentrations of Cr (VI) respectively, m (g) is the mass of dry adsorbent used, and V (L) is the volume of solution, q_e (mg/g) is the equilibrium uptake capacity of Fe-RM, η is the removal efficiency of Cr (VI).

The effect of calcinations temperature of RM, concentration of hydrochloric acid and $FeCl_3$ on the efficiency η was examined and the results were shown in Fig.2 (a), (b) and (c). It can be seen from Fig. 2 (a) that there was a maximum η value at 600 ℃ and then a sharp decrease with the temperature exceed to 600 ℃. From Fig. 2 (b), the highest η was corresponding to the 3.0 mol/L of concentration of hydrochloric acid. It can be also observed from Fig.2 (c) that the η decreased and the q_e increased with the initial Cr (VI) concentration increasing from 20 to 120 mg/L. The subsequent adsorption experiments were carried out using red mud activated with a 0.3 mol/L$FeCl_3$ solution.

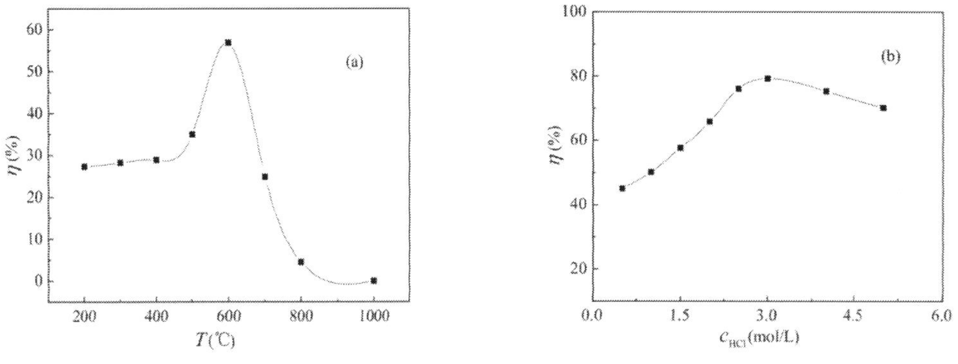

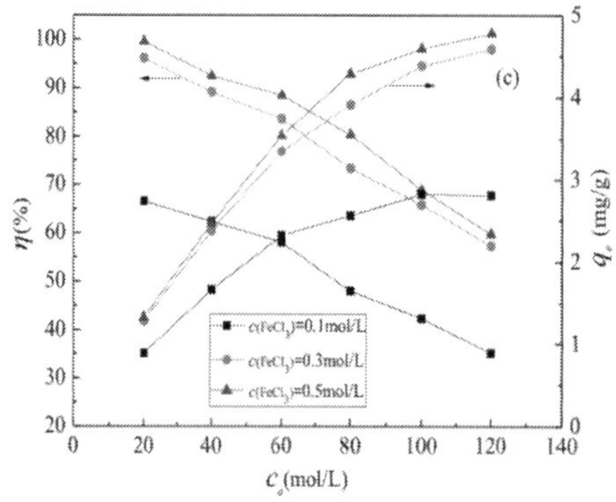

Fig. 2 Effect of calcination temperature (a), concentration of HCl (b) and $FeCl_3$ (c) on the η and q_e

The effect of Fe-RM concentration (a), pH value (b) and contact time (c) on the removal efficiency η and uptake capacity q_e were investigated by single factor experiments at room temperature and the results were showed in Fig.3 (a), (b), (c).

From Fig. 3 (a), it can be seen that the η increased from 23.4 to 69.5 % when the Fe-RM concentration increased from 1 to 8 g/L. Then the efficiency did not increase significantly when the concentration of Fe-RM was exceeded to 8 g/L. It was also observed from Fig. 3 (b) that there was a maximum η at pH = 5 and a sharp decrease when the pH exceeded to 5. According to the Fig.3 (c), the η increased from 24.6 % to 68.7 % within 50 min, as the contact time was extended to 180 min, little change of η indicating the sorption equilibrium state was achieved.

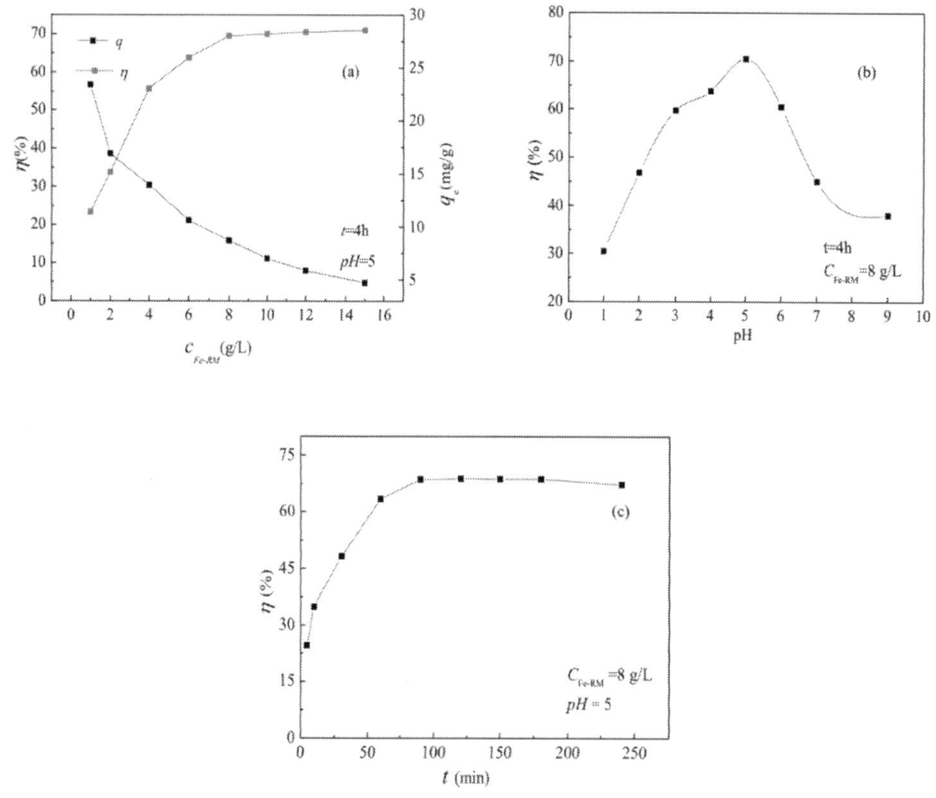

Fig. 3 Effect of sorbent concentration (a), pH (b) and contact time (c) on the removal efficiency

The effect of adsorption temperature on the sorption capacity were also examined under the conditions of

Fe-RM concentration 8 g/L, the initial Cr (VI) concentration 100 mg/L, pH = 5, and contact time 4 h. The experimental results were showed in Fig. 4.

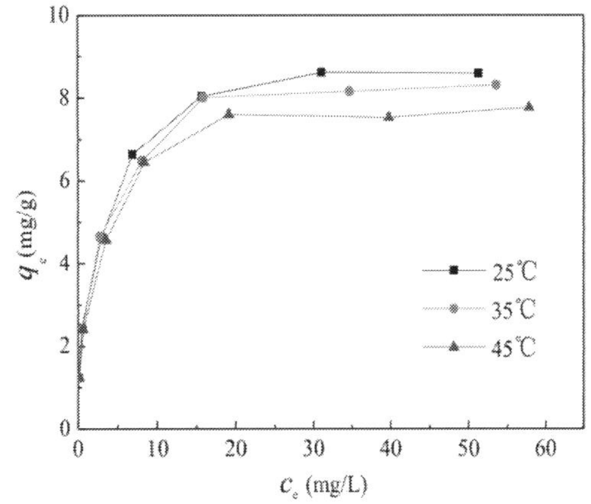

Fig.4 Effect of adsorption temperature on the sorption capacity q_e of Fe-RM

It was found that there is a negative effect of sorption temperature on the uptake capacity q_e. The optimum temperature was selected as 25 ℃ for further sorption experiments.

3.2 Adsorption thermodynamics

Thermodynamic behavior of Cr (VI) sorption onto Fe-RM was investigated using the thermodynamic parameter[5] (see Table1).

Tab.1 Thermodynamic parameters

T(K)	K_D (L/g)	ΔG (kJ/mol)	ΔH (kJ/mol)	ΔS (J/(mol • K))
298	3.177	-2.867		
308	2.897	-2.722	-7.182	-14.473
318	2.648	-2.577		

The thermodynamic parameters (ΔG、ΔH、ΔS) in above table showed that the sorption process of Cr (VI) onto Fe-RM was feasible, spontaneous and exothermic in nature. The K_D is the distribution coefficient.

There are several frequently used models like Langmuir, Freundlich, and Tempkin[5] isotherm models for analyzing the experimental data and the results were listed in Table 2.

Tab.2 Isothermal adsorption parameters

Model	Equation parameters		R^2
Langmuir	K_L=5.311 L/mg;	q_m=5.81 mg/g	0.8738
Freundlich	K_F=14.319mg$^{0.701}$L$^{0.299}$g^{-1};	$1/n$= 0.299	0.9720
Tempkin	K_T=2.7610^3 L/mg;	B=1.197	0.9453

3.3 Adsorption kinetics

Several well-known kinetic models, Lagergren[6], Morris-Weber , Elovich and pseudo-second-order models have been applied to fit these experimental data obtained for sorption of Cr (VI) onto Fe-RM (Table 5).

Tab.3 Kinetic models parameters

Model	Equations	Parameter	R^2
Lagergren	y = -0.0126x + 0.5959	k_1=0.0291 (min^{-1})	0.7223
pseudo-second-order	y = 0.1093x +1.2286	k_2=0.0097 (g/(mg·min^{-1}))	0.9983
Elovich	y = 1.5820x + 0.7608	α=2.5590, β=0.6321	0.9486

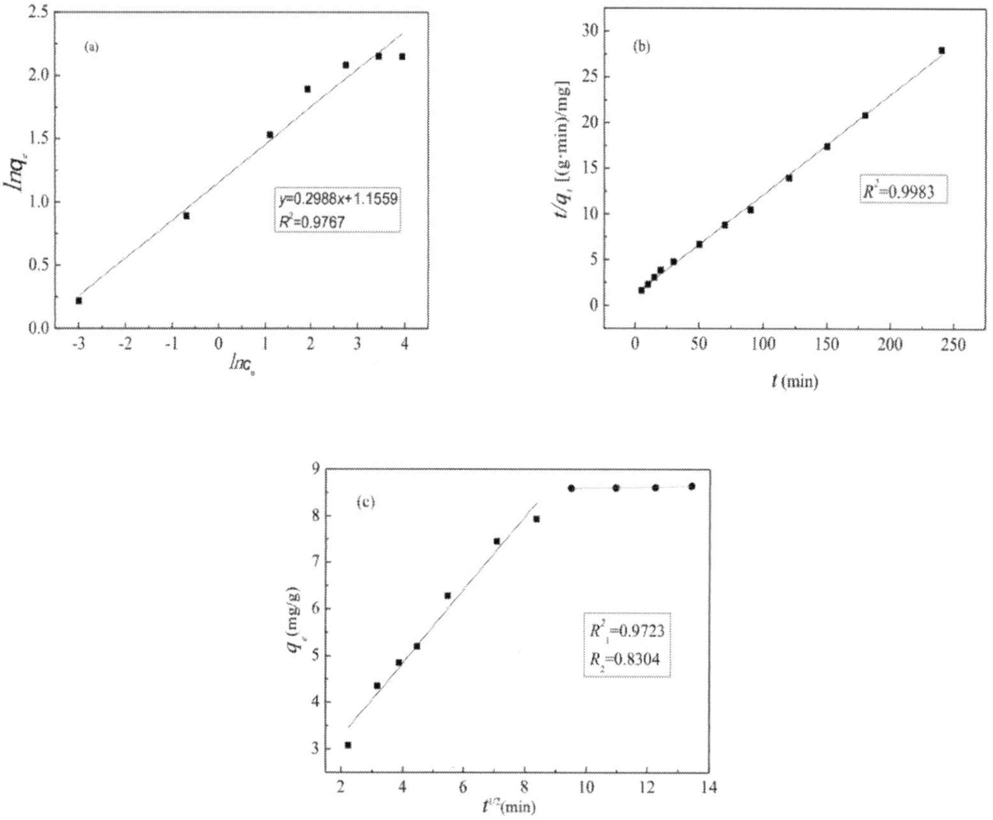

Fig. 5 Freundlich isotherm (a), pseudo-second-order (b) and intra-particle diffusion kinetic plot (c)

From Fig. 5 (a), the adsorption behavior of Fe-RM-Cr (VI) system was consistent with the Freundlich model and Tab.2), which indicated that adsorption process existed in multi-molecular physical adsorption. The correlation coefficient for the pseudo-second-order model was 0.9983 (Fig. 5 (b) and Tab. 3), which showed that the kinetic model provided a best correlation for sorption of Cr (VI) onto Fe-RM in contrast to other models. It also revealed that the sorption process happened through surface sorption and intra-particle diffusion (Fig. 5 (c)). As seen from Tab. 2 and Tab. 3, there is chemical sorption from Tempkin and Elovich model. Thus, the sorption process for this system was completed through physisorption and chemisorption.

4. Conclusions

The red mud from alumina industry was activated via calcination, HCl soaking and $FeCl_3$ modification, respectively in this study. The $FeCl_3$ modification method was the more efficient than the other two. The maximum removal efficiency of Cr (VI) was 99.6% at the optimum conditions of pH = 5, contact time 180 min, sorbent concentration 8 g/L, and initial Cr (VI) concentration 20 mg/L. The calculated thermodynamic parameters (ΔH, ΔG, ΔS) showed the sorption to be feasibility, exothermic and spontaneous at the solid-liquid interface. The experimental data were correlated by Langmuir, Freundlich and Tempkin isotherm models, and it was proved that the Freundlich isotherm model offered the better correlation results than other two models. The pseudo-second-order model provided the good description of kinetic behavior. The sorption process was completed through physisorption and chemisorption. Based on all above results, Fe-RM can be used as an efficient and alternative sorbent for the treatment of wastewaters containing Cr (VI) ions due to the advantages of being natural, low-cost and high sorption capacity.

5. References

1 D. Mohan, C.U. Pittman, Activated carbons and low cost adsorbents for remediation of tri- and hexavalent chromium from water, J. Hazard. Mater., 137 (2006) 762-811.

2 X.W. Wu, H.W. Ma, Y.R. Zhang, Adsorption of chromium(VI) from aqueous solution by a mesoporous aluminosilicate synthesized from microcline, Appl. Clay Sci., 48 (2010) 538-541.

3 E. Pehlivan, S. Cetin, Sorption of Cr(VI) ions on two Lewatit-anion exchange resins and their quantitative determination using UV-visible spectrophotometer, J. Hazard. Mater., 163 (2009) 448-453.

4 H.S. Altundogan, S. Altundogan, F. Tumen, M. Bildik, Arsenic adsorption from aqueous solutions by activated red mud, Waste Manage, 22 (2002) 357-363.

5 H. Javadian, Application of kinetic, isotherm and thermodynamic models for the adsorption of Co(II) ions on polyaniline/polypyrrole copolymer nanofibers from aqueous solution, J. Ind. Eng. Chem., 20 (2014) 4233-4241.

6 S.S. Baral, S.N. Das, P. Rath, Hexavalent chromium removal from aqueous solution by adsorption on treated sawdust, Biochem. Eng. J., 31 (2006) 216-222.

493649 Ultrasound Assisted Synthesis of Zirconium Impregnated Activated Carbon Nanocomposite and Its Effective Use for Defluorination of Water

Abstract:

Abstract

The scarcity for potable water over a wide region of Rajasthan, Gujarat and Andhra Pradesh in India and some other countries like China, United States and Mexico is quite prominent owing to excessive fluoride contamination in the ground water and surface water of those regions [1]. Fluoride is a necessary element of human body helping to construct healthy teeth and calcification of the dental enamel among infants, inhibiting dental caries when present within a concentration range of 0.5-1 mg/L. But long exposure and intake of water with high fluoride concentration leads to physiological, chronic and toxic effects. The permissible limit of fluoride in drinking water ranges from 0.5–1.5 mg/L as per the norms of World Health Organization (WHO) [2]. The well known water defluorination techniques include precipitation, membrane separations, and adsorption processes. Precipitation is the most frequently used technique for fluoride decontamination, but it has several drawbacks such as formation of fluoride precipitates as well as waste disposal issues. Comparatively, membrane separation is a smarter and cleaner way of fluoride decontamination by the use of ion exchange, reverse osmosis or electrodialysis. On the other hand, adsorption is a method that has been under the limelight for several decades as one of the most popular and economically sound technologies having dynamic applications in pollution prevention and water decontaminations. In this regard, research attention has been given to both utilizing bioresources as the raw materials for synthesizing novel adsorbents very inexpensively. Furthermore research effort is underway to improve the efficiency of the commercially available adsorbents by physical or chemical functionalization. Present study reports the feasibility of fluoride (F^-) adsorption by use of zirconium (Zr(VI)) impregnated powdered activated carbon (PAC) nanocomposite synthesized through the assistance of ultrasound. Because of high electronegativity, fluoride has a strong affinity for rare earth metals such as zirconium, lanthanum and impregnating the zirconium metal ions on the activated carbon multiplies the fluoride adsorption efficiency of activated carbon by 3 to 5 times [3]. During synthesis, ultrasound was found to have significant impact on reducing the particle size with high specific surface area as well as substantial reduction in energy consumption. The nanocomposite was characterized by using scanning electron microscopy (SEM), energy dispersive X-ray (EDX), X-ray diffraction (XRD), N_2 adsorption analysis (BET) and Fourier Transform Infrared Spectroscopy (FTIR) to study the physical and chemical properties. The pH_{pzc} (Point of zero charge) of the adsorbent was found to be 5.03; hence all experiments were carried out at an optimum pH of 4 for adsorption of strong electronegative fluoride ions with an optimum adsorbent dosage of 2 g/L. The initial fluoride concentration was varied from 2.5 to 20 mg/L and the adsorption capacity was obtained upto 5 mg/g with equilibrium reaching within 3 hours. The adsorption kinetics was found to follow pseudo-second order kinetics and the isotherm to follow Langmuir isotherm model. The thermodynamic study showed the endothermic and spontaneous nature of the process within the studied temperature range. The effects of various co-existing ions (HCO_3^-, NO_3^-, SO_4^{2-}, Cl^-) on the process performance were also studied and found to have no significant effects on fluoride adsorption capacity. The adsorbent maintained high adsorption efficiency after the regeneration studies carried out for five adsorption-desorption cycles which signifies its potential to be a good defluorinating agent.

Keywords: Ultrasonication; Fluoride; Nanocomposite

Figure 1: Effect of solution pH on fluoride removal (F^-: 10 mg/L)

Figure 2: Effect of adsorbent dosage on fluoride adsorption (F^-: 10 mg/L)

Figure 3: Effect of coexisting ions on fluoride adsorption

(F^-: 5 mg/L, HCO_3^-, NO_3^-, SO_4^{2-}, Cl^- :10 mg/L; pH 4)

Figure 4: Fluoride removal after regeneration of the adsorbent (F^-: 10 mg/L)

Reference:

[1] P. Miretzky, A.F. Cirelli, Fluoride removal from water by chitosan derivatives and composites: a review, Journal of Fluorine Chemistry, 132(4) (2011) 231–240.

[2] X. Yu, S. Tong, M. Ge, J. Zuo, Removal of fluoride from drinking water by cellulose@ hydroxyapatite nanocomposites, Carbohydrate polymers, 92(1) (2013) 269-275.

[3] R.S. Sathish, N.S.R. Raju, G.S. Raju, G. Nageswara Rao, K.A. Kumar, C. Janardhana, Equilibrium and kinetic studies for fluoride adsorption from water on zirconium impregnated coconut shell carbon, Separation Science and Technology, 42(4) (2007) 769-788.

496308 Lithium Selective 14-Crown-4 Ethers: Synthesis, Polymerization and its Application for the Recovery of Lithium from Dilute Solutions

Abstract:

The demand to recover lithium (Li^+) from alternative aqueous resources that contain diluted Li^+ with abundant competing alkali metal ions (Na^+, K^+ and Cs^+) has renewed research interests to develop highly selective Li^+ sequestrants. Crown ethers (CEs) have flexible structures whose cavity sizes or metal ion affinities can be tailored to impart selectivity towards Li^+. Among the CEs, 12-14 membered CE rings are known to form stable Li^+ complexes in both organic and aqueous solutions, in the presence of other alkali metal ions. Specifically, dibenzo-14-crown-4 ether (DB14C4) and its derivatives are known as Li^+ complexants due to their ideal cavity dimensions (1.2-1.52 Å for 4-6 coordination number with Li^+). Despite this ideal cavity size-match relationship, complexation of DB14C4 and its derivatives with larger M^+ remains a challenge. The sandwich type 2:1 crown ether-metal ion (CE-M^+) complex is still remarkable in currently known 14-membered CE derivatives. This problem limits their application as Li^+selective chelates in separation process.

A new type of lithium selective CEs containing rigid and bulky moieties were developed having both rigid and bulky subunits. The rigid aromatic groups enhance the rigidity of 14-crown-4 ether cavity while the bulky subunits provide a blocking mechanism to prevent bigger metal ions to from a complex as well as preventing formation of higher ordered sandwich type complexes. All of the CEs synthesized showed high Li^+ distribution but 3d with bulky tetramethyl showed the highest Li^+ selectivity in the presence of other alkali metals. Furthermore, no D_{K+} and D_{Cs+} were analyzed for 3d. Selectivity of Î±$_{Li/Na}$=3344 for 3d is among the highest in the literature.

Since the resulting 14-membered CEs are terminated with hydroxyl groups, they can be further polymerized. Like for instance via epoxy polymerization in the presence of a porogen (PEG 400) to obtain a porous hyper-crosslinked CE-epoxy resin. In particular, the CE (3d) epoxy monomer was synthesized by reacting the dihydroxy CE with allyl bromide to give CE-diene intermediate followed by epoxidation of the terminal dienes with an oxidant m-CPBA. Epoxy polymerized CE (3d) with different linear, cyclic and aromatic diamines (1,5-diaminopentane, 4,4â€™-methylenebis(cyclohexylamine), and 4,4â€™-diaminodiphenylmethane) in the presence of PEG 400 resulted in different CE epoxy polymer resins. The CE epoxy polymer resins were characterized by TGA-DSC, BET surface area analysis to determine the glass transition temperature (Tg) and porosity, and SEM to show the morphology of the synthesized polymer resin. Batch and continuous lithium extraction experiments were done to evaluate the performance of the CE-epoxy polymer resins.

This work was supported by Basic Science Research Program through the National Research Foundation of Korea (NRF) funded by the Ministry of Science, ICT and Future Planning (2015R1A2A1A15055407) and by the Ministry of Education (No. 2009-0093816 and 2017R1D1A1B03028102).

496981 Defluoridation of Water Using Amine Functionalized Cellulose Nanofibers

Abstract:

Fluoride content in potable water that is higher than the World Health Organization (WHO) limit of 1.5 mg L^{-1}, can cause fluorosis [1]. As groundwater in several countries is affected by excess fluoride content, there is a great need for the treatment of fluoride contaminated water [2]. Several defluoridation techniques such as adsorption, ion exchange, coagulation-precipitation, and membrane process have been reported in the literature. However, adsorption technique is widely studied because of its economic viability, easy operation, low maintenance cost and efficiency of fluoride ion removal [3-5].

In this study, we have investigated the fluoride removal capacity of amine-functionalized cellulose nanofibers derived from modified electrospun cellulose acetate fibers under batch conditions. Electrospun cellulose acetate nanofibers were modified by the deacetylation process to get the cellulose fibers. Further, the surface of cellulose fibers was functionalized with amine functional groups using ethylenediamine to improve its binding capacity for fluoride ions. Thus prepared adsorbent was characterized with field emission scanning electron microscopy (FESEM), Fourier transform infrared spectrophotometer (FT-IR) and Brunauer-Emmett-Teller (BET) surface analyzer.

Figure 1: SEM images of electrospun cellulose acetate fibers at different magnifications

Batch adsorption experiments were performed as the function of contact time, the dosage of adsorbent, initial fluoride concentration, temperature, and pH. Langmuir, Freundlich, and Dubinin-Radushkevich isotherm models were applied to the equilibrium data and adsorption kinetics were tested by pseudo-first-order and pseudo-second-order kinetic models to determine the best-fit of sorption kinetics. To regenerate the adsorbent, desorption experiments were carried out for fluoride adsorbed nanofibers at various pH values by adding 0.1 M HCl or 0.1 M NaOH. Finally, the prepared adsorbent was successfully used for the field water sample collected from a nearby fluoride affected village.

References:

1. World Health Organization. Guidelines for drinking-water quality, First addendum to third addition, Vol. 1, Recommendations.-3rd ed.

2. Jagtap, S., Yenkie, M.K., Labhsetwar, N., and Rayalu, S. Fluoride in drinking water and defluoridation of water, Chem. Rev. **2012**, 112, 2454-2466.

3. Meenakshi, Maheshwari, R.C. Fluoride in drinking water and its removal, J. Hazard. Mater. **2006**, B137, 456-

463.

4. Gupta, V.K., Ali, I., Saini, V.K. Defluoridation of wastewaters using waste carbon slurry, Water Res. **2007**, 41, 3307-3316.

5. Chen, N., Zhang, Z., Feng, C., Li, M., Zhu, D., Chen, R., Sugiura, N. An excellent fluoride sorption behavior of ceramic adsorbent, J. Hazard. Mater. **2010**, 183, 460-465.

501065 Uranium Adsorption on Organophosphorus-Derivitized Extractive Scintillating Resins

Abstract:

Nuclear nonproliferation efforts and treaty verification require portable, robust radiation detectors capable of detecting trace-levels of radionuclides in environmental matrices. A recent development in environmental sensing is a portable, flow cell detector that utilizes extractive scintillating resin. Extractive scintillating resin serves the dual purpose of (1) concentrating the radionuclide of interest through adsorption and/or ion-exchange and (2) serving as a radiation transducer through scintillation. At the heart of this analytical method lies a separations challenge, as the incorporation of analyte-selectivity can enhance the sensitivity of the radiation measurement.

Uranium solution chemistry has a strong dependence on pH, ionic strength, solution composition (i.e. the presence of other ions) and the redox potential of the system. Due to its complex chemistry, there is no one material that is universally selective for uranium. Designing a uranium-selective separation material requires a careful consideration of the behavior of both ligand and uranium at a given set of conditions.

This contribution describes our efforts to synthesize extractive scintillating resins that selectively concentrate and detect uranium from near-neutral pH ground water. Organophosphorus-derivitized resins were synthesized by two approaches: (1) synthesis of a 4-vinylbenzyl chloride containing polymer followed by solid-phase synthesis techniques to add the phosphonic acid moiety and (2) one-pot polymerization that directly incorporates an alkyl phosphate moiety. Resins were compared on the basis of capacity for uranium, detection efficiency, and volume to detection. The volume to detection was evaluated in neutral pH simulated ground waters (pH 4-8) using a Shewart-3Ï*f* alarm statistic applied to data collected in real-time in a flow-cell detector. Resins exhibited similar binding capacities (0.18 mmol g^{-1}) and detection efficiencies (~40%); however, the alkyl phosphate resins more rapidly achieve the Shewart-3Ï*f* alarm criteria than the phosphonic acid resins. Simulations utilizing Density Functional Theory (DFT) are used in conjunction with thermodynamic models in Visual MINTEQ software to understand how the binding mechanisms for both functional groups change as a function of pH. The data imply that as pH increases from 4 to 8, the binding mechanism shifts from *ion-exchange* to *ligand-exchange*. Understanding the binding mechanisms of uranium with organophosphorus moieties will contribute to the rational design of future uranium-adsorbent materials for environmental systems.

499974 Evaluation of Sugar Beet Processing Lime Cake for the Removal of Synthetic Dyes from Aqueous Solutions

Abstract:

Waste management is one of the most important issues in industries. Their (re)-utilization provide additional revenue that has potential to decrease the process costs. In traditional sugar beet processing, the sugar is extracted from plants and non-sugar components are removed by precipitation. During the process, the impurities are flocculated via sequential addition of limewater (milk of lime and calcium hydroxide suspension) and carbon dioxide. The resultant precipitate contains a significant amount of calcium carbonate and also minerals, nitrogenous- and phosphoric compounds. The content of the sludge slightly varies with the technology and the raw material utilized. In the literature, the waste material has several names such as filter cake, spent lime or sugar beet processing lime waste, etc. Due to its content, the vast amount of sugar beet processing lime waste (SBPLW) is difficult to be disposed and cause severe environmental and ecological problems.

(Efficient management of the waste materials is one of the most important issues in industries. Their (re)-utilization provide additional revenue is also directly decreasing the production costs. In traditional sugar beet processing, the product is extracted from plants and non-sugar components are removed by precipitation. During the process, the impurities are flocculated via sequential addition of limewater (milk of lime and calcium hydroxide suspension) and carbon dioxide. The resultant precipitate contains a significant amount of calcium carbonate and also minerals, nitrogenous- and phosphoric compounds. The content of the sludge slightly varies with the technology and the raw material utilized. In the literature, the waste material has several names such as filter cake, spent lime or sugar beet processing lime waste, etc. Due to its content, the vast amount of sugar beet processing lime waste (SBPLW) is difficult to be disposed and cause severe environmental and ecological problems.)

Previously, several types of applications have been proposed for SBPLW. It was evaluated in desulphurization of fuel gases, animal feed preparations and soil amending. Due to the high amount of calcium carbonate (CC), SBPLW was also used to replace CC to save the natural reserves. It was tested to be used in paper mills and composite preparations. Currently, most of the sugar companies send the waste material to the other industries to be used as a filling material in cement, breeze block and foam concrete productions. However, SBPLW may be used in a relatively more critical objective prior to this low-value handling.

In this study, SBPLW was evaluated to be used in the removal of synthetic dyes from aqueous solutions. Synthetic dyes in aqueous streams are poisonous to human life as well as habitat. Before their discharge, the toxic dyes in aqueous streams and solutions should be removed. Methylene Blue (MB) was chosen as the model synthetic dyestuff in experimental studies. Firstly, the basic characterization of SBPLW was performed and then equilibrium, kinetic and thermodynamic studies were conducted. Experiments were carried out in a broad range of MB concentration. The results were compared with those obtained with commercially available active carbon.

The characterization of SBPLW was carried out by Fourier Transform Infrared Spectroscopy (FT-IR), Scanning

Electron Microscope (SEM), Energy-dispersive X-ray Spectrophotometry (EDX). The removal process reached equilibrium within 240 minutes and the kinetic parameters were calculated. The influence of initial pH of aqueous MB solution on adsorption was examined in the pH range of 3â€“10 and the maximum removal was obtained at pH 4.6 as 76.1% using an MB solution having 200 ppm. Isotherms were applied to the results to figure out which model describes better among the ones tested in this study. Consistent with the literature, adsorption efficiency decreased with the increase in MB concentration and increased with adsorbent dose. Temperature negatively affected the MB adsorption and the thermodynamic parameters were calculated for the adsorption process using SBPLW. According to the values, this adsorption process is a physical process.

497848 The Accuracy of Pore Size Distribution Obtained from Non-Local Density Functional Theory in Amorphous Microporous Materials - Polymers and Large Organic Molecules

Abstract:

Typically, amorphous microporous materials have attractive characteristics such as a large surface area and high porosity. Although the industrial utilization of these materials is promising, the accurate survey of their pore size distribution (PSD) is still a challenging task. Pore size distribution (PSD) is an important structural property that facilitates the suitable selection of these materials for specific processes. However, the accurate characterization of their PSD remains a challenging task. The experimental PSD is usually obtained indirectly through N_2 adsorption at 77 K by calculating the PSD using the non-local density functional theory (NLDFT). In this work, we investigated the accuracy of PSD by constructing N_2 isotherms using Gibbs ensemble Monte Carlo simulations. NLDFT-generated PSDs were compared against reference PSDs that were obtained by direct insertion of simulation probes with various sizes. We studied 11 different amorphous materials with surface areas between 5 - 1698 m^2/g. We concluded that the NLDFT indeed produced various artifacts in the micropore region resulting in main peaks that did not truly represent the dominant pore size. Most importantly, we propose an alternative method for choosing the smoothing parameter that can produce PSDs that are significantly more consistent with the reference distribution.

495733 Gas Sorption and Swelling in Flexible Metal-Organic Frameworks

Abstract:

Gas Sorption and Swelling in Flexible Metal-Organic Frameworks
Sahar Bakhshian and Muhammad Sahimi

Metal organic frameworks (MOFs) have recently attracted considerable attention as new nanoporous materials with applications in separation, catalysis, gas capture and storage and drug delivery. MIL-53 (Al or Cr) family as a subclass of the MOFs displays a remarkable structural phase transition, often referred to as the breathing transition, due to the adsorption of guest molecules in its morphology. In the breathing phenomenon, the structure oscillates between two distinguishable phases called the large-pore (lp) phase and the narrow-pore (np) phase.
While molecular simulations can be used to compute the sorption isotherms of gas molecules in a flexible structure, such computations are expensive and applicable to a limited number of unit cells. We have developed a model based on the energetics of the system that couples guest adsorption with the host-guest interaction and structural deformation of the MOFs. The total energy of the system â€“ the MOF plus the gas â€“ is written as the sum of three terms: The elastic energy of the material, the fluid-phase energy, and the interaction energy between the fluid (gas) and the solid matrix. By minimizing the total energy with respect to both the fluidâ€™s density and the displacement field of the solid we derive the governing equations for both. The equations are then solved by the finite-element method. Adsorption/desorption isotherms, along with the spatial distribution of the induced strain in the MIL-53 framework, are then computed. The proposed model can be applied to the structural transitions in materials, such as adsorption-induced structural deformation, at large scales, such as crystal scale. The model is completely general and may be used in studying of the adsorption-induced swelling of porous media by any type of gas.
When the model was applied to adsorption of CO2 in MIL-53 (Al) samples, all the reported experimental features of the phenomenon were reproduced. We computed the adsorption/desorption isotherms and volumetric strains various at different temperatures. At low pressures the system experiences contraction in which the structure transforms form the lp to the np phase, whereas at high pressures the system undergoes swelling and reverses the structural transformation from the np to the lp phase.

491156 Hydrogen Sulphide Absorption Performance in Various Amines Solution for Combined Desulphurization and Dehydration

Abstract:

Hydrogen sulphide absorption performance in various amines solution for combined desulphurization and dehydration

Usman Shoukat, Diego D. D. Pinto, Hanna K. Knuutila*

*Corresponding author e-mail: hanna.knuutila@ntnu.no

Department of Chemical Engineering, Norwegian University of Science and Technology (NTNU), Trondheim, Norway. INtroduction

Almost quantitative removal of hydrogen sulphide (H_2S) from natural gas is necessary prior to use or pipeline transport. The most commonly used method for H_2S removal is scavengers. However, these are usually non-regenerative substances and because of cost and handling they cannot be used for high H_2S concentrations in the gas stream. Alkanolamines are commonly used for regenerative H_2S removal processes (Kohl and Nielsen 1997). Natural gas is commonly saturated with water. The presence of water in the gas accelerates corrosion, and potentialize the formation of solid gas hydrates, which cause problems in gas flow in pipelines. One common way to avoid hydrate formation and to achieve problem free continuous gas transportation operations is to add inhibitors in gas pipelines (Kohl and Nielsen 1997). A system which can removed H_2S and control hydrate simultaneously could reduce equipment size and help to reduce the installation and operational costs for both subsea and platform operations.

The objective of this study is to identify regenerative solvents suitable for combined selective H_2S removal in high H_2S content feed gas and hydrate formation control. Several amine-based solvents will be screened in this study and the effect of solution composition, inlet H_2S concentration and temperature on H_2S loading in the solutions will be discussed. The amines chosen for this work were systematically chosen so that insight into the influence of amine alkanol groups, alkyl chain length, hydroxyl group and hindrance of the tertiary amines at H_2S absorption are given based on their molecular structure. MEthodology

In this study, various amines solutions (20 wt.%) with water, ethylene glycol (MEG) and triethylene glycol (TEG) are studied for H_2S absorption in a custom built experimental setup at temperatures relevant to subsea conditions (e.g., 5 °C). The apparatus design is based on the CO_2 screening apparatus used by (Ma'mun et al. 2007) as shown in Figure 1. As for the CO2 screening apparatus, the H2S screening is operated at atmospheric conditions. The H_2S inlet partial pressure was chosen to be 0.03 kPa and 1 kPa and the experiments were run for 120 minutes for all solutions at 5 and 40 °C. The tests were performed in amine-water and amine-glycol (MEG or TEG) solutions. ICP-MS was used for the measurement of total sulfur in the liquid phase to obtain accurate absorbed H_2S in amine solution.

Figure 1: Schematic flow diagram of screening apparatus.

Preliminary Results

H_2S absorption in various tertiary amines in presence of water, MEG and TEG was performed. The preliminary results show that the effect of MEG and TEG on the absorption of H_2S into the amine solutions is significant. For example, comparing the H_2S absorption at 5°C and 0.03 kPa pH_2S into 20wt% aqueous N-Methyl diethanolamine (MDEA) solution with glycols (MEG/TEG)-MDEA solutions shows that replacing of glycols with water in amine solutions restricts the H_2S absorption significantly as shown in Figure 2. MDEA-MEG solution restricts the H_2S absorption by 30% as compared to MDEA-H_2O solution while in case of MDEA-TEG restriction is 50% as compared to MDEA-H_2O and 27% as compared to MDEA-MEG solution.

Figure 2: Restriction in H_2S absorption in amine solutions caused by addition of glycols (MEG / TEG)

Key words: H_2S absorption, alkanol amines References

Kohl, A. L. and R. B. Nielsen (1997). Gas Purification (Fifth Edition). Houston, Gulf Professional Publishing: 40-186.

Ma'mun, S., H. F. Svendsen, K. A. Hoff and O. Juliussen (2007). "Selection of new absorbents for carbon dioxide capture." Energy Conversion and Management **48**(1): 251-258.

495039 An Investigation on Chemical Absorbents for the Effective Removal of Hydrogen Sulfide from Crude Oils

Abstract:

An Investigation on Chemical Absorbents for the Effective Removal of Hydrogen Sulfide from Crude Oils

Tracy J. Benson, Obakore Agbroko, and Karishma Piler

Sour crudes (containing > 0.5% sulfur) are of low quality and cause harm to equipment, the environment, and to humans. Sustainable measures for crude oil sweetening are of significant importance. Removing of H_2S at the wellhead before transporting via pipeline or railway increases the value of crude oil and may even be necessary to conform to legal transport laws. Chemical agents, known as scavengers, are used to scrub H_2S from crude oils. Scavengers are either of the regenerative or non-regenerative type. The type of scavenger used depends upon the size of the wellhead facility as well as the pH or water content of the crude stream. The aim of this work is directed towards smaller wellhead facilities where onsite scrubbing and stripping operations are not feasible, making non-regenerative scavengers a better option.

From a survey of the literature, non-regenerative scavengers, chiefly of the triazine type, vary by constituent side chain moieties that allow the scavenger to be more soluble in one medium versus another. Yet, little information is provided about the true solubility of these scavengers in various crude mediums. Therefore, a series of triazine-type scavengers with differing side chains have been tested in a series of solvents ranging from hydrophilic to hydrophobic with the purpose of understanding the solubility phenomenon as well as the appropriate liquid activity coefficient models for these systems. The chemical processing, including recovery of scavenger or end use of products, will be discussed for decision-making at the confluence of chemistry and chemical engineering for hydrogen sulfide removal. This work seeks how and why some scavengers perform better for some crudes and environmental conditions than others with the intent to develop designer scavengers and processing conditions for the crude oil industry.

492791 Hydrophobic Deep Eutectic Solvents: Their Discovery and Design for Separations

Abstract:

In 2003 a new innovative category of solvents was discovered, namely deep eutectic solvents (DESs).[1] DESs are stated as the next generation of solvents, overcoming most of the disadvantages of ionic liquids (ILs).[2] These advantages of DESs are a simple preparation of the solvent by mixing the two components of the DES and the ability to produce sustainable solvents by the choice of components.

The first DESs presented in literature were all produced from hydrophilic components. In 2015 a number of us reported, for the first time, the preparation of hydrophobic DESs by combining decanoic acid with quaternary ammonium salts.[3] Their hydrophobicity was tested via measuring the amount of quaternary ammonium salt that leached to the water phase in combination with the water uptake of the DES.

The application tested upon the discovery of hydrophobic DESs was their use for the removal of volatile fatty acids from water.[3]All the tested DESs show better extraction efficiency in comparison with the industrial standard trioctylamine.

Next to their use for the extraction of volatile fatty acids from water, hydrophobic DESs were applied for the removal of metal ions from water.[4] Transition metal ions could be extracted with high distribution coefficients (D), higher than 0.99, while alkali metal ions only show low to moderate D. The extraction of Co^{2+} from water was used as a model system to investigate the variation of metal concentration, the solvent-to-feed ratio and an investigation into the extraction time. The results show that with an increase of the initial concentration of Co^{2+}, the D drops to approximately 0.83. A D of 0.80 was measured when the solvent-to-feed ratio was decreased to 1:10. The results into the extraction time showed that within 5 s 99% of the Co^{2+}could be removed.

Another application tested with hydrophobic DESs for the first time, is their use for the capture of carbon dioxide (CO_2). To the best of our knowledge, the results that will be shown report the highest solubilities in DESs measured until now. The solubilities are even similar to the best ones measured in ILs.

Finally, the first results on the formation and use of hydrophobic DESs to form DES-H_2O emulsions, including an application, will also be shown.

References

[1] A. P. Abbott, G. Capper, D. L. Davies, R. K. Rasheed, and V. Tambyrajah, *Chemical Communications*, 2003, 70–71.

[2] M. Francisco, A. van den Bruinhorst, and M. C. Kroon, *Angew. Chem. Int. Ed.*, 2013, **52**, 3074–85.

[3] D. J. G. P. van Osch, L. F. Zubeir, A. van den Bruinhorst, M. A. A. Rocha, and M. C. Kroon, *Green Chemistry*, 2015, **17**, 4518–4521.

[4] D. J. G. P. van Osch, D. Parmentier, C. H. J. T. Dietz, A. van den Bruinhorst, R. Tuinier, and M. C. Kroon, *Chemical Communications*, 2016, **52**, 11987–11990.

499310 Thickening of Liquid Digestate: Integration of Vacuum Evaporator into a Biogas Plant

Abstract:

More than 17 thousand biogas plants (BGP), with a total installed capacity of more than 8,200 MWel, were established in Europe in 2015. The rise of number of BGP can be associated with increased amounts of available digestate. Digestate is a by-product of anaerobic fermentation and is predominantly used as a fertilizer in agricultural areas close to the BGP. Digestate in liquid form contains little nutrients, and its storage and application in the fields bring high costs to the operator. Promising method to improve nutrient concentrations in the liquid digestate is water separation via vacuum evaporation. Integration of vacuum evaporator (VE) into BGP should decrease the amount of fuels necessary for liquid digestate transport. The integration should further increase efficiency of waste heat utilization. Plausible payback period is a prerequisite of a successful integration of the technology on the large scale.

This paper evaluates the integration of VE in BGP. For this purpose a mathematical model was developed which provides details on mass and heat flows in the process and calculates the payback period too. The model reflects real operational and technological limitations. Limitations of liquid digestate thickening in VE include maximum admissible dry matter content which still allows for transport via common pumps.

It can be concluded that the integration of VE in BGP makes sense with the increase of transport distances, BGP capacity, thermal power consumption and dry matter content in the concentrate and with the decrease of feed-in tariffs and electrical power consumption. Integration of VE technology into BGP must be preceded by analysis of unitâ€™s capacity in terms of produced digestate, and quality and amount of waste heat. Further, amounts of auxiliary chemical components and subsequent elimination of residual ammonia from the distillate must be also considered before the actual integration.

494434 Process Optimisation in the Retreatment of Gold-Bearing Sand Dumps: A Case Study of Zimbabwe

Abstract:

In this paper, we apply process optimisation to identify opportunities of maximizing gold recovery in a small scale heap leaching plant. Less work on improving leaching gold ore, specifically targeting optimizing most process variables in heap leaching, is available as yet. Bottle rolls gave the total leach result ~ 95 % to use as a base line whilst column tests 81% gave a good indication of the "relative" response of ore to different operating parameters. The choice and blend of test methods is about maximising information return within available time and money. Preliminary pilot results currently underway prove that utilizing heap leaching would potentially reduce capital costs and increase in recoverable gold.

The investigation reported in this paper is not intended to set out ideal operating parameters for industrial processes, but aims to show how process optimisation can be used to develop ways of improving recovery. The anticipated benefits of our research are that it would reduce reagent consumption, create background knowledge and greatly improve overall mineral recoveries.

Research on Separation Performance of Supersonic Separator with a Forward Helical Guide Blade

Huirong Liang[1,2], Shuai Zhang[2], Yong Kang[2], Kegang Ling[1], Sai Wang[1]
(1) Petroleum Engineering Department, University of North Dakota, Grand Forks, ND,
(2) School of Mechanical Engineering, Xi'an Shiyou University, Xi'an, China

Abstract: The supersonic separation is a new approach to dehydrate the natural gas in recent years. In the conventional supersonic separator, a cyclone or a delta wing is employed to generate the swirling flow. Unfortunately, the separation efficiency of conventional supersonic separator is low. In order to overcome the disadvantages, helical guide blade was designed to incorporate into the swirling device which is fixed at the upstream of the Laval nozzle in this paper. The computational fluid dynamics modeling was performed to investigate the flow characteristics in the supersonic separator and optimize the structure of the blade. The swirling flow generated from the blade promoted the separation of the droplets from the gas. The optimization results showed that a pitch of 30 mm and a blade length of 120 mm are the best geometry structure for the swirling device in the supersonic separator in the dehydration process, which can create the stable flow field and benefit the separation effect.

1 Introduction

The natural gas collected from the gas wells is generally saturated with large volumes of water vapor, which inevitably causes clogging and corrosion in natural gas transportation pipelines, valves, and other auxiliary equipment when water vapor condenses out. It also seriously affects the power supply and quality of the natural gas, and reduces the utilization efficiency of the natural gas. Therefore, it is imperative to effectively remove the free water or saturated water in the natural gas treatment process. There are various methods for the liquid separation from the gas[1-3]. Supersonic separator is a new type of gas-liquid separation equipment with the advantages of low energy consumption, safe and efficient operation, and pollution-free to environment, which has attracted the attention in the separation sector[4-6]. Jassim et al.[7] and Karimi and Abdi[8] researched the single-phase flow of natural gas in the Laval nozzle under a high pressure condition. Yang et al.[9-10] investigated the characteristics of the pressure recovery and real gas using CFD method. Malyshkina[11] studied the effect of the Mach number on the liquid component and obtained the optimum Mach number of 1.4-1.8 for a supersonic separator. Wen et al.[12] optimized the geometry structure of the diffuser in the supersonic separator under the conditions of single phase flow.

In a conventional supersonic separator, the shockwave usually occurs at the expanding section of the Laval nozzle, which results in a short low-temperature section and a high flow resistance, consequently a low separation efficiency[13-15]. In order to improve the separation performance of the supersonic separator, a new type of the swirling device was designed and the installation site was also changed from the downstream to upstream of the Laval nozzle in the new design. The flow characteristics were then investigated in our newly designed supersonic low-temperature gas-liquid separator using three-dimensional numerical simulation.

2 Supersonic separators

The conventional supersonic gas-liquid separator is mainly composed of a Laval nozzle, a separation straight tube, an annular drainage tube and a diffuser as illustrated in Fig.1. The Laval nozzle is a key part including a converging tube, a throat tube, and a diverging tube. In the conventional structure, the straight tube is usually integrated with a cyclone to generate a strong vortex flow. Because of the strongly negative effect of the shockwave, the separation efficiency is relatively low. To overcome this disadvantage, instead of the cyclone a swirling device, which uses a helical guide blade, was designed and installed at the entrance of the Laval nozzle in the separator, as shown in Fig.2.

The design of the supersonic separator is based on the theories of supersonic wind tunnels, thermodynamics, and fluid mechanics. The saturated natural gas experiences an adiabatic expansion when flowing through a special variable-section tube of the Laval nozzle. The temperature drops rapidly to the dew point of the gas and the pressure energy can be converted into kinetic energy. With that a flow field forms with low temperature, low pressure, and high velocity. The liquid droplets start to condense in the diverging section of the Laval nozzle and are centrifuged onto the wall when the mixture of liquid and gas flows through the swirling device. They are collected and drained through the annular drainage tube. Finally, the dry gas exits through the diffuser.

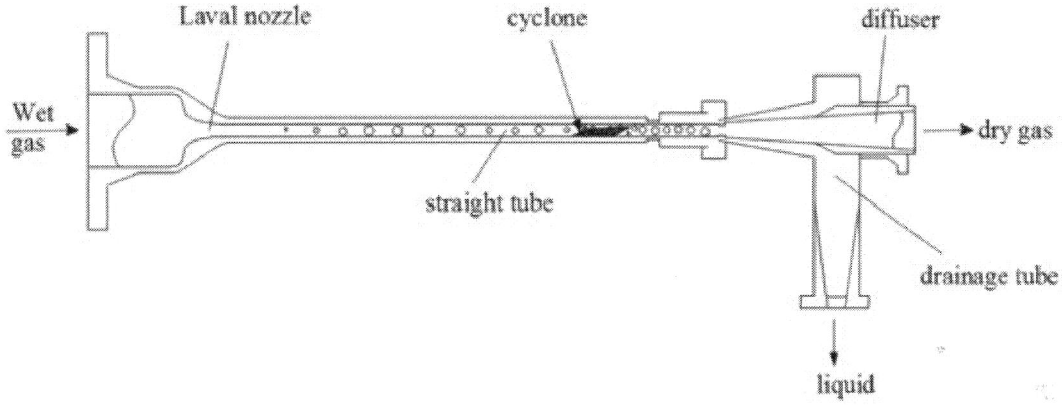

Fig 1. Structure schematic diagram of the conventional supersonic separator

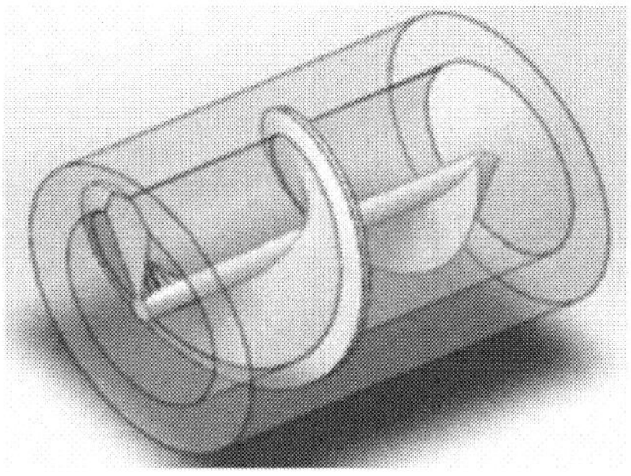

Fig 2. Schematic diagram of the designed helical guide blade

3 Calculation model

3.1 Conservation equations

The internal flow in the supersonic separator follows the laws of the fluid dynamics, i.e. three governing equations, including the continuity, momentum and energy equations[9]. Because natural gas is compressible, the energy conversion of the gas flow is involved.

Continuity equation:

$$\frac{\partial \rho}{\partial t}+div(\rho\vec{u})=0$$

Momentum equation:

$$\frac{\partial \rho u}{\partial t}+div(\rho u\vec{u})=-\frac{\partial p}{\partial x}+\frac{\partial \tau_{xx}}{\partial x}+\frac{\partial \tau_{yx}}{\partial y}+\frac{\partial \tau_{zx}}{\partial z}+F_x$$

$$\frac{\partial \rho \upsilon}{\partial t}+div(\rho \upsilon\vec{u})=-\frac{\partial p}{\partial y}+\frac{\partial \tau_{xy}}{\partial x}+\frac{\partial \tau_{yy}}{\partial y}+\frac{\partial \tau_{zy}}{\partial z}+F_y$$

$$\frac{\partial \rho \omega}{\partial t}+div(\rho \omega\vec{u})=-\frac{\partial p}{\partial z}+\frac{\partial \tau_{xz}}{\partial x}+\frac{\partial \tau_{yz}}{\partial y}+\frac{\partial \tau_{zz}}{\partial z}+F_z$$

Energy equation:

$$\frac{\partial(\rho T)}{\partial t}+div(\rho\vec{u}T)=div\left[\frac{\kappa}{c_p}\cdot gradT\right]+S_T$$

where ρ is the density of the gas; t is the time; $\vec{u}$ is the velocity vector; p is the pressure; T is the temperature; F_x, F_y and F_z are the volume force on the micro-body; τ_{xx}, τ_{xy} and τ_{xz} are the component of the viscous stress acting on the surface of the micro-body.

3.2 Turbulence model

In the supersonic separator, the flow after the swirling device is strongly turbulent so that the anisotropic effect is very significant especially when the gas flow occurs at supersonic speed. Therefore, it is very important to select an appropriate model to simulate the turbulent flow. According to the flow characteristics of the supersonic separator, RNG k-ε model is used to predict the complex swirling flow in this paper. RNG k-ε model involves turbulent vortices and improves the accuracy of the simulation for the swirling flow. It can satisfy the requirements for the calculation accuracy and speed.

3.3 Numerical Schemes

In this study, a three-dimensional physical model of supersonic separation was establishedas Fig.3. The flow field is discretized using different structure grid in different sections. The finite difference method is utilized to solve the governing equations. The second-order upwind scheme is adopted to discretize the flow equations and the appropriate boundary conditions are set in the model. The pressure boundary conditions are specified at inlet and outlet of the supersonic separator, respectively. The walls of the flow field and the surface of the blade are all set as the solid walls, and moreover, no-slip and adiabatic boundaries. The standard wall function method is used to deal with the boundary turbulence.

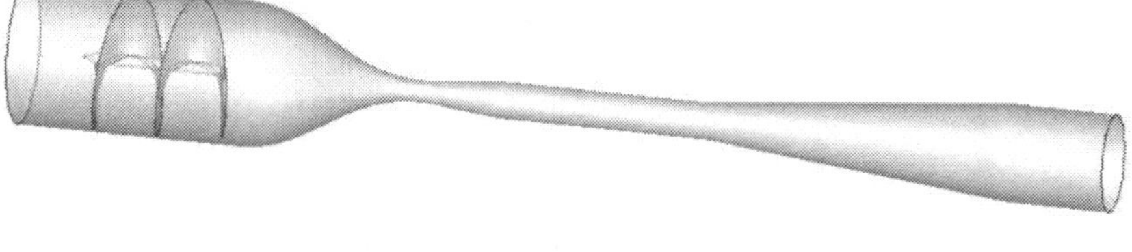

Fig 3. Schematic diagram of the physical model

4 Results and discussion

4.1 The characteristics of the flow field

In this physical model, the swirling device is installed at the entrance of the Laval nozzle with a blade number of 1, a pitch size of 60 mm, and a blade length of 120 mm. The cloud map distributions of temperature, velocity and pressure are shown in Fig.4, Fig.5 and Fig.6, respectively. From these maps, it can be seen that the pressure gradually decreases, the temperature drops rapidly, and the velocity greatly increases when the gas flows into the Laval nozzle through the helical guide blade. The Mach number in the throat is close to 1, which is consistent with the mechanism of the expansion refrigeration in the Laval nozzle. The flow regime transition from subsonic to sonic then to supersonic flow has been achieved within the Laval nozzle. The fluid flows at supersonic speed in the separation straight tube, consistent with the designed Mach number (*Ma*=2.4). It can be concluded that the process of air flow through the Laval nozzle is adiabatic expansion, and the temperature and the pressure significantly decreased in the diffuser tube. The extremely low temperature makes it possible that the gas phase condenses into liquid droplets. The long low-temperature field can reduce the re-evaporation of the droplets. The outlet temperature of the nozzle is far lower than the dew point temperature of the gas, indicating that the saturated gas has been transited into the oversaturated gas and a large number of droplet nuclei begin to grow into the liquid drops. In the separation straight tube, the liquid droplets are thrown to the wall by the centrifugal force with separation from the gas. In the diffuser, the velocity of the main gas flow gradually decreases, and the temperature and the pressure is picking up.

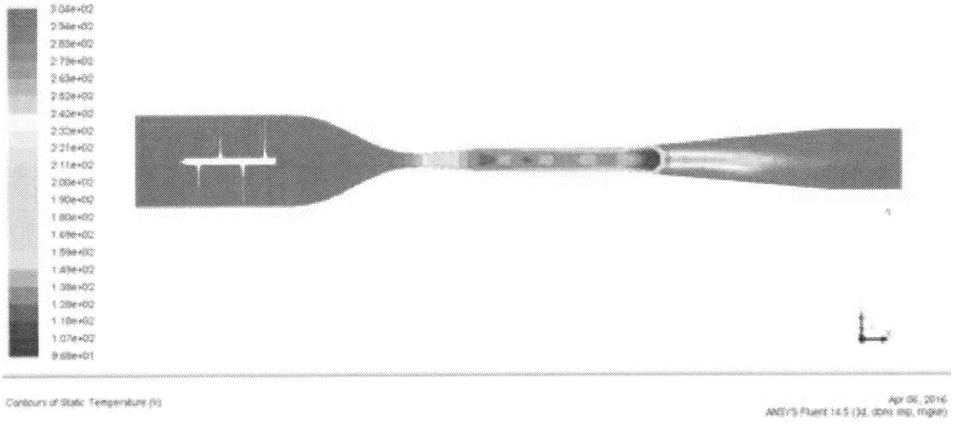

Fig 4. The distribution of the temperature along the axis (z=0)

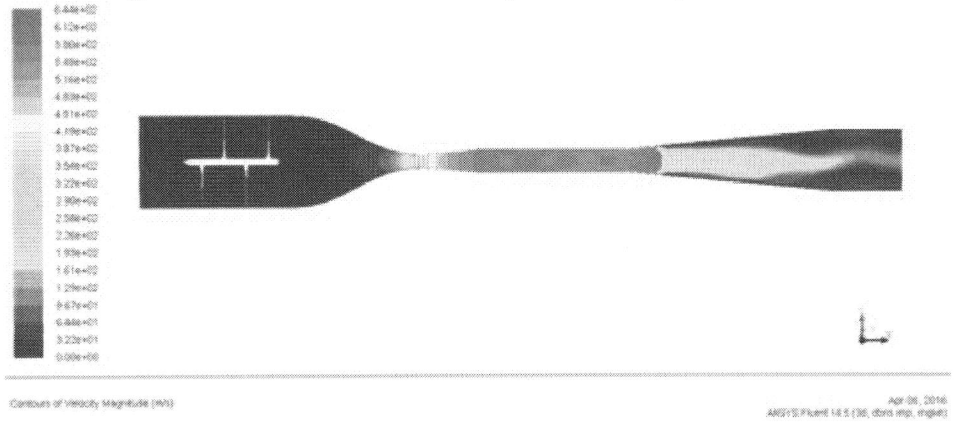

Fig 5. The distribution of the velocity along the axis (z=0)

Fig 6. The distribution of the pressure along the axis (z=0)

Fig.7 shows that the vector diagram of the tangential velocity at different cross sections. It can be found that in (a), the secondary flow generates at the front edge of the helical guide blade; and the tangential velocity is low with a messy distribution. The swirling flow becomes strong and uniform at the tail edge of the blade, as shown in (b). From (c) and (d), we can see that the inlet tangential velocity rises in the separation straight tube, but the uniformity of the swirling field is reduced and the right half of the fluid field is mainly axial flow. A high tangential speed is conducive to the efficient separation of the droplets.

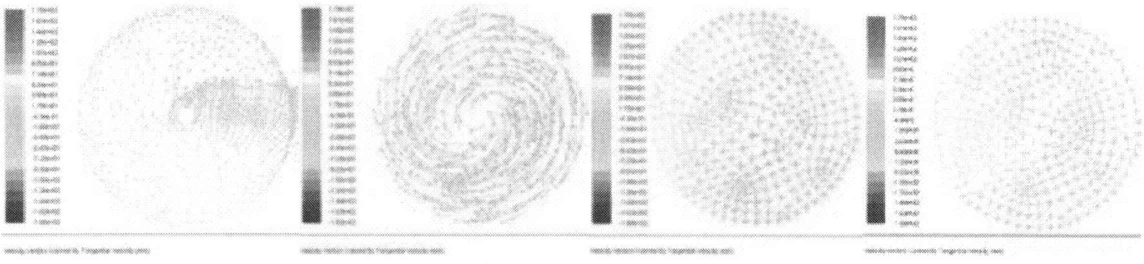

(a) x=0.07m **(b) x=0.2m** **(c) x=0.46m** **(d) x=0.66m**

Fig 7. The vector diagram of the tangential velocity at different cross sections

4.2 The effect of the geometrical parameter of the guide blade

Two parameters including the pitch and the blade length were considered in order to research the effect of the guide blade on the separation performance.

Fig.8 shows that the contrast curve of the temperature, the pressure and the velocity based on three different pitches. It can be seen that the temperature change is gentler in (a), the rate of pressure drop decreases in (b), and the flow rate increases slightly in (c) and with the increase in pitch when the air flows through the spiral guide blade. In the Laval nozzle, the temperature drops rapidly, reaching the valley at the outlet of the diverging tube. The temperature drop is inversely proportional to the pitch. In the upstream of the nozzle, the increase rate of the velocity goes down with the increase of the pitch. When the fluid passes through the straight tube, because the temperature is always low, the re-evaporation of the condensed droplets is better suppressed in the process of the centrifugal separation. The temperature and pressure rise suddenly near the entrance of the diffuser because the transition from the supersonic flow to the subsonic flow generates a shock. The flow velocity, the temperature, and the pressure change become gentler after the shock with the decrease of the pitch. In other words, the flow performance for the pitch of 30 mm and 60 mm are more stable than 90 mm. The tangential velocity distributions of supersonic separator with three kinds of pitch are compared in Fig.8 (d). With the increase of the pitch, the vortex of the tail edge of the guide blade gradually deviates from the centerline of the straight tube, and the outlet tangential velocity increases and then decreases in the radial direction. The slope of the tangential velocity change diminishes and the maximum tangential velocity near the wall fades with the increase of the pitch. When the pitch is 30 mm, the swirling vortex is centered, the tangential velocity and the centrifugal acceleration are the highest, and the tangential velocity is symmetrical in the radial direction, which can ensure the separation straight tube a better centrifugal force. Based on the above discussions, it was found that the model with the pitch of 30 mm is the best in both stability and the separation performance.

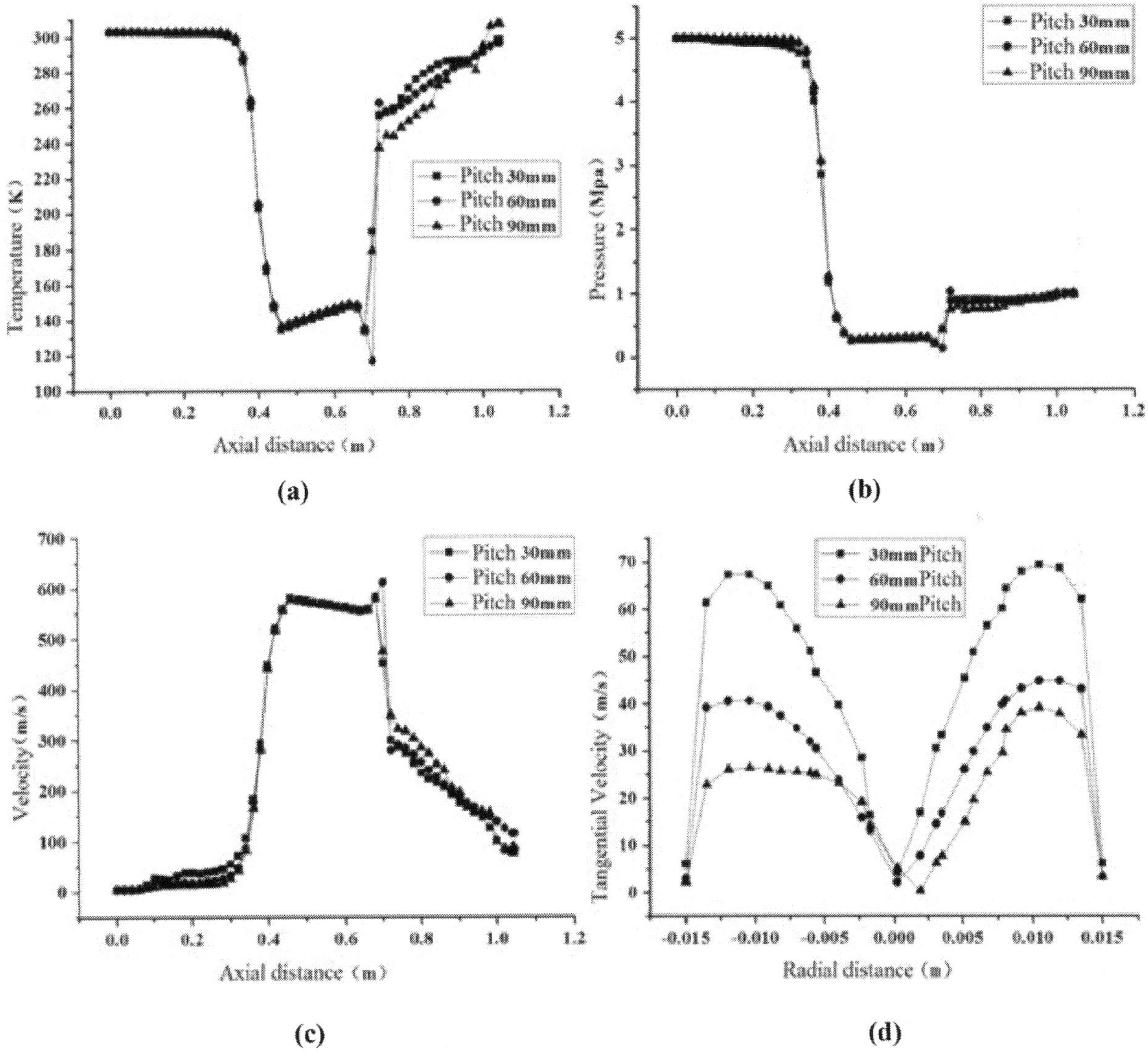

Fig 8. The effect of the pitch on (a) the temperature, (b) the pressure, (c) and (d) the velocity

With other parameters fixed, the length of the blade of the basic model was changed to study the effect of helical guide blade on the flow field performance. On comparison of the temperature, velocity and pressure changes along the axial direction based on three different blade length of 90 mm, 120 mm and 150 mm shown in Fig.9, it indicated that the changes of the velocity and temperature are not obvious with the increase of blade length. In the separation straight tube, the pressure drop goes up with the increase of the length of the blade because the collision area between the fluid and the wall increases. When the fluid flows into the diffuser, there is a shock near the entrance that creates a distinct change in the flow field. When the blade length is 150 mm, there are many jumps in the curves of the flow velocity, temperature and pressure in the diffuser tube, which demonstrated that the excess length of the blade enlarges the instability of the flow field in the diffuser. In contrast, for both models with 90 mm and 120 mm blades, their flow field changes are gentle and similar. The outlet tangential velocity variation along the radial direction in the separated straight tube among three kinds of guide blades are compared, as shown in Fig.9 (d). With the increase of the length of the blade, the tangential velocity of the same radial point increases, and the changes in the tangential velocity for both 120 mm and 150 mm models are very close. In conclusion, the model with a blade length of 120 mm has a relatively more stable flow field and a more appropriate swirling strength than others.

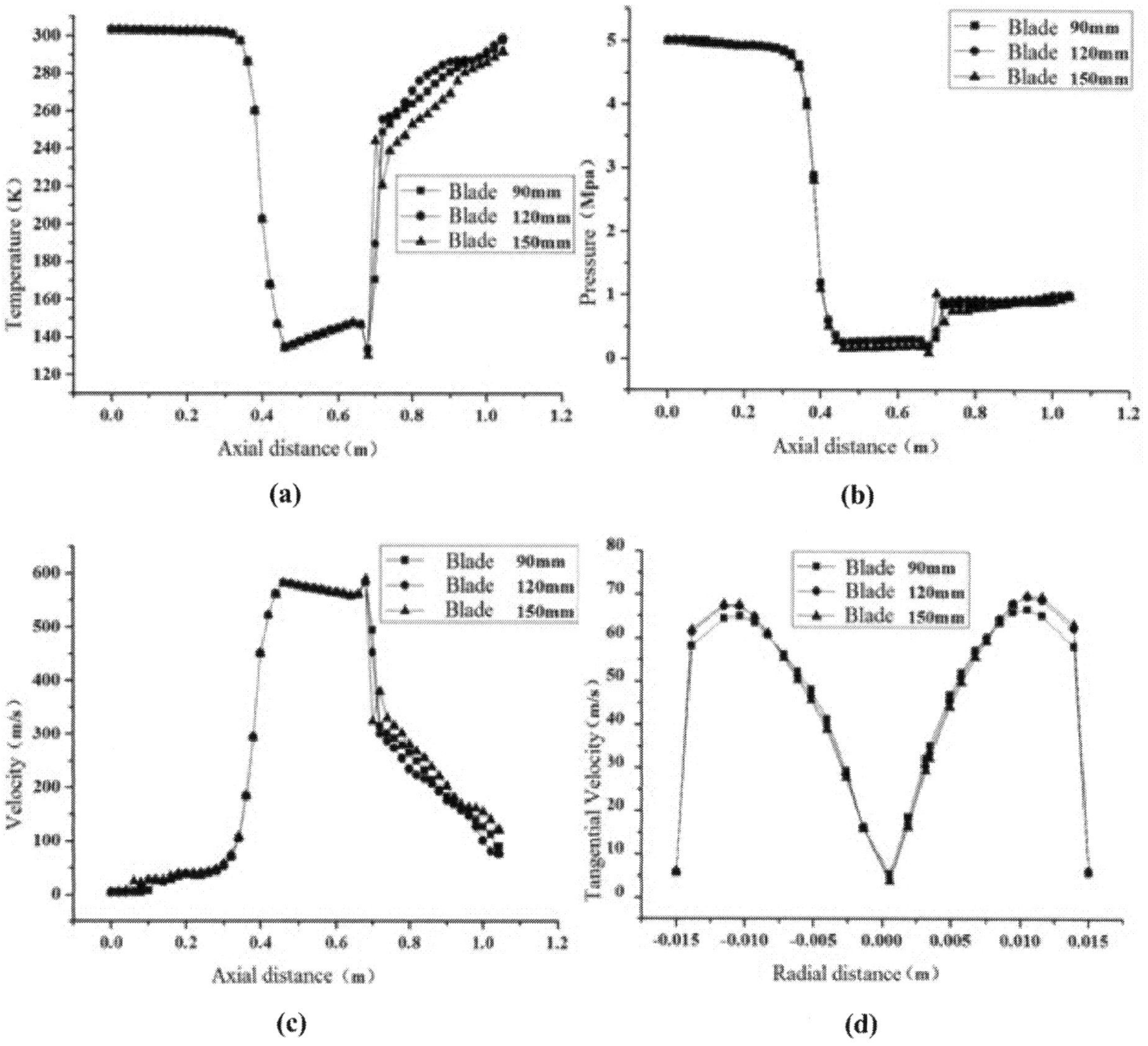

Fig 9. The effect of the blade length on (a) the temperature, (b) the pressure, (c) and (d) the velocity

5 Conclusions

The guide blade is fixed in the upstream of the Laval nozzle in this paper. The physical model has been established and analyzed by CFD method. The swirl flow field with a low tangential velocity was firstly formed in the swirl tube. When the fluid passes through the Laval tube, the axial velocity of the fluid rapidly accelerates to supersonic velocity and the tangential velocity increases as well. The fluid can keep the diffuser and the separation straight tube in a low temperature state, which can avoid the re-evaporation of the droplets and reduce the loss of the blade shock. The flow field of the supersonic separator was analyzed by adjusting the magnitude of the pitch. With the decrease of the pitch, the centrifugal acceleration in the straight tube increased, and the separation performance of the outlet of the separation tube was improved, and the flow field in the downstream expansion tube became stable. According to the comparison, the separation performance of the model with 30 mm pitch is best, and the centrifugal acceleration is highest among three kinds of pitches. It was found that the model with 120 mm blade length had a more stable flow field and better swirl separation ability than others.

Acknowledgement

The authors thank the Science and Technology Department of Shaanxi Province for providing the financial support of the Shaanxi Province Science and Technology Co-ordination Innovation Project (2012KTCG01-12).

References

1. I. Folgueira, I. Teijido, A. García-Abuín, D. Gómez-Díaz, A. Rumbo, Carbon dioxide absorption behavior in 2-(ethylamino) ethanol aqueous solutions, Fuel Process.Technol. 131 (2015) 14–20.
2. N. Álvarez-Gutiérrez, M.V. Gil, F. Rubiera, C. Pevida, Adsorption performance indicators for the CO2/CH4 separation: application to biomass-based activated carbons, Fuel Process. Technol. 142 (2016) 361–369.
3. G.Q. Chen, S. Kanehashi, C.M. Doherty, A.J. Hill, S.E. Kentish,Water vapor permeation through cellulose acetate membranes and its impact upon membrane separation performance for natural gas purification, J. Membr. Sci. 487 (2015) 249–255.
4. P.B. Machado, J.G. Monteiro, J.L. Medeiros, et al., Supersonic separation in onshore natural gas dew point plant, J. Nat. Gas Sci. Eng. 6 (2012) 43–49.
5. V.I. Alferov, L.A. Baguirov, L. Dmitriev, V. Feygin, S. Imaev, J.R. Lace, Supersonic nozzle efficiently separates natural gas components, Oil Gas J 103 (2005) 53–58.
6. C. Wen, X. Cao, Y. Yang, W. Li, An unconventional supersonic liquefied technology for natural gas, Energy Educ. Sci. Technol. Part A Energy Sci. Res 30 (1) (2012) 651–660.
7. E. Jassim, M.A. Abdi, Y. Muzychka, Computational fluid dynamics study for flow of natural gas through high-pressure supersonic nozzles: part 1. Real gas effects and shockwave, Pet. Sci. Technol. 26 (2008) 1757–1772.
8. A. Karimi,M.A. Abdi, Selective dehydration of high-pressure natural gas using supersonic nozzles, Chem. Eng. Process. 48 (2009) 560–568.

9. Y. Yang, C. Wen, S. Wang, Y. Feng, Theoretical and numerical analysis on pressure recovery of supersonic separators for natural gas dehydration, Appl. Energy 132 (2014) 248–253.
10. Y. Yang, C.Wen, S.Wang, Y. Feng, Numerical simulation of real gas flows in natural gas supersonic separation processing, J. Nat. Gas Sci. Eng. 21 (2014) 829–836.
11. M. Malyshkina, The procedure for investigation of the efficiency of purification of natural gases in a supersonic separator, High Temp. 48 (2010) 244–250.
12. C.Wen, X. Cao, Y. Yang,W. Li, Numerical simulation of natural gas flows in diffusers for supersonic separators, Energy 37 (2012) 195–200.
13. W.M. Jiang, Z.L. Liu, H.W. Liu, J. Zhang, X.J. Zhang, Y.X. Feng, Two dimensional simulation and analysis of the flow in supersonic separator, J. Eng. Therm. 29 (2008) 2119–2121.
14. X.W. Liu, Z.L. Liu, Y.X. Li, Numerical study of the high speed compressible flow with non-equilibrium condensation in a supersonic separator, J.Clean Energy Technol. 3 (2015) 360–366.
15. X.W. Liu, Z.L. Liu, H.Q. Wu, Flow behavior analysis of the cyclone back-placed supersonic separator, J. B. Univ. Technol. 40 (2014) 1394–1401.

491847 The Influence of Microstructure on Membrane Distillation: Accurate 3-D Reconstructions for Analysis of Pore-Scale Phenomena

Abstract:

Water filtration technologies play a central role in addressing the water-energy climate nexus through applications such as desalination, contaminant removal, and demineralization. While reverse osmosis is the most common method for such applications, it requires significant power input, and is unable to handle highly-concentrated brines, making waste brine management/chemical discharge a challenge. Membrane distillation (MD) has emerged as a promising approach to desalinate high-concentration brines, brackish waters, produced waters, and seawater to near-saturated conditions, which allows for a smaller physical footprint than other current techniques. MD is a thermal process that can operate at comparatively low temperatures, which enables running on waste heat from industrial or renewable sources. Though attracting considerable attention, several technological barriers must be solved before MD sees significant industrial application. Membrane pore-structures play a dominant role in determining conductive heat losses and membrane permeability to vapor, but are poorly understood to date. Most studies focus on MD hydrodynamics and few attempts have been made to understand pore-scale phenomena within membrane pores. Without sufficient understanding of MD processes at the pore scale, appropriately designed MD systems (membranes and feed spacers) cannot be developed to optimize MD for industrial applications.

This poster will describe initial attempts to determine design criteria for the development of new membranes via predictive pore-scale modeling of heat and vapor transport. Current transport models struggle to predict microstructure influence due to inadequate membrane pore-structure representation. Typically, effective membrane parameters are determined by fitting to bench-scale vapor flux measurements, which immediately removes any predictive capability, neglects the link between specific membrane microstructural properties and the resulting mass and heat flows, and prevents a true understanding of the coupling between local hydrodynamic characteristics and the pore fluxes that act as a boundary condition for these flows.

We address this issue by directly measuring the 3-D pore-structures of commercial membranes using focused ion beam scanning electron microscopy (FIB-SEM). Via FIB-SEM, structures can be numerically reconstructed for accurate 3-D pore-scale simulations of heat and vapor transport. These 3-D pore scale simulations are then used as a "numerical experiment," for validation of reduced-order, low-cost simulations. These reduced-order models can then be incorporated into system-level, hydrodynamic models to directly link membrane microstructural properties with MD module performance and identify optimal membrane properties. This poster will demonstrate FIB-SEM as an appropriate technique for developing accurate 3-D transport models at a pore-scale level. The reconstructed 3-D structures are analyzed to determine microstructure parameters such as porosity, tortuosity and mean pore diameter that are important in understanding transport phenomena. We will also discuss initial efforts to couple the 3-D reconstructed microstructures with 1-D reduced-order heat- and mass-transfer simulations.

Mitigation of Thin Film Composite Membrane Biofouling via Immobilizing Nano-Sized Biocidal Reservoirs in the Membrane Active Layer

Drexel UNIVERSITY

Alireza Zirehpour, Ahmad Rahimpour, Ahmad Arabi Shamsabadi, Mohammad Sharifian Gh., and Masoud Soroush

17AIChE

Introduction

Forward osmosis (FO) is a high water-recovery and low-cost membrane-based technology that has potential applications in desalination and wastewater treatment. However, biofouling has limited the use of FO, in particular, for feeds containing microorganisms. Applying silver-based materials has been recognized as an effective approach to decreasing membrane biofouling. One method is direct blending, which has several disadvantages. First, the biofouling mitigation does not last long. Second, membrane selectivity deteriorates as a result of low compatibility between the inorganic nanomaterials and organic membrane. In this work, we investigate the use of silver-based metal-organic framework (MOF) nanocrystals to mitigate the biofouling in FO membranes. The MOF nanocrystals have a good compatibility with the selective layer because of their organic ligand. Consequently, the membrane active layer is improved without any detrimental effects on its selectivity. The results highlight the great potential of the silver-based MOF in improving the biofouling resistance of membranes.

Silver-based MOF Characterization

The silver-based MOF was characterized using TEM, IR, and energy-dispersive X-ray (EDX). The TEM image in Fig.1a shows the morphology of the MOF nanocrystals prepared by the ultrasonic method. The chemical composition of the MOF was characterized via EDX analysis (Fig.1b). The EDX analysis indicates that the oxygen-to-silver atom ratio in the MOF building block is 2.52. The DLS measurement shows that the size distribution of nanocrystals has a sharp and narrow peak with an approximately 33 nm mean size (Fig.1d), which is in the range of the MOF nanocrystal sizes measured using the TEM image. The organic ligands and the small size of the MOFs lead to no defects in the membrane active layer when MOFs are present in the active layer.

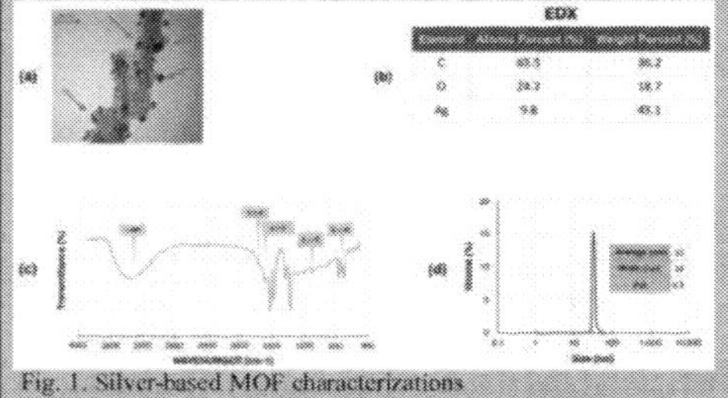

Fig. 1. Silver-based MOF characterizations

Thin-film composite (TFC) and nanocomposite (TFN) Membranes: Results

TFC and TFN Membranes Characterization

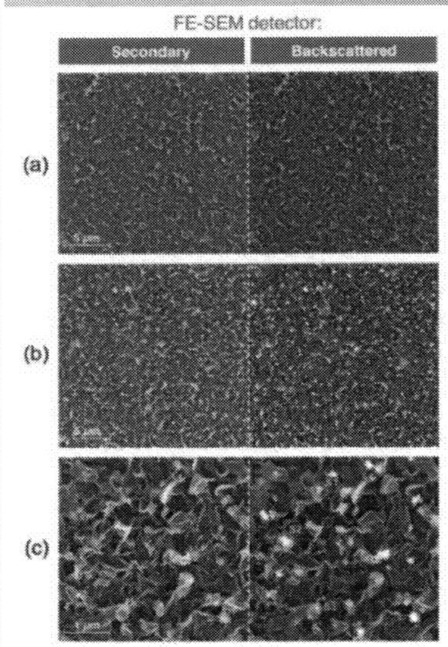

Fig. 2. FE-SEM images of the membranes using different imaging detectors. (a) TFC membrane, (b) TFN membrane, and (c) a higher-magnification image of the TFN membrane surface. The left-hand images were taken using a secondary detector, while the right-hand image were taken using a backscattered imaging detector. The backscattered image of TFN membrane shows numerous bright spots on the surface, attributed to the silver element as a MOF nanocrystal characteristic. The orange arrows in the right-side image show the positions of some of the bright spots in the backscattered image.

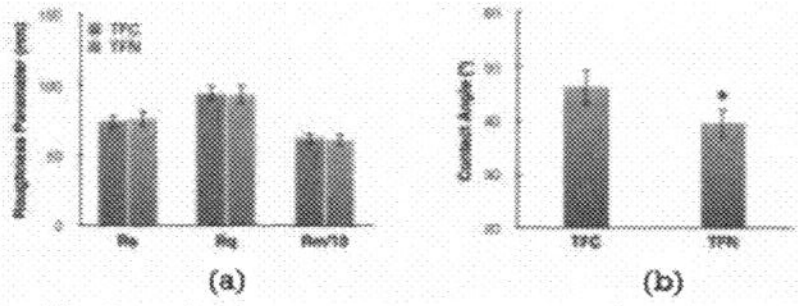

Fig. 3. (a) Surface roughness parameters of membranes, and (b) water contact angles of membrane surfaces

Antibacterial Activity of the Membranes

Fig. 4. Biocidal activity of the TFC and TFN membranes, the normalized number of attached live bacteria on the TFC and TFN membranes for *E. coli* and *S. aureus* bacteria. The values were normalized by the number of attached live bacteria on the TFC. The orange and violet bars illustrate the antibacterial activity of the TFN membranes after 24 days and 6 months of storing the TFN membranes in water, respectively. Each bar indicates the standard deviation of three independent test, each with a fresh membrane. Asterisks above the bars represent the statistical significance (P-value < 0.05), determined by a student's t-test.

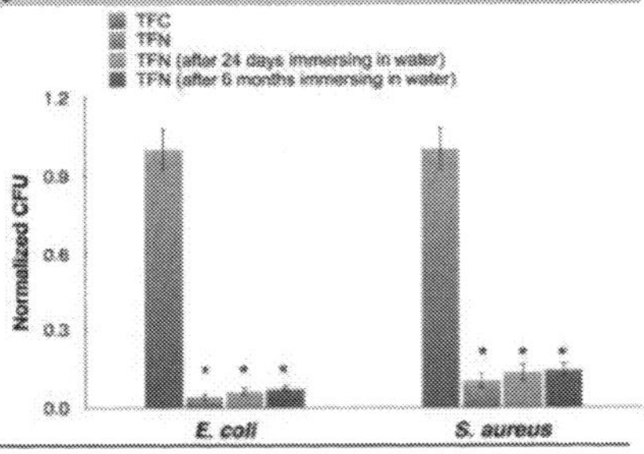

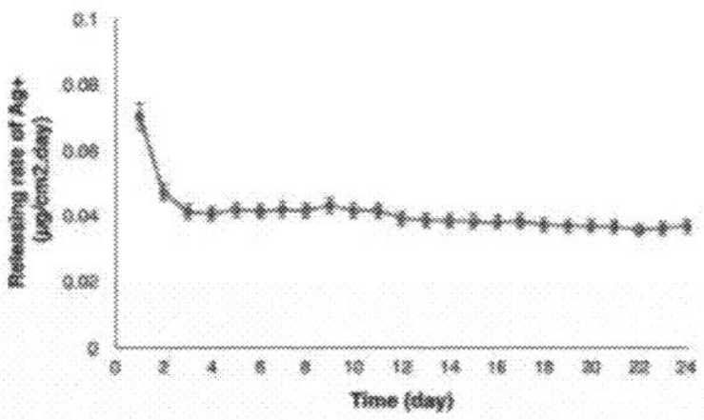

Fig. 5. Release rate of Ag^{+} from the TFN membrane. The silver-ion release from the TFN membrane decreased sharply from the initial value of about 0.07 $\mu g.cm^{-2}.day^{-1}$ during the first 2 days. The release rate then decreased very slowly afterwards. This implies that the MOF can work as an Agreservoir immobilized in the membrane active layer, providing a controlled sustained Ag release. These results indicate that the MOF imparts a long-lasting antibacterial activity to the membrane surface to mitigate biofouling in a prolonged period of the membrane operation.

Performance Evaluation of the Membranes

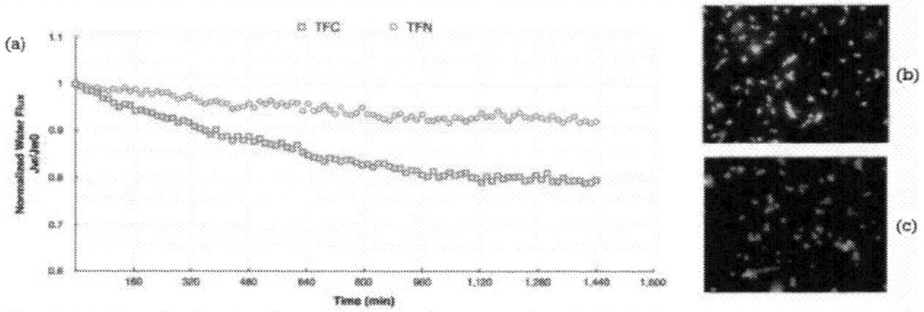

Fig. 6. (a) Normalized water fluxes of the membranes vs. time in biofouling experiments using *E. coli*. Fluorescence microscopy image of (b) TFC membrane surface, and (c) TFN membrane surface using the live/dead staining experiment.

Conclusion

This study showed that the MOF nanocrystals endow the membranes with TFN membranes antibacterial activity. It demonstrated that it is possible to improve both the biocidal activity and the hydrophilicity of the membrane active layer, without any adverse effects on the membrane selectivity. The MOFs are very attractive materials for developing high-performance FO membranes. They have great potential in improving the biofouling resistance of FO membranes that require both antibacterial and anti-adhesion features.

495469 Ethanolamine Separation By Nanofiltration: A Molecular Simulation Study

Abstract:

Biomass is an abundant resource of renewable feedstock on the Earth and there has been increasing interest to convert biomass into fuels and valuable chemicals (e.g. *C*, *H*, and *O*-based bioethanol, sugar alcohals/polyols, furan and organic acid). Nevertheless, nitrogen-containing chemicals are often more valuable and widely used in pharmaceuticals, textiles and many more. Particularly, ethanolamine (ETA) is extensively being used for scrubbing acidic gases and as an important feedstock in the production of detergents, emulsifiers, pharmaceuticals, chemical intermediates, etc. After the conversion of biomass, however, the solution usually contains ETA, ammonia (used as a nitrogen source during conversion), as well as a large amount of water. Traditional separation using thermal distillation is energy intensive and may decompose ETA. In this context, the technically feasible and economically viable separation of ETA is highly desirable. Herein, we report a molecular simulation study to investigate the separation of ethanolamine (ETA) by nanofiltration (NF) through a polymer of intrinsic microporosity (PIM-1) and two activated carbon membranes namely curved corannulene (CRNL) and curved corannulene functionalized with hydroxylic groups CRNL-$(OH)_2$. The hierarchy of water flux and permeability through the membranes are CRNL-$(OH)_2$ > PIM-1 > CRNL. Due to the hydrophilic nature, CRNL-$(OH)_2$ possesses the highest water flux and permeability compared to the other two. Along with higher water permeation, 100% ETA retention is obtained through CRNL-$(OH)_2$, indicating its suitability for ETA separation. The survival time-correlation function (STCF) of water around the non-accessible atom of the PIM-1 membrane shows a remarkably faster decay than around the easily accessible atom. The STCF of water in hydrophilic membrane shows a slower decay than in hydrophobic counterpart. Further, the life time of water hydrogen bonds within the membranes follow the reverse trend to water flux. The simulation study provides microscopic insight into the structural and separation properties of ETA through the NF membranes, and suggests that CNRL-$(OH)_2$ membrane is an interesting candidate for ETA separation.

495615 Separation of Hexavalent Chromium Cr(VI) from Wastewater through Supported Liquid Membrane Using Environmentally Benign Solvent

Abstract:

Heavy metals are carcinogenic and not biodegradable. Among the various heavy metals hexavalent chromium (Cr(VI)) is considered one of the most deleterious pollutant. Tanning, electroplating, dyeing, paints/pigments and mining are the major industries that are responsible for chromium (Cr) contamination to natural water body (Chakravarti et. al., 1995; Chiha et. al., 2006). Separation or removal of hexavalent chromium (Cr(VI)) through liquid membrane (LM) is very promising technology. Among the various LM techniques supported liquid membrane (SLM) technique can be used as an effective substitute to traditional solvent extraction methods (Chakrabarty et. al., 2010; Bhatluri et. al., 2014). Some of the noted advantages of SLM are the low volume of solvent required, low operating cost and energy consumed. It can be seen that most of the researchers, dealing with Cr(VI) transport through SLM, used conventional hazardous solvent. Djane et al. (1999) have reported separation of total chromium, Cr (III) & Cr (VI), through SLM using two serially connected membrane units with kerosene as a solvent. They have collected samples from the drainage path of the tannery industry to extract total chromium. Venkateswaran et al. (2005) have used tri-n-butyl phosphate (TBP) as a carrier to separate Cr(VI) through SLM where polytetrafluoroethylene (PTFE) membrane used as a membrane support. They have optimized parameters to maximize the Cr(VI) separation and tried applicability of those parameters with real plating effluent. There are very few researchers who have explored the environmentally benign solvent in supported liquid membrane (SLM) to separate the heavy metal. The present study explores the potential of environmentally benign solvent with SLM technique for simultaneous extraction and recovery of Cr(VI). With the context of environmentally benign solvent, sunflower oil used as a green solvent and n-methyl-n,n,n-trioctylammonium chloride (a.k.a Aliquat 336) as an extractant/carrier, where Aliquat 336 showed very good extraction capabilities for transport of Cr(VI). Various physico-chemical parameters are tuned to optimize the process, which would be more relevant for commercial operation. Di-sodium ethylene-di-amine-tetraacetic acid (Na2-EDTA) was selected as stripping agent for its affinity towards metal. SLM method has proved very effective method over the other traditional solvent extraction methods. Polyvinylidene fluoride (PVDF) was used as a support in order to hold and contain the membrane liquid in the SLM structure. The size of the pores in the support is 0.22 μm. The transport performance of Cr(VI) was affected by various parameters such as pH of feed phase, pH of strip phase, concentration of strip phase, concentration of carrier, stirring speed and period of operation (i.e. run time). The role of Na2-EDTA as a stripping agent is very crucial for recovery of Cr from membrane phase. Various strip concentration ranges from 0.005 - 0.04 M had been experimented to check the effect of concentration of Na2-EDTA on the recovery efficiency. Both extraction and recovery are maximum at 0.03 M Na2-EDTA. However, beyond 0.03 M concentration both extraction and recovery start declining. This is due to crowding effect of excess Na2-EDTA in strip phase which inhibits the transport of Cr(VI) from strip-membrane interface to bulk of strip phase. The pH of strip phase plays a vital role on extraction and recovery of Cr(VI). The kinetics of extraction and recovery were examined in the pH range of 4 to 9. The recovery of Cr(VI) was increased with increase in pH and reached its optimum value at pH 6.5.

Again beyond pH 6.5 or at higher pH the recovery slowly decreased. Aqueous solution of Na2-EDTA is stable at lower acidic condition, but very low pH, Cr ions compete with H+ ions and create problem to form Cr-EDTA complex. On the other hand, at very high pH metal ions are prone to form insoluble hydroxide and it makes metal ions less accessible to EDTA. The period of operation or time is an important parameter for any chemical process viability in case of industrial scenario. The experiment was performed for 48 h and samples were periodically taken. Maximum extraction and recovery of Cr(VI) after 48 h are over 80% and 70% respectively. The effect of carrier (i.e. Aliquat 336) concentration in organic phase on extraction and recovery were studied at concentration ranging from 0-3%. The extraction and recovery of Cr(VI) increased with increase in carrier concentration and reached its maximum at 1 volume% Aliquat 336 concentration. With further increase in carrier concentration the extraction and recovery of Cr(VI) are decreased. Carrier helped to form solute-carrier complex at the feed-membrane interface which enhance the diffusive mass transfer rate, but beyond 1 volume% carrier concentration the extraction rate of Cr(VI) decreased because of increase in viscosity of membrane phase which resist the diffusion rate of solute-carrier complex. Over 80 % extraction and over 70 % recovery were obtained with following operating conditions: concentration of carrier = 1% (v/v) aliquat 336, pH of feed phase = 4.56, pH of strip phase = 6.5, concentration of strip phase = 0.03 M, stirring speed =120 rpm and run time = 48 h. The results of SLM study can further be used to design the three phase hollow fibre membrane (HFM) based separation process.

REFERENCES

Bhatluri, K. K., Manna, M. S., Saha, P., Ghoshal, A. K., 2014. Supported liquid membrane-based simultaneous separation of cadmium and lead from wastewater. Journal of Membrane Science; 459: 256–263.

Chakrabarty, K., Saha, P., Ghoshal, A. K., 2010. Separation of mercury from its aqueous solution through supported liquid membrane using environmentally benign diluent. Journal of Membrane Science; 350 (1): 395–401.

Chakravarti, A. K., Chowdhury, S. B., Chakrabarty, S., Chakrabarty, T., Mukherjee, D. C., 1995. Liquid membrane multiple emulsion process of chromium (VI) separation from wastewater. Colloids and Surfaces A: Physicochemical and Engineering Aspects; 103: 59-71.

Chiha, M., Samar, M. H., Hamdaoui, O., 2006. Extraction of chromium (VI) from sulphuric acid aqueous solutions by a liquid surfactant membrane (lsm). Desalination; 194: 69 – 80.

Djane, Nii-K., Ndungu, K., Johnsson, C., Sartz, H., Tornstrom, T., Mathiasson, L., 1999. Chromium speciation in natural waters using serially connected supported liquid membranes. Talanta; 48(5): 1121 – 1132.

Monteiro, M. I. C., Fraga, I. C. S., Yallouz, A. V., de Oliveira, N. M. M., Ribeiro, S. H., 2002. Determination of total chromium traces in tannery effluents by electrothermal atomic absorption spectrometry, flame atomic absorption spectrometry and uv-visible spectrophotometric methods. Talanta; 58: 629 – 633.

Venkateswaran, T., Palanivelu, K., 2005. Studies on recovery of hexavalent chromium from plating wastewater by supported liquid membrane using tri-n-butyl phosphate as carrier. Hydrometallurgy; 78: 107-115.

500818 Water Stable Metal-Organic Framework Based Adsorbent and Membrane for Precious Metal (Silver) Separation from Wastewater

Abstract:

Wastewater treatment techniques have gained significant importance nowadays because the water scarcity is threating all people around world. Adsorption based membrane purification techniques have an exciting potential owing to their economic and environmental advantages. In this study, water and acid stable UiO-66 was mixed with â€˜green polymerâ€™ cellulose to produce mixed matrix membranes for silver adsorption applications. Highly stable membranes were obtained because the desirable properties of cellulose (biodegradable, cheap, and natural polymer with highly stable structure due to â€“OH bonding in its structure) and consistency UiO-66 (water and acid stable porous filler with very high surface area) were combined successfully.

Firstly, zirconium metal-organic framework (UiO-66) is synthesized. Then, silver adsorption capacity of UiO-66 is investigated in detail at several varying parameters such as pH conditions, contact time and adsorption isotherms. Afterwards, pure cellulose and cellulose/UiO-66 mixed matrix membranes are synthesized, and their silver adsorption capacities are investigated in batch and continuous conditions. UiO-66 MOFs and cellulose/UIO-66 mixed matrix membranes are also characterized using XRD, FTIR, and SEM analysis techniques.

To the best of our knowledge, this is the first application of UiO-66 in silver pollutant removal from water supplies. Pure UiO-66 powders exhibited very promising performance for silver adsorption across a broad pH range from very acidic (1) to neutral (7) conditions, with the best adsorption capacity at pH 2. It achieved a remarkable silver uptake capacity of 77 mg/g, and it reached the equilibrium just after 1 hour of contact time. This remarkable performance of rapid silver adsorption and high stability of UiO-66 in water provides promising insights to the water treatment applications.

Cellulose/UiO-66 membranes showed better silver and arsenic adsorption capacity than pure cellulose membranes. Incorporation of MOF particles in cellulose resulted in highly porous green membranes with promising adsorption performances for silver. Cross-flow filtration geometry improved their efficiency further due to penetration of pollutants through the membrane due to applied positive pressure across the membrane. Finally, 4 grams/m^2 silver adsorption rate was achieved with the cellulose/UiO-66 membrane in a 1-hour experiment.

501963 Fabrication of Dense ZSM-5 and Fe-ZSM-5 Membranes for High Throughput Desalination

Abstract:

Seawater desalination has been a hot topic worldwide due to the increasing need of fresh water. Zeolite membranes with uniform and molecular-sized pores have excellent thermal, mechanical and chemical stabilities. They keep stable when being applied in a long-term desalination process. The pore size of MFI-type zeolites is about 0.55 nm, continuous MFI-type zeolite membranes could be used for efficient desalination processes with a high rejection rate of ions.

The seed layers of silicalite-1 crystal grains sized at ca. 450 nm were prepared on porous alpha-alumina substrates using the spin-coating technique. Dense and defect-free ZSM-5 and Fe-ZSM-5 membranes were fabricated over porous alpha-alumina supported seed layers by secondary growth. The synthesized MFI membranes were characterized by using SEM, XRD, FT-IR, UV-Vis, TG, BET and contact angle measurements. According to the permporometry measurements, the defect size of ZSM-5 membranes could be as large as 3.5 nm after template removal at 450 °C. However, the defect size of Fe-ZSM-5 membranes was less than 0.6 nm when calcined under low-temperature.

The synthesized ZSM-5 and Fe-ZSM-5 membranes were tested with respect to pervaporation and reverse osmosis processes. The results indicated that defects inside zeolite membranes, the surface hydrophilicity of membranes, and the composition of feeding solutions were key factors in desalination processes. Compared with ZSM-5 membranes, Fe-ZSM-5 membranes possessed more satisfactory desalination performance. When dealing with 0.1 mol/L NaCl solution, the ion rejection rate of Fe-ZSM-5 membrane could be up to 98.6% for pervaporation and 93.9% for reverse osmosis, the corresponding fluxes were 0.76 and 1.90 kg/($m^2 \cdot h$), respectively. With respect to multi-component solutions such as 0.1 mol/L NaCl_0.1 mol/L KCl_0.1 mol/L $CaCl_2$, the rejection rate of ZSM-5 and Fe-ZSM-5 membranes would be closely related to their hydrated ion diameters.

Acknowledgements

This research was supported by the National Natural Science Foundation of China (Grant No. 21136008) and the State Oceanic Administration of China (Grant No. 201105026)

497545 Quantifying Bacterial Adhesion to Polymeric Membranes By Single-Cell Force Spectroscopy

Abstract:

Biofouling constitutes a major technical problem in membrane-based processes for water treatment. The permeate flux, selectivity and useful life of the membrane deteriorate as the biofilm develops. Understanding bacterial adhesion, the crucial first step in biofilm formation, is essential for the formulation of biofouling-resistant membranes. In this study, we use AFM-based single-cell force spectroscopy to measure the adhesion force of the bacterium *Pseudomonas fluorescens* on polymeric ultrafiltration membrane surfaces. Our investigation examines the effects of membrane and solution chemistry, and cell-polymeric membrane contact time, on the adhesion of single bacterial cells. This study yields insight into the formulation of biofouling-resistant membranes for water treatment, as well a better understanding of the interaction between bacterial surface structures mediating adhesion to the membrane surface.

495667 Ceramic Membrane Based Technology for the Clarification of Mosambi Znd Orrange Juice

Abstract:

Ultrafiltration (UF) and microfiltration (MF) of mosambi, orange, lemon, grape, apple, carrot, water melon, blood orange and pineapple juice were reported well. Amongst several beverage processing sectors, citrus fruits constituting orange, lemon, pineapple and mosambi are prominent due to their wider availability, low cost as well as high nutrition value to human health. Citrus fruits primarily constitute both lower molecular weight compounds (such as sugar, acid, salt, flavour, aroma compounds, etc.) as well as higher molecular weight polysaccharides (such as pectic material, cellulose, hemicellulose etc.) in addition to haze causing proteins and microorganisms. The presence of pectic material and protein in fruit juice is responsible for cloudiness and post bottling haze formation and their fermentation during long storage. The objective of clarification of fruit juice using membrane filtration is to eliminate the high molecular weight pectic material and its derivatives and retain low molecular weight solutes (valuable for human health) such as sucrose, acid, salt, aroma and flavour compounds in the clarified juice.

Recently, much work has been reported with respect to the clarification of citrus fruits using membrane technology. A critical insight into the literatures infers that mainly polymeric membranes have been studied for mosambi juice and other citrus juice clarification. This is due to the fact that polymeric membranes are inexpensive in comparison with the ceramic membrane and are very easy to use. However, the major drawback of polymeric membrane is their low corrosion resistance in acidic media and hence they possess lesser life cycle times for juice processing applications. For any industrial application, the lifespan of membrane is always a major issue. Many varieties of fruit juices are acidic in nature with pH in the range of 3 to 5. Henceforth, though experimental studies indicate the successful applicability of polymeric membranes for juice processing applications, industrialization of membrane technology for juice processing has not been promising so far. For these circumstances, the usage of ceramic membranes with high corrosion resistance in both acidic as well as basic media is rendered to be more beneficial, especially in the citrus juice sector. One of the striking limitations of ceramic membranes is their higher cost (1600 $\$/m^2$) in comparison with the polymeric membranes (50 to 200 $\$/m^2$ (2010)). The higher cost of ceramic membranes is due to the utilization of expensive inorganic precursors such as alumina and zirconia and higher sintering temperature (more than 1100 °C) during membrane fabrication. Therefore, the development of low cost ceramic membranes is envisaged by utilizing low cost inorganic precursors and lower sintering temperatures.

Microfiltration (MF) of mosambi and orange juice using low cost ceramic membrane prepared from locally available inorganic precursors will be discussed. The average pore diameter, total porosity and hydraulic resistance of the prepared membrane were evaluated as 0.285 μm, 23.6 % and 9.26×10^{11} m^2/m^3, respectively. Dead-end MF experiments were conducted for both centrifuged (CJ) and enzyme treated centrifuged (ETCJ) mosambi and orange juices. It was observed that after MF, important properties like TSS, pH, acidity and density of both CJ and ETCJ were almost unaffected. However, significant improvement in juice colour, clarity and AIS was observed. It was also noticed that the clarified juice can be stored in refrigerated condition for more than 30 days without significant change in juice quality. Different membrane pore blocking models were used to analyze the observed permeate flux decline.

494440 Nanoporous Crystals Channeled Two-Dimensional-Material Membranes with Highly-Enhanced Water Purification Performance

Abstract:

Two-dimensional (2D) materials have been received increasing attention in various fields such as physics, material science, chemistry and engineering. The unique feature of atomic thickness enables 2D materials to develop high-performance separation membranes with maximum permeate flux because of the minimum membrane thickness that can be achieved. As a representative class of 2D materials, graphene-based materials have been used to fabricate membranes for selectively transporting water, ions and gases. In particular, graphene oxide (GO) membranes show great potential in water purification and gas separation due to the facile processing property and versatile functionality of GO. While for practical implementation, the productivity and stability of current GO membranes for water purification still need to be improved. Intercalation and crosslinking approaches were employed to address the challenges, however, selectivity and flux were often reduced, respectively.

In this work, we report nanoporous crystals embedded graphene laminate membranes for water purification. Reduced graphene oxide (rGO) nanosheets obtained from a solution-chemical process act as building blocks to construct the 2D channels through a pressure-driven filtration process. By incorporating the 3D nanoporous crystals with sub-nano sized aperture size (e.g., UiO-66, Prussian blue) into the 2D graphene laminates, both the inter-layer spacing and numbers of nanofluidic channels are increased, leading to a highly-enhanced water transport property. The optimized 3D/2D membranes exhibit 15 times higher water permeability (30 L m^{-2} h^{-1} bar^{-1}) than that of the rGO membrane with similar high dye rejection rate (>95% for rhodamine B). The significance of such 3D nanoporous structure and transport mechanism through the 3D/2D membranes are systematically studied. This general approach of enhancing the molecular transport through 2D nanofluidic channels proposed here may also find application in gas separation and battery membranes.

Acknowledgements

This work was financially supported by the National Natural Science Foundation of China (Nos. 21490580, 21476107, 21406107).

488918 Two-Dimensional Mxene Membrane for Water Purification

Abstract:

Two-dimensional MXENE MEMBRANE for WATER PURIFICATION

Li Ding, Yanying Wei, **Haihui Wang***

South China University of Technology, School of Chemistry and Chemical Engineering, No. 381 Wushan Road, Guangzhou 510640, Guangdong, China

Corresponding author: hhwang@scut.edu.cn Introduction

Two-dimensional (2D) materials are considered to be promising candidates for advanced membrane separation process. Several 2D materials, such as graphene and graphene oxide (GO),[1] exfoliated nanosheets of metal-organic frameworks (MOFs)[2] and zeolite nanosheets,[3] and the transition metal dichalcogenides (TMDs),[4] have attracted increasing attention in preparing for future functional separation membranes due to their outstanding mechanical properties, excellent thermal stability and superior flexibility. Nowadays, a novel kind of 2D layered material named MXenes - a family of early transition metal carbides, has received increasing attention. MXene is normally produced by selectively etching the A layer from MAX phases. Owing to its flexibility, superior structural stability, high electrical conductivity and hydrophilic surfaces, MXene has been widely used in super capacitors,[5] Lithium-ion batteries,[6] oxygen-evolution reaction,[7] and heavy metal adsorption.[8] However, there so far no report on the inorganic MXene-based membranes till now, except the paper by Gogotsi *et al.* for ion sieving.[9] Herein, we propose a kind of 2D lamellar membranes with $Ti_3C_2T_X$ MXene nanosheets, and its application in water purification.

Keywords: Two-dimensional • MXene • Ti3C2TX • membrane • separation • water purification
Experimental

The preparation of the MXene membrane for water purification is as follows. First, Ti_3AlC_2 particles were first etched by HF solution to generate $Ti_3C_2T_X$ powder. By extracting Al as AlF_3, the interaction between the layers is weakened. The MXene nanosheets can be obtained by sonication-assisted exfoliation. Second, the positively charged $Fe(OH)_3$ colloidal solution was chosen to intercalate the negatively charged MXene nanosheets to create expanded nanochannels. Subsequently, after a simple vacuum filtration process and hydrochloric acid solution (HCl) treatment to remove the $Fe(OH)_3$ nanoparticles, the ultimate MXene membrane can be obtained. Results and discussion

The MXene membrane supported on anodic aluminum oxide (AAO) substrate shows excellent water permeance (more than 1000 L m^{-2} h^{-1} bar^{-1}) and favorable rejection rate (over 90 %) for molecules with sizes larger than 2.5 nm. The water permeance through the MXene membrane is much higher than that of the most membranes with similar rejections. Additionally, the long-time operation reveals the outstanding stability of the MXene membrane for water purification.

Conclusions

In summary, a new kind of a 2D lamellar membrane based on stacks of Mxene nanosheets are prepared successfully by filtration deposition on AAO substrates. During the on-filtration, colloidal $Fe(OH)_3$ has been used as distance holder followed by HCl dissolution. The MXene membrane exhibits an excellent water permeance (more than 1000 L m^{-2} h^{-1} bar^{-1}) and a high rejection rate (90 %) for molecules with sizes larger than 2.5 nm when applied in water purification. Mxenes, as a new kind of 2D materials, opens a door for the development of highly efficient membranes for water treatment, selective gas separation and other practical separation applications.

Acknowledgment

We gratefully acknowledge the funding from by the Sino-German center for Science Promotion (GZ 911), the Natural Science Foundation of China for Distinguished Young Scholars of China (no. 21225625), Natural Science Foundation of China (21536005, 51621001) and the Natural Science Foundation of the Guangdong Province (2014A030312007).

References

[1] a) C. Lee, X. Wei, J. W. Kysar, J. Hone, *science* **2008**, *321*, 385-388; b) R. Joshi, P. Carbone, F. Wang, V. Kravets, Y. Su, I. Grigorieva, H. Wu, A. Geim, R. Nair, *Science* **2014**, *343*, 752-754.

[2] Shen, W. Jin, *Angew. Chem. Int. Ed.* **2014**, *53*, 6929-6932; d) J. Shen, G. Liu, K. Huang, W. Jin, K. R. Lee, N. Xu, *Angew. Chem.* **2015**, *127*, 588-592.

[3] K. Varoon, X. Zhang, B. Elyassi, D.D. Brewer, M. Gettel, S. Kumar, J.A. Lee, S. Maheshwari, I.A. Mittal, C.Y. Sung, M. Cococcioni, L.F. Francis, A.V. McCormick, K.A. Mkhoyan, M. Tsapatsis, *Science*, **2011**, *334*, 72-75.

[4] A. Castellanos-Gomez, M. Poot, G. A. Steele, H. S. van der Zant, N. Agraït, G. Rubio-Bollinger, *Adv. Mater.* **2012**, *24*, 772-775.

[5] a) M. R. Lukatskaya, O. Mashtalir, C. E. Ren, Y. Dall'Agnese, P. Rozier, P. L. Taberna, M. Naguib, P. Simon, M. W. Barsoum, Y. Gogotsi, *Science* **2013**, *341*, 1502-1505.

[6] M. Naguib, J. Halim, J. Lu, K. M. Cook, L. Hultman, Y. Gogotsi, M. W. Barsoum, *J. Am. Chem. Soc.* **2013**, *135*, 15966-15969.

[7] a) T. Y. Ma, J. L. Cao, M. Jaroniec, S. Z. Qiao, *Angew. Chem. Int. Ed.* **2015**, *54*, 1-6; b) D. Kong, H. Wang, Z. Lu, Y. Cui, *J. Am. Chem. Soc.* **2014**, *136*, 4897-4900.

[8] Q. Peng, J. Guo, Q. Zhang, J. Xiang, B. Liu, A. Zhou, R. Liu, Y. Tian, *J. Am. Chem. Soc.* **2014**, *136*, 4113-4116.

[9] C. E. Ren, K. B. Hatzell, M. Alhabeb, Z. Ling, K. A. Mahmoud, Y. Gogotsi, *J. Phys. Chem. Lett.* **2015**, *6*, 4026-4031.

488714 Nano-Structuring of UF and RO Membranes with Hydrophilic Polymers â€“ Scalability of Membrane Synthesis Via Atmospheric Pressure Plasma-Induced Graft Polymerization

Abstract:

Surface nano-structuring of membranes with hydrophilic polymer brush layers via atmospheric pressure plasma (APP)-induced graft polymerization (APPIGP) is an effective approach for reducing mineral scaling and biofouling propensity of RO membranes. However, production of large membrane surface area needed for the fabrication of spiral-wound elements requires uniformity of surface activation prior to graft polymerization. Accordingly, in the present study, the application of surface activation by APP, over a large membrane surface area, via a robotic dispensing system was investigated with the goal of optimizing surface activation with respect to the target membrane performance. The plasma treatment time processing conditions were systematically evaluated to arrive at the optimal conditions necessary to achieve uniformity of surface activation. Two different types of APP sources, air and helium-oxygen, were investigated for activating polyamide (PA) based RO and polysulfone (PSf) UF membranes at various plasma operating conditions (i.e., source-substrate distance, plasma exposure time). The optimal plasma operating conditions were quantified via the attainable increase in surface wettability (via contact angel measurements with multiple solvents) which was shown to correlate with the efficiency of subsequent surface graft polymerization. Subsequently, surface graft polymerization was accomplished using a suitable water soluble vinyl monomer to form a polymer brush layer. The uniformity of surface topography was assessed via AFM and XPS analyses, in addition to quantifying membrane permeability and salt rejection. Results of the current study revealed that the APP and graft polymerization conditions can be tailored to provide uniform surface structuring for both RO and UF membranes, which is critical for the scale up of the APPIPG process for producing high performance membranes for water treatment and desalination.

499823 Effect of Membrane Surface Chemistry on Water Permeance and Antifouling Properties

Abstract:

Fouling is one of the major challenges for the adoption of membrane technology for water purification. To mitigate the fouling, membrane surface can be chemically modified to reduce favorable interactions between the membrane and foulants, such as the coating with polydopamine (PDA). In this study, we systematically investigate the effect of various chemistries on the membrane surface on water permeance and antifouling properties. A versatile platform of membrane surface modification using PDA as a bio-glue is adopted. Specifically, membranes are co-deposited with dopamine and functional materials with different hydrophilicity. The functional materials containing thiol and/or acrylate groups can react with PDA through Michael addition to be grafted onto the surface. We have evaluated a series of acrylate-containing macromonomers including decyl acrylate, poly(ethylene glycol) methylether acrylate (PEGMEA), poly (1,3 dioxolane) acrylate (PDXLA), hexafluoroisopropyl acrylate (HFIPA), and a zwitterionic polymer (p($PBMA_{160}$-*co*-$DTMAEL_{42}$)) on UF membranes. Effect of coating solution composition on the membrane surface hydrophilicity is studied. For example, when p($PBMA_{160}$-*co*-$DTMAEL_{42}$), PDXLA, and PEGMEA are grafted, the water contact angle decreases because of the hydrophilic nature of these compounds. The coating layer is also validated using XPS. Effect of the surface modification on the pure water permeance is evaluated. The antifouling properties in the membranes are determined using a crossflow system and model foulants such as bovine serum albumin (BSA) with negative charge, lysosome with positive charge, and alginate. For example, the p($PBMA_{160}$-*co*-$DTMAEL_{42}$)/PDA modified membrane exhibits less flux decline (32% reduction) that those coated by PDA (45% reduction) and unmodified one (53% reduction) after 4-h filtration with a BSA solution. We will systematically compare the effect of the surface chemistry on the water permeance and antifouling properties, aiming to derive the structure/property relationship.

Fluoride removal from Antarctic Krill (Euphausia superba) by Donnan Dialysis

Guojia Yan[1,2], Yan Bao[1], Xiaolan Lu[2], Ming Tan[1]*, Yang Zhang[1]*

[1]Qingdao Institute of Bioenergy and Bioprocess Technology, Chinese Academy of Sciences, Qingdao, 266101, PR China

[2] Ocean University of China, Qingdao 266100, PR China

Introduction

Krill has great health benefits to human being. However, a high fluorine level in Antarctic krill has attracted attention which is the reason why krill has not been widely consumed. In this study, Donnan Dialysis (DD) was used to remove fluoride from different synthetic solutions and real Antarctic krill solution in batch / semi-batch mode. Fluoride content after treatment was expected to fulfill the standard of US FDA in food and Canadian Food Inspection Agency standard in fish and fish products. Effect of receiver NaCl concentration, experiment stages and flow rate were compared. Afterwards, the optimum conditions were applied in a real solution.

Experimental

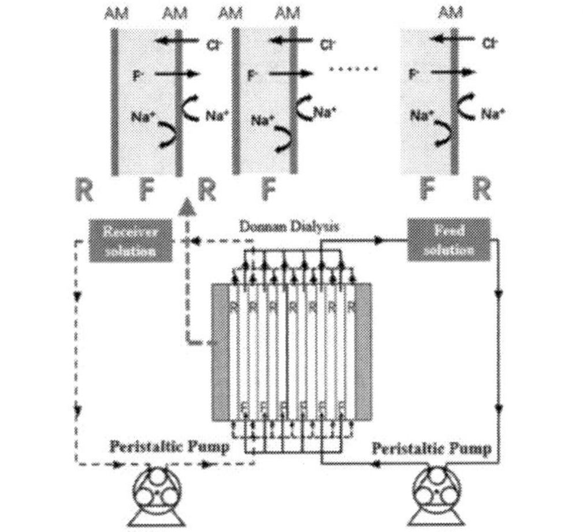

Fig. 1 The mechanism of DD and schematic diagram of the stack configuration

Table 1 Water compositions and experiment conditions of the synthetic solution

	Feed/mg.L^{-1}	Receiver/mol.L^{-1}	Stage	Flow rate/cm.s^{-1}
1	19F$^-$+71Cl$^-$	0.1	1	0.01
2	19F$^-$+71Cl$^-$	0.5	1	0.01
3	19F$^-$+71Cl$^-$	1	3	0.01
4	19F$^-$+71Cl$^-$	1	1	0.03
5	19F$^-$+71Cl$^-$	2	1	0.01

Condition Optimization

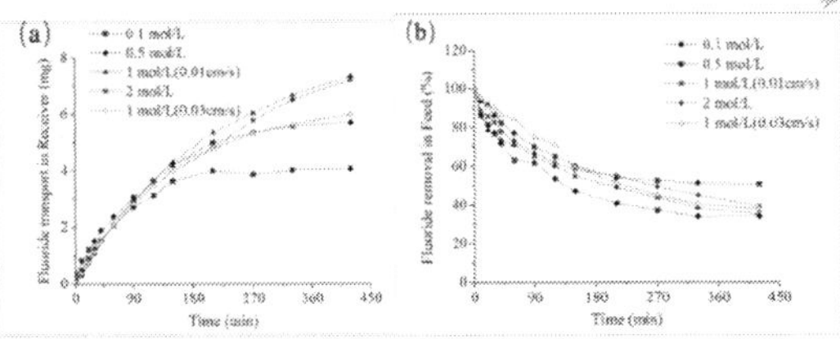

Fig. 2 Fluoride transport with receiver concentration of 0.1, 0.5, 1 and 2 mol L^{-1} (flow rate of 0.01 cm s^{-1} and 0.03 cm s^{-1}). (a) presence of fluoride (weight) in receiver cell, (b) presence of fluoride (percentage) in feed cell.

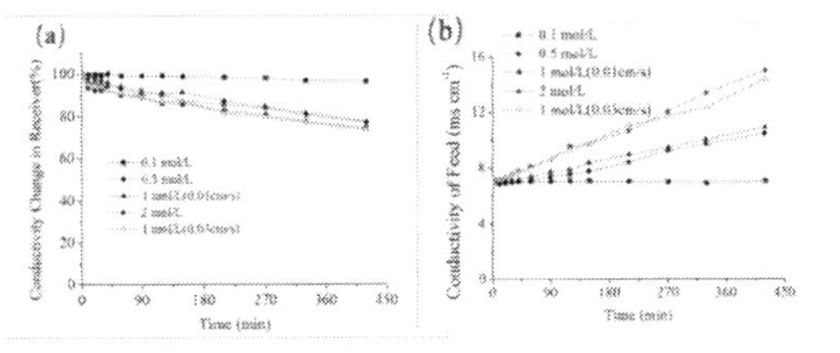

Fig. 6 Correlations between conductivity decline in the receiver cell (a) and conductivity increment (b). 0.1 mol L^{-1}, 0.5 mol L^{-1}, 1 mol L^{-1} (with flow rate of 0.01 cm s^{-1} and 0.03 cm s^{-1})

Results

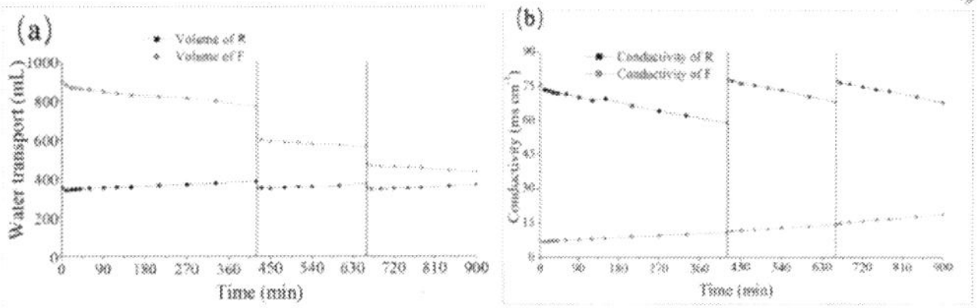

Fig. 8 Water volume (a) and conductivity change (b) of synthetic water with receiver concentration of 1 mol L^{-1} (with flow rate of 0.01 cm s^{-1}).

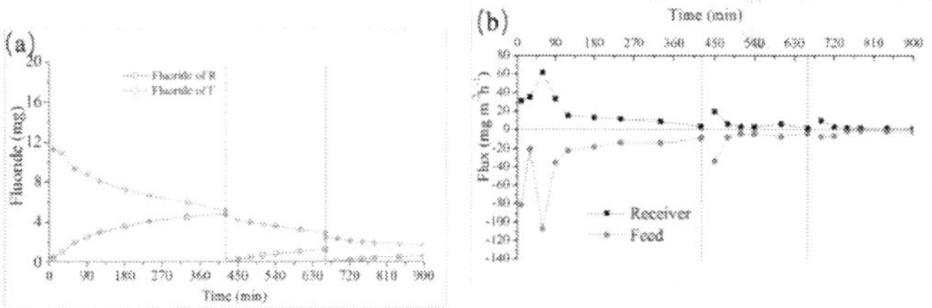

Fig. 9 Fluoride transport of krill solution with receiver concentration of 1 mol L^{-1} (flow rate: 0.01 cm s^{-1}). (a) presence of fluoride (weight) in both cell, (b) presence of flux in both cell

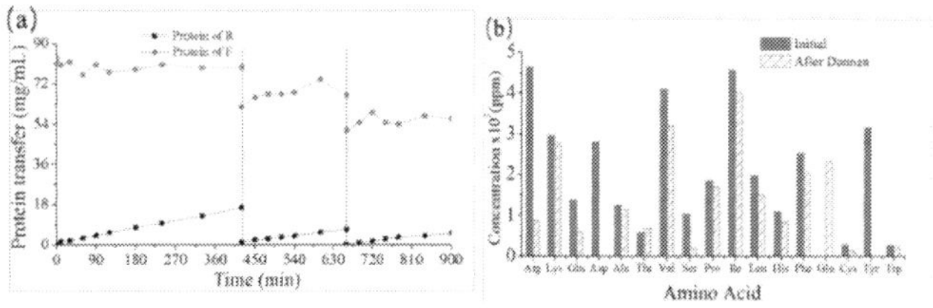

Fig. 11 Protein(a) and amino acids changes(b) in krill solution with receiver concentration of 1 mol L^{-1} (with flow rate of 0.01 cm s^{-1})

Conclusion

- Receiver concentration and flow rate play important roles in fluoride removal by Donna Dialysis. 1 mol L^{-1} NaCl gets higher removal percentage than 0.1, 0.5 and 2 mol L^{-1} NaCl in receiver chamber. Flow rate of 0.01 cm s^{-1} promotes the transport of fluoride. 93.66%±0.81% and 89.34% ± 0.55% fluoride have been removed in synthetic water and real solution by three stages experiment.
- Sulfate moved slower than fluoride in the krill broth, this resulted in a higher removal rate of fluoride than that of sulfate. Even though protein and amino acids decreased 32.57% and 35.58% respectively.
- Owning to a very low energy consumption of DD, the process may treat one cubic of Krill solution with 4.0 USD. if seawater (or concentrated seawater) is used as the receiver solution (0.5 – 1.0 M), the cost of NaCl is negligible.

Acknowledgement

Financial support by the National Science and Technology Major Project of the Ministry of Science and Technology (2016ZX05040-003), the National Natural Science Foundation of China (51508548), Key Research and Development Program of Shandong (2016CYJS07A02) and the Instrument Functional Exploitation and Technical Innovation fund.

http://www.qibebt.ac.cn, **E-mail:** zhangyang@qibebt.ac.cn, **Tel.: +8653280662728**

Contents lists available at ScienceDirect

Journal of Membrane Science

journal homepage: www.elsevier.com/locate/memsci

A new superior competitor for exceptional propylene/propane separations: ZIF-67 containing mixed matrix membranes

Heseong An[a,1], Sunghwan Park[b,1], Hyuk Taek Kwon[c], Hae-Kwon Jeong[c,d], Jong Suk Lee[a,*]

[a] Department of Chemical and Biomolecular Engineering, Sogang Universty, Seoul 121-742, Republic of Korea
[b] Institute of Convergence Chemical Engineering System, Korea University, 5-1 Anam-dong, Seongbuk-gu, Seoul 136-713, Republic of Korea
[c] Artie McFerrin Department of Chemical Engineering, Texas A&M University, 3122 TAMU, College Station, TX 77843-3122, USA
[d] Department of Materials Science and Engineering, Texas A&M University, 3122 TAMU, College Station, TX 77843-3122, USA

ARTICLE INFO

Keywords:
ZIF-67
6FDA-DAM
Mixed matrix membranes
Propylene/propane separations
Physical aging resistance

ABSTRACT

Recently, the ZIF-67 molecular sieve has emerged as an excellent substitute for the ZIF-8 counterpart due to its potentially high C_3H_6/C_3H_8 separation performance. Here, for the first time, we investigated the effect of ZIF-67 in mixed matrix membranes (MMMs) for C_3H_6/C_3H_8 separations by integrating them into 6FDA-DAM polymer matrix. The 6FDA-DAM/ZIF-67 (80/20 wt/wt) MMMs achieved an ideal selectivity of 29.9 and a separation factor of 19 under single and equimolar mixed gas conditions, respectively, at 35 °C. The intrinsic C_3H_6/C_3H_8 separation performance of ZIF-67 was estimated from our experimental transport results based on the Maxwell model. The calculation demonstrated its superior C_3H_6/C_3H_8 selectivity of 251, which is twice of that of ZIF-8, supporting the high molecular sieving capacity of ZIF-67 containing MMMs. Moreover, a thorough investigation on the temperature-dependent gas transports elucidated that the energetic selectivity is a major contributor for the high C_3H_6/C_3H_8 diffusivity selectivity in ZIF-67 containing MMMs. However, the minimum aperture size of ZIF-67 (4.5 Å), derived by the cuboid dimensions of C_3H_6 and C_3H_8, substantially sacrificed the C_3H_6/C_3H_8 entropic selectivity of ZIF-67 containing MMMs. Lastly, the defect-free incorporation of ZIF-67 nanoparticles into 6FDA-DAM polymer matrix effectively retarded physical aging process, maintaining the excellent separation performance of the associated MMMs for 75 days.

1. Introduction

Propylene/propane separation is one of the most important processes in the petrochemical industry due to the highly utile applications of both gases in our everyday life [1]. Propylene is the second most important feedstock next to ethylene with its annually increasing demands: the 90 million tons of global demand in 2014 was expected to continually increase, exceeding 130 million tons in 2023 based on IHS Chemical North American Propylene Supply Study [2]. Currently, propylene/propane (i.e., C_3H_6/C_3H_8) separation is implemented mainly by highly energy-intensive cryogenic distillation due to the similar physical properties between C_3H_6 and C_3H_8 (i.e., T_b of C_3H_6 and C_3H_8 is −47.6 °C and −42.1 °C, respectively) [3]. Therefore, the energy-efficient alternatives have been actively sought-out for the C_3H_6/C_3H_8 separations.

Membrane-based separation technique is promising due to its good energy efficiency as well as low capital cost. Conventional polymeric membranes, however, have been yet to reach and/or surpass the performance limit *so-called* "the upper bound" [4], albeit their great advantage of high processability. A number of different approaches have been attempted to achieve high C_3H_6/C_3H_8 separation performances. While facilitated transport membranes showed prominent performances for C_3H_6/C_3H_8 separations, their poor long-term stability severely depressed the practical utilization [5,6]. Among the diverse strategies, using molecular sieve materials has drawn a great attention due to their excellent and stable C_3H_6/C_3H_8 separation performances. Nevertheless, the inorganic membranes composed of molecular sieves such as zeolite [7] and metal organic framework (MOF) [8] have yet to control the formation of non-selective grain boundary microstructures [9], making it difficult to use those materials solely. Instead, by embedding selective inorganic molecular sieves into polymeric materials, the composite membranes referred to as mixed matrix membranes (MMMs) have demonstrated a promising potential for C_3H_6/C_3H_8 separations [10].

Despite many advantages of MMMs, the non-ideal morphology issues of MMMs are still challenging impediments including particle

* Corresponding author.
E-mail address: jongslee@sogang.ac.kr (J.S. Lee).
[1] These authors contributed equally.

http://dx.doi.org/10.1016/j.memsci.2016.12.053
Received 24 October 2016; Received in revised form 23 December 2016; Accepted 24 December 2016
Available online 26 December 2016
0376-7388/

agglomerations or formation of interfacial voids caused by poor wetting of particles by polymer matrix in the development of MMMs [11]. Also, selection of appropriate molecular sieves whose transport properties are matched up with those of polymer matrix is an arduous task to achieve high separation performance of MMMs [12]. Among many previous studies of MMMs for C_3H_6/C_3H_8 separations, Zhang et al. [13] successfully demonstrated the excellent C_3H_6/C_3H_8 separation performance of MMMs by using 6FDA-DAM polyimide and zeolitic imidazolate framework (ZIF-8). Due to the flexibility of organic linkers, the effective pore size of ZIF-8 was evaluated to be larger than its apparent pore size (i.e., 3.4 Å) and suitable for C_3H_6/C_3H_8 separations [14].

Very recently, Kwon and colleagues substantially improved C_3H_6/C_3H_8 separation performance by heteroepitaxially growing ZIF-67 (Co-substituted ZIF-8) on ZIF-8 seed layers [15]. They suggested that ZIF-67 might be intrinsically more efficient than ZIF-8 for C_3H_6/C_3H_8 separation due to the stiffer Co-N bond as compared to the Zn-N bond in ZIF-8. Also, Krokidas and co-workers [16] further confirmed that ZIF-67 is more efficient for C_3H_6/C_3H_8 separations based on their newly-developed force fields and subsequent molecular dynamics simulations. It is interesting that besides more-commonly observed linker-substitution, metal-substitution also affects the molecular sieving efficiency of the associated ZIF crystals.

In this work, we demonstrated the high C_3H_6/C_3H_8 separation performance of ZIF-67 containing MMMs by using defect-free 6FDA-DAM/ZIF-67 MMMs for the first time. The MMMs showed superior C_3H_6/C_3H_8 separation performances for single gas as well as mixed gas, far exceeding the upper bound limit for polymeric membranes. Also, our thorough investigation on the separation performance of ZIF-67 containing MMMs allows to estimate the inherent separation performance of ZIF-67 for C_3H_6/C_3H_8 separations. Moreover, the temperature-dependent gas transport characterizations were performed to systematically analyze the molecular sieving mechanism. Finally, the influence of the ZIF-67 nanoparticles on the physical aging of MMMs was carefully investigated.

2. Theory and background

Gas separation membranes have two important parameters to evaluate their separation performance: permeability and permselectivity. Permeability is defined as a flux of gas penetrants through membranes normalized by given pressure and membrane thickness.

$$P_i = \frac{N_i \times l}{p_2 - p_1} \quad (1)$$

Where P_i and N_i are the permeability and the flux of gas i, respectively. l is the thickness of membrane and p_1 and p_2 are the pressure of downstream and upstream, respectively. The common unit for permeability is Barrer and defined as:

$$1\ Barrer = 10^{-10} \frac{cm^3(STP)\ cm}{cm^2\ s\ cmHg} \quad (2)$$

The gas transport of dense polymeric membranes is based on the solution-diffusion mechanism. Permeability can be expressed by the product of diffusivity (D_i) for kinetic factor and solubility (S_i) for thermodynamic factor [17].

$$P_i = D_i \times S_i \quad (3)$$

Sorption isotherms in glassy polymers are described by the dual-mode model. This model can be represented by the Henry's law mode and the Langmuir mode: the equilibrium and non-equilibrium structures [18].

$$S_i = \frac{C_i}{p_i} = k_{D,i} + \frac{C'_{H,i} \times b_i}{1 + b_i \times p_i} \quad (4)$$

Where C_i is the concentration of gas i in polymer, p_i is the partial pressure. $k_{D,\ i}$ is the Henrry's constant for Henry's law mode, $C'_{H,\ i}$ is the Langmuir sorption capacity, and b_i is the Langmuir affinity constant. Also, for microporous molecular sieves with a finite number of sorption sites, the solubility is expressed only by the Langmuir mode as follow.

$$S_i = \frac{C_i}{p_i} = \frac{C'_{H,i} \times b_i}{1 + b_i \times p_i} \quad (5)$$

For single gas separations, the permselectivity ($\alpha_{i/j}$) is simply defined as the ratio of the permeability of gas penetrant i over that of j. Also, the permselectivity can be expressed in terms of diffusivity selectivity and solubility selectivity as follows:

$$\alpha_{i/j} = \frac{P_i}{P_j} = \frac{D_i}{D_j} \times \frac{S_i}{S_j} \quad (6)$$

Since diffusion of gas penetrants in polymer is a thermally activated process, it requires the energy barrier, the *so-called* activation energy for diffusional jumps. The temperature dependence of diffusivity is expressed in Arrhenius equation form.

$$D_i = D_{o,i} \exp(-\frac{E_{D,i}}{RT}) \quad (7)$$

Where $D_{o,\ i}$ is a pre-exponential factor of diffusivity, $E_{D,\ i}$ is an activation energy for diffusion, R is the universal gas constant, and T is the temperature. Temperature dependence of solubility can be expressed by van't Hoff equation.

$$S_i = S_{o,i} \exp(-\frac{H_{S,i}}{RT}) \quad (8)$$

Where $S_{o,\ i}$ is a pre-exponential factor of solubility, $H_{S,\ i}$ is a heat of sorption, generally negative. From Eqs. (3), (7), and (8), temperature dependence of permeability is described as:

$$P_i = P_{o,i} \exp(-\frac{E_{P,i}}{RT}) \quad (9)$$

where

$$P_{o,i} = D_{o,i} \times S_{o,i} \quad (10)$$

$$E_{P,i} = E_{D,i} + H_{S,i} \quad (11)$$

3. Experimental

3.1. Materials

6FDA-DAM with a molecular weight of ca. 220,000 g/mol was purchased from Akron Polymer Systems. N-Methyl-2-pyrrolidone (NMP, extra pure, Daejung Chemical Co. LTD) was used as a solvent for the membrane fabrications. Cobalt nitrate hexahydrate ($Co(NO_3)_2 \cdot 6H_2O$, 98%, Sigma-Aldrich) and 2-methylimidazole ($C_4H_6N_2$, 99%, Sigma-Aldrich) were used for the synthesis of ZIF-67 nanoparticles without further purification.

3.2. Preparation of ZIF nanoparticles

Commercially available ZIF-8 particles (Basolite Z1200, Sigma-Aldrich) were used in this work. ZIF-67 nanoparticles were synthesized by modifying a literature procedure [15]. ZIF-67 particles were synthesized at room temperature by intensively mixing a cobalt metal source solution containing deionized (DI) water of 8g and cobalt nitrate hexahydrate of 1.17g with a ligand source solution containing DI water of 80g and 2-methylimidazole of 22.70g for 5 min. Then, an aging step was followed sequentially at room temperature for 12 h. The homogeneous purple mixture was centrifuged at 8000 rpm for 30 min and decanted. To remove all the residual solvent and unreacted speices, the prepared nanoparticles were washed in fresh methanol at least three times and then dried at 100 °C under vacuum for 12 h before use.

3.3. Fabrication of dense films

6FDA-DAM powders were dried at 100 °C under vacuum for 12 h to remove any residual solvent. They were dissolved in NMP with an appropriate concentration. The polymer solutions were rolled for 12 h to obtain completely homogeneous solutions. Neat polymeric membranes were fabricated by using a knife-casting method on the flat and clean glass plate and then immediately dried at 180 °C under vacuum for 12 h. The formed thin films were peeled off from the glass plate and further dried at the same conditions for 12 h. Molecular sieve-containing polymer solutions were prepared by the "priming" method, which can be found elsewhere [19]. Each particle was dispersed in 70 wt% of the total NMP by water bath sonicator (MUJIGAE, South Korea) with a power of 100 W for 1 h. It should be noted that the interaction between nanoparticles and solvent plays a role in uniformly dispersing nanoparticles in solvent [20]. Our previous work exhibited that commercial ZIF-8 particles were well-dispersed in NMP while they were agglomerated in both chloroform ($CHCl_3$) and tetrahydrofuran (THF) [20]. Since ZIF-67 involves the same organic ligands as those for ZIF-8, NMP was chosen as the solvent for the MMMs. Also, a horn type sonicator was avoided because ZIF particles can form agglomerates by the Ostwald ripening effect at this step [21]. The dilute "priming" dope containing 30 wt% of the total NMP and 15 wt% of the total 6FDA-DAM powders was added into suspensions. Remaining 85 wt% of the total 6FDA-DAM was added into the mixture after the dope turned homogeneous. The prepared composite solution was treated by the same procedure as the one for neat polymeric membranes to fabricate mixed matrix membranes.

3.4. Transport characterization

Single gas permeation measurements were preceded by a standard isochoric technique [22] at 2 atm and at three different temperatures (i.e., 35, 50, and 70 °C) using the gases with purity of 99.95%. The permeation test for each type of sample was carried out by using at least four different films from at least four individual fabrications. Each permeation test was conducted after verifying the air leakage rate of permeation systems was less than 1% of the steady state permeation rates. More detailed procedures were described in elsewhere [23]. Also, all the permeation tests were performed as soon as membranes were fabricated except those for the aging study. Samples for the aging study were stored in vacuum condition at 25 °C for the considered aging time. Mixed gas permeation measurements were performed for C_3H_6/C_3H_8 binary gas mixture (50/50 mol/mol) at C_3H_6 partial pressure of 2 atm and 35 °C. The stage cut, which is the fraction of the feed that permeates through the membrane, was controlled less than 1% in order to prevent concentration polarization [24]. The composition of permeates was analyzed by gas chromatography (GC, 7890B (Agilent)). Equilibrium sorption isotherms of C_3H_6 and C_3H_8 in ZIF-67, ZIF-8 nanoparticles, 6FDA-DAM, and ZIF-67 containing MMMs were evaluated at 35 °C by using the pressure decay method [25].

3.5. Supplementary characterization

Field emission gun scanning electron microscopy (FEG-SEM, Inspect F (FEI)) was employed with 10 kV to characterize the morphology of our synthesized ZIF-67 nanoparticles as well as MMMs. Fourier transform infrared spectroscopy (FT-IR, FT-IR frontier (Perkin-Elmer)) was used to evaluate interactions between 6FDA-DAM polymer and ZIF-67 in MMMs over the range of 4000–400 cm^{-1}. Wide angle X-ray diffraction (WAXD, D/Max 2500/PC (Rigaku)) was used to verify the phase and crystallinity of ZIF-67 nanoparticles with Cu-Kα radiation (λ=1.5406 Å). The glass transition temperature (T_g) of membranes was analyzed by using differential scanning calorimetry (DSC, Q20 TA instruments) to investigate change of polymer chain mobility with the incorporation of ZIF-67. Each sample was heated from 25 °C to 100 °C at a ramp rate of 5 °C/min under N_2 purge of 50 ml/min. The samples were heated at 100 °C for 30 min to remove any moisture and then were heated to 410 °C at the same condition. Thermo gravimetric analysis (TGA, TGA1 (METTLER TOLED)) was performed to determine the actual content of ZIF nanoparticles in MMMs. More specifically, TGA was run from 25 °C to 800 °C with a heating rate of 5 °C/min under air atmosphere (50 cm^3/min), fully transforming the samples to cobalt oxide (CoO). The residual cobalt (Co) weight percentage of ZIF-67 was comparable with the theoretical weight percentage of Co in ZIF-67 nanoparticles (i.e., 26.4%). As such, the actual weight percentages of ZIF-67 in MMMs were calculated by the ratio of Co wt% of MMM to that of ZIF-67 [13].

4. Results & discussion

4.1. Characterization of ZIF-67 nanoparticles

The house-made ZIF-67 nanoparticles showed a uniform particle size of 300 – 400 nm with dodecahedral shapes (Fig. 1a). As shown in Fig. 1b, the FT-IR spectra of ZIF-67 indicated covalent bonds between cobalt and nitrogen of 2-methylimidazole (2-mIM) ligands at the designated wavenumber of 426 cm^{-1} [15]. The XRD patterns of ZIF-67 exhibited a typical sodalite (SOD) topology (Fig. 1c) with theoretical aperture size of ~ 3.3 Å [16]. Nevertheless, due to the flexible 2-methylimidazolate (2-mIM) ligands in ZIF-67, the effective aperture size is anticipated to be larger than the theoretical one similarly to the case of ZIF-8, whose theoretical pore size is 3.4 Å but the effective pore size is found between 4.0 and 4.2 Å [14]. Also, the equilibrium sorption isotherms of C_3H_6 and C_3H_8 for ZIF-67 nanoparticles were characterized at 35 °C (Fig. 1d). As represented in Fig. 1d, the sorption capacities of C_3H_6 and C_3H_8 for ZIF-67 nanoparticles are analogous; therefore, the C_3H_6/C_3H_8 sorption selectivity was 1.1 at 29.3 psia. The equilibrium sorption isotherm data for both C_3H_6 and C_3H_8 in ZIF-67 were fitted well with the Langmuir model in Eq. (5) and the associated parameters are summarized in Table S1. Both Langmuir sorption capacity, C_H' and Langmuir affinity constant, b for each penetrant in ZIF-67 were very similar to those for ZIF-8 within the experimental uncertainties. It reflects that the accessible micro-voids and affinities of ZIF-67 for C_3H_6 and C_3H_8 are comparable with those of ZIF-8 (Table S1), despite the different transition metals connected with 2-mIMs (i.e., Co for ZIF-67 vs. Zn for ZIF-8).

4.2. Physical properties of ZIF-67 containing MMMs

ZIF-67 nanoparticles were well-dispersed in the 6FDA-DAM (PI) polymer matrix up to nominal concentration of 20 wt% with the assistance of proper ultra-sonication treatments while mitigating the Ostwald ripening [21]. As shown in Fig. 2, the uniformly distributed ZIF-67 nanoparticles showed good adhesion with PI. It should be noted that ZIF-67 particles at the concentration of 30 wt% started to agglomerate in MMMs with a selectivity loss of 10%. This is likely due to the fact that the van der Waals interactions between ZIF-67 particles are stronger than the interaction of ZIF-67 particles with either polymer or solvent (Fig. S1) [11,26]. Such non-ideal morphology is detrimental to the separation performance because gas penetrants prefer to bypass through the non-resistant void volumes formed within the agglomerates regardless of their sizes. Nevertheless, our 6FDA-DAM/ZIF-67 MMMs with up to 20 wt% of ZIF-67 exhibited that the formation of ZIF-67 clusters was efficaciously mitigated probably due to a combination of the selection of an appropriate solvent (i.e., NMP) and the exclusion of direct sonication. For the sake of convenience, the ZIF-67 containing MMMs used in this work were named as a function of ZIF-67 contents: PZ67-10 for PI/ZIF-67 (90/10 wt/wt) and PZ67-20

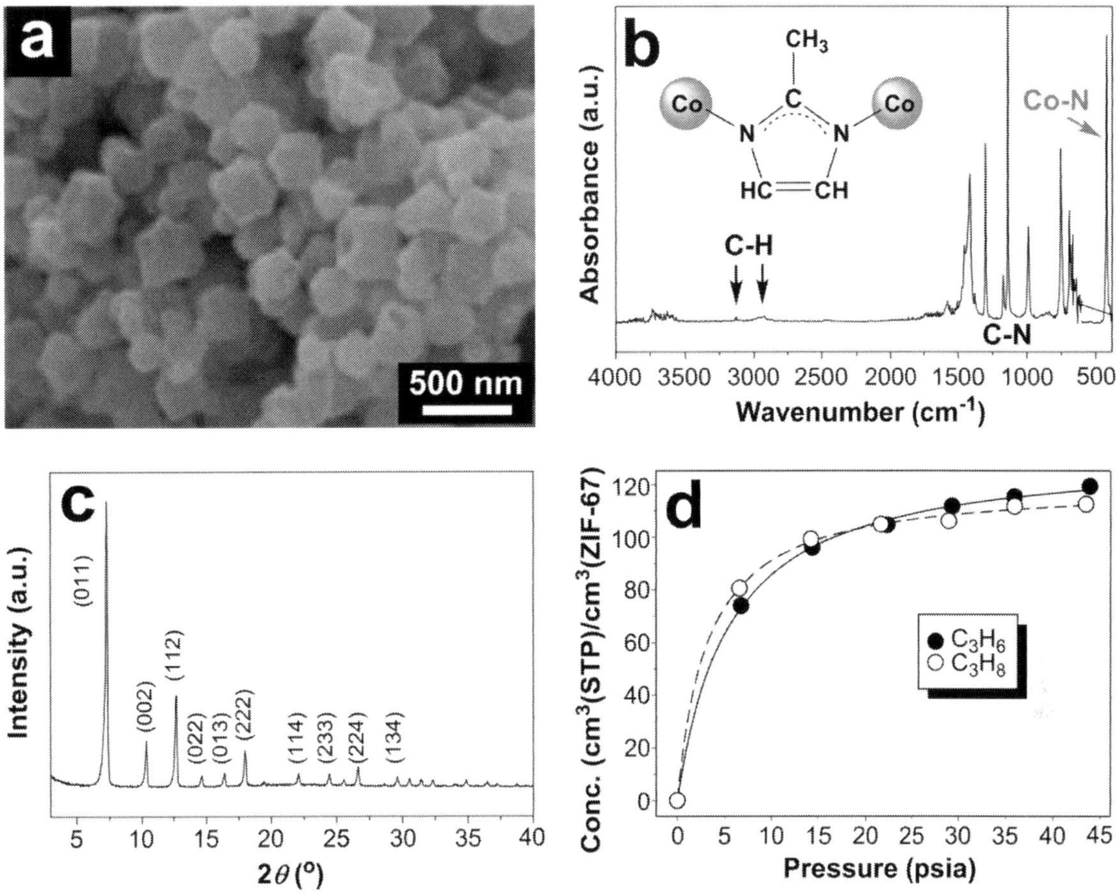

Fig. 1. ZIF-67 nanoparticle characterizations: SEM (a), FT-IR (b), XRD (c), and sorption isotherm for C_3H_6 and C_3H_8 (d) at 35 °C.

for PI/ZIF-67 (80/20 wt/wt).

Specific interaction between 6FDA-DAM and ZIF-67 nanoparticles was investigated by conducting FT-IR analysis. The FT-IR spectra of ZIF-67 MMMs were almost identical to those of the neat 6FDA-DAM (Fig. S2), confirming that no specific interaction exists between 6FDA-DAM and ZIF-67. This is not surprising since 6FDA-DAM/ZIF-8 MMMs also showed no specific interaction between 6FDA-DAM and ZIF-8 which involves the same organic ligand of 2-mIM [27]. Nevertheless, the 2-mIM-induced hydrophobic properties of ZIF-67 nanoparticles allow a good affinity with the polymer matrix, suppressing the formation of interfacial voids [13]. Despite the lack of specific interaction, the strong affinity between polymer matrix and dispersed particles tends to rigidify the polymer matrix around the embedded particles [27]. As shown in Fig. S3, the glass transition temperature (T_g) of PI (i.e., 371 °C), which is consistent with the reference value [28], increased upon the addition of ZIF-67 nanoparticles into the polymer matrix. While the rigidification of polymer chains itself can induce changes in separation performance (typically decreasing permeability with increasing selectivity), its effect on the inherently rigid polymer with high T_g is generally negligible [29].

As the gas separation performances of MMMs depend highly on the amount of embedded particles, the actual contents of ZIF-67 in MMMs are important for the accurate characterization of gas transport in MMMs [30]. The residual weight percentages of ZIF-67 and their MMMs were characterized using TGA, (Fig. S4) and the corresponding percentages of ZIF-67 in MMMs are shown in Table S2. The actual ZIF-67 concentrations in MMMs are nearly comparable with the nominal ones as shown in Table S2.

4.3. Gas transport properties of ZIF-67 containing MMMs

4.3.1. C_3H_6/C_3H_8 separation performances

Both C_3H_6 and C_3H_8 single gas transport behaviors in pristine 6FDA-DAM (PI) and ZIF-67 containing MMMs were characterized at 2 atm and 35 °C and the corresponding results are summarized in Table 1. The C_3H_6 and C_3H_8 permeabilities in PI are 16.63 and 1.47 Barrer, respectively, consistent with the reference values within the experimental uncertainties (i.e., C_3H_6 permeability: 15.7 Barrer and C_3H_8 permeability: 1.27 Barrer) [13]. In MMMs, the C_3H_6/C_3H_8 separation performance was continually improved with increasing the content of ZIF-67, overcoming the polymeric upper bound limit for C_3H_6/C_3H_8 separations (Fig. 3) [4]. Both C_3H_6 permeability and C_3H_6/C_3H_8 permselectivity of PZ67-20 increased by as much as 105.3% and 164.6%, respectively, comparing with those of PI. Interestingly, the C_3H_8 permeability decreased with increasing ZIF-67 concentrations in MMMs, which is attributed to the realization of molecular sieving effect of ZIF-67 for C_3H_8 gas molecules, significantly increasing the C_3H_6/C_3H_8 permselectivity. It should be noted that the realization of molecular sieving effect of ZIF-67 in MMMs was due to a proper match with respect to the permeable behaviors between PI and ZIF-67 as well as a good interfacial adhesion between them.

Furthermore, C_3H_6/C_3H_8 separation performances for PI/ZIF-67 MMMs were compared to those for PI/ZIF-8 counterparts, since ZIF-8 nanoparticles are known to be the most appropriate molecular sieve candidate for C_3H_6/C_3H_8 separations [31,32]. The ZIF-8 containing MMMs, which were also named PZ8-10 for 6FDA-DAM/ZIF-8 (90/10 wt/wt) and PZ8-20 for 6FDA-DAM/ZIF-8 (80/20 wt/wt), were

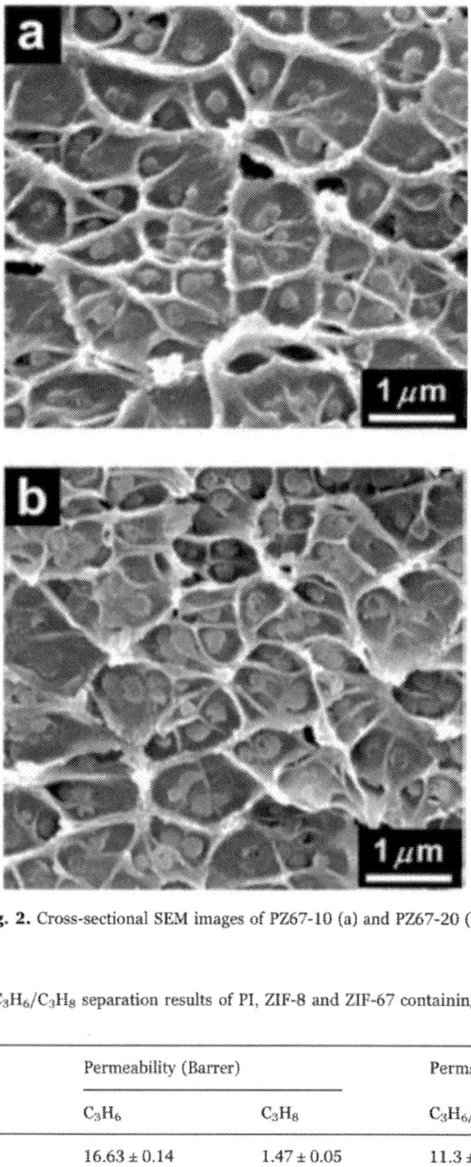

Fig. 2. Cross-sectional SEM images of PZ67-10 (a) and PZ67-20 (b).

Table 1
Single gas C_3H_6/C_3H_8 separation results of PI, ZIF-8 and ZIF-67 containing MMMs at 35 °C.

Sample	Permeability (Barrer)		Permselectivity
	C_3H_6	C_3H_8	C_3H_6/C_3H_8
PI	16.63 ± 0.14	1.47 ± 0.05	11.3 ± 0.3
PZ67–10	24.52 ± 0.11	1.25 ± 0.02	19.6 ± 0.2
PZ67–20	34.14 ± 0.04	1.14 ± 0.01	29.9 ± 0.2
PZ8–10	26.10 ± 0.03	1.60 ± 0.03	16.3 ± 0.3
PZ8–20	37.73 ± 0.56	1.82 ± 0.06	20.7 ± 0.4

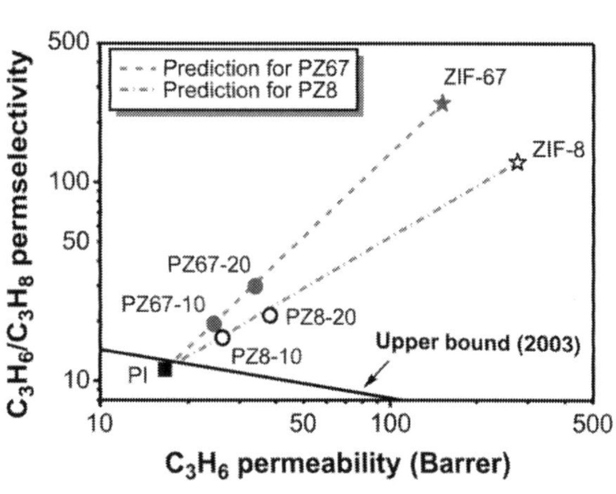

Fig. 3. Single gas C_3H_6/C_3H_8 separation performances of PI and MMMs and the predictions for MMMs and ZIFs with the upper bound curve [4].

fabricated by the same process as the one for PI/ZIF-67 MMMs and their gas permeation measurements were performed at the same conditions. The permeation results of our ZIF-8 containing MMMs were very close to those of ZIF-8 containing MMMs demonstrated by Zhang et al. [13] (Fig. S5). As shown in Fig. S5, the PZ67 MMMs exhibited a significant improvement in C_3H_6/C_3H_8 selectivity compared to PZ8 MMMs, but rather lower C_3H_6 permeability. Especially, the C_3H_6/C_3H_8 permselectivity of PZ67-20 which involves the actual ZIF-67 content of 21.1 wt% was comparable with that of PZ8 MMMs containing the actual ZIF-8 content of 48.0 wt% by Zhang et al. [13]. As indicated in Table S3, such difference was caused mainly by diffusivity selectivity rather than solubility selectivity. As discussed in the Section 4.1, the sorption capacities of ZIF-67 for both C_3H_6 and C_3H_8 gas molecules are almost identical, which is similar to ZIF-8 nanoparticles. The ZIF-67 containing MMMs, however, exhibited 25 ~ 32% greater diffusivity selectivity than ZIF-8 containing analogs due to the enhanced molecular sieving behavior of ZIF-67 (Table S3).

The intrinsic C_3H_6/C_3H_8 separation performances of ZIF-67 and ZIF-8 were predicted from our single gas permeation results by using the Maxwell model [13] and are represented in Table S4. The validity of Maxwell model-based C_3H_6 and C_3H_8 permeabilities of ZIFs was verified by confirming that our predicted permeation results for ZIF-8 are almost identical with those for ZIF-8 by Zhang et al. [13] (See Table S4). It should be noted that the direct comparison of the predicted separation performance of ZIF-8 with the experimental counterpart from neat polycrystalline ZIF membranes is far-fetched due to the difficulty of controlling grain boundaries of polycrystalline ZIF membranes. The intrinsic C_3H_6 permeability and C_3H_6/C_3H_8 selectivity of ZIF-67 are 150.7 Barrer and 251, respectively (vs. C_3H_6 permeability of 275.6 Barrer and C_3H_6/C_3H_8 selectivity of 125 for ZIF-8) (Fig. 3). ZIF-67 provided 45% lower C_3H_6 permeability, but 100% higher C_3H_6/C_3H_8 permselectivity than ZIF-8, thereby affecting the tendency of C_3H_6/C_3H_8 separation performance for the respective MMMs. As Kwon et al. demonstrated based on FT-IR spectra [15], the coordination bonding between cobalt and nitrogen in ZIF-67 is stiffer than that between zinc and nitrogen in ZIF-8, consequently leading to the more restricted flipping motion of 2-mIM organic ligands and thereby the reduction in the effective aperture size. This results in improved discrimination between C_3H_6 and C_3H_8 penetrants while leading to reduced permeabilities of C_3H_6 and C_3H_8. It is noteworthy that improving C_3H_6/C_3H_8 selectivity of polymer membranes is challenging due to their similar molecular size as well as condensable properties [33].

Furthermore, the mixed gas permeation measurements were performed for PI and ZIF-67 containing MMMs by using a binary gas mixture of C_3H_6/C_3H_8 (50/50 mol/mol) at C_3H_6 partial pressure of

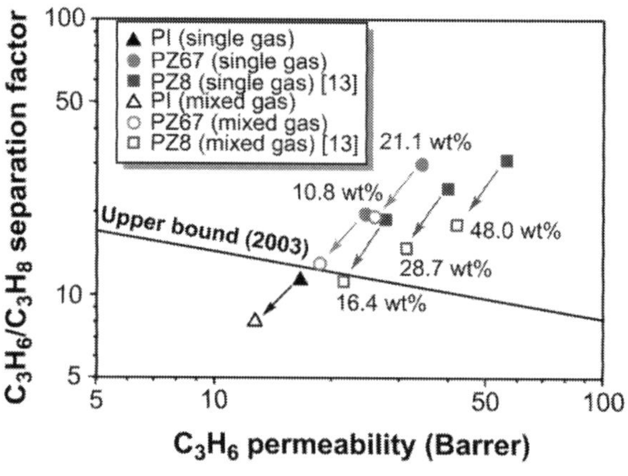

Fig. 4. C_3H_6/C_3H_8 separation performances of PI and PI-based MMMs under single as well as mixed gas conditions. Note: the data of PZ8 were taken from the Zhang et al.'s work [13] for comparison and the numbers next to the symbols of MMMs represent the actual weight fractions of ZIFs.

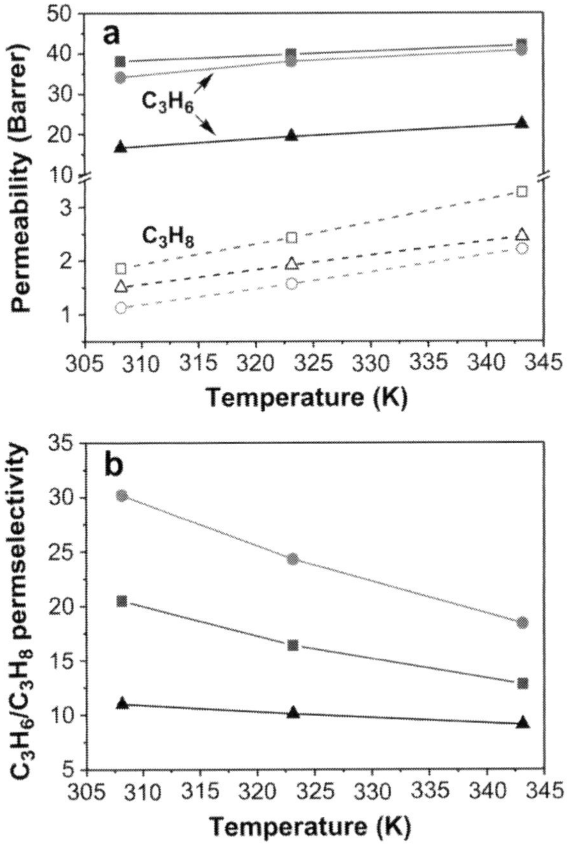

Fig. 5. Temperature dependence of permeability (a) and permselectivity (b) for PI (triangle), PZ67-20 (circle), and PZ8-20 (square) at 2 atm.

2 atm and 35 °C with the mixed gas separation performance for PI/ZIF-8 MMMs from Zhang et al.'s work [13] for reference (Fig. 4). As represented in Fig. 4, the mixed gas separation performance was depreciated compared to the corresponding single gas counterpart due to the competition effect between two gas molecules as well as C_3H_6-induced bulk effect. The strong negative sorption coupling effects between C_3H_6 and C_3H_8 owing to the similarity of their condensabilities may diminish C_3H_6 permeability and C_3H_6/C_3H_8 separation factor [34]. Also, the high permeability of C_3H_6 for PI and ZIF-67 containing MMMs induces a bulk effect, which accelerates movement of larger gas molecules (C_3H_8) by being carried by smaller gas molecules (C_3H_6), abating the separation factors in mixed gas conditions [35]. Nevertheless, it should be noted that a high separation factor can be still realized at a practical concentration level (i.e., 20 wt%) of molecular sieves by utilizing ZIF-67 nanoparticles.

4.3.2. Temperature-dependent characteristics of gas transports

Temperature dependence of C_3H_6/C_3H_8 separation performance was evaluated by characterizing transport behaviors at three different temperatures of 35, 50, and 70 °C in order to further investigate the molecular sieving behavior of ZIF-67 and ZIF-8 nanoparticles in MMMs. Both C_3H_6 and C_3H_8 permeabilities of all samples (i.e., PI, PZ67-20, and PZ8-20) were gradually enhanced with increasing operating temperature (Fig. 5a). The activation energy for permeation is decided by the combination of two competitive components: the activation energy of diffusion (positive) and the heat of sorption (negative) (See Eq. (11)). Axiomatically, the permeability augmentation with increasing temperature indicated that the diffusivity is a more dominant contributor than the solubility for all three samples. Also, Fig. 5b exhibits that the C_3H_6/C_3H_8 selectivity decreased continually for all samples with increasing temperature, since the larger penetrant, C_3H_8, has higher diffusional activation energy than the smaller penetrant, C_3H_6 (Table 2).

Table 2
Activation energies for permeation (E_P) and diffusion (E_D) and heats of sorption (H_S) for PI and MMMs.

Sample	Gas	E_P (kJ/mol)	E_D (kJ/mol)	H_S (kJ/mol)
PI	C_3H_6	7.3	20.4	−13.1
	C_3H_8	12.2	24.4	−12.2
PZ67−20	C_3H_6	4.9	16.8	−11.9
	C_3H_8	16.7	31.6	−14.9
PZ8−20	C_3H_6	2.4	11.6	−9.2
	C_3H_8	14.2	22.4	−8.2

Overall, the permeation activation energy of C_3H_6 decreased while that of C_3H_8 increased with the addition of molecular sieves (ZIF-67 and ZIF-8 nanoparticles) compared to that of respective penetrant in PI. Comparing with the activation energy of permeation for PI, the extent of reduction in the activation energy of permeation of C_3H_6 for PZ67-20 was less than that for PZ8-20, but the extent of increase in the permeation activation energy of C_3H_8 was greater for PZ67-20 than that for PZ8-20, which is consistent with our single gas permeation results (See Fig. 3). These contrasting changes in the activation energy for permeation through PZ67-20 and PZ8-20 are construed in terms of diffusivity and solubility for each MMM. The diffusional activation energies of C_3H_6 for both PZ67-20 and PZ8-20 decreased compared to that for PI due to the free diffusion of C_3H_6 through the large cages of the molecular sieves. Interestingly, the diffusional activation energy of C_3H_8 through PZ8-20 still decreased compared to that for PI while that through PZ67-20 rather increased. This observation strongly supports that ZIF-67 performs better for the C_3H_6/C_3H_8 separation based on molecular sieving principle than ZIF-8, even though both PZ67-20 and PZ8-20 substantially increased the absolute values of $\Delta E_{D,\ C3H6,C3H8}$ (i.e., $|E_{D,\ C3H6}-E_{D,\ C3H8}|$) as compared to PI. The heat of sorption of the MMMs (PZ67-20 and PZ8-20) also exhibits different trends depending on the molecular sieves (Table 2). In general, the sorption process of a gas in a polymer involves an exothermic process and the more negative the heat of sorption, the stronger the interactions between the gas and the polymer. Both PI and PZ8-20 show a negative value of $\Delta H_{S,\ C3H6,C3H8}$ (i.e., $H_{S,\ C3H6}-H_{S,\ C3H8}$), implying that the sorption of C_3H_6 is more favorable than that of C_3H_8 in both PZ8-20 and PI. The PZ67-20, however, exhibits a positive value of $\Delta H_{S,\ C3H6,C3H8}$, indicating a more favorable sorption of C_3H_8 than C_3H_6 in PZ67-20, which may consequentially make it difficult for C_3H_8 to desorb from the sorption site on diffusional jump.

4.3.3. Energetic selectivity vs. entropic selectivity

For C_3H_6/C_3H_8 separation, hitherto, ZIF-67 and ZIF-8 nanoparticles in MMMs showed the different extents of molecular sieving effects, which strongly affect the diffusivity selectivity of MMMs. Diffusivity selectivity can be separated into the energetic selectivity associated with the size of gas penetrants and membrane pores and the entropic selectivity associated with the shape of gas penetrants and membrane pores [36]. There have been a few studies considering the energetic and entropic selectivity of polymeric membranes for gas separations and most of them are focused on O_2/N_2 separations. Especially, the Koros group intensively studied energetic and entropic contributions to the diffusivity selectivity of O_2/N_2 separations for several 6FDA-based polyimide membranes [37]. They concluded that the entropic selectivity of those polymers is negligible or favorable to N_2 than O_2 (i.e., the entropic selectivity of polymeric membranes for O_2/N_2 separations was less than unity). Nevertheless, the diffusivity selectivity of most 6FDA-based polyimides in their work was greater than 4 due to the high energetic selectivity. The relatively large size of N_2 requires more activation energy than O_2 for diffusional jump which restricts its movements through locally ordered chain segments [37]. Although

Table 3
The energetic, entropic, and diffusivity selectivity for C_3H_6/C_3H_8 separations in PI, PZ67-20 and PZ8-20.

Sample	Energetic selectivity	Entropic selectivity	Diffusivity selectivity
PI	4.8 ± 0.002	2.1 ± 0.01	10.1 ± 0.1
PZ67–20	322.7 ± 0.4	0.08 ± 0.0001	24.8 ± 0.2
PZ8–20	67.8 ± 0.3	0.28 ± 0.001	18.7 ± 0.2

we confirmed that the 6FDA-DAM exhibits the O_2/N_2 entropic selectivity less than unity (Table S5), the entropic contribution for C_3H_6/C_3H_8 separations was 2.1 (Table 3) which is probably associated with the nature of polyatomic molecules.

Previous research demonstrated that a high separation performance of inorganic molecular sieves including zeolite 4A or carbon molecular sieve (CMS) membranes results from a considerable contribution of entropic selectivity [33,36,38]. Unlike CMS membranes possessing slit-like rigid pore structures, the high C_3H_6/C_3H_8 separation performances of both PZ8-20 and PZ67-20 were realized by a substantial enhancement of energetic selectivity (Table 3). ZIFs in MMMs significantly improved the energetic selectivity with sacrificing entropic selectivity in increasing diffusivity selectivity for C_3H_6/C_3H_8 separations, which may be attributed to the well-defined cylinder-like pore structures of the ZIFs by the six membered rings. Such high energetic contribution to diffusivity selectivity of ZIF-containing MMMs was consistent with the results from polycrystalline ZIF-8 membranes; the ZIF-8 membranes with C_3H_6/C_3H_8 diffusivity selectivity of 125 showed an extremely high energetic selectivity of 1.3E+11 and a low entropic selectivity of 1.0E-10 [39]. Moreover, the energetic selectivity of C_3H_6/C_3H_8 for PZ67-20 was 375% greater than that for PZ8-20, while the entropic selectivity of C_3H_6/C_3H_8 for PZ67-20 was 71% less than that for PZ8-20 (Table 3). A significant improvement in the energetic selectivity of C_3H_6/C_3H_8 for PZ67-20 results in higher C_3H_6/C_3H_8 diffusivity selectivity compared to PZ8-20 albeit lower C_3H_6/C_3H_8 entropic selectivity.

The higher C_3H_6/C_3H_8 energetic selectivity of PZ67-20 over PZ8-20 is attributed to the efficient size sieving effect of ZIF-67. When gas penetrants execute diffusional jumps from a normal state to another normal state, they have to pass through the activated state by overcoming the repulsion energy as well as the desorption energy [40]. In molecular sieving media such as ZIF-67, the normal state indicates the sorbed gas molecules within their large cavities, while the activated state corresponds to the gas molecules passing through constricted passages between large cavities [37]. Although both ZIF-67 and ZIF-8 involve the same organic ligand, 2-mIM, the higher electronegativity of cobalt in ZIF-67 than that of zinc in ZIF-8 induces stiffer ionic bond with nitrogen in 2-mIM, possibly restricting the flipping motion of organic ligands [15,16]. Consequentially, the more restricted flipping motion increased the efficiency of molecular sieving effect of ZIF-67 comparing with that of ZIF-8, requiring more repulsion energy for the diffusional jump of C_3H_8. Also, the more negative heat of sorption of C_3H_8 than that of C_3H_6 in ZIF-67 (Table 2) corresponds to more desorption energy of C_3H_8, contributing to increase in the energetic selectivity of C_3H_6/C_3H_8.

Although the smaller entropic selectivity of PZ67-20 than PI is counter-intuitive (Table 3), this can be explained by applying the transition state theory (TST) to C_3H_6/C_3H_8 transport behaviors in ZIF-67. Based on the TST, the entropic selectivity can be estimated by dividing the ratio of partition functions of the fast gas for the activated and normal states by that for the slow gas (Equation S2) [36]. The partition function is the product of the translational, vibrational, and rotational partition functions (Equation S3). It should be noted that the vibrational partition functions can be simply withdrawn for the consideration of entropic selectivity in TST due to the unrestricted movements of gas molecules in both normal and activated states [36]. The partition functions for polyatomic gas molecules such as C_3H_6 and C_3H_8 were calculated by using the Equations S4, S5, and S6 represented in Supporting Information (Table S6).

As shown in Fig. 6, both C_3H_6 and C_3H_8 are simplified as cuboids with the corresponding dimensions [33]. In order for both C_3H_6 or C_3H_8 to penetrate through ZIF-67, the minimally required aperture size of ZIF-67 could be conjectured to be at least 4.5 Å based on their cuboid dimensions (Fig. 6). Nevertheless, it should be noted that the effective pore size of ZIF-67 could be smaller than 4.5 Å. Also, the penetrants slightly larger than 4.5 Å might be still accessible to ZIF-67, due to the flexible characteristics of 2-mIM in ZIF-67 [41]. The large cavities of ZIF-67 where gas molecules stay at the normal state were presumed to have the same size as those of ZIF-8 (i.e., 12.5 Å) [39], since the sorption capacity of ZIF-67 was analogous to that of ZIF-8 as discussed in the Section 4.1. In these regards, both C_3H_6 and C_3H_8 penetrants in the large cavities of ZIF-67 can freely translate into x, y, and z directions coupled with all the possible rotations; both C_3H_6 and C_3H_8 penetrants have 3 translational modes as well as 3 rotational modes at the normal state of ZIF-67 as depicted in Fig. 5. When those gas molecules, however, are in the activated state, the translation of C_3H_8 would be allowed to the directions of y and z within a very limited space of 0.2 Å, which is the difference between the presumed window size of ZIF-67 (4.5 Å) and the cross-sectional dimensions of C_3H_8 (4.3 Å), while the translation of C_3H_6 would be possible only for y direction (Fig. 5). Interestingly, although C_3H_8 exhibited one more translational degree of freedom compared to C_3H_6 at the activated

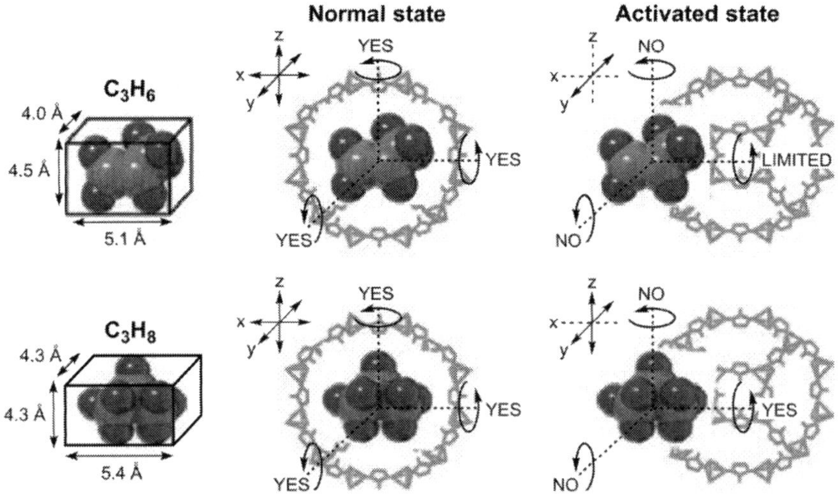

Fig. 6. Illustration of translational and rotational degree of freedom in the normal and activated state for C_3H_6 and C_3H_8 in passing through ZIF-67.

state, the translational entropic selectivity of C_3H_6/C_3H_8 for ZIF-67 was close to 2 (Table S6). It is because the y-directional translational movements of planar C_3H_6 might be more mobile than the movements of bulky C_3H_8. The significantly declined entropic selectivity of PZ67-20 was mainly ascribed to the considerable rotational limitation of C_3H_6 compared to C_3H_8 in the activated state of ZIF-67, albeit the impeded rotations in the direction of y and z for both penetrants in the activated state. While C_3H_8 gas molecules were able to rotate along the x axis in the activated state, the vertical rotation of C_3H_6 was very limited, since the height of C_3H_6 of 4.5 Å was admittedly stuck in the expected minimum window size (i.e., 4.5 Å) of ZIF-67. The low entropic selectivity of C_3H_6/C_3H_8 (i.e., 0.08) for ZIF-67 essentially resulted from the remarkably low rotational entropic selectivity of 0.04 (Table S6).

4.3.4. Physical stability of ZIF-67 containing MMMs

Substantial reduction in the permeability of glassy polymeric membranes over time, *so-called* physical aging, is a critical drawback for their gas separation applications. It is, therefore, of great interest to arrest their physical aging. In general, this physical aging is more critical for glassy polymers with high fractional free volume (*FFV*) since their non-equilibrium excessive free volume tends to become equilibrium states via the time-dependent chain relaxation process or densification [42].

The C_3H_6 permeability variations for both PI and PZ67-20 were monitored up to 75 days at 35 °C and 2 atm while the thickness of the membranes was maintained similar (i.e., 10 ± 1 μm) in order to eliminate any thickness-induced complication [42–44]. As shown in Fig. 7a, the permeability of C_3H_6 for PI substantially decreased in the early stage, especially within 10 days, and reached 3.82 Barrer in 75 days, which is only 25% of the initial permeability. Changes in C_3H_6 permeability for PZ67-20 were, however, relatively marginal over time. A closer analysis reflects that a substantial reduction in the C_3H_6 permeability for PI can be attributed likely to the substantial reduction in the C_3H_6 diffusivity. On the contrary, the PZ67-20 exhibits gradual, but still marginal reduction in both C_3H_6 diffusivity and C_3H_6 solubility (Fig. 7b and c).

To further investigate the reduced physical aging rate of ZIF-67 containing MMMs compared to PI, the dual-mode model parameters of C_3H_6 and C_3H_8 for pristine and aged PI and PZ67-20 were obtained by fitting the experimental sorption isotherm data represented in Fig. 8 (Table 4). The Langmuir affinity constant, *b* values of C_3H_6 and C_3H_8 for both PI and PZ67-20 remain almost constant over the test period and also the slight deviations are still within error ranges considering that all samples were still being aged during characterization. Langmuir affinity constant of a specific gas depends on the chemical structure and therefore, it is not expected to change over time as long as the chemical structure is conserved. The PI, however, exhibits severe reduction in Henry's law constant, k_D and Langmuir sorption capacity, C_H' for both C_3H_6 and C_3H_8 over time, while the PZ67-20 shows trivial reduction in those parameters despite the fact that the actual percentage of polymer in PZ67-20 is substantially lower (i.e., 78.9 wt% and 64.9 vol%) compared to neat polymer matrix. The parameters, k_D and C_H', have to do with the densely packed equilibrium regions and the non-equilibrium micro-voids in polymer, respectively [45]. The trivial reduction in these parameters implies that the shrinkage of overall free volumes of PI was definitely hampered by the presence of ZIF-67 nanoparticles.

Interestingly, there was a similar observation with our results for the deceleration of physical aging rates induced by the introduction of nanoparticles, polyhedral oligomeric silsesquioxane (POSS), in poly(1-(trimethylsilyl)-1-propyne) (PTMSP) [46]. The permeability of PTMSP, well-known for a high permeability due to its high free volume, severely decreased over time even after crosslinking. However, the PTMSP containing 10 wt% POSS showed the significantly improved the physical aging resistance by occupying some of the large free volume elements, despite the permeability loss due to the impermeable characteristic of POSS [46]. The embedded nanoparticles in polymer matrix can play a role of buttress between polymeric chains, preventing decrease in free volume (Fig. 9a and b). More specifically, ZIF-67 nanoparticles rigidify polymer chains of the large free volume elements of PI around them through hydrophobic nature-induced strong affinity, verified by increase in T_g (Fig. S3), retarding their rapid relaxation. Such beneficial anti-aging effect by nanoparticles, however, might be infeasible in the absence of a good adhesion between nanoparticles and polymer (Fig. 9c) since the polymer chains still have enough freedom to be relaxed. It should be noted that the PZ8-20 also showed the same physical stability as PZ67-20 did (Fig. S6). Such beneficial features of mitigating permeability reduction over time assure a potential application of ZIF-67 containing MMMs for gas separations.

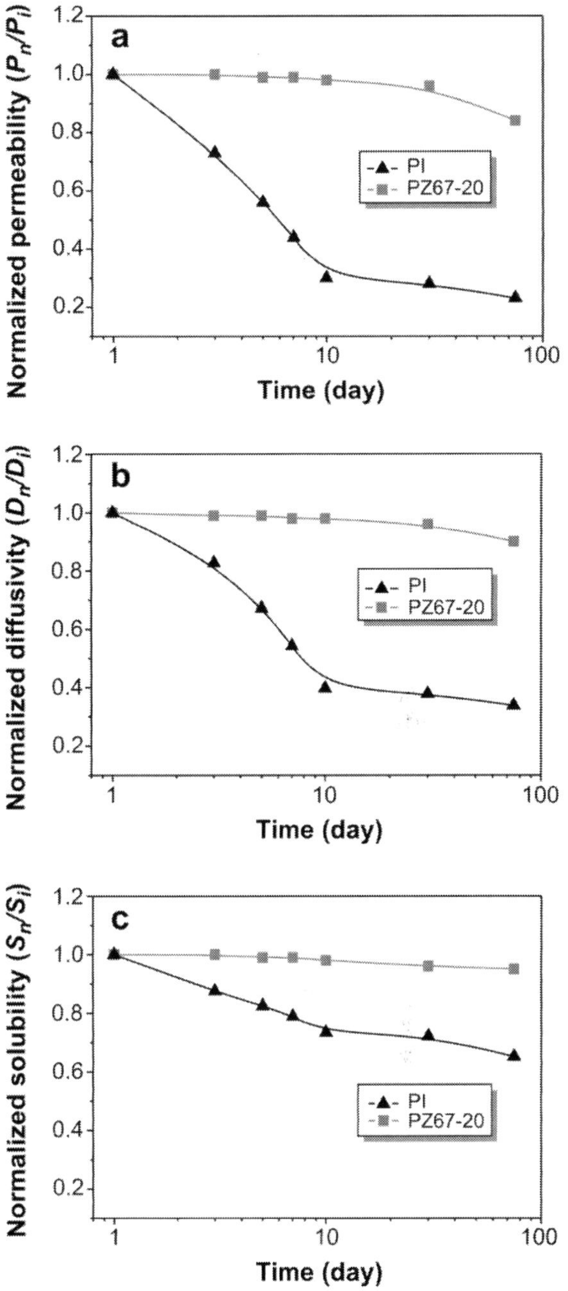

Fig. 7. Normalized permeability (a), diffusivity (b), and solubility (c) of C_3H_6 for PI and PZ67-20 at 35 °C as a function of time.

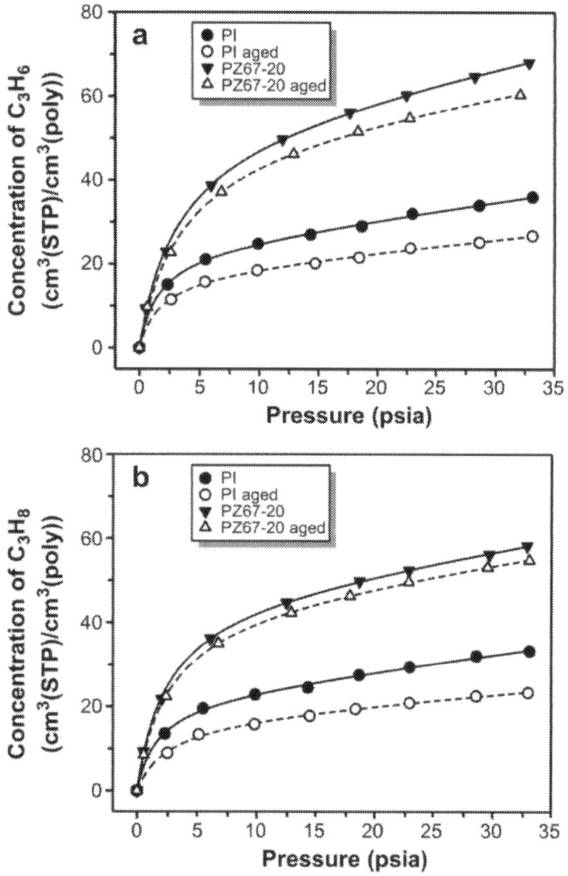

Fig. 8. Sorption isotherms of C_3H_6 (a) and C_3H_8 (b) at 35 °C for PI and PZ67-20 before vs. after aging of 30 days.

Table 4
Dual-mode model parameters for C_3H_6 and C_3H_8 in PI and PZ67-20 at 35 °C.

Sample	Aging (days)	Gas	k_D (cm³(STP)/ cm³(poly)/psia)	C_H' (cm³(STP)/ cm³(poly))	b (1/psia)
PI	–	C_3H_6	0.39 ± 0.02	24.11 ± 0.79	0.51 ± 0.07
		C_3H_8	0.37 ± 0.03	22.46 ± 1.03	0.57 ± 0.08
	30	C_3H_6	0.28 ± 0.03	18.20 ± 0.85	0.57 ± 0.09
		C_3H_8	0.22 ± 0.03	17.47 ± 1.00	0.38 ± 0.06
PZ67–20	–	C_3H_6	0.58 ± 0.06	53.76 ± 2.28	0.32 ± 0.03
		C_3H_8	0.44 ± 0.03	46.80 ± 1.08	0.41 ± 0.02
	30	C_3H_6	0.45 ± 0.06	50.92 ± 2.10	0.30 ± 0.03
		C_3H_8	0.41 ± 0.02	44.65 ± 0.60	0.38 ± 0.01

5. Conclusions

A new promising molecular sieve, ZIF-67, for C_3H_6/C_3H_8 separations was successfully incorporated into 6FDA-DAM polymer matrix with a good dispersion as well as good adhesion with polymer matrix. 6FDA-DAM/ZIF-67 (80/20 wt/wt) MMMs (i.e., PZ67-20) involving a practically feasible ZIF-67 content level of 21.1 wt% achieved a C_3H_6/C_3H_8 separation factor of 19 under the equimolar binary mixture conditions at 35 °C. Temperature-dependent transport studies enabled thorough investigation on the molecular sieving behavior of ZIF-67 nanoparticles in MMMs. The higher C_3H_6/C_3H_8 selectivity of PZ67-20 MMMs over that of PZ8-20 counterparts was attributed to a higher diffusivity selectivity induced by a combination of much higher repulsion energy and more negative heat of sorption for C_3H_8 penetrants in ZIF-67. Also, the high C_3H_6/C_3H_8 diffusivity selectivity of PZ67-20 MMMs was achieved by a significantly high energetic selectivity despite loss in the entropic selectivity. A loss in the C_3H_6/C_3H_8 entropic selectivity of PZ67-20 compared to that of PI was successfully demonstrated by using the transition state theory. The Maxwell model was used to estimate the intrinsic separation performance of ZIF-67, showing a C_3H_6 permeability of 150.7 Barrer, which is 45% less than that of ZIF-8, but a C_3H_6/C_3H_8 selectivity of 251, which is 100% higher than that of ZIF-8. In the light of the fact that both C_3H_6 and C_3H_8 are similarly sized as well as condensable, ZIF-67 can be a promising replacement for ZIF-8 molecular sieves for C_3H_6/C_3H_8 separations. In addition to the excellent separation performance of ZIF-67 containing MMMs, the membranes showed substantially improved physical aging resistance due to a buttress role of ZIF-67 between polymer chains in the large free volume elements, further supporting their practical application for C_3H_6/C_3H_8 separations.

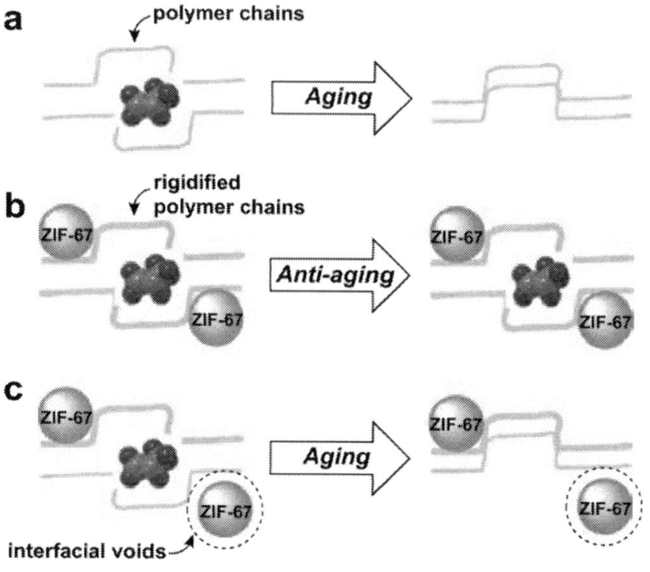

Fig. 9. Schematic of free volume changes of polymer by physical aging for PI (a), ZIF-67 containing MMMs with strong adhesion (b), and ZIF-67 containing MMMs with weak adhesion (c).

Acknowledgement

This research was supported by R & D Convergence Program of MSIP (Ministry of Science, ICT and Future Planning) and NST (National Research Council of Science & Technology) (CRC-14-1-KRICT), and the Korea CCS R & D Center (KCRC) (No. 2014M1A8A1049315). H.-K.J. acknowledges the financial support from the National Science Foundation (CBET-1510530) and in part from the Qatar National Research Fund (NPRP 7-042-2-021). The authors would like to thank to K.Y. Cho for help on the revision process.

Appendix A. Supporting information

Supplementary data associated with this article can be found in the online version at doi:10.1016/j.memsci.2016.12.053.

References

[1] J.J. McKetta, Petroleum Processing Handbook, Marcel Dekker, Inc, New York, 1992.
[2] M. Manning, Natural Gas Liquids Challenging Oil as Petrochemical Feedstock in North America, Increasing Global Demand for On-purpose Production of Propylene, IHS Says (in:), IHS Inc., 2014.
[3] J.R. Alcantara-Avila, F.I. Gomez-Castro, J.G. Segovia-Hernandez, K. Sotowa, T. Horikawa, Optimal design of cryogenic distillation columns with side heat pumps for the propylene/propane separation, Chem. Eng. Process. 82 (2014) 112–122.
[4] R.L. Burns, W.J. Koros, Defining the challenges for C_3H_6/C_3H_8 separation using polymeric membranes, J. Membr. Sci. 211 (2003) 299–309.

[5] S.U. Hong, J.H. Jin, J. Won, Y.S. Kang, Polymer-salt complexes containing silver ions and their application to facilitated olefin transport membranes, Adv. Mater. 12 (2000) 968–971.
[6] M. Fallanza, A. Ortiz, D. Gorri, I. Ortiz, Polymer-ionic liquid composite membranes for propane/propylene separation by facilitated transport, J. Membr. Sci. 444 (2013) 164–172.
[7] I.G. Giannakopoulos, V. Nikolakis, Separation of propylene/propane mixtures using faujasite-type zeolite membranes, Ind. Eng. Chem. Res. 44 (2005) 226–230.
[8] Y.C. Pan, T. Li, G. Lestari, Z.P. Lai, Effective separation of propylene/propane binary mixtures by ZIF-8 membranes, J. Membr. Sci. 390 (2012) 93–98.
[9] D.A. Newsome, D.S. Sholl, Molecular dynamics simulations of mass transfer resistance in grain boundaries of twinned zeolite membranes, J. Phys. Chem. B 110 (2006) 22681–22689.
[10] R.J. Lin, L. Ge, H. Diao, V. Rudolph, Z.H. Zhu, Propylene/propane selective mixed matrix membranes with grape-branched MOF/CNT filler, J. Mater. Chem. A 4 (2016) 6084–6090.
[11] G.X. Dong, H.Y. Li, V.K. Chen, Challenges and opportunities for mixed-matrix membranes for gas separation, J. Mater. Chem. A 1 (2013) 4610–4630.
[12] J.Q. Liu, T.H. Bae, W.L. Qiu, S. Husain, S. Nair, C.W. Jones, R.R. Chance, W.J. Koros, Butane isomer transport properties of 6FDA-DAM and MFI-6FDA-DAM mixed matrix membranes, J. Membr. Sci. 343 (2009) 157–163.
[13] C. Zhang, Y. Dai, J.R. Johnson, O. Karvan, W.J. Koros, High performance ZIF-8/6FDA-DAM mixed matrix membrane for propylene/propane separations, J. Membr. Sci. 389 (2012) 34–42.
[14] C. Zhang, R.P. Lively, K. Zhang, J.R. Johnson, O. Karvan, W.J. Koros, Unexpected molecular sieving properties of zeolitic imidazolate framework-8, J. Phys. Chem. Lett. 3 (2012) 2130–2134.
[15] H.T. Kwon, H.K. Jeong, A.S. Lee, H.S. An, J.S. Lee, Heteroepitaxially grown zeolitic imidazolate framework membranes with unprecedented, J. Am. Chem. Soc. 137 (2015) 12304–12311.
[16] P. Krokidas, M. Castier, S. Moncho, D.N. Sredojevic, E.N. Brothers, H.T. Kwon, H.K. Jeong, J.S. Lee, I.G. Economou, ZIF-67 Framework: a promising new candidate for propylene/propanepropylene/propane separation.separation. experimental data and molecular simulations, J. Phys. Chem. C. 120 (2016) 8116–8124.
[17] W.J. Koros, G.K. Fleming, Membrane-based gas separation, J. Membr. Sci. 83 (1993) 1–80.
[18] D.R. Paul, Gas sorption and transport in, Ber. Bunsenges. Phys. Chem. 83 (1979) 294–302.
[19] S. Husain, W.J. Koros, Mixed matrix hollow fiber membranes made with modified HSSZ-13 zeolite in polyetherimide polymer matrix for gas separation, J. Membr. Sci. 288 (2007) 195–207.
[20] S. Park, W.R. Kang, H.T. Kwon, S. Kim, M. Seo, J. Bang, S.H. Lee, H.K. Jeong, J.S. Lee, The polymeric upper bound for N-2/NF3 separation and beyond; ZIF-8 containing mixed matrix membranes, J. Membr. Sci. 486 (2015) 29–39.
[21] J.A. Thompson, K.W. Chapman, W.J. Koros, C.W. Jones, S. Nair, Sonication-induced Ostwald ripening of ZIF-8 nanoparticles and formation of ZIF-8/polymer composite membranes, Micro. Mesopor. Mat. 158 (2012) 292–299.
[22] W.G. Kim, J.S. Lee, D.G. Bucknall, W.J. Koros, S. Nair, Nanoporous layered silicate AMH-3/cellulose acetate nanocomposite membranes for gas separations, J. Membr. Sci. 441 (2013) 129–136.
[23] S. Park, A.S. Lee, Y.S. Do, S.S. Hwang, Y.M. Lee, J.H. Lee, J.S. Lee, Rational molecular design of PEOlated ladder-structured polysilsesquioxane membranes for high performance CO_2 removal, Chem. Commun. 51 (2015) 15308–15311.
[24] K.C. Obrien, W.J. Koros, T.A. Barbari, E.S. Sanders, A new technique for the measurement of multicomponent gas-transport through polymeric films, J. Membr. Sci. 29 (1986) 229–238.
[25] W.J. Koros, D.R. Paul, Design considerations for measurement of gas sorption in polymers by pressure decay, J. Polym. Sci. Pol. Phys. 14 (1976) 1903–1907.
[26] R. Mahajan, W.J. Koros, Factors controlling successful formation of mixed-matrix gas separation materials, Ind. Eng. Chem. Res. 39 (2000) 2692–2696.
[27] Q.L. Song, S.K. Nataraj, M.V. Roussenova, J.C. Tan, D.J. Hughes, W. Li, P. Bourgoin, M.A. Alam, A.K. Cheetham, S.A. Al-Muhtaseb, E. Sivaniah, Zeolitic imidazolate framework (ZIF-8) based polymer nanocomposite membranes for gas separation, Energ. Environ. Sci. 5 (2012) 8359–8369.
[28] J.H. Kim, W.J. Koros, D.R. Paul, Physical aging of thin 6FDA-based polyimide membranes containing carboxyl acid groups. Part I. Transport properties, Polymer 47 (2006) 3094–3103.
[29] D.Q. Vu, W.J. Koros, S.J. Miller, Mixed matrix membranes using carbon molecular sieves - I. Preparation and experimental results, J. Membr. Sci. 211 (2003) 311–334.
[30] S. Park, J. Bang, J. Choi, S.H. Lee, J.H. Lee, J.S. Lee, 3-Dimensionally disordered mesoporous silica (DMS)-containing mixed matrix membranes for CO_2 and non-CO_2 greenhouse gas separations, Sep. Purif. Technol. 136 (2014) 286–295.
[31] O. Shekhah, R. Swaidan, Y. Belmabkhout, M. du Plessis, T. Jacobs, L.J. Barbour, I. Pinnau, M. Eddaoudi, The liquid phase epitaxy approach for the successful construction of ultra-thin and defect-free ZIF-8 membranes: pure and mixed gas transport study, Chem. Commun. 50 (2014) 2089–2092.
[32] H.T. Kwon, H.K. Jeong, In situ Synthesis of thin zeolitic-imidazolate framework ZIF-8 membranes exhibiting Exceptionally high propylene/propane separation, J. Am. Chem. Soc. 135 (2013) 10763–10768.
[33] M.L. Chng, Y.C. Xiao, T.S. Chung, M. Toriida, S. Tamai, Enhanced propylene/propane separation by carbonaceous membrane derived from poly (aryl ether ketone)/2,6-bis(4-azidobenzylidene)-4-methyl-cyclohexanone interpenetrating network, Carbon 47 (2009) 1857–1866.
[34] R.J. Swaidan, X.H. Ma, E. Litwiller, I. Pinnau, Enhanced propylene/propane separation by thermal annealing of an intrinsically microporous hydroxyl-functionalized polyimide membrane, J. Membr. Sci. 495 (2015) 235–241.
[35] M. Das, W.J. Koros, Performance of 6FDA-6FpDA polyimide for propylene/propane separations, J. Membr. Sci. 365 (2010) 399–408.
[36] A. Singh, W.J. Koros, Significance of entropic selectivity for advanced gas separation membranes, Ind. Eng. Chem. Res. 35 (1996) 1231–1234.
[37] A. Singh-Ghosal, W.J. Koros, Energetic and entropic contributions to mobility selectivity in glassy polymers for gas separation membranes, Ind. Eng. Chem. Res. 38 (1999) 3647–3654.
[38] M. Rungta, L.R. Xu, W.J. Koros, Carbon molecular sieve dense film membranes derived from Matrimid (R) for ethylene/ethane separation, Carbon 50 (2012) 1488–1502.
[39] K.H. Li, D.H. Olson, J. Seidel, T.J. Emge, H.W. Gong, H.P. Zeng, J. Li, Zeolitic imidazolate frameworks for kinetic separation of propane and propene, J. Am. Chem. Soc. 131 (2009) 10368–10369.
[40] X. Ning, W.J. Koros, Carbon molecular sieve membranes derived from Matrimid (R) polyimide for nitrogen/methane separation, Carbon 66 (2014) 511–522.
[41] K. Zhang, R.P. Lively, C. Zhang, R.R. Chance, W.J. Koros, D.S. Sholl, S. Nair, Exploring the framework hydrophobicity and flexibility of ZIF-8: from biofuel recovery to hydrocarbon separations, J. Phys. Chem. Lett. 4 (2013) 3618–3622.
[42] Y. Huang, D.R. Paul, Effect of film thickness on the gas-permeation characteristics of glassy polymer membranes, Ind. Eng. Chem. Res. 46 (2007) 2342–2347.
[43] L.L. Cui, W.L. Qiu, D.R. Paul, W.J. Koros, Physical aging of 6FDA-based polyimide membranes monitored by gas permeability, Polymer 52 (2011) 3374–3380.
[44] R.R. Tiwari, Z.P. Smith, H.Q. Lin, B.D. Freeman, D.R. Paul, Gas permeation in thin films of "high free-volume" glassy perfluoropolymers: part I. Physical aging, Polymer 55 (2014) 5788–5800.
[45] W.H. Lin, T.S. Chung, Gas permeability, diffusivity, solubility, and aging characteristics of 6FDA-durene polyimide membranes, J. Membr. Sci. 186 (2001) 183–193.
[46] S.D. Kelman, B.W. Rowe, C.W. Bielawski, S.J. Pas, A.J. Hill, D.R. Paul, B.D. Freeman, Crosslinking poly[1-(trimethylsilyl)-1-propyne] and its effect on physical stability, J. Membr. Sci. 320 (2008) 123–134.

506543 High Performance Ultrafiltration Membrane with Cysteine-Functionalized Graphene Oxides for Sustainable Water Production

Abstract:

Ultrafiltration membranes based on Polyethersulfone (PES) nanocomposites with cysteine-modified graphene oxides (CysGO) were prepared and their separation performance was investigated. Graphene oxides were functionalized with cysteine, zwitterionic amino acid, to enhance hydrophilicity. The modification of graphene oxides carried out through thiol-ene click reaction. PES nanocomposites membranes containing different concentration of CysGO were prepared by non-solvent induced phase separation (NIPS) method. Scanning electron microscope (SEM) images indicated all membranes showed a typical thin active layer and a finger-like support layer. Water contact angle was measured to characterize the surface hydrophilicity of prepared membranes. As CysGO was added, a compact and tight water layer adjacent to the surface of membrane was formed, which showed enhanced hydrophilicity. The filtration performance of obtained membranes was evaluated by a dead-end filtration cell test. Bovine serum albumin (BSA) solution was used for model foulants. The presence of CysGO tended to enhance fouling resistance of membrane since zwitterionic moieties of CysGO inhibited absorption of BSA protein on membranes. The flux of membranes decreased with the introduction of CysGO since the extra wall of CysGO was induced on the membrane.

490198 Unique Design Considerations for Maximum-Boiling Azeotrope Via Extractive Distillation System: Acetic Acid/N,N-Dimethylacetamide Separation

Abstract:

Extractive distillation system is commonly used in industry to separate azeotropic or close boiling mixtures. Adding a suitable heavy entrainer into the system can enhance the relative volatility of the original two components so that the lighter component can be obtained from the distillate of an extractive distillation column, while another component can be obtained from the distillate of a second entrainer recovery column. The bottoms of the second column containing mostly the entrainer can be recycled back to the extractive distillation column. For minimum-boiling azeotropic mixture, the enhancement capability of relative volatility is most important for the selection of suitable entrainer. As for maximum-boiling azeotrope, another important feature of the ternary system needs to be taken into consideration also for the determination of suitable entrainer. This is because the type of ternary diagram/RCM for maximum-boiling azeotrope is different than that of the minimum-boiling azeotrope. A distillation boundary will be formed for the ternary diagram/RCM of maximum-boiling azeotrope with a heavy entrainer. Depending on the degree of curvature of the distillation boundary, the feasible minimum entranier-to-feed ratio can be estimated with given product purity specifications. Thus together with the enhancement of relative volatility, the estimated minimum value of entranier-to-feed ratio can be used as another screening tool for selecting a suitable entrainer of maximum-boiling azeotrope systems.

An extractive distillation system for separating acetic acid and N,N-Dimethylacetamide is used as an illustrative example to elaborate the new findings for determining the most suitable entrainer for maximum-boiling azeotorpe systems. Optimal design flowsheets of two candidate entrainers, triethylene glycol (TEG) and tetraethylene glycol (TTEG), are obtained to compare the total annual cost and the steam cost of the two candidate extractive distillation systems. It is demonstrated that the estimated minimum value of entranier-to-feed ratio is crucially important in the determination of the most suitable entrainer. Using TTEG as entrainer, although having greater enhancement of the relative volatility of acetic acid over N,N-Dimethylacetamide, the feasible minimum entranier-to-feed ratio is much larger than that of the TEG system. Thus, the economics of the overall system using TTEG as entrainer is inferior to that of the system using TEG as entrainer. This minimum value of each candidate entrainer can easily be estimated by the ternary diagram/RCM of the candidate system without needing of time-consuming procedure in obtaining the optimal design flowsheet of the extractive distillation system.

494197 Pilot Deposition of Zeolite-Y Nanoparticles on Polyethersulfone Substrate for Composite Membrane Fabrication in CO_2 Separation

Abstract:

A vacuum-assisted deposition setup was developed in pilot scale to continuously deposit zeolite-Y (ZY) nanoparticles on a 14-inch wide nanoporous polyethersulfone (PES) substrate. The deposited substrate was used for composite membrane fabrication for carbon capture from flue gas in coal- and/or natural gas-fired power plants. The effects of operating parameters, including ZY dispersion concentration, pressure differential, and web speed, on the ZY layer thickness and uniformity were studied. A ZY dispersion concentration of 0.08 wt.%, a pressure differential of 3.7 inch Hg, and a web speed of 3 feet/min were identified as the optimal operating conditions for depositing a thin and uniform ZY layer with a thickness of approximately 160 nm. A mathematical model with two adjustable parameters was derived from Darcyâ€™s law to correlate the operating parameters to the ZY layer thickness. The model was validified by the parametric study, and the adjustable parameters were determined by fitting the experimental data. After coating an amine-containing polymer selective layer on top of the ZY layer, the composite membrane showed a CO_2 permeance of 800 GPU (1 GPU = 10^{-6} cm^3 (STP)·cm^{-2}·s^{-1}·$cmHg^{-1}$) with a CO_2/N_2 selectivity over 150 at the typical flue gas temperature of 57 °C, which was superior to the performance of the membrane coated on the bare PES substrate. This improvement was due to the reduced selective layer penetration, resulting from the smaller interparticle pore size on the ZY layer. The scale-up vacuum-assisted deposition process developed in this study is also suitable for the scale-up of the deposition of other fine particles onto a porous polymeric substrate. The developed roll-to-roll fabrication process can be applied to the mass production of the nanoparticle layers, which could potentially reduce the fabrication cost of inorganic membranes in a variety of applications.

494204 Carrier Saturation Phenomenon in Facilitated Transport Membrane for CO_2 Separation from Low Concentration Sources

Abstract:

Amine-containing facilitated transport membranes exhibit superior CO_2 separation performance because the CO_2 transport is enhanced by the reversible reaction between CO_2 and the active carrier, amine, incorporated in the membrane. Continuously increasing the CO_2 partial pressure leads to the constant CO_2 flux, which does not increase with CO_2 partial pressure, i.e., the carrier saturation phenomenon. At the carrier saturation, all available carrier molecules have been reacted with the high pressure CO_2. However, this carrier saturation phenomenon can be beneficial for CO_2 capture from low CO_2 concentration (~1% CO_2) sources, e.g., the residual flue gas after the primary CO_2 capture system (with 90% CO_2 removal) and coal mine gas streams. For CO_2 separation from low concentration sources, an improved membrane separation performance can be obtained. With reducing the CO_2 concentration in the feed gas, both CO_2 permeance and CO_2/N_2 selectivity increased. These were mainly due to more available amine carriers for CO_2 molecular transport at lower CO_2 concentration conditions. In this study of CO_2 separation from low concentration feed, the membrane used consisted of polyvinylamine serving as the fixed-site carrier and piperazine glycinate acting as the mobile carrier. A CO_2 permeance of 785 GPU (1 GPU = 10^{-6} cm^3 (STP)$\cdot cm^{-2}\cdot s^{-1}\cdot cmHg^{-1}$) and a CO_2/N_2 selectivity of 150 were obtained with a feed of 20% CO_2 and 80% N_2 on a dry basis (16.6% CO_2 and 66.2% N_2 along with the saturation water vapor at 17.2%) at the typical flue gas temperature of 57°C. By reducing the CO_2 concentration in the feed, a significant CO_2 permeance increase was observed when the CO_2 concentration was below 5%. For instance, the membrane showed a CO_2 permeance of 950 GPU and a CO_2/N_2 selectivity of 190 with a feed of 1% CO_2 and 99% N_2 on a dry basis. The observed performance increase was due to the carrier saturation phenomenon as more carrier molecules were available for CO_2 transport at a lower CO_2 concentration. Additionally, it was found that the effect of SO_2 was less significant when the CO_2 concentration was low. The CO_2 permeance dropped by approximately 2.5% with an introduction of 3 ppm SO_2 in the feed gas of 20% CO_2 and 80% N_2 on a dry basis at 57°C. However, no performance reduction was observed after exposing the membrane to 3 ppm SO_2 with the feed gas of 1% CO_2 and 99% N_2 on a dry basis at the same temperature. Therefore, the studied membrane can be applied for CO_2 separation from low concentration sources with an exceptional separation performance.

494207 Improved Polyethersulfone Substrates for Composite Membranes in CO_2 Separation

Abstract:

Pilot-scale nanoporous polyethersulfone (PES) substrates were fabricated by vapor- and non-solvent-induced phase inversion steps successively. N-methyl-2-pyrrolidone (NMP) and water were employed as solvent and non-solvent, respectively. 2-Methoxyethanol (2-ME) was incorporated into the casting solution as a pore-forming additive because of its high affinity with water, leading to an interconnected spongy structure in the bulk of the substrate. The 14-inch wide PES substrates were fabricated successfully by using a continuous roll-to-roll casting machine. Two hydrophilic polymers, hydroxylated PES (PES-OH) and polyvinylpyrrolidone (PVP), were incorporated for modifying the pristine substrate both hydrophilically and morphologically. The surface morphologies and gas permeation resistances of the PES membranes were characterized by scanning electron microscopy (SEM) and a gas permeation unit, respectively. The pristine PES substrate, with a pore size of 41 nm and a surface porosity of 12%, showed a CO_2 permeance of about 9200 GPU (1 GPU = 10^{-6} cm^3 (STP)$\cdot cm^{-2}\cdot s^{-1}\cdot cmHg^{-1}$) and a CO_2/N_2 selectivity ~ 0.9 (Knudsen diffusion). The addition of hydrophilic polymers reduced the gas permeation resistances of PES substrates because of the more open morphology. By incorporating appropriate amounts of PES-OH and PVP in the casting solutions, modified PES substrates showed CO_2 permeances of 14100 GPU and 21000 GPU, respectively, indicating a significant reduction of gas permeation resistance. The developed PES substrates were applied for the preparation of composite membranes for CO_2 separation. Amine-containing selective layers, with a thickness of approximately 175 nm, were coated on the developed pristine PES substrate, PES-OH modified substrate, and PVP-modified substrate to prepare 3 composite membranes, respectively. From the gas transport measurements by using a simulated flue gas, these 3 composite membranes showed CO_2 permeances of 785 GPU, 836 GPU, and 843 GPU, and CO_2/N_2 selectivities of 150, 148, and 160, respectively. Based on the resistance-in-series model, the separation performance improvements could be attributed to the reduced mass transfer resistances of the modified substrates. In addition, a thinner selective layer can be coated on the hydrophilic polymer modified PES substrates due to the enhanced hydrophilicity, which can further improve the separation performances of the composite membranes in the future. The developed fabrication process has demonstrated the potential for the mass production of the suitable PES substrate for composite membranes in CO_2 separation.

494912 Polymeric Membrane Systems for CO_2 Capture from Flue Gas: A Techno-Economic Analysis

Abstract:

The increasing consumption of fossil fuels has resulted in immense CO_2 emissions after the Industrial Revolution. Membrane processes have been suggested as an alternative to amine absorption for post-combustion carbon capture. Despite of the tremendous research efforts, CO_2-selective polymeric membranes typically exhibit a trade-off between CO_2 permeance and CO_2/N_2 selectivity. This presentation will show how to overcome the limitations as a result of this trade-off by noval membrane process designs. In the first membrane process, called the hybrid air sweep process, the flue gas flows to a vacuum membrane stage using a membrane with high CO_2/N_2 selectivity and moderate CO_2 permeance. The retentate gas then flows to an air-sweep membrane stage employing another membrane with high CO_2 permeance but low CO_2/N_2 selectivity. The CO_2-laden sweep air is fed to a power plant boiler as combustion air. A highly selective amine-containing membrane with a CO_2 permeance of 1100 GPU (1 GPU = 10^{-6} cm^3 (STP)·cm^{-2}·s^{-1}·$cmHg^{-1}$) and a CO_2/N_2 selectivity >140 is used for the vacuum stage to enrich the CO_2 to >95% purity whereas a highly permeable poly(ethylene oxide)-based membrane with a CO_2 permeance of 2000 GPU and a CO_2/N_2 selectivity of 20 is employed for the air sweep stage. This hybrid membrane configuration yields a capture cost of \$40.2/tonne CO_2 in the 2007 dollar basis for 90% CO_2 recovery, which nearly meets the target of the Department of Energy (DOE) set for 2025. In the second membrane process, i.e., retentate recycle process, the flue gas is passed to two stages employing the membrane with high CO_2/N_2 selectivity; these two stages are placed in an enriching cascade. A portion of the CO_2-depleted retentate of the first membrane stage is recycled as its own sweep gas. It is identified that the highly selective amine-containing membrane is needed for both stages and a 20% retentate recycle renders a capture cost of \$43.1/tonne CO_2. These cost calculations in the 2007 dollar basis have included both membrane module installation cost and 20% process contingency.

494913 Membranes for CO_2 Capture from Low Concentration Sources: A Technical and Economic Feasibility Study

Abstract:

In the past decade, membranes and membrane processes have been studied extensively for carbon capture from various gas sources in the energy sector, such as syngas (~35% CO_2) from an integrated gasification combined cycle, flue gas (11 â€" 15% CO_2) in coal-fired power plants, and gas turbine flue gas (~4% CO_2) of a natural gas combined cycle. However, there are other more challenging candidates for carbon capture where the CO_2 concentration is low but the total CO_2 emission is large. In this presentation, a first-of-a-kind membrane process is proposed for carbon capture from low CO_2 concentration (~1%) sources. The low CO_2 concentration sources have included the treated flue gas in coal-fired power plants in which 90% CO_2 has been removed by a primary carbon capture unit, e.g., membrane or aqueous amine absorption, as well as coal mine gas streams. The low CO_2 concentration feed gas is passed to a 2-stage enriching membrane cascade where the CO_2 is sequentially enriched to 95% purity. Vacuum is pulled on the permeate sides of both membrane stages to provide the transmembrane driving force. The techno-economic analysis shows that a high CO_2/N_2 selectivity is crucial to achieve the 95% CO_2 purity. If the CO_2 permeance is higher than 1000 GPU (1 GPU = 10^{-6} cm^3 (STP)$\cdot cm^{-2}\cdot s^{-1}\cdot cmHg^{-1}$), a minimum selectivity of 80 is required. A high selectivity of about 200 is preferred from a cost and energy standpoint. Further increasing the selectivity shows little benefit. By using the state-of-the-art amine-containing facilitated transport membrane with a CO_2 permeance of 1100 GPU and a CO_2/N_2 selectivity of 200, a capture cost of \$294.7/tonne CO_2 (in the 2011 dollar basis) can be achieved for 90% CO_2 recovery. This corresponds to an increase in the cost of electricity of 21.2%. The membrane process is feasible for CO_2 mitigation with a feed CO_2 concentration spectrum of 0.6 â€" 2.0%. If the feed CO_2 concentration falls within this range, 90% CO_2 recovery and 95% purity can be achieved with similar total capital and operational expenditures. It is also found that the 90% CO_2 removal may not be the most economic option for low CO_2 concentration sources. Reducing the CO_2 recovery leads to a smaller energy penalty, and a lowest capture cost is reached at 70% recovery. These cost calculations in the 2011 dollar basis have included both membrane module installation cost and 20% process contingency.

489682 Polyelectrolyte Modified Graphene Oxide/Polypropylene Composite Membranes for Organic Solvent Nanofiltration

Abstract:

Polyelectrolyte Modified Graphene Oxide/Polypropylene Composite Membranes for Organic Solvent Nanofiltration

Dan Hua and Tai-Shung Chung*

Department of Chemical & Biomolecular Engineering, National University of Singapore, Singapore 117585, Singapore

* Corresponding author. Tel.: (+65) 6516 6645; fax: (+65) 6779 1936.

E-mail address: chencts@nus.edu.sg (T. S. Chung).

Abstract:

In this work, we have molecularly designed graphene oxide (GO) /polypropylene (PP) composite membranes by three steps: (1) grafting the PP surface with polyethylene glycol (PEG) with the aid of argon plasma (TPP), (2) a pressure-assisted filtration of GO dispersion and (3) functionalization by various polyelectrolytes. The fabricated polyelectrolyte modified GO/PP composite membranes not only show excellent adhesion between the GO layer and PP support, but also exhibit satisfying separation performance for organic solvent nanofiltration (OSN) of dye/ethanol mixtures. It is interesting to find that the TPP/GO membrane with hyperbranched polyethylenimine (HPEI) modified on its outmost layer has high rejections toward cationic dyes. It has a rejection of 95% toward Alcian blue with a high permeance of 14.9 L m^{-2} h^{-1} bar^{-1}. Similarly, it exhibits excellent anionic dye/ethanol separation with poly(styrenesulfonate) (PSS) modified on its outmost layer. It shows a rejection up to 97% toward Rose bengal with a total permeance of 3.1 L m^{-2} h^{-1} bar^{-1}. Experimental results indicate that both molecular sieving and electrostatic repulsion are critical factors for GO based membranes to separate dye/organic solvent mixtures. This work may both widen the selection of membrane support materials for GO deposition and provide useful insights for developing GO based membranes for OSN.

Acknowledgements

The authors are grateful to National Research Foundation, Prime Minister¡¯s Office, Singapore for funding this research under its Competitive Research Program for the project entitled, ¡°Development of solvent resistant nanofiltration membranes for sustainable pharmaceutical and petrochemical manufacture¡±; (CRP Award NO. NRF-CRP14-2014-01 (NUS grant number: R-279-000-466-281)). Special thanks are due to Dr. Gui Min Shi, Ms. Yu Zhang, Miss Anastasia Frances Frederica and Miss Jie Gao for their valuable suggestions and help.

492171 The Effects of Chemical Structure on Gas Transport Properties in a Family of Polyethersulfones Polymers

Abstract:

The effects of chemical structure on gas transport properties in a family of polyethersulfones polymers

Ali Naderi[1], Wai Fen Yong[1], Youchang Xiao[1] and Tai-Shung Chung[1,*]

Martin Weber[2], Christian Maletzko[3],

[1]Department of Chemical and Biomolecular Engineering, National University of Singapore, Singapore 117585, Singapore

[2]Advanced Materials & Systems Research, BASF SE, GMV/W - B001, 67056 Ludwigshafen, Germany

[3]Performance Materials, BASF SE, G-PM/PU, 67056 Ludwigshafen, Germany

*Correspondence author. Tel: +65 6516 6645; fax: +65 67791936.

E-mail address: chencts@nus.edu.sg(T.-S. Chung).

Abstract

Polymeric membranes are considered as a promising technology for gas separation. Polysulfone (PSF) and polyethersulfone (PES) have been used in the membrane applications due to their thermal stability, chemical resistance, processability and intrinsic properties. In this work, a series of BASF polyethersulfones polymers containing of different backbone structures have been studied. Their permeation, sorption properties and physical properties were investigated. The gas permeabilities of the dense films including oxygen, nitrogen, methane, and carbon dioxide were measured as a function of pressure. The sorption of methane and carbon dioxide were assessed via the dual-volume pressure decay method using a XEMIS microbalance system and the diffusion coefficients of these gases through the membranes were determined by solution-diffusion model. The permeability of the membranes decreases slightly with the increase in pressure for all gases which can be due to the saturation of Langmuir sites in high pressure. To elucidate the microstructural behavior of different PES, positron annihilation life time spectroscopy (PALS) and dynamic mechanical analyses (DMA) have been examined. The results of PALS and DMA reveal that the polyphenylsulfone (PPSU) with a higher segmental motion in the polymer chain offers a higher free volume which

subsequently lead to a higher CO_2 and CH_4 diffusion rates through the membrane. Among all the PES, the trimethyl polyethersulfone (TPESU) shows the highest CO_2 and CH_4 sorption. The high CO_2 and CH_4 sorptions in the TPESU could be attributed to the high polymer chain-penetrant interactions of this polymer compared to the others. The diffusivity coefficients of CO_2 of each membrane in the Henry and Langmuir sites have been calculated according to the dual mode sorption and the immobilization models. The results reveal that a higher mobility of the Langmuir site than the Henry site in the PPSU compared to the others. These results are in accordance with the Î³-transition results that have been obtained via DMA analyses.

493201 High-Resolution Scalable Propylene/Propane Separation for ZIF-8 Polycrystalline Membranes on Ceramic Tubular Supports

Abstract:

High-resolution Scalable Propylene/Propane Separation for ZIF-8 Polycrystalline Membranes on Ceramic Tubular Supports

Jingze Sun[1] and Hae-Kwon Jeong*[1,2]

[1]Artie McFerrin Department of Chemical Engineering and [2]Department of Materials Science and Engineering, Texas A&M University, College Station, TX 77843-3122

* Corresponding author: hjeong7@tamu.edu

Zeolite-imidazole frameworks (ZIFs) have been extensively studied as a new membrane material for gas separations due to their robust synthesis protocols, chemical and thermal stability, and well-defined ultra-micropores (~ 2 to 5 Å).[1, 2] ZIFs are comprised of metal centers (Zn, Co, and Cd) and bridging imidazole-based ligands. ZIF-8, composed of zinc nodes and 2-methylimidazole linkers, has been shown very promising for propylene/propane separations since its effective aperture size lies in between propylene (~ 4 Å) and propane (~ 4.2 Å).[3] In fact, some of the polycrystalline ZIF-8 membranes synthesized by various methods have been reported to exhibit impressive propylene/propane separation factors up to 200.[4, 5] These highly propylene-selective ZIF-8 membranes were supported mostly on alumina disks or occasionally on hollow fibers (polymer or ceramic).

For their large-scale industrial applications, however, the productivities of ZIF-8 membranes must be significantly improved.[6] One way to increase the membrane productivity is to increase membrane area per volume. As such, current disk supports are not desirable. Though hollow fiber supports (either polymer or ceramic) are most desirable, they are unlikely to be used for large-scale applications in near future. This is because ceramic hollow fibers are highly fragile while polymer hollow fibers require much more development. While hollow fiber supports are transformative, ceramic tubular supports are readily available for large-scale applications. For example, ceramic tubular supports have been used for commercial polycrystalline zeolite membranes.[7] Though ZIF-8 membranes on ceramic tubular supports were reported, none of those membranes showed promising propylene/propane separation capabilities.[8, 9]

Here, we would like to present polycrystalline ZIF-8 membrane supported on alumina tubes showing high propylene/propane separation performances. The tubular membranes were synthesized using our microwave seeding and secondary growth method. The effects of the synthesis parameters on the microstructures of tubular ZIF-8 membranes will be presented. Finally, the propylene/propane separation performances of ZIF-8 membranes supported on alumina tubes will be discussed.

References and Notes:

(1) Park, S.; Kang, W. R.; Kwon, H. T.; Kim, S.; Seo, M.; Bang, J.; Lee, S. h.; Jeong, H. K.; Lee, J. S., The polymeric upper bound for N2/NF3 separation and beyond; ZIF-8 containing mixed matrix membranes. *J. Membr. Sci.* **2015,** 486, 29-39.

(2) Chen Zhang, R. P. L., Ke Zhang, Justin R. Johnson, Oguz Karvan, William J. Koros, Unexpected molecular sieving properties of zeolitic imidazolate framework-8. *The journal of physical chemistry letters* **2012,** 3, (16), 2130-2134.

(3) Li, K.; Olson, D. H.; Seidel, J.; Emge, T. J.; Gong, H.; Zeng, H.; Li, J., Zeolitic imidazolate frameworks for kinetic separation of propane and propene. *J. Am. Chem. Soc.* **2009,** 131, (30), 10368-9.

(4) Lee, M. J.; Kwon, H. T.; Jeong, H. K., Defect-dependent stability of highly propylene-selective zeolitic-imidazolate framework ZIF-8 membranes. *J. Membr. Sci.* **2017,** 529, 105-113.

(5) Pan, Y. C.; Li, T.; Lestari, G.; Lai, Z. P., Effective separation of propylene/propane binary mixtures by ZIF-8 membranes. *J. Membr. Sci.* **2012,** 390, 93-98.

(6) Baker, R. W., Future directions of membrane gas separation technology. *Ind. Eng. Chem. Res.* **2002,** 41, (6), 1393-1411.

(7) Kosinov, N.; Gascon, J.; Kapteijn, F.; Hensen, E. J., Recent developments in zeolite membranes for gas separation. *J. Membr. Sci.* **2016,** 499, 65-79.

(8) Hara, N.; Yoshimune, M.; Negishi, H.; Haraya, K.; Hara, S.; Yamaguchi, T., Diffusive separation of propylene/propane with ZIF-8 membranes. *J. Membr. Sci.* **2014,** 450, 215-223.

(9) Zhang, X.; Liu, Y.; Kong, L.; Liu, H.; Qiu, J.; Han, W.; Weng, L.-T.; Yeung, K. L.; Zhu, W., A simple and scalable method for preparing low-defect ZIF-8 tubular membranes. *J. Mater. Chem. A* **2013,** 1, (36), 10635.

496991 Cross-Linked Polyimides for Membrane H_2/CO_2 Separation at Elevated Temperatures

Abstract:

Cross-linked Polyimids for Membrane H_2/CO_2 Separation at Elevated Temperatures

Maryam Omidvar, Mark Swihart, and Haiqing Lin

Department of Chemical and Biological Engineering

University at Buffalo, The State University of New York)

Membrane technology has attracted significant interests for H_2 purification and CO_2 capture in the pre-combustion processes. The key to the success of this technology is membrane materials with high H_2 permeability and H_2/CO_2 selectivity at elevated temperatures of syngas processing such as 100 °C or above. Polyimides such as P84 have been explored for H_2/CO_2 separation due to their strong size-sieving ability and thus high H_2/CO_2 diffusivity selectivity. In this study, we demonstrate that these polymers can be cross-linked chemically to further improve the size sieving ability and thus H_2/CO_2 selectivity. We systematically evaluate the effect of the cross-linking in P84 using amines such as propane diamines, butane diamines and p-xylene diamine on the H_2/CO_2 separation properties. The cross-linking can significantly increase H_2/CO_2 selectivity while retaining moderate H_2 permeability.Â For example, cross-linking of P84 with butane diamine with 42% conversion rate increases H_2/CO_2 selectivity from 2.4 to 8.0, while decreasing H_2 permeability from 40 to 24 Barrers at 100 °C. The properties of 24 Barrers for H_2 permeability and 8.0 for H_2/CO_2 selectivity are very close to the Robesonâ€™s upper bound for H_2/CO_2 separation, demonstrating the potential of this approach for membrane development. These cross-linked polymers are thoroughly evaluated using FTIR, DSC, TGA, WAXD, and XPS. The effect of cross-linking degree on the physical properties will be discussed, including glass transition temperature, fractional free volume and pure- and mixed-gas H_2/CO_2 separation properties at various temperatures. This presentation will also discuss the fundamental structure/property relationship in these highly cross-linked polymers with promising H_2/CO_2 separation properties.

496374 Preparation and Characterization of Zeolite-Polymer Mixed Matrix Membranes Filled with KFI and Rho Type Zeolites

Abstract:

Separation of CO_2 from gas mixtures is a popular subject of investigation. The tailoring of membranes for this purpose includes the development of suitable polymeric and zeolitic membranes. Fast permeability through polymeric membranes combined with high selectivity of crystalline molecular sieve zeolites may lead to improved performances in the separation of CO_2 from CH_4 as well as other applications. Up to now different types of zeolite-polymer mixed matrix membranes have been prepared by using zeolites, such as A, X/Y, silicalite/ZSM-5, SSZ-13, SAPO-34, BEA, sodalite and L. Although zeolite NaA has been determined to be very suitable for CO_2/CH_4 and CO_2/N_2 separations, the very hydrophilic nature of this zeolite results in adsorption of water which complicates the separation process.

In this study, zeolite-polymer mixed matrix membranes containing KFI and RHO type zeolites (20% by weight) were prepared. These zeolites have pore sizes comparable with that of zeolite NaA but are less hydrophilic and thus may be very useful in the separation of CO_2 from various gas mixtures. Polyvinyl acetate was selected as the polymer phase due to its compatibility with zeolites in the mixed matrix membranes. The zeolites were synthesized by using suitable conditions reported in the literature. The zeolites as well as the membranes obtained were characterized by using X-ray diffraction (XRD), scanning electron microscopy (SEM), adsorption and integrated gravimetric analysis (IGA) techniques.

The zeolites were identified to be crystalline without impurity phases according to the XRD analyses performed. The SEM pictures indicated a homogeneous distribution of the zeolite particles, of about 1-3 μm, in the mixed matrix membranes. Adsorption of CO_2 and N_2 in the pure zeolite and polymer as well as mixed matrix membrane samples assured the high selectivity of both KFI and RHO zeolites for CO_2. The zeolite-polymer interactions were also investigated by using adsorption data for pure and mixed phases. The results implied that the composite materials exhibited properties that could not be explained by considering the pure zeolite and polymer phases, alone. This may support the view of the existence of an interphase between the zeolite and polymer phases in the mixed matrix membranes. According to the gravimetric sorption measurements, it was determined that polymeric membranes filled with KFI and RHO type zeolites presented quite different behaviors for the sorption of CO_2 and N_2, when compared to the pure polymeric membrane, which might provide significant improvements for the separation of CO_2 from gas mixtures.

Permeability measurements should also be performed to confirm the advantages of using mixed matrix membranes filled with KFI and RHO zeolites. Different polymer types and various zeolite loadings may also be tested in order to determine the extent of the possible improvements.

490289 Gas Separation from Intrinsic Defects of Single Layer Graphene

Abstract:

Block-copolymer derived nanoporous carbon membranes for high throughput gas separation
Atom thick graphene membrane, the thinnest possible molecular barrier, is an ideal membrane for the gas separation.[1–3] However, route to fabricate a large-area, crack-free graphene membrane has not been demonstrated. Cracks and tears eclipses the intrinsic separation performance from nanoporous graphene, therefore till-date, gas mixture separation through single-layer graphene film has not been reported. Herein, by developing, a novel transfer technique, we report gas separation from chemical vapor deposition (CVD) synthesized atom thick graphene membrane from an area as large as 1 mm^2. No cracks or tears were observed in the graphene membrane by SEM. To the best of our knowledge, before this work, gas separation studies from single-layer graphene membranes have been limited to the length scale of a few micron.

The gas permeation experiments with four gases (H_2, He, CH_4, and CO_2) at variable temperatures (25-200 °C) were conducted to study the transport properties of the intrinsic defects of atom-thick graphene film. Resulting permeance of CH_4 and CO_2 were 6 and 16 GPU, respectively at 200 °C. Moreover, He/CH_4 ideal gas selectivity was as high as 29 with He permeance of 183 GPU at 200 °C. Interestingly, we also observed H_2/CO_2 selectivity (8.8) with H_2 permeance of 137 GPU at 200 °C, indicating that average size of nanopores in graphene was somewhat smaller than that of the kinetic diameter of CO_2. The permeance increased with temperature, indicating molecular transport was in the activated transport regime. This high gas selectivity observed from a 1 mm^2 large graphene membrane is first proof-of-principle report on gas separation from large-scale single-layer graphene film.

References:

(1) Koenig, S. P.; Wang, L.; Pellegrino, J.; Bunch, J. S. Selective Molecular Sieving through Porous Graphene. *Nat Nano* **2012**, *7* (11), 728–732.

(2) Celebi, K.; Buchheim, J.; Wyss, R. M.; Droudian, A.; Gasser, P.; Shorubalko, I.; Kye, J.-I.; Lee, C.; Park, H. G. Ultimate Permeation across Atomically Thin Porous Graphene. *Science* **2014**, *344* (6181), 289–292.

(3) O'Hern, S. C.; Stewart, C. a; Boutilier, M. S. H.; Idrobo, J.-C.; Bhaviripudi, S.; Das, S. K.; Kong, J.; Laoui, T.; Atieh, M.; Karnik, R. Selective Molecular Transport through Intrinsic Defects in a Single Layer of CVD Graphene. *ACS Nano* **2012**, *6* (11), 10130–10138.

494314 Relationship between pK_a of Amines in Microgel Particle Membranes and CO_2 Permeance

Abstract:

Technology development of CO_2 capture technology from mass CO_2 emission sources such as thermal power plants is needed to decrease CO_2 concentration in the atmosphere. One of the most realistic strategies to reduce the CO_2 emission is capturing (separating) CO_2 from exhaust gases and storing them under ground. Membrane separation method is theoretically the most energy-saving separation method driven only by transmembrane pressure difference. However, there is a trade-off between selectivity and permeability.[(1)] Development of materials to overcome the trade-off and to realize reduction of CO_2emission is highly required.

We have developed CO_2 selective membranes using amine-containing gel particles (GPs) as the separation function layer on a support. The GPs consist of *N*-Isopropylacrylamide (NIPAm), *N,N'*-Methylene-bis-acrylamide (BIS) and amine-containing monomer. Properties of GPs such as particle size, amount of amine, swelling behavior, and pK_a of amines can be controlled by tuning polymerization procedures.[(2)] In this paper, we report the effects of pK_a on CO_2separation efficiency.

GPs with variety of pK_a were prepared by free radical polymerization of monomers. Concentration and pK_a of amines were quantified by acid base titration. CO_2 separation membranes were prepared by spray coating of GPs solution to the porous support membranes. Membranes whose thickness of prepared GPs layer ranged from 2 μm to 130 nm determined by scanning electron microscope are prepared. Humidified gas mixture of CO_2/N_2(10/90 by vol.) was fed to the membranes, and the composition of the permeate was quantified by gas chromatography equipped with a thermal conductivity detector.

CO_2 permeance of amine-containing GPs membranes is in reverse proportion to thickness of the GPs layer when the thickness of GPs layer was more than 500 nm. Meanwhile, CO_2 permeance didn't change dramatically when the thickness of GPs layer was less than 500 nm. It is suggested that surface reaction at the membrane surfaces was a rate-limiting step of CO_2 permeation when the thickness of GPs layer was less than 500 nm. What's more, CO_2 permeance depended on pK_a of amines in GPs when the thickness of GPs layer was 250 nm. The GP with pK_a of 6.6 showed the CO_2 highest permeance of 195 GPU, while 61 GPU for the GP with pK_a of 8.3. It has been reported that CO_2 absorption rate depends on pK_a.[(3)] Thus the CO_2 absorption and desorption rates were equal at p$K_a$of 6.6, and permeation resistance was the smallest, which resulted in giving the highest permeance.

In conclusion, we developed amine-containing GPs membranes with tunable pK_a and found that

CO_2 permeance was the fastest at pK_a of 6.6. It can be for the first time to clarify that pK_a value determines the CO_2 permeance. This finding would provide an insight for further development of CO_2 separation membranes.

References

1. Robeson, L. M., J. Membr. Sci, 320, 390-400 (2008).

2. Yue, M., Hoshino, Y., Ohshiro, Y., Imamura, K., and Miura, Y., Angew. Chem. Int. Ed. 126, 2692-2695 (2014).

3. Littel, R. J., Swaaij, W. V., and Versteeg, G. F., Aiche. J. 36, 1633-1640 (1999).

500061 Effect of Moisture on Mechanical Properties of an Amphiphilic Block Copolymer Membrane

Abstract:

Applications such as CO_2 capture and flue gas dehydration require membranes that are mechanically robust at elevated temperatures and that have high selectivity for CO_2 and/or water over other gases. Polystyrene-*b*-poly(ethylene oxide) is an amphiphilic block copolymer. In the phase-separated state (e.g. for high molecular weight polymer) glassy polystyrene (PS) domains act as mechanical crosslinks resulting in exceptional toughness. Hydrophilic poly(ethylene oxide) (PEO) domains confer selectivity for polar compounds such as water and carbon dioxide. An important consideration that has not been investigated is the impact of moisture on the mechanical properties of this material. We report our findings on the effect of moisture on the glass transition temperature of PS and the resulting impact on the mechanical properties of the block copolymer using humidity-controlled differential scanning calorimetry and dynamic mechanical analysis. In order to understand how mechanical properties are transmitted from the length scale of block copolymer phase separation to macroscopic lengths important for membrane performance, we have investigated local measurements of block copolymer dynamics using a time-resolved scattering technique, x-ray photon correlation spectroscopy.

494225 The Effects of SiO_2 and Zeolite 4A/SiO_2 on the Transport Behavior of CO_2 and CH_4 through Polydimethylsiloxane Nanocomposite Membranes

Abstract:

The Effects of SiO_2 and Zeolite 4A/SiO_2 on the Transport Behavior of CO_2 and CH_4 through Polydimethylsiloxane Nanocomposite Membranes

Abstract

Emmanuel O. Ogbole, Jianzhong Lou, Shamsuddin Ilias

Polymeric membrane-based gas separation provides energy-efficient unit operation in the process -industry. Polydimethylsiloxane (PDMS) is one of the common polymers used for developing mixed matrix membranes (MMMs) for gas separation as it offers one of the highest permeability coefficients for a wide range of gas species, modest selectivity, good material compatibility with inorganic materials, etc. Single-filler and double-fillers mixed matrix membranes were prepared with polydimethylsiloxane (PDMS) and different weight percent of SiO_2 and zeolite 4A/SiO_2. The effects of the single filler (SiO_2) and double-fillers (zeolite 4A/SiO_2) on the transport behaviors of CO_2 and CH_4 through the nanocomposite membranes were investigated. Permeability of CO_2 and CH_4 through the neat PDMS membrane and the PDMS mixed matrix membranes (MMMs) were experimentally measured. CO_2 and CH_4 permeability were theoretically predicted using Maxwell model. Scanning electron microscopy (SEM) images and visual observation showed that SiO_2 enhanced the uniform dispersion of zeolite 4A in the double-filler-PDMS membrane. The incorporation of SiO_2 nanoparticles and zeolite 4A/SiO_2 disrupted the PDMS polymer chains packing arrangement; thereby, altering the performance of MMMs and the transport behavior of the gases in the single filler (SiO_2-PDMS) MMMs and double filler (zeolite 4A/SiO_2-PDMS). The experimental permeability measurements of CO_2 and CH_4 through the (SiO_2-PDMS) MMMs showed a measure of deviation from the theoretical model. The 2.5wt% zeolite 4A/2.5wt%SiO_2-PDMS MMM exhibited significant average CO_2/CH_4 selectivity improvement over the neat membrane and other MMMs from 3.09 to 4.89.

496023 Composite Ionic Liquid and Dense Polymeric Membranes for CO_2/N_2 and CO_2/CH_4 Gas Separation at Elevated Pressures

Abstract:

Different types of ionic liquids (ILs) were blended with polysulfone (PSF) to produce functional dense polymer-supported IL membranes (DPSILMs). The main aim was to investigate the applicability of the DPSILMs in high-pressure CO_2 separation from N_2 and CH_4 streams with less or no IL loss. The DPSILMs showed a clear chemical and physical change in the PSF structure and a good IL distribution in PSF. CO_2/N_2 and CO_2/CH_4 binary mixtures (5 mol% CO_2) were used.

CO_2/N_2 separation results obtained from both systems showed that the addition of [C_4mim][NTf_2], [DIP-C_4mim][NTf_2], and [N_{4441}][formate] to PSF increased the separation efficiency of CO_2. However, CO_2/N_2 selectivity decreased with the addition of [P_{4441}][formate]. The optimum IL concentrations for CO_2/N_2 separation are 0.5 wt% [C_4mim][NTf_2] and [DIP- C_4mim][NTf_2], 50 wt% for [P_{4441}][formate], and 25 wt% for [N_{4441}][formate].

On the other hand, all of the used ILs showed increased CO_2/CH_4 selectivity compared to that of pure PSF. The highest CO_2/CH_4 selectivities were obtained with DPSILMs having low IL concentrations hence the optimum IL concentrations are 2.5 wt% for [C_4mim][NTf_2] and [DIP- C_4mim][NTf_2], 5 wt% for [P_{4441}][formate], and 0.5 wt% for [N_{4441}][formate].

Stability measurements on DPSILMs were conducted with regards to IL loss. DPSILMs incorporated [N_{4441}][formate] and [P_{4441}][formate] exhibited small loss of IL compared to that reported in literature. Conversely, no loss of [C_4mim][NTf_2] and [DIP- C_4mim][NTf_2] was observed. Thus, the synthesized DPSILMs are stable at high pressures for long durations.

AUTHOR INDEX

AUTHOR INDEX

AUTHOR INDEX

AUTHOR INDEX

AUTHOR INDEX

AUTHOR INDEX

AUTHOR INDEX

AUTHOR INDEX

AUTHOR INDEX

AUTHOR INDEX

AUTHOR INDEX

AUTHOR INDEX

AUTHOR INDEX

Separations Division 2017

Core Programming Area at the 2017 AIChE Annual Meeting

Minneapolis, Minnesota, USA
29 October - 3 November 2017

Volume 2 of 2

ISBN: 978-1-5108-5791-9

Printed from e-media with permission by:

Curran Associates, Inc.
57 Morehouse Lane
Red Hook, NY 12571

proceedings.com

Some format issues inherent in the e-media version may also appear in this print version.

Printed by Curran Associates, Inc. (2018)

For permission requests, please contact AIChE
at the address below.

AIChE
120 Wall Street, FL 23
New York, NY 10005-4020

Phone: (800) 242-4363
Fax: (203) 775-5177

www.aiche.org

Additional copies of this publication are available from:

Curran Associates, Inc.
57 Morehouse Lane
Red Hook, NY 12571 USA
Phone: 845-758-0400
Fax: 845-758-2633
Email: curran@proceedings.com
Web: www.proceedings.com

TABLE OF CONTENTS

VOLUME 1

VOLUME 2

Some abstracts in this book refer to images that were not made available by the publisher

17AIChE
Annual Meeting, Minneapolis, MN

METAL-ORGANIC FRAMEWORK/GRAPHENE OXIDE COMPOSITE FILLERS IN MIXED-MATRIX MEMBRANES FOR CO_2 SEPARATION

Stavroula Anastasiou, Nidhika Bhoria, Jeewan Pokhrel,
K. Suresh Kumar Reddy, Georgios N. Karanikolos*

Department of Chemical Engineering,
Petroleum Institute, Khalifa University of Science & Technology, Abu Dhabi, UAE

KHALIFA UNIVERSITY OF SCIENCE AND TECHNOLOGY جامعة خليفة للعلوم والتكنولوجيا

جامعة خليفة
KHALIFA UNIVERSITY

Masdar مصدر

CO_2 Problem - Opportunity

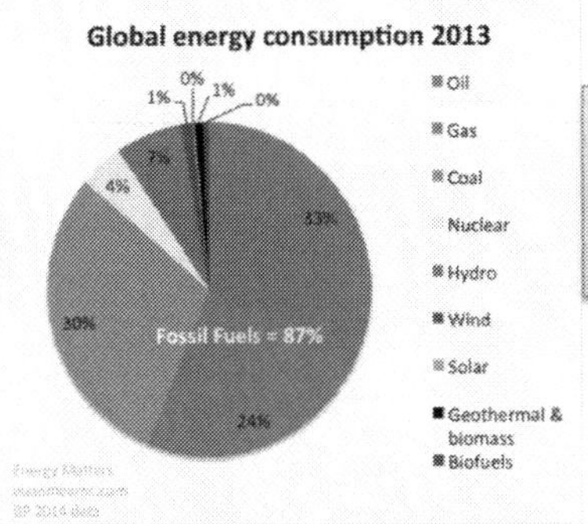

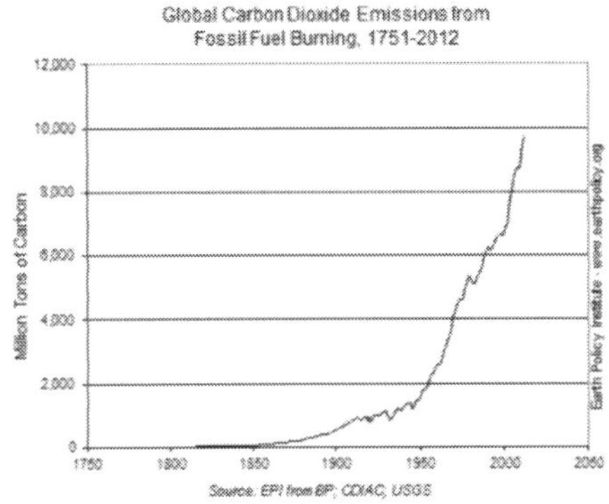

Major greenhouse gas → Global warming → Climate change

- Extreme weather effects: droughts, floods
- Increasing of sea levels
- Extinction of plants and animals species, unable to follow rapid climate change

Challenge for the fossil fuel industry

- Reduction of CO_2 release into the atmosphere
- CO_2 removal from natural gas before further processing or selling

Captured CO_2 - EOR

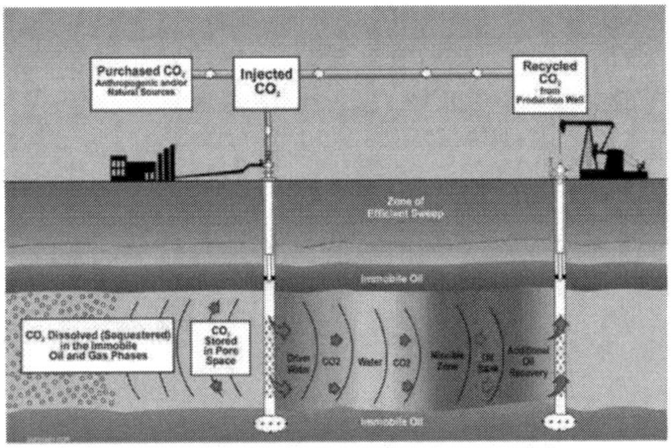

Source: Advanced Resources International and Melzer Consulting, Optimization of CO_2 Storage in CO_2 Enhanced Oil Recovery Projects, prepared for UK Department of Energy & Climate Change, November 2010.

Major EOR medium: CO_2

➢ ***High extraction cost***

Injection of captured CO_2 in depleted oil reservoirs

↓

Increase in recovery through the EOR process

Dual Gain for oil industry by:

- Substantial reduction of CO_2 emissions
- Increase of recovery from oil fields

CO_2 capture techniques

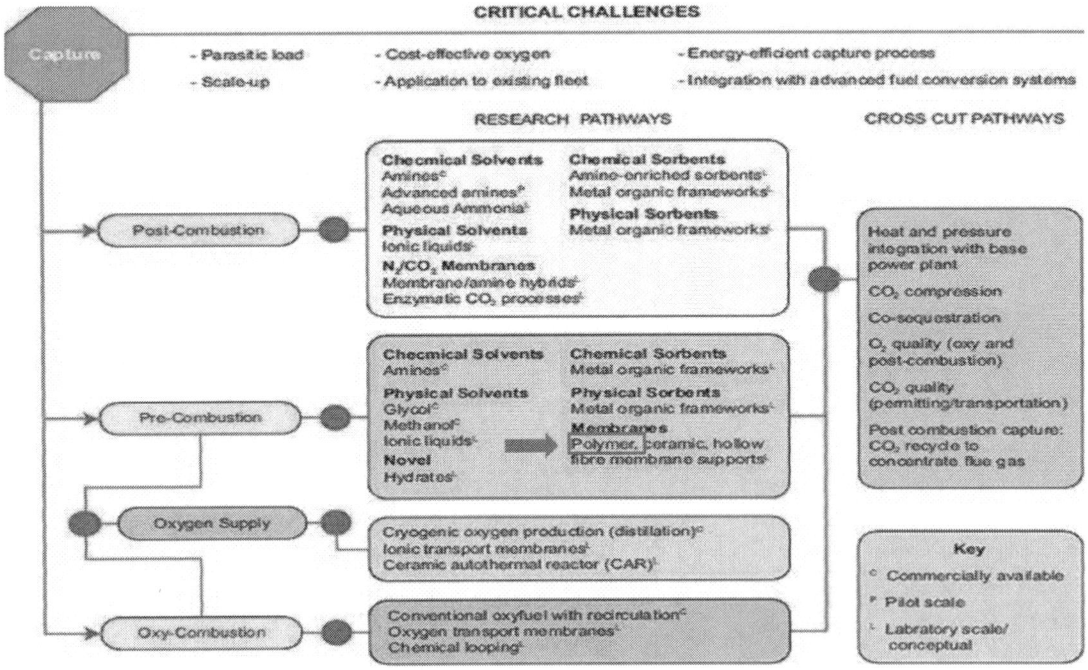

Membrane Separation

The Problem: 15% of the global energy (~60×10^{15} Btu/yr) is currently used for separation and purification, out of which, only a small fraction is dedicated to the actual separation due to thermodynamic limitations (e.g. 10% efficiency in distillation columns)

Action is needed: Energy-conservation strategies relying on replacing or combining thermally driven separation processes with more efficient ones

Promising option: Energy-efficient high-resolution molecular separation with membranes based on preferential adsorption and/or sieving of molecules with minute size and shape differences

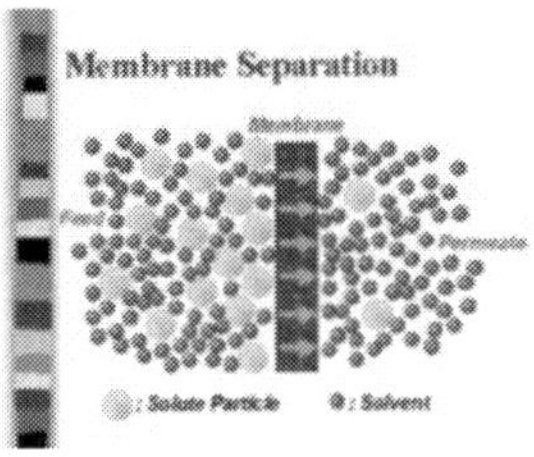

CO_2 and H_2S capture
Olefin/paraffin separations **(ethylene/ethane, propylene/propane, styrene/ethyl benzene)**
Isomer separations **(p-xylene/mixed xylenes, n-paraffins/isoparaffins**
Removal of organics/dilute organics from water **(ethanol, isopropanol, acetic acid, ethylene glycol)**
Other hydrocarbon separations
Inorganic chemicals

Bottom-up Design of Novel Separation Media

Improving current technologies (increase efficiency, reduce cost, reduce energy demand)

Focus on materials → Hierarchical, Nanotechnology-based approaches in design and development of novel materials and hybrids

meter	m	10^{0}	1 m
centimeter	cm	10^{-2}	0.01 m
millimeter	mm	10^{-3}	0.001 m
micrometer	µm	10^{-6}	0.000001 m
nanometer	nm	10^{-9}	0.000000001 m

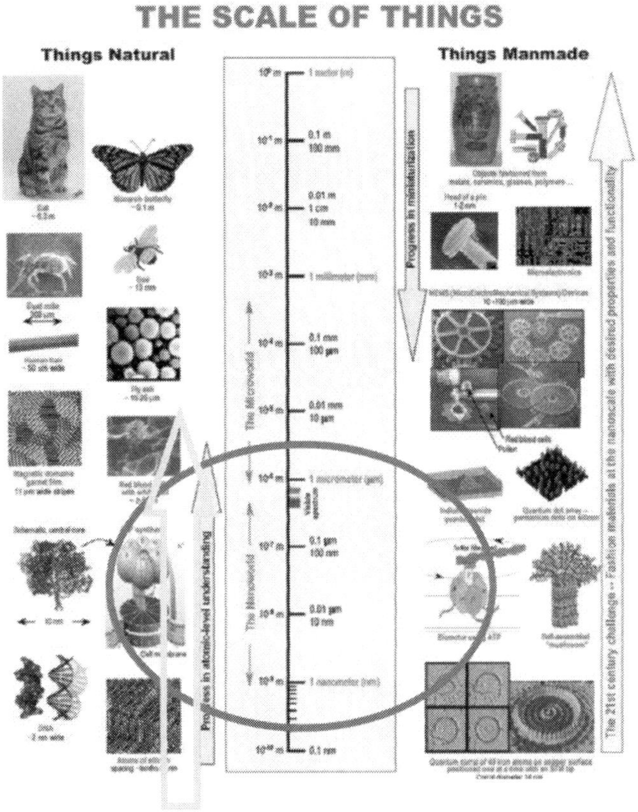

Bottom-up Materials Design

- High Resolution
- High Functionality
- Increased Flexibility
- Nature-mimetic

Mixed-Matrix membranes (MMMs)

Modification of polymers with inorganic fillers (zeolites, mesoporous silicas, activated carbons) for MMMs synthesis to enhance permeability/selectivity/robustness:

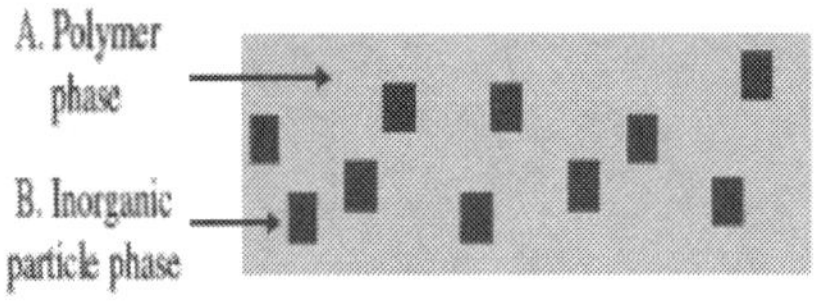

Metal-Organic Frameworks as new potential fillers in the polymer matrix:

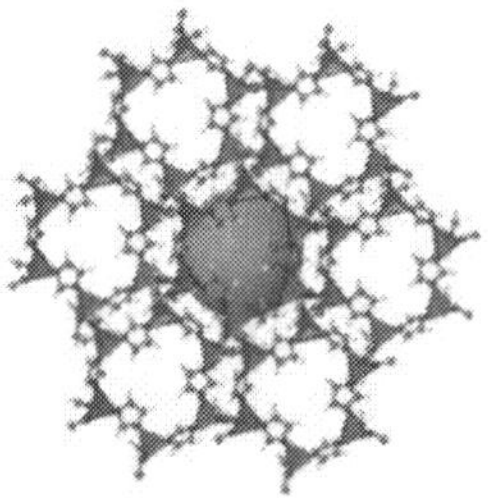

- New class of microporous materials
- Properties:
  - High surface area
  - High pore volume
  - Enhanced sorption capacity
  - Selective affinity for gases
  - Tunable pore structure/structural flexibility

Metal ions
Organic ligands

Mixed-Matrix Membrane Components

ZIF-8:

- Thermal stability up to 400 °C
- Chemical stability in the presence of water, benzene, and other aromatic hydrocarbons
- Main pore cavity: 11.6 Å, windows: 3.4 Å

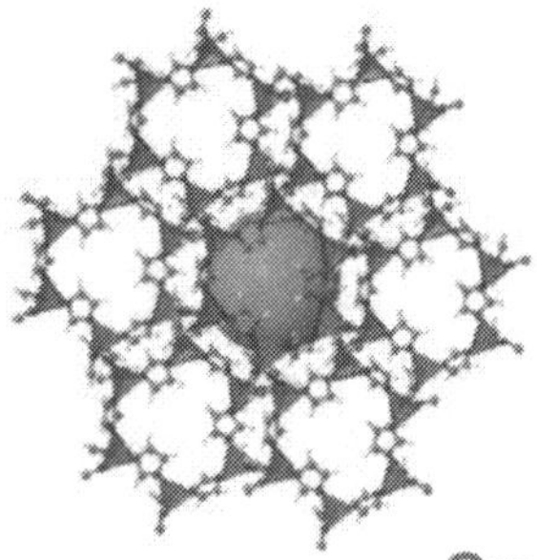

Polysulfone (PSF):

- Thermally and chemically stable thermoplastic polymer
- Easy processing and membrane fabrication

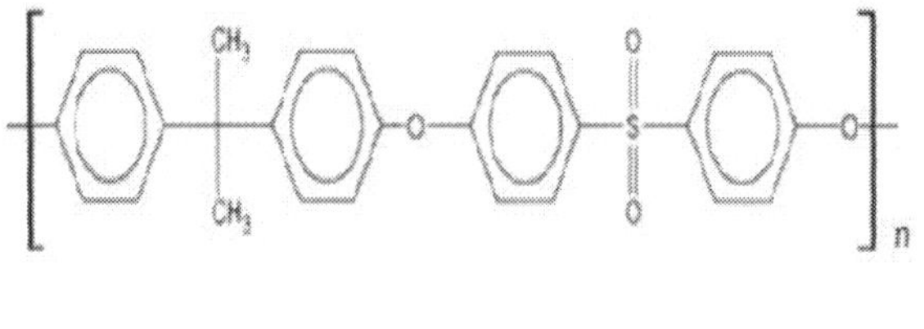

Graphene Oxide (GO):

- Support for ZIF/GO composite fillers

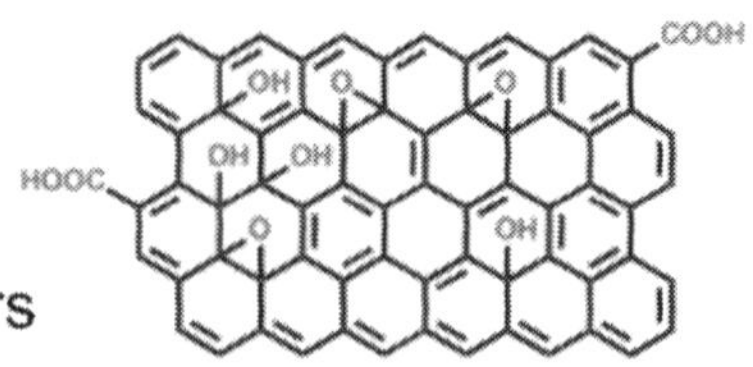

Membrane Fabrication

GO synthesis

Oxidation of graphite with H_2SO_4, H_3PO_4 and $KMnO_4$ → addition of H_2O_2/H_2O to stop oxidation → washing, centrifugation, freeze-drying to obtain the GO

Fillers synthesis

ZIF-8

- Solutions of $Zn(NO_3)_2 6H_2O$, 2-methylimidazole and methanol
- Centrifugation and washing
- Vacuum drying of obtained crystals

ZIF-8/GO

- Mixing of $Zn(NO_3)_2 6H_2O$,
- 2-methylimidazole and DMF
- Addition of dissolved GO
- Reaction, filtration, washing
- Vacuum drying

Membrane Fabrication

PSF+ZIF-8 and MMM
(5% w/w ZIF-8, 95% w/w PSF)

- Solutions of PSF and ZIF-8 with DMF
- Mixing of two solutions with stirring and probe-sonication
- Solvent evaporation and MMM casting
- Vacuum drying

Pure PSF membrane

- Dissolution of PSF in DMF
- Solvent evaporation and MMM casting
- Vacuum drying

PSF+ (ZIF-8 /GO) MMM
(5% w/w ZIF-8,
5% w/w GO, 90% w/w PSF)

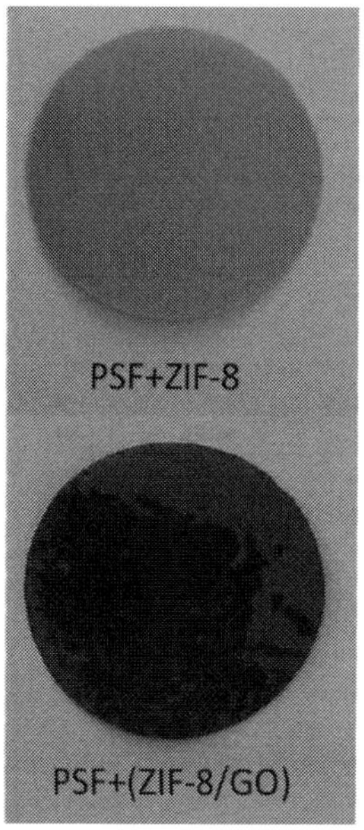

Filler Morphology

SEM

GO

ZIF-8

ZIF-8/GO

X-Ray Diffraction

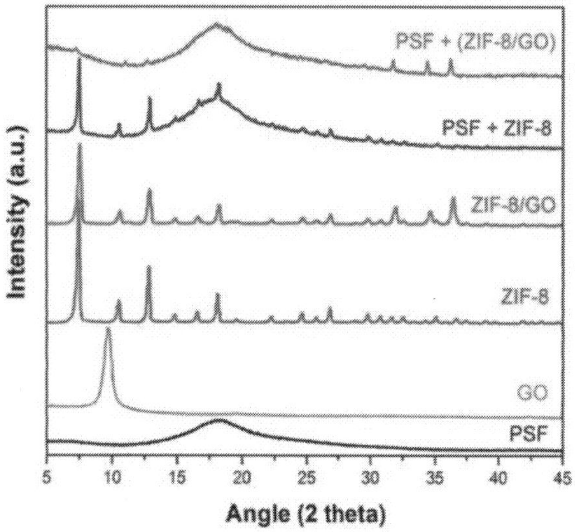

- The ZIF-8/GO pattern is similar to that of ZIF-8
  - ✓ Absence of GO peak → Complete GO exfoliation
- In both MMMs the filler reflection peaks are present.
  - ✓ Slight move of PSF peak from 18.2 ^{0} to 18.4 ^{0}

FTIR Spectroscopy

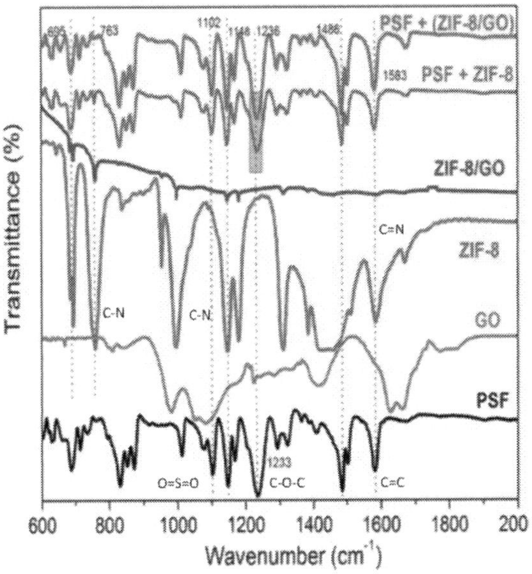

- The MMMs PSF+ZIF-8 and PSF+(ZIF-8/GO) spectra show peaks that are assigned to both the polymer and the filler
- A slight shift of PSF peak (1233 cm^{-1}) to higher wavenumbers in both MMMs.

CO_2 Adsorption

CO_2 adsorption isotherms at 25 ^oC

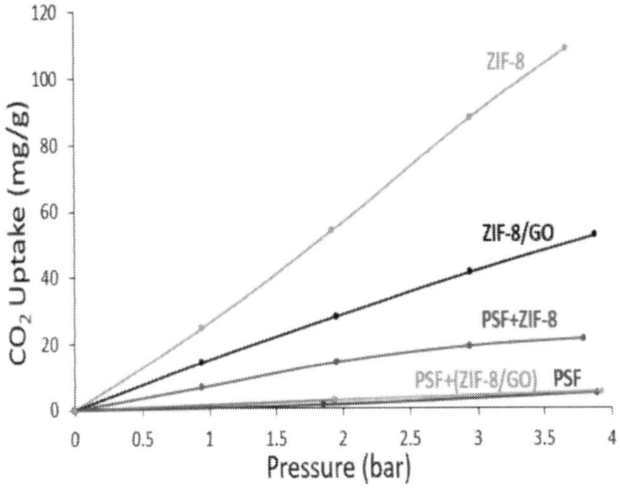

BET Surface Area for **ZIF-8**: 1318 m^2/g

& for **ZIF-8/GO**: 689 m^2/g

CO_2 adsorption

Sample	Max*.CO_2 Uptake (mg/g)
ZIF-8	109.1
ZIF-8/GO	52.7
PSF+ZIF-8	21.5
PSF+(ZIF-8/GO)	5.5
PSF	5.2

* at 4 bar

Gas Permeation

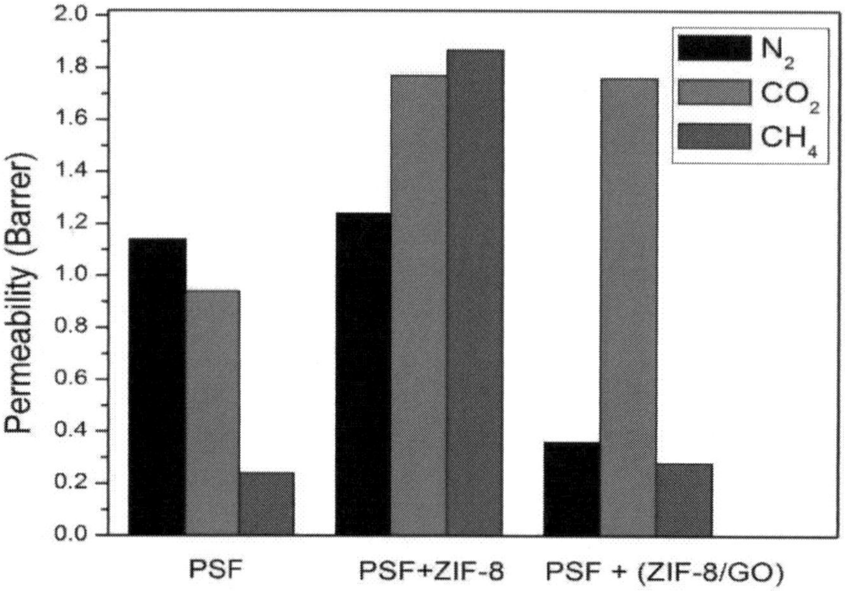

(1 Barrer = $1 \cdot 10^{-10}$ cm^3 (STP)·cm/(cm^2·s·cmHg))

CO_2/CH_4 Separation

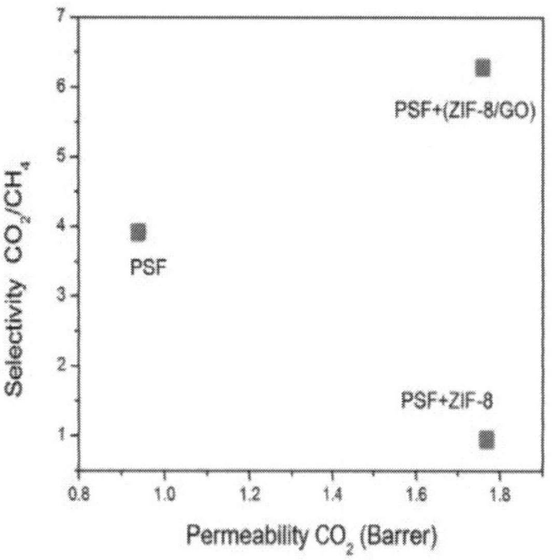

- Both MMMs:

Higher CO_2 permeability compared to pure PSF

- PSF+ZIF-8 MMM:

CO_2 permeability increased at the expense of CO_2/CH_4 selectivity

- PSF+(ZIF-8/GO) MMM:

increased permeability and CO_2/CH_4 selectivity

G.N. Karanikolos et al., Separ. Pur. Tech. 2017

CO_2/N_2 Separation

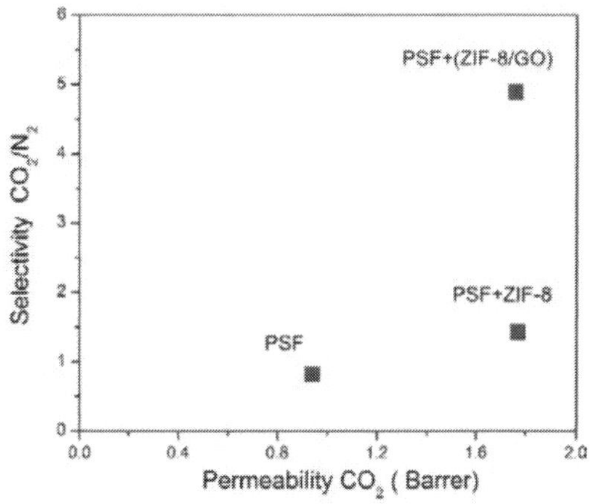

• Both MMMs:

Higher CO_2 permeability compared to pure PSF

• PSF+ZIF-8 MMM:

CO_2 permeability increased with slight increase in CO_2/N_2 selectivity

• PSF+(ZIF-8/GO) MMM:

Increased permeability and CO_2/N_2 selectivity

Conclusions

- PSF MMMs with ZIF-8 fillers exhibited enhanced CO_2 permeabilities.
- The MMMs with ZIF-8/GO composite fillers exhibited both enhanced CO_2 permeability and CO_2 selectivity for CO_2/CH_4 and CO_2/N_2.
- MOF/GO "dual-action" fillers: MOF enhances affinity of the membrane for CO_2. GO flakes create hindered tortuous diffusion pathways in the matrix allowing diffusion of smaller molecules, such as CO_2, while limiting diffusion of larger molecules, such as N_2 and CH_4, thus enhancing selectivity. The latter is facilitated by the interaction of the gas molecules with the GO polar functional groups leading to decreased solubility of the non-polar gases in contrast to the polar ones.
- Composite fillers combining the functionality of MOFs with GO, and highly versatile matrices have the potential to enhance membrane performance for large scale CO_2 separation from hydrocarbon, flue gas, and related mixtures.

THANK YOU

جامعة خليفة
KHALIFA UNIVERSITY

Masdar مصدر

Funding: Gas Research Center, Abu Dhabi

496226 Adsorption Rate Constant and Equilibrium Constant in Chiral Separation for Trans-Stilbene Oxide, Linalool and Ibuprofen By Supercritical Fluid Chromatography

Abstract:

The enantiomers of chiral molecules have many different biological activities in R and S forms, and their separation is essential in the pharmaceutical, nutraceutical, and food industries [1]. Supercritical fluid chromatography (SFC) has been attracting attention for separating enantiomers with scCO2 by adding the small amount of an organic solvent. The mobile phase properties can be tuned by adding organic species as a modifier [2] as well as by changing temperature and pressure. However, while the elution time of an analyte in SFC is significantly affected by the analytical conditions such as temperature, pressure, modifier species and the composition, the effects of analytical conditions on retention factors have not been clarified well. It is difficult to find the optimum analytical conditions. A predictive method or a predictive correlation of retention factors as a function of analytical conditions is required. In this study, SFC separation was performed for the three kinds of enantiomers, i.e., trans-stilbene oxide (tSO), linalool, and ibuprofen, and the chromatograms were analyzed using the moment method [3].

As a result, the adsorption equilibrium constant K_A, defined as the rate constant ratio of desorption k_d to adsorption k_a were almost constant, independent of the flow rate of the mobile phase. On the other hand, the adsorption rate constant k_a can be approximated by a linear curve passing through the origin and it can be said that there is a tendency to depend strongly on the flow velocity with correlation coefficients $R^2 = 0.99$ or higher. As the flow velocity increases, since the number of analyte molecules adsorbing and desorbing passing through the silica gel surface increased, this resulted in increasing k_a and k_d. Thus, K_A did not depend on the flow rate and were almost constant for each component.

[1] Min, L. *et al. J. Sep. Sci.,* **36**, 3093 (2013).

[2] Horikawa, Y. *Chromatogr.*, **32**, 153 (2011).

[3] Wakao, N., Kaguei, S., Heat & Mass Transfer in Packed Beds, Gordon & Breach,

New York (1982).

500445 Prediction of Water Uptake in Ion Exchange Membranes Based on Gel Swelling Models Combined with Ion Sorption Model

Abstract:

Ion sorption and diffusion in ion exchange membranes are significantly influenced by water uptake in the membrane. Therefore, fundamental understanding of the relationship between polymer structure and water uptake contributes better design of the membranes. However, in spite of its huge effect on membrane performance, there have been a smaller number of studies about water uptake prediction model describing the relationship in an ion exchange membrane field compared with ion sorption model.

In this work, prediction models for water uptake are discussed using gel swelling models combined with ion sorption model. Flory-Rehner model was firstly introduced to predict swelling of the polymer networks which have ionic functional group and Gaussian distribution polymer chains. [1] This model has been modified by Tanaka [2] and Peppas [3] in order to incorporate the effects of diluted polymer solution and non-ideality of polymer chains for hydrophilic gels. Recently, Donnan and Manning model was developed to predict coion concentration in ion exchange membranes. [4]

Strongly charged cation and anion exchange membranes were synthesized in this study. Water uptake in those membranes were measured as a function of external NaCl concentration within 0.01 M to 1.0 M, and interpreted using the gel swelling models. Water uptake predicted by Flory-Rehner model which accounted for the effect of diluted polymer solution and non-Gaussian effect showed good agreement with experimentally obtained water uptake. In this presentation, coion concentration was also predicted by Donnan and Manning model, and water uptake prediction based on gel swelling model combined with ion sorption model is also discussed.

[1] P. J. Flory, CRC press, 1953.

[2] T. Tanaka et al., Phys. Rev. Lett., 45(20), 1636-1639, 1980.

[3] Peppas and Lucht, Chem. Eng. Commun., 30.3-5, 291-310, 1984.

[4] J. Kamcev et al., Phys. Chem. Chem. Phys., 18, 6021-6031, 2016.

497805 Design of Ionic Liquid Epoxy Functionalized Ion Exchange Resin Wafers for Low Energy Electrodeionization

Abstract:

Current improvement in and development of new separation technologies is targeted at simplifying operation methods, high product recovery, minimizing waste production, reducing costs & energy requirements while generally ensuring that the process is environmentally friendly. One widely used technology in the chemical industry is electrodeionization; a charged-based separation technique that utilizes both ion exchange and electrodialysis for efficient removal of ions. Researchers have immobilized resins into wafers using common binding polymer agents. (e.g. polyethylene) which has helped prevent ion leakage and uneven flow in EDI, yet poor current efficiencies make the process energy intensive. There has yet to be a study on the use ionic liquid cured epoxy in resin wafers for improved performance in EDI system. This research aims at designing and examining the impact of ionic liquid epoxies on performance indicators in resin wafer electrodeionization focusing on the effect on energy consumption.

495406 Research on CO_2 Sorption Capacity of Two Silica-PEI Samples in the Bubbling Fluidized-Bed Reaction System

Abstract:

In this study, the CO_2 sorption capacity of two silica-PEI (molecular weight of PEI: 800 and 5,000 g/mol) samples, which have been supplied by the University of Nottingham (UNOTT), has been measured in the bubbling fluidized-bed reaction system. At first, the CO_2 sorption and desorption behaviours of silica-PEI samples were investigated in a bubbling fluidized-bed reactor (90 cm tall bed with 0.5 cm i.d.). Those experiments were carried out by the cyclic tests (5 times of sorption step and 5 times of desorption step) at 65 to 70°C of sorption temperature and 115 to 125°C of desorption temperature. Secondly, two bubbling fluidized-beds CO_2 capture system has been operated continuously in order to evaluate the dynamic sorption capacity of the samples by varying several operating variables such as moisture contents in the feed gas, operating temperatures of each reactor, etc. The CO_2 removal performance has also been evaluated according to the several operating variables.

502481 Separation of CO_2/N_2 mixture Using MIL-101(Cr)/PVA Pellets By PSA Process

Abstract:

In the past two decades, Metal Organic frameworks (MOFs) are being developed as promising candidates for CO_2 capture due to their high porosity and selective CO_2 capture ability. In this study, the metal organic framework MIL-101(Cr) was synthesized by hydrothermal method in a batch to yield about 15 gm product. The PVA/H_2O polymer binder was used to make pellets (of about 2 mm diameter) of MIL-101(Cr) to be used in column studies. The ratio of MOF powder to PVA/H_2O was varied to optimize the beads preparation. The synthesized MOF powder and MOF/PVA pellets were characterized by BET surface area analysis and XRD analysis. The pure component equilibrium isotherms of CO_2 and N_2 were measured on MIL-101(Cr) powder and MIL-101(Cr)/PVA pellets. A 2 cm diameter column of 30 cm length containing about 11 gm of adsorbent was used in the column studies. A simulated dry flue gas (~15% CO_2 and balance N_2) was prepared in laboratory and used as feed mixture. Using data acquisition system, the feed flowrate, column pressure were recorded during breakthrough and process experiments. Based on the breakthrough time, cycle times were chosen for the PVSA (Pressure/Vacuum Swing Adsorption) process. The PVSA cycle studies have been carried out between 1.3 bar (adsorption) and 0.1 bar (desorption) at a flowrate of 150 SCCM. The effect of various steps, such as purge with the product (N_2) and rinse with CO_2 have been studied.

495693 Adsorption and Desorption Breakthrough Behaviors of Carbon Dioxide, Nitrogen, Water Mixture over PEI-Silica Solid Sorbent in a Temperature Programmed Oven

Abstract:

TPAD (Temperature Programmed Adsorption & Desorption) system has been developed for evaluating adsorption and desorption behaviors using under 1 g amount of sorbents. In order to control the system temperature precisely, all the equipments, such as fixed bed, pre-heater, 4 ports valve and filters were installed in a temperature programmed oven (Agilent 6890 GC oven). And all the tube lines, valves, and regulators were insulated by heating tapes with temperature controllers. The effect of temperature, feed concentration, retention time and cyclic operation on CO_2/N_2 and $CO_2/H_2O/N_2$ breakthrough behaviors over PEI-Silica sorbent were studied in a fixed bed with 1/4 inch outside diameter. From the experimental data, adsorption kinetics and equilibrium parameters could be estimated. One of the most important features of TPAD system is that the kinetic and the equilibrium information can be obtained by using very small amount of sorbents (under 1g).

490232 Separation and Purification of Cyclopentadiene and Methyl CyclopentadieneÂ from Pyrolysis Carbon 9

Abstract:

Nowadays, along with the expansion of ethylene production, the efficient use of by-product of Pyrolysis C9 (Carbon 9) in ethylene production is an important subject to petroleum refining industry. Currently, C9 Distillate Fraction is mainly used as fuel and in making petroleum resin, which causes serious waste because in C9 Distillate Fraction, dicyclopentadiene and methyl cyclopentadiene, with high value-added dimer, cannot be efficiently recycled.

Nowadays, highly pure dicyclopentadiene and methyl ring dicyclopentadiene are mainly obtained from extraction of the Pyrolysis in C5. It's difficult to attain highly pure dicyclopentadiene and methyl ring dicyclopentadiene dimer because the components of the boiling point of C5 are fairly similar to those of the cyclopentadienyl and methyl ring cyclopentadienyl. At the same time, the price of Pyrolysis C9 is lower than that of pyrolysis C5. If the highly pure dicyclopentadiene and methyl ring dicyclopentadiene dimer are separated from Pyrolysis C9, it will meet the rising requirement of CPD, MCPD and their dimers and bring huge economic value.

In this paper, the author analyzes the highly different boiling points of the component in the C9 and cyclopentadiene and methyl cyclopentadiene monomer, and then obtains highly pure cyclopentadienyl and methyl cyclopentadiene by use of the depolymerization of dicyclopentadiene and methyl cyclopentadiene dimer though two consecutive reactive distillation columns. With C9 fraction as raw material, it first enter the depolymerization reaction distillation tower 1, DCPD in the distillation column continuous depolymerized and high-pure CPD obtained from the top of distillation tower, at the same time, the MCPD is enriched in tower kettle in the form of dimer. Then liquid from the tower 1 kettle enters the depolymerization reaction distillation column 2, DMCPD depolymerized in this tower and high-pure MCPD is obtained from the top of distillation tower 2. High-pure cyclopentadiene and methyl cyclopentadiene were obtained individually at the top of two reactive distillation columns. This method will provide a new direction of the comprehensive utilization of C9.

The calculation and designation of cyclopentadienyl and methyl dicyclopentadiene separation columns are based on the small scale experiments of the MCPD and CPD separated from Pyrolysis C9. On the basis of the small scale experiments, by verifying the process of cyclopentadienyl and methyl ring cyclopentadiene produced by depolymerization rectification, the author obtains optimum conditions, and ensures the process of industrialization and technological conditions. Eventually by adopting two reactive distillation columns, the author achieves high purifying products of dicyclopentadiene and methyl ring dicyclopentadiene dimer.

Experiments proved that it is feasible to use byproduct C9 fraction as the raw material to extract high-pure CPD and MCPD in the process of two depolymerized distillation columns. It has promising prospects for industrial development and great economic benefits. CPD>99% of cyclopentadienyl which is obtained from the top of depolymerization distillation column 1, converts into highly pure DCPD product after simple dimerization. And MCPD>97% of methyl cyclopentadienyl which is obtained from the top of depolymerization distillation column 2, converts into highly pure

DMCPD product after simple dimerization. CPD yield reaches to 95%, and MCPD yield reaches to 92%.

502691 Tailoring Porous Polymers for Adsorption and Catalytic Applications

Abstract:

Microporous activated carbon (AC) and zeolites are industrially applied for air pollution control, but pore blockage, irreversible adsorption, and co-adsorption of moisture reduce their contaminant removal efficiencies and/or prevent reuse. Porous polymers are gaining traction in environmental engineering because they can have tunable physical and chemical properties that allow tailoring for specific applications. Our previous studies show that hydrophobic styrene-divinylbenzene (St-DVB) polymers maintain their catalytic NO oxidation performance when moisture is present, while AC fails due to competitive water adsorption caused by increased hydrophilicity and surface oxidation during use. It was also observed that polymers provide faster adsorption/desorption kinetics and higher desorption efficiency than AC at low temperatures. However, commercial St-DVB polymers have low micropore (d_p < 2 nm) volume, reducing adsorption capacity for low concentration contaminants. To overcome, the Friedel-Crafts reaction is applied to increase micropore volume by hyper-cross-linking polymer segments and reducing the network mesh size. The Friedel-Crafts reaction can also be directly used to synthesize microporous polymers from aromatic compounds such as benzene, phenol, or methylbenzene using external cross-linkers. This method not only produces cost-effective microporous polymers, but also allows user-defined surface functional groups by changing the aromatic building blocks or by post-synthesis modifications.

In this work, commercial St-DVB polymers were first hyper-cross-linked with dichloroalkanes to increase micropore volume by up to 333%, without sacrificing the inherent hydrophobicity of the material. Consequently, hexane adsorption capacity increased by up to 218% at relative pressure (p/p_0) = 0.05 and, in a separate series of experiments, catalytic NO oxidation rate increased by 100% ([NO] = 500 ppm_v). Next, microporous polymers were directly prepared using the Friedel-Crafts reaction with benzene or phenol monomers, formaldehyde dimethyl acetal or 4,4â€²-Bis(chloromethyl)-1,1â€²-biphenyl (BCMBP) cross-linkers, and $FeCl_3$ as the reaction catalyst. Maximum surface area and micropore volume were 1600 m^2/g and 0.43 cm^3/g, respectively, when BCMBP self-condensed (reacted as monomer and cross-linker). The surface area and micropore volume decreased by 50% and 95% when 0.2 ml and 2 ml benzene was used as a co-monomer, respectively. Using co-solvents of dichloroethane and hexane reduced surface area, micropore volume, and total pore volume by 50%, 40%, and 70%, respectively. Overall, this work describes novel polymer modification and production processes that highlight opportunities to tailor the productâ€™s physical and chemical properties. The overarching goal is to exploit this flexibility, in an effort to present microporous polymers as competitive materials for use in air pollution control applications.

491754 Chemical Separations for Improved Catalytic Upgrading of Fast Pyrolysis Bio-Oils

Abstract:

Chemical Separations for Improved Catalytic Upgrading of Fast Pyrolysis Bio-oils

Mi (Aimee) Lu and Michael Z. Hu

Oak Ridge National Laboratory, Oak Ridge, TN 37831

Selective removal of target component(s) from streams of thermochemical conversion pathways for bio-oil liquids or vapors was investigated by *selective separations* and *reactive conversions* to enhance bio-refinery processing economics through *catalyst preservation* and *process intensification* strategies. In one task on pre-processing of condensed bio-oil liquids, we develop molecular removal technology to address two major barriers: (1) Mitigation of catalyst deactivation by highly reactive and thermally unstable foulants, such as carbonyl molecules (e.g., carboxylic acid and aldehydes), to enhance the stability of bio-oils and thereby enhance the lifetime of downstream upgrading catalysts. (2) Improvement of carbon and hydrogen efficiency by removing low-molecular-weight C1-C4 oxygenates from raw bio-oils to prevent the H_2-consuming conversion of these small molecules into low-value hydrocarbon gasses. The recovered carbonyls are considered for production of high-value chemicals. The effectiveness of adsorption and membrane technologies are experimentally and economically evaluated for their separation performance and impacts on bio-oil stability, hydrotreatment catalyst lifetime, carbon and hydrogen efficiency, and cost reduction in both materials and operational maintenance. In another task on in-situ chemistry tailoring of hot pyrolysis vapors, we develop catalytic hot gas filter to process (deoxygenate) vapors before they are condensed into liquid and upgraded to hydrocarbon fuels. Besides the filtration removal of inorganic mineral (alkali/alkaline metals), catalyst-loaded filters are being developed to enable selective target conversions (such as carboxylic acids to ketones or alcohols), producing low-carbonyl bio-oils for better performance and lifetime of upgrading catalysts.

489440 Determination of Adsorption Equilibrium Constants from Experimental Chromatograms for Liquid Adsorption

Abstract:

Equilibrium and kinetic parameters are very important factors for the design of absorbers. Determination techniques of intraparticle diffusivities and fluid-to-solid mass transfer coefficients from experimental result(s) are already proposed for liquid adsorption. Some of them require very small amount of sample mixture.

For determination of equilibrium parameters, on the other hands, â€œBatch Bottle Techniqueâ€ is usually employed. This technique is very simple but some amount of fluid volume is required. The other environment-friendly determination procedure of the parameters employs chromatogram technique. The procedures are roughly divided into tow determination methods. First, chromatograms are estimated numerically under conditions of various equilibrium constants and the curves are compared with the experimentally obtained chromatograms. Then, the optimal equilibrium parameters are determined from the comparison results. Second, derivatives of the isotherm are evaluated from chromatograms experimentally obtained by stepwise technique with the aide of moment analysis and amounts adsorbed are estimated by use of numerical integration. Then, the equilibrium parameters are determined from equilibrium relationships between amounts adsorbed and fluid concentrations by use of traditional technique. The former technique requires calculation time and the constants obtained by the latter technique might have larger experimental error.

In the presentation, new simple determination technique of the isotherm constants directly from the derivatives of the isotherm without any integration procedures will be shown.

504852 Recovery of Lactic Acid from the Pretreated Fermentation Broth Based on a Novel Hyper-Cross-Linked Meso-Micropore Resin

Abstract:

Lactic acid (2-hydroxypropionic acid) is one of the most important organic chiral compound, containing both the hydroxyl and carboxylic acid groups, which is widely used in food, pharmaceutical, chemicals, cosmetics and other industries. Recently, as a monomer for the synthesis of polylactic acid, the demand of lactic acid has a significant increasing. Production of lactic acid by fermentation is the main production method in industry, because the fermentation can make full use of cheap biomass resources, and can produce one of the two stereoisomers of lactic acid or racemic mixture with high yield and productivity. However, the separation of high purity lactic acid from fermentation broth has been a difficult and hot spot.

In this work, an innovative benign process for recovery lactic acid from its fermentation broth is proposed using a novel hyper-cross-linked meso-micropore resin. No environmentally harmful wastes are produced in this process, since deionized water (eluent) is the only substance added to the separation process. This work focuses on modeling the competitive adsorption behaviors of glucose, lactic acid and acetic acid ternary mixture and explosion of the adsorption mechanism. The characterization results showed the resin had a large BET surface area and specific pore structure with hydrophobic properties. By analysis of the physicochemical properties of the solutes and the resin, the mechanism of the separation is proposed as hydrophobic effect and size exclusion. Subsequently three chromatographic models, i.e., GR (General Rate Model), TDM (Transport Dispersive Model) and LDF (Linear Driving Force) models were applied to predict the competitive breakthrough curves of the ternary mixture under different operating conditions. The pore diffusion was the major limiting factor for the adsorption process, which was consistent with the BET results. In the end of this work, the selected model (TDM) was validated by comparison of the chromatographic peaks of the ternary-mixture with pretreated real fermentation broth used as a feed solution. According to this work and comparison with other literature, the novel HD-06 resin can be a good potential adsorbent for the future SMB continuous separation process.

505144 Development of CuCl-Supported Nanoporous Adsorbent Exhibiting High Performances (Adsorption Capacity and Selectivity) of Carbon Monoxide Separation, and Strong Resistance to Oxidation under Atmospheric Condition

Abstract:

Carbon monoxide (CO) is produced in various industrial fields including steam reforming process, partial oxidation and steelmaking process [1]. Especially in the steelmaking process, CO was emitted within off-gas streams such as coke oven gas, blast furnace gas and Linz-Donawitz gas. The concentration of CO in those off-gas is 20~80%, and the amount of annul emission of CO in Republic of Korea is approximately three million ton. The CO gas generated in the steelmaking field is generally burned off in the form of CO_2 to recover energy. However, if CO can be separated from the off-gas, CO can be better utilized as a precursor of high-valued chemicals such as acetic acid, poly-urethane, polycarbonate, formic acid, acrylic acid and phosgene, instead of emitting a greenhouse gas, CO_2. In addition, CO is recently expected to be highly used in the form of syn-gas to synthesize gasoline via Fischer-Tropsch catalytic process. There have been various methods to separate CO, such as cryogenic distillation, absorption method, and adsorption method [2,3]. Cryogenic distillation is commonly used in large-scale CO separation in various fields. However, this process is highly-energy demanding for maintaining cryogenic condition. In addition, it is quite challenging to separate CO and nitrogen which show quite similar boiling point. In the case of absorption method, $CuAlCl_4$ dissolved in toluene is normally used as absorbent, and this process is called as COSORB process. This method is highly advantageous for the generation of high-purity CO gas and recovery rate. However, toluene can be sometimes disposed to harmful species and Cu^+ ion can be easily oxidized during the process. Adsorption method such as pressure swing adsorption is quite promising, and this process is especially suitable for medium-scale CO separation process. This process requires low energy consumption and low cost for process operation. In this method, the separation performance of adsorbent is very important. Development of new adsorbent can make innovation of the adsorption process.

We have developed nanoporous adsorbent supporting CuCl, which exhibits notable performance in selective adsorption of CO. The adsorbent was prepared impregnating CuCl on nanoporous materials via thermal dispersion or impregnation method. In this work, we could find new supporting material which exhibits unexpected characteristics. This supporting material shows high affinity of CuCl, so that exhibiting high adsorption capacity for CO, and extremely low affinity to CO_2. In these reasons, new CO adsorbent synthesized using this supporting material exhibits high adsorption capacity of CO (32 $cm^3\ g^{-1}$) and extremely low adsorption capacity of CO_2 (2.8 $cm^3\ g^{-1}$) under 100 kPa of each gas at 293 K. Thus, this adsorbent exhibits excellent CO/CO_2 separation coefficient higher than 10. In addition, we have developed simple method of pelletizing new adsorbent to show reasonable rigidity. More interestingly, new adsorbent is highly resistant to oxidation of CuCl under atmospheric condition. This result is quite notable, as considering the fact that CuCl is easily oxidized in air. The oxidation of CuCl can cause the dramatic decrease of the adsorption capacity of CO. Our new adsorbent can maintain CO adsorption capacity after six-month exposure to air. More detail results of the material synthesis and CO adsorption analysis will be presented in 2017 AICHE annual meeting. New developed material will be promising adsorbent for selective separation of CO from combustion exhaust and off-gas stream of various industry fields. Furthermore, it can be used as highly-performing adsorbent in separation processes of Ï€-bond-containing species such as olefin and sulfur.

[1] Y. Xie, J. Zhang, J. Qiu, X. Tong, J. Fu, G. Yang, H. Yan, Y. Tang, Adsorption 3 (1996) 27.

[2] F. Kasuya, T. Tsuji, Gas Sep. Purification 5 (1991) 242.

[3] F. Gao, Y. Wang, S. Wang, Chem. Eng. J. 290 (2016) 418.

505145 Synthesis of Highly Performing Nanoporous Carbon Adsorbent for Separation of Siloxane and Ammonia Impurities from Land-Fill Gas

Abstract:

Land-fill gas or bio-gas is composed of large portion of methane and carbon dioxide, and small amount of impurities such as nitrogen, oxygen, hydrogen sulfide, siloxane and ammonia. For example, in Metropolitan Landfill area in Republic of Korea, land-fill gas is recently composed of 47% of methane, 38% of carbon dioxide, 13% of air, 1.8% of hydrogen sulfide, 50ppm of siloxane, and 1ppm of ammonia. Land-fill gas is sometimes separated to form high-purity methane gas by removing other species [1,2]. Hydrogen sulfide can be removed by absorption method using sodium hydroxide solution, and carbon dioxide can be removed by several methods such as water scrubbing absorption, pressure-swing adsorption, and membrane separation. Tiny amount of siloxane and ammonia can be separated via adsorption process, but non-recyclable adsorbents have been so far used for the separation of these species [3,4]. Most challenging part is the separation of air. Cryogenic distillation method is commonly used for this step, but this process is highly energy demanding and highly cost.

Recently, we have been developing low-cost separation process to upgrade land-fill gas to medium-grade-purity (~70%) methane gas which can be still utilized to fuel for various industrial uses such as sewage drying. This process is quite economic, because separation of air is not included. In addition, we developed separation process of siloxane and ammonia residues using recyclable and highly-performing adsorbents. For the synthesis of suitable adsorbent of each component, it was quite necessary to consider the chemical properties of siloxane and ammonia. Siloxane species are known to be polymerized in acidic or basic condition to form bulkier species. The polymeric bulky siloxane can cause adsorbent deactivation due to pore blocking and difficult regeneration. The ammonia gas is well known as basic molecules which have strong affinity to acidic species. In these reasons, siloxane separation, we synthesized neutral carbon materials with large surface area by various methods. In addition, we developed carbon-based acidic ammonia-adsorbent by simple methods. In parallel, adsorption-desorption isotherms and breakthrough curve of siloxane and ammonia were measured for thus synthesized adsorbents. Regeneration tests of those adsorbents were also performed. Our new adsorbents for siloxane and ammonia species exhibited higher adsorption capacity and breakthrough amount than previous commercial adsorbents. More notably, the adsorption performance of new siloxane adsorbent could be completely regenerated by heat treatment. In the case of ammonia adsorbent, we are modifying the ammonia adsorbent to be easily regenerated by simple heat treatment. And then, we will develop adsorption-regeneration process of siloxane and ammonia residue. Detail results for synthesis and the adsorption measurement of the synthesized adsorbents will be presented in the conference.

[1] S. Cavenati, C.A. Grande, A.E. Rodrigues, Energy&Fuels 6 (2005) 2545.

[2] E. Ryckebosch, M. Drouillon, H. Vervaeren, Biomass and Bioenergy 35 (2011) 1633.

[3] A. Cabrera-Codony, M.A. Montes-Moran, M. Sanchez-Polo, M.J. Martin, R. Gonzalez-Olmos, Environ. Sci. Technol. 48 (2014) 7187.

[4] A. Qajar, M. Peer, M.R. Andalibi, R. Rajagopalan, H.C. Foley, Micropor. Mesopor. Mater. 218 (2015) 15.

505466 Synthesis of Various Water Adsorbents with Controllable Adsorption Properties for Application of Water Adsorption Chiller

Abstract:

Adsorption chiller is a type of refrigeration system, operating on the basic principle of heat removal by evaporation of liquid refrigerants (e.g., water, ammonia, and alcohol) and consequent adsorption in porous materials. It has been considered and gaining great interest as an environment-friendly and energy-efficient system, owing to its several advantages such as utilization of renewable heat, low electric capacity, and the use of non-Freon-gas-type refrigerants [1-3]. Many research institutes and private companies, including Mycom, Nishiyodo, Mitsubishi Plastics, Zeopower, Invensor, SorTech, and Shanghai Jiao Tong University, have developed adsorption chillers. The key to the development of adsorption chillers with high efficiency and good performance is the development of suitable adsorbents that can adsorb as much refrigerant as possible at room temperature and desorb the refrigerant easily using low-temperature heat during the regeneration step. When water is used as refrigerant in this chiller, nanoporous materials with high water uptake under 9 ~ 12 Torr at room temperature and with remarkable desorbing capability under 353 K are highly desirable to generate chilled water (~283 K). However, it is hard to control the chemical properties of the adsorbent materials in such a way as to achieve the desired performances in both sides. Furthermore, the feasibility of mass production and low production costs for the adsorbent are also required.

In this work, we have synthesized various types of water adsorbent such as zeolite derivatives, mesoporous silica and activated carbon which show large difference of water adsorption amount (so called, â€˜dynamic sorption amountâ€™) between conditions of adsorption step and desorption-regeneration step of water adsorption chiller. Water sorption isotherms of the synthesized adsorbents were measured using volumetric sorption analyzer, and dynamic sorption amounts of the samples were also measured using gravimetric sorption analyzer for several cycles of conversion between adsorption and regeneration steps. According to the results of the sorption analysis, all the synthesized adsorbent exhibited significantly larger dynamic sorption amount than previous commercial water adsorbents such as regular-density silica gel and AQSOA^{TM} FAM series adsorbents. Synthesis methods of the present water adsorbents are quite simple and low-cost, so that we could synthesize the water adsorbents in large-scale batch. In addition, we developed precise methods for pelletizing the adsorbents and coating the adsorbents on aluminum metal surface for the application of water adsorption chiller. Detail results for the synthesis, the adsorption measurement, the palletization, and the coating of the synthesized adsorbents will be presented in the conference.

[1] F. Meunier, Journal of Heat Recovery Systems 6 (1986) 491.

[2] K.C. Ng et al., Applied Thermal Engineering 21 (2001) 1631.

[3] K. Cho et al., J. Nanosci. Nanotechnol. 17 (2017) 5869.

506936 Cross-linked Highly Sulfonated Poly(arylene ether sulfone) Prepared by Thiol-ene Reaction for Fuel Cell Application

Abstract:

End group capped highly sulfonated poly(arylene ether sulfone) (SPAES) was cross-linked by thiol-ene click reaction with allyl polysulfone (APSf). The end group cross-linking effects such as chemical and mechanical properties and proton conductivity of cross-linked SPAES (CSPAES) membranes were investigated by changing the contents of polymeric cross-linker, APSf. The C-SPAES membranes exhibited remarkably enhanced chemical and mechanical properties compared with pristine SPAES. Moreover, while the proton conductivity of pristine SPAES membrane couldnâ€™t be measured due to the low stability, C-SPAES membranes showed high proton conductivities in various humidity condition (up to 190.6 mS/cm2 in 100 %RH).

507003 Graphene Oxide-doped Polyethersulfone Hollow Fiber Membranes for Bioartificial Kidney Application

Abstract:

There is a great commercial significance of developing efficient hollow fiber membrane (HFM) materials for bioartificial kidney (BAK). The idea is to develop a single HFM-based unit, which can perform the dual functions: (1) growth of kidney cells on outer surface of HFMs, and (2) uremic toxin clearance, so that kidney functions desirable for survival of chronic kidney disease patients can be achieved. In this context, polyethersulfone (PES) HFMs containing optimized amount of graphene oxide (GO) were successfully developed. The uniform dispersion of GO in PES matrix was achieved with remarkably enhanced thermally stability. GO-doping also improved the hydrophilicity of the developed membranes. This resulted in low hemolysis (0.37 ± 0.15 %), prolonged coagulation times, and low SC5b-9 marker level (6.84 ± 1.7), i.e., significantly better compatibility with human blood in the lumen-side of the developed HFMs. The density of kidney cells attached and proliferated on the outer surface of the developed was also considerably higher than that on the undoped PES HFMs. The separation performance also remarkably enhanced with the addition of GO as high pure water permeability (154 Â± 3 mL/m2/h/mmHg) was measured, and solute rejection profile was found to be similar to that of the commercial dialyzer membranes. The uremic toxins (urea, creatinine and phosphorous) clearance (in simulated blood) was also almost two-fold better than that of the undoped PES HFMs. Thus, these results indicated that the GO-doped PES HFMs, developed in this study, are a potential membrane material for BAK application.

References:
1. Modi, A., Verma, S.K. and Bellare, J., 2017. Graphene oxide nanosheets and d-Î±-Tocopheryl polyethylene glycol 1000 succinate (TPGS) doping improves biocompatibility and ultrafiltration in polyethersulfone hollow fiber membranes. Journal of Colloid and Interface Science, 504, pp.86-100.
2. Verma, S.K., Modi, A., Singh, A.K., Teotia, R. and Bellare, J., 2017. Improved hemodialysis with hemocompatible polyethersulfone hollow fiber membranes: In vitro performance. Journal of Biomedical Materials Research Part B: Applied Biomaterials. (Accepted).
3. Bellare, J.R., Modi, A., and Verma, S.K., 2017. Biocompatible polymerbased nanocomposite hollow fiber membranes for simultaneous/sequential cell culture and filtration and a process for manufacturing the same. Indian Patent Application No. 201721012547.

507289 Multi-stage Electrodialysis for Coal Chemical Industry Wastewater Treatment

Abstract:

During coal mining, about 0.5 tonnes of water are consumed to produce 1-2 tonnes of coal, the coal chemical industry in China is facing a severe water shortage problem. Zero liquid discharge is required with the stricter regulations for wastewater discharge. Comparing with the other membrane concentrating techniques such as forward osmosis, disc tube reverse osmosis, membrane distilliation, electrodialysis (ED) with ion exchange membranes are considered as a promising technology for salt concentrating for a moderate salinity concentration. Matching between cation and anion exchange membrane is especially important in ED for salt concentrating. Results in this study indicate that cation exchange membrane has a greater significance for the ED concentrating performance compared with anion exchange membrane. A three-stage ED is also developed for salt concentrating (Figure 1). It is estimated that $Na+$ and $Cl-$ ions carry about 9 hydrated water ions at high salt concentration, comparing with about 15-18 molecules at low salt concentration. It is feasible to concentrate the salt to a concentration higher than 20wt% with a three-stage ED. The energy consumption can be as low as 4.9-12.1 kWh/m3. It is concluded that ED is a very effective technology to minimize the wastewater capacity for evaporation in the zero liquid process.

Keywords: Ion exchange membrane; electrodialysis, zero liquid discharge, coal chemical wastewater

Acknowledgements: This research is supported by the National Natural Science Foundation of China (21676259) and National High Technology Research and Development Program 863 (No. 2015AA021001).

Figure 1. A three-stage electrodialysis for salt concentrating

509783 Ethanol water separation using biomimetic aquaporin embedded membranes

Abstract:

Aquaporin based biomimetic membranes has been an active topic of recent research due to the selectivity and permeability of aquaporin to water. Most efforts have been focused on using the protein incorporated membrane for desalination purposes. However, there is potential for ethanol water separation as well. While several methods for ethanol water separation exist, they are often costly due to high temperature or pressure requirements, and are limited by azeotropic conditions. In using aquaporin, it may be possible to perform this separation at low temperatures and pressure as well as around the azeotrope. While some researchers have speculated on the permeability of ethanol to aquaporin, no studies exist to support or refute these claims. It is hypothesized that an ethanol water mixture will exhibit some form of separation across aquaporin membranes. Aquaporin maintains selectivity in two ways, having charged residues to reject ionic species, and having a narrow pore diameter to reject large molecules. Since ethanol and water have similar polarities, and ethanol is larger than water, it is anticipated that water will transport across the aquaporin protein leaving behind an ethanol rich solution.

509837 The Effect of the Chemical Structure of Four Dianhydride Precursors on the Gas Separation of Thermally Rearranged Polybenzoxazole for Natural Gas Separation

Abstract:

Natural gas purification via membrane separation is a promising method that could potentially result in a considerably cost-effective and sustainable purification process. A number of polymers have been studied; however, the two polymers that stand out are polyimide and polybenzoxazole. Among the two polymers, polybenzoxazole has better separation properties mainly because of the size and distribution of its free volume elements. Polybenzoxazole is made through the thermal rearrangement of a polyimide which is formed via the condensation reaction of diamine and dianhydride precursors. Everything from its permeation properties to the path of its formation have been studied. However, a glaring paucity within these studies exists: there are no studies that directly compare the dianhydride precursors of the polyimides and their effects on the resulting polybenzoxazole. The aim of this research is to compare the permeability and selectivity properties of thermally rearranged polybenzoxazole membranes formed by the condensation reaction of a diamine and four different dianhydride precursors. The four precursors are presented below. Out of the four dianhydride precursors to be examined, the BPDA is expected to have the optimum separation properties due to its bridging group.

509892 BN-Carbon-Membrane as a Novel/Alternative Separator for Li-Sulfur Batteries with Excellent Performance

Abstract:

Lithium-sulfur (Li-S) batteries have gained considerable attention as next-generation batteries because of their high theoretical capacity of 1672 mAh g-1, lower environmental impact, and reduced cost. However, several critical limitations produce poor electrochemical performance of Li-S batteries, especially those associated with polysulfide shuttling effect.

The introduction of a carbon-modified separator has been tested in order to alleviate these systemic challenges. However, this approach was focused on the suppression of dissolved polysulfides only in the cathodic side, i.e. without considering additional entrapment of polysulfides in the anodic side. Herein, we first constructed a multifunctional tri-layer membrane constituted by a carbon layer and a boron nitride (BN) layer, in order to improve the electrochemical performances of Li-S batteries and protect a Li anode from undesired side-reactions. The use of BN-Carbon-separator, the sulfur cathode exhibited stable capacity over 250 cycles and an exceptional specific capacity (702 mAh g-1) at a high current density of 4 C. Moreover, the BN-Carbon-separator presented a synergic effect on the uniform plating/stripping of Li and thus restraining critical dendritic lithium growth on the anode; this derived in a considerably stable operation of Li anode with high Coulombic efficiency and enhanced cycling performance.

495669 Purification of Styrene from the Styrene/Xylene Mixture By a New Technique Combining Distillation and Crystallization

Abstract:

Stripping crystallization (SC) is a new separation technique which combines distillation and crystallization. Basically, SC is operated at a triple-point condition, in which the liquid is simultaneously vaporized and crystallized due to the three-phase equilibrium. In essence, the process is continued until the liquid phase is completely eliminated and only the pure solid crystals remain in the feed. Thus, crystal washing is not required since only pure solid crystals remain in the feed and no impurities are adhered on the crystal surfaces at the end of SC. In the previous research, SC has been successfully applied to separate p-xylene from the mixed xylenes.

SC is introduced in this work to purify styrene (ST) from the multi-component system consisting of ST, ethylbenzene (EB), p-xylene (PX), m-xylene (MX) and o-xylene (OX). Styrene is one of the most important aromatic compounds and is used extensively in the manufacture of polystyrene. Due to the close boiling points of ST, EB, PX, MX and OX, it is rather complicated and energy-intensive to separate them by conventional distillation. A thermodynamic model is developed to determine the three-phase equilibrium conditions and direct the SC experiments. The experiments show that SC can be effectively applied to purify ST from the mixture. However, as EB is more easily incorporated into the ST crystal lattice compared to PX, MX and OX, it becomes more difficult to further purify ST for a higher concentration of EB in the mixture.

References:

1. Shiau, L. D.*, Wen, C. C. and Lin, B. S. “Separation and purification of p-xylene from the mixture of m-xylene and p-xylene by distillative freezing”, *Ind. Eng. Chem. Res*, **44(7)**, p2258-2265, 2005.

2. Shiau, L. D.*, Wen, C. C. and Lin, B. S. “Separation of p-xylene from the multicomponent xylene system by stripping crystallization”, *AIChE Journal*, **54(1)**, p337-342, 2008.

3. Shiau, L. D.*; Liu, K. F. "Investigations into the effects of the cooling rate on stripping crystallization", *Ind. Eng. Chem. Res*, **52**, p1716-1722, 2013.

496164 Performance Evaluation of Long Chain Alkyl Ionic Liquids and Their Mixtures for CO_2 Solubility at Elevated Temperature and Pressure

Abstract:

Performance Evaluation of long Chain Alkyl Ionic Liquids and Their Mixtures for CO_2 Solubility at Elevated Temperature and Pressure

Ruh Ullah, Tausif Altamash and Majeda Khraisheh*

Department of Chemical Engineering, Qatar University, Doha, Qatar

E-mail: m.Khraisheh@qu.edu.qa

Technological Development worldwide has significantly intensify the energy demand and per capita energy consumption for last two decades which has subsequently threatened the globe with the biggest environmental issue of global warming associated with greenhouse gas emissions from numerous industries including energy and cement sectors[1]. On the other hand aqueous mono-ethanol amine (MEA) scrubbing techniques have technological issues of corrosion, degradation and economically penalizing the energy sector to mitigate CO_2 emission[2]. To overcome the issues of multiple applications and material degradation other compounds like large number of solid sorbents (such as metal organic framework, covalent organic polymers, silica, activated carbon and nanoclays) and other liquids such as ionic liquid and deep eutectic solvents (DES) have been under investigation since long for pre- and post-combustion CO_2 capture and separation. Although various solid sorbents such as covalent organic polymers and polybenzimidazole have shown promising results as long as capturing capacity and repeatability of the sorbent are concerns[3, 4]. However, chemisorption, lower selectivity, porosity dependent adsorption and slow adsorption kinetics are among the major issues which hinder the large scale applications of solid sorbents. Alternatively, studies have shown that ionic liquids are very promising [5, 6] capture solvents for CO_2 and other toxic gases since; their physical and chemical properties at various temperatures could be tailored through modifying their cationic and anionic moieties[7]. Additionally, extremely low vapor pressures, high thermal and chemical stability, non-flammable nature, high solvation capacity, and promising gas solubility features, ionic liquids have emerged as attractive compounds since the last decade. Moreover, modifying solvent properties at room temperature through hydrogen bond donors, ionic liquids mixtures may have great potential for efficient CO_2 absorption.

Since, mixing two or three components capable of intermolecular interactions through the hydrogen bonding may generate channels to absorb more CO_2. Thus, there is a possibility of designing specific ionic combinations which possess suitable properties and can be tested for high pressure CO_2 capture at various temperatures. Based on the advantageous features ionic liquids, we report testing and preparation of novel mixtures of ionic liquids based on their long alkyl chain. CO_2 capture of three ionic liquids and their mixtures were obtained with the state of art equipment magnetic suspension balance (MSB). Mole fraction CO2 solubility of three different ionic liquids at 25 °C and maximum of 50 bars clearly shows differences in the absorption capaity indicating effect of alkyl chain. The initial findings have indicated that, the novel mixtures have favorable strong thermo physical stability and appreciable CO_2 capturing capacity. TGA, FTIR, viscosity, density, pH, and thermal conductivity analysis have also indicated the suitability of these ionic mixtures for CO2 and other gases solubility. Moreover, the repeated use of material at various temperature and pressure without any decrease in CO_2 capturing capacity make these compounds capable to be considered for further investigation on larger scale applications.

References

1. Štreimikienė, D., *Residential energy consumption trends, main drivers and policies in Lithuania.* Renewable and Sustainable Energy Reviews, 2014. **35**(0): p. 285-293.

2. Mondal, M.K., H.K. Balsora, and P. Varshney, *Progress and trends in CO2 capture/separation technologies: A review.* Energy, 2012. **46**(1): p. 431-441.

3. García, G., et al., *Deep Eutectic Solvents: Physicochemical Properties and Gas Separation Applications.* Energy & Fuels, 2015. **29**(4): p. 2616-2644.

4. Ullah, R., et al., *A detailed study of cholinium chloride and levulinic acid deep eutectic solvent system for CO2 capture via experimental and molecular simulation approaches.* Physical Chemistry Chemical Physics, 2015. **17**(32): p. 20941-20960.

5. Huang, K., et al., *Facilitated separation of CO2 and SO2 through supported liquid membranes using carboxylate-based ionic liquids.* Journal of Membrane Science, 2014. **471**(0): p. 227-236.

6. Karadas, F., M. Atilhan, and S. Aparicio, *Review on the Use of Ionic Liquids (ILs) as Alternative Fluids for CO2 Capture and Natural Gas Sweetening.* Energy & Fuels, 2010. **24**(11): p. 5817-5828.

7. Atilhan, M., et al., *Double Salt Ionic Liquids Based on Ammonium Cations and Their Application for CO2 Capture.* The Journal of Physical Chemistry C, 2016. **120**(31): p. 17829-17844.

496762 Capture of Nitric Oxide in Simulated Flue Gas By a Metallic Functional Ionic Liquid

Abstract:

Capture of nitric oxide in simulated flue gas by a metallic functional Ionic Liquid

Ying Sun[1], Shuhang Ren[1], Yucui Hou[2], Kai Zhang[1], Weize Wu[1,*]

[1]State Key Laboratory of Chemical Resource Engineering, Beijing University of Chemical Technology, Beijing 100029, China

[2]Department of Chemistry, Taiyuan Normal University, Taiyuan 030031, China

Nitric oxide (NO) is one of the main air pollutants threatening human health and environment. Currently, selective catalytic reduction (SCR) is the most widely used technology in industry. However, the catalyst poisoning, equipment corrosion and NH_3 escape make the SCR technology to be challenged. In last decade, ionic liquids (ILs) have been widely studied on efficient and reversible removal of CO_2 and SO_2 in fue gas due to their benign physicochemical properties, such as negligible vapor pressure, high stability and tunable structure. However, there is very little research on NO absorption by ILs. In this work, a metallic functional ionic liquid $[Bmim]_2FeCl_4$ was designed and applied to efficiently and reversibly capture NO in a simulated flue gas. The solubility of 2000 ppm NO in the IL was high up to 0.238 mol of NO / mol of IL at 30 °C, and the absorbed NO could be easily desorbed from the IL. NO solubility increased with decreasing temperature and increasing NO partial pressure. Meanwhile, the absorption capacity of IL did not change after 4 absorption/desorption cycles. Furthermore, the absorption mechanism was studied by FT-IR, and the result demonstrated that there was a chemical interaction between Fe^{2+} and NO.

The project is financially supported by the National Natural Science Foundation of China (No. 21176020 and 21306007)

Keywords: Absorption; Nitric oxide; Simulated flue gas; Functional ionic liquid

* Corresponding author; E-mail: wzwu@mail.buct.edu.cn; Tel./Fax: +86 10 64427603.

489305 Vaporization of a Single N-Pentane Liquid Drop in a Flowing Mmicsible Liquid Media

Abstract:

Heat exchange in a direct contact process that involves bubbles, drops or particles as a dispersed phase is a complex phenomenon. This is due to the wide range of variables such as phase velocities, phase temperatures, the physical and thermodynamic properties of both phases and system holdup or void fraction.

Vaporization of a single drop in immiscible liquid represents the basis of the three-phase direct contact heat exchanger which has several advantages over surface or classical type heat exchangers. Generally, can operate at very low temperature differences between the fluid streams, even as low as . The absence of the internal separating wall between fluid streams leads to the elimination of fouling and corrosion problems, as well as an increase in the heat transfer rate between the fluids. It has very high heat transfer coefficient (about 20 â€“ 100 times more than a single phase heat exchanger) because of the large heat transfer area and the latent heat effect due to phase change.

The direct contact process simply occurs when bubbles or drops are injected into a continuous phase. There is a great similarity in the heat transfer mechanism for drops and bubbles in a direct contact process. In both cases, a two-phase bubble/drop is formed just after the dispersed phase first contacts the continuous phase. This configuration, (two-phase bubble/drop) arises from two instantaneously separating parts: a vapour phase, which normally concentrates at the top of the structure, in thermal equilibrium with a liquid phase at the bottom, because of the effect of gravity.

In this paper, vaporization of a single n-pentane drop in a direct contact with another flowing immiscible liquid (warm water) has been experimentally investigated. The experiments were carried out utilising a cylindrical Perspex tube of 10 cm diameter and 150 cm height. Saturated liquid n-pentane and warm water at 45°C were used as the dispersed phase and continuous phase, respectively. Photron FASTCAM SA 1.1, high speed camera (75,000f/s) with software V. 321 was implemented during the experiments to measure the rise and growth velocity of the two-phase bubble-droplet from the point detached the nozzle tip to the end of evaporation was recorded. Each run was studied frame by frame to give the exact time of evaporation. . Five different continuous phase flow rates (10, 20, 30, 40 and 46 Lâ„h) were used in the study. The results indicated that the as evaporation proceeded, the vapour accumulated at the top and the liquid part seemed to stay at the bottom of the bubble-droplet system. The liquid part could be recognized only up to about 8%, after that it occupied a relatively small volume of the system. Before reaching this stage, the system oscillated and the liquid sloshed from side to side. In addition, increase in flow rate of the continuous phase results in increasing the two-phase bubble.

494851 Extraction of Phenolics from Coal By Deep Eutectic Solvents

Abstract:

Deep eutectic solvents (DES) are some combination of components that result in a solution with a lower melting point than the individual compounds. DES's have been looked at recently as a tool to make processes more "green". One such example is the extraction of sulfur containing compounds in fuel to decrease the amount of sulfur containing gases produced from combustion [1]. DES's have shown promise in their capacity to better dissolve organic compounds than other solvents such as methanol or water [2]. Coal is a material that can benefit from the use of DES's. Traditional solvents have already been researched in an attempt to dissolve select compounds form coal [3]. DES's, which have shown to be effective solvents for a wide variety of organic compounds, can be used with coal to extract select compounds, or a set of compounds with similar characteristics.

DES's have already been used to extract phenolic compounds from olive oil [4].One potential use of this particular DES is the solvation of phenolic substances from coal. Several different DES's were examined and tested qualitatively to see if they were able to extract phenolic compounds from coal. GCMS was used to identify extracted species. It was found that of the examined compounds ethanolamine hydrochloride and ethylenediamine showed the most promise. Various ratios of ethanolamine hydrochloride and ethylenediamine where then used to extract phenolic chemicals from Clarion 4a coal dust that had particle sizes of 35-75 micron. GCMS grade hexane was then used to extract the compounds of interest from the DES to identify. Qualitative GCMS was performed to evaluate species extracted, and quantitative mass balances were used to determine the effect of DES ratios.

[1]. Li, Changping, et al. "Extraction desulfurization process of fuels with ammonium-based deep eutectic solvents." *Green Chemistry* 15.10 (2013): 2793-2799.

[2]. Dai, Yuntao, et al. "Natural deep eutectic solvents as new potential media for green technology." *Analytica chimica acta* 766 (2013): 61-68.

[3]. Darlage, Larry J., and Maurice E. Bailey. "Solvent effects in the depolymerization of coal." *Fuel* 55.3 (1976): 205-210.

[4]. García, Aránzazu, et al. "Extraction of phenolic compounds from virgin olive oil by deep eutectic solvents (DESs)." *Food chemistry* 197 (2016): 554-561.

504944 Efficient Decolorization of Citric Acid Fermentation Broth using Carbon Materials Prepared from Phosphoric Acid Activation of Hydrothermally Treated Corncob

Abstract:

Citric acid, as an important organic acid, is largely used in aspects of food, beverage, chemical and pharmaceutical fieldsï¼Œand plays increasingly important role in the society. However, currently the downstream aspect of separating the citric acid is facing a tremendous problemâ€"â€"the decolorization of the citric acid fermentation broth. In order to costly and efficiently remove the pigments, a novel biomass-based carbon material, hydrochar, is put forward. During the preparation process, by adjusting the hydrothermal conditions (temperature, time) and feedstock type (food residue, agricultural, etc), we obtained an excellent hydrochar to be used in the decolorization of citric acid fermentation broth. Additionally, for further improving its adsorption performance, the post chemical activation is applied to optimize it textural characteristics. Meanwhile, the pigments adsorption process on the hydrochar-based carbon material (HBCM) was fully investigated. Meanwhile, the material was fully characterized by SEM, BET, FTIR, XPS and pH_{pzc}, and several mathematic models were used to analyze the adsorption mechanism involved in the adsorption process. Finally, the tailor-made HBCM had the excellence performance to achieve 99% decolorization ratio and no loss of citric acid at the optimum operating conditions. It could be a potential adsorbent for the removal of pigments in the citric acid fermentation broth.

504943 Recovery of Monosaccharides from Dilute Acid Corncob Hydrolyzate based on Nanofiltration Technologyï¼šModeling and Optimization

Abstract:

Lignocellulosic biomass such as woody materials and agricultural residues is an abundant, environmentally feasible, easily available, and renewable feedstock for the biofuel production. However, utilization of such biomass requires a pre-treatment process, through which polymeric carbohydrates can be decomposed to monosaccharide. Dilute acid pretreatment is the most commonly used method . However, this way produces inhibitors like the furfurals, hydroxymethyl furfural (HMF), phenolic compounds and several other acids such as acetic acid and formic acid . It has been widely conformed that the presence of these by-products during fermentation of sugars may seriously impact the productivity of bio-based products and the growth of bacterial strain. For example, phenolics are lethal to Clostridium which is widely used to produce butanol and butyric acid even at low concentrations. So it is important to remove these inhibitors from hydrolysates.

Nanofiltration is a high-efficiency technology to remove inhibitors and recover monosaccharides from the dilute acid lignocellulose hydrolyzates. In this study, the effect of feed pH, permeate flux and Na_2SO_4 concentration on the rejection of monosaccharides and inhibitors was firstly investigated. The results showed that the separation performance of carboxylic acids and furans from monosaccharides was optimal at pH 3. All the rejection of carboxylic acids and furans decreased with increasing Na_2SO_4 concentration, especially for carboxylic acids. Subsequently, donnan steric pore flow model coupled with mass balance (DSPM-MS model) was successfully applied to predict the rejection of solutes at different permeate flux. Based on the DSPM-MS model a parallel multi-objective optimization strategy was developed to select a reasonable permeate flux j_v and operating time t, in which maximum total inhibitor remove efficiency Prinhibitor, maximum recovery rate of monosaccharides Ysugar and minimum water consumption EC were set as objective functions. The results showed that in the case of maximum Ysugar, more monosaccharides were recovered and less water was consumed as well. The optimal operating condition was consequently verified with the practical hydrolysate solutions. It had consumed about 35 min and 13L water to remove 95% inhibitors, meanwhile, 93.55% glucose, 90.75% xylose and 90.53% arabinose were recovered.In the end of this work, a batch column equipped with strong acid cation-exchange resins Sa-2 were used to recover monosaccharides from the hydrolysate. Using water as eluent, 95.37% of sulfuric acid and 94.87% of monosaccharides was recovered. This work demonstrated that nanofiltration combined with electrolyte exclusion chromatography could be a potential integrated process to recover monosaccharides and inorganic acids in the practical dilute acid corncob hydrolyzate.

493678 Organic-Inorganic Composite Membranes: Fundamental Study and Engineering Application

Abstract:

Organic-inorganic Composite Membranes: Fundamental Study and Engineering Application

Wanqin Jin

State Key Laboratory of Materials-Oriented Chemical Engineering, Nanjing Tech University,

Nanjing, 210009, P.R.China. wqjin@njtech.edu.cn

Abstract

By combining advantages of organic and inorganic materials, organic-inorganic composite membranes are regarded as a hot research interest in fields of membrane technology and chemical engineering. This presentation will review our recent progress in organic-inorganic composite membranes, covering from fundamental study to engineering application. It aims to give an overview of design, preparation and application of the organic-inorganic composite membranes used for molecular separation. Various membrane types with different materials will be involved in the talk, including ceramic-supported polymer composite membranes, mixed matrix membranes, and graphene-based membranes. Meanwhile, the presentation will illustrate typical examples of the membranes applied in fields such as bio-fuels production, volatile organic compounds (VOCs) recovery, carbon dioxide capture and so on. In addition, large-scale production and industrial implementation of the organic-inorganic composite membranes will also be introduced.

This work was financially supported by the National Basic Research Program of China (No. 2009CB623400), the National Natural Science Foundation of China (Nos. 21406107, 2149580015), the Innovative Research Team Program by the Ministry of Education of China (No.IRT13070), the Top-notch Academic Programs Project of Jiangsu Higher Education Institutions (TAPP).

***Keywords*:** organic-inorganic composite membranes; molecular separation; industrial implementation

495668 Modeling of Structural Defects in MFI Zeolite Membranes

Abstract:

Modeling of Structural Defects in MFI Zeolite Memabranes

Sungwon Hong[1], Dongjae Kim[2], Jaewook Nam[2] and Jungkyu Choi[1]*

[1] *Department of Chemical & Biological Engineering, Korea University, Seoul 02841, Republic of Korea*

[2] *School of Chemical Engineering, Sungkyunkwan University, Seoul 02841, Republic of Korea*

hsw1110@korea.ac.kr, ssdj91@skku.edu, jaewooknam@skku.edu, jungkyu_choi@korea.ac.kr*

Introduction

Defects of zeolite membranes often lower their separation performance. Thus, the investigation of the defects is highly critical in achieving high separation performance. While general characterization methods (e.g. scanning electron microscopy; SEM) that examine the membrane surface cannot detect defects, the FCOM measurement is able to identify the defective structure inside the zeolite membrane using dye molecules of appropriate size [1]. In this work, various dyeing conditions (times and concentrations) were applied to a MFI zeolite membrane in an attempt to investigate the defective structure. Furthermore, the quantitative analysis is practiced to measure the defects in numerical form.

Experimental

A MFI membrane was synthesized on an ¥á-alumina disc support by employing hydrothermal growth of MFI seed layers. Then, zeolite membranes were calcined to remove organic templates in the zeolite pores. Calcined MFI membranes were further contacted with a dye molecule (~ 1 mM aqueous fluorescein). The size of fluorescein molecules (~0.9 nm) is smaller than that of intercrystalline microporous defects (~1-2 nm), but larger than that of zeolitic pores (0.5~0.6 nm). This size difference allows for the selective dyeing of defects in membranes. The dyed membranes were measured with Carl Zeiss LSM 700 confocal microscope in 488 nm wavelength. Quantitative analysis of the measured data was followed, investigating the bright sections in the FCOM images.

Results and discussion

The result images of FCOM analysis with different dyeing times are shown in Fig 1. From Fig. 1a, the appearance of bright broad lines is observed. At the same time, a number of isolated bright spots shown in Fig. 1a were pronounced in Fig. 1b and they were interconnected with the increased dye time (Fig. 1c). The dye molecules seem to prefer to access larger defects (here, crack) first and then approach smaller defects (here, grain-boundary). Interestingly, the black broad line, which was bright in both Fig. 1a and 1b, results from a quenching phenomenon, which happens when the dye molecules are stacked up and concentrated. Considering this phenomenon, crack defects could be selectively extracted among the grain-boundary defects for quantitative analysis. FCOM characterizations with different dye concentrations showed a similar trend to those with the variation in the dyeing time. Using these data, each of the defects (grain-boundary and crack) was quantified analyzed.

Fig. 1. FCOM images of MFI membranes, which were contacted with the dye solution in various times. (Fig. 1a is the shortest time condition and Fig. 1c is the longest time condition.)

Conclusions

FCOM measurement of zeolite membranes is a simple and effective technique to observe the defects in zeolite membranes. The experiments using FCOM measurement were performed systematically to elucidate the dyeing mechanism, and accordingly, the properties of the defects. In addition, a quantitative analysis was conducted to show the amount of the defects in MFI membranes, which can be associated with the corresponding separation performance.

References

[1] G. Bonila et al., J. Membr. Sci. 182 (2001), 103-109

487952 Filling of Clay Nanoparticles into Thin Film Nanocomposite Membranes to Improve Their Efficiency Toward Brackish Water Desalination

Abstract:

Clay nanoparticles, ≥ 50nm, were successfully prepared by solvothermal method and filled into thin film nanocomposite (TFN) membranes used in reverse osmosis application. Membranes were prepared by the interfacial polymerization (IP) of *m*-phenylenediamine (MPD) aqueous solution and trimesoyl chloride (TMC) isooctane solution [1] on a porous layer of polysulfone (PSU). Different percentages, 0.01, 0.02, 0.05, 0.07, 0.1, and 0.15, of the NPs were filled into the membranes and an improvement in terms of performance was recorded. Different techniques were employed to characterize the NPs and membranes properties to give better understanding on observations. At the optimum addition, 0.15%, NaCl rejection and water flux increased from 96.25% and 45.6 L/m^2 hr to 97.45% and 59.0 L/m^2 hr, respectively, in operational conditions of 2000 ppm NaCl, 300 psi, and 25 °C.

References

1- Mohammed Kadhom, Jun Yin, Baolin Deng, "A Thin Film Nanocomposite Membrane with MCM-41 Silica Nanoparticles for Brackish Water Purification", *Membranes*, 6, 50, (2016) 1-12.

482718 Advanced Membrane Separation to Improve Efficiency of Thermochemical Conversion of Biomass

Abstract:

Thermochemical conversion pathways include fast pyrolysis, catalytic fast pyrolysis, and hydrothermal liquefaction of biomass. The objective of the work is to develop and employ robust advanced membranes to improve the efficiency of bio-oil processing. These porous membranes take advantage of surface tunability (from superhydrophilicity to superhydrophobicity) to achieve high permeability while maintaining high selectivity. This effort is relevant to bioenergy technology barrier areas (1) Aqueous Phase Utilization and Wastewater Treatment and (2) Liquefaction of Biomass and Bio-Oil Stabilization. Key challenges are enabling higher flux (at desirable separation factor), improving chemical/thermal stability and fouling resistance, and developing a path forward to a small unit volume but high-productivity membrane module. Our team is working toward demonstration of a separation technology to meet performance targets (>1 liter/m^2/h liquid dewatering flux at separation factor >20) and achieve a >8% reduction in MFSP ($/gge). The speaker will present recent advancements in vapor-phase dewatering and liquid-phase separation technologies. For example, a >5x improvement in the permeation flux for dewatering membranes was demonstrated which will lead to a more efficient process for recovering carbon from aqueous waste streams. Our work has led to an industry-funded commercialization effort that is seeking to integrate advanced emulsion separation technologies with biofuel conversion reactors to increase overall production yield.

497724 Antifouling Membranes By Surface Modification Using Hydrophilic Polymers

Abstract:

Antifouling Membranes by Surface Modification Using Hydrophilic Polymers

Nima Shahkaramipour, Chong Cheng, and Haiqing Lin*

[1]Department of Chemical and Biological Engineering, University at Buffalo,

The State University of New York, Buffalo, NY 14260, USA

*Corresponding author. Tel: +1-716-645-1856, Email: haiqingl@buffalo.edu

Abstract

Polymeric membranes for wastewater reuse are often fouled by suspended solids and dissolved organic matters, resulting in dramatic decrease in water flux. Fouling can be mitigated by enhancing hydrophilicity of membrane surface to avoid favorable interactions between the membranes and foulants. However, the modification of membrane surface is often complicate and difficult to integrate into industrial membrane production processes. Herein we demonstrate a facile one-step coating of superhydrophilic zwitterions on the surface of ultrafiltration (UF) and nanofiltration (NF) membranes by coating with dopamine and sulfobetaine methacrylate (SBMA). In the presence of oxygen, dopamine forms polydopamine (PDA) adhering onto the membrane surface and anchors SBMA via Michael addition to form a robust thin superhydrophilic layer, as confirmed by contact angle measurement, thickness measurements, and x-ray photoelectron spectroscopy (XPS). The effect of coating solution composition and coating time on the water flux through the membranes with time was thoroughly studied. The modified UF membranes demonstrate up to 80% higher water flux than the uncoated ones, when tested with water containing bovine serum albumin (BSA) as a model foulant in a crossflow system. This presentation will also discuss the effect of zwitterion chemistry on the antifouling properties of the modified membranes for the wastewater reuse. This facile approach of membrane modification is also adapted for post-modification of a commercial nanofiltration membrane module, which demonstrates superior antifouling properties when tested with real wastewater at a wastewater treatment plant.

498489 Chlorine-Tolerant Block Polymer Nanofiltration Membranes

Abstract:

The design of next-generation membranes with highly-ordered, well-defined nanostructures is critical to meeting the increasing demand for highly selective separation devices in the water treatment arena. Furthermore, the design of the membrane chemistry is critical to advancing the use of these novel devices. Specifically, chlorine is a commonly used in water treatment systems as a disinfecting agent to reduce the effects of biofouling yet state-of-art nanofiltration membranes, based on polyamide chemistries, cannot tolerate extended periods of exposure to chlorine concentrations above 0.1 ppm. This limited resistance to chlorine necessitates the implementation of costly de-chlorination steps. Therefore, a chemically robust, high-performance nanofiltration device is urgently required to reduce the cost of nanofiltration processes. Due to the ability to self-assemble into a membrane with a well-defined nanostructure, block polymers are ideal candidates to generate highly-selective nanofiltration membranes. In this poster, the careful design of a block polymer chemistry tailored to enable a robust separation profile and to withstand the presence of chlorine will be discussed. In particular, the fabrication of an ultra-thin, self-assembled polyisoprene-*b*-polystyrene-*b*-poly(2-acrylamido-ethane-1-1-disfulonic acid) (PI-PS-PADSA) coating atop of a poly(vinylidene fluoride) hollow fiber support generated a filtration device with an extremely high solute selectivity, which is capable of fully fractionating solutes with only 8 Å difference in diameter. The performance of this nanofiltration device upon exposure to sodium hypochlorite solutions for extended periods of time will be discussed. Preliminary results provide strong evidence for the utility of block polymer nanofiltration membranes in chlorine-rich environments.

491871 Mixed Matrix Membranes with Improved Interfacial Morphologies Via Supramolecular Interactions

Abstract:

Mixed matrix membranes (MMMs) are highly attractive for gas separation applications since they can integrate superior sieving properties of nano-filler particles and excellent processability of polymer matrix. However, preparation of MMMs with homogeneously dispersed fillers and defect-free interfacial morphology of ideal interphase contact is challenging due to intrinsically different physical and chemical properties of different phases. Here, we report novel nanocomposite membranes with desirable interfacial morphologies tailored by supramolecular interactions between the filler and polymer matrix. These new MMMs are prepared from single-walled carbon nanotubes (SWCNTs) and a triptycene-containing polyimide. SWCNTs show ultrahigh transport rates of gas molecules, conferring on SWCNT-based membranes the potential for high gas permeability. The paddle-wheel like triptycene moieties in polymer matrix induce strong Ï€-Ï€ supramolecular interactions with the graphene wall of nanotubes, which effectively improves the interfacial contact between phases and assists homogenous dispersion of SWCNTs via nano-confinement effect.

MMMs was prepared using commercial SWCNTs by solution casting method, which were systematically varied in surface functionality of SWCNTs, metal impurities, aspect ratio of nanotubes and filler loading. Good adhesion between SWCNTs and triptycene-polyimide and homogeneous dispersion of SWCNTs in polymer matrix without aggregation are verified by cross-sectional SEM images for all membranes. Fluorescence quenching phenomenon provides a direct evidence for the Ï€-Ï€ stacking interaction between SWCNTs and the polyimide. The improved interfacial properties lead to excellent separation performance in these MMMs. For example, a composite membrane with only 2 wt% filler loading showed nearly an order of magnitude increase in gas permeabilities from the pure polymer without sacrificing CO_2/CH_4 and CO_2/N_2 selectivities.

We will also demonstrate the design of triptycene modified ZIF-90 particles and triptycene-based polyimide MMMs with improved interfacial interaction via Ï€-Ï€ stacking interaction and interlocking phenomenon associated with the hook-like structure of triptycene moieties. Detailed discussion will be provided regarding the structures of triptycene used for modification, degree of surface modification and filler loading, with a focus on relating microscopic structures with macroscopic transport properties.

496234 A Systematic Investigation of Ionic Liquids As Effective Draw Solutes for Forward Osmosis

Abstract:

The growing concern on energy shortage has driven research to find energy-efficient technologies including those which mitigate water scarcity and environmental pollution. Among the alternative technologies, forward osmosis (FO) has emerged as one of the most promising candidates due to its versatile application in power generation, wastewater reclamation, and clean water production. This led to intensive research about FO which has an advantage of not using high hydraulic pressure as compared to reverse osmosis. Two elements of FO system are of importance to realize its feasibility as a practical and merchantable technology: FO membranes and draw solutes (DS). Despite the progress in membrane technology, draw solutions are still limited in terms of its reusability thereby hampering its economic viability.

In this study, fourteen ionic liquids (ILs) were investigated as DS for FO. These materials are coined as green substances due to their high ionicity, low vapor pressure and high thermal stability. Advancement in IL research has led to the development of polymerizable ILs or ILs incorporated in hybrid materials. All of these known properties of ILs suggest their strong potential as precursors for the development of next generation DS solutes. All IL candidates were thoroughly characterized and certain constants unavailable in literature were gathered. Initially, their Vanâ€™t Hoff factors (*i*) were determined by freezing point depression method. Their ability to generate osmotic pressures (p) were calculated and compared with NaCl as a conventional osmotic agent. The ILs were ranked and screened; their ability to generate osmotic pressures were correlated well with their ionicity (Vanâ€™t hoff), volume properties, melting points and diffusion coefficients. Ionic liquids which exhibited the highest Î Â were further investigated in FO operation conducted under FO and PRO modes. Results reveal the potential of ILs as alternatives to NaCl, especially under PRO mode wherein the internal dilutative concentration polarization (DICP) is much less than in FO mode. One significant benefit of ILs over NaCl is their negligible solute reverse flux and higher reverse flux selectivity. However, it is noted that some ILs were unsuitable given their high affinity to the selective layer of the CTA-FO membrane used. These results suggest that IL-membrane affinity played a critical role in the potential of certain ILs as DS. Thus, the best ILs are those which generate high osmotic pressures but have low affinity to the selective layer of the FO membrane. Suitable ILs can be recovered and recycled through membrane distillation (MD). A final hybrid system of FO-MD can be used for water purification and desalination application using ILs as DS.

This research was supported by the National Research Foundation of Korea (NRF) funded by the Ministry of Science, ICT & Future Planning (No. 2016R1A2B1009221) and the Ministry of Education (2009-0093816).

497047 Development of Robust, Ion-Selective Anion Exchange Membranes through Incorporation of Ionic Liquid Materials for Water Purification Via Electrodialysis

Abstract:

The onset of climate change and rising global population has caused greater water scarcity, a considerable problem which threatens our world. Electrodialysis (ED) is an electrically-driven separation process aimed at water treatment through ion removal, which is achieved through the selective control and transport of ionic species. ED is a popular water treatment method; however, ED is limited by membrane scaling, lack of ion selectivity, and high energy consumption for high salinity feeds. This research is focused on the synthesis of unique imidazolium-based anion exchange membranes in order to obtain enhanced selectivity for specific ions.

498322 Sorption Enhanced Mixed Matrix Materials Comprising Palladium Nanoparticles and Polybenzimidazole for H_2/CO_2 Separation

Abstract:

Polymeric membranes have been widely explored for energy-efficient and low-cost CO_2 capture and H_2 purification in the integrated gasification combined cycle (IGCC) processes. Conventional approaches are focused on designing rigid polymers with strong size sieving ability achieving high H_2/CO_2 diffusivity, such as poly[2,2â€™-(m-phenylene)-5,5â€™-bisbenzimidazole] (PBI). In contrast, we investigate mixed matrix materials (MMMs) containing PBI and palladium (Pd) nanoparticles, which have strong affinity towards H_2 and thus exhibit extremely high H_2/CO_2 solubility selectivity. Pd nanoparticles with uniform diameters of 6 - 8 nm are prepared via hot-injection approaches, which show H_2/CO_2 solubility selectivity of 840, compared with < 0.1 for typical polymers. The effect of Pd loading (0 - 70 wt%) and temperature on the pure- and mixed-gas H_2/CO_2 separation properties is thoroughly evaluated. The loading of Pd nanoparticles dramatically increases H_2 solubility and H_2/CO_2 solubility selectivity, resulting in a significant increase in H_2 permeability and H_2/CO_2 selectivity at temperatures of 100 - 200 °C. For example, adding 70 wt% Pd in PBI increases H_2 permeability from 25 to 70 Barrers, and H_2/CO_2 selectivity from 13 to 29 at 150 °C. Such performance is above the upper bound of the Robesonâ€™s plot for H_2/CO_2 separation, demonstrating their potential for industrial H_2/CO_2 separation. This presentation will also examine morphology of the MMMs and describe a model to elucidate the relationship of the structure and H_2 transport properties in these sorption enhanced MMMs.

501223 Perfluoropolymers and Hydrocarbon Polymer Analogs with Pendant Rings for Gas Separation Membranes

Abstract:

Amorphous glassy perfluoropolymers are attractive for membrane gas separation, such as He/CH_4 and CO_2/CH_4 separation. These perfluoropolymers often have pendant rings to suppress crystallinity and achieve high free volume and thus high gas permeability. However, their applications are hampered by the high cost due to the difficulty in synthesis. We hypothesize that the pendant rings can be incorporated in hydrocarbon polymers to achieve similarly superior gas separation properties. In this study, we design and synthesize hydrocarbon polymers with structures similar to perfluoropolymers, aiming to elucidate the effect of pendant rings on gas separation properties. Specifically, we synthesize a perfluoropolymer of poly(perfluoro-4-methyl-2-methylene-1,3-dioxolane) (PPMMD) and its hydrocarbon polymer analogue, poly(4-methyl-2-methylene-1,3-dioxolane) (PMMD). These polymers with pendant rings are thoroughly characterized using FTIR, NMR, GPC, WAXD, TGA and DSC for physical properties such as molecular weight, crystallinity, and glass transition temperature. PPMMD shows a H_2 permeability of 236 Barrers and H_2/CO_2 selectivity of 4.1 at 35 °C, while the PMMD shows a H_2 permeability of 9.6 Barrers and H_2/CO_2 selectivity of 7.3 at 35 °C. Gas sorption results will be described using the dual mode sorption model. A series of poly(cyclic ketene acetals) with structures similar to PMMD are synthesized and characterized for gas separation. This presentation will also discuss the fractional free volume in these polymers and the use of a free volume model to describe the gas transport properties in these hydrocarbon polymers and perfluoropolymer analogs.

491059 A Computational Study of Water Sorption and Its Effect on CO_2 Separation Performance in Graphene-Oxide Based Membranes

Abstract:

Membrane technologies have been actively studied as a potentially viable energy-saving alternative for CO_2 capture in comparison to conventional absorption-based amine scrubbing. Recently, graphene oxide (GO) has gained much attention as an advanced material for gas-separation membranes by virtue of its ultra-thin laminate structure suitable for use as a separation membrane. Moreover, water vapor can be absorbed into the hydrophilic GO membrane in humid conditions. Because of the solubility difference between CO_2 and other nonpolar gases in water, the intercalated water can increase the selectivity of CO_2 over nonpolar gases such as N_2. Hence, a detailed understanding of the water sorption behavior of GO-based membranes is important in order to better evaluate and also find an effective way to improve their CO_2 separation performance.

This poster will present some results of our recent computational studies of water sorption and its effect on CO_2 separation performance in GO-based membranes, especially the influences of relative humidity, OH group content and distribution, and GO geometry. This presentation will specially touch on changes in the amount and dynamics of intercalated water under varying conditions of humidity, temperature, and GO structure, along with the effects of intercalated water on the separation of CO_2 from a mixture of gases. The improved understanding helps to better understand the properties and performance of GO-based membranes for CO_2 separation, and will also contribute to the optimal design and development of more efficient GO-based membrane systems.

502585 Novel Mixed Matrix Membrane Using Metal Organic Framework and Graphene Oxide for CO_2 separation

Abstract:

The reduction in CO_2 emissions is very important to mitigate the problem of global warming and climate change. Membrane technology for CO_2 separation has shown positive results. Solution diffusion and facilitated transport mechanism are the most important in membrane gas separation. Among the different types of membranes, mixed matrix membranes (MMM's) have shown great potential because they are expected to combine the selectivity of ceramic membranes with the low cost and ease of manufacture of polymer membranes. Moreover, the performance of the membranes is governed by two characteristics, permeability; that is the flux of a specific gas through the membrane, and selectivity; the membrane's preference to pass one gas species over the other.

The overreaching goal of this study is to develop and explore approaches to enhance the performance of Mixed Matrix Membrane (MMM) by properly modifying the Metal Organic Frame work (MOF). In this work a octahedral metal-organic framework UIO-66 with surface area 1680 m^2g^{-1} is synthesized and polyetherimide (PEI) has chosen as a matrix .The asymmetric flat sheet of mixed matrix membrane (MMM) has been fabricated via a dry/wet phase inversion technique and characterized by thermogravimetric analysis (TGA), differential scanning calorimetry (DSC), field emission scanning electron microscopic (FESEM) and pure gas permeation testing. In addition, graphene oxide has synthesized from graphite powder and have been used as a filler of mixed matrix membrane to improve the mechanical properties and enhance the sustainable CO_2 capture properties.

493286 Sorption of Organic Acids on Weak-Base Anion Exchange Resins with Different Basicities

Abstract:

There has been a rapid shift in recent years in the utilization of resource base from fossil fuel feedstocks to renewable feedstocks for the production of value added chemicals. By the year 2025, bio-based production of these chemicals is expected to contribute to at least 25% of the total production. Production methods based on fossil fuels are cost-effective and efficient due to highly optimized processes in the chemical manufacturing sector. Emerging technologies in metabolic engineering and bio refineries are helping biobased industries to compete more effectively with the chemical industries to bring products to the market. Among value added chemicals, organic acids belonging to the carboxylic acid group are considered important platform chemicals. In particular, acetic acid and lactic acid have large market potentials due to their numerous applications in the food and chemical industries. However, to achieve market viability, the downstream processing costs of separation and purification of the acids from the fermentation broth, a major cost component of the total product cost, should be significantly reduced.

Ion exchange using weak-base anion exchange resins has proven to be an economic and efficient method for the separation of organic acids. In this research, the uptakes of acetic and lactic acids on several resins of different basicities were studied. Experiments were conducted at different pH values to determine the effects of the pH of the solution, pKa of the acid, and the pKa of the resin on acid sorption from pure solutions, mixtures of lactic and acetic acids, and from fermentation broths. The resins studied include Reillex 425 and a more basic resin Purolite A835. Sorption data for pH range from 2 to 10 for lactic and acetic acids along with equilibrium models and mechanistic analysis of resin basicity will be presented.

497667 N_2 Selective Cation Exchanged Clinoptilolite - Adsorption Separation of N_2 from CH_4 Containing MixturesÂ

Abstract:

In this study, a new modified N_2 selective adsorbent material was developed for the application N_2 removal from unconventional natural gas and biogas upgrading processes where inert nitrogen contamination is a concern. Adsorption based separations have the potential to achieve a more economically efficient separation of N_2 from hydrocarbon mixtures containing CH_4 than current separation processes such as cryogenic distillation. Most commercially available adsorbents are only moderately selective for CH_4 over N_2; however, inverse selectivity is desired for this separation for applications such as pressure swing adsorption in order to maintain the CH_4 gas in the higher pressure raffinate stream. Currently, the only N_2 selective adsorbent that has shown commercial success for this application is titanosilicate ETS-4.

The potential of cation exchanged clinoptilolite adsorbents were studied for the selective separation of N_2 from CH_4 containing gas mixtures. Single component equilibrium adsorption isotherms and uptake measurements of CH_4 and N_2 confirm the promising N_2 selective properties for both cation exchanged powders and pellets. Adsorption breakthrough curve experiments were used to show the effects of column dynamics and competitive adsorption of CH_4 and N_2 containing mixtures on the new materials.

498246 Fundamentals of Competitive Adsorption Phenomena within Dilute, Multi-Component Aqueous Mixtures

Abstract:

Increasing global energy demand combined with decreasing fossil fuel supplies serve as the impetus for the development of alternative sources of fuels and chemical intermediates. Toward this goal, many biologically based processes have been recently invented to convert biomass into short chain oxygenated molecules that can be subsequently converted to fuels and chemicals. The primary advantage of these biological processes are their high selectivity and mild operating conditions. However, the primary drawback is that the final product is typically produced as a very dilute aqueous solution. The main objective of this study is to investigate the major factors that affect the performance of solid adsorbent materials for the extraction of small chain alcohols, aldehydes, and ketones from dilute aqueous solutions that contain a wide range of molecules and ions, similar to those derived from fermentation processes. To understand the fundamentals of adsorption using microporous, aluminosilicate materials, adsorption isotherms were collected for alcohols and aldehydes on proton and Na-exchanged forms of BEA, MOR, FER, FAU and MFI zeolites. Langmuir and Freundlich adsorption models were used to develop a quantitative understanding of the relationship between adsorbent structure and performance by determining the effect of solution composition and adsorbent structure on the parameters of each model (equilibria constants, saturation capacities, and interaction parameters.)

The effects of different zeolite structure on the parameters for the Langmuir model were initially probed using butanol as the adsorbate. Maximum adsorption capacities were similar for all zeolites tested (2.5 mmol/g on FAU, 1.8 mmol/g on MOR, 2.2 mmol/g on MFI, and 2.2 mmol/g on BEA). The adsorption equilibrium coefficient parameter was lower for FAU and MOR (30 L/mol) compared to BEA and MFI (140 L/mol). For the Freundlich model, the adsorption equilibrium coefficient was inversely proportional to zeolite pore size whereas the adsorption capacity parameter was directly proportional to molecule size.

Multi-component mixtures were used to probe competitive adsorption on BEA zeolite. These systems were analyzed using the Langmuir competitive adsorption model. For an equimolar binary system of butanol with smaller alcohols (propanol or ethanol) on BEA zeolite, two phenomena were observed, competitive adsorption between butanol and propanol and co-adsorption of ethanol and butanol. Maximum capacities for propanol were unaffected by the presence of butanol (~2.2 mmol/g), however, the maximum capacity for ethanol increased compared to its value in a single component system (from 0.9 to 2.5 mmol/g). The value of the adsorption equilibrium coefficient was the same for butanol in both binary (with propanol) and single component systems (134 mol/g), however competitive adsorption with propanol (with an adsorption coefficient of 22 L/mol) caused a decrease in its maximum capacity. In the presence of ethanol, the adsorption equilibrium coefficient for butanol increased from 134 to 241 L/mol, which led to a corresponding increase in butanol uptake.

487890 Evaluation of Solid Adsorbents As Buffer Materials for Indoor Air CO_2 Control

Abstract:

In this study, we evaluated solid sorbents for their ability to passively control indoor CO_2 concentration in buildings or rooms with cyclic occupancy (e.g. offices, bedrooms). Silica supported amines were identified as suitable candidates and systematically evaluated in the removal of CO_2 from indoor air by equilibrium and dynamic techniques. In particular, sorbents with various amine loadings were synthesized using tetraethylenepentamine (TEPA), poly(ethyleneimine) (PEI), and a silane coupling agent 3-aminopropyltriethoxysilane (APS). TGA analysis indicates that TEPA impregnated silica not only displays a relatively high adsorption capacity when exposed to ppm level CO_2 concentrations, but also is capable of desorbing the majority of CO_2 by air flow (e.g. by concentration gradient). In 10 L flow-through chamber experiments, TEPA based sorbents reduced outlet CO_2 by up to 5% at 50% RH and up to 93% of CO_2 adsorbed over 8 h was desorbed within 16 h. In 8 m^3 flow-through chamber experiments, 18 g of the sorbent powder spread over a 2 m^2 area removed approximately 8% of CO_2 injected. By extrapolating these results to real buildings, we estimate that meaningful reductions in the CO_2 can be achieved, which may help reduce energy requirements for ventilation and/or improve air quality.

499173 Adsorption Modeling for Selective Capture of Uranium from Seawater

Abstract:

One of the worldâ€™s rising global issues is how to meet the necessary energy demands of tomorrow without destroying the resources and environment of today. This challenge is being met by research into solar and wind energy, biofuels, and nuclear power. To facilitate the use of nuclear power in place of more traditional fossil-energy sources may require the utilization of the ocean as a new source of uranium. Adsorption technology can be employed to farm uranium from seawater in order to provide a new, untapped source of fuel for nuclear power. The ocean has the potential to provide a vast amount of uranium to add to the nuclear fuel cycle, as it contains nearly 1000 times more uranium than conventional mines. However, the average concentration of that resource is roughly 3.3 ppb, which presents a technological challenge. Recently, a significant effort has been focused on the utilization of amidoxime ligands for the selective removal of uranium from seawater, but there are still many other environmental factors such as pH, temperature, competing ions, and current flows that further complicate the design of a capture system. The realization of the ocean uranium recovery system will require an advanced understanding of the physical-chemical processes involved, as well as a set of engineering modeling tools to aid in the process design. An advanced modeling framework coupled with detailed, high-resolution adsorption process models is being developed in order to give engineers a reliable design tool for the implementation of a seawater-capture system. These process models are coupled with reaction schemes developed using *ab initio* methods and have been shown capable of predicting uranium uptake under laboratory conditions for specific adsorbents.

499994 Nanostructured Conducting Polymers for Electrochemically Mediated Separations of Organic Pollutants from Water

Abstract:

Many stimuli-responsive polymeric adsorbents with chemical (e.g., CO_2, pH, oxidizing/reducing agents) or thermal responsiveness have been developed for separations of organic compounds from water. Electrochemical stimuli, however, have rarely been used to modulate the affinity between target hydrophobic organic molecules and polymeric adsorbents. Here, we report an approach for the separation of organic pollutants from water using an intrinsically conducting polymer, polypyrrole (PPy), doped with surfactant AOT as an adsorbent. We have discovered a new approach to fabricate PPy that results in nanoporous structures and can undergo hydrophobic-hydrophilic transitions under electrochemical control. In its hydrophobic state, the PPy polymer can adsorb organic uncharged species from water, and can subsequently be regenerated by application of an electrical potential to switch the polymer to its hydrophilic state and induce the release of adsorbed pollutants, or their electrochemical decomposition. The PPy polymer has been taken through multiple adsorption-regeneration cycles without decay in its adsorption capacity, showing the robustness and reusability of the adsorbent. Our fabrication method also creates hierarchical porous structures in the PPy polymer which allows for increased adsorption capacity of greater than 100 mg pollutant/g polymer for a model organic pollutant molecule.

500180 Understanding Gravimetric and Volumetric Hydrogen Cryo-Adsorption Trade-Off in Metal-Organic Frameworks (MOFs) and its Link to Material Properties

Abstract:

With the recent release of hydrogen-powered fuel cell vehicles that store hydrogen at 700 bar, alternatives are being explored to reduce storage pressure, including the use of an adsorbent-filled tank with operating adsorption/desorption cycles between 100 bar/77 K and 5 bar/160 K. Metal-organic frameworks (MOFs) are promising adsorbents for cryo-adsorbed hydrogen storage due to their highly tunable structure and chemistry, which can be potentially optimized to obtain the highest possible material performance. MOF volumetric and gravimetric hydrogen capacities are key factors that need to be maximized to attain a MOF-filled tank that is neither too large nor too heavy for practical application. Because the goals of maximizing MOF volumetric and gravimetric adsorption individually are incompatible, an in depth understanding of the trade-off between MOF volumetric and gravimetric loadings is necessary to achieve the best compromise between these two properties.

Here we first study, both experimentally and computationally, the trade-off between volumetric and gravimetric cryo-adsorbed hydrogen deliverable capacity in a set of well-known MOFs, including the highly stable zirconium MOFs, NU-1101, NU-1102, and NU-1103, among which we identified the best performers. On the grounds of excellent agreement between simulated and measured hydrogen isotherms, we use molecular simulation to elucidate the mechanism of hydrogen cryo-adsorption and its connection to materials properties such as volumetric and gravimetric surface areas, and pore volume and void fraction. We found that the higher variability in gravimetric deliverable capacity, in contrast to volumetric capacity, occurs due to the proportional relation between gravimetric surface area and pore volume in contrast to the inverse relation between volumetric surface area and void fractions. Using an "inverse molecular design" approach, we explored textural property combinations that could potentially allow attaining specific hydrogen adsorption targets.

Continuing with the inverse design philosophy, then we explored the impact of pronounced alterations in MOF chemistry on hydrogen cryo-adsorption performance. Specifically, we studied the introduction of "blanket" metal catecholate sites in sixty yet-to-be-synthesized MOFs based on a dozen topologies, which were constructed using a topologically-based crystal constructor (ToBaCCo) code. Using molecular simulation, we explored to what extent tuning the interaction strength between hydrogen and MOF metal catecholate sites could enhance volumetric and gravimetric hydrogen cryo-adsorption, depending on the operating conditions.

References:

1. DA Gómez-Gualdrón, TC Wang, P García-Holley, RM Sawelewa, E Argueta, RQ Snurr, JT Hupp, T Yildirim, OK Farha, ACS Appl. Mater. Interfaces, DOI:10.1021/acsami.7b01190

495349 Modeling of Multicomponent Sorption of Fermentation Products on Resin Sorbents

Abstract:

Industrial-scale production of fuels and chemicals from renewable feedstocks via fermentation technologies has rapidly developed in recent years. Ethanol is one such biofuel that is being produced on a large scale using yeast fermentation. The broth from fermentation contains about 15% ethanol along with a variety of compounds such as carboxylic acids, glycerol, and sugars which are generated as byproducts during fermentation. Separation of ethanol from such complex broths using conventional distillation and sorbent drying, as currently practiced, is highly energy intensive. Packed bed adsorbers using various resins have been studied as more efficient alternatives, due to their selectivity and ability to operate in a continuous mode. Using mathematical models for the analyses of these systems offers the flexibility to study and understand the effect of different operational parameters on the separation process. The aim of this research is to develop a model of the sorption process for the separation of ethanol from fermentation broth using COMSOL Multiphysics® software. The PDEs representing multicomponent diffusion and equilibrium sorption of organic acids and ethanol were developed and solved in the Coefficient form PDE interface available in COMSOL. The effects of initial concentration, resin affinity, and other parameters on fixed-bed breakthrough curves including roll-over effects will be presented.

502470 Pervaporation and Vapor Permeation Separation of Xylene Mixture By Randomly Oriented MFI Â Zeolite Membranes

Abstract:

MFI zeolite membranes have attracted much attention for separation of p-xylene from xylene isomer mixture. However, all studies reported were either focused on vapor-permeation or pervaporation separation of xylene separation properties. Furthermore, the vapor permeation studies were conducted under unpractical condtions ofÂ quite low xylene activity (<3%). There is lack of a comparative study on xylene separation characteristics of the MFI membranes of different structures in the different operation modes. In this work, randomly oriented MFI zeolite membranes synthesized with and without using organic template on Î±-alumina support were studied for the separation of xylene isomers by pervaporation and vapor-permeation methods. Vapor permeation selectivity and permeance of binary mixture of p-xylene and o-xylene through MFI zeolite membranes were measured in various p-xylene activity in a wide temperature range.Â It was found that the separation factor of p-xylene/o-xylene decreases with increasing p-xylene activity. Membranes obtained from template-free precursor exhibit much higher separation factors of p-xylene over o-xylene than the membrane synthesized with template by using both separation methods due to less intercrystalline defects in the former membrane than the latter. The p-xylene/o-xylene separation factor of both membranes in vapor-permeation (with highest of 162) is found to be much higher than that of pervaporation (highest 34) because the latter was operated at highest activity.

492851 Suppressing Substructure Collapse in Carbon Molecular Sieve (CMS) Hollow Fibers

Abstract:

Carbon Molecular Sieve (CMS) membranes exceed the so-called polymer permeability/selectivity trade-off upper bound for most challenging gas-pairs separations. Asymmetric hollow fibers are preferred membrane formats, due to the high surface area to module volume ratios this format enables. During pyrolysis to form CMS, the porous asymmetric polymeric precursor support tends to collapse, thereby reducing productivity. Previous work showed that this support collapse could be overcome for Matrimidâ polyimide precursors by pretreating precursor fibers with vinyltrimethoxysilane (VTMS) (henceforth to be referred as V-treatment). Such V-treated CMS fibers show little support collapse, with much higher permeances as compared to the untreated fibers. CMS fibers pyrolyzed from 6FDA BPDA: DAM (henceforth to be referred as 6F), a fluorinated copolyimide with higher intrinsic permeability than Matrimidâ , shows less sub-structure collapse than Matrimidâ even without V-treatment. The fundamentally different response between these two polyimide precursors reflects a smaller temperatures interval between the glass transition temperature and temperature of the onset of carbonization for the 6F vs. Matrimidâ. Optimization of the V-treatment process under varied pyrolysis conditions will be discussed to illustrate approaches to achieve improved performance relative to the process conditions for Matrimidâ. Fundamental factors controlling performance will be discussed to show how the versatile V-treatment process can be optimized for different precursors, temperatures and advanced pyrolysis options, such as doping with trace O_2 during pyrolysis.

497124 Chabazite SAPO-34 Zeolite Membranes for Krypton/Xenon Separation: Enhanced Separation Performance and Process Modeling

Abstract:

Nuclear power is considered to be an economical means of carbon-free energy, but it requires effective ways to deal with radioactive waste products. Separation of radioactive ^{85}Kr from a Kr/Xe mixtures is one of the off-gas processing steps in the used nuclear fuel (UNF) recycling scheme proposed by the US DOE in reducing the volume of radioactive waste for storage. Instead of conventional cryogenic distillation, membrane-based separation could be an attractive separation method. Zeolite membranes offer an opportunity to exploit the size difference of Kr (kinetic diameter 0.36 nm) and Xe (0.396 nm) to achieve preferential permeation of the minority component (10 vol % of ^{85}Kr in gaseous mixture) from the mixture, resulting in an efficient separation with very low energy requirements. SAPO-34 is a silicoaluminophosphate zeolite molecular sieve with chabazite framework topology with uniform micropores, intermediate in size (0.38 nm) between the kinetic diameters of Kr and Xe. Our previous work indicated the potential for engineering the separation properties of these membranes, but revealed the need to considerably improve the membrane performance and evaluate the expected performance of a multistage membrane process for ^{85}Kr recovery. In this study, our main goals were to obtain SAPO-34 membranes with enhanced performance *via* thickness reduction and ion exchange, and to conduct process modeling that is useful for future technoeconomic analysis of zeolite membrane systems for Kr/Xe separation.

In the first part of this work, we demonstrate considerable thickness reduction of SAPO-34 membranes. Porous Î±-Al_2O_3 supports were seeded using steam-assisted conversion (SAC) methods to make a continuous seed layer followed by a conventional hydrothermal reaction for the secondary-growth process. The synthesized membranes showed dense membrane layers with thicknesses as low as 2.4 Âµm. The reduction of membrane thickness led to a large increase in Kr permeance from the previously reported [1] 7.5 GPU to 26.3 GPU with ideal selectivity of 21 at 298 K. Cation-exchanged membranes after thickness reduction showed moderate permeance reduction for both Kr and Xe, but showed large increases (more than 50%) in selectivity. The effects of the type and valency of the exchanged cations is discussed. These membranes were also shown to have excellent Kr selectivity in mixed-gas permeation during ambient and sub-ambient (-18°C) operation. In order to estimate the size of the separation system, process modeling of multistage membrane process was conducted using experimentally obtained permeances and selectivities. The required Xe purity in the retentate of each membrane stage is fixed as 99.9% and the permeate from each stage is sent to the next stage as feed till a Kr purity of 90+% (suitable for waste storage) is reached. Among the cation-exchanged SAPO-34 membranes, K-SAPO-34 membranes showed the best potential which requires a small membrane area (~7 m^2) as well as the fewest number (~3) of stages for handling a 1 L/min Kr/Xe feed mixture. Based upon these findings, we conclude that the development of a compact, low cost membrane system for Kr separation is quite feasible.

[1] Y. H. Kwon, C. Kiang, E. Benjamin, P. Crawford, S. Nair, and R. Bhave, "Krypton-xenon separation properties of SAPO-34 zeolite materials and membranes". *AIChE J.* **63**, *2*, 761 (2017).

501537 Synthesis of MOF-Based Membranes through Chemical Vapor Growth

Abstract:

Metal organic framework (MOF) membranes possess unique properties for gas separation. To make MOF membranes a viable separation technology requires development of cost-effective synthesis of thin and high quality MOF membranes on scalable polymer support, or embedding nano-sized MOF crystals into polymer matrix to form defect-free mixed matrix membranes (MMMs). It was recently reported that atomic layer deposited ZnO can be transformed to ZIF-8 by exposure to 2-methylimidazole vapor [*Nat. Mater.*2016, 15:304â€“310]. Here, we developed a modified chemical vapor growth method as novel techniques for the synthesis of MOF-based supported membranes and MMMs. In this method, ZnO nanocrystals of ~5 nm in diameter were synthesized and used as metal source to react with vapor of organic ligands forming microporous ZIF structures.

For supported MOF membranes, this chemical vapor growth eliminates the homogeneous nucleation/growth of MOFs and the re-dissolution of membrane materials during membrane synthesis, enabling the formation of conformal ZIF-8 membranes as thin as 300 nm on porous polyethersulfone substrate through vapor treatment of a thin coating of ZnO nanocrystals. The membrane exhibits a molecular sieving effect towards gas molecules with the He/N_2 and CO_2/N_2 ideal selectivities close to those of ZIF-8 membranes made by the conventional liquid-phase synthesis. For MOF-based MMMs, the non-selective MOF-polymer interface remains a major challenge to overcome. In this work, a novel approach was developed to in situ form MOFs inside the polymer matrix by the chemical vapor growth method. Polymer/ZnO nanocomposite membranes were first prepared, followed with a ligand vapor treatment process to convert the embedded ZnO nanocrystals to MOFs inside the polymer matrix. The volume expansion of ZnO after being converted to MOF (~17 times for complete conversion to ZIF-8) is expected to occupy the ZnO-polymer interface, producing defect-free MOF-based MMMs. SEM and XRD confirmed the successful in situ phase transformation of ZnO to ZIF-8 inside the polymer matrix. The mechanical properties, gas permeation and separation properties of these interesting MMMs will also be discussed.

493156 Zeolitic-Imidazolate Framework ZIF-8 Membranes on Ceramic Tubular Supports for Scalable High-Resolution Propylene/Propane Separation

Abstract:

Zeolite-imidazole frameworks (ZIFs) have been extensively studied as a new membrane material for gas separations due to their robust synthesis protocols, chemical and thermal stability, and well-defined ultra-micropores (~ 2 to 5 Å).[1, 2] ZIFs are comprised of metal centers (Zn, Co, and Cd) and bridging imidazole-based ligands. ZIF-8, composed of zinc nodes and 2-methylimidazole linkers, has been shown very promising for propylene/propane separations since its effective aperture size lies in between propylene (~ 4 Å) and propane (~ 4.2 Å).[3] In fact, some of the polycrystalline ZIF-8 membranes synthesized by various methods have been reported to exhibit impressive propylene/propane separation factors up to 200.[4, 5] These highly propylene-selective ZIF-8 membranes were supported mostly on alumina disks or occasionally on hollow fibers (polymer or ceramic).

For their large-scale industrial applications, however, the productivities of ZIF-8 membranes must be significantly improved.[6] One way to increase the membrane productivity is to increase membrane area per volume. As such, current disk supports are not desirable. Though hollow fiber supports (either polymer or ceramic) are most desirable, they are unlikely to be used for large-scale applications in near future. This is because ceramic hollow fibers are highly fragile while polymer hollow fibers require much more development. While hollow fiber supports are transformative, ceramic tubular supports are readily available for large-scale applications. For example, ceramic tubular supports have been used for commercial polycrystalline zeolite membranes.[7] Though ZIF-8 membranes on ceramic tubular supports were reported, none of those membranes showed promising propylene/propane separation capabilities.[8, 9]

Here, we would like to present polycrystalline ZIF-8 membrane supported on alumina tubes showing high propylene/propane separation performances. The tubular membranes were synthesized using our microwave seeding and secondary growth method. The effects of the synthesis parameters on the microstructures of tubular ZIF-8 membranes will be presented. Finally, the propylene/propane separation performances of ZIF-8 membranes supported on alumina tubes will be discussed.

References and Notes:

(1) Park, S.; Kang, W. R.; Kwon, H. T.; Kim, S.; Seo, M.; Bang, J.; Lee, S. h.; Jeong, H. K.; Lee, J. S., The polymeric upper bound for N2/NF3 separation and beyond; ZIF-8 containing mixed matrix membranes. *J. Membr. Sci.* **2015**, 486, 29-39.

(2) Chen Zhang, R. P. L., Ke Zhang, Justin R. Johnson, Oguz Karvan, William J. Koros, Unexpected molecular sieving properties of zeolitic imidazolate framework-8. *The journal of physical chemistry letters* **2012**, 3, (16), 2130-2134.

(3) Li, K.; Olson, D. H.; Seidel, J.; Emge, T. J.; Gong, H.; Zeng, H.; Li, J., Zeolitic imidazolate frameworks for kinetic separation of propane and propene. *J. Am. Chem. Soc.* **2009**, 131, (30), 10368-9.

(4) Lee, M. J.; Kwon, H. T.; Jeong, H. K., Defect-dependent stability of highly propylene-selective zeolitic-imidazolate framework ZIF-8 membranes. *J. Membr. Sci.* **2017**, 529, 105-113.

(5) Pan, Y. C.; Li, T.; Lestari, G.; Lai, Z. P., Effective separation of propylene/propane binary mixtures by ZIF-8 membranes. *J. Membr. Sci.* **2012**, 390, 93-98.

(6) Baker, R. W., Future directions of membrane gas separation technology. *Ind. Eng. Chem. Res.* **2002**, 41, (6), 1393-1411.

(7) Kosinov, N.; Gascon, J.; Kapteijn, F.; Hensen, E. J., Recent developments in zeolite membranes for gas separation. *J. Membr. Sci.* **2016**, 499, 65-79.

(8) Hara, N.; Yoshimune, M.; Negishi, H.; Haraya, K.; Hara, S.; Yamaguchi, T., Diffusive separation of propylene/propane with ZIF-8 membranes. *J. Membr. Sci.* **2014**, 450, 215-223.

(9) Zhang, X.; Liu, Y.; Kong, L.; Liu, H.; Qiu, J.; Han, W.; Weng, L.-T.; Yeung, K. L.; Zhu, W., A simple and scalable method for preparing low-defect ZIF-8 tubular membranes. *J. Mater. Chem. A* **2013**, 1, (36), 10635.

497032 Gas Mixture Separation through Nanoporous Graphene Membranes

Abstract:

Graphene sheets perforated with nanometer-scale pores have the potential to become high performance gas mixture separation membranes. Gas molecule translocation through molecularly-sized graphene pores can be an activated process, and differences in activation energy between different gases could give rise to high separation factors. These membranes could also achieve extremely high permeance due to graphene's single-atom thickness. Finally, graphene's outstanding chemical, thermal, and mechanical stability could enable these membranes to withstand the pressure differentials and elevated temperatures necessary for gas separations.

To date, very few experimental measurements of gas mixture permeation through graphene membranes have been reported. Some prior experimental works have investigated the transport of single gases through graphene membranes with nanometer-scale pores,[1] larger rips and tears,[2] or capillaries between multilayered graphene sheets.[3, 4] In these studies, gas separation factors were estimated from single gas permeation measurements, which may not accurately account for phenomena such as competitive adsorption and diffusion that could influence mixture separations with these membranes.[5-8] Park and coworkers provided the first demonstration of gas mixture separation through graphene membranes with pores of 7.6 nm to 1 µm in diameter.[9] While these studies have provided highly valuable insights into of the properties of graphene membranes, further direct measurements of gas mixture separations are critical for understanding graphene membrane separation performance.

For the first time, we have measured the temperature dependence of H_2, He, CH_4, CO_2, and SF_6 permeance from 22 – 208 °C through intrinsic defects in three suspended graphene membranes. These measurements reveal surprising insights about the mechanisms of gas permeation through the graphene membranes. The three membranes we tested show substantially different behavior. Membrane A exhibits permeance on the order of 10^{-5} to 10^{-4} mol m^{-2} s^{-1} Pa^{-1} with minimal variation as a function of temperature and decreasing flux with increased gas molecular weight, suggesting that transport through this membrane is likely dominated by the direct gas impingement pathway. Membrane B shows complex behavior, with the permeance increasing as a function of temperature from 27 to 200 °C, and a CO_2/SF_6 separation factor of 12 at 200 °C, likely indicating a mix of several gas permeation mechanisms, potentially including activated translocation through molecularly sized nanopores. Membrane C shows no measureable permeance of any gas above the detection limit of our technique, indicating that it may be a molecularly impermeable barrier. Overall, this work demonstrates that graphene membranes can provide high mixture separation selectivity, and temperature-dependent permeance measurements are an important strategy for understanding complex physical phenomena involved in these separation processes.

References

1. S. P. Koenig, et al., *Nature Nanotechnology*, **7**, 728 (2012).

2. M. S. H. Boutilier, et al., *ACS Nano*, **8**, 841 (2014).

3. R. K. Joshi, et al., *Science*, **343**, 752 (2014).

4. H. W. Kim, et al., *Science*, **342**, 91 (2013).

5. L. W. Drahushuk, et al., *Langmuir*, **28**, 16671 (2012).

6. D. E. Jiang, et al., *Nano Letters*, **9**, 4019 (2009).

7. J. Schrier, *ACS Applied Materials & Interfaces*, **3**, 4451 (2011).

8. H. J. Liu, et al., *Nanoscale*, **5**, 9984 (2013).

9. K. Celebi, et al., *Science*, **344**, 289 (2014).

499589 A General Adsorption/Reaction Framework for Modular and Multi-Functional Process Design

Abstract:

Modular chemical processes are easily deployable for small scale operations in remote locations to produce value-added chemicals from unconventional resources and distributed feedstocks such as stranded natural gas, CO_2 from flue gas and associated shale gas. The advantages of modular processes include compactness, prevention of inefficiencies of on-site equipment manufacturing, freedom of constructing high-quality equipment in specialized locations, and faster production time. However, modular processes often fail to achieve economies of scale. One way to address this challenge is to design *multifunctional* modular processes using existing single-functional materials. To this end, we first propose a multi-material design approach which integrates multiple phenomena within a single unit and thereby leads to multifunctional modular processes using existing materials. We use a systematic framework to select and arrange materials (e.g., adsorbent and catalyst blocks) with unique tasks (e.g., adsorption, reaction, storage) in a sequential block-by-block arrangement inside a packed column, and configure operational cycles for integrated, periodic and multifunctional processes. The task-integrated column operates in a cyclic or periodic manner, where each cycle consists of several steps. We use a single superstructure that embeds different design alternatives using blocks, where each block represents a specific material performing a specific task for a specific time step. The functional block-based superstructure is general such that it contains numerous cyclic process configurations for separation processes including pressure/temperature swing adsorption (PSA/TSA), as well as for integrated separation-conversion processes including sorption-enhanced reaction processes (SERP).

Next, in order to accurately predict the performance of different processes generated via multi-material design, we develop Generalized Reaction-Adsorption Model (GRAM) â€“ a high-fidelity process model for dynamic adsorption-reaction systems. GRAM is a one-dimensional, pseudo-homogeneous, non-isothermal, non-adiabatic, and non-isobaric model that captures the reaction and adsorption dynamics in hybrid adsorption-reaction systems. The model is validated using different experimental data from literature. Specifically, the model predictions show agreement with experimental observations for sorption-enhanced steam-methane reforming (SMR)[1,3], sorption-enhanced reverse water gas shift (RWGS)[2], and conventional SMR[4,5]. Computationally, the model handles multi-feed, multi-functional and multi-product processes well.

The GRAM model is further incorporated within the multi-material design framework to design several novel multifunctional modular processes for cost-effective carbon capture and utilization using unconventional energy sources such as stranded and distributed natural gas, CO_2-contaminated shale gas, biogas, landfill gas, fuel gas and flare gas. The objective is to produce valuable products such as hydrogen and syngas. Computational studies are performed for several configurations of the column containing (i) pure catalyst, (ii) pure adsorbent, (iii) heterogeneously-compartmentalized adsorbent and catalyst, and (iv) homogeneously-distributed uniform mixture of adsorbent and catalyst. These intensified processes use sections of adsorbent and catalyst for simultaneous capture and

conversion of reactants.

References:

[1] Carvill, B.T., Hufton, J.R., Anand, M., Sircar, S. Sorption-Enhanced Reaction Process. AIChE Journal. 1996.

[2] Jang, H.M., Lee, K.B., Caram, H.S., Sircar, S. High-purity hydrogen production through sorption enhanced water gas shift reaction using K_2CO_3-promoted hydrotalcite. Chemical Engineering Science. 2012.

[3] Hufton, J.R., Mayorga, S., Sircar, S. Sorption-enhanced reaction process for hydrogen production. AIChE Journal. 1999.

[4] Shu, J., Grandjean, B.P.A., Kaliaguine, S. Methane steam reforming in asymmetric Pd- and Pd-Ag/porous SS membrane reactors. Applied Catalysis. 1994.

[5] Gallucci, F., Paturzo, L., Basile, A. A simulation study of the steam reforming of methane in a dense tubular membrane reactor. International Journal of Hydrogen Energy. 2004.

496493 Ultrasonic Tuning of TiO_2 Based Mixed Oxides Structural Properties and Catalytic Activity

Abstract:

Ultrasound (US) has a potential to impact every stage of the preparation of a material in liquid phase. US can produce nanomaterials in amorphous state (due to the very high cooling rates), which is relevant to fields such as catalysis, magnetism and coating processes [1]. US generates shock waves that facilitate coating and insertion/intercalation processes and improves the distribution of the active phase on a support. We pursue ultrasound powder synthesis as a greener method to lower solvent amount and catalyst precursor, as well as to accelerate lengthy preparation method such as sol-gel and impregnation. US-assisted synthesis produces catalysts with better morphology and activity. Simple tuning of operating conditions increases the range of obtainable features for different purposes [2,3].

We report the ultrasound assisted sol-gel and impregnation synthesis of sulphated ZrO_2-TiO_2 and Mn oxides-TiO_2 systems. We characterized the catalysts with N_2 adsoption, XRD, SEM-EDX and TEM. We tested the sulphated ZrO_2-TiO_2 in the free fatty acids (FFA) esterification reaction and the sulphated ZrO_2-TiO_2 and the Mn_xO_y-TiO_2 systems in the ultrasound-assisted photocatalytic degradation of recalcitrant water pollutants (pharmaceutical products). We adopted a ultrasound probe with a tip of 13 or 20 mm diameter. The ultrasound operating frequency was fixed at 20 KHz. The nominal power was 500 W and we varied the amplitude from 20 to 60 % adopting continuous ultrasound or ultrasound pulses (0.5 on and 0.5 off). In the sol-gel synthesis of sulphated ZrO_2-TiO_2 systems, ultrasound pulses increased the surface area up to 220 m^2g^{-1} for the lowest solvent: precursors ratio of 15:1 mol, corresponding to the highest power density. Solvent: precursors molar ratios of 30:1 and 60:1 gave a SSA of 150 m^2g^{-1} and 30 m^2g^{-1}, respectively. The samples obtained with pulses were also the ones exhibiting the highest Brønsted surface acidity. The catalysts with the highest acidity/SSA ratio were the most active in the free fatty acids esterification reaction.

We deposit Mn_xO_y on TiO_2 in one step with the ultrasound-assisted sol-gel synthesis or via ultrasonic impregnation in a second step. We prepared Mn/TiO_2 catalyst in one step with a classical sol–gel method, starting from butyl-titanate, deionized water, nitric acid, anhydrous ethanol, and manganese nitrate [4]. After mixing at room temperature to yield a yellowish transparent sol, we dried the precursor at 80°C for 24h, converting it into a xerogel. The calcination to obtain the Mn/TiO_2 catalyst was standard at 500°C for 5h. We prepared the same catalyst applying US, either maintaining the same calcination parameters or decreasing the calcination temperature to 400°C for 2 hours to trigger the formation of anatase instead of rutile as an active phase.

Mn/TiO_2 two steps synthesis consisted of the US-assisted surface deposition of Mn-oxides on TiO_2 (both nanometric or micrometric), synthesized in a preceding stage [5]. Generally, US promote homogeneous nucleation in a first step, and enhance the crystals growth in a second stage. Acoustic cavitation, in particular for the two-steps Mn deposition, generates high-speed micro jets which impact the TiO_2 support with a cleaning and de-agglomerating effect. Ultrasound increases the dispersion of the Mn species over both supports and has a role in stabilizing Mn_xO_y species that samples synthesized with traditional sol-gel don't normally exhibit (preliminary XRD results) [6]. The modulation of the main sonication operating parameters produces a mixed-size distribution of Mn and Mn-Ox species on TiO_2 rather than a narrow distribution of small crystallites, which often turns out to be better for the photocatalytic activity. US also improves the distribution of Mn_xO_y over micro-metric TiO_2 (SEM-EDX analysis), which is meaningful from a safety point of view since body barriers may not "recognize" nano-powders. The photocatalytic activity of Mn_xO_y over nano and micro-size TiO_2 is comparable. Considering instead the comparison between Mn and other metals, Mn is cheaper and one of a kind, considering the variety of oxides that can form. By means of US we can also modulate and enhance the formation of specific oxidation states [6,7].

References

[1] T.J. Mason, Ultrasound in synthetic organic chemistry, *Chem. Soc. Rev.*, 26 (1997), pp. 443–451.

[2] S.J. Musevi, A. Aslani, H. Motahari, H. Salimi, Offer a novel method for size appraise of NiO Nanoparticles by PL analysis: synthesis by sonochemical method, *J. Saudi Chem. Soc.*, 20 (2016), pp. 245–252.

[3] C. Pholnak, C. Sirisathitkul, D.J. Harding, Characterizations of octahedral zinc oxide synthesized by sonochemical method, *J. Phys. Chem. Solids*, 72 (2011), pp. 817–823.

[4] Qi-lin Chen, Rui-tang Guo, Qing-shan Wang, Wei-guo Pan, Ning-zhi Yang, Chen-zi Lu, Shu-xian Wang, *J. Taiwan Inst. Chem. Eng.*, 64 (2016) 116–123.

[5] M. Stucchi, C.L. Bianchi, C. Pirola, G. Cerrato, S. Morandi, C. Argirusis, G. Sourkouni, A. Naldoni, V. Capucci, *Ultrason. Sonochem.*, 31 (2016) 295–301.

[6] M.I. Díez-García, V. Manzi-Orezzoli, M. Jankulovska, S. Anandan, P. Bonete, R. Gómez, T. Lana-Villarreal, Effects of ultrasound irradiation on the synthesis of metal oxide nanostructures, *Phys. Procedia*, 63 (2015), pp. 85–90.

[7] K.D. Bhatte, D.N. Sawant, D.V. Pinjari, A.B. Pandit, B.M. Bhanage, One pot green synthesis of nano sized zinc oxide by sonochemical method, *Mater. Lett.*, 77 (2012), pp. 93–95.

497703 Multi-Model Operability Approach for Process Design, Intensification and Modularity: Application to Nonlinear and High-Dimensional Membrane Reactors

Abstract:

Membrane reactors (MRs) have arisen as alternatives for conventional reactors, enabling previously economically unfavorable processes. MRs are also intensified processes that are candidates for modular applications. Such reactors provide a built-in membrane separation layer along the entire reactor length, where one or more reaction products can be selectively removed. This removal shifts the reaction equilibrium towards the products, increasing reactants conversion and production rate. However, the process design of MRs is intrinsically challenging. MRs combine chemical reactions with separation transport phenomena, which results in models of complex and nonlinear nature for their representation. MRs are also commonly placed in highly constrained environments, which poses an additional difficulty to the process design task, especially if process integration is needed for modularity. In this presentation, a multi-model operability approach is proposed to address these challenges. Operability was first introduced as a tool for the design and control interface to calculate the inputs that are needed to achieve specified desired outputs in a process [1]. Here, the operability pathway is extended to provide a framework able to tackle nonlinear high-dimensional and constrained systems. This framework attains process intensification, ensuring high efficiency units, suitable for modular systems applications.

In the past decades, process operability has been traditionally applied to low-dimensional nonlinear systems. The implementation of operability for high-dimensional nonlinear systems was successfully performed by employing optimization algorithms based on nonlinear programming (NLP) [2,3]. The novelty of this work derives from the fully utilization of linear programming (LP) tools to handle systems of the same complexity, with the advantage of computational time reduction. In particular, the original nonlinear system is divided into several subsystems and a linear model is calculated for each subsystem. The properties of all linear models are then analyzed, such as the eigenvalues (or singular values), so that the minimum number of model sets that adequately represents the process nonlinearity is obtained. A multi-level optimization algorithm is then formulated based on LP concepts, aiming intensification and control targets, while considering process requirements and constraints.

The developed framework is applied to a membrane reactor for the direct methane aromatization conversion (DMA-MR) to hydrogen and benzene. DMA-MR corresponds to a candidate system for modular natural gas utilization onsite at stranded gas fields without the need of building expensive pipelines [2]. Preliminary results for a low-dimensional system show that benzene production can be optimized and the membrane reactor footprint reduced by 60% in terms of volume (catalyst zone) and 80% in membrane area. These results indicate the potential of the proposed framework to facilitate process intensification towards modularity. In this presentation, a comparison will be established between this approach and previously developed NLP-based approaches [4] to analyze the trade-off between accuracy and

computational expense. The expansion to a higher dimensional system will then be discussed, contemplating membrane parameters, design and operating conditions that are critical for process intensification and modularity.

References:

1. Vinson D. R. and Georgakis C. "A new measure of process output controllability". J. Proc. Cont. 2000, 10(2-3), 185-194.
2. Carrasco J. C. and Lima F.V. "An optimization-based operability framework for process design and intensification of modular natural gas utilization systems". Accepted for publication in Comput. Chem. Eng. 2017; DOI: CACE-5645.
3. Carrasco J. C. and Lima F. V. "Novel operability-based approach for process design and intensification: application to a membrane reactor for direct methane aromatization". AIChE J. 2017; 63(3): 975-983.
4. Carrasco J.C. and Lima F. V. "Operability-based approach for process design, intensification, and control: application to high-dimensional and nonlinear membrane reactors". In proceedings of the FOCAPO/CPC, 2017.

493441 A Modeling Methodology for Predicting Intensified CO_2 removal Efficiency with Rotating Packed Bed

Abstract:

The opportunity in process intensification (PI) of carbon dioxide (CO_2) capture has drawn much attention recently. Since last decade, rotating packed bed (RPB) has been considered as an intensified alternative to conventional packed bed (PB) in CO_2 chemical absorption processes which require excessive absorber volume resulting from mass transfer resistance across vapor-liquid interface during the absorption process [1-3].

Reliable predictions of intensified CO_2 removal efficiency in RPB are essential to design the intensified CO_2 absorption/stripping process. In this study, we present a modeling methodology with sound theoretical basis for predicting the CO_2 removal efficiency in RPB. The methodology assumes the modeling of RPB and that of PB are identical if the gravitational force presented in RPB can be properly taken into account in modeling the mass transfer resistance. The methodology consists of three steps. First, we validate the PB model with PB experimental data [4]. The PB model is based on Aspen rate-based PB model with liquid film discretization (Aspen RateSep) [5]. Second, we predict the CO_2 removal efficiency of RPB experiments [1,4] with the validated PB model. Through this step, we identify the difference in CO_2 removal efficiency with and without the gravitational effect. Third, we account for the gravitational effect in RPB with the adjustable mass transfer factors in the validated PB model. Specifically, the liquid mass transfer coefficient factor and the interfacial area factor are chosen to account for the RPB gravitational effects with the validated PB model. The absolute average deviation (AAD) between the prediction results and the experimental RPB data [1,4] are found to be less than 10% when lean loading of MEA is in the typical operating range of 0.16 ~ 0.44 mol CO_2/mol amine, suggesting a robust predictive capability of our modeling methodology.

This modeling methodology not only provides new insights on RPB, but also avails chemical engineers a practical and theoretically sound modeling capability to estimate the intensified CO_2removal efficiency when replacing PB with RPB. Furthermore, by taking advantages of proven modeling tools in commercial simulators, the methodology could be widely practiced by engineers worldwide to examine the process intensification potential of replacing PB with RPB.

[1] Yu, C.H., Chen, M.T., Chen, H., Tan, C.S., "Effects of Process Configurations for Combination of Rotating Packed Bed and Packed Bed on CO_2 Capture", *Applied Energy*, 175, 269-276 (2016).

[2] Yu, C.H., Lin, Y.X., Tan, C.S., "Effects of Inorganic Salts on Absorption of CO_2 and O_2 for Absorbents Containing Diethylenetriamine and Piperazine", *Int. J. Greenh. Gas Cont.*, 30, 118-124 (2014).

[3] Yu, C.H., Cheng, H.H., Tan, C.S., "CO_2 Capture by Alkanolamine Solutions Containing Diethylenetriamine and Piperazine in a Rotating Packed Bed", *Int. J. Greenh. Gas Cont.*, 9, 136-147 (2012).

[4] Thiels, M., Wong, D.S.H., Yu, C.H., Kang, J.L., Jang, S.S., Tan, C.S. Modeling and Design of Carbon Dioxide Absorption in Rotating Packed Bed. *11th IFAC Symposium on Dynamics and Control of Process Systems*.895-900 (2016).

[5] Zhang, Y., Chen, H., Chen, C.C., Plaza, J.M., Dugas, R., Rochelle, G.T. Rate-Based Process Modeling Study of CO_2 Capture with Aqueous Monoethanolamine Solution. *Ind. Eng. Chem. Res.*48, 9233-9246 (2009).

Keywords: Process Intensification; Rotating Packed Bed; Aspen Plus; CO_2 Chemical Absorption; Mass Transfer Coefficient; Interfacial Area

482122 Carbon Dioxide Capture By Sodium Hydroxide-Glycerol Aqueous Solution in a Rotating Packed Bed

Abstract:

Carbon dioxide capture by chemical absorption has been one of the most popular approaches. However, low solubility of CO_2 in water and high energy demand for absorbent regeneration hinder its feasibility. In this study, glycerol, a biodiesel waste by-product, was added to sodium hydroxide (NaOH) aqueous solution to relax these drawbacks, and to serve as an efficient absorbent in a rotating packed bed (RPB) for CO_2 capture. The experimental results suggested that the addition of glycerol in the NaOH aqueous solution boosted the overall CO_2 absorption efficiency to more than 90% easily. Under the high G (rotating speed) condition, the (chemical) enhancement factor could go up to more than 50. Furthermore, at same NaOH concentration, the mass transfer coefficients (K_Ga) could be doubled even when the viscous glycerol was added because of intensified centrifugal field. It is clear indicated that, in an RPB, glycerol addition in NaOH solution showed very positive effect on CO_2 capture not only for better mass transfer performance but also for the potential utilization of biodiesel by-product.

484778 Graphic Synthesis Method for Multi-Technique Integration Separation Sequences of Multitudinous Refinery Gases

Abstract:

Graphic Synthesis Method for Multi-technique Integration Separation Sequences of Multitudinous Refinery Gases

Xuehua Ruan, Hongyan Xiao, Xiaobin Jiang, Xiaoming Yan, Yan Dai, Gaohong He

School of Petroleum and Chemical Engineering, State Key Laboratory of Fine Chemicals, Dalian University of Technology, Panjin, Liaoning, P.R. China, 124221

Corresponding Author Email: hgaohong@dlut.edu.cn

In petroleum and chemical industries, considerable crude oil, i.e., 3~8 %, is converted into refinery gases and used as furnace fuels. Many valuable species in the refinery gases, e.g., H_2, C_2H_6, C_2H_4, C_3H_8 and C_4H_{10}, are depleted. Currently, this wastage is going on increasing because more and more hydro-processes are built up for refining heavy sour crude oils to high quality products. Accordingly, the effective separation of valuable species from refinery gases has been one of the most potential approach to improve crude oil utilization percentage.

Nowadays, some separation unit operations have been attempted to fractionate and reclaim refinery gases, e.g., pressure swing adsorption, glassy polymer membrane permeation and cryogenic condensation for recovering H_2, shallow condensation, gasoline absorption and rubbery polymer membrane permeation for concentrating light hydrocarbons, and distillation for further separating hydrocarbons. Nevertheless, any simplex technique cannot fractionate all the valuable components efficiently and sufficiently from multitudinous refinery gases. It is necessary to integrate various techniques together as the hybrid processes to achieve the recovery task. In this instance, a shortcut method is established in this work to construct the multi-technique integration processes, in which highly efficient separation unit operations are coupled together within an optimal sequence to improve efficiency.

This shortcut synthesis method is a graphic means based on expressing separation units by vectors in composition space. Three supporting technical measures are summarized:

1. After ranking components with their separation characteristics and belongings in products, refinery gases are reduced to be pseudo-ternary mixtures of H_2, light hydrocarbons and fuel gases, meanwhile the composition space is visualized into triangular coordinate system.

2. Based on the efficiency analysis, dominant feed ranges for adsorption, condensation, glassy and rubbery membrane separation are determined, in which the represented technique could be more efficient than the others for the

composition matched refinery gases.

3. Separation operations, usually acting with the selectivity unique to a certain species group, are estimated quickly by the shortcut calculations and then represented with vectors in the triangular coordinate system.

The shortcut synthesis method and its application for construct multi-technique integration systems are sketched out in Figure 1.

Figure 1. Graphic synthesis of a multi-input cascade coupling separation sequence.

The synthesis is launched by positioning the feedstock in the triangular coordinate system according to the composition. According to the location of the feed stream, the separation unit suitable to deal with this stream can be determined facilely by matching with the dominant ranges. Afterwards, the shortcut calculations are conducted by vector analysis to orientate the residue from a separation unit and select the next separation unit.

Thanks to running the separation unit operations in their preferred ranges and avoiding the stream mixing with great composition difference, the multi-technique systems designed by the expert synthesis approach can reclaim refinery gases sufficiently and efficiently. For multitudinous refinery gases in large-scale petrochemical complex, the overall recovery ratio for valuable species is higher than 95 %, while the specific energy consumption is usually lower than 0.50 kWh/Nm^3 refinery gases. The integration processes designed by this means have been successfully industrialized in Sinopec and PetroChina.

Acknowledgement:

The authors are grateful to the financial support from National Natural Science Foundation of China (No. 21606035, U1663223, 21527812), Education Department of the Liaoning Province of China (LT2015007), and Changjiang Scholars Program (T2012049).

497879 Pinducer Analysis for the Design of a Nucleation Subsystem for Continuous Slug-Flow Crystallization

Abstract:

The reliable and convenient control of crystallization processes has received increased academic and industrial interest. Ultrasonication has been used to aid crystallization, such as accelerating nucleation, reducing crystal size, reducing aggregation, or changing polymorphic ratios [1]–[3]. Most bench-scale continuous sonocrystallization studies are based on configurations involving an ultrasonic probe directly inserted into the solution/slurry or a tube placed within a bath. An alternative multiphase millifluidic tubular crystallizer design combined the advantages of using a probe (focused energy) and a bath (indirect ultrasonication) for continuous nucleation, with the nuclei size tuning by changing the ultrasonic amplitude [4].

To better understand and improve the design of a continuous crystallization process, we employ a needle-tip pressure transducer ("pinducer") to measure and analyze the "primary effect" of ultrasonication, which is the generation of pressure waves (the term "primary effect" is used as in Ref. [5]). This analysis is used to gain understanding on the crystallization outcome (the "secondary effect"). This separation of the analysis of the primary effect from the secondary effect provides greater insight and understanding than past studies that only reported the crystallization results of applying ultrasonication. The pressure waves of ultrasonication are measured for different equipment configurations, and different positions and orientations of the pinducer relative to the ultrasonication probe. The quantification of the pressure waves as a function of frequency provides insights into bubble dynamics and is used to guide the design of the nucleation subsystem and its positioning relative to other subsystems of the continuous crystallization apparatus. The analysis of the spatial localization of energy dissipation that occurs during ultrasonication informs the design of the focused indirect ultrasonication zone to enhance reproducibility and the control of nuclei size distribution [6]. Although the focus of this study was on a particular continuous crystallizer design, the same procedures for the analysis of physical phenomena via pinducers employed in this study could be applied to provide insights and guidance into other ultrasonication-facilitated crystallization designs.

References

[1] J. R. G. Sander, B. W. Zeiger, and K. S. Suslick, "Sonocrystallization and sonofragmentation," *Ultrason. Sonochem.*, vol. 21, no. 6, pp. 1908–1915, 2014.

[2] O. Narducci, A. G. Jones, and E. Kougoulos, "Continuous crystallization of adipic acid with ultrasound," *Chem. Eng. Sci.*, vol. 66, no. 6, pp. 1069–1076, 2011.

[3] R. J. P. Eder, S. Schrank, M. O. Besenhard, E. Roblegg, H. Gruber-Woelfler, and J. G. Khinast, "Continuous sonocrystallization of acetylsalicylic acid (ASA): Control of crystal size," *Cryst. Growth Des.*, vol. 12, no. 10, pp. 4733–4738, 2012.

[4] M. Jiang, C. D. Papageorgiou, J. Waetzig, A. Hardy, M. Langston, and R. D. Braatz, "Indirect ultrasonication in continuous slug-flow crystallization," *Cryst. Growth Des.*, vol. 15, no. 5, pp. 2486–2492, 2015.

[5] V. S. Sutkar and P. R. Gogate, "Design aspects of sonochemical reactors: Techniques for understanding cavitational activity distribution and effect of operating parameters," *Chem. Eng. J.*, vol. 155, no. 1–2, pp. 26–36, 2009.

[6] M. Jiang, C. Gu, and R. D. Braatz., "Analysis of focused indirect ultrasound via high-speed spatially localized pressure sensing and its consequences on nucleation," submitted.

499788 Understanding Metastable Zone Width without Nucleation Rate

Abstract:

Traditionally, the metastable zone width (MZW) in solution crystallization is thought to originate from the path-dependence of the steady-state nucleation rate of a solute. However, new experimental findings appear to go against this popular notion. In this work, we propose an alternate viewpoint to understand the MZW, and show that MZW may not necessarily be a function of the nucleation rate as conventionally understood.

Our work presents experimental data on the MZWs of various solutes and show that these data can be explained without invoking the concept of a nucleation rate. We use data from our own work as well as that from the recent literature. The applicability of the new concept of MZW to both evaporative and cooling crystallizations is shown. The stochastic view point of nucleation is contrasted against a more â€˜deterministicâ€™ view point of the classical nucleation theory. The pitfalls of obtaining nucleation kinetics from MZW experiments are discussed. Both induction time and MZW measurements are recommended as complementary experimental techniques to characterize the kinetics of nucleation in solution crystallization.

499208 Molecular Interactions between Nanocellulose and Crystallizing Pharmaceuticals

Abstract:

Cellulose is a highly abundant natural polymer that, being a plant-based material, exhibits renewability and biodegradability. It is frequently used in the form of microcrystalline cellulose as a filler and binder in pharmaceutical tablets. A nanoscale form of cellulose can be prepared by removing much of the amorphous cellulose material and leaving small crystalline particles with a large concentration of hydrogen bonding groups on the surface. Due to its strong hydrogen bonding tendency and the presence of charges on the nanocellulose surface, it has the potential to bind with drug molecules and act as a release inhibitor, a stabilizer or a site for heterogeneous crystallization of the drug itself. Beyond hydrogen bonding and charge, however, the nanocellulose surface can be readily altered to display a variety of functional groups, leading to flexibility for rational design of nanocellulose-drug composites. We use nanocellulose functionalization, such as carboxylation and sulfonation, to design composites containing nanocellulose and crystallized pharmaceutical products, with an aim towards using the molecular interactions between nanocellulose and drugs to direct formation of desired polymorphs and crystal sizes. The high surface area of the nanocellulose provides a high concentration of interaction sites and the small size of the nanocellulose excipient particles has interesting performance in promoting assembly and packing of the composite in drug products such as compressed tablets and dense free standing films.

500635 A New Model for Solubility Prediction to Guide Solvent Selection for Process Development

Abstract:

Solvent selection to obtain adequate yield and impurity rejection is a critical activity in the development of crystallization processes for pharmaceutical molecules. This activity requires the determination of solubilities for a solute molecule (moreover a polymorph of a solute molecule) in a variety of different solvents and solvent combinations. This is referred to as solvent screening. There are roughly fifty different solvents commonly used in synthetic chemical manufacturing. Although experimentally determining the solubility of a single solute in each of these solvents is an approachable (be it impractical) problem, experimental screening becomes unapproachable for mixtures of two or more solvents. Conventionally, intuition has driven the reduction of the complete set of binary blends to a more tractable set of mixed solvent systems on a case-by-case basis. As such the majority of solvent combinations remaining unexplored for a given molecule. A predictive model was developed to better facilitate the exploration of single solvent systems and blends.

This model was developed starting from a mechanistic perspective for crystal growth and dissolution, which is most applicable at temperatures far from the melt. The model can be implemented based on either the regression of a limited set of solubility data and the subsequent prediction of solute characteristics (magnitude of polar, dispersive and hydrogen bonding interactions, size) or using molecular modelling to approximate these characteristics directly. The regression approach has proven to be efficient (requiring minutes) and accurate for single solvent systems. This model has been extended to mixtures of two solvents, and coupled with crystallization process design calculations to predict yield, and provide a rough process outline to the end user. The intention being that once a â€œgoodâ€ system has been found (determined using the model) it can be verified experimentally. This model has been applied to several systems (~20 solutes and 200 solute/solvent combinations) including the modelling of a solubility data set for paracetamol and caffeine collected within the Enabling Technologies Consortium.

Heterogeneous Nucleation of Small Molecule Crystals in the Presence of Nanoparticles

Lucrèce H. Nicoud, Leia M. Dwyer, Allan S. Myerson*

Chemical Engineering Department, Massachusetts Institute of Technology, Cambridge, MA, U. S. A.
***Corresponding author**: myerson@mit.edu

Abstract
In this study, we evaluated the possibility to use nanoparticles as a means to obtain fundamental insights on the heterogeneous nucleation of organic molecules. To do so, we investigated the crystallization of diphenhydramine hydrochloride in ethanol in the presence of silica coated magnetic nanoparticles. We performed crystallization experiments at two different scales, namely in 1 mL vials and in 100 mL vessels. At both scales, crystallization was induced by two different ways, either by cooling or by anti-solvent (heptane) addition. It was found that the presence of nanoparticles promotes crystallization, in particular at low supersaturation. However, the high variability of induction times and the aggregation of the nanoparticles makes it difficult to extract quantitative information regarding the impact of the nanoparticles on the crystallization kinetics.

Introduction

Surfaces are ubiquitous in crystallizers, in particular due to the walls of the vessel, the impeller or the presence of dust. Surfaces generally lower the energy barrier to nucleation, so that heterogeneous nucleation commonly dominates over primary homogeneous nucleation in most laboratory or plant settings. However, the physical mechanisms underlying heterogeneous nucleation remain poorly understood. Because the early stages of solution crystallization play a decisive role in determining the final crystal properties, such as the crystal size, shape and polymorphism, a fundamental understanding of nucleation is a key requisite to improve control over crystallization processes [1, 2].

Experimental studies aiming at investigating crystallization in the presence of surfaces employed various types of templates including crystals [3], polymers [4] and self-assembled monolayers (SAMs) [5, 6]. The latter are highly ordered and oriented molecular assemblies that form spontaneously by the adsorption of a surfactant on a solid surface [7]. SAMs have been reported to influence crystal size [8], enantiopurity [9], morphology [10], and polymorphism [11].

In this work, we evaluate the possibility to use nanoparticles to investigate the mechanisms of heterogeneous nucleation. Nanoparticles seem particularly suitable for this purpose due to their large surface area and tunable properties. Nanoparticles are indeed available in a wide range of sizes (from few nanometers to few micrometers), shapes (most often spheres, but also cubes, rods and nanostars) and functionalities (mainly thanks to the large portfolio of thiols and silanes, which allow the surface functionalization of gold and silica nanoparticles, respectively). One may therefore envision performing studies with various types of well characterized nanoparticles in an attempt to understand how the uncontrolled presence of foreign objects such as dust may impact nucleation in regular crystallizers. Nanoparticles have been shown to promote the formation of protein crystals at the microliter scale [12, 13, 14], but their quantitative impact on the crystallization kinetics still needs to be elucidated.

In this context, we investigate the impact of silica coated magnetic nanoparticles on the crystallization behaviour of diphenhydramine hydrochloride (DPH), which is a commercial pharmaceutical ingredient. Silica coated magnetic nanoparticles were selected as they combine two main advantages: *(i)* the magnetic core allows performing magnetic decantation, which is very practical for several preparation steps (e.g. to wash, concentrate) and for a potential reuse after crystallization, *(ii)* the silica shell allows functionalizing the surface with a wide variety of silanes. We start our study with experiments performed at the milliliter scale and then increase the crystallization volume to 100 mL. At both scales we induce crystallization by two different ways, namely anti-solvent (heptane) addition and cooling.

1. Materials and Methods

1.1 Materials

1.1.1 Suppliers

DPH was purchased from BeanTown Chemical at a purity $\geq 99\%$. Anhydrous heptane 99% was purchased from Sigma-Aldrich and 200 proof ethanol from Deacon Labs. Silica coated magnetic nanoparticles with a total diameter of 200 nm were purchased from Spherotech Inc (catalog number SIM-025-10H). N-decyltriethoxysilane was purchased from Gelest (catalog number SID2665.0) to perform functionalization of the silica surface of the nanoparticles with decane.

1.1.2 DPH solution preparation

DPH solutions were prepared by dissolving DPH in ethanol at 35 °C under magnetic stirring. The solutions were then filtered with a PTFE syringe filter prior to performing crystallization experiments. Filters with a 0.2 µm and 0.45 µm cut-off were used for crystallization studies at the small scale (1 mL) and large scale (100 mL), respectively. DPH solutions were prepared at a concentration of 315.8 g/L^{EtOH} so as to reach 300 g/L^{EtOH} after addition of the nanoparticle solution (95 % DPH solution and 5 % nanoparticle solution in volume). For control experiments performed in the absence of nanoparticles, the corresponding quantity of pure ethanol was added to reach the same final DPH concentration. In the following, the nanoparticle concentrations are given in grams of nanoparticles per liter of DPH solution (i.e. DPH dissolved in ethanol before a potential addition of heptane). The units are abbreviated as g/L^{DPH} for the sake of simplicity.

1.1.3 Nanoparticles solution preparation

Before usage, the nanoparticles (Nps) were washed, first with water and then with ethanol. To do so, a given volume of nanoparticle solution was placed in a tared vial on top of a magnet. After magnetic decantation, the solvent was removed and a volume of water equal to twice the initial volume of the nanoparticle solution was placed on top of the nanoparticles. The nanoparticles were then redispersed by sonicating the solution for 2 min (Branson 3510 ultrasonicator bath). This wash procedure was then repeated with ethanol. After magnetic decantation and solvent removal, the last traces of ethanol were eliminated by drying the nanoparticles in a vacuum oven at around 60 °C. The vial was then weighted to determine the mass of nanoparticles and an appropriate volume of ethanol was added to reach a target nanoparticle concentration. The nanoparticles were redispered by placing the vial in the bath sonicator for 10 min.

It was checked with dynamic light scattering (DLS) that the nanoparticles are properly dispersed after sonication. Measurements were performed with a DynaPro NanoStar instrument (Wyatt Technology) at a 90° angle using a laser whose wavelength was 658 nm.

1.2 Nanoparticles functionalization

Before functionalization, the silica coated magnetic nanoparticles were washed as described in the previous section. A 5 g/L solution of nanoparticles in a 95/5 % v/v ethanol/water mixture was prepared. The water was acidified with acetic acid at a concentration of 10^{-2} M. The silane was added to the solution to reach a 2 % v/v concentration. The solution was sonicated in the bath sonicator during 2 min. The functionalization reaction was then performed at 25 °C under shaking at 500 rpm during 24 h. The temperature was subsequently increased to 70 °C during 2 h. The solvent was removed by magnetic decantation and the nanoparticles were dried in the vacuum oven at 90 °C during 4 h. Finally, the nanoparticles were washed twice with ethanol using sonication to redisperse the nanoparticles at each wash step.

The extend of functionalization was assessed by Fourier transform infrared (FTIR) spectroscopy (IdentifyIR Smiths). To do so, few microliters of a concentrated nanoparticle suspension were deposited on the sample holder and 256 scans were recorded after ethanol evaporation.

1.3 Solubility

Saturated solutions of DPH were prepared by magnetically stirring an excess of DPH crystals in contact with a selected solvent (either pure ethanol or a mixture of ethanol and heptane) during 24 h. The temperature was maintained at a set value thanks to a water bath. After 24 h, the solution was filtered with a 0.2 µm PTFE syringe filter and the filtrate was placed in tared vials. The vials were weighted immediately after introducing the filtrate and after evaporation of the solvent. The solvent was first evaporated under the hood and the last traces were removed by drying the crystals in a vacuum oven at around 60 °C. Once the mass of the samples was constant over time, the solubility was computed as the mass of crystals divided by the mass of solvent that had evaporated.

1.4 Crystallization likelihood (1 mL)

The tendency of a solution to crystallize was assessed by preparing at least 5 samples per investigated condition and visually inspecting after 24 h whether crystals formed. To do so, 1 mL samples were placed in glass vials (ThermoScientific, C4000-1W) incubated in a thermoshaker (MultiTherm, heat/cool, Benchmark Scientific) at a selected temperature and under 500 rpm shaking. For anti-solvent crystallization, the temperature was set to 25 °C. The formed crystals were observed with an optical microscope (Nikon Eclipse ME 600).

1.5 Crystallization kinetics (100 mL)

Batch crystallization kinetic experiments were performed using an EasyMax 102 worksation (Mettler Toledo). The solution was stirred at 250 rpm with a pitched-blade impeller (Alloy-C22, 38 mm diameter, downwards). The total solution volume was 100 mL. For anti-solvent crystallization, the reactor temperature was maintained at 25 °C and heptane was poured at once on top of the DPH solution to reach a heptane concentration of 33 % v/v^{tot}. For cooling crystallization, the reactor temperature was decreased from 25 °C to 15 °C at a rate of 1 °C/min.

The crystallization kinetics was monitored by focused beam reflectance measurements (FBRM) using a lasentec probe (S400Q).

In few experiments, particle vision measurements (PVM) were performed to visually observe the nanoparticle aggregates and the crystals (ParticleView V19, Mettler Toledo).

2. Results and Discussion

2.1 Choice of nanoparticles

The core-shell nanoparticles selected for this study combine the advantages of a practical handling and of a possible functionalization thanks to the magnetic core and to the silica surface, respectively. Indeed, the magnetic behaviour of the nanoparticles is very convenient for all the preparation steps such as washing, concentration and surface functionalization. The silica shell instead prevents the nanoparticles from extensive aggregation and allows the grafting of a wide variety of silanes, which then confer various surface properties to the nanoparticles.

A transmission electron microscopy picture of the nanoparticles is shown in Figure 1, where it is seen that the magnetic core is around 160 nm and the silica shell approximately 20 nm thick. The total diameter of 200 nm was confirmed by DLS analysis.

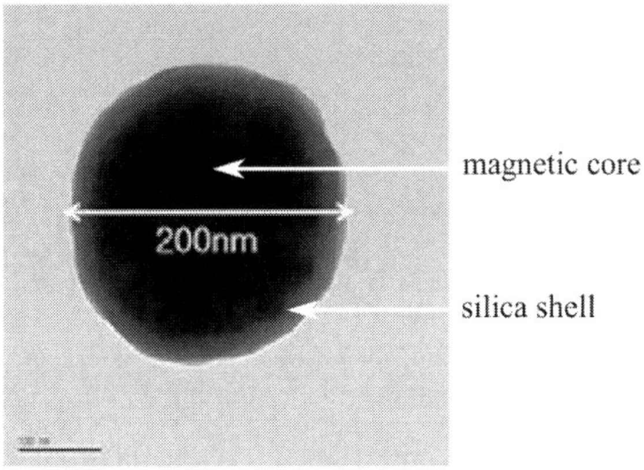

Figure 1. Transmission electron microscopy of the silica coated magnetic nanoparticles. Provided by Spherotech Inc.

To give an order of magnitude, a 1 g/L solution of such nanoparticles corresponds to a concentration of 6×10^{13} nanoparticles per liter and to a volume fraction of 0.03 %. Therefore, even a quite dilute solution of nanoparticles provides a large number of foreign objects, which are as many potential nucleation sites.

2.2 Nanoparticle functionalization

The silica shell of the nanoparticles allows functionalizing the surface with various silanes. As an example, the FTIR spectrum of nanoparticles functionalized with n-decyltriethoxysilane, which provides a functionalization with decane molecules, is shown in Figure 2. As a comparison, the spectra of the non-functionalized (bare) nanoparticles and of the silane are also shown. In contrast to the bare nanoparticles, the functionalized nanoparticles exhibit three small additional peaks in the range from 2800 to 3000 cm^{-1}. These peaks coincide with those of the silane, thus suggesting that the functionalization reaction was successful. The low intensity of the peaks corresponding to the grafted decane as compared to the pure silane can be attributed to the small mass of the surface groups with respect to the bulk of the nanoparticles.

2.3 Solubility

Figures 3(a) and (b) show experimental data of DPH solubility as a function of the heptane volumetric fraction and temper-

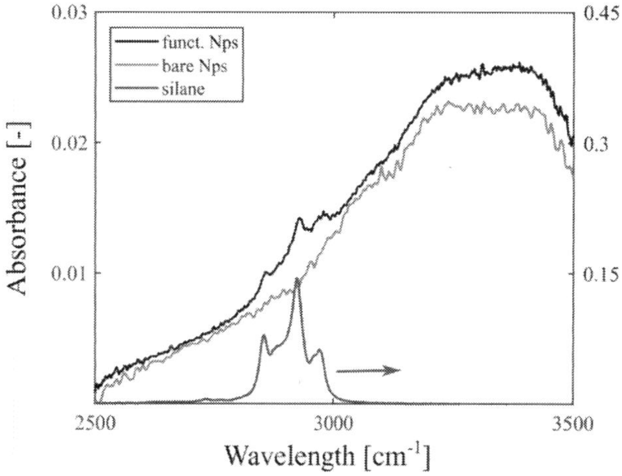

Figure 2. FTIR spectrum of nanoparticles functionalized with decane as compared to the spectra of the bare nanoparticles and of the corresponding silane (whose intensity is read on the right axis).

ature, respectively. The solid line in Figure 3(a) represents the dilution line upon heptane addition starting from a DPH solution at 300 mg/mL^{EtOH} = 382.2 mg/g^{EtOH}. It is observed that crystallization is expected to occur from a thermodynamic point of view at heptane concentrations above approximately 10 % in volume. In the case of cooling, the DPH concentration is kept constant, as indicated by the flat solid line in Figure 3(b). It is observed that crystallization is expected to occur at temperatures roughly below 23 °C.

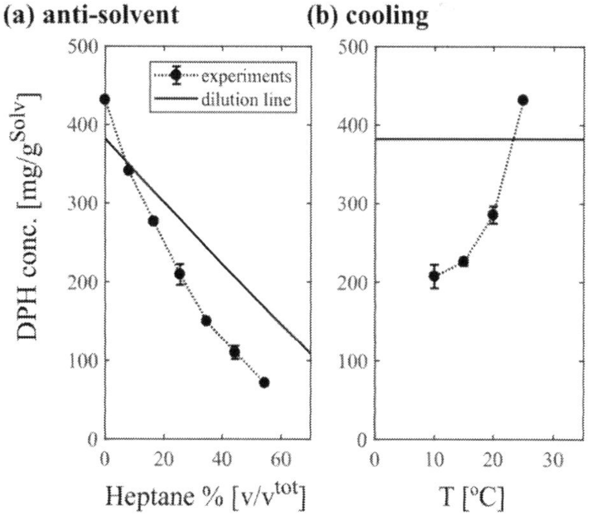

Figure 3. DPH solubility as a function of (a) heptane volumetric fraction and (b) temperature. Symbols correspond to experimental data obtained by gravimetry, whereas solid lines indicate the operating conditions, i.e. a DPH solution at 300 mg/mL^{EtOH} = 382.2 mg/g^{EtOH} diluted with heptane or cooled down.

2.4 Crystallization likelihood (1 mL)

2.4.1 Non-functionalized nanoparticles

In a first set of experiments, the impact of the nanoparticles (Nps) on the crystallization behaviour of DPH was investigated at the milliliter scale in glass vials. For each condition under investigation, at least five samples were prepared and visually inspected after 24 h of shaking at a target temperature to determine whether crystals formed during the experiment time frame. The results for anti-solvent crystallization and cooling crystallization are shown in Figure 4(a) and (b) respectively.

(a) anti-solvent

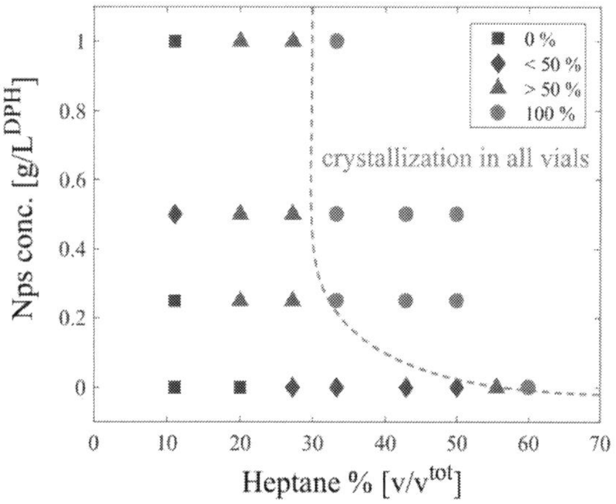

(b) cooling

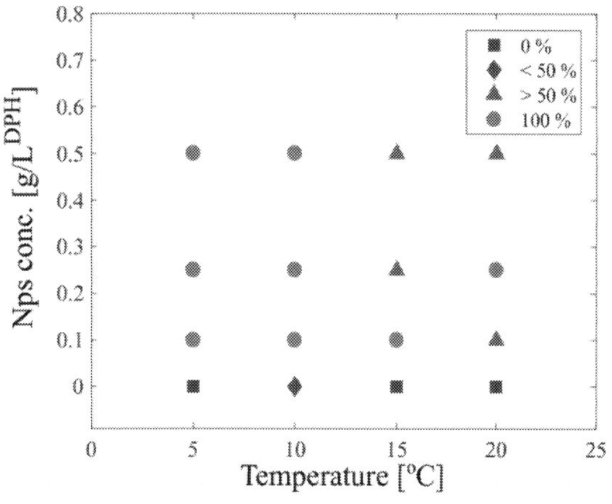

Figure 4. Crystallization likelihood at the milliliter scale in (a) anti-solvent crystallization and (b) cooling crystallization. For each condition, at least five samples were prepared and visually inspected after 24 h of shaking to determine whether crystals formed. The percentages in the caption indicate the fraction of samples that crystallized. In the case of anti-solvent crystallization, the temperature was set to 25 °C. Non-functionalized nanoparticles were used.

Let us first consider the samples without the nanoparticles, i.e. the row of symbols at the bottom of both diagrams. According to the solubility data shown in Figure 3, the selected conditions of heptane concentration and temperature were expected to induce crystallization in all samples. However, it is observed in Figure 4 that most of the samples did not crystallize within 24 h. For example, none of the samples whose heptane concentration was below 20 % crystallized (blue squares). At heptane concentrations between 25 and 55 %, only some of the samples crystallized (purple symbols) and it was necessary to reach 60 % heptane so as to observe crystallization in all the vials (red circles). To verify that there was no mistake in sample preparation, a seed was added in the samples that did not crystallize within 24 h. This resulted in rapid crystallization, thus confirming that all the solutions were supersaturated. This means that, although all the samples were expected to crystallize from a thermodynamic point of view, a significant kinetic barrier prevented crystallization to occur during the time frame of the experiment.

Interestingly, it is observed in Figure 4 that the addition of non-functionalized nanoparticles resulted in the crystallization of many more samples as compared to the situation without nanoparticles. For instance, all the samples at 33 % heptane crystallized in the presence of nanoparticles, whereas less than half crystallized in the absence of nanoparticles. Similarly, none of the samples at 15 °C crystallized in the absence of nanoparticles whereas the vast majority of the samples with the nanoparticles crystallized.

Overall, these data show that nanoparticles promote crystal formation, in particular at low supersaturation values, i.e. in the metastable zone.

2.4.2 Functionalized nanoparticles

Following the idea of a previous study [15], crystallization experiments were performed with nanoparticles functionalized with decane, which is an anti-solvent for DPH. As a reference, the addition of 50 % decane in volume reduces the solubility of DPH from 432 to 137 mg/g^{Solv}. The functionalization of the silica coated magnetic nanoparticles with decane was discussed in Section 2.2.

According to the original work, the addition of nanoparticles functionalized with anti-solvent molecules was able to induce the crystallization of slightly under-saturated solutions of sodium chloride, D-mannitol and fenofibrate. Here, we investigated whether the addition of decane functionalized nanoparticles could induce the crystallization of DPH solutions at 25 °C with heptane concentrations varying from 0 to 10 % in volume. According to the solubility data shown in Figure 3(a), such conditions lead from slightly under-saturated up to saturated solutions. Several nanoparticle concentrations ranging from 0.25 g/L^{DPH} to 1 g/L^{DPH} were tested and several repetitions were performed for each condition. However, none of the samples crystallized within 24 h.

After observing that decane functionalized nanoparticles were not able to induce the crystallization of DPH solutions in the range 0-10 % heptane, larger heptane volume fractions were investigated. As in the case of the bare nanoparticles, crystallization was observed in all the samples whose heptane concentration was above 33 %. Nevertheless, one noteworthy difference was observed in terms of crystal color. Figures 5(a) and (b) show pictures of the crystals formed with non-functionalized nanoparticles and with decane functionalized nanoparticles, respectively. The top pictures were taken with a regular camera, whereas the bottom pictures were taken with an optical microscope. In both samples, the heptane concentration was 43 % v/v^{tot} and the nanoparticle concentration 0.25 g/L^{DPH}. Due to the red/brown colour of the nanoparticles, it is possible to visually assess the entrapment of nanoparticles in the crystal lattice. It is seen in Figure 5 that the crystals formed in the presence of the bare nanoparticles are darker than those formed in the presence of the nanoparticles functionalized with decane. This might suggest that the surface functionalization with decane reduced nanoparticle entrapment.

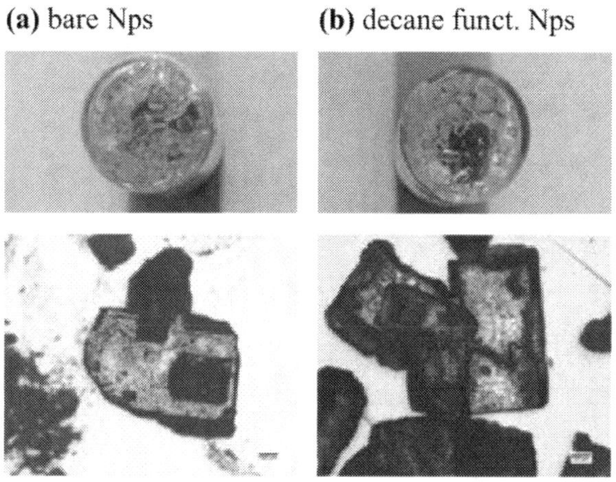

Figure 5. Pictures of DPH crystals formed in the presence of (a) bare and (b) decane functionalized nanoparticles. The heptane concentration was 43 % v/v^{tot} and the nanoparticle concentration 0.25 g/L^{DPH}. After crystallization, the crystals were washed with heptane. The top pictures were taken with a regular camera, whereas the bottom pictures were taken with an optical microscope.

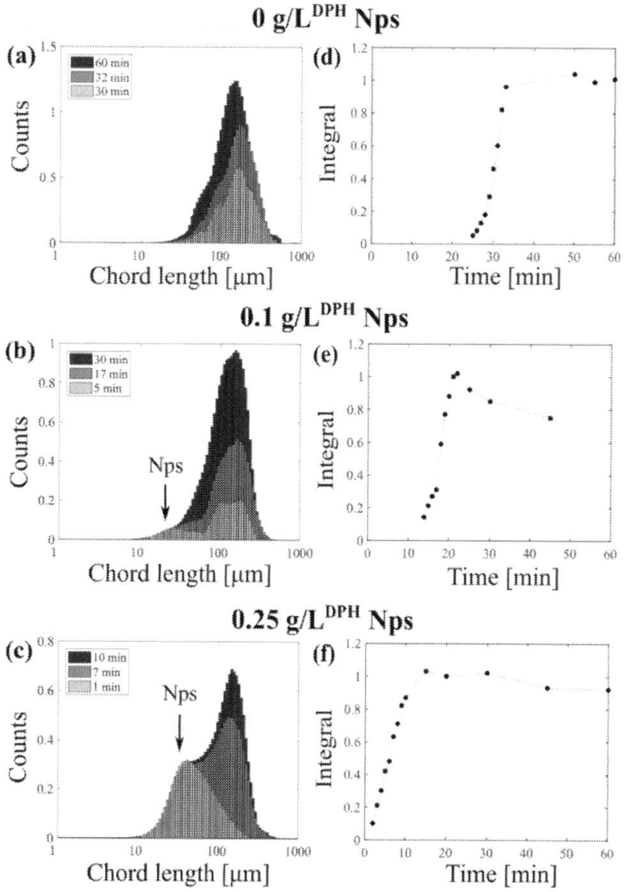

Figure 6. (a-c) Cube weighted chord length distributions monitored by FBRM at different time points and nanoparticles concentrations, as indicated in the legend. (d-f) Integral of the chord length distribution as a function of time for the same three nanoparticle concentrations. The peak corresponding to the nanoparticles (centered at around 20 µm) was substracted before computing the integral and the curves were normalized between 0 and 1. Bare nanoparticles were used.

2.5 Crystallization kinetics (100 mL)

After investigating the crystallization likelihood at the milliliter scale, batch crystallization kinetics were performed in a 100 mL stirred vessel equipped with a FBRM probe. For anti-solvent crystallization, a heptane concentration of 33 % v/v^{tot} was selected, whereas for cooling crystallization a temperature of 15 °C was chosen.

Figure 6(a) shows cube weighted chord length distributions recorded in the absence of nanoparticles at three different time points during anti-solvent crystallization. It is observed that the crystal size is in the order of 100 µm. The number of crystals increases with time, but their size remains relatively constant. This indicates that crystal growth is significantly faster than crystal nucleation.

Figures 6(b) and (c) show the chord length distributions at three different time points in the presence of 0.1 g/L^{DPH} and 0.25 g/L^{DPH} bare silica coated magnetic nanoparticles, respectively. As compared to the situation without the nanoparticles, an additional peak centered at around 20 µm is observed. This peak does not correspond to crystals but is due to the presence of nanoparticles. This was evidenced by an experiment performed below the saturation (25 °C, no heptane), during which a similar peak was observed, as further discussed in Section 2.6.

In order to evaluate the mass of crystals formed as a function of time, the peak corresponding to the nanoparticles was substracted and the integral of the cube weighted chord length distribution was computed. The results are presented in Figures 6(d), (e) and (f) for 0, 0.1 and 0.25 g/L^{DPH} nanoparticles, respectively. In the absence of nanoparticles, a lag time of approximately 25 min is observed before crystallization starts. Then, the mass of crystals increases sharply until it reaches a plateau. In few experiments, a slight decrease in the number of counts was observed at large reaction times (see Figures 6(e)). This might be due to crystals intermittently sticked to the FBRM probe falling off the probe window.

Several repetitions of these anti-solvent experiments were performed and the results are reported in Figure 7(a) in terms of lag times. It is seen that the average lag time decreases with the concentration of nanoparticles.

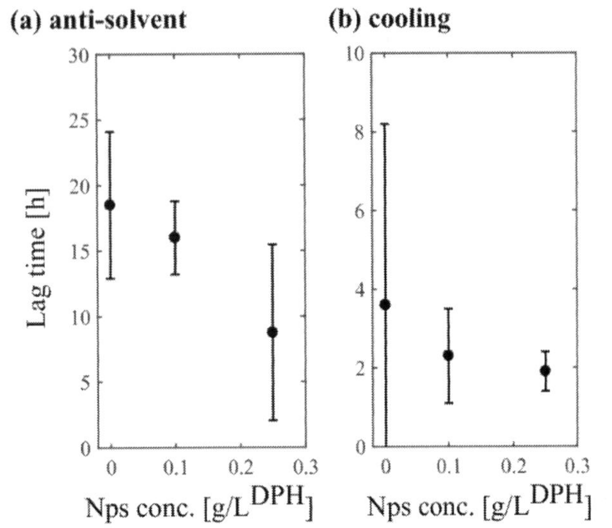

Figure 7. Lag time as a function of the concentration of nanoparticles in the case of (a) anti-solvent and (b) cooling crystallization. Non-functionalized nanoparticles were used.

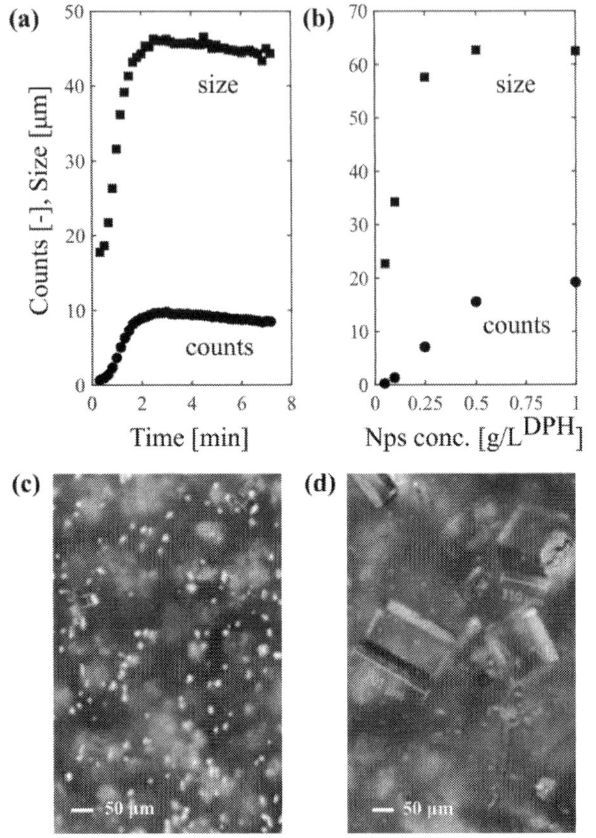

Figure 8. (a) Total number of counts (circles) and average size (squares) of the cube weighted distribution measured with FBRM as a function of time in a 300 g/L^{EtOH} DPH solution at 25 °C in the presence of 0.25 g/L^{DPH} of bare nanoparticles. (b) Total number of counts (circles) and average size (squares) of the cube weighted distribution measured with FBRM at the early stages of an anti-solvent crystallization experiment. (c) PVM images recorded after 8 min of the experiment reported in panel (a). (d) PVM images recorded after 3 h of cooling at 15 °C following the experiment reported in panel (a).

Similar studies were carried out in the case of cooling crystallization. Chord length distributions comparable to those obtained during anti-solvent crystallization were obtained. The lag time is shown as a function of the nanoparticle concentration in Figure 7. Once more, it is seen that the average lag time decreases upon nanoparticle addition. However, the poor reproducibility makes it difficult to properly quantify the impact of the nanoparticles.

2.6 Nanoparticle aggregation

In this section, we investigate in more details the aggregation behaviour of the nanoparticles. Figure 8(a) shows the total number of counts (circles) and the average size (squares) of the cube weighted distribution measured with FBRM in an experiment performed in the presence of 0.25 g/L^{DPH} of bare nanoparticles below the supersaturation, i.e. where no crystals can form. In this experiment, the temperature was maintained at 25 °C and no heptane was added. It is seen that both the number of counts and the average size increase during the first two minutes and then become almost constant with time. Note that an average size close to 45 μm is reached, which is more than two order of magnitude larger than the initial nanoparticle diameter (200 nm as measured by DLS).

The nanoparticle aggregates were visualized by PVM and a representative picture recorded after 8 min is shown in Figure 8(c). It is seen that particulates in the order of few tens of microns are observed, in agreement with the FBRM data shown in Figure 8(a).

After performing particle vision measurements, the temperature was decreased to 15 °C so as to induce crystal formation. Figure 8(d) shows a representative picture recorded after 3 h of cooling, where both nanoparticle aggregates and crystals are observed. The crystals appear much larger than the nanoparticle aggregates, in line with the bimodal distribution shown by the FBRM data in Figure 6.

The impact of the concentration of nanoparticles on their aggregation behaviour was studied under supersaturated conditions in the case of anti-solvent crystallization. It was already observed in the context of Figure 6 that an increase in the nanoparticle concentration from 0.1 to 0.25 g/L^{DPH} leads to an increase in the height of the peak corresponding to the nanoparticles. Figure 8(b) shows how the total number of counts and the average size related to the peak of the nanoparticles increase as a function of the nanoparticle concentration. A sharp rise in the size of the nanoparticle aggreates is observed at nanoparticle concentrations below

0.25 g/L^{DPH}. Above that value, the nanoparticle aggregates reach an average size of around 60 μm.

Nanoparticle aggregation represents a major issue in this study. Indeed, aggregation reduces both the the surface area and the number of particles in solution. Assuming that the aggregates are compact spheres with a uniform size distribution, the surface area and the number of particles are inversely proportional to the aggregate diameter and to the cube of their diameter, respectively. Considering that in the present case aggregates are approximately 100 times larger than the primary particles, aggregation roughly reduces the surface area by a factor 100 and the number of particles by 10^6.

Besides reducing the surface area and the number of particles, aggregation also complicates the interpretation of experimental data. Indeed, aggregates are non regular structures, which are difficult to characterize with accuracy in terms of shape and size distribution. In addition, aggregation is affected by a number of parameters, such as time and nanoparticle concentration as shown in Figure 8, but also potentially temperature, solvent type, solute concentration, nanoparticle functionalization, etc. It is therefore difficult to rationalize the quantitative impact of nanoparticles on the crystallization kinetics monitored under various experimental conditions. For instance, the observation that the aggregate size increases with nanoparticle concentration prevents a detailed analysis of the impact of the nanoparticles concentration on the kinetics of crystallization. Similarly, studies aiming at investigating different surface functionalization would require the decoupling between the impact of surface functionalization on the aggregation of nanoparticles and on the kinetics of crystallization.

3. Conclusion

In this study, we assessed the possibility to use nanoparticles to gain fundamental insights into heterogeneous nucleation mechanisms. This was motivated by the fact that nanoparticles offer a large surface area and are tunable in terms of various properties, such as size, shape and surface chemistry.

At the milliliter scale, it was shown that silica coated magnetic nanoparticles promote the crystallization of diphenhydramine, which was induced either by heptane addition or cooling. Then, the crystallization kinetics was monitored with FBRM in 100 mL stirred reaction vessels and it was found that the presence of nanoparticles reduces the lag time before crystallization occurs. However, the effect of the nanoparticles seems less pronounced than at the milliliter scale and the large variability in the lag time measurements makes it difficult to draw solid conclusions.

Finally, it was shown that the nanoparticles significantly aggregate under the considered experimental conditions, which reduces the surface area and the number of particles in solution. This complicates the interpretation of experimental results and hinders the quantitative analysis of the impact of the nanoparticles on the crystallization kinetics, at least for the system under investigation.

Acknowledgments

Financial support from the Swiss National Foundation is gratefully acknowledged (grant P2EZP2_168909). The support of DARPA for a portion of this project is gratefully acknowledged.

References

[1] A. S. Myerson. *Hanbook of industrial crystallization, 2nd Edition*. Butterworth-Heinemann, 2002.

[2] P. G. Vekilov. Nucleation. *Cryst Growth Des*, 10(12):5007–5019, 2010.

[3] M. D. Ward. Bulk crystals to surfaces: Combining x-ray diffraction and atomic force microscopy to probe the structure and formation of crystal interfaces. *Chem Rev*, 101(6):1697–1725, 2001.

[4] C. P. Price, Ad. L. Grzesiak, and A. J. Matzger. Crystalline polymorph selection and discovery with polymer heteronuclei. *J Am Chem Soc*, 127(15):5512–5517, 2005.

[5] A. Singh, I. S. Lee, K. Kim, and A. S. Myerson. Crystal growth on self-assembled monolayers. *CrystEngComm*, 13(1):24–32, 2011.

[6] J. Aizenberg, A. J. Black, and G. M. Whitesides. Control of crystal nucleation by patterned self-assembled monolayers. *Nature*, 398(6727):495–498, 1999.

[7] J. C. Love, L. A. Estroff, J. K. Kriebel, R. G. Nuzzo, and G. M. Whitesides. Self-assembled monolayers of thiolates on metals as a form of nanotechnology. *Chem Rev*, 105(4):1103–1170, 2005.

[8] A. Y. Lee, I. S. Lee, S. S. Dette, J. Boerner, and A. S. Myerson. Crystallization on confined engineered surfaces: A method to control crystal size and generate different polymorphs. *J Am Chem Soc*, 127(43):14982–14983, 2005.

[9] J. Chen and A. S. Myerson. Pasteur revisited: Chiral separation by crystallization on self-assembled monolayers. *CrystEngComm*, 14(24):8326–8329, 2012.

[10] A. Y. Lee, A. Ulman, and A. S. Myerson. Crystallization of amino acids on self-assembled monolayers of rigid thiols on gold. *Langmuir*, 18(15):5886–5898, 2002.

[11] R. Hiremath, J. A. Basile, S. W. Varney, and J. A. Swift. Controlling molecular crystal polymorphism with self-assembled monolayer templates. *J Am Chem Soc*, 127(51):18321–18327, 2005.

[12] Y. Takeda and F. Mafuné. Induction of protein crystallization by platinum nanoparticles. *Chem Phys Lett*, 647(Supplement C):181–184, 2016.

[13] S. Ko, H. Y. Kim, I. Choi, and J. Choe. Gold nanoparticles as nucleation-inducing reagents for protein crystallization. *Cryst Growth Des*, 17(2):497–503, 2017.

[14] F. Hodzhaoglu, F. Kurniawan, V. Mirsky, and C. Nanev. Gold nanoparticles induce protein crystallization. *Cryst Res Technol*, 43(6):588–593, 2008.

[15] S. A. Kulkarni and A. S. Myerson. Reversible control of solubility using functionalized nanoparticles. *Chem Comm*, 53(8):1429–1432, 2017.

487832 The Influence of Solution Conditions on the Self-Assembly of Pre-Nucleation Clusters

Abstract:

The concept of pre-nucleation clusters has been established to, successfully, explain phenomena, in nucleation kinetics, that they were inconsistent with the classical nucleation theory.[1] The key idea is that pre-nucleation clusters are formed providing a comfortable niche for nuclei to be formed.[2] Phenomenological theories exist describing this two-step nucleation process.[3] However, there is a lack of qualitative understanding of the fundamental phenomena underpinning the formation of pre-nucleation clusters both in a supersaturated and an understaturated environment.

Using the models proposed in more recent works is being shown that bigger pre-nucleation clusters provides an environment giving more chances of survival to the nucleus inside it. These findings were tested experimentally using a number of industrially relevant molecules (arginine, glycine, glutamic acid). These molecules are particularly attractive as they have relatively large nucleation times. In addition the formation of pre-nucleation clusters as nucleation precursors for them has been reported in literature. Using DLS it was shown that in aqueous systems, the size of the pre-nucleation clusters scales with the supersaturation. These findings are in accordance with the findings proposed from previous works. Upon filtration, both in a supersaturated and an undersaturated regime, the pre-nucleation clusters recover their initial size in a few hours, in the absence of any agitation.

Experiments were conducted in systems containing water and ethanol to investigate the influence of solutionâ€™s chemical potential on the size of the pre-nucleation clusters. It is shown that, in both the undersaturated and the supersaturated regime, the formation of the pre-nucleation clusters follows the fundamentals of self-assembly in solution.[4] The findings suggest that the influence of the solvent can overcome that of supersaturation, for example an aqueous system containing arginine at Sâ‰ˆ0.5 (0.8 g of L-Arginine) gives bigger pre-nucleation clusters than a solution containing 30 % v/v ethanol at Sâ‰ˆ0.9.

The findings of this work verify the majority of the findings reported in literature. However, it reports different kinetics of recovery of the pre-nucleation clusters size upon filtration. It highlights the applicability of the fundamentals of self-assembly in solution. The findings suggest that in depth understanding of the self-assembly phenomena are required for the development of efficient strategies for the control of crystallization kinetics via stirring[5] and for the formulation of amorphous solid dispersions.

References

1. P.G. Vekilov. Nucleation. *Crystal Growth & Design*. **2010**, 10, 5007.
2. B. Peters. On the coupling between slow diffusion transport and barrier crossing in nucleation. *The Journal of Chemical Physics*. **2011**, 135, 044107.
3. W. Pan, A.B. Kolomeisky, P.G. Vekilov. Nucleation of ordered solid phases of proteins via a disordered high-density state: Phenomenological approach. *The Journal of Chemical Physics*. **2005**, 122, 174905.
4. J.N. Israelachvili. Intermolecular and Surface Forces. 3rd ed. Oxford, Academic Press; **2011**.
5. A. Jawor-Baczynska, B.D. Moore, H.S. Lee, A.V. McCormick, J. Sefcick. Population and size distribution of solute-rich mesospecies within mesostructured aqueous amino acid solutions. *Faraday Discussions*, **2013**, 167, 425.]
6. M. Diwan, H. Li, F. Nordstrom. Primary Nucleation Under High Shear Conditions. *AICHE Conference Proceedings*, **2015**

490516 Simulations and Experiments Validation on the Membrane Based Response Technology for Nucleation Detection and Crystallization Control

Abstract:

Simulations and Experiments validation on the Membrane based Response Technology for Nucleation Detection and Crystallization Control

Xiaobin Jiang, Wu Xiao, Dapeng Lu, Rui Zhao, Gaohong He*

State Key Laboratory of Fine Chemicals, School of Chemical Engineering, Petrochemical Energy-efficient Separation Technology Engineering Lab of Liaoning Province, Dalian University of Technology, Dalian, P. R. China

*Corresponding author: Email: hgaohong@dlut.edu.cn

Abstract

The mass transfer flux decline and the crystallization on the membrane interface is often considered as a fouling behavior [1,2]. For now, non-attention had been paid on applying this sensitive nucleation response phenomenon for the crystal nucleus detection. While, the inevitable nucleation on the membrane interface under certain operational conditions inspired us that the membrane interface could perform as the response medium of the nucleation detection. The responding mechanism is constructed base on the chain effect of ¡®nucleation on the pores of the membrane interface-transmembrane flux sharply decrease¡¯. The inflection point of the transmembrane flux confirms the nucleation timing on the membrane interface. The accurate recording of the transmembrane flux data can be utilized to calculate the solution concentration to determine the metastable zone width (MSZW), which is key data to determine the crystallizaiton operative window [3,4].

Obviously, without relying on the optical signal, membrane based nucleation response method would be easy and more meaningful in the case of opaque solutions that laser intensity response method is invalid [5]. This advantage will improve the controlling accuracy and flexibility during MSZW measurement. Moreover, with the proposed method, it is easy to enhance control over the process. The real operative conditions during crystallizaiton can be achieved in this membrane based technology. Detailed simulations and experiments on the typical binary aqueous solution systems were implemented to represent the role of T_f, T_p and A/V_0 on the supersaturation degree adjustment.

In addition, it also reported that the polymer interface with nanometer scale pores had significant impact of on inducing crystallization and polymorphism control [6,7]. The foundational work can promote the theoretical research and application of the MDC based nucleation response and crystallization control by providing the qualified microporous membrane as desired nucleation detection and response interface.

Acknowledgment

We acknowledge fnancial contribution from from National Natural Science Foundation of China (Grant No. 21527812, 21676043, U1663223), Changjiang Scholars Program (T2012049), the Fundamental Research Funds for the Central Universities (DUT16TD19) and Education Department of the Liaoning Province of China (No. LT2015007).

References

[1] Meng, S.; Ye, Y.; Mansouri, J.; Chen, V., Fouling and crystallisation behaviour of superhydrophobic nano-composite PVDF membranes in direct contact membrane distillation. *Journal of Membrane Science* **2014,** 463, 102-112.

[2] Chen, G.; Lu, Y.; Krantz, W. B.; Wang, R.; Fane, A. G., Optimization of operating conditions for a continuous membrane distillation crystallization process with zero salty water discharge. *Journal of Membrane Science* **2014,** 450, 1-11.

[3] Ulrich, J.; Jones, M. J., Industrial Crystallization. *Chemical Engineering Research and Design* **2004,** 82, (12), 1567-1570.

[4] Ulrich, J.; Strege, C., Some aspects of the importance of metastable zone width and nucleation in industrial crystallizers. *Journal of Crystal Growth* **2002,** 237, 2130-2135.

[5] Jiang, X.; Ruan, X.; Xiao, W.; Lu, D.; He, G., A novel membrane distillation response technology for nucleation detection, metastable zone width measurement and analysis. *Chemical Engineering Science* **2015,** 134, 671-680.

[6] Curcio, E.; L¨®pez-Mej¨ªas, V.; Di Profio, G.; Fontananova, E.; Drioli, E.; Trout, B. L.; Myerson, A. S., Regulating Nucleation Kinetics through Molecular Interactions at the Polymer¨CSolute Interface. *Crystal Growth & Design* **2014,** 14, (2), 678-686.

[7] Diao, Y.; Helgeson, M. E.; Siam, Z. A.; Doyle, P. S.; Myerson, A. S.; Hatton, T. A.; Trout, B. L., Nucleation under Soft Confinement: Role of Polymer¨CSolute Interactions. *Crystal Growth & Design* **2012,** 12, (1), 508-517.

495322 Global Minimization of Multicomponent Distillation Configurationsâ€™ Total Cost

Abstract:

Separation is the most universal process in the chemical industries for separation of multicomponent mixtures. Multicomponent mixtures are separated in sequential steps in multiple distillation columns. The sequence that a feed mixture is separated into products, referred to as distillation configuration, is considered as a candidate configuration for achieving the separation task. As the number of components in the feed mixture increases, the number of candidate configurations increases rapidly, thus, it is challenging for practitioners in the early stages of the process design to find the most economic configuration(s) among the vast search space of configurations.

Shah and Agrawal[1] proposed a method to enumerate the basic configurations for separation of a given n-component feed mixture into n nearly-pure product streams using n-1 columns. Although for a given number of components the configurations have the same feed and final products, installation and operation costs vary depending on the vapor duties, vapor and component flows, etc. Nallasivam et al.[2]developed a global minimization algorithm (GMA) to find the optimal flows that result in the minimum vapor duty for each configuration, and proposed several strategies to improve the convergence and speed of the algorithm. However, the vapor duty as an objective function is loosely proportional to the capital cost of a given configuration and a reliable formulation is needed to identify the optimal configuration with minimum total cost.

In this work, we propose a new objective function formulation and extend the GMA method to minimization of the total cost. The new objective function includes calculation of column height and diameter, tray size and diameter, and heat exchanger capital and annualized operating costs. The proposed method identifies the globally optimal configuration from a complete ranklist of all configurations, including partially and completely thermally coupled basic configurations. Finally, the efficacy of the method is demonstrated by considering three case-studies where the effect of separation difficulty and balancing weight of capital and operating costs are studied for five-component feed mixture.

1. Shah, V. H., Agrawal, R. A Matrix Method for Multicomponent Distillation Sequences. AIChE J., 56(7), 1759-1775, 2010.
2. Nallasivam, U., Shah, V. H., Shenvi, A. A., Tawarmalani, M., Agrawal, R. Global Optimization of Multicomponent Distillation Configurations: 2. Enumeration based Global Minimization Algorithm. AIChE J., 62(6), 2071-2086, 2016.

496961 A New Minimum Reflux Calculation Method for Multiple-Feed Distillation Columns Distilling Ideal Multicomponent Mixtures

Abstract:

To separate a multicomponent mixture, a series of distillation columns are usually required. Each sequence or arrangement of distillation columns generates a distillation configuration. In many distillation configurations, one or more columns have more than one feed streams. These multiple feed columns are very common in many industrial applications, such as multi-effect distillation systems. On the other hand, the minimum reflux ratio of a column is related to its energy consumption and capital cost, and thus is a key parameter in distillation design. Therefore, to design and build energy efficient and cost effective distillation configurations, it is practically very important to have an efficient and accurate method to determine the minimum reflux ratio for multiple feed columns.

Despite its practical significance, this problem is intellectually difficult to solve. Most existing approaches involve either rigorous tray-by-tray calculations or iterative guessing of reflux ratio which are computationally expansive. Here, we present a simple and easy-to-use algorithmic method to determine the minimum reflux ratio for a multiple feed distillation column distilling ideal multicomponent mixtures. This method uses and extends the classic Underwoodâ€™s method, which is a short-cut method originally developed to describe simple columns with exactly one feed. The geometric interpretations of the Underwoodâ€™s equations are exploited, which provide some valuable insights in calculating the minimum reflux ratio for a multiple feed column. Compared with other existing short-cut based approaches, our method does not require guessing of pinch location or controlling split, which may lead to incorrect solutions.

Through several case studies for ternary separation involving columns with two and three feed streams, we demonstrate the accuracy and effectiveness of our new approach. Furthermore, we show that this method can be easily incorporated into a global optimization framework to facilitate the calculation of the minimum heat duty requirement for a multicomponent distillation configuration under minimum reflux. Finally, we envision that this approach can be further extended to a general complex column with multiple feeds and multiple side-product withdrawals.

500242 An Easy-to-Use Rule Synthesizes Numerous New Dividing Wall Columns for Thermally Coupled Distillations

Abstract:

We present a new, easy-to-use rule, which we derive in this work within a formal framework, to synthesize dividing wall columns (DWCs) for multi-component thermally coupled distillation. As a result of the availability of this new, easy-to-use rule we derive, we observe a drastic expansion in the search-space of possible DWCs to implement any attractive thermally coupled distillation in the form of a DWC. We demonstrate the application of this easy-to-use rule, and the multitude of options it leads to, through examples. To get a measure of the magnitude of expansion in the search space of DWCs as a result of application of our rule, we present simple formulae to determine their number. Among them, we identify a subset of DWCs in which the vapor flow through every section of the DWC can be controlled during operation by means external to the column. We believe any thermally coupled distillation can be implemented and optimally operated in the form of a DWC because of the availability of numerous options borne out of this work.

492902 Novel Solvent Exchange Distillation System- a Comparison Study with Batch Separations

Abstract:

A novel solvent exchange distillation system for simple and azetropic systems has been modeled by Aspen Plus. The results including compositions, flow rate and energy have been compared with constant volume batch distillation system results obtained with spreadsheets and with Aspen Batch Modeler. Operating conditions such as reflux ratio, number of stages, heating parameters and mole fractions have changed. Control parameters and design specifications have been perfectly regulated. The economic analysis including total capital and utilities' cost have been presented and compared. The results show the novel solvent exchange method gives high purity with lower energy consumption than the batch systems. The advantages of the new exchange system are most evident when the initial solvent is the less volatile component.

EXPERIMENTAL AND CFD ANALYSIS OF A SIEVE TRAY

Nadine Zandoná Rafagnim, University of Blumenau, Blumenau, SC – Brazil
Vitória Arend Castamman, University of Blumenau, Blumenau, SC - Brazil
Jaci Carlo Schramm Camara Bastos, University of Blumenau, Blumenau, SC - Brazil
Dr. Dirceu Noriler, University of Campinas, Campinas, SP - Brazil
Dr. Henry França Meier, University of Blumenau, Blumenau, SC - Brazil
Dr. Marcela Kotsuka da Silva, University of Blumenau, Blumenau, SC - Brazil

Introduction

In chemical industry, processes generate products that differ in composition from the raw material through chemical reactions with subsequent separation steps. Among them, distillation is widely used and makes components of different boiling points migrate between vapor and liquid phases as consequence of heat addition.[1]

Models used for distillation columns projects are based on equilibrium and nonequilibrium stages and assume perfect mixture between phases. Although the results provided by such models are useful, the flow pattern is of great importance on a sieve tray[2], since it is associated with the process efficiency. Therefore, tools such as Computational Fluid Dynamics (CFD), that provide a detailed description of fluid dynamic behavior, can be very useful.

CFD is the solution of mathematical models by means of numerical methods, which allows studies of complex problems by partially eliminating simplifications.[7] These models, however, need validation with experimental data to demonstrate their ability to represent real physical phenomena. In this regard, pressure drop values are commonly employed. According to Bennett et al.,[8] the total tray pressure drop is result of three contributions: the vapor flow through perforations (dry tray pressure drop), the liquid over the plate (clear liquid height) and the surface tension.

In the present work, physical (experimental) measures were obtained for total pressure drop, dry pressure drop and clear liquid height. The data are compared with numerical results from a transient three-dimensional model, with two Eulerian phases, using k-epsilon turbulence model for both phases.

Materials and Method

Experimental Section

Experimental data was obtained in a 0.350 m diameter acrylic column, which contains three sieve trays with 0.500 m space between them. The lower plate distributes the gas phase and the upper plate the liquid phase. This creates real conditions of entry in the central plate, where the experimental measurements take place. As materials, water is used to represent the liquid phase and air the vapor phase. The sieve tray apparatus is shown in Figure 1.

The trays have 68 squared holes distributed in triangular arrangement. In addition to the total pressure drop, dry pressure drop, and (global) clear height data, local clear liquid heights are also measured. Hence, the global clear height is understood as average, while local clear liquid height are discrete quantities that vary because of gas momentum changes in different positions. In this study, these measures are taken in 17 points demonstrated by black dots in Figure 2. The indicated x and y coordinates will be further used in the Results and Discussion section and are positioned in the origin ($x = y = 0$).

Figure 1 – Sieve tray apparatus

Figure 2 – Sieve tray with 68 squared holes

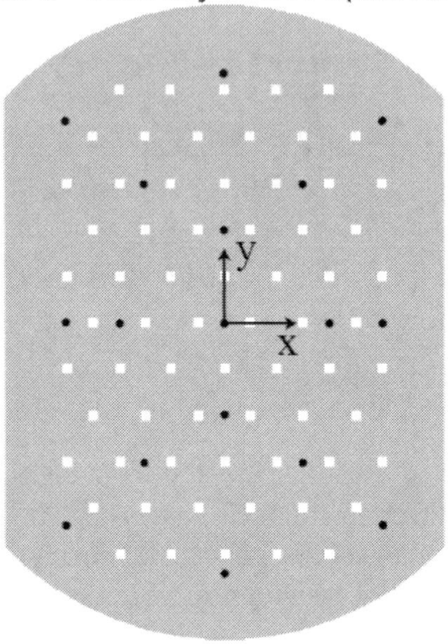

Numerical Section

The Eulerian-Eulerian framework was applied for the multiphase model, where phases are treated as an interpenetrating continuum and have separate transport equations. This model is widely employed in distillation flow studies.[2 – Erro! Fonte de referência não encontrada.]. The main Unsteady Reynolds Averaged Navier-Stokes equations (URANS) for the kth phase of the model are displayed below:

- Mass Continuity Equations:

$$\frac{\partial}{\partial t}(f_k \rho_k) + \boldsymbol{\nabla} \cdot (f_k \rho_k \boldsymbol{v}_k) = 0 \qquad (1)$$

f_k, ρ_k, and $\boldsymbol{v}_k$ represent, respectively, the volume fraction, density and velocity vector of the kth phase.

- Momentum Equations:

$$\frac{\partial}{\partial t}(f_k \rho_k \boldsymbol{v}_k) + \nabla \cdot (f_k \rho_k \boldsymbol{v}_k \boldsymbol{v}_k) = -f_k \nabla p + \nabla \cdot \left[f_k \mu_k^{eff} \left(\boldsymbol{\nabla v}_k + \boldsymbol{\nabla v}_k{}^T\right)\right] + \rho_k f_k \boldsymbol{g} + \boldsymbol{F}_{kj}^D \qquad (2)$$

p_k is the pressure field (which is the same for both liquid and gas phases) and $\boldsymbol{g}$ is the gravity vector. The terms μ_k^{eff} and $\boldsymbol{F}_{kj}^D$ represent the effective viscosity and the drag force, respectively, and are obtained from eq. 3:

$$\mu_k^{eff} = \mu_k + \mu_k^t \qquad (3)$$

μ_k^t is the turbulent viscosity, which is calculated as a function of the turbulent kinetic energy (k) and turbulence dissipation rate (ε) from the k-epsilon model.[2] The drag force is calculated as follows:

$$F_{kj}^{D} = \frac{3}{4}\frac{f_\beta \rho_\alpha}{d_p} Cd\left|\boldsymbol{v}_\beta - \boldsymbol{v}_\alpha\right|\left(\boldsymbol{v}_\beta - \boldsymbol{v}_\alpha\right) \quad (4)$$

where the subscripts α and β represent the continuous and dispersed phase, respectively. d_p is the bubble diameter and Cd the drag coefficient, which was estimated using the correlation proposed by Krishna et al.[3] for the rising of a swarm of large bubbles in the churn turbulent regime:

$$Cd = \frac{4}{3}\frac{\rho_\alpha - \rho_\beta}{\rho_\alpha}\boldsymbol{g} d_p \frac{1}{\left|V_s/\overline{f_\beta}\right|^2} \quad (5)$$

V_s is the superficial velocity and $\overline{f_\beta}$ is the gas average holdup calculated from Bennett et al.[8] as:

$$\overline{f_\beta} = 1 - exp\left[-12{,}55\left(V_s\sqrt{\frac{\rho_\beta}{\rho_\alpha - \rho_\beta}}\right)^{0{,}91}\right] \quad (6)$$

The use of eq. 5 eliminates the dependence of the drag force on the bubble diameter, which is convenient, since its value is hard to determinate for a sieve tray.

The simulations were performed using Fluent 15.0, a commercial code based on the finite volume method. The simulated geometry is a section containing the middle sieve tray of the experimental apparatus. Its boundary conditions are schematized in Figure 3. For the calculations, the geometry is divided into smaller control volumes by a hexahedral mesh of approximately 1.4 million elements with refinement in the perforations for better velocity profiles in the holes representation.

Figure 3 – Simulated geometry and boundary conditions

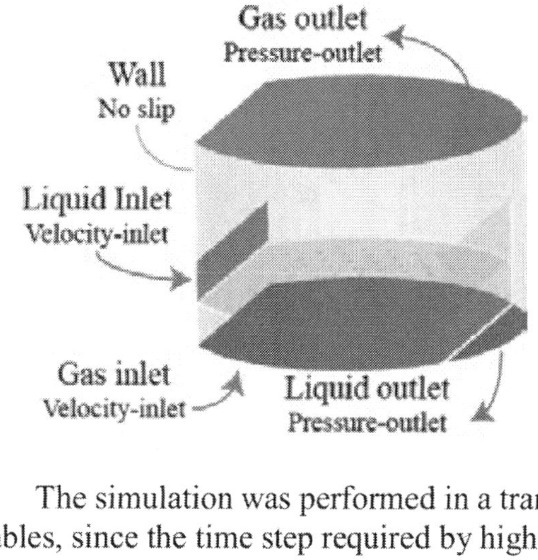

Figure 4 – Hexahedral mesh with refinement in the perforations

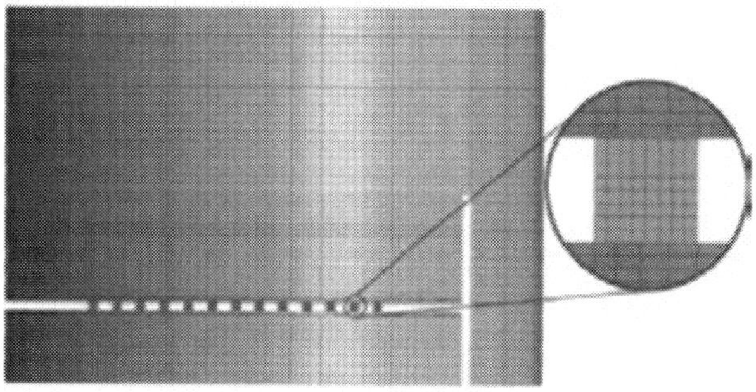

The simulation was performed in a transient regime with first order interpolation schemes for all variables, since the time step required by higher order schemes was too small (~1e-05) and would make the solution computationally expensive. Three different gas velocities at the inlet were simulated: 0.36, 0.43, and 0.50 m/s, and liquid flow rate equals to 25 L/min for all cases.

Results and Discussion

Experimental and Numerical clear liquid heights along the x-coordinate (see Figure 2) and y = 0 for the three gas superficial velocities (0.36, 0.43 and 0.50 m/s) are presented in Figure 5.

Figure 5 – Clear liquid height behavior in a central plane along the liquid path (x-coordinate with y = 0) for 0.36, 0.43 and 0.50 gas superficial velocities

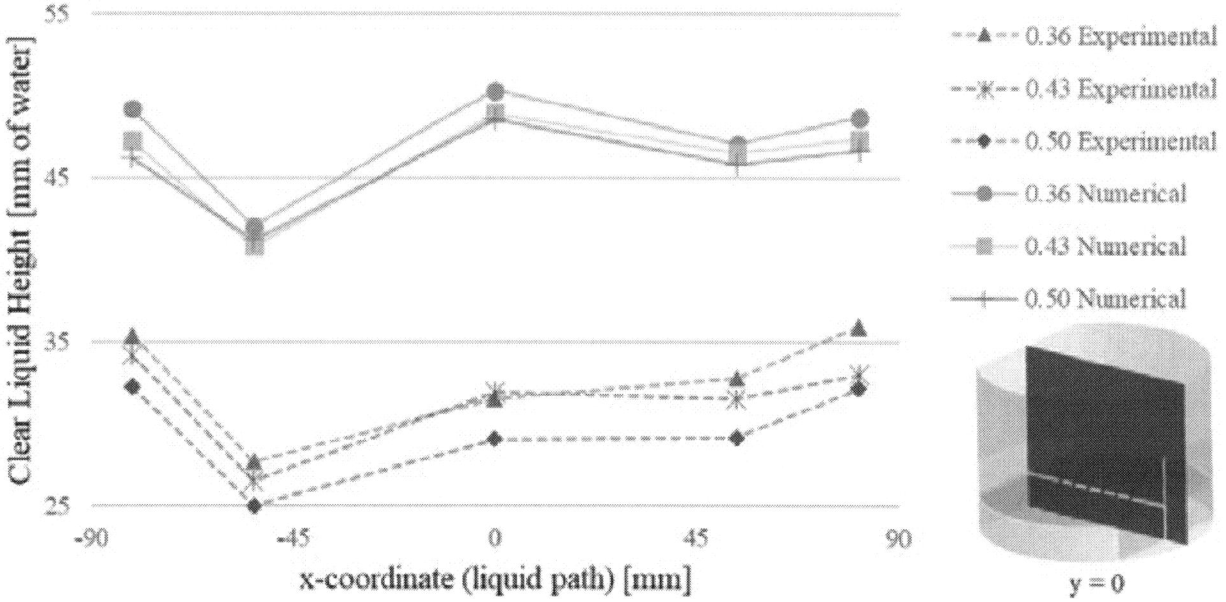

Similar behavior is observed for experimental and numerical results, but the latter present higher values. Clear liquid height tends to decrease with higher gas superficial velocities, which might be associated with the increased expulsion from the liquid when the gas velocities rises as consequence of larger drag force. This assumption is also justified by Figure 6, where numerical volume fraction profiles in the same central plane show that the increase in the gas velocity makes the gas-liquid contact region expand while the global clear liquid height decreases, as demonstrated in Table 1.

Figure 6 – Numerical liquid volume fraction profiles in a central plane for gas superficial for 0.36, 0.43 and 0.50 gas superficial velocities.

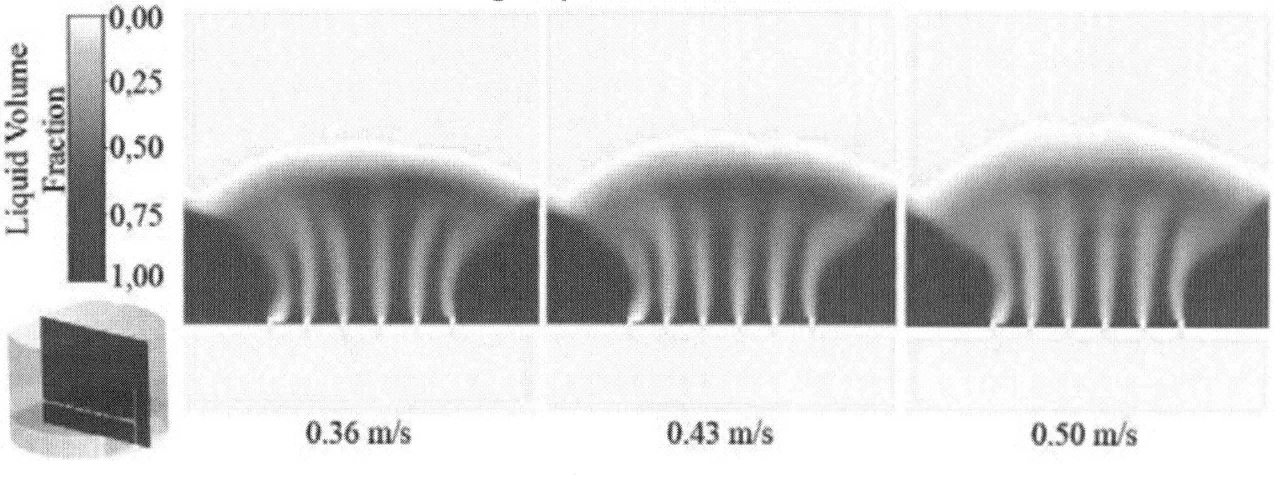

Table 1 – Experimental and numerical results for global clear liquid height

Gas Superficial Velocity [m/s]	Global Clear Liquid Height [mm of water column]		
	Experimental	Numerical	Bennet et al.[8]
0.36	35.57	53.37	54.10
0.43	34.61	52.21	52.10
0.50	31.04	52.17	50.20

Table 1 also shows there is an overprediction of global clear liquid height values by Bennet et al.[8] correlation, which is noted as well in numerical results that use also the authors' liquid holdup correlation in the drag model (eq. 6)

In Figure 7 and Figure 8 clear liquid height distributions over the tray are shown as a result of experimental and numerical data collected in the seventeen points presented in Figure 2. In both numerical and experimental color maps it is possible to see regions of smaller liquid heights near the middle of the plate and higher heights in the peripheries. The numerical model was not able to capture the small asymmetries observed experimentally, effect that might be associate with the uneven way the gas entries the sieve tray apparatus.

Figure 7 – Experimental clear liquid height distribution, interpolated from values of seventeen points

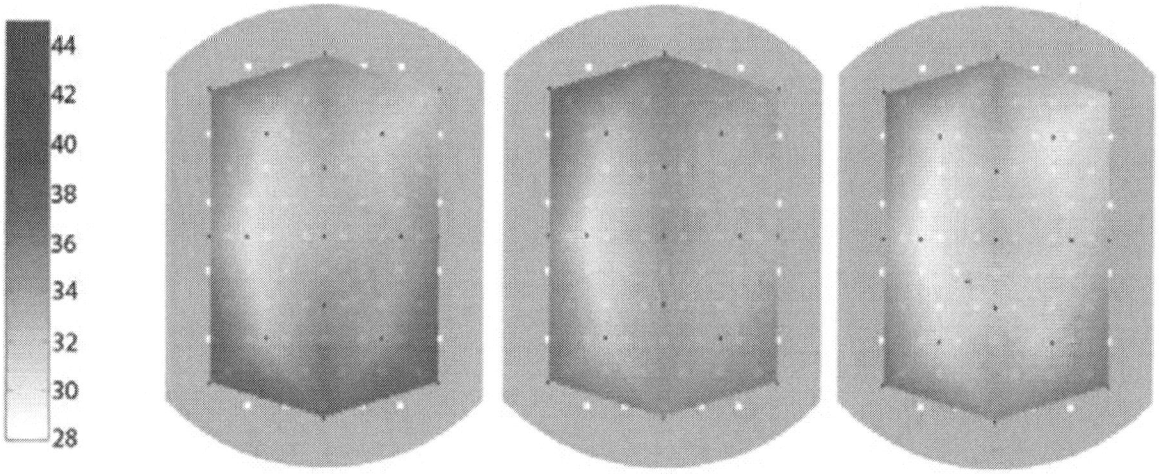

Figure 8 – Numerical clear liquid height distribution, interpolated from values of seventeen points

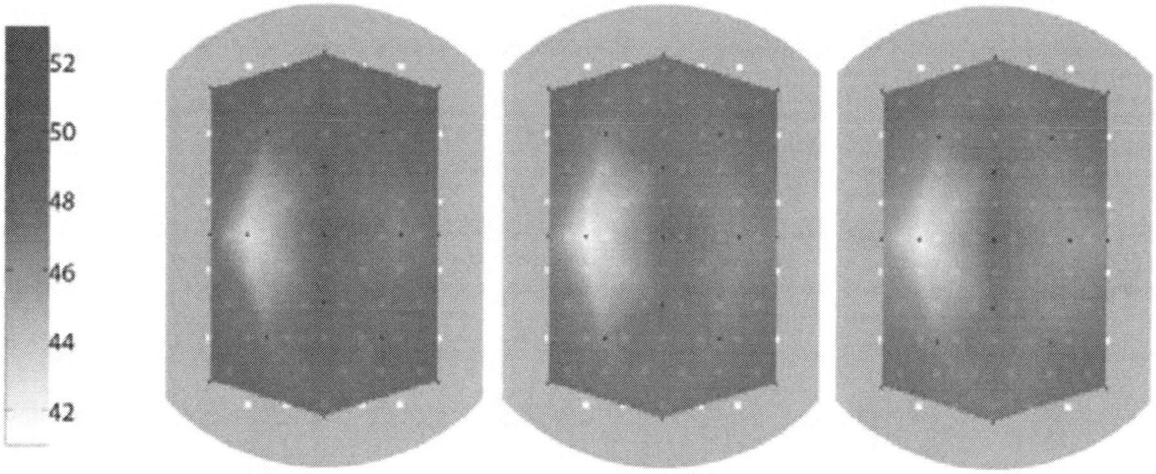

Figure 9 – Total and dry pressure drops for Vs = 0.36, 0.43, and 0.50 m/s

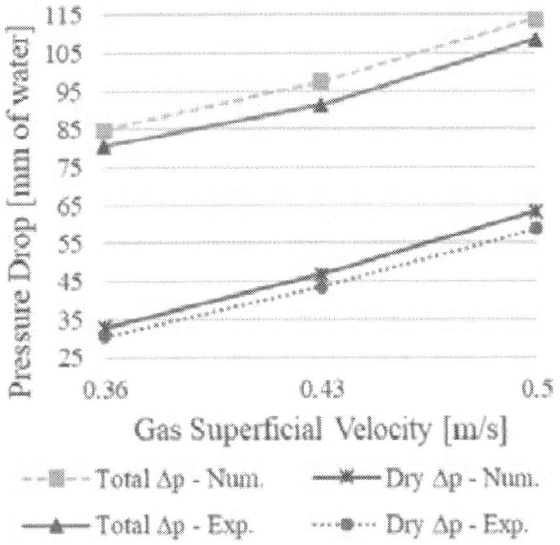

Figure 10 – Isosurface for liquid volume fraction = 0.5

In Figure 9, total and dry pressure drop measures are compared. Numerical and experimental results present the same behavior showing that with raise in gas velocity, increases also the dry and total pressure drops. The numerical values are greater than the experimental once again, with mean relative error of 6% for total pressure drop and 8% for dry pressure drop.

Mesh refinement in the perforation region was fundamental to obtain good results for dry pressure drop, since this variable has a strong dependency on the gas velocity profile in the holes, which can only be well represented by sufficiently refined mesh.[9]

Despite the Eulerian-Eulerian framework inability to reconstruct interface, in Figure 10 an isosurface for liquid volume fraction equals to 0.5 is used to capture it. This image evidence the model capability of reproducing the irregular flow on a sieve tray, with different bubble sizes and shapes rising and bursting over the liquid surface.

In Figure 11 two-dimensional stream lines for gas and liquid phases are shown in two different planes: one from the inlet to the weir and other cross-sectional, 10 mm above the tray. The gas rises from the perforations almost without recirculating, while the liquid has two main recirculating zones: one close to the liquid inlet and other to the weir. According to Noriler et al.,[2] these two regions will also present higher efficiencies compared to others. From the cross-sectional plane, the liquid flow from the wall to the center of the tray is observed.

Figure 11 – Stream lines for (a) gas phase in a central plane from the inlet to the weir; (b) liquid phase in a central plane from the inlet to the weir; (c) liquid phase in a cross-sectional plane at 10 mm above the tray

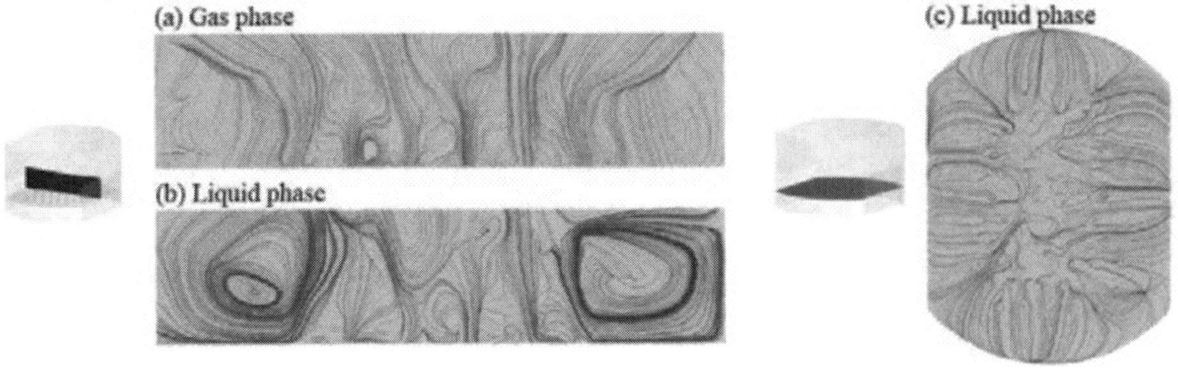

Conclusions

Distillation is a separation process of major importance in chemical industry. Its efficiency is highly affected by flow patterns not taken under consideration by the conventional equilibrium and nonequilibrium models. In this regard, Computational Fluid Dynamics can be a useful tool by allowing the description of fluid dynamic phenomena with detail, which has been evidenced in the present work.

Numerical dry and total pressure drop results are in good agreement with experimental data. The flow behavior described by the CFD model is also very similar to that observed physically. However, the numerical representation of the clear liquid heights still needs improvement, considering it was only qualitatively validated. These differences might be associated with the drag model, given that the Bennett et al.[8] gas holdup correlation underpredicts the gas holdup for the distillation apparatus. Thus, there is indication that improving the drag force model is necessary. Since changes in clear liquid height would probably affect the total pressure drop, the uncertainty about the former affects the reliability of both results.

References

1. Seader, J. D.; Henley, E. J.; Roper, D. K. Separation process principles: chemical and biochemical operations. Hoboken : Wiley, 2011. 821p.
2. Noriler, D.; Barros, A. A. C.; Maciel, M. R. W.; Meier, H. F. Simultaneous Momentum, Mass, and Energy Transfer Analysis of a Distillation Sieve Tray Using CFD Techniques: Prediction of Efficiencies. Ind. Eng. Chem. Res., v. 49, p. 6599-6611, 2010.
3. Krishna, R.; van Baten, J. M.; Ellenberger, J.; Higler, A. P.; Taylor, R. Chem. Eng. Res. Des. v. 77, p. 639-646, 1999.
4. Rahimi, R.; Sotoodeh, M. M.; Bahramifar, E. The effect of tray geometry on the sieve tray efficiency. Chem. Eng. Sci., v. 76, p. 90-98, 2012.
5. Sun, L.; Xubo, L.; Jiang, S.; Wang, W.; Lyu, H.; Wang. P.; Gao, H. Computational Fluid Dynamics Hydrodynamic Analysis of a Cross-Orthogonal Fixed-Valve Tray;
6. Muniz, M.; Decker, R. K.; Wiggers, V. R.; Noriler, D.; Meier, H. F. Application of nonlinear analysis techniques (chaos) on sieve tray multiphase and turbulent flows as a model validation instrument. In: AIChE Annual Meeting, 2014, Atlanta. 2014 AIChE Annual Meeting, 2014.
7. Maliska, C. R. Transferência de Calor e Mecânica dos Fluidos Computational [Computational Heat Transfer and Fluid Flow]. Rio de Janeiro, Livros Técnicos e Científicos Editora, 2004.
8. Bennett, D. L.; Agrawal, R.; Cook, P. J. New Pressure Drop Correlations for Sieve Tray Distillation Columns. AIChE J., v.26, p.432-442, 1983.
9. Rafagnim, N. Z.; Castamann, V. A.; Meier, H. F.; Noriler, D.; Silva, M. K. Avaliação de Aspectos de Modelos Numérico Empregado para Simulação de Escoamento em Pratos Pefurados de Destilação [Assessment of a numerical model applied for distillation sieve tray simulation]. Proc. of the XXXVIII Brazilian Crongress of Particulate Systems, 2017.

495335 Optimal Design of Advanced Distillation Schemes for Enhancing Energy Efficiency of Close-Boiling Ethylbenzene/Styrene Separation in Plastics Industry

Abstract:

Distillation is by far the most widely used separation technology in chemical process industry, which is commonly utilized in separating close-boiling component in industry. Considering the capital and energy intensity of the task, schematic selection of optimal distillation strategies becomes a significant decision. The importance of designing distillation systems of both sustainability and feasibility is continuously challenging chemical engineers.

Styrene monomer (SM) is the simplest and by far the most important member of a series of aromatic monomers that is used in the plastics industry. Currently, catalytic dehydrogenation of ethylbenzene (EB) is an overwhelming technology for SM production. However, the unreacted EB must be separated from SM through distillation in order to guarantee downstream processes. The normal boiling points (NBPs) of EB and SM are 136.2 and 145.7°C, respectively. Obviously, they form a close-boiling system, which requires energy intensive distillation operation for the separation. For example, a typical 500,000 metric tonnes annually (MTA) styrene production plant contains an EB/SM distillation column with about 70 equilibrium stages, which operates at a reflux ratio of 7.1 to achieve an EB impurity level in the final SM product of 100 ppm. This column accounts for 75-80% of the total energy consumptions in the current production process. In this sense, this study introduces a reliable shortcut method through optimizing the target of total annualized cost (TAC). Two energy-saving distillation candidates â€" double-effect distillation (DED) and mechanical vapor recompression (MVR) â€" are applied to potentially reduce energy requirement in conventional EB/SM column. In the first round evaluation (shortcut method), MVR stands out to be a more attractive option than DED alternative. In addition, the shortcut calculation results are used as initial values for the rigorous simulations. In the second round comparison (rigorous method), the advanced distillation schemes (ADSs) using DED and MVR are proved that they can save by up to 30% and 40% of operating costs, respectively, compared to conventional distillation schemes (CDSs). In the view of TAC, the ADSs can cut a corner of ~35-40% from the CDSs. Specifically, the ADS using MVR slightly outperforms its DED counterpart in TAC comparison.

495054 Effect of Ions on Coalescence in Liquid Two-Phase Systems

Abstract:

Liquid/liquid processes, e.g., extraction processes or polymerization, play an important role in industrial applications. The efficiency of these processes is mainly determined by the drop size distribution (DSD), which is the result of the competing phenomena drop breakage and coalescence. These phenomena were investigated extensively in the last decades. However, due to its complexity, especially coalescence is by no means completely understood. To get a comprehensive understanding of the coalescence process considering influencing parameters such as fluid dynamics and ion addition and to evaluate/optimize model descriptions of the DSD, a systematic experimental analysis of the influence of ion addition and fluid dynamics on coalescence was conducted.

A bottom-up approach was used to systematically quantify the impact of the described parameters in the system toluene/water separately: first the fundamental behavior of single drop coalescence was investigated in a fully automated test cell using high speed imaging [1]. With this setup, serial and detailed examinations (e.g. coalescence efficiency and time) of binary droplet collisions under variable system conditions are possible. In the second step of the approach, lab-scale experiments in a stirred tank and a Kühni-column (both DN150) were conducted, measuring the drop size distribution and the height of the dense-packed zone, respectively, for the addition of different ions in varying concentrations.

In the single drop experiments, a significant coalescence inhibition was observed for increasing ion concentrations (see Figure 1). Additionally, drop size and ion species had a major impact on the coalescence efficiency and coalescence time (hence the large error bars in Figure 1). The dependence of the coalescence behavior on the ion species could also be observed in lab-scale stirred tank experiments, as can be seen in Figure 2 for different ions. At concentrations above 10^{-3} mol/L all ions decrease the Sauter mean diameter to a specific degree, especially at a prompt stirrer speed decrease the strong impact of ion species on coalescence behavior becomes apparent. A first approach to model the coalescence behavior with addition of OH^- ions (=increasing pH) and the implementation into the population balance equation [2] showed promising results and is now tested for application for the investigated ions in our system.

The comprehensive experimental and numerical results of the systematic analysis will be presented and discussed.

Figure 1. Coalescence efficiency for single drops with varying drop diameter as a function of ionic strength.

Figure 2. Transient behavior of the Sauter mean diameter in a stirred tank for the addition of different ions.

Financial support within the DFG project â€œCoalescence efficiency in binary systemsâ€ is gratefully acknowledged.

[1] J. Villwock, F. Gebauer, J. Kamp, H.-J. Bart, M. Kraume, Chem. Eng. Technol. 37(7), 1103-1111 (2014)

[2] J. Kamp, M. Kraume, Chem. Eng. Sci. 126, 132-142 (2015)

487979 Turbidity Control

Abstract:

In biobased processes it is a common task to recover valuable components from the fermentation broth by liquid-liquid extraction. Despite the advantages of this process like processing of effluents at low temperature etc., a few drawbacks may cause severe problems. A problem often encountered in biobased processes is the formation of emulsions. This behavior is caused by surface active constituents which may be byproducts from metabolizing the substrate or may be added intentionally for other purposes. Examples for these components are proteins or surfactants.

Splitting of these emulsions can be performed with several splitting methods like thermal, chemical, mechanical or electrical splitting. Due to the limited performance of thermal, chemical and mechanical splitting technologies electrical splitting was investigated. This project focused on the investigation and modelling of turbidity control in the electrical field.

The emulsions were prepared by mixing ShellSol® K (SSK), deionized water and *4*-dodecylbenzenesulfonic acid (4DBSA). Prior to the experimental investigation of splitting stable oil in water emulsions had to be prepared. The stirring intensity and the amount of dispersed phase were therefore varied in a mixing cell. Afterwards the emulsion was analyzed in a Rhosonics SUSS 2008 ultrasonic scanner. After phase separation the aqueous phase still contained residual turbidity. The turbidity (oil in water emulsion) contained 0.5 wt% of solvent phase consisting of SSK and 0.1 wt% of 4DBSA. For investigating turbidity control solvent mixtures according to this composition were prepared by homogenizing the "turbidity-mixture" with a rotor stator device IKA T18 with the dispersing element S18 N – 19 G. The separation progress for different electric field intensity and different electrode geometries was then monitored by time lapse photography. The cell was illuminated from the backside of the test cell to detect the brightness (intensity I) of the sample. A Matlab routine was used to compare the brightness I_0 before, during and after complete separation (complete separation $I/I_0 = 1$) and the brightness at the beginning (separation $I/I_0 = 0$).

The resulting separation (splitting) trends were then correlated with the field intensity.

486815 Scale-up Rules for Taylor-Couette Disc Contactor Design

Abstract:

In separation and isolation of value added products from low grade aqueous feed, liquid-liquid extraction has established as a basic unit operation. The Taylor-Couette Disc Contactor (TCDC) has shown suitability for applications in bioseparations due to the simple design of internals. The internals of the TCDC are similar to the design of the Rotating Disc Contactor (RDC) but without stator rings and with increased shaft and disc diameter [1]. The design prevents from formation of hydrodynamic dead zones, crud accumulation and fouling. The increased shaft diameter and the increased disc diameter engender a similar flow pattern compared to banded two-phase flow of Taylor-Couette reactors (TCR), making the TCDC particularly suitable for the processing of biogenic raw materials and harsh operation conditions.

The industrial scale implementation of processes for mass transfer with and without chemical conversion demands reliable scale up rules. Therefore, it is important to elaborate basic experiments for validation of CFD-simulation and for the dimensional analysis [2]. For liquid-liquid phase contactors the reliable prediction of hydrodynamics is a critical issue for the design and scale-up. Since design rules for the prediction of hydrodynamic data for the TCDC are still not available, correlations for the determination of the drop size distribution, the Sauter mean diameter and the dispersed phase hold-up where derived via dimensionless analysis [3]. The correlations have been validated with experimental data from 0.1 and 0.3 m diameter pilot plant scale operation with 1 m active column height. The results of mass transfer experiments have confirmed that the concentration profile and the separation efficiency can simply be modeled with CSTR cascade design [4]. The determination of the residence time distribution has confirmed CSTR cascade operation when exceeding a critical rotational speed. The outcome of CFD-analysis and the experiments provides simple tools for the basic design of the TCDC.

Literature

[1] E. Aksamija, C. Weinländer, R. Sarzio, M. Siebenhofer, *Sep. Sci. Technol.* **2015**, 50 (18), 2844–2852. DOI: 10.1080/01496395.2015.1085406

[2] M. Zlokarnik, *Chem. Eng. Technol.* **2004**, 27 (1), 23–27. DOI: 10.1002/ceat.200403158

[3] A. Grafschafter, M. Siebenhofer, *Chemie Ing. Tech.* **2017**, 4, DOI: 10.1002/cite.201600142

[4] A. Grafschafter, E. Aksamija, M. Siebenhofer, *Chem. Eng. Technol.* **2016**, 39 (11), 2087–2095. DOI: 10.1002/ceat.201600191

501405 Modeling Internal Recycle within a Mixer-Settler to Aid Extraction Performance

Abstract:

Abstract

Precise models for solvent extraction have become necessary as the demand for rare earth elements (REE) continues to rise. Similarly, improved solvent extraction methods are required to ensure the supply of REE is maintained in the future. This study employed and modeled internal recycle â€" remixing the organic phase during extraction â€" to develop a precise model of and an improved method for solvent extraction.Â The recycled organic phase consisted of the cationic extractant (bis[2-ethylhexyl] phosphate [HDEHP]) in Isopar-L and the aqueous phase consisted of yttrium (Y) in dilute sulfuric acid.

Although internal recycle within a mixer-settler has been shown to minimize emulsions (Ryon et al, 1963) and reduce entrainment (Rowden et al, 1974), it has the added benefit of increasing extraction efficiency when performed in the correct proportions. Treybal indicated that only by recycling the dispersed organic phase during solvent extraction would the mixer-settler efficiency increase (Treybal, 1964), whereas Hoh determined that recycling either the dispersed or continuous phase was beneficial as long as the organic to aqueous phase ratio (O:A) in the mixer remained equal to 2:1 for their system (Hoh et al, 1989). Although Hoh's model suggests that the change in mixer O:A caused the change in efficiency, a thorough model relating the internal recycle parameters, mixer O:A, organic phase residence time and mixer efficiency is needed.

The current study modeled the effect of internal recycle on mixer O:A and mixer residence time, then determined the corresponding efficiency for each of the internal recycle rates. To model the mixer O:A and residence time, a mathematical relation was developed. To determine the efficiency for a specific recycle rate, Y was extracted from a surrogate phosphate-derived byproduct stream. The quantity of Y extracted in a single stage employing recycle was compared against batch equilibrium data. Using both the mathematical relation and the extraction results, the final model related extraction efficiency with mixer residence time and mixer O:A. When internal recycle was not employed, the mixer O:A was equal to the external O:A (1:10), the mixer residence time was approximately 45 seconds, and the efficiency was 55%. When 80% of the of the organic phase was internally recycled, the mixer O:A was 1:2, the mixer residence time was approximately 155 seconds, and the efficiency was 85%. Experimental results confirmed the trends predicted by the model and show a strong correlation between mixer O:A, mixer residence time and mixer-settler efficiency.

References

1. Ryon, A. D. and Lowrie, R.S., 1963. U.S. Atomic Energy Comm., ORNL-3381

2. Rowden, G.A., Scuffham, J.B. and Warwich, G.G.I., 1974. The effect of change in operating organic/aqueous ratio on the operation of a mixer-settler. Proc. ISEC'74, 1:81

3. Treybal, R.E., 1964. Recycle in liquid extraction. I&EC Fundamentals, 3(3):185

4. Hoh, Y.C., Ju, S.J., Chiu, T.M., 1989. Effect of internal recycle on mixer-settler performance. Hydrometallurgy, (23): 105-118

Acknowledgements

This research is supported by the Critical Materials Institute, an Energy Innovation Hub funded by the U.S. Department of Energy, Office of Energy Efficiency and Renewable Energy, Advanced Manufacturing Office. This research is also supported by the Graduate Assistance in Areas of National Need Fellowship funded by the U.S. Department of Education. Special thanks to Natasha Ghezawi and Samantha Thorpe for their contributions and Dr. David DePaoli of Oak Ridge National Laboratory for his expertise.

496447 Liquid Membrane Separation Process for the Removal of As (III) & As (V) from Drinking Water

Abstract:

For the past two decades arsenic contamination in drinking water around the globe has led to the research and development of several removal technologies due to its high toxicity. The liquid membrane separation technique being highly selective, simple design and requires low energy has been employed here for the removal of arsenic from its aqueous solution for the purification of drinking water. This work is intended to identify environmentally benign organic solvent, receiving phase and extractant for the extraction and recovery of arsenic ions from its aqueous solution through two-phase equilibrium study. Sodium arsenite and sodium arsenate salts have been used for feed phase solutions of As (III) and As (V), respectively. Amongst the conventional organic chemicals and environmentally benign solvents used as liquid membrane, sesame oil with Aliquat as the extractant increased the extraction of arsenic ions from 50% (without extractant) to 90% (with extractant).

The main aim of two phase study is to identify LM solvent and extractant to remove arsenic from water. Equal volume of aqueous solution of sodium arsenite and arsenate was taken separately and added to environmentally benign solvent in a conical flask. The mixture was continuously stirred by a mechanical shaker. Then it was kept undisturbed for some time until the two phase separated and the aqueous phase was carefully separated from organic phase. The concentration of As (III) and As (V) in the aqueous phase after extraction was measured by using AAS (at wave length 193.7 nm using VGA mode).

Two phase experiments were carried out with various green solvents. Maximum extraction for As (III) 53.3% and As (V) 64% was obtained using sesame oil. Sesame oil mainly consists of linoleic acid, oleic acid, palmitic acid and stearic acid. Saturated fatty acids like palmitic acid and stearic acid reacts with arsenic forming arsenolipid of the type (CH3)2As(O)-(CH2)nCOOH (n=14 and 16) with a dimethylarsinoyl group, (CH3)2As(O)-, replacing the methyl group in palmitic and stearic acids (Rumpler et al. (2008)). Arsenic reacts with unsaturated fatty acids like oleic acid forming C19H37AsO3 (Rumpler et al. (2008)).

Based on literature review, two phase experiments were conducted with four extractants, viz. trioctylamine (TOA), tributyl phosphate (TBP), di-2-ethylhexyl phosphoric acid (D2EHPA) and N- N-dimethyloctyl amine (Aliquat). In order to improve the percentage of extraction, extractants were added to the selected solvent (sesame oil) to identify the suitable extractant for arsenic ions. Though all four extractants showed a high percentage of extraction, aliquat being a basic extractant reacts with both dissociated and undissociated forms of arsenic (Pancharoen et al. (2009)). Aliquat in sesame oil has shown best extraction of 92%. (Figure 2).

The concentration of the selected extractant, Aliquat (NR4+Cl-) was varied from 1-10% (v/v). In neutral pH range of about 6-8 which is found in groundwater, As (III) exists mostly as H3AsO3 while As (V) occurs mostly as H2AsO4- and HAsO42- (Pancharoen et al. (2009)). As reported by Pancharoen et al. (2009) aliquat reacts with both undissociated H3AsO3 and dissociated forms H2AsO4- and HAsO42-, 6% (v/v) aliquat showed maximum extraction of As (III) (93.6%) while 10% (v/v) of aliquat was required for extraction of As (V) (95.2%) (Figure 3).

In subsurface environments, researchers have found that arsenic contamination might occur due to two reasons, one could be arsenic mobilization due to oxidation of pyrite minerals containing arsenic and the other could be reduction of iron oxy-hydroxides leading to the release and dissolution of arsenic in groundwater (Bose and Sharma (2002)). Based on this interaction of iron and arsenic, iron salts viz. ferric nitrate, ferric sulphate, ferric chloride, ferrous sulphate and ferrous chloride are selected as the receiving phase for recovery of As (III) and As (V). The idea is to implement the reverse mechanism that leads to contamination. Ferric nitrate for As (III) and ferrous chloride for As (V) were found to give a high recovery percentage.

The concentration of the selected receiving phases (ferric nitrate and ferrous chloride) was varied from 1.25-5ppm. It was observed that 5ppm of ferric nitrate and ferrous chloride showed maximum recovery of As (III) (69%) and As (V) (75%), respectively.

Based on the results of two phase studies, over 90% of As (III) and As (V) have been extracted by incorporating Aliquat in sesame oil. Ideal condition for maximum extraction of arsenic was found to be 200rpm at 25â°C for a duration of 12 hours. Different extractant concentration was observed to give maximum extraction for As (III) and As (V). 6% and 10% (v/v) Aliquat showed maximum extraction of As (III) (93.6%) and As (V) (95.2%), respectively. Around 70% recovery was possible with 5ppm of ferric nitrate and ferrous chloride for As (III) and As (V), respectively. Further experiments for three phase studies is being carried based on the performance of two phase studies.

REFERENCES

Bose P, Sharma A. Role of iron in controlling speciation and mobilization of arsenic in subsurface environment. Water Res. 2002; 36(19):4916-26.

Guell R, Fontas C, Antico E, Salvado V, Crespo J.G, Velizarov S. Transport and separation of arsenate and arsenite from aqueous media by supported liquid and anion-exchange membranes. Sep Purif Technol. 2011; 80(3):428-34.

Hylton K, Mitra S. A microfluidic hollow fiber membrane extractor for As (V) detection. Anal Chim Acta. 2008; 607(1):45-9.

Lothongkum A.W, Suren S, Chaturabul S, Thamphiphit N, Pancharoen U. Simultaneous removal of arsenic and mercury from natural-gas-co-produced water from the Gulf of Thailand using synergistic extractant via HFSLM. J Memb Sci. 2011; 369: 350-8.

Mukherjee A, Sengupta M.K, Hossain M.A, Ahamed S, Das B, Nayak B, Lodh D, Rahman M.M, Chakraborti D. Arsenic contamination in groundwater: A global perspective with emphasis on the Asian scenario. J Health Popul Nutr. 2006; 24(2): 142-63.

Pancharoen U, Poonkum W, Lothongkum A.W. Treatment of arsenic ions from produced water through hollow fiber supported liquid membrane. J Alloys Compd. 2009; 482(1-2): 328-34.

Perez M.E.M, Reyes-Aguilera J.A, Saucedo T.I, Gonzalez M.P, Navarro R, Avila-Rodriguez M. Study of As(V) transfer through a supported liquid membrane impregnated with trioctylphosphine oxide (Cyanex 921). J Memb Sci. 2007; 302(1-2):119-126.

Prapaswat T, Ramakul P, Satayaprasert C, Pancharoen U, Lothongkum A.W. Separation of As(III) and As(V) by hollow fiber supported liquid membrane based on the mass transfer theory. Korean J Chem Eng. 2008; 25(1): 158-63.

Rumpler A, Edmonds J.S, Katsu M, Jensen K.B, Goessler W, Raber G, Gunnlaugsdottir H, Francesconi K.A. Arsenic-containing long-chain fatty acids in cod-liver oil: a result of biosynthetic infidelity. Angew Chem Int Ed. 2008; 47: 2665-7.

490284 Extraction of Dyes Using an Aqueous Two Phase System in Stratified and Slug Flow Regimes of a Microchannel

Abstract:

In this work analysis of an Aqueous two phase (polymer-salt) system for extraction of sunset yellow dye is carried out. A polymer-salt ATPS i.e.; Polyethylene glycol-600 and anhydrous sodium sulfate is used for the extraction. Conditions are chosen to ensure that the extraction results in a concentration of the dye in one of the phases. The dye has a propensity to come to the Polyethylene glycol-600 phase. This extracted sunset yellow dye is degraded photo catalytically into less harmful components .The cloud point method was used to obtain the binodal curve of ATPS. From the binodal curve, the composition of salt and Polyethylene glycol -600 was chosen such that the volume of Polyethylene glycol-600 rich phase is low. This was selected to concentrate the dye from a dilute solution in a large volume of contaminated solution into a small volume. This pre-concentration step provides a high reaction rate for photo catalytic degradation reaction. Experimentally the dye is extracted from the salt phase to Polyethylene glycol -600 phase in batch extraction. This was found to be very fast and all dye was extracted .The concentration of sunset yellow dye in salt and polymer phase is measured at 482nm by Ultraviolet- visible spectrophotometry. The extraction experiment in micro channels under stratified flow is analyzed to determine factors which effects the dye extraction. Focus will be on obtaining slug flow by adding nanoparticles in micro channel. The primary aim is to exploit the fact that slug flow will help improve mass transfer rate from one phase to another through internal circulation in dispersed phase induced by shear.

Keyword---- Aqueous two phase system, binodal curve, Extraction ,sunset yellow dye

498369 N-Nitrosodimethylamine (NDMA) Removal By Thin Film Composite Polyamide Reverse Osmosis Membranes

Abstract:

Water reuse and desalination are the most feasible ways to alleviate water scarcity in an increasingly drier planet. Although thin film composite polyamide (PA) membranes are capable of high ion rejection (above 99%), removal of small neutral molecules, such as disinfection byproducts (DBPs), is still challenging. DBPs (e.g., N-nitrosodimethylamine, NDMA), formed during water treatment, are insufficiently retained by commercial nanofiltration (NF) and reverse osmosis (RO) PA membranes, prepared by interfacial polymerization. In this paper, we explore NDMA removal by TFC membranes prepared via molecular layer-by-layer (mLbL) deposition. The mLbL approach enables the formation of a PA layer whose chemistry can be controlled at the single monomer level, thereby facilitating fundamental relations linking polymer chemistry to membrane morphological and transport properties.

495399 Reaction Kinetics for Microcystin -LR Degradation By Ozonation in a Basic Medium

Abstract:

Microcystin-LR (MC-LR) is a cyclic heptapeptide released by some cyanobacteria during algal blooms in eutrophic waters that can bio-accumulate in fish and be transferred to higher organisms. Awareness of the fatal effect of MC-LR in humans began in 1996, when thirty eight patients undergoing dialysis at a renal diseases institute in Brazil died of acute hepatic failure as a result of contamination of the water used for hemodialysis. The cause of the deaths was a mystery initially, until it was reported that the water used for hemodialysis at the institute was contaminated with the toxin MC-LR. In 1988, The World Health Organization set the provisional concentration limit of MC-LR for potable water at 1Î¼g/L. Generally, conventional water treatment methods, such as precipitation, coagulation, flocculation, sand filtration can be effective for removal of intracellular toxins, as intact cells are physically removed, but these processes are ineffective for extracellular toxins released when algal cells lyse. Ozonation can effectively destroy extracellular MC-LR from water at a high pH but it is necessary to predict the extent of degradation after a specified duration of treatment. Studies have shown that pH largely determines the rate of most ozonation reactions. In this study, the rate constants, k_{03} and k_{OH}, values will be determined. The probable degradation pathway of MC-LR will be examined using liquid chromatography coupled with mass spectrometry. Membrane filtration using hydrophobic membranes have also been shown to be effective for the removal of MC-LR at high pH, but problems that may arise from this choice of treatment include accumulation of the toxin on the surface of the membrane and cost for disposal of the contaminated membranes. The combination of membrane filtration and ozonation will be further examined to create a treatment process that can be effectively implemented.

500487 Novel Cylindrical Cross-Flow Hollow Fiber Membrane Module for Direct Contact Membrane Distillation-Based Desalination

Abstract:

Direct contact membrane distillation (DCMD) process for desalination is attractive especially for high salt concentrations if low-cost steam/waste heat is available and waste brine disposal cost for inland desalination is factored in. We recently developed successfully a hollow fiber membrane potted in a rectangular module with hot brine in cross-flow over the fibers and obtained high and stable water vapor flux under demanding scaling conditions. However, the current rectangular membrane module design having a low membrane surface area/unit device volume appears to be inadequate for larger-scale plants. Here we have described a novel cylindrical cross-flow module containing high-flux composite hydrophobic hollow fiber membranes for membrane surface areas: 0.15 and 0.6 m^2. These two modules used respectively 316 or 1266 porous fluorosiloxane-coated porous hollow fibers of PP (ID/OD: 330, 630 μm). The pipe-like module design is simple, easily scalable and packs about four times larger membrane surface area per unit equipment volume compared to the earlier rectangular cross-flow module design. The distilled water production rates from 1 wt% salt-containing feed brine has been studied over a hot brine temperature of 60-91^{0}C. A model has been developed to describe the observed water production rates of such devices in the dead-end brine feed introduction configuration. The model appears to describe the observed water vapor production rates well for different feed brine temperatures at various brine flow rates. The model flux predictions have been explored over a range of hollow fiber lengths to compare the present results with those obtained earlier from rectangular modules which had significantly shorter hollow fibers.

A Novel Process for Concurrent Desalination and Boron Removal (CDBR)

Süer Kürklü[1], Sadiye Velioglu[1,2,3], M. Göktug Ahunbay[1], S. Birgül Tantekin-Ersolmaz[1], William B. Krantz[3,4]

[1] Department of Chemical Engineering, Istanbul Technical University

[2]Department of Chemical & Biomedical Engineering, Nanyang Technological University

[3]Singapore Membrane Technology Center (SMTC), Nanyang Technological University

[4]Department of Chemical & Biological Engineering, University of Colorado

Annual Meeting of the AIChE
Minneapolis Minnesota
November 1, 2017

SMTC
Member of the NEWRI Ecosystem

1

Plants must Walk the Boron Tightrope!

Insufficient Boron

Excessive Boron

2

The Challenge!

- Boron in excess of a few ppm is toxic for many agricultural crops and can be deleterious to human health
- Desalinated water is used both for irrigating agricultural crops and for providing potable water
- The WHO recommends a maximum boron concentration of 0.5 ppm for both irrigation and potable water
- The boron concentration in seawater averages 4.6 ppm but is as high as 9.6 ppm in the Eastern Mediterranean
- Current technologies for desalination and boron removal are problematic

3

Outline

- Boron tolerance limits for agricultural crops
- Overview of current methods for desalinating and removing boron from seawater
- Objectives of this study
- Overview of the novel Concurrent Desalination and Boron Removal (CDBR) process
- Performance metrics for the CDBR process
- Summary

4

Boron Levels at which the Yield of Agricultural Crops is Affected

Affected at levels < 0.5 ppm	
Blackberry	Lemon

Affected at levels of 0.5 – 0.75 ppm						
Apricot	Avogado	Cherry	Cowpea	Grape	Grapefruit	Kadota Fig
Onion	Orange	Peach	Pecan	Persimmon	Plum	Walnut

Affected at levels of 0.75 – 1.0 ppm						
Barley	Garlic	Jerusalem Artichoke	Kidney Bean	Lima Bean	Mung Bean	Lupine
Peanut	Sesame	Strawberry	Sunflower	Sweet Potato	Wheat	

Affected at levels of 1.0 – 2.0 ppm				
Carrot	Pea	Red Pepper	Potato	Radish

The yield of several other crops is affected at boron levels of 2 – 15 ppm

5

Membranes can do it all!

…… or can they?!

"To combat the rising sea levels we construct thousands of desalination plants to suck up the water."

- Boron is in seawater primarily as boric acid
- Boric acid & water have similar diameters (~2.7 Å)
- Hydrophilicity effects permit some degree of boron rejection using RO membranes
- At near neutral pH conditions, the boron rejection of commercial RO membranes is at most 90%

6

Concurrent Desalination and Boron Removal using two SSROs in Series

35000 ppm salt and 10 ppm boron in feed
RO membranes with 99.7% salt rejection and 90% boron rejection
< 100 ppm salt and 0.5 ppm boron in product water
85% pump efficiency, 90% energy-recovery-device efficiency

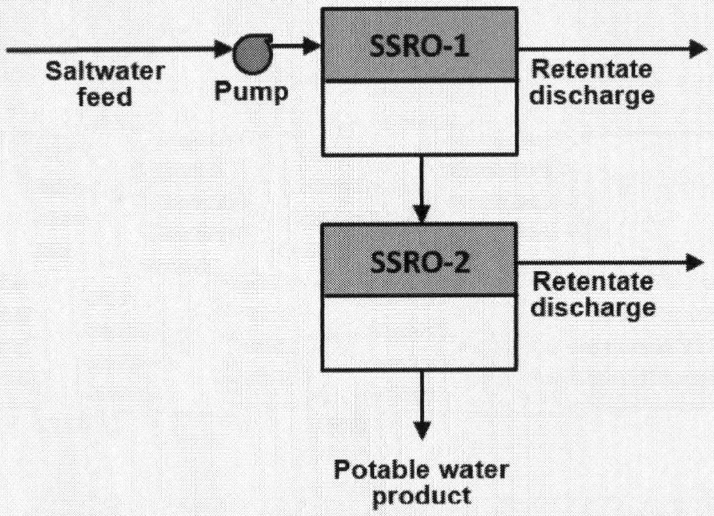

For 55% recovery, OPD = 66 bar, SEC_{net} = 2.68 kWh/m^3
For 65% recovery, OPD = 89 bar, SEC_{net} = 3.38 kWh/m^3
Safety factor for SSRO-1 is between 0.46 and 0.90

7

Current Commercial Methods for Desalination and Boron Removal

Two methods for boron removal from water are used on an industrial scale: reverse osmosis (RO) at high pH conditions and ion exchange:†

RO Method

- At a pH above the pKa of 9.15 boric acid dissociates into borate ions that fully hydrate and are highly rejected both because of their size and charge
- This method requires adding some base to raise the pH that must be neutralized after the boron is removed
- Raising the pH causes precipitation of calcium and magnesium hydroxides that result in membrane scaling
- This is a multi-step process that involves high pressure desalination, boron removal, neutralization and scaling mitigation

Ion-Exchange Method

- Uses ion-exchange chelating resins with ligands containing hydroxyl groups in the *cis* position capable of forming complexes with borates
- Requires resin regeneration using HCl or H_2SO_4 and neutralization with NaOH
- Requires a separate high pressure desalination step and incurs high costs owing to resin regeneration
- Multi-step process involving desalination, boron removal & resin regeneration

† Bodzek, M., The removal of boron from the aquatic environment – state of the art, *Desalination and Water Treatment* 57(3) (2016) 1107-1131.

8

Objectives of this Study

(On Target with 3 Darts – 3 Objectives – 3 Process Configurations!)

- Develop novel process technology for concurrent desalination and boron removal using currently available commercial membranes
- Develop mathematical model for this new process technology
- Access performance metrics of this novel process technology

9

The 3-3 Concurrent Desalination & Boron Removal (CDBR) Process

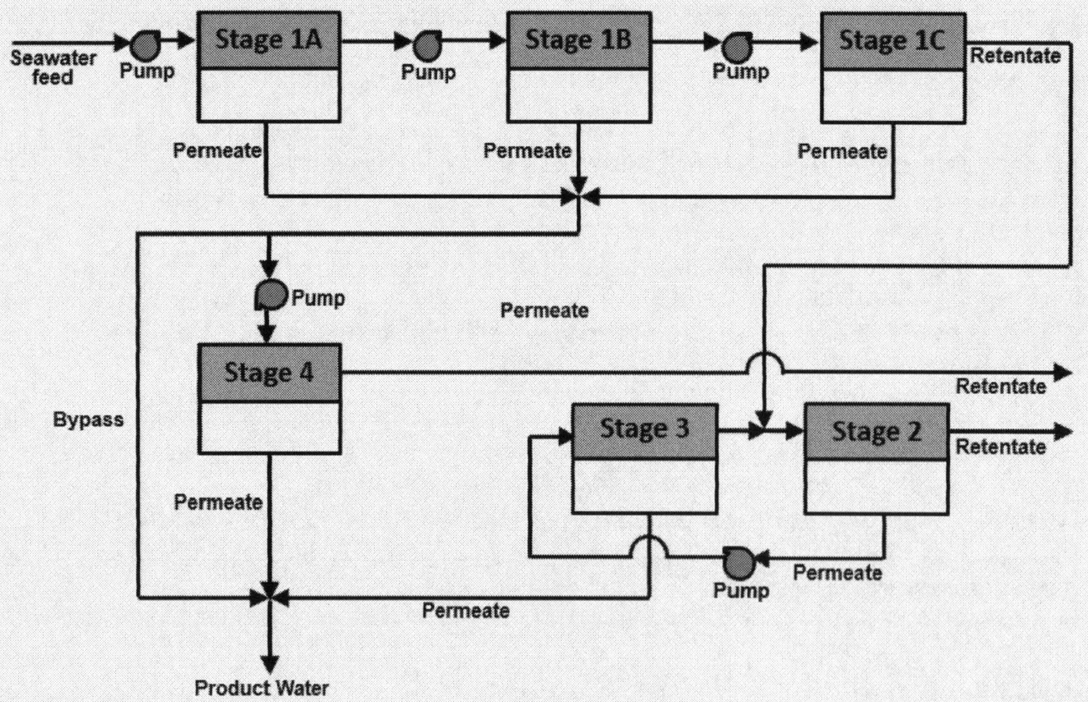

10

Features of the CDBR Process

- Capitalizes on SSRO performance by using its retentate brine as the feed to a CMCR and using its permeate as the feed to a LPMS for additional boron removal
- Reduces the OPD by employing countercurrent retentate and permeate flow in the CMCR
- Reduces the OPD by employing permeate recycling in the CMCR
- Reduces the OPD by employing retentate recycling via one or more NF stages in the CMCR
- Operates all stages at same OPD to avoid interstage pumping on retentate side and entropy-of-mixing effects
- Maintains safety factors ≥ 1 in all stages to mitigate concentration polarization and scaling

SSRO: Single-stage reverse osmosis
CMCR: Countercurrent membrane cascade with recycle
LPMS: Low pressure membrane stage
OPD: Osmotic pressure differential
NF: Nanofiltration

11

Modeling the CDBR Process

- Involves component & overall mass balances on each stage & mixing point(s)
- 1-3 CDBR involves 19 equations in 35 unknowns ⇒ 16 degrees of freedom
- 2-3 CDBR involves 25 equations in 44 unknowns ⇒ 19 degrees of freedom
- 3-3 CDBR involves 28 equations in 50 unknowns ⇒ 22 degrees of freedom
- Resulting set of coupled algebraic equations can be solved analytically

12

Assumptions in Modeling the CDBR Process

- Operation at the thermodynamic limit
- Concentration polarization, scaling and fouling not considered
- Pressure loss in flow channels and owing to permeation neglected
- Use of an energy-recovery device (ERD) for high pressure stages
- 85% efficiency for high pressure pumps
- 90% efficiency for energy-recovery devices
- Linear relationship between osmotic pressure and concentration
- Osmotic pressure determined only by the salt concentration
- Availability of membranes with the required salt & boron rejections
- Fixed cost reduction at lower pressure operation ignored
- Maintenance cost reduction at lower pressure operation ignored

13

Osmotic Pressure Differential for Concurrent Desalination and Boron Removal

35000 ppm salt & 10 ppm boron feed; < 100 ppm salt & 0.5 ppm boron product

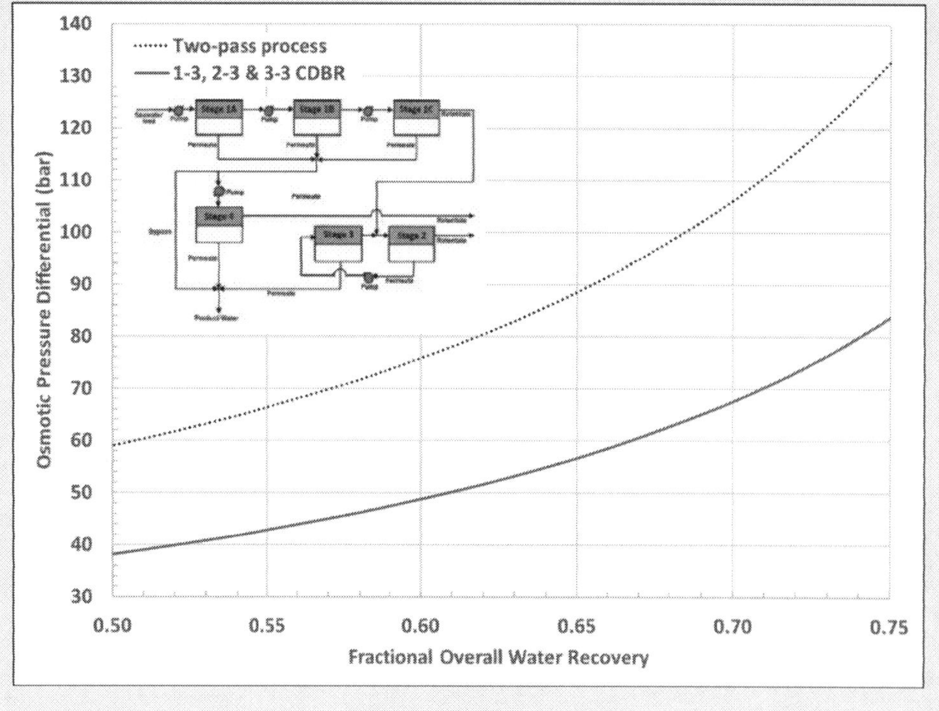

14

Net Specific Energy Consumption for Concurrent Desalination and Boron Removal

35000 ppm salt and 10 ppm boron feed; < 100 ppm salt and 0.5 ppm boron product

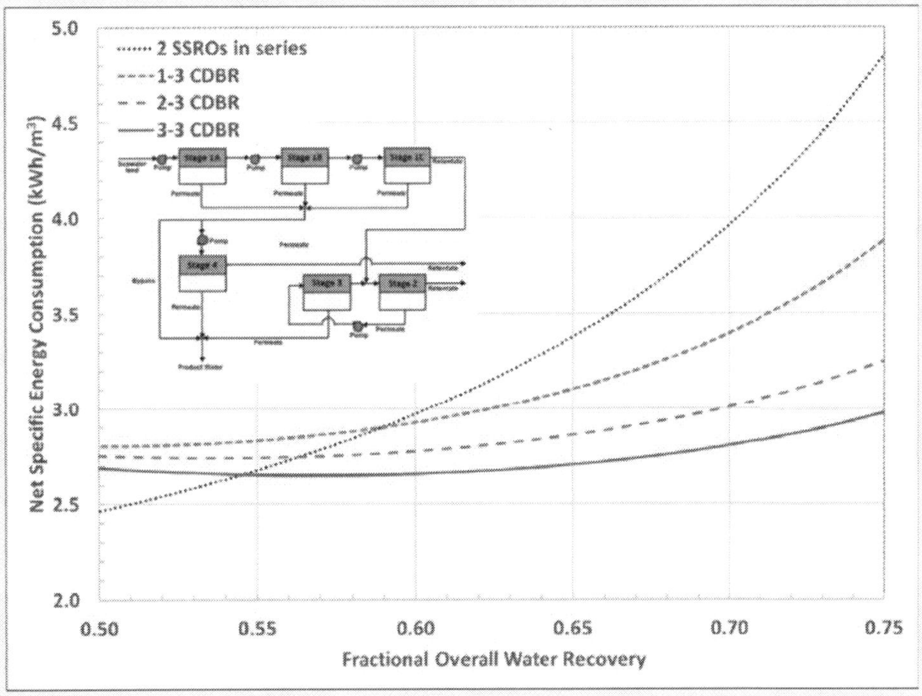

15

Summary of Performance Metrics

35000 ppm salt and 10 ppm boron feed; < 100 ppm salt and 0.5 ppm boron product
99.7% salt rejection and 90% boron rejection RO membranes
85% pump efficiency, 90% energy-recovery-device efficiency

Process	50% Recovery		60% Recovery		70% Recovery	
	OPD (bar)	SEC_{net} (kWh/m^3)	OPD (bar)	SEC_{net} (kWh/m^3)	OPD (bar)	SEC_{net} (kWh/m^3)
2 SSROs in series	59.0	2.46	75.8†	2.97	106†	3.96
1-3 CDBR‡	38.1	2.80	48.8	2.93	67.7‡	3.40
2-3 CDBR	38.1	2.75	48.8	2.78	67.7	3.01
3-3 CDBR	38.1	2.69	48.8	2.66	67.7	2.81

† Maximum allowable pressure for conventional RO modules is 69 bar
‡ Safety factor less than 1 in SSRO stage for recoveries greater than 65%

16

Summary

- The CDBR process can achieve concurrent desalination to < 100 ppm and boron removal to 0.5 ppm using currently available commercial membranes
- The CDBR process allows a high water recovery at an OPD and SEC_{net} considerably lower than conventional RO
- The CDBR process has a very competitive net specific energy consumption owing to operation at a lower OPD
- The CDBR process can significantly reduce the fixed and maintenance costs owing to operation at a lower OPD
- The high recovery of the CDBR process produces a highly concentrated brine that could be used in a hybrid CDBR-PRO process to harvest its osmotic potential energy
- CDBR technology can be retrofit onto existing desalination plants

17

For More Information

- Krantz, W.B., S. Velioğlu, S. Kűrklű, M.G. Ahunbay, S.B. Ersolmaz, "Concurrent Desalination and Boron Removal (CDBR) Process," Patent Application PCT/TR2016/050387 filed on October 19, 2016.
- Kűrklű, S., S. Velioğlu, M.G. Ahunbay, S.B. Ersolmaz, W.B. Krantz, "A Novel Energy-Efficient Concurrent Desalination Boron-Removal (CDBR) Process," Desalination, 423 (2017) 79-94.

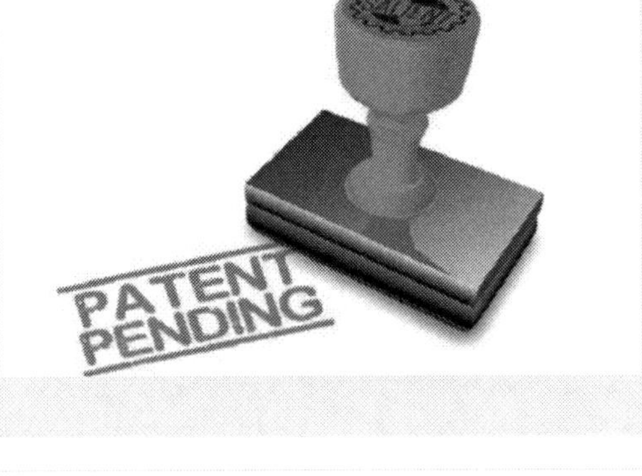

18

Acknowledgment

➢ This research has been supported by the Scientific and Technological Research Council of Turkey (TUBITAK) through Grant No. 114Y165.

19

Questions?

krantz@colorado.edu

20

Specification of the Degrees of Freedom

Specification	Number	Case
Feed flow rate and concentrations	3	
Salt and boron rejections in stage 1A	2	
OPD in stage 1A	1	
Stage 2 recovery plus stage 3 recovery equal to one	1	
Recoveries in stages 2, 3 and 4	3	
OPDs equal in stages 2 and 3	1	1-3 CDBR
Boron rejection scaled to predicted salt rejection in stage 2	1	
Boron rejection scaled to predicted salt rejection in stage 3	1	
Salt rejection scaled to predicted boron rejection in stage 4	1	
Bypass split fraction	1	
Permeate product boron concentration	1	
	16	
Salt and boron rejections in stage 1B	2	2-3 CDBR
OPD in stage 1B	1	
	19	
Salt and boron rejections in stage 1C	2	3-3 CDBR
OPD in stage 1C	1	
	22	

21

Specifying the Degrees of Freedom

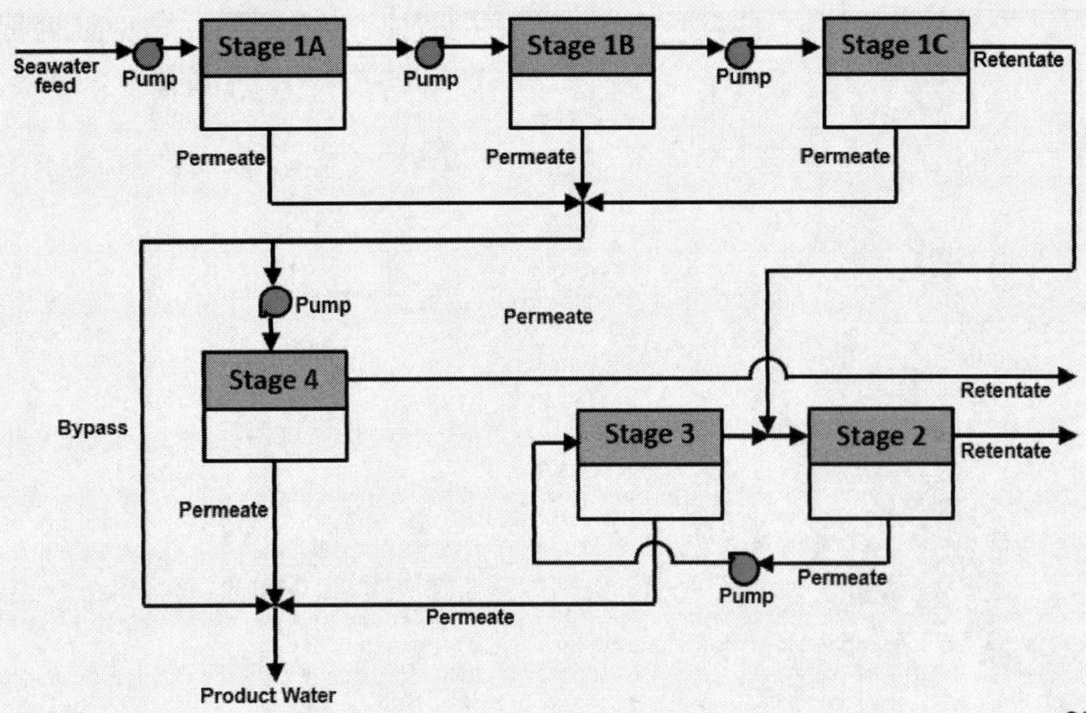

22

487951 Condition Optimization for the Fabrication of Desalination Thin Film Composite Membranes

Abstract:

Reverse osmosis is one of the most spread technologies all over the world for salty water purification, in which thin film composite (TFC) is the common form of membranes used. In this paper, we examined the effects of various membrane fabrication conditions on the TFC membrane performance including 1) m-phenylenediamine (MPD) and trimesoyl chloride (TMC) exposure times; 2) different ways for the support PSU sheets phase inversion; 3) the support layer sheet thickness and 4) drying temperature for the fabricated membranes. The operation conditions used in this work were: 300 psi, 25 °C, 2000 ppm NaCl, and the samples were run for at least 8 hours. It was found that the best salt rejection and water flux results were obtained when the reaction time was short during the interfacial polymerization and when the plate was immersed vertically during the phase inversion for PSU support preparation. Increasing the PSU sheet thickness increased proportionally the salt rejection and decreased the water flux. And increasing the drying temperature led to an increase in the salt rejection and water flux for short reaction time. Based on the study, a TFC membrane with salt rejection of 98.8 ± 0.5 % and water flux of 76.1 ± 2.7 L/m^2hr was fabricated using 25 sec MPD contacting time and 5 sec TMC reaction time and 110 °C drying temperature. To the best of our knowledge, this result is the best ever reported for simple prepared TFC membranes.

References

1- Mohammed Kadhom, Jun Yin, Baolin Deng, "A Thin Film Nanocomposite Membrane with MCM-41 Silica Nanoparticles for Brackish Water Purification", *Membranes*, 6, 50, (2016) 1-12.

488743 Scale up of RO and UF Membrane Surface Nano-Structuring with Hydrophilic Polymer Brush Layers - Evaluation of Uniformity of Membrane Performance

Abstract:

A novel process of membrane surface nano-structuring with hydrophilic polymer brush layers was evaluated for the scaled-up synthesis of high performance polyamide (PA) RO and polysulfone (PSf) UF membranes. The process is based on surface activation via treatment with atmospheric pressure plasma (APP) followed by free-radical graft polymerization. Surface nano-structuring was evaluated with two types of atmospheric pressure plasma (APP) sources based on air and a helium-oxygen mixture. In order to enable uniform level of surface treatment a robotic system was utilized that can translate the plasma source over a large surface area. Surface graft polymerization was subsequently carried out with acrylic acid (AAc) at various graft polymerization conditions (i.e., initial monomer concentration, graft polymerization time) so as to optimize the synthesized a poly(acrylic acid) (PAAc) polymer brush layer which was confirmed via X-ray photoelectron spectroscopy (XPS). The uniformity of surface activation was evaluated via characterization of surface wettability and topography. At the optimal conditions, membrane sheets of size suitable for 21â€ spiral-wound RO elements were synthesized. Performance of the nano-structured PA RO and PSf UF membranes, with respect to water permeability and salt rejection, was then quantified for different sections of the large sheet membranes. The results of the current study demonstrate that APPIGP approach can be scaled-up to produce commercial scale surface nano-structured RO and UF membranes.

502514 Polybenzimidazole Based Membranes for Organic Solvent Nanofiltration (OSN)

Abstract:

Polybenzimidazole Based Membranes for Organic Solvent Nanofiltration (OSN)

Ruiyi Liu, Irina Valtcheva, Piers Gaffney, Patrizia Marchetti, Andrew Livingston

Barrer Centre, Department of Chemical Engineering, Imperial College London

Organic liquids are ubiquitous in chemical science based industries, which range in scale from refining to pharmaceutical production. It is generally accepted that 40-70% of capital and operating costs in these industries are dedicated to separations; and a substantial fraction of this cost is related to processing of organic liquids. Membrane technology utilised in Organic Solvent Nanofiltration (OSN) has the potential to provide game changing alternatives to conventional concentration and purification technologies such as evaporation, liquid extraction, adsorption and chromatography. The membranes must offer resistance to organic environments, attractive selectivities and permeabilities. Ideally they should also be resistant to physical aging under use.

A good polymer candidate which could overcome these challenges is 2,2–(*m*-phenylene)–5,5–bibenzimidazole (PBI), which belongs to the class of polybenzimidazoles and possesses thermal, mechanical and chemical stability towards organic solvents, acids and bases. The imidazole ring of PBI provides a reaction site which can be used for chemical modification (crosslinking) of the polymer chains. The polymer used for this study is commercially available Celazole PBI (~30 000 Da) in *N,N*-dimethylacetamide (DMAc) solution. Membranes were prepared by casting solutions of PBI with different concentrations on polypropylene non-woven fabric, followed by phase inversion in a water bath. In order to improve the chemical resistance, the polymer was modified in a post treatment step using crosslinking with 3 wt% *p*-α,α'-dibromoxylene (DBX) in organic solvent. The reactions were carried out at 70-80 °C for 24 h, and the chemical modification was verified by FTIR, microanalysis and x-ray photoelectron spectroscopy (XPS).

The chemical conditions of three commercial applications were chosen to demonstrate the performance of the obtained PBI membranes in extreme pH conditions – peptide and oligonucleotide synthesis, and carbon capture and storage (CCS). The challenging conditions are as follows: a) 20 % v/v piperidine/DMF (peptides); b) 3 % v/v dichloroacetic acid/acetonitrile (oligonucleotides); and c) 20 % v/v monoethanol amine/water (CCS). In order to assess the membrane stability in such media, first a dip test was performed. Small pieces of membrane were placed in the acidic and basic solutions for several months. No visible degradation of the polymer or colour change of the transparent solutions was observed. Filtration experiments were conducted in a crossflow set up using a range of polystyrene oligomers dissolved in solvent to determine MWCO. The polystyrene solution was then replaced by one of the basic or acidic solutions which were used to treat the membranes for several hours. Finally the polystyrene solution was filtered again through the membranes in order to evaluate whether the extreme pH change degraded them or not.

PBI membranes showed good performance after the acid and basic treatment in terms of rejection. However for the oligonucleotides application, residual cross-linking agent decorating the membrane with unreacted functional groups interfered with the oligo synthesis. This was overcome by a post-crosslinking functionalization which also improved permeance and allowed tailoring of the permeation properties.

498990 Organic Solvent Nanofiltration with NovelÂ Perfluoropolymer Membranes and Other Novel Membrane Processes

Abstract:

A very large variety of membranes polymeric or otherwise have been investigated over the last two decades for organic solvent nanofiltration (OSN). The materials and structures used to make OSN membranes include among others the following: poly (dimethylsiloxane) (PDMS); mixed matrix membranes (MMMs) of PDMS with zeolites and other fillers; asymmetric integrally skinned polyimide (PI) membrane crosslinked with aromatic or aliphatic diamines; polyaniline; polypyrrole; interfacially polymerized polyamide with or without functionalized carbon nanotubes; carbon-based membranes including graphene. Most of the studies involved polymeric membranes. One of the weaknesses of most polymeric membranes is their varying tendencies for swelling with demanding organic solvents aprotic or otherwise; this gets reflected in the solute rejection behaviors of such polymers. Diamine-crosslinked PI membranes show excellent resistance to organic solvents but have some swelling in the presence of water. To this end we have started studying membranes from particular classes of fluoropolymers which are extremely inert. Initially these membranes were studied for pervaporation-based selective removal from organic-organic mixtures and aqueous-organic mixtures. Next they were studied with a variety of solvents and solvent mixtures for OSN over a considerable pressure range. We report results of solvent flux and solute rejection for such membranes using dye solutes. These studies were made with dense flat membranes of different thicknesses supported on appropriate porous supports. We will also present totally new results on novel membrane processes for organic solvent-based processing.

510034 High-performance Ceramic Supported Thin Film Composite Membrane for Organic Solvent Nanofiltration

Abstract:

The commercial success of membrane-based separations processes related to water treatment have helped launch an emergent field of membrane based organic solvent nanofiltration (OSN). Polyamide, a mainstay selective layer found in reverse osmosis (RO) membranes, has demonstrated some excellent performance in OSN. RO membranes, however, are built on solvent intolerant polymers (polysulfone), thus necessitating the use of polyamides on other types of supporting materials. One such support material is a ceramic ultrafiltration membrane. In this work, we demonstrate the formation of a polyamide selective barrier layer *in-situ* onto a ceramic UF membrane and evaluate the think film composite for use in OSN. We have selected ceramic tubular UF membranes provided by Fraunhofer IKTS as substrates for this work. The resulting membranes, after activation with dimethylformamide, have demonstrated a pure methanol permeance of 26.3 $Lm^{-2}h^{-1}bar^{-1}$, and dye rejection for methyl orange (327.3g/mol), acid fuchsin (585.5g/mol), chlorophyllin (724.2g/mol) and brilliant blue (825.6g/mol) of 30 %, 90%, 95%, and 91%, respectively. This work represents the first time that polyamide selective layers have been directly formed onto a tubular ceramic membrane and tested for OSN.

488984 Defect-Free PIM-1-Based Hollow Fiber Membranes

Abstract:

Microporous polymers are one of the newest classes of materials investigated for membrane applications. Polymers of intrinsic microporosity (PIMs) have rigid backbones that lead to high permeabilities and good selectivities along with solution processability. Current research on PIMs has been exclusively performed on flat sheet membranes to understand fundamental membrane properties. To bridge the gap between the development of new high performance polymers and industrially attractive technologies, more efficient membrane units such as hollow fibers are needed. To date, only dilute PIM-1 containing blends have been fabricated into fibers, which exhibit large surface area to volume ratios and are highly scalable. This work is the first to demonstrate the fabrication of defect-free, integrally skinned, asymmetric PIM-1 hollow fiber membranes utilizing an immiscible liquid protective layer. PIM-1-based fibers have been utilized to demonstrate various organic liquid molecular separations.

489162 Extraction and Phase Separation of Api's in Low Interfacial Tension Mixtures

Abstract:

Membranes are widely used in the biotechnology, food, petroleum, and water treatment industries, to name a few. Currently, they have not seen widespread use is in the pharmaceutical industry. This is due to the often harsh or chemically incompatible nature of many of the solvents used when producing pharmaceuticals. Additionally, nearly all pharmaceuticals are still produced using batch or semi-batch unit operations. However, the relatively new movement towards continuous production of pharmaceuticals presents an exciting opportunity for membrane based applications. Although many industries, such as petroleum refining and food production, have been using continuous processing for decades, the approach is relatively new to the pharmaceutical industry. Chemically inert membranes, such as polytetrafluoroethylene (PTFE), can be used for phase separation, purification, and extraction during or after synthesis of a pharmaceutical by utilizing membrane wettability. The organic phase will selectively wet the PTFE surface and permeate through, while the aqueous phase will be retained. This is a powerful tool during synthesis because it allows for continuous purification or extraction, instead of using semi-batch unit operations. Fully continuous production of a pharmaceutical allows for smaller scale and dead volumes, increased heat and mass transfer, faster process control, safer operation, and allows for the exploration of reactions that are too dangerous or difficult to perform in batch.

In this work, we utilize a unique membrane module with integrated pressure control that decouples the separation from downstream disturbances. This unique module removes the need for the user to individually control the pressure of the aqueous and permeate sides of the membrane. The pressure is controlled using a thin perfluoroalkoxy (PFA) polymer diaphragm that acts to modulate the pressure between the aqueous and organic sides of the membrane. The upper and lower bounds for operating this system while achieving complete separation correspond to the capillary and permeation pressure, respectively. The capillary pressure must be greater than the diaphragm pressure and the diaphragm pressure must be greater than the permeation pressure. These pressures can be controlled by varying the pore size and type of membrane and the thickness of the diaphragm.

This system has been successfully applied to the final synthesis steps of the production of several pharmaceuticals, including lidocaine and diazepam. These complex mixtures contained several organic solvents as well as aqueous salt solutions. The goal was to extract the active pharmaceutical ingredients (API's) into the organic phase; hexanes for lidocaine and ethyl acetate for diazepam. In general, this type of separation would not be possible using a membrane, since the interfacial tension is too low, leading to breakthrough of the aqueous phase into the organic phase. We present a method to modulate the interfacial tension so that complete separation of the two phases is achieved. In addition to completely separating the two phases, 100% of the API's were extracted into their respective organic phases. This was not previously possible and is a step towards the reliable extraction and phase separation of API's in a fully continuous operation.

490848 Tunable Mesoporous Films from Graft Copolymers with Degradable Side Chains

Abstract:

Most membrane separations function through the use of a steric exclusion mechanism to effect the separation of target solutes from solution based on molecular size. Consequently, the development of polymer materials that enable the development of membranes with easily tunable pore sizes has been a major focus of recent membrane research. To this end, self-assembled block polymers have been used to prepare ultrafiltration membranes (pore diameters 10 nm - 100 nm) with tunable pore sizes, but the low molecular weight block polymers required to achieve smaller pore sizes and the poor mechanical properties of these low molecular weight materials limits their utility for many membrane separations from organic solvents. Here, to overcome this challenge a series of poly(styrene-graft-poly(lactide) methacrylate) copolymers have been prepared by a â€œgrafting throughâ€ approach with styrene and poly(lactide) methacrylate macromonomers in a solution free radical polymerization reaction. Characterization of the copolymer materials by size-exclusion chromatography and H^1 NMR confirm good control over composition and molecular weights above 200 kDa. Films cast from these copolymers were characterized by scanning electron microscopy, small angle X-ray scattering, and dead-end filtration testing demonstrating that the pore size of the resultant membranes is easily varied from 15 nm to 5 nm by changing the molecular weight of the sacrificial, pore-templating poly(lactide) methacrylate group. Further exploration of structure-property relationships, membrane performance, and crosslinked materials shows potential utility of this materials platform for organic solvent separations.

495843 Effects of Pore Size in Polybenzimidazole Membranes on the Performance of Organic Solvent Nanofiltration

Abstract:

Organic solvent nanofiltration (OSN) through a membrane is an economically viable technology for the separation and recovery of organic solvents. The pore size in the membrane is crucial to the membrane performance (based on solvent flux and solute rejection). In this work, we report a molecular simulation study to investigate the effects of pore size (centred at 0.75, 0.85 and 0.95 nm, respectively) in three polybenzimidazole (PBI) membranes. Quantitative relationships between the pore size and the microscopic properties of solvent (methanol) in the membranes are established. Methanol in the membranes is revealed to possess two states: a bound state that strongly interacts with the polymer matrix and a bulk-like state that can move freely in the membrane. The movement of methanol can induce the formation of pores and the variation of pore size distribution in the membranes. Methanol flux and solute rejection are estimated at different pore sizes. It is unravelled that methanol flux is primarily governed by the percentage of the bulk-like methanol, while solute reject is largely determined by the pore size. This simulation study highlights the crucial effects of pore size on the performance of OSN membranes, and would facilitates the development of new membranes for high-performance OSN.

482851 Recovery of Inorganic Phosphorus Via Metal-Exchanged ZSM-5

Abstract:

Interest is growing in the development of separation processes for efficient removal and regeneration of P from wastewaters. Dissolved P, including phosphate (PO_4^{3-}), hydrogen phosphate (HPO_4^{2-}) and dihydrogen phosphate ($H_2PO_4^{-}$), is a primary source of water contamination originating from industrial wastewater or agricultural runoffs. The presence of high levels of P in ponds, rivers and lakes can lead to eutrophication, threatening the existence of fish and other aquatic life. On the other side, P is broadly used as fertilizers in agriculture to promote the growth of crops, whose demand escalates as the population continues to grow. The world's supply of P today, however, relies on depleting feedstocks such as phosphate rocks. It thus becomes imperative, not only for mitigation of pollution, but also for sustainable production of fertilizers, to develop efficient and cost-effective methods for separation of P from aqueous solutions.

The most common method of phosphorus separation is chemical precipitation, typically in the form of struvite. Although feasible for implementation, this method consumes valuable feedstocks such as ammonia and/or magnesium salts. Meanwhile, struvite precipitation requires operation under high-pH conditions and precise control over the molar ratio of ammonium (NH_4^+), magnesium (Mg^{2+}), and phosphates, which add challenges to the process design and incur additional costs for scale-up applications. The use of other cations, such as Al^{3+} and Fe^{3+} that are capable of precipitating phosphate under more robust conditions, however, generates products undesired for agricultural purposes. In light of the challenges present in chemical precipitation, efforts have been devoted to the development of new separation processes, such as using sorbents for capture of the phosphate anions.

Zeolites have attained great attention as ion-exchange materials due to their chemical and structural stability, tunable and selective adsorption properties, employment of naturally abundant elements, and environmental compatibility. Natural zeolites such as clinoptilolite and gibbsite have previously been reported for capture and release of phosphates, but their limited capacity and specificity have inspired the exploration of artificial zeolitic materials as more robust sorbents. Here we report on the use of copper-substituted Zeolite Socony Mobil-5 (Cu-ZSM-5) as sorbents for efficient recovery of P. It has been demonstrated that phosphate (inorganic P) can be captured from synthetic solutions and released in brine solutions. By tailoring the synthesis of the zeolites, the capacity and efficiency of P capture/release were systematically studied. The captured inorganic P was released from the adsorbents when they were added to regenerative solutions and exhibited high-performance, stability, and recyclability after multiple successive treatments. Variations in Cu content and alterations to the Cu oxidation state in the adsorbents have enabled us to determine the active site for capture and to elucidate a potential mechanism for inorganic P capture and release. Fast capture and release of phosphate anions are demonstrated with >90% efficiency of recovery using synthetic solutions of Na_2HPO_4. The zeolite sorbents are also found to be recyclable and sustain high recovery efficiency after multiple capture-release cycles.

483502 Metal Organic Framework for Subsequent/Simultaneous Removal of CO_2 and H_2SÂ from Natural Gas and Biogas

Abstract:

Metal Organic Framework for Subsequent/simultaneous removal of CO_2 and H_2S from natural gas and biogas

Youssef Belmabkhout, Prashant M. Bhatt, Amandine Cadiau, Åukasz J. WeseliÅ„ski and Mohamed Eddaoudi*

Functional Materials Design, Discovery & Development (FMD3), Advanced Membranes & Porous Materials Center (AMPM), King Abdullah University of Science and Technology, Saudi Arabia

Hydrogen sulfide (H_2S) and Carbon dioxide (CO_2) are among the major impurities found in natural gas. The level of these acid gases need to be reduced to fulfil specifications necessary to guaranty optimal corrosion free operations during transportation through pipelines. Enhancing the fraction of hydrocarbons, particularly CH_4, by removing CO_2 and H_2S is also crucial to enhance the calorific value of natural gas. In this work, we prepared various Metal Organic Frameworks (MOFs) and evaluated their tolerance to H_2S, using powder X-ray diffraction (PXRD) and cyclic adsorption. The performance of these MOFs for subsequent or simultaneous H_2S/CO_2 removal from CH_4 was evaluated using dynamic mixed gas breakthrough adsorption column testing using $H_2S/CO_2/CH_4$: 05/05/90 gas mixture. Based on these results we unveil a range of H_2S/CO_2 MOF adsorbents capable for either selective H_2S or CO_2 adsorption or for simultaneous H_2S/CO_2 adsorption from natural gases. These MOFs can provide an alternative to existing technologies for removal of acid gases (H_2S, CO_2) with different compositions.

Metal-Organic Frameworks with High Volumetric Hydrogen Storage Capacity

November 1st, 2017

Justin Purewal, Mike Veenstra.
Ford Motor Company, Research & Advanced Engineering

Don Siegel, Adam Matzger, Alauddin Ahmed, Antek Wong-Foy, Saona Seth
University of Michigan, Department of Chemistry and Mechanical Engineering Department

Go Further

Hydrogen Storage in Fuel Cell Vehicles: Advantages and Challenges

Hydrogen storage influences the key factors of fuel cell vehicles:

- ***Driving Range*** is 1.5x to 3x greater than battery electric vehicles and near conventional gasoline vehicles
- ***Fueling Time***: is in 3 to 5 minutes similar to conventional gasoline vehicles rather than hours for BEVs
- ***Specific Energy:*** is significantly better than a battery electric vehicle as the on-board energy increases

Hydrogen storage is one of the most significant cost items on a fuel cell vehicle:

- ***Storage System Cost*** is a close 2nd to the fuel cell system based on 5.6 kg usable and 80 kW net fuel cell

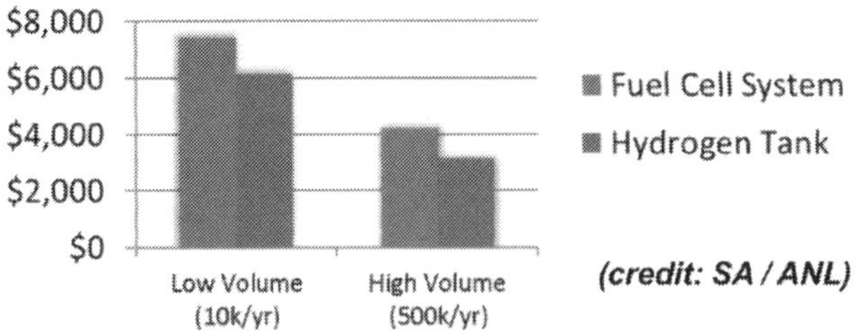

(credit: DOE record: https://www.hydrogen.energy.gov/program_records.html)

Hydrogen storage is a key enabler for the commercialization of FCVs

Challenge of Volumetric Energy Density

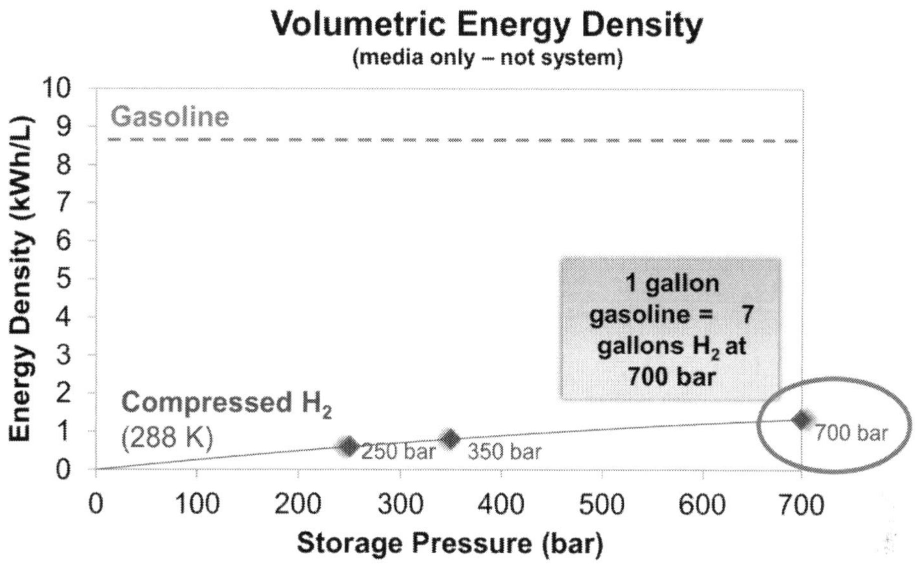

Challenge of Volumetric Energy Density

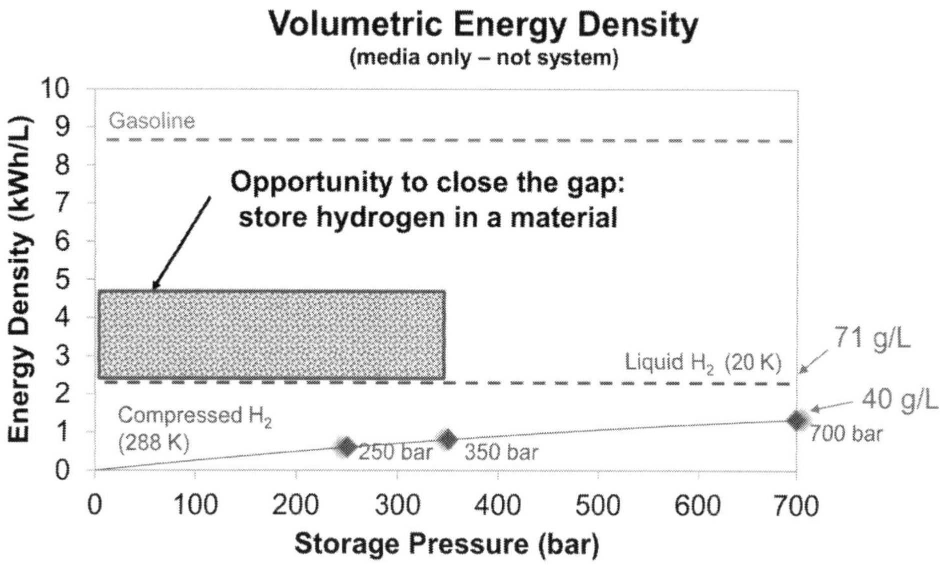

Hydrogen Storage Targets

DOE Target *(updated May 2017)*	2020	Ultimate	700 bar (typical)
System Gravimetric Density	4.5 wt.%	6.5 wt.%	4.4 wt.%
System Volumetric Density	30 g/L	50 g/L	25 g/L
Storage System Cost	$333/kg	$266/kg	$560/kg

Usable hydrogen capacities

-Increase system volumetric density (*e.g., improve storage material properties*)
-Reduce the system cost (*e.g., reduce use of high strength carbon fiber*)

Cryo-Adsorbent Hydrogen Storage

Taking the cryo-compressed concept one step further:
Fill the tank with high-surface area adsorbent (e.g., MOF-5)

HexCell (MOF-5)
Flow Through Chilled H_2 Cooling

MATI (MOF-5)
Isolated LN2 Flow Cooling

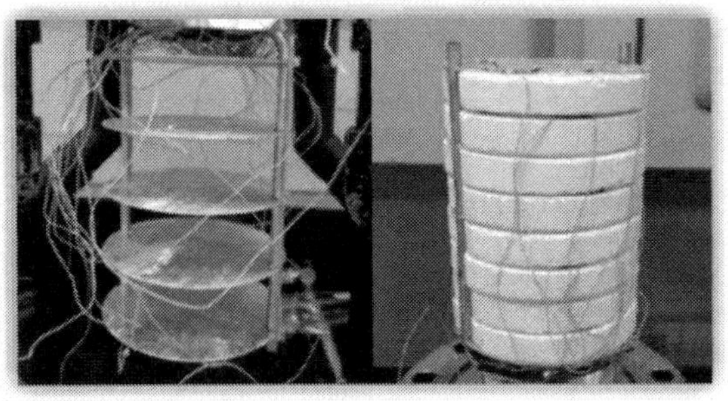

HSECoE (2010-2015) tested this concept with MOF-5 as adsorbent (pictured above)

- High volumetric H_2 densities can be achieved at $p < 100$ bar
- Enables the use of lower-cost type-1 aluminum tanks

Cryo-Adsorbent *vs.* 700 bar

System Comparison (*useable* 5.6 kg of H_2)

Internal HX and Media:	HexCell + powder MOF-5	MATI + 0.32 g/cc MOF-5 pucks	700 bar* Compressed H_2
System Mass (kg)	161	159	128
System Volume (L)	304	263	224
Estimate System Cost at 500K units	\$2,720**	\$2,897**	\$3,134
Gravimetric Capacity (g-H_2/g-system)	3.5 %	3.5%	4.4 %
Volumetric Capacity (g-H_2/L-system)	18.5 g/l	21.3 g/l	25.0 g/l

- Full tank: P = 100 bar, T = 80 K
- Empty tank: P = ~5 bar, T = ~140 K
- Single, Aluminum (6061-T6) Type 1
- LN_2 vessel wall chilling channels
- * 2015 AMR references (ref)

Lightly packed MOF-5 powder ~0.18 g/cc

*2013 AMR references
ANL Project ID: ST001
SA Project ID: ST100

**PNNL cost analysis

ADVANTANGES
- ✓ 7x lower pressure
- ✓ Lower tank cost
- ✓ Lower infrastructure capital cost
- ✓ Enhanced dormancy (?)

CHALLENGES
- ❖ Volumetric capacity
- ❖ Cryogenic certification
- ❖ Insulation robustness
- ❖ Liquid hydrogen pathway penalties

Adsorbent systems offer advantages using low pressure with challenges

Go Further

Material-Level H_2 Volumetric Capacity (for MOF-5)

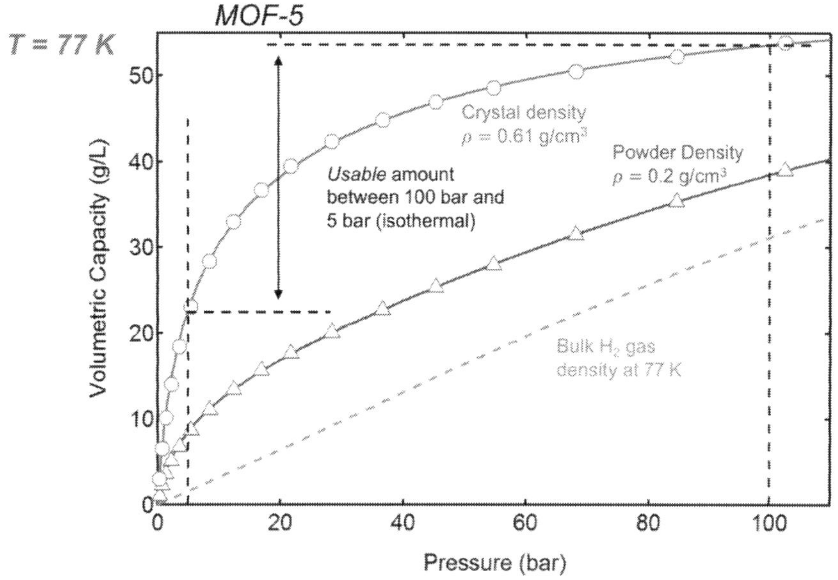

Our goal has been to maximize the usable hydrogen storage amount at 77 K. Packing density of MOF-5 powder has large effect on volumetric capacity.

Go Further

Experimental Screening of MOFs

Approach: computational screening of known MOF structures (e.g., CSD database). Then optimize synthesis.

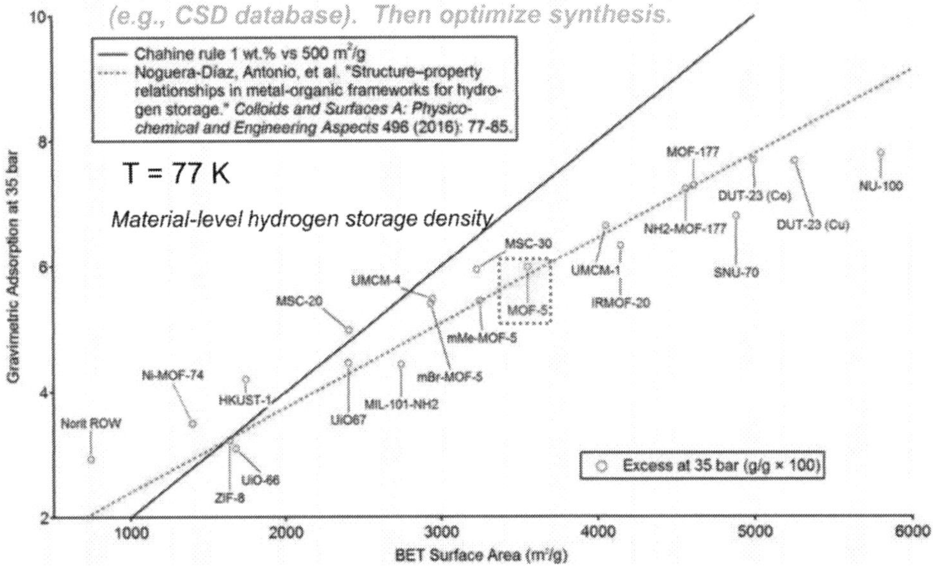

Correlation between BET surface area (m^2/g) and *gravimetric* H_2 adsorption at 35 bar, 77 K

Go Further

Experimental Screening of MOFs

Volumetric H_2 Adsorption versus volumetric BET surface area

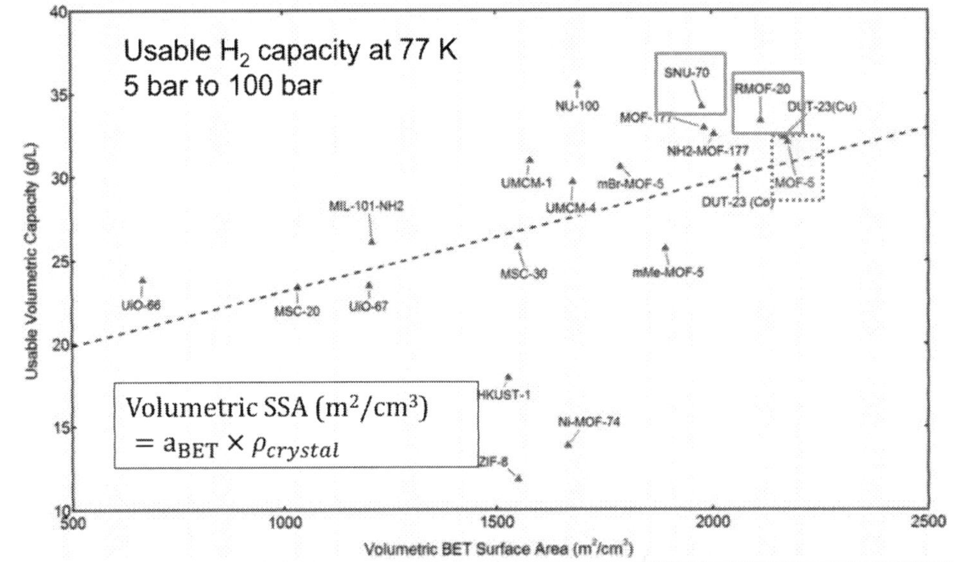

Stronger correlation between volumetric BET surface area (m^2/g) and volumetric hydrogen storage at 77 K

Go Further

Synthesis of IRMOF-20

$$H_2TTDC + Zn(NO_3)_2 \cdot 4H_2O \xrightarrow[100\ ^\circ C]{N,N\text{-diethylformamide}} Zn_4O(TTDC)_3$$

H_2TTDC =

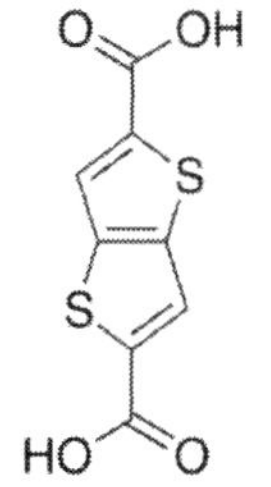

Thieno[3,2-*b*]thiophene-2,5-dicarboxylic acid

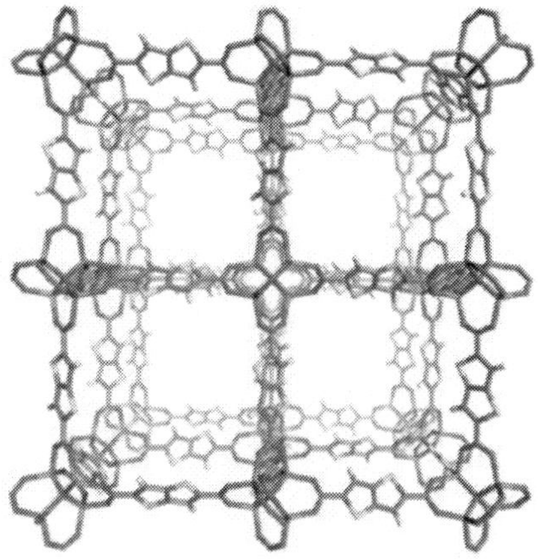

Activated by:
1) Multiple solvent exchanges
2) RT vacuum drying

BET S.A. =	4073 m^2/g (94% of calc'd)
Calculated =	4324 m^2/g
Literature =	3409 m^2/g

Rowsell, J. L. C.; Yaghi, O.M. *J. Am. Chem. Soc.* **2006**, *128*, 1304.

Synthesis of SNU-70

SNU-70

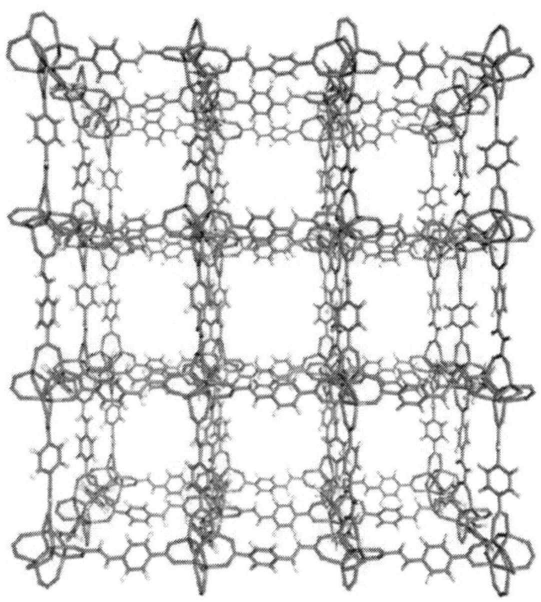

Same metal cluster (SBU) as IRMOF-20 and MOF-5 and same cubic topology

Different, but related linker

SNU-70 **IRMOF-20** **MOF-5**

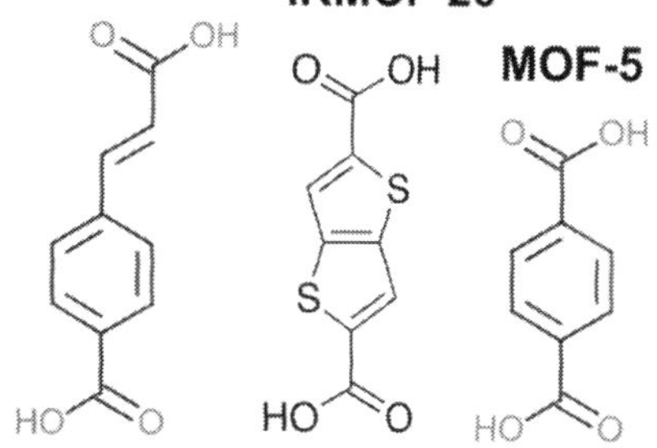

SNU Linker: 4-(2-carboxy-vinyl) benzoic acid (H_2CVB)

Improvement in H_2 Storage Compared to MOF-5 Benchmark

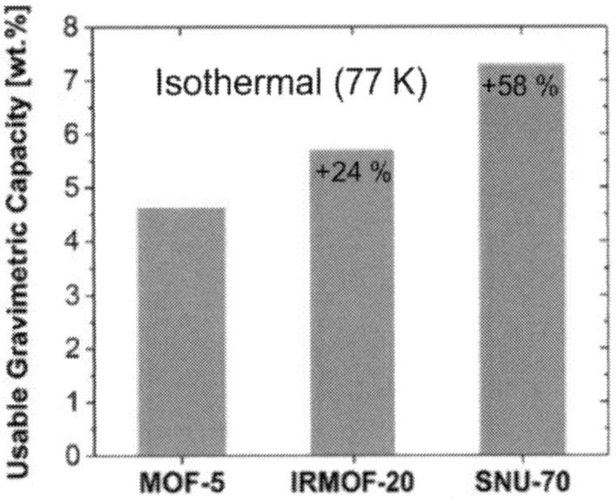

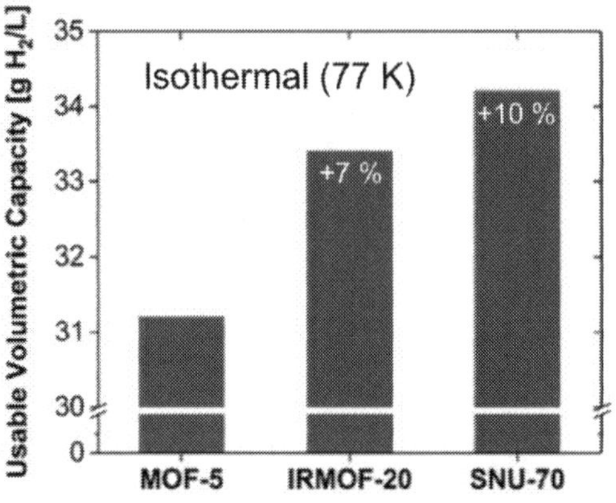

Compound	Single Crystal Density [g/cm³]	Pore Volume [cm³/g]	Gravimetric Surface Area [m²/g]	Volumetric Surface Area [m²/cm³]	Void Fraction	Pore Diameter/ Aperture [Å]
MOF-5	0.60	1.36	3,563	2,172	0.81	15.1/7.9
IRMOF-20	0.51	1.65	3,913	2,000	0.84	17.3/9.3
SNU-70	0.40	2.14	4,756	1,905	0.86	16.8/10

Go Further

Hydrogen Storage Model (Material Level)

Measured isotherms at 77 K, 103 K, 118 K, 138 K, 200 K and 295 K

Modified D-A model parameters (MOF-5)	
nMax	252.794 g/kg
a	2239 J/mol
b	19.5 J/mol*K
p0	16920 bar
vA	2.01 L/kg
m	2

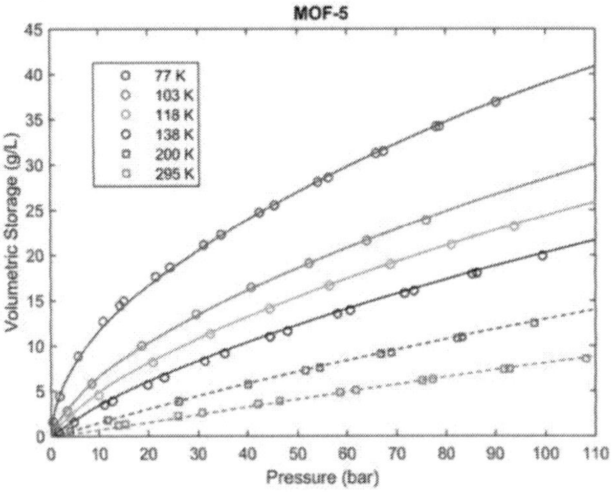

Fit H_2 adsorption data (multiple temperatures) to modified Dubinin-Astakhov model

Go Further

Hydrogen Storage Model (Material Level)

Estimated isotherms at 80 K and 160 K

Modified D-A model parameters (MOF-5)	
nMax	252.794 g/kg
a	2239 J/mol
b	19.5 J/mol*K
p0	16920 bar
vA	2.01 L/kg
m	2

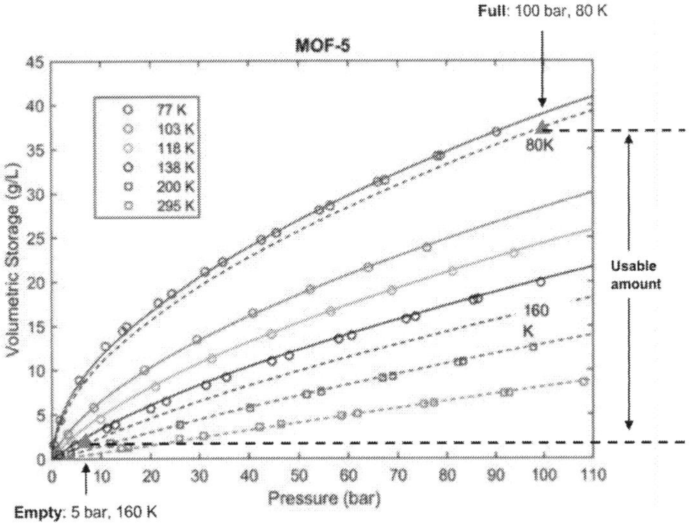

Usable hydrogen is calculated at "Full" and "Empty" states for system

Go Further

System-Level Hydrogen Storage Capacity

Assumed powder density of 0.2 g/cm³ (which we can achieve for all four MOFs by simple tapping or vibrating.

System Properties

Initial/Full Pressure:	100 bar
Initial/Full Temp:	80 K
Final/Empty Pressure:	5 bar
Final/Empty Temp:	160 K
Useable Hydrogen:	5.6 kg
Heat Exchanger:	HexCell
MOF Density:	0.2 g/cc
Pressure Vessel:	Type 1 Al
Insulation Thickness:	23 mm
Outer Shell Thickness:	2 mm

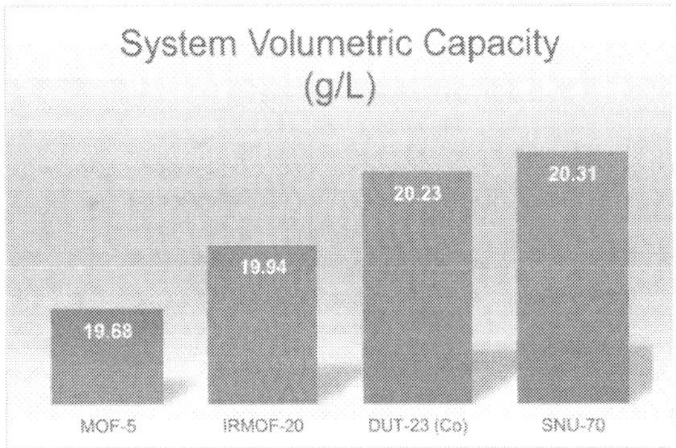

Improved material-level capacity yields improvement to system-level capacity...

But still below 700 bar compressed (25 g/L)

Go Further

Increasing SNU-70 packing density

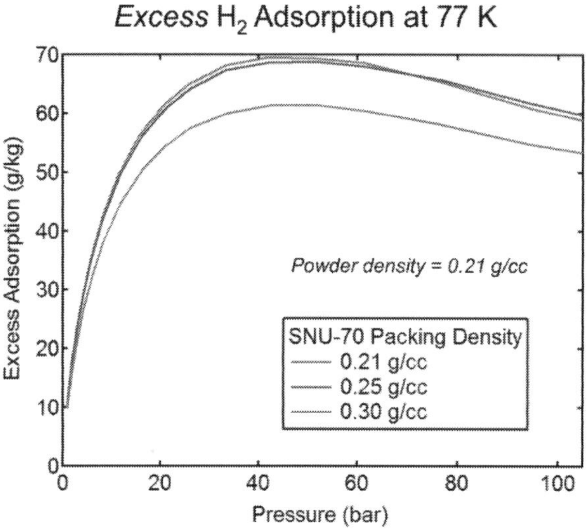

SNU-70 can be compacted to a density of 0.25 g/cc without decrease in excess H_2 adsorption at 77 K. Further work needed

Go Further

Pathway to Meeting DOE 2020 Targets

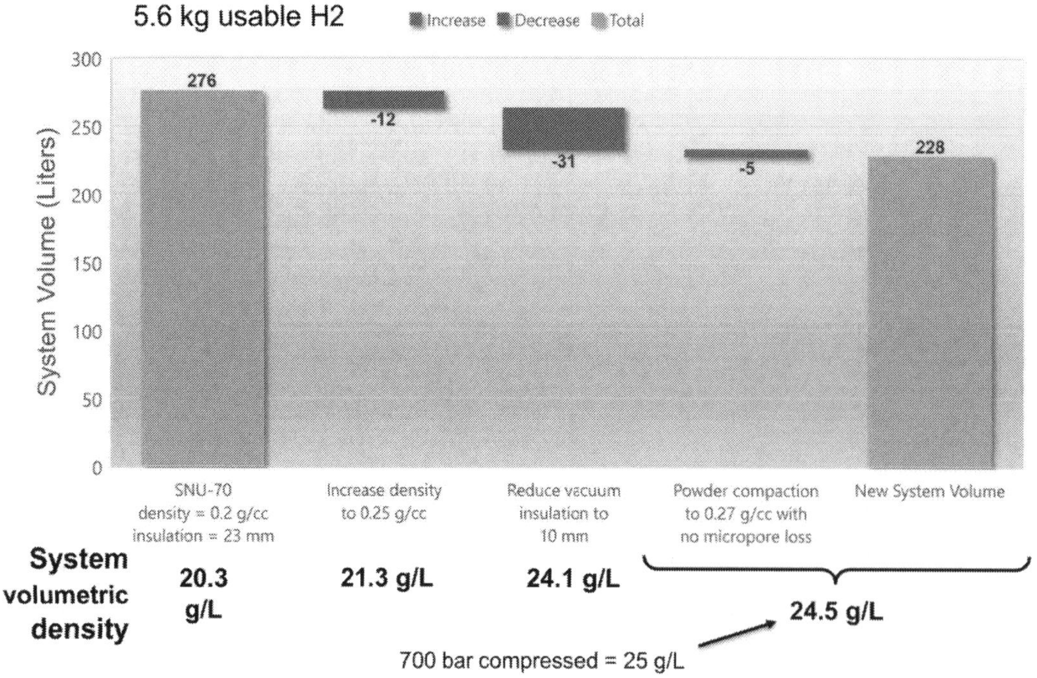

Go Further

Conclusions

1. Cryo-adsorbent based H_2 storage has the potential to be competitive with 700 bar compressed at 7x lower pressure
2. Earlier cryo-adsorbent system prototypes and models established MOF-5 as the benchmark sorbent
3. We have identified sorbents which exceed MOF-5 in terms of *usable volumetric H_2 capacity* at 77 K: SNU-70, IRMOF-20, NU-100
4. SNU-70 can be compacted to 0.25 g/cc without decrease in excess H_2 adsorption
5. Analyzed strategies for surpassing volumetric capacity of 700 bar compressed, meeting 2020 DOE targets with cryo-adsorbent system models

Extra Slides

Synthesis of MOF-5

Performed air-free synthesis[1] of the benchmark compound MOF-5

$$H_2BDC + Zn(NO_3)_2{\bullet}6H_2O \xrightarrow[80\ ^\circ C]{\text{N,N-diethylformamide}} Zn_4O(BDC)_3$$

H_2BDC =

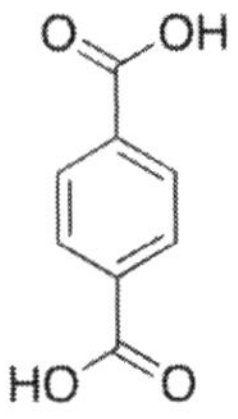

Benzenedicarboxylic acid

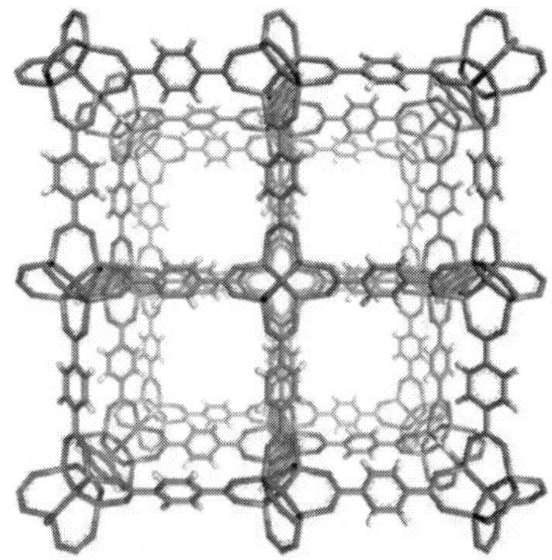

Activated by:

1) Multiple solvent exchanges
2) RT vacuum drying

BET S.A. =	3512 m^2/g
Calculated =	3563 m^2/g
Literature =	3800 m^2/g [1]

[1]Kaye, Dailly, Yaghi, and Long, 2007. JACS, 129,14176:

MOF-5 Benchmark Data

- Measured performance of air-free synthesis of MOF-5
- Hydrogen uptake & BET surface area are essentially identical to pre-commercial MOF-5 supplied by BASF for HSECoE sub-scale tank
- Data adopted as benchmark

Excess H_2 Adsorption

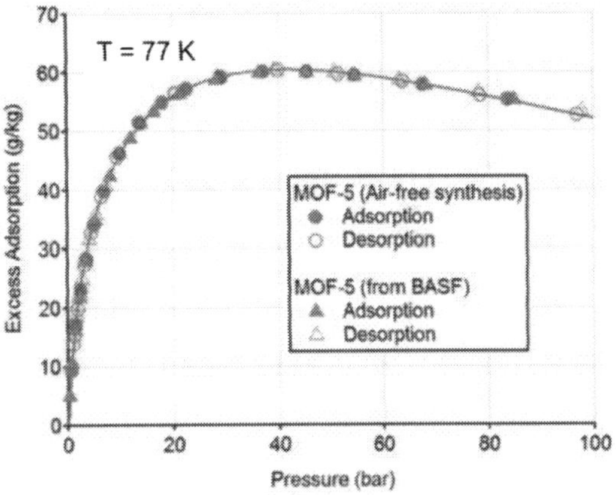

Hydrogen ***Total*** *Storage*

	Total	
p (bar)	Volumetric (g/L)	Gravimetric (wt.%)
5	22.2	3.5
35	44.4	6.8
50	47.8	7.3
100	53.3	8.0

Hexcell Design: Type-1 Al tank with MOF5 and honeycomb HX

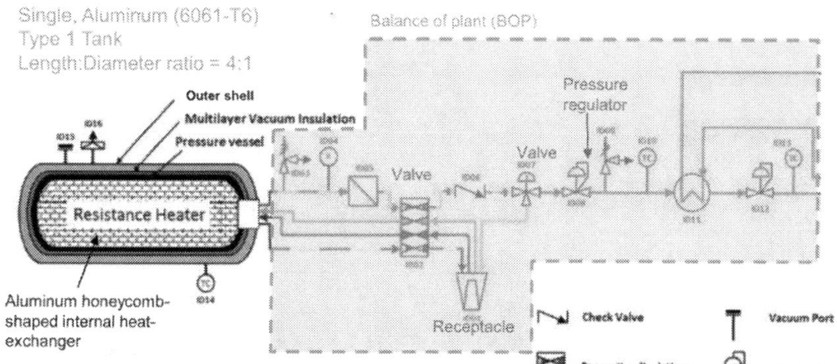

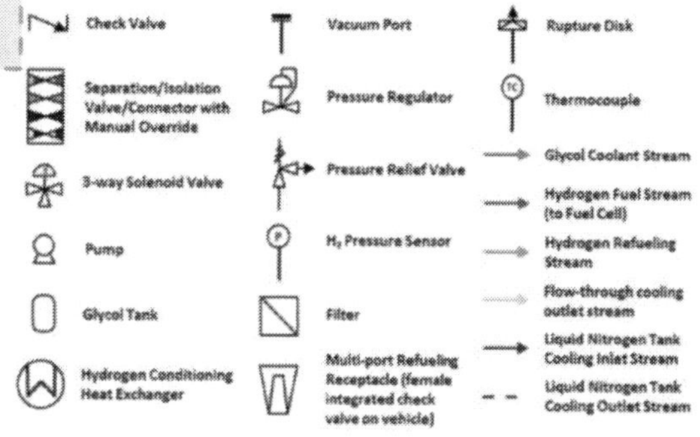

System Properties	
Initial/Full Pressure:	100 bar
Initial/Full Temp:	80 K
Final/Empty Pressure:	5.5 bar
Final/Empty Temp:	160 K
Useable Hydrogen:	5.6 kg
Heat Exchanger:	HexCell
MOF Density:	~0.2 g/cc
Pressure Vessel:	Type 1 Al
Insulation Thickness:	23 mm
Outer Shell Thickness:	2 mm

Go Further

Details of System-Level Mass and Volumes

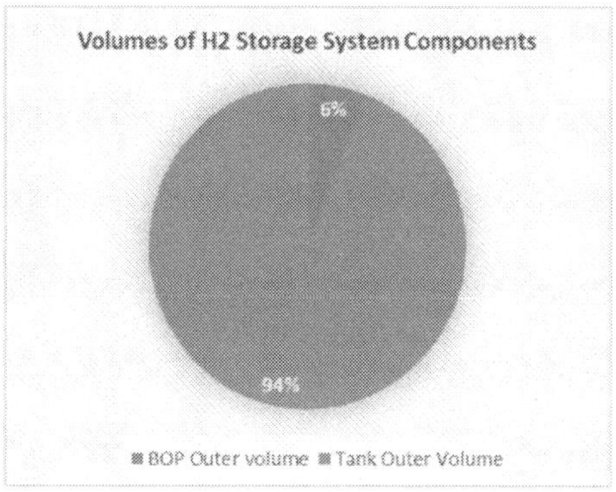

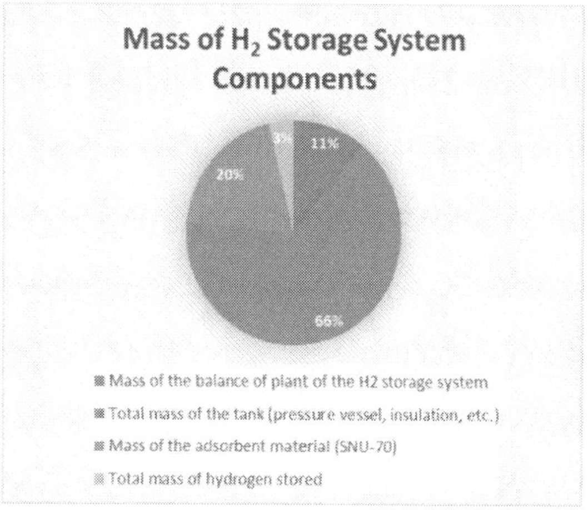

Majority of system mass is from the tank + insulation + BOP (not MOF or hydrogen). System volume is primarily tank.

Go Further

Cryogenic Hydrogen Storage: *Advantages and Drawbacks*

Advantages

1. At high volumes (>500 kg/day) more efficient and cost-effective to transport H_2 in liquid tankers

2. An LH2 trailer can transport up to 6 times as much hydrogen load as conventional compressed gas trailer

3. Infrastructure footprint at station is smaller for for LH2 and CCH2 compared to 700 bar
   a) No multi-stage compressor and high-power chiller

Challanges

1. Insulation robustness
   a) Multi-layer vacuum super insulation
   b) Hydrogen permeation, VOC outgassing from carbon-fiber composite tanks, degrades vacuum
2. Driving range uncertainty
   a) Final fill level depends on initial temperature and fuel level
3. Liquefaction energy and CO_2 penalty

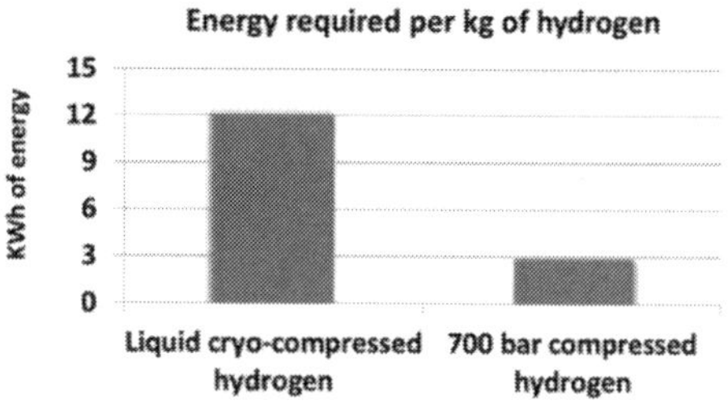

494606 Three-Dimensional Lignin-Based Smart Aerogel for Controllable Oil/Water Separation

Abstract:

Three-dimensional lignin-based smart aerogel for controllable oil/water separation

Jingxian Jiang, Qinghua Zhang, Xiaoli Zhan, Fengqiu Chen

College of Chemical and Biological Engineering, Zhejiang University, Hangzhou 310027, P. R. China

ABSTRACT

Recent years, oil/water separation has become a worldwide challenge. Among all these approaches to deal with this severe problem, adsorption separation shows promising prospects due to its high efficiency, easy manipulation, and practicability under harsh circumstances. Oil sorbents derived from biomass which is a kind of available renewable resource, also have the advantages of cost-effectiveness, environmental-friendliness and convenient enrichment of multi-functionality. Herein, we present a facile method to fabricate a self-standing compressible three-dimensional lignin-based aerogel with porous structure, low density and a smart surface that can be responsive to medium pH. Based on the hydrolytic condensation effect of silane coupling agent, the copolymer P(DMAEMA-SA-KH570) synthesized by free radical polymerization was covalently bonded onto the lignin-based aerogel through simple dip-coating method. The integrated aerogel exhibited superhydrophobic and superoleophilic properties and could be served as effective oil sorbent when medium pH was 7 with an adsorption capacity for oil and organic solvents of 20-40 times of its own weight. And it could realize oil desorption when medium pH was changed to 1. Furthermore, the cycling stability of aerogel was evaluated by both mechanical squeezing and pH controlling. This work presents a method to fabricate robust smart oil-absorbing aerogel for controllable oil/water separation based on lignin and offers a new viable sustainable approach for oil recovery.

TUNABLE FABRICATION OF CARBON NANOFIBERS FROM RENEWABLE AND WASTE RESOURCES

(Abstract ID: 496751)

E. Svinterikos[1], M. Al-Marzouqi[1], A.M. Soliman[2,3], I. Zuburtikudis[4]*

[1]Department of Chemical and Petroleum Engineering, United Arab Emirates University (U.A.E.U), Al Ain, U.A.E

[2]Department of Chemistry, United Arab Emirates University (U.A.E.U), Al Ain, U.A.E

[3]Nuclear Fuel Technology Department, Hot Laboratory Center, Atomic Energy Authority, Cairo, Egypt

[4]Department of Chemical Engineering, Abu Dhabi University, Abu Dhabi, U.A.E., E-mail: ioannis.zuburtikudis@adu.ac.ae

Abstract

The research presented here reports the utilization of biorenewable and waste resources for the development of versatile adsorbents for wastewater remediation. Lignin was blended with recycled poly(ethylene terephthalate) and spun into nanofibers of controlled diameter via electrospinning. The nanofibers were carbonized and activated under N_2 and CO_2 respectively at 600 °C, in order to transform them into activated carbon nanofibers (ACNFs). These ACNFs have a mesoporous structure with BET surface area around 314 m^2/g and average pore width 6.7 nm. After treating them with HNO_3, the ACNFs were used for the successful adsorption of Pb^{2+} ions from aqueous solutions. The fitting of the Langmuir adsorption isotherm reveals that they exhibit a theoretical maximum adsorption capacity of 23.75 milligrams of Pb^{2+} per gram of ACNFs.

1. Introduction

The utilization of inexpensive renewable and waste resources for producing high added-value products can have positive implications in the sustainable economy of the future. This concept has been the main motivation behind our research, which focuses on the development of versatile adsorbents from inexpensive bioresources and waste feedstock.

Lignin is the second most abundant natural polymer on earth behind cellulose [1]. It is a major by-product of the paper industry, while it can also be extracted from waste by-products of agriculture activities, such as date pits and date palm pruning woods [2,3]. Despite its large availability, only a very small percentage of its worldwide production is utilized for the manufacture of commercial products, mainly adhesives and dispersants [2]. However, various research attempts have exhibited its potential for producing a variety of products, such as carbon fibers, thermoplastic elastomers and membranes among others [4]. Its advantages include the relatively high carbon content (>60 wt%) and the presence of functional groups, whereas its structural complexity and its relatively poor mechanical properties hinder its more widespread use [4].

* Corresponding author: Tel: 00971564913992, Email: ef_svinterikos@uaeu.ac.ae (Efstratios Svinterikos)

Poly(ethylene terephthalate) (PET) is one of the most widely used commodity plastics and the main feedstock for the manufacture of water bottles. In addition, it is the most widely recycled plastic in the USA in terms of weight of plastic material recycled [5]. Therefore, new potential applications for recycled PET would contribute to sustainable growth.

Carbon fibers have been the focus of research due to an array of interesting properties such as high tensile strength, low density and good flexibility [6,7]. When their diameter is reduced in the submicron range, their surface area is maximized and they are termed as carbon nanofibers (CNFs). They are currently being used as fillers for composite materials in the aerospace and automotive industries, while they have potential application in adsorption, catalysis and energy storage among others [6,7]. One of the methods for producing carbon fibers is the carbonization of suitable precursor polymer fibers.

Electrospinning is a well-established technique for the manufacture of submicron polymeric fibers from a wide variety of polymers. Its advantage lies in the easy adjustment of the process conditions, which creates on-demand fibrous structures with controlled diameters. Briefly, the technique produces fibers by applying an electrostatic field between the tip of a syringe that contains the polymer solution and a grounded collector; this field stretches the droplets of the polymer solution. More details on the technique can be found in ref. 8. Minimizing the average fiber diameter to the nano-scale increases their surface area and enhances their applicability for various industrial and medical applications.

The treatment of wastewater for the removal of organic, inorganic, and biological pollutants is of major importance for the protection of the environment and the public health. Heavy metals, such as Pb, Cr, Cd, etc., tend to accumulate in the soil, and therefore create severe health risks after they are transferred in the food chain [9]. Among the various methods that have been developed for the removal of heavy metals from wastewater, the metal ion adsorption carries significant advantages due to its high efficiency and the availability of low-cost adsorbents [9].

In the present study, the removal of Pb^{2+} ions from wastewater was accomplished using activated mesoporous carbon nanofibers, which were prepared from carbonizing precursor electrospun fibers from a blend of lignin with recycled PET. Therefore, a road-map for the effective preparation of value-added products from inexpensive resources is presented.

2. Materials and methods

Kraft lignin (alkali lignin, low sulfonate content, Mw ~ 10,000) was obtained from Sigma-Aldritch and was used as received. The source of recycled PET was waste water bottles. All the bottles that were used came from the same water-bottling company in the U.A.E. and were chosen for their availability (popular water brand). They were dried and cut into pieces for the preparation of the spinnable solutions. Trifluoroacetic acid (TFA, 99%) purchased from Merck was used as solvent for the electrospinning. Solutions of lignin/PET with various concentrations were prepared and left under magnetic stirring at room temperature for 6 h until the PET pieces had been completely dissolved.

The precursor fibers were carbonized and activated by heating them inside a temperature-programmed tubular furnace (Tube Furnace GSL-1500X-50, MTI). Initially the fibers were heated under N_2 at a rate of 4 °C/min until they reached 600 °C, where they were held for 1h. Then the atmosphere was switched to CO_2 for 1h in order to activate them. Finally, the fibers were cooled under N_2 until they reached room temperature.

The morphology of the fibrous mats was examined using Scanning Electron Microscopy (JEOL Neoscope JCM-5000) after they were gold-coated. The mean fiber diameter was measured using an image analyzer (ImageJ, National Institutes of Health, U.S.A.). Around 100 measurements in each sample were used for the calculation of the mean diameter and the standard deviation. The functional groups in the structure of the precursor as well as of the activated carbon fibers were investigated using Attenuated Total Reflection-Fourier Transform Infrared Spectroscopy (ATR-FTIR), with an IRPrestige-21/Shimadzu spectrophotometer and a MIRacle ATR tool by PIKE Technologies. The porosity characteristics and the BET surface area of the carbon fibers were measured through N_2 physisorption using a Micromeritics, Tristar II Plus instrument.

The adsorption capacity of the carbon fibers was assessed through batch adsorption experiments at room temperature. The initial solutions of Pb^{2+} ions were prepared by dissolving the appropriate amount of $Pb(NO_3)_2$ in deionized water. 20 mg of activated carbon nanofibers were added to 25 mL of water solution, with the initial concentration of Pb^{2+} ions ranging between 10 to 100 ppm. The flasks were sealed and shaken at 200 rpm for 24 h in order to let them reach the equilibrium concentration. The concentration of Pb^{2+} ions in the water solutions before and after adsorption was measured using ICP with a Varian 710-ES instrument.

3. Results and discussion

3.1 Preparation of the precursor polymer nanofibers

The first stage of the research involved the preparation of the precursor carbon nanofibers from the blend of lignin with PET through electrospinning. The target here was the continuous fabrication of bead-free mats with the minimum average fiber diameter. In order to achieve this, five process and material parameters were selected (solution concentration, spinning distance, voltage, lignin/PET ratio and flow rate) and their effect on the average fiber diameter was examined. Following a design-of-experiments statistical methodology, the process was optimized and a minimum average fiber diameter of 191 ± 60 nm was achieved. Details on the methodology and the statistical analysis can be found in ref. 10. In the research presented here, all the carbonized and activated nanofibers were produced from precursor fibers consisting of a lignin/PET ratio of 1/1.

3.2 Carbonization and activation

After the electrospinning step, the precursor nanofibers were carbonized and activated in order to obtain a carbon fibrous structure of high surface area and porosity, following the procedure described in section 2. The carbonization is generally achieved with heating under inert atmosphere at elevated temperatures (>600 °C), in a process during which the volatiles including the non-carbon elements are removed, and the material is transformed into a carbon structure with regions of graphite and disordered carbon [11]. Further processing with CO_2 increases the surface area by opening pores on the material surface, while some oxygen-containing functional groups are introduced [11]. Attempts to activate the precursor fibers at temperatures higher than 600 °C, led to oxidation and burning of the fibers. A possible explanation is the occurrence of the so-called reverse Boudouard reaction ($CO_2 + C_{(solid)} \leftrightarrow 2CO$), an endothermic reaction, which is favorable at high temperatures (>700 °C) and consumes solid carbon in the presence of CO_2 [12,13].

The precursor nanofibers of minimum diameter were carbonized and activated, but it was observed that the fibrous structure was not preserved. Instead, there was extensive sintering and merging of the fibers, which compromised the surface area (figure 1). In contrast, fibrous mats of larger average diameter (>400-450 nm) exhibited a well-preserved fibrous structure, with only a minor degree of merging, and some curling, which however didn't destroy the fibrous network (figure 2). It is obvious in figure 2 that the fiber diameters after carbonization and activation remain at the same levels as the precursor fibers. Figure 2d shows that the activated carbon fibrous network resembles the structure of a sponge. The results presented here and the adsorption experiments are based on these activated carbon nanofibers (ACNFs).

Figure 1. Carbonization and activation of precursor nanofibers of minimum diameter; a) the precursor nanofibers (the average fiber diameter here is 145 ± 54 nm), b) the nanofibers after carbonization and activation

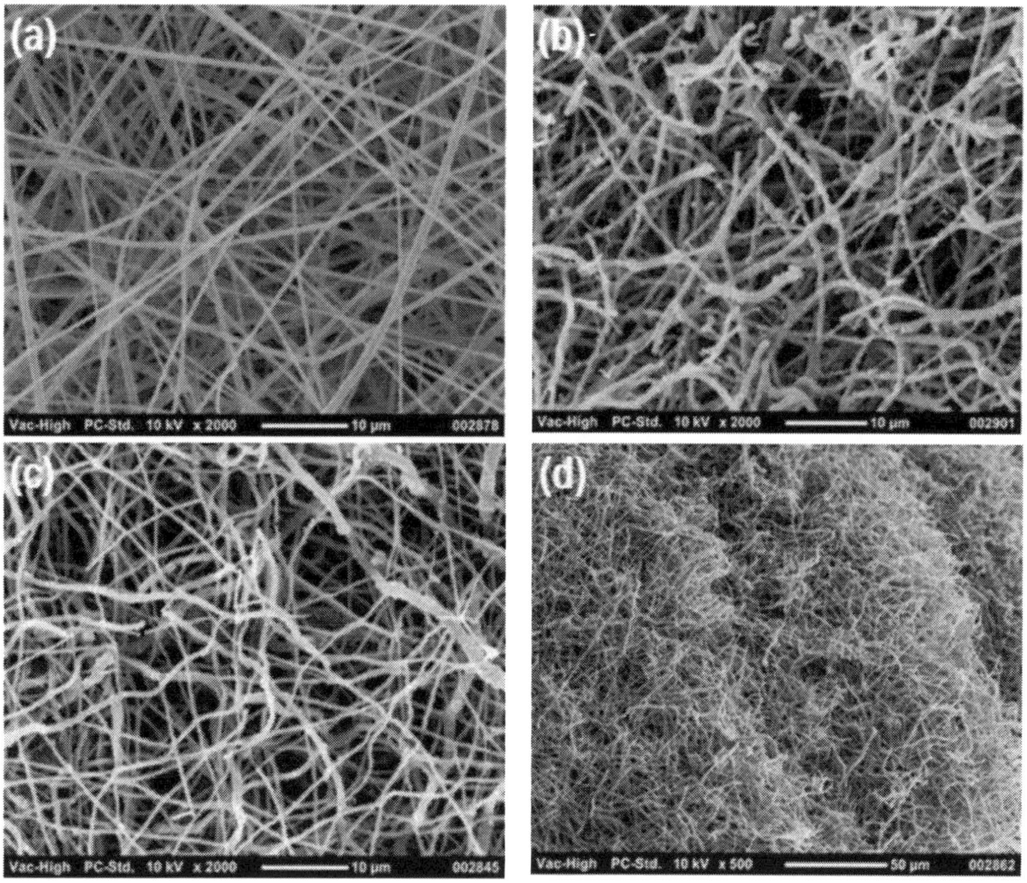

Figure 2. Carbonization and activation of a fibrous mat of larger average diameter. a) the precursor nanofibers (average diameter is = 586 ± 183 nm), b) the fibers after carbonization (average diameter = 605 ± 140 nm), c) the carbonized fibers after activation (average diameter = 546 ± 147 nm), d) the same fibers as in (c) but at lower magnification

3.3 Characterization of the carbon nanofibers

Table 1. Pore characteristics of the carbonized and activated nanofibers

CNFs (N_2 @ 600 °C)	ACNFs (N_2/CO_2 @ 600 °C)
BET surface area: 28 m^2/g	BET surface area: 314 m^2/g
Total pore volume: 0.028 cm^3/g	Total pore volume: 0.182 cm^3/g
BJH Adsorption average pore width: 28.8 nm	BJH Adsorption average pore width: 6.7 nm

Both carbonized and activated fibers were produced from the same initial precursor mat, with average diameter ~ 500 nm. The weight yield after carbonization was ~ 33 %, while after activation it dropped to ~ 30 % compared to the initial weight of the precursor polymer nanofibers. Table 1 shows the results of the N_2 adsorption measurements for the nanofibers which were carbonized with N_2 at 600 °C (CNFs), and for those which were carbonized and activated with N_2/CO_2 at 600 °C (ACNFs). Figure 3 shows the N_2 adsorption isotherms for both the carbonized and the activated nanofibers. Table 1 reveals that the activation with CO_2 greatly enhances the surface area and the pore volume of the nanofibers. At the same time, the average pore width is reduced from an average value of 28.8 nm for the carbonized fibers, to 6.7 nm after activation. Moreover, figure 3 shows that the isotherms are typical of mesoporous materials with hysteresis loop between adsorption and desorption, while the almost horizontal slope is typical of materials which contain slit-shaped rather than round-shaped pores.

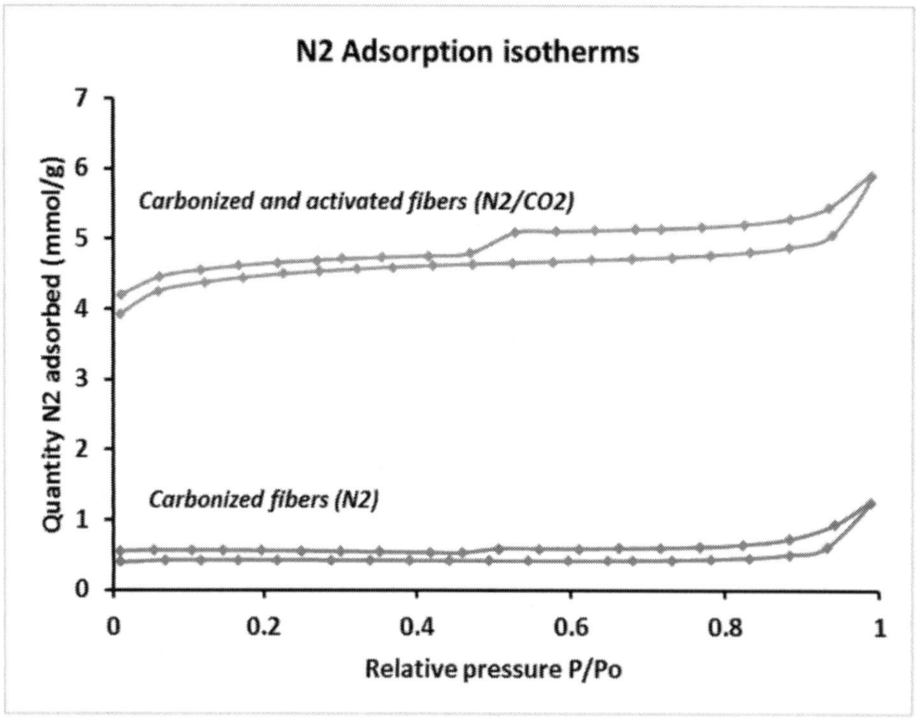

Figure 3. N_2 adsorption isotherms for the carbonized fibers at 600 °C (bottom) and for the activated fibers at the same temperature (top)

In order to investigate the presence of functional groups in the ACNFs, their infrared spectrum was recorded, both for CNFs and ACNFs, and the results are presented in figure 4, in the range of 850 – 2010 cm^{-1}. In the same figure, the infrared spectra of the starting precursor polymer nanofibers and of pure graphite are presented.

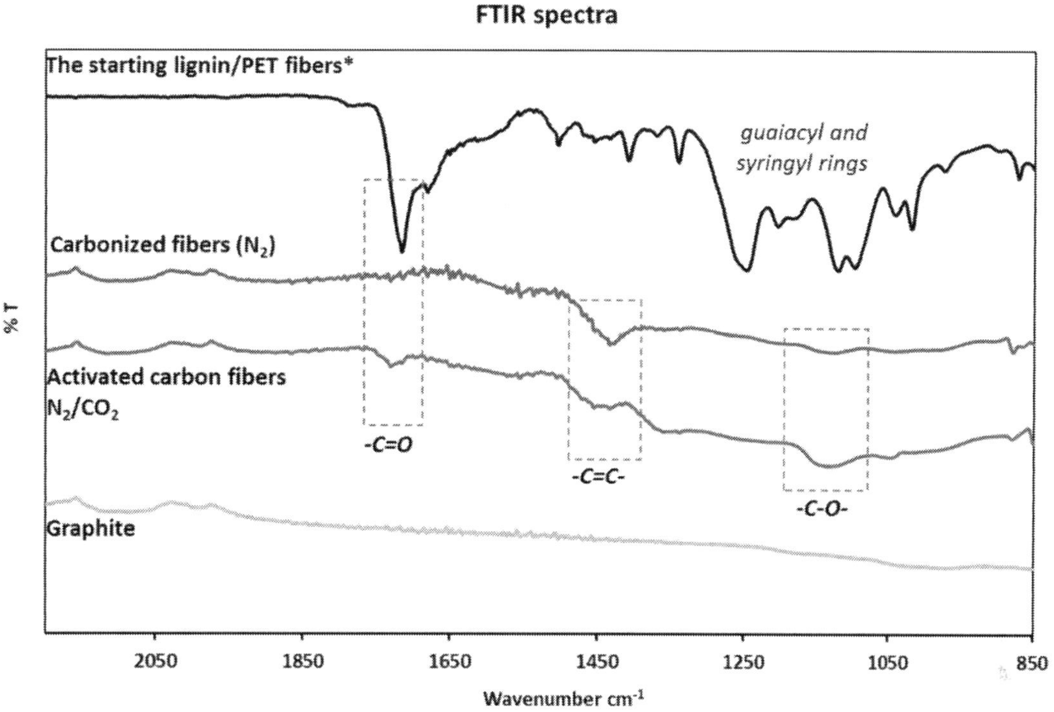

Figure 4. The infrared spectra of the carbonized (CNFs) and activated carbon nanofibers (ACNFs) in comparison with the spectrum of the precursor polymer nanofibers and of the spectrum of pure graphite

The spectrum of the precursor nanofibers (consisting on lignin/PET ratio of 1/1) reveals the presence of various functional groups, which are described in detail in ref. 10. The spectrum of pure graphite (which is usually prepared at much higher carbonization temperatures, of more than 1000 °C) is a straight line without peaks as it doesn't contain any functional groups. In contrast, the spectra of the carbonized and the activated nanofibers presented here indicate the presence of oxygen containing functional groups. These are located in the region around 1700-1750 cm^{-1} (-C=O stretching vibration) and in the region around 1130 cm^{-1} (-C-O- stretching vibration) [9,14]. The absorbance at 1450 cm^{-1} (C=C aromatic nucleus) indicates the attachment of functional groups to the aromatic rings. These oxygen-containing functional groups probably derive from the presence of an amount of lignin which wasn't transformed into carbon, and also by the treatment with CO_2 during the activation process. These groups are desirable for the purpose of adsorption, as they represent active sites on which ions can be attached due to electrostatic interactions.

3.4 Adsorption of Pb^{2+} ions

In monitoring and modeling the adsorption of metal ions from aqueous solutions, the pH is of high significance. Depending on the pH, different ionic species prevail in the solution. As indicated by diagrams of electrode potential vs. pH (Pourbaix diagrams), which are available for almost all metal ions (see ref. 15), as the pH increases and the solution becomes alkaline, the prevalence of OH^- ions leads to hydrolysis and to the formation of metal-hydroxide ions ($Me(OH)_x$) which precipitate [9]. Therefore, if the adsorption is attempted at alkaline pH, there will possibly be some precipitation simultaneously with adsorption, and when measuring the concentration of ions before and after, there will be dubious conclusions on the actual quantity of ions adsorbed. Therefore, the adsorption of metal ions should preferably be attempted in acidic regions of pH, where metal ions are stable in the solution.

Regarding the activated carbon nanofibers prepared in the previous steps, we observed that when we added them in the aqueous solutions of Pb^{2+} ions (which initially exhibited acidic pH), the pH was rising to around 10 – 10.5. This is probably due to some amount of untransformed lignin remaining in the carbon materials. Lignin contains OH^- groups and thus its aqueous solutions have an alkaline character. So, in order to proceed with the adsorption experiments, we decided first to neutralize the activated carbon nanofibers with a proper treatment. Two different treatment routes were employed; the first one was washing the ACNFs several times with de-ionized water until the supernatant reached an almost neutral pH; the second one was boiling them with HNO_3 1M at elevated temperature (~90 °C). In both cases, the ACNFs were dried after treatment. In the case of washing with de-ionized water, it was observed that the surface area increased to around 511 m^2/g, indicating the washing away of some lignin which remained in the ACNFs. Figure 5 shows SEM images of the ACNFs before treatment, and after treating them either with de-ionized H_2O or with HNO_3. It is clear that the fibrous structure is not influenced by the treatment.

Figure 5. The ACNFs before treatment (a), after washing with de-ionized water (b), and after boiling with HNO_3 (c)

Figure 6 shows the infrared spectrum of the untreated ACNFs in comparison with the spectrum of the ACNFs after washing with de-ionized water, and their spectrum after treatment with HNO_3. In the case of the fibers washed with de-ionized water, the spectrum appears to be more flat, probably because some of the oxygen-containing functional groups were attached on residual lignin and were washed away with it after the treatment. In contrast, the spectrum of the fibers treated with HNO_3, includes a very broad region of infrared adsorption, around 1100 cm^{-1}, and another broad peak at around 1450 cm^{-1}. This indicates the presence of oxygen-containing functional groups attached to aromatic rings present in the ACNFs. The treatment with HNO_3 is expected to oxidize some double bonds in the aromatic structure of the carbon material, and the infrared spectrum gives a strong indication of this action.

The adsorption capacity of the ACNFs treated with de-ionized water and of those treated with HNO_3 was assessed through batch adsorption experiments, with the initial concentration of Pb^{2+} ions ranging from 10 to 100 ppm. The measurement of the equilibrium concentration after adsorption allowed the construction of the isotherms and the fitting of a suitable model. It was found that the model which best fits the data is the Langmuir isotherm. The adsorption isotherms for the linear form of the Langmuir model are presented in figure 7.

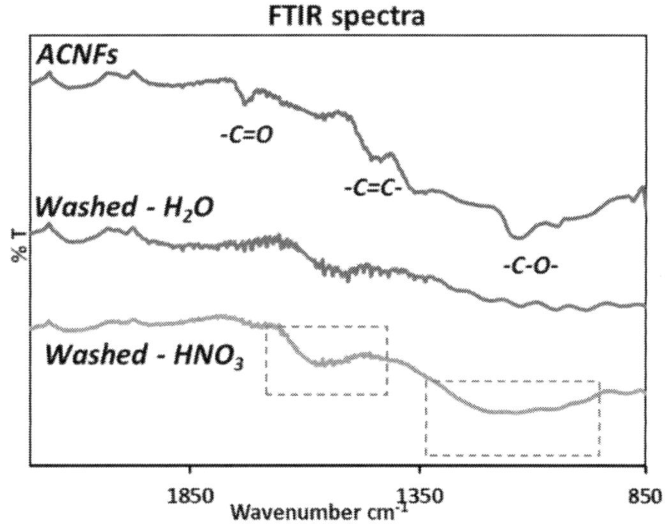

Figure 6. The infrared spectra of the untreated ACNFs, of the ACNFs treated with de-ionized H_2O and of the ACNFs treated with HNO_3

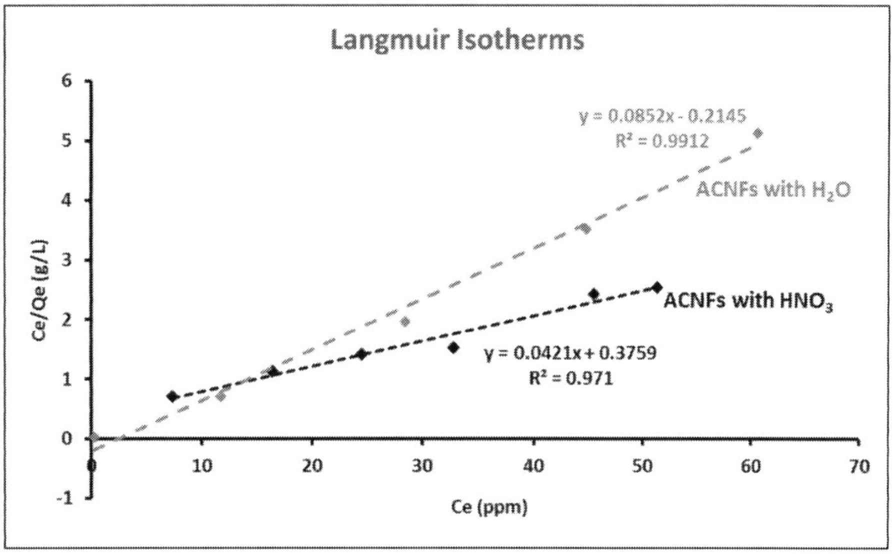

Figure 7. The linearized Langmuir adsorption isotherms for Pb^{2+} adsorption on ACNFs treated with de-ionized H_2O (red line) and on ACNFs treated with HNO_3 (black line)

The linear form of this model is $Ce/Qe = (1/K*Q_{max}) + (Ce/Q_{max})$, where Ce (mg/L) is the equilibrium concentration of the adsorbate, Qe (mg/g) is the amount of metal ions adsorbed per unit mass of adsorbent at equilibrium, Qmax (mg/g) is the theoretical maximum adsorption capacity and K (L/mg) is the Langmuir adsorption constant. The ACNFs treated with HNO_3 exhibited higher adsorption capacity and their theoretical maximum adsorption capacity (Qmax) was calculated to be 23.75 mg of Pb^{2+} per g of ACNFs. In contrast, the theoretical Qmax for the ACNFs washed with H_2O was much lower, 11.73 mg/g. The superior adsorption capacity of ACNFs treated with HNO_3 is expected, and an indication for this was provided by the infrared spectra. Washing the nanofibers with de-ionized water removed the residual lignin and together some functional groups, which would provide active sites for the ion adsorption. In contrast, after the treatment with HNO_3 some additional functional groups were probably introduced due to the oxidizing action of the strong acid. Hence, the larger adsorption capacity exhibited by the acid-treated ACNFs. The presence of these functional groups will be investigated in more details with standard titration methods in the next steps of our study.

4. Conclusions

In this research, the successful preparation of adsorbents from renewable and waste resources was demonstrated. Lignin was blended with recycled PET and precursor carbon nanofibers of controlled average diameter were prepared via electrospinning. After their carbonization and activation at 600 °C, mesoporous activated carbon nanofibers (ACNFs) were produced, with average fiber diameters similar to the that of the corresponding precursor nanofibers. The BET surface area of the ACNFs was measured to be 314 m^2/g, while the presence of oxygen-containing functional groups was confirmed with FTIR. After these ACNFs were treated with either de-ionized water or HNO_3, they were used for the adsorption of Pb^{2+} ions from aqueous solutions. The ACNFs treated with HNO_3 exhibited a higher adsorption capacity (23.75 mg/g) due to the presence of functional groups acting as active sites.

References

1. Duval, A.; Lawoko, M. React. Funct. Polym. (2014), 85, 78.
2. Alekhina M, Erdmann J, Ebert A, et al (2015) J. Mater. Sci. 50, 6395–6406
3. M.H Abdul Latif et al., J. Nat. Sci. Res. (2015), Vol.5, No.4
4. Thakur VK, Thakur MK, Raghavan P, et al. ACS Sustain Chem Eng (2014) 2:1072–1092
5. NAPCOR - National Association for PET Container Resources (USA), http://www.napcor.com/PET/landing_petrecycling.html, accessed on 14/11/2017
6. Xu X, Zhou J, Jiang L, Lubineau G, et al. Carbon 80 (2014) 91-102
7. Frank E, Steudle LM, et al. Angew. Chem. Int. Ed. (2014), 53, 5262 – 5298
8. Ko FK, Wan Y. Introduction to nanofiber materials, Cambridge University Press, 2014, p. 223-227.
9. A.M. Soliman, H.M. Elwy, T. Thiemann, Y. Majedi, F. Labata, N.Al-Rawashdeh, J. Taiwan Inst. Chem. Eng., 58 (2016), 264 – 273
10. E. Svinterikos, I. Zuburtikudis, J. Appl Polym. Sci. (2016), 133, 43936
11. Frank, E.; Hermanutz, F.; Buchmeiser, M. R. Macromol. Mater. Eng. (2012), 297, 493
12. R.D. Parmar, A. Kundu, K. Karan, J. Power Sources 194 (2009) 1007–1020
13. Y. Liu, L. Liu, L. Hong, Catal. Today 281 (2017) 352–359
14. Y. Lin, Y. Liao, Z. Yu, S. Fang, X. Ma, Energy Convers. Manage. 151 (2017) 190–198
15. N. Takeno, Atlas of Eh-pH diagrams, Geological Survey of Japan Open File Report No.419, National Institute of Advanced Industrial Science and Technology, May 2005

497611 Size-Controllable MOF Composite Particles for Bio-Butanol Recovery By Adsorption: Formulation and Mass Transfer Resistances

Abstract:

Biobutanol is regarded as a promising substitute for fossil-based chemicals and fuels. It can be produced from renewable resources by the so-called ABE fermentation. Its separation from the aqueous mixture in which it is produced (i.e. fermentation broth) is very challenging. Adsorption has been identified as an efficient alternative for the traditional energy-intensive distillation technique. Among various porous materials, including zeolites, active carbons and metal-organic frameworks (MOFs), we found that the ZIF-8 MOF material shows interesting properties giving perspectives for an economical and low energy-intensive production of biobutanol [1].

The formulation of the ZIF-8 MOF material into a resistant structure is a key aspect towards its implementation into an actual separation process. Therefore, resistant MOF composite beads were produced by combining polymers with ZIF-8 crystals under different conditions. The formulation method, which easily allows varying the particle size, will be presented. In addition, the adsorptive properties, studied under dynamic and static conditions in liquid and vapor phase, as well as the stability of the composite particles under chemical and mechanical stress will be discussed. Mass transfer resistances related to the formulation will be highlighted. Additional results obtained by various characterization techniques, such as argon and mercury porosimetry will also be presented. The results of these different experiments demonstrated that the formulation method allows to produce resistant ZIF-8 MOF composite particles with controllable size and with limited change in properties of the original material, which are important for the efficient adsorptive recovery of bio-butanol. The formulation method is, moreover, expected to be applicable to any other nanoporous material.

[1] Cousin Saint Remi, J., Remy, T., Van Hunskerken, V., Van de Perre, S., Duerinck, T., Maes, M., De Vos, D., Gobechiya, E., Kirschhock, C.E.A., Baron, G.V. and Denayer, J.F.M. Biobutanol Separation with the Metal-Organic Framework ZIF-8. *ChemSusChem* **4**, 1074-1077 (2011).

498158 Synthesis and Granulation of a Ca-X Zeolite-Based Molecular Sieve and Adsorption Equilibrium of Oxidative Coupling of Methane Gases

Abstract:

During last decades, some developments on the methane chemistry have brought the attention to its use as as feasible source of value-added chemicals [1]. In particular, the exploitation of methane from biogas is of interest for current biorefineries, and waste treatment plants. Nowadays, Biogas is mainly used as fuel for energy generation or in some other cases is just sent to the flares for final treatment [2], [3]. Developing of new products from biogas might help improving the overall economics of its production.

The Oxidative Coupling of Methane (OCM) allows the conversion of the methane contained in the natural gas into ethylene, which is a major building block for the chemicals and plastics industry. While significant efforts have been put into the development of new catalyst and reactors for the OCM, there is still a challenge to overcome in the separation of the reactor effluent gases. As reaction conversion is around 30%, it is expected a large amount of methane in the final product. The low concentration of ethylene and the large amount of byproducts, make the recovery process quite expensive[4].

Current industrial processes for isolation of high purity olefins generally involve high pressure cryogenic distillation in the olefin/paraffin separations[5]. This process is energy intensive and requires large capital and operating costs. In order to reduce the energy consumption, and to improve the economic potential of the process, novel separation techniques are required. Among different alternatives, selective adsorption and pressure swing can be used for selective removal of ethylene from the gas mixture. Despite the process is well known, the development of selective materials suitable for industrial operation is necesary. Most research efforts in this topic has been focused on the use of activated carbons and crystalline silicates (zeolites) [6].

Taking into account the aforementioned, A Na-X zeolite type was synthesized via hydrothermal treatment. The synthesis was carried out in stirred batch reactors at different gel composition (Varying Na_2O/Al_2O_3, H_2O/Al_2O_3 and SiO_2/Al_2O_3Ratios) and temperatures. Experimental design followed a simplex optimization method using as objective goal the crystallinity of the adsorbent material (characterized by XRD). Afterwards, the synthetized material was subjected to an Ion-exchange process with calcium, in order to improve the yield and selectivity for ethylene adsorption. Once obtained, the Ca-X zeolite, was put through an agglomeration process to obtain granulated molecular sieves (ca. 3 mm diameter). For this purpose an inorganic binding material was used. The obtained particles were characterized by measuring surface area, pore distribution, adsorption capacity and adsorption isotherms with OCM gases. Results indicate that the material is suitable to be used as adsorbent in pressure swing separation systems for OCM gases.

References:

[1] K. Skutil y M. Taniewski, "Some technological aspects of methane aromatization (direct and via oxidative coupling)", *Fuel Process. Technol.*, vol. 87, núm. 6, pp. 511–521, jun. 2006.

[2] British Petroleum Co, "Statistical Review of World Energy | Energy economics | BP Global", *bp.com*. [En línea]. Disponible en: http://www.bp.com/en/global/corporate/energy-economics/statistical-review-of-world-energy.html. [Consultado: 26-oct-2016].

[3] J. Tollefson, "'Flaring' wastes 3.5% of world's natural gas", *Nature*, ene. 2016.

[4] X. S. Nghiem, "Ethylene production by oxidative coupling of methane: New process fow diagram

[5] J. A. Hugill, F. W. A. Tillemans, J. W. Dijkstra, y S. Spoelstra, "Feasibility study on the co-generation of ethylene and electricity through oxidative coupling of methane", *Appl. Therm. Eng.*, vol. 25, núm. 8–9, pp. 1259–1271, jun. 2005.

[6] D. M. Ruthven, *Principles of Adsorption and Adsorption Process*, First. New York: Wiley-Interscience, 1984.

498895 Development of Aminopolymer-Based Sorbents for CO_2 Capture with Improved Capacity and Oxidative Stabilit

Abstract:

Sustainable and regenerable carbon capture materials, such as poly(ethyleneimine) (PEI) and other aminopolymers supported in mesoporous solids, have been widely studied as they can effectively capture CO_2 from both flue-gas (10% CO_2) and from ambient air, via direct-air-capture (400 ppm CO_2). Primary and secondary amines act as the adsorptive/reactive centers for the capture of CO_2; however, the secondary amines are susceptible to oxidative degradation at elevated temperatures, which results in loss of CO_2 capture capacity. This oxidative stability is especially vital for air-capture adsorbents due to the presence of high concentrations of oxygen in the dilute CO_2 feed stream.

Preliminary work on small oligomers and dendrimers with ethyl versus propyl backbones indicated amines separated by propyl linkers have increased CO_2 capture capacity and are more stable under oxidation conditions [1]. We extend this work to synthesize a propyl analog of commercially available branched PEI, i.e. poly(propyleneimine) (PPI) from cationic ring opening polymerization of azetidine. Amine oxide sorbents are prepared through impregnation of mesoporous silica (SBA-15) with the polymer and the CO_2 capture capacity and oxidative stability are investigated.

[1] S.H. Pang, L.Lee, M.A. Sakwa-Novak, R.P. Lively, C.W. Jones, J. Am. Chem. Soc. 139 (2017) 3627â"3630

500851 Experimental Studies of 4.6-Dimethyldibenzothiohene Adsorption on Metal-Exchanged Mesoporous Y Zeolites

Abstract:

Experimental Studies of 4.6-Dimethyldibenzothiohene Adsorption on Metal-exchanged Mesoporous Y Zeolites

Kevin X. Lee, George Tsilomelekis and Julia A. Valla

Department of Chemical & Biomolecular Engineering, University of Connecticut, 191 Auditorium Road, Unit 3222, Storrs, CT 06269-3222, USA,

Phone: +1-860-486 4602, e-mail: ioulia.valla@.uconn.edu

According to the latest transportation fuel regulations, the maximum allowable concentration of sulfur in diesel is 15 ppmw.[1] This new limit corresponds to a 97% reduction of sulfur since 2005. Owing to the strict administrative standard, a majority of industries and research institutions face considerable challenges as further reduction and removal of sulfur from diesel fuels require more energy-intensive conditions and active catalysts. Alternatively, adsorptive desulfurization using metal-exchanged mesoporous Y zeolites has shown to be very effective in removing refractory sulfur compounds from transportation fuels.[2] The introduction of mesoporosity improves mass transfer and minimize diffusion limitations of bulky sulfur compounds. The incorporation of metals such as Ce or Cu via the ion-exchanged method has shown to increase the adsorptive active sites, consequently improving the capacity and selectivity of sulfur compounds. The adsorption of thiophenic rings on metal cations can proceed either via the p-complexation or a direct S-M s bond.[3,4] The objective of this work is to extend our studies toward the desulfurization of a much larger thiophenic compound such as the 4,6-dimethyldibenzothiophene (DMDBT) using our novel catalyst. To the best of our knowledge, DMDBT removal by metal-exchanged mesoporous Y zeolite has yet to be reported, which is expected due to the bulky and unreactive nature of such molecule. This paper will highlight the remarkable desulfurization performance of DMDBT using metal-exchanged mesoporous Y zeolites, supported by fundamental IR spectroscopic experiments.

Metal-exchanged mesoporous Y zeolites were first prepared by introducing mesoporosity via a top-down surfactant-assistant technique.[5] The mesoporous zeolite was subjected to an ion-exchange procedure to replace the protons with either Cu or Ce metals. The modified zeolites were characterized using various techniques. The crystallinity and porosity, as well as the presence of metal cations/nanoparticles were imaged using scanning transmission electron microscopy (TEM). The pore size distribution and crystal structure were identified using N_2 adsorption/desorption isotherm and x-ray diffraction (XRD), respectively. Surface acidity and metal distributions were determined using pyridine and CO adsorptions by diffuse reflectance infrared spectroscopy (DRIFT-FTIR), respectively. DRIFTS-FTIR experiments were also used to study the adsorption mechanism of thiophenic compounds on various modified zeolites. The metal loadings on the zeolite were quantified using ICP.

Breakthrough experiments were conducted in a fixed-bed column to study the dynamic adsorption of model fuel (DMDBT in octane). The desulfurized effluent was collected periodically until saturation had been reached and analyzed using a gas-chromatograph-sulfur chemiluminescence detector (GC-SCD). Figure 1 shows the breakthrough curves of DMDBT adsorbed on parent and modified Y zeolites. The parent Y shows an extremely early breakthrough of DMDBT, suggesting diffusion limitations toward the internal active sites. This is confirmed when the use of mesoporous Y zeolite (ie. SAY) extended the breakthrough point from 0 to 150 mL/g of sulfur-free fuel produced. Metal-exchanged mesoporous Y zeolite (ie. CuSAY) has shown to improve the removal of DMDBT by about 20 mL/g. The best desulfurization sorbent was reported by bimetallic mesoporous Y zeolite as nearly 200 mL/g of DMDBT-free fuel was produced. These results suggest that in combination with the mesopores, the synergistic effects of Ce and Cu metals significantly improve the capacity and selectivity of DMDBT.

Figure 1: Breakthrough Curves of DMDBT on various Y-derived zeolites.

The adsorption mechanism of sulfur on Y zeolites can be revealed by FTIR studies. About 25 mg of zeolite was weighed into the DRIFT cell and activated in-situ with either H_2 or N_2 for 1 hour. Upon activation, DMDBT solids were added into the cell and allowed to adsorb at 200 °C. Temperature programmed desorption (TPD) studies were carried out to investigate the sites at which DMDBT is adsorbed. Figure 2 shows the DMDBT adsorption spectra on the parent Y zeolite. The OH region on the left shows that DMDBT adsorbs on the acidic hydroxyl sites (3630 cm^{-1} and 3550 cm^{-1}) more preferentially than the silanol sites (3741 cm^{-1}). TPD profile shows that the hydroxyls begin to restore at 350 °C. On the CC region, peaks at 1640, 1570, 1480, 1445, 1413, and 1375 cm^{-1} are apparent. The peaks at 1570 cm^{-1} and 1480 cm^{-1} are a consequence of a shift to higher wavenumbers, implying an increase in electron density. The positive shift is typical when new s-bonds are formed, suggesting a direct interaction (ie. H-bonding) of the proton in the zeolite with the sulfur in the DMDBT ring. Meanwhile, the peaks located at 1445 cm^{-1}, 1413 cm^{-1} and 1375 cm^{-1} represent a shift to lower wavenumbers upon adsorption of DMDBT on the parent Y. This negative shift implies that the proton can also interact with the DMDBT ring, resulting in a loss in electron density. The same IR study can be conducted on the other modified Y zeolites to fully understand the adsorption mechanism, as well as the internal active sites at a fundamental level.

Figure 2: O-H and C-C vibration regions of DMDBT adsorption parent Y zeolite.

References

1. U.S. Environmental Protection Agency. EPA sets Tier 3 motor vehicle emissions and fuel standards. (2014).

Available at: https://www.epa.gov/regulations-emissions-vehicles-and-engines/final-rule-control-air-pollution-motor-vehicles-tier-3. (Accessed: 2nd March 2016)

2. Lee, K. X. & Valla, J. A. Investigation of bifunctional zeolites for the adsorptive desulfurization of fuels. *Appl. Catal. B Environ.* **201,** 359–369 (2017).

3. Hernández-Maldonado, A. J., Yang, F. H., Qi, G. & Yang, R. T. Desulfurization of transportation fuels by p-complexation sorbents: Cu(I)-, Ni(II)-, and Zn(II)-zeolites. *Appl. Catal. B Environ.* **56,** 111–126 (2005).

4. Velu, S., Ma, X. & Song, C. Selective Adsorption for Removing Sulfur from Jet Fuel over Zeolite-Based Adsorbents. *Ind. Eng. Chem. Res.* **42,** 5293–5304 (2003).

5. García-Martínez, J., Johnson, M., Valla, J., Li, K. & Ying, J. Y. Mesostructured zeolite Y—high hydrothermal stability and superior FCC catalytic performance. *Catal. Sci. Technol.* **2,** 987 (2012).

484139 Scaling up Dividing Wall Distillation Columns

Abstract:

Scaling up Dividing Wall Distillation Columns

Bailee Roach, Bruce Eldridge

James R. Fair Process Science and Technology Center

The University of Texas at Austin

Dividing wall distillation columns (DWC) exhibit reduced energy consumption and lower capital cost when compared to traditional multi-column separation schemes. Several open literature articles address the modeling of a DWC but very limited information is available containing both model development and experimental validation. In addition, investigation of the hydraulic impact of the non-circular geometry and heat transfer across the dividing wall is lacking.

The presentation will report results from a comprehensive analysis of DWC operation performed at the Separations Research Program (SRP) facility. Air/water hydraulic tests were performed to determine the impact of the dividing wall geometry on the hold-up, pressure drop, and wetted area. The results from the hydraulic study were incorporated into a rigorous process simulation which was used to match pilot DWC data from a 6 inch diameter column. The rigorous simulation captured both heat loss to the atmosphere and heat transfer across the wall. The simulation was used for the development of an industrial column design and to provide guidance for industrial applications.

484141 Optimizing Dividing Wall Distillation Columns

Abstract:

Optimizing Dividing Wall Distillation Columns

Colton Andrews, Bailee Roach, Bruce Eldridge

James R. Fair Process Science and Technology Center

The University of Texas at Austin

A dividing wall distillation column (DWC) is an energy efficient configuration capable of producing a high purity tertiary separation within a single column. A DWC includes additional degrees of freedom, making modeling and optimization more complex than standard distillation columns. This presentation covers the application of a rigorous simulation and optimization approach to DWC studies reported in the open literature.

Three DWC column configurations (one four-product DWC and two three-product DWCs) were reconstructed using a process simulator. Each cases was optimized using multivariable optimization software that can tests hundreds of design parameters. From a base case, the optimization software varied the DWC design parameters (number of stages, feed location, reboiler duty, etc.) across a specified range. Using the SHERPA optimization method, the objective function of the optimization software was set to minimize/maximize the key process parameters used to design a DWC. For example, the product stream purities were maximized whereas the reboiler duty was minimized. From the simulations, the â€œbestâ€ design was determined and a scale-up for each optimized design was conducted.

The optimization software in combination with the process simulator forms a powerful and efficient tool for the optimization of DWC simulations and designs. The enhanced efficiency and ease of use allows for many opportunities to test the various complicated design characteristics of the DWC. In the future, as more simulations and optimizations are performed using the optimization software in conjunction with the process simulator technique, the trends in design and operation of a DWC can be studied and, ultimately, a wider acceptance and understanding of the DWC can be achieved.

Introduction to Dividing Wall Columns

Manish Bhargava
Director – Advanced Separation Systems
GTC Technology
AIChE Annual Meeting, November 2017

GTC Technology
• Engineered to Innovate®

Agenda

GTC Technology • Engineered to Innovate

GT-DWC® as part of -

- Standalone grassroots / revamp applications
  - Case study 1 : Side cut naphtha splitter revamp
  - Case study 2 : Two-column naphtha splitter sequence revamped to DWC
  - Case study 3 : High purity products with overhead heat integration

- Optimization blocks in complex gasoline production processes
  - Case study 4: NHT Unit
  - Case study 5: Isomerization
  - Case study 6: CCR gas plant

Page 2

Conventional Sidecut Column Revamps
Naphtha / Reformate Splitters

Side cut above Feed

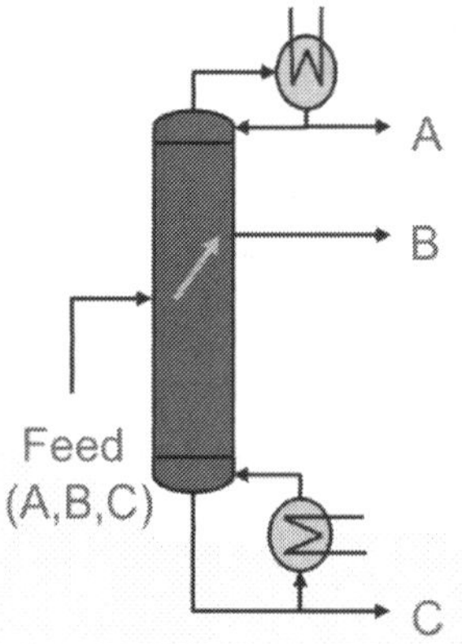

- Lighter components mixes with side cut
- Problems in meeting Initial Boiling Point

Side cut below Feed

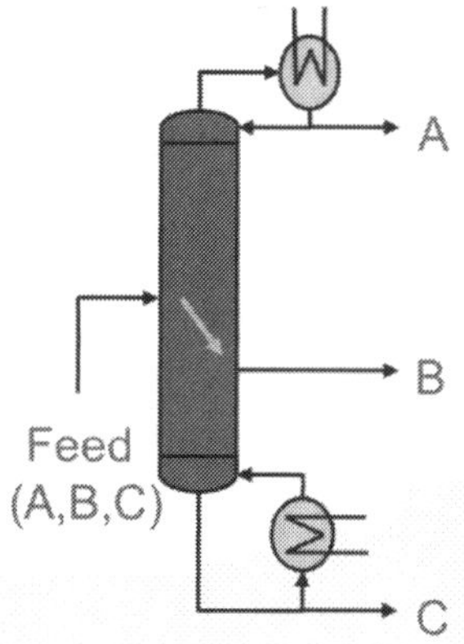

- Heavier components in feed mixes with side cut
- Problems in meeting Final Boiling Point

Page 3

GT-DWC℠ in Side Cut Column Revamps
When High-purity Side Product Required

Conventional Sidedraw Column → **GT-DWC℠**

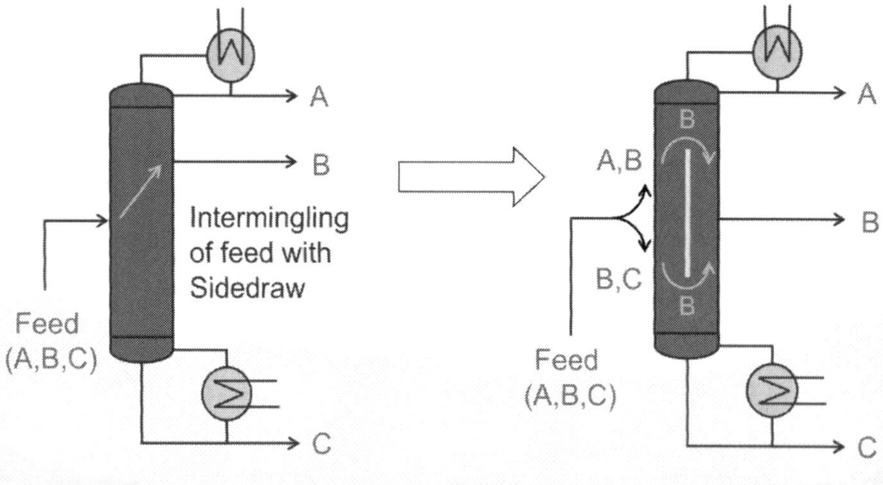

When **high purity middle product** is desired, GT-DWC should be considered instead of a conventional side draw column.

Case Study 1: Commercial Column Retrofit

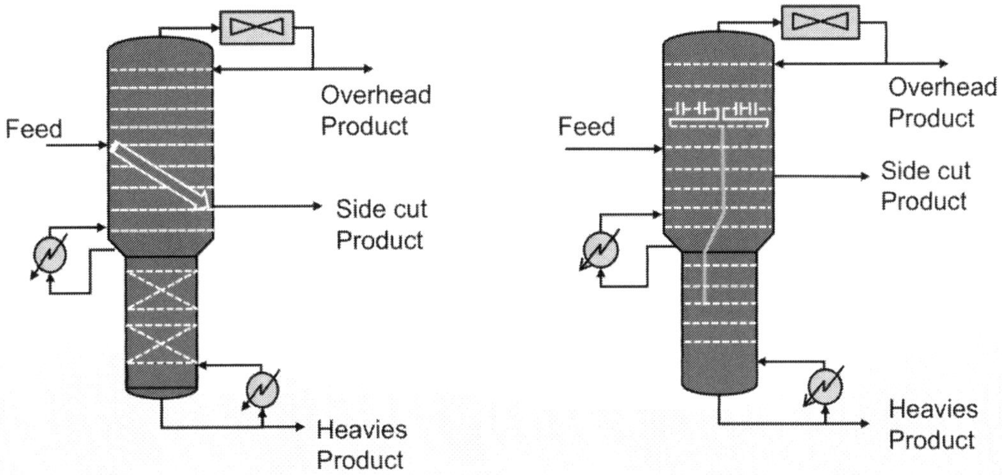

- Intermixing of feed with side cut
- Smaller diameter puts a restriction on vapor/liquid loading

- Side cut separated from feed
- Better quality

GTC Internal Liquid Split Distributer

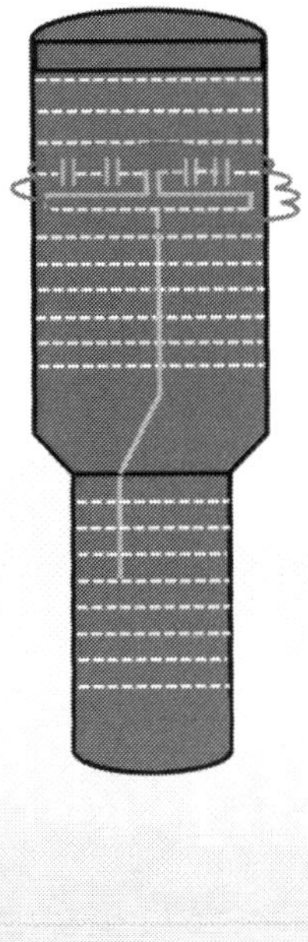

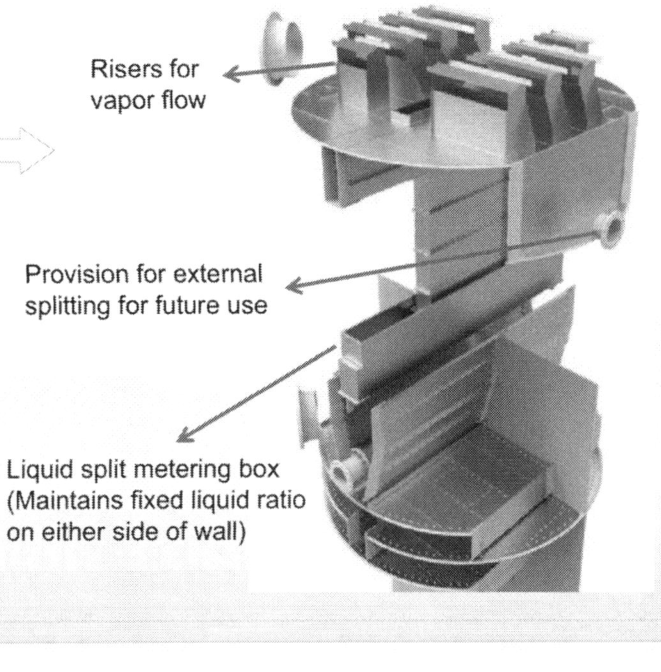

Dividing Wall Column Trays

GTC Technology
- Engineered to Innovate

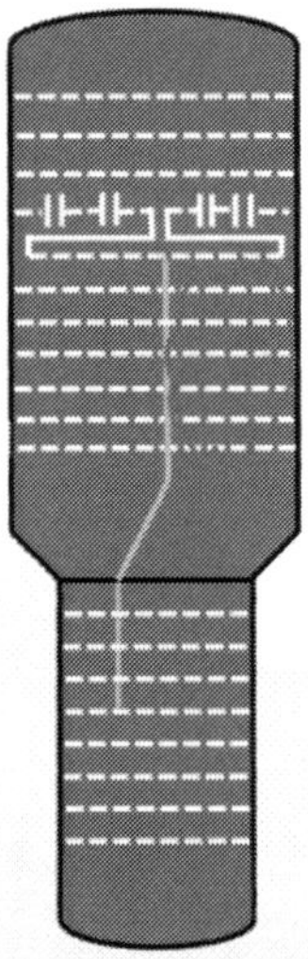

Page 7

Installation of Dividing Wall

GTC Technology
- Engineered to Innovate

Page 8

Product Specs after Revamp to GT-DWC®

GTC Technology – Engineered to Innovate

Parameter	Units	Naphtha Splitter before Revamp	Napththa Splitter after Revamp to GT-DWC®
Side Draw Flow Rate	lb/hr	398,000	353,000
D86 (IBP/FBP)	Deg° F	223 to 376	231 to 356
Overlap (Heart Cut Naphtha D86 95%-Heavy Cut Naphtha D86 5%)	Deg° F	99	27

- Client was planning to install a 2nd column in sequence with the existing Naphtha Splitter.
- A spare column was available for this retrofit.
- Cost of revamp to GT-DWC® was 1/4 of the alternate two-column design being explored by the client.

Case 2: Original 2-Tower System Revamped to GT-DWC®

GTC Technology – Engineered to Innovate

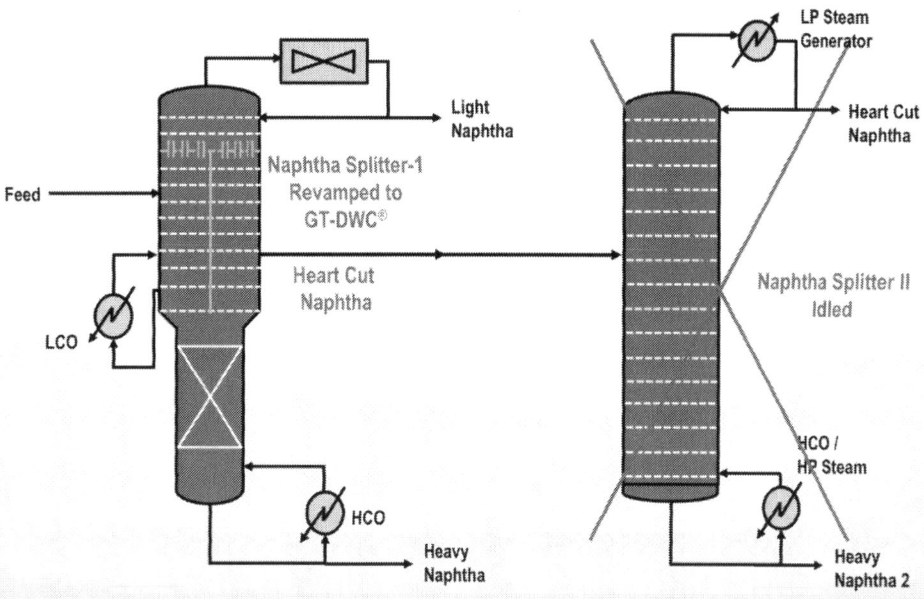

Comparison : Product Recoveries Original Configuration vs GT-DWC®

GTC Technology – Engineered to Innovate

Parameter	Units	Original Design (NS-I and NS-II)	Naphtha Splitter-I after Revamp to GT-DWC®
Feed	lb/hr	679,000	679,000
Mid Cut	lb/hr	363,700	363,500
D86 (IBP-FBP)	˚F	231-338	231-337
Overlap (Side Cut D86 95%- Heavy Cut Naphtha D86 5%)	˚F	8.1	7.0
Total heating duty	MMBtu/hr	151	111

Benefits of GT-DWC

- Heating duty reduced by ~ 26%
- HP steam usage eliminated
- Naphtha Splitter-II is idled
- 3% higher product recoveries
- Heart Cut Naphtha obtained in Naphtha Splitter-I
- Project payback (based on energy reduction benefits only) = 10 Months

Case 3: GT-DWC® for High-Purity Products with Overhead Heat integration (TonenGeneral, Japan)

GTC Technology – Engineered to Innovate

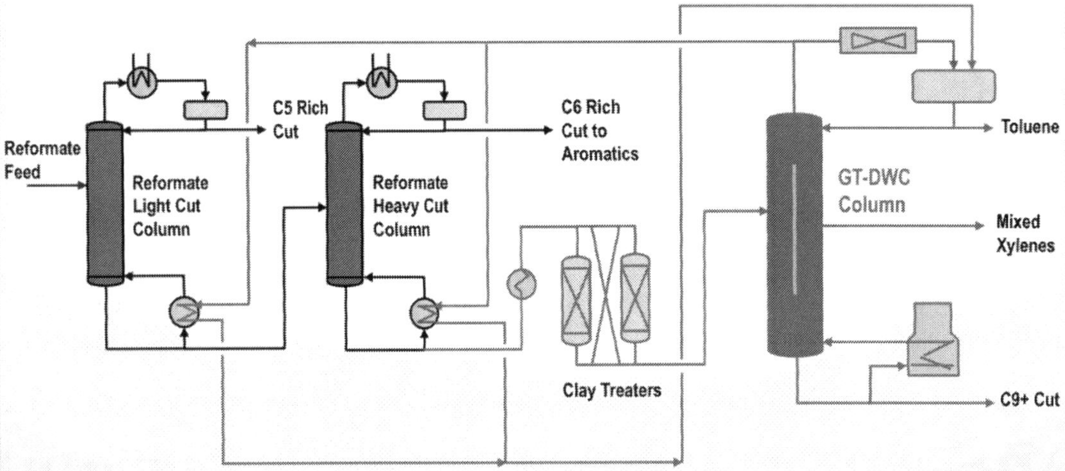

Page 12

GT-DWC® Column Internals

Off Center
Dividing Wall

TonenGeneral Mixed Xylenes Recovery Unit

- Installation of GT-DWC® Column at Chiba Refinery, Japan

Bird's-Eye View of TG Xylene Recovery Unit

GTC Technology – Engineered to Innovate

Mixed Xylene Recovery Unit Operational since March 2016

Page 16

Advanced Distillation Technologies in Gasoline Complex

GTC Technology – Engineered to Innovate

- Gasoline Production is energy intensive
- GTC's Advanced Distillation technologies can make substantial reductions in energy consumption and improve product quality.

Page 17

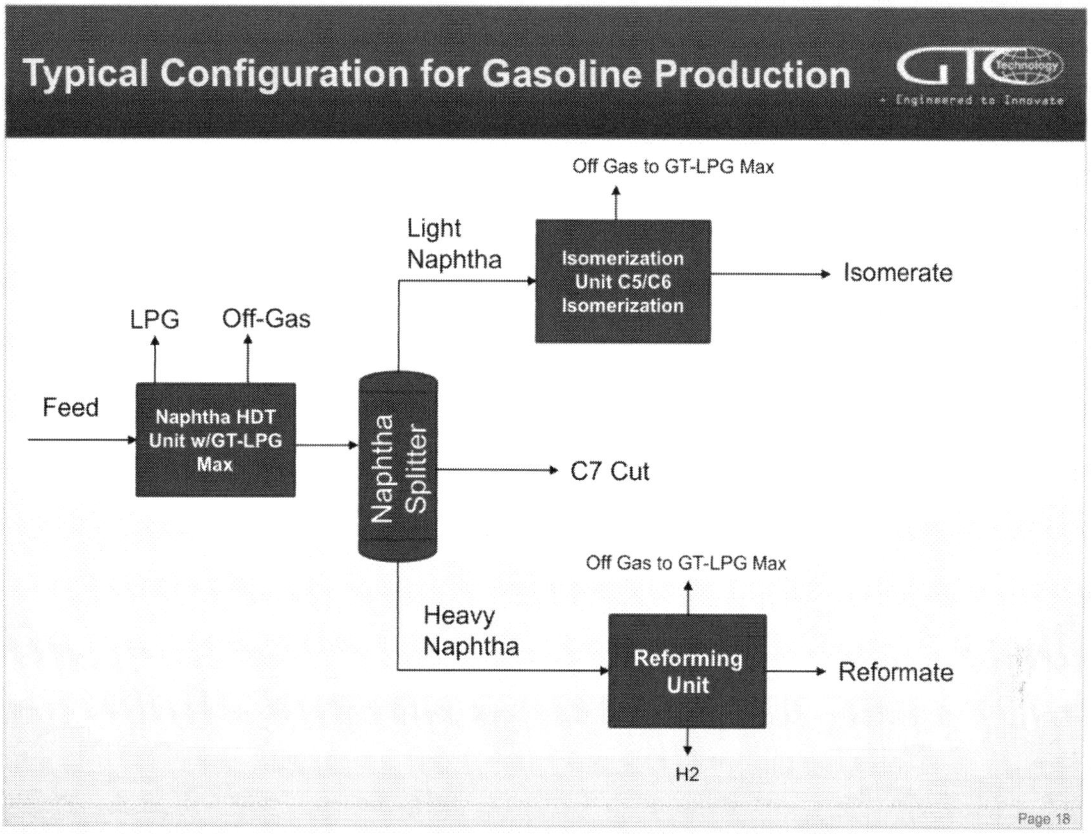
Typical Configuration for Gasoline Production
GTC Technology
Engineered to Innovate
Off Gas to GT-LPG Max
Light Naphtha
Isomerization Unit C5/C6 Isomerization
Isomerate
LPG
Off-Gas
Feed
Naphtha HDT Unit w/GT-LPG Max
Naphtha Splitter
C7 Cut
Off Gas to GT-LPG Max
Heavy Naphtha
Reforming Unit
Reformate
H2
Page 18

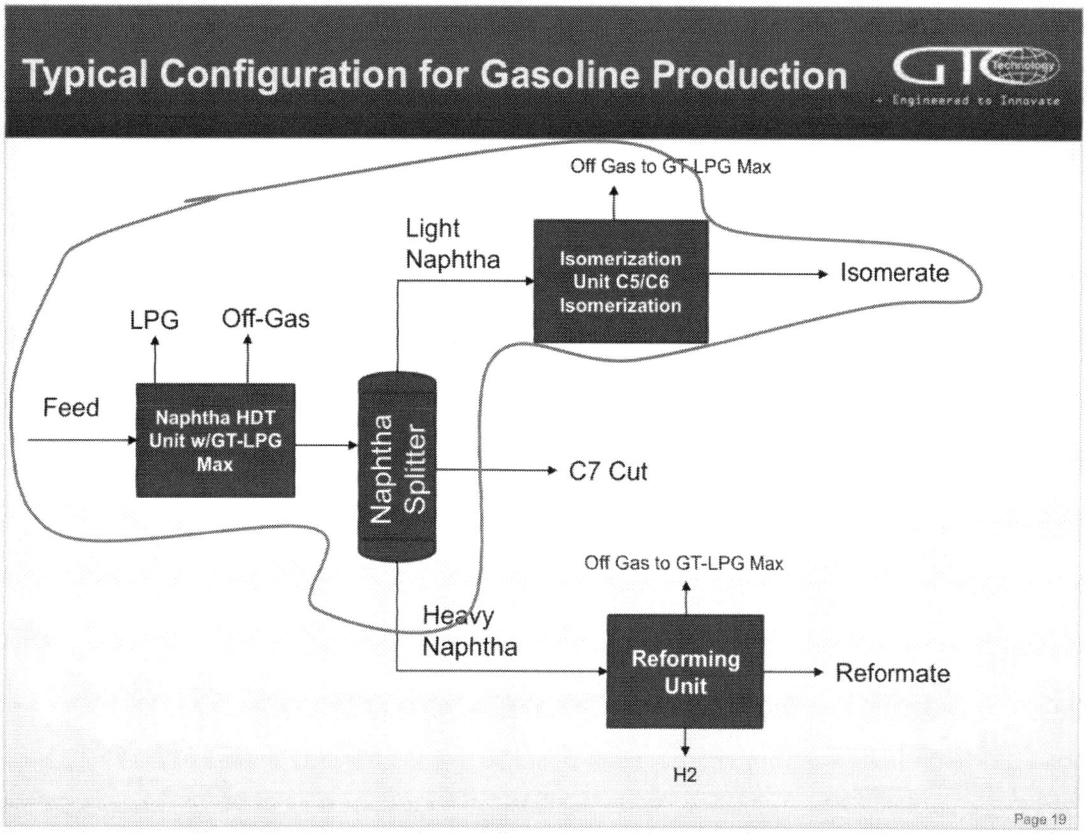
Typical Configuration for Gasoline Production
GTC Technology
Engineered to Innovate
Off Gas to GT-LPG Max
Light Naphtha
Isomerization Unit C5/C6 Isomerization
Isomerate
LPG
Off-Gas
Feed
Naphtha HDT Unit w/GT-LPG Max
Naphtha Splitter
C7 Cut
Off Gas to GT-LPG Max
Heavy Naphtha
Reforming Unit
Reformate
H2
Page 19

NHT + Naphtha Splitter + Isomerization Unit

GTC Technology – Engineered to Innovate

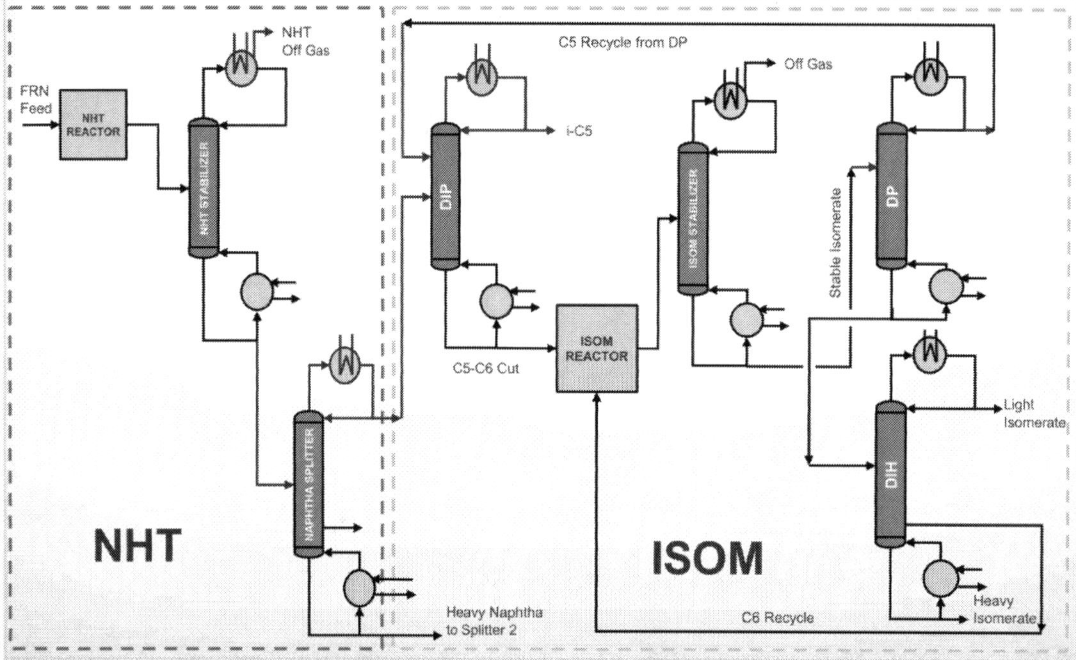

GTC Confidential

Page 20

Process Scheme with GTC's Advanced Distillation Techniques

GTC Technology – Engineered to Innovate

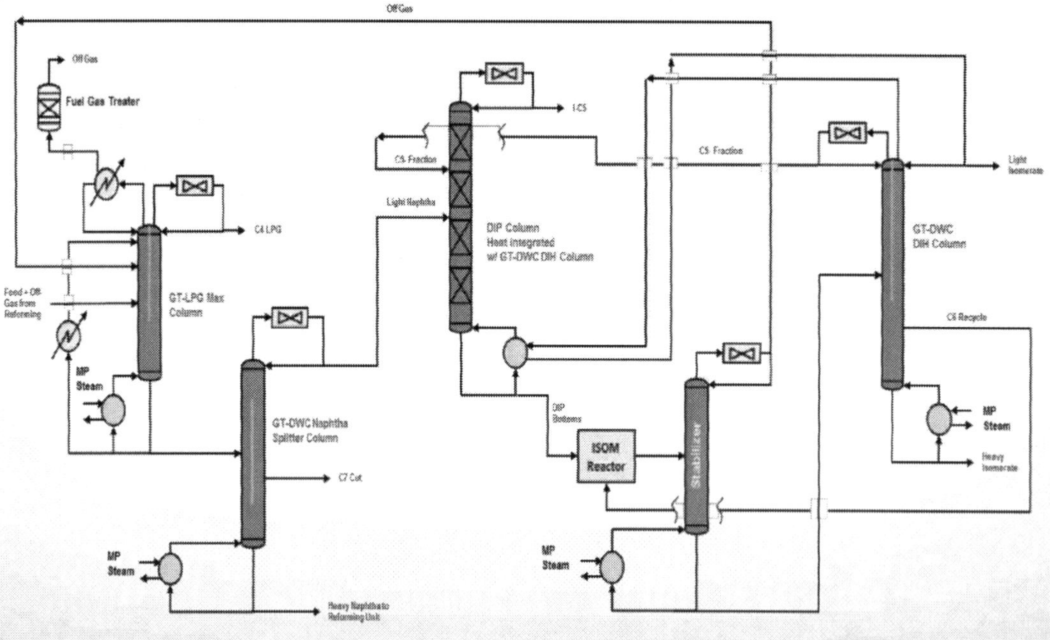

Page 21

GTC's Advanced Distillation Application for Gasoline Production

GTC Technology – Engineered to Innovate

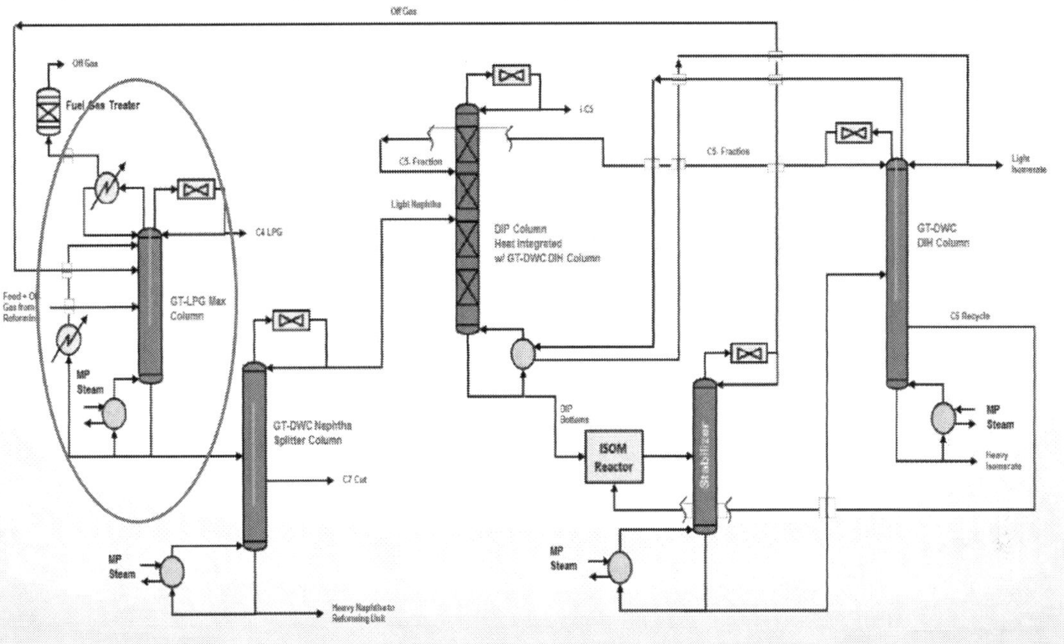

Conventional Stabilizer Design

GTC Technology – Engineered to Innovate

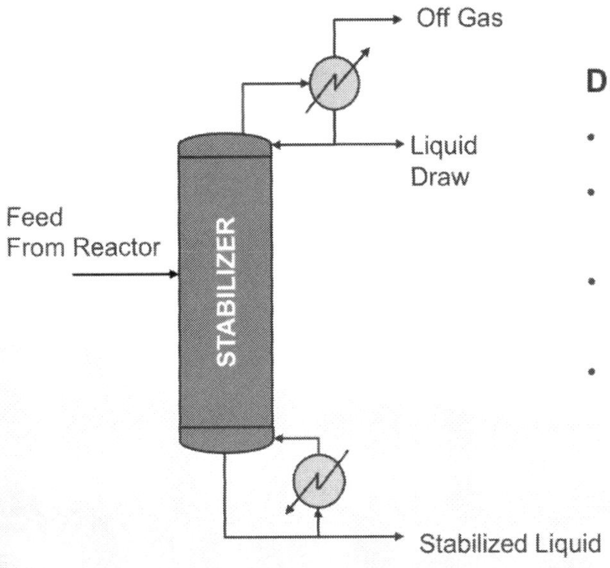

Drawbacks

- Operates at higher pressure
- Liquid losses to off gas due to partial condensation.
- Lights contamination of liquid overhead product
- High bottoms temperature

GT-LPG MAX® Using Top Dividing Wall; Absorption and Distillation in the Same Column

GTC Technology – Engineered to Innovate

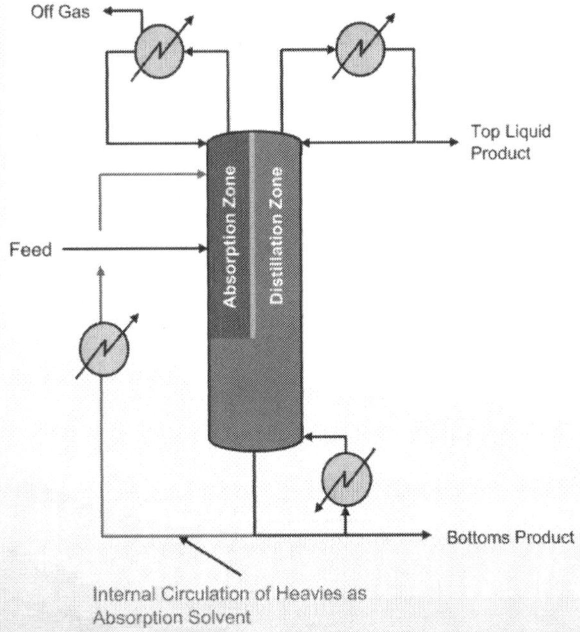

Benefits

- Performs two unit operations (absorption & distillation) in a single column.
- Lights are removed on feed side of dividing wall by reboiled absorption.
- Top liquid product is separated from heavies on the other side of dividing wall via distillation.

GTC's Advanced Distillation Application for Gasoline Production

GTC Technology – Engineered to Innovate

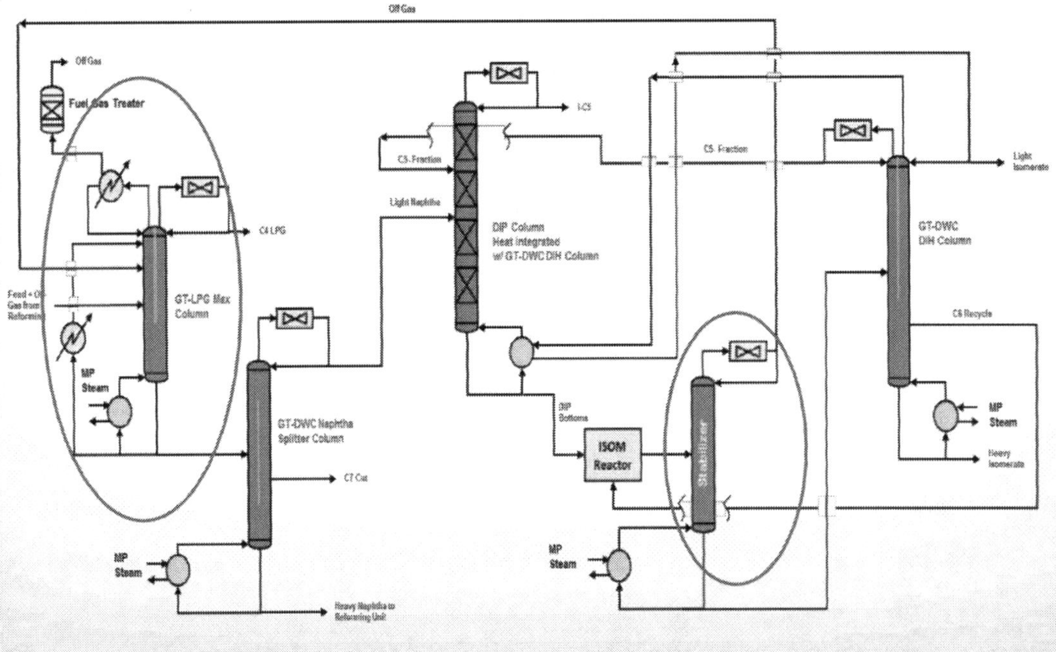

GTC's Advanced Distillation Application for Gasoline Production

GTC Technology – Engineered to Innovate

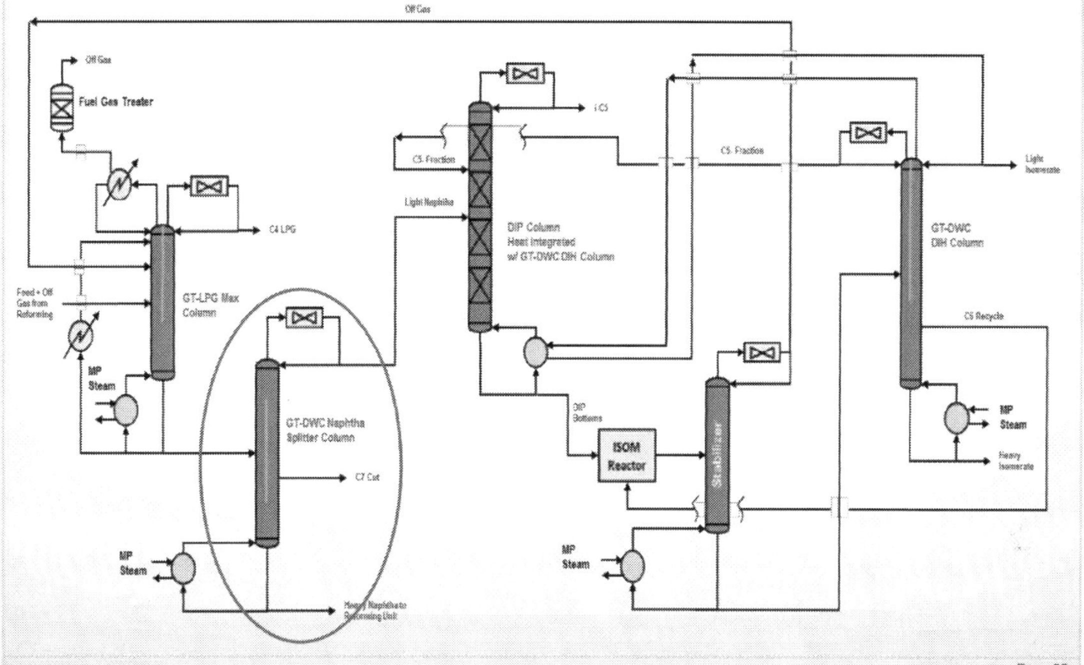

Page 26

GTC's Advanced Distillation Application for Gasoline Production

GTC Technology – Engineered to Innovate

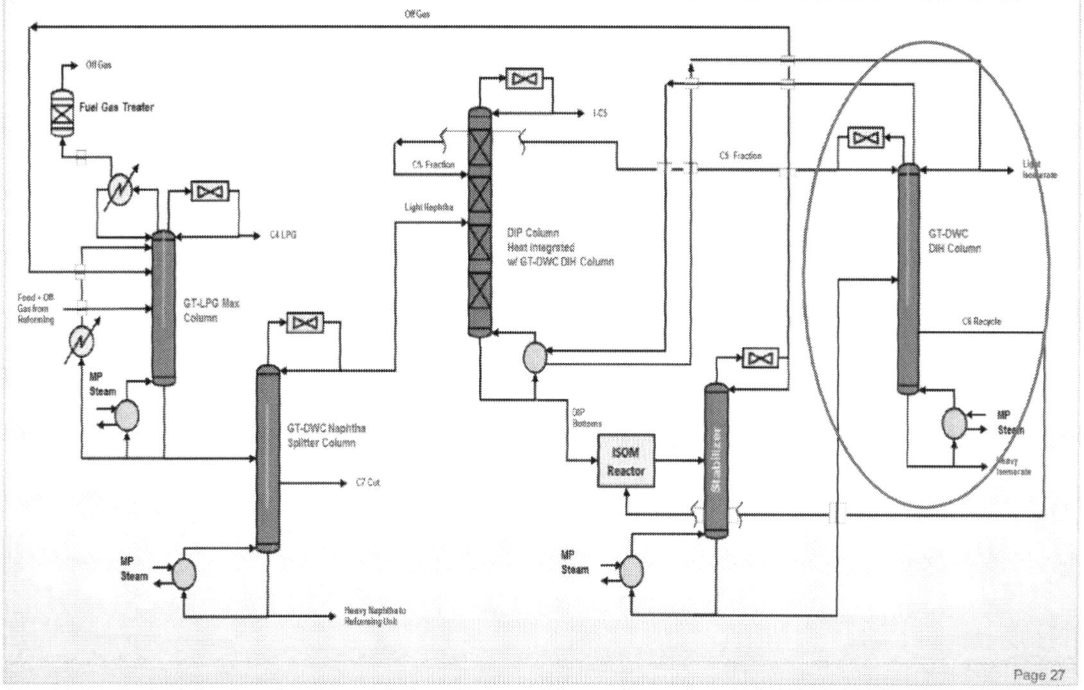

Page 27

Heat Integrated GT-DWC®

GTC Technology – Engineered to Innovate

- Conventional side draw column provides limited opportunity for heat integration
- Top dividing wall column gains the thermodynamic efficiency, plus affords meaningful heat integration.

Conventional Sidedraw Column **GT-TDWC℠**

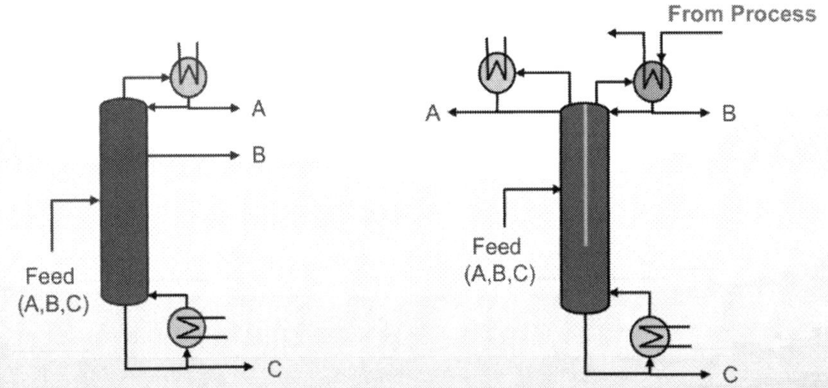

Heat Integration using GT-TDWC℠

GTC Technology – Engineered to Innovate

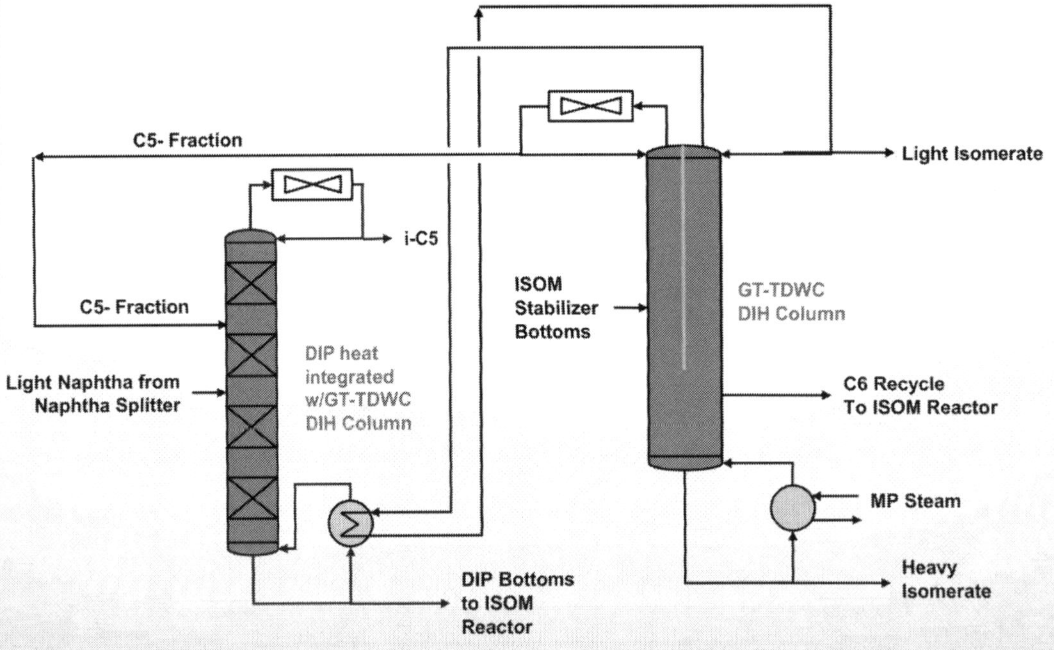

Top Dividing Wall Column Installation at BPCL, India

GTC Technology
+ Engineered to Innovate

Summary

GTC Technology
+ Engineered to Innovate

	Conventional NHT + ISOM	NHT + ISOM with GT-DWC℠
No. of columns	7	4
OPEX Savings (%)	-	30
CAPEX Savings (%)	Base	70 % of Base

Typical Configuration for Gasoline Production

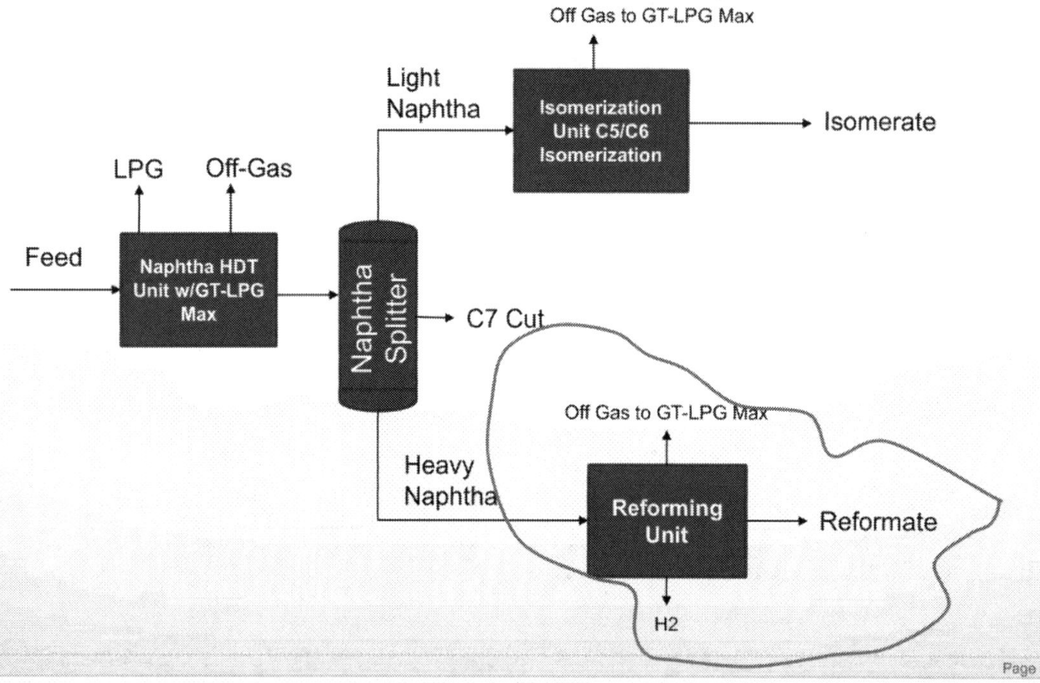

Page 32

Reformer Unit Gas Plant – Conventional Scheme

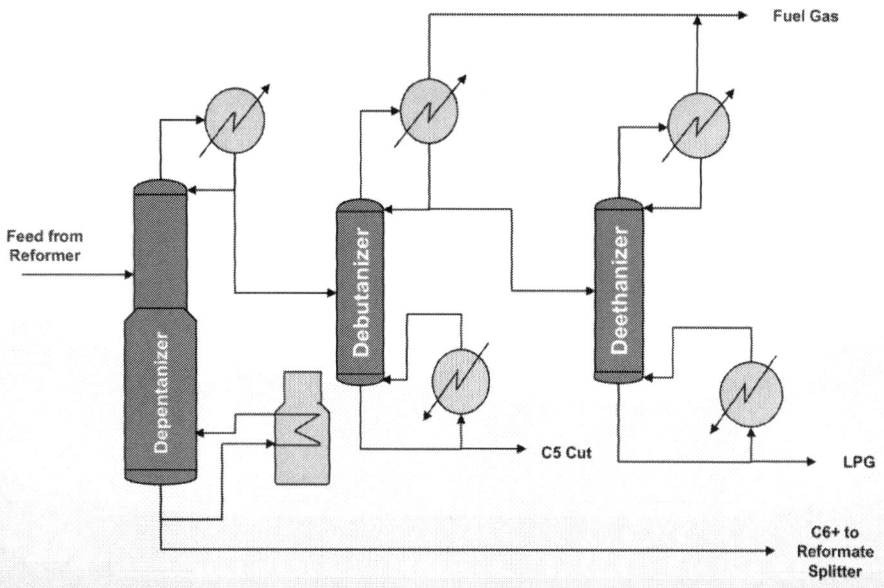

GTC Confidential

Page 33

Reformer Unit Gas Plant – Conventional Scheme

GTC Technology
- Engineered to Innovate

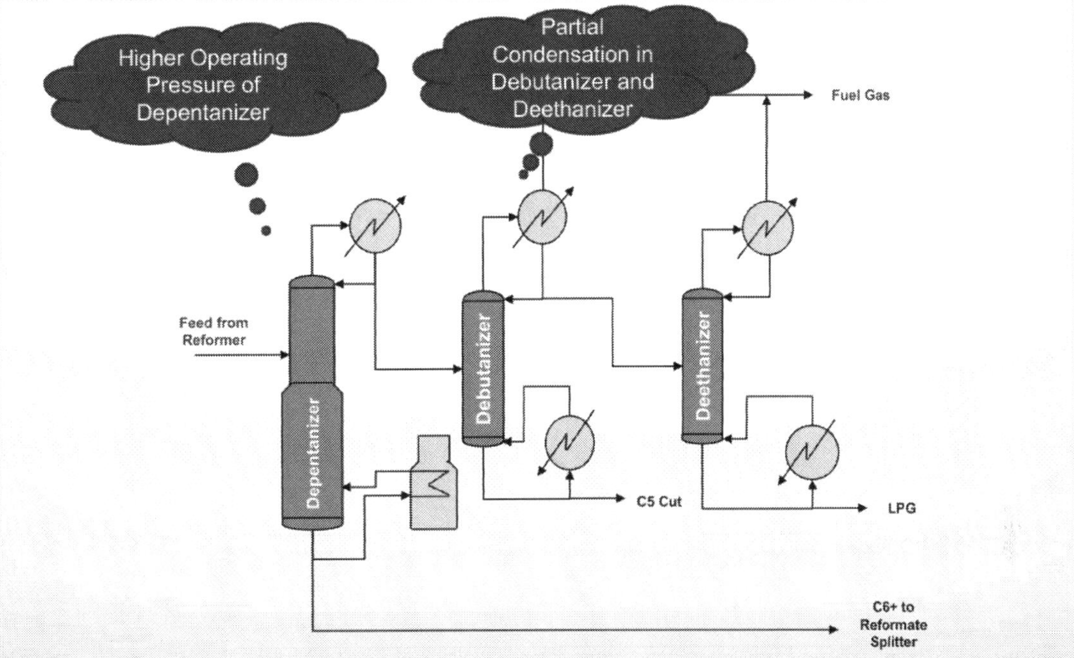

Reformer Unit Gas Plant : GT-DWC® Configuration

GTC Technology
- Engineered to Innovate

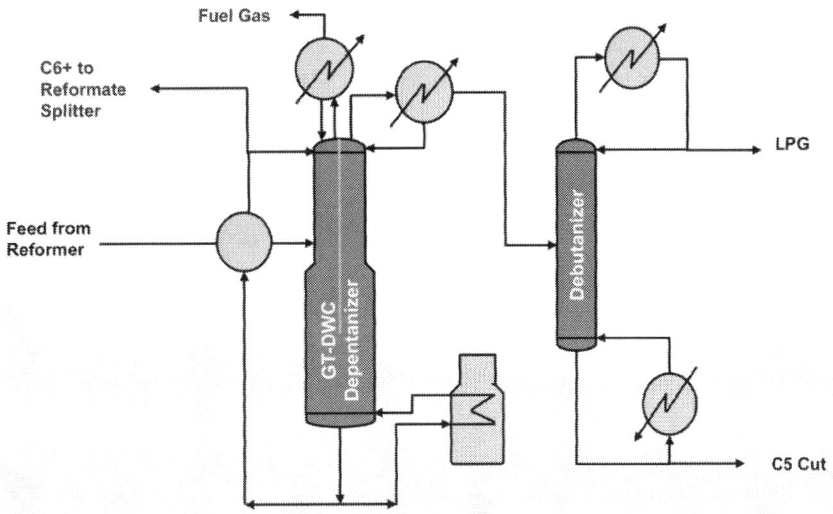

GT-DWC® vs Conventional Scheme

GTC Technology – Engineered to Innovate

Items	Units	Conventional Design			GT-DWC Design	
Columns		DeC5 Column	DeC4 Column	DeC2 Column	GT-TDWC DeC5	DeC4 Column
Column diameter	ft	8.0/10.5	4.3	2.3	11.5	5
No. of Trays		40	33	27	40	40
Operating pressure	psig	110	154	445	75	165
Total Heating Duty	MMBtu/hr	63.1	9.1	1.6	63.5	10.3
		73.8			73.8	
Total Installed Cost	MM$	10.7			8.5	
Feed	lb/hr	852,642			852,642	
C6+ Cut	lb/hr	805,721			805,721	
C5- Cut	lb/hr	17,906			17,992	
Off Gas	lb/hr	12,705			3,761	
LPG	lb/hr	16,310			25,168	

GTC Confidential Page 36

Summary

GTC Technology – Engineered to Innovate

- Unconventional distillation applications are an overlooked means to reduce refinery energy consumption
  - GT-DWC℠ reduces 20–30% OPEX through energy savings
  - GT-DWC℠ reduces 20–30% CAPEX by requiring a single column for multi-component separation
  - ➢ GT-TDWC℠ allows new solutions for off gas recovery using absorption/stripping
- Advanced Distillation schemes can offer CAPEX and OPEX reduction of more than 40% by combining multiple unit operations and heat integration opportunities.
- After successful standalone installations, complex processes are utilizing DWCs for further optimization.

488947 Improving Themal Efficiency and CO_2Â Emission of the Dividing Wall Column

Abstract:

This work aims at analyzing the performance enhancement of a dividing wall column for the separation of a ternary system. It is true that to fractionate a ternary mixture, at least two conventional distillation columns are required, which are arranged in a sequence. In the light of improving its energetic and economic potential, and at the same time to reduce space requirement, two towers are proposed to merge into one column with a dividing wall and it is named as dividing wall column (DWC). The dividing wall is placed at an intermediated position inside the cylindrical shell and the mathematical model of this column is developed by considering heat transfer through that metal wall. The DWC model is simulated and verified using the Aspen Plus flowsheet simulator. For a ternary hydrocarbon mixture, the dominance of this proposed heat integrated structure over a conventional distillation sequence (CDS) is evaluated in the context of mainly two performance indicators, namely thermal efficiency and carbon dioxide emission. It is investigated that the proposed dividing wall column improves the thermal efficiency by 37.5%. The attractiveness of the proposed DWC can also be measured by its potential in reducing CO_2 emission level that is obtained 22.59% for the sample case.

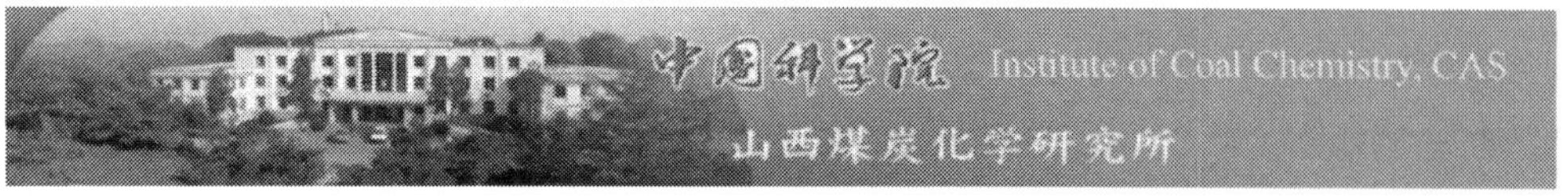

Design, Optimization and Control of Energy-Saving Dividing-Wall Column for Separating Azeotropes

Ming Xia

2017 AIChE Annual Meeting, Nov. 1, Minneapolis, MN

State Key Laboratory of Coal Conversion,
Institute of Coal Chemistry, Chinese Academy of Sciences,
Taiyuan 030001, China
E-mail:mxia@sxicc.ac.cn; Tel: +86 0351 4121793

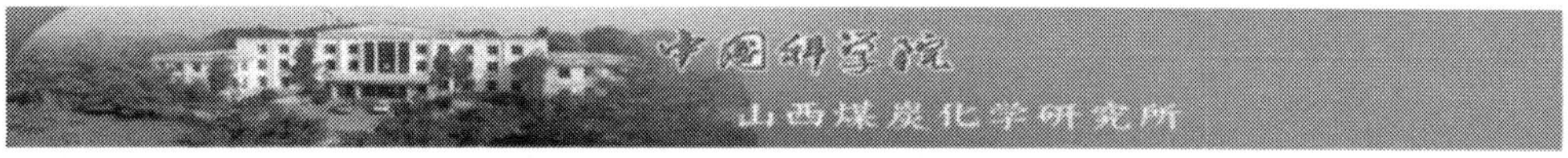

Outline

1. Problem Statement & Solutions
2. Ordinary Dividing-Wall Column Review
3. Dividing-Wall Column for Separating Azeoptropes
4. Conclusions

中国科学院
山西煤炭化学研究所

1. Problem Statement & Solutions

中国科学院
山西煤炭化学研究所

1.1 Energy Consumption Status

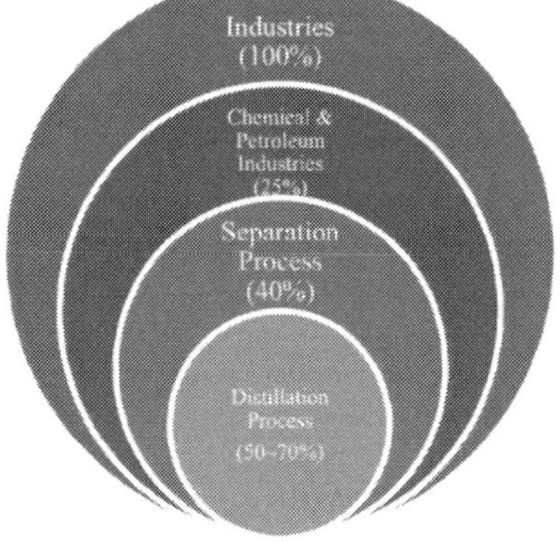

Chemical & Petroleum Industries' energy consumption possesses 25% of the entire industries' (100%).

And separation process consumes 40% energy of the chemical & petroleum industries.

Most importance, around 50~70% energy in separation process are consumed in distillation process.

Finding solutions to energy conservation/-saving in distillation is one of main tasks for chemical engineer to improve entire industries.

1.2 Solutions

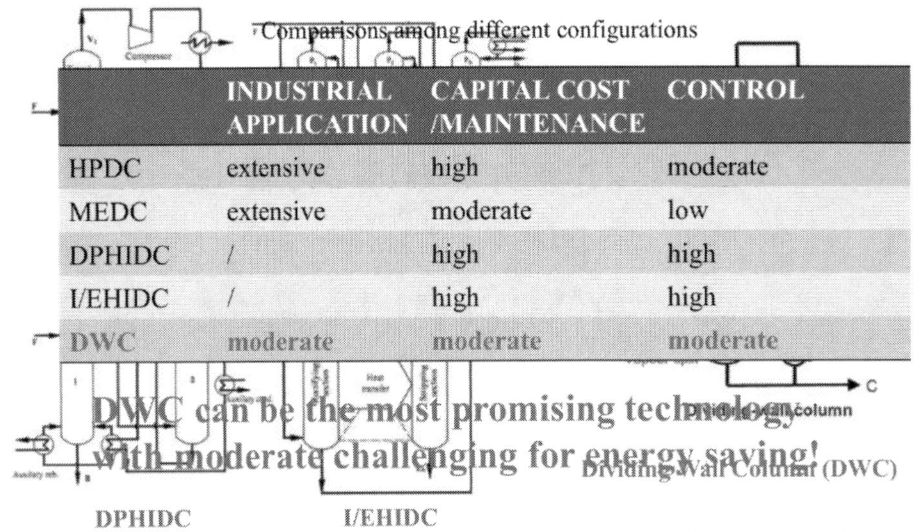

Comparisons among different configurations

	INDUSTRIAL APPLICATION	CAPITAL COST /MAINTENANCE	CONTROL
HPDC	extensive	high	moderate
MEDC	extensive	moderate	low
DPHIDC	/	high	high
I/EHIDC	/	high	high
DWC	moderate	moderate	moderate

DWC can be the most promising technology with moderate challenging for energy saving!

2. Ordinary Dividing-Wall Column Review

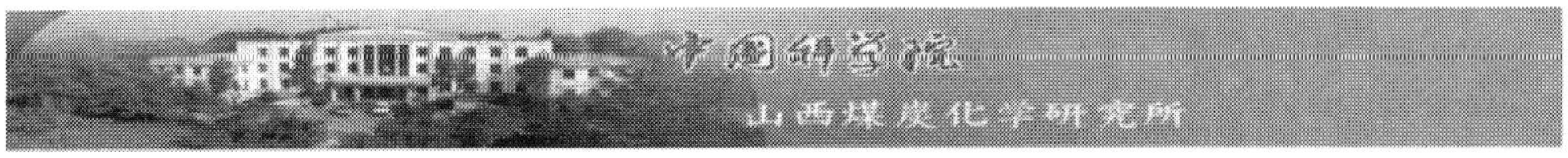

2.1 Intro: Short History

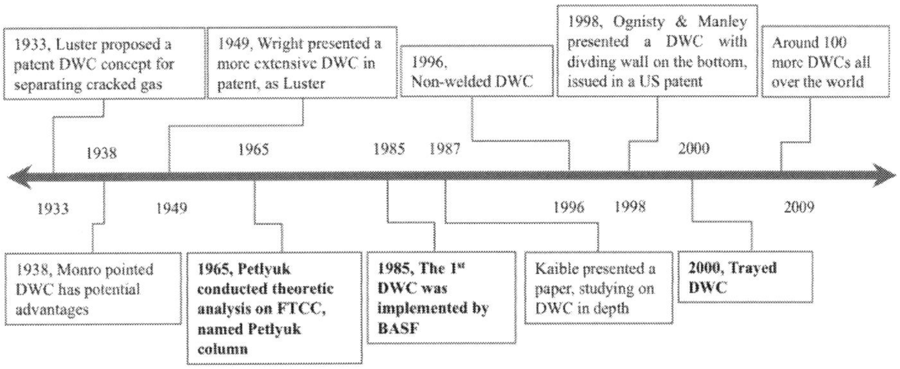

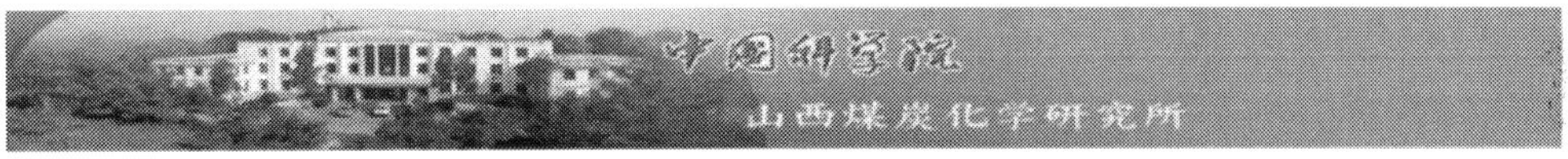

2.1 Intro: Petlyuk & DWC

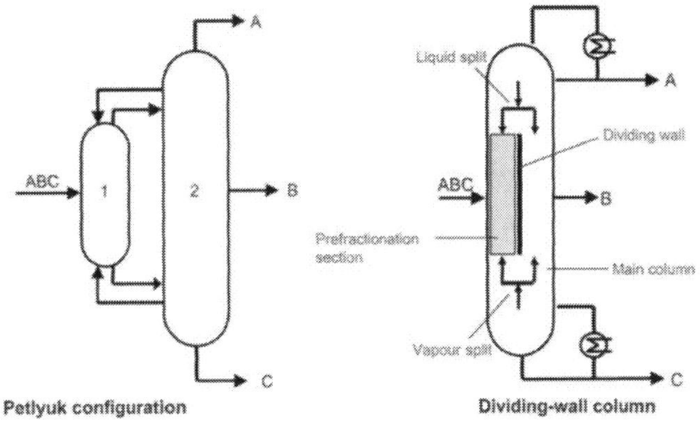

Petlyuk Column, proposed by Petlyuk 1965

DWC, engineering implementation of Petlyuk column

2.1 Intro: Applications & Industrial Cases

- Separating ternary mixture with low fractions of light & heavy components, and with almost equal relative volatility of light to middle and that of middle to heavy components.
  - Hydrocarbons
    - ✓BTX
    - ✓hypothetical systems with various constant relative volatilities
  - Alcohols
    - ✓ethanol/propanol/butanol,
    - ✓methanol/2-propanol/butanol,
    - ✓butanol/pentanol/hexanol,
    - ✓ethanol/1-propanol/1-butanol
  - Aldehydes
  - Ketones
  - Acetals
  - Amines, and others

2.2 Design/Opt.: Problems & Parameters

- Total stages N_t in 6 sections;
- Overhead reflux ratio RR;
- Liquid split ratio $L_{II}:L_{III}$;
- Vapor split ratio $V_{IV}:V_V$;
- Reboiler duty QR;
- Sidestream flowrate S

- divided wall vertical location:
- top, middle, or bottom
- divided wall horizontal location:
  - pass through axis or not?
- total stages at the left and right of the divided wall:
  - equal, or not?

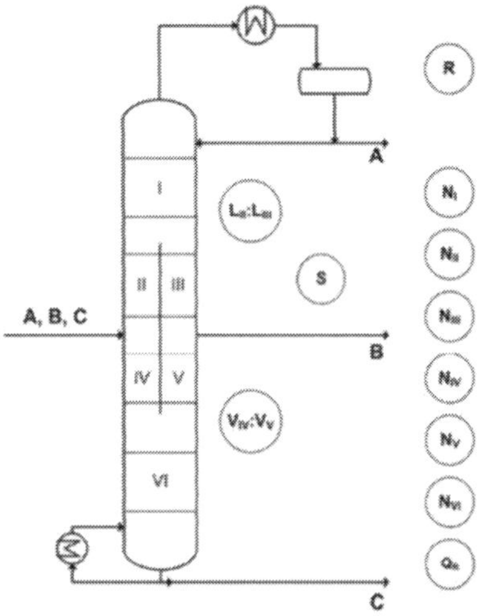

DWC Parameters

2.2 Design/Opt.: Economic Optimization Function

- It is widely acknowledged that Douglas suggested screening the optimum design on the basis of minimizing the total annual cost (TAC) including annual capital costs and operating costs.

min TAC = OC + CI/3

Operation Costs involve steam, cooling water, electricity, etc

Capital Investment Costs involve columns, heat exchangers, compressors and other vessels

2.2 Design/Opt.: Modeling

- The existing or equivalent models in Aspen+/Hysys can be used for design/optimization.

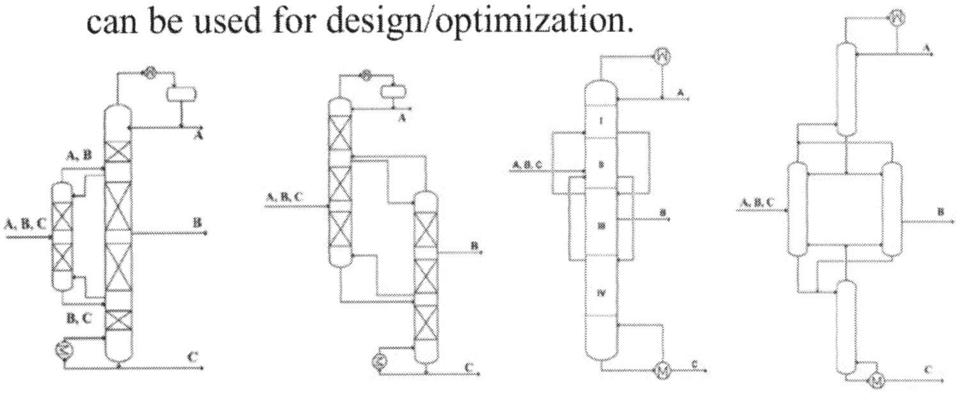

Prefractionator-Main Column Model 1 | Operatable Column Model 2 | Pump-around Column Model 3 | Four Column Model 4

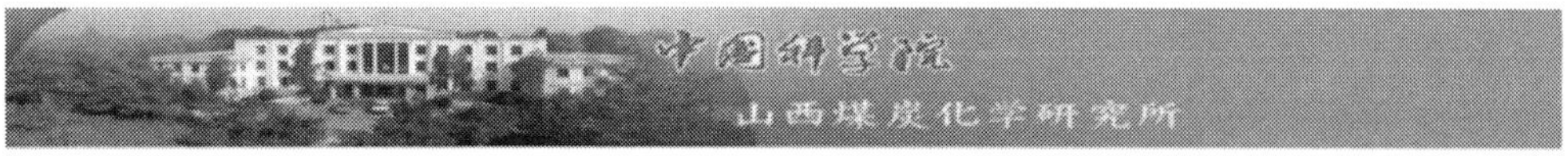

2.2 Design/Opt.: Optimization Sequences

1. Given P;
2. Given NTi;
3. Vary NFi, NSi to minimize QR;
4. Calculate TAC to answer if TAC is minimized. If No, back to 2, if Yes, go to next;
5. The minimized TAC gives optimum NTi, NFi, NSi, and thereby the optimum design is determined.

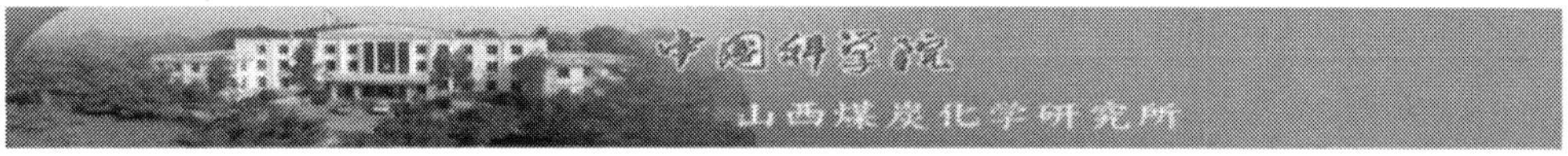

2.3 Control: Control Structures

- Besides design, control problem is one of the main issue that needs to be discussed in depth.
- Degree of freedom:
  - Overhead reflux ratio RR, or D; --Y
  - Liquid split ratio L_{II}:L_{III}; --Y
  - Vapor split ratio V_{IV}:V_{V}; -- N
  - Reboiler duty QR, or B; --Y
  - Sidestream flowrate S. --Y

2.3 Control: Three-point Control--LVS

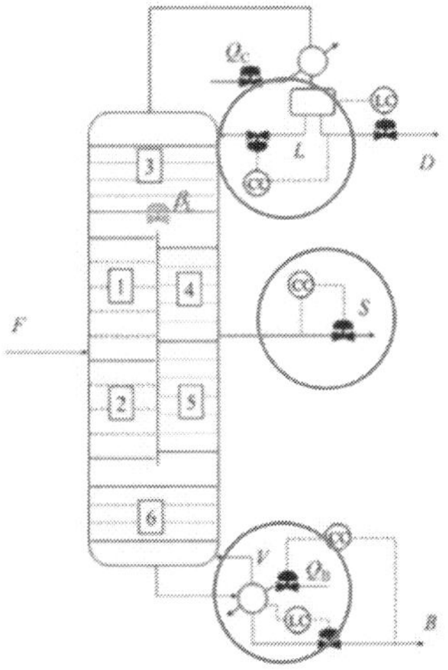

LVS by Wolff & Skogestad, 1995

- Pairings: L-x_{D1}, V-x_{B3}, S-x_{S2}
- Dynamic simulations: this structure effectively handle feed changes and setpoint value changes. However, it cannot reject even small changes in setpoint of product purity.

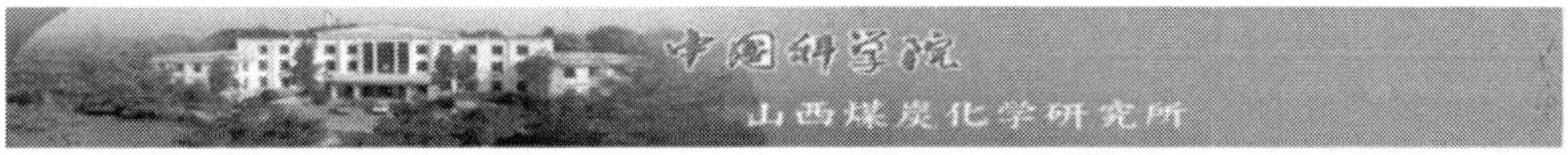

2.3 Control: Three-point Nested Control--LSV

- Nested Pairings: L-x_{D1}, V-x_{s2}, S-x_{B3}
- Linear analysis revealed LSV gives not bad control. However, LSV cannot effectively reject moderate disturbances.

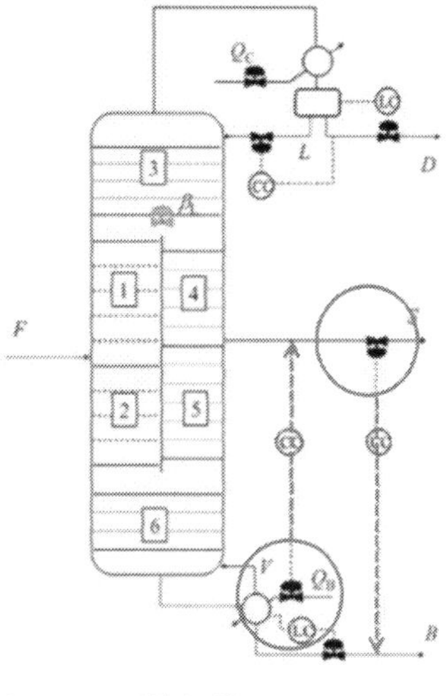

LSV by Wolff & Skogestad, 1995

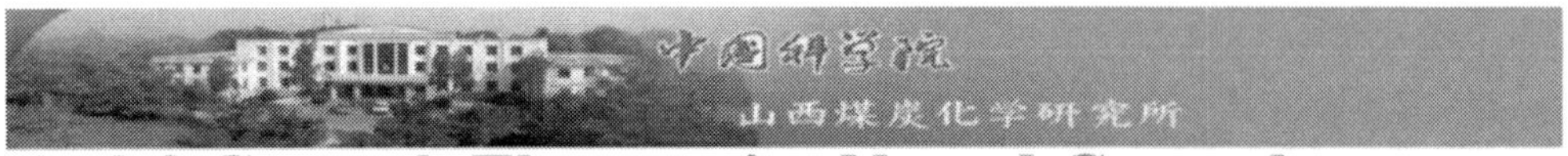

2.3 Control: Three-point Nested Control--DSV

- Nested Pairings: D-x_{D1}, V-x_{s2}, S-x_{B3}
- Linear analysis revealed LSV gives not bad control. However, LSV cannot effectively reject moderate disturbances.

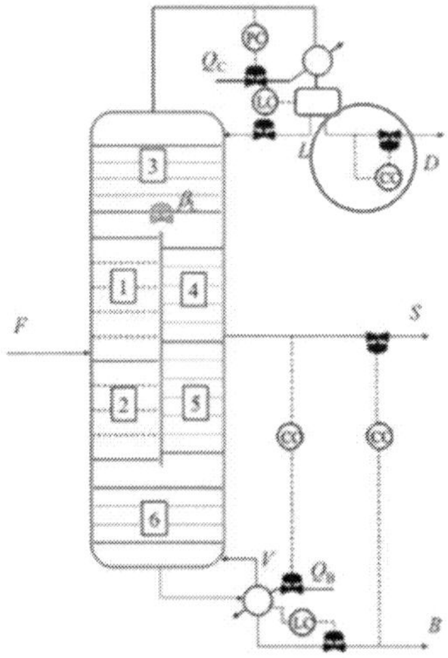

DSV by Abdul Mutalib et al., 1999

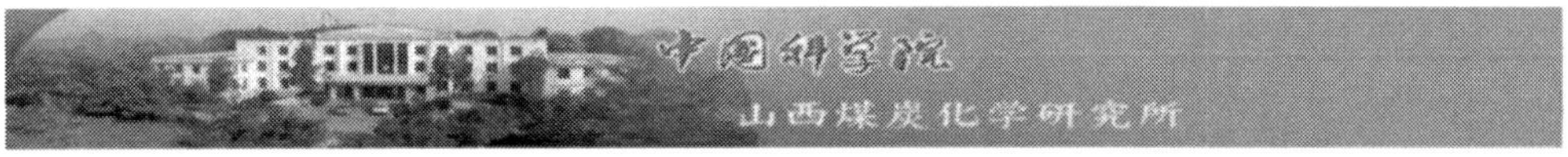

2.3 Control: Four-point Control--LVSβ_L

- On the basis of LVS, a new MV β_L is activated.
  - to control x_{s1}/x_{s3}; (Woff & Skogestad, 1995)
  - to control x_{s3}; (Ling & Luyben)

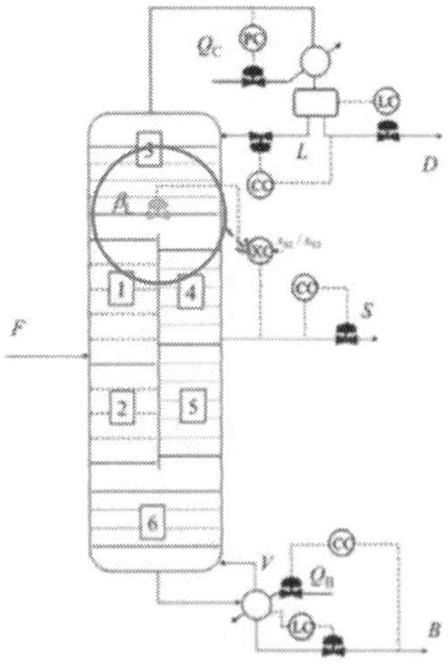

LVSβ_L by Wolff & Skogestad, 1995

2.3 Control: Other Controls

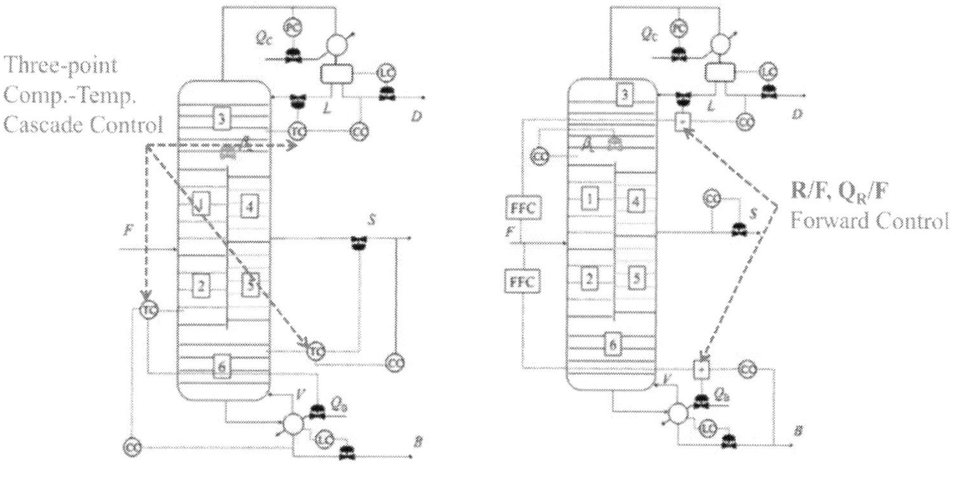

Other Control Structures

中国科学院
山西煤炭化学研究所

2.4 Summary

- Before 2012, Ordinary DWC has been extensively studied, with focus on design and control.
- Up to 2013, rare researches focused on application of DWC in special distillation such as extractive, heterogeneous azeotropic, reactive distillations, although extractive dividing-wall column has been applied in industry.

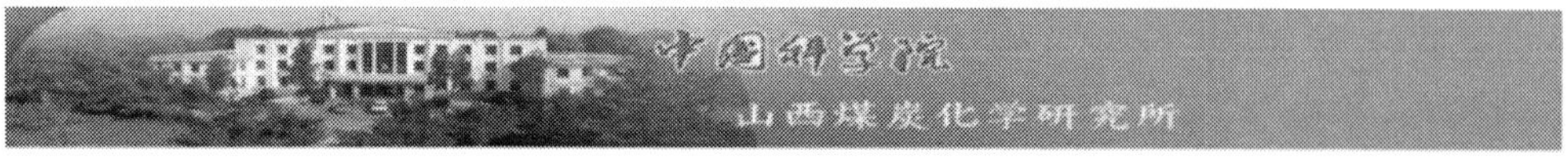

3. Dividing-Wall Column for Separating Azeoptropes

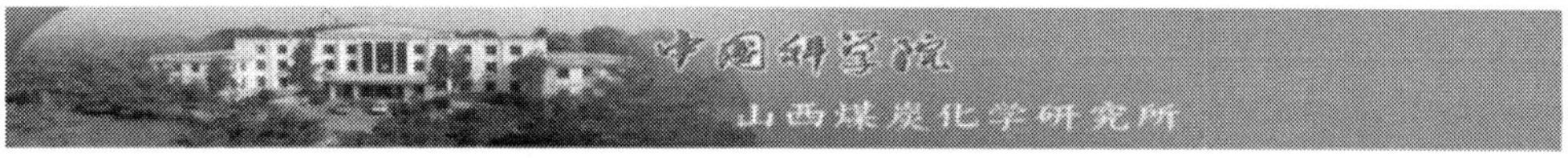

3.1 Intro: Special DWC (SDWC)

ODWC, DWC with a dividing wall in the middle & generally for separating zeotropes

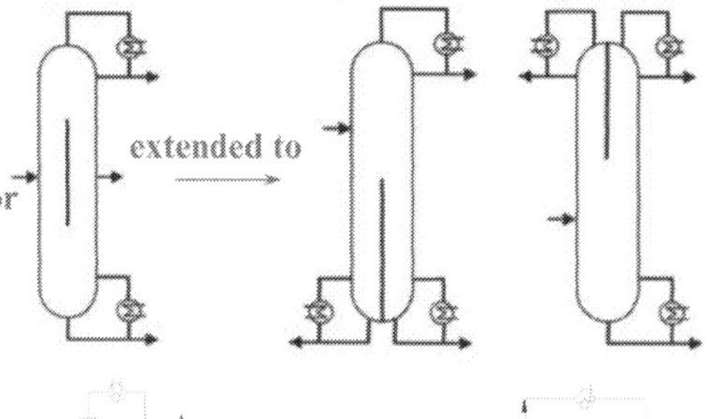

SDWC, DWC with dividing wall in the bottom/top & gene -rally for azeotropes

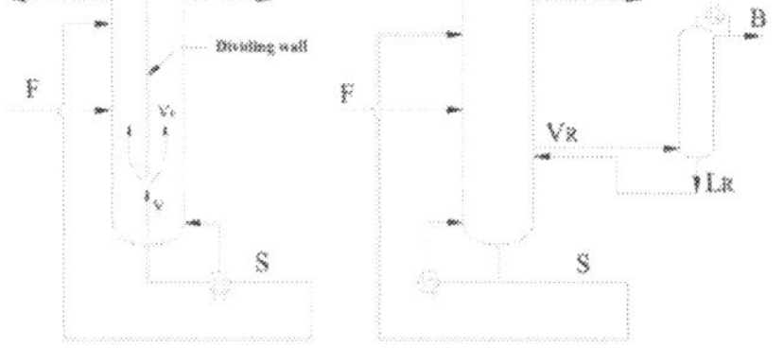

Extractive Dividing-Wall Column

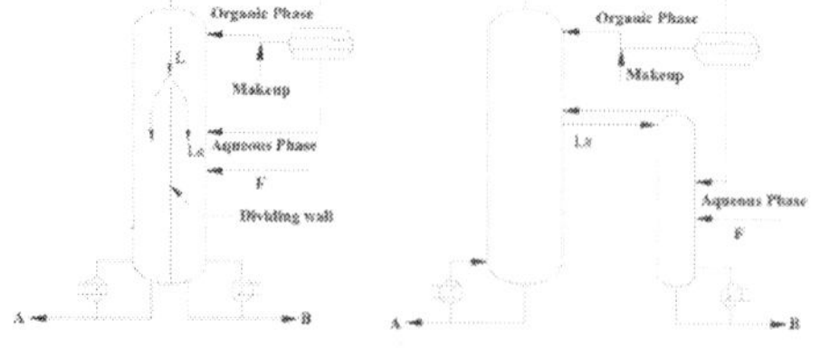

Azeotropic Dividing-Wall Column

3.2.1 EDWC Design/Opt.: Sequence

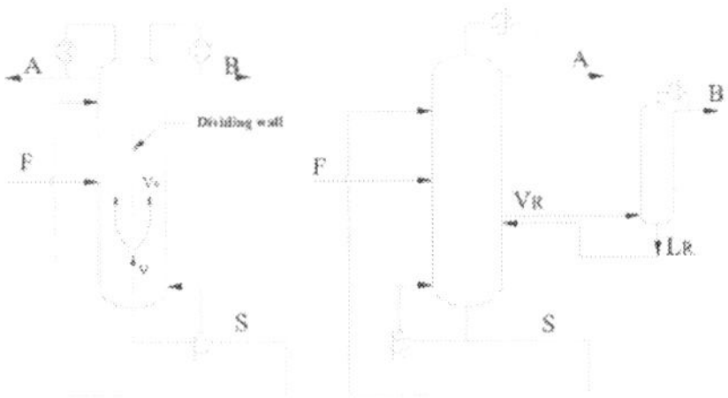

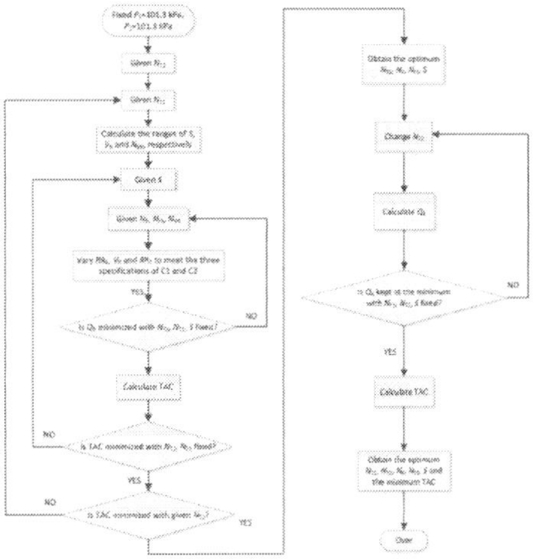

Base on previous optimization sequence for conventional extractive distillation column(Chien & Luyben) and ODWC, we proposed a optimization sequence in terms of TACmin for EDWC to facilitate the optimization.

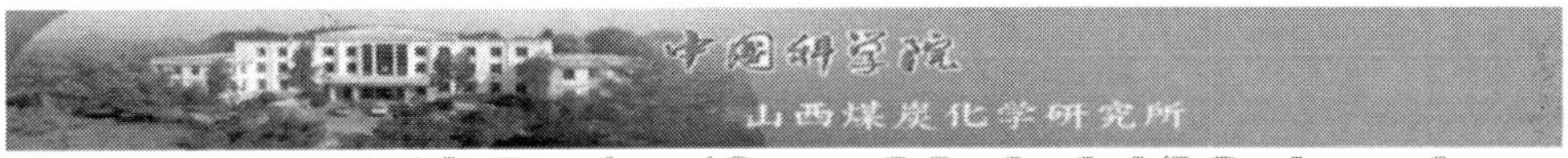

3.2.1 EDWC Design/Opt.: Methylal/Methanol System

The Azeotrope

01	Number Of Components: 2	Temperature 41.38 C	
	Homogeneous	Classification: Unstable Node	
		MOLE BASIS	MASS BASIS
	METHYLAL	0.8708	0.9412
	METHANOL	0.1292	0.0588

Conventional Extractive Distillation Column

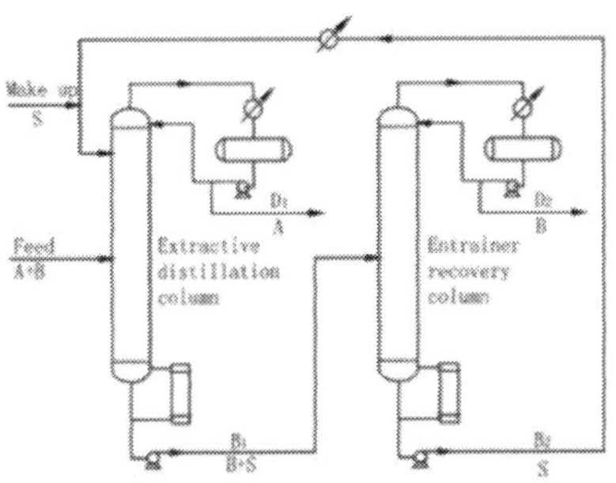

Partially Thermal Coupled

Extractive Dividing-Wall Column

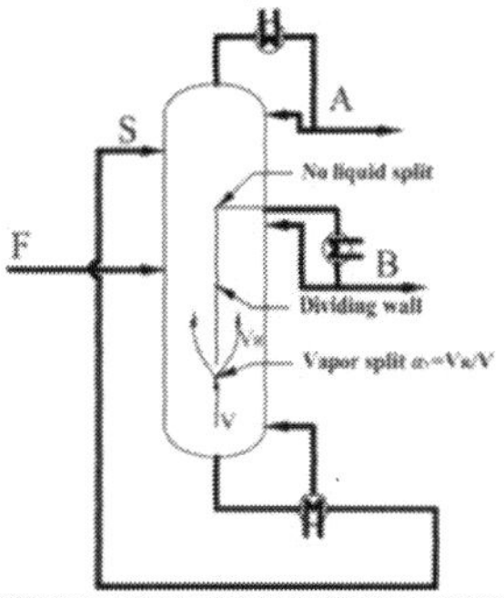

3.2.1 EDWC Design/Opt.: Econo-comparisons

	optimum EDWC design	optimum two-column design	
Total stages			
MC/C_1	52	52	
RC / C_2	12	12	
S (kg/h)	3000.000	3000.000	
Column diameters (m)	0.893	0.776, 0.560	
Tray space (m)	0.4	0.4, 0.3	
$Q_{R,\ total}$ (kW)	**983.194**	**1072.100**	Energy Saving ~8.3 %
Economical data (10^3\$/year)			
CI	**1158.428/(953.945)[a]**	**1430.918**	
OC	**411.782**	**429.407**	
TAC	**759.310/(715.591)[a]**	**858.863**	TAC Reduction 11.6% (16.7%)

a: The values in the brackets are the values of CI and TAC if the manufacturing difficulty and proficiency of the EDWC design and that of the two column design are equal.

3.2.2 EDWC Control: CCS1 & CCS2

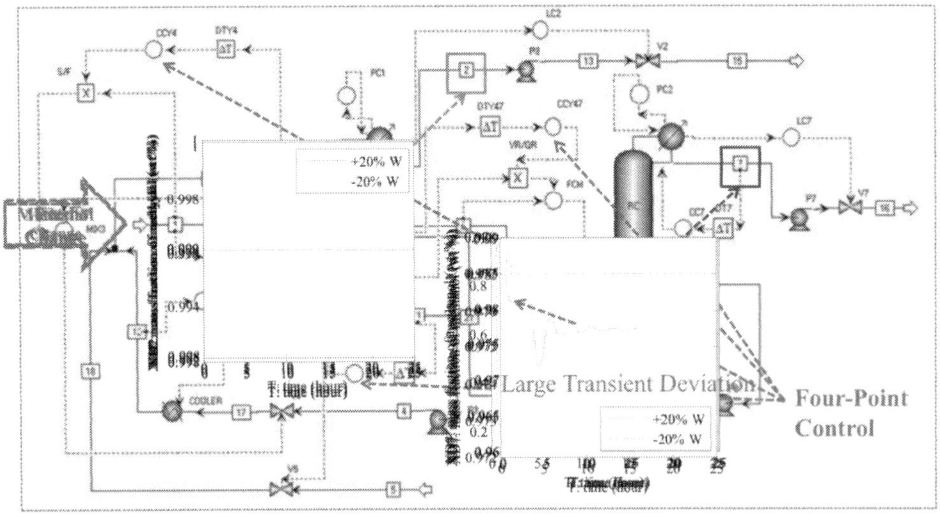

3.3.1 ADWC Design/Opt.: Sequence

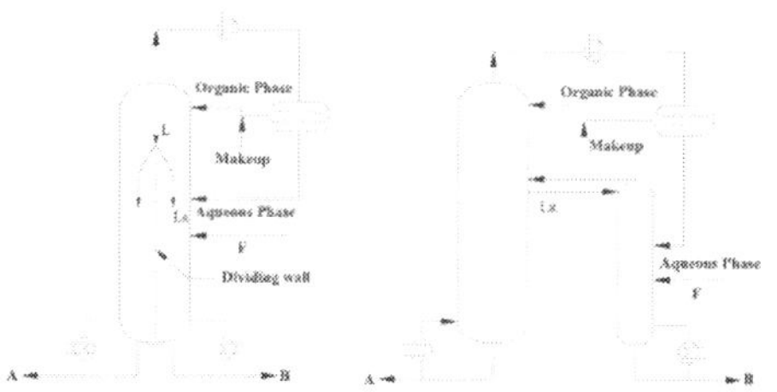

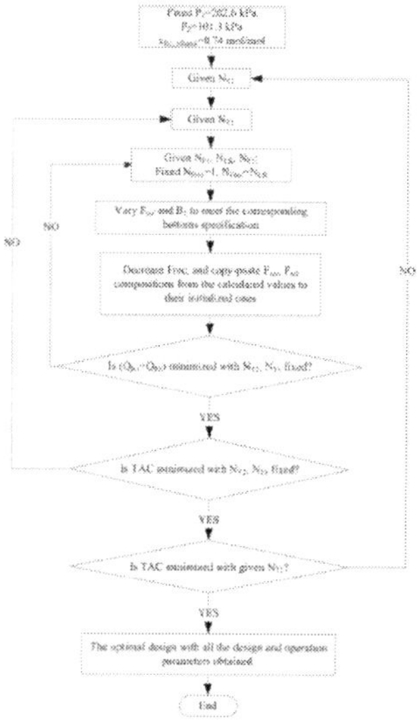

We proposed a optimization sequence in terms of TACmin for conventional heterogeneous azeotropic distillation column to facilitate the optimization. This sequence can be modified and further used for A1, A2, and A3.

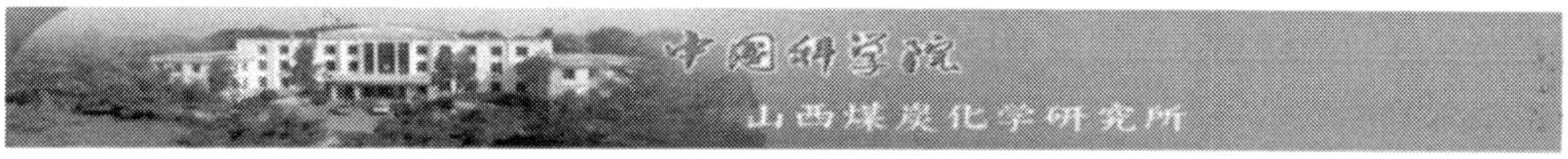

3.3.1 ADWC Design/Opt.: Ethanol/Water System

The Azeotrope

01	Number Of Components: 2 Homogeneous		Temperature 78.14 C Classification: Unstable Node	
			MOLE BASIS	MASS BASIS
		ETHANOL	0.8921	0.9549
		WATER	0.1079	0.0451

Azeotropic Column-Recovery Column Sequence A1

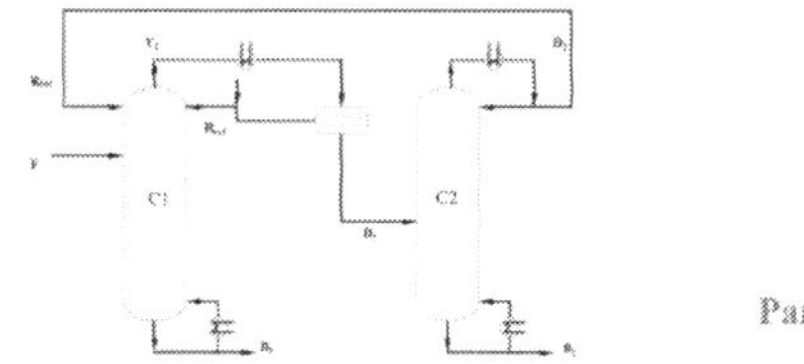

Azeotropic Column-Stripping Column Sequence A2

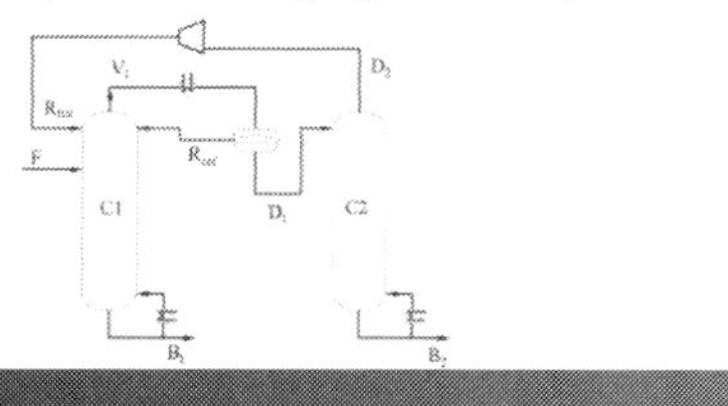

Azeotropic Dividing-Wall Column A3

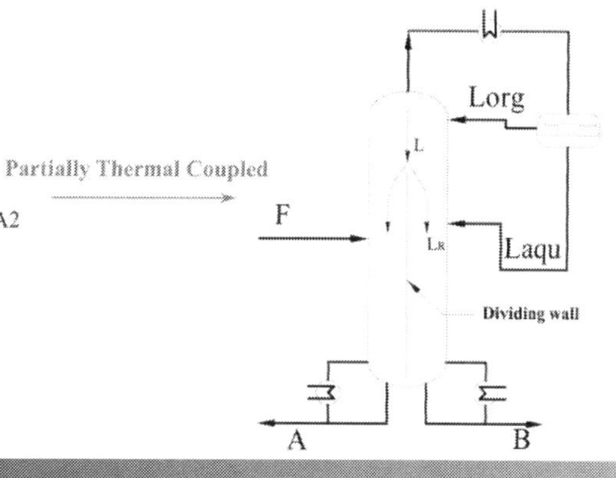

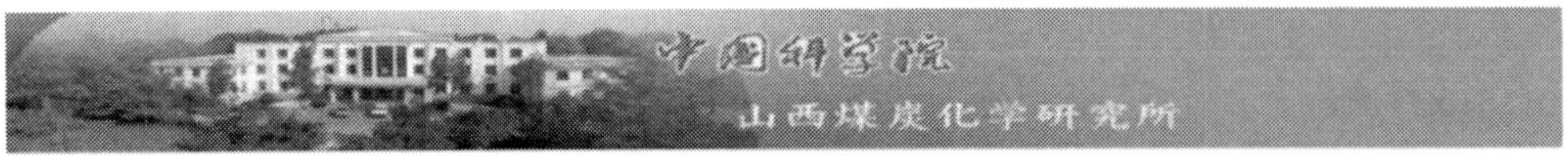

3.3.1 ADWC Design/Opt.: Econo-comparisons

Items\sequences	A1	A2	A3	A2~A1/A3~A1 saving gain[c]/%
$Q_{R,\ total}$ (kW)	1095.51	978.68	861.56	**-10.7/-21.4**
OC (10^3\$/year)	846.591	948.872	727.303	**-9.4/-14.1**
CI (10^3\$/year)	429.894	398.846	339.313	**+12.1/-21.1**
TAC (10^3\$/year)	683.871	683.508	557.504	**0/-18.5**

c: saving gain = $(item_{Ai,\ total} - item_{A1,\ total})/item_{A1,\ total} \times 100$ %,
where i = 2, 3; item = $Q_{R,\ total}$, OC, CI, TAC.

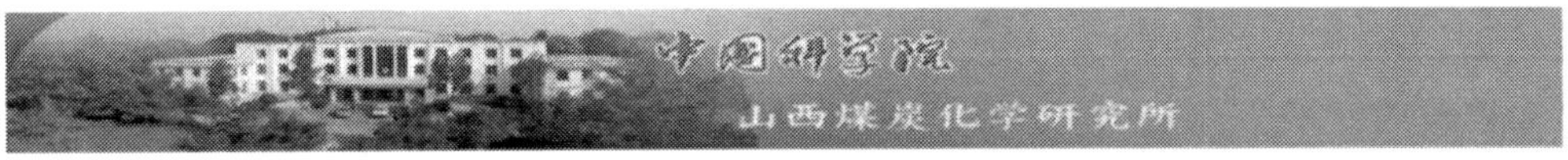

3.3.2 ADWC Control: Controlled-Tray Location

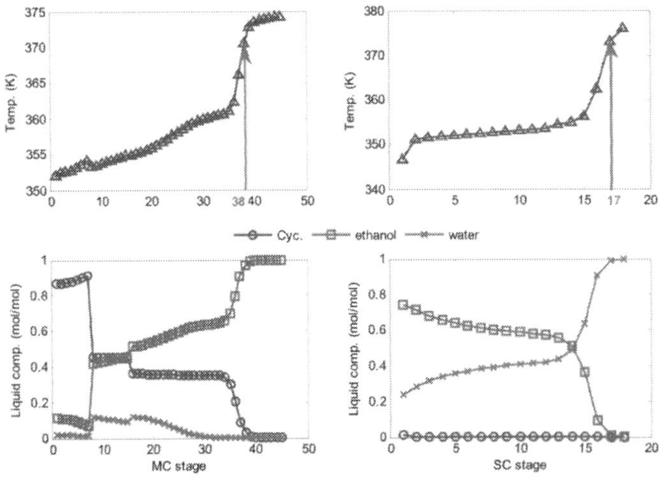

3.3.2 ADWC Control: TCS1

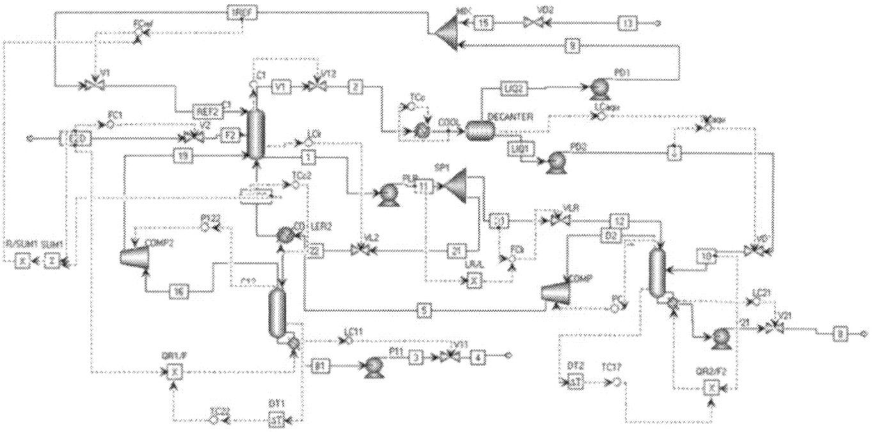

中国科学院
山西煤炭化学研究所

3.3.2 ADWC Control: TCS1 Feed Changes

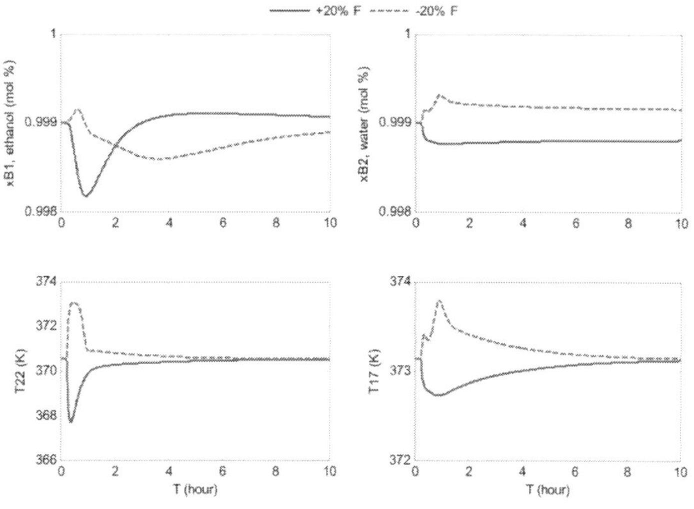

3.3.2 ADWC Control: TCS1 Composition Changes

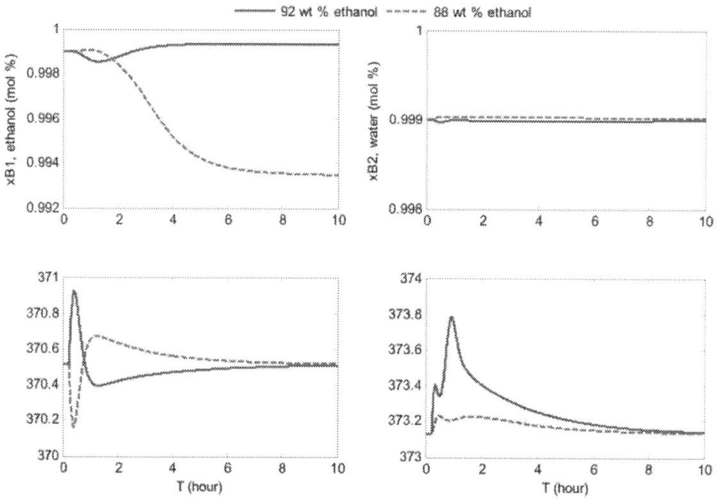

中国科学院
山西煤炭化学研究所

3.3.2 ADWC Control: TCS2 Feed Changes

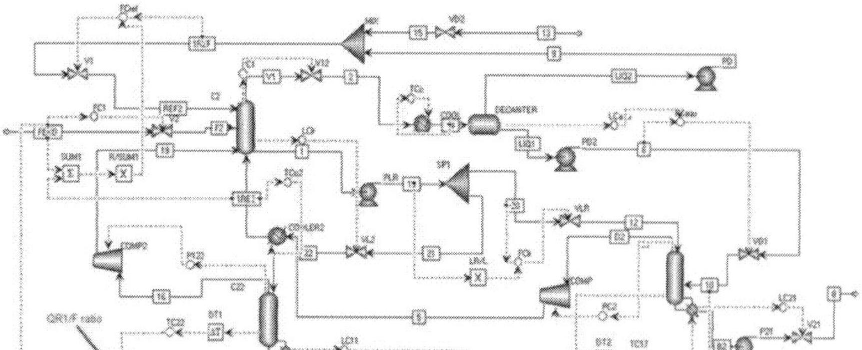

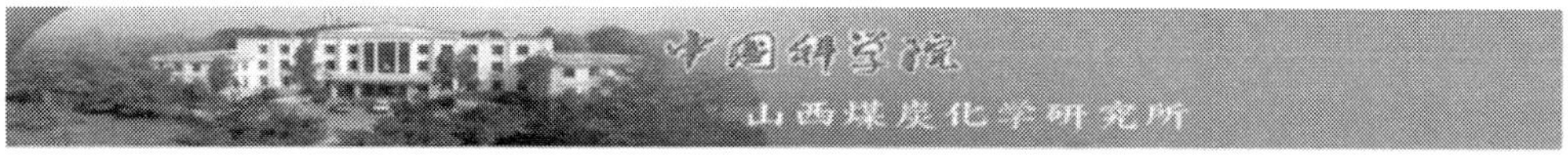

3.3.2 ADWC Control: TCS2 Feed Changes

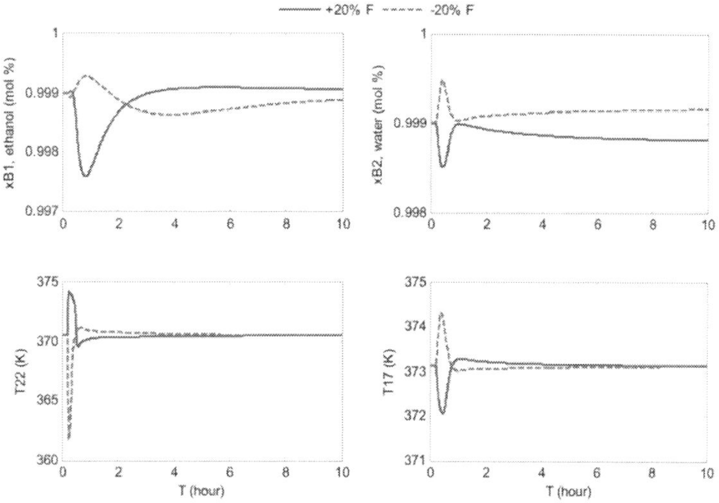

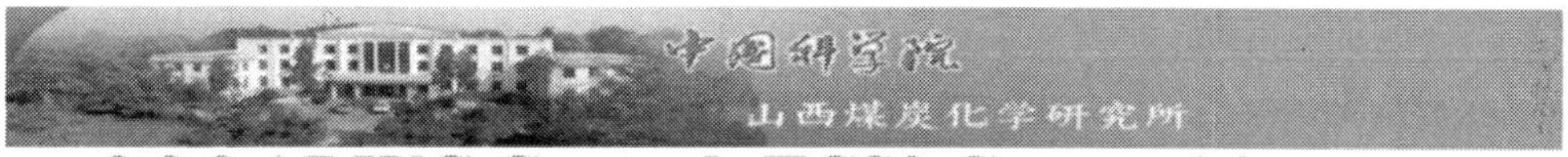

3.3.2 ADWC Control: TCS2 Composition Changes

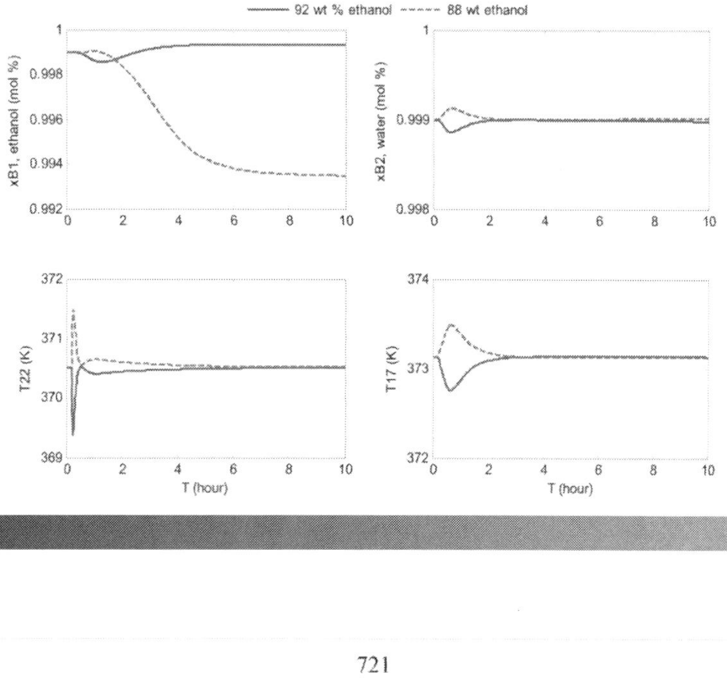

4. Conclusions

中国科学院
山西煤炭化学研究所

4. Conclusions

- DWC can be effectively extended to EDWC and ADWC for separating azeotropes. A considerable saving in energy consumption is obtained.
- There is some control problems in EDWC when key component disturbances is encountered since the vapor split at the bottom of the dividing wall cannot be adjusted.
- As for ADWC, there are almost no control problems since the liquid split at the top of the dividing wall can be practically implemented, with the vapor split being fixed.

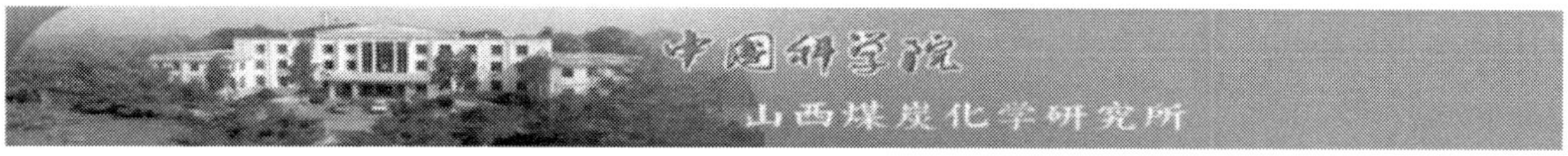

Thank you for your Attention!

510024 Equation-Based Model for Dividing Wall Column Simulations

Abstract:

The literature on Dividing Wall Columns (DWCs) has been growing extremely rapidly of late. Indeed, it is already sufficiently large that more than one review article has been published (Dejanović, I., Matijašević, L. & Olujić, Ž. Dividing wall column—a breakthrough towards sustainable distilling. *Chem. Eng. Process. Process Intensif.* **49,** 5pp 59–580, 2010; Yildirim, Ö, Kiss, A.A., Kenig, E.Y., Dividing wall columns in chemical process industry: A review on current activities, *Separation and Purification Technology*, **80**, pp 403-417, 2011; Kaibel, B. Dividing-Wall Columns, in *Distillation: Equipment and Processes* pp 183–199, Academic Press, 2014).

With apparently few exceptions, simulations of DWCs are done using an interlinked multi-column model in a commercial simulator and it is not hard to find statements lamenting the degree of difficulty of simulating DWCs in this way. For example: Dejanovic et al. (2010) wrote:

Carrying out DWC performance simulations requires great experience and these are more or less computationally very demanding. ... well established commercial software packages still do not contain a DWC as a standard model. This however will occur sooner or later, most probably as a simultaneous, equation based model.

The *sooner or later* mentioned by Dejanovic et al., in fact, means *now*. An imminent release of the freely available column simulation program ChemSep will include an equation based equilibrium-stage model for simulating DWCs of arbitrary complexity (this means with multiple walls in any configuration).

In fact, it has long been recognized that a simultaneous equation based model is probably the smart tool to use. Many investigators cite a paper by Becker et al. (Becker, H., Godorr, S., Kreis, H. & Vaughan, J. Partitioned Distillation Columns–Why, When & How. *Chem. Eng.* **108,** 68–68, 2001) who make such a recommendation, but they provide no evidence at all of having done so. Nor do any of the many other writers that cite Becker's work. Thus, one goal of this work is provide definitive evidence that a simultaneous column model can converge quickly and easily, even in cases where the multi-column models fail.

In addition, the model in ChemSep is compared to experimental data from the University of Texas at Austin. Finally, we demonstrate that the ChemSep standalone DWC model can be used as a single unit operation model within commercial process simulation tools via the CAPE-OPEN mechanism, thereby greatly simplifying the task of building DWC models.

494836 Understanding and Controlling Organic Molecule and Metal Organic Framework Crystallization

Abstract:

Small organic molecules have had a dramatic impact on our health and daily life over the past century. Small molecule pharmaceuticals have increased human lifespans, and organic pigments have expanded in use in textiles and displays. This impact is set to accelerate in the near future, as small molecules are explored for novel applications such as organic electronics, or metal-organic frameworks for chemical separations, catalysis and sensing. One of the major barriers for using small organic molecules for new applications is the limited understanding we possess on how molecules aggregate together to form different crystal habits and phases. Different crystal structures and morphologies can have wildly varying physical, chemical and physiological properties. Thus, if we do not control the crystallization organic molecules, we cannot predict its behavior for the aforementioned applications. Understanding the crystallization process can also help form metastable phases. These metastable phases can be more useful than the equilibrium phase for many applications. Metastable phases permit tunable optical bandgap for optoelectronics, control over pore size and shape in metal organic frameworks (MOF), and increased bioavailability in pharmaceuticals. General methods used to create metastable phases, like confinement or rapid cooling, require small length scales and extreme rates of heat and mass transfer. Moreover, these processes need precise control to get reliable results. This talk will focus on flow coating and microfluidic methods of controlling organic molecule and MOF crystallization characteristics, and the use of these materials for various applications.

502410 Design of API Amorphous Precipitation Processes Via the Use of an FBRM Based Method

Abstract:

This presentation will provide an FBRM based method to aid in the design of amorphous precipitation processes for active pharmaceutical ingredients. The attributes of amorphous particles formed during precipitation from solution are dependent on the glass transition temperature of the system. Glass transition temperature measurements of amorphous wetcakes and slurries are difficult to obtain using differential scanning calorimetry. This FBRM based method provides a â€˜softening temperatureâ€™ of amorphous particles in suspension. The softening temperature is shown to correlate with the glass transition temperature of amorphous solids (as determined by differential scanning calorimetry). By determining the softening point of a precipitation system as a function of solvent composition, a process can be designed to deliver desired particle properties. A case study will be provided where a precipitation process was developed to provide a narrow range of powder specific surface areas to meet formulation requirements. Another case study will be provided where the filtration rate of an amorphous slurry was drastically improved by adjusting the temperature of a precipitation. The optimal temperature to maximize filtration rate was determined by softening point measurements obtained via the FBRM based method.

494690 Facile Synthesis of Soft-Templated Carbon Nitrides with High Surface Area and Porosity

Abstract:

With high physicochemical stability, a band gap in the visible region, and basic site moieties, carbon nitrides are attractive metal-free semiconductor and catalytic candidates for a variety of applications. However, the tendency of carbon nitrides to form extended sheets limits their surface area (~10 m^2/g) and porosity (<0.1 cm^3/g), thus impeding their performance as catalysts. Hard templating approaches can effectively introduce porosity and increase the surface area of carbon nitrides, but they require multiple steps and hazardous reagents (e.g., HF) for template removal. In this talk, we present a facile soft-templating synthesis of high surface area (140-270 m^2/g) carbon nitrides containing both meso- and macropores (total 0.4-0.8 cm^3/g), while maintaining an overall low C/N ratio (0.70-0.78 at/at). Interesting morphologies like tubular and sponge-like arise, in contrast to the sheet-like structures that are more commonly observed for carbon nitrides. These textural and morphological features are achieved by taking advantage of the interaction between the melamine cyanurate adduct and surfactant (Triton X-100, Pluronic F-127, and Pluronic P-123), during both the synthesis and the thermal treatment in N_2 at 500 C. Compared to bulk melamine-derived carbon nitride, the synthesized carbon nitrides show improved light absorption and enhanced photocatalytic performance in a Rhodamine B dye degradation under visible light irradiation.

500552 Production of Large Quantities of ZIF-8 through Batch and Semi Continuous Processes

Abstract:

Zeolitic imidazolate framework-8 (ZIF-8) is a nanoparticle, which contains a large enough surface area for it to be applicable in many modern-day transport operations. Since transport operations are a large factor in many innovative fields, such as sustainable energy and gas purification, production and utilization of particles in new applications is the key to the future growth of increasingly efficient heat and energy transfer operations. Due to the fact that the current laboratory synthesis method for ZIF-8 is extremely lengthy and inefficient, the research conducted sought to design a large-scale method to produce the large quantities of ZIF-8 required for testing new applications. By taking the unit operations perspective, both a batch process and a semi-continuous process were designed and implemented to produce large quantities of ZIF-8 through the use of a stirred tank reactor and a spray dryer. Several methods of characterization show that the ZIF-8 produced from the designed bulk synthesis method hold the same properties as that produced from the traditional lab scale processes.

Thermodynamic study of silica precipitation in a hydrometallurgical process

Christian Manfoumbi[1, 2], Martine Meireles[1], Kevin Roger[1], Pierre Roblin[1], Denis, Beltrami[2], Bruno Courtaud[2], Philippe Ribagnac[2]

[1]Laboratoire de Génie Chimique, Université de Toulouse, CNRS, INPT, UPS, 31062 Toulouse, France.
[2]ERAMET Research, 1 Avenue Albert Einstein 78190 Trappes, France.

Abstract

ERAMET Research, research center of the French mining group ERAMET, has developped a hydrometallurgical process to recover several value elements (Nb, Ta, U and rare earths) in a polymetallic deposit. The value elements are recovered from the rest of the ore by filtration after a leaching step in an acid medium that aim to dissolve them. The efficiency of the process according to the yield of recovery for the value elements depends on this filtration step which remains difficult to operate due to the presence of silica precipitated in the suspensions. Indeed, it is dissolved during the leaching step, few soluble in these conditions, it precipitates in the form of silicon oxide which polymerizes in high acidity conditions to form a gel extremely detrimental for the filtration step.

In this study we investigate the composition effect of the leaching solution on the sol-gel time transition as well as the morphology and the filterability of the silica gel. We are looking by small-angle X-ray scattering (SAXS) the morphological evolution of the solid phase during silica precipitation. Our results show that under the process conditions in terms of temperature and pH, the composition has not effect on the shape of the particles and the gel structure. When it appears its structure remains the same whatever the conditions. That is why it is not possible to use some strategies based on a change in the composition of the liquor to improve filterability. To improve the filterability we have considered another strategy based on a modification of the silica gel structure by a pH change from 0 to 9 and then 9 to 0 after a consolidation step by precipitation of sodium silicate at pH 9.

Introduction

Filtration is one of the main operations used in the field of hydrometallurgy where most of the solid-liquid separation steps are realized by this technique. The efficiency of the hydrometallurgical processes for the recovery of the value elements depends primarily on the productivity of this step. Silicon is the second most abundant element of the earth's crust and is therefore present in the majority of deposits, as amorphous form, quartz or embeded in silicates crystallites. The leaching conditions of these ores lead sometimes to its dissolution and its precipitation under the form of gels which are responsible of the difficulties during the filtration.

These problems related to the presence of silicon have recently been encountered on a hydrometallurgical process developed by ERAMET Research, to recover niobium Nb, tantalum Ta, uranium U and rare earths elements REE in a polymetallic deposit. In this one as described by these references [1] [2] the value elements are dissolved during an acidic leaching step of a solid concentrated in value elements called "**Valuable elements dissolution**". During this step, silicon is dissolved in the form of silica and precipitates to form a gel. In addition, it was found that a slight variability in the chemical composition of the ore could lead to different filtration rates under the same leaching conditions. Phenomenologically, it has recently shown that the filtration rate of precipitated silicas depends directly of the morphology of silica gels [3] [4].

Our study investigates the effect of the composition of leaching solutions on the precipitation and morphology of silica gels formed during this step and their filterability. We examine after the role of ferric iron Fe (III), a major cation in these solutions, on the structure of silica gel to know if the ionic composition can modify the structure and changes the filterability of the gel. From the considerations about the mechanisms discussed in this paper we describe a procedure to change the structure of silica gel formed in acid medium. The filterability of this gel is compared with the gel obtained under the pH and temperature conditions of the leaching process.

I. Composition effect on filterability and silica gel structure at pH 0

I.1 Experimental procedure

The tests are made from two types of solutions which concentrations can be modulated according to the chosen conditions. The first is an acidic solution at 1 mol/L of H^+ prepared from a 95% by weight of concentrated sulfuric acid solution diluted in deionized water. To study the effect of Fe (III), a ferric sulphate solution at 150 g / L of Fe (III) is added to this solution to reach a final concentration of 60 g/L of Fe (III). The second solution is prepared by diluting a sodium silicate solution (27.5% by weight of SiO_2, density 1.39 g / mL, Sigma Aldrich) in 100 mL of deionized water.

The precipitation experiments are carried out by adding 900 mL of the acid solution into the reactor heated at 85 ° C. The sodium silicate solution at pH 11 is then quickly injected. During the course of precipitation, samples are collected and "quenched" at pH 2 and 30 ° C to stop the reaction as described in the literature by *Iler* and *Alexander* [5] [6] [7]. The quenched samples are analyzed by Small Angle X-ray Scattering (SAXS) on the XEUSS 2.0 platform available in house. For the analyses, volumes of 30 μL are introduced into borosilicate capillaries and placed in a vacuum chamber. The sample-detector distance used was 1 to 2.5 m giving scattering vectors q in the range of 0.005 and 0.5 $Å^{-1}$. Each analysis takes 20 minutes and the sample is analyzed three times at 20 ° C. For samples sufficiently diluted, the scattered intensity depends mainly on the volume fraction of precipitated silica and on the shape of the objects in the suspension.

At the end of the precipitation tests, the suspensions are filtered on a filtration set-up to study the influence of the gels structure on the filterability. The filtrate mass collected as a function of time is recorded on an analytical balance with a sensittivity of ± 0.01 g. The Ruth relationship (equation 1) is used to calculate the specific resistance from the filtrate volumes by dividing the filtrate mass acquired over time by the density of filtrate measured at the end of each filtration test [8] [9]. Specific resistance α is the macroscopic quantity that will be used to evaluate the filterability of the gels. The higher it is slower is the rate of filtration.

$$\frac{t_{i+1} - t_i}{V_{i+1} - V_i} = \frac{\mu \alpha M}{2\Delta P A^2}(V_{i+1} - V_i) + \frac{\mu R_m}{\Delta P A} \quad (1)$$

In equation 1, A is the surface of the filter medium (in m^2), ΔP (in Pa) is the pressure difference applied to the fluid that passes through the cake and the filter, μ is the fluid viscosity (in Pa.s), R_m and R_c (in m^{-1}) are respectively the hydraulic resistances of the media and the filter cake. At the end of each test, the solid is recovered and washed twice at mass ratios water/solid of 5. It is then dried at 80 ° C, until the dry mass become constant, this mass corresponds to the mass of solid deposited at the media surface during the filtration.

I.2 Silica gels structures and filterability

Figure 1 represents the log-log scattered intensity of the objects in solution as a function of the scattering vector for the samples collected at different times of the precipitation on both the reference system and the system with Fe (III). Figure 1 shows that at the beginning of precipitation, typically for durations less than 60 minutes, the scattered intensity is constant for the low values of the scattering vector, between 0.01 and 0.1 $Å^{-1}$. This asymptote, called the

Guinier plateau, is related to the presence of the finite size objects [10]. The intensity corresponding to this plateau increases over time between 2 and 60 minutes due to the increase in volume fraction of precipitated silica. The evolution of the plateau is illustrated by the dotted lines in Figure 1. Beyond 60 minutes, the intensity increases continuously with the scattering vector, the disappearance of the Guinier plateau is characteristic of the formation of connected network [11].

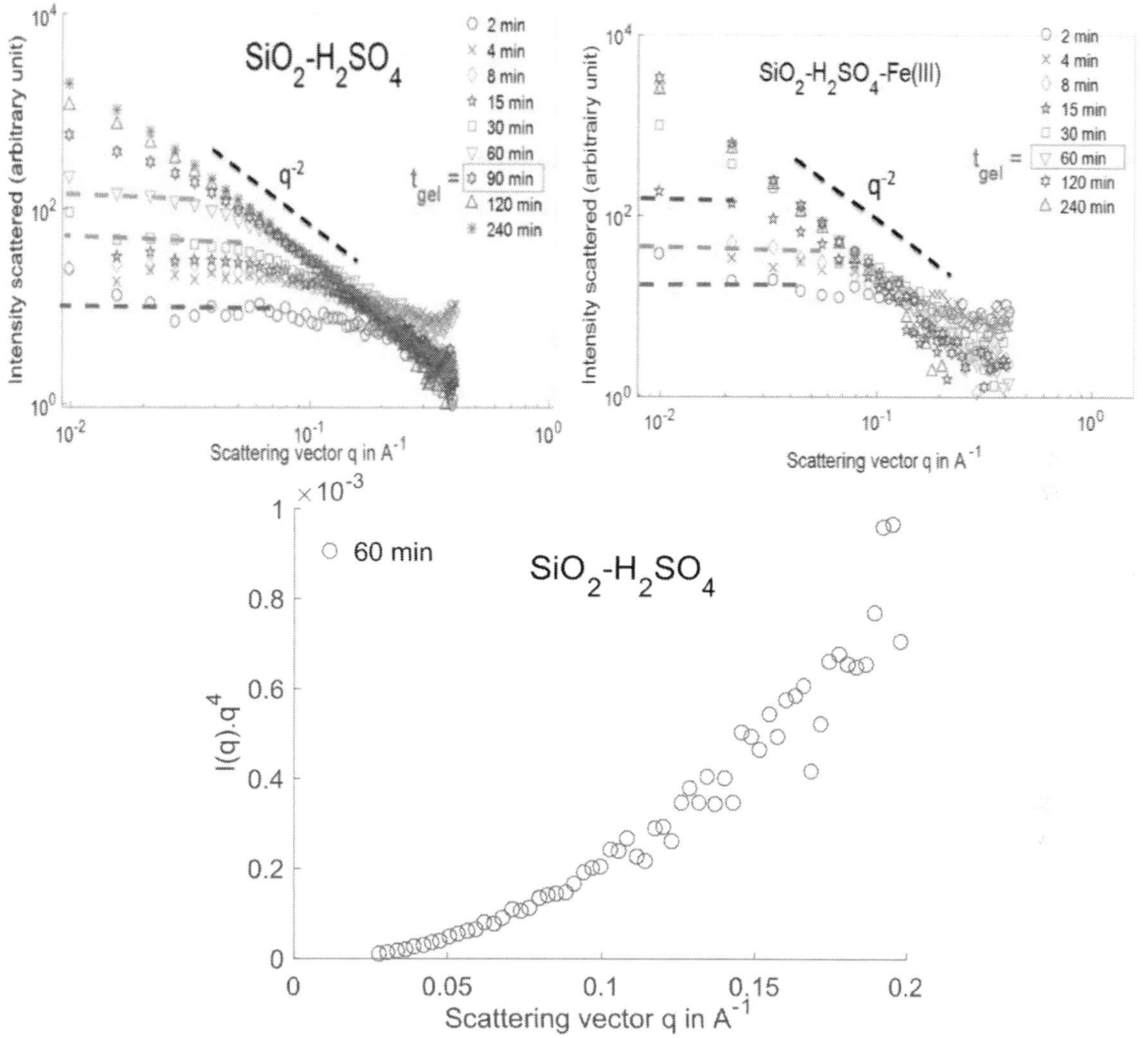

Figure 1 : *In the top, intensity scattered by the objects as a function of the scattering vector for different times of precipitation during the experiment at 1 mol/L of* H^+ *and 85°C for a silica concentration of 13 g/L. At the bottom, Porod representation I(q).q4= f(q) for the sample collected at 60 minutes on the reference system* SiO_2-H_2SO_4.

At the large values of the scattering vector (0.1 and 0.3 $Å^{-1}$) the superposition of all spectra means that the objects have the same shape and approximately the same size whatever the precipitation time. If the objects had a spherical shape with smooth surfaces the intensity would be expected to decrease according to a power law of -4, and as consequence in the Porod representation we would observed a plateau for the large values of the scattering vector [12] [13].

Indeed, in this case, the objects have the form of polymers because if we use the Porod representation which consists to plot I (q).q^4 as a function of q, the intensity decreases following a power law of -2 q^{-2} representative of polymeric objects (Figure 1). Thus, we show that the

objects formed during the precipitation at 1 mol/L of H^+ are not spherical. During the silica precipitation oligomers precipitate as finite size objects before sticking to each others to form a network. This percolation mechanism leads to the appearance of a gel after 90 minutes on the reference system and 60 minutes in the presence of Fe (III). These times are called t_{gel} in Figure 1. We can see that the spectra obtained on the two systems are quite similar. Precipitation leads to similar objects in the form of polymers. The presence of salt does not change the nature of the gel. However we can note that Fe (III) accelerates the formation rate of the gel, as an illustration of its effect on the gelation kinetics.
The filtration curves in representation t / V versus V on the reference system are shown in Figure 2. The results correspond to the suspension collected after 2, 4 and 6 hours of precipitation. The results show that filtration becomes difficult as soon as the gel appears after 4 hours of precipitation. The specific resistance increases by a factor of 222 between 2 and 6 hours and by a factor of 6 between 4 and 6 hours. The pictures of suspensions corresponding to the different precipitation times illustrate the texture of different silica gels. We can see that after 2 hours, the solution is blue but remains transparent. The resistance for the test at 2 hours corresponds to the resistance of the filter media R_m since the mass of deposited solid is negligible. This is due to the fact that at the microscopic scale the gel has already formed as we can see in Figure 1 for this precipitation time, but all connections between the microscopic structures are not yet established to form the macroscopic gel that we can observe from 4 hours. If we compare the case at 4 and 6 hours we can observe that, once the gel has formed, the specific resistance does not increase so much, the structure of the gel evolves very slightly after the formation of the macroscopic gel.

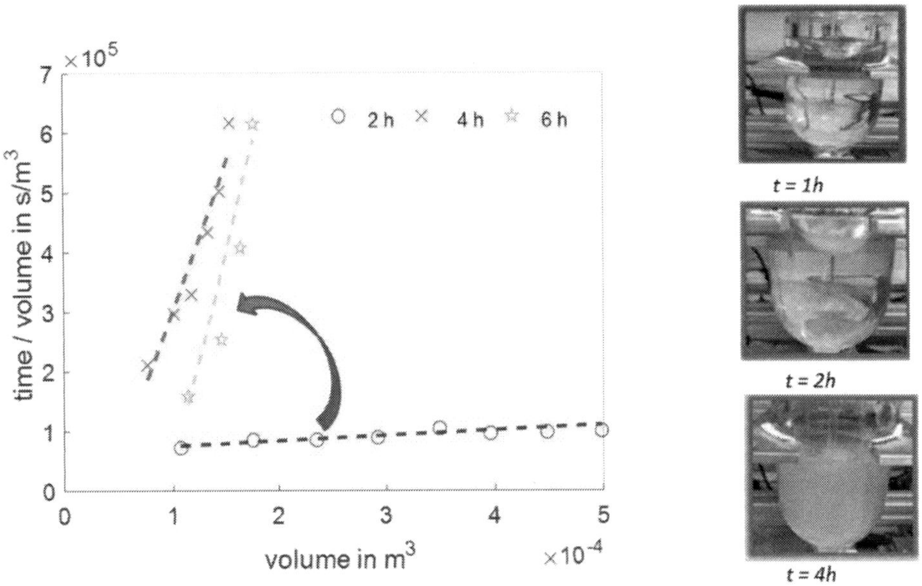

Figure 2 : *Filtration curves t/V versus V for the suspensions obtained at different residence time and filtered at 7 bars. The suspensions are made by precipitation of the sodium silicate solution at 1 mol/L of* H^+ *and 85° C for an initial silica concentration of 13 g/L.*

The comparison of the filtration tests between the suspensions obtained after 4 hours of precipitation on the reference system and that in the presence of Fe (III) shows that the specific resistance is 12 times higher than that of the reference test for the same precipitation time. These results are explained by the fact that in the presence of Fe (III) the kinetics of the gel formation is faster, the number of connections between polymers is larger than on the reference system. The specific resistances calculated from the slopes in Figure 2 and for the test in the presence of Fe (III) are given in Table 1 with the parameters which allowed their determination.

Table 1 : *Influence of precipitation time and Fe(III) on the specific resistance values for the suspensions collected during the precipitation experiments at 1 mol/L of H^+ and 13 g/L of silica. The operating pressure is 7 bars, the surface of the filter media is 0.0038 m^2.*

Precipitation time in h	**2**	**4**	**6**	**4 with Fe(III)**
Specific resistance α in m/kg	_	3.4×10^{12}	1.4×10^{13}	4.3×10^{13}
Media resistance R_m	6.3×10^{10}	_	_	_
Slope of curves t/V= f(V)	9×10^7	4×10^9	7×10^9	1.4×10^{11}
Filtrate mass in kg	0.821	0.225	0.208	0.117

I.3 Discussion about the precipitation mechanisms and the mesostructure of silica gels as a function of pH

The work of *Jolivet* [14] have shown that the precipitation in a very acid medium (below pH 3) and between pH 3 and 9 consist in an oxolation mechanism, the structural differences of the silica gels as a function of the pH are then related to the silanols groups reactivities. The mechanisms explained by *Jolivet* [14] are in agreement with those reported by *Iler* [6] and *Scherer* [15]. Below pH 2 silica polymerized to form small particles which aggregates and lead to a **polymeric gel** [15]. In the paragraph I.2 of this study we have validated this for the structure of the silica gel in very acidic medium. *Jolivet* also shown that between pH 3 and 9, the mechanism involved consists in a first nucleophilic attack by the anionic forms of silica SiO^- or OH^- followed by the formation of a siloxane bond. The silicic anions attack specifically the sites where the partial charge is the highest and that occurs in the middle of the chain. Several points of condensation are then generated and allow a three-dimensional 3D increase of the structure leading to gels consisting of dense particles denoted as **particulate gels** [12]. The filtration of these gels is not difficult provided that there is a consolidation step which will be described below. In acidic medium, it is the hydroxyl functions at the end of the olygomeric chain that are concerned by the condensation, particularly those with the most negative partial charge. This leads to the growth of two- dimensional 2D chains in preferential directions.

By considering the mechanisms explained previously, we have imagined that by shifting the polymeric gel from 0 to 9, the charge on the surface of objects would increase. The salinity of the environment would then promote aggregation and the occurence of flocs in solution. Since particulate gels are easier to filter than polymeric gels, we have decided to modify the structure of the polymeric gel into a gel consisting of much denser particles by changing the pH from 0 to 9 and then from 9 to 0 aftter a consolidation step at pH 9. The shift of pH 9 to 0 is imposed by the fact that the value elements in the deposit and the metallic salts are only soluble in a very acid medium (at pH 0). In our strategy we have chosen to considere a time of 1 hour allowing to dissolve the value elements before to change the pH to improve the filterability.

II. Modification of silica gel structure by changing the pH

II.1 Experimental procedure

The conventional precipitation protocol that consists in precipitating the sodium silicate solution at 1 mol/L H^+ and 85 ° C was used. After 1 hour, a first change of pH from 0 to 9 occurs by adding 8 mol/L of sodium hydroxide solution. A **consolidation** step is then performed to change the structure of the gel. The consolidation is done at pH 9 by adding in the agitated tank a solution of sodium silicate under the most favorable conditions for the precipitation at the surface of aggregates in suspension, especially in the empty spaces leaved by the particles aggregated. This step is ensured by injecting at a low flow rate a sodium silicate solution at 70 g/L. The pH is kept constant during the addition of silicate by adding small volumes of sulfuric acid solution at 1 mol/L of H^+. After this consolidation step the transition from pH 9 to 0 is carried out by adding a sulfuric acid solution at 8 mol/L.

II.2 Study of silica gel sturcture during the pH changing procedure by SAXS

Figure 3a shows the SAXS spectrum after 1 hour of precipitation at 1 mol/L of H^+ and 85 ° C as well as the spectra during the pH changing and the consolidation stage. Before the first pH changing (from 0 to 9) the objects have the polymer shape as shown in section I.2. The change from pH 0 to 9 causes the disappearance of the Guinier plateau, the intensity decreases continuously when the scattering vector increases according to a power law of q^{-2} characteristic of the of a polymeric gel structure. The disappearance of the Guinier plateau during this pH change is due to the aggregation of the polymers to form the gel. On the other hand, we can observe that at the end of the consolidation step at pH 9, the intensity decreases according to a power law of q^{-3} for the scattering vector ranges between 0.07 and 0.2 $Å^{-1}$. The network appears as formed from dense particles of spherical shape with rough surfaces. The same type of spectrum is obtained during the change of pH from 9 to 0, the intensity also decreases according to a power law in q^{-3}, which means that the change of pH from 9 to 0 after the consolidation step at pH 9 does not involve a structure modification.

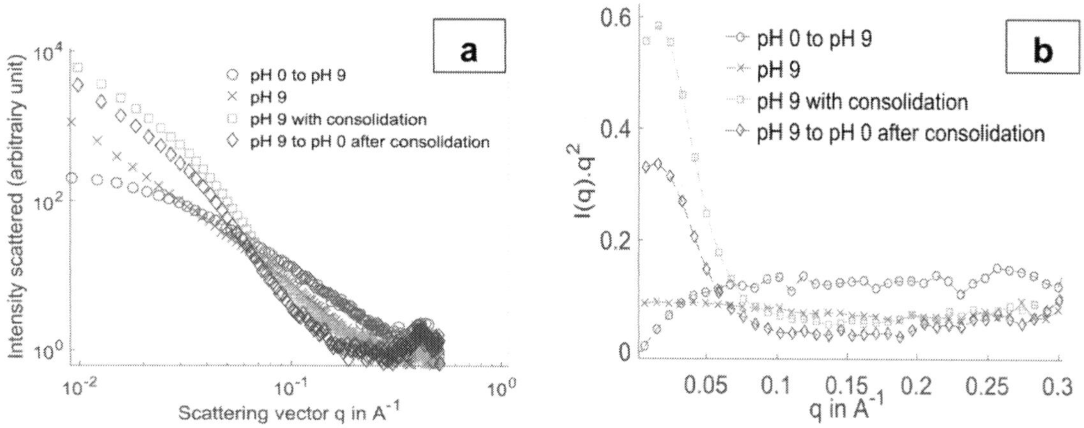

Figure 3 : SAXS spectra of the silica gel during the experiment starting from the reference system at 1 mol/L of H^+ after 1 hour of precipitation at 85°C.

Kratky's representation which consists to plot I (q).q^2 as a function of q allows as the Porod's representation to have some informations about the structure of the gel (Figure 3b).In this representation, the intensity scattered by the particle-like polymers leads to a plateau for the large values of the scattering vector. Figure 3b shows that the intensity scattered in Kratky's representation evolves differently for the consolidated samples. For the intermediates values of scattering vector those between 0.07 and 0.2 $Å^{-1}$ corresponding to the aggregate scale, the scattered intensity for the consolidated samples increases initially to reach a maximum and gradually decreases when the scattering vector increases to form a plateau. For the large values of the scattering vector, a plateau is observed for all samples, which means that the smaller objects have the polymer shape whatever the sample. The decrease in plate values during the different pH change and the consolidation stage is due to the dilution during the adding of the reagents. Thus, these results show that the structural modification occurs at the aggregate scale. The consolidation step causes the precipitation of silica at the surface of aggregates leading to a change structure of the polymeric gel.

II.3 Filterability of the gel after the shifting pH

The filtration tests realized on the gel at pH 0 after the pH shifting protocole show a significant improvement in filterability compared to the reference system (Figure 4). The polymeric gel obtained under the standard precipitation conditions (1 mol/L of H^+ and 85 ° C) allows an important mass transfer of liquid during the first moments. The collapse of its structure or its crushing on the filter surface causes a loss of permeability of the deposit. The gel obtained by pH change is filtered more slowly if we compare the rates before the gel collapse. However, this gel has a mechanical resistance to compression larger than the polymer gel obtained at pH 0, its filtration rate is much larger. The solids texture at the end of the filtration tests is also shown in Figure 4 which shows the two solids are macroscopically different.

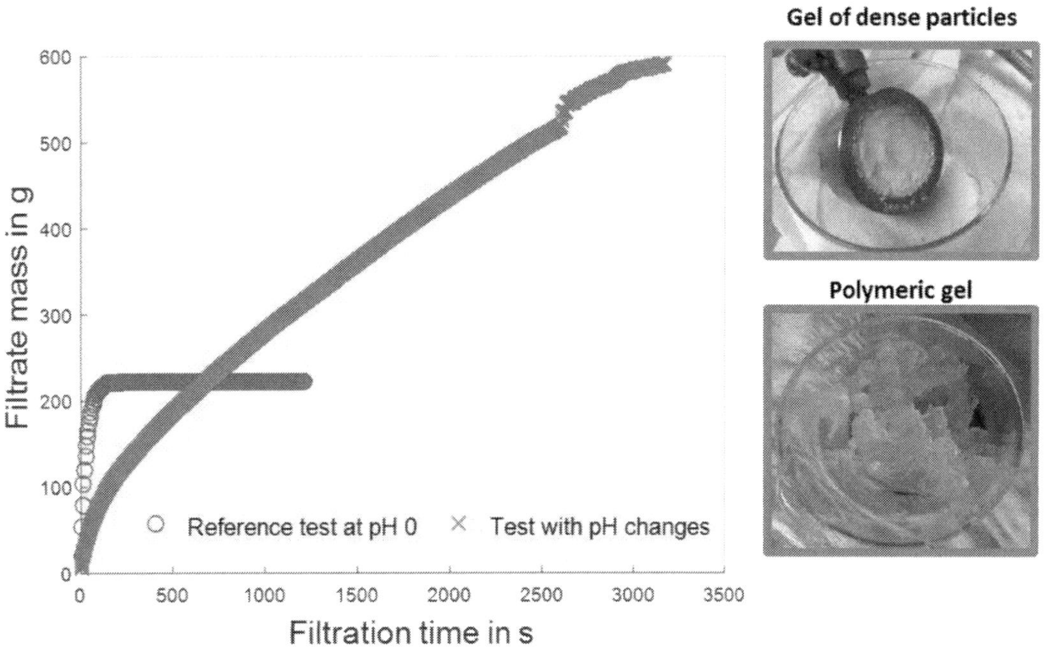

Figure 4 : Comparison of the filtration rates between the gel obtenained after 4 hours of precipitation at 1 mol/L of H^+ and 85 °C and the gel obtenained by pH changing.

III. Conclusion

We studied the composition effect of hydrometallurgical solutions on the structure of silica gels and their filterability. The morphological evolution of the objects during the precipitation has been monitored by SAXS. We have shown that under these process conditions, silica precipitates in the form of polymeric objects and subsequent percolation leads to the gel formation responsible of bad performances at the filtration step. The ionic composition neither changes the structure of the gels nor the shape of the precipitated objects, however, it has a kinetics effect on the gel time transition. To improve the filterability of the pulp in the process we have propose to change the structure of the silica gels by a pH shifting protocole. From scattering experiments, we show that this protocole contributes to a modification of the structure of the gel and allows significant improvement of filterability.

Bibliographic references

[1] D. Beltrami, P. Ribagnac, and G. J. Deblonde, "Leaching of niobium- and REE-bearing iron ores : significant reduction of H 2 SO 4 consumption using SO 2 and activated carbon .," *Sep. Purif. Technol.*, 2017.

[2] D. Beltrami, G. J. P. Deblonde, and V. Weigel, "Selective recovery of niobium and tantalum from low-grade concentrates using a simple and fluoride-free process," *Sep. Purif. Technol.*, vol. 162, pp. 180–187, 2016.

[3] Moussa, Deka, M. Meireles, B. Cabane, and J. Gummel, "Filtration of precipitated silica aggregates: Length scales, percolation threshold and yielding behaviour," *Sep. Purif. Technol.*, vol. 156, pp. 2–11, 2015.

[4] D. M. Ragueh, "Filtration de silices précipitées: mise en évidence des relations entre propriétés macroscopiques et échelles locales caractéristiques dans les dépôts," Thèse de l'Université Paul Sabatier, 2011.

[5] R. K. Iler, "Polymerization of polysilicic acid derived from 3.3 ratio sodium silicate," *J. Phys. Chem.*, vol. 57, no. 6, pp. 604–607, 1953.

[6] R. K. Iler, "The chemistry of silica: solubility, polymerization, colloid and surface properties, and biochemistry," *Lavoisier*, p. 892 pp, 1979.

[7] G. B. Alexander, "The Polymerization of Monosilicic Acid," *J. Am. Chem. Soc.*, vol. 76, no. 8, pp. 2094–2096, 1954.

[8] K.-J. Kim, V. Chen, and A. G. Fane, "Ultrafiltration of Colloidal Silver Particles: Flux, Rejection, and Fouling," *J. Colloid Interface Sci.*, vol. 155, no. 2, pp. 347–359, 1993.

[9] D. Antelmi, B. Cabane, M. Meireles, and P. Aimar, "Cake collapse in pressure filtration," *Langmuir*, vol. 17, no. 22, pp. 7137–7144, 2001.

[10] F. A., Guinier, G., *Small Angle Scattering of X-Rays*, John Wiley. New York, 1955.

[11] D. W. Schaefer, T. Rieker, J. S. Lin, D. Fischer, C. Chen, and G. Beaucage, "conference papers Multilevel Multilevel structure Structure of of reinforcing Reinforcing silica and carbon Silica and Carbon," *J. Appl. Crystallogr.*, pp. 587–591, 2000.

[12] P. W. J. G. Wijnen, T. P. M. Beelen, K. P. J. Rummens, H. C. P. L. Saeijs, and R. A. Van Santen, "Silica gel from water glass: a SAXS study of the formation and ageing of fractal aggregates," *J. Appl. Crystallogr.*, vol. 24, no. pt 5, pp. 759–764, 1991.

[13] D. W. Schaefer and A. J. Hurd, "Growth and structure of combustion aerosols: Fumed silica," *Aerosol Sci. Technol.*, vol. 12, no. 4, pp. 876–890, 1990.

[14] J.-P. Jolivet, *De la solution à l'oxyde*, Savoirs Ac. PARIS, 1994.

[15] C. Brinker and G. Scherer, "Sol-Gel Science: The Physics and Chemistry of Sol-Gel Processing," *Advanced Materials*, vol. 3, no. 10. p. 912, 1990.

489046 Separation of Isomers

Abstract:

In chemical industry there are several reactions in which a linear product and a number of branched side products are simultaneously produced. One example is the hydroesterification of long-chain unsaturated esters of fatty acids where linear and branched diesters are produced. For application only the linear ones are of interest, so that there is a demand to separate these isomers. This separation is a challenging task as distillation is very elaborative because of small separation factors and extraction often does not meet the high purity demands. Possible unit operations for the separation of isomers are the solvent crystallization and the adsorption. Solvent crystallization is promising because linear and branched molecules show different solubilities in a solvent. The challenge is the appearance of a superposition of a liquid-liquid equilibrium (LLE) and a solid-liquid equilibrium (SLE) during the separation process. Since this superposition can affect the product properties it has to be understood in order to build up suitable crystallization processes. To design a suitable adsorption process, the adsorption isotherms must be estimated. But in both cases, it is challenging purchasing the pure isomers to analyze the solid-liquid-liquid equilibrium (SLLE) or the adsorption isotherm. Here a thermodynamic model can be used to overcome these limitations.

The basic idea of the developed methodology is to only perform experiments with linear molecules and use the gained results for the prediction of phase equilibria of systems containing linear and branched molecules. For this purpose, it is mandatory to use a thermodynamic model that considers the architecture of the branched molecules. The lattice cluster theory (LCT) was developed to consider the architecture of the molecules without any additional adjustable parameter. The LCT therefore allows for the prediction of thermodynamic properties of branched molecules based on pure component parameters of linear molecules. In the case of SLLE calculations, the investigated solution consists of an alcohol and an alkane. To consider the self-association of the alcohol the LCT is combined with the chemical association lattice model (CALM).Â To calculate the adsorption isotherms of alkanes, the LCT is combined with the density functional theory (DFT).

In both cases the developed methodology can be divided into two steps. First, experiments with linear molecules are required. Regarding the solvent crystallization, binary LLE of a linear alkane dissolved in an alcohol were estimated. For the adsorption, pure component isotherms of linear alkanes were used. These experimental data are then used to fit the required model parameters. In the framework of the LCT applied to incompressible systems for the SLLE calculation, there is only one adjustable parameter for a binary system, namely the interaction energy Î"Îµ. Additionally, there are two adjustable parameters to consider the self-association of the alcohol in case of LLE. The LCT parameter for the compressible version to calculate the adsorption isotherms of alkanes were fitted to their vapor-liquid equilibria and the external potential of the DFT is fitted to pure substances adsorption isotherm. In the second step the fitted parameters are used for the prediction of the LLE of binary systems including a branched alkane and an alcohol and to predict the adsorptions isotherms of the branched isomer. This methodology will be applied and discussed for the design of solvent crystallization and adsorption processes.

How Reliable Are Isotherm Measurements In Metal-Organic Frameworks?

Jongwoo Park, Joshua D. Howe, David S. Sholl*

Nov. 1, 2017

Georgia Institute of Technology
School of Chemical and Biomolecular Engineering

Acknowledgements

NATIONAL ENERGY TECHNOLOGY LABORATORY
Grant DE-FE0026433

UNCAGE ME
Award DE-SC0012577

U.S. DEPARTMENT OF ENERGY

Any opinions, findings, conclusions or recommendations expressed herein are those of the author(s) and do not necessarily reflect the views of the DOE

Stimulating discussions with **R. Lively**, **K. Walton**, and **M. Realff** are gratefully acknowledged

Reproducibility in Science

- Reproducibility *crisis* rocking science

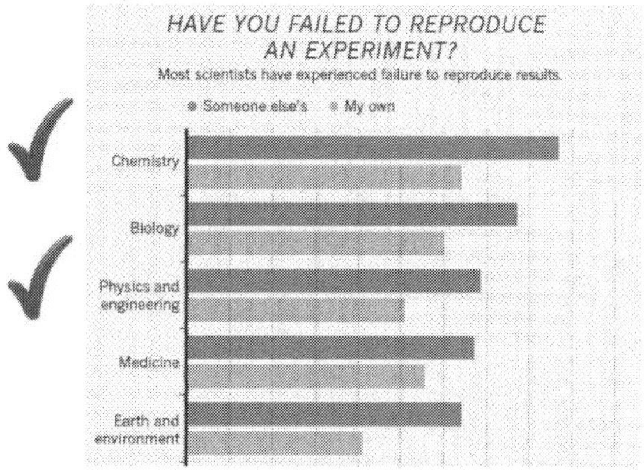

Significant "crisis" of reproducibility ...

No metrics on "what reproducibility is or should be" in materials chemistry

[Baker, M., *Nature*, **2016**, *533*, 452.]

Reproducibility in Science

- Reproducibility in materials chemistry

Reproducibility in DFT calculations of solids

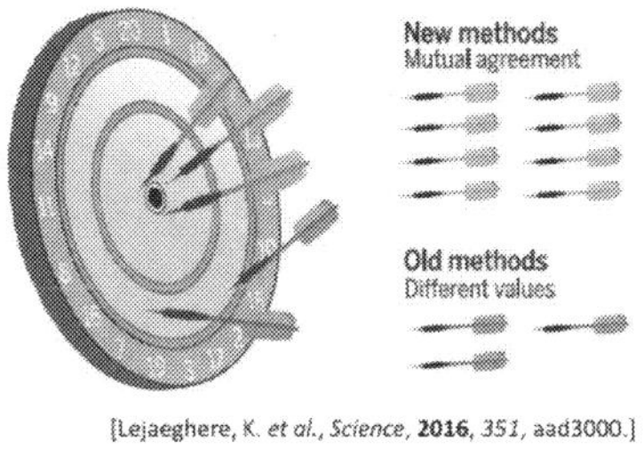

[Lejaeghere, K. *et al.*, *Science*, **2016**, *351*, aad3000.]

Reproducibility in predictions of the EOS-derived observables of crystals depending on DFT codes

Irreproducibility in H_2 storage materials

[Broom, D. P. *et al.*, *Energy Environ. Sci.*, **2016**, *9*, 3368.]

Reproducibility of H_2 isotherm measurements using porous materials *via* interlaboratory test

Identify the contributing factors causing irreproducibility ...

Yield a "consensus" in reproducible experiment

CO_2 Adsorption in MOFs

- **What are reproducible CO_2 adsorption quantities at similar adsorption conditions?**
  - ✓ Adsorptive CO_2 capture and separation
  - ✓ Metal-organic frameworks (MOFs)
    - Emerging class of crystalline porous adsorbent for adsorption applications
  - ✓ Molecular adsorption equilibrium
    - Uncertainty quantification for CO_2 single-component adsorption isotherm measurements
    - Yield consensus bounds across CO_2 adsorption quantities in MOF materials

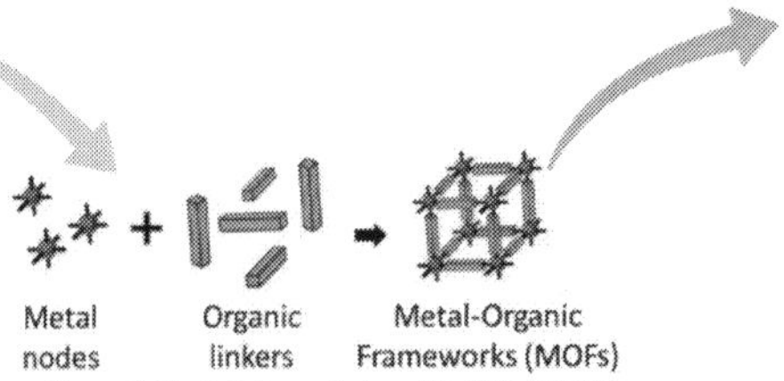

[Simon, C. M. *et al.*, *Energy Environ. Sci.*, **2015**, *8*, 1190.]

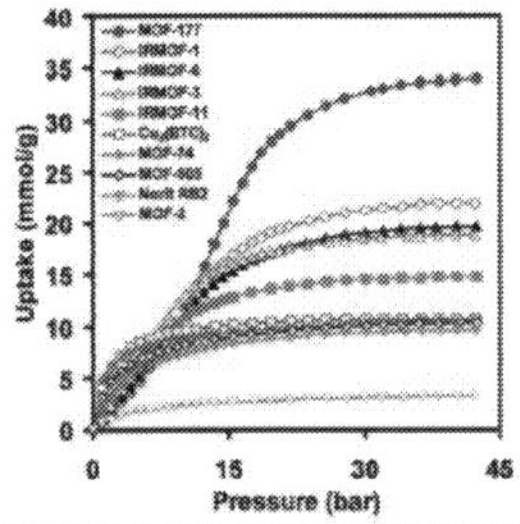

[Millward, A. R. *et al.*, *JACS*, **2005**, *127*, 17998.]

Meta-Analysis: Open Literature Database

- NIST/ARPA-E database of novel and emerging adsorbent materials
  - Systematically collects single-component adsorption data for adsorbents from the peer-reviewed literature

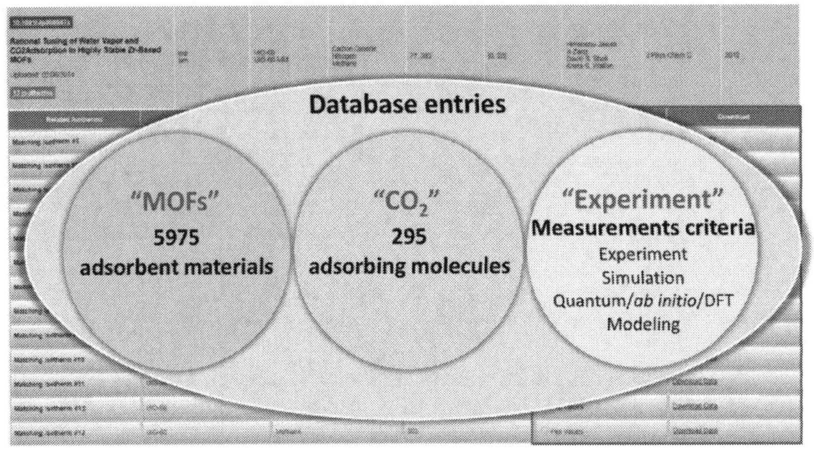

[NIST/ARPA-E Adsorption Database, http://adsorbents.nist.gov.]

Reproducibility Analysis Metrics

- Complementary statistical metrics to access the consensus bounds of isotherms in MOFs

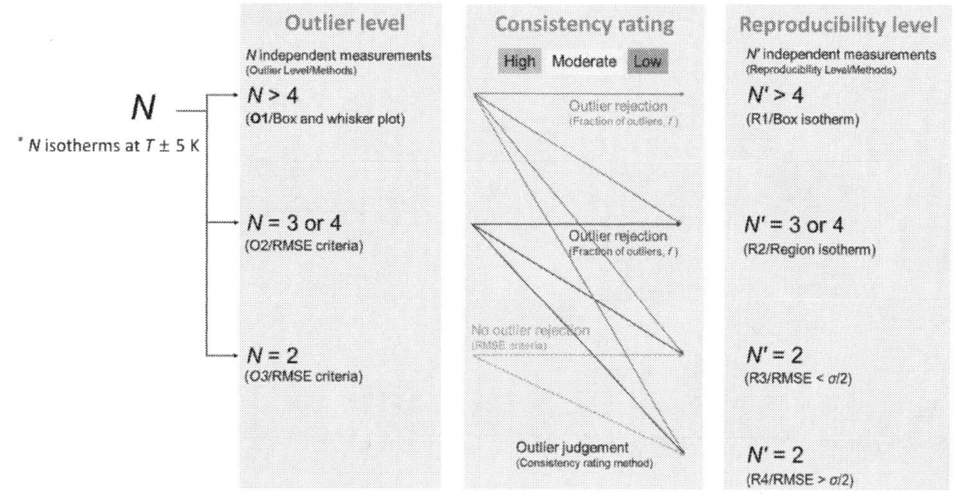

Taxonomy for Reproducibility Analysis

- CO_2 adsorption in HKUST-1 at 298 ± 5 K

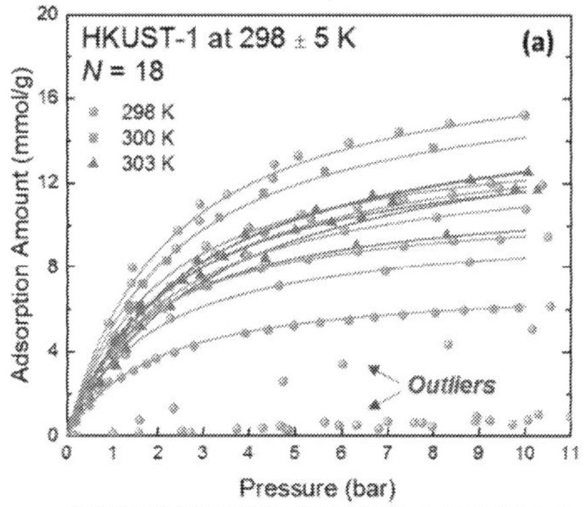

Moderate
$0.25 < f \leq 0.4$

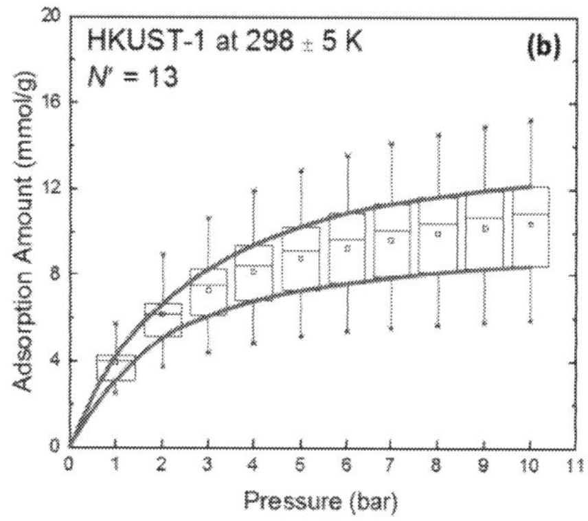

O1/Box and whisker plot **Consistency** **R1/Box isotherm**

- Consensus bounds to the variations between isotherms assigned *via* IQR of the measured quantities

* f: fraction of outliers, IQR: interquartile range

Taxonomy for Reproducibility Analysis

- CO_2 adsorption in UiO-66 at ambient temperatures

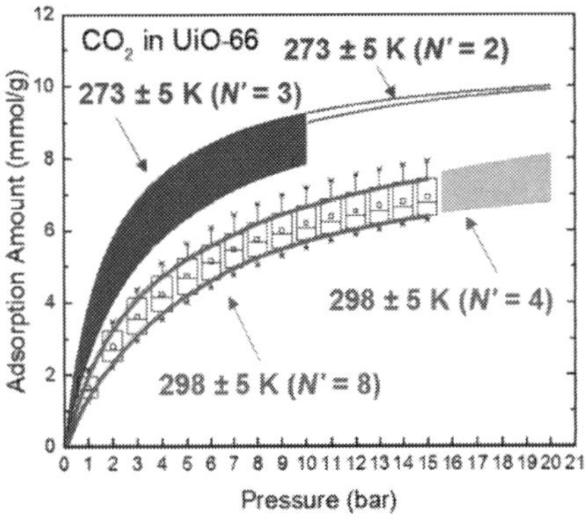

R1/Box isotherm

R2/Region isotherm

R3, R4/Pair of fitted isotherms

- "Reproduced" adsorption isotherms with consensus bounds given *via* IQR/shaded region/pair of isotherms
- **Encapsulates the different and useful facets of the overall reliability of a set of experimental measurements**

* RMSE: root mean square error

Reproducibility of BET Surface Area

- Reproducibility analysis of N_2 adsorption isotherms at 77 K in HKUST-1

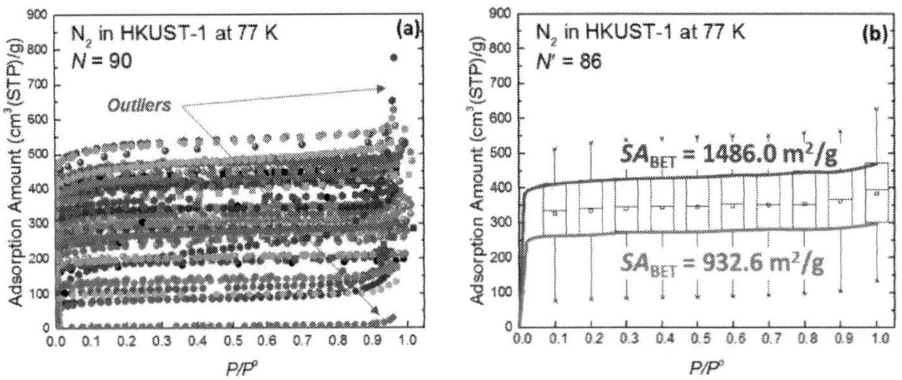

- Consensus bounds to the BET surface area* (SA_{BET}) estimated from consensus N_2 isotherms at 77 K

* BET plots for pseudo-experimental isotherms fitted over $0.05 < P/P^o < 0.15$

Reproducibility of Isotherms in MOFs

- Reproducibility analysis of CO_2 adsorption isotherms in *all* MOFs with $N > 1$

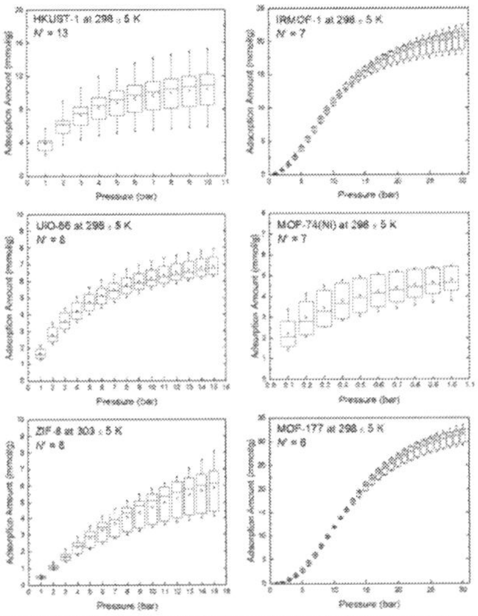

R1 level MOFs

UiO-66 **MOF-74(Ni)**
MIL-101(Cr) **ZIF-8**
MIL-53(Al)

IRMOF-1 **MOF-177**
HKUST-1

Georgia Tech | School of Chemical & Biomolecular Engineering

Reproducibility of Isotherms in MOFs

- Reproducibility analysis of CO_2 adsorption isotherms in *all* MOFs with $N > 1$

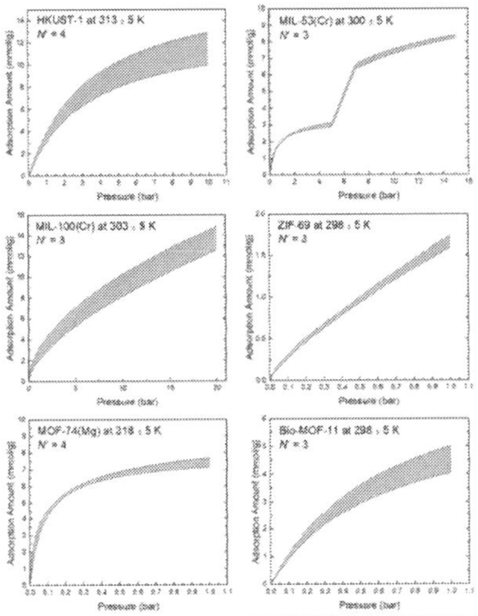

R2 level MOFs

MOF-74(Mg) Bio-MOF-11
MIL-100(Cr) ZIF-69
MIL-53(Cr) ZIF-7
PCN-200

HKUST-1

Georgia Tech | School of Chemical & Biomolecular Engineering

Reproducibility of Isotherms in MOFs

- Reproducibility analysis of CO_2 adsorption isotherms in *all* MOFs with $N > 1$

MOFs	*T* range	*P* range
CAU-1	273 ± 5 K 298 ± 5 K	[0, 1] bar
CPM-5	273 ± 5 K 298 ± 5 K	[0, 1] bar
IRMOF-3	298 ± 5 K	[0, 30] bar
MOF-508	303 ± 5 K 323 ± 5 K 343 ± 5 K	[0, 5] bar
NU-100	298 ± 5 K	[0, 1] bar
NU-1000	273 ± 5 K 293 ± 5 K	[0, 1] bar
SDU-1	273 ± 5 K	[0, 1] bar
UMCM-1	298 ± 5 K	[0, 15] bar
ZIF-11	273 ± 5 K	[0, 1] bar
ZIF-71	298 ± 5 K	[0, 1] bar

R3 level MOFs

CPM-5, IRMOF-3, MOF-508, CAU-1, NU-100, NU-1000, UMCM-1, SDU-1, ZIF-71, ZIF-11

R4 level MOFs

CPM-5

* Table curates MOFs with $N = N' = 2$ CO_2 isotherms and corresponding adsorption conditions

Georgia Tech | School of Chemical & Biomolecular Engineering

Reproducibility of Isotherms in MOFs

- Statistically quantified CO_2 adsorption range in MOFs with $N' \geq 3$

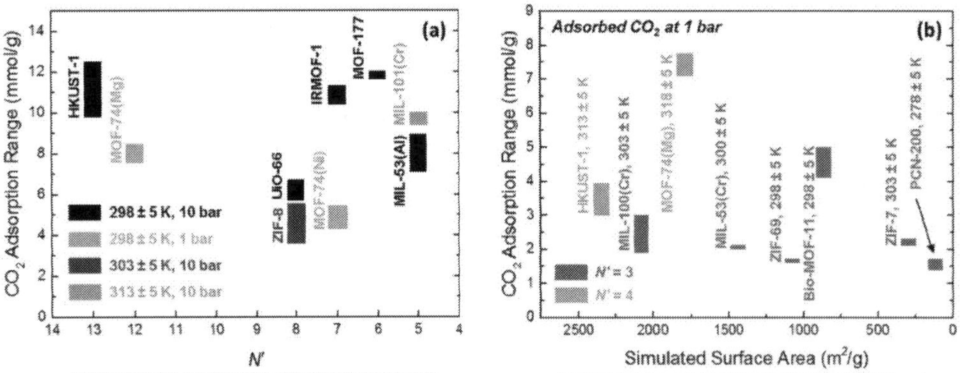

R1 level MOFs **R2 level MOFs**

- Firm conclusions about the reproducibility of CO_2 isotherms can be drawn for a small set of MOFs
- **Around 20% of adsorption isotherms are labeled as outliers**

Georgia Tech | School of Chemical & Biomolecular Engineering

Reproducibility of Isotherms in MOFs

III. Results and Discussion

- Implications for atomistic classical simulations of adsorption in MOFs

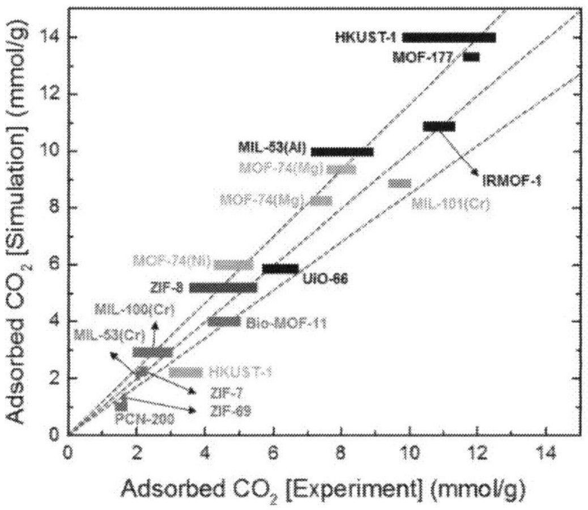

[*] GCMC simulation details
van der Waals: UFF/TraPPE force fields
Electrostatics: EQeq/TraPPE charges
Rigid structure approximation

[Park, J. *et al.*, *J. Mater. Chem. A*, **2017**, *5*, 12258.]

- Generic force fields are often inaccurate for
  (i) open metal site MOFs
  (ii) MOFs with marginal stability
- Simulations overlap with experiments when a 15% relative error is allowed
- ***Standard force field* enables reliable modeling of the extant reproducible isotherms within acceptable experimental uncertainty**

Georgia Tech | School of Chemical & Biomolecular Engineering

[*] GCMC: Grand Canonical Monte Carlo

Reproducibility of Isotherms in MOFs

III. Results and Discussion

- Reproducibility map for CO_2 isotherms in MOFs
  - All known MOFs for which *any* conclusions can be drawn about the reproducibility of CO_2 at $T \pm 5$ K

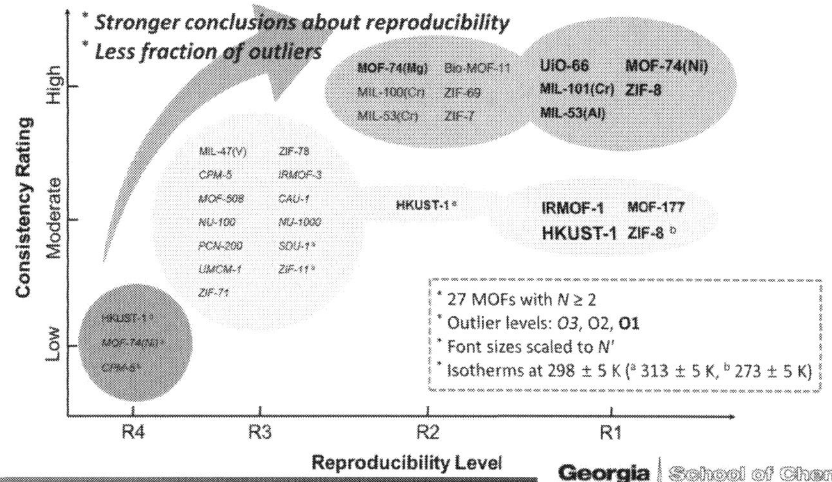

Georgia Tech | School of Chemical & Biomolecular Engineering

Conclusions

Meta-analysis of CO_2 isotherms in MOFs

- Exhaustive and unbiased compilation of single-component adsorption isotherms in MOFs

Establish uncertainty bounds across reported adsorption isotherm measurements

- Statistical quantification of different level of discrepancies for existing adsorption isotherms

Comprehensive summary of the state of knowledge regarding CO_2 adsorption in MOFs

- 20% of isotherms in the literature classified as outliers
- Metrics could aid the reproducibility of adsorption in MOFs
- Utilization of metrics with minimal changes would be possible

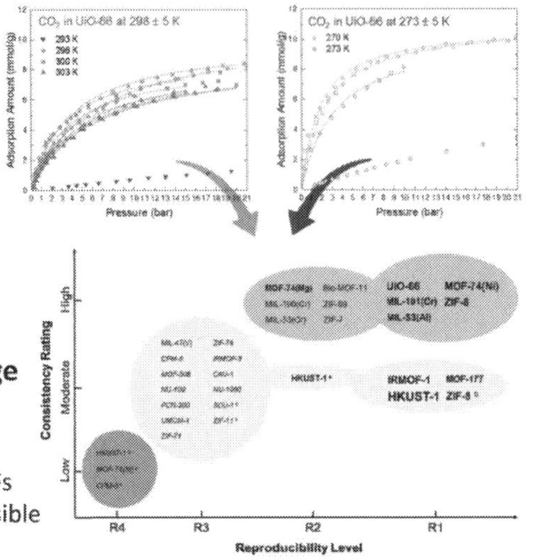

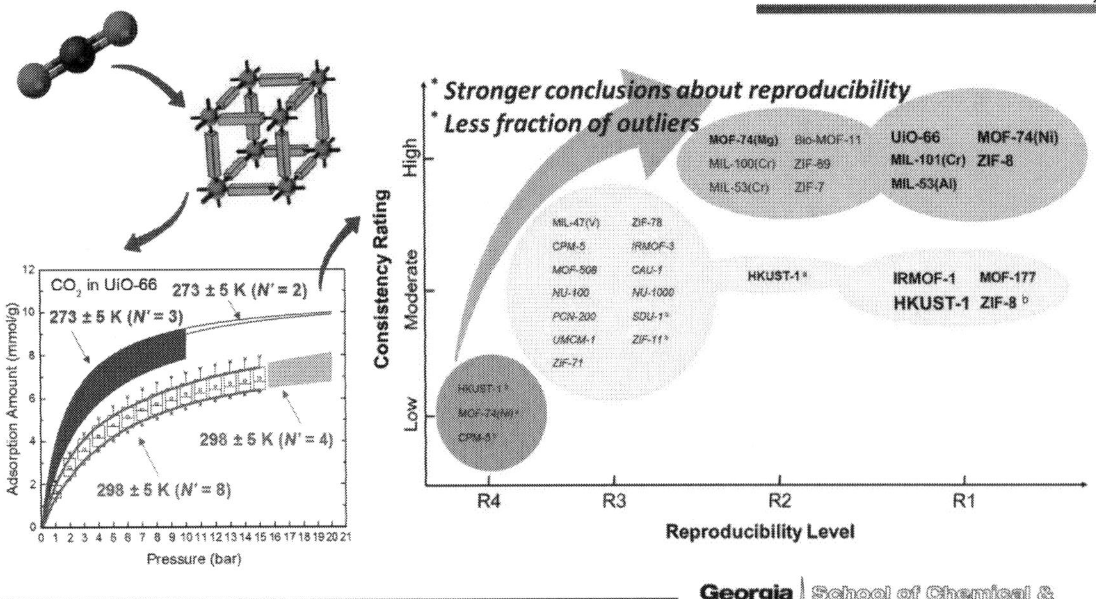

2017 AIChE Annual Meeting

Characterization of Adsorbent Materials

Supporting Information

Jongwoo Park, Joshua D. Howe, David S. Sholl*

Nov. 1, 2017

Georgia Institute of Technology
School of Chemical and Biomolecular Engineering

Georgia Tech | School of Chemical & Biomolecular Engineering

Meta-Analysis: Analysis Metrics Platform

Supporting Information

Explore adsorption isotherms

- Multiple, unbiased collection of isotherms in MOFs
- Temperature range of 10 K ($T \pm 5$ K)

Fit isotherms to classical models

- Coefficient of determination (R^2) metric
- Identify comparable pressure range

Outlier isotherm detection

- Number of data set dependent metric
- Reject outlier isotherms
- Consistency rating

Isotherm reproducibility assessment

- Number of data set dependent metric
- Isotherm bounds: Box/Region isotherms

Quantification of *reproduced* adsorption range

- Absence of the absolute reference adsorption isotherm
- Adsorbate-adsorbent independent algorithmic analysis platform

Georgia Tech | School of Chemical & Biomolecular Engineering

Outlier Level, Consistency Rating

Table S1. Outlier isotherm detection and consistency rating for adsorption isotherms in MOFs.

Outlier Level	Outlier Detection Criteria and Methods	Consistency Rating	Consistency Rating Criteria and Methods
O1	Tukey's method: $N > 4$ *Label outlier for adsorption amount outside the bounds defined by box and whisker plot*	High	Fraction of outliers, f $f \leq 0.25$
		Moderate	Fraction of outliers, f $0.25 < f \leq 0.4$
		Low	Fraction of outliers, f $f > 0.4$
O2	Error index statistics: $N = 3$ or 4 *Label outlier for isotherm curves with RMSE larger than $\sigma/2$ relative to a comparison measurement*	High	Fraction of outliers, f $f \leq 0.25$
		Moderate	Fraction of outliers, f $0.25 < f \leq 0.4$
		Low	Fraction of outliers, f $f > 0.4$
O3	$N = 2$ *Inadequate N to label outlier*	Moderate	RMSE $< \sigma/2$
		Low	RMSE $> \sigma/2$

- Outlier detection: enables meaningful statistical comparison for normal range of data measured in reasonably well qualified parent MOFs

Reproducibility Level

Supporting Information

Table S2. Isotherm reproducibility assessment for adsorption isotherms in MOFs.

Reproducibility Level	Criteria and Methods
R1	Box isotherm: $N' > 4$ *Isotherm bounds assigned by interquartile range of adsorbed amount as a function of pressure*
R2	Region isotherm: $N' = 3$ or 4 *Isotherm bounds assigned by the upper and lower measurement over a common pressure range*
R3	$N' = 2$: *Pair of fitted isotherms with RMSE $< \sigma/2$*
R4	$N' = 2$: *Pair of fitted isotherms with RMSE $> \sigma/2$*

- Exploration of adsorption behavior in parent MOFs enables yielding qualitative agreement in principle
- Graphical representation of reproducibility depending on different levels of information available

Georgia Tech | School of Chemical & Biomolecular Engineering

Box Isotherm: R1 Level

- Temperature dependent isotherm-like band diagram

✓ Consensus bounds to the isotherm quantities that are consistent with the underlying experimental data

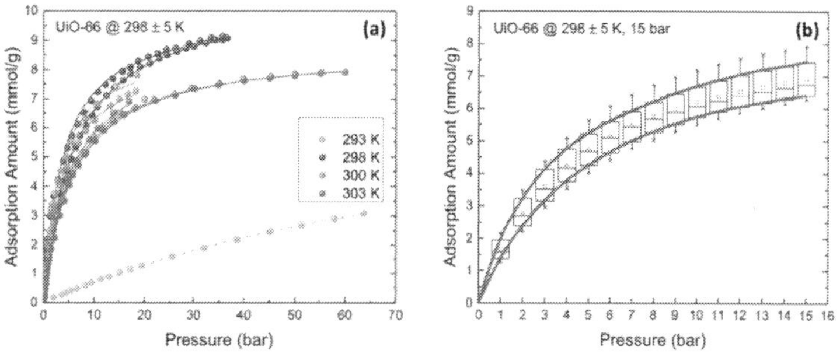

- **(a)** Raw experimental adsorption isotherms fitted with Langmuir model
- **(b)** At high consistency rating, consensus bounds to the isotherm quantities are assigned: solid curves connecting IQRs of the adsorbed amount as a function of *P*

Box Isotherm: R1 Level

Supporting Information

- Temperature dependent isotherm-like band diagram

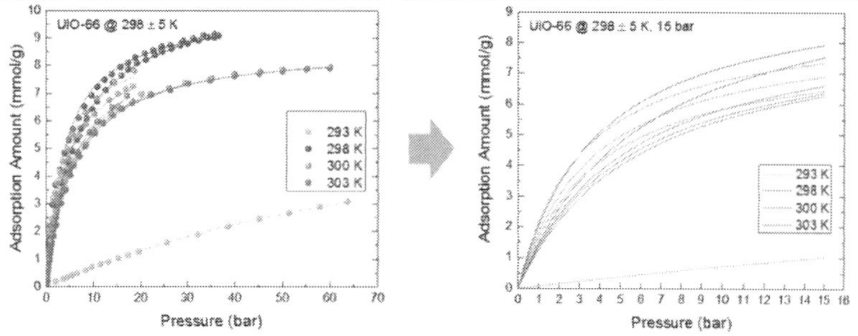

- Explore *N* independent experimental adsorption isotherms at a temperature $T \pm 5$ K
- Fit isotherms to classical Langmuir adsorption model ($R^2 \geq 0.99$)
- Identify comparable pressure range between fitted isotherms

Box Isotherm: R1 Level

- Temperature dependent isotherm-like band diagram

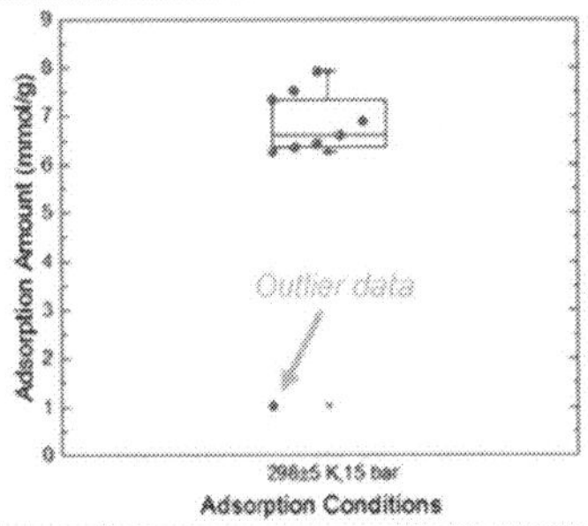

- O1 outlier level ($N > 4$)
- Outlier isotherm detection by Tukey's method
- Box and whisker plot for adsorption amount at any pressure point
- Reject isotherm containing outlier data
- High consistency rating ($f \leq 0.25$)

Box Isotherm: R1 Level

- Temperature dependent isotherm-like band diagram

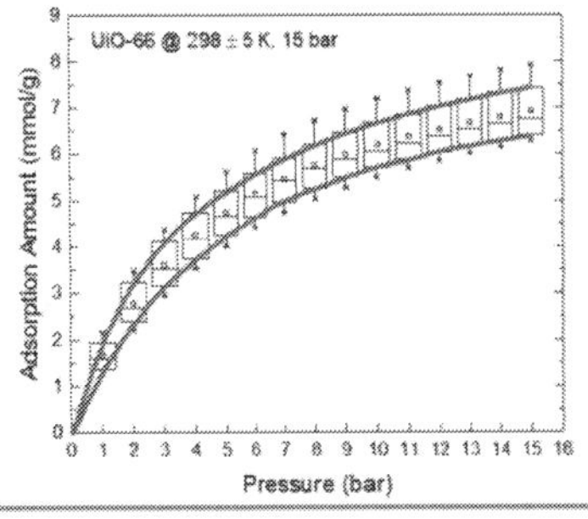

- R1 reproducibility level ($N' > 4$)
- Band diagram represented by box isotherm
- Isotherm bounds assigned by upper and lower quartile of adsorption amount at each pressure
- Quantification of adsorption range (mmol/g): [6.4, 7.4] @ 298 ± 5 K, 15 bar

Region Isotherm: R2 Level

- Temperature dependent isotherm-like band diagram

✓ Consensus bounds to the isotherm quantities that are consistent with the underlying experimental data

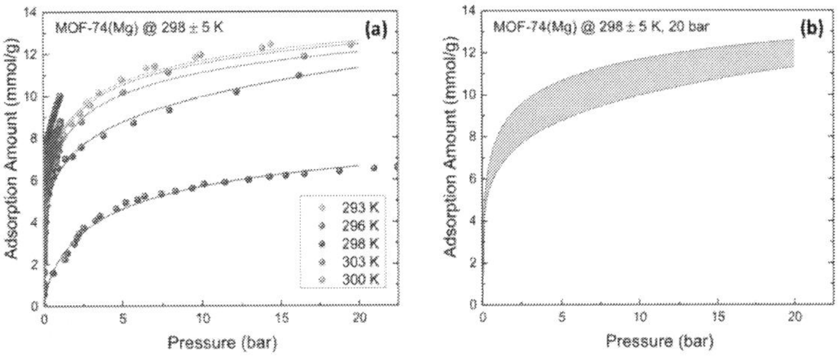

- **(a)** Raw experimental adsorption isotherms fitted with Langmuir-Freundlich model
- **(b)** At high consistency rating, consensus bounds to the isotherm quantities are assigned: encompass all of the measured data into a shaded region

Region Isotherm: R2 Level

- Temperature dependent isotherm-like band diagram

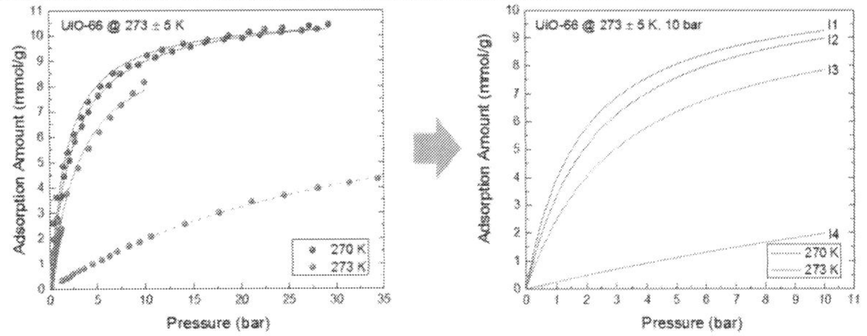

- Explore *N* independent experimental adsorption isotherms at a temperature $T \pm 5$ K
- Fit isotherms to classical Langmuir adsorption model ($R^2 \geq 0.99$)
- Identify comparable pressure range between fitted isotherms

Region Isotherm: R2 Level

Supporting Information

- Temperature dependent isotherm-like band diagram

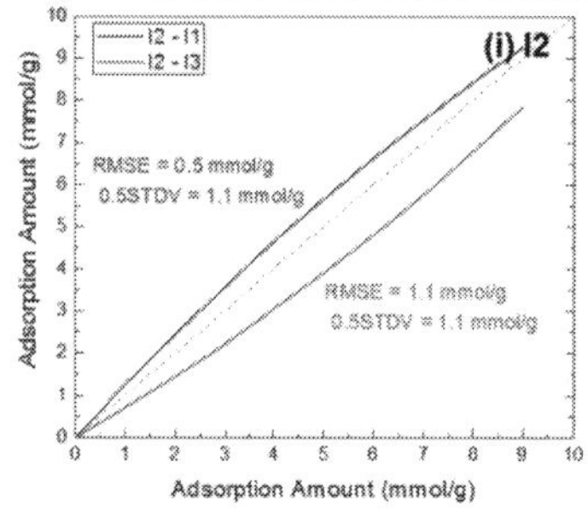

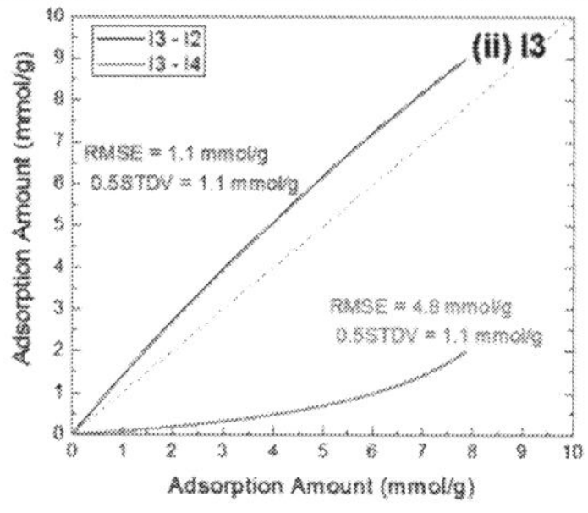

- O2 outlier level (N = 4)
- Outlier isotherm detection by error index statistics
- RMSE for adsorption amount within comparable pressure range
- Reject isotherm with RMSE larger than half of STDV of the comparison isotherm*

* Comparison isotherm: **(i) I2**, **(ii) I3**

- High consistency rating ($f \leq 0.25$)

Georgia Tech | School of Chemical & Biomolecular Engineering

Region Isotherm: R2 Level

Supporting Information

- Temperature dependent isotherm-like band diagram

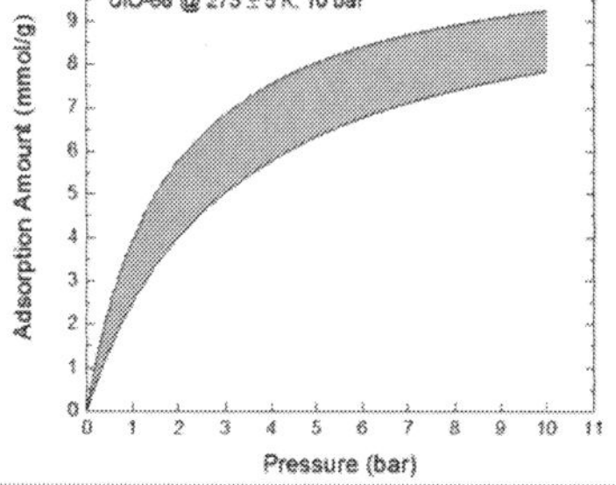

- R2 reproducibility level (N' = 3)
- Band diagram represented by region isotherm
- Isotherm bounds assigned by upper and lower adsorption amount within pressure range
- Quantification of adsorption range (mmol/g): [7.9, 9.3] @ 273 ± 5 K, 10 bar

Georgia Tech | School of Chemical & Biomolecular Engineering

Pair of Fitted Isotherms: R3/R4 Level

- Temperature dependent isotherm-like band diagram *not applicable*
  - ✓ Inadequate N to label outlier (N = 2)
  - ✓ Assign consensus bounds to the isotherm quantities only based on pair of fitted isotherms (N' = 2)

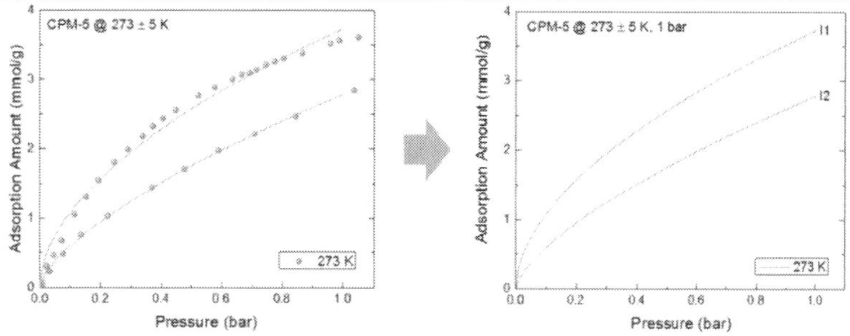

- Explore N independent experimental adsorption isotherms at a temperature $T \pm 5$ K
- Fit isotherms to classical Freundlich adsorption model ($R^2 \geq 0.99$)
- Identify comparable pressure range between fitted isotherms

Pair of Fitted Isotherms: R3/R4 Level

- Temperature dependent isotherm-like band diagram *not applicable*
  - ✓ Inadequate N to label outlier (N = 2)
  - ✓ Assign consensus bounds to the isotherm quantities only based on pair of fitted isotherms (N' = 2)

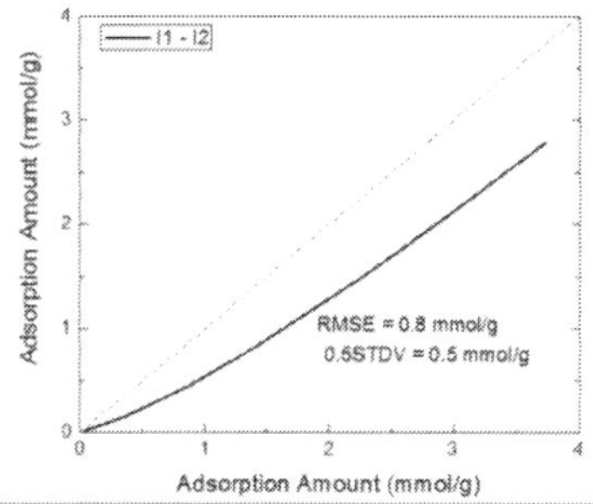

- O3 outlier level (N = 2)
- Unable to label outlier isotherm
- Error index: RMSE for adsorption amount within comparable pressure range
- Consistency rating depending on RMSE criteria
- Low consistency rating (RMSE > 0.5STDV)

Pair of Fitted Isotherms: R3/R4 Level

Supporting Information

- Temperature dependent isotherm-like band diagram *not applicable*

✓ Inadequate *N* to label outlier (*N* = 2)

✓ Assign consensus bounds to the isotherm quantities only based on pair of fitted isotherms (*N'* = 2)

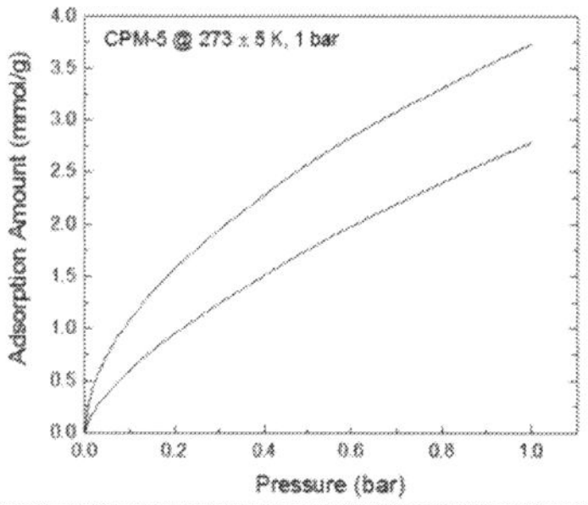

- R4 reproducibility level (*N'* = 2)
- Band diagram not applicable
- Isotherm bounds assigned by upper and lower adsorption amount within pressure range
- Quantification of adsorption range (mmol/g): [2.8, 3.7] @ 273 ± 5 K, 1 bar

Georgia Tech | School of Chemical & Biomolecular Engineering

Pair of Fitted Isotherms: R3/R4 Level

Supporting Information

- Temperature dependent isotherm-like band diagram *not applicable*

✓ Assign consensus bounds to the isotherm quantities only based on pair of fitted isotherms (*N'* = 2)

✓ Lowest level of confidence with respect to reproducibility of isotherm measurements

Table S4. List of MOFs for *N* = *N'* = 2 CO_2 isotherms at given temperature and pressure ranges identified from the NIST/ARPA-E Adsorption Database. MOFs are ordered alphabetically.

MOFs	*T* range	*P* range	MOFs	*T* range	*P* range
CAU-1	273 ± 5 K	[0, 1] bar	NU-100	298 ± 5 K	[0, 1] bar
	298 ± 5 K		NU-1000	273 ± 5 K	[0, 1] bar
CPM-5	273 ± 5 K	[0, 1] bar		293 ± 5 K	
	298 ± 5 K		SDU-1	273 ± 5 K	[0, 1] bar
IRMOF-3	298 ± 5 K	[0, 30] bar	UMCM-1	298 ± 5 K	[0, 15] bar
MOF-508	303 ± 5 K	[0, 5] bar	ZIF-11	273 ± 5 K	[0, 1] bar
	323 ± 5 K		ZIF-71	298 ± 5 K	[0, 1] bar
	343 ± 5 K				

Georgia Tech | School of Chemical & Biomolecular Engineering

Surface Area Scaling

Supporting Information

- Effect of surface area discrepancy: defects in real material

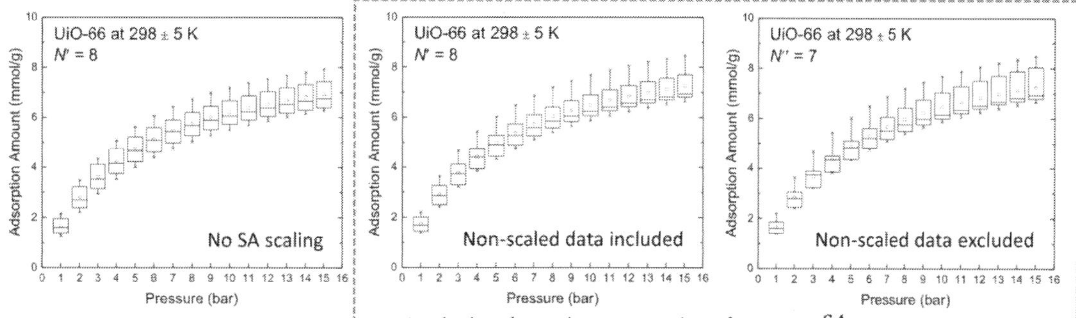

Intrinsic adsorption properties of defect-free material*: $q_{exp} \cdot \frac{SA_{simulated}}{SA_{experimental}} = q_{exp,scaled}$

- Quality of MOFs can vary depending on details of synthesis, activation, and stability
- Tightens the consensus bounds provided that uncertainty only comes from surface defect
- Variations in CO_2 adsorption quantities cannot be ascribed simply to variations in the surface areas of the reported materials

* [Demir, H., *J. Mater. Chem. A*, **2015**, *3*, 23539.]

Analysis of N_2 Adsorption Isotherms

Supporting Information

- Reproducibility analysis metrics for N_2 isotherms: BET surface area analysis

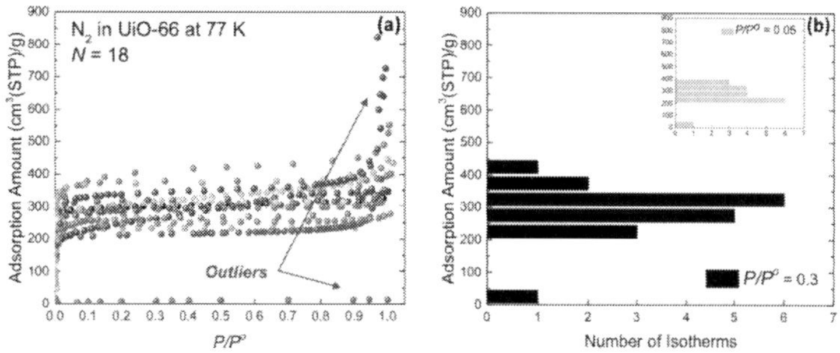

- N_2 isotherm at 77 K: material properties (*e.g.* BET surface area, pore volume) regardless of application
- Wide distribution of adsorption quantity: discrepancy between qualities of experimental samples despite they purport to have synthesized the same material in principle
- Uncertainty in N_2 adsorption reveals challenges in material characterization linked with deviation in consensus bounds of CO_2 adsorption isotherms

Analysis of N_2 Adsorption Isotherms

- Reproducibility analysis metrics for N_2 isotherms: BET surface area analysis

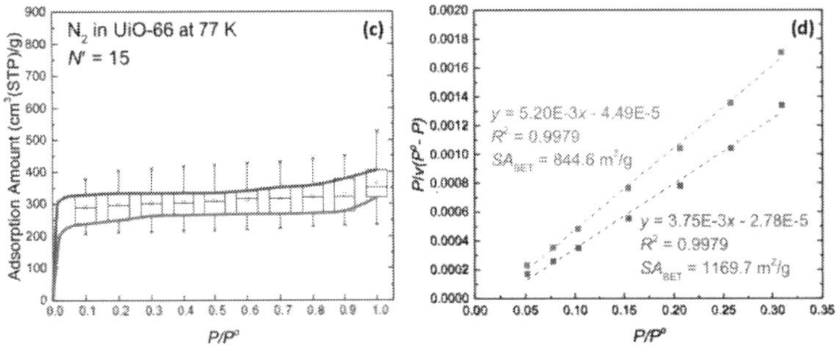

- Utility of metrics: consensus bounds in N_2 adsorption amount measurements with R1 reproducibility level
  : generation of pseudo-experimental isotherms by connecting Q_U's and Q_L's
- BET theory applied for pseudo-experimental isotherms over $0.05 < P/P^o < 0.15$
- Quantification of consensus bounds in BET surface area based on statistical metrics
- Expect more reproducible CO_2 isotherm measurements provided more consistent surface areas

Analysis of N_2 Adsorption Isotherms

- Reproducibility analysis metrics for N_2 isotherms: BET surface area analysis

Table S3. Surface area analysis using reproducibility analysis metrics for N_2 adsorption isotherms at 77 K enabling statistical quantification of consensus bounds in BET surface area (SA_{BET}).

MOFs	SA_{BET} [a] [m^2/g]		SA_{BET} range [b] [m^2/g]
	$P/P^o \leq 0.01$	$0.05 < P/P^o < 0.15$	$0.05 < P/P^o < 0.15$
HKUST-1	1909.5	1667.1	[932.6, 1486.0]
UiO-66	1180.4	1041.7	[844.6, 1169.7]

- BET theory applied over different linear regions of BET plot using simulated N_2 isotherms* at 77 K: variation ~ 10% using different P/P^o ranges indicating minor effect for these materials
- SA_{BET} range: SA_{BET}s calculated from pseudo-experimental isotherms over conventional P/P^o range
- HKUST-1: larger discrepancy in SA_{BET} range results in larger deviation in CO_2 adsorption range
- UiO-66: smaller difference in SA_{BET} range results in smaller variation in CO_2 adsorption range

* Simulated N_2 adsorption isotherms at 77 K obtained by GCMC

Unknown Reproducibility in MOFs

Supporting Information

- Molecules for which no conclusions can be drawn about reproducibility of adsorption in MOFs

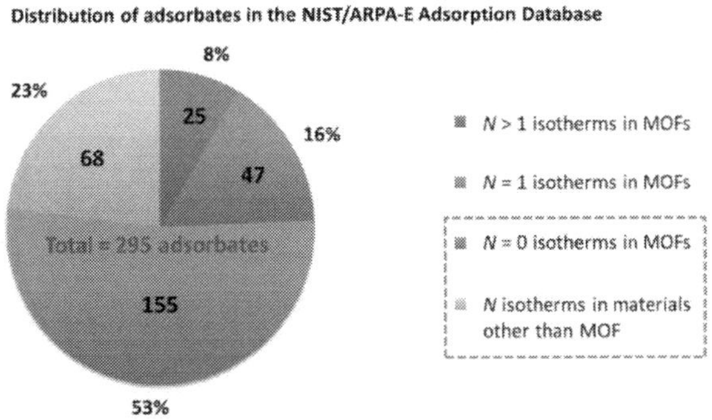

- 76% of adsorbing molecules in the NIST/ARPA-E Adsorption Database for which no conclusions can currently be drawn about the reproducibility of experimental data in MOFs
- 25 molecules for which any information about reproducibility is currently available for adsorption in MOFs
- 12 sorbates have a material/temperature range pair for which $N > 2$

Georgia Tech | School of Chemical & Biomolecular Engineering

Temperature Dependency in Analysis

Supporting Information

- Reproducibility analysis metrics for CO_2 adsorption isotherms at single T K

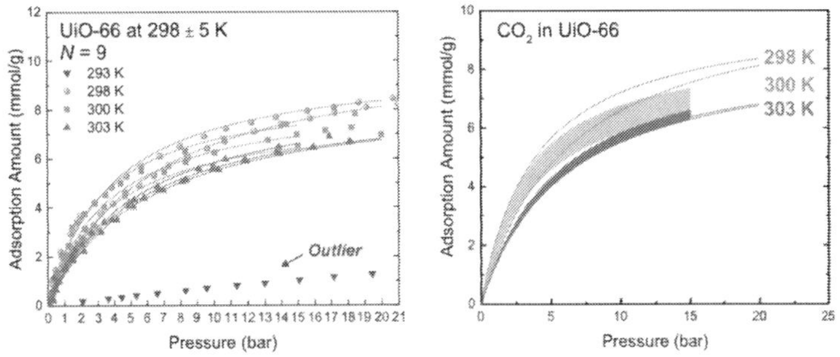

- CO_2 adsorption in MOFs over a range of P from N independent measured isotherms at $T \pm 5$ K: assumption is that regarding $T \pm 5$ K as a same adsorption condition
- Presence of uncertainty even in single T K and variation in adsorption amount is not clearly consistent with what might be expected from the heat of adsorption

Georgia Tech | School of Chemical & Biomolecular Engineering

Reproducibility In DFT Calculations

Supporting Information

- Property predictions of crystals and molecules *via* computational chemistry

✓ Reproducibility depends strongly on implementation of the governing equation of quantum physics

Calculations of equilibrium lattice parameter for Si within the DFT-PBE framework

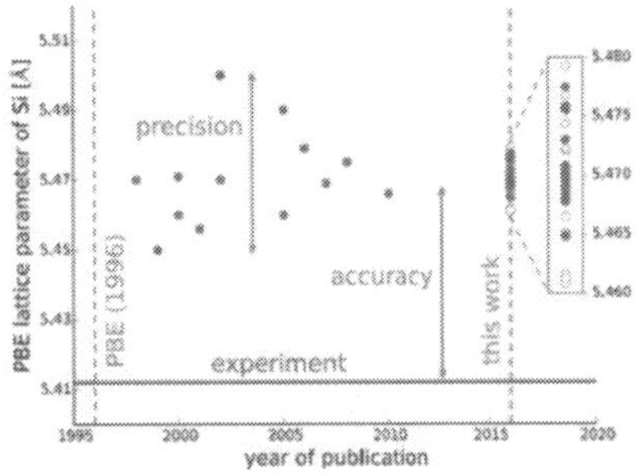

Quadratic energy difference between two EOS with distinguishable implementations

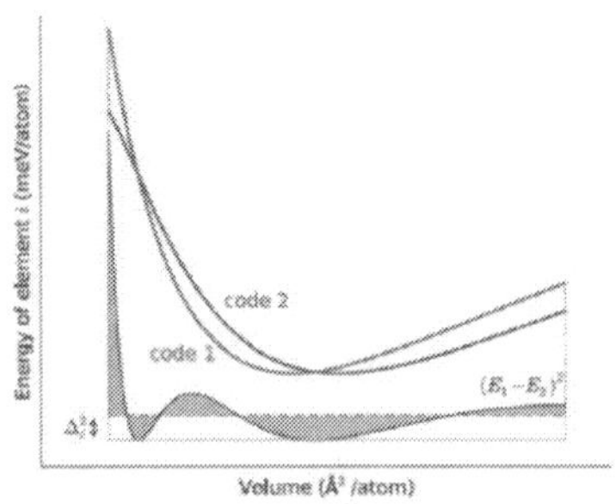

- Access the precision of DFT methods and demonstrate reproducibility among widely used DFT codes
- Large-scale pairwise code comparison with respect to the quality of the PBE functional

[Lejaeghere, K. *et al.*, *Science*, **2016**, *351*, aad3000.]

Georgia Tech | School of Chemical & Biomolecular Engineering

Reproducibility In DFT Calculations

Supporting Information

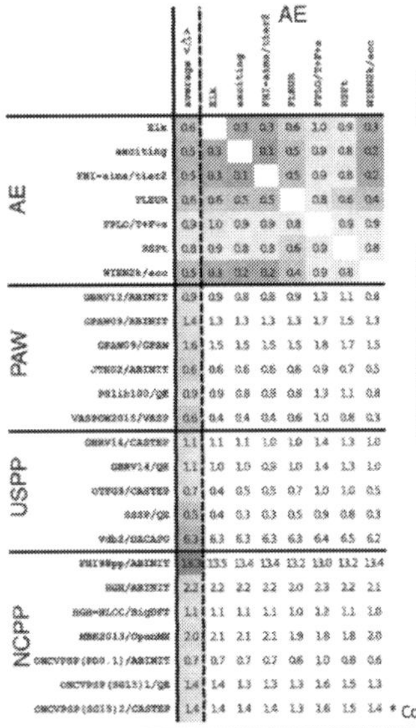

* Comparing all-electron (AE) methods [potential set/code]

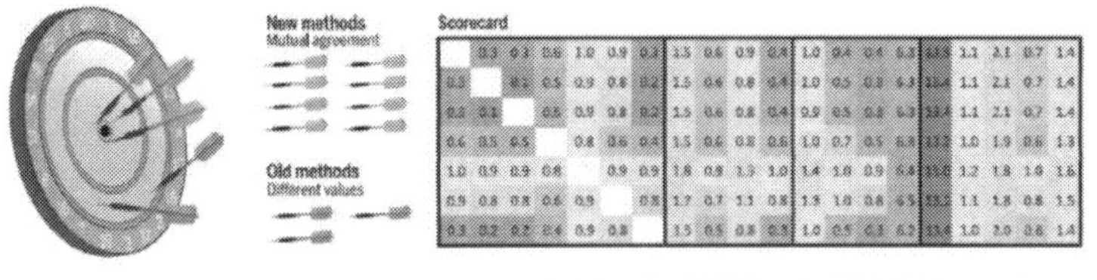

Δ matrix to compare EOS-derived observables:

$$\Delta_i(a,b)=\sqrt{\frac{\int_{0.94V_{0,i}}^{1.06V_{0,i}}\left(E_{b,i}(V)-E_{a,i}(V)\right)^2 dV}{0.12V_{0,i}}}$$

... RMS [meV/atom] difference between the EOS of methods *a* and *b* over *a* ± 6% interval round the equilibrium volume for every element *i*

- Precision of DFT implementations (*codes and pseudopotentials*) can be determined in the absence of one absolute reference code
- Recent methods yield nearly indistinguishable EOS, with the associated error bar comparable to that between different high-precision experiments
- Validity of recent DFT methods: correctly converged calculations yield reliable predictions

[Lejaeghere, K. *et al.*, *Science*, **2016**, *351*, aad3000.]

Georgia Tech | School of Chemical & Biomolecular Engineering

Round Robin Characterization

Supporting Information

- A Round Robin characterization of H_2 sorption properties of carbon based material

✓ Round Robin Test (RRT) protocol: avoid bias between experimentally measured values and expected capacity values from already published results

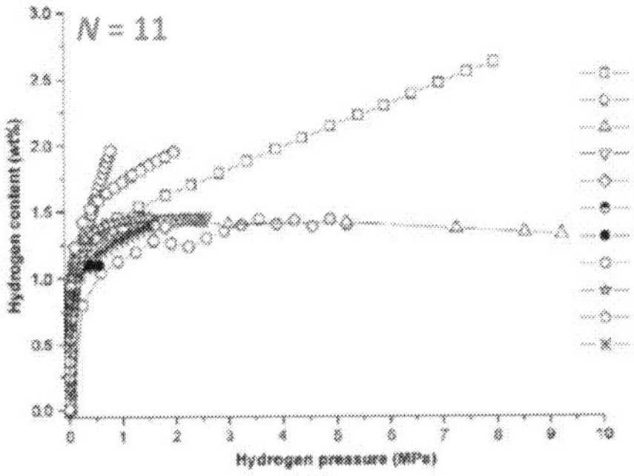

Fig. 4 – The adsorption isotherms at 77 K from eleven RRT participants.

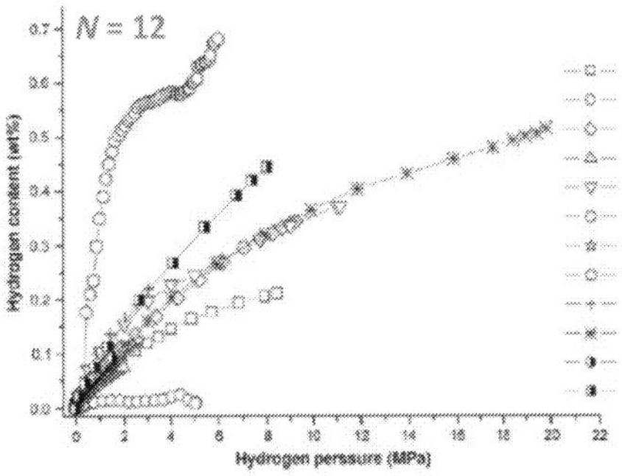

Fig. 8 – The adsorption isotherms recorded at ambient temperature from twelve RRT participants.

- Interlaboratory consistency for H_2 sorption *via* pressure-composition-isotherm (PCI) measurements
- Overall picture of the dispersion of results conducted in different participant laboratories

Round Robin Characterization

Supporting Information

- A Round Robin characterization of H_2 sorption properties of carbon based material

✓ H_2 physisorption: statistical evaluation for a need of a corrective action

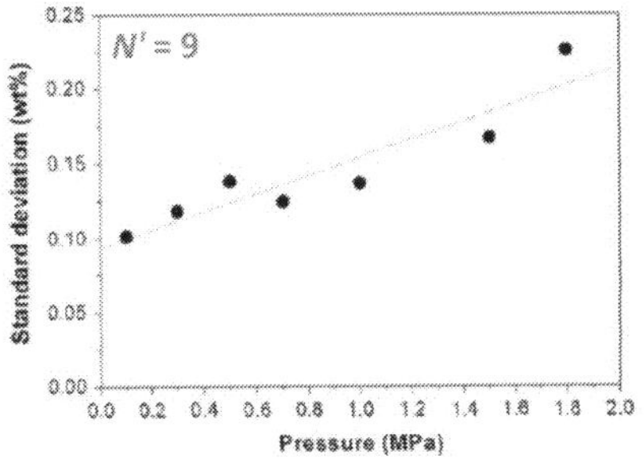

Fig. 6 – The standard deviation of the hydrogen content calculated from adsorption isotherms at 77 K as function of pressure. The dotted line is guide for the eye.

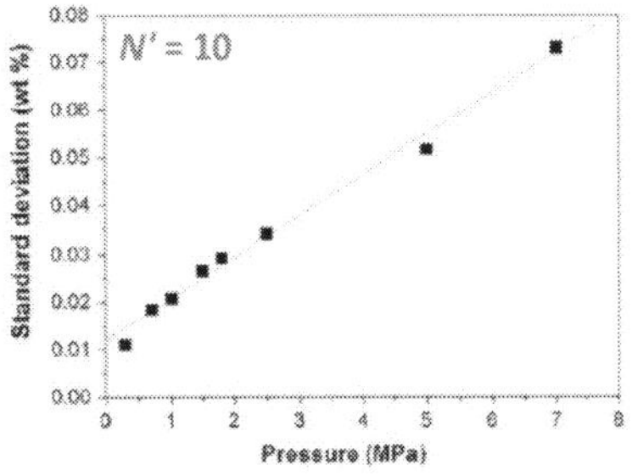

Fig. 10 – The standard deviation of hydrogen content calculated from adsorption isotherms at ambient temperature as function of pressure. The dotted line is guide for the eye.

- Statistical analysis performed for *selected isotherms* that show reversibility of adsorption and desorption
- Maximum relative standard deviations are 14% and 36% at 77 K and ambient temperature, respectively

Round Robin Characterization

Supporting Information

- A Round Robin characterization of H_2 sorption properties of carbon based material

✓ H_2 physisorption: statistical evaluation for a need of a corrective action

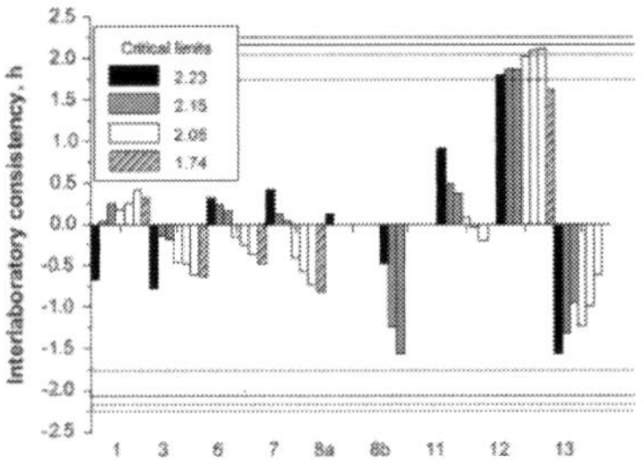

Fig. 7 – The interlaboratory consistency for adsorption isotherms at 77 K as function of participant code number. The bars correspond from left to right to the following pressures: 0.1, 0.3, 0.5, 0.7, 1.0, 1.5 and 1.8 MPa. The critical values of interlaboratory consistency are also plotted as straight lines.

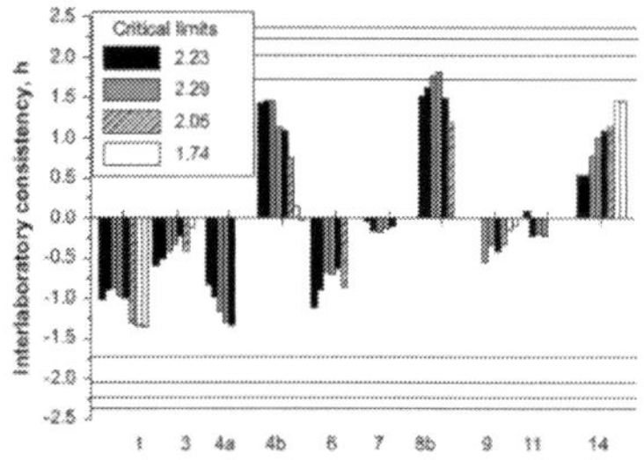

Fig. 11 – The interlaboratory consistency for adsorption isotherms at ambient temperature as function of participant code number. The bars correspond from left to right to the following pressures: 0.3, 0.7, 1.0, 1.5, 1.8, 2.5, 5.0 and 7.0 MPa. The critical values of interlaboratory consistency are also plotted as straight lines.

- Participants approaching or exceeding the critical *h* value need to take a corrective action
- 4 similar isotherms from different laboratories proposed as reference H_2 sorption behavior at both *T*'s

[Zlotea, C. *et al.*, *Int. J. Hydrogen Energy*, **2009**, *34*, 3044.]

Georgia Tech | School of Chemical & Biomolecular Engineering

Towards Branches of Materials Chemistry

Supporting Information

ACS APPLIED MATERIALS & INTERFACES — Letter

www.acsami.org

Improving Efficiency and Reproducibility of Perovskite Solar Cells through Aggregation Control in Polyelectrolytes Hole Transport Layer

ABSTRACT: Here, we report that the performance of perovskite solar cells (PSCs) can be improved by aggregation control in polyelectrolytes interlayer. Through counterions tailoring and solvent optimization, the strong aggregation of polyelectrolytes P3CT-Na can be broken up by P3CT-CH_3NH_3. When using P3CT-CH_3NH_3 to replace P3CT-Na as hole transport layer, the average efficiency is greatly improved from 16.9 to 18.9% (highest 19.6%). Importantly, efficiency over 15% is obtained in 1 cm^2 devices with P3CT-CH_3NH_3, ~50% higher than that with P3CT-Na (10.3%). Our work demonstrates the important role of aggregation control in polyelectrolytes interlayer, providing new opportunities to promote its application in PSCs.

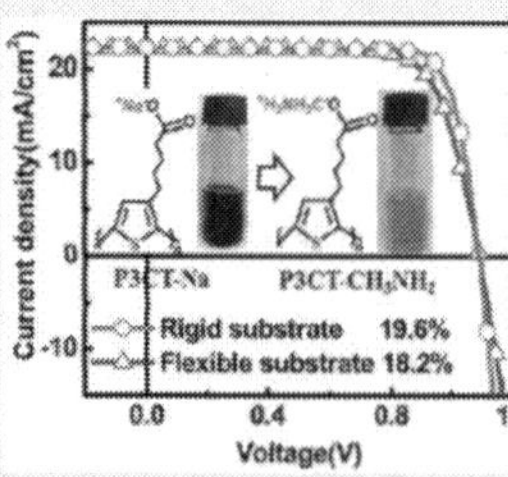

- Reproducibility of electrical performance of solar cells is crucial for scale-up (*i.e.* large area devices)
- In perovskite solar cells, the device efficiency and reproducibility can be impressively improved by controlling the aggregation of the polyelectrolyte interlayer

[Li, X. *et al.*, *ACS Appl. Mater. Interfaces*, **2017**, *9*, 31357.]

Georgia Tech | School of Chemical & Biomolecular Engineering

Towards Branches of Materials Chemistry

Supporting Information

- Improving reproducibility of perovskite solar cells: efficiency distribution

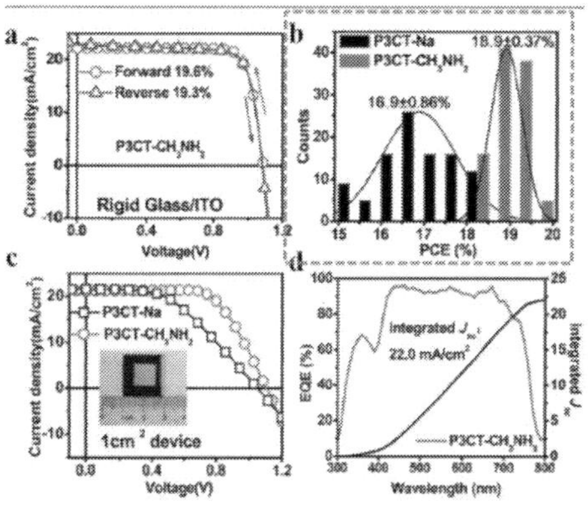

Figure 4. (a) J–V curves of perovskite solar cells with P3CT-CH$_3$NH$_2$ on rigid glass/ITO; (b) efficiency distribution among 100 separated devices; (c) J–V curves of large area devices with 1 cm^2; (d) EQE of devices with P3CT-CH$_3$NH$_2$.

Box plot of device efficiency

* PCE: power conversion efficiency

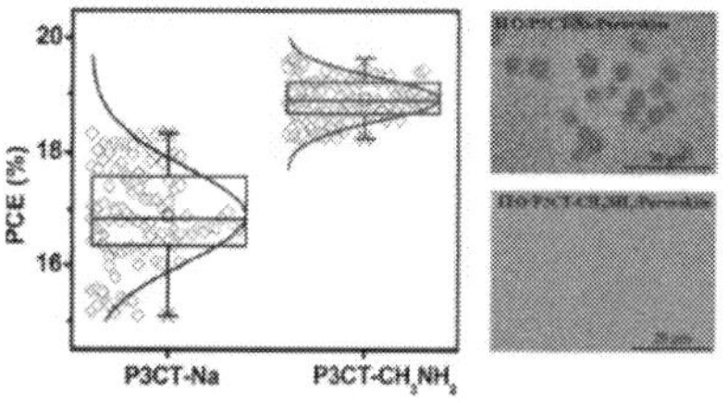

- PCE is determined by a variety of factors, *i.e.* J_{sc}, V_{oc}, FF
- Device reproducibility evaluation based on PCE

- Superior reproducibility (P3CT-CH$_3$NH$_2$) demonstrated by narrower efficiency distribution relative to conventional sample
- Consensus bounds for validating PCE and contributing factors:
  - reliability of novel system (*i.e.* self-testing/interlaboratory)
  - further scale-up estimation *etc.*

[Li, X. *et al.*, *ACS Appl. Mater. Interfaces*, **2017**, *9*, 31357.]

Georgia Tech | School of Chemical & Biomolecular Engineering

Is There A Reproducibility Crisis?

Supporting Information

- Reproducibility concerns in various science fields

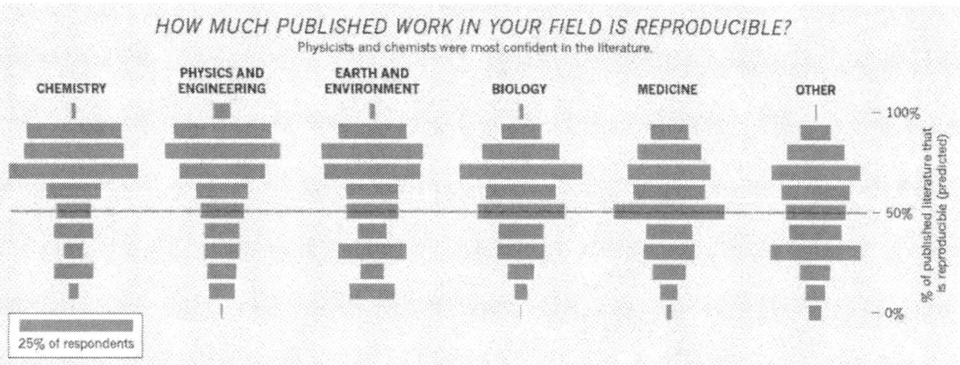

- Chemists were most confident about reproducibility, but showed the highest portion to failure in reproduction of their results and does not have well-established metrics to test reproducibility

[Baker, M., *Nature*, **2016**, *533*, 452.]

Georgia Tech | School of Chemical & Biomolecular Engineering

Is There A Reproducibility Crisis?

Supporting Information

- Contributing factors to irreproducible research

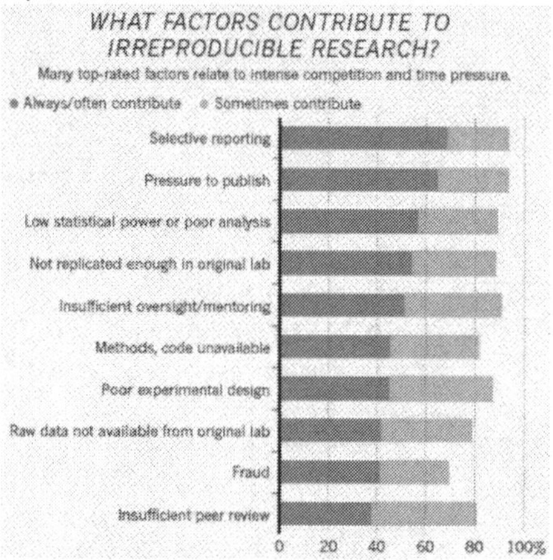

  - Small portion pointed to obstacles such as variability in reagents or the use of specialized techniques that are difficult to repeat

[Baker, M., *Nature*, **2016**, *533*, 452.]

Georgia Tech | School of Chemical & Biomolecular Engineering

Is There A Reproducibility Crisis?

Supporting Information

- Check list to improve reproducibility

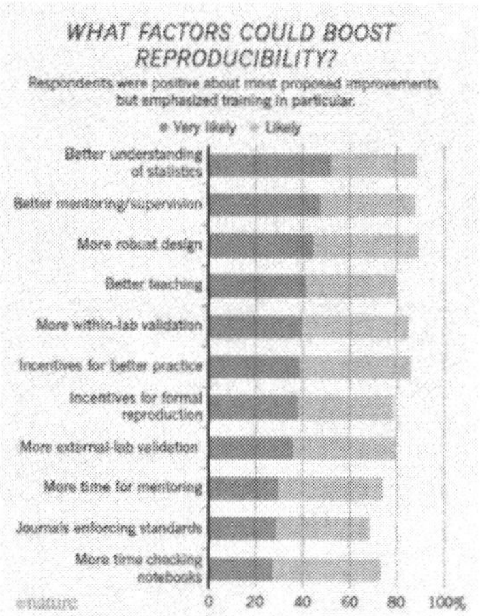

  - Mostly focusing on "human resources" for reproducible results

[Baker, M., *Nature*, **2016**, 533, 452.]

Georgia Tech | School of Chemical & Biomolecular Engineering

Modeling Adsorption Properties

Supporting Information

- Indirect adsorption modeling for deformed/amorphous MOFs

✓ Structure-property map extended to detect anomaly experimental data points

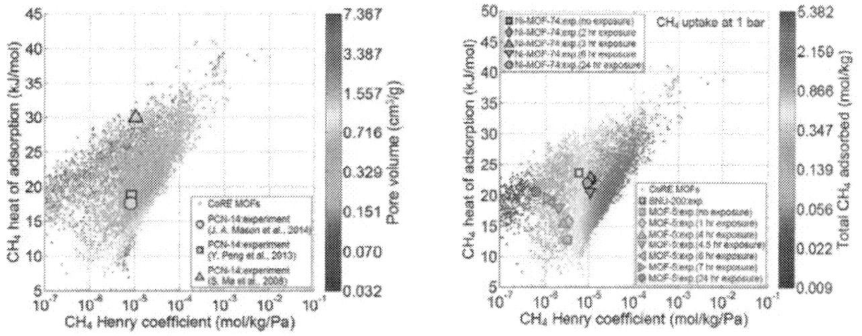

- Potential use as a detection method for experimental errors that might not be apparent without the proper context of large amounts of data
- Structurally deformed MOFs fundamentally share similar sets of properties compared with crystalline MOFs located nearby in the map

[Jeong, W. *et al.*, *PNAS*, **2017**, *114*, 7923.]

Georgia Tech | School of Chemical & Biomolecular Engineering

Atomistic Classical Simulation

Supporting Information

- Adsorption modeling in porous materials

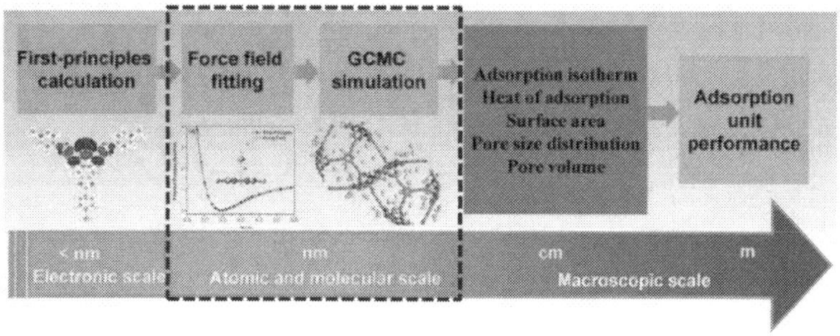

- **MC simulation in Grand Canonical ensemble**: adsorption equilibrium

⬇

Atomic level insight of adsorption behavior

[Xiang, Z. *et al.*, *Energy Environ. Sci.*, **2010**, 3, 1469.]

Georgia Tech | School of Chemical & Biomolecular Engineering

502345 Toolkit for a Reliable Characterization of Hierarchical Structured Nanoporous Materials By Physical Adsorption and Mercury Porosimetry

Abstract:

In recent years, major synthetic efforts have been focused on the introduction of hierarchical pore structures into many different materials such as zeolites, carbons, silicas, MOFs, hybrid layered and pillared structures. Nanoporous materials which consist of hierarchical pore networks with pore widths spanning the entire micro-meso- and even macropore range allow one to overcome accessibility and diffusion limitations associated with classical microporous materials for catalysis and separations.

In this talk, recent developments in the structural characterization of these materials will be described. An accurate textural characterization is crucial not only within the discovery process of novel hierarchically ordered nanoporous materials but also contributes to advance their application in a variety of areas including heterogenous catalysis, separation, battery devices, drug delivery. Among a variety of techniques, gas adsorption is widely applied because it assesses a wide range of pore sizes, spanning the entire micro- and mesopore range. Within the last two decades major progress has been achieved in physical adsorption characterization also because of the development of advanced approaches based on statistical mechanics such as molecular simulation and density functional theory (DFT) [1]. This progress, coupled with the availability of high resolution experimental methodologies for the adsorption of various subcritical fluids (also allowing one to couple adsorption techniques with complimentary techniques such as small angle scattering), has led to major advances in the structural characterization of nanoporous materials. However, major challenges still exist concerning an in-*depth* characterization of complex pore networks. Application of characterization techniques to hierarchical materials has revealed new phenomena, such as cavitation processes during desorption, and led to significant refinement of these techniques [2]. Here, we propose a "toolkit" of recommended experimental and state-of-the-art data reduction methods for a comprehensive study of the pore architecture of hierarchically structured materials by gas adsorption but also emphasize the importance of combining advanced physical adsorption with other complementary techniques such as mercury porosimetry [2].

[1] M. Thommes, K. Kaneko, A.V. Neimark, J.P. Olivier, F. Rodriguez Reinoso, J. Rouquerol and K.S.W Sing, (IUPAC Technical Report), *Pure Appl. Chem.* **87**, 1051 (2015)

[2] K. A. Cychosz, R. Guillet-Nicolas, J. Garcia-Martinez, J., M. Thommes, M. *Chem. Soc. Rev.* **46**, 389, (2017)

502361 Temperature Dependence of CO_2 Sorption in Micro-Mesoporous Carbons

Abstract:

Temperature Dependence of CO_2 Sorption in Micro-Mesoporous Carbons

F. Silvio P. Dantas[1], Richard Cimino[1], Katie A. Cychosz[2], Matthias Thommes[2], Alexander V. Neimark[1]

[1]Rutgers University, Department of Chemical and Biochemical Engineering, 98 Brett Road, Piscataway, NJ 08854, USA

[2]Quantachrome Instruments, 1900 Corporate Dr., Boynton Beach, FL 33426, USA

Carbon dioxide has proven to be a suitable probe molecule to characterize microporous carbons due to its ability to be adsorbed at ambient temperatures, allowing for faster diffusion rates and penetration into narrowest micropores not normally accessible by cryogenic adsorbates. Until recently, the limit in resolution of CO_2 adsorption @ 273 K was ~ 2nm due to the high pressures (~35 bar) necessary at 273 K to fill mesopores. However, recent advances in instrumentation have allowed CO_2 isotherms to be measured up to the saturation pressure. At this temperature, high pressure CO_2 experiments in micro-mesoporous carbons have shown that CO_2 isotherms are reversible and do not show hysteresis in the mesopore region. However, upon temperature reduction, the hysteresis phenomenon does occur. This feature has important implications for the application of theoretical methods for determining the pore size distribution (PSD) by CO_2 adsorption. In this work, we investigate the influence of temperature on the adsorption of CO_2 as it pertains to the characterization of micro-mesoporous carbons, with the goal of developing a reliable method of assessing the PSD for these systems. Using Grand Canonical Monte Carlo (GCMC) and Quenched Solid Density Functional Theory (QSDFT), theoretical adsorption isotherms were calculated in model carbon pores in the temperature range of 205 K to 273 K. The adsorbent was modeled using simple homogeneous slit and cylindrical shaped pores while the adsorbate was represented either as a 3-center molecule (GCMC) or its LJ representation (QSDFT). The theoretical isotherms were combined into adsorption "kernels", which were then used to solve the integral adsorption equation to obtain the surface area, pore volume and pore size distribution of several micro-mesoporous carbon samples. The results were compared to the conventional QSDFT methods for both argon and nitrogen adsorption. The theoretical results are compared to experimental adsorption isotherms on reference micro-mesoporous carbons.

497725 Pore Size Distribution in Hierarchical Materials: Insights from Molecular Simulations

Abstract:

Hierarchical materials, such as mesoporous zeolites, comprise of a distribution of pore sizes ranging from micropores (<2 nm) to mesopores (2â€“50 nm). Knowledge of pore size distribution is critical in designing and optimizing materials for various applications including, catalysis, separation, and gas storage and sequestration. Among the various available techniques for pore characterization, gas adsorption allows for determining a wide range of pore sizes in a non-destructive manner. Deducing the pore size distribution from the measured adsorption isotherms for argon, nitrogen, and/or carbon dioxide relies on theoretical procedures that utilize classical density functional theory (DFT) for the prediction of isotherms. These DFT predictions are limited to specific pore geometries (spherical, cylindrical, and slit pores) and represent the pore walls either as uniform surfaces or incorporating some degree of heterogeneity using interaction parameters tailored to entire classes of adsorbents, such as carbons, silicas, and zeolites.

In this work, large-scale Monte Carlo simulations are carried out to probe the adsorption in hierarchical materials. An attempt is made to understand the dependence of adsorption on confinement in mesoporous zeolites. By investigating large number of pore dimensions, a correlation between the capillary condensation pressure and the geometry of the pores is established. In addition to pore geometry, influence of adsorbate/adsorbent interactions on the adsorption behavior are also investigated.

489013 Investigating Imine-Based Porous Organic Cage Formation Mechanisms Using Time-Resolved Mass Spectrometry and Quantum Chemical Calculations

Abstract:

The synthesis of porous organic cages represents an important synthetic puzzle in self-sorting. Herein, we explored the formation of imine based porous organic cages using time-resolved electrospray ionization mass spectrometry, density functional theory and *ab initio* calculations. Specifically, the iminization reactions of triformylbenzene with (1R,2R)-1,2-diaminocyclohexane, and triformylbenzene with 1,5-pentanediamine were studied. We found that the typical [4+6] cycloimine cage between triformylbenzene and (1R,2R)-1,2-diaminocyclohexane, CC3-R, proceeds through a [2+3] intermediate, followed by rearrangement of the imine bonds towards the larger [4+6] cage. On the other hand, the iminization reaction between triformylbenzene and 1,5-pentanediamine first formed insoluble polymeric species that was later transformed to thermodynamically stable [2+3] cages. The presence of the low-energy small cage transition state in CC3-R synthesis is supported by theoretical calculations. The combination of DFT and *ab initio* calculations also indicated that common DFT methods could not capture the accurate energies of the electron-rich cage structures. Single point energy calculations at MP2 level is required for theoretical interpretation of the cycloimination reactions.

483647 Thermodynamic Studies and Process Modeling for the Separation of Aromatics from Aliphatics with Ionic Liquids

Abstract:

The separation of aromatics from aliphatics is a routine step in oil refinery processes. Current methods for this separation include extraction, extractive distillation, and azeotropic distillation. The main problems with these methods are: 1) suitability only for aromatic concentration of the feed mixture higher than 20%, 2) low aromatic/aliphatic selectivities, and 3) low capacities. All of these problems call for the development of new technology for aromatic/aliphatic separation, with the design of an ionic liquid (IL) solvent for the extraction process being a promising research direction.

Well-known for their favorable properties including negligible vapor pressure, good thermal stability, large liquid range, and versatility for creation, ILs are highly attractive as potential substitutes for current commercial solvents [1,2,3]. Process modeling can guide the design of ILs. For example, through fundamental studies of the thermodynamic models for liquid-liquid-equilibrium (LLE) and vapor-liquid-liquid-equilibrium (VLLE), we can better understand the thermodynamic limits of the product purities one can achieve with various ILs, and we can evaluate energy and economic costs by building process models to compare the use of IL extraction solvents to traditional solvents for aromatic/aliphatic separation like sulfolane.

To study the aromatic/aliphatic extraction process, toluene-heptane mixtures are used as a model, with a current commercial solvent (sulfolane) used as the benchmark for comparison. We evaluate two methods of recovering the aromatics from the IL-based extract. The first takes advantage of the nonvolatility of ILs, by simply evaporating the aromatics out of the aromatic-IL mixture, thus recovering the IL solvent by flash evaporation. To reduce the loss of solvent, ILs with good thermal stabilities are preferred. We compare this to using a Gas-Antisolvent Process (GAP), where gaseous CO_2 is used to facilitate a liquid-liquid phase split between an aromatic-rich and an IL-rich phase [4,5].Â We focus on ILs with the bis(trifluoromethylsulfonyl)imide ($[Tf_2N]^{â^{\wedge}}$) anion paired with 1-hexyl-3-methylpyridinium ($[hmpy]^+$), 1-hexyl-3-methylimidazolium ($[hmim]^+$), 1-*n*-butylthiolanium ($[bthiol]^+$), and triethyloctyl phosphonium ($[P_{2228}]^+$) cations.

For the thermodynamic modeling of the heptane-toluene-IL ternary LLE, COSMO-SAC and COSMO-RS are investigated for all four ILs.Â The ability of these models to fit the experimental ternary LLE data [6] are compared.Â On this basis, COSMO-SAC is chosen to model both the flash evaporation and GAP IL recovery methods.

Process models are built for the separation of toluene from heptane using an IL extraction solvent.Â Theoretical product purities are calculated for each of the four ILs considered, and the sensitivity of product purities to variation in the number of extractor stages, solvent flow rate, temperature and pressure of flash separation, etc. is analyzed. Capital

cost and energy cost are also calculated for each IL process. Results are compared with the current commercial sulfolane process in both product purities and cost analysis. The four types of ILs are compared in terms of the desired physical properties, based on the results of the thermodynamic modeling and the process modeling.

[1] Lin, W.-C., Tsai, T.-H., Lin, T.-Y., Yang, C.-H. J. Chem. Eng. Data 2008; 53: 760-764.

[2] Cháfer, A., de la Torre, J., Font, A., Lladosa, E. J. Chem. Eng. Data 2015; 60: 2426-2433.

[3] Brennecke. J. F., Maginn. E. J. AIChE J. 2001; 47: 2384-2389.

[4] Mellein, B. R., Brennecke, J. F. J. Phys. Chem. B 2007; 111: 4837-4843.

[5] Blanchard, L. A., Brennecke, J. F.. Ind. Eng. Chem. Res. 2001; 40: 287-292.

[6] Lubben, M. J., Canales, R., Gonzalez-Miquel, M., Lyu, Y., Stadtherr, M. A., Brennecke, J. F., to be submitted.

495361 Equilibrium Data Determination for Sucrose Esters Separation

Abstract:

Sucrose esters are biobased emulsifiers used in high value niche markets such as cosmetics and food [1]. They are produced via basic or enzymatic transesterification of sucrose and fatty acid methyl esters. With the aim of improve reactants incompatibility, traditional production processes use a large quantity of solvents (near 60%wt of reactive media) as Dimethyl Sulfoxide and Dimethyl Formamide [2]. The use of these regulated solvents in the reaction process obligates an exhaustive downstream separation process to fulfill the product specifications. The energy intensive solvent removal process is unavoidable because solvent in the product must be in the traces level [3]. For this reason, a solvent-free process using emulsifiers would be preferred [4]. Regardless the process, the sucrose ester purification is a complex task owing to the complexity of the reactive mixture.

Different purification processes to obtain commercial sucrose ester have been proposed [5–10]. Mainly, an organic solvent (non-polar) is used to extract residual fatty acid methyl esters (and the solvent if used). Then water or other solvent (polar) is used to wash the precipitated sucrose esters and to separate them from residual sucrose and catalyst. Despite this downstream processing has been described in the literature, there is a lack of information regarding the phase equilibrium of the process. This data are required for further modeling and design of the separation process.

In this work, the downstream purification of sucrose esters produced by a solvent-free reaction was studied. Taking into account that the main goal of the solvent-free process is to reduce de use of toxic solvents in the process, green solvents were selected as extracting agents in the product purification. The phase equilibria in mixtures of ethyl acetate, methyl palmitate and sucrose palmitate; and water, sucrose and sucrose palmitate have been evaluated. The solid-liquid equilibria of both systems were evaluated at different temperatures (between 5 and 30°C), and the characterization of samples was performed by HPLC method. In both cases, sucrose palmitate was a mixture of sucrose esters with different degree of substitution, so that selectivity of sucrose esters towards the extracting solvent was also evaluated. This is important since it will determine separations yields and esterification degree of final product. Obtained data was used to evaluate performance of different thermodynamic predictive methods. Results may be used in the design and optimization of purification operations for sucroesters purification.

References

[1] B.A.P. Nelen, J.M. Cooper, Sucrose esters, in: R.J. Whitehurst (Ed.), Emuls. Food Technol., Backwell Publishing Ltd, 2004: pp. 131–162.

[2] H.B. Hass, N.J. Summit, F.D. Snell, W.C. York, L.I. Osipow, Process for producing Sugar Esters, 2893990, 1959.

[3] L.I. Osipow, W. Rosenblatt, S. Valley, Esterification of polyhydric compounds in the presence of transparent emulsifying agent, 3480616, 1969.

[4] J. Fitremann, Y. Queneau, J. Maitre, A. Bouchu, Co-melting of solid sucrose and multivalent cation soaps for solvent-free synthesis of sucrose esters, Tetrahedron Lett. 48 (2007) 4111–4114. doi:10.1016/j.tetlet.2007.04.015.

[5] K. James, Purification of Sucrose Esters, US 4104464, 1978.

[6] S. Kea, C.E. Walker, Separation and purification of sugar esters synthesized from both aqueous and nonaqueous systems, US 4710567, 1987.

[7] S. Matsumoto, Y. Hatakawa, A. Nakajima, Process for purifying sucrose fatty acid esters, US 5008387, 1991.

[8] S. Matsumoto, Y. Hatakawa, A. Nakajima, Process for recovering unreacted sucrose from reaction mixture synthesis of sucrose fatty acid esters, US 4995911, 1991.

[9] S. Matsumoto, Y. Hatakawa, A. Nakajima, Method of producing powdery high hlb sugar fatty acid ester, 5144022, 1992.

[10] Y. Koyama, N. Kawase, H. Yamamoto, S. Kawata, Y. Kasori, Process for producing sucrose fatty acid ester, EP 0659760A2, 1994.

495650 A Molecular Design Method Based on the COSMO-SAC Model for Ionic Liquid in Extractive Distillation

Abstract:

A Molecular Design Method Based on the COSMO-SAC

Model for Ionic Liquid in Extractive Distillation

FANG Jing, Li Chunli, Wang Yijing, Zhao Rui

(School of Chemical Engineering, Hebei University of Technology, Tianjin, China)

Extractive distillation is widely used for azeotropic systems because of its superiority. At present, most organic solvents or salt are used universally as extraction solvents. Comparing with traditional organic solvents, the radio of solvent to feed using ILs could decrease 20% and the heat duty on reboilers in extractive distillation could decrease about 17%. However it is difficult to select appropriate ILs for one certain azeotropic system.

In this article, a COSMO-SAC model was used to screen for prospective solvent from a wide variety of ILs for extractive distillation. In 1983, UNIFAC model which was applied in molecular design of solvent was based on the group contribution method (GCM). Because of absence of group interaction parameters, it is not suitable for selecting ILs. In 1995, COSMO-RS which was based on the solvent molecular orbital continuum model was proposed by Klamt. This method can provide promising results but the equation for chemical potential does not converge for boundary conditions and final expression for activity coefficient fails to satisfy the thermodynamic consistency relationships. The COSMO-SAC model was introduced by Lin and Sandler, which has an obvious advantage is that the interaction parameters are too limited to perform the UNIFAC calculation.

In order to reduce the calculation workload, dividing the ILs into cations and anions. Applying materials studio to drawn the structure of ILs and optimizing its

structure and energy. Based on the COSMO-SAC model, the Ïƒ-profile database of ILs

was established. Selectivity and solubility were used as the indexes for solvent

screening. Solvents can change the relative volatility of the components but it is difficult to calculate the activity coefficient of each component. Turning to the other aspect, the infinite dilution activity coefficient of component is easy to obtain. The larger the value, the better the separation results. Thus, selecting the solvent which has high infinite dilution activity coefficients will obtain better separation results. According to the molecular design method, three suitable extractive distillation were determined. Figure 1 shows structure of three ILs.

There are two systems-acetonitrile-water and ethanol-cyclohexane are selected to test the results of chosen ILs. The first system is a typical polar binary azeotropic system while the other is a binary azeotropic system composed of one polar component and one nonpolar component. Adding various weight fractions (15 and 30%) into two chosen systems to obtain the vapor-liquid equilibrium database. The database of acetonitrile-water system are plotted in Figures 2-4 while ethanol- cyclohexane in Figures 5-7.

From these figures, we can conclude that the azeotropic point vanished when the mass fractions of the ILs reached 30%, the accuracy of the chosen ILs are proved again. In this article, the Wilson equation was used to predict the vapor composition for molecular design. Comparing the experimental and calculated relative volatilities of two systems with three types of ILs. Relative volatility for compared of experiment data and calculated results were shown in figure 8 and figure 9. Through calculation, for acetonitrile-water-IL systems the overall average relative deviation was 4.92%, while

ethanol-cyclohexane-IL was 5.45%.

499066 Replacement of Toxic Organic Solvents with Natural Non-Toxic Diluents for the Recovery of Pyruvic Acid

Abstract:

Lately, reactive extraction (RE) has been shown to be one of the most promising techniques for the recovery of bio-molecules, e.g. carboxylic acids, from biological production media. It is a modified version of traditional solvent extraction (SE) process. Different from SE, the organic phase in RE involves a chemical extraction using an extraction agent that has the ability to form complex structures with the target molecule, carboxylic acids. The technique is preferred due to its high efficiency, low energy demand and simplicity of the approach. It is also favored for its availability to be used for heat-sensitive chemicals since the process is generally carried out at room temperature. However, high toxicity levels of the organic solvents used during the process is the most critical disadvantage of the method. Various alternatives have been used to replace these solvents; however, low efficiency and high cost still prevent the commercially use of the process.

Pyruvic acid (PA) is an important biochemical, mostly utilized in the production of amino acids and pharmaceuticals. Following to its biological production; an efficient, low-cost and environmentally-friendly recovery process is required. Several methods have been tested for the purpose; however, none of them could be widely accepted yet. As previously stated, RE is shown to be a promising candidate for the aim due to its several advantages. In addition, the use of non-toxic diluents to replace the traditional organic solvents will make the process environmentally-friendly and will significantly reduce the hazard level.

In this study, 12 different vegetable oil types (sunflower, soybean, corn, almond, canola, hazelnut, olive, sesame, linseed, safflower, castor and black cumin (black seed) oils) were tested with various extractants for the purpose. Low-cost of these non-toxic solvents will add a significant advantage to the commercialization of the operation. Equilibrium studies were performed to determine the most effective non-toxic diluent and extractant type; and investigate the effects of the process parameters on the recovery. Results were compared with the ones obtained with 1-octanol, which is accepted as the state of art solvent for the technique. For all types of extractants, highest extraction efficiencies were obtained with almond oil. Among the extractants, tertiary amines (TAs) provided the highest recovery rates, and the quaternary ammonium salts and organophosphorus extractants followed TAs. Extraction efficiencies with vegetable oils+trioctylamine were about 94.5-96% when [PA]/[TA]=1.0-1.7 while it was 95-97% with 1-octanol. Consistent with the literature, distribution ratio increased with extractant concentration while decreased with the initial aqueous pH, PA concentration and temperature. Back extraction of the target molecule was performed using various back extraction agents. Firstly, several types of sodium salts were tested for the purpose and highest recovery was reached as 88% with sodium nitrate. Alternatively, water soluble amines were used and over 95% recovery was achieved. Results of the related experiments also exhibited that the organic phase can be successfully utilized for the consecutive forward reactive extraction operations.

The results obtained in this study showed that toxic organic solvents widely preferred in industrial extractive separations of bio-molecules (e.g. carboxylic acids) can be replaced with non-toxic and natural vegetable oils. Efficient back extractions resulted in an aqueous solution of PA and allowed the organic phase to be re-used in the following RE, which significantly reduce the costs. Therefore, an efficient, low-cost and environmentally-friendly recovery of PA was achieved in the present study.

496691 Efficient Extraction of Phenolic Compounds from Oils with Dicationic Ionic Liquids Via Forming Deep Eutectic Solvents

Abstract:

Efficient extraction of phenolic compounds from oils with dicationic ionic liquids via forming deep eutectic solvents

Youan Ji[1], Yucui Hou[2], Shuhang Ren[3], Congfei Yao[4], Weize Wu[1,*]

[1]State Key Laboratory of Chemical Resource Engineering, Beijing University of Chemical Technology, Beijing 100029, China

[2]Department of Chemistry, Taiyuan Normal University, Taiyuan 030031, China

Phenolic compounds, which are widely applied in producing synthetic fibers, engineering plastics, pesticides, medicines, explosives and so on, are necessary to be extracted from oil mixtures before further applications. In this work, six dicationic ionic liquids (DILs, including 1,2-bis[N-(N'-methylimidazolium)] ethane 2Br$^-$, DIL1; 1,3-bis[N-(N'-methylimidazolium)] propane 2Br$^-$, DIL2; 1,4-bis[N-(N'-methylimidazolium)] butane 2Br$^-$, DIL3; N,N,N, N',N',N'-hexaethyl-ethane-1,2-diammonium dibromide, DIL4; N,N,N,N',N',N'-hexaethyl - propane-1,3-diammonium dibromide, DIL5; N,N,N,N',N',N'-hexaethyl-butane-1,4- diammonium dibromide, DIL6) were found and used to extract phenolic compounds from model oil and coal tar. We have studied the effect of time, temperature, DIL:phenol mole ratio, and initial phenol concentration on the extraction efficiency (*EE*) of phenol. It was found that the DIL:phenol mole ratio was only 0.3 or 0.4 for the highest *EE* of phenol. The extraction process could be finished within 5 min, and *EE* did not significantly change with increasing temperature. Also, the ultimate phenol concentration remained constant despite the difference in initial phenol concentrations. Of these DILs, DIL5 showed the lowest ultimate phenol concentration of 3.9 g/dm^3, and *EE* of phenol could reach as high as 97.0%. We also reused these DILs and the results showed that these DILs could repeatedly extract phenol from model oil. In addition, *EE* of phenolic compounds from real coal tar using DIL5 was found to be 92.7%. The mechanism was analyzed by FT-IR, and the results showed that there was a hydrogen bond between phenol and DIL. Moreover, a comparison was made between DILs and mono-ILs. It was found that the DILs were much more thermally stable, and showed smaller solubility in oil mixtures than mono-ILs.

This work is financially supported by the National Basic Research Program of China (2011CB201303), and Specialized Research Fund for the Doctoral Program of Higher Education (20120010110005).

Keywords: Extraction; Phenolic compound; Model oil; Dicationic ionic liquid; Deep eutectic solvent

* Corresponding author; E-mail: wzwu@mail.buct.edu.cn; Tel./Fax: +86 10 64427603.

Supercritical extraction of essential oil from *Achillea wilhelmsii*: Experiments and modelling

Bizhan Honarvar[a,*], Davood Cheraghi[b], Ali Bagheri[a], Milad Boostani[c]

[a] Department of Chemical Engineering , Marvdasht Branch , Islamic Azad University, Marvdasht , Iran

[b]Young Researchers and Elite Club, Marvdasht Branch , Islamic Azad University, Marvdasht , Iran

[c]Young Researchers and Elite Club, Yasooj Branch, Islamic Azad University, Yasooj, Iran

Tel. / fax: +989173145175

Email: honarvar2@gmail.com

Abstract

In this study, supercritical carbon dioxide (SC-CO_2) extraction was used to extract the essential oil from *Achillea wilhelmsii*. In this way, a series of experiments were designed to investigate the effect of different operating conditions such as temperature (35, 40 and 45 °C), pressure (150, 170 and 190 bar), mean particle size (0.04, 0.06 and 0.08 cm) and dynamic extraction time (0-90 min) on extraction yield. The extracts obtained using SC-CO_2 were analyzed using gas chromatography-mass spectrometry (GC-MS). In chromatographic analyses of essential oil from *Achillea wilhelmsii*, Carvacrol, Camphor, α-Pinene, Borneol, Thymol and Pinocarvone were identified. The best yield of *Achillea wilhelmsii* (3.63%) was reported at 45 °C and 190 bar and 40 min. In the second part of the paper, a mathematical modeling for extraction of oil from *Achillea wilhelmsii* using SC-CO_2 was performed. The model was solved by finite element method (FEM). The results of the model showed a good agreement with experimental data.

Keywords: Supercritical extraction, *Achillea wilhelmsii*, Mathemathical modeling, gas chromatography-mass spectrometry

Nomenclature

C_f	oil concentration in the supercritical fluid phase (kg m^{-3})
C_s	solute concentration in the supercritical fluid phase on the surface of solid (kg m^{-3})
d_p	particle diameter (m)
D_{ax}	axial dispersion coefficient (m^2 s^{-1})
D_{ext}	extractor diameter (m)
D_{AB}	molecular diffusion coefficient (m^2 s^{-1})
K	extract equilibrium constant between the solid and fluid phases
k_f	external mass transfer coefficient (m s^{-1})
L	extractor length (m)
M	Molecular weight (g mol^{-1})
P	pressure (bar)
Pe_b	Péclet number for the bed, ($L\ \upsilon\ D_{ax}^{-1}$)
Q	solvent flow rate (kg s^{-1})
q	solute concentration in the solid phase (kg m^{-3})
R_p	particle radius (m)
Re	Reynolds number ($\rho\ \upsilon\ d_p\ \mu^{-1}$)
Sc	Schmidt number ($\mu\ \rho$-1 D_{AB}^{-1})
Sh	Sherwood number ($d_p\ k_f\ D_{AB}^{-1}$)
T	temperature (K)
t	time (min)
T_{cB}	critical temperature of the solvent (K)
v_B	molar volume of the solvent (cm^3)
v_{cB}	critical volume of the solvent (cm^3 mol^{-1})
v_{rB}	residue volume of the solvent (cm^3 mol^{-1})
V	molar volume (cm^3 mol^{-1})
v	superficial fluid velocity (m s^{-1})
Y	extraction yield
z	dimensionless axial coordinate along the bed, $x\ L^{-1}$
Greek letters	
μ	Viscosity (kg m^{-1} s^{-1})
ρ	Density (kg m^{-3})
ε	extractor void fraction
τ	dimensionless time, ($t\ v\ L^{-1}$)
Subscripts	

c	Critical
ext	Extractor
f	Fluid
p	Particle
s	surface of particle
0	Initial

1. Introduction

Supercritical carbon dioxide (SC-CO_2) extraction is a powerful developing technique in separation process that produces compounds by pure technology, under mild conditions. Carbon dioxide (CO_2) is the most widely used solvent in SC-CO_2, since it is physiologically harmless, environmentally safe, non-explosive, and readily available and it can be easily removed from products. Nowadays, supercritical fluid extraction is widely used as an attractive alternative extraction method to conventional liquid extraction in wide variety areas including the industries of food, pharmacy and environmental engineering, chemical and oil industries [1, 2].

Achillea wilhelmsii is an umbelliferae species in the flora of Iran full of flavonoid compounds. Flavonoids are highly effective natural antioxidant that control the blood cholesterol and triglyceride and are used especially in kidney problems. To date, there are no publications found on essential oil extraction from *Achillea wilhelmsii* by supercritical fluid extraction. In this study supercritical carbon dioxide was employed to extract compounds from *Achillea wilhelmsii*.

Mathematical models are powerful and useful tools for the design and scale up of SC-CO_2 processes. Since the laboratory scale up to pilot and industrial scales and the cost evaluation of processes cannot be done without mathematical modeling, several attempts at mathematical modeling of supercritical fluid extraction processes have been proposed [3-10].

In these models, the mass balance equations can be solved analytically or numerically, depending on the assumptions made [11, 12]. In any case, when the model describes the phenomena occurring in the extractor vessels well, it can safely be used for process designing and process scale-up. In order to solve the differential mass balance equations and to analyze the adequacy of the model it is necessary to compare the calculated overall extraction curves (OECs) with the experimental ones. The objective of this research was the application of a theoretical model for describing SC-CO_2 extraction from *Achillea wilhelmsii* and to study the influence of some process parameters such as supercritical fluid pressure, temperature, mean particle size and extraction time on the extraction yield. For verification of the model our experimental data have been used.

2. Experimental

2.1. Material

The plant used in this study named *Achillea wilhelmsii* was collected close to the city of Marv-dasht in Iran ($29^0$52'27" North and $52^0$48'09" East). It was then dried in a dark place at room temperature (around 25 °C). The sample was ground in a Panasonic blender (model MX-J225G) to produce a powder. Carbon dioxide with the purity of 99.95% was obtained from Abu-Qaddareh Company (Shiraz). Dichloromethane with the purity of 99.99%, purchased from Merk Kga A, was used as solute recovery.

2.2. Experimental procedure

The experiments were carried out in bench scale apparatus. Fig. 1 shows the schematic diagram of the extractor apparatus. Carbon dioxide fed from tank A was liquified by a condenser (B) (a refrigerator at temperature of -20 °C). The liquid CO_2 was pumped into a surge tank (D) (where the pressure oscillation is dampened) by a reciprocating pump (C) (Haskel pump, MS-110, USA). The pressurized CO_2 was then flown into a coil which emerged in a warm water bath (H_1). In order to control the temperature of the extraction process, the surge tank and extraction column (E) were surrounded by a regulating hot water jacket (shell and tube). At the outlet of the surge tank, a pressure gauge (J) (Indumart, EN 837-1, Canada) within the range of 0-250 bar by a division of 1 bar was placed. All temperatures were measured using Pt100 RTD sensors (accuracy of ± 0.1 °C). For this study, 6 g of dried *Achillea wilhelmsii* and glass beads were loaded in a high pressure vessel (E) with an internal volume of 250 ml. Glass beads prevent channeling of the flow in packed beds and the dead volume. The SC-CO_2 was in contact with the sample for 50 min and during this time, it solved the essential oil of the sample. This situation is called "static time". The dynamic extraction was started by opening the exit valve for the SC-CO_2 extraction system. The static extraction allows the sample of *Achillea wilhelmsii* to soak in the CO_2 in order to equilibrate the mixture at the desired pressure and temperature. During the dynamic extraction time, CO_2 carrying the crude extract was allowed to flow through a heated backpressure valve (F). The valve was heated using an element to avoid the Joule-Thomson effect. The SC-CO_2 was then entered into a U-shaped tube (G) in which dichloromethane was at 0°C. The essential oil was solved in dichloromethane. Then, CO_2 passed through a flow meter (I) and was released into the atmosphere. In the end, the essential oil+dichloromethane mixture was collected as a sample which was used in GC analysis. [13, 14] are given similar procedure for SC-CO_2 extraction from plants.

The extracted essential oil was weighed by an analytical balance (RADWAG, MX-J225G) with an accuracy of 10^{-4} g. The extraction yield can be calculated as:

$$\text{Yield (\%)} = \frac{\text{g of extracted oil}}{\text{g of dried plant powder}} \times 100 \qquad (1)$$

2.3. Gas chromatography-mass spectrometry (GC-MS) analysis

In the present study, the oil was analyzed in an Agilent 7890A system (Agilent Technologies, USA) including a split/splitless injector, a flame ionization detector, a 5975C triple-axis mass spectrometer detector and Chemstation software. The column used was an Agilent HP-5MS capillary column (30 m × 0.25 mm (ID), 0.25 μm film thickness). The oven temperature started at 60 °C, then increased to 210 °C at 3 °C/min, then increased to 240 °C at 20 °C/min and finally kept at 240 °C for 8.5 min. Sample injections (0.2 μL) were performed in split mode. Helium (purity > 99.99 mole %) was used as carrier gas (flow rate of 1 mL/min) and the split ratio was 1:50. The injector temperature was 280 °C and the mass spectrometer ion source and interface temperatures were 230 and 280 °C respectively. The ionization energy was 70 eV with a mass range of 50 to 480 AMU.

Around 50 peaks were detected in the *Achillea wilhelmsii* supercritical extracts. The peak identification of compounds within the extracts was performed by comparing their mass spectra and retention indices with standards. Mass spectra were also compared with those in the NIST (National Institute of Standards and Technology).

3. Mathematical modeling

All the tests and measurements were performed in a fixed bed containing the *Achillea wilhelmsii* as the stationary phase with flowing SC-CO_2 as the mobile phase. The purpose of the experiments was extraction of essential oils from the *Achillea wilhelmsii* and determining the extraction yield. The extraction yield of essential oils from the *Achillea wilhelmsii* can be mathematically described by a set of partial differential equations. The assumptions of the model are listed as follows:

1- All the solid particles are spherical and mono-sized
2- The solid particles are considered as a lumped system duo to the small size.
3- Concentration and temperature changes during the extraction process are negligible.
4- Concentration gradients of extractor in radial direction are negligible.
5- Bed porosity remains constant during the extraction process.

6- The physical properties of supercritical CO_2 are constant along the bed and during the extraction time.

According to the above assumptions, the governing differential equation for the fluid phase can be written in the following dimensionless form:

$$\frac{\partial C_f}{\partial \tau} - \frac{1}{Pe_b}\frac{\partial^2 C_f}{\partial z^2} + \frac{\partial C_f}{\partial z} + \frac{1-\varepsilon}{\varepsilon}\frac{3L}{R_p}\frac{k_f}{v}\left(C_f - C_s\right) = 0 \qquad (2)$$

where C_f is the solute concentration in the bulk phase, C_s is the solute concentration in the surface of the solid phase, Pe_b is Péclet number for the bed, z is dimensionless axial coordinate along the bed, L is extractor length, R_p is particle radius, ε is extractor void fraction, τ is dimensionless time v is superficial fluid velocity and k_f is external mass transfer coefficient.

The boundary and initial conditions for this equation are as follows:

$$\text{At } \tau = 0 \rightarrow C_f = 0 \qquad (3)$$

$$\text{At } z = 0^- \rightarrow C_f = 0 \qquad (4)$$

$$\text{At } z = 0^+ \rightarrow \frac{\partial C_f}{\partial z} = Pe_b\left(C_f - 0\right) \qquad (5)$$

$$\text{At } z = 1 \rightarrow \frac{\partial C_f}{\partial z} = 0 \qquad (6)$$

Similarly, the mole balance for the solid phase can be expressed in a dimensionless differential equation as the following:

$$\frac{dq}{d\tau} = -3\frac{k_f}{v}\frac{L}{R_p}\left(C_s - C_f\right) \qquad (7)$$

where q is solute concentration in the solid phase, k_f is external mass transfer coefficient and v is superficial fluid velocity.

With this initial condition:

At $\tau = 0 \rightarrow q = q_0$ (8)

In the above equations, q and C_s can be related by the following equilibrium:

$$q = KC_s \quad (9)$$

In which, K is equilibrium constant.

4. Numerical solution

The differential equations of (2) to (9) were simultaneously solved using finite element method FEM with FlexPDE software. The FlexPDE is software based on the standard Galerkin FEM which is created by PDE Solutions Inc. This software is simple to use and solves partial differential equations without needing to know the details of programming or finite element implementation. The two dimentional (2-D) domain for numerical solution in FlexPDE software and corresponding boundary conditions is shown in Fig. 2. Using this model the change of solute concentration in the fluid phase and solid phase can be expressed as a function of time and axial coordinate along the bed.

5. Model parameters

Critical pressure, critical temperature, acentric factor and average molecular weight of *Achillea wilhelmsii* extract were obtained from our experiments.

The residual viscosity correlation of Jossi et al [15] was used for viscosity calculations. Density of CO_2 was calculated using the Soave/Redlich/Kwong equation of state [15]. Several correlations have been presented for diffusion coefficients of solutes in supercritical fluids. He and Yu correlation was used as follows [15]:

$$D_{AB} = \alpha + 10^5 \left(\frac{T}{M_A}\right)^{0.5} exp\left(\frac{0.3887}{0.23 - \upsilon_{rB}}\right) \quad (10)$$

where A is solute, B is solvent and $\upsilon_{rB} = \frac{\upsilon_B}{\upsilon_{cB}}$, $\upsilon_B = \frac{ZRT}{P}$.

$$\alpha = 14.882 + 0.005908 \frac{T_{cB} V_{cB}}{M_B} + 2.0821 \times 10^{-6} \left(\frac{V_{cB} T_{cB}}{M_B}\right)^2 \quad (11)$$

The mass transfer coefficient was estimated using Eq. (12) and correlation of Tan et al. [16] was selected, which is valid over *Re* range from 2 to 40 and *Sc* from 2 to 40 for determination of *Sh* correlation.

$$k_f = \frac{Sh.D_{AB}}{d_p} \tag{12}$$

$$Sh = 0.0306\, Re^{0.83} Sc^{0.33} \tag{13}$$

where *Re* and *Sc* are Reynolds and Schmidt numbers, respectively, and calculate as below:

$$Re = \frac{2R_p v \rho_f}{\mu_f} \tag{14}$$

$$Sc = \frac{\mu_f}{\rho_f D_{AB}} \tag{15}$$

where μ_f is fluid viscosity and ρ_f is fluid density.

The axial dispersion coefficient (D_{ax}) was given by correlation of Funazukuri et al. [17] as shown below:

$$\frac{\varepsilon D_{ax}}{D_{AB}} = 1.317\left(\varepsilon Re Sc\right)^{1.392} \tag{15}$$

6. Calculation of equilibrium constant [9]

Equilibrium constant should be determined so the model data must fit with the experimental data. On the other hand, the value of *K* is obtained by the sum of the squares of the difference between the experimental yield (Y_{exp}) and the predicted yield (Y_{model}), which must be minimized. The criterion is described by Eq. (17) as:

$$\mathrm{OBJ} = \Sigma(Y_{model} - Y_{exp})^2 \tag{17}$$

In this case a value for K was guessed, then the model was solved and OBJ was calculated for constant velocity data only (small scale and constant residence time data were used to test the model). Operation stops if the value of OBJ is lower than a specified small number, otherwise the procedure continues [9]. Therefore, there are three eqs. which can be solved simultaneously.

7. Results and discussion

In this study, the developed mathematical model and the experimental results were compared and discussed. The results obtained from the model were presented along with the experimental data to optimize the process parameters such as pressure, temperature. The extraction bed and the particles showed the following characteristics: particle true density of 977 kgm^{-3}; particle mean diameter of 0.06 cm, respectively; and bed-plus-particle porosity of 0.728. Table 1 shows the process parameters; it can be noted that the quantity of extractable oil varies with pressure and temperature.

Table 2 shows the chemical composition of the *Achillea wilhelmsii*-extract and lists the fractions of compounds that were identified. The following six compounds were identified in each sample: Carvacrol, Camphor, α-Pinene, Borneol, Thymol and Pinocarvone. Over 29 wt% of the extract in each sample could not be identified. Initially, Camphor and Borneol showed the maximum and minimum concentration in the extract, respectively. However, the composition significantly changed by the end of the experiment, and the compounds with maximum and minimum concentrations in the extract were as follows: Camphor and Pinocarvone for all experiments.

Table 3 shows the modeling results obtained for the following parameters in this study: molecular diffusion coefficient (D_{AB}), axial dispersion coefficient (D_{ax}), fluid viscosity (v), fluid density (ρ), Reynolds number (Re), Schmidt number (Sc), Sherwood number (Sh) and external mass-transfer coefficient (k_f). As can be seen, the maximum extraction yield (3.63 %) was achieved at 45°C and 190bar.

Figs. 3 and 4, illustrates the comparison between model and constant velocity experimental data at 170 bar and 40°C, and 190 bar and 45°C, respectively. As can be seen, there is a good agreement between model and experimental data at constant residence time condition. The difference between the experimental extraction yield and the predicted extraction yield at the initial time is higher than that at the other times. The comparison between Fig. 3 and Fig. 4 shows the extraction yield increases with increasing pressure and temperature.

Variation of extraction yield with temperature and pressure has been shown in Fig. 5. It is clear from this figure that extraction yield increases with increase of temperature; but if pressure is too high, extraction yield is a strong function of pressure.

Effect of extractor void fraction (ε) on extraction yield is shown in Fig. 6. It is clear that extraction yield decreases with increasing extractor void fraction at constant pressure (190 bar), temperature (40°C) and particle diameter (0.06cm).

Variation of extraction yield with particle diameter is shown in Fig. 7. It is clear from this figure that extraction yield decreases with increasing. The results in Fig. 7 demonstrate that the extraction yield is increased by decreasing particle diameter at constant pressure (190 bar) and temperature (45°C).

Fig. 8 shows the influence of extraction time and pressure on the yield of extract. Yield of extract is a weak function of extraction time at low value of pressure and is also a strong function of pressure at large value of extraction time.

Fig. 9 presents the predicted dimensionless concentration (C_f/q_0) at different heights of the extractor as a function of time. This figure shows a very sharp increase of desorption rate at the beginning of the extraction process. At the beginning there is an appropriate extraction rate which can be duo to a high driving force for mass transfer. After about 35 min, the solute concentration tends towards zero which means that there is not enough driving force for mass transfer.

In Fig. 10 the dimensionless concentration is plotted vs the bed length at different times. This figure shows that after the maximum concentration in Fig. 9, as the extraction time increases, the outlet concentration from the bed decreases. At a fixed time, as the bed length increases, the concentration also increases linearly.

8. Conclusion

In this study, extraction of essential oil from *Achillea wilhelmsii* using SC-CO_2 at different pressures and temperatures were carried out. The best efficiency of the plant was obtained 3.63% at pressure of 190 bar, temperature of 45°C, dynamic extraction time of 40 min and particle diameter of 0.06 cm. The essential oil obtained using the experiments were analyzed using gas chromatography-mass spectrometry (GC-MS). In chromatographic analyses, Carvacrol, Camphor, α-Pinene, Borneol, Thymol and Pinocarvone were identified. In the second part of this work, a mathematical model based on mass balance was used to simulate extraction yield. The differential equations of the extraction process were solved using the numerical method of FEM. Finally, the results were compared with the experimental data. It was found that the results are in good agreement with the experimental data.

9. Acknowledgements

The financial and laboratory support of Dr. Honarvar in the department of chemical engineering, Marvdasht Branch, Islamic Azad University, Marvdasht, Fars, Iran, is gratefully acknowledged.

References

[1] M. Herreroa, J.A. Mendiola, A. Cifuentes, E. Ibáñeza, Supercritical fluid extraction: recent advances and applications, *J. Chromatogr. A*. 1217 (16) (2010) 2495–2511.

[2] T.M.Takeuchi, P.F. Leal, R. Favareto, L. Cardozo-Filho, M.L. Corazza, P.T.V. Rosa, M.A.A. Meireles, Study of the phase equilibrium formed inside the flash tank used at the separation step of a supercritical fluid extraction unit, *J. of Supercritical Fluids*. 43 (3) (2008) 447-459.

[3] T. Hatami, R. N. Cavalcanti, T.M. Takeuchi, M.A.A. Meireles, Supercritical fluid extraction of bioactive compounds from Macela (Achyrocline satureioides) flowers: Kinetic, experiments and modeling, *J. of Supercritical Fluids*. 65 (2012) 71–77.

[4] M. Pilavtepea, M. Yucel, S.S. Helvaci, M. Demircioglub, O. Yesil-Celiktas, Optimization and mathematical modeling of mass transfer between Zostera marina residues and supercritical CO_2 modified with ethanol, *J. of Supercritical Fluids*. 68 (2012) 87–93.

[5] O. Boutin, A.D. Nadaï, A.G. Perez, J.-H. Ferrasse, M. Beltran, E. Badens, Experimental and modelling of supercritical oil extraction from rapeseeds and sunflower seeds, *Chem. Eng. Res. Des*. 89 (11) (2011) 2477–2484.

[6] E. Rahimi, J.M. Pradob, G. Zahedi, M.A.A. Meireles, Chamomile extraction with supercritical carbon dioxide: Mathematical modeling and optimization, *J. of Supercritical Fluids*. 56 (1) (2011) 80–88.

[7] T. Hatami, R.N. Cavalcanti, T.M. Takeuchi, M.A.A. Meireles, Supercritical fluid extraction of bioactive compounds from Macela (Achyrocline satureioides) flowers: Kinetic, experiments and modeling, *J. of Supercritical Fluids*. 65 (2012) 71–77.

[8] F. Bashipour, S.M. Ghoreishi, Experimental optimization of supercritical extraction of β-carotene from Aloe barbadensis Miller via genetic algorithm, *J. of Supercritical Fluids*. 72 (2012) 312–319.

[9] M.A.A. Melrelesa, G. Zahedi, T. Hatami, Mathematical modeling of supercritical fluid extraction for obtaining extracts from vetiver root, *J. of Supercritical Fluids*. 49 (1) (2009) 23–31.

[10] I. Goodarznia, M.H. Eikani, Supercritical carbon dioxide extraction of essential oils: Modeling and simulation, *Chem. Eng. Sci*. 53 (7) (1998) 1387-1395.

[11] H. Sovová, Rate of the vegetable oil extraction with supercritical CO_2—I. Modelling of extraction curves, *Chem. Eng. Sci*. 49 (3) (1994) 409-414.

[12] M. Goto, M. Sato, T. Hirose, Extraction of peppermint oil by supercritical carbon dioxide, *J. Chem. Eng. Japan*. 26 (4) (1993) 401-407.

[13] G. Sodeifian, S.A. Sajadian, Investigation of Essential oil extraction and antioxidant activity of Echinophora platyloba DC. using supercritical carbon dioxide, *J. of Supercritical Fluids*. 121 (2017) 52-62.

[14] G. Sodeifian, S. Ghorbandoost, S.A. Sajadian, N.S. Ardestani, Extraction of oil from Pistacia khinjuk using supercritical carbon dioxide: experimental and modeling, *J. of Supercritical Fluids*. 110 (2016) 265-274.

[15] B.E. Poling, J.M. Prausnitz, J.P. O'Connell, The Properties of Gases and Liquids, Vol. 5, Mcgraw-hill.: New York, (2001).

[16] C.-S. Tan, S.-K. Liang, D.-C. Liou, Fluid–solid mass transfer in a supercritical fluid extractor, *Chem. Eng. J.* 38 (1) (1988) 17-22.

[17] T. Funazukuri, C. Kong, S. Kagei, Effective axial dispersion coefficients in packed beds under supercritical conditions, *J. of Supercritical Fluids*. 13 (1-3) (1998) 169-175.

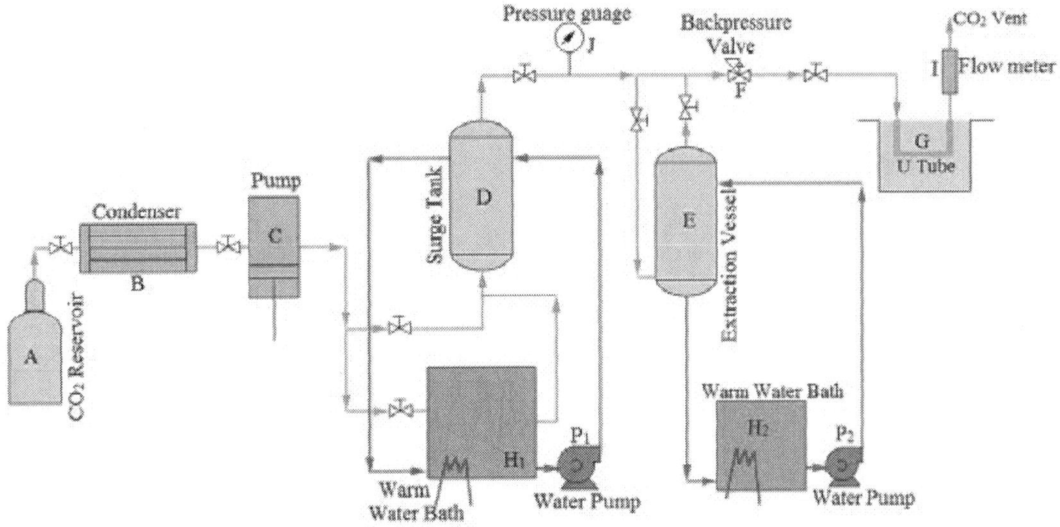

Fig. 1. *Schematic diagram of supercritical extraction system: A: CO_2 Tank; B: Condenser;*

C: Pump; D: Surg Tank; J: Pressure guage; H_1, H_2: Warm Water baths; P_1, P_2: Water pumps+; E: Extraction column, F: Backpressure valve; G: U Tube Separator; I: Flow meter.

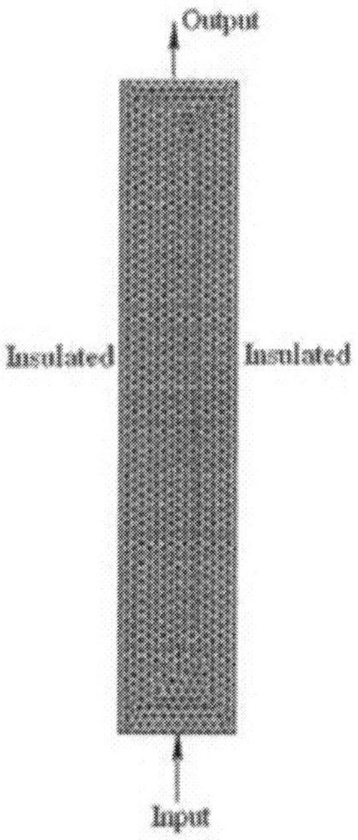

Fig .2. *Solution domain and corresponding boundary condition*

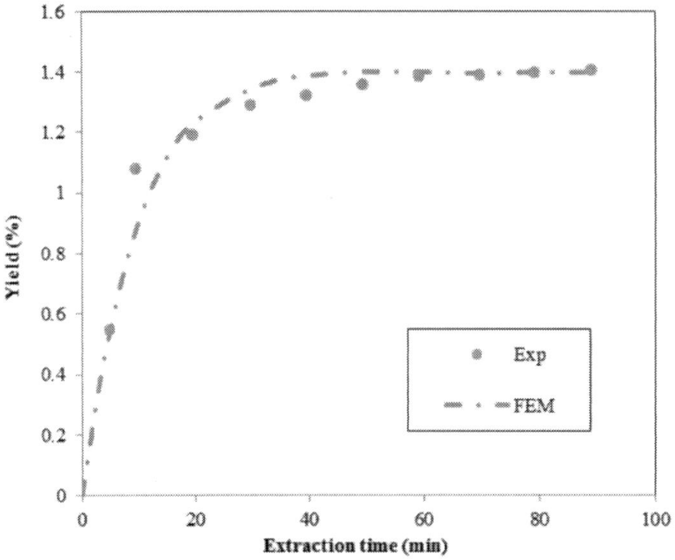

Fig. 3. *Comparison of FEM with experimental data at P = 170bar, T = 40°C and* d_p *= 0.06cm*

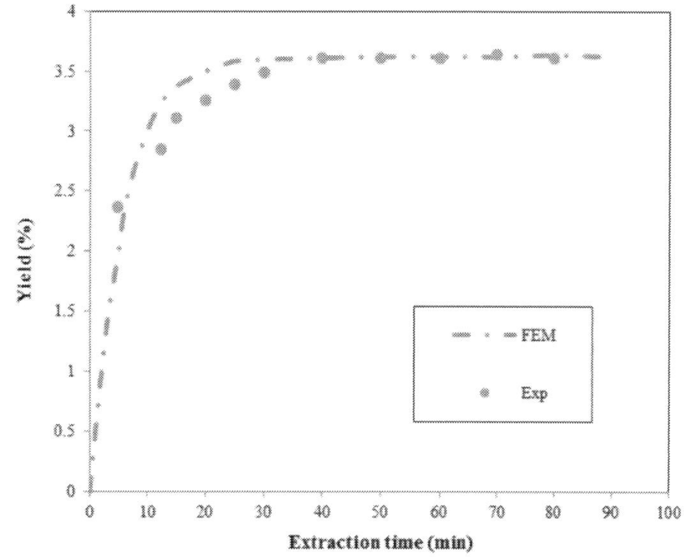

Fig. 4. *Comparison of FEM with experimental data at P = 190bar, T = 45°C and* d_p = *0.06cm*

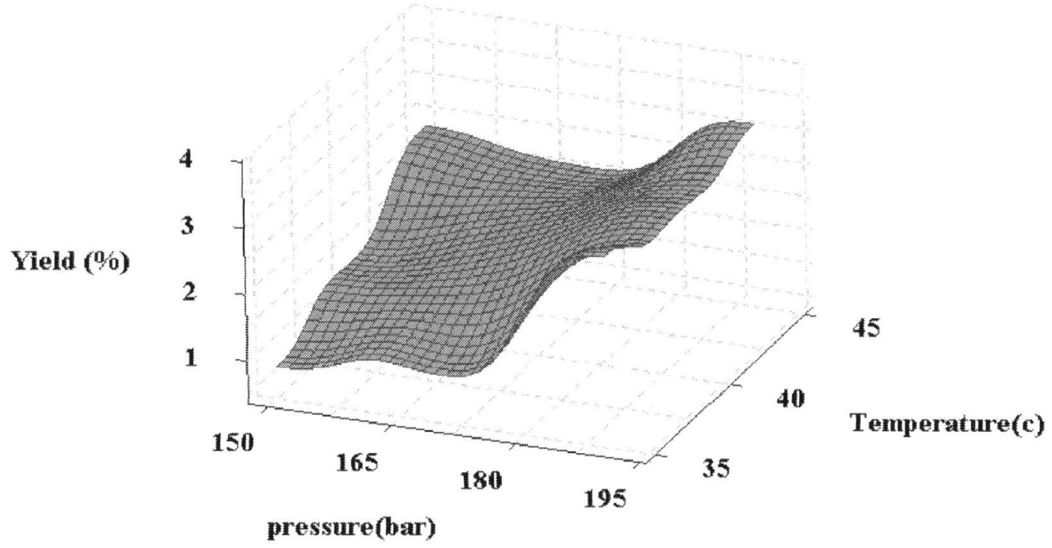

Fig. 5. *Influence of pressure and temperature on the yield of extraction at* $d_p = 0.06cm$, $\varepsilon = 0.728$.

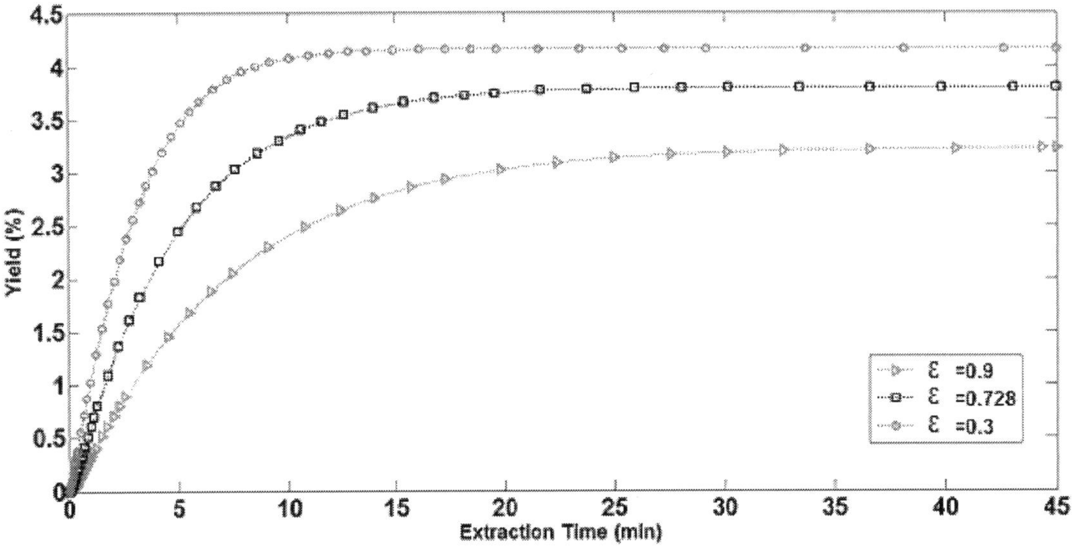

Fig. 6. *Influence of extractor void fraction ε on the yield of extraction at P = 190bar, T = 40°C and d_p = 0.06cm*

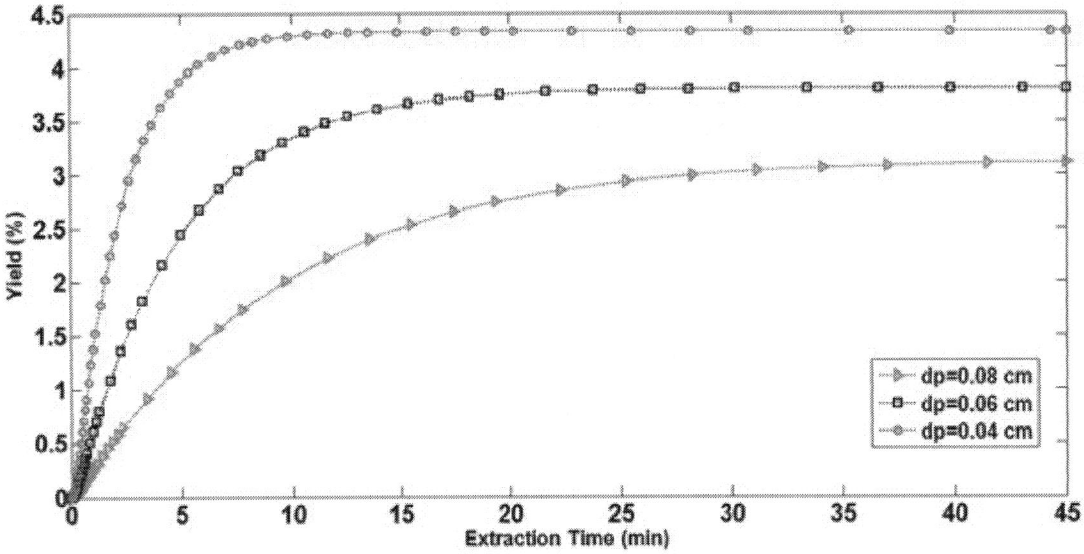

Fig. 7. *Influence of particle diameter on the yield of extraction at P = 190bar, T = 45°C, ε = 0.728.*

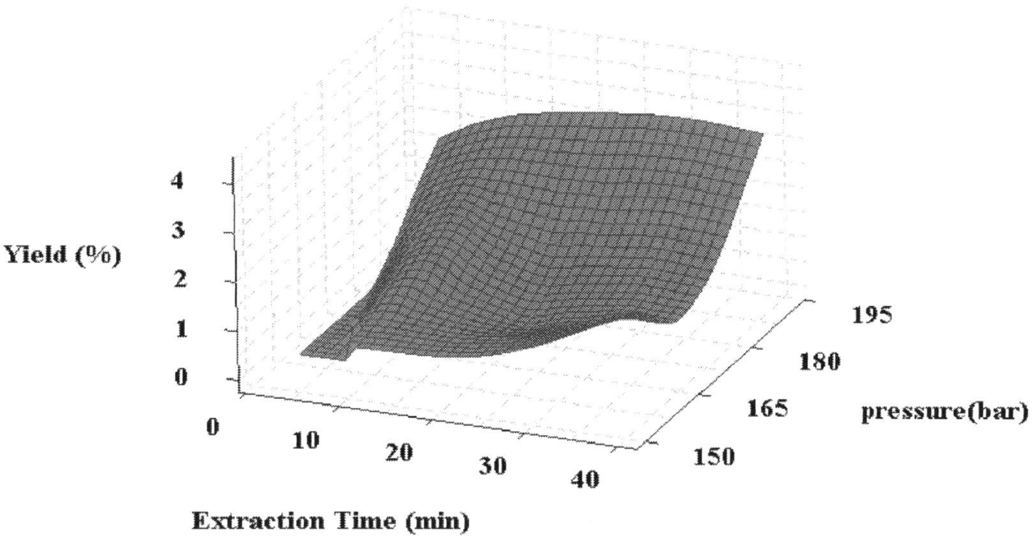

Fig. 8. Influence of pressure and extraction time on the yield of extraction at $d_p = 0.06$ cm, $\varepsilon = 0.728$

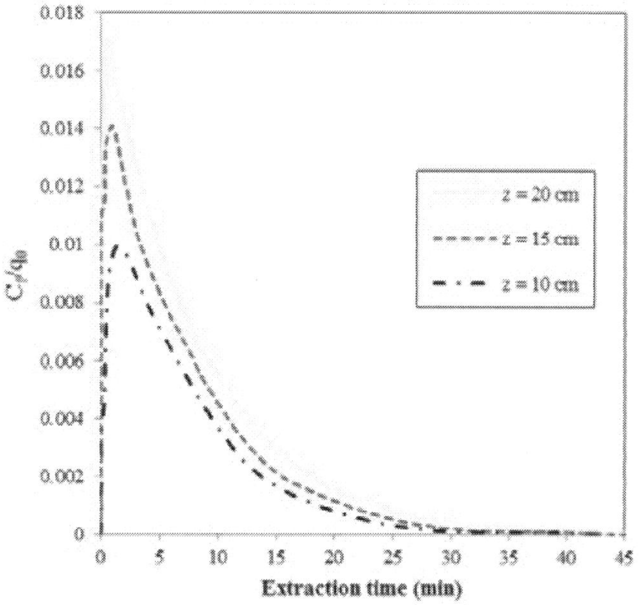

Fig . 9. Dimensionless concentration changes in different parts of the extractor

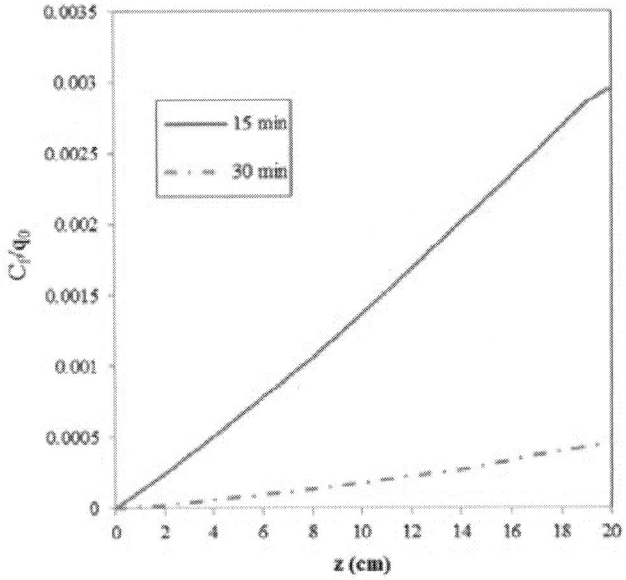

Fig . 10. Dimensionless concentration changes in different times

Table 1

Process parameters for SC-CO_2 extraction from *Achillea wilhelmsii*.

Conditions	Exp.1	Exp.2	Exp.3	Exp.4	Exp.5	Exp.6
P (bar)	190	170	190	150	150	170
T (°C)	45	35	35	45	40	45
d_p(cm)	0.06	0.06	0.06	0.06	0.06	0.06
Q ($\times 10^4$ kg s^{-1})	5.34	5.64	5.68	4.94	5.20	5.03
ε	0.728	0.728	0.728	0.728	0.728	0.728
Static time (min)	45	45	45	45	45	45
ρ (kg m^{-3})	977	977	977	977	977	977
D_{ext}(cm)	1.2	1.2	1.2	1.2	1.2	1.2
L (cm)	20	20	20	20	20	20
M_{feed}(g)	6	6	6	6	6	6

Table 2

The composition of extracted oil from *Achillea wilhelmsii* as a function of time for the four experimental conditions described in Table 1.

	T(min)	Carvacrol (%)	Camphor (%)	α-Pinene	Borneol (%)	Thymol (%)	Pinocarvone (%)	Not identified
Exp. 1	10	18.37	37.11	16.51	4.23	8.03	4.56	11.19
	20	15.82	33.37	11.65	3.50	2.84	1.26	31.56
	40	10.52	34.26	6.32	3.63	3.68	0.63	41.80
	60	3.57	19.40	5.75	0.36	1.53	0.09	69.31
Exp. 2	10	12.61	26.37	10.21	7.63	9.42	3.82	29.93
	20	14.75	19.84	9.26	9.73	7.71	1.62	37.09
	40	3.23	15.48	6.68	5.27	6.81	2.33	60.20
	60	4.11	10.06	8.67	6.73	5.94	0.85	63.64
Exp. 3	10	21.40	35.04	14.35	3.36	6.43	4.11	15.31
	20	13.56	36.5	10.9	4.61	9.5	2.64	22.29
	40	9.7	27.21	11.75	2.55	5.36	0.25	43.18
	60	11.56	29.5	7.83	3.72	3.16	0.18	44.05
Exp. 4	10	16.55	36.60	15.8	4.74	8.42	3.85	14.04
	20	15.82	27.61	10.3	3.43	9.57	4.18	29.09
	40	4.33	29.73	6.22	3.30	7.14	0.61	48.67
	60	5.41	22.78	8.64	1.57	3.25	0.09	41.74

Table 3

Physical properties of supercritical fluid and estimated parameters of model.

Conditions	Run 1	Run 2	Run 3	Run 4	Run 5	Run 6
P (bar)/T (°C)	190/45	170/35	190/35	150/45	150/40	170/45
v ($\times 10^4$ m s^{-1})	5.34	5.64	5.68	4.94	5.20	5.03
ρ (kg m^{-3})	801.53	846.14	852.70	742.24	780.11	755.57
$\mu(\times 10^5 kgm^{-1}s^{-1})$	7.14	7.51	8.75	6.21	6.77	6.71
D_{AB} ($\times 10^9 m^2 s^{-1}$)	9.54	8.43	8.32	10.86	9.91	10.06
Re	54.83	54.71	47.43	58.07	56.02	56.17
Sc	9.38	10.57	12.33	7.75	8.77	8.65
Sh	22.14	23.17	21.45	21.74	22.36	21.91
D_{ax} ($\times 10^5 m^2 s^{-1}$)	6.51	6.83	6.91	6.21	6.42	6.37
$k_f(\times 10^4 m\ s^{-1})$	3.51	3.22	3.01	3.92	3.62	3.74
Yield (%)	**3.63**	1.39	2.96	2.14	1.36	2.56

484353 a Comprehensive Analysis of Gas Sorption and Transport in Thermally Rearranged Polymers

Abstract:

A comprehensive analysis of gas sorption and transport in thermally rearranged polymers

Michele Galizia‡, Kevin A. Stevens, Zachary P. Smith , Donald R. Paul, Benny D. Freeman

J.McKetta Department of Chemical Engineering, 200 E. Dean Keeton Street, 78712 Austin,

and Center for Energy and Environmental Resources, 10100 Burnet Road, 78758 Austin,

The University of Texas at Austin, Texas (USA)

‡ *Permanent address: Department of Chemical, Biological and Materials Engineering, 100 E. Boyd St., 73019 Norman, University of Oklahoma, OK (USA)*

Permanent address: Department of Chemical Engineering, 25 Ames St., 02142 Cambridge, Massachusetts Institute of Technology, MA (USA)

Thermally Rearranged polymers based on PBOs (poly benzoxazoles) [1] are a class of materials endowed with outstanding separation performance, which abundantly surpass the 2008 Robeson upper bound [2]. As recently reported by Robeson et al. [3], the outstanding behavior exhibited by TR polymers, uncommon among glassy polymers, derives from a special combination of high gas diffusivity, high gas solubility and high diffusivity-selectivity. In this study, these experimental findings were rationalized on theoretical basis using the *Non-equilibrium thermodynamics of glassy polymers* [4]. First, solubility, diffusivity and permeability coefficients for H_2, N_2, CH_4, and CO_2 in several TR polymers and their polyimide precursor, HAB-6FDA, were measured in a wide temperature (-10 to 50°C) and pressure (up to 32 atm) range. At fixed temperature, gas solubility, diffusivity and permeability coefficients increased with increasing extent of thermal conversion [5,6].

To investigate the molecular origin of the change in gas sorption behavior with increasing TR conversion, the enthalpic and the entropic contributions to the solubility coefficient were calculated. The increase in gas solubility with increasing TR conversion is essentially due to an increase in the entropic contribution, with the enthalpic counterpart being barely affected by thermal conversion. Thus, the increase in gas transport properties observed in TR samples relative to HAB-6FDA polyimide is essentially due to an increase of excess, non-equilibrium fractional free volume. Such analysis justifies the high gas solubility, diffusivity and permeability coefficients exhibited by TR polymers [6].

A theoretical analysis of gas permeability and diffusivity was also performed. Experimental permeability data were described using the solution-diffusion model. Specifically, solubility coefficients were predicted using the *non-equilibrium lattice fluid* (NELF) model, and diffusion coefficients were expressed as the product of a mobility factor, *L*, which was assumed to vary exponentially with concentration, and a thermodynamic factor, which was calculated using the NELF model [4,7]. Interestingly, the mobility factor correlates with the penetrant critical volume, V_C , according to the equation:

$$L = k/(V_C)^\eta$$

Polymers having large values of η exhibit stronger size sieving ability than polymers with smaller values of η. TR polymers exhibit greater values of η than other glassy polymers, which suggests that the former have stronger size sieving ability than conventional polymers. This result fully agrees with the experimental findings reported by Robeson et al. [3]. The high size sieving ability shown by TR polymers derives from a favorable size and distribution of free volume cavities. This conclusion is supported by PALS analysis and Monte Carlo simulations [8]. So, coalescence of smaller free volume cavities to form larger cavities is believed to take place during conversion of polyimides to PBOs.

[1] H.B. Park et al., Science 2007 **318** 254

[2] L.M. Robeson, J. Membr. Sci. 2008 **320** 390

[3] L.M. Robeson et al., J. Membr. Sci. 2017 **525** 18

[4] F. Doghieri et al., Macromolecules 1996 **29** 7885

[5] K.A. Stevens et al., J. Membr. Sci. 2017, accepted

[6] M. Galizia et al., Macromolecules 2016 **49** 8768

[7] G.C. Sarti et al., J. Membr. Sci. 2013 **43** 176

[8] Y. Jiang et al., Polymer 2011 **52** 2244

499838 Ionic Polyimides: New Dimensions in the Design of Polymer Gas Separation Membranes

Abstract:

Ionic liquids (ILs) and polyimides (among several other classes of materials) have been at the forefront of gas separation membranes in recent years. Although ILs and polyimides are seemingly disparate classes of materials, ILs and polyimides can be covalently attached via condensation polymerization reactions to form "ionic polyimides". Based on rational selection of precursors and reagents, there are 10,000's of feasible ionic polyimide structures along with the potential to incorporate elements from PIMs and TR polymers. We have already developed several unique ionic polyimide materials based on monomers formed from amino-tethered imidazoles with various aromatic dianydrides. These diimide monomers are then reacted with dihalides through the imidazole moeities via SN2 reactions to form the ionic polyimides. Ionic polyimides show excellent processability and display increased ordering in the presence of ILs. This presentation will discuss our design strategies, polymer characterization and performances as membranes in the separation of CO2, H2, N2 and CH4.

501257 Tailoring Backbone Rigidity of Triptycene-Containing Polymers for Enhanced Gas Transport Propertie

Abstract:

Gas transport in dense polymer membranes is a pressure driven, solution-diffusion process, which involves sorption of a species into the membrane surface, diffusion across the membrane through a series of â€˜jumpsâ€™ through the free volume elements, and desorption at the downstream. In this process, the transient free volume spaces are largely generated by inefficient chain packing and dictated by the segmental motions of polymer chains to produce gaps that open and close within the polymer. Therefore, polymer backbone rigidity, which governs the local segmental motions and chain packing plays a critical role in constructing low-barrier diffusion pathways enabling fast and selective gas transport. In this talk, we report two new strategies in tailoring polymer backbone rigidity of triptycene-containing polymers to further improve their overall gas separation performance. Specifically, the first strategy is to incorporate a super rigid triptycene-based diamine as a comonomer in polyimide synthesis. The resulting copolymers are composed of a rigid element with the triptycene skeleton directly linked to imide rings and a relatively flexible element containing ether linkage. The ratio of the two components was systematically varied to finely tune the chain rigidity and thus gas transport properties. It is found that the copolymer composition of 50% of the rigid structure caused a 42 °C increase in glass transition temperature and a more than 2-fold increase in gas permeability. Moreover, the much improved gas permeability was accompanied with almost constant gas selectivities due to the unique internal free volume of the triptycene moiety, leading to the overall separation performance laterally approaching the upper bound. A second strategy is to investigate the integration of highly size-sieving triptycene moieties in polybenzoxazole (PBO) structure, which is arguably a more rigid backbone structure with heterocyclic ring than polyimides. A series of triptycene-containing PBOs with systematically varied backbone rigidity were prepared, all of which displayed superior gas separation performance that is far beyond the 2008 upper bound. Strikingly, via increasing the backbone rigidity by adjusting the content of triptycene units, the highly permeable triptycene-PBO copolymers showed a vertically increasing trend in permselectivities, leading to the overall separation performance of several films surpassing the most recently reported 2015 upper bounds for H_2/CH_4 separations. In this talk, synthesis and characterization of these new highly rigid triptycene-containing polymers will be presented. Fundamental physical and transport properties will be discussed to illustrate the fundamental relationship between microscopic structures with macroscopic transport properties for these new triptycene-containing polymer membranes.

500792 Ultrathin and Pinhole-Free Gas-Selective Polymeric Films Synthesized Via Chemical Vapor Deposition

Abstract:

Gas separation has gained enormous interest due to the increasing industrial demand for pure gas species and the urgency for carbon capture. Among various gas separation techniques, polymer membrane based separation process is considered to be an energy-efficient alternative for conventional cryogenic distillation and sorbents based separation. In order to simultaneously achieve high flux and high selectivity, a typical membrane has an anisotropic structure consisting of an ultra-permeable support and an ultrathin gas-selective skin layer, since the flux is inversely proportional to the skin layer thickness. However, avoiding the pinhole formation in sub-100 nm membranes has proven to be difficult for solution based fabrication techniques, such as phase inversion. In this regard, we introduced chemical vapor deposition (CVD) as a powerful technique to synthesize ultrathin and pinhole-free gas-selective polymeric films. CVD was chosen for several reasons: 1) depositing pinhole-free films with precise thickness control at nanometer level; 2) large library of starting monomers; 3) substrate independency; 4) in situ covalent crosslinkability; and 5) scalability. In this talk, sub-100 nm pinhole-free gas-selective membranes as large as 175 cm^2 were successfully deposited on top of ultrapermeable poly[1-(trimethylsilyl)-propyne] (PTMSP) cast membranes. The large library of starting monomers has allowed us to fine tune the polymer structure to meet different gas separation purposes. For instance, the robust porphyrin derived sub-100 nm microporous films led to high CO_2/N_2 selectivity (ca. 85) and a negligible H_2/CO_2 selectivity (< 2). In contrast, an ionic monomer acrylic acid (AA) derived sub-10 nm membrane displayed a significant H_2/CO_2 selectivity (ca. 45) but a moderate CO_2/N_2 selectivity (ca. 15). Such CVD synthesized gas-selective films were also successfully cross-linked by employing a vaporized cross-linker to enhance the film durability. Future work will focus on designing new gas-selective CVD polymers and exploring new substrate options that are less prone to physical aging. The work reported in this talk should provide insight to the membrane community on the possibility of employing CVD to fabricate gas-selective polymeric films.

492182 The Effects of Backbone Chemical Structure on Gas Transport Properties in a Family of Polyethersulfones Polymers

Abstract:

The effects of chemical structure on gas transport properties in a family of polyarylethersulfones polymers

Ali Naderi[1], Wai Fen Yong[1], Youchang Xiao[1] and Tai-Shung Chung[1], *

Martin Weber[2], Christian Maletzko[3]

[1]Department of Chemical and Biomolecular Engineering, National University of Singapore, Singapore 117585, Singapore

[2]Advanced Materials and Systems Research, BASF SE, RAP/OUBGMV/W - B1, 67056 Ludwigshafen, Germany

[3]Performance Materials, BASF SE, GE-PMF/SUKTE/NE-F206, 67056 Ludwigshafen, Germany

*Correspondence author. Tel: +65 6516 6645; fax: +65 67791936.

E-mail address: chencts@nus.edu.sg (T.-S. Chung).

Abstract

Polymeric membranes are considered as a promising technology for gas separation. Polysulfone (PSF) and polyethersulfone (PESU) have been used in the membrane application due to their thermal stability, chemical resistance, processability and gas separation properties. In this work, a series of BASF polyarylethersulfone polymers (PAES) containing different backbone structures have been studied. Their permeation, sorption properties and physical properties were investigated. The gas permeabilities of the dense films including oxygen, nitrogen, methane, and carbon dioxide were measured as a function of pressure. The sorption of methane and carbon dioxide were assessed via the dual-volume pressure decay method using a XEMIS microbalance system and the diffusion coefficients of these gases through the membranes were determined by solution-diffusion model. The permeability of the membranes decreases slightly with the increase in pressure for all gases which can be due to the saturation of Langmuir sites in high pressure. To elucidate the microstructural behavior of different PES, positron annihilation life time spectroscopy (PALS) and dynamic mechanical analyses (DMA) have been examined. The results of PALS and DMA reveal that the polyphenylsulfone (PPSU) with a higher segmental motion in the polymer chain offers a higher free volume which

subsequently lead to a higher CO_2 and CH_4 diffusion coefficient through the membrane. The diffusivity coefficients of CO_2 of each membrane in the Henry and Langmuir sites have been calculated according to the dual mode sorption and the immobilization models. The results reveal a higher mobility of the Langmuirâ€™s population by Henryâ€™s population in the PPSU compared to the others. These results are in accordance with the Î²-transition results that have been obtained via DMA analyses.

494170 Hydrophilic and Morphological Modification of Polyethersulfone Substrates for Composite Membranes in CO_2 Separation

Abstract:

Pilot-scale nanoporous polyethersulfone (PES) substrates were fabricated by vapor- and non-solvent-induced phase inversion steps successively. N-methyl-2-pyrrolidone (NMP) and water were employed as solvent and non-solvent, respectively. 2-Methoxyethanol (2-ME) was incorporated into the casting solution as a pore-forming additive because of its high affinity with water, leading to an interconnected spongy structure in the bulk of the substrate. The 14-inch wide PES substrates were fabricated successfully by using a continuous roll-to-roll casting machine. Two hydrophilic polymers, hydroxylated PES (PES-OH) and polyvinylpyrrolidone (PVP), were incorporated for modifying the pristine substrate both hydrophilically and morphologically. The surface morphologies and gas permeation resistances of the PES membranes were characterized by scanning electron microscopy (SEM) and a gas permeation unit, respectively. The pristine PES substrate, with a pore size of 41 nm and a surface porosity of 12%, showed a CO_2 permeance of about 9200 GPU (1 GPU = 10^{-6} cm^3 (STP)·cm^{-2}·s^{-1}·$cmHg^{-1}$) and a CO_2/N_2 selectivity ~ 0.9 (Knudsen diffusion). The addition of hydrophilic polymers reduced the gas permeation resistances of PES substrates because of the more open morphology. By incorporating appropriate amounts of PES-OH and PVP in the casting solutions, modified PES substrates showed CO_2 permeances of 14100 GPU and 21000 GPU, respectively, indicating a significant reduction of gas permeation resistance. The developed PES substrates were applied for the preparation of composite membranes for CO_2 separation. Amine-containing selective layers, with a thickness of approximately 175 nm, were coated on the developed pristine PES substrate, PES-OH modified substrate, and PVP-modified substrate to prepare 3 composite membranes, respectively. From the gas transport measurements by using a simulated flue gas, these 3 composite membranes showed CO_2 permeances of 785 GPU, 836 GPU, and 843 GPU, and CO_2/N_2 selectivities of 150, 148, and 160, respectively. Based on the resistance-in-series model, the separation performance improvements could be attributed to the reduced mass transfer resistances of the modified substrates. In addition, a thinner selective layer can be coated on the hydrophilic polymer modified PES substrates due to the enhanced hydrophilicity, which can further improve the separation performances of the composite membranes in the future. The developed fabrication process has demonstrated the potential for the mass production of the suitable PES substrate for composite membranes in CO_2 separation.

500273 Comparison between Perfluoropolymers and Hydrocarbon Polymer Analogs with Pendant Rings for Membrane Gas Separation

Abstract:

Amorphous glassy perfluoropolymers are attractive for membrane gas separation, such as He/CH_4 and CO_2/CH_4 separation. These perfluoropolymers often have pendant rings to suppress crystallinity and achieve high free volume and thus high gas permeability. However, their applications are hampered by the high cost due to the difficulty in synthesis. We hypothesize that the pendant rings can be incorporated in hydrocarbon polymers to achieve similarly superior gas separation properties. In this study, we design and synthesize hydrocarbon polymers with structures similar to perfluoropolymers, aiming to elucidate the effect of pendant rings on gas separation properties. Specifically, we synthesize a perfluoropolymer of poly(perfluoro-4-methyl-2-methylene-1,3-dioxolane) (PPMMD) and its hydrocarbon polymer analogue, poly(4-methyl-2-methylene-1,3-dioxolane) (PMMD). These polymers with pendant rings are thoroughly characterized using FTIR, NMR, GPC, WAXD, TGA and DSC for physical properties such as molecular weight, crystallinity, and glass transition temperature. PPMMD shows a H_2 permeability of 236 Barrers and H_2/CO_2 selectivity of 4.1 at 35 ^oC, while the PMMD shows a H_2 permeability of 9.6 Barrers and H_2/CO_2 selectivity of 7.3 at 35 ^oC. Gas sorption results will be described using the dual mode sorption model. A series of poly(cyclic ketene acetals) with structures similar to PMMD are synthesized and characterized for gas separation. This presentation will also discuss the fractional free volume in these polymers and the use of a free volume model to describe the gas transport properties in these hydrocarbon polymers and perfluoropolymer analogs.

487948 Metal-Organic Frameworks Uio-66 and MIL-125 Nanoparticles Enhance the Performance of Thin Film Nanocomposite Membrane for Water Desalination

Abstract:

Two metal-organic frameworks (MOFs) UIO-66 and MIL-125 nanoparticles (NPs) were filled into thin film nanocomposite (TFN) membranes to increase their performance in reverse osmosis application. Membranes were prepared by the interfacial polymerization of 2% *m*-phenylenediamine (MPD) in aqueous solution and 0.15% trimesoyl chloride (TMC)-isooctane [1] solution. Polysulfone (PSU) sheets were prepared by phase inversion and employed to support the synthesized membranes. Different NPs weight ratios, 0.05, 0.1, 0.15, 0.2, and 0.3 %, were used to study the impacts of these additives on membrane efficiency. The as-prepared membranes and NPs were characterized using different techniques for better understanding of the observations. At the optimum addition, the water flux was enhanced from 62.5 L/m^2 h to 74.9 L/m^2 h by adding 0.15% UIO-66 and to 85.0 L/m^2 h by adding 0.3% MIL-125. The NaCl rejection was not significantly affected by embedding UIO-66 and even slightly improved by MIL-125, always at â‰¥ 98.5% as tested at 2000 ppm salt concentration and 300 psi transmembrane pressure.

References

1- Mohammed Kadhom, Jun Yin, Baolin Deng, â€œA Thin Film Nanocomposite Membrane with MCM-41 Silica Nanoparticles for Brackish Water Purificationâ€, *Membranes*, 6, 50, (2016) 1-12.

492605 The Time Evolution of Fouling Development upon Design Aspects of Direct-Flow Filtration

Abstract:

Key words: hollow fibre membrane, direct-flow filtration, packing density, membrane fouling, filtration resistance

Current membrane manufactures are prone to particle removal and membrane cleaning technologies because of the very challenge fouling issue that they are trying to address. It is that fouling decreases the filtration flux over time during constant pressure filtration and contrastively increases the transmembrane pressure difference (TMP) during constant flux filtration, which both restrain the productivity of the filtration process. In literature, most fouling studies have been directed to the constant pressure cross-flow filtration performances, however filtration operations with constant fluxes are more preferred in practice. This study explores the time evolution of fouling development upon design parameters of constant flux direct-flow filtration (DFF), which permits the use of smaller diameter fibres and generates less tangential flow compared with classical cross-flow filtration (CFF). DFF is usually achieved by capping one end of a CFF hollow fibre module to maintain no outlet flow from the module.

Fouling formation and its effect on filtration are influenced by various parameters, such as filtration resistance, backwash cleaning frequencies and module design parameters, out of which resistance is mostly emphasised. As reported in the literature, the rate of change of resistance is proportional to the instantaneous filtration resistance raised to a power, which varies from filtration to filtration. Hence, our investigation started with a DFF model developed by dividing the module into thousands of sections along the axial distance, which allows us to adapt a positional-changing resistance for each filtration circle. The model was validated by comparing the results with the mathematical model resulting from solution of the Navier-Stokes equations, and one group of results are given in Fig.1, Fig.2 and Fig.3. Whilst there is little evolution in the TMP profiles (Fig.1), the evolution of the resistance profile (Fig.2) results in the changing profiles of local fluxes (Fig.3). These results reflect a particular fouling model and will be contrasted with two other fouling models in the presentation. The models will be manipulated with module design parameters aiming to explain the fouling mechanism and approach for optimization of DFF.

Fig. 1. The transmembrane pressure differences (TMP) upon the operating circles versus axial distance at constant flux 70 LMH. The module had a length of 1.6m and 55% packing density of 1.3mm outer diameter fibres.

Fig. 2. The local resistance in log scale upon the operating circles versus axial distance at constant flux 70 LMH. The module had a length of 1.6m and 55% packing density of 1.3mm outer diameter fibres.

Fig. 3. The local flux upon the operating circles versus axial distance at constant flux 70 LMH. The module had a length of 1.6m and 55% packing density of 1.3mm outer diameter fibres.

495089 Overcoming Operating Pressure Barrier in High Recovery Membrane Desalination Via Hybrid RO-NF Processes

Abstract:

In the application of reverse osmosis (RO) membrane technologies for inland brackish water desalination, achieving high product water recovery (up to 90 % or higher) is critical. Even a small enhancement in water recovery has significant impacts on concentrate volume reduction, which in turn directly affect the capacity requirements and thus overall costs of concentrate management and disposal solutions. RO water recovery in inland water desalination is often limited due to two major challenges: a) membrane mineral scaling, and b) mechanical limit of operating pressure. The challenge of membrane mineral scaling can be overcome via antiscalant treatment, as well as the integration of intermediate concentrate demineralization technologies to remove mineral scale precursors from primary RO concentrate. Given that osmotic pressure rises rapidly with water recovery, mechanical limit of operating pressure in membrane systems imposes an upper limit to product water recovery, in addition to the high capital cost of high pressure pumps. New methods are needed for addressing such pressure-imposed water recovery limit in extending the operating range (and thus applicability) of pressure-driven membrane technologies. The present study addresses the issue of concentrate volume reduction in inland brackish water desalination via high recovery hybrid RO-NF process integration. The hybrid RO-NF processes can operate at higher water recovery levels relative to conventional RO processes, while keeping the required operating pressures within the limits imposed by process components (e.g., pumps, membrane elements, and pressure vessels). For example, process analysis indicates that a two-stage hybrid RO-NF process for desalting inland brackish water (e.g., subsurface agricultural drainage water, mine drainage water, produced water, etc.) at up to 90% water recovery can have up to 54% lower operating pressure (e.g., 700 psig vs. 1,500 psig for 15,000 mg/L feed TDS) relative to a conventional two-stage RO process. However, in order to maintain high product water quality, the reduction in the required applied pressure necessitates recycling of NF permeate to RO feed, which increases the required volumetric feed capacity of pumping (and thus energy consumption). Notwithstanding, the above increase in energy requirement remains modest (i.e., 17% higher than a two-stage RO process). Based on process modeling, analysis, and proof-of-concept experimental data, a set of fundamental principles for process design and optimization of hybrid RO-NF processes will be presented with the aim of demonstrating the promising applicability of pressure-driven membrane technologies for desalination of highly saline waters.

495868 Development and Performance of a Silver Nanoparticle Impregnated Reusable Anti-Biofouling Membrane in a Continuous Cross Flow Membrane Module

Abstract:

It is well-known that biofouling decreases membrane-life significantly with simultaneous decrease in permeate flux. To resolve this issue, silver nanoparticle (Ag-NP) - as a biocidal material - has been impregnated in a polyethersulfone (PES) membrane; the latter being typically used in water treatment applications. The main challenges in using such a membrane would be to: (i) prevent membrane biofouling, (ii) achieve zero bacterial cell (*E. coli, MTCC 1302*) count in the permeate water, (iii) maintain silver (Ag) concentration in the permeate stream within the permissible limit of drinking water and (iv) obtain a high tensile strength of the membrane for continuous performance. Addressing these factors would ensure a long and productive service-life of the membrane.

Therefore, the goal of the present research was to develop and test the anti-biofouling efficacy of an Ag-NP incorporated, sulfonated, polyethersulfone (SPES) membrane and reuse it over multiple cycles. We call it as the Ag-SPES membrane and tested it in a laboratory-scale cross-flow membrane module. To this end, the PES membrane having an overall thickness of 150 μm was first sulfonated and then Ag-NPs were selectively impregnated only on the upper, external surface (having a thickness of 1.74 μm) of the membrane by UV-reduction. This leaves the bulk of the membrane and its lower surface free of any Ag-NP. Cross-section FEG-SEM images of the Ag-SPES membrane confirm the selective presence of Ag-NPs only on the upper surface of the otherwise thicker (150 μm) membrane. This ensures better anti-biofouling performance with lower Ag loading, since most of the Ag on the upper surface can come into contact with the *E. coli* in the inlet, contaminated water; whereas there is no extra Ag in the bulk or lower surface of the membrane, which otherwise does not come into contact with *E. coli*, and is hence of no consequence. Such selective Ag-NP impregnation as a thin layer in the post-membrane formation stage, ensures superior anti-biofouling performance with a much lower amount of Ag itself.

Now, in actual operation, contaminated, inlet water contains a low concentration of microorganisms (like *E. coli*), which overt time attaches to the membrane surface and slowly leads to biofouling after a somewhat long duration. However in current experiments, a very high *E. coli* cell concentration (10^4 CFU/ml of *E. coli*) was purposefully used for reducing the timescale of biofouling. This facilitates an easy check on anti-biofouling efficacy of the Ag-SPES membrane.

To this end, *E. coli* cell suspension at 10^4 CFU/ml of *E. coli* concentration was dispersed in PBS (phosphate saline buffer) and was passed through both PES (as control) and Ag-SPES (as test) membranes. However, *E. coli* cells having diameter 0.8-1 μm and length 1.5-2.5 μm are much larger than the 200 nm pore diameter of the membrane. Therefore, cells do not pass through the pores and instead attaches on the membrane surface. This was confirmed by both confocal microscopy and cryo-FEG-SEM images of PES membrane used in the membrane module, showing presence of only live bacterial cells on the outer surface of the membrane. This leads to biofouling and hence as expected, PES membrane showed a continuous and significant decrease in permeate water flow rate, from 64.5 to 12.4 ml/min, after the passage of 240 L of water (in 4 h). In contrast, due to the presence of Ag-NPs on the upper surface of Ag-SPES membranes, cells attached on this membrane are completely killed, showing only a marginal decrease in permeate flow rate (from 58.3 to 50.2 ml/min in 4 h).

In addition, a very low Ag concentration (0.01 mg/L) was also maintained in the permeate stream, which is well below the recommended limit of 0.1 mg/L (WHO) for use as drinking water. Finally, due to sulfonation and impregnation of Ag-NPs, the PES membrane showed an increase in maximum tensile strength from 2.78 MPa to 3.92 MPa, which is beneficial from a mechanical standpoint of membrane integrity in a water treatment process.

Therefore, we could achieve all the four targets set out by us in the beginning. Furthermore, we also found that the Ag-SPES membrane can be synthesized with different, controlled amounts of Ag loading (e. g. 4, 8.8 and 15 wt.% Ag), by tuning the degree of sulfonation of the Ag-SPES membrane. On capturing time-dependent confocal microscopy (using SYTO-9 and propidium iodide) images it turns out that the Ag-SPES membrane with 8.8 wt.% Ag loading (used in the previous results of this study) shows maximum permeability and best anti-biofouling performance. In contrast, the membrane with 4 wt.% Ag is sub-optimum and hence takes a longer time to kill the attached *E. coli* cells; whereas the membrane with 15 wt.% Ag shows a significant decrease in permeability due to pore blockage from excess Ag.

During long usage over repeated cycles, eventually the Ag content in the membrane may decrease, which might affect the anti-biofouling performance of the membrane. To simulate this condition, the Ag-NPs were purposefully and completely removed from the membrane by acid wash. Subsequently, the membranes were again reloaded with the same amount of Ag loading. It was observed that this does not affect the membrane performance. For example, Ag-SPES membrane reloaded with the optimum 8.8 wt.% Ag (after 4 cycles of operation and reloading), shows an almost identical permeate flow rate of 49.71 ml/min after 4 h, compared to 50.2 ml/min (after 4 h) obtained for the fresh, new membrane.

Thus this work demonstrates the viability of the reusable, anti-biofouling, Ag-SPES membrane, to produce decontaminated water over a long period. This was achieved without compromising the mechanical strength of the membrane and maintaining a permissible Ag concentration in the treated water, which is safe for potable purposes. Additionally, after certain periods of operation, Ag-NP could be selectively impregnated as a thin layer on the upper surface of the membrane repeatedly, gaining back the same degree of performance in continued operation.

497918 Performance of a Cost-Effective Micro-Scale Biochar/Psf Mixed Matrix Hollow Fiber Membrane in the Treatment of Methylene Blue Simulated Wastewater

Abstract:

Colorful dye containing water has increasingly become an important issue with the increasing public concerns on environmental protection. A micro-scale biochar particles/PSF mixed matrix hollow fiber membrane (MMM) was successfully prepared for the removal of methylene blue from water. The static and dynamic adsorption of methylene blue by MMM was investigated. The results revealed that the MMM exhibited a high removal efficiency of methylene blue under a wide pH range of 4 to 10. The adsorption process on biochar and MMM obeyed the intraparticle surface diffusion model and Langmuir isotherm model. At neutral pH, the maximum adsorption capacity was for 544.459 mg/g for biochar and 165.808 mg/g for MMM, respectively. Regeneration could be achieved by using 1 M NaCl in 90 % ethanol aqueous solution, with the desorption rate above 92 %.Furthermore, the MMM displayed good performance in treating simulated methylene blue wastewater through continuous filtration. More importantly, the MMM showed an excellent reusability for methylene blue removal, achieving81 % of the permeate yield of the fresh MMM after 3 regeneration cycles. Finally, the adsorption mechanism studies indicated that the adsorption of methylene blue was associated with electrostatic interaction, hydrogen bonding as well as hydrophobic interaction.

499081 Treating Poultry Processing Wastewaters By Ultrafiltration

Abstract:

Commercial poultry processors use an average of 26.5 L of water per bird during processing of live birds to meat. It has been estimated that in 2007, US poultry processors slaughtered 8.9×10^9 broilers, producing 16.0×10^6 metric tons of ready-to-eat meat. Poultry processing water (PPW) is an example of an oily wastewater. These wastewaters contain large amounts of proteins, fats and carbohydrates from meat blood, skin and feathers. Before discharge, the majority of the soluble and organic particulate matter must be removed.

Use of membrane filtration to treat PPW is attractive as one can avoid the use of chemical additives as are often used in processes such as dissolved air flotation. In addition, appropriately selected membrane pore sizes can provide an absolute barrier to pathogens. In this study different ultrafiltration membranes were evaluated for treatment of PPW. Pretreatment of the feed wastewater is critical in order to suppress membrane fouling. Here pretreatment using electrocoagulation has been investigated.

Our results indicate that electrocoagulation followed by sedimentation could be a viable pretreatment technology prior to ultrafiltration. The evolution of the floc size distribution after sedimentation has been determined using laser diffraction particle sizing. Optimizing the sedimentation time is critical when developing a viable process. In this work we have investigated different sedimentation times. Flux decline, and rejection of a number of different ultrafiltration membranes in tangential flow filtration mode have been investigated. The BOD, COD, TDS in the feed, retentate and permeate have been determined. In addition membrane surface properties have been characterized using scanning electron microscopy, contact angle and zeta potential measurements. Membrane performance has been related to membrane surface properties. Finally the membranes were systematically cleaned and regenerated between runs. Our results indicate that while ultrafiltration is feasible for treating PPW.

500862 Ceramic Hollow Fiber Supported Metal-Organic Framework Uio-66 for Enhanced Adsorptive Separations

Abstract:

Ceramic Hollow Fiber Supported Metal-Organic Framework Uio-66 for Enhanced Adsorptive Separations

Chenghong Wang,[a, b] Dr. Melanie Lee,[a] Dr. Xinlei Liu,[a] Dr. Bo Wang,[a]

Professor J. Paul Chen,[b] and Professor Kang Li*[a]

[a] The Barrer Center, Department of Chemical Engineering, Imperial College London, London SW7 2AZ, United Kingdom

[b] NUS Graduate School for Integrative Sciences & Engineering (NGS), Department of Civil and Environmental Engineering, National University of Singapore, Singapore, 117456, Singapore

ABSTRACT

A novel ceramic hollow fiber was prepared through controllable spinning and sintering to form a two-distinct-layer geometry: one very thin barrier layer containing 3D-pore network structure at the lumen, and one unique layer containing a plurality of conical micro-channels towards the shell side.[1] These micro-channels not only reduce the mass transfer resistance as a transport network, giving rise to competitive permeation fluxes, but also form pockets within which functional materials can be readily deposited.

Recently, metal-organic frameworks (MOFs), a new class of porous materials, has exhibited a great potential in various applications, owing to their unique properties like exceptionally high porosity, large surface area and customizable chemical functionality.[2] One of the representative examples is UiO-66, a zirconium-based MOF, which has shown the best adsorption capacity for arsenic removal from wastewater.[3] However, the current discussion regarding MOF adsorbent is majorly based on the particle samples. In order to be industrially applicable, the powder materials have to be specifically shaped or supported, since in practical dispersed particles would easily leak through the application compartment, resulting in challenging spent-particles-separation issues and severe safety concerns.[4]

Herein, we developed and optimized a functional adsorptive membrane by incorporating UiO-66 into the ceramic hollow fiber. The fast adsorption kinetics provided by UiO-66 allows an efficient adsorptive separation process to be realized within the micro-channels, and the thin barrier layer can prevent any loss of the active adsorbents from leaking out. As a proof of concept, this adsorptive membrane was found to be capable of effectively remediating

arsenic contaminated water and producing potable water recovery (less than 10 ppb – required by World Health Organization). In comparison, to achieve a similar performance, the conventional adsorption packing column required eight times amount of active UiO-66 adsorbents. Hence, this adsorptive membrane has exhibited a promising potential to be applied for industrial wastewater decontamination. Looking forward, different functional membranes can be prepared based on specific adsorptive applications, as a wide range of adsorbents can be loaded into this novel ceramic hollow fiber, of which the micro-channel size and barrier layer pore size can be easily manipulated during fabrication.

Figure 1. Schematic diagram of arsenic contaminated water remediation by the adsorptive membrane. The inset demonstrates an enlarged cross-sectional view of the adsorptive membrane. Blue molecule: water; green molecule: arsenic pollutant.

KEYWORDS

MOF, ceramic hollow fiber, adsorption, arsenic, water remediation

REFERENCES

[1] M. Lee, B. Wang and K. Li, *J. Membr. Sci.*, 2016, **503**, 48.

[2] H. Furukawa, K. E. Cordova, M. O'Keeffe and O. M. Yaghi, *Science*, 2013, **341**, 1230444.

[3] C. Wang, X. Liu, J. P. Chen and K. Li, *Sci Rep*, 2015, **5**, 16613.

[4] J. He, T. S. Siah and J. Paul Chen, *Water Res.*, 2014, **56**, 88.

ACKNOWLEDGEMENTS

This work was supported by the research funding provided by Engineering and Physical Sciences Research Council in the United Kingdom (Grant no. EPSRC, EP/J014974/1). C. Wang gratefully acknowledges the National University of Singapore Graduate School for Integrative Sciences and Engineering scholarship.

489582 Apparatus Design for Liquid-Liquid Extraction Combined with Heterogeneously Catalyzed Esterification

Abstract:

In biorefinery processes downstream processing often suffers from poor economics. Target constituents do not turn up in broths in appropriate concentration for simple separation technologies. For instance, black liquor evaporation condensate from pulping contains carboxylic acids at low concentration. Isolation of these constituents from the aqueous effluent fails because economic feasibility cannot be achieved by state of the art downstream processes. The gap between expectations and needs may be solved by combining simple solvent extraction with proper chosen chemical conversion by esterification due to changes in the substance properties. The esterification of carboxylic acids is a well investigated chemical reaction, and esters have a wide applicability. Unfavorably esterification reactions are slow chemical equilibrium reactions and catalytic acceleration is recommended. High water content of the reaction broth shifts the chemical equilibrium composition to the reactant side. Via liquid-liquid extraction the carboxylic acids can be transferred into the solvent phase prior to or after esterification, and the limitation of conversion because of high water load is avoided. Due to the limited solubility in the solvent phase the byproduct water is transferred to the aqueous phase. Conversion of the carboxylic acids is either accelerated by catalysis of esterification in the carrier phase or in the solvent phase.

However, the application of reactive separation processes accelerated with heterogeneous catalysis does induce new demands to apparatus design, when implemented in a continuous process. The Taylor-Couette Disc Contactor (TCDC), a hybrid of the Rotating Disc Contactor (RDC) and the Taylor-Couette Reactor (TCR) satisfies the requirements necessary for intensive liquid-liquid-solid phase contact. In operation the main advantage of the TCDC over the RDC is, that the TCDC does not need stator rings to provide sufficient phase contact. Accumulation of the catalyst in dead zones of stator rings along the reactor can therefore be avoided. Shaft and rotor discs induce a similar flow pattern to banded two phase flow in Taylor-Couette reactors. Banded two phase flow provides proper mixing and sufficient residence time for solid catalysts in the rotor compartments.

Several experiments were carried out to quantify liquid-liquid flow in presence of a solid phase in the TCDC. Beyond the critical rotational speed the catalyst is dragged by the toroidal vortexes. Even in the presence of the dispersed solvent phase the catalyst particles remain within the single compartment for a reasonable residence time. The results confirm applicability of the three-phase reactor TCDC in intensification of mass transfer unit operations by chemical conversion of phase transfer constituents.

497171 Recovery of Both Low and High Molecular Weight Lignin Fractions By Extraction with Hot Aqueous Organic Solvent Systems

Abstract:

The current technologies being used to recover lignins from the black liquor by-product streams of both pulp-and-paper mills and lignocellulosic biorefineries all generate lignins that are "low-ash" compared to the parent black liquors, which are 50% ash and 20% Na. These so-called Kraft or alkali lignins have been proposed for low- and medium-value materials applications such as the partial substitution of phenol in phenol–formaldehyde resins and of polyols in polyurethane. However, for many higher-value applications, this level of metal content renders the lignin unfit for use. For example, if lignin is to be considered for applications such as high-strength carbon fibers and electrodeposited coatings, it must be much cleaner or "ultrapure", with a total metals content of no more than 100-150 ppm. Furthermore, the molecular weight (MW) of the lignin should be controllable within well-defined limits.

The Aqueous Lignin Purification using Hot Acids (ALPHA) process was invented with this requirement in mind, as ALPHA uses a biorenewable solvent system for simultaneously cleaning and fractionating Kraft lignins. Lignins of controlled MW with a total metals content of less than 100 ppm (and <20 ppm Na) can be obtained in only 1-2 extraction steps. The essence of the discovery is that when Kraft lignins are combined with hot acetic acid–water mixtures at the appropriate conditions, two equilibrium liquid phases are formed, with the metal salts being extracted into the solvent phase and a purified, solvated lignin-rich liquid phase that is remarkably free of metals being obtained as the desired product. Very low metals content in the solvated lignin phase is achieved because the metals can easily diffuse out of the liquefied lignin phase and into the solvent phase. With other extraction methods, the lignin is processed as a solid, so diffusion is far slower.

In this work, the fractionation of lignin into low and high MW fractions was investigated. High MW fractions are of interest as precursors for carbon fibers, and low MW fractions are being considered for coatings applications. In both cases, the lignins must also be "ultrapure", with a Na content less than 50 and total metals content less than 100 ppm being present. Two-three stages were used in ALPHA to obtain the desired lignin products. For the high MW lignin, an acetic acid/water (AA/H2O) solvent ratio of 50/50 was initially used to isolate a relatively pure fraction of lignin containing the highest 50% MW cut of the lignin. That fraction was then fractionated yet again with a 70/30 AA/H2O solvent mixture, and the resulting lignin contained less than 25 ppm Na and contained the highest 10% MW cut of the lignin. That lignin cut was then converted into carbon fibers, and the best fiber properties reported to date were obtained. In an analogous manner, ALPHA was operated as three stages so as to obtain a low MW cut of lignin. This time, a solvent composition rich in AA was started with, and water was then added to the solvent mixture to weaken the solvent power of the solvent mixture, precipitating out lower and lower MW cuts of lignin. One of the lowest MW cuts obtained in the aforementioned ALPHA setup contained less than 20 ppm Na and <100 ppm total metals, and is currently undergoing product testing in a coatings application.

493988 Reactive Extraction of Lactic Acid with Trioctylamine/Octanol/n-Undecane

Abstract:

The pulp and paper industry serves as backbone in the biorefinery. Black liquor from pulping contains, besides lignin which is mainly used for heat production, different carboxylic and hydroxycarboxylic acids in considerable amount.These acids are not recovered at present [1]. Especially formic acid, acetic acid and lactic acid are of interest, as they may serve as platform chemicals in the future.Isolation of lactic acid from industrial process streams is a promising way to cover the increasing demand of lactic acid. Due to its use as monomer for the poly lactic acid production lactic acid receives priority attention in the present work. The conventional lactic acid production and isolation from fermentation broths requiresa high amount of chemicals, andthe process generatesa high amount of unwanted, low value by-products.

Liquid-liquid extraction is a promising alternative separation technology to state of the art isolation of lactic acid [2]. The present work provides the basis for selective separation of lactic acid by using reactive liquid/liquid extraction. Phase equilibrium measurements were performed in dependence on the composition of the solvent phase starting with physical extraction with n-Undecane and Octanol, followed by reactive extraction with mixtures of Trioctylamine/Octanol/Undecane. Trioctylamine (TOA) is well known for its ability to extract carboxylic acids [3]. The modifier Octanol affects the distribution characteristic and, together with the diluent n-Undecane, it affects the physical properties of the solvent to avoid third phase formation at higher acid concentration. The optimized solvent phase containing 40w% TOA/40w% Octanol/20w% Undecane provides highly efficient extraction of lactic acid.The influence of the pH-value on distribution and hence on separation was considered. Back extraction was performed with $NaHCO_3$.

As mentioned, black liquor does contain several carboxylic acids. Isolation of lactic acid with amines must consider co-extraction of carboxylic acids. The effect of acetic acid and formic acid on yield and selectivity of lactic acid extraction was therefore investigated too.

The results underline the high potential of reactive liquid/liquid extraction for the recovery of carboxylic and hydroxycarboxylic acids from black liquor.

[1] E. Sjoestroem, "The behavior of wood polysaccharides during alkaline pulping processes," *Tappi*, vol. 60, no. 9, pp. 151–154, 1977.

[2] S. Basetty and B. R. V V, "Equilibrium Studies on Reactive Extraction of Lactic Acid using Tri- butyl amine in Chloroform," vol. 38, pp. 99–104, 2012.

[3] C. I. Sci, C. Engimeer, N. York, A. Review, F. Mechanics, A. Reviews, P. Alto, F. Mech, M. N. York, A. Sci, P. Press, N. York, A. Sci, and T. R. The, "Amine Extraction of Hydroxycarboxylic Acids . 1 . Extraction of Citric," pp. 1554–1562, 1992.

499956 Using Deep Eutectic Solvents (DESs) to Extract Lignin from Black Liquor

Abstract:

Keywords: Lignin, Black liquor, Deep Eutectic Solvents, Extraction

Corresponding author: J.A. Lalman, Telephone: +1 519 253 3000 ext 2519; Fax: +1 519 971 3686, lalman@uwindsor.ca

Abstract

Recent developments in renewable technologies to convert low-value agriculture residues into value-added products is rapidly emerging in Canada, the United States, the European Union and other nations (Stigsson, 2017; Arkell et al., 2014; Humpert et al., 2016). The global annual lignin production is estimated at 50 million tonnes with 1.1 million tonnes being used to produce lignosulphonates and the remaining used for energy production in pulping mills (Edgar, 2015). However, this data excludes lignin production from corn and wheat biorefineries. The demand for products produced from lignin and other renewable energy feedstocks such as cellulose and hemicellulose is expected to increase with depleting fossil fuel supplies (Edwards, 2000; Orr, 2006).

Utilizing cellulose and hemi-cellulose to produce products such as methane, ethanol, diesel and hydrogen has focused mainly on using non-woody and woody biomass (Lalman et al., 2017). In biorefineries such as in the pulp and paper industry, cellulose is used to manufacture paper products while in other similar industries such as sugar cane, the sugar component, a disaccharide, is extracted and converted into value-added materials such as consumer sugar products and ethanol. In the pulp and paper industry, the feedstock is mainly composed of cellulose, a glucose polymer, hemicellulose, a polymer consisting of xylose, mannose, glucose, galactose and lower levels of arabinose and rhamnose, plus lignin. In the sugar cane industry, the feedstock composition used to produce sugar products and ethanol is lignin plus sucrose, a disaccharide. The lignin component, designated as black liquor, is an aqueous soluble fraction produced in pulp and paper mills while in the sugar cane industry, the lignin component is a solid and labelled as bagasse. A common process employed by these industries is to add value by converting lignin into heat and energy by employing combustion as the terminal process (Lalman et al., 2017).

Black liquor, a waste stream from the pulp and paper industry, has a lignin content ranging between 29 – 54% (Humpert et al., 2016). The high calorific energy value of black liquor (Demirbas, 2003), based on the dry solid content, is a driving factor for considering it as an excellent candidate for upgrading into value-added chemicals. In the pulp and paper industry, approximately 66% of electricity and heat required to remove water from wood pulps is derived from black liquor combustion (Humpert et al., 2016). A few reports have examined converting lignin model compounds and black liquor photocatalytic by-products into value-added chemicals and electricity (Lalman and Ray, 2009; Shewa and Lalman, 2014). These researchers employed TiO_2 photocatalysis to produce short chain biodegradable carbon chemicals.

Lignin in black liquor can be recovered by employing precipitation under acid or alkaline conditions as well as membrane separation (Keyoumu et al., 2004; Jonsson and Wallberg, 2009). Membrane processes such as ultrafiltration is primarily used to recover lignosulphonates from sulphite pulp mill spent liquors while Kraft lignin is mainly recovered by precipitation. Major disadvantages during lignin precipitation include operating under elevated temperatures and pressures, using hazardous chemicals and plugging of the filter with a subsequent reduction in process efficiency (Tomani, 2010; Lake, 2011).

Recently, solvent extraction has been examined to recover lignin from a variety of renewable pretreated feedstocks. Deep eutectic solvents (DESs) are recognized as an alternative to traditional solvents and ionic liquids (Hou et al., 2008). DESs are generally composed of two or three components which are capable of associating with each other, through hydrogen bonding, to produce a eutectic mixture (Abbott et al., 2003). The resulting eutectic mixture has a melting point lower than that of each individual component. Desirable characteristics of DESs include low cost, high solute solubility, wide potential window, able to operate under low temperature conditions, and having negligible impact on the environment (Abbott et al, 2003). DESs are produced by mixing an organic salt plus a HBD (Yu et al., 2008).

Over the past few years, a number of studies have focused on using DESs to recover lignin from lignocellulosic biomass (Jablonsky et al., 2015; Kumar et al., 2016). These researchers reported delignification of rice straw and wheat straw using choline chloride (ChCl):Lactic acid (LA) (1:5) and ChCl:oxalic acid (OXA) (1:1) DESs and concluded they were able to recover approximately 60% of the lignin fraction. With recent developments in utilizing DESs to recover lignin from lignocellulosics, expanding and developing the technology to include black liquor is expected to add-value to lignin waste streams from pulping mills and other biorefineries. Hence, the objective of the current study is to extract lignin from black liquor using DESs.

Black liquor was provided by a pulp and paper mill located in Ontario, Canada. The study examined the impact of varying the black liquor to DES ratio, operating temperature, and the salt:HBD ratio on lignin recovery. The eutectic mixtures were prepared using methods reported by Ghareh Bagh et al. (2013, 2014, and 2015). The organic salt plus HBD were placed in a jacketed vessel and vigorously mixed at a specific temperature and atmospheric pressure to produce a homogenous colorless liquid. The list of DESs is presented in Table 1. ChCl was used as the salt and the selected HBDs included LA, OXA, malic acid (MA) and urea.

Table 1. DESs used in this study

DESs	**Salt**	**HBD**	**Ratio**	
ChCl:LA	ChCl	LA	1:9	1:10
ChCl:OXA.2H_2O		OXA·2H_2O	1:2	1:3
ChCl:MA		MA	1:1	1:2
ChCl:Urea		Urea	1:1	1:2

Notes: LA = lactic acid; OXA = Oxalic acid; MA = Malic acid

The DESs were synthesized at temperatures less than 100°C under atmospheric pressure. Black liquor plus DESs were mixed in a jacketed vessel at 300 rpm for 6 hours using a specified DES/black liquor ratio at 25, 60, 75, 90°C, respectively. Milli-Q (MQ) water was added to precipitate the lignin fraction. The MQ plus lignin mixtures were filtered and the recovered solids was analysed spectrophotometrically to determine the acid soluble lignin [TAPPI UM 250]. The insoluble lignin (Klason lignin) was determined gravimetrically using TAPPI T 222 om-02.

The temperature, black liquor:DES ratio, and DES utilized for extraction were optimized based on the mass fraction (percent) of lignin recovered. The optimization design was based on a one-factor approach. The lignin recovered and purity was a function of the temperature, type of DES and DES:black liquor ratio. The optimum was observed at 90°C for a 1:3 OXA:DES ratio and a 1:3 black liquor:DES ratio. An approximately 850 mg lignin/g of solid extracted and the recovery was approximately 80%. For the MA-DES and LA-DES combinations under same temperature and black liquor:DES ratio, the optimum recovery range was approximately 73% and 78% with 770 and 800 mg lignin/g solid extracted, respectively. However, at ambient, a 1:3 black liquor:DES ratio, the lignin recovery for ChCl:LA (1:9) was 60% and approximately 380 mg lignin/g solid extracted.

References

Abbott, A.P., Capper, G., Davies, D.L., Rasheed, R.K., Tambyrajah, V. (2003) Chemical Communications. 70–71.

Arkell, A., Olsson, J., Wallberg, O. (2014) Chemical Engineering Research and Design. 92, 1792-1800.

Demirbas, A. (2003) Energy Sources. 25(7), 629-635.

Edgar, C. 2015. Global Lignin Products Market-Segmented By Product Type, Source, Application, and Geography-Trends and Forecasts (2015-2020)-Reportlinker Review. http://www.prnewswire.com/news-releases/global-lignin-products-market-segmented-by-product-type-source-application-and-geography-trends-and-forecasts-2015-2020-reportlinker-review-300145371.html

Edwards, J. D. (2000) Part II conference, petroleum provinces of the 21st century, San Diego, CA.

Ghareh Bagh, F.S., Mjalli, F. S., Hashim, M. A., Hadj-Kali, M.K.O., AlNashef, I.M. (2013) Journal of Chemical and Engineering Data. 58, 2154-2162.

Ghareh Bagh, F.S., Hadj-Kali, M.K.O., Mjalli, F. S., Hashim, M. A., AlNashef, I.M. (2014) Journal of Molecular Liquids. 199, 344-351.

Ghareh Bagh, F. S., Shahbaz, K., Mjalli, F. S., Hashim, M. A., AlNashef, I.M. (2015) Journal of Molecular Liquids. 204, 76-83.

Hou, Y., Gu, Y., Zhang, S., Yang, F., Ding, H., Shan, Y. (2008) Journal of Molecular Liquids. 143, 154–159.

Humpert, D., Ebrahimi, M., Czermak, P., (2016) Membranes 6, 42-54.

Jablonskey, M., Skulcova, A., Kamenska, L., Vrska, M., Sima, J. (2015) BioResources. 10, 8039-8047.

Jönsson, A.S., Wallberg, O., (2009) Desalination. 237, 254–267.

Keyoumu, A., Sjödahl, R., Henriksson, G., Ek, M., Gellerstedt, G., Lindström, M. E. (2004) Industrial Crops and Products. 20, 143–150.

Kumar, A.K., Pharik, B. S., Pravakar, M. (2016) Environmental Science and Pollution Research. 23, 9265–9275.

Lake, M.A., Blackburn, J.C. (2011) International Patent Application. PCT/US2010/049773.

Lalman, J.A., Ray, S. (2009) Provisional Patent Application Serial Number 61/272,567.

Lalman, J.A., Shewa, W.A., Gallagher, J., Ravella, S. (2017) Biomass and Bioenergy *In* Quality living through chemurgy and green chemistry, Lau, P.C.K. (editor), Springer, New York, NY.

Orr, L. (2006) Global and Energy Project. Stanford University. (after John Edwards, American Association of Petroleum Geologists).

Shewa, W.A., Lalman, J.A. (2014) Proceedings of the Water Environment Federation. 20, 2484-2503.

Stigsson, L., Arkell, A., Olsson, J. (2017) EP3004455 A4.

TAPPI UM 250. (1991) TAPPI Useful Methods. Atlanta, GA, USA.

TAPPI T 222 om-02. (2002) TAPPI Test Methods. Atlanta, GA, USA.

Tomani, P. (2010) Cellulose Chemical Technology. 44 , 53-58.;

Yu, Y. H., Lu, X. M., Zhou, Q., Dong, K., Yao, H. W., Zhang. S. J. (2008) Chemistry - A European Journal. 14, 11174-11182.

499970 Efficient and Selective Recovery of Formic Acid and Acetic Acid from Their Mixed Solutions

Abstract:

Formic and acetic acids are among the low molecular weight carboxylic acids that have wide applications in various industries. They co-exist in the media of several biological productions and wastewater streams. Their high production amounts and costs make their low-cost and efficient recovery very significant and beneficial.

Selective recovery of carboxylic acids from their binary acid aqueous solutions with high efficiency is so important. One of the most advantageous methods for the purpose is reactive extraction, a modified version of traditional liquid-liquid extraction process. It is superior to various types of separation techniques with its high distribution coefficients (K_D), low energy requirements, thermal stability and ease of process. Despite there are several studies on reactive extraction in the literature, most of them were carried out using single acid synthetic solutions. On the other hand, several types of aqueous solutions contain carboxylic acids, e.g., biological production media and wastewater streams are multi-acid solutions.

In the present study, selective recovery of formic and acetic acid from their mixed solutions by reactive extraction was aimed. In order to have a successful operation, system variables should be operated at optimum values. Firstly, both acids were recovered from their single acid solutions using tertiary amines dissolved in 10 different types of organic diluents. These diluents or solvents were selected from different types of chemical classes. The K_D and extraction efficiency (E%) values were compared and most efficient solvents for the operation were determined. In addition, separation factor values were calculated using the K_D values of each acid obtained in their separate solutions. For both acid types, the highest K_D and E% values were attained with intermediate molecular weight alcohols. In addition, highest separation factors were also obtained with them, too. Furthermore, separation factors increased with the increase in initial concentration of the extractant.

In the second part of the experimental part, formic and acetic acids were recovered via reactive extraction from their binary acid solutions. Highest separation factors were obtained with 0.5 M TOA in alcohols while the lowest were with 0.1 M TOA level, which is consistent with the results, obtained using single acid solutions. The effect of initial pH on the selective recovery was also probed. The highest recovery rates were obtained with the solutions; pH was not adjusted (pH~2.2) and with the organic phases, 1-hexanol was used as the diluent. Moreover, effects of the acid ratio and total acid concentration (where the acid ratio is 1) were also investigated. Highest selective extraction efficiencies (separation factor~8) were obtained with total acid concentration of 0.5 M (0.25 M each acid in aqueous solution). According the results, selective recovery was not influenced from the acid ratio at this total acid level.

498325 Simulation and Comparative Evaluation of Different Methods of Microalgal Lipid Extraction and Conversion to Biodiesel Using Aspen HYSYS

Abstract:

Microalgal lipids are prized commodity for its tremendous potential to be used directly as a bio-diesel equivalent. Microalgal derived biofuels can be used as it is in modern day transportation vehicles, without much engine modification. Foreseeing the trend of mindless use of exhaustible fossil fuel reserves, increased demand and insufficient supply sources, it is imperative to devise efficient systems of bio-diesel production. Microalgae is one such green and relatively cost-effective option for biodiesel production. Biodiesel is essentially similar to diesel but with some biological characteristics. Better processing of biodiesel could give same value as that of diesel with less environmental damage, thus it should be explored as a viable alternative. However, the major bottleneck associated with production of biodiesel from microalgae at commercial scale lies in high costs associated with various unit operations of the microalgal bio-oil production such as cultivation, harvesting and downstream cost mainly in product recovery steps. There are several methods which have been used for the lipid extraction from microalgal biomss like: bead beating, homogenizing, sonication, solvent extraction, supercritical fluid extraction etc or simultaneous extraction and transesterification process for the conversion of microalgal lipids into fatty acid methyl esters or FAME (bio-diesel).

Generally, lipids are triglycerides which react with alcohol to produce ester by transesterification reaction in the presence of acid/base catalyst. Every method involves use of different types of solvent, physical condition (eg. temperature, pH etc) electricity consumption (cell disruption). The paper presents an overall overview of process modeling of different methods of lipid extraction from the microalgal biomass and its conversion into biodiesel using Aspen Plus software. Using Aspen, each unit operation, and flow stream, physical properties of each reactant, product and intermediate is run sequentially until convergence and careful mass and energy balance is designed. Every aspect of the process is taken into account such as cell disruption, extraction, transesterification etc and the final biodiesel yields compared from different processes. Additional, other properties such as cloud point, cetane number and heating value of the algae biodiesel was assessed and compared. The possibility of any by-product generation such as glycerol was considered.

The above discussed comparisons of technologies and processing conditions allow to select the best options, for each process stage from an environmental perspective. After the oil extraction and conversion, the economy results is dependent on the final use (and quality) of the microalgae oil free cake. It could be used as a substrate for biogas production or as animal feed, fertilizer etc. It was observed through Aspen modeling that rather than individual processes of oil extraction and conversion, simultaneous extraction and transesterification significantly affects overall performance of the process. The simulation results demonstrate promising results in terms of product recovery, yield, and product quality and could be used to select best choice for microalgal biodiesel production.

AIChE 2017
Annual Meeting

北京化工大学
Beijing University of Chemical Technology

Operation of Reactive Distillation Columns Disproportionating Trichlorosilane to Silane: Dynamic Impact of the number of Reactive Sections

Xinxiang Zang, Yang Yuan, Haisheng Chen, Liang Zhang, Shaofeng Wang and Kejin, Huang

College of Information Science and Technology
Beijing University of Chemical Technology, P.R. China

CONTENTS

1. Introduction

Case 1: A + B ↔ C + D
$\alpha_C > \alpha_A > \alpha_B > \alpha_D$

Case 2: A + B ↔ C + D
$\alpha_A > \alpha_C > \alpha_D > \alpha_B$

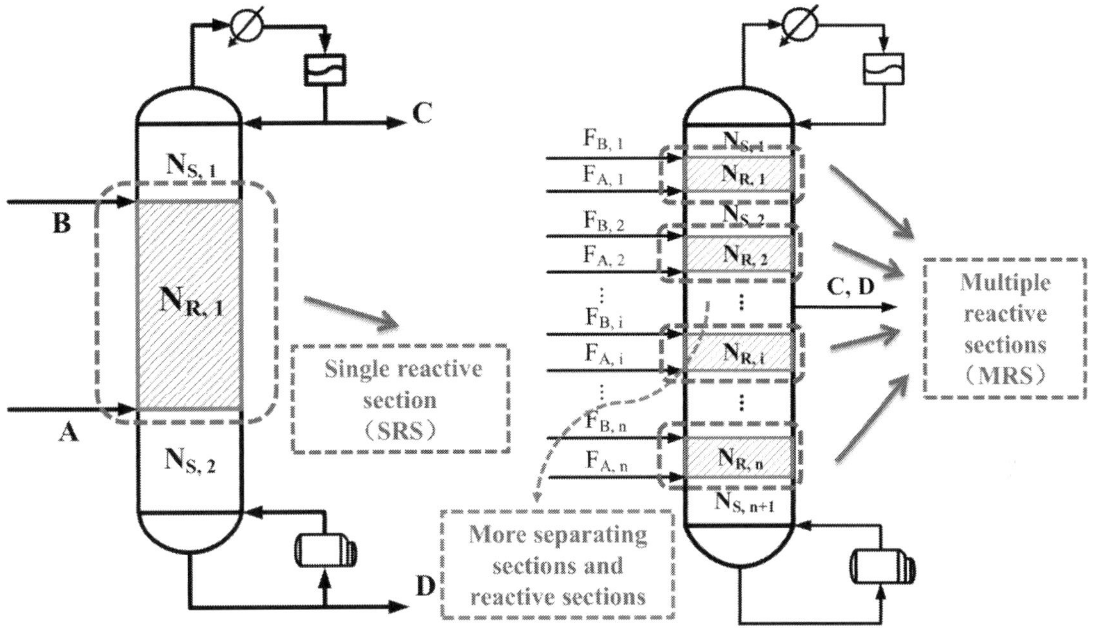

2. A Selected Reaction

A Typical Reaction

- Disproportionation of Trichlorosilane to Silane

- Reaction rates for the first-, second-, and third-stage reactions

$r_1 = k_1(x^2_{TCS} - xSTCxDCS/K_1)$

$r_2 = k_2(x^2_{DCS} - x_{TCS}x_{MCS}/K_2)$

$r_3 = k_3(x^2_{MCS} - x_{DCS}x_{SIL}/K_3)$

Three Consecutive Reversible Reactions

Bottom product

① 2TCS ↔ STC + DCS

$\Delta Hr = 6402$ kJ/kmol

② $2DCS$ ↔ TCS + MCS

$\Delta Hr = 2226$ kJ/kmol

③ $2MCS$ ↔ DCS + $silane$ → Top product

$\Delta Hr = -248$ kJ/kmol

Arrangement of relative volatility of components:

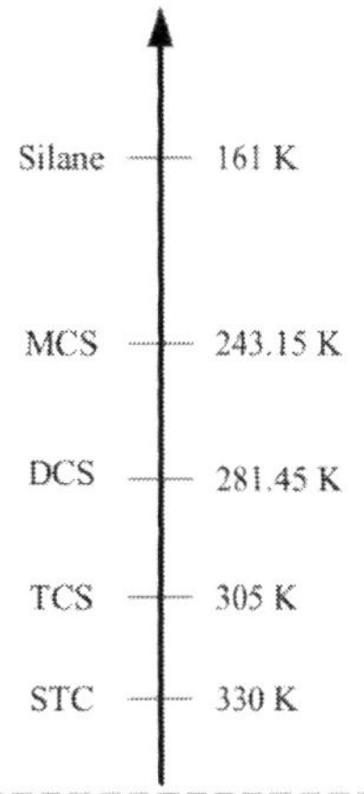

2. Design of RDC-SRS, RDC-DRS and RDC-TRS

Prerequisites for RDC-SRS, RDC-DRS and RDC-TRS

- Share the same total number of stages
- Share the same total number of reactive stages.
- Share the same total amount of catalyst employed.
- All the columns are operated at the same pressure profile.

2. Design of RDC-SRS, RDC-DRS and RDC-TRS

Optimal Design of RDC-SRS

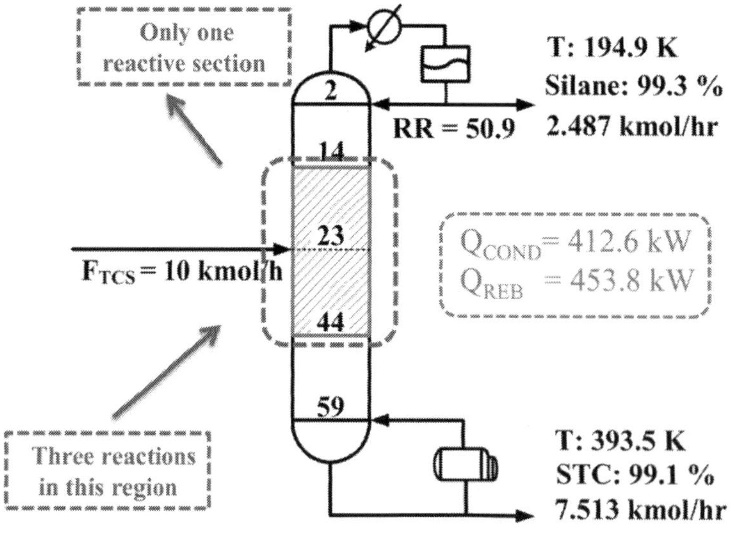

2. Design of RDC-SRS, RDC-DRS and RDC-TRS

Optimal Design of RDC-DRS

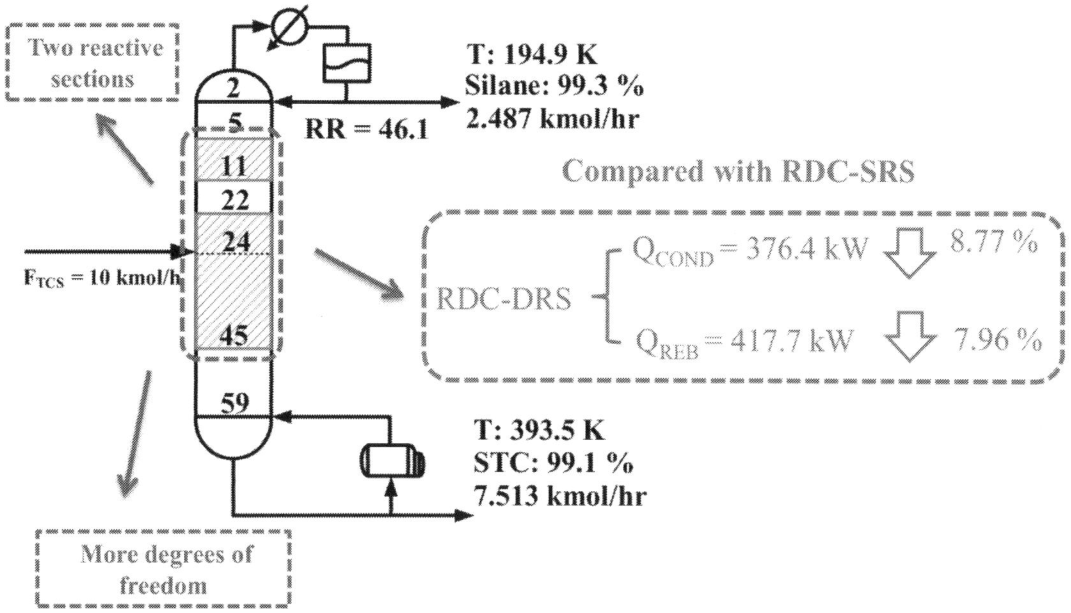

2. Design of RDC-SRS, RDC-DRS and RDC-TRS

Optimal Design of RDC-TRS

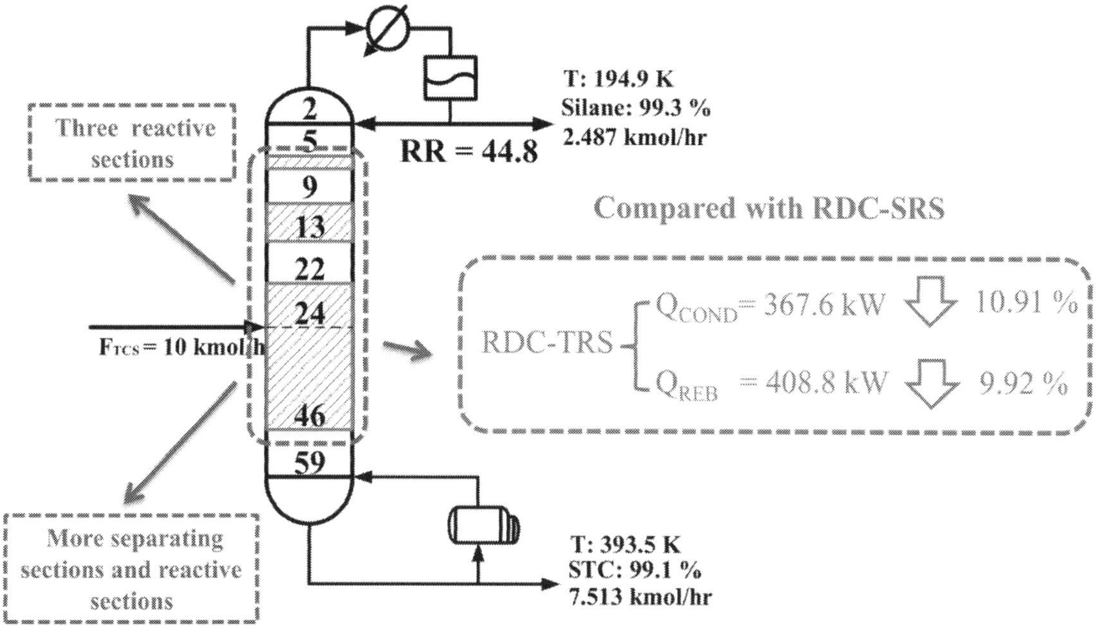

AN IMPORTANT FINDING

Arranging the multiple reactive sections brings more degrees of freedom

⇩

Interaction between reactions will be greatly alleviated

⇩

RDC-MRS can reduce the heat duties of condenser and reboiler greatly.

⇩

With the incremental arrangement of reactive sections, from SRS to TRS, the steady-state performance is getting better and better.

4. Dynamic Analysis of the RDC-SRS and RDC-MRS

Open-loop Controllability Evaluation

- Analysis of Two Regulation Paths

  ⇩

  RR – Top and Bottom Product Composition Path
  Q_{REB} – Top and Bottom Product Composition Path

- Analysis of the Disturbance Path

  ⇩

  Feed Flow Rate – Top and Bottom Product Composition Path

4. Dynamic Analysis : Open-loop Controllability Evaluation

Analysis of Two Regulation Paths:

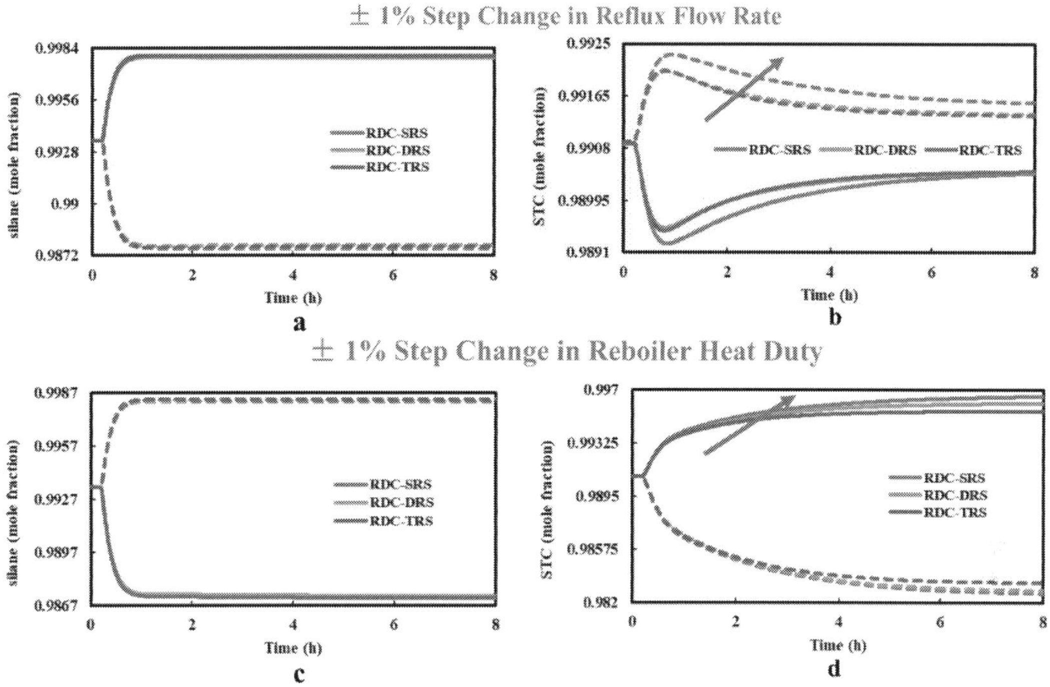

4. Dynamic Analysis: Open-loop Controllability Evaluation

Relative Gain Array (RGA) at 1 x 10^{-4} and 1 x 10^{4} rad/h

Process design		w = 1×10^{-4} rad/h		w = 1×10^{4} rad/h	
		X_{silane}	X_{STC}	X_{silane}	X_{STC}
RDC-SRS	RR	1.0945	-0.0945	1.0515	-0.0515
	Q_{REB}	-0.0945	1.0945	-0.0515	1.0515
RDC-DRS	RR	1.0856	-0.0856	1.0478	-0.0478
	Q_{REB}	-0.0856	1.0856	-0.0478	1.0478
RDC-TRS	RR	1.0923	-0.0923	1.0479	0.0479
	Q_{REB}	-0.0923	1.0923	-0.0479	1.0479

Implying **that** arranging the multiple reactive sections **presents essentially** favorable net effects **to process controllability in the regulation path.**

4. Dynamic Analysis: Open-loop Controllability Evaluation

Analysis of the Disturbance Paths:

± 1% Step Change in Feed Flow Rate

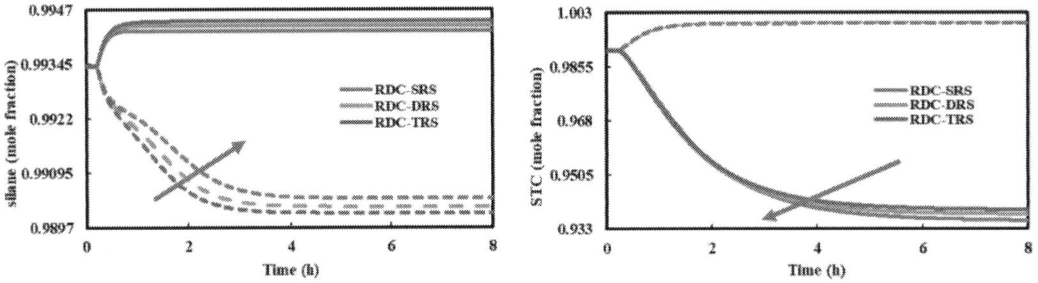

a

b

± 5% Step Change in Feed Composition (5% silane and 95 % TCS)

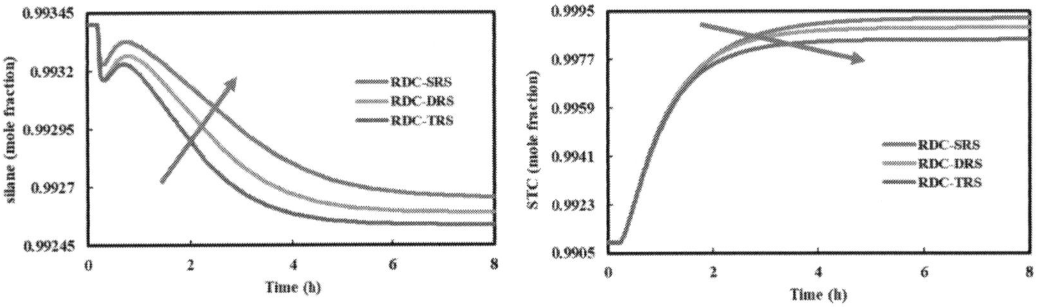

c

d

4. Dynamic Analysis: Closed-loop Controllability Evaluation

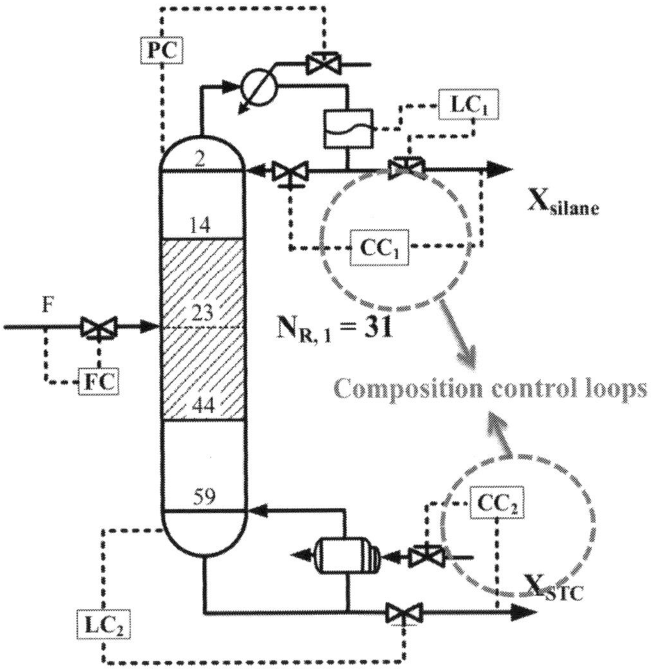

Control Scheme

- One flowrate control loop.
- Column pressure control loop.
- Two level control loops.
- One pressure control loop
- Two composition control loops.

Closed-loop Responses of RDC-SRS and RDC-MRS

± 5% Step Change in Feed Flow Rate

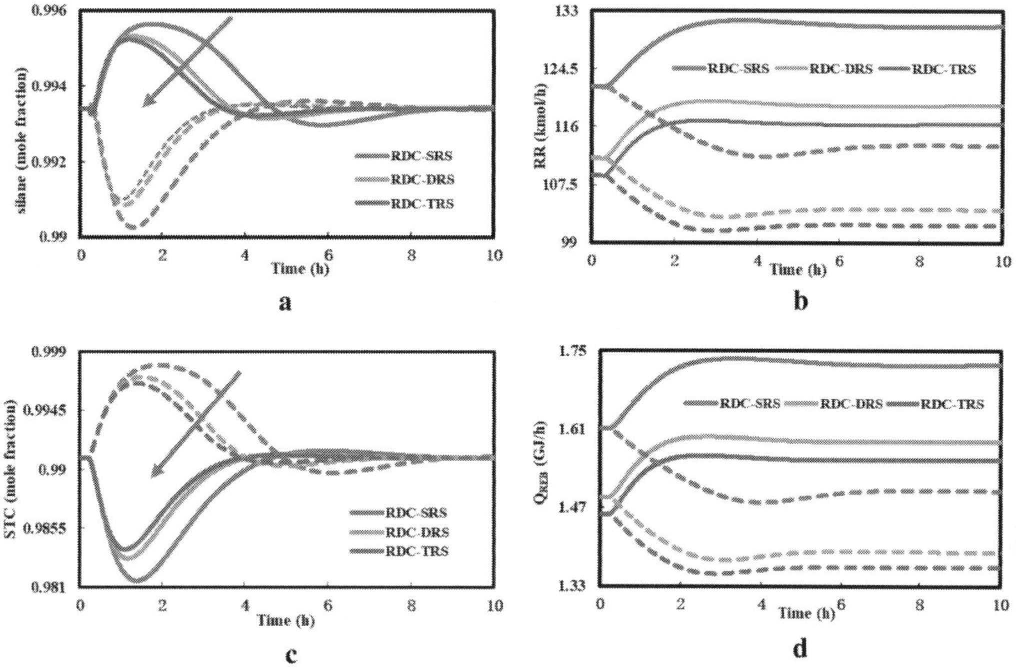

Closed-loop Response of RDC-SRS and RDC-MRS

± 5% Step Change in Feed Component (5% silane and 95% TCS)

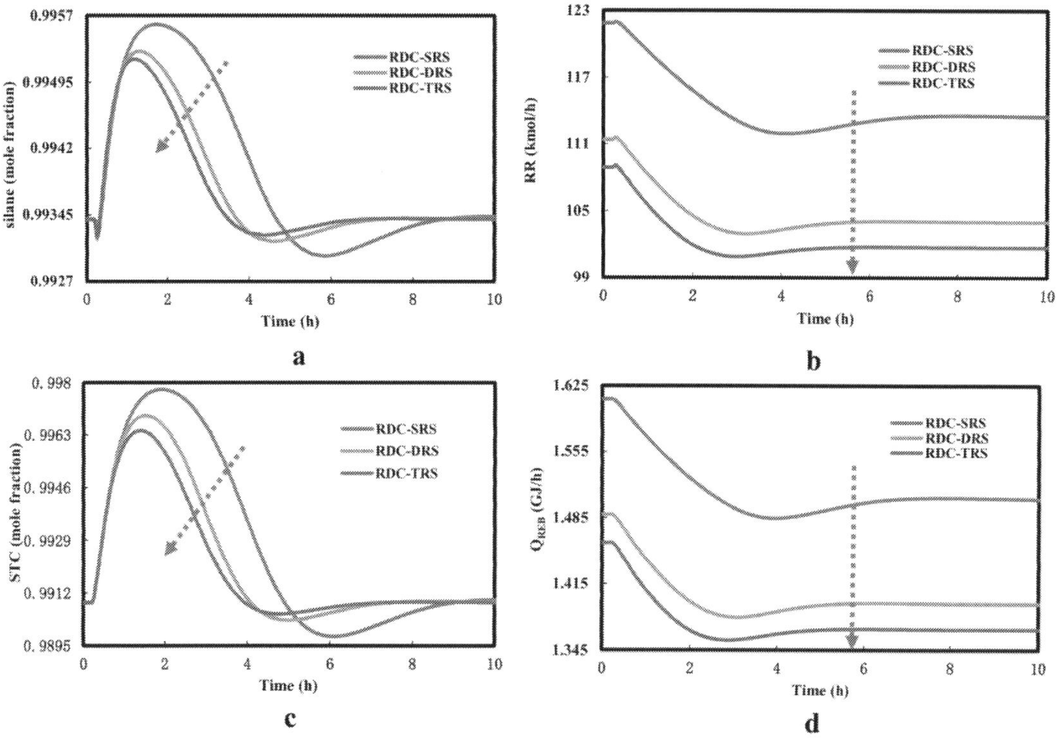

Closed-loop Responses of RDC-SRS and RDC-MRS

Servo Responses of the RDC-SRS, RDC-DRS and RDC-TRS to a ± 0.5% Set point Change in the Top Product Purity

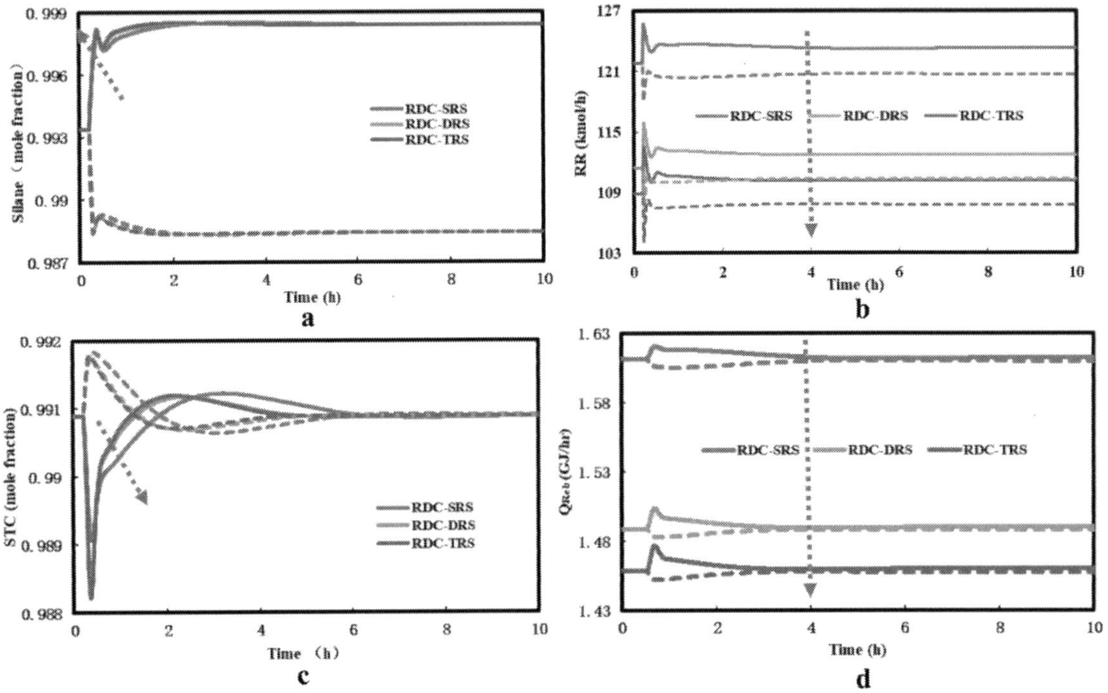

Closed-loop Responses of RDC-SRS and RDC-MRS

Servo Responses of the RDC-SRS, RDC-DRS and RDC-TRS to a ± 0.5% Set-point Change in the Bottom Product Purity

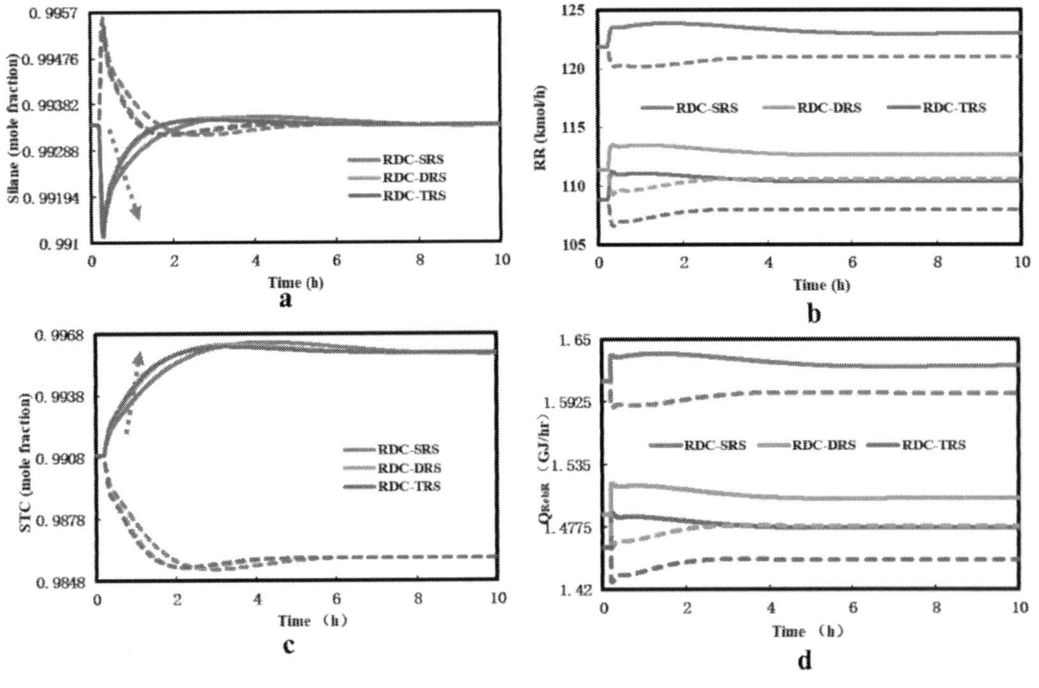

Summary

(1) A novel configuration, RDC-MRS, especially RDC-TRS, is proposed.

(2) A case study on the disproportionation of trichlorosilane to silane is studied to evaluate the steady-state performance and dynamic characteristics.

- Arranging multiple reactive sections **renders more degrees of freedom for process synthesis and design.**
- Arranging multiple reactive sections **reinforces the internal mass and energy integration**
- Arranging multiple reactive sections **reduces energy consumption and operating cost significantly.**
- **Compared with RDC-SRS,** RDC-MRS **has better performance in steady state and dynamic characteristics**
- RDC-MRS **should be considered to be great significant for design and control in separating complicated reacting mixture.**

Beijing University of Chemical Technology

Thanks for your attention !

493751 A Novel Potential Application of SiC Ceramic Foam Material to Distillation: Foam Monolithic Tray

Abstract:

Applications using foams material as distillation packing or tray have demonstrated higher performance than classical column internal except for the operation range. To overcome this disadvantage, a series of novel foam column internal have been developed, such as Foam Monolithic Tray, Structured Corrugation Foam Packing and so on. In this paper, the hydrodynamics performance and mass transfer efficiency of SiC foam monolithic tray is to be examined, with special emphasis focused on the effect of foam cell size and tray thickness. The hydraulic performance parameters and the mass transfer efficiency of the foam monolithic tray were measured in 600 mm i.d. and 300 mm i.d. diameter, respectively. The experimental values are compared with last trends and traditional column trays. In general, the comparison results showed that the hydrodynamics and mass transfer performances of Foam monolithic tray is acceptable as the mass transfer elements in the distillation column. The results also showed that the Foam monolithic tray has higher operation flexibility and significantly reduce the capital cost at industrial scale.

493330 Experimental and Simulation Studies of High-Temperature Ethane Dehydrogenation in Microporous Zeolite Membrane Reactor

Abstract:

Experimental and Simulation Studies of High-Temperature Ethane Dehydrogenation in Microporous Zeolite Membrane Reactor

*Shailesh Dangwal, Ruochen Liu, Seok-Jhin Kim**

School of Chemical Engineering, Oklahoma State University,

420 Engineering North, Stillwater, OK 74078, USA

ABSTRACT

Ethylene is an important chemical engineering ingredient and used in polymerization, oxidation, and alkylation. The increasing demand for ethylene has stimulated substantial research into the development of new processes to reduce energy consumption. Catalytic ethane dehydrogenation using a membrane reactor is an attractive solution because the cracking equilibrium can be shifted in favor of ethylene by selective removal of hydrogen. We reported the intensification of ethane dehydrogenation reaction in MFI zeolite membrane reactors operating with a Pt/Al_2O_3 catalyst. The effects of the reaction conditions and the membrane separation properties on the ethane conversion in the membrane reactor were investigated experimentally and by simulations using a simple one-dimensional model which was validated by the experimental data. It was demonstrated that the rate of H_2 membrane permeation significantly enhances the ethane conversion. MFI membrane reactors allow the equilibrium limit of ethane conversion to be surpassed at high temperatures. The model calculations have shown that near-completion ethane conversion (>99.5%) may be achieved under realistic operating temperature, pressure, space velocity and catalyst load even for membranes with moderate H_2 selectivity, ($aH_2/C_2H_4<10$) and H_2 permeance ($<1.8\times10^{-7}$ mol.s.m^2.Pa).

*Corresponding author: E-mail: seokjhin.kim@okstate.edu

501253 Experimental and Numerical Study of an Intensified Water-Gas Shift (WGS) Reaction Process Using a Membrane Reactor (MR)/Adsorptive Reactor (AR) Sequence

Abstract:

Reactive separation processes, such as membrane reactors (MRs) and adsorptive reactors (ARs), have been attracting increased interest for industrial applications. Such a hybrid system combining a membrane reactor (MR) and an adsorptive reactor (AR) in tandem (with the AR following the MR, and the MRâ€™s reject stream serving as the ARâ€™s feed) can produce an ultra-pure H_2 product (without the need for using a post-processing step), and can deliver WGS reaction conversion exceeding equlibrium. The proposal configuration (AR following MR) provides significant flexibility in addition to high-purity hydrogen and carbon dioxide production.

The primary objective of this work is to first develop and simulate an advanced, and detailed mathematical model for the MR/AR sequence and then validate this model using experimental data obtained at the bench-scale. The MR system is composed of a reaction zone packed with catalyst and quartz pellets, and a water/hydrogen permeation zone. In the AR, both catalyst and adsorbent materials are used for simultaneous reaction and separation. The AR is a dynamically operated process. Both steady-state and dynamic experiment are carried out in a laboratory setting to capture the individual MR/AR performances. Experiments are carried out for a broad range of temperatures (up to 300 ^{o}C) pressures (up to 25 bar) relevant to realistic application conditions. The velocity, temperature and species concentration profiles in the MR reaction and permeation zones are captured by momentum/species transport models accounting for convection/reaction /diffusion/conduction mechanisms. The rigorous Maxwell-Stefan and dusty gas models are applied to describe mass diffusion fluxes. Finally, the performances of the combined system (the membrane reactor (MR) followed by an adsorptive reactor (AR) is simulated for a broad range of the operating conditions and design parameters. The simulation results are compared with the experimental findings.

[HC1]and pressures
[HC2]experimental

493066 Synthesis & Performance Study of Pd-Au/PSS Membrane Reactor to Produce Highly Pure Hydrogen Via Natural Gas Steam Reforming

Abstract:

One possible use of hydrogen, without direct greenhouse gas emissions, is as feed for a fuel cell (FC), with the most readily available technology being a proton exchange membrane FC (PEMFC). In order to avoid poisoning of PEMFCâ€™s Pt-based catalyst due to the presence of ppm levels of CO, the hydrogen feed needs to be ultra-pure. The industrial process for hydrogen production, which is a multi-step energy-intensive process followed by further separation/purification, can be a potential source [1]. However, as an alternative method, a Pd-based membrane reactor (MR) may be used due to its ability to provide high-purity hydrogen directly. Moreover, the MR works at milder operating conditions compared to a traditional system.

In recent years, Pd-based composite membranes, i.e., thin metallic layers supported on porous materials such as ceramics or stainless steel, have been considered due to their lower cost (thin Pd layer) and higher mechanical resistant (porous support) compared to dense Pd-based membranes [2,3]. Moreover, the use of Pd-alloys, in particular, Ag, Au, Pt, has grown in the recent years given their ability to resist sulfur poisoning in addition to their enhanced hydrogen permeability [4]. Therefore, the aim of this study was to fabricate a Pd-Au membranes supported on porous stainless steel (PSS) with the intent to produce high-purity hydrogen from natural gas steam reforming. The membrane used in this study was fabricated via an electroless plating technique for depositing Pd and electro-plating for the deposition of Au.

The initial characterization of the membrane by way of ideal selectivity took place at 400 °C with H_2, He and N_2, and with DP in the range of 1.5 - 3.0 bar. After ideal selectivity characterization of the Pd-Au/PSS membrane, the natural gas steam reforming reaction was carried out and the MR performance in terms of CH_4 conversion, hydrogen production and hydrogen permeate purity has been analyzed and will be discussed.

References

[1] Rostrup-Nielsen, J.R., Catalytic steam reforming. 1984: Springer.

[2] S. Liguori, A. Iulianelli, F. Dalena, P. Pinacci, F. Drago , M. Broglia, Y. Huang, A Basile, "Performance and long-term stability of Pd/PSS and Pd/Al_2O_3 membranes for hydrogen separation", *Membranes*, 4 (2014) 143-162.

[3] B. Anzelmo, J. Wilcox, S. Liguori, "Natural gas steam reforming reaction at low temperature and pressure conditions for hydrogen production via Pd/PSS membrane reactor", *Journal of Membrane Science*, 522 (2017) 343-350.

[4] Chen, C.-H. and Y.H. Ma, "The effect of H_2S on the performance of Pd and Pd/Au composite membrane". *Journal of Membrane Science*, 362 (2010) 535-544.

497163 Non-Isothermal CFD Study of Ethanol Steam Reforming in a Catalytic Membrane Reactor

Abstract:

Non-isothermal CFD Study of Ethanol Steam Reforming in a Catalytic Membrane Reactor

Rui Ma[1], Bernardo Castro-Dominguez[2], Anthony G. Dixon[1], Yi Hua Ma[1]

[1.] Center for Inorganic Membrane Studies, Department of Chemical Engineering, Worcester Polytechnic Institute, 100 Institute Road, Worcester, MA 01609, USA.

[2.] University of Limerick, Sreelane, Castletroy, Co. Limerick, Ireland

Single unit hydrogen production and separation achieved by conducting ethanol steam reforming (ESR) in a catalytic membrane reactor (CMR) has attracted more interest due to the carbon neutral property of bio-ethanol, and the high reactant conversion due to process intensification.[1] In ESR process, temperature control helps to achieve higher conversions. Compared with traditional packed bed reactors (PBR), heat supply for CMR needs to compensate for the enthalpy that hydrogen carried across the membrane during the permeation process, in addition to the heat loss from endothermic reactions. Furthermore, hydrogen flux through the membrane is temperature dependent as the hydrogen permeability follows Arrhenius law. In this work, a comprehensive heat and mass transfer study of an ESR process in a membrane reactor is carried out using an experimentally-validated non-isothermal CFD model. Retentate and permeate domains were modeled by introducing a negative and positive hydrogen flux term respectively, using Sieverts' law. In order to increase the hydrogen permeation driving force, a purge gas was applied to the permeate side. Effective thermal conductivity and diffusivity were introduced to the simulation in order to take into account the catalyst bed.

A "cold spot" adjacent to the membrane was observed (Fig 1) since this highly endothermic process benefits from hydrogen removal adjacent to the membrane. From the reaction rate study of the ESR process, the process takes place very rapidly, and a slight reverse reaction for methane steam reforming was observed due to the decrease in equilibrium conversion at lower temperature. This phenomenon is most noticeable within the "cold spot". A clear hydrogen generation and depletion along the reactor on the retentate side was observed. In addition, the reactions reach equilibrium shortly after reactants enter the reactor which leads to an inefficient use of the reactor and the membrane surface area; thus, counter-current flow of the purge gas is studied, which helps with shifting the reaction equilibrium throughout the entire reactor. Using this model, the effect of temperature and pressure on hydrogen recovery and yield are studied in both co-current and counter-current cases, and a process optimization study was carried out.

Fig 1. Temperature distribution in the reactor at LHSV=3.77h-1, Pret =6 bar, S/E=5 and Tin=723K on both permeate and retentate side.

1) Ma, R., Castro-Dominguez, B., Mardilovich, I.P., Dixon, A.G. and Ma, Y.H., "Experimental and Simulation Studies of the Production of Renewable Hydrogen through Ethanol Steam Reforming in a Pilot-Scale Catalytic Membrane Reactor," Chem. Eng. J., 303, 302-313 (2016).

511513 Heat Integrated Steam Reforming Membrane Reactor

Abstract:

A heat integrated steam reforming membrane reactor was investigated for development of a gasoline fuel processor for fuel cell vehicles. This reactor utilizes Le Chatelierâ€™s Principle for driving thermodynamically equilibrium limited steam reforming reactions at higher pressures. Steam reforming at higher pressures also reduces the palladium alloy membrane area required. Coupling steam reforming with a hydrogen selective membrane provides essentially pure hydrogen for optimal fuel cell performance.

Engineering models were used extensively to help define and refine the fuel processor design. The catalytic heat exchange reactor was coupled to a separate palladium-alloy membrane and used recycle of hydrogen depleted reformate to achieve high efficiencies at moderate operating temperatures and pressures (650Â°C, 20 Bar). Recycle was achieved by using a steam ejector. Catalytic combustion of the hydrogen depleted reformate with lean air fuel cell cathode exhaust (11% O2) supplied the heat required for steam reforming. Our modeling suggested that our steam reforming approach could have thermodynamic efficiencies of up to 40%.

Screening studies were undertaken to identify high activity noble metal steam reforming catalysts. A bimetallic Ru-Ir catalyst was found to have exceptionally high activity and stability. Order-of-magnitude, improvements in catalyst performance and reactor geometry were made.

In laboratory studies excellent heat transfer was observed when conducting catalytic combustion and steam reforming co-currently on opposing walls of the wash coated reactor tubes. Several membrane geometries were used. Hydrogen permeate was obtained from a variety of feeds, including isooctane, naphtha and methane, at reaction temperatures of 600-700°C. The experimental results were in close agreement with the modeling predictions.

While considerable research and development would be required to advance this technology, these results demonstrate the potential of the heat integrated steam reforming membrane reactor system.

486015 Improving Dehydrogenation Conversion and Selectivity By Utilizing Thermally Stable Membranes

Abstract:

In an equilibrium-limited alkane dehydrogenation reaction, due to its endothermic nature, high temperature is demanded to reach high conversion. Removal of product H_2in a dehydrogenation reaction shifts the equilibrium forward to realize a higher conversion at mild conditions. Thermodynamic computation demonstrates that with the assistance of a membrane separation unit, equilibrium limit can be exceeded in alkane dehydrogenation by selective removal of hydrogen. By operating at a lower temperature, high catalyst performance can be achieved; while catalyst coking is less likely to occur. In this research, dehydrogenation of methylcyclohexane to toluene is studied as a model reaction.

To study the impact of membrane utilization on the dehydrogenation process, blend polymer materials with high thermal stability, chemical resistance and transport properties has been selected as membrane materials. The permeabilities of dense films of Polybenzimidazole (PBI) and Matrimid® blends were evaluated from room temperature to 300 C. Permeabilities increased exponentially with temperature and the selectivities decreased. All blend materials were stable over the multi-day testing.

The potential impact of selective hydrogen removal via thermally-stable polymer blends was calculated from finite element modelling. The model incorporated the measured permeation properties, measured reaction equilibrium information, and catalyst performance (kinetics and coking). As the process can be operated at a range of temperatures, the influence of system temperature on conversion, permeation rate, and catalyst deactivation is analysed. The use of a single membrane unit inter-stage between two packed beds has the potential to increase reactant conversion by 35% compared to operation without inter-stage conversion.

494458 Integration of Synthesized Catalysts and Membrane in Packed and Fluidized Bed Membrane Reformer for Production of Hydrogen through Steam Reforming

Abstract:

Hydrogen is the only energy carrier which can cater to the rising power demands across the world without compromising with the environmental impact. Its high efficiency to generate power comes in conjunction with fuel cells that can produce a desired output with absolutely no carbon emissions. However, the hydrogen fed to fuel cells must be of high purity to ensure stable performance of Pt-based catalysts of fuel cell anode. Therefore, production of hydrogen with high purity is considered crucial. In the current work, hydrogen generation is targeted with methanol steam reforming using synthesized Cu-based catalysts. To achieve high hydrogen purity, Pd-based dense supported membranes were employed after their synthesis, characterization and testing. The target of this work is to suitably integrate the catalysts with the membranes in one single unit. Such integration will cause continuous permeation of hydrogen through a H_2 selective Pd membrane. Disturbed reaction equilibrium will therefore cause an increase in the forward reaction rate according to Le Chatelierâ€™s rule and hence improve methanol conversion. This system of simultaneous generation and separation of hydrogen in one single unit is termed as a membrane reformer (MR). Catalyst placing relative to the membrane determines the reaction residence time in a membrane reformer and thereby plays a key role. Further, this factor is inter-related with reaction kinetics, hydrogen mass transfer rate from reaction zone to membrane, and lastly membrane permeation flux and perm-selectivity. In order to observe the effect of these parameters on total system performance, the membrane reformer study was paralleled between a packed bed and a fluidized bed MR. Packed bed consists of a catalyst bed (~bed ht. 2 cm) in the shell of the reactor with membrane in its lumen separated by a distance of approximately 15 cm. In contrast to that, fluidized bed gives the advantage of catalyst being fluidized in the same shell that brings it in close contact with the membrane.

The current study would determine performance of packed bed MR with varying catalyst positions and parameters such as temperature, flow rate and weight hour space velocity (WHSV). Based on these operating conditions, fluidization regime was characterized by u/u_{mf} in the range of 1.4â€"3.6 (where u_{mf} corresponds to the minimum fluidization velocity). All testing was performed in an in-house designed test rig. Dense Pd-based membranes were prepared using modified electroless deposition approach on porous SS (PSS) with yttria stabilized zirconia (YSZ) as the intermediate layer. YSZ minimizes the inter-metallic diffusion between Pd and SS. Further, catalysts with varying bi-metallic compositions were synthesized by incipient wetness impregnation. Optimization of standalone membrane as well as catalyst performance was performed prior to the integration.

Once integrated, the packed bed membrane reformer (MR) showed >95% methanol conversion and 45% H_2 recovery at 723 K and 3 bar. Further, a comparison of packed bed MR was made with a fluidized bed MR. It was observed that with increasing u/u_{mf} greater contact between the catalyst and the membrane enhances the hydrogen yield and thereby leads to better separation in comparison to packed bed. Moreover, a complete description of these results will be elaborated in terms of effectiveness factor, reaction kinetics and hydrogen permeation flux through the membrane.

496102 Facile Fabrication of Ultrathin Molecular Sieving ZIF-8 Hollow Fiber Membranes on Polydopamine-Polyethyleneimine Coated Microporous PVDF Supports

Abstract:

Zeolitic Imidazolate Frameworks (ZIFs)-based membrane emerges as a potential and attractive material for molecular sieving gas separation process. ZIF-8 with sodalite topology is one of the most ZIFs studied recently due to its relatively simple fabrication process and environmentally friendly solvent used during its synthesis. However, scaleable formation of thin, defect free films remains challenging. Many ZIF-8 membranes were synthesized by coating ZIF-8 seeds on the surface of inorganic support. However, this process is relatively high cost, complex and the membranes are relatively brittle. Polymeric porous support is one alternative to overcome such problems for inorganic support. The growing of defect free ZIF layers on polymeric support often requires complex surface functionalization as the support frequently lacks in heterogeneous site and ligands needed for the growing process of ZIFs.

To overcome these problems, we chemically-functionalized PVDF hollow fiber porous supports before the deposition of ZIF-8 particles using a simple coating process. Co-deposition of polydopamine (PDA) and polyethyleneimine (PEI) was conducted to induce the crystallization process of ZIF-8 on the support by the presence of cathecol group and produces more uniform coating of PVDF surface compared to traditional PDA deposition. Thin molecular sieving membrane was successfully synthesized simply by 5 hours deposition of support in ZIF-8 precursor solution. Thin layer of ZIF-8 (< 1 Î¼m in thickness) produced very high hydrogen permeance of approximately 105 x 10^{-7} mol m^{-2} s^{-1} Pa^{-1}, which is one of the highest permeance for ZIF-8 membranes, and H_2/CO_2 selectivity of 7.5. Compared to our previous results utilizing TiO_2/APTES modified PVDF support, the results from this study show slightly lower gas permeances but provides a easier fabrication route.

500171 Effect of Graphene Oxide on Gas Transport and Sorption in Poly(dimethylsiloxane)-Based Membranes

Abstract:

We report the gas transport properties of poly(dimethylsiloxane)/graphene oxide (PDMS/GO) thermoset composites synthesized from graphite and amine-terminated telechelic PDMS. The material was prepared by a simple method using a thermally-activated reaction between the amine terminal end groups of the telechelic PDMS and epoxides on the surface of GO. A systematic reduction in pure gas permeability was observed for various gases, such as N_2, O_2, CH_4, and CO_2, by adding different amounts of GO into the PDMS matrix. At 3.55 vol. % (8 wt. %) GO loading, the composites resulted in a more than 99.9% reduction in gas permeability for all gases. Moreover, with increasing GO concentration, the composites showed a systematic increase in gas permselectivity for several pairs of gases. For instance, the permselectivity of CO_2/N_2 and CO_2/CH_4 at the maximum GO loading of 3.55 vol. % (8 wt. %) increased by approximately a factor of 2 and 3, respectively, compared to the neat PDMS membrane. The solubility of N_2, O_2, CH_4, and CO

effect of GO filler on gas transport, specifically in terms of the individual effects on the gas solubility and diffusivity. In general, the gas solubilities in PDMS/GO composites were not affected by the addition of GO at concentrations of up to 3.55 vol. % (8 wt. %). In this regard, these results suggest that the reduction of gas permeabilities in PDMS/GO composites is mainly attributed to the dispersed phase of impermeable GO, which provides longer tortuous paths for gas diffusion. The permselectivity of PDMS/GO composites is controlled by the diffusivity selectivity, and both appear to increase with increasing GO concentration, which clearly illustrates that increasing the concentration of GO in the PDMS matrix enhances the size-sieving capability. Theoretical models for predicting the tortuosity of mixed matrix membranes (MMM) are discussed in detail to provide further understanding of the correlation between gas diffusion and morphology in PDMS/GO composites.

491518 Tailoring Interfacial Properties in Mixed Matrix Membranes Via Supramolecular Interactions

Abstract:

Mixed matrix membranes (MMMs) are highly attractive for gas separation applications since they can integrate superior sieving properties of nano-filler particles and excellent processability of polymer matrix. However, preparation of MMMs with homogeneously dispersed fillers and defect-free interfacial morphology of ideal interphase contact is challenging due to intrinsically different physical and chemical properties of different phases. Here, we report novel nanocomposite membranes with desirable interfacial morphologies tailored by supramolecular interactions between the filler and polymer matrix. Specifically, these new MMMs are prepared from single-walled carbon nanotubes (SWCNTs) and a triptycene-containing polyimide. SWCNTs show ultrahigh transport rates of gas molecules, conferring on SWCNT-based membranes the potential for high gas permeability. The paddle-wheel like triptycene moieties in polymer matrix induce strong Ï€-Ï€ supramolecular interactions with the graphene wall of nanotubes, which effectively improves the interfacial contact between phases and assists homogenous dispersion of SWCNTs via nano-confinement effect.

A series of MMMs was prepared using commercial SWCNTs by solution casting method, which were systematically varied in surface functionality of SWCNTs, metal impurities, aspect ratio of nanotubes and filler loading. In all cases, good adhesion between SWCNTs and triptycene-polyimide and homogeneous dispersion of SWCNTs in polymer matrix without aggregation are verified by cross-sectional SEM images.Â Fluorescence quenching phenomenon provides a direct evidence for the Ï€-Ï€ stacking interaction between SWCNTs and the polyimide. The improved interfacial properties lead to excellent separation performance in these MMMs. For example, a composite membrane with only 2 wt% filler loading showed nearly an order of magnitude increase in gas permeabilities from the pure polymer without sacrificing CO_2/CH_4 and CO_2/N_2 selectivities. The significant improvement in permeability is the synergistic result of improved diffusivity due to the incorporation of SWCNTs and improved interface morphology due to supramolecular interactions between phases. In this talk, the synthesis, fabrication and characterization of these new MMMs will be presented. Detailed discussion will be provided regarding the dependence of gas transport properties on surface functionality, impurities, aspect ratio of nanotubes and filler loading, with a focus on relating microscopic structures with macroscopic transport properties.

495684 Chemical Vapor Deposition on Chabazite (CHA) Zeolite Membranes for Improving CO_2/N_2 Separation Performance

Abstract:

Chemical Vapor Deposition on Chabazite (CHA) Zeolite Membranes for Improving CO_2/N_2 Separation Performance

Yanghwan Jeong[1], Jungkyu Choi[1]*

[1]Department of Chemical & Biological Engineering, College of Engineering, Korea University, Anam-dong, Seongbuk-gu, Seoul, 02841, Republic of Korea

*E-mail: jungkyu_choi@korea.ac.kr

Chabazite (CHA) zeolites with a pore size of 0.37 * 0.42 nm^2 are expected to separate CO_2 (0.33 nm) from larger N_2 (0.364 nm) in post-combustion flue gases by recognizing their minute size differences. Furthermore, the hydrophobic siliceous constituent in CHA membranes can allow for maintaining the CO_2/N_2 separation performance in the presence of H_2O in contrast with the CO_2 affinity-based membranes. In an attempt to increase the molecular sieving ability, the pore mouth size of all silica CHA (Si-CHA) particles was reduced via the chemical vapor deposition (CVD) of a silica precursor (tetraethyl orthosilicate). Accordingly, an increase of the CVD treatment duration decreased the penetration rate of CO_2 into the CVD-treated Si-CHA particles. Furthermore, the CVD process was applied to siliceous CHA membranes in order to improve their CO_2/N_2 separation performance. Compared to the intact CHA membranes, the CO_2/N_2 maximum separation factor (max SF) for CVD-treated CHA membranes was increased by 2 fold under dry conditions. More desirably, the CO_2/N_2 max SF was increased by 3 fold under wet conditions at 50 ¡ÆC, a representative temperature of the flue gas stream. In fact, the presence of H_2O in the feed disfavored the permeation of N_2 more than that of CO_2 through CVD-modified CHA membranes and thus, contributed to the increased CO_2/N_2 separation factor.

495524 Molecular Dynamics Simulation of Mixed Matrix Membrane Formed By Porous Organic Cage and Polymer with Intrinsic Microporosity

Abstract:

Membrane based gas separation is an energy efficient and environmentally friendly technology that is widely applied in areas[1-2] such as CO_2 capture and sequestration, air separation, and natural gas sweetening. However, the trade-off between permeability versus selectivity still hinders the performance of most polymer based membrane[3]. A promising strategy to tackle this trade-off is mixed matrix membranes (MMMs) in which a solid porous framework is embedded into a polymer host. Recently, porous organic cages (POC)[4-5], organic molecules with intrinsic cages, were in situ crystalized in a polymer of intrinsic microporosity (PIM-1) to form MMMs with enhanced permeability and minimal selectivity sacrifice[6]. In this work, we study the MMMs formed by POC and PIM-1 using all atom molecular dynamics simulation. By analyzing structural properties (such as free void volume, pore size distribution), thermal properties (such as gas adsorption), and dynamical properties (such as gas diffusivity), we tried to elucidate the reason for the enhanced performance of POC/PIM-1 MMMs. We hope that results from these simulations will provide insights into MMM design and optimization.

References

1. Haszeldine, R. S., Carbon Capture and Storage: How Green Can Black Be? *Science* **2009**, *325*, 1647-1652.

2. Vinh-Thang, H.; Kaliaguine, S., Predictive Models for Mixed-Matrix Membrane Performance: A Review. *Chemical reviews* **2013**, *113*, 4980-5028.

3. Robeson, L. M., The Upper Bound Revisited. *Journal of Membrane Science* **2008**, *320*, 390-400.

4. Tozawa, T.; Jones, J. T.; Swamy, S. I.; Jiang, S.; Adams, D. J.; Shakespeare, S.; Clowes, R.; Bradshaw, D.; Hasell, T.; Chong, S. Y., Porous Organic Cages. *Nature materials* **2009**, *8*, 973-978.

5. Jones, J. T.; Hasell, T.; Wu, X.; Bacsa, J.; Jelfs, K. E.; Schmidtmann, M.; Chong, S. Y.; Adams, D. J.; Trewin, A.; Schiffman, F., Modular and Predictable Assembly of Porous Organic Molecular Crystals. *Nature* **2011**, *474*, 367-371.

6. Bushell, A. F.; Budd, P. M.; Attfield, M. P.; Jones, J. T.; Hasell, T.; Cooper, A. I.; Bernardo, P.; Bazzarelli, F.; Clarizia, G.; Jansen, J. C., Nanoporous Organic Polymer/Cage Composite Membranes. *Angewandte Chemie International Edition* **2013**, *52*, 1253-1256.

488912 High Performance Composite Membrane for Olefin/Paraffin Separation

Abstract:

High Performance Composite Membrane for Olefin/Paraffin Separation

Zhong Tang, Bo Lu, Xiaojuan Hu, Kristy Pham, Xueliang Dong, Lin-feng Li

Bettergy Corp., Peekskill, NY 10566

Abstract

Bettergy Corp. is developing a novel, robust, olefin/paraffin membrane separation technology with high selectivity and flux and excellent stability. The membrane can be used as a more efficient replacement for the energy-intensive cryo-distillation processes that are currently being used. Bettergy's membrane can meet the stringent performance and cost requirements for olefin/paraffin separation, significantly reducing the cost of olefin production. Light olefins such as ethylene and propylene are produced in great quantities in petrochemical plants worldwide. The global light olefins market size was 178.6 million tons in 2008 and is expected to reach 267.9 million tons in 2020. In the United States, ethylene and propylene are the most commonly produced light olefins, with yearly production exceeding 25 million tons for ethylene and 16 million tons for propylene, respectively, representing a market over $50 billion. [1] An important process in the manufacture of olefins is large-scale separation of the olefin from corresponding paraffin. In the petrochemical industry, this separation is carried out by cryo-distillation, which is highly energy-intensive due to the cryogenic temperatures required for the process. [2]Membrane separation for olefin/paraffin mixtures offers an appealing energy-efficient alternative to the cryo-distillation process.

There have been previous attempts to develop membranes for olefin/paraffin separation. Although substantial effort has been dedicated in developing a high flux, high selectivity olefin separation membrane, the state-of-the-art membrane cannot meet the stringent requirements of real life olefin/paraffin separation. Technical barriers include (1) a trade-off between selectivity and flux; (2) poor stability especially under the practical operating environment; (3) fouling or poisoning in the presence of contaminants (olefin, alkynes, dienes, olefin sulfide, and other sulfur species); and (4) high cost of the membranes. Given the significant costs associated with current olefin/paraffin separation technologies, the membrane technology developed by Bettergy has the potential to find a huge market in olefin/paraffin separation. Bettergy estimates that by utilizing its novel membrane technology to replace current cryo-distillation process, the cost of production can be reduced significantly, with the potential for savings of tens of billions of dollars annually on a global basis.

Bettergy has experimentally proved that our unique composite membrane can separate propylene/propane gas mixture achieving propylene selectivity of ~ 50-100, and propylene permeance up to 0.1-1.0×10^{-8} mol/m^2.s.Pa. The membrane has exhibited very stable performance during more than 1500 hours of test. A

comparison of the membrane performance between Bettergy's novel composite membrane and several kinds of other membranes reported in the literature shows that Bettergy's membrane exhibits the best performance in terms of selectivity to propylene, permeance. In addition, this composite membrane shows excellent stability, can be fabricated in spiral-wound module with low cost.

The simulation of propylene/propane membrane separation has been performed. Only a one-stage membrane process is considered in this work. The feed gas – propylene 88.0% and propane 12.0%, target permeate propylene purity – 99.5%, feed rate – 1000 lb/hr. The trans-membrane pressure is set to 325 psi and operation temperature is 28°C. Results show that in order to meet the target permeate propylene concentration of 99.5%, the maximum propylene recovery rate possible is about 72%. The membrane area required for this operation with feed rate of 1000 lb/hr is about 500 m^2.

Bettergy has a number of patents based on this membrane technology that have been either granted or pending. [3]The scale-up modules are also being developed and showing reproducible performance. Bettergy is actively seeking industrial partners to conduct pilot plant testing and further commercialize this advanced membrane technology.

---

[1] 2009 World Petrochemical Production Report, http://www.petrochemistry.net/statistics-western-european-market-review.html

[2] T. Ren, M.K. Patel, K. Blok, Steam cracking and methane to olefins: Energy use, CO_2 emissions and production costs, *Energy*, **2008**, 33, 817-833

[3] US Patent Pub No. 20150321141, WO2015171970, Novel Composite Membrane for Olefin/Paraffin Separation; US Patent Appl No. 15/482,942, Mixed Matrix Membrane for Olefin/Paraffins Separation and Method of Making Thereof.

Comparative Study of Two Nano-Composite Membranes for Efficient CO_2 Removal

Ahmad A. Shamsabadi[1], Farzad Seidi[2], Ehsan Salehi[3], Mohammad Nozari[4], Ahmad Rahimpour[5] ,and Masoud Soroush[1]

[1]Chemical and Biological Engineering, Drexel University, Philadelphia, USA
[2]Vidyasirimedhi Institute of Science and Technology, Royong, Thailand
[3]Arak University, Arak, Iran
[4]Chemistry, Drexel University, Philadelphia, USA
[5]Chemical Engineering, Babol University of Technology, Babol, Iran

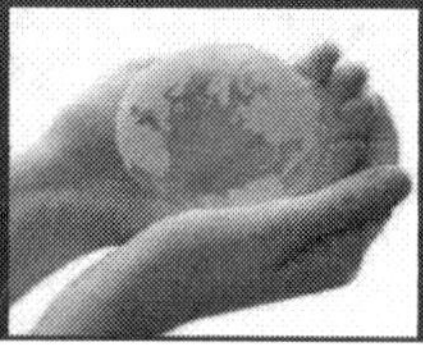

Current Level of CO_2>400 ppm

Average global atmospheric CO_2 level:
February 2016: 402.59 ppm
October 2017: 403.44 ppm

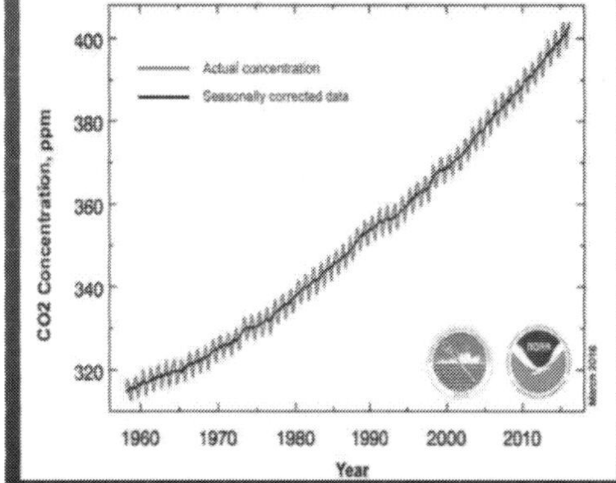

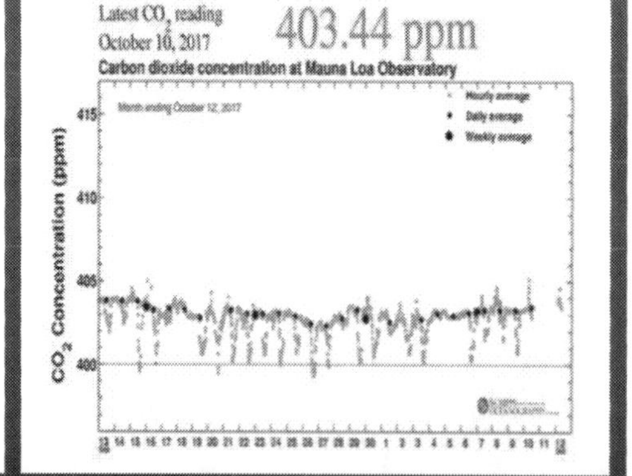

US National Oceanic and Atmospheric Administration's (NOAA)

2

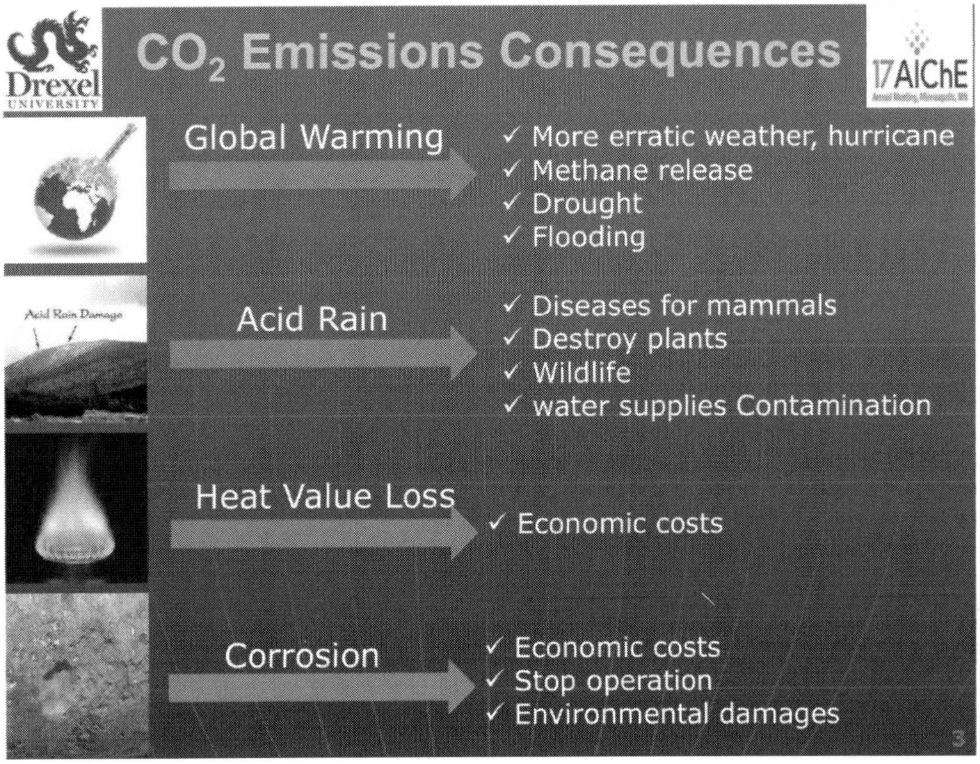
CO2 Emissions Consequences
Drexel UNIVERSITY
17AIChE
Global Warming
✓ More erratic weather, hurricane
✓ Methane release
✓ Drought
✓ Flooding
Acid Rain Damage
Acid Rain
✓ Diseases for mammals
✓ Destroy plants
✓ Wildlife
✓ water supplies Contamination
Heat Value Loss
✓ Economic costs
Corrosion
✓ Economic costs
✓ Stop operation
✓ Environmental damages
3

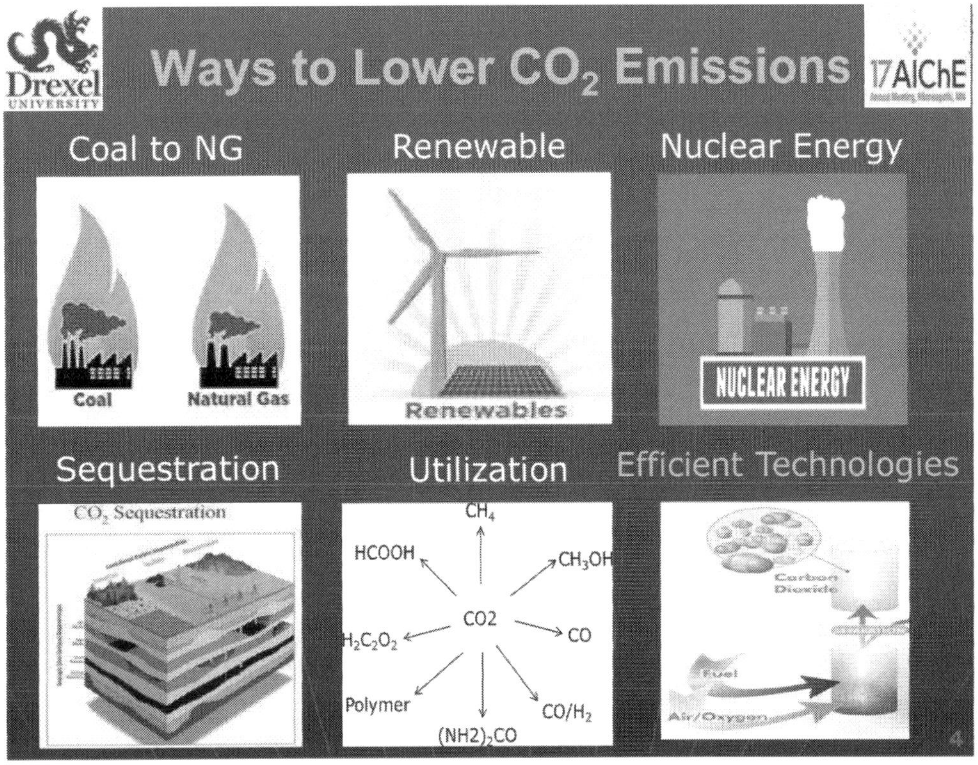
Ways to Lower CO2 Emissions
Drexel UNIVERSITY
17AIChE
Coal to NG
Coal
Natural Gas
Renewable
Renewables
Nuclear Energy
NUCLEAR ENERGY
Sequestration
CO2 Sequestration
Utilization
CH4
HCOOH
CH3OH
CO2
H2C2O2
CO
Polymer
CO/H2
(NH2)2CO
Efficient Technologies
Carbon Dioxide
Fuel
Air/Oxygen
4

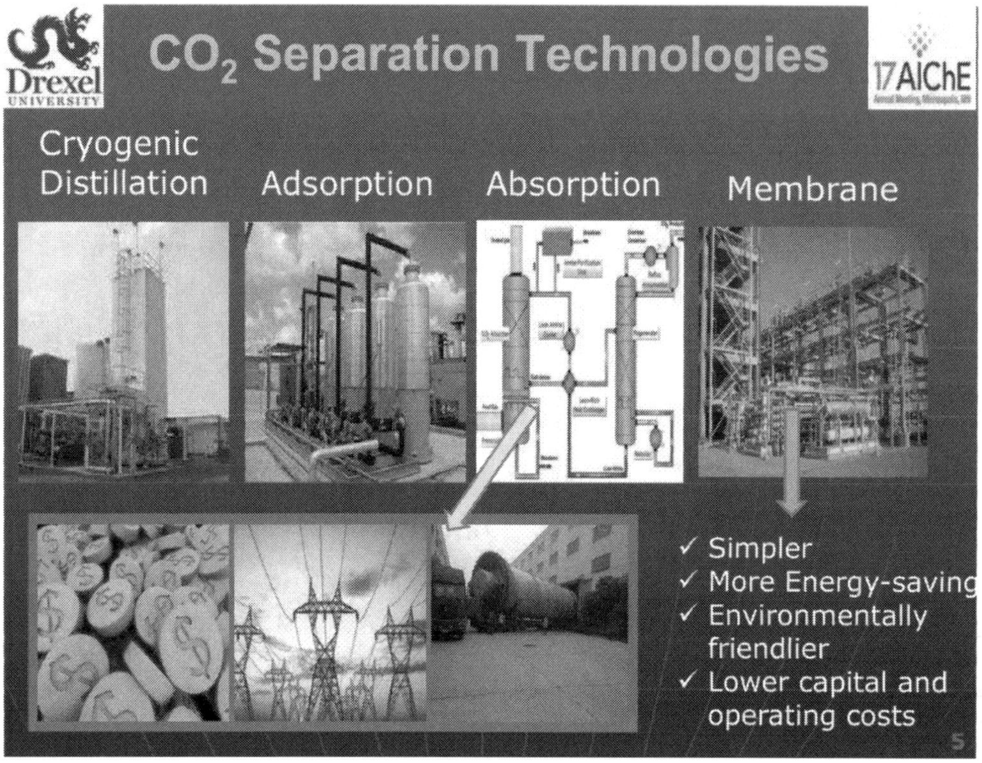
Drexel UNIVERSITY
CO_2 Separation Technologies
17 AIChE
Cryogenic Distillation
Adsorption
Absorption
Membrane
✓ Simpler
✓ More Energy-saving
✓ Environmentally friendlier
✓ Lower capital and operating costs
5

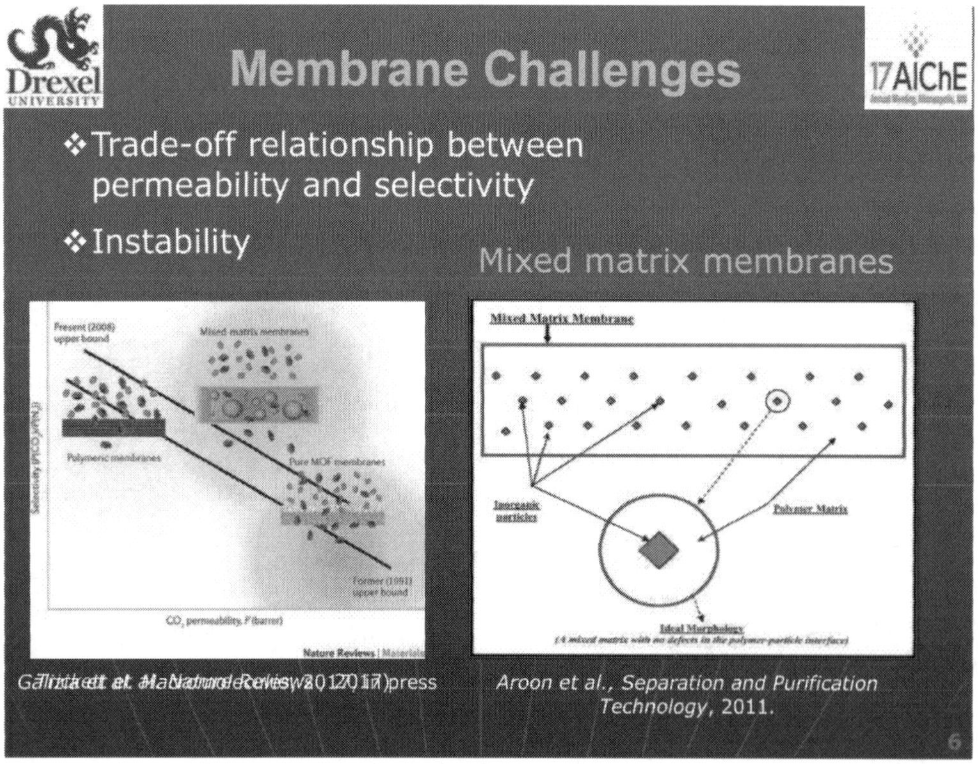
Drexel UNIVERSITY
Membrane Challenges
17 AIChE
❖ Trade-off relationship between permeability and selectivity
❖ Instability
Mixed matrix membranes
Present (2008) upper bound
Mixed matrix membranes
Polymeric membranes
Pure MOF membranes
Former (1991) upper bound
CO_2 permeability, P (barrer)
Nature Reviews | Materials
Mixed Matrix Membrane
Inorganic particles
Polymer Matrix
Ideal Morphology
Aroon et al., Separation and Purification Technology, 2011.
6

Nanocomposite Fabrication Challenges

Agglomeration | Interfacial defects | Incompatibility

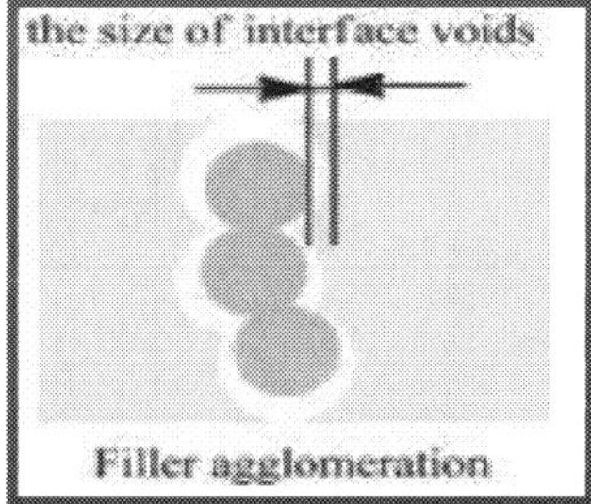

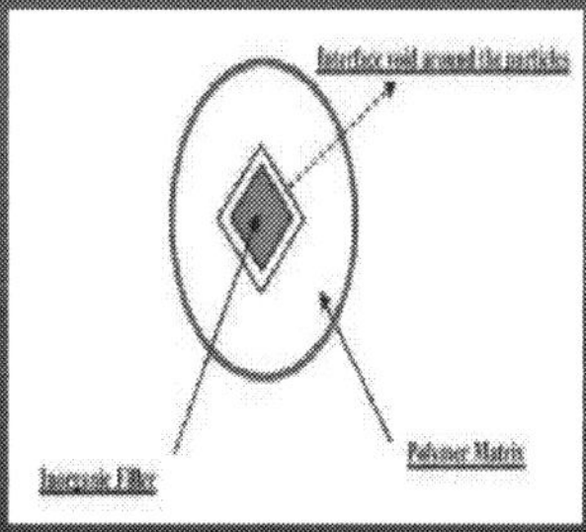

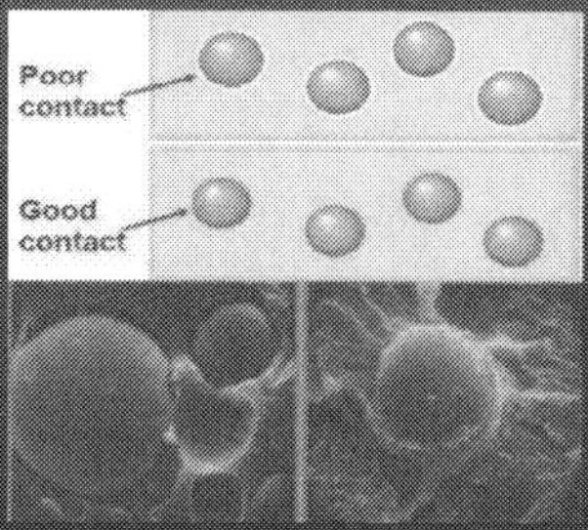

Wang et al. J Membrane Science,2015

Aroon et al., Separation and Purification Technology,2011

http://slideplayer.com/slide/7448844/

7

This Research

Addressing the trade-off relationship between permeability and selectivity, and nanocomposite fabrication challenges through:

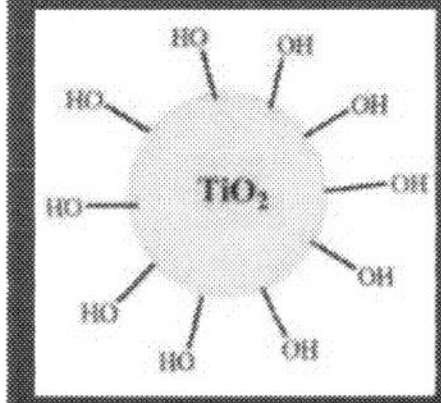

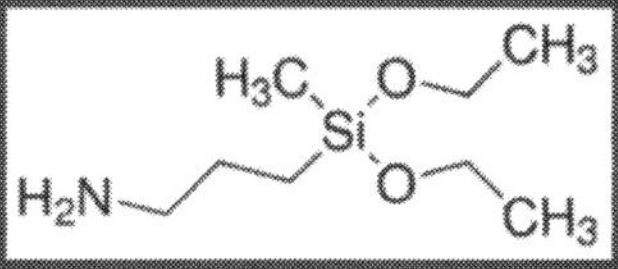

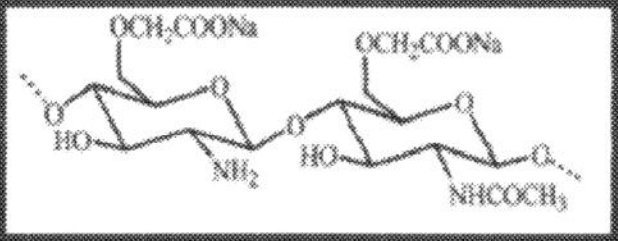

Pebax-1657

8

Surface Modification of TiO_2

Drexel University | 17 AIChE

- TiO_2 Nanoparticles (NPs)

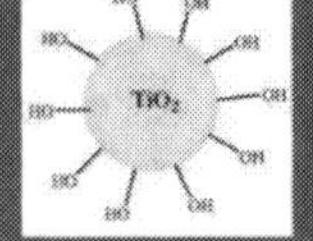

- (3-aminopropyl)-diethoxy-methylsilane (APDEMS)

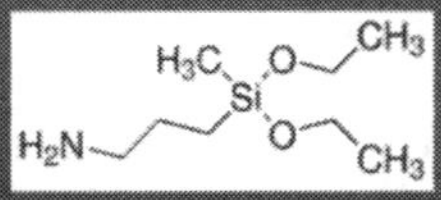

- Carboxymethyl Chitosan (CMC)

- Modification conditions

TiO_2 Concentration (g/L)	Medium	Modifier	Modifier/TiO_2 Ratio	Temperature (°C)	Time (h)
25	Ethanol	APDEMS	1	80	24
2	DI Water (pH=2.5)	CMC	1	25	24

11

Characterization of Modified-TiO_2

Drexel University | 17 AIChE

TGA thermographs of the TiO_2, TiO_2-AS, and TiO_2-CMC nanoparticles

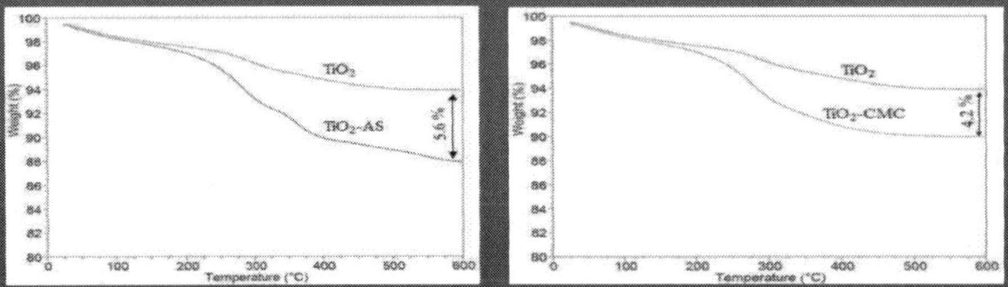

(a) N_2 Adsorption (b) pore size distribution (NLDFT method)

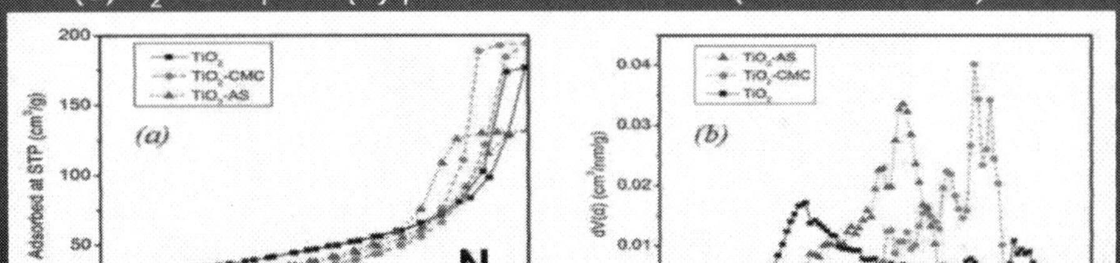

10

TiO_2-based/Pebax-1657 Nanocomposite Membranes

Drexel UNIVERSITY

17 AIChE

Pebax-1657

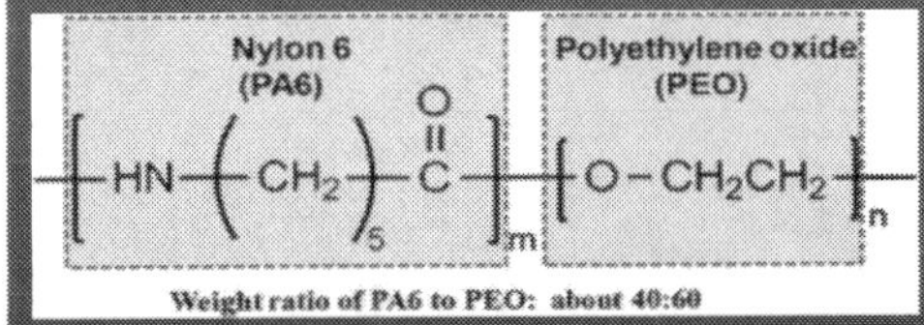

- Commercial polymer
- Mechanically robust
- Good thermal resistance
- Specific structure

FTIR-spectra of nanoparticles and membranes

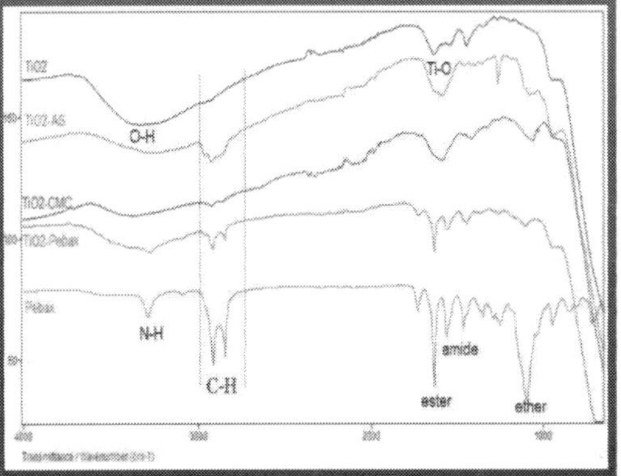

11

Pebax/TiO_2-AS and Pebax/TiO_2-CMC Interactions

Drexel UNIVERSITY

17 AIChE

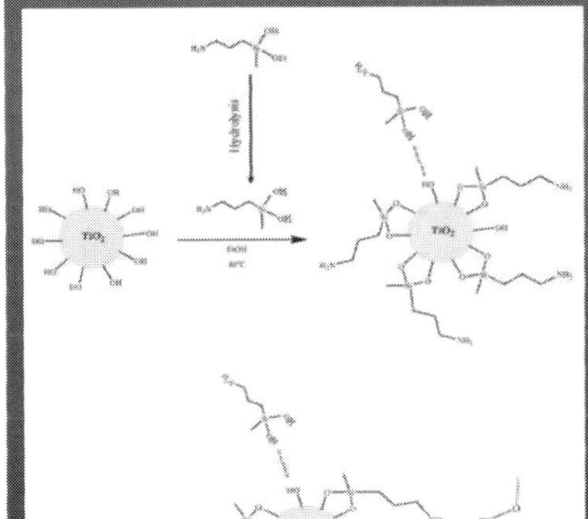

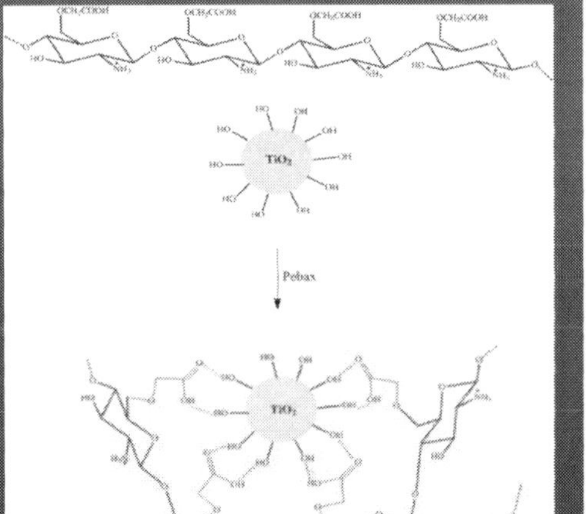

12

TGA and T_g of Membranes

TGA

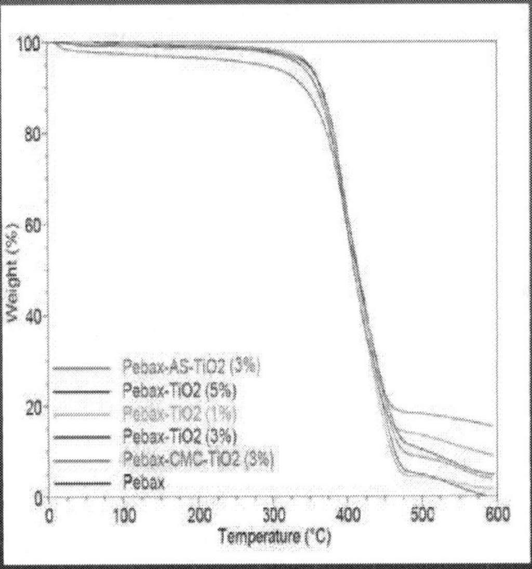

Membrane	T_g (°C)
Pebax	-52.2
Pebax-TiO2-1%	-55.9
Pebax-TiO2-3%	-51.1
Pebax-AS-TiO2-3%	-50.4
Pebax-CMC-TiO2-3%	-49.5
Pebax-TiO2-5%	-48.1

13

SEMs of Membranes

Top surface | Cross section

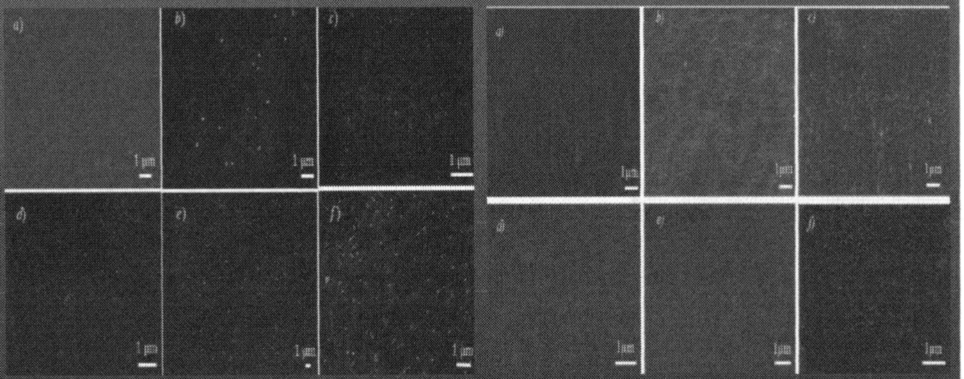

(a) Pebax, (b) Pebax-TiO_2 (1%), (c) Pebax-TiO_2 (3%), (d) Pebax-CMC-TiO_2 (3%) (e) Pebax- AS-TiO_2 (3%) (f) Pebax-TiO_2 (5%).

14

Membranes Performance

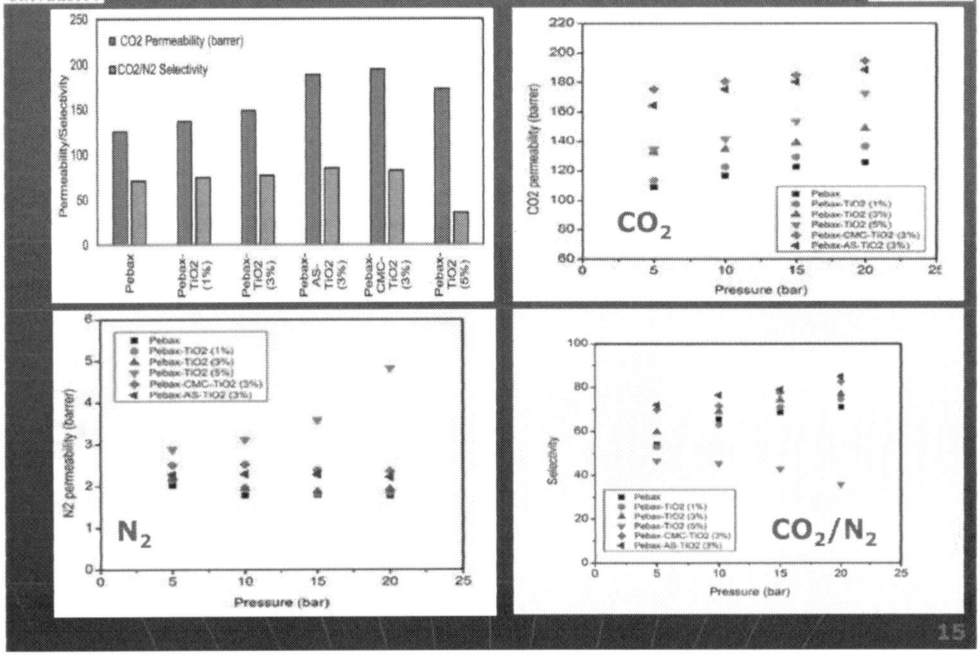

15

CO_2 and N_2 diffusivity and Solubility (5 bar, 25 °C)

Membrane	D (×10⁶ cm²/s)		S (×10⁴ cm³(STP)/cm³ cmHg)	
Gas	CO_2	N_2	CO_2	N_2
Pebax	1.47	0.74	73.96	2.71
Pebax-TiO_2-(3%)	1.73	0.8	76.56	2.75
Pebax-AS-TiO_2-(3%)	2.26	0.84	83.28	2.71
Pebax-CMC-TiO_2-(3%)	2.39	0.87	81.12	2.88

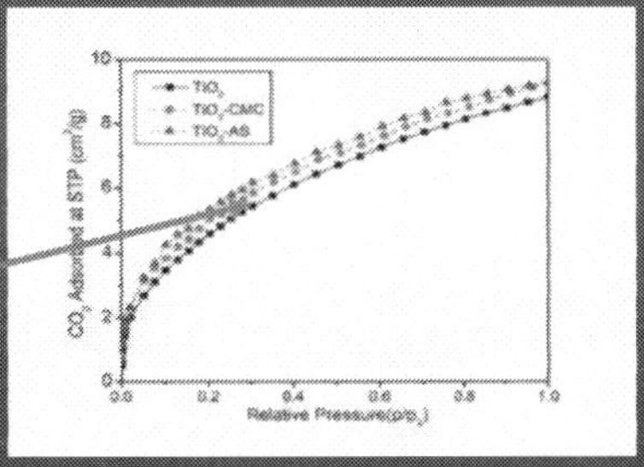

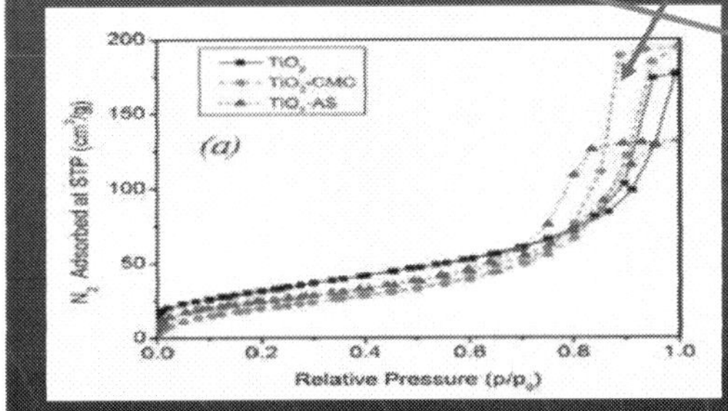

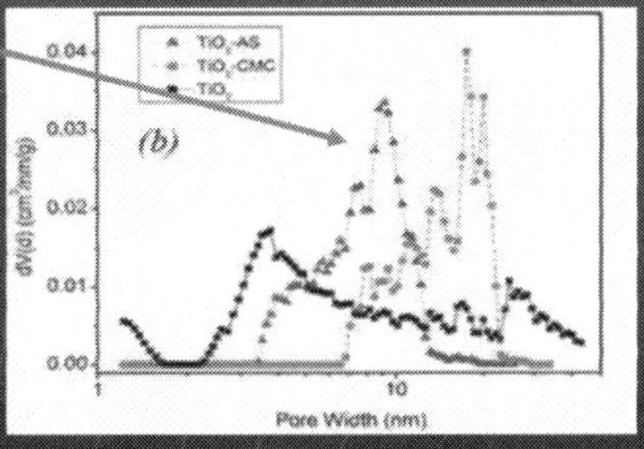

16

Membranes Performance and Stability

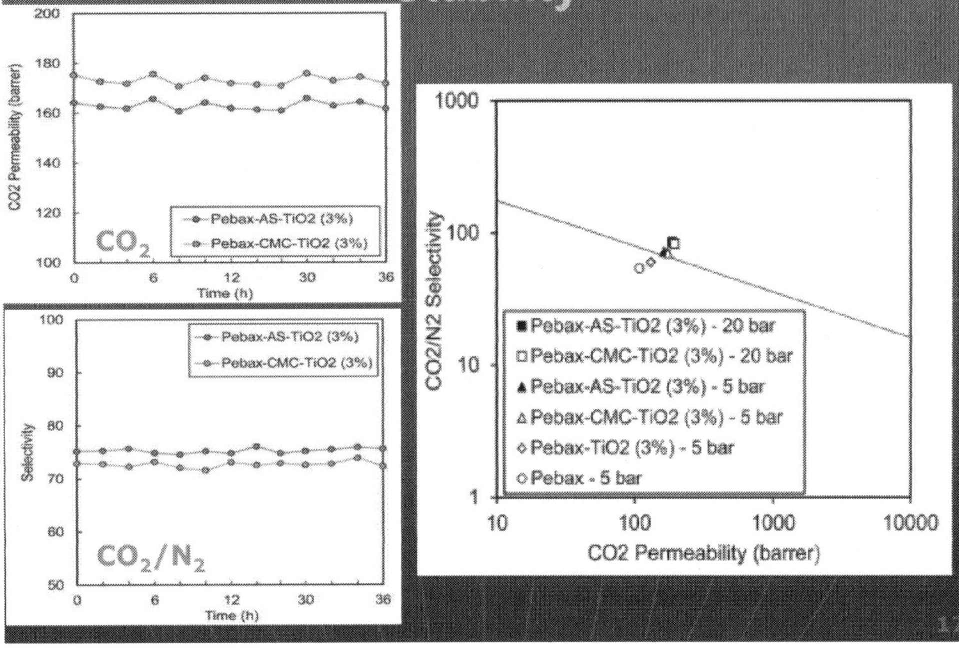

17

Concluding Remarks

- Loading Pebax with TiO_2 modified with CMC and APDEMS provides an appealing combination of fixed facilitated transport and molecular sieving trait of inorganic nanoparticles
- Considerable enhancement of permeability and selectivity was achieved for both modifications
- The highest selectivity and permeability were obtained for Pebax-AS-TiO_2 and Pebax-CMC-TiO_2 membranes, respectively
-

18

Concluding Remarks (Cont.)

- The performance of the mixed matrix membranes exceeds the 2008 CO_2/N_2 Robeson upper bound
- Modifications led to better polymer/ particles interactions
- Increasing pressure improves the performance of the membranes
- TiO_2 nanoparticles have great potential for chemical modifications and are easy to embed into polymers for CO_2 removal

19

497783 Calculation of Free Energy Barriers for Attachment of Molecules during Crystal Growth and Nucleation

Abstract:

Physical properties of crystalline materials depend upon their molecular structure and shape. Following the adoption of Quality by Design approaches by the Food and Drug Administration, the theoretical and computational approaches are increasingly implemented to predict structure (polymorphs), shape and properties of the crystalline materials. The most common method to predict crystal structure is the lattice energy minimization technique, which calculates the configuration of molecules in the lattice and its energy using the state-of-the-art optimization algorithms. Whereas the theoretical methods to predict crystal shape relies on the calculation of growth rates using either the interfacial energy or the kink energy. Such methods for the crystal structure and shape prediction depend on the estimation of crystal energies and hence to fail to capture the underlying dynamics of nucleation and growth, where solute molecules continuously attach to and detach from the crystal surfaces in a supersaturated solution. In order to effectively model the dynamics of crystallization processes, it is essential to know the free energy barriers for the attachment and detachment of solute molecules. Here we propose a method to calculate free energy barriers for the attachment and detachment of molecules using the double-well potential method. This approach consists of finding the configuration of molecules in the solvation shell, followed by the construction of the double-well potential diagrams to calculate free energy barriers. Application of the proposed method to calculate growth rates and nucleation rates of Glutamic acid will be shown.

492558 Population Balance Modeling and Optimization of an Integrated Batch Crystallizer – Wet Mill System for Crystal Size Distribution Control

Abstract:

Batch crystallization is one of the most efficient and economic separation and purification technique of solid crystalline materials, which has particularly important role in the fine chemical and pharmaceutical industry. From the point of view of particle formation is also significant as during the process the crystal size distribution (CSD) can be adjusted, within a certain domain.

Through the crystal size is often the most critical property of the end-product (keeping in mind that the purity, yield and crystal form specifications must be respected), the control of crystallization process in many practical case consists on controlling the product CSD. In crystallization systems' engineering, from the perspective of product size, the general objective is the production of large, uniform crystals [1] in order to facilitate the downstream operations. However, the production of small, uniform crystals is also of great interest due to the increased specific surface and bioavailability.

Since the crystallization is governed by the simultaneously ongoing nucleation and growth, which are non-linear functions of concentration, controlling the relative rate of these generally raises complex control issues. Generally, model predictive control (MPC) is employed to overcome the process nonlinearity's while respecting various constraints. In contrast with the continuous tank systems, the control of batch operation is sometimes more challenging due to the fact that the initial and final system states are far from each other, during which transition the process nonlinearity's has more significant effects [2]. Out of the model based techniques, with the quick spread of process analytical technologies the model free control methods, such as the direct nucleation control (DNC) and supersaturation control (SSC) [3]. The batch crystallization control, regardless on the way of implementations, in most cases relies in the manipulation of applied temperature profile, which influences the crystallization kinetics through the supersaturation. In crystallization practice the attainable particle size domain is always delimited by the crystallization kinetics [4].

In order to widen the attainable crystal size domain, other manipulated variables affecting crystal size are also implied, such as using antisolvent or, more often, applying combined antisolvent-temperature variations for supersaturation generation. Out of the nucleation and growth, the secondary operations, like the agglomeration and growth are also promising sup-processes to be controlled from the point of view of crystal size control [5]. The seed addition is a very popular startup procedure of batch crystallizers, which also presents high potential of tailoring the product CSD. Growth rate modifier additives were also successfully applied to adjust the crystal size and shape during the crystallization.

If the CSD cannot be obtained directly in the crystallization, secondary operations, such as granulation and milling are required to adjust the size of crystals properly. This however is not an advantageous solution as involves extra operations in the production line and, moreover, the quality of crystals not reaches those that obtained in direct crystallization. The straightforward question arises if the crystallization and milling could be coupled with each other for the sake of production uniform crystals in lower size domains? There is a forward looking experimental work analyzing a continuous crystallizer-wet mill cascade system [6] which was also successfully extended to continuous wet mill based direct nucleation control for cooling crystallization [7]. Others attempted to model and simulate the attainable crystal size distribution in a very similar setup.

For the mathematical description of particulate systems the population balance modelling (PBM) is the widely accepted and applied approach [8]. Since the introduction of PBM's a large number of publications was addressed to this topic. The PBM's, from the most basic form of nucleation and growth assuming one crystal dimension have been extended to multidimensional cases and involving additional crystallization mechanisms such as fragmentation, agglomeration, crystal-solution heat transfer [9] *etc*. From process engineering perspective, the numerical solution of the PB equations (PBE) is the bottleneck of crystallization simulation and is often more challenging that the derivation of the PBM. Numerous methods have been addressed to PBE solution from the moment based methods, through Monte Carlo simulations and method of characteristics to discretization based techniques. To improve the real time applicability of full PBM based simulations, the solution time is usually improved by various techniques from using adaptive grid, through application of parallel and GPU calculations to coordinate system transformations.

Despite of the fact that PBM modelling was successfully applied for various crystallization systems and that using a recycle wet-mill can considerably widen the attainable crystal size domain of batch crystallizers a study for the optimization of such a promising integrated system has not been published yet. The first objective of this work is to develop a PB based model for a batch crystallizer-wet mill system, involving primary nucleation, growth and dissolution for the crystallizer and fragmentation and attrition for the wet mill and to solve the equations efficiently and accurately [10]. The second objective of the study is to analyze and optimize the system through numerical experimentation. The authors address the current study as a quality-by-design (QbD) approach driven deeper understanding effort of simultaneous role of system integration and hidden potential of process optimization.

The optimization results indicate that the target crystal size distribution can be successfully achieved in the cases when the pure crystallizer cannot even approximate the shape of target CSD. The optimized system in all cases presents similar operation pattern, which can be summarized in 4 point:

1) The crystallizer is cooled quickly to produces some crystals

2) The pump is turned on to transport the crystals to the external wet mill (the mill is not running yet)

3) The pump is turned off (the crystallizer and wet mill is decoupled from each other). During this, the crystallizer is heated up to the solubility temperature (complete dissolution occurs) while the wet mill operates with high rotation speed (the existing crystals are milled).

4) When the crystals are sufficiently milled, the system is started to be cooled in controlled manner while the seeds are added dynamically through the recirculation stream.

Consequently, the optimal operation of integrated crystallizer-external wet mill system utilizes the strengths of optimal dynamic seeding and in situ seed generation, making unnecessary the time and high accuracy demanding seed preparation step.

Nevertheless, the system model was not validated for a particular system, in which context the results are not valid for any practical crystallization. Although, real crystallization kinetics and a realistic breakage model was used. Indeed, if the nucleation, growth, dissolution and breakage parameters would be change the attainable crystal size domain would certainly be modified but the contribution of this work, the optimal trajectories of temperature, flowrate and wet mill rotation speed, would remain the same.

References

[1] A. Mersmann, *Crystallization Technology Handbook*. Marcel Dekker Inc., New York, Basel, 2001.

[2] Z. K. Nagy and R. D. Braatz, "Robust nonlinear model predictive control of batch processes," *AIChE J.*, vol. 49, no. 7, pp. 1776–1786, Jul. 2003.

[3] Z. K. Nagy, G. Fevotte, H. Kramer, and L. L. Simon, "Recent advances in the monitoring, modelling and control of crystallization systems," *Chem. Eng. Res. Des.*, vol. 91, no. 10, pp. 1903–1922, Oct. 2013.

[4] T. Vetter, C. L. Burcham, and M. F. Doherty, "Regions of attainable particle sizes in continuous and batch crystallization processes," *Chem. Eng. Sci.*, vol. 106, pp. 167–180, Mar. 2014.

[5] B. Szilágyi and B. G. Lakatos, "Model-based analysis of stirred cooling crystallizer of high aspect ratio crystals with linear and nonlinear breakage," *Comput. Chem. Eng.*, vol. 98, pp. 180–196, 2017.

[6] Y. Yang, L. Song, T. Gao, and Z. K. Nagy, "Integrated Upstream and Downstream Application of Wet Milling with Continuous Mixed Suspension Mixed Product Removal Crystallization," *Cryst. Growth Des.*, vol. 15, no. 12, pp. 5879–5885, 2015.

[7] Y. Yang, L. Song, Y. Zhang, and Z. K. Nagy, "Application of wet milling-based automated direct nucleation control in continuous cooling crystallization processes," *Ind. Eng. Chem. Res.*, vol. 55, no. 17, pp. 4987–4996, 2016.

[8] H. M. Hulburt and S. Katz, "Some problems in particle technology. A statistical mechanical formulation," *Chem. Eng. Sci.*, vol. 19, pp. 555–574, 1964.

[9] B. G. Lakatos and B. Szilagyi, "Modeling Crystallization from Solution with Heat Effects," *Cryst. Growth Des.*, vol. 15, no. 12, pp. 5726–5737, 2015.

[10] B. Szilágyi and Z. K. Nagy, "Graphical Processing Unit (GPU) Acceleration for Numerical Solution of Population Balance Models Using High Resolution Finite Volume Algorithm," *Comput. Chem. Eng.*, 2016.

500509 A Cellular Automata Approach for Simulation of Crystal Growth

Abstract:

Cellular Automata (CA) provides a computational platform for combining micro-scale fluctuation theory with macro-scale field models to predict discontinuities, sharp transitions and moving boundaries in solidification of impure systems. The CA system can consider solute and heat conservation equations subject to complex boundary conditions at the solid-liquid interface. Constitutional and curvature undercooling on the equilibrium interface tare easy to incorporate. These variables are needed to predict the boundary between stable crystal growth and the unstable dendritic growth that lead to defect formation in many applications. The stochastic procedures in place for nucleation and dendritic growth can also be applied without making strong assumptions about the underlying rate processes. The price one has to pay can be significant computational cost due to the detailed modeling of the micro-structure.

The aim of the current paper is to consider the application of CA to the Horizontal Ribbon Growth (HRG) process. HRG was proposed by Shockley for producing single crystalline silicon sheets [1]. It is a continuous process and it has potential to overcome the limitations of traditional batch-scale wafer processes based on Czochralski crystallization and wire sawing. In HRG, a thin silicon solid sheet is produced and extracted continuously from a molten silicon pool, minimizing in this way material losses. In order to achieve fast production speed, the pulling rate has to be determined in conjunction with proper cooling set up and the presence of impurities may introduce defects and other instabilities that must be controlled.

Many technical challenges have been encountered in the effort to stable and fast ribbon production in the HRG process. Experiments show that sudden dendritic growth occurs at the crystal front when the pulling speed exceeds a threshold which depends on a number of parameters in the system including the accumulation of impurities [2]. Experimental results have shown cases of non-smooth and unstable solid-liquid interface and sharp wedge has occurred [3].

In the current paper we develop the CA approach for thin film crystal growth and show how the models can be used to derive stability/instability boundaries and control strategies to suppress these. We show results that demonstrate computational tractability and we present parametric tests to show how different cooling conditions, impurity concentrations, and other process related parameters relate to process stability. A method to control the process is proposed and it shown how model results relate to experimental results reported in the open literature.

References:

1. William, S. (1962). *U.S. Patent No. 3,031,275*. Washington, DC: U.S. Patent and Trademark Office.
2. Kudo, B. (1980). Improvements in the horizontal ribbon growth technique for single crystal silicon. *Journal of Crystal Growth*, *50*(1), 247-259.
3. Kellerman, P., Kernan, B., Helenbrook, B. T., Sun, D., Sinclair, F., & Carlson, F. (2016). Floating Silicon Method single crystal ribbonâ€"observations and proposed limit cycle theory.Â *Journal of Crystal Growth*,Â *451*, 174-180.

Optimal Crystal Size Control using a Plug Flow Crystallizer Configuration with Recycle

Xiaodong Xu, Yuan Yuan, Stevan Dubljevic

Department of Chemical and Materials Engineering
University of Alberta, Edmonton, Canada

1Nov 2017

American Chemical Engineering Conference
Minneapolis

Outline

A. Introduction

B. Modeling of Plug Flow Crystallizer with Recycle

C. Problem Statement

D. Steady State

E. Case Study and Simulation Study

F. Conclusion

Introduction

Batch process

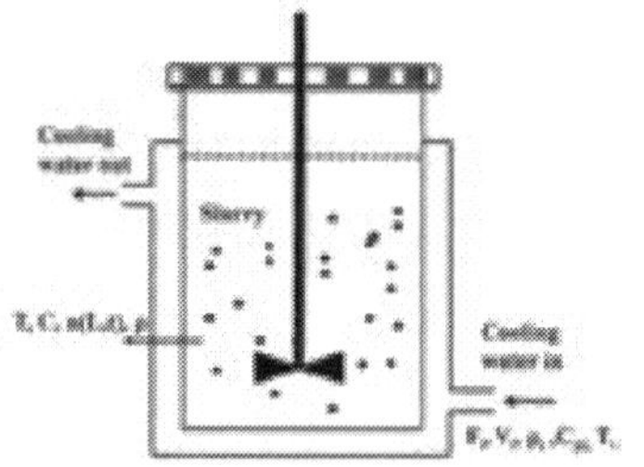

MSMPR

Plug flow process

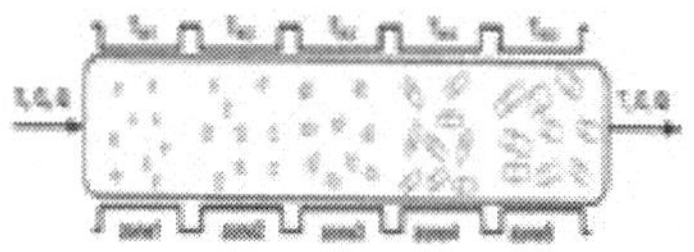

- More manipulated variables: T, v_z;
- Control of crystal size and shape distribution;
- Produce narrow size and shape distribution.
- Contributions:
  - Myerson, A.S, 2002, Handbook of Ind. Crystal., B.-H. Elsevier Science, Boston
  - Debasis *et all*, 2006, Chem. Eng. Sci
  - Alvarez, A.J. *at all*, 2010, Cryst. Growth Des.
  - Nagy, Z.K and R.D. Braatz, 2012, Advanced and new directions in crystallization control, Ann. Rev. Chem. Bio. Eng.
  - Vetter *et. all.*, Chem. Eng. Sci. 2014

Introduction

PFC process with recycle

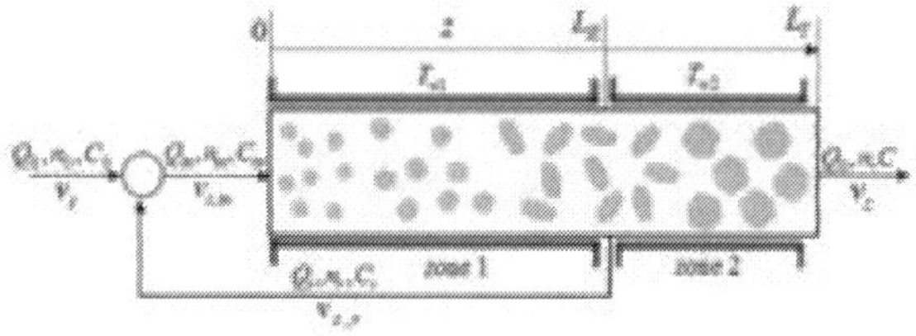

- $r = \frac{Q_r}{Q_0}$
- $v_{z,in} = (1+r)v_z$
- $C_{in} = \frac{rC(L_E)+C_0}{1+r}$
- $n_{in} = \frac{r}{1+r}n(L_E)$

Manipulated variables:

- Jacket temperature: T_{w1}, T_{w2}
- Flow velocity: v_z
- Extract position: L_E
- Recycle ratio: r

Mass balance

First part $0 \le z \le L_E$

Solution concentration:

$$\frac{\partial C_1}{\partial t} = -(1+r)v_z \frac{\partial C_1}{\partial z} - \rho_c \int_0^\infty G_v(\sigma_1) n_1(V,z,t)dV$$

$$C_1(0,t) = \frac{C_0 + rC_1(L_E,t)}{1+r}$$

$$C_1(z,0) = C_0$$

Second part $L_E \le z \le L_T$

Solution concentration:

$$\frac{\partial C_2}{\partial t} = -v_z \frac{\partial C_2}{\partial z} - \rho_c \int_0^\infty G_v(\sigma_2) n_2(V,z,t)dV$$

$$C_2(L_E,t) = C_1(L_E,t)$$

$$C_2(z,0) = C_0$$

- Solubility: $s(T) = 2.88 \times 10^{-4}T^3 - 1.65 \times 10^{-3}T^2 + 4.62 \times 10^{-2}T + 6.01 \times 10^{-1}$
- Supersaturation: $\sigma = \ln\left(\frac{C}{s}\right)$
- Growth rate: $G_v(\sigma) = 2G_{110}h_{110}h_{101} + G_{101}h_{110}^2$

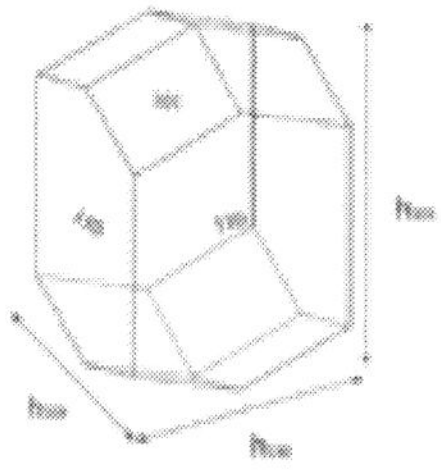

G. Cogni *et. all* Chemical Engineering Science, 138 (2015), 592–599

Energy balance model

First part $0 \le z \le L_E$

Temperature:

$$\frac{\partial T_1}{\partial t} = -(1+r)v_z \frac{\partial T_1}{\partial z} - \frac{\rho_c \Delta H_c}{\rho C_p} \int_0^\infty G_v(\sigma_1) n_1(V,z,t)dV - \frac{U_c A_c}{\rho C_p}(T_1 - T_{w1})$$

$$T_1(0,t) = \frac{T_0 + rT_1(L_E,t)}{1+r}$$

$$T_1(z,0) = T_0$$

Second part $L_E \le z \le L_T$

Temperature:

$$\frac{\partial T_2}{\partial t} = -v_z \frac{\partial T_2}{\partial z} - \frac{\rho_c \Delta H_c}{\rho C_p} \int_0^\infty G_v(\sigma_2) n_2(V,z,t)dV - \frac{U_c A_c}{\rho C_p}(T_2 - T_{w2})$$

$$T_2(L_E,t) = T_1(L_E,t)$$

$$T_2(z,0) = T_0$$

- Solubility: $s(T) = 2.88 \times 10^{-4}T^3 - 1.65 \times 10^{-3}T^2 + 4.62 \times 10^{-2}T + 6.01 \times 10^{-1}$
- Supersaturation: $\sigma = \ln\left(\frac{C}{s}\right)$
- Growth rate: $G_v(\sigma) = 2G_{110}h_{110}h_{101} + G_{101}h_{110}^2$
- Super-density: $\rho = 1 + C(z,t)$ [g/cm^3]

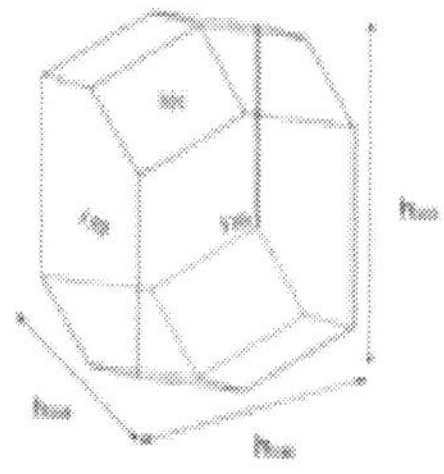

Population balance model

First part $0 \le z \le L_E$

NDF:

$$\frac{\partial n_1}{\partial t} = -(1+r)v_z\frac{\partial n_1}{\partial z} - \frac{\partial (G_v(\sigma_1)n_1)}{\partial V} - B_{seed}\delta(V - V_0, z - z_0)$$

$$n_1(V,0,t) = \frac{r}{1+r}n_1(V,L_E,t)$$

$$n_1(0,z,t) = 0$$

$$n_1(V,z,0) = 0$$

Second part $L_E \le z \le L_T$

NDF:

$$\frac{\partial n_2}{\partial t} = -v_z\frac{\partial n_2}{\partial z} - \frac{\partial (G_v(\sigma_2)n_2)}{\partial V} - B_{seed}\delta(V - V_0, z - z_0)$$

$$n_2(V,L_E,t) = n_1(V,L_E,t)$$

$$n_2(0,z,t) = 0$$

$$n_2(V,z,0) = 0$$

- Crystal volume: $V(z,t) = (h_{110}(z,t))^2 h_{101}(z,t)$
- Growth rate (110): $G_{110} = 0.1843 \times \sigma^3 - 1.1699 \times \sigma^3 + 2.8885 \times \sigma^3 - 2.5616$
- Growth rate (101): $G_{101} = 0.1893 \times \sigma^3 - 1.2264 \times \sigma^3 + 2.9887 \times \sigma^3 - 2.5348$
- Heights: $\frac{dh_{110}(z,t)}{dz} = \frac{G_{110}}{v_z}$, $\frac{dh_{101}(z,t)}{dz} = \frac{G_{101}}{v_z}$

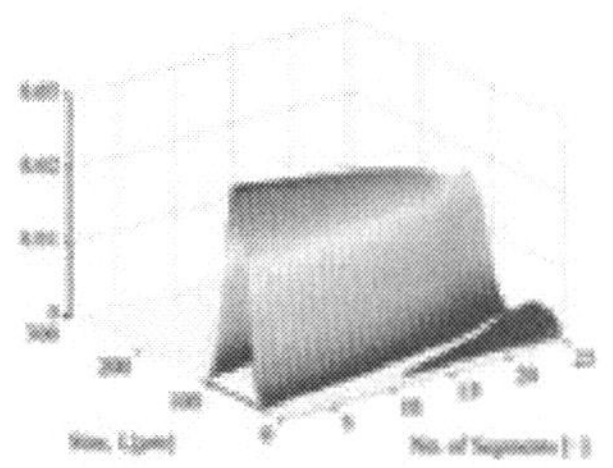

7 / 22

Problem statement

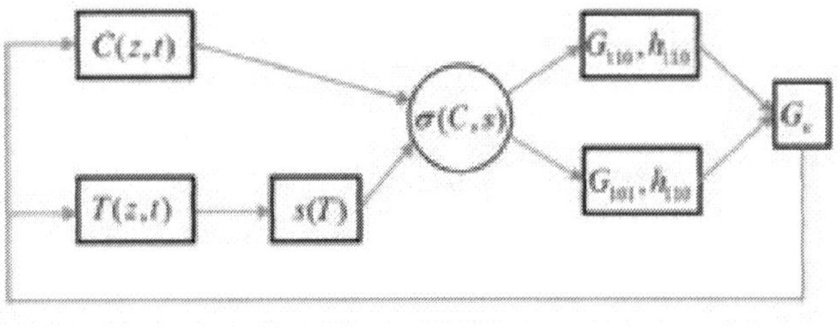

- Nonlinearity
- Coupling

- Design problem:

$$\min_{T_{w1},T_{w2},L_E} \gamma_1 L_T - \gamma_2\eta, \eta = \left[1 - \frac{C(L_T)}{C_{IN}}\right] \times 100$$

- Control problem:

$$\min_{T_{w1},T_{w2},r,L_E} \int_0^{t_f}\int_0^{L_T} \left[w_1(\alpha(z,t) - \alpha_{set})^2 + w_2(h_{110}(z,t) - h_{set})^2\right] dzdt$$

8 / 22

Method of moments

- Define the kth moment as:

$$M_k(z,t) = \int_0^\infty V^k n(V,z,t)dV$$

$M_0(z,t)$: Total number of crystals; $M_1(z,t)$: Total volume of crystals.

First part $z \in [0, L_E]$

$$\frac{\partial M_{0,1}}{\partial t} = -(1+r)v_z \frac{\partial M_{0,1}}{\partial z}$$
$$\frac{\partial M_{k,1}}{\partial t} = -(1+r)v_z \frac{\partial M_{k,1}}{\partial z} + kG_v(\sigma_1)M_{k-1,1}$$
$$M_{0,1}(0,t) = \frac{B_{seed}}{(1+r)^2 v_z} + \frac{r}{1+r} M_{0,1}(L_E,t)$$
$$M_{k,1}(0,t) = \frac{B_{seed}V_0^k}{(1+r)^2 v_z} + \frac{r}{1+r} M_{k,1}(L_E,t)$$

- B_{seed}: Seeding rate

Second part $z \in [L_E, L_T]$

$$\frac{\partial M_{0,2}}{\partial t} = -v_z \frac{\partial M_{0,2}}{\partial z}$$
$$\frac{\partial M_{k,2}}{\partial t} = -v_z \frac{\partial M_{k,2}}{\partial z} + kG_v(\sigma_2)M_{k-1,2}$$
$$M_{0,2}(L_E,t) = M_{0,1}(L_E,t)$$
$$M_{k,2}(L_E,t) = M_{k,1}(L_E,t)$$

- V_0: Seed size

Steady state model

- Moments

$$\frac{\partial M_{0,1}}{\partial z} = 0, \quad \frac{\partial M_{k,1}}{\partial z} = \frac{kG_v(\sigma_1)}{(1+r)v_z} M_{k-1,1}$$
$$M_{0,1}(0) = \frac{B_{seed}}{(1+r)^2 v_z} + \frac{rM_{0,1}(L_E)}{1+r}$$
$$M_{k,1}(0) = \frac{B_{seed}V_0^k}{(1+r)^2 v_z} + \frac{rM_{k,1}(L_E)}{1+r}$$

$$\frac{\partial M_{0,2}}{\partial z} = 0, \quad \frac{\partial M_{k,2}}{\partial z} = \frac{kG_v(\sigma_2)}{(1+r)v_z} M_{k-1,2}$$
$$M_{0,2}(L_E) = M_{0,1}(L_E)$$
$$M_{k,2}(L_E) = M_{k,1}(L_E)$$

- Concentration

$$\frac{\partial C_1}{\partial z} = \frac{\rho_c}{(1+r)v_z} G_v(\sigma_1) M_{0,1}$$
$$C_1(0) = \frac{C_0}{1+r} + \frac{r}{1+r} C_1(L_E)$$

$$\frac{\partial C_2}{\partial z} = \frac{\rho_c}{(1+r)v_z} G_v(\sigma_2) M_{0,2}$$
$$C_2(L_E) = C_1(L_E)$$

- Temperature

$$\frac{\partial T_1}{\partial z} = -\frac{\rho_c G_v(\sigma_1) M_{0,1}}{\rho C_p (1+r) v_z} - \frac{U_c A_c (T_1 - T_{w1})}{\rho C_p (1+r) v_z}$$
$$T_1(0) = \frac{T_0}{1+r} + \frac{r}{1+r} T_1(L_E)$$

$$\frac{\partial T_2}{\partial z} = -\frac{\rho_c G_v(\sigma_2) M_{0,2}}{\rho C_p v_z} - \frac{U_c A_c (T_2 - T_{w2})}{\rho C_p v_z}$$
$$T_2(L_E) = T_1(L_E)$$

Optimization for steady state model

$$\min_{T_{w1},T_{w2},v_z} \int_{L_E}^{L_T} w_1(\alpha(z)-\alpha_{set})^2 + w_2(h_{110}(z)-h_{set})^2 dz$$

$s.t.$

$$G_{110} = 0.1843 \times \sigma^3 - 1.1699 \times \sigma^3 + 2.8885 \times \sigma^3 - 2.5616$$

$$G_{101} = 0.1893 \times \sigma^3 - 1.2264 \times \sigma^3 + 2.9887 \times \sigma^3 - 2.5348$$

$$G_v = 2G_{110}h_{110}h_{101} + G_{101}h_{110}^2$$

$$\frac{dh_{110}}{dz} = G_{110}, \frac{dh_{101}}{dz} = G_{101}, \alpha(z,t) \approx \frac{h_{110}(z,t)}{h_{101}(z,t)}$$

Manipulated variables:

- The superficial velocity v_z has to be a constant
- T_{w1} and T_{w2} can be spatially varying or constants along the space

Case study

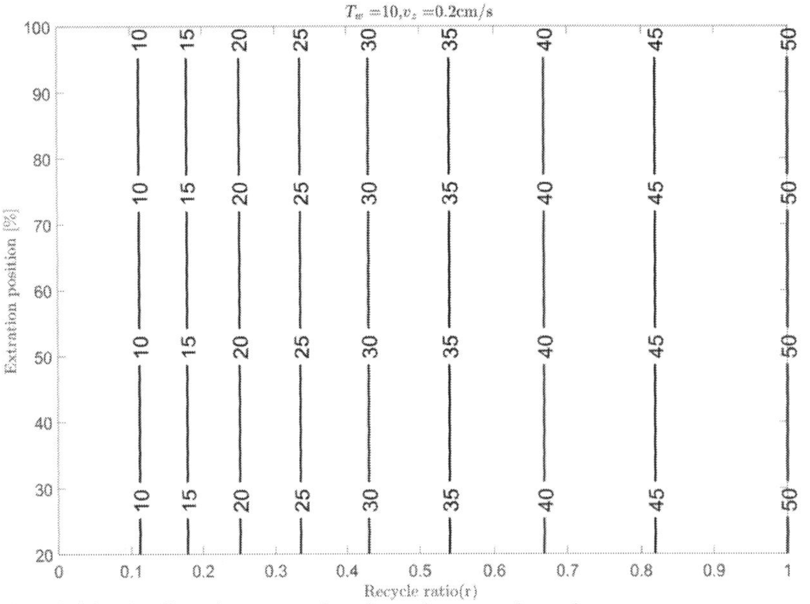

- The yield η is directly proportional to the recycle ratio r.

Case study

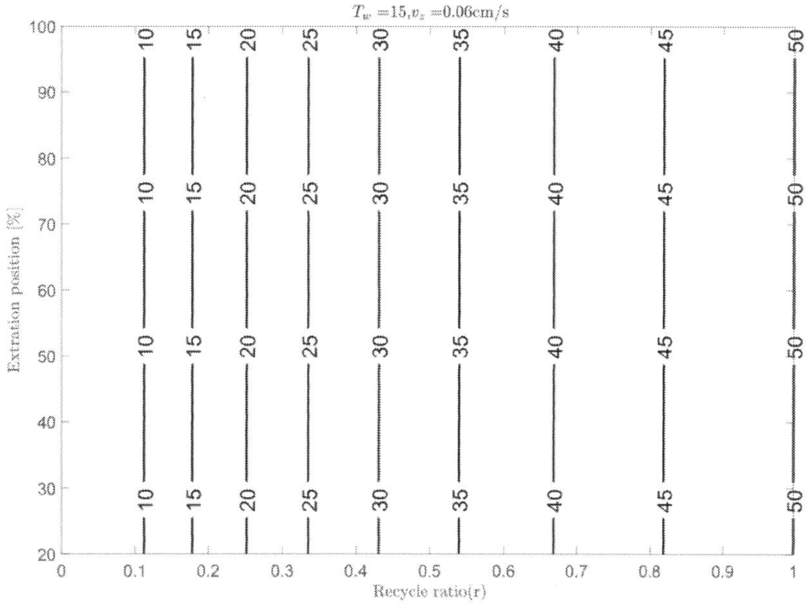

- The yield η is slightly influenced by the extraction position.

Case study

The shape of crystal

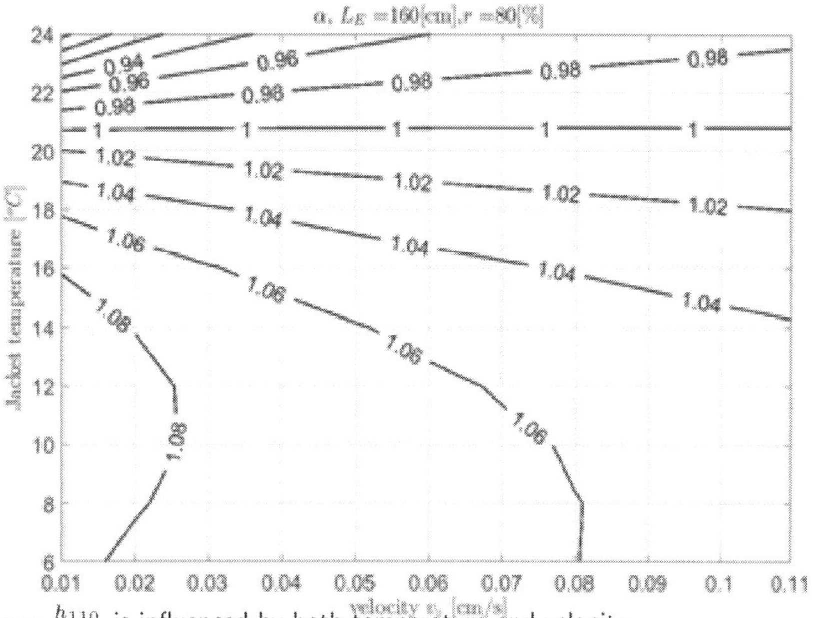

- $\alpha = \frac{h_{110}}{h_{101}}$ is influenced by both temperature and velocity.

Case study

The height of crystal (h_{110})

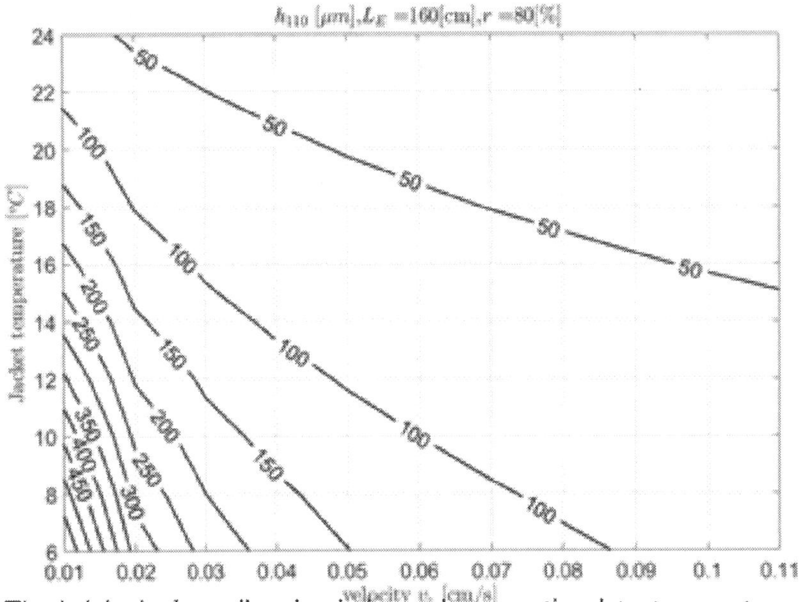

- The height in h_{110} direction is inversely proportional to temperature and flow velocity.

- The recycle ratio and the extraction position can be chosen to maintain appropriate yield.
- Temperature and superficial flow velocity are chosen to obtain desired crystals shape and size distribution.

Multivariable optimization problem

The resulting optimization problem:

$$\min_{T_{w1},T_{w2},vz} \int_0^{t_f} \int_{L_E}^{L_T} w_1(\alpha(z,t)-\alpha_{set})^2 + w_2(h_{110}(z,t)-h_{set})^2 dzdt$$

$s.t.$

$$G_{110} = 0.1843\times\sigma^3 - 1.1699\times\sigma^3 + 2.8885\times\sigma^3 - 2.5616$$

$$G_{101} = 0.1893\times\sigma^3 - 1.2264\times\sigma^3 + 2.9887\times\sigma^3 - 2.5348$$

$$G_v = 2G_{110}h_{110}h_{101} + G_{101}h_{110}^2$$

$$\frac{dh_{110}}{dz} = G_{110},\ \frac{dh_{101}}{dz} = G_{101}, \alpha(z,t) \approx \frac{h_{110}(z,t)}{h_{101}(z,t)}$$

$$\frac{\partial M_{0,1}}{\partial t} = -(1+r)v_z\frac{\partial M_{0,1}}{\partial z},\ \frac{\partial M_{k,1}}{\partial t} = -(1+r)v_z\frac{\partial M_{k,1}}{\partial z} + kG_v(\sigma_1)M_{k-1,1}$$

$$\frac{\partial M_{0,2}}{\partial t} = -v_z\frac{\partial M_{0,2}}{\partial z},\ \frac{\partial M_{k,2}}{\partial t} = -v_z\frac{\partial M_{k,2}}{\partial z} + kG_v(\sigma_2)M_{k-1,2}$$

$$\frac{\partial C_1}{\partial t} = -(1+r)v_z\frac{\partial C_1}{\partial z} - \rho_c G_v(\sigma_1)M_{0,1}$$

$$\frac{\partial C_2}{\partial t} = -v_z\frac{\partial C_2}{\partial z} - \rho_c G_v(\sigma_2)M_{0,2}$$

$$\frac{\partial T_1}{\partial t} = -(1+r)v_z\frac{\partial T_1}{\partial z} - \frac{\rho_c\Delta H_c G_v(\sigma_1)M_{0,1}}{\rho C_p} - \frac{U_c A_c}{\rho C_p}(T_1 - T_{w1})$$

$$\frac{\partial T_2}{\partial t} = -v_z\frac{\partial T_2}{\partial z} - \frac{\rho_c\Delta H_c G_v(\sigma_2)M_{0,2}}{\rho C_p} - \frac{U_c A_c}{\rho C_p}(T_2 - T_{w2})$$

Shape: $\alpha = 0.88$

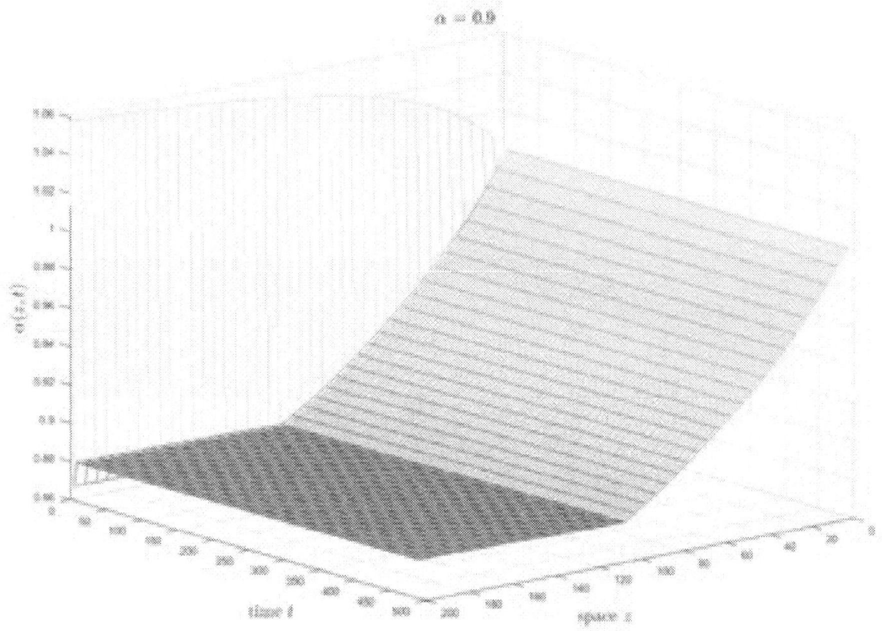

Height: $h_{110} = 130\mu m$

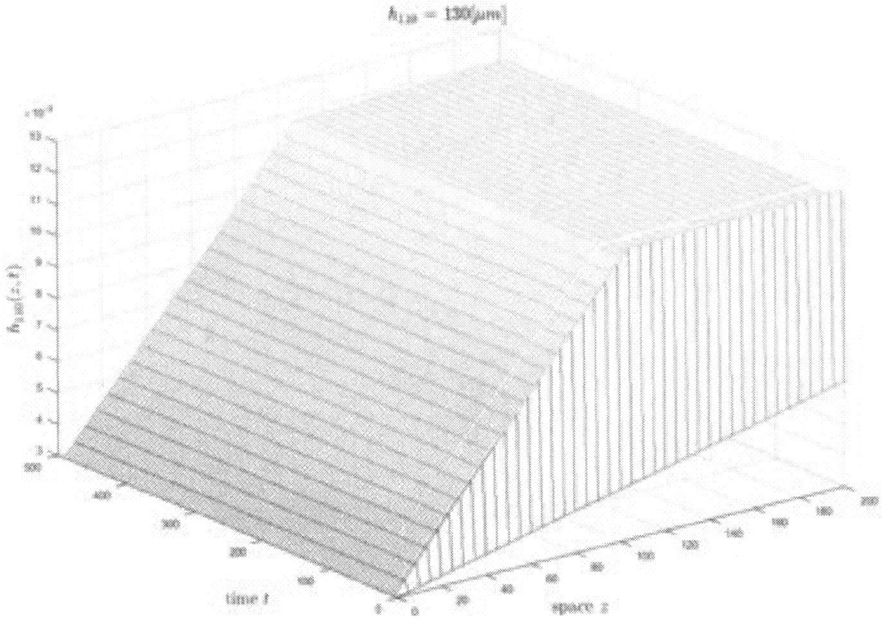

Temperature

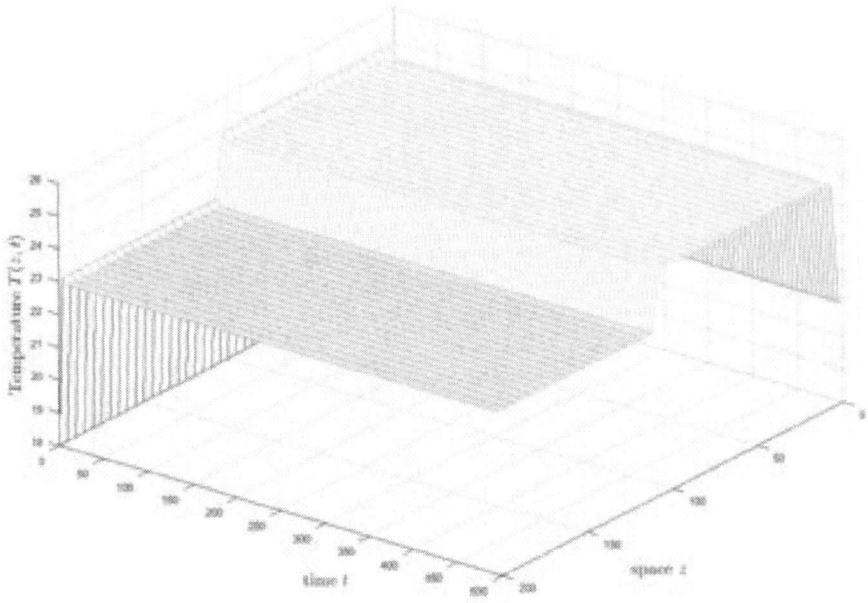

Concentration

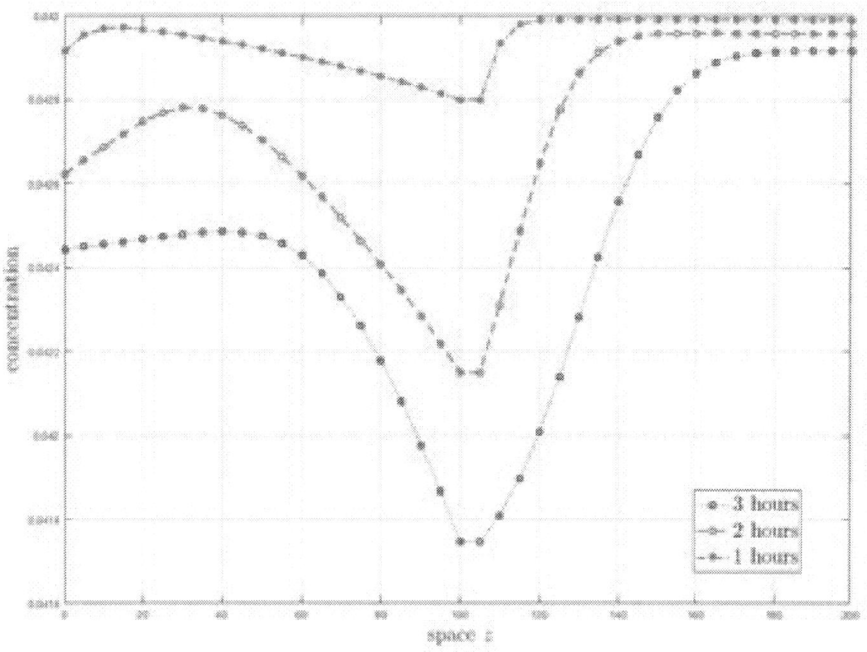

Conclusion:

- Model of PFC process with recycle and crystals growing from seeds
  - Mass balance
  - Energy balance
  - Population balance
- Reduced-order moments model: method of moment:

$$M_k(z,t) = \int_0^\infty V^k n(V,z,t)dV$$

- Manipulated variables: r,L_E, T_{w1}, T_{w2} and v_z.
  - yield
  - shape and size
- MOP for desired shape and size distributions

497753 Multiphysics Modeling and Simulation of Microfluidic Platforms for Screening of Pharmaceutical PolymorphsÂ

Abstract:

Polymorphs are the crystalline materials with different molecular arrangements and hence varying physicochemical properties such as bioavailability, chemical stability, and mechanical strength. Almost 60% of marketed oral drugs are produced in the solid form, whose quality is mostly determined by their polymorphic structure. Due to the regulatory issues, it is essential to develop a platform to effectively screen these polymorphs to identify stable and metastable forms of drugs. In the pharmaceutical industry, the commonly employed technique to screen polymorphs is a microtiter plate method which uses 96-well, 384-well, and 1586-well plates. Such high-throughput technique can evaluate up to 10,000 compounds per day. However, setbacks are often encountered during scale-up due to differences in the environmental conditions of a microtiter plate and a real crystallizer. Here we propose a design of a novel microfluidic crystallizer that can potentially overcome the drawbacks of the microtiter plate method by inducing hydrodynamically controlled nucleation and growth of single crystals. Our microfluidic platform can control the supersaturation at the stagnation point where single crystal can nucleate and grow. In this talk, I will present the results from the Multiphysics simulations of microfluidic crystallization and propose an optimal design to enable high-throughput screening of polymorphs.

497114 Motion-Based Multiple Object Tracking of Ultrasonic-Induced nucleation: A Case Study of L-Glutamic Acid

Abstract:

Motion-Based Multiple Object Tracking of Ultrasonic-Induced Nucleation: A Case Study of L-glutamic Acid

Zhenguo Gao [1,2] Dan Zhu [1,2] Yuanyi Wu [1] Sohrab Rohani [1]* Junbo Gong [2], Jingkang Wang [2]

1. The University of Western Ontario, Department of Chemical and Biochemical Engineering, London, Ontario N6A 5B9, Canada

2. Tianjin University, School of Chemical Engineering and Technology, Tianjin, 300072, P. R. China

Keywords: Ultrasonic irradiation, Nucleation, Motion-based multiple object tracking <>Introduction

Nucleation from solution has aroused great interest over the past decades. Scientists have considered atomistic levels on the order of 10^{-10} m and time scales in the order of 10^{-13} s to unravel the nucleation mechanism with the help of advanced instruments[1] . However, in industrial crystallization, there is still a limitation in monitoring nucleation at the initial stage of critical nucleus formation[2] . Existing measuring techniques used to monitor nucleation include the focused beam reflection measurement (FBRM), turbidity, ultrasonic velocity, electrical conductivity and light transmittance measurement of bulk solution[3] . The accuracy of the monitoring devices affects the measurement of nucleation kinetics. In order to improve the accuracy, progress has been made based on high-speed imaging, multivariate statistical process monitoring charts, etc[4,5] . Recently, motion-based multiple object tracking (MMOT) has been widely used in computer vision field. The MMOT method detects moving objects in a video stream, predicts their locations in the next frame, and records the object count automatically, which shows its potential application in monitoring the occurrence and disappearance of nuclei in a bulk solution.

External factors such as ultrasound, microwave, magnetic field and electric field have been studied to control crystal nucleation, which exhibit effectiveness in some cases[6,7] . Ultrasonic irradiation creates sequential compression and expansion of the solution, which leads to bubble formation and growth. Finally, the bubbles collapse and release energy, promoting nucleation within a short induction time and at a lower supersaturation level. Parameters, e.g. ultrasound power, frequency, with or without pulse, have significant influence on nucleation process and crystal qualities[8,9] . In this study, the MMOT was introduced in the crystallization area, for the first time, to monitor the nucleation of L-glutamic acid. Results showed that ultrasonic irradiation shortened the induction time dramatically. The MMOT-based device showed higher accuracy and precision compared to the FBRM technology as the bubbles were transparent in the MMOT technology, but considered as nuclei by the FBRM. <>Materials and Experiments

L-glutamic acid (LGA) was used as model compound. An ultrasonic processor was inserted into a double jacketed crystallizer to generate ultrasound. During nucleation process, a waterproof USB-based micro-camera (Magnification: 1-300X) was used to record a video stream to monitor the nucleation process as shown in Figure 1(a). A home-designed vial adaptor with a solution flow channel (Figure 1b) was printed by a 3D printer and connected to the end of the micro-camera to allow capture crystal images as the slurry flowed through the adaptor. Figure 2 shows the experimental setup. In the first step, clear solutions were prepared in the crystallizer at 80 . Afterwards, nucleation temperature was reached as soon as possible by switching the 3-way valve to the cooling refrigerated circulator. The temperature was kept constant to study the nucleation kinetics until the crystals appeared in the video stream. The images recorded by micro-camera were parsed by MMOT in MATLAB.

Figure 1. (a) Portable USB-based micro-camera setup; (b) 3D print vial adaptor.

Figure 2. (a) Schematic of experimental setup; **(b)** Experimental setup picture. Results and Discussions

Different conditions of solution concentration (30 g/L and 40 g/L) and nucleation temperature (35 and 45) were tested. The tracking model is solely based on motion, in which the background is subtracted and a Kalman filter is used to predict the assigned track's location in next frame. The nuclei count in MMOT was smoothed by a moving average (MA) model as shown in equations 1 and 2.

$$\text{MA} = \qquad = \qquad (1)$$

$$\text{MA}_{\text{current}} = \text{MA}_{\text{prev}} + \qquad (2)$$

Where c indicates the nuclei counts along the time series and MA_{prev} is the previous averaged value. The moving window n was optimized to be 30 which tightly fitted the nuclei counts. The time at which a continuous increase in the micro-camera counts was sensed was taken as the onset of nucleation. The induction time was calculated as the period between the time the ultrasound processor was switched on and the onset of nuclei occurrence. Under the ultrasonic irradiation at 40 g/L and 35 , Figure 3 and Figure 4 show the results at 38 s and 95 s of same process after the ultrasonic processor was started.

Figure 3. Detection results at 38 s with ultrasonic irradiation (14 W). **(a)** In-situ plot of nuclei counts versus time based on MMOT; **(b)** Plot of nuclei counts against time of the whole process. The blue line and black line are raw data and smoothed data. **(c)** Detected objects marked in an original video. **(d)** Detected objects marked in the video after background subtraction.

Figure 4. Detection results at 95 s with ultrasonic irradiation (14 W). **(a)** In-situ plot of nuclei counts versus time based on MMOT; **(b)** Plot of nuclei counts against time of the whole process. The blue line and black line are raw data and smoothed data. **(c)** Detected objects marked in an original video. **(d)** Detected objects marked in the video after background subtraction.

As shown in Figure 3c and 4c, only one object had been detected, which indicates no nucleation occurred before 38 s, while large quantities of objects had been detected after 95 s. A continuous increase of nuclei counts was detected after 72 s in Figure 4a, with the constant nucleation occurring in raw image (Figure 4c), which confirmed the onset of nucleation at 72 s. Comparing Figure 4a with Figure 3a, it can be concluded that the threshold of background noise is around 5, which can be ignored before the continuous increase of nuclei count. At least six experiments were repeated under the same condition and average results were reported in Table 1.

Table 1. Comparison of induction time and crystal form of LGA with/without ultrasonic irradiation.

Ultrasonic power (W)	Concentration (g/L)	Nucleation temperature ( )	S_0, a (C_0/C^*)	S_0, ß (C_0/C^*)	Induction time (s)	Standard deviation (s)	Crystal form
14	30	35	1.91	2.51	181.3	8.2209	a
14	40	35	2.54	3.35	77.8	4.0249	a
14	30	45	1.31	1.78	585.6	22.1088	a, ß
14	40	45	1.74	2.37	113.0	7.7136	a, ß
0	30	35	1.91	2.51	1824.3	100.8806	a
0	40	35	2.54	3.35	381.3	40.8084	a
0	30	45	1.31	1.78	3600.0	169.7056	ß
0	40	45	1.74	2.37	737.3	68.3002	ß

Note: Standard deviation was calculated by , and *X* means average of all (i=1, 2, 3..., n).

There is a sharp decrease in induction time compared with the nucleation kinetics without ultrasonic irradiation [10]. For a rising of nucleation temperature, the induction time is longer. Ultrasonic irradiation shows potential application to promote nucleation in crystallization process design. Experimental results show the measurement accuracy based on MMOT is much better than FBRM as the latter does not distinguish between the nuclei and the bubbles generated by ultrasound. However, limitation exists in MMOT method, which is similar to the PVM (Particle Vision and Measurement) technology, the detection limit of objects count is around 100 and the occurrence of multiple crystals can affect the accuracy. Conclusions

In this study, the MMOT method was, for the first time, introduced in the crystallization field to monitor nucleation process. Results show better accuracy compared with other methods such as the FBRM technology. The nuclei count measured by MMOT, which is different from the FBRM count, can be used as a parameter to control nucleation in designing crystallization processes. Based on the MMOT technology, potential applications can be developed to measure solubility, metastable zone width, particle size and even solution-mediated polymorphic transformation.

<>Reference

1. J. Zhang, P. Chen, B. Yuan, W. Ji, Z. Cheng, and X. Qiu, *Science*, 2013, **342**, 611–4.

2. A. S. Myerson, *Handbook of industry crystallization*, Butterworth-Heinemann, 2002.

3. C. Brandel and J. Horst, *Faraday discussions*, 2015, **179**, 199–214.

4. S. Bauerecker, P. Ulbig, V. Buch, L. Vrbka, and P. Jungwirth, *J Phys Chem C*, 2008, **112**, 7631–7636.

5. L. Simon, K. Oucherif, Z. Nagy, and K. Hungerbuhler, *Chem Eng Sci*, 2010, **65**, 4983–4995.

6. S. Jiang, University of Leeds, 2012.

7. D.-C. Yin, *Prog Cryst Growth Ch*, 2015, **61**, 1–26.

8. S. Bhangu, M. Ashokkumar, and J. Lee, *Cryst Growth Des*, 2016, **16**, 1934–1941.

9. M. Jiang, C. Papageorgiou, J. Waetzig, A. Hardy, M. Langston, and R. Braatz, *Cryst Growth Des*, 2015, **15**, 2486–2492.

10. Schöll, Vicum, Müller, and Mazzotti, *Chem Eng Technology*, 2006, **29**, 257–264.

494505 Understanding the Removal of Pharmaceuticals from Water Effluents By Adsorption in Activated Carbons – a Molecular Simulation Approach

Abstract:

There is an increased concern in recent years on the removal of pharmaceuticals from water effluents. Many of these compounds are at concentrations of mg/L or below, are hardly biodegradable even at the outlet of wastewater treatment plants and much on their potential impact on environment and human health are still not completely known. Among the main removal technologies reported in the literature, removal by adsorption is one of the most promising techniques because of its convenience when applied in current water treatment processes. Adsorption using activated carbon is ideal for removing organic compounds due to the availability of high surface area and combination of well-developed pore structure and surface functional group properties [1]. Hence, the identification of the mechanisms governing the adsorption phenomena of pharmaceuticals in aqueous solutions is of fundamental importance.

The main objectives of this study was to evaluate, by molecular simulations, the competitive adsorption of pharmaceutical compounds as pollutants contented in water on activated carbons of varied characteristics, and to compare the removal capacities from equilibrium simulations on single component and aqueous mixtures. Molecular simulations using the Grand Canonical Monte Carlo (GCMC) method have been performed in order to understand the interaction of pharmaceutical molecules (ibuprofen, diclofenac, naproxen, paracetamol and amoxicillin, all with similar dimensions, except amoxicillin) on activated carbons in both, pure and aqueous mixtures. "Realistic" nanoporous carbon models based on units of polyaromatic molecules with different number of rings, defects and polar-oxygenated sites have been employed for this purpose [2].

The modeling approach we have used for the activated carbons, inspired by the early work of Gubbins and co-authors (see, for instance, reference [3]) represents a middle ground between the slit pore model and the fully atomistic models of carbon; it is an extension of the model packing idealized structures first developed by Segarra and Glandt (SG) [4]. We have modeled graphene planes as platelets, but including polar groups and defects, not explicitly considered in the SG model, and in a similar fashion of the extended Liu-Monson model [5]. Furthermore, in order to create imperfections on the surface, increased surface roughness were created by deleting up to 10–20% of the atoms in the inner carbon layers. Details on the calculations can be found in [2].

Our methodology allowed us to obtain a successful prediction of the drug adsorption properties. Comparison of the simulated capacities with experimental reported values is highly consistent. The model is sufficiently flexible to tune its parameters in a systematic way and introduce additional features to reflect specific properties of the material of interest, *i.e.*, effects of platelet sizes and geometries, surface area, concentration and nature of surface groups, among others. Results confirm the role of surface sites in the enhancement of water adsorption and reduction of drug adsorption as the oxygenated sites exceeds

10%wt. Reduction of the surface functionality gives a higher adsorption capacity of pharmaceuticals in aqueous mixture, since there is not competition with water molecules.

This work is a further step in order predict macroscopic performance of activated carbons in drug removal applications, providing useful molecular-level insights into the drug/water adsorption mechanism and allowing finding the most favorable material.

References:

1. L. Nielsen, T. J. Bandosz, Analysis of the competitive adsorption of pharmaceuticals on waste

derived materials, Chemical Engineering Journal 287 (2016) 139–147.

2. D. Bahamon, L. Carro, S. Guri, L.F. Vega, Computational study of ibuprofen removal from water by adsorption in realistic activated carbons, Journal of Colloid and Interface Science, 498 (2017), 323–334.

3. J. Pikunic, C. Clinard, N. Cohaut, K.E. Gubbins, J.M. Guet, R.J.M. Pellenq, I. Rannou, J.-N. Rouzaud, Structural modeling of porous carbons: constrained reverse Monte Carlo method, Langmuir 19 (2003) 8565–8582.

4. E.I. Segarra, E.D. Glandt, Model microporous carbons: microstructure, surface polarity and gas adsorption, Chem. Eng. Sci. 49 (1994) 2953–2965.

5. J.C. Liu, P.A. Monson, Monte Carlo simulation study of water simulation study of water adsorption in activated carbon, Ind. Eng. Chem. Res. 45 (2006) 5649–5656.

504009 Insights into the Adsorption and Phase Behaviour of Fluids in Nanoporous Materials with Hierarchical Pore Structure:Â Towards an Advanced Textural Characterization

Abstract:

In recent years, major synthetic efforts have been focused on the introduction of hierarchical pore structures into many different materials such as zeolites, carbons, silicas, MOFs, hybrid layered and pillared structures. Nanoporous materials which consist of hierarchical pore networks with pore widths spanning the entire micro-mesopore range (i.e. up to 50 nm) allow one to overcome accessibility and diffusion limitations associated with classical microporous materials for catalysis and separations.

In this talk recent developments in the structural characterization of these materials will be described. An accurate textural characterization is crucial not only within the discovery process of novel hierarchically ordered nanoporous materials but also contributes to advance their application in a variety of areas including heterogenous catalysis, separation, battery devices, drug delivery. Among a variety of techniques, gas adsorption is widely applied because it assesses a wide range of pore sizes, spanning the entire micro- and mesopore range. Within the last two decades major progress has been achieved in physical adsorption characterization also because of the development of advanced approaches based on statistical mechanics such as molecular simulation and density functional theory (DFT). This progress, coupled with the availability of high resolution experimental methodologies for the adsorption of various subcritical fluids (also allowing one to couple adsorption techniques with complimentary techniques such as small angle scattering), has led to major advances in the structural characterization of nanoporous materials [1]. However, major challenges still exist concerning an in-*depth* characterization of complex pore networks. Application of characterization techniques to hierarchical materials has revealed new phenomena, such as cavitation processes during desorption, and led to significant refinement of these techniques [2].

[1] M. Thommes, K. Kaneko, A.V. Neimark, J.P. Olivier, F. Rodriguez Reinoso, J. Rouquerol and K.S.W Sing, (IUPAC Technical Report), *Pure Appl. Chem.* **87**, 1051 (2015)

[2] K. A. Cychosz, R. Guillet-Nicolas, J. Garcia-Martinez, J., M. Thommes, M. *Chem. Soc. Rev.* **46**, 389, (2017)

504010 Molecular Studies of Supercapacitors: Ionic Liquids Adsorbed into Porous Carbon Electrodes

Abstract:

Energy storage continues to attract the attention of the scientific and engineering research communities. Renewable energy sources, such as solar and wind, require so-called grid-based energy storage so that energy generated during periods of high energy production can be made available during periods of low or zero energy production. Mobile energy storage (for everything from mobile phones to automobiles and beyond) is clearly increasing exponentially in importance. Among the many choices for energy storage devices, electrical double layer capacitors (EDLCs), also called supercapacitors, are attracting considerable attention. Supercapacitors store electrical energy via ion electrosorption directly in the EDLs at the electrolyte-electrode interface, suggesting that such liquid-solid interfaces play a dominant role in the underlying energy storage mechanism and the resulting device performance. Because electrical energy in supercapacitors is stored based on physical phenomena rather than chemical reaction (as in batteries), supercapacitors have fast rates of charge/discharge and a virtually limitless number of charge cycles (unlike batteries, which are often limited to 10^4or less cycles). Much of the goal of supercapacitor research is aimed at increasing the amount of energy stored (energy density is the strong point in favor of batteries), which in turn focuses attention on the electrolyte, the nature of the electrode, and the electrode-electrolyte interactions.

To date, ionic liquids (ILs) have become emerging candidates for electrolytes used in supercapacitors, due to their exceptionally wide electrochemical window, excellent thermal stability, nonvolatility, and relatively inert nature; meanwhile carbons are the most widely used electrode materials in supercapacitors, due to their high specific surface area, good electrical conductivity, chemical stability in a variety of electrolytes, and relatively low cost. To improve the energy density and the transport properties of the charge carriers in supercapacitors, carbons have been developed in diverse forms such as activated carbons, carbon nanotubes (CNTs), onion-like carbons (OLCs), carbode-derived carbons and graphene. Using molecular modeling combined with molecular experimental probes, such as SAXS, SANS, NMR, and AFM, we report on our investigations into the interfacial phenomena occurring between the IL electrolytes and electrodes of varying geometries to understand the energy storage mechanism of supercapacitors that rely on EDLs established at IL-electrode interfaces.

504011 Structure of Ice in Confinement; Water in Mesopores

Abstract:

We report X â€“rays diffraction studies of water adsorbed in nanoporous activated carbon fibres (ACFs) and CMK-3 and CMK-8 carbon mesopores of different pore sizes. The fibres are built of turbostratic nanoparticles separated by quasi two-dimensional voids, forming narrow slit-shaped pores; CMK-3 and CMK-8 are the reverse carbon replica of silica SBA-15 and KIT-6 porous matrices. In order to determine the structure of water within the pores and its influence on the fibresâ€™ structure, mean interatomic and intermolecular distances have been estimated from the positions of the maxima of the normalized angular distribution functions obtained by X-ray diffraction [1]. We observe significant changes in the interlayer distance of the carbon nanoparticles; the results suggest that very high pressures arise within the pores,of the order a few hundreds MPa, as has been observed in molecular simulations [1,2]. Such results are confirmed by observation of high pressure forms of ice in cylindrical nanocarbons , using neutron and X â €“ray diffraction [3]. For water in multi-walled carbon nanotubes (MWCN), below the pore melting point cubic ice was observed. This structure of ice can be obtained during re-crystallisation from high pressure phases at low temperature for bulk water. For CMK-3 and CMK-8 mesopores we observed the existence of a stacking-disordered ice, I_{sd} [4,5]. This metastable ice is neither cubic nor hexagonal, and is not a simple mixture of the two, but a combination of cubic sequences intertwined with hexagonal sequences, which was identified as having the space group p3m1 [4]. Moreover, the stacking disorder can vary in complexity depending on the way the ice is formed and on the prevailing thermal conditions during this process. An analysis of the kind of I_{sd} formed in CMK-3 and CMK-8 of different pore sizes will be presented. These crystal forms, which occur in bulk water only at temperatures below 180 K in the case of cubic ice, and at pressures of hundreds or thousands of MPa are stabilized by the confinement.

References

[1] M.Sliwinska-Bartkowiak, M.Drozdowski, M.Jazdzewska, Y.Long, J.Palmer, K.E.Gubbins, Phys. Chem. Chem. Phys., **14**, 71454, 2012.

[2] Y. Long, J.Palmer, B.Coasne, M.Sliwinska-Bartkowiak, K.E.Gubbins, Phys. Chem. Chem. Phy*s* , **14** 17163, 2011 .

[3] M.Jazdzewska , M. Sliwinska-Bartkowiak, A.I. Beskrovnyy, S.G. Vasilovskiy, S.W. Ting, K.Y. Chan, L.L. Huang, K.E. Gubbins, Phys. Chem. Chem. Phys., **13** ,9008 ,2011

[4] Malkin, T. L., Murray, B. J., Salzmann, C., Molinero, V., Pickering, S. J., & Whale, T. F. Phys. Chem. Chem. Phys., **17(1)**, 60-76, 2014; A.Haji-Akbari, P.G.Debenedetti, PNAS, **112**, 10582, 2015

[5]K.Domin,K.Y.Chan,H.Yung,A.Sterczynska,M.Jurek,K.E.Gubbins, M.Sliwinska-Bartkowiak, J.Chem..Eng. Data, 61, 4252-4260, 2016

504012 Towards Understanding the Role of Microstructure in Energetic Material Response: Coarse-Grain Modeling and Simulation

Abstract:

Mechanical and thermal loading of energetic material (EM) composites can incite responses over a wide range of spatial and temporal scales due to inherent microscale features. Many energy transfer processes within these materials are atomistically governed, yet the overall material response is manifested at the micro- and mesoscale â€“ scales beyond those amenable to atomistic simulation techniques. Moreover, continuum level approaches rely on field-based formulations that are empirically based, which lack sufficient fidelity to capture microstructure effects.

Particle-based microscale simulation methods that utilize coarse-grain models offer a promising route for extending the attributes of atomistic modeling toward the mesoscale. We have developed such microscale methods based upon the constant-energy dissipative particle dynamics method that includes chemical reactions. Coarse-grain models of EMs have been built using the Multiscale Coarse-Graining approach. Time-resolved depictions of an EM responding to insult has been simulated for a variety of microstructures, including samples with intra- and inter-granular voids of varying shape, size and relative orientation, varying grain shapes and sizes, and polymer binder. A sampling of these results will be presented.

504014 Adsorption and Transport in Multiscale Porous Media

Abstract:

Hierarchical porous materials such as hierarchical zeolites, which combine several porosity scales, are widely used in industry (adsorption, separation, catalysis) to overcome slow diffusion in microporous solids (< 2 nm) and enhance access to their large surface area. Available modeling approaches for adsorption and transport in such multiscale porous media are limited to empirical parameters which cannot be derived from molecular coefficients. In particular, existing approaches do not offer the ground for a bottom up model of adsorption/transport in multiscale materials as (1) they describe empirically the adsorption/transport interplay and (2) they do not account for the breakdown of hydrodynamics at the nm scale.

In this talk, I will present a multiscale model of adsorption and transport in hierarchical materials obtained by adding mesopores (~ few nm) and macropores (> 10 nm) to existing microporous crystals [1]. I will first show how adsorption, permeance, and transport in such media can be described without having to rely on macroscopic concepts such as hydrodynamics [2,3]. Using fundamental parameters and coefficients available to simple experiments, we will see how transport coefficients can be rigorously obtained from simple models in the framework of Statistical Mechanics. Then, I will present a multiscale model of adsorption and transport in hierarchical materials [4]. This approach consists of upscaling accurate molecular simulations in a lattice model. Thanks to the use of atom-scale simulations, which capture the different adsorption and transport regimes upon varying the temperature, pore size, pressure, etc. this bottom-up model does not rely on hydrodynamics and, hence, does not require assuming a given adsorption or flow type. I will also discuss NMR experimental results on transport in hierarchical zeolites [5].

[1] B. Coasne et al., Langmuir 29, 7864 (2013).

[2] K. Falk et al. Nature Comm. 6, 6949 (2015).

[3] T. Lee et al. Nature Comm. 7, 11890 (2016).

[4] A. Botan et al. Phys. Rev. E 91, 032133 (2015).

[5] A. Galarneau et al., J. Phys. Chem. C 120, 1562 (2016).

501637 Regeneration of Ammonia-Loaded Metal Halide Absorbents

Abstract:

With more than 150 million tons production per year, ammonia is one of the most commonly used chemical in the world. The conventional Haber-Bosch process for ammonia synthesis requires high pressure to ensure suitable conversion.[1] Such high pressure makes the ammonia synthesis costly, energy- and capital-intensive.[2] We are developing an innovative process to synthesize green ammonia at lower pressure, with hydrogen from water, nitrogen from air, and wind-generated electricity.[3] We seek to develop a process for lower pressure ammonia synthesis, because low pressure process is safer, cheaper, more robust, and more amenable to downscaling for distributed production from renewables.

In our proposed reaction-absorption process, we are synthesizing ammonia using a conventional catalyst and a column of solid absorbents.[4] These absorbent materials – alkaline earth metal halides - have attractively high equilibrium absorption for ammonia by forming an ammine complex. These complexes have various coordination numbers, some of which stable even at temperatures as high as 400 °C (close to reaction temperature).[5,6] Metal halides can be reversibly loaded and unloaded with ammonia. We have already proved and reported the viability of high temperature ammonia absorption into metal halides, enabling low pressure ammonia synthesis.

In this talk, we are reporting some of our most recent findings regarding our search for a practical regeneration protocol to unload ammonia from the absorbent materials. Pressure and temperature swing desorptions were studied to figure out the best operating conditions. Results indicate that depending on the coordination number of the ammine complex, which is formed in the absorber, a combination of pressure and temperature swing is required. Vacuum desorption showed to be fast and reliable. Currently, we believe a full regeneration is not necessary. Thus, cyclical partial desorptions at high ammonia loadings is probably the solution for sustainable regeneration of absorbent materials.

References

(1) Vojvodic, A.; Medford, A. J.; Studt, F.; Abild-Pedersen, F.; Khan, T. S.; Bligaard, T.; Nørskov, J. K. Exploring the Limits: A Low-Pressure, Low-Temperature Haber–Bosch Process. *Chem. Phys. Lett.* **2014**, *598*, 108.

(2) Jennings, J. R. *Catalytic Ammonia Synthesis*, 1st ed.; Plenum Press, 1991.

(3) Reese, M.; Marquart, C.; Malmali, M.; Wagner, K.; Buchanan, E.; McCormick, A.; Cussler, E. L. Performance of a Small-Scale Haber Process. *Ind. Eng. Chem. Res.* **2016**, *55* (13), 3742.

(4) Malmali, M.; Wei, Y.; McCormick, A.; Cussler, E. L. Ammonia Synthesis at Reduced Pressure via Reactive Separation. *Ind. Eng. Chem. Res.* **2016**, *55* (33), 8922.

(5) Sørensen, R. Z.; Hummelshøj, J. S.; Klerke, A.; Reves, J. B.; Vegge, T.; Nørskov, J. K.; Christensen, C. H. Indirect, Reversible High-Density Hydrogen Storage in Compact Metal Ammine Salts. *J. Am. Chem. Soc.* **2008**, *130* (27), 8660.

(6) Wagner, K.; Malmali, M.; Smith, C.; McCormick, A.; Cussler, E. L.; Zhu, M.; Seaton, N. C. A. Column Absorption for Reproducible Cyclic Separation in Small Scale Ammonia Synthesis. *AIChE J.* **2017**.

500302 Development of a TSA Process for Metabolic CO_2 Removal from Spacecraft Cabins Using a Structured 13X Adsorbent

Abstract:

NASA has a great deal of interest in the use of structured sorbents in environmental systems for space applications for a variety of reasons. The most important reason perhaps is to minimize or even eliminate the formation of sorbent dust. Sorbent dusting most recently caused the failure of a TSA unit on board the International Space Station (ISS). The use of a Catacel structured 13X adsorbent may alleviate the dusting problem associated with the TSA unit in the CDRA and possibly provide even better performance characteristics.

The Catacel structured 13X adsorbent is constructed of corrugated metal sheets rolled into a cylindrical shape in such a way that it forms parallel triangular channels with all three walls coated with zeolite crystals. One of the Catacel structured 13X adsorbents available for testing at USC contains 741 cells per square inch (CPSI)), has a 13X coating thickness of about 31 m, a bulk density of 171 kg/m^3, and a low pressure drop of < 12 kPa/m at 20 m/s (a very high velocity). These properties make it ideal for use in high flow rate, low pressure drop applications where low attrition (i.e., minimal dusting) is a requirement. Moreover, since it is made of 50 m thick metal foil sheets, it should also have a very high thermal conductivity. This property may give rise to shorter cycle times in a TSA process, thus making the TSA beds much smaller than those in the current CDRA TSA unit.

The overall goal of this project is to design a temperature swing adsorption (TSA) process that utilizes a structured 13X adsorbent. This TSA cycle design would a drop-in replacement for the TSA unit on board the ISS. This design is based on simulations being carried out with the USC dynamic adsorption process simulator (DAPS) and validated experimentally using the USC single bed apparatus in a TSA mode. The latest results of this project will be presented.

502851 PSA Based CO_2 Capture Above the Dew Point of Synthesis Gas for IGCC Power Plants

Abstract:

TDA is developing a novel sorbent based PSA process that removes CO_2 via physical adsorption from synthesis gas. The relatively strong affinity of the sorbent to CO_2 enables effective operation at temperatures up to 300°C (well above the dew point of synthesis gas stream generated by commercial gasifiers). However, because the sorbent and the CO_2 do not form a true covalent bond, the energy needed to regenerate TDAâ€™s sorbent (5.4 kcal per mol of CO_2) is much lower than that observed for either chemical absorbents (e.g., 29.9 kcal/mol CO_2 for sodium carbonate) or amine-based solvents (e.g., 14.2 kcal/mol CO_2 for monoethanolamine). TDAâ€™s sorbent can be regenerated isothermally and CO_2 can be recovered at pressure (as high as 150 psia). Thus, the energy needed to regenerate the sorbent and compress the CO_2 for sequestration is significantly lower than that for any other technology reported to date. The high surface area and favorable porosity of the sorbent also provides a unique platform to introduce additional functionality, such as active groups to catalyze the water-gas-shift (WGS) reaction or to remove trace metals (e.g., Hg, As). Hence, TDAâ€™s CO_2 sorbent could be used to do a warm gas capture of CO_2 above the dew point of the synthesis gas in an Integrated Gasification Combined Cycle (IGCC) power plant. TDA's pre-combustion CO_2 capture system achieves a net plant efficiency of 34% on a coal high heating value basis when operated on bituminous coal, which is significantly higher than that can be achieved with the Selexol™ scrubber at 31.6% (state-of-the-art pre-combustion carbon capture technology). In a current DOE sponsored project, TDA is carrying out field tests with the warm gas CO_2 capture unit at 0.1 MWe scale at the National Carbon Capture Center (NCCC) in Wilsonville, AL and has so far completed over 500 hours of field tests achieving greater than 90% carbon capture. In this paper we present the results from the 8-bed PSA cycle performance in the field test with coal derived synthesis gas at NCCC.

500569 On the Development of a PSA Process for Natural Gas Purification

Abstract:

The objective of this work was to evaluate the separation of CH_4 and CO_2 gas mixtures via simulation using the USC Dynamic Adsorption Process Simulator (DAPS) and using single and multi-bed bench scale pressure swing adsorption (PSA) systems. A variety of conditions were studied, including the effects of the feed composition, high to low pressure ratio, high pressure, low pressure, cycle time, light reflux ratio, heavy reflux ratio, and feed flow rate (feed throughput) on the purity and recovery of CH_4. The latest single and multi-bed experimental and DAPS simulation results will be presented.

500573 Limits of Rapid Pressure Swing Adsorption Processes: Does a Minimum Bed Size Factor Exist?

Abstract:

The goal of most rapid pressure swing adsorption processes (RPSA) is to increase the process productivity to permit process size reduction for portable applications. In literature and industry, productivity is typically represented by the bed size factor (BSF), which is inversely related to productivity. Significant BSF reduction is possible with small diameter particles and fast cycling. The tradeoff for operating with a higher cycling frequency is concentration front spreading resulting in a loss in product recovery. Front spreading is commonly attributed to 1) an increase in pressure drop 2) resistance to mass transfer and 3) resistance to heat transfer. A minimum BSF is thought to exist when the product loss becomes significant. For RPSA design, there is interest in what limits BSF reduction, if such a limit exists. However, conclusive experimental evidence that demonstrates what causes a minimum BSF has not been previously presented.

In this study, a two column PSA system was used to determine if a minimum BSF exists and what may be causing it. The process used a commercial LiLSX zeolite (~0.5 mm in diameter) and a traditional 4-step feed pressurization cycle in an experimental system capable of achieving cycles as low as 3 seconds. The results from the study demonstrate a definite minimum in BSF at a cycle time ~ 4 seconds for a product oxygen purity ~ 96% (argon free air feed). This represents the first known study to experimentally demonstrate a minimum BSF using a two column process. Experiments were conducted at various pressure ratios and column dimensions to determine their effect on the minimum BSF. It will further be postulated why the minimum is happening and how to predict process conditions where it will occur.

489750 Thermal Modulation in High-Capacity Pressure-Swing Adsorption Via Incorporation of Microencapsulated Phase-Change Material

Abstract:

Post combustion separation of CO_2 from flue-gas will be a key component in the portfolio of technologies that enable reduction of global carbon emissions. Metal-organic-frameworks (MOFs) have shown promise as possible adsorbent materials for CO_2 capture due to their large capacity driven by both functionalized surfaces and high pore volume. This large capacity for CO_2 can be unlocked for high recovery separations using a small pressure swing at sub-ambient conditions. These high capacities are necessary to drive down post-combustion CO_2 capture costs; consequentially, large amounts of heat are released during the adsorption process, leading to significant thermal fronts propagating through the adsorption column. For sub-ambient adsorption processes, these fronts induce significant inefficiency in the sorbent utilization due to the elevation of local adsorption temperatures. Managing these thermal fronts can be achieved via incorporation of phase change materials, which adds both sensible thermal mass and latent heat storage into the system, dampening the thermal fronts created during rapid adsorption and desorption cycles. In this work, we demonstrate the ability to incorporate microencapsulated phase change materials (Î¼PCM) into polymeric fiber contactors via solution spinning techniques at loadings up to 75 weight percent and in the presence of the MOF UiO-66. Moreover, we show that these composite monolithic contactors are effective in mitigating the thermal fronts created during the CO_2 adsorption process. Thermal gravimetric analysis and differential scanning calorimetry demonstrate that the Î¼PCM remain intact through the spinning process and are operable for thermal modulation at the target temperature of -30 Â°C. Sub-ambient breakthrough experiments with simulated flue gas are performed to show the effect of thermal modulation on the breakthrough behavior and sorbent capacity for CO_2. In addition, temperature profiles in the bed under breakthrough of pure CO_2 are collected to better understand the effectiveness of Î¼PCM for thermal front mitigation relative to fibers with no Î¼PCM. Our findings suggest that further optimization of the relative loadings of sorbent to thermal modulator can result in more efficient pressure swing adsorption.

496155 The Exact Equivalence between Diffusion and LDF Models

Abstract:

When designing and optimizing pressure swing adsorption or temperature swing adsorption processes the equivalence between the linear driving force model and the diffusion model is often invoked to reduce the computational effort by an order of magnitude. This is very effective for relatively slow cycles where Glueckauf's original approximation [1] provides sufficient accuracy.

Over the past 35 years, from Nakao and Suzuki's [2] contribution, there has been a significant effort by several groups to extend this approach to fast cycles, with some success but no formal exact equivalence has been accepted by the adsorption community.

The starting point of our analysis is the fact that any adsorption process consists of a cyclic sequence of steps and therefore the basis for establishing the correct equivalence comes from studying the response of an adsorption column to a sinusoidal input in concentration, ie the method to be used is based on the frequency response analysis [3]. We show that the typical approach of simulating both models and finding an LDF correction factor [2] is in fact based on matching the amplitude ratios obtained from the two models, but that the corresponding phase lags are not correct. We revisit the equivalence criterion proposed by Rouse and Brandani [4] and prove that this can be presented in the form of explicit relationships, ie no approximate solution is needed, and that the two parameters obtained have physical meaning. Matching the amplitude ratio and the phase lag of the response at cyclic steady state is strictly valid only for linear systems, but the proposed method reduces to Gluekauf's equivalence for longer cycles, which we know if effective also for non-linear systems [5].

Single particle kinetics, kinetics of a perfectly mixed adsorber and the kinetics of a full adsorption column are shown to yield the exact match between the two models and the basis for the equivalence of non-sinusoidal inputs is discussed. The equivalence is shown to provide the means to accelerate convergence to cyclic steady state for fast cycles, but if the aim is to model also the approach to cyclic steady state, then no exact equivalence is possible and the actual solution of the diffusion equation with appropriate numerical grids [6] should be used.

References

[1] Gluekauf E. Trans. Faraday Soc.; 1955, 51, 1540-1541.

[2] Nakao S., Suzuki M. J. Chem. Eng. Japan; 1983, 16, 114-199.

[3] Stephanopoulos G. *Chemical Process Control. An Introduction to Theory and Practice*, Prentice-Hall, Englewood Cliff, NJ 1984

[4] Rouse A.J. and Brandani S. A New LDF Approximation for Cyclic Adsorption Processes, In AIChE Separation Technology Topical Conference Vol. 2, P. Bryan and A. Serbezov Eds., 2001, p 739–743.

[5] Sircar S. and Hufton J.R. Adsorption; 2000, 6, 137–147.

[6] Ahn H. and Brandani S. Adsorption; 2005, 11, 113–122.

496409 Laboratories in the Chemical Engineering Curriculum: Current State and Best Practices

Abstract:

Hands-on experiences with laboratory and pilot-scale equipment are one of the hallmarks of the undergraduate chemical engineering experience. Lab experiences require significantly more space, infrastructure, and instructional time than do classroom sections or computational studios, and this combined with recent growth in enrollment means labs are getting a hard look at many institutions.

This session focuses on the undergraduate laboratory experience and will present results from the most recent AIChE Survey on Unit Operations lab as well as invited speakers on standard and innovative approaches to this important experience.

486368 Towards the Control of Fouling Behavior of Hollow-Fiber Membranes Via High-Throughput Screening

Abstract:

Hollow-fiber (HF) membranes are the preferred configuration for various water and wastewater treatment applications. The fouling of HF membranes is a complex phenomenon that is known to depend on a multitude of parameters and have a severe effect on membrane filtration performance. An ongoing challenge is the significant effort and time that is required to develop and test the effect of different operating parameters. We have recently developed a high-throughput (HT) platform for HF membranes [1] that allows for six filtration tests to be conducted in parallel with each test using a single HF membrane - this design is hereafter referred to as 'high-throughput hollow fiber' (HT-HF) module. As a proof of concept, preliminary experiments were conducted with un-modified polyvinylidene fluoride (PVDF) membranes in constant flux mode to evaluate the fouling tendency for different humic acid (HA) types and concentrations, solution conditions (pH, conductivity), and filtrate flux. It was observed that increasing the pH and decreasing the conductivity results in higher transmembrane pressures and lower apparent sieving coefficients. For example, the final transmembrane pressure (TMP) value decreased from 2.2 to 1.3 psi as the HA solution conditions were changed from 5% phosphate buffered saline (PBS) to 100% PBS while this change in solution conditions has increased the apparent sieving coefficient from 20% to 54%.

Furthermore, we have extended the application of the HT-HF module by performing experiments with a 'library' of potential foulants (humic acids, proteins, polysaccharides), testing different membrane pre-treatment and cleaning strategies, and using surface modified PVDF membranes. Surface modification of PVDF membranes have been studied by different groups in conventional format using different modifying agents such as polydopamine [2–4], polyvinyl alcohol [5], polyvinylpyrrolidone [6] and chitosan [7]. Furthermore, previous studies have shown the usefulness of developing a HT strategy to study the effect of surface modification on polyethersulfone flat sheet membrane performance using a monomer library in 96 well filter plate format towards natural organic matter [8] and protein fouling [9]. However, none of the previous works have studied the surface modification of HF membranes in HT format. In this study, PVDF HF membranes were modified using multiple approaches and modification conditions by using nanoparticles, polyelectrolytes and other polymers deposition in HT format using the HT-HF module. The modified membranes were characterized using contact angle measurements, Fourier transform infrared spectroscopy and scanning electron microscopy. The effect of surface modification condition on membrane hydraulic permeability, TMP profiles and separation performance were studied and the results were compared statistically to each other and to those of un-modified HF membranes. The HT platforms and the results introduced here will be used to develop new strategies for controlling membrane fouling and to better understand membrane fouling behavior.

References

[1] A.S. Kazemi, L. Boivin, S. Mi Yoo, R. Ghosh, D.R. Latulippe, Elucidation of filtration performance of hollow-fiber membranes via a high-throughput screening platform, J. Membr. Sci. 533 (2017) 241–249.

[2] H. Zhu, S. Zhu, A versatile and facile surface modification route based on polydopamine for the growth of MOF films on different substrates, Can. J. Chem. Eng. 93 (2015) 63–67.

[3] L. Shao, Z.X. Wang, Y.L. Zhang, Z.X. Jiang, Y.Y. Liu, A facile strategy to enhance PVDF ultrafiltration membrane performance via self-polymerized polydopamine followed by hydrolysis of ammonium fluotitanate, J. Membr. Sci. 461 (2014) 10–21.

[4] B.D. McCloskey, H.B. Park, H. Ju, B.W. Rowe, D.J. Miller, B.D. Freeman, A bioinspired fouling-resistant surface modification for water purification membranes, J. Membr. Sci. 413–414 (2012) 82–90.

[5] J.R. Du, S. Peldszus, P.M. Huck, X. Feng, Modification of poly(vinylidene fluoride) ultrafiltration membranes with poly(vinyl alcohol) for fouling control in drinking water treatment, Water Res. 43 (2009) 4559–4568.

[6] Z. Xu, L. Li, F. Wu, S. Tan, Z. Zhang, The application of the modified PVDF ultrafiltration membranes in further purification of Ginkgo biloba extraction, J. Membr. Sci. 255 (2005) 125–131.

[7] S. Boributh, A. Chanachai, R. Jiraratananon, Modification of PVDF membrane by chitosan solution for reducing protein fouling, J. Membr. Sci. 342 (2009) 97–104.

[8] M. Zhou, H. Liu, J.E. Kilduff, R. Langer, D.G. Anderson, G. Belfort, High-throughput membrane surface modification to control NOM fouling., Environ. Sci. Technol. 43 (2009) 3865–71.

[9] M. Zhou, H. Liu, J.E. Kilduff, R. Langer, D.G. Anderson, G. Belfort, High Throughput Synthesis and Screening of New Protein Resistant Surfaces for Membrane Filtration, AIChE J. 56 (2010) 1932–1945.

486100 Influence of Backwashing on the Pore Size of Hollow Fiber Ultrafiltration Membranes

Abstract:

Backwashing is a common method for fouling mitigation. However, its impact on the pore-size distribution (PSD) of hollow fiber (HF) membranes has not been studied to date. This study quantitatively assessed the effects of filtration and backwashing cycles on the PSDs of polyacrylonitrile (PAN) and polyvinylidene difluoride (PVDF) HF membranes by evapoporometry (EP) characterization. The membranes were characterized before and after repeated cycles of filtration and backwashing in the absence of any foulants, and for a feed solution of bentonite and humic acid that caused fouling both on and within the membrane pores. Firstly, in the absence of any foulants, backwashing caused the appearance of larger pores, the effect of which was greater for the rubbery PVDF membrane than for the glassy PAN membrane. Secondly, backwashing was more effective in removing the fouling within the larger pores, but could not remove all the deposits within the smaller pores, which provides a mechanistic explanation for the progressive increase in the transmembrane pressure (TMP) with each backwashing cycle. Thirdly, the membranes that did not undergo the 10th backwashing at the end of 10 cycles of filtration and backwashing displayed a marked shift of the PSD towards smaller pores due to the deposition of foulants on and within the largest pores, whereas those that underwent the 10 complete backwashing cycles achieved nearly complete recovery of the larger pores accompanied by an irreversible increase in the diameter of the largest pores. Fourthly, a higher backwashing flux led to similar average pore diameters of the fouled and virgin membranes due to the increased effectiveness in restoring the smallest pores, but the corresponding higher filtration flux negated the benefits due to a greater fouling extent particularly for the larger pores. Finally, in order to achieve the desired permeation and rejection properties, possible enlargement of the pores needs to be taken into consideration when choosing a UF membrane and when specifying the backwashing intensity and protocol.

494413 A Simple, Reactive Approach to Mitigate Fouling and Concentration Polarization in Reverse Osmosis Systems

Abstract:

A Simple, Reactive Approach to Mitigate Fouling and Concentration Polarization in Reverse Osmosis systems

Rajarshi Guha[a], Boya Xiong[b], Michael Geitner,[a] Tevin Moore[a], Darrell Velegol[a], **Manish Kumar[a]**

[a]Department of Chemical Engineering, The Pennsylvania State University, University Park, PA 16802, United States

[b]Department of Civil and Environmental Engineering, The Pennsylvania State University, University Park, PA 16802, United States

Abstract

Fouling of high pressure membranes is a recalcitratnt problem, particularly in reverse osmosis (RO) and nanofiltration (NF) systems. Colloidal, organic and biological fouling are the most commonly encountered challenges in industrial desalination and wastewater treatment processes and there is no convenient way to clean the membrane without interrupting operation or resorting to expensive fouling mitigation infrastructure. Therefore, significant energy savings can be achieved by real time fouling elimination and concentration polarization mitigation. Although several important technological improvements were achieved in membrane design and fouling control in past few years, for example- TiO_2 embedded composite antifouling membranes, photothermal nanoheater membranes or carbon nanotube containing composite electrolytic membranes, which are promising in reducing different types of fouling. However, all the reported methodologies suffer from requiring extensive new infrastructure, lack of scalability and most importantly, efficiency without loss in productivity.

Here, we have demonstrated a simple scalable approach of coating RO/ NF membrane with bioinspired adhesive polymer polydopamine and copper oxide nanoparticles, which decomposed H_2O_2 (0.006% - 0.024%) to oxygen bubbles and thereby, swept foulants away from the surface and at the same time reduced concentration polarization by enhancing mass transfer co-efficient. The polydopamine layer embedded the nanoparticles and acted as the reactive catalytic surface. Additionally, polydopamine layer helped to protect the underlying polyamide layer from hydroxyl radicals by acting as free radical scavenger. Such composite membrane eliminated flux decline with colloidal foulants in repeated runs and even increased the flux level beyond the baseline value with organic humic acid foulants. Enhanced catalytic activity of humic acids in presence of H_2O_2 and reduction of concentration polarization were found to be responsible for such flux enhancements. Embedded copper oxide nanoparticles also retarded *E. Coli* attachment to the membrane surface in stirred cell experiments. On the other hand, such composite membrane was found to eliminate concentration polarization upon H_2O_2 addition by bubble induced mixing. Therefore, such scalable facile approach holds promise to mitigate fouling in industrial membrane systems, reduce concentration polarization and thereby, saving considerable energy.

Figure. Schematic of catalytic/ reactive membrane assembly and realization of mechanism of action.

495199 Graphene Oxide Membranes in Extreme Operating Environments: Concentration of Kraft Black Liquor

Abstract:

Kraft weak black liquor (WBL) is a hot (80-95°C), corrosive (pH ~12), high total solids (15-18 wt%), and high-volume (~500 gal/min in a typical pulp and paper mill) by-product of biomass pretreatment in papermaking and other processes. WBL contains lignin, hemicellulose, organic acids, inorganic salts, and water. Dewatering of WBL is currently performed by a six-stage evaporator train to produce strong black liquor (SBL, up to 80 wt% solids). More than 500 million tons/year of WBL are generated worldwide, and the concentration of WBL to SBL by evaporation is highly energy-intensive (about 0.2 Quads/year in the US alone). Membrane-based concentration of black liquor is attractive as an energy-efficient alternative, but challenging due to the harsh operating conditions and high fouling potential of BL.[1]

This talk will discuss our pursuit of chemically robust, low-cost, and scalable membranes for black liquor concentration, focusing in particular on the development of graphene oxide (GO) nanofiltration membranes for lignin rejection. Thin (< 250 nm), uniform, and defect-free GO membranes were fabricated on macroporous poly(ethersulfone) (PES) supports and shown to possess excellent stability in real mill-derived BL during prolonged operation (1 month) at realistic temperature and feed conditions. Furthermore, high lignin rejection (~99%) and flux (~20 kg/m^2/hr) was obtained when concentrating BL feeds at transmembrane pressures (TMPs) up to 50 bar and temperatures up to 85°C. We discuss in the detail the characterization of the GO membrane structure, morphology, and permeation properties.[2] A preliminary discussion of the economics and techniques for scale-up of the present membranes will be included.

1. Kevlich, N. S.; Shofner, M. L.; Nair, S., Membranes for Kraft Black Liquor Concentration and Chemical Recovery: Current Progress, Challenges, and Opportunities. *Separation Science and Technology* **2017**, *52* (6), 1070-1094.

2. Rashidi, F.; Kevlich, N. S.; Sinquefield, S. A.; Shofner, M. L.; Nair, S., Graphene Oxide Membranes in Extreme Operating Environments: Concentration of Kraft Black Liquor by Lignin Retention. *ACS Sustainable Chemistry & Engineering* **2017,** *5* (1), 1002-1009.

500398 Water Permeation through Structural Defects of Single-Layered Graphene Oxide Membranes

Abstract:

Comparing with traditional methods, membrane-based water treatment technologies are essential due to the low energy consumption and high efficiency in removing contaminants or salts. As the flux cross the membrane decreases with membrane thickness, to balance the pay off between permeability and selectivity, membrane should possess pores large enough to allow the passage of water molecules but small enough to block the large solutes, and the membrane thickness should be as thin as possible to provide high flux, which makes the "ideal" membrane might be in the form of one atomic thickness and with suitable pores. Materials like graphene and its derivatives just meet this requirement and have been considered as a promising candidate to prepare ultrathin membranes which mostly closing to the "ideal" state. However, the harsh conditions to prepare macroscopic single-layer graphene with sufficient densities and size controllable pores have hindered its practical applications. Herein, by clarifying two distinct water transportation mechanisms for membranes with sub-monolayer and multilayer GO coverage, a methodology to fabricate nominal single-layered GO membrane was proposed for the first time. The Water flux exhibited a transition from linear to exponential decrease as the GO coverage increase, the condition to prepare nominal single-layered GO membrane could be extrapolated from the turning point of the transition, the as-fabricated membrane which with thickness closing to one-carbon-atom exhibits high water permeance around 64 $L\cdot m^{-2}\cdot h^{-1}\cdot bar^{-1}$, and defects on GO flakes provide major contribution for the membrane sieving property. By evaluating the separation performance of this membrane with rigid molecules, the effective defect size of GO was determined to be ~1.2-1.7 nm. This membrane with nominal single-layer GO cover also show great potential in protein separation.

502718 Braided-Reinforced Thin Film Composite (TFC) Nanofiltration Hollow Fiber Membranes

Abstract:

Hollow fiber membranes are attractive for use in water treatment processes because of their high packing density relative to flat sheet and spiral wound configurations. They are also easily submersed in tanks for bioreactor applications. For water purification in particular, they are commonly use in microfiltration (MF) an ultrafiltration (UF) applications. To enable low pressure drop in fiber lumens, inner diameters of these fibers can be rather large, thus limiting the pressure tolerance of these membranes. Low pressure tolerance of larger capillary fibers has largely excluded the use of hollow fiber from higher pressure membrane processes reverse osmosis (RO) and nanofiltration (NF) except for cases where integrated asymmetric membranes are considered. Use of the thin-film composite (TFC) platform with polyamide based selective layers, while common with flat sheet membranes for RO an NF, have not found commercial success and have largely been relegated to the academic community for studies on forward osmosis (FO). In this work, we show that hollow fiber TFC membranes can be made for higher pressure applications. To address the pressure tolerance issues, we build our TFC using a commercial hollow fiber PVDF UF membrane that is reinforced with an embedded braid as a support. Onto this support is formed the selective barrier layer *in-situ* from piperazine (PIP) an trimesoyl chloride (TMC) monomers. Through careful optimization the monomer concentration and reaction conditions, we have demonstrated softening with a 89 % $MgSO_4$ rejection and a permeance of 8 LMH/bar. This work represents the first time such a reinforce hollow fiber membrane has been used as a support for a TFC membrane and could broaden opportunities for hollow fiber membranes in processes where salt selection (RO, pressure retarded osmosis) is required.

497318 Naphthenic Acids Removal from High Tds Produced Water By Persulfate Mediated Iron Oxide Functionalized Catalytic Membrane, and By Nanofiltration

Abstract:

Increasing stress on the existing water resources necessitates advancement in water treatment technology. During oil extraction process, oil industries generate large amounts of produced water containing organic contaminants, such as naphthenic acids (NA) and very high concentrations of inorganic salts. Recovery of potable water from such water is highly energy intensive due to its high salt concentration. Here, we explored catalytic properties of iron oxide (Fe_xO_y) functionalized membranes in oxidizing NA from water containing high concentrations of total dissolved solids (TDS) using persulfate as an oxidizing agent. Catalytic decomposition of persulfate by Fe_xO_y functionalized membranes followed pseudo-first order kinetics with an apparent activation energy of 18 Kcal/mol. Fe_xO_y functionalized membranes were capable of lowering the NA concentrations to less than discharge limits of 10 ppm at 40 ^{O}C. Oxidation state of iron during reaction was quantified. Membrane performance was evaluated for extended operating period. Coupled nanofiltrationâ€"functionalized membrane was evaluated. Commercially available nanofiltration membranes were investigated to selectively retain NA from water containing high concentrations of dissolved salts. Commercial NF membranes, NF270 (Dow) and NF8 (Nanostone) had NA rejection of 79% and 82%, respectively. The developed Fe_xO_yÂ functionalized membrane can then treat the lower volume concentrated NF stream (retentate) for efficient removal of NA.

483735 Measuring 3D Gas Adsorption Isotherms By X-Ray Computed Tomography

Abstract:

Imaging techniques constitute a novel paradigm in characterization methods of porous media; X-ray computed tomography is of particular interest because it allows for the non-destructive determination of spatially resolved properties such as macropore space[1], and distribution of binder content[2] and minerals[3]. While the resolution is being continuously increased moving from medical scanners towards monochromatic synchrotron micro-CT and nano-CT setups, bench scale laboratory systems are still limited to voxel sizes above 1μm^3, and therefore do not allow visualising micro- and meso-porosity (less than 50 nm) that is ubiquitous in both natural (e.g., clay, coal, shale) and most porous materials used in the chemical industry (e.g. adsorbent materials, catalysts, membranes). The most widespread technique for the characterization of nanoporous solids indeed is gas adsorption. While the interpretation of sub-critical adsorption isotherms in terms of pore size, pore structure and connectivity is being continuously refined[4], gas adsorption remains a technique that only provides bulk (macroscopic) information and can therefore not provide any information regarding the distribution of these properties.

The images obtained from X-ray CT scans contain spatially resolved quantitative information of the linear attenuation coefficient: the gray scale value of each voxel is the volume weighted average of the attenuation of each component and/or phase present in that specific voxel. Based on this observation, Pini[5] showed in a recent communication that it is possible to extract the excess adsorption from the subtraction of a scan of a porous solid taken in the presence of an adsorbate, and a scan of that same solid taken under vacuum or in the presence of an inert (e.g. Helium). This work has been extended here by developing the relevant experimental and analysis protocols to measure full excess adsorption isotherms by X-ray CT imaging for medical and micro-CT setups. By applying the devised protocols to fixed beds of commercial zeolite 13X pellets (ZeoChem, Uetikon Switzerland) and activated carbon rods (Norit, USA), we measured spatially distributed adsorption isotherms of CO_2 -- one isotherm for each cubic millimeter -- by means of X-ray CT imaging with a medical CT scanner (Universal Systems), as illustrated in Fig. 1. Pellets can be imaged in a similar way at a larger resolution by means of micro-CT imaging (Zeiss Versa XRM500): this enables obtaining adsorption isotherms that are spatially resolved in a sub-millimeter scale.

Fig. 1 X-ray CT image of a slice positioned at the interface between a Zeolite 13X and an activated carbon layer at 30 bar CO_2. From difference images between CO_2 scans taken at various pressures and a helium scan, excess adsorption isotherms can be reconstructed for every voxel. Two examples are shown for the rod-like particles (activated carbon, voxels 1 and 3, blue) and for the spherical particles (zeolite 13X, voxels 2 and 4, red).

Because this technique ultimately allows to quantitatively probe and visualize the spatial distribution of typical metrics,

e.g. pore volume and pore size distribution, within individual pellets and across packed beds, it is anticipated that it can support the development of formulation processes of adsorbent powders into shaped particles by providing direct structural information of the properties of interest, i.e., the adsorption isotherms. Moreover, this work provides a novel way to probe nanoporosity of heterogeneous porous media in multiple dimensions and in a non-destructive way. As such, it is expected to bear significant importance in the characterization of nanoporosity within natural porous media, such as shales or reservoir rocks, where three-dimensional information on the pore size, pore volume and pore inter-connectivity are essential for the accurate description of gas transport.

References:

1. Blunt, M. J; Bijeljic, B; Dong, H; Gharbi, O; Iglauer, S; Mostaghimi, P; Paluszny, A; Pentlan, C. Pore scale imaging and modelling. *Advances in Water Resources* **2013**, *51*, 197-216.

2. Mitchell, S; Michels, N; Kunze, K; perez-Ramirez, J. Visualization of hierarchically structured zeolite bodies from macro to nano length scales. *Nature Chemistry* **2012**, *4*, 825-831.

3. Lai, P; Moulton, K; Krevor, S. Pore-scale heterogeneity in the mineral distribution and reactive surface area of porous rocks. *Chemical Geology* **2015**, *411*, 260-273.

4. Thommes, M; Kaneko, K; Neimark, A.V; Olivier, J. P; Rodriguez-Reinoso, R; Rouquerol, J; Sing, K. S. W. Physisorption of gases, with special reference to the evaluation of surface area and pore size distribution. *Pure and Applied Chemistry* **2015**, *87*, 1051-1069.

5. Pini R. Multidimensional Quantitative Imaging of Gas Adsorption in Nanoporous Solids. *Langmuir* **2014**, *30*, 10984-10989.

497037 Measurement of Water Adsorption Equilibrium and Kinetics Using the ZLC Technique

Abstract:

We present a new configuration of the zero length column apparatus designed specifically to measure adsorption equilibrium and kinetics for vapours at near saturation conditions. We demonstrate its use using water on different adsorbents including mesoporous materials showing non-type I isotherms.

The initial study was shown to reproduce with high accuracy the desorption curves generated using a Quantachrome Aquadyne gravimetric system, including scanning desorption curves in the hysteresis region. The significant improvement gained by using the ZLC lies in the time required for generating the desorption curves as well as the fact that thousands of equilibrium points are obtained from a single experiment. The full characterisation of a sample is reduced to one week compared to 6 to 8 weeks on the gravimetric system, therefore indicating a significant potential in using this approach to characterize the properties of nanoporous materials.

Results on various commercial pellets will be presented, both below and inside the hysteresis region. Experiments at higher flowrates reveal that the kinetic behaviour in the hysteresis region is quite complex and further analysis is required

503251 Novel Approach for Kinetic Measurements for Low Capacity Adsorbents

Abstract:

The key advantage of the Zero Length Column (ZLC) technique is the capability to characterise an adsorbent, both from the equilibrium and kinetic point of view, by using less than 10 mg of sample, more rapidly and accurately than other conventional methods. Determination of the kinetics and transport mechanism may be prevented in some cases when the adsorption capacity of the sample is relatively low. In these cases, the decoupling of the kinetics of adsorption from the kinetics of the system becomes quite difficult due to the fact that the ZLC response curve is too close to the response of the system with no adsorbent. This means that it is not possible to rely on the traditional asymptotic method to extract any meaningful kinetic information in such systems.

Here we present a novel ZLC approach specifically developed to overcome this issue. The method is based on the measurement of the adsorption capacity at different adsorption times and relates this to the approach of the system to saturation. The new methodology was validated by measuring the kinetics of adsorption of N_2 in 4A at very low temperatures (down to -20 °C). A cryogenic cell, purposely designed for ZLC tests at very low temperatures, was used for these experiments. The study was also extended at different temperatures to extract the activation energy of the diffusion process. The new methodology represents an easy alternative to the traditional ZLC approach which allows to overcome the limitations experienced when working with systems characterised by low adsorption capacity.

COLORADOSCHOOLOFMINES
EARTH • ENERGY • ENVIRONMENT

Vibrational Gravimetric Analysis of Capillary Condensation in Nanoporous Rocks

Younki Cho, Ryan Lo, Keerthana Krishnan, Xiaolong Yin
ADVISOR: DR. Xiaolong Yin
COLORADO SCHOOL OF MINES

American Institute of Chemical Engineers
ANNUAL MEETING
2nd NOVEMBER 2017

COLORADOSCHOOLOFMINES
EARTH • ENERGY • ENVIRONMENT

Outline

- Objectives
- Theoretical Background
- Overview of other methods used
- System Calibration
- Results and Discussion
- Conclusion
- Acknowledgement
- References

Objectives

- Measure change in mass of a sample due to capillary condensation by monitoring the natural frequency of sample using oscilloscope
- Perform the measurement at high-pressure and high-temperature conditions
- Apply the method to:
  - Berea sandstone (macroporous)
  - Niobrara shale (crushed, macroporous and mesoporous)

IUPAC definitions:

- **Micropore**: Pore with width not exceeding about 2.0 nm
- **Macropore**: Pore with width exceeding about 0.05 μm (50nm)
- **Mesopore**: Pores of intermediate size between macropores and micropores. $2\text{nm} \leq mesopores \leq 50nm$

Background: Adsorption and Condensation

(QuantachromeCorp, 2010)

Background: Adsorption and Condensation

COLORADOSCHOOLOFMINES
EARTH • ENERGY • ENVIRONMENT

Adsorption:

- Ability of solid materials to attract molecules of gasses or solutions with which they are in contact.

  -Monolayer adsorption

  -Multilayer adsorption

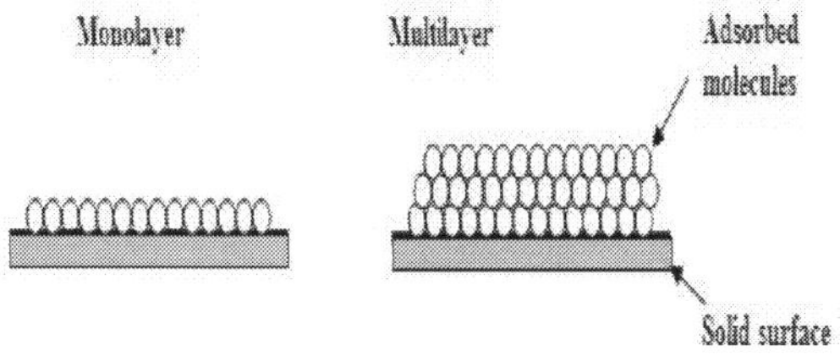

Figure 1: Source: nptel.ac.in

Capillary Condensation:

- Multilayer adsorption to fill the entire pore space
- Vapor phase condenses at lower pressure than bulk saturation pressure
- Condensation first appears in smallest pores

For a cylindrical pore, the kelvin equation is given by:

$$\ln\left(\frac{P_v}{P_{v0}}\right) = -\frac{2\sigma cos\theta}{r} * \frac{V_m}{RT}$$

Overview of methods used to determine gas solid interactions

COLORADOSCHOOLOFMINES
EARTH • ENERGY • ENVIRONMENT

▪Manometric methods (Sieverts' method)

-Instruments work with same principle as BET surface area and pore size distribution

-More widely used other than volumetric methods

▪Other volumetric method

-Flowing volumetric method: amount of hydrogen adsorbed by a sample is a function of flow rate at isobaric conditions.

-Can be used only for low pressure measurements

▪Differential volumetric method

-Uses differential pressure transducers

-Advantageous over Sieverts' method, cannot be used for thermal variations

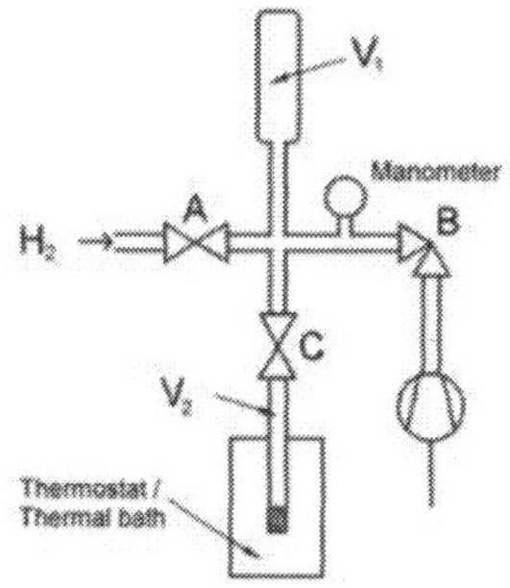

Figure 2: Manometric method

Gravimetric methods

Magnetic suspension

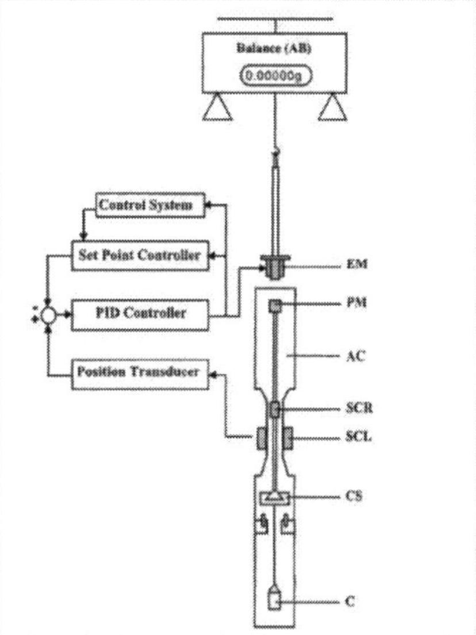

Figure 3: Magnetic Suspension method

Microbalance

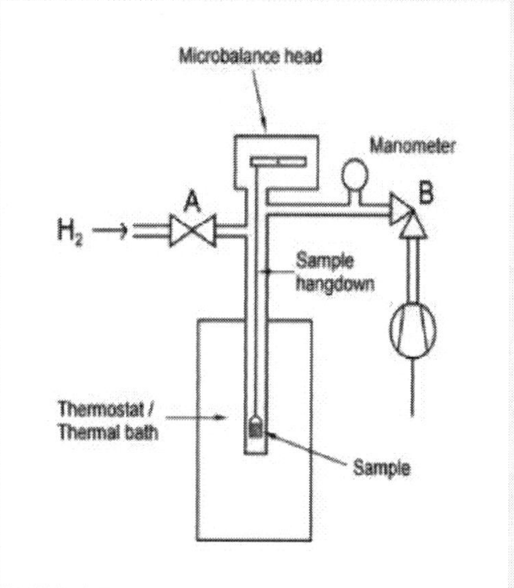

Figure 4: Microbalance method

Source: Rubotherm.org

Vibrational gravimetric method

Quartz-crystal microbalance

- Detect µg-level mass changes to crystal and coating polymer material [1]
- Addition of polymeric materials reduces frequency of crystal
- Gas sorption onto coating material causes mass change and further reduces frequency

Beam is used to generate a frequency of 200 Hz.[2]

$$\Delta m = \frac{k}{2\pi^2 f^3}\Delta f$$

PVT cell

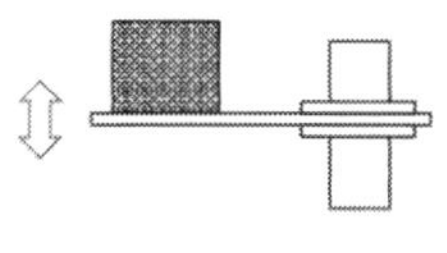

Figure 5:

1. Bonner and Cheng 1975, J. Polym. Sci. C 13:259-264.

2. Biscoe and Mahgereteh 1983, J. Phys. E 17:483-487.

Our Vibration Based Method

COLORADO SCHOOL OF MINES
EARTH • ENERGY • ENVIRONMENT

Change in Frequency of a Porous Solid

Change in the density of gas co-accelerated outside of porous solid

ΔM_a
a = added mass

Change in the density of co-accelerated gas inside of porous solid

ΔM_p
p = pore gas

Change in the mass due to adsorption and condensation

$\Delta M_{Adsorption\ and\ Condensation}$

Spring

Mass

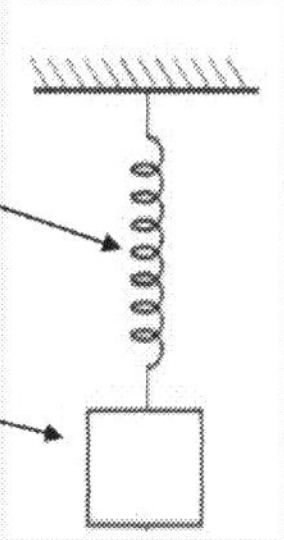

Calibration Methodology

COLORADO SCHOOL OF MINES
EARTH • ENERGY • ENVIRONMENT

Change in frequency is related to change in mass

$$f = \frac{1}{2\pi} * \sqrt{\left(\frac{k}{M_{Total}}\right)}$$

$$\Delta f = \Delta M_{Total}$$

Where, $\Delta M_{Total} = \Delta M_a + \Delta M_p + \Delta M_{Adsorption+Condensation}$

Calibration is done following the two steps below:

1. **Calibration for added mass (using a non porous mass of the same shape as the porous solid):**

$$\Delta M_a = \alpha P$$

Where, $\alpha = added\ mass\ coefficient$ (unit: g/psi)

2. **Calibration for pore gas (using a macroporous solid):**

$$\Delta M_p = (Pore\ Volume) * \frac{P}{zRT} * MW$$

Where, $MW = Molecular\ weight$

$$\Delta M_{Total} - \Delta M_a - \Delta M_p = \Delta M_{Adsorption+Condensation}$$

Non Porous Solid: Frequency Vs Pressure

COLORADO SCHOOL OF MINES
EARTH • ENERGY • ENVIRONMENT

- Calibration for α using non-porous solid.
- Different masses considered: 106.338g, 90.956g, 66.275g and 56.434g

$$M + M_a = \frac{k}{4\pi^2 f^2}$$

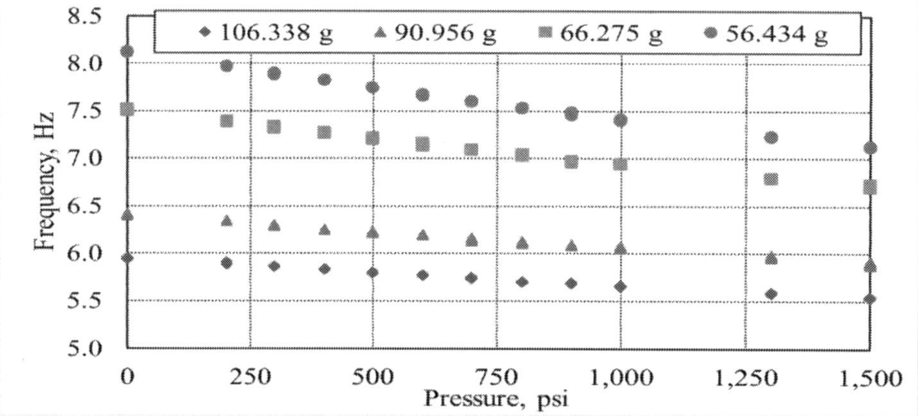

Figure 6: System Calibration for different mass
Larson et al. (2017). Experimental technique to measure mass under high pressure conditions using oscillatory motions of a spring-mass system

Non Porous Solid: Added mass coefficient

COLORADO SCHOOL OF MINES
EARTH • ENERGY • ENVIRONMENT

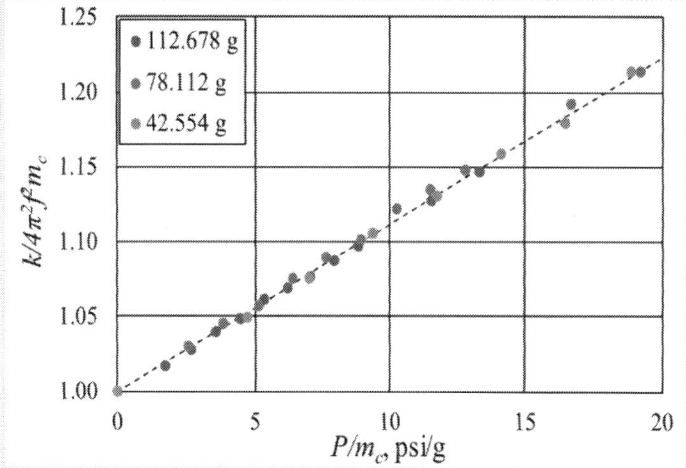

$$\frac{k}{4\pi^2 f^2} = m + m_a = m + \alpha P$$

$$\frac{k}{4\pi^2 f^2 m_C} = 1 + \alpha\left(\frac{P}{m_C}\right)$$

Figure 7: Determining added mas coefficients
Larson et al. (2017). Experimental technique to measure mass under high pressure conditions using oscillatory motions of a spring-mass system

Non Porous Solid: Frequency Vs Pressure at high temperature

COLORADOSCHOOLOFMINES
EARTH • ENERGY • ENVIRONMENT

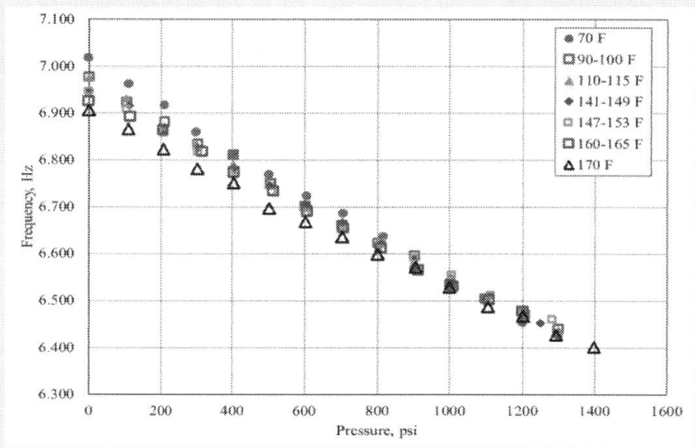

- Gas = Nitrogen
- Mass = 80.06 g
- Measured till 170°F
- α is reduced at high temperature as nitrogen becomes less dense.
- Initial frequency drops as rising temperature due to thermal effect on spring.

Figure 8: Frequency Vs Pressure data at high temperatures

Added mass coefficients

COLORADOSCHOOLOFMINES
EARTH • ENERGY • ENVIRONMENT

Temperature (°F)	Added Mass coefficients (g/psi) Niobrara
Gas: Propane	
70	0.02605 ± 0.00170
115	0.02602 ± 0.00079
150	0.02658 ± 0.00014
Gas: Nitrogen	
70	0.00841
Temperature (°F)	**Added Mass coefficients (g/psi) Berea**
Gas: Propane	
70	0.01341
Gas: Nitrogen	
70	0.00554

Macroporous Solid: Understanding ΔM_p

COLORADOSCHOOLOFMINES
EARTH • ENERGY • ENVIRONMENT

Gasses used:

-Nitrogen (0-1500 psi)

-Propane (0-350 psi)

▪**Berea sandstone core (Macroporous solid)**: obtained from Cleveland Quarries

-1.5 in diameter x 1.25 in length
- Porosity = 19.7%
- Permeability = 175 mD
- Mass = 75.9 gram

▪Sample was stored in oven before and after experiment

Berea Sandstone

COLORADOSCHOOLOFMINES
EARTH • ENERGY • ENVIRONMENT

▪Both Nitrogen and Propane were used

▪Mass = 93 g (mass of rock+weight holder), Temperature = 70 (°F)

▪Added mass coefficient of Propane is higher than Nitrogen as Propane is denser than Nitrogen

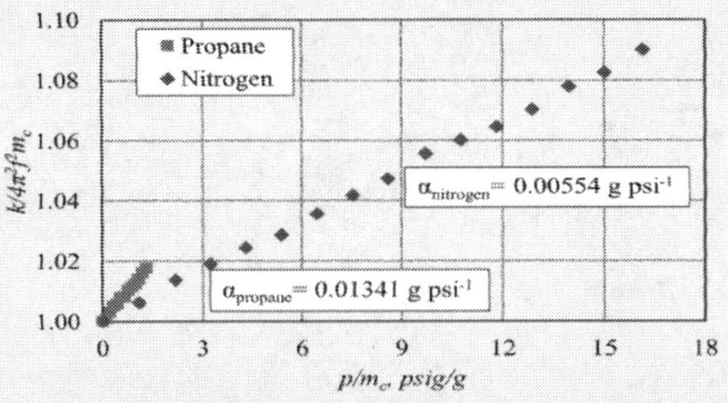

Figure 10: **$k/4\pi^2f^2m_c$ as a function of p/m_c of Propane and Nitrogen gas at 70 F**

Berea Sandstone: Need for ΔM_P

COLORADO SCHOOL OF MINES
EARTH • ENERGY • ENVIRONMENT

- In macroporous solids, both N_2 and Propane have little adsorption and they do not condense until pressure becomes extremely close to P_{v0} (bulk vapor pressure)
- A- no noticeable difference between added mass correlation and measured frequencies.
- B-Compressed nitrogen gas has influence of frequencies at high pressure

At 1500 psi, 0.9 g of compressed nitrogen occupies the pores. Deflection in B due to pore gas.

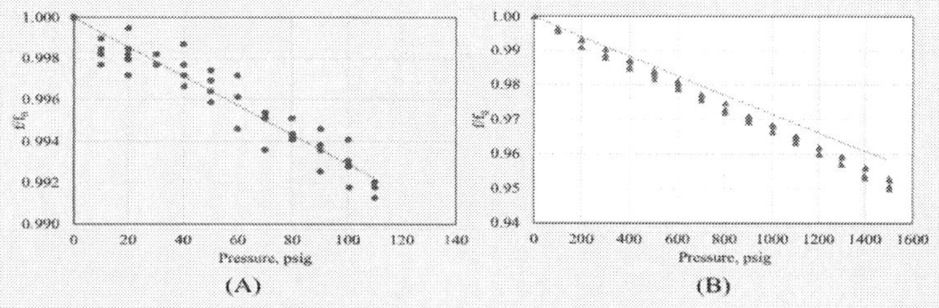

Figure 11: Normalized frequency measurements on Berea sandstone in a function of pressure for (A) propane and (B) nitrogen. The dot lines represent the added mass correlation calculated from the added mass coefficient.

Correlation for ΔM_p

COLORADO SCHOOL OF MINES
EARTH • ENERGY • ENVIRONMENT

Mass of nitrogen in pores determined using:

$$\Delta m_p = \left(\frac{PV}{TRz}\right) mw_{N_2}$$

Figure shows that the measured Δm_p (horizontal axis) are in good agreement with calculated Δm_p, proving that Δm_p can be calculated using pore volume and gas EOS

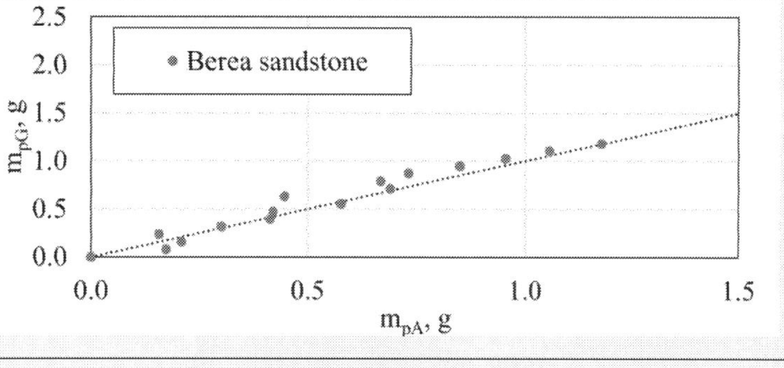

Figure 12: Δm_p in Berea Sandstone

Mesoporous Niobrara Shale

COLORADO SCHOOL OF MINES
EARTH • ENERGY • ENVIRONMENT

Gasses used:

Propane: A condensable gas at pressure and temperature of interest

Niobrara shale(Mesoporous solid): fresh from nearby quarry

- -cleaned using Toluene and Methanol
- -crushed to 20/40 mesh size
- -Porosity = 44.5 %
- -Permeability = 0.7-1.6 µD
- -20/40 crushed sample
- -Mass = 98.963g.

▪Sample was stored in oven before and after experiment

Niobrara Shale- Frequency Vs Pressure

COLORADO SCHOOL OF MINES
EARTH • ENERGY • ENVIRONMENT

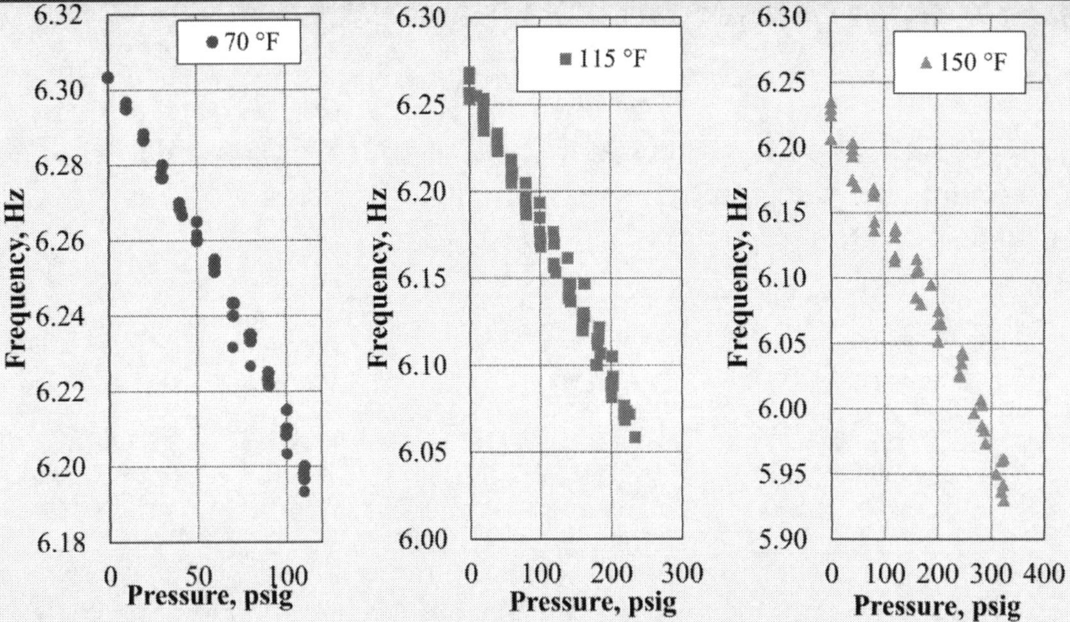

Figure 13: Frequencies of Niobrara shale sample as a function of pressure at 70°F. The dot lines represent the added mass correlation calculated from the added mass coefficient

Niobrara Shale- Final Results

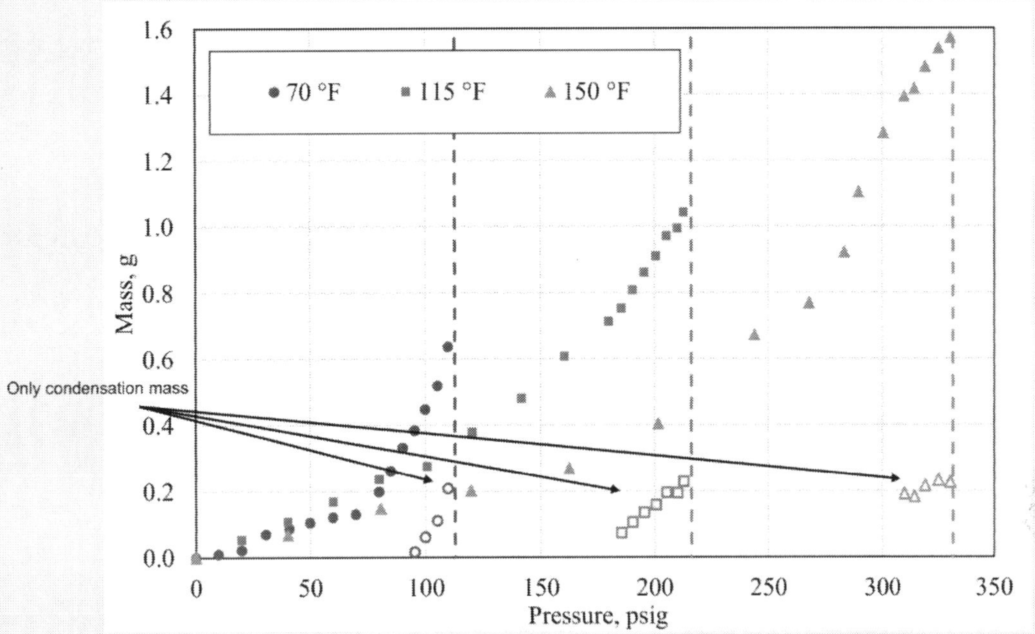

Conclusion

- We designed a simple device and methodology to measure capillary condensation in mesoporous rocks:
- Vibration (oscillation) based measurement principle.
- Effect of added mass corrected by calibration using non-porous samples
- Effect of pore gas, as shown by Berea sample can be determined using pore volume and gas density
- Additional mass gained in Niobrara, a mesoporous rock is due to capillary condensation was measured
- Mass of condensation analyzed using Kelvin equation
  - Smallest pore size: 2-4 nm (agrees with literature)

Acknowledgement

COLORADOSCHOOLOFMINES
EARTH • ENERGY • ENVIRONMENT

This work was supported by:

- ACS PRF
- Colorado School of Mines Undergraduate Research Fellowship
- Unconventional Natural Gas and Oil Institute
- Unconventional Reservoir Engineering Project Consortium

References

COLORADOSCHOOLOFMINES
EARTH • ENERGY • ENVIRONMENT

- D.P. Broom, Hydrogen Storage Materials, Green Energy and Technology, 2011
- Agarwal and Schwarz Carbon Volume 26, No.6, pp 873-887, 1988
- Larson, Zachary, Younki Cho, and Xiaolong Yin. "Experimental technique to measure mass under high pressure conditions using oscillatory motions of a spring-mass system." Measurement Science and Technology 28, no. 6 (2017): 065902
- Prasad, Manika, Kenechukwu C. Mba, Tracy Sadler, and Mike L. Batzle. "Maturity and impedance analysis of organic-rich shales." SPE Reservoir Evaluation & Engineering 14, no. 05 (2011): 533-543
- Petrophysical rock typing of unconventional shale plays: A case study for the Niobrara formation of the Denver-Julesburg (dj) basin, Asm Kamruzzaman, 2015

COLORADO SCHOOL OF MINES
EARTH • ENERGY • ENVIRONMENT

Thank you

11/19/2017

Vibrational Gravimetric Analysis of Capillary Condensation in Nanoporous rock-Keerthana Krishnan

Process Flow Chart

COLORADO SCHOOL OF MINES
EARTH • ENERGY • ENVIRONMENT

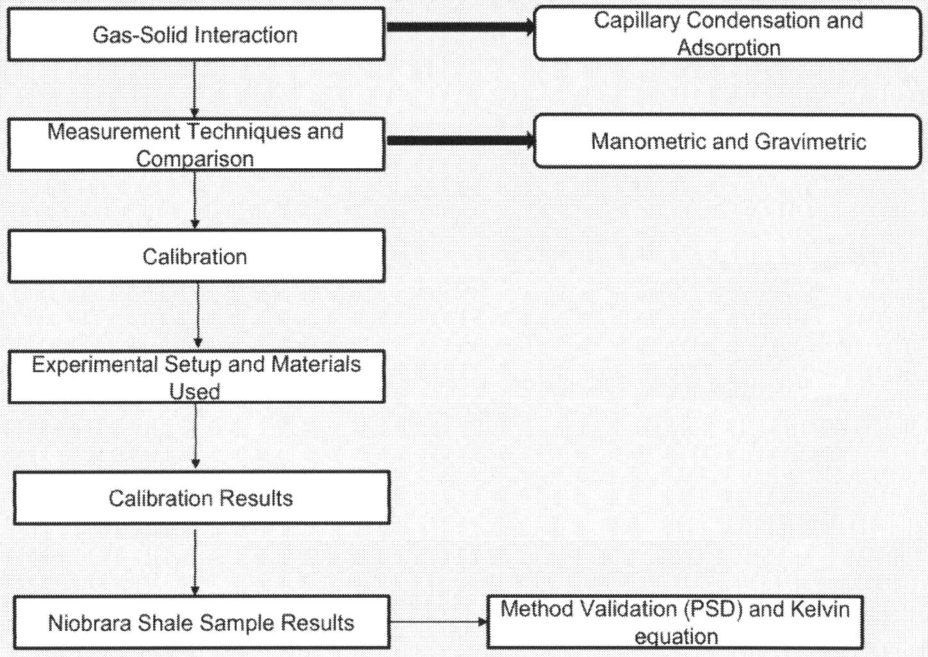

11/19/2017

Vibrational Gravimetric Analysis of Capillary Condensation in Nanoporous rock-Keerthana Krishnan

26

Experimental Design

COLORADO SCHOOL OF MINES
EARTH • ENERGY • ENVIRONMENT

- Determination of spring constant
- Accounting for surrounding fluid
- Determination of unknown mass

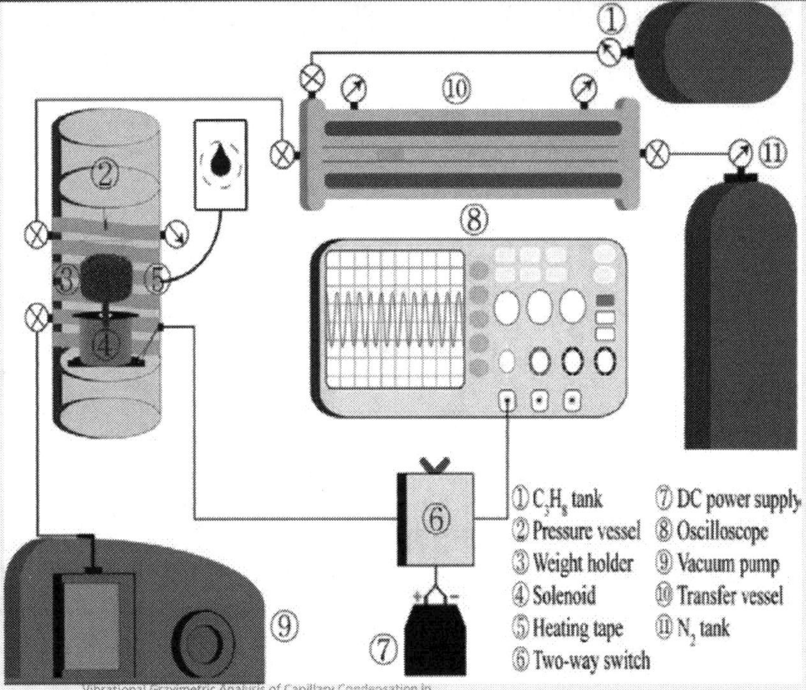

Figure 4: Experimental Setup
Cho et al., Vibrational Gravimetric Analysis of Capillary Condensation in Nanoporous Rock, 2017

PSD

COLORADO SCHOOL OF MINES
EARTH • ENERGY • ENVIRONMENT

- Kelvin equation is used to determine the PSD of pores filled with the condensates
- Kelvin equation- improved by adding density and Interfacial Tension(IFT) from PREOS and Parachor method

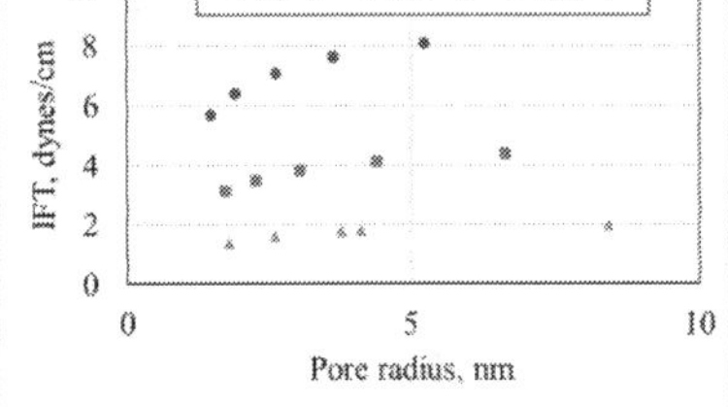

Figure 15: Interfacial tension as a function of pore radius at different temperatures

PSD cont..

COLORADO SCHOOL OF MINES
EARTH • ENERGY • ENVIRONMENT

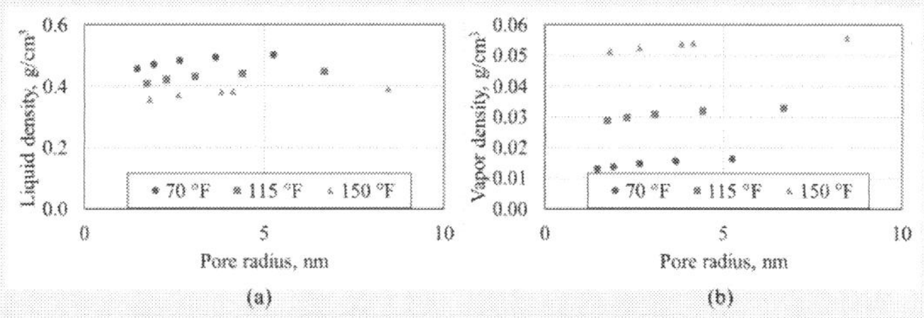

Figure 16: Densities of (a) liquid and (b) vapor as a function of pore radius at different temperatures

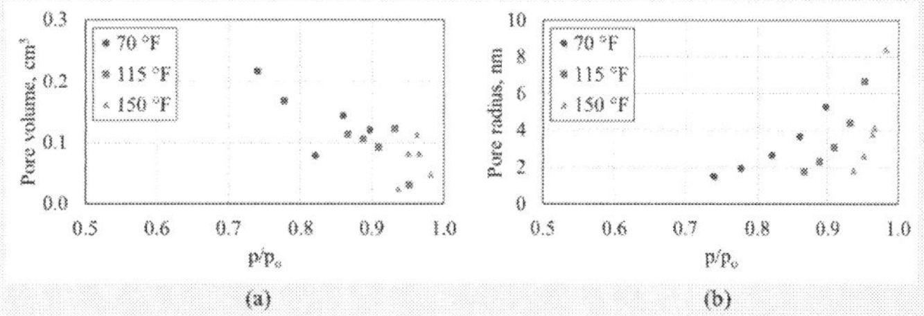

Figure 17:(a) Pore volume and (b) pore radius as a function of normalized pressure (p/p_o) at different temperatures. The normalized pressure is a ratio of vapor pressure to corresponding saturation pressure at the temperature

Calibration Cont..

COLORADO SCHOOL OF MINES
EARTH • ENERGY • ENVIRONMENT

•**Part 2: Estimation of PSD**

-Only the pores with condensate liquid is estimated

-Estimate interfacial tension σ

$$\sigma = [P_r(\rho_L - \rho_V)]^e$$

Where e = 3.88

$P_r = 155.33\ (for\ propane)$ from National Institute of Standards

$\rho_L and\ \rho_V$: estimated using PR EOS

-Pore radius is estimated using Kelvin equation:

$$r = \frac{-2\sigma}{\rho_L RT ln\left(\frac{p_v}{p_v^0}\right)}$$

Where, p_v is vapor pressure and p_v^o is saturation pressure

▪Liquid pressure p_L is determined using Young-Laplace equation:

$$p_c = p_v - p_L = \left(\frac{2\sigma cos\theta}{r}\right)$$

$\theta = 0$ (to maximize surface coverage)

▪Minimum pressure corresponds to metastable state

▪Molar volume of gas and liquid is calculated using PREOS

▪Convergence criteria : $(\mathrm{v_t} - \mathrm{v_{t-1}}) \leq 1 \times 10^{-7}$

Where, $v = \frac{m(r_{i+1}) - m(r_i)}{\rho_L(r_{i+1}) - \rho_v(r_i)}$

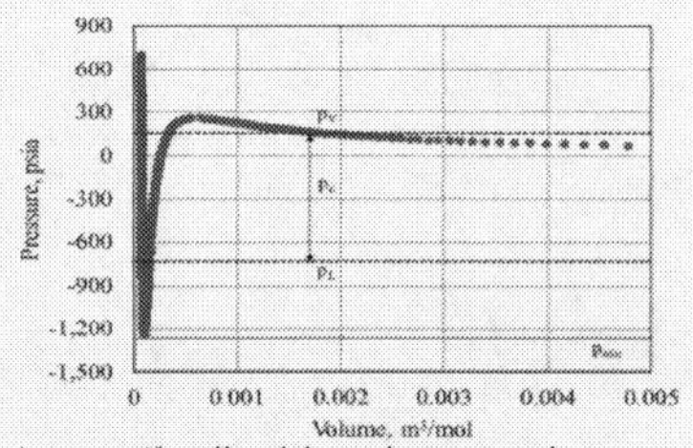

Figure 7: The pressure difference between the liquid and vapor phases represents the capillary pressure in an isothermal pressure-volume curve of propane at 70 °F from PR EOS. The red line is the minimum liquid pressure

Determination of Porosity

▪Volumetric method:

-change in volume is related to total pore volume

-crushed sample and deionized water

▪Vibrational gravimetric method:

-Nitrogen isotherm at 70F and 2000 psig using vapor density from NIST

Result:

-52 % porosity from volumetric technique and 54.9 % from vibrational gravimetric method

Calibration for Ma

COLORADO SCHOOL OF MINES
EARTH • ENERGY • ENVIRONMENT

Determination of mass gained

- $k = 4\pi^2 f^2 m_c$ Eq1

Effect of viscous damping is neglected

- $k = 4\pi^2 f^2 (m + m_a)$ Eq2

-To determine added mass, mc = m

-As added mass is proportional to density of fluid- proportional to pressure,

-the equation above becomes,

$$\frac{k}{4\pi^2 f^2 m_c} = 1 + \left(\frac{\alpha P}{m_c}\right) \quad \text{Eq3}$$

-Substituting $\alpha P = m_a$, we get

$$m = \frac{k}{4\pi^2 f^2} - (\alpha P) \quad \text{Eq4}$$

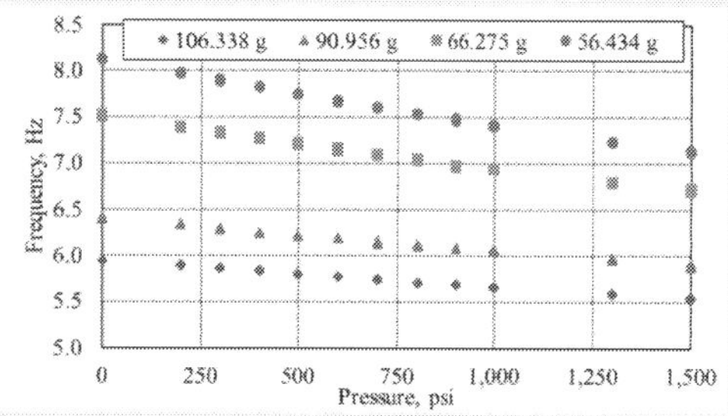

Figure 5: System Calibration for different mass
Larson et al. (2017). Experimental technique to measure mass under high pressure conditions using 1 oscillatory motions of a spring-mass system

Calibration for Ma

COLORADO SCHOOL OF MINES
EARTH • ENERGY • ENVIRONMENT

- Three different weights: 42.554 g, 78.112 g and 112.678 g.
- Nitrogen-70F; 0 upto 1500 psig
- 5 reading at each pressure- standard deviation: 0.13%
- $\alpha = 0.01118 \frac{g}{psi}$

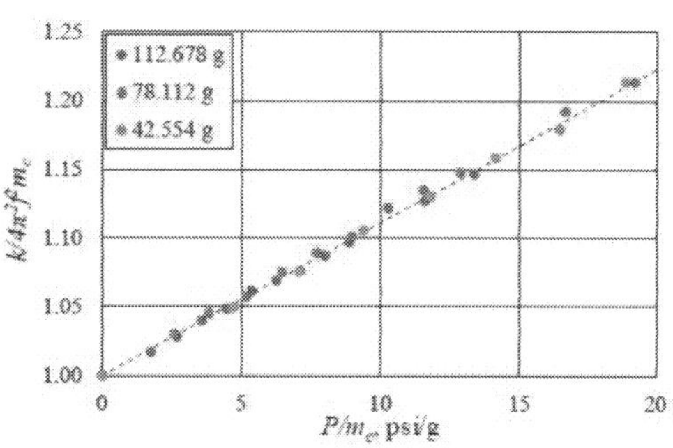

Figure 6: $\frac{k}{4\pi^2 f^2 m_c}$ as a function of $\frac{P}{m_c}$ at 70 °F. Data collected using nitrogen in pressure range of 0−1500 psig. Larson et al. (2017). Experimental technique to measure mass under high pressure conditions using 1 oscillatory motions of a spring-mass system

Non Porous Solid remove

COLORADOSCHOOLOFMINES
EARTH • ENERGY • ENVIRONMENT

-Test temperature: 70 °F, 115 °F and 150°F using Propane

-Average spring constant: 0.154, 0.151 and 0.150 $N\ mm^{-1}$

-Added mass coefficients calculated with two different weights

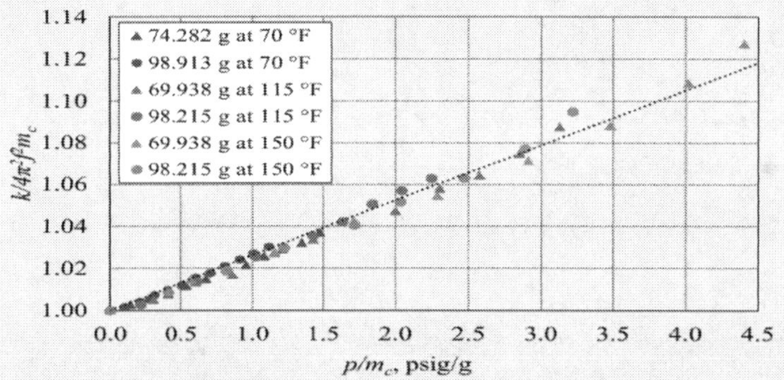

Figure 9: $k/4\pi^2 f^2 m_c$ as a function of p/m_c at 70, 115, and 150 °F. Data are collected using propane up to saturation pressure for the crushed Niobrara sample.

Calibration data – focus on the linear part

COLORADOSCHOOLOFMINES
EARTH • ENERGY • ENVIRONMENT

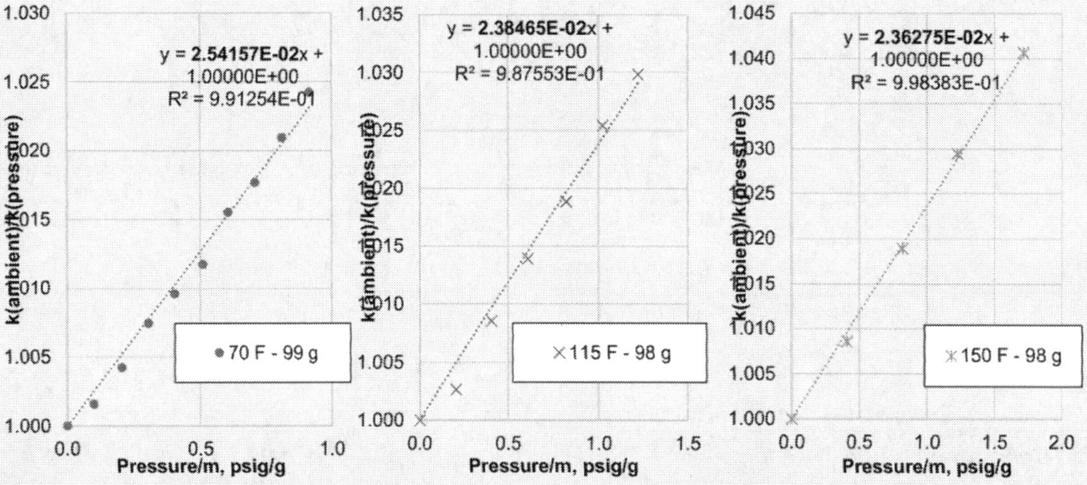

COLORADOSCHOOLOFMINES
EARTH • ENERGY • ENVIRONMENT

1	2	3	4	5	6	7	8	9	10
Temp, F	Pressure, psig	mass, g	sample mass, g	Bulk density, g/cc	Porosity, %	Pore volume, cc	Propane density from NIST, g/cc	Gas in pore, g	Condensation mass, g
70	80.05	1.98E-01	79.795	2.7	44.53	23.72	0.01294	0.31	-1.09E-01
70	85.00	2.61E-01	79.795	2.7	44.53	23.72	0.013747	0.33	-6.54E-02
70	90.13	3.31E-01	79.795	2.7	44.53	23.72	0.0146	0.35	-1.50E-02
70	95.30	3.84E-01	79.795	2.7	44.53	23.72	0.015477	0.37	1.64E-02
70	99.90	4.47E-01	79.795	2.7	44.53	23.72	0.016274	0.39	6.12E-02
70	105.20	5.19E-01	79.795	2.7	44.53	23.72	0.017211	0.41	1.10E-01
70	109.85	6.36E-01	79.795	2.7	44.53	23.72	0.018052	0.43	**2.08E-01**
115	185.50	7.53E-01	79.817	2.7	44.53	23.73	0.028649	0.68	7.34E-02
115	190.40	8.08E-01	79.817	2.7	44.53	23.73	0.029615	0.70	1.05E-01
115	195.40	8.61E-01	79.817	2.7	44.53	23.73	0.030624	0.73	1.35E-01
115	200.60	9.10E-01	79.817	2.7	44.53	23.73	0.031699	0.75	1.58E-01
115	205.30	9.71E-01	79.817	2.7	44.53	23.73	0.032696	0.78	1.95E-01
115	210.00	9.94E-01	79.817	2.7	44.53	23.73	0.033718	0.80	1.94E-01
115	212.80	1.04E+00	79.817	2.7	44.53	23.73	0.034339	0.81	**2.28E-01**
150	310.00	1.40E+00	79.917	2.7	44.53	23.76	0.050641	1.20	1.92E-01
150	314.75	1.42E+00	79.917	2.7	44.53	23.76	0.051941	1.23	1.83E-01
150	319.57	1.48E+00	79.917	2.7	44.53	23.76	0.053306	1.27	2.16E-01
150	325.40	1.54E+00	79.917	2.7	44.53	23.76	0.055026	1.31	2.33E-01
150	330.60	1.57E+00	79.917	2.7	44.53	23.76	0.056632	1.35	2.27E-01

Conclusion

COLORADOSCHOOLOFMINES
EARTH • ENERGY • ENVIRONMENT

- Vibrational gravimetric analysis and capillary condensation in Niorbrara shale suing Berea sandstone as baseline
- Calibration done using non porous sample, nitrogen upto 170 F.
- Can be extended to multicomponent gas mixtures as well
- No adsorption observed at 70F for nitrogen in both samples regardless of pore size
- Capillary condensation observed in Niobrara at different temperatures
- Pore radius of Niobrara: 1.5 nm – 8.4 nm
- Porosity is lower whereas permeability agrees well with condensate pores

Determination of anthracene, phenanthrene and carbazole in crude anthracene by capillary GC

*LIU Wei, JIANG Su-yu, ZHANG Dan, Li Hui-ping**
School of Chemical Engineering and Energy, Zhengzhou University, Zhengzhou 450001, China
** Corresponding author phone: (+86) 0371-67781807; fax: (+86) 0371-67781807*
E-mail: huipingli@zzu.edu.cn

Abstract

A method was established for the determination of anthracene, phenanthrene and carbazole in crude anthracene by capillary GC. The qualitative analysis of main components in crude anthracene was analyzed by GC-MS, where N,N-dimethylformamide (DMF) was used as solvent and phenoxazine as internal standard substance (ISS). The results showed that the components of the crude anthracene could be separated well and twenty-two components were identified. The average contents of three main components were 39.59% for anthracene, 17.93% for phenanthrene, 15.48% for carbazole respectively. The relative standard deviations (*RSD*s) of the measurements are less than 2%. The recovery ratio is over 98%. The method is of high accuracy and precision, also, it is quick, sensitive, simple and suitable for industrial production analysis.

Key word: capillary column GC; anthracene; phenanthrene; carbazole

1. Introduction

Crude anthracene (CA), yellow-green crystal[1-2], is a distillation fraction from coal tar with the boiling point range of 300°C~360°C. The main components in CA such as anthracene (A), phenanthrene (P) and carbazole (C) are of highly value-added [3], and are not or cannot be economically obtained from the petroleum industry. It is significant to analyze the content of CA accurately, which is the basis for the separation and purification of A, P, C from CA. Therefore, it is very important to establish a method which can analyze the content of main components in CA accurately and quickly.

CA consists of many components, differing only slightly in physical properties [4], therefore, it is difficult to analyze and separate them. For example, Wang, J. G. [5] studied the content of A in CA using GC with internal standard method. Li, C. [6] obtained the contents of A and P in CA using GC with external standard method. Here, CA was analyzed by GC-MS qualitatively and the content of A, P and C were determined quantitatively by the internal standard curve method in this work, which can eliminate the influence of some operations, also, reduce the trouble of weighing and calculating, and is more accuracy, reliable than external standard method.

2. Experiment Sections

2.1 Instruments and reagents

Shimadzu GC-MS QP2010, Tianmei GC-7900, Anthracene (A), phenanthrene (P), carbazole (C), and phenoxazine are GC grade reagents, DMF (A.R.). Crude anthracene (CA) was from Henan Baoshun Chemical Technology Co., Ltd.

2.2 Chromatographic conditions

GC: SE-54 capillary column (30 m×0.32 mm×0.5 μm) was selected. The temperature-programmed route as follows: staring from 140°C (0.1 μL injection volume) for 4 min, then increased to 180°C by 20°C/min, held for 5 min, and then further increased to 300°C by 20°C/min, held for 5 min. The temperature of the vaporization chamber and FID detector: 300°C, the flow rate of H_2: 40 mL/min, the pressure of air: 0.36 MPa, carrier gas N_2 (purity>99.999%): 0.5 MPa.

GC-MS: SE-54 capillary column (30 m×0.32 mm×0.25 μm) was used, the temperature were: 300°C (inlet), 250°C (FID), 230°C (ion source) and 250°C (surface). The temperature-programmed route was the same as the above GC (0.2 μL injection volume). And the components were qualitatively analyzed by NIST 08 library. The flow rate was 1.04 mL/min and the split ratio was 30: 1.

2.3 Method

A amount of CA was fully dissolved in DMF. The samples were analyzed by GC-MS qualitatively (Fig. 1) and quantitatively analyzed by GC with the internal standard curve method. The qualitative results of the main components are shown in Tab. 1.

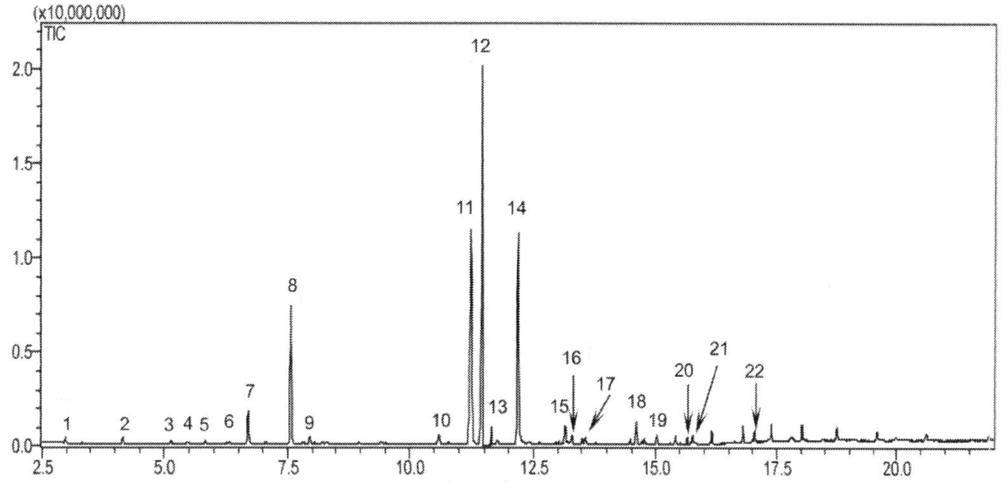

Fig. 1 GC-MS spectra of CA

Tab. 1 Qualitative results of the components in CA

sequence	retention time/min	component	sequence	retention time/min	component
1	2.974	Naphthene	12	11.467	Anthracene
2	4.159	1-Methylnaphthalene	13	11.767	Naphtho[2,1-b]thiophene
3	5.135	Biphenyl	14	12.184	Carbazole
4	5.451	2-Ethylnaphthalene	15	13.150	2-Methylanthracene
5	5.826	2,3-Dimethylnaphthalene	16	13.300	9-Methylanthracene
6	6.315	Acenaphthene	17	13.567	3-Methylcarbazole
7	6.697	Dibenzofuran	18	14.592	Fluoranthene
8	7.558	Fluorene	19	15.017	Pyrene
9	7.951	4-Methylbiphenyl	20	15.649	1,2-Benzfluorene
10	10.590	Dibenzothiophene	21	15.767	2,3-Benzfluorene
11	11.234	Phenanthrene	22	17.040	Benz(a)anthracene

Fig. 1 and Tab. 1 showed that the composition of CA is complex, of which twenty-two components were identified. However, the main components are A, P, C, fluorene and dibenzofuran.

3. Results and discussion

3.1 Selection of chromatographic column

SE-54 column was used to analyze CA, as shown in Fig. 2.

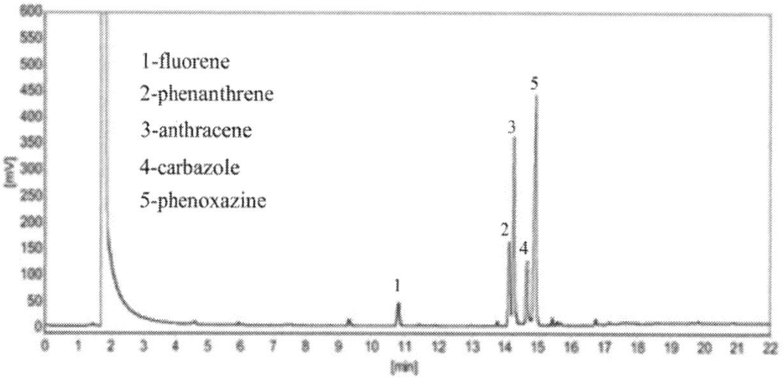

Fig. 2 Chromatogram of CA with the ISS added

It can be seen from Fig. 2 that the peaks of A (No. 3), P (No. 2), C (No. 4), fluorene (No. 1) and phenoxazine (No. 5) were well separated. Also, the peaks of A, P, C were close to each other, and the retention time of P was shorter.

3.2 Standard curves of A, P and C

The standard curves of A, P and C were obtained by using $m_i/m_s \sim A_i/A_s$ (shown in Fig. 3, m_i and A_i -- the mass and peak area of the component tested, m_s and A_s -- the mass and peak area of the ISS respectively). Where y is the ratio of peak area and x is the mass ratio of component tested to the mass of the ISS. The results showed that there was a good linear relationship between m_i/m_s and A_i/A_s.

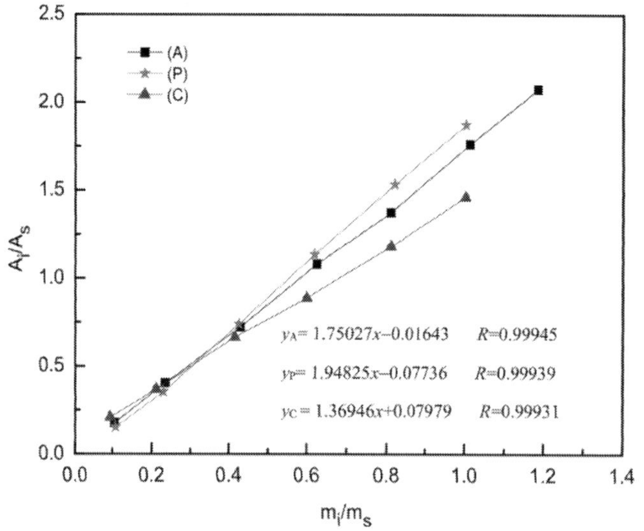

Fig. 3 Standard curves for A, P and C

3.3 Precision test

The precision of the determination results was verified by measuring relative standard deviations (*RSD*) [7]. The formula is as follows:

$$S = \sqrt{\frac{\sum (X_i - \overline{X})^2}{n-1}} \qquad RSD = \frac{S}{\overline{X}}$$

Where S is the standard deviation of the results, X_i is the content of components tested, *RSD* is the relative standard deviation of the results, and $\overline{X}$ is the average of the results tested, n is the number of measurements.

The *RSD*s were calculated and shown in Tab. 2.

Tab. 2 Results of precision of CA (unit /%)

Sample No.		X_i /%						$\overline{X}$ /%	S /%	RSD /%
1	A	41.13	40.18	40.23	40.00	40.11	41.25	40.48	0.553	1.366
	P	19.70	19.33	19.28	19.30	19.28	19.32	19.37	0.162	0.836
	C	14.61	14.54	14.84	15.26	14.90	14.68	14.64	0.262	1.766
2	A	36.76	37.20	37.35	37.62	37.24	37.80	37.33	0.363	0.971
	P	16.99	17.28	17.49	17.27	17.35	17.76	17.36	0.257	1.482
	C	14.32	14.40	14.73	14.80	14.54	14.94	14.59	0.241	1.650
3	A	38.39	38.18	37.63	37.23	37.88	37.31	37.77	0.465	1.232
	P	16.94	16.74	17.08	16.72	17.07	16.67	16.87	0.184	1.088
	C	14.66	14.88	14.80	14.39	14.51	14.55	14.63	0.261	1.803
4	A	42.46	42.66	42.25	41.41	41.08	41.46	41.89	0.651	1.554
	P	18.63	18.81	18.64	18.49	18.18	18.48	18.54	0.212	1.142
	C	17.28	17.03	17.13	16.69	16.73	17.31	17.03	0.266	1.563

Tab. 2 showed that the *RSD*s were as follows: A: 0.971%~1.554%; P: 0.836%~1.482%; C: 1.563%~1.803%, which are less than 2%.

3.4 Recovery test

Here, the accuracy of the determination results was verified by measuring the recovery ratio of the sample based on the following formula [8]:

$$R = \frac{(m_2 - m_1)}{\Delta m} \times 100\%$$

Where R is the recovery ratio of the sample, m_1 is the measured mass value of the original sample, m_2 is the measured mass value of the sample with the ISS added, and Δm is the mass of the ISS added.

The measurement results are listed in Tab. 3.

Tab. 3 Results of recovery ratio of CA

Sample No.	Object	m_1 /g	Δm /g	m_2 /g	R /%
1	A	0.0245	0.0098	0.0342	98.98
			0.0132	0.0376	99.30
	P	0.0115	0.0076	0.0191	99.57
			0.0130	0.0244	98.85
	C	0.0103	0.0067	0.0170	99.26
			0.0133	0.0235	99.31
2	A	0.0332	0.0080	0.0412	99.54
			0.0123	0.0456	100.85
	P	0.0147	0.0070	0.0216	99.02
			0.0130	0.0274	98.05
	C	0.0134	0.0090	0.0222	98.81
			0.0146	0.0279	99.50
3	A	0.0417	0.0071	0.0487	99.39
			0.0120	0.0538	100.90
	P	0.0181	0.0085	0.0265	98.29
			0.0125	0.0305	99.08
	C	0.0166	0.0082	0.0249	101.27
			0.0145	0.0311	100.22

From Tab. 3, it can be seen that the relative error is within ±2%, therefore, the method can be considered as being of high accuracy.

3.5 Determination of CA

The percentage (X_i %) of A, P, C was calculated according to the following formula:

$$X_i\% = \left(m_i/m_s\right) \times \left(m_s/m_0\right) \times 100\ \%$$

Where m_s/m_0 is the mass ratio of the ISS to CA, and m_i/m_s was calculated by the standard curve. The results were shown in Tab. 4.

Tab. 4 Measurement results of CA

Object	X_i /%						$\bar{X}$ /%
A	42.91	41.67	40.00	37.80	36.76	38.39	39.59
P	18.47	18.11	19.30	17.76	16.99	16.94	17.93
C	17.29	16.58	15.26	14.94	14.12	14.66	15.48

4. Conclusions

Twenty-two components in crude anthracene were identified by GC-MS in this study, of which anthracene, phenanthrene, carbazole, fluorene and dibenzofuran were the main components. The average contents were 39.59% for anthracene, 17.93% for phenanthrene, 15.48% for carbazole respectively. The relative standard deviations (*RSDs*) of the measurements were 0.971%~1.544% for anthracene, 0.836%~1.482% for phenanthrene and 1.563%~1.803% for carbazole. The recovery ratio were 98.98%~100.90% for anthracene, 98.05%~99.57% for phenanthrene and 98.81%~101.27% for carbazole. The results revealed that the analysis method is of high accuracy and great significance to analyze and purify the main components from crude anthracene, also, it is helpful to make full use of resources and reduce environmental pollution.

5. References

[1] He, J.P.; Li H. Recycling technology of coking chemical products [M]. Beijing: Metallurgical Industry Press, 2006: 264-268.

[2] Xiao, R.H. Coal tar chemical industry [M]. Beijing: Metallurgical Industry Press, 2009: 207-211.

[3] Pan, B.; Wang, B.; Wang, Y.; Xu, P.; Wang, L.; Chen, J.; Ma, D. A simple carbazole-N-benzimidazole bipolar host material for highly efficient blue and single layer white phosphorescent organic light-emitting diodes. J. Mater. Chem. C 2014, 2 (14), 2466-2469.

[4] Ye, C.P.; D, X.X.; Li, W.Y.; Wu, T.T.; Fan, M.M.; Feng, J. Highly efficient solvent screening for separating carbazole from crude anthracene. Energy Fuels 2016, 30, 3529-3534.

[5] Wang, J.G.; Liu, C.X. et al. Determination of anthracene in coal tar crude anthracene by capillary gas chromatography [J]. Journal of Wuhan University of Science and Technology, 2006, 29 (6): 567-568.

[6] Li, C.; Chen, G.P. et al., et al. Determination of anthracene and phenanthrene in crude anthracene by capillary gas chromatography [J]. Metallurgical Analysis 2015, 35 (1): 72-76.

[7] Xu, G.W. et al. Modern practical gas chromatography [M]. Beijing: Chemical Industry Press, 2004: 218-221.

[8] Qi, M.L. et al. Analysis and application of gas chromatography [M]. Beijing: Science Press, 2012: 66-69.

501228 Characterization of Chitosan Hydrogel with Improved Acid Stability Switched on By Carbon Dioxide

Abstract:

I. Introduction

Along with the development of chemical technology comes the obvious drawback of industrial processes: wastewater pollution. There have been many examples of water contamination around the world, which all seem to deliver the same message: it is now urgently necessary to develop more advanced, effective and sustainable technologies for wastewater treatment.

Amongst the wastewater treatment methods currently available, adsorption has been the focus of many studies for its effectiveness and relatively straightforward operation. There are various types of adsorbents, such as zeolite, modified-alumina, activated-clay, activated-carbon to bentonite, fly ash, pine leaves and chitosan [1]. As the world is switching its focus to "sustainable technology", adsorbents should be cheap, environmentally-friendly and widely abundant. One candidate that can meet those requirements is chitosan hydrogel. Chitosan, a derivation from the N-deacetylation of chitin which can be extracted from the shells of crustacean animals, is one of the most abundant natural bio-polymer [2]. In many studies, this material has been reported to possess excellent adsorption capacities towards many synthetic dyes and heavy metals. The adsorption capacity of chitosan hydrogel for anionic dye pollutants is particularly high at low pH due to the protonation of amino groups on chitosan molecules. In this process, NH_2 groups are protonated, which would allow chitosan hydrogel to strongly and quickly adsorb negatively-charged ions on anionic dye molecules.

In spite of the promising results, there are two existing problems with chitosan adsorbent materials. First of all, at acidic conditions, chitosan hydrogel is protonated and thus can be easily dissolved. This is a conflict for many researchers, as using chitosan at low pH conditions for anionic pollutants would require making the material stable in acids. To achieve this, chitosan is usually cross-linked using glutaraldehyde (GLA), epichlorohydrin (ECH) or ethylene glycol diglycidyl ether (EDGE). However, although these cross-linkers can increase chitosan's acid stability, they have been regularly avoided due to their high toxicity and the negative impact on chitosan's adsorption capacity [3]. The second problem is that, even if chitosan is successfully modified to survive in acids, the use of conventional acidic compounds to lower the pH (usually HCl or H_2SO_4) would lead to further wastewater treatment such as using basic compounds, for example NaOH, to neutralise the mixture afterwards. Furthermore, HCl or H_2SO_4 are known to corrode process equipment, which may result in malfunctioning and extra maintenance costs.

To tackle these problems, we have devised a new technique, CO_2-switchable system. CO_2-switchable system is a concept originally proposed by Jessop's group where carbon dioxide can be deployed as a trigger for modifications of various solvents' properties, such as degree of viscosity, polarity or hydrophilicity [4] [5]. In our case, we decided to utilise carbon dioxide to decrease the solution's pH, which is a much "greener" method compared with using inorganic acids such as HCl or H_2SO_4. With our results up to now, it can be shown that in CO_2-switchable system, chitosan undergoes protonation to become a cationic polymer, which can significantly speed up the adsorption mechanism between chitosan hydrogel and anionic dyes. Another interesting result that was obtained is the enhancement of chitosan hydrogel's stability in acid when the temperature of CO_2-switchable system was increased to 45°C or higher. At this condition, the material became mechanically stable, retained its shape and weight and most importantly, it did not dissolve throughout the whole adsorption process. Nevertheless, to achieve a detailed comprehension of the reaction between chitosan and carbon dioxide, further characterization was urgently needed.

Based on that motivation, in this study the effects of carbon dioxide on chitosan hydrogel at temperatures higher than 45°C were investigated. To understand the changes inside chitosan's structure, FT-IR spectra were studied. Dynamic light scattering method was also employed to investigate the zeta-potential of chitosan hydrogel after making contact with CO_2 at different temperatures. Last but not least, the stability of chitosan hydrogel in acid after pre-treatment with CO_2 was tested with acetic acid aqueous solution

II. Experiment

Chitosan flakes were dissolved in CH_3COOH solution to form a hydrogel mixture. This hydrogel mixture was left under stirring for 5 hours before being collected into a syringe. From this syringe, chitosan hydrogel was dropped into NaOH solution for coagulation. The coagulated beads were left overnight before being washed extensively with DI water until a neutral pH was reached.

After that, chitosan hydrogel beads were put in an aqueous solution. The solution was heated in 65°C with continuous CO_2 bubbling for 3 h. After that, the hydrogel beads were taken out and oven-dried at 50°C.

Chitosan hydrogel after pre-treatment by CO_2 and oven-drying was analyzed by Fourier-Transform Infrared Spectroscopy (FT-IR). Zeta-potential was measured by Dynamic Light Scattering (DLS) method.

For analysis of acid stability, these hydrogel beads were put in an aqueous solution of acetic acid. The weight of chitosan hydrogel was measured and recorded at different time intervals.

III. Results and discussion

The FT-IR spectra revealed noticeable changes to represent the formation of a carbamate cross-link, which is produced from the reaction between a primary amine and CO_2 as reported by literatures [6,7].

DLS was employed as a useful tool for inspecting the role of CO_2 on turning chitosan hydrogel into a cationic polymer. The result of the treated chitosan hydrogel shows positive zeta potentials. It is important to be reminded again that at 65°C, chitosan hydrogel was found to possess a higher acid stability than at 25°C. This, coupled with DLS results, further confirms the assumption from FT-IR spectra that a carbamate cross-link has been formed.

The results from the acid stability experiment show the changes in chitosan hydrogel's weight for both chitosan hydrogel treated by CO_2 switchable system and chitosan hydrogel that is not pre-treated. For the treated chitosan, after 3 hours the weight of chitosan was largely unchanged and in fact, slightly increased due to swelling.

IV. References

1 M. T. Yagub, T. K. Sen, S. Afroze and H. M. Ang, *Adv. Colloid Interface Sci.*, 2014, **209**, 172–184.

2 M. Li, S. Cheng and H. Yan, *Green Chem.*, 2007, **9**, 894.

3 R. Huang, Q. Liu, J. Huo and B. Yang, *Arab. J. Chem.*, 2013.

4 P. P. G. Jessop, D. J. D. Heldebrant, X. Li, C. A. C. Eckert and C. C. L. Liotta, *Nature*, 2005, **436**, 1102.

5 P. G. Jessop, S. M. Mercer and D. J. Heldebrant, *Energy Environ. Sci.*, 2012, **5**, 7240.

6 D. Nagai, A. Suzuki, Y. Maki, H. Takeno, *Chem. Commun.*, 2011, **47**, 8856.

7 V. Stastny and D. M. Rudkevich, *J. Am. Chem. Soc.*, 2007, **129**, 1018–1019.

491702 Multiscale Modeling of the Breakthrough Behavior of Adsorption Columns

Abstract:

The present work is associated with the development of a comprehensive theoretical approach to model the breakthrough behavior of adsorption columns used to separate gaseous components. The adsorption of gaseous species onto the surface of a porous solid is a stochastic process and is simulated using kinetic Monte Carlo methods. The molecules attach and detach form the sites of the surface through three micro-processes: adsorption or deposition, desorption, and migration of surface diffusion. Adsorption columns usually operate under transient, non-steady state conditions. Analysis of the transient behavior of a gas mixture flowing through a column filled with adsorbent yields the so-called breakthrough time which is the time required to achieve a desired degree of separation. Modeling of the column comprises conservation statements for the species (mass balances) that contain the amounts adsorbed onto the porous material. The conservation equations can be solved numerically using a finite difference/element technique in which the equations are discretized in space. The amount deposited onto the adsorbent is calculated from the kinetic Monte Carlo simulations. The process is repeated recursively until breakthrough behavior is achieved. The results of the kinetic Monte Carlo simulations are compared against experimental adsorption isotherms for methane, carbon dioxide, and nitrogen as well as for binary mixtures of these gases on microporous activated carbon. Furthermore, the calculated breakthrough times are verified experimentally using a custom-made dynamic column experiment to determine the outlet concentration. The multiscale model will be used in the sweetening (separation of sour gases) and inert rejection (separation of nitrogen) of natural gas.

490592 Impurity Rejection in Crystallization â€“Quantification and Mechanisms

Abstract:

Other than isolating the product as a solid, the perhaps most important objective in crystallization of pharmaceuticals is purification. Nevertheless, most of the engineering research in academia and industry has been focused on particle size and polymorph control. Notwithstanding their importance, crystallization as a unit operation is relied on not just in the API step but across the entire synthetic scheme to reject a number of different impurities. These can be reagents, reaction by-products, process impurities, degradants, chiral impurities, inorganic impurities and even genotoxic impurities. Purification via crystallizations is thus critical in producing a drug substance that is suitable for pre-clinical and clinical trials and eventually for market demands. A stringent regulatory oversight is also in place to ensure that the API meets all purity and quality attributes prior to administration to humans.

In Pharma industry, most of the impurity control in crystallization has traditionally been carried out by synthetic chemists who mainly rely on a before and after approach that is highly empirical in nature. Presented herein is the use of engineering principles to accurately quantify rejection of impurities throughout the crystallization. Various mechanisms of impurity rejection are identified and linked to thermodynamic and/or kinetic effects. Case studies from pharmaceutical development are presented that provide a clear understanding on the impurity purge mechanisms and how they are affected by process conditions. This work thus enables simple control measures to be introduced in the crystallization process that are capable of producing materials with targeted purity profile, whether that is for toxicity studies or for clinical trials.

492027 Bi-Directional Control of Glycosylation of Antibodies

Abstract:

Monitoring and control of critical quality attributes (CQAs) is important for biologics manufacturing. CQAs need to be controlled within desirable ranges in order to manufacture a clinically effective product. These ranges can be particularly tight for biosimilars, because similarity to the innovator molecule must be demonstrated. To address these challenges, new flexible methods of monitoring and modulating CQAs are required. Multivariate model predictive control (MPC) can be used to target CQAs within desirable ranges to ensure the delivery of high quality biologic products. In this study, MPC was applied to control the CQA of interest (CQA1) of a recombinant antibody by using two media additives and a cell culture parameter as modulation levers. A partial factorial experiment (n=25) was first conducted in bench-top bioreactors to understand the impact of these levers on CQA1. Both media additives were able to modulate CQA1 but in opposite directions. The cell culture parameter alone was able to modulate CQA1 in both directions. After these levers were identified and characterized, an empirical multivariate model was created to predict and control CQA1 during the cell culture process. Next, the robustness of this control strategy was tested in 3 L bench-top and 100 L pilot-scale bioreactors. For the test, CQA1 was purposefully deviated in either direction (high or low) away from the target using the cell culture parameter as a lever. Our results indicate that the model effectively predicted the amount of media additive required to bring CQA1 back to the target. The target CQA1 range was met for both scales at the end of the cell culture. Thus, the use of media additives to achieve bi-directional model predictive control of CQA1 has been achieved. This study demonstrates that MPC is a robust tool to modulate CQAs within desired targets for cell culture processes.

492107 Bioprocess Optimization for Production of Recombinant Proteins in Chinese Hamster Ovary Cells in Consideration of Metabolic Shifts By Shear Stress

Abstract:

Biopharmaceuticals such as monoclonal antibody drugs have attracted considerable interest in recent years for currently untreatable diseases. A series of large production plants have been built in different parts of the world. Chinese hamster ovary (CHO) cells are widely used as a host cell line and cultured in fed-batch cultivation in a large stirred bioreactor for commercial production of biopharmaceuticals. When a bioreactor is designed for commercial-scale production, it is indispensable to evaluate the impact of shear stress on cell growth and cell viability due to their mechanical fragility. In this study we focused on the impacts of shear stress on cellular metabolism to optimize fed-batch cultivation for production of tissue plasminogen activator (tPA) in CHO cells.

We went through the two steps to optimize a fed-batch cultivation process. The first step is the quantitative evaluation of the effects of shear stress. In the case of a stirred bioreactor, a quantitative evaluation is difficult to be performed due to the distribution of shear stress in a tank. Therefore we introduced a flow chamber which can impart a steady laminar shear stress on the cells for the evaluation of the effects of shear stress and found that metabolism shift occur with increased shear stress. We determined the appropriate range of shear stress for the production from the results. The second step is the optimization of an impeller configuration and feed medium in consideration of the metabolism shift. We designed 3L bioreactors which sustain appropriate shear stress in a tank using computational fluid dynamics (CFD) and feed medium in consideration of metabolism shift. CHO cells producing tPA protein were cultured with our optimized bioprocess. As a result, the production of tPA increased by 1.5 times compared to the conventional fed-batch process.

We found that metabolism shift occur by increasing shear stress and the optimization of a bioprocess taken into account the metabolism shift contributes to increasing productivity.

Using machine learning tools in bioprocess scale up when the number of batches is small

Viktor Konakovsky
Minneapolis, MN, US
Nov 2nd 2017

FUJIFILM Diosynth biotechnologies | BioRAPID | Newcastle University | DTU Danmarks Tekniske Universitet | BIOPRO World Talent Campus

Topics

- What was happening in my project BioRapid so far
- Today I will talk about using PCA in combination with TOST time series
- The case study will focus on quantifying equivalence between 2L and 200L

FUJIFILM Diosynth biotechnologies | BioRAPID | Newcastle University | DTU Danmarks Tekniske Universitet | BIOPRO World Talent Campus

My project BioRapid

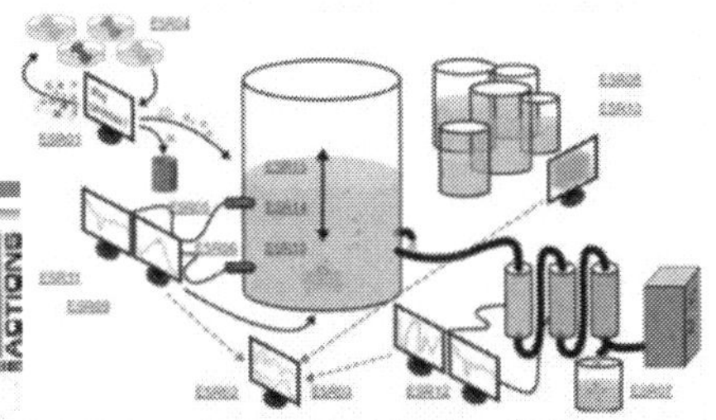

Meet my Team

Bio RAPID

MARIE CURIE ACTIONS

http://www.bio-rapid.eu/

Financial support from the EU-Horizon 2020 Marie Skłodowska-Curie actions (MSCA) ITN project BIORAPID (no. 643056).

FUJIFILM Diosynth biotechnologies | Bio RAPID | Newcastle University | DTU Danmarks Tekniske Universitet | BIOPRO World Talent Campus

Motivation

- This work was initiated while writing a report...

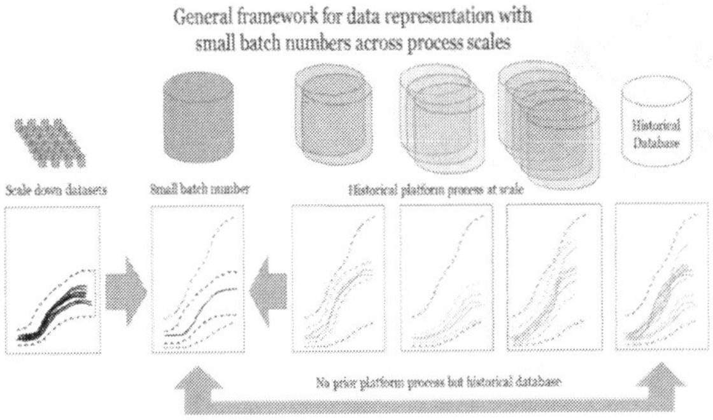

MARIE CURIE ACTIONS

Bio RAPID

"The Rapid Bioprocess Development ITN, effective training framework in rapid development of novel bioactive molecules from the very early stages of potency and efficacy testing to the bio-manufacturing process characterization and monitoring"

EU-Horizon 2020 - Marie Curie ITN project BIORAPID (Grant Agreement n° 643056)

Deliverable D2.5: General framework for data representation with small batch numbers, across process scales

- ...and we are writing up to publish the results soon in Biotechnology Journal

FUJIFILM Diosynth biotechnologies | Bio RAPID | Newcastle University | DTU Danmarks Tekniske Universitet | BIOPRO World Talent Campus

Scientific output

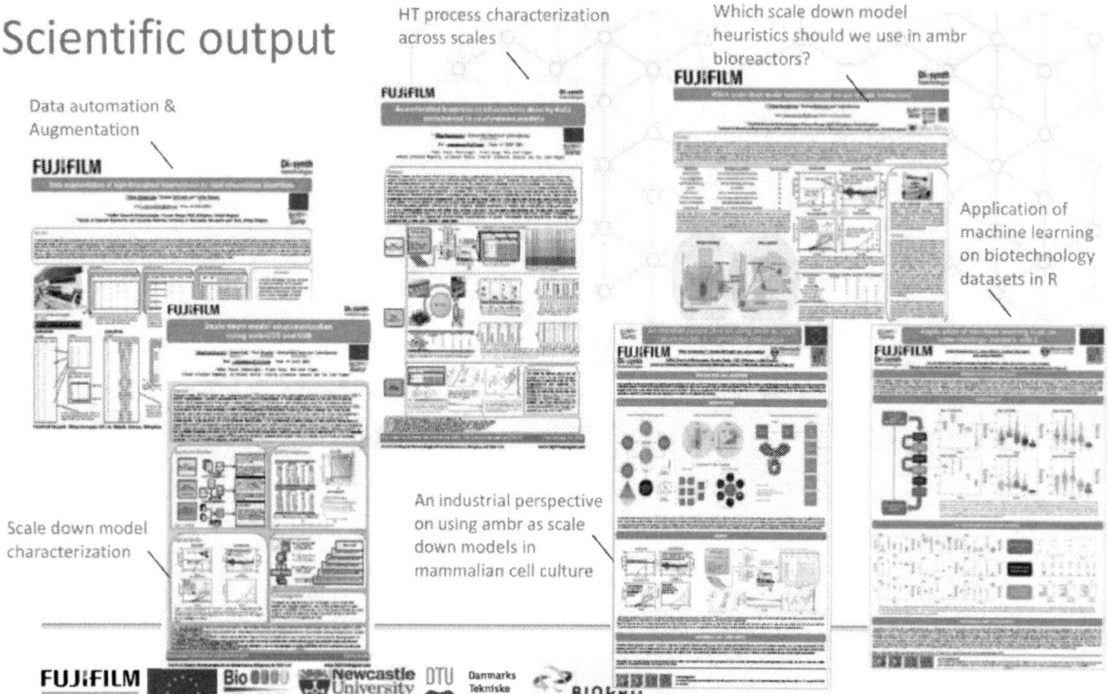

FUJIFILM Diosynth biotechnologies · BioRAPID · Newcastle University · DTU Danmarks Tekniske Universitet · BIOPRO World Talent Campus

Introduction

My part is to get value out of databases, and to develop predictive modeling and machine learning tools on dedicated industrial projects. Those projects were mostly focused on high throughput upstream data sources in mammalian cell culture but also extend to CLD and to microbial manufacturing

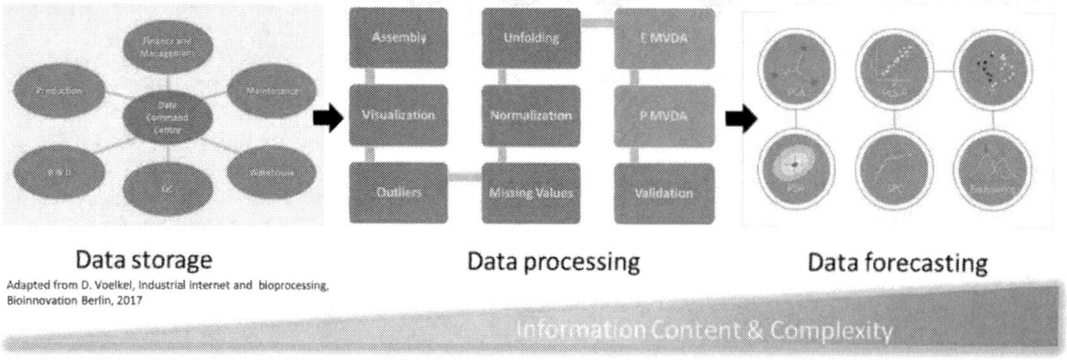

FUJIFILM Diosynth biotechnologies · BioRAPID · Newcastle University · DTU Danmarks Tekniske Universitet · BIOPRO World Talent Campus

6

The case

- In HT (high throughput) we have a lot of data, but as we scale up this is not the case any more
- Often at this stage we want to know whether batches are equivalent, or if not, what is different?
- This dataset is a CHO cell culture fed-batch

Scale	***Regime***	***n***
2L	C	6
2L	D	9
10L	E	2
200L	D	1
sum		**18**

FUJIFILM Diosynth biotechnologies · Bio RAPID · Newcastle University · DTU Danmarks Tekniske Universitet · BIOPRO World Talent Campus

The TOST

- Question: Is data from group A equivalent to group B?
- Typically would use TOST (two one sided t-test) to quantify a significant difference
- Statisticians in bioprocessing often use 2 standard deviations (SD) of a reference group to calculate delta

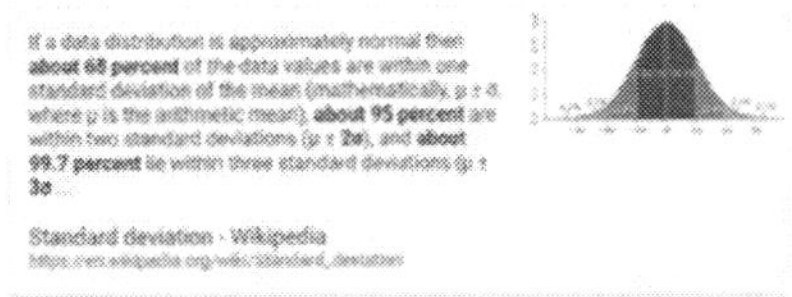

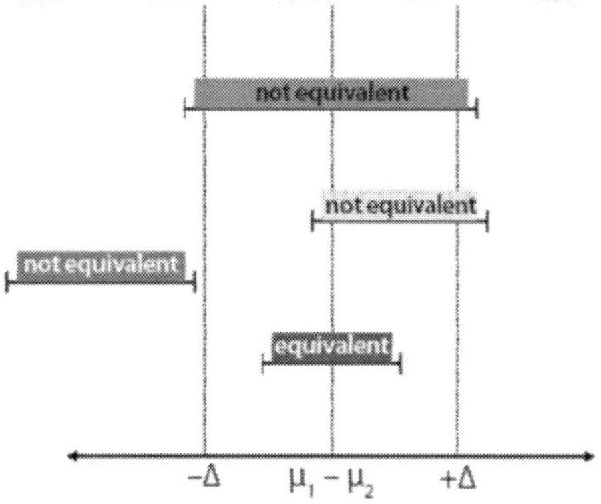

Ref: C. DiCesare et al.: Development, Qualification, and Application of a Bioreactor Scale-Down Process: Modeling Large-Scale Microcarrier Perfusion Cell Culture." *BioProcess International, 2016*

FUJIFILM Diosynth biotechnologies · Bio RAPID · Newcastle University · DTU Danmarks Tekniske Universitet · BIOPRO World Talent Campus

The challenge

- What if the number in one group is just 1? -> Time series TOST testing
- Definition of how many SD are used to form the equivalence interval
- Time series could be problematic, rules for passing/failing equivalence are required
- What if one variable is equivalent, another is not, the next is, the nth variable again is not...

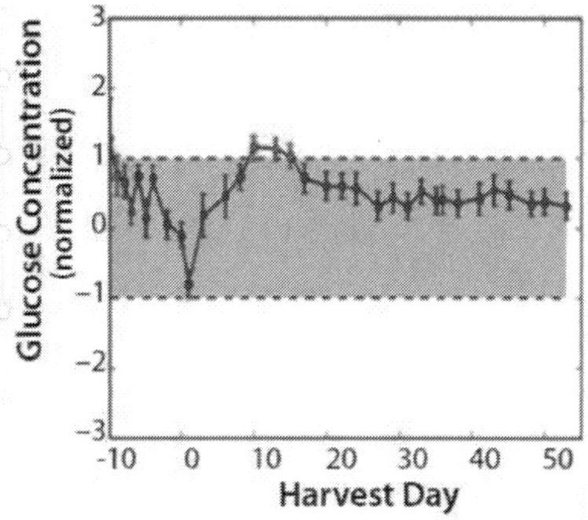

DiCesare used a rule of ‚three in a row' as maximum acceptable equivalence outliers

Ref: C. DiCesare et al.: Development, Qualification, and Application of a Bioreactor Scale-Down Process: Modeling Large-Scale Microcarrier Perfusion Cell Culture." *BioProcess International, 2016*

FUJIFILM Diosynth biotechnologies | Bio RAPID | Newcastle University | DTU Danmarks Tekniske Universitet | BIOPRO World Talent Campus

The challenge

- Especially relevant for CQAs or CPPs (critical quality attributes or critical process parameters)
- If it was critical to be equivalent, it becomes relevant how close equivalence was missed
- Sometimes those differences can be explained, but often this is means lengthy investigations have to be done

Variable	***Equivalent?***	***Was it a CQA or CPP?***	***P value debate***
V1	Pass	Yes	OK
V2	Pass	Yes	OK
V3	Pass	No	OK
V4	Pass	Yes	OK
V5	Fail	Yes	Clear fail
V6	Pass	Yes	OK
V7	Fail	Yes	Borderline fail
V8	Pass	Yes	OK
V10	Pass	Yes	OK
...			

FUJIFILM Diosynth biotechnologies | Bio RAPID | Newcastle University | DTU Danmarks Tekniske Universitet | BIOPRO World Talent Campus

The opportunity

- Are there any tools that allow us to capture most of the variance in the process?
- Can we get a quantitative measure of how equivalent the runs are in %, including which variables are equivalent and which are not?
- Today I present how PCA can be combined with a TOST test on time series to do exactly that using the TOSTER package in R

Ref: D Lakens, Two One-Sided Tests (TOST) Equivalence Testing, Package 'TOSTER', 2017

Goal

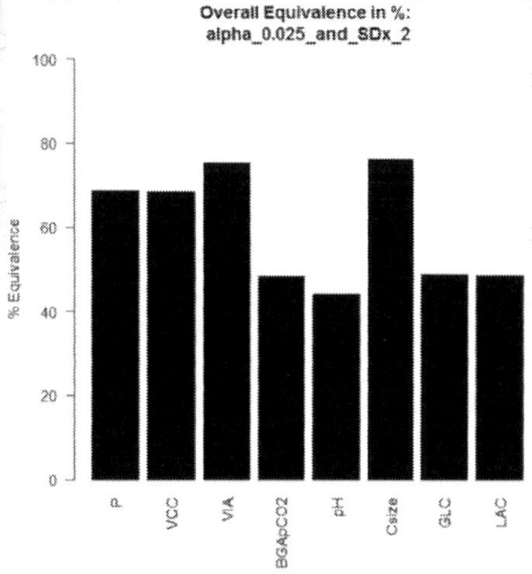

Total Equivalence (TEQ): 59.78%

Data

- Prepare dataset by filtering the data into subsets of all batches per day
- Data was mean centered and divided by the SD before extracting the scores with PCA

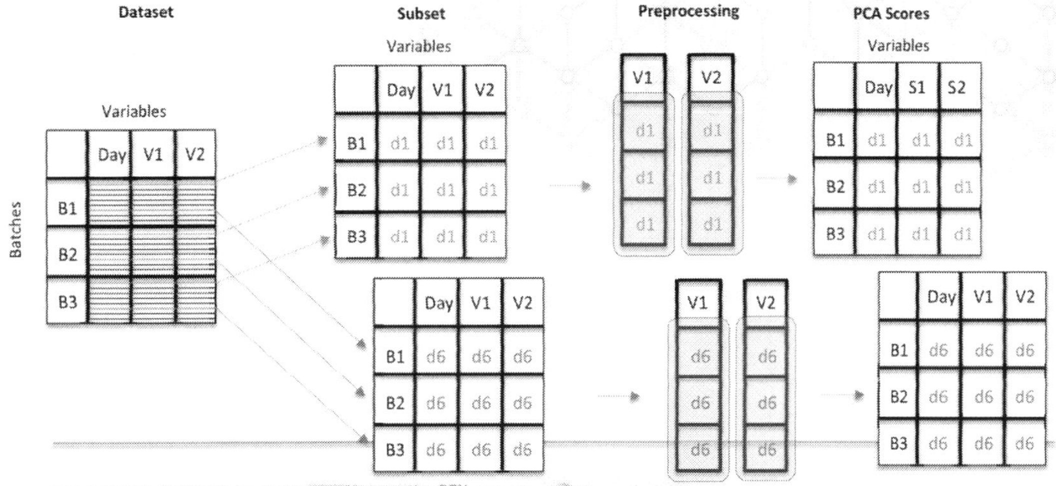

Workflow

- Take the scores and run a TOST test on it using the parameters alpha and SD multiplier to define whether these PCs on that day are equivalent or not

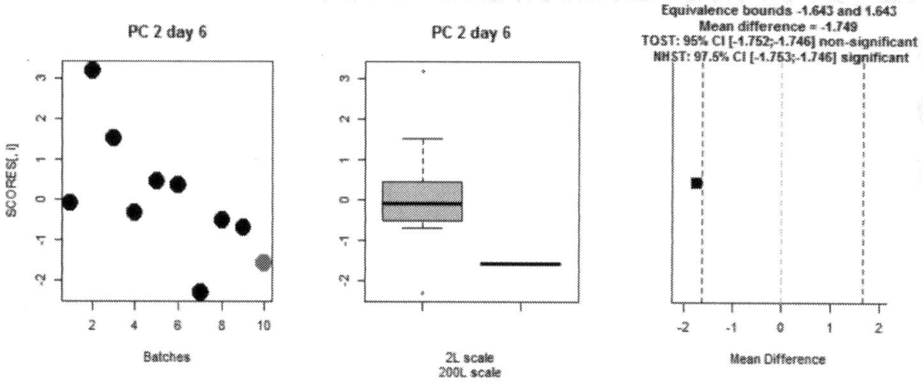

- alpha_0.025_and_SDx_1

FUJIFILM Diosynth biotechnologies Bio RAPID Newcastle University DTU Danmarks Tekniske Universitet BIOPRO World Talent Campus

Workflow

- Take the scores and run a TOST test on it using the parameters alpha and SD multiplier to define whether these PCs on that day are equivalent or not

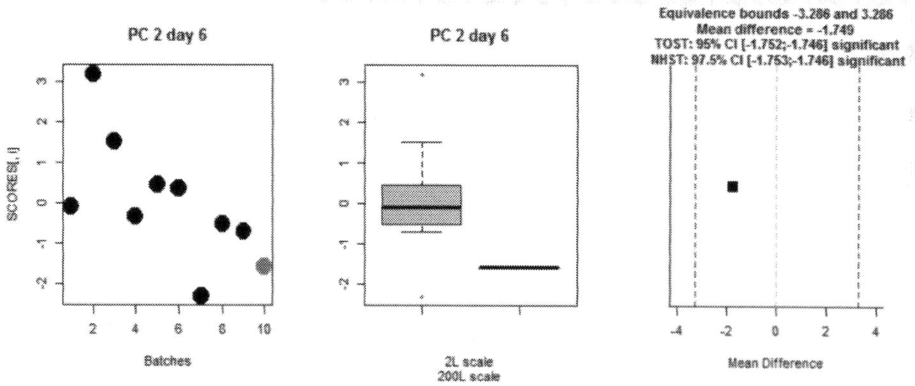

- alpha_0.025_and_SDx_2

FUJIFILM Diosynth biotechnologies Bio RAPID Newcastle University DTU Danmarks Tekniske Universitet BIOPRO World Talent Campus

Workflow

- Take the scores and run a TOST test on it using the parameters alpha and SD multiplier to define whether these PCs on that day are equivalent or not

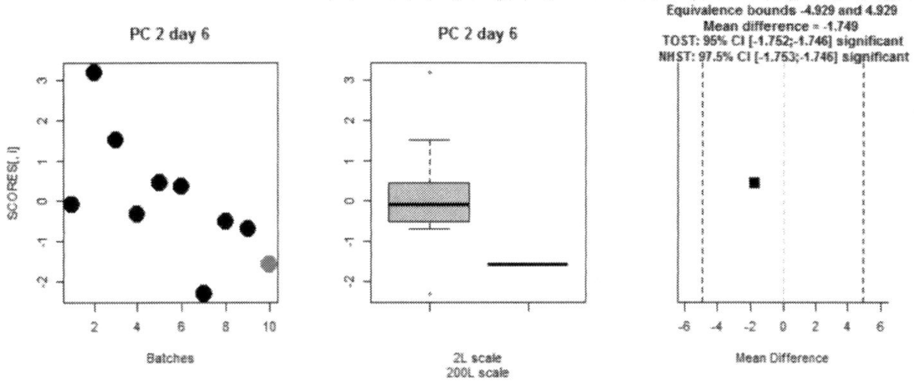

- alpha_0.025_and_SDx_3

Workflow

- The parameter alpha is not relevant in this example as there is only one observation in one of the groups. If would determine if the data point lies within or outside of equivalence.

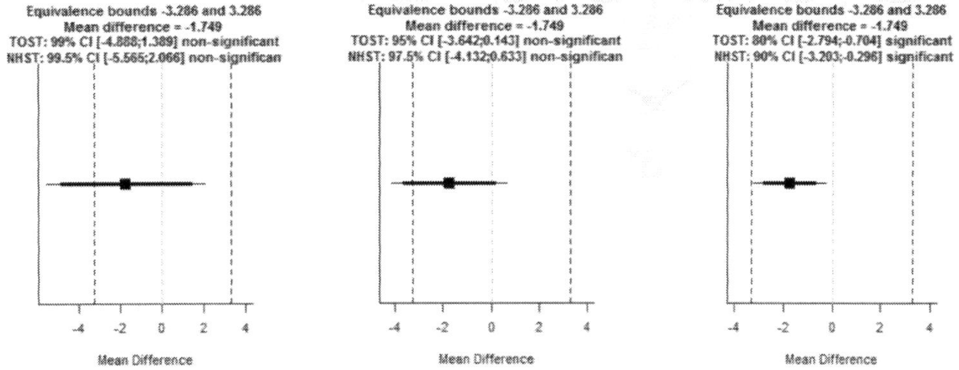

- alpha_0.005_and_SDx_2 alpha_0.025_and_SDx_2 alpha_0.1_and_SDx_2

Workflow

- Different combinations of conditions, alpha and SDmultiplier, are evaluated in a grid search because they have a strong impact on TOST

alpha = 0.005, 0.010, 0.025, 0.050, 0.100
SDmultiplier = 1.0, 1.5, 2.0, 2.5, 3.0

- Create a weight matrix: which PCs were equivalent or not

Name	PC1	PC2	PC3	PC4	PC5	PC6	PC7	PC8
alpha_0.005_and_SDx_1	1	0	0	1	0	0	0	0
alpha_0.01_and_SDx_1	1	0	0	1	0	0	0	0
alpha_0.025_and_SDx_1	1	0	0	1	0	0	0	0
alpha_0.05_and_SDx_1	1	0	0	1	0	0	0	0
alpha_0.1_and_SDx_1	1	0	0	1	0	0	0	0
alpha_0.005_and_SDx_1.5	1	1	0	1	0	0	1	0
alpha_0.01_and_SDx_1.5	1	1	0	1	0	0	1	0
alpha_0.025_and_SDx_1.5	1	1	0	1	0	0	1	0
alpha_0.05_and_SDx_1.5	1	1	0	1	0	0	1	0
alpha_0.1_and_SDx_1.5	1	1	0	1	0	0	1	0
alpha_0.005_and_SDx_2	1	0	0	1	0	1	1	0
alpha_0.01_and_SDx_2	1	1	0	1	0	1	1	0
alpha_0.025_and_SDx_2	1	0	0	1	0	1	1	0
alpha_0.05_and_SDx_2	1	1	0	1	0	1	1	0
alpha_0.1_and_SDx_2	1	1	0	1	0	1	1	0
alpha_0.005_and_SDx_2.5	1	1	0	1	0	1	1	0
alpha_0.01_and_SDx_2.5	1	1	0	1	0	1	1	0
alpha_0.025_and_SDx_2.5	1	1	0	1	0	1	1	0
alpha_0.05_and_SDx_2.5	1	1	0	1	0	1	1	0
alpha_0.1_and_SDx_2.5	1	1	0	1	0	1	1	0
alpha_0.005_and_SDx_3	1	1	0	1	0	1	1	0
alpha_0.01_and_SDx_3	1	1	0	1	0	1	1	0
alpha_0.025_and_SDx_3	1	1	0	1	0	1	1	0
alpha_0.05_and_SDx_3	1	1	0	1	0	1	1	0
alpha_0.1_and_SDx_3	1	1	0	1	0	1	1	0

FUJIFILM Diosynth biotechnologies · BioRAPID · Newcastle University · DTU Danmarks Tekniske Universitet · BIOPRO World Talent Campus

17

Workflow

- Let us have a look what those PCs represent - weighting of all PCs, on day 6

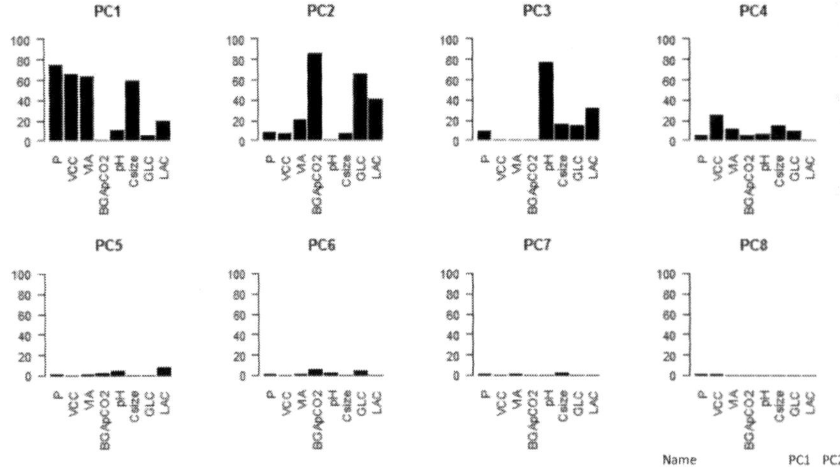

Name	PC1	PC2	PC3	PC4	PC5	PC6	PC7	PC8
SD	1.72	1.54	1.23	0.88	0.41	0.38	0.2	0.13
Var%	37	30	19	10	2	2	1	0
CumVar	37	67	86	95	98	99	100	100

FUJIFILM Diosynth biotechnologies · BioRAPID · Newcastle University · DTU Danmarks Tekniske Universitet · BIOPRO World Talent Campus

18

Results

- Sum up PCs passing the TOST test

- For every condition
- For every variable
- For every day

- The variance covered in the passing PCs gives a measure of overall equivalence in %

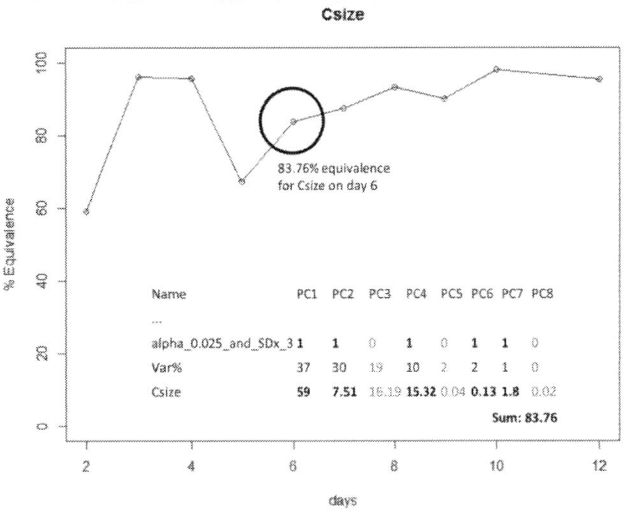

FUJIFILM Diosynth biotechnologies | Bio RAPID | Newcastle University | DTU Danmarks Tekniske Universitet | BIOPRO World Talent Campus

Results

- Visualize explained equivalence per variable as time profiles, for all days and all variables

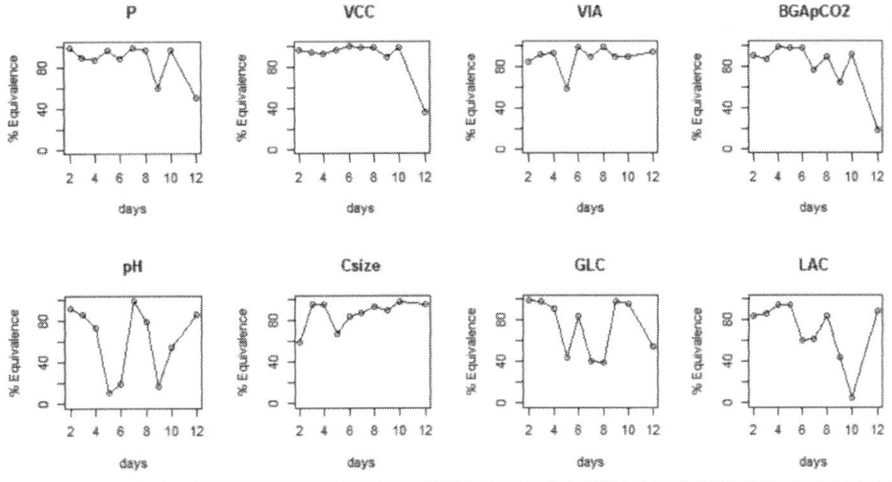

FUJIFILM Diosynth biotechnologies | Bio RAPID | Newcastle University | DTU Danmarks Tekniske Universitet | BIOPRO World Talent Campus

Results

- Get an overall metric for equivalence with a certain condition of alpha and SDmultiplier

Mean of all equivalent variables over the whole process time:

Total Equivalence (TEQ): 59.78%

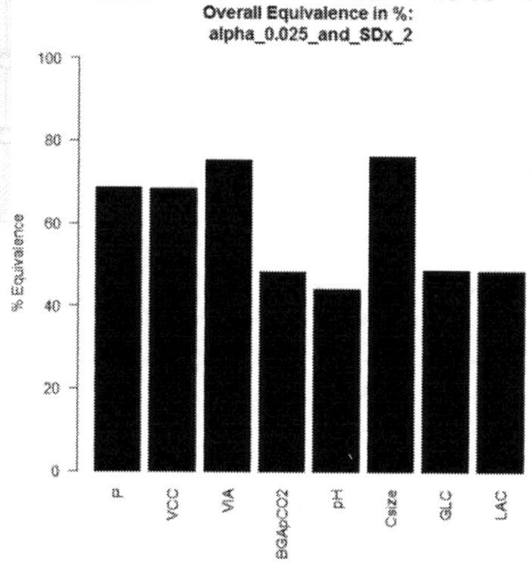

FUJIFILM Diosynth biotechnologies | Bio RAPID | Newcastle University | DTU Danmarks Tekniske Universitet | BIOPRO World Talent Campus

Results

- Visualize main differences per variable as time profiles

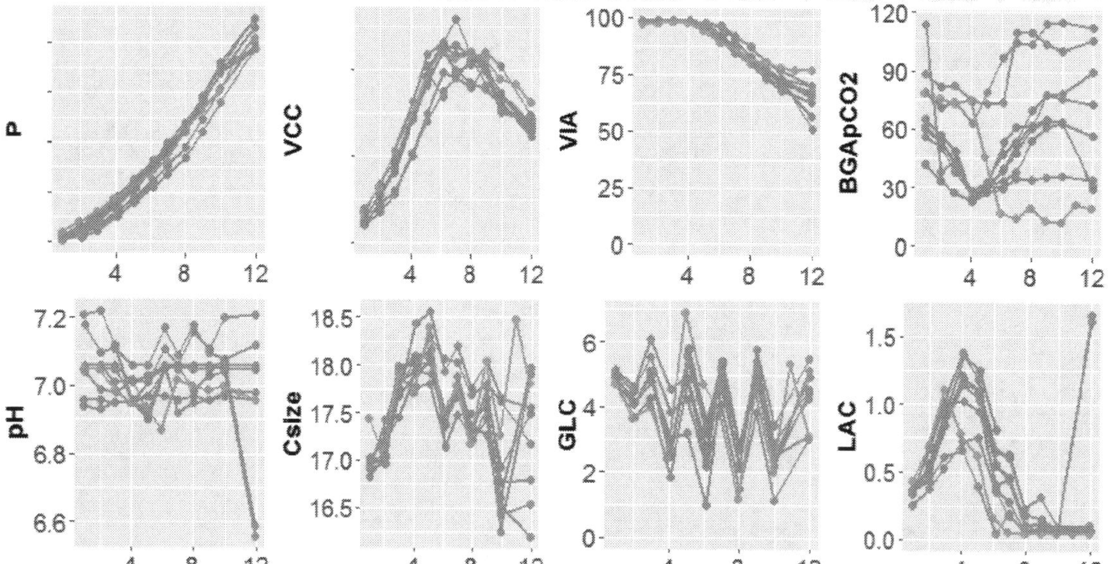

Results

- Get an overall metric for conditions tested at different levels of alpha and SD multipliers

TEQ: 28.17%

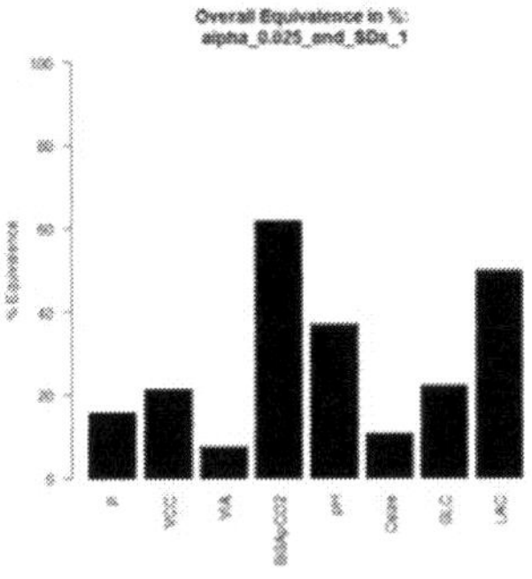

- alpha_0.025_and_SDx_1

TEQ: 59.78%

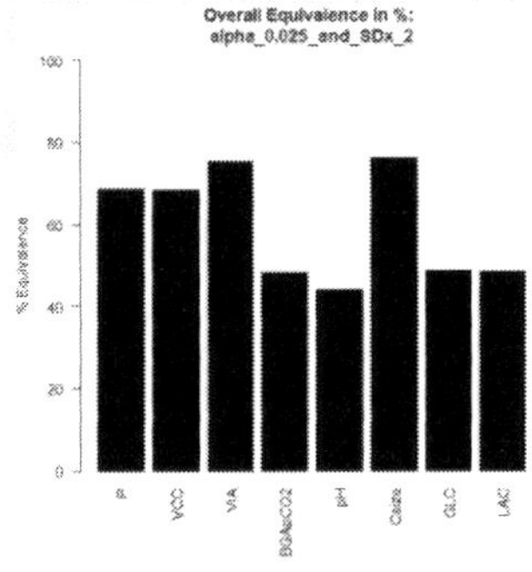

alpha_0.025_and_SDx_2

TEQ: 79.80%

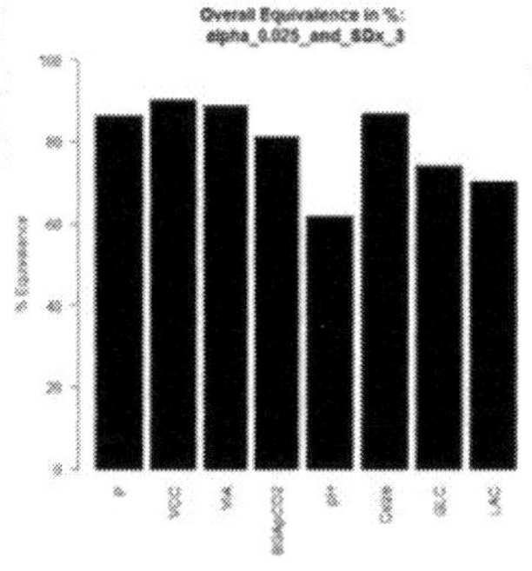

alpha_0.025_and_SDx_3

FUJIFILM Diosynth biotechnologies | BioRAPID | Newcastle University | DTU Danmarks Tekniske Universitet | BIOPRO World Talent Campus

Conclusions

- A method was presented that enables us to quantify equivalence with a novel metric TEQ
- The metric is based on PCs, which inherently explain the most variance in a dataset, and could be better suited than regular variables to compare groups holisticly using time series TOST
- Two major factors influencing the TOST (alpha and SDmultiplier) were evaluated for the new metric to allow for a better understanding of the TOST results
- This method summarizes which variables were most and least equivalent, which may help to find the root cause of undesireable process behaviour faster
- It is designed for HT datasets, but also worked for small batch numbers of n = 1, when data was scarce during scale up

FUJIFILM Diosynth biotechnologies | BioRAPID | Newcastle University | DTU Danmarks Tekniske Universitet | BIOPRO World Talent Campus

Acknowledgements

Thanks to my Mentors
Graham McCreath, Head of Process Design, Fujifilm Diosynth Biotechnologies
Jarka Glassey, Professor, University of Newcastle

Thanks for continued support at Fujifilm Diosynth Biotechnologies!

Thanks to my colleagues and friends especially at NCL and DTU!

This project has received funding from the European Union's Horizon 2020 research and innovation programme under the Marie Curie Skłodowska-Curie grant agreement No 643056.

25

Results

- A different (non equivalent) case with more data: Regime C vs Regime D

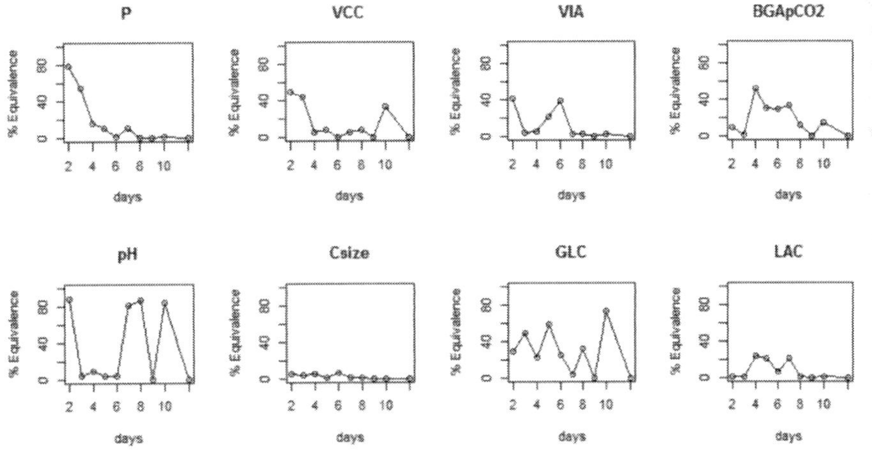

FUJIFILM Diosynth biotechnologies
Bio RAPID
Newcastle University
DTU Danmarks Tekniske Universitet
BIOPRO World Talent Campus

Results

- A different (non equivalent) case with more data: Regime C vs Regime D

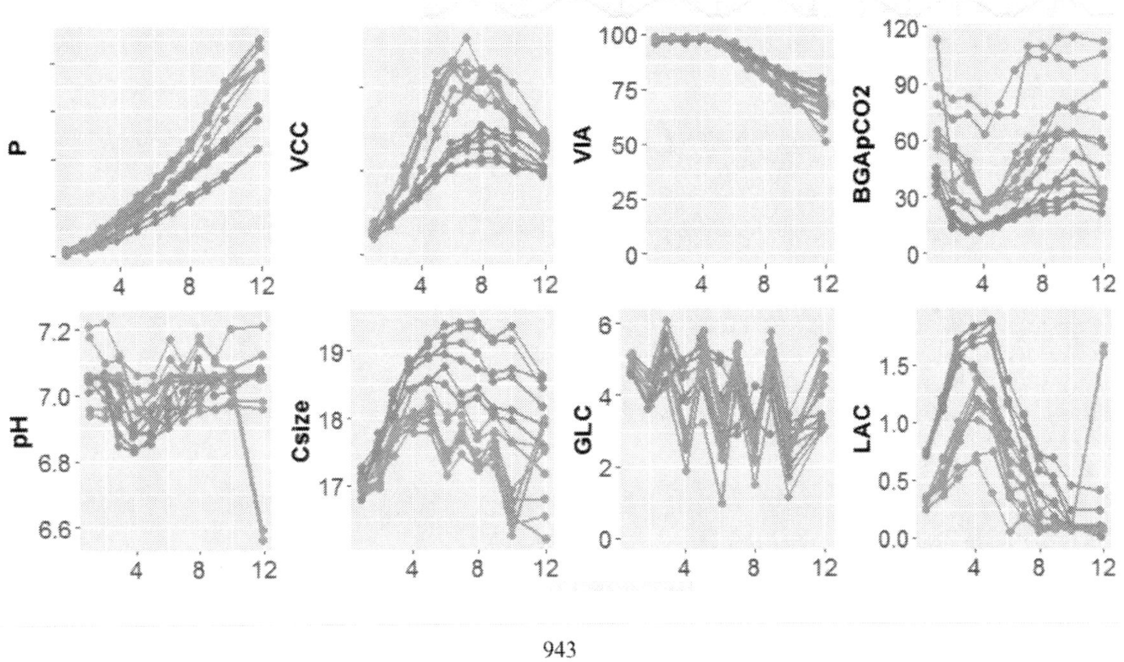

Results

- A different (non equivalent) case with more data: Regime C vs Regime D

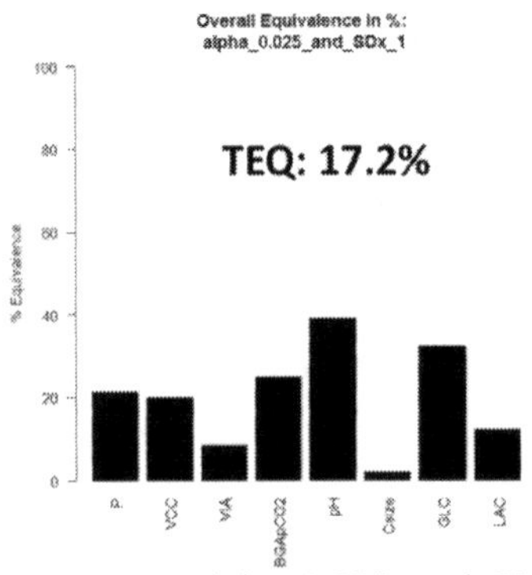

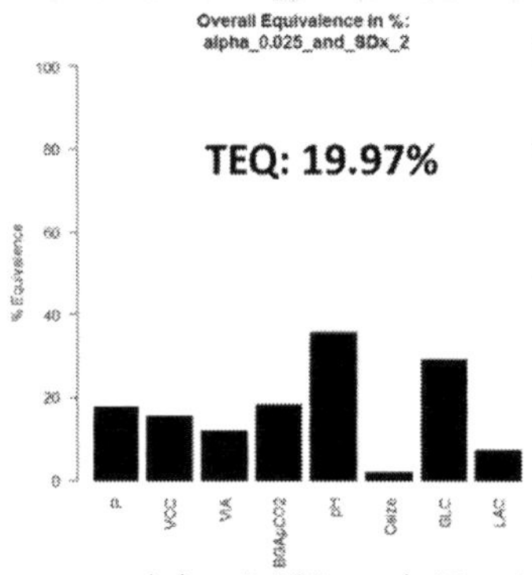

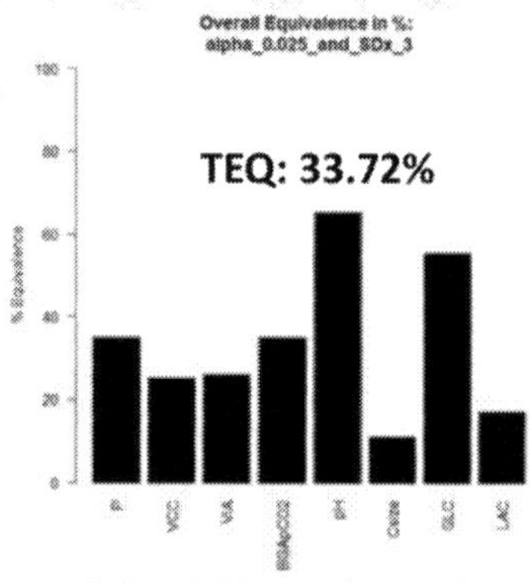

- alpha_0.025_and_SDx_1 alpha_0.025_and_SDx_2 alpha_0.025_and_SDx_3

FUJIFILM Diosynth biotechnologies | Bio RAPID | Newcastle University | DTU Danmarks Tekniske Universitet | BIOPRO World Talent Campus

Workflow

- Combine weighted equivalence matrix with variance covered by individual variables.
- Not all PCs will be used, only those that passed the TOST test.

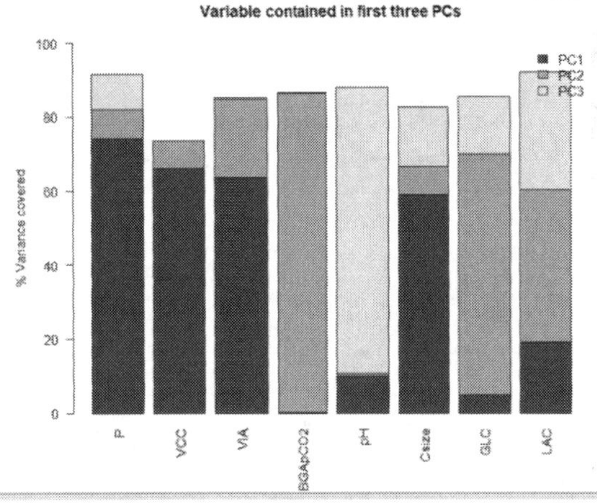

Name	PC1	PC2	PC3	PC4	PC5	PC6	PC7	PC8
...								
alpha_0.025_and_SDx_3	1	1	0	1	0	1	1	0
SD	1.72	1.54	1.23	0.88	0.41	0.38	0.2	0.13
Var%	37	30	19	10	2	2	1	0
CumVar	37	67	86	95	98	99	100	100
P	74.16	7.83	9.67	5.32	1.36	0.57	0.57	0.52
VCC	66.2	7.23	0.07	25.7	0.01	0.02	0.18	0.59
VIA	63.69	21.1	0.45	11.11	1.17	1.07	1.27	0.15
BGApCO2	0.18	86.35	0.12	4.95	2.57	5.81	0	0.01
pH	9.98	0.7	77.25	6.34	4.05	1.66	0.01	0.01
Csize	59	7.51	16.19	15.32	0.04	0.13	1.8	0.02
GLC	4.73	65.23	15.54	9.14	0.22	4.84	0.14	0.16
LAC	19.06	41.18	32.02	0.24	7.27	0.01	0.08	0.15

FUJIFILM Diosynth biotechnologies | Bio RAPID | Newcastle University | DTU Danmarks Tekniske Universitet | BIOPRO World Talent Campus

A workflow-based framework for managing product analytical data and statistical results for lot release

G. Joglekar[1], Q. Cai[2], P. DeLaurentis[1], L. Mockus[1], K. Morris[2], G.V. Reklaitis[1]
[1]Davidson School of Chemical Engineering, Purdue University
[2]Lachman Institute for Pharmaceutical Analysis, Long Island Univ

AIChE Annual Meeting
Paper 665(e)
November 2, 2017

Outline

- Open data requirements
- KProMS system
- The collaborative research project
- KProMS benefits
- Conclusions

Objective

- Demonstrate the use of a workflow-based knowledge management system to support a collaborative research project
  - ✓ Data gathering
  - ✓ Data curation
  - ✓ Data analysis
  - ✓ Multiple groups
  - ✓ Open data

Open Data

- Directive to funding agencies to ensure that public has access to the results of government-funded research (Office of Sc and Tech Policy, 2013)
- The directive requires that data are
  - ✓ Richly described with machine-actionable metadata
  - ✓ Complete, self-explanatory, and accurate
  - ✓ Accessible and preserved beyond the life of a grant
  - ✓ Protected for confidentiality and privacy
- Private funders require that data be released with some degree of openness
- Companies would have similar requirements for internal use of confidential data

Data Sharing

- Even though more data are becoming available, they are not used widely because
  - ✓ Full context is not available
  - ✓ Meta-data is not available
- Data sharing among academic researchers very sporadic
- Tech transfers often time consuming
- The need for well curated data is expected to grow in the future
- There are no guidelines for good practices, or standards for publishing data
  - ✓ Putting it on the cloud is not sufficient

KProMS: Knowledge Provenance Management System

- KProMS framework produces shareable and reusable data by
  - ✓ Capturing the context of data
  - ✓ Streamlining data submission into a repository
  - ✓ Organizing data for machine and manual access over the web

KProMS Architecture

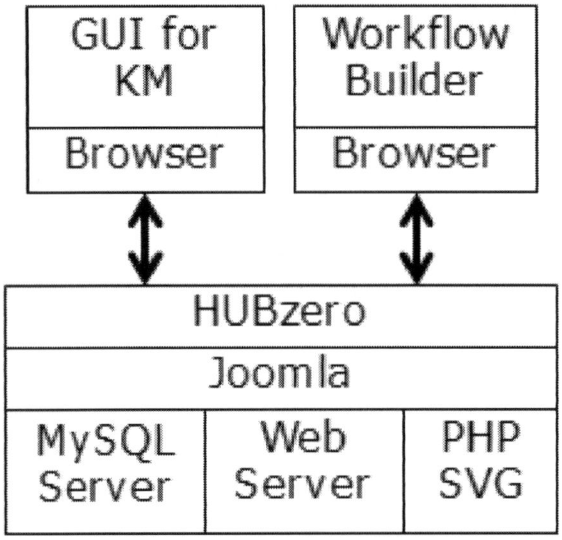

Joglekar, G., Giridhar, A., Reklaitis, G.V. "A Workflow Modeling System for capturing data provenance", Computers & Chemical Engineering (DOI: 10.1016/j.compchemeng.2014.04.006)
Joglekar, G., Giridhar, A., Reklaitis, G.V. "A Novel Knowledge Management System Based on Workflows", Computer Aided Chemical Engineering Series, Vol 33, Elsevier.

Collaborative Project

- Perform analytical tests associated with lot release on drugs purchased in open market
  - ✓ Dissolution
  - ✓ Content uniformity
  - ✓ Physical properties: Weight, Dimension, Mechanical strength
- Perform Bayesian analysis for hypothesis testing
- Funded by the FDA
  - ✓ Long Island University, Purdue University

Project Execution

- Model each experimental procedure as a workflow
- Procure drug to be studied
- Record each experiment as an instance of the associated workflow model
  - ✓ Parameter values
  - ✓ Upload instrument data
- Perform appropriate data checks
- Perform data analysis

Interactions of Working Groups

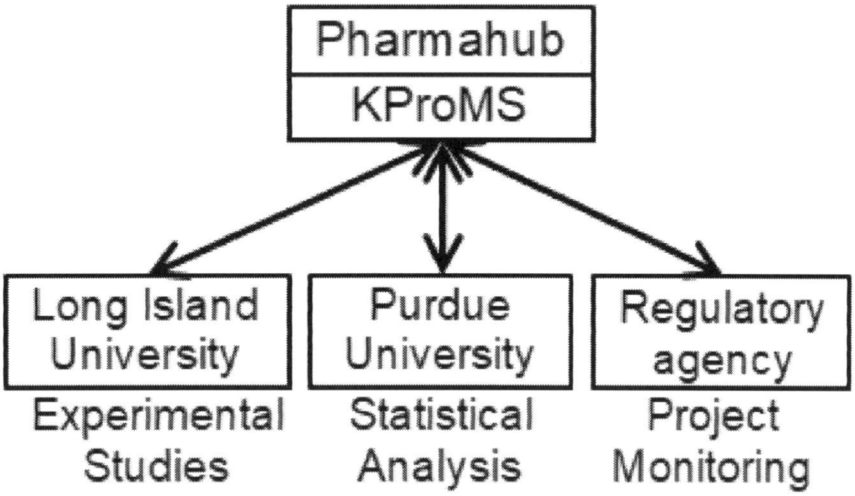

Dissolution Workflow

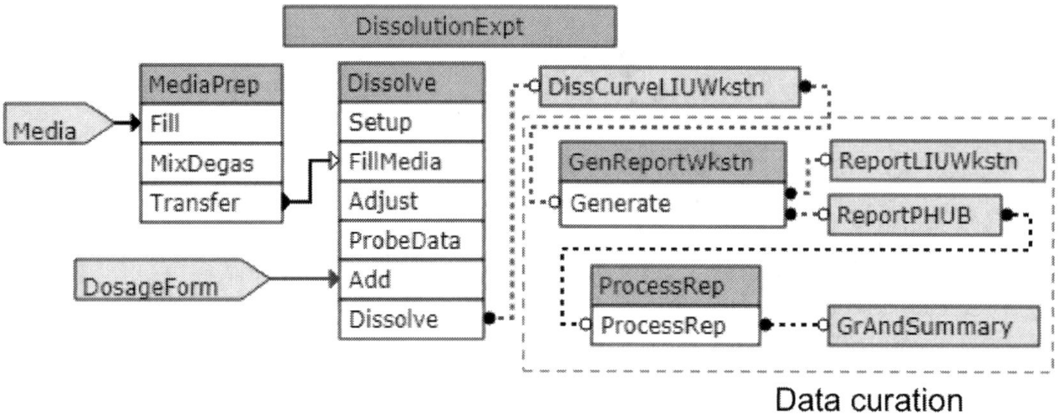

Parameters

		Parameter
Material:Media		Amount, liter
Material:DosageForm		Material lot, ID
Task:MediaPrep	*Subtask*:Fill	pH, number
	Subtask:MixDegas	Degas, Yes/No
Task:Dissolve		Number of vessels, integer
	Subtask:Setup	Bath temperature, C Accessory, Paddle/Basket
	Subtask:FillMedia	Amount per vessel, ml
	Subtask:ProbeData	Exposure Autoset, Yes/No Standard A blank, Yes/No Standard B blank, Yes/No Sample blank, Yes/No Standard A, g/l Standard B, g/l Response factor, Number Wavelength max, nm Path length, mm Calibration curve, Yes/No Baseline correction, Yes/No
	Subtask:Add	Number of units per vessel, number
Task:GenReportWkstn	*Subtask*:Generate	Program name, OPT-DISS
Task:ProcessRep	*Subtask*:ProcessRep	Library program, DigData Input file type, Dissolution report Output data, Pct diss and graph/Zip data

GrAndSummary Data Node

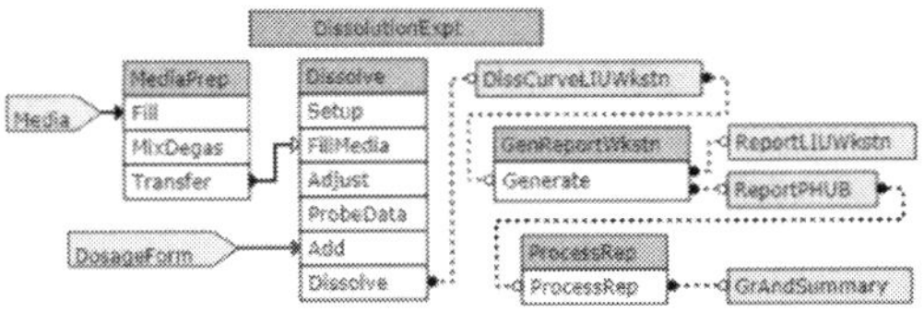

Number of tablets	6					
Dissolution end time	00:30:00					
Pct dissolved at end time	Ch1	Ch2	Ch3	Ch4	Ch5	Ch6
	104.95	100.99	99.90	102.41	102.74	97.26
Av Pct dissolved	101.3750					
Std dev	2.6447					

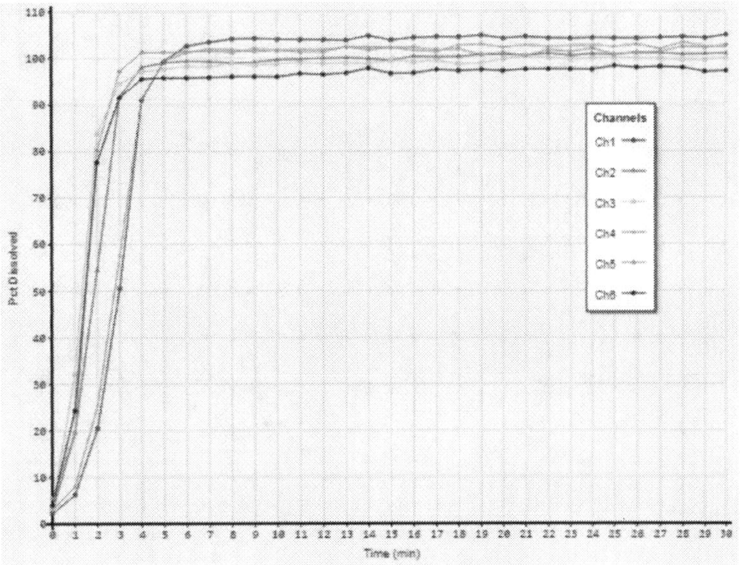

Workflows

Dissolution

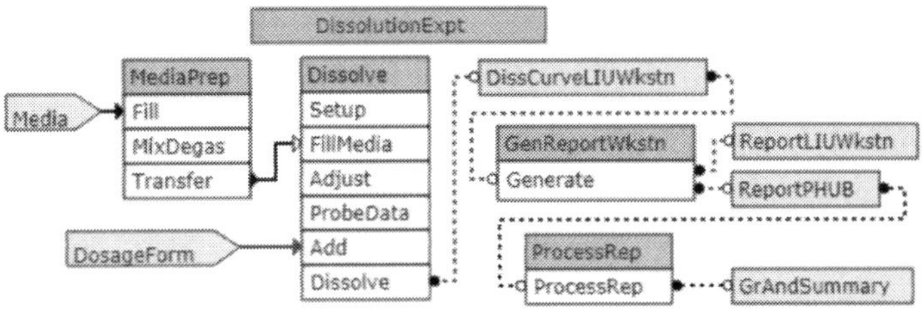

Content uniformity

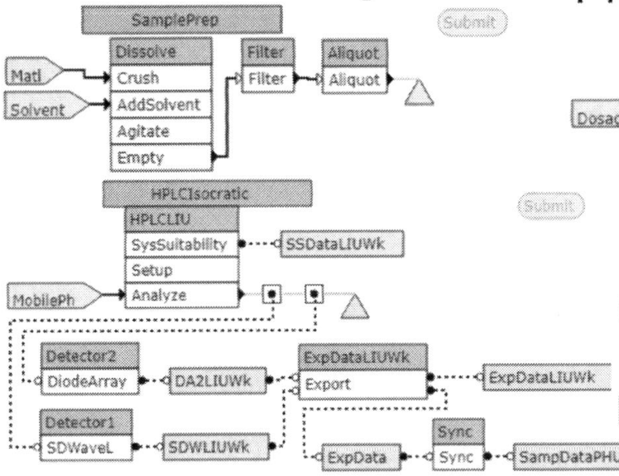

Physical Properties

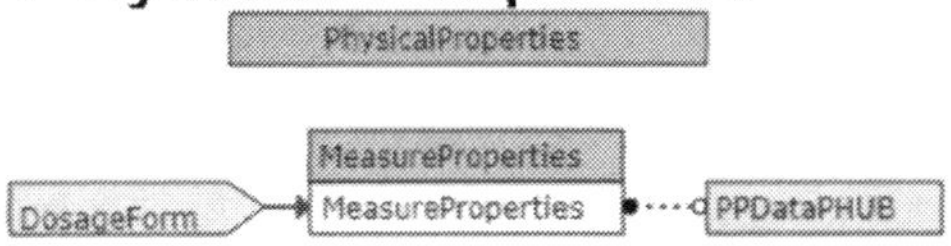

Bayesian Analysis

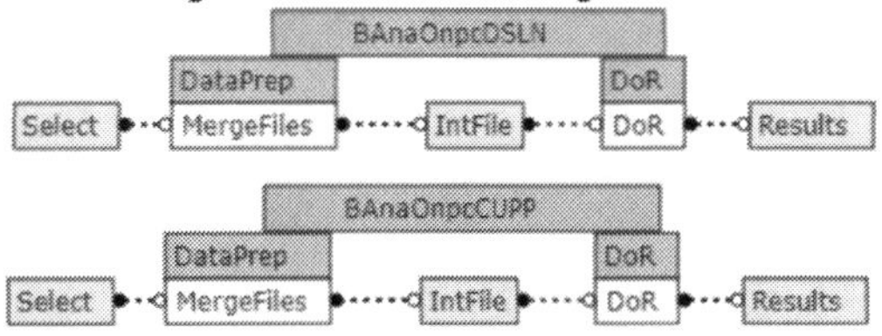

Bayesian Analysis

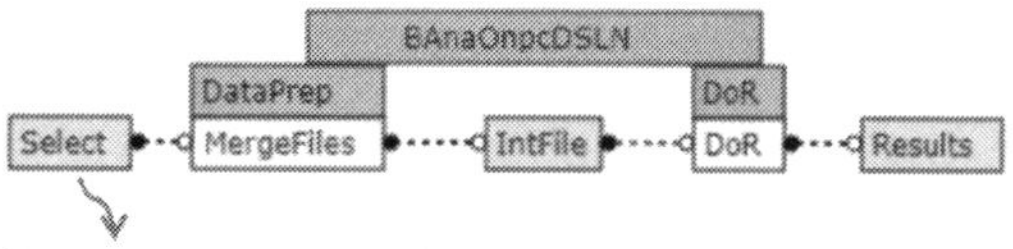

IntFile – Downloaded to PC

Results – Uploaded to repository

DoR – Only records the R package used. Execution done on PC

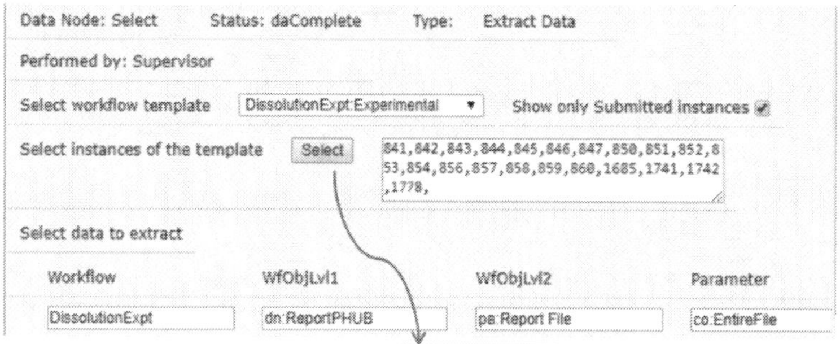

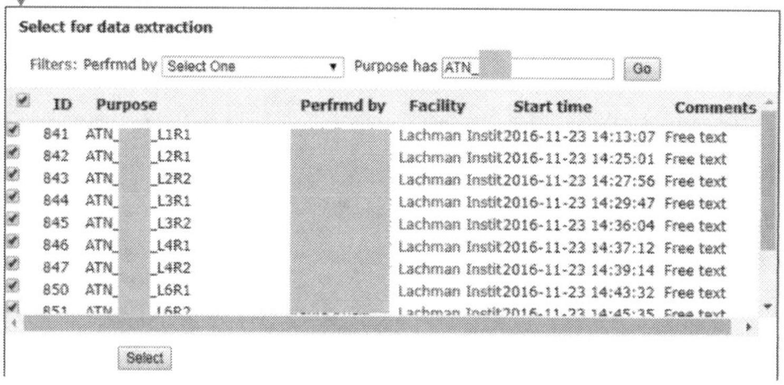

Benefits of Using KProMS

- Workflows tailored to the underlying procedure
- All steps and parameters explicitly defined
- Workflow developed based on input from all collaborators
- Improved quality of data
- Data available on the web as soon as it is submitted
- Facilitates discovery
- Uploaded data is protected

Summary

- KProMS framework provides complete context to the underlying data
- KProMS satisfies the requirement of Open Data archival system

Questions?

495398 Evaluation of Recovery Pathways for Improved Bioavailability of Clofazimine Nanoparticles

Abstract:

Clofazimine is an anti-mycobacterial agent used as part of a multidrug treatment for leprosy for over four decades. Recently clofazimine has shown promising activity against multidrug resistant tuberculosis. Due to its extremely poor aqueous solubility and high lipophilicity, the formulation development has focused on the solubility enhancement. Nanoparticles enhance bioavailability by two mechanisms. First, relative to larger drug crystals, nanoparticles have higher surface-to-volume ratios, which enhances dissolution rates. Second, rapid nanoparticle precipitation can trap drug molecules in higher energy, amorphous states, which have higher solubility. Here we use Flash Nanoprecipication (FNP) to form clofazimine-loaded nanoparticles with hypromellose acetate succinate (HPMC-AS) as polymer stabilizers, and investigate how to recover the clofazimine-loaded nanoparticles from the aqueous dispersions. Dry nanoparticle powders that can be readily reconstituted would enable delivery of drug dispersions to children and infants. Three approaches were used to produce dry powders: direct salt precipitation, spray drying and lyophilization. The yield efficiency, particle morphology, drug loading, dissolution rate and the amorphous/crystalline state of clofazimine in the dry powders are determined and compared among the processes. Based on these results, we evaluate the advantages and disadvantages of each process regarding the maintenance of the bioavailability after drying.

497815 Assessment of the Impact of Interfacial and Shear Stress on Biologics Drug Product Via Mini-Piloting Tools

Abstract:

Assessment of the Impact of Interfacial and Shear Stress on Biologics Drug Product via Mini-piloting Tools

During biologics development, processes such as membrane filtration, filling, and mixing are critical in the drug product manufacturing. However, these processes, as well as turbulent flow conditions, and multiple passes through pumps, valves and â€œpinchâ€ points; impose various stresses on the protein. These stresses can lead to protein instability. Shear, cavitation, and interfacial stress are often attributed as the main causes of protein unfolding and aggregation.

Proper risk analysis needs to be in place to understand the susceptibility of the protein to unfold and aggregate in presence of interfacial and shear stress. There are certain techniques, such as agitation/shaking studies that have been traditionally used to understand the impact of these stresses on the protein physical stability. However, the stresses applied in these systems are convoluted, making it difficult to define the control strategy, (i.e. adjustment in process parameters to reduce foaming/bubble formation, etc.). In this study, we have developed two mini-piloting tools that allow for the isolation of interfacial and shear stress respectively. The interfacial stress is applied to the sample by generating uniform bubbles when recirculating air from the headspace of the vial in a closed loop. The bubbles generated are passed through the solution for finite periods of time. For the shear stress, an estimation of the shear rates present during processing is made by fundamental transport relations and CFD modeling. The small scale shear stress tool simulates the shear experienced by the protein during processing, using a small-scale high pressure pump connected to a fine stainless steel tubing.

These systems help simulate the normal operating ranges as well as proven acceptable ranges for unit operation such as TFF, mixing and filling. Different protein formulations were subjected to controlled interfacial and shear stresses. The quality attributes evaluated after the stress were focused on size variant (particulate matter and HMW species), which is most likely the mechanism associated with interfacial and shear stress. It was possible to identify the susceptibility of the protein to aggregate depending of the stress that was applied. In addition, these tools contributed in formulation screening by evaluating which formulation would better mitigate the impact of the imposed stress.

During development, it is important to identify the protein formulations that have a tendency to aggregate in order to implement appropriate risk mitigation strategies. And also, to have scale down models to validate the design space for the process parameters of key unit operations.

497373 CO_2-Philic Polymer Membranes for High Flux CO_2 Separation

Abstract:

The vast majority of the worldâ€™s energy is presently derived from the burning of fossil fuels, which releases vast quantities of carbon dioxide (CO_2) into the environment and results in undesirable climate change. Practical and cost efficient methods of CO_2 separation and capture would thus solve one of the most challenging problems today. This presentation summarizes our effort on the development of novel polymer membranes functionalized with CO_2-philic groups for high flux CO_2 separation. Our strategy focuses on tuning solubility selectivity in addition to diffusivity selectivity for achieving high permeability membranes combined with good selectivity. Various synthetic techniques including ROMP and post functionalization were used and the careful design permits to prepare well-defined novel high permeable polymers containing CO_2-philic groups. This study demonstrated the addition of CO_2-philic groups (e.g. amidoxime and PEO) significantly increased the solubility selectivity of CO_2 over N_2. The membrane performance is also highly dependent on the balance of gas/functional group interaction, intra/inter- molecular interaction (H-bonding etc.) of membrane, packing, and polymer dynamics. Tuning the balance of the interaction enables to achieve the CO_2 separation performance over the Robeson upper bound, e.g. CO_2 permeability 3000-8000 Barrer and CO_2/N_2 selectivity 18-20. The structure-property relationships especially on CO_2 uptake, CO_2 and N_2 permeability, CO_2/N_2 selectivity to the polymer structure will be discussed.

496766 Highly Cross-Linked Polymers for Membrane H_2/CO_2 Separation at Elevated Temperatures

Abstract:

Highly Cross-linked Polymers for Membrane H_2/CO_2 Separation at Elevated Temperatures

Maryam Omidvar, Mark Swihart, and Haiqing Lin

Department of Chemical and Biological Engineering

University at Buffalo, The State University of New York)

Membrane technology has attracted significant interest for pre-combustion H_2 purification and CO_2 capture. The key to the success of this technology is membrane materials with high H_2 permeability and H_2/CO_2 selectivity at elevated temperatures of syngas processing (>100 °C). Polyimides such as P84 have been explored for H_2/CO_2 separation due to their strong size-sieving ability and resulting high H_2/CO_2 diffusivity selectivity. In this study, we demonstrate that these polymers can be cross-linked chemically to further improve H_2/CO_2 selectivity. We systematically evaluate the effect of cross-linking P84 using amines such as propane diamines, butane diamines and p-xylene diamine on the H_2/CO_2 separation properties. The cross-linking can significantly increase H_2/CO_2 selectivity while retaining moderate H_2 permeability. For example, cross-linking of P84 with butane diamine with 42% conversion increases H_2/CO_2 selectivity from 2.4 to 8.0, while decreasing H_2 permeability from 40 to 24 Barrers at 100 °C. The combination of 24 Barrers for H_2 permeability and 8.0 for H_2/CO_2 selectivity is very close to the Robesonâ€™s upper bound for H_2/CO_2 separation, demonstrating the potential of this approach for membrane development. These cross-linked polymers are thoroughly evaluated using FTIR, DSC, TGA, WAXD, and XPS. The effect of cross-linking degree on the physical properties will be discussed, including glass transition temperature, fractional free volume and pure- and mixed-gas H_2/CO_2 separation properties at various temperatures. This presentation will also discuss the fundamental structure/property relationship in these highly cross-linked polymers with promising H_2/CO_2 separation properties.

499468 High-Throughput Direct Synthesis of Cubdc MOF Nanosheets and Their Application in CO_2 Separation

Abstract:

Membranes based on metal organic frameworks (MOFs) have attracted increased research attention especially for their application in separation of carbon dioxide[1,2]. High aspect ratio MOF nanosheets, having thickness on the order of few nanometers and large lateral dimensions are ideal for fabrication of high performing membranes for gas separations.

CuBDC MOF nanosheets first introduced by the Gascon group were synthesized via a gradient synthesis strategy[3]. They show promising performance in the separation of carbon dioxide from natural gas streams.

In this work, the bottom-up synthesis strategy for CuBDC nanosheets was modified to eliminate the need for gradient synthesis. Instead CuBDC nanosheets were obtained by a direct mixing scheme consisting of homogeneous mixing of the metal and linker solutions. Nanosheets with an average thickness of 30 nm and average lateral dimensions of 2.5 micron were obtained. The aspect ratio of CuBDC nanosheets was optimized by varying parameters such as synthesis time, temperature and concentration of the synthesis solutions. It was found that lower temperature favors higher aspect ratios. The direct synthesis scheme developed herein is advantageous as it is scalable and high-throughput.

Mixed matrix membranes were fabricated by incorporating the CuBDC nanosheets in a polymer matrix and tested for CO_2/N_2 separation performance. A significant increase in the ideal selectivity of CO_2/N_2 was obtained reiterating the promising separation performance of CuBDC nanosheets.

[1] J. E. Bachman, J. R. Long, *Energy Environ. Sci* **2016**, *9*, 2031–2036.

[2] Y. Peng, Y. Li, Y. Ban, H. Jin, W. Jiao, X. Liu, W. Yang, *Science* **2014**, *346*, 1356–1359.

[3] T. Rodenas, I. Luz, G. Prieto, B. Seoane, H. Miro, A. Corma, F. Kapteijn, F. X. Llabrés i Xamena, J. Gascon, *Nat. Mater.* **2015**, *14*, 48–55.

483115 New Negative Emissions Technology: Indirect Ocean Capture Separating CO_2 from Air

Abstract:

Negative emissions technologies (NETs), which result in the removal of CO_2 from the atmosphere, will likely prove crucial to limiting global warming to 2 °C. Few viable NET options exist, and detailed, transparent cost assessments are scarce, hindering policy development and strategic investment.

We propose a new platform of membrane-based negative emissions technologies, which we term Indirect Ocean Capture, and provide detailed experimental results demonstrating real-world feasibility. Our research presents the design, pilot-scale construction, testing, and fully integrated techno-economic analysis of a new platform of negative emissions technologies for removing CO_2 from the atmosphere. The technological platform has two unique implementations but both rely on bi-polar membrane electrodialysis in series with hydrophobic membranes used for water-gas separations. The fully automated system extracts CO_2 in the following manner: A custom build bi-polar membrane electrodialysis system is used to efficiently produce HCl and NaOH in low concentrations. In the first implementation the continuous production of HCl at a stable concentration is pumped into degassed seawater. The seawater is acidifed to pH 4.5 and pumped across hydrophobic membrane contactors under vacuum pressures. High purity CO_2 is evolved from the seawater and extracted across the membrane. The continuous production of NaOH at a stable concenrtaion is pumped into the acidified seawater to return the seawater to its original alkalinity, and then returned to the oceans. In the second implementation, NaOH is added to seawater such that the seawater achieves a $9.3 < pH < 9.6$. $CaCO_3$ is precipitated from solution while basic seawater is removed. HCl is pumped into the basic seawater to return the seawater to its original pH, and then returned to the oceans.

We built a pilot-scale platform of this technology and quantified the performance of this previously untested negative emissions platform. Four different membranes from two different companies were analyzed to optimize acid and base production during electrodialysis. Further membranes for CO_2 separation from water were evaulated for efficiency. Finally, we used our experimental measurements as input into a fully integrated techno-economic quantitative model that calculates the cost of indirect ocean capture, including the effects of co-location with other industrial processes that use seawater, and the resultant costs of equivalent emissions in dollars per ton of captured CO_2. Our analysis takes the important step of determining the optimized design in pilot-scale, as well as quantifying the economics for a real-world implementation verified through industrial partner. This represents the first and only detailed process design and cost assessment for indirect ocean capture.

495079 Study of Carrier Saturation Phenomenon in Facilitated Transport Membrane for CO_2 Capture from Low CO_2 Concentration Sources

Abstract:

Amine-containing facilitated transport membranes exhibit superior CO_2 separation performance because the CO_2 transport is enhanced by the reversible reaction between CO_2 and the active carrier, amine, incorporated in the membrane. Continuously increasing the CO_2 partial pressure leads to the constant CO_2 flux, which does not increase with CO_2 partial pressure, i.e., the carrier saturation phenomenon. At the carrier saturation, all available carrier molecules have been reacted with the high pressure CO_2. However, this carrier saturation phenomenon can be beneficial for CO_2 capture from low CO_2 concentration (~1% CO_2) sources, e.g., the residual flue gas after the primary CO_2 capture system (with 90% CO_2 removal) and coal mine gas streams. For CO_2 separation from low concentration sources, an improved membrane separation performance can be obtained. With reducing the CO_2 concentration in the feed gas, both CO_2 permeance and CO_2/N_2 selectivity increased. These were mainly due to more available amine carriers for CO_2 molecular transport at lower CO_2 concentration conditions. In this study of CO_2 separation from low concentration feed, the membrane used consisted of polyvinylamine serving as the fixed-site carrier and piperazine glycinate acting as the mobile carrier. A CO_2 permeance of 785 GPU (1 GPU = 10^{-6} cm^3 (STP)$\cdot cm^{-2}\cdot s^{-1}\cdot cmHg^{-1}$) and a CO_2/N_2 selectivity of 150 were obtained with a feed of 20% CO_2 and 80% N_2 on a dry basis (16.6% CO_2 and 66.2% N_2 along with the saturation water vapor at 17.2%) at the typical flue gas temperature of 57°C. By reducing the CO_2 concentration in the feed, a significant CO_2 permeance increase was observed when the CO_2 concentration was below 5%. For instance, the membrane showed a CO_2 permeance of 950 GPU and a CO_2/N_2 selectivity of 190 with a feed of 1% CO_2 and 99% N_2 on a dry basis. The observed performance increase was due to the carrier saturation phenomenon as more carrier molecules were available for CO_2 transport at a lower CO_2 concentration. Additionally, it was found that the effect of SO_2 was less significant when the CO_2 concentration was low. The CO_2 permeance dropped by approximately 2.5% with an introduction of 3 ppm SO_2 in the feed gas of 20% CO_2 and 80% N_2 on a dry basis at 57°C. However, no performance reduction was observed after exposing the membrane to 3 ppm SO_2 with the feed gas of 1% CO_2 and 99% N_2 on a dry basis at the same temperature. Therefore, the studied membrane can be applied for CO_2 separation from low concentration sources with an exceptional separation performance.

502281 Recent Applications of Separex Membranes Beyond CO_2 Removal

Abstract:

Gas separation membranes have been used commercially for over 35 years to process natural gas, most commonly with the goal of removing CO_2. This presentation will explore two recent applications of Honeywell UOP's Separex membranes: (1) Condensing mode operation of the membrane and (2) Modular design for fuel gas conditioning.

The operation of membranes in condensing mode has commonly included carbon dioxide enhanced oil recovery (CO_2-EOR), natural gas liquids (NGL) recovery and operations targeting reduction of pretreatment costs. We have developed a new membrane element, "NGL–Extract", for these condensing mode applications. NGL-Extract elements replace older generation membrane elements in commercial plants that produce NGL from a feed with a high CO_2 content. The results showed up to 70% increase in gas production while utilizing less membrane area than the older elements. In addition, results from field and laboratory tests confirmed that NGL–Extract elements are able to operate robustly under condensing conditions even with the presence of common contaminants.

The other application under discussion is fuel gas conditioning. Here the target is to use raw field gas as a substitute for purchased fuel. Raw field gas typically includes contaminants such as CO2 and H2S. Gas engines need a consistent supply of gas with a high methane number and low contaminants and heavy hydrocarbons to have higher efficiencies and prevent engine knocking. Honeywell UOP's Fuel Gas Conditioning unit with Separex membranes recovers C2+ to provide a lean engine fuel for power applications. The rich gas stream as the byproduct from fuel gas conditioning is sent back to the main gas pipeline for NGL recovery. In the approach presented here a fit for purpose fuel-gas conditioning system was developed. It utilizes Separex membranes to perform the separation and delivers high pressure treated gas. The units are pre-engineered and consist of a catalog of pre-treatment options, membrane and project-specific value blocks that can be arranged to deliver specific solutions. This flexible design approach covers a wide range of operating pressures, feed and product flowrates.

493661 a New Experimental Technique for the Study of Gas Permeation of Binary Mixture

Abstract:

It is well known that the gas permeation follows the mechanisms of 'sorption-diffusion' in microporous inorganic membranes and 'solution-diffusion' in polymeric membranes. [1,2] However, the traditional experimental technologies for investigating those diffusion/permeation processes, such as diffusion cell, are costly, difficult to operate (e.g. using GC or MS for the composition analysis of permeate), and prone to experimental errors (e.g. the use of helium as the sweeping gas as down-stream, which may cause back-diffusion to upstream).

In this study, we proposed and validated a new experimental technique for effective study of such processes. Binary gas permeation (Helium and CO_2) were measured on a Carbon Molecular Sieve Membrane (CMSM) derived from Kapton polyimide at 700. The transient permeation time lag and steady state permeation fluxes were measured for the binary mixture as well as for each pure gas component.

It was found that the proposed method, which can readily condense the CO_2 component into solid phase via a physical/nondestructive method, offers the good advantages over the traditional methods, by revealing the 'true' permeation behavior of each component in the binary mixture. The method can provide us with deep insight into the gas permeation processes together with the Maxwell-Stefan theory. The figure below shows the permeation data of one set of He-CO_2 mixture.

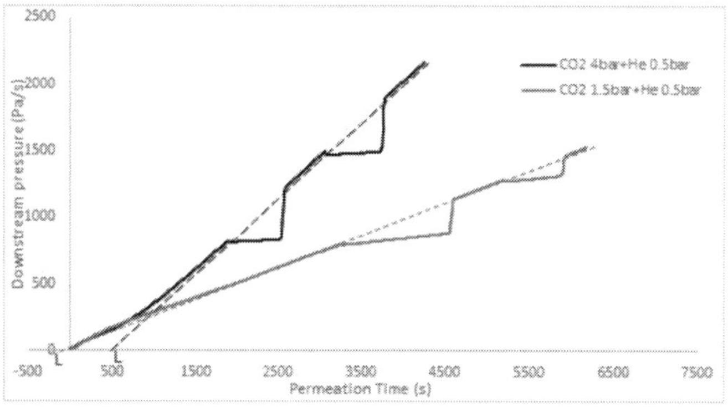

Figure 1. The pressure of binary permeated side against time by using a new method.

Key words: Time lag; adsorption; diffusion; permeation; gas mixture

1. Koresh, J. E.; Soffer, A., Mechanism of permeation through molecular-sieve carbon membrane. Part 1.- The effect of adsorption and the dependence on pressure. *Journal of the Chemical Society, Faraday Transactions 1: Physical Chemistry in Condensed Phases* **1986,** *82* (7), 2057-2063.

2. W.J. Koros, et al, Polymeric membrane materials for solution-diffusion based permeation separations, Progress in Polymer Science, 13, 1988, 339-401

502747 Insights and Rational Design of Metal-Organic Frameworks for Enantiomers Separations

Abstract:

Recent advances in the design and synthesis of metal-organic frameworks have demonstrated the potential of this material class. This research paper analyzes the existing literature on chiral metal-organic frameworks as gathered in the CoRE MOF database [2]. The CoRE MOF database contains most of the reported MOF structures from literature. The crystal structure and synthesis condition are known for each represented network. In many cases, only specific properties in the domains of adsorption, catalysis, electrochemistry... are reported. The extend of reported properties is not only limited by the scope of the original author(s) but in some cases also by the technological advance at the time of reporting. A combination proper chemical and topological analysis with selective computational screening of adsorptive properties was performed.

Chiral frameworks are characterized by the presence of either chiral topology (e.g. srs, unc, lcy...) or enantiomers as linkers, or both. Chiral topologies are identified using structural analysis of their 3D net. Interestingly, only 19 out of the 362 chiral nets in the Reticular Chemistry Structure Resource (RCSR) database are found within reported MOF structures. MOF structures with chiral linkers (or introduced by post synthetic modification) are identified by data mining of article contents. Analysis of the reported frameworks gives insight in the organic and inorganic building blocks needed to form a chiral frameworks. There are 2 topological factors mentioned in literature that contribute to chirality: asymmetric metal node or helix formation. Interestingly, our analysis shows that in many cases chirality is the result of symmetry distortion at the metal node by chelation or mixed O- and N-terminated organic linkers. This essentially means that no chiral building blocks are required to result in some degree of chirality in the framework. Several asymmetric metal clusters are identified for Zn/Cu, In, Co, Ni and Cd that are of interest in the design of novel frameworks (no need for chiral linkers). The formation of helixes or warped pore architecture is more likely if a three-connecting organic building block is present rather than 2 or 4 connecting linkers. On the basis thereof, several new hypothetical structures are proposed and evaluated for chiral separations. The analysis of structural components is complemented with molecular simulations on the adsorption for R- and S-isomers (organic acid, alcohol, amine, amino acids) to establish relationships and compare impact of key features (linker, node, helix) on the separation potential. A one-to-one comparison between framework types is not straightforward as pore size, topological net type and chemical composition are typically linked but select cases where only one aspect is different indicate synergetic effects.

Particular attention is given to cyclodextrin Metal-Organic structures as a subset to illustrated the effect of pore size and topology towards chiral separation potential. The three different ring sizes in alpha, beta and gamma cyclodextrin lead to different pore dimensions irrespective of the packing in a crystal structure. Our results on the CD-MONT metal-organic cage structures [2] indicate that a more confined pore enhances the chiral separation potential. This finding is in line with observations for stereoisomers in shape selective nanoporous materials. The differences in free energy between R- and S- isomers are partly enthalpic and

partly entropic. The effect of topology is illustrated by another subset of gamma cyclodextrin [3] metal-organic frameworks with different chiral net. Both the number of cyclodextrin rings that a probe enantiomer simultaneously interacts with as the relative pore dimension (resulting from tubular or cage type pore formation) have significant effects.

[1] Chung et al. Chem. Mater., 2014, 26 (21), pp 6185–6192.

[2] Wei et al. Chem. Sci., 2012, 3, pp 2282-2287.

[3] Forgan et al. JACS, 2012, 134, 406, Hartlieb et al. J. Am. Chem. Soc., 2016, 138 (7), pp 2292–2301.

494097 Selective Carbohydrate Adsorption in Solvated Nu-1000

Abstract:

As the efforts for green chemistry and sustainability continue to grow, production of biofuels becomes more important. During the production of bioethanol, cellulose-based feedstocks are sent through physicochemical pretreatments to enhance the amount of extractable sugar per biomass processed. The drawback of the pretreatment process is that besides obtaining sugar, furanic species, 5-hydroxymethylfurfural (HMF) and furfural, are also formed as a result of side reactions. The selective removal of these furanic compounds is crucial since HMF and furfural have inhibiting effects on fermentation and effective separation of them will allow more economical production of cellulosic ethanol.

Metal-organic frameworks (MOFs) can provide solutions to many separation problems due to their ordered, highly porous, chemically versatile structures. MOFs are comprised of metal nodes and organic linkers and considering various organic linkers and metal types that can be selected, it is possible to create structures which can have favorable interactions for a particular species and an unfavorable interaction for another species. For the selective capture of furanic species over sugars (glucose, fructose, xylose), NU-1000, a Zr-based MOF, has been experimentally shown to have almost infinite selectivity in aqueous-phase.[1]

The aim of this computational study is to elucidate the reasons behind why NU-1000 can specifically adsorb some carbohydrates (HMF, furfural) but not others (glucose, fructose, xylose). To have a fundamental understanding about this highly selective separation, classical molecular dynamics (MD) simulations are employed to calculate relative free binding energies where solvation effects are considered. Using MD results, the high adsorption selectivity for furanic compounds over sugars will be explained by describing the interaction characteristics of compounds in NU-1000.

References 1. Yabushita, M. *et al.* Complete furanics-sugar separations with metal-organic framework NU-1000. *Chem. Commun.* **52,** 11791â€"11794 (2016).

495665 Molecular Simulation of Combined Chemi- and Physi-Sorption of Carbon Monoxide on Cobalt in the Presence of Supercritical Hexane

Abstract:

The use of supercritical fluid (SCF) solvents to enhance heterogeneous catalysis is growing, with several examples focused on energy and fuel applications. One such application is the use of supercritical hexane to enhance the overall catalytic chemistry of Fischer-Tropsch (FT) synthesis for the production of liquid hydrocarbon fuels. Catalytic reaction engineering experiments have demonstrated that FT synthesis in SCF hexane has several advantages over conventional FT processing, including improved heat management, reduced catalyst deactivation via coking, faster reaction rates, and better product selectivity. More recent molecular simulation studies have revealed unique and relevant aspects to adsorption in/with SCF fluids, including crossover behavior (from adsorption to depletion) for SCF hexane physisorption on a cobalt FT catalyst [1] and a density dependence for the free energy of carbon monoxide chemi-sorption on cobalt which correlates to the bulk SCF hexane isothermal compressibility [2].

This work explores the combined chemical and phase equilibrium present during chemisorption within a SCF solvent environment via reactive Monte Carlo simulations. A previously developed Morse potential for carbon monoxide interaction with cobalt [2] has been added into an existing Monte Carlo code [3], to allow for the proper energetics associated with carbon monoxide chemisorption onto cobalt. Grand canonical Monte Carlo simulations are conducted to explore the adsorption loadings and energetics of carbon monoxide and SCF hexane solvent in cobalt slit pores of varying dimensions under a range of pressure/density conditions, spanning from the ideal gas limit to SCF hexane conditions. Simulations reveal the larger loading and heat of adsorption associated with the chemisorbed CO, versus the smaller loadings and heat of adsorption associated with the physisorbed hexane solvent.

[1] Benjamin, K. M., Asiaee, A., Veer, C., Losinski, C., Gunderson, S., and Larson, T., "Monte Carlo Simulation of Near- and Supercritical Hexane Fluid and Physisorption Phase Behavior", *J. Phys. Chem. C*, **2016**, 120, 21336-21343.

[2] Asiaee, A. and Benjamin, K. M., "Molecular Simulation of CO Chemisorption on Co(0001) in Presence of Supercritical Fluid Solvent: A Potential of Mean Force Study", *J. Chem. Phys.*, **2016**, 145, 084506.

[3] Martin, M. G., "MCCCS Towhee: a tool for Monte Carlo molecular simulation", *Mol. Simulat.*, **2013**, 39, 1212-1222.; http://towhee.sourceforge.net

497713 Atomistic Understandings of the CO_2 Uptake Difference in Photo Responsive Metal-Organic Frameworks

Abstract:

Global climate change owing to anthropogenic CO_2 emissions has prompted nationwide actions from academia to industries to look for strategies to reduce atmospheric CO_2 concentrations. To this end, carbon capture and sequestration has been regarded as a promising technology to substantially reduce the CO_2 emission from power plants, one of the largest point emission sources. Conventionally, amine absorbers are used to separate CO_2, but the energy requirement of a plant can be increased by 25-40%.[1-2] Recently, metal-organic frameworks (MOFs) have drawn significant attention as a promising alternative because of their desirable properties (e.g., large surface area, adjustable chemical functionality, high selectivity of CO_2 over N_2, etc.).[3] Moreover, experimentists have further developed photo-switchable (azobenzene functionalized) MOFs whose affinity for CO_2 adsorption can be modified by light irradiation (transformation from trans to cis configurations), resulting in a CO_2 uptake difference of more than 50%.[4] This has potentially led to a more energy-efficient CO_2 capture process as compared to traditional materials. Despite the success, the atomistic understandings of the root cause behind the observed uptake difference remain limited. In this study, by using state-of-the-art molecular simulations and density functional theory calculations, we investigate the adsorption mechanisms in azobenzene functionalized MOF-5 (azo-MOF5) that leads to the experimentally observed CO_2 uptake difference between cis and trans configurations. The free energy landscape of CO_2 molecules as well as minimum energy configurations will be computed and analyzed in detail to identify preferential adsorption sites in azo-MOF5 for CO_2. The effect of the azobenzene density on the uptake difference will also be studied. The outcomes of this study are anticipated to provide insights into the design of this stimulus-responsive MOFs to push the development of CO_2 capture and sequestration.

Reference

1. D'Alessandro, D. M.; Smit, B.; Long, J. R., Carbon Dioxide Capture: Prospects for New Materials. *Angew. Chem. Int. Ed.* **2010**, *49*, 6058-82.

2. Huck, J. M.; Lin, L. C.; Berger, A. H.; Shahrak, M. N.; Martin, R. L.; Bhown, A. S.; Haranczyk, M.; Reuter, K.; Smit, B., Evaluating Different Classes of Porous Materials for Carbon Capture. *Energ. Environ. Sci.* **2014**, *7*, 4132-4146.

3. Millward, A. R.; Yaghi, O. M., Metal-Organic Frameworks with Exceptionally High Capacity for Storage of Carbon Dioxide at Room Temperature. *J. Am. Chem. Soc.* **2005**, *127*, 17998-17999.

4. Park, J.; Yuan, D. Q.; Pham, K. T.; Li, J. R.; Yakovenko, A.; Zhou, H. C., Reversible Alteration of Co2 Adsorption Upon Photochemical or Thermal Treatment in a Metal-Organic Framework. *J. Am. Chem. Soc.* **2012**, *134*, 99-102.

497937 Sorption-Relaxation Behavior in Polymers of Intrinsic Microporisity during Gas Separation from Molecular Simulations

Abstract:

Polymers of intrinsic microporosity (PIMs) are high free volume polymers that can function as molecular sieves, making PIMs alternative materials for separation applications. These polymers are unable to pack efficiently due to their rigidity and contorted shapes, which results in their desirable porosity. Our group has been investigating the permeability and selectivity of gases in PIMs using atomistic molecular simulations. In this work, we further improve our predictive capability by considering the motion of the polymer framework during adsorption (sorption-relaxation scheme) by repetitively implementing a hybrid simulation of molecular dynamics and Monte Carlo simulation (MD/MC). A python package, *pysimm*, was used to facilitate the workflow between MD and MC simulators (LAMMPS and Cassandra). Structural properties, such as radial distribution function, were compared in the presence and the absence of the sorption-relaxation scheme. The permeability and selectivity of important gas pairs such as CO_2/CH_4 and O_2/N_2 were evaluated at typical industrial gas conditions including the presence of H_2S and water. Our results suggested that hybrid MD/MC simulations can provide important insights for future adsorption studies. This procedure can be applied to other polymeric membranes, composite membranes, and flexible porous materials to account for possible swelling, relaxation, gate-opening, or breathing phenomena within the frameworks.

498482 Framework Flexibility Driven Adsorptive Separation of C_8Â Aromatic Isomers in Metal Organic Frameworks: A Computational Exploration

Abstract:

Separation of xylene isomers from mixtures of C_8 aromatic compounds using metal organic frameworks is of potential interest for practical production of these species. In this work, we have used computational methods to explore adsorption of xylenes in several MOFs. It was found that use of rigid framework approximation in molecular simulations to describe adsorption in MOFs gives drastically different results than experiments. To address this issue, new methods combining ab-initio and MD calculations were developed to study the flexibility of these materials and its effect on adsorption of C_8 aromatic compounds in MOFs. In addition to the well-known structural transitions that can occur for materials like MIL-53, rotation of organic linkers and changes in torsion angle between metal-oxygen bonds have strong influence on the highest capacity and selectivity of adsorbates. We have applied these methods to C_8 aromatic adsorbates in MIL-53 series to address the question of how much significant the effect of flexibility in these MOFs are.

498988 Molecular Simulation Studies Probing Transport and Adsorption of C8 Aromatics through MFI Nanosheet Membranes

Abstract:

This work focuses on the separation of mixture of C8 aromatic compounds using MFI-type zeolite membrane. The model membrane system comprises of the 5nm-thick zeolite nanosheet with periodic replications in the *a* and *c* directions. The free surfaces normal to the *b* direction are saturated with silanol groups. We have utilized both first principles and force-field-based molecular dynamics (MD) simulations to understand the diffusion barriers for transport across the zeolite channels. We use an ensemble of framework structures obtained from MD simulations to generate adsorption isotherms for mixtures containing the C8 molecules using Gibbs Ensemble Monte Carlo simulations. We report the adsorption and transport selectivities of p-xylene to other aromatics in the mixture and compare them with experimental observations.

500742 Computational Screening of Zeolites for Gas Separation Applications from Multi-Component Mixtures

Abstract:

There is abundant literature on high-throughput *in-silico* screening of micro-porous materials such as zeolites and metal organic frameworks for various gas separation applications.[1–5] Recent efforts have been directed towards the combined computational screening of candidate materials and process optimization for various gas separation processes, e.g., CO_2 capture[6], natural gas purification[7], and hydrogen sulfide removal[8]. To properly evaluate the applicability of a process, it is important to not just perform process level optimization, but also simultaneously determine candidate materials that are most suitable for different process configurations. Though simultaneous materials and process optimization is useful for optimizing well-established processes, it may not be feasible for quick evaluation of novel and emerging technologies such as integrated carbon capture and conversion[9], new sorption-enhanced reaction processes[10], and combined separation and storage[11]. To develop novel and intensified processes to improve productivity and efficiency, it is necessary to leverage on properties (i.e. storage, selectivity) of different gases on different materials. For separation of the desired gas from a multi-component gas mixture, the working capacity of the material depends on the relative selectivity parameters, nature of isotherms and the concentrations of the gases.

To this end, we have performed Grand-Canonical Monte Carlo simulations for gases such as CO_2, CH_4, CO, H_2, H_2S, H_2O and N_2 to obtain gas (adsorbate) loading on existing pure-silica zeolites (adsorbents) from the IZA-SC database for different temperatures (298, 323 and 373 K) and pressures (1.3, 5.3, 10.7, 21.3, 42.7, 85.3, 101.3, 266.6 15, 580, 6500 kPa). These compute-intensive simulations are performed on a high performance computing cluster named Ada. Based on the data obtained from these molecular simulations, selectivity parameters are then obtained for different combination of gases (CO_2/N_2, CH_4/N_2, CO_2/CH_4, H_2/CO, H_2S/CH_4, H_2/CH_4, $CO_2/N_2/H_2O$, $CO_2/H_2S/CH_4$, etc.) and their concentrations in feed mixtures. The data from molecular simulations is fitted to single, dual, multi-site Langmuir, extended Langmuir, Freundlich and Toth adsorption isotherm models to determine the best fitting model for each gas-zeolite data set. The isotherm parameters are obtained by solving the minimization problem of the least-square error between the data and the model prediction to global optimality. The adsorption isotherm model is then incorporated into the process model.

The top zeolites for each application of gas separation from a multi-component mixture of different concentrations are then rank ordered based on different materials-centric metrics (adsorption selectivity, saturation loadings, working capacity, heat of adsorption, Henry's constant, etc.). In addition, zeolites which would not be suitable for a particular application are also determined. The information gathered will enable the design of novel process configurations for different gas separation applications.

REFERENCES:

(1) Martin, R. L.; Willems, T. F.; Lin, L.-C.; Kim, J.; Swisher, J. A.; Smit, B.; Haranczyk, M. Similarity-Driven Discovery of Zeolite Materials for Adsorption-Based Separations. *Chemphyschem A Eur. J. Chem. Phys. Phys. Chem.* **2012**, *13* (16), 3595–3597.

(2) Colon, Y. J.; Snurr, R. Q. High-Throughput Computational Screening of Metal-Organic Frameworks. *Chem. Soc. Rev.* **2014**, *43* (16), 5735–5749.

(3) Lin, L.-C.; Berger, A. H.; Martin, R. L.; Kim, J.; Swisher, J. A.; Jariwala, K.; Rycroft, C. H.; Bhown, A. S.; Deem, M. W.; Haranczyk, M.; et al. *In silico* Screening of Carbon-Capture Materials. *Nat Mater* **2012**, *11* (7), 633–641.

(4) Krishna, R.; van Baten, J. M. *In silico* Screening of Metal-Organic Frameworks in Separation Applications. *Phys. Chem. Chem. Phys.* **2011**, *13* (22), 10593–10616.

(5) Wu, D.; Wang, C.; Liu, B.; Liu, D.; Yang, Q.; Zhong, C. Large-Scale Computational Screening of Metal-Organic Frameworks for CH_4/H_2 Separation. *AIChE J.* **2012**, *58* (7), 2078–2084.

(6) Hasan, M. M. F.; First, E. L.; Floudas, C. A. Cost-Effective CO_2 Capture Based on *in silico* Screening of Zeolites and Process Optimization. *Phys. Chem. Chem. Phys.* **2013**, *15* (40), 17601–17618.

(7) First, E. L.; Hasan, M. M. F.; Floudas, C. A. Discovery of Novel Zeolites for Natural Gas Purification through Combined Material Screening and Process Optimization. *AIChE J.* **2014**, *60* (5), 1767–1785.

(8) Liu, T.; First, E. L.; Hasan, M. M. F.; Floudas, C. A. A Multi-Scale Approach for the Discovery of Zeolites for Hydrogen Sulfide Removal. *Comput. Chem. Eng.* **2016**, *91*, 206–218.

(9) Iyer, S. S.; Bajaj, I.; Balasubramanian, P.; Hasan, M. M. F. Modular Process Intensification of Carbon Capture and Conversion to Syngas. *Submitted.*

(10) Carvill, B. T.; Hufton, J. R.; Anand, M.; Sircar, S. Sorption-Enhanced Reaction Process. *AIChE J.* **1996**, *42* (10), 2765–2772.

(11) Iyer, S. S.; Hasan, M. M. F. A Novel Plug-and-Store Technology for Natural Gas Purification and Strage. In *CAMX 2015 - Composites and Advanced Materials Expo*; 2015.

500013 Unraveling of Pristine and Defective Metal-Organic Framework (MOF) Structures through Molecular Simulation

Abstract:

Metal-organic frameworks (MOFs) are extremely attractive porous materials for applications in gas storage, chemical separations, catalysis, and others. The appeal of MOFs for these applications is in great part due to the chemical and structural tunability of MOFs, which arises from the virtually unlimited combination of constituent organic and inorganic building blocks that these materials can be made of. However, the unlimited diversity of MOFs that could be synthesized for a given application raises the challenge of identifying the most promising structures among so many candidates. To address this challenge, the vast MOF structure space can be navigated in "high throughput" fashion—typically done computationally—or in "rational exploration" fashion, where the latter approach relies on a solid, reliable knowledge about the links between synthesis, structure, property and performance for currently synthesized MOFs, and the use of this knowledge to design higher-performance materials. Therefore, a thorough understanding of the structure of currently synthesized and tested MOFs is critical as to not to mislead rational MOF design. In many instances, attaining this understanding can be difficult, partly due to advances in MOF synthesis leading to increasingly more complex MOF structures, and partly due to the realization that MOF structural defects can significantly affect performance.

Molecular simulation and auxiliary computational methods are valuable—and relatively accessible—tools to help unravel the structure of existing and newly synthesized MOFs, especially when used in synchrony with experimental characterization. Here, we first present the utilization of automated algorithms for construction of MOF computational models, nitrogen adsorption simulations and other auxiliary tools to elucidate the network structure of a new uranium-based MOF, **NU-1301** that synthesizes into the largest crystallographic unit cell for any non-biological material to date. Despite the simplicity of the constituent building blocks, we found this MOF to present a complex hierarchical structure based on a previously unknown network topology, *nun*, which has now been added to the Reticular Structure Chemistry Resource (RSRC) database.

Then, we present the utilization of MOF computational models and nitrogen adsorption simulations to elucidate structural details of the defective structure of the well-known, zirconium-based MOF, **UiO-66**. Specifically, we illustrate how features in nitrogen adsorption isotherms are highly dependent on MOF structural features, to a level that can be used to discern the local MOF structure around defective regions, if sufficiently accurate structural MOF models are provided to be used to obtain "standard" simulated isotherms for comparison with experimental ones. Using this method, we elucidated how the structure of **UiO-66** varies as the synthesis method is varied, for instance, through the change of crystallization modulators.

493867 Entropy-Driven, Two-Step Crystallization of Colloidal Clathrate Crystal

Abstract:

Non-classical crystallization, such as two-step crystallization, is the primary mechanism of solidification for many substances. Despite its importance, aspects of the system responsible for determining the crystallization mechanism are generally poorly understood. Identifying idealized model systems that exhibit non-classical crystallization is therefore desirable. Here, we report two-step crystallization in a three-dimensional system of hard shapes. Unlike previous reports of two-step nucleation and growth, in our system it is driven by entropy alone. Monte Carlo simulations reveal that crystallization is preceded by the appearance of a dense amorphous phase.The thermodynamics of crystallization is investigated by determination of the equation of state and via free energy calculations.

499352 Unraveling the Coupling between Demixing and Crystallization in Mixtures

Abstract:

Crystallization has remained an incredibly complex and mysterious process. This is for instance the case on the nanoscale where our ability to control nanoparticle properties is limited by our lack of understanding of how nanoparticles nucleate and grow. This is however key for many applications as metallic nanoparticles are increasingly used as protein markers, in high-density magnetic recording and in catalysis. Using molecular simulation, we shed light on the mechanisms underlying crystal nucleation and growth in metallic and bimetallic systems. In particular, we unravel the competition between the two nonequilibrium processes of phase separation and crystallization in bimetallic systems and discuss its impact on structure and composition on the nanoscale.

500651 La Mer Burst Nucleation and Growth: Assumptions, Models, and Data

Abstract:

Recent efforts to synthesize monodispersed nanoparticles have employed in situ reactions to steadily and homogeneously generate a mounting solute supersaturation. As the supersaturation climbs past a critical level in solution, the nucleation rate spikes giving birth to a large population of nuclei over a very short interval of time. Subsequent growth of these nuclei lowers the supersaturation and thereby suppresses further nucleation. The result is a large population of nearly monodisperse particles. Classic experiments by T. Sugimoto and others have been modeled with deterministic models having separate stages of supersaturation build-up, nucleation, growth, and sometimes ripening. We examine assumptions in these models within the context of chalcogenide nanoparticle synthesis (Prof. J. S. Owen) and gold nanoparticle synthesis (Prof. S. L. Scott). We present a fully non-equilibrium stochastic model for LaMer burst nucleation with comparisons to results from simpler models. The results should clarify whether fit parameters from simple models accurately resolve the microscopic nucleation and growth kinetics.

489027 A Stochastic Model of Primary Nucleation of Polymorphs

Abstract:

Polymorphism is a crucial feature of crystals, with particular importance in the fine chemical and pharmaceutical industry, since different polymorphs have different physical and chemical properties, such as solubility and crystal structure. Crystal nucleation is known to be a stochastic phenomenon; however, the implications of its statistical nature for the order of formation of the polymorphs (Ostwaldâ€™s rule of stages) have not yet been explored. Similarly, the role of stochastic nucleation in polymorphs transformation has not been thoroughly investigated either.

Based on large experimental evidence, and building upon previous theoretical work, we have developed a stochastic model of primary nucleation of multiple polymorphs from a clear solution, accounting also for growth, dissolution, and secondary nucleation. With this model, one can simulate and qualitatively study the interplay of all phenomena acting in the system - often coupled and not directly measurable - and their role in determining detection times and polymorphs transformation. Assessing such relationships and the mechanisms underlying them would allow to not only gain a deeper understanding of polymorphism, but also to better design and optimise the production processes of polymorphs.

492589 Step Velocity of a Crystal Edge with Alternating Rows of Growth Units

Abstract:

The morphology of a crystal impacts bioavailability, catalytic activity, processability, etc. Mechanistic growth models have accurately predicted crystal shapes for a number of organic molecules. [1-3] A critical parameter in these models is the step velocity, i.e., the rate at which a spiral or 2D nucleus edge spreads across the surface. The step velocity depends on the density of kink sites along an edge and the net attachment rate to a kink site; kinks are favorable surface sites for attachment of growth units (atoms, molecules, etc.). [4] Determining these quantities is complicated for non-centrosymmetric crystals, as their multiple growth units can lead to complex step edge configurations.

In this work, we examine an edge with alternating rows of A and B growth units (A-B edge), which can exhibit stable and unstable behavior depending on the supersaturation and anisotropic interactions between A and B. [5] For example, the B row can actually dissolve under supersaturated conditions. We capture these phenomena in a stochastic 1D nucleation model which employs a Fokker-Planck equation to treat formation of an AB nucleus (new A and B rows). This model is used to determine the supersaturation-dependent kink density by balancing kink creation (1D nucleation) with kink annihilation (kink collisions). The resulting step velocity predictions yield great agreement with kinetic Monte Carlo simulations for various supersaturations and anisotropic interactions. We find that the step velocity undergoes a cross-over in supersaturation dependence as the edge changes from unstable to stable (via increasing supersaturation). This suggests the possibility of a supersaturation-dependent aspect ratio of spirals and 2D nuclei with an A-B edge. The step velocity model captures important changes in growth at higher supersaturations and will improve crystal shape predictions.

[1] J. Li, C.J. Tilbury, S.H. Kim, M.F. Doherty, *Prog. Mater. Sci.* **2016**, *82*, 1-38.

[2] M.A. Lovette, A.R. Browning, D.W. Griffin, J.P. Sizemore, R.C. Snyder, M.F. Doherty, *Ind. Eng. Chem. Res.* **2008**, *47*, 9812-9833.

[3] P. Dandekar, Z.B. Kuvadia, M.F. Doherty, *Annu. Rev. Mater. Res.* **2013**, *43*, 359-386.

[4] P.G. Vekilov, *Cryst. Growth Des.* **2007**, *7*, 2796-2810.

[5] C.J. Tilbury, M.N. Joswiak, B. Peters, M.F. Doherty, *Cryst. Growth Des.* **2017**, *17*, 2066-2088.

497948 Multiscale, Multiphysics, Mechanistic Model for Computation of Face-Specific Growth Rates

Abstract:

The morphology or shape of the crystalline material is a critical determinant of their physical properties such as stability, flowability, wettability, bulk density, and dissolution rates. For example, the plate- or needle-like pharmaceutical crystals are difficult to granulate and compress into a tablet form. Identification of theoretical models to predict the effect of crystallization conditions on crystal morphologies is imperative to develop control strategies for preventing such undesired morphologies to grow. The dynamic and steady-state prediction of crystal morphology requires the knowledge of face-specific growth rates of crystals. The current models to predict crystal growth rates assumes the regimes of crystal growth mechanism (spiral, 2D nucleation and rough) based on the empirical calculations of interfacial free energies of solute and solvent molecules. The growth rate is then estimated using the height, spacing, and velocity of layers spreading on the crystal surface. However, a single face of the crystal can have multiple growth mechanisms on the surface and there need not be a strict barrier between different crystal growth mechanisms. This could be due to different processes â€" diffusion, desolvation and surface integration, and topography evolution - occurring simultaneously on the nucleated crystal surface. There is a need for a multiscale model which accounts for all these processes. In this talk, I will describe the multi-scale model for determining crystal growth rate using Kinetic Monte-Carlo method. This multi-scale model captures the most fundamental mechanism of crystallization â€" the association and dissociation of molecules â€" and find the most probabilistic realization of the surface. Our method considers probabilities of events such as nucleation of 2D islands and screw dislocations and spreading of the layer based on the energy barrier for integration of molecules onto the crystal surface. The prediction of growth rates of Glutamic acid crystals will be shown as an example.

486860 Anisotropic Growth Kinetics of Triblock Janus Colloids

Abstract:

Anisotropic colloidal particles have been shown to exhibit equilibrium crystal structures of vastly different density and coordination number depending upon surface geometry and thermodynamic conditions. However, due to very small free energy differences between accessible structures, it is not known whether self-assembly into large single crystals is feasible. We use molecular dynamics simulations to calculate interface propagation velocities for different faces of these anisotropic crystals. Our results show that the growth kinetics are strongly influenced by geometric factors which are difficult to predict from equilibrium properties. Developing a better understanding of these effects could unlock a new set of tools for directed self-assembly by leveraging growth kinetics to select between structures with similar free energies.

Systematic Analysis and Optimization of Power Generation in Pressure Retarded Osmosis: Effect of Multi-Stage Design

Mingheng Li

November 2, 2017

Outline

Motivation

Global Grand Challenges in Energy and Water

Humanity's Top Ten Problems for next 50 years

1. **ENERGY**
2. WATER
3. FOOD
4. ENVIRONMENT
5. POVERTY
6. TERRORISM & WAR
7. DISEASE
8. EDUCATION
9. DEMOCRACY
10. POPULATION

2003	**6.3**	**Billion People**
2050	**8-10**	**Billion People**

[1] Professor Richard Smalley, Nobel Laureate

Membrane-Based Processes

- Reverse Osmosis (RO)

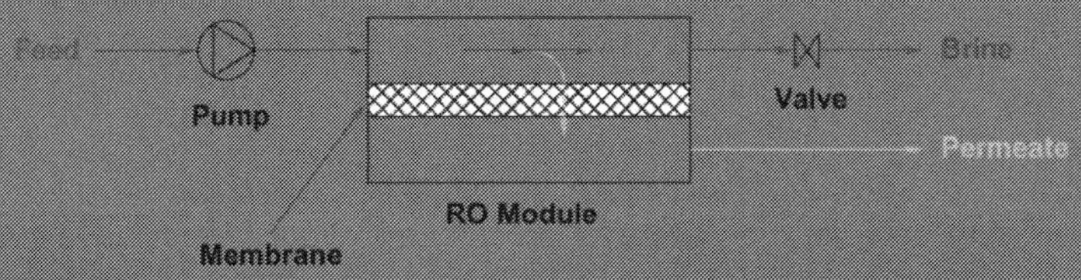

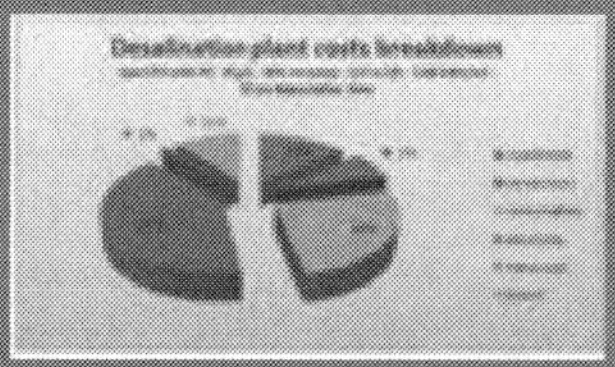

- Pressure Retarded Osmosis (PRO)

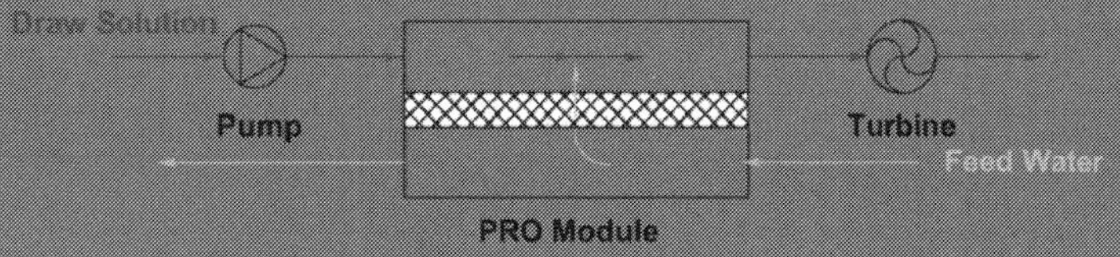

[1] https://www.lenntech.com/processes/desalination/energy/general/desalination-costs.htm

Need for System-Level Understanding of PRO

- There is a lack of computational framework to address scale-up issues in PRO
  - Optimal applied pressure much lower than $\Delta\pi/2$ is observed in some experiments (Xu et al., JMS, 2009; She et al., JMS, 2012; Sharif et al., Membranes, 2014).
  - Power generation in PRO is much lower than theoretical prediction.
- Mathematical models helps elucidate the understanding of energy issues in PRO.

Methodology

Characteristic Equation of RO and PRO

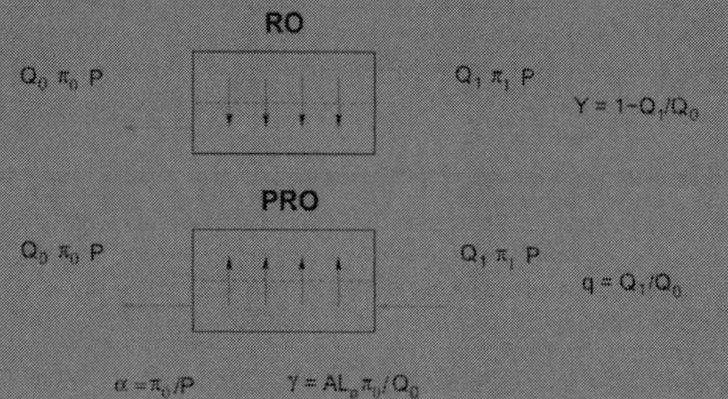

Assumptions:

- Negligible pressure drop
- Negligible concentration polarization
- Perfect transverse mixing of salt

$\Rightarrow$

γ:	Dimensionless membrane area	$\rightarrow$ Design parameter
α^{-1}:	Dimensionless applied pressure	$\rightarrow$ Operating parameter
q:	Dimensionless flow rate ($Y = 1 - q$)	$\rightarrow$ Performance parmater

[1] Li, IECR, 2010; AIChE J., 2015

Model Accounting for PRO Efficiency

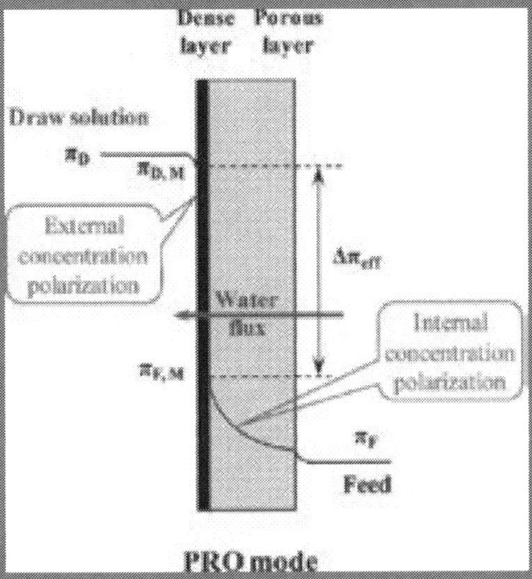

- Detrimental Effects in PRO
  - External Concentration Polarization (ECP)
  - Internal Concentration Polarization (ICP)
  - Reverse Salt Leakage
- Equation accounting for PRO efficiency

[1]Wang, Ong, and Chung, IECR, 2010
[2]Li, AIChE J., 2017

Energy Production in PRO

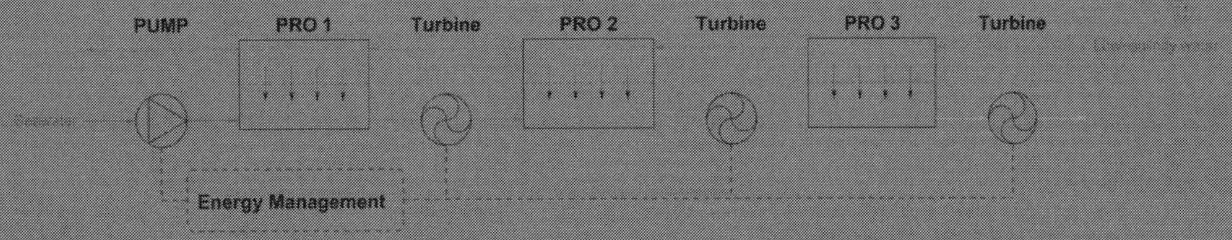

- Normalized Specific Energy Production (NSEP) in PRO
- NSEP in Single- or Multi-stage PRO

[1]Li, AIChE J., 2017

Optimization Model for PRO

[1]Li, AIChE J, 2017

Results & Discussion

Results of NSEP in PRO

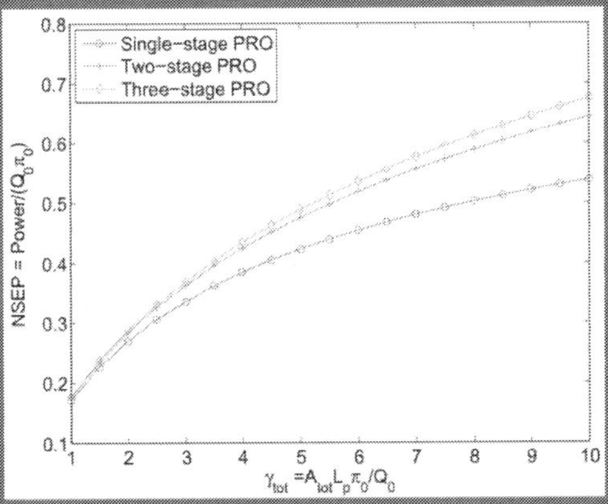

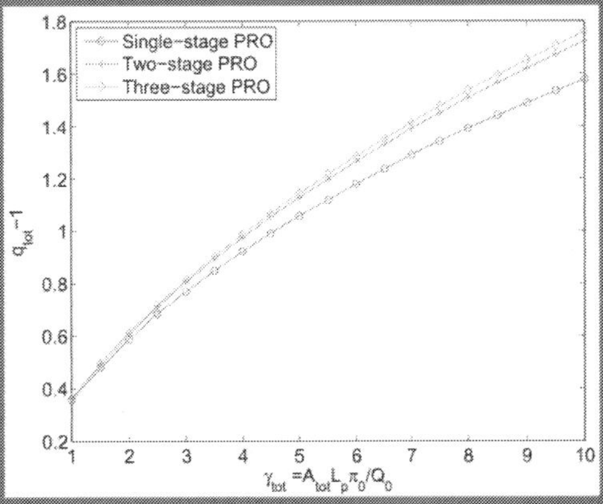

- The increased volume gain is not the sole cause to the boosted power generation in multi-stage design.
- $NSEP = \sum_{j=1}^{k}(q_j - 1)/\alpha_j$ (if ignoring equipment efficiencies), implying pressure profile has a contribution to NSEP as well.

Mingheng Li | Multi-stage PRO | AIChE 2017 | 13 / 24

Pressure Profile $\gamma_{tot} = 4$ (top) and 10 (bottom)

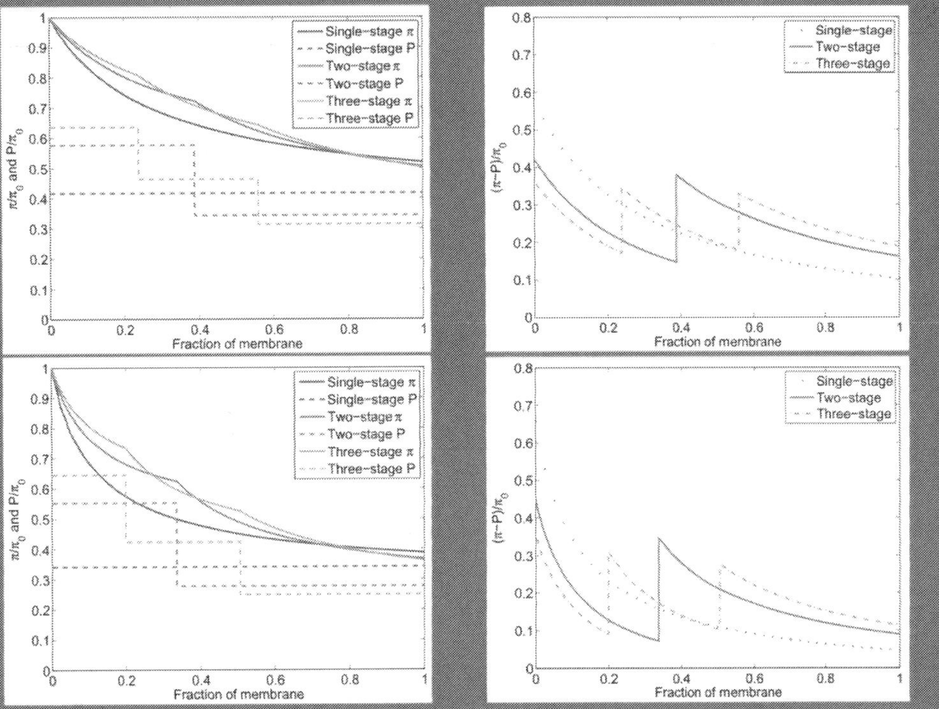

Mingheng Li | Multi-stage PRO | AIChE 2017 | 14 / 24

Power Density

- Relationships between NSEP and Power Density (PD)

- PD vs L_p PD vs π_0 PD vs Q_0/A

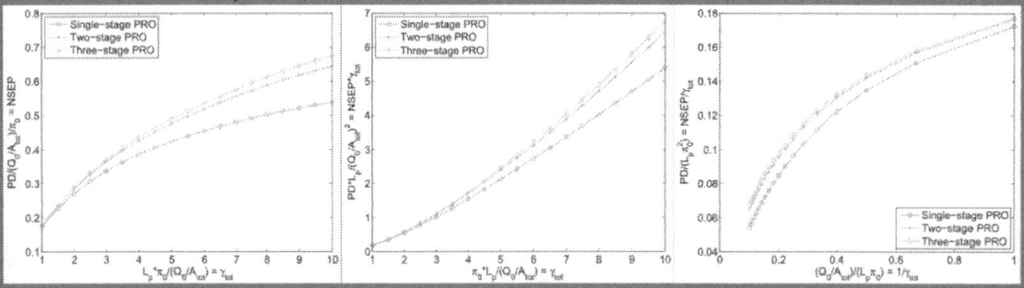

Optimization at the Thermodynamic Limit

- As $\gamma_j \to \infty$, driving force $\to 0$. i.e. $\alpha_j = q_j$
- Optimization model

- Solution

Optimization at the Thermodynamic Limit

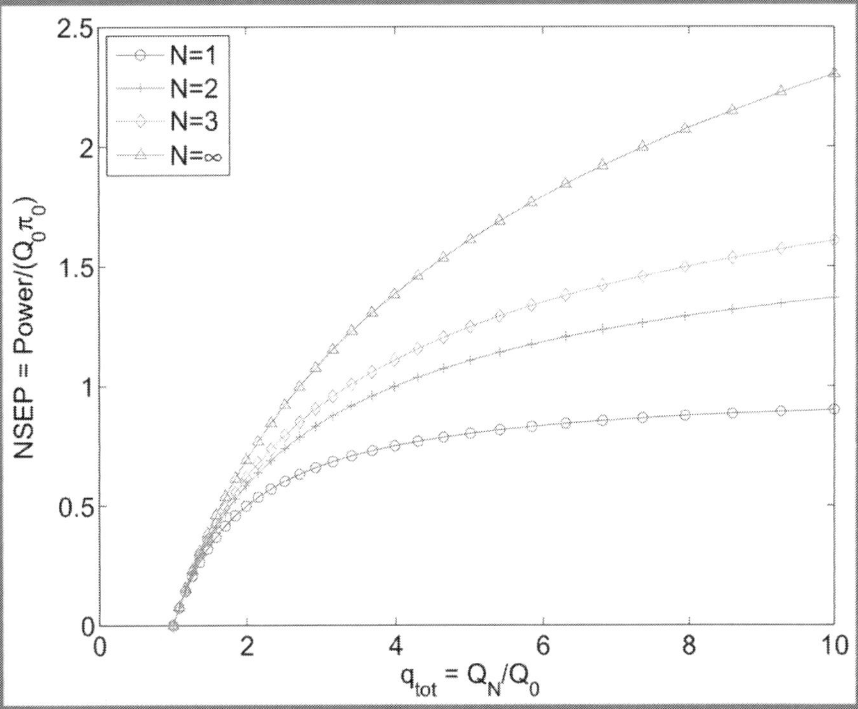

Infinite Stages at the Thermodynamic Limit

- Infinite PRO stages at the thermodynamic limit
- Fully reversible PRO

How Far from the Thermodynamic Limit

γ_{tot}=4	Single-Stage	Two-Stage	Three-Stage
q_{tot}	1.9241	1.9774	1.9895
NSEP	0.3852	0.4251	0.4347
NSEP (eq)	0.4803	0.5777	0.6147
NSEP(γ_{tot}=4)/NSEP(eq)	80%	74%	71%

γ_{tot}=10	Single-Stage	Two-Stage	Three-Stage
q_{tot}	2.5784	2.7234	2.7595
NSEP	0.5388	0.6440	0.6751
NSEP (eq)	0.6122	0.7881	0.8612
NSEP(γ_{tot}=10)/NSEP(eq)	88%	82%	78%

Optimization Accounting for Process Efficiencies

- η_{PRO} = 80%, η_{turb} = 80%, and η_{pump} = 80%

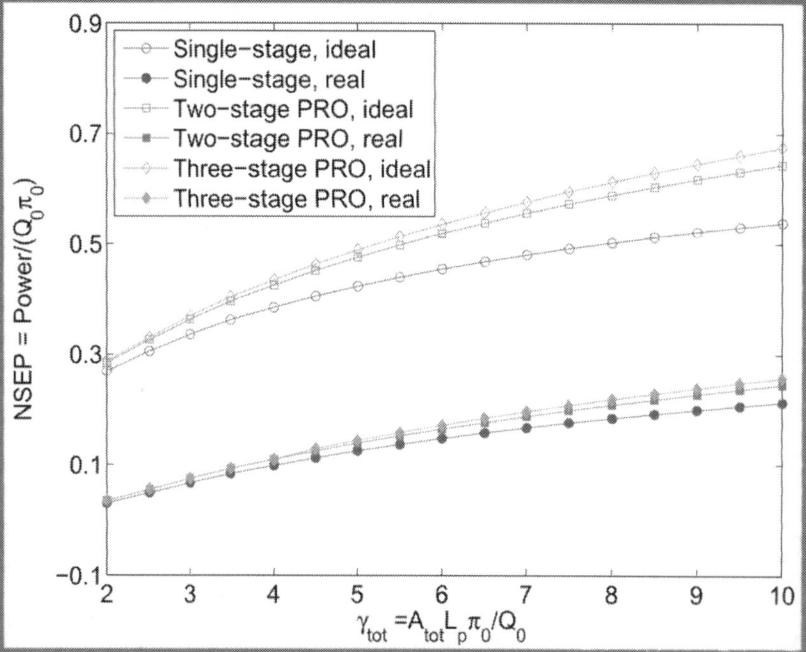

Conclusions

Conclusions

- NSEP increases with number of stages and $\gamma_{tot} = A_{tot} L_p \pi_0 / Q_0$.
- Multistage not only improves volume gain ratio, but also allows a stage-dependent, progressively decreasing pressure, both enhancing power production.
- At the thermodynamic limit where γ_{tot} goes to infinity, the theoretical maximum NSEP by an N-stage PRO system is $N\left(1 - q_{tot}^{-1/N}\right)$. For one stage, NSEP is no more than 1.
- For infinite number of stages, the NSEP at the theoretical limit becomes $\ln q_{tot}$, which is equivalent to that of a fully reversible PRO.
- Process efficiencies have a significant effect on NSEP.

Acknowledgment and References

- Acknowledgement
  - American Chemical Society Petroleum Research Fund (No. 55347-UR9)

- References

  - Li, M. "Systematic Analysis and Optimization of Power Generation in Pressure Retarded Osmosis: Effect of Multi-Stage Design," *AIChE Journal*, in press, 2017.
  - Li, M. "Reducing Energy Consumption of Seawater Desalination: Staged RO or RO-PRO? " *Desalination*, **422**, 124-133, 2017.
  - Li, M. "Analysis and Optimization of Pressure Retarded Osmosis for Power Generation, " *AIChE Journal*, **61**, 1233-1241, 2015.

499267 Removal and Prevention of Biofouling in Forward Osmosis Membrane Bioreactors

Abstract:

With increasing populations and declining clean water supplies, treatment of municipal wastewater for reuse is becoming more viable. Membrane treatment processes will be an integral part in reuse treatment, however membrane fouling results in decreased process efficiencies. Forward osmosis (FO) is a membrane-based separation process that uses the osmotic pressure resulting from a salt gradient rather than a hydraulic pressure as the driving force, resulting in energy savings over traditional membrane treatment processes. The objectives of this project were to 1) design and build a submerged FO membrane bioreactor (MBR) with three independent membrane loops, and 2) to evaluate alternative membrane cleaning mechanisms to improve and maintain flux performance.

As part of the first objective, the hardware and programming logic code for the FO MBR system were developed and integrated to create a system that was able to measure and compare membrane flux over time for three independent FO membrane cassettes where all three were exposed to the same biological fouling conditions. The MBR contained *E. coli* JW3818 Î"*rfaH* grown in LB broth with 50 mg/L kanamyacin, and half strength LB with kanamycin was used to replace volume lost from the MBR through the FO membrane. As part of the second objective, various thermal osmotic backwashing methods were evaluated for flux recovery and propensity for re-fouling. It was found that osmotic backwashing at 70 Â° C for 30 min resulted in 15% greater flux recovery compared to the baseline case of 60 min at ambient temperature. During this process, the hot water removes the adhered biofilm allowing for increased flux recovery. It is hypothesized that the elevated temperature of the thermal backwash results in cell death at the membrane surface, thereby enhancing the osmotic backwashing efficiency. Confocal Scanning Laser Microscopy (CSLM) confirmed that less biofilm was present after thermal backwashing than with ambient backwashing. Based on these preliminary results, hot osmotic backwashing is a promising membrane cleaning technique to remove and decrease reoccurrence rate of biofouling in FO MBR operations.

502721 Understanding the Performance of Forward Osmosis Hollow Fiber Membranes at the Module Scale

Abstract:

This study evaluates an approach to make thin film composite (TFC) hollow fiber membranes for forward osmosis (FO) by simply employing commercial ultrafiltration (UF) membranes as support material. A thin polyamide film with excellent selective properties was synthesized on the inner surface of hollow fibers via interfacial polymerization. Besides demonstrating good performance as TFC FO membrane supports, the commercial hollow fiber UF membranes also provide a systematic platform with consistent properties to study structure-performance relationship. A series of commercial UF membranes were used to evaluate how support layer surface pore size (MWCO) impacted the properties of the selective layer and overall performance of the TFC membranes.

Aside from using commercial hollow fibers as FO membrane supports at bench scale, we also demonstrate that the TFC hollow fiber FO membranes can be made on existing modules at pilot scale. A series of commercial modules with different fiber size were used to make TFC hollow fiber modules. The resultant modules were evaluated under various operating conditions. While we evaluate how performance was impacted by module operating conditions, overall the modules demonstrate impressive FO results.

With the availability of reproducible membranes and modules, we developed a computational fluid dynamics (CFD) model with COMSOL Multiphysics to optimize module design for FO application. The FO process in a hollow fiber membrane was simulated in a 2D axis-symmetry geometry and described by fluid dynamics coupled with mass transfer. We verified the model accuracy by constructing modules with the same dimensions and fibers and testing under pertinent conditions. The agreement between the simulation and experimental results provided insight into how independent variables would impact module performance. To allow our colleagues to take the advantage of this model, we also built it into an application and will release it for the community to use for free.

497043 Understanding the Impact of Membrane Properties and Transport Phenomena on the Energy Efficiency of a Membrane Distillation Desalination System

Abstract:

Direct contact membrane distillation (DCMD) is a promising thermal desalination process that can treat high salinity waters using low-grade heat. DCMD has several advantages from the perspective of the water-energy nexus. These include the utilization of waste heat (below 100ºC), perfect rejection of nonvolatile solutes, low areal footprint, and high scalability. However, currently, the energy efficiency of DCMD is relatively low compared to other work-based and thermal desalination processes. In this study, we aim to quantify how membrane properties and process conditions affect the exergy or second-law efficiency (η_{II}) of a DCMD desalination system with external heat recovery. We begin by developing a fundamental understanding of mass and heat transport through an MD membrane, before modeling how membrane properties and transport phenomena affect entropic losses in an MD system. In particular, we analyze how the membrane permeability coefficient (B) and thermal conduction coefficient (K) impact MD performance. We show that increasing the B value of a membrane by reducing its thickness, initially leads to an increase in η_{II} before conductive heat loss through the membrane causes η_{II} to fall. For a typical MD membrane with a porosity of 0.90, material thermal conductivity of 0.20 W m^{-1} K^{-1}, and a nominal pore diameter of 0.6 µm, we find that the optimal permeability coefficient is 1.59×10^{-6} kg m^{-2} s^{-1} Pa^{-1} (572 kg m^{-2} h^{-1} bar^{-1}). This value corresponds to an optimal membrane thickness of around 95 µm. Our analysis stresses the importance of effective heat recovery in DCMD. We show that an external heat exchanger with a minimum approach temperature of 10ºC reduces energy consumption by 63%. Finally, we demonstrate that increasing the ratio B/K, rather than just the B value, is key to increasing the exergy efficiency of DCMD desalination. For example, increasing membrane porosity from 0.70 to 0.90, which yields a 160% increase in B/K, leads to a 42% increase in η_{II} from 5.3% to 7.6%. The advantages of reducing the bulk pressure (P) in the membrane pores are also explored. For a typical membrane, halving P from 1.0 bar to 0.5 bar, results in a 21% increase in η_{II} from 7.0% to 9.2%. We conclude by identifying that the maximum exergy efficiency achievable as membrane porosity tends to unity is 10% for a bulk membrane pressure of 1.0 bar and 12% for a bulk membrane pressure of 0.5 bar, given perfect heat recovery.

498547 Development of a Multi-Stage Membrane Distillation - Electrocoagulation Process for Treatment of Hydraulic Fracturing Flow Back Waters

Abstract:

There is an urgent need to develop innovative and more effective technologies for treatment of wastewaters. Many of these wastewaters such as hydraulic fracturing flow back waters, contain very high dissolved solids (TDS). Treatment of hydraulic fracturing flowback waters can be very challenging as they often contain TDS in excess of 200,000 ppm as well as dissolved surfactants and small organic compounds. Pressure driven membrane processes such as reverse osmosis are impractical for treating very high salinity wastewaters due to the high osmotic back pressure that must be overcome. Membrane distillation has been proposed as a new unit operation for treatment of very high TDS wastewaters. One of the advantages of membrane distillation is the fact that low grade waste heat may be used.

Here we have developed a multi-staged membrane distillation system for maximizing water recovery from actual hydraulic fracturing flowback waters. The multi-stage system has been designed to different on-site applications for wastewaters with specific TDS levels. We show that ideal membrane properties vary as the TDS of the fee increases above 300,000. At lower TDS values membranes should be selected that maximize flux. However at TDS values above 300,000, membrane surface properties such as roughness becomes important. At theses higher TDS values leakage of the feed through the membrane pores and into the permeate side is a major concern. In addition, appropriate pretreatment of the feed to remove dissolved organic compounds is essential prior to membrane distillation. Here we have developed an efficient electrocoagulation process for pretreatment of the hydraulic fracturing flow back waters.

Here different membranes have been used in the different stages and performance of the system in terms of water recovery potential and specific energy consumption has been evaluated. Significant concentration of the feed has been conducted and over 60% water recovery has been achieved for actual hydraulic fracturing flowback waters from Marcellus shale containing over 100,000 ppm TDS.

490626 Permeability and Selectivity Limits of Biomimetic Desalination Membranes

Abstract:

Membrane-based desalination is increasingly applied to alleviate water scarcity through the purification of nontraditional water sources, such as seawater, brackish groundwater, and municipal wastewater. Due to fundamental material limitations, performance improvements for the industry standard thin-film composite (TFC) membranes have been incremental. As such, there has been highly active research in exploiting biological water channels, such as the membrane protein aquaporin, and designing synthetic water channels to produce â€œnext-generationâ€ desalination membranes. Most of these design strategies would incorporate the channels within an amphiphilic lipid or block copolymer bilayer to form the membrane selective layer. Based on published single-channel water permeability measurements, 96-98% the membrane surface area would likely comprise just the bilayer to achieve adequate water permeability. As such, the intrinsic permeability characteristics of the bilayer will play a crucial role in determining the overall water/solute permselectivity of the resulting membrane. In this study, solution-based analytical methods assessing large unilamellar vesicles are used to measure the permeability of water and several model solutes through lipid and block copolymer bilayers . Results are combined with published single-channel permeabilities to yield the permeability and selectivity limits of defect-free biomimetic desalination membranes. Comparison with the performance of a commercial TFC membrane shows that biomimetic desalination membranes may be advantageous for some desalination applications, but disadvantageous for others.

495606 Dipole Oriented Water Wires Confined in Artificial Chiral Membrane Channels

Abstract:

Aquaporins (AQPs) are known for their highly selective water transport through cell membranes. Oriented water-wires spanning the AQPs pore is of considerable importance for the selective translocation of water over ions. We recently discovered that water translocation through artificial I-quartets (self-assembled imidazole structures) increases when water molecules display dipolar orientation. Here, we show that water confined inside self-assembled I-quartets integrated into supported lipid bilayers (SLB) form oriented water-wires exhibiting a chiral super-structure, as visualized with static solid state X-ray single-crystal structures and sum-frequency generation (SFG) spectroscopy. The integration and assembly of I-quartets into SLB were also monitored with a Quartz Crystal Microbalance with Dissipation (QCM-D) and modeled with molecular simulations in order to provide physical insight into dipolar water orientation within artificial I-quartet channels. This study provides the first experimental proof of dipole orientated water wires within artificial water channels inserted in bilayer membranes and the characterization of their hydrogen-bond connectivity and chiral alignment. Uncovering the interplay between the hydrogen-bonded water structure and water transport through the self-assembled i-quartets is critical to understanding the function of natural membrane channels and will guide the design principles for developing artificial water channels for water purification.

487911 Bioinspired One-Step Co-Deposition PVDF Membrane Toward Multifunctional Applications: Oil/Water Emulsions Separation and Soluble Contaminants Adsorption

Abstract:

Bioinspired One-Step Co-Deposition PVDF Membrane toward Multifunctional Applications: Oil/Water Emulsions Separation and Soluble Contaminants Adsorption

Guangfa Zhang, Jingxian Jiang, Qinghua Zhang, Xiaoli Zhan, Fengqiu Chen

College of Chemical and Biological Engineering, Zhejiang University, Hangzhou 310027, P. R. China

ABSTRACT:

Bioinspired materials with special wettability towards water and oils have been proven very efficient in removal of insoluble organic pollutants in water. However, these materials cannot remove water-soluble organic contaminants (e.g., dyes and heavy metal ions), and often suffer some problems such as easy contamination by oils, low water tolerance and complicated preparation methods. Herein, we present a simple approach for the fabrication of fish scale and mussel-inspired underwater superoleophobic PVDF membranes via co-deposition of polydopamine (PDA) and chitosan mediated by tyrosinase. The facile modification procedure makes the membranes underwater superoleophobic with high contact angles of 160.1°and ultra-low adhesion force for oils. The as-prepared membranes not only can separate surfactant-stabilized oil/water emulsions with high fluxes, separation efficiency, and excellent antifouling properties, but also can effectively absorb heavy metal ions and dyes (copper ion and methyl blue) from water with very high removal efficiencies in a short time by simply passing through these membranes. Therefore, the simple and facile method has excellent potential in practical multifunctional wastewater treatment applications.

KEYWORDS: bioinspired PVDF membranes, underwater superoleophobic, oil/water emulsion separation, contaminant adsorption

491692 Synthesis of High-Performance Biologically-Inspired Nanofiltration Membranes for Water Treatment

Abstract:

Aquaporins are membrane protein water channels present in cells, and they restrict the passage of contaminants and small molecules such as urea and boric acid, without preventing the passage of water. Biomimetic membranes, an innovation in water filtration technology attempts to replicate this natural process of highly selective and efficient transport of different molecules across a semi-permeable membrane occurring at the cell level. Therefore, aquaporins have received worldwide attention because of their potential to form biomimetic membranes with high flux and selectivity for water reuse. However, challenges involved in the incorporation of aquaporin proteins in membranes limit their applicability. One of them is to attach aquaporins to the membranes without chemically altering or damaging the aquaporins during the binding to the membrane. The second challenge is to design and prepare an assembly that allows biomimetic membranes with aquaporins to sustain hydraulic water pressure gradients without losing their integrity and performance. The overarching objective of this project is to form a biomimetic membrane made of unaltered aquaporins dispersed in a polymeric membrane selective layer and capable of operation under high hydraulic pressure. Polybenzimidazole (PBI) membranes were surface modified with treated aquaporins in order to achieve higher water flux and selectivity. Membranes modified with aquaporins displayed lower flux declines and higher flux recovery values after backwash as compared to unmodified PBI membranes. Also, modified membranes showed improved rejection values for both protein and salt solutions of different concentrations. However, there was leakage of ions between the channels. Therefore, in order to improve the rejection of protons, ions and other impurities, aquaporin channels were aligned with the direction of water flow. Functional groups were installed on AqpZ for covalent attachment to the polymer matrix so that the proteins could be immobilized to the membranes and aligned in the direction of the flow.

495240 Artificial Water Channelsâ€•Bioinspired and Energy-Efficient Filtration Materials

Abstract:

Artificial water channels¯Dbioinspired and energy-efficient filtration materials

Yue-xiao Shen and Manish Kumar*

Department of Chemical Engineering, The Pennsylvania State University, University Park, PA 16802

Water purification is emerging as an important challenge in the 21st century,[1] globally, and even in our own backyard with accidents of unsafe and scarce drinking water in Milwaukee, Flint and California in recent years. Membrane-based technologies that have been extensively used to produce fresh water from seawater and to purify microbiologically and chemically contaminated water are energy intensive. Nature provides excellent examples for energy-efficient desalination and water filtration. Mangrove trees purify saline water through its root systems with minimal energy input (Fig. A). In cell membranes, including those in the mangrove roots (Fig. B), biological water channel proteins aquaporins (AQPs) (Fig. C) conduct single channel water transport while excluding all other molecules. This mechanism has inspired us to study on the design of artificial structures that mimic AQPs and led to the exciting development of biomimetic membranes for energy-efficient desalination using these structures.[2]

Here we report that peptide-appended pillar[5]arene (PAP) artificial water channels (Fig. D) combine the advantages of biological water channels Aquaporins (AQPs) and carbon nanotubes (CNTs).[3] These channels were functionally inserted into stable poly(butadiene)-*b*-poly(ethylene oxide) (PB-PEO) amphiphilic block copolymers.[4] We showed that the water conductance of PAP channels in BCPs was in the range of AQPs¡¯ (~10^8-10^9 water molecules/s) (Fig. E). They could also be vertically aligned in 2 dimensional block copolymers membranes (Fig. F), with much higher packing density (~2´10^5/¦Ìm^2) than current synthetic CNTs-based membranes (0.1~2.5´10^3/¦Ìm^2). The channel based 2D materials were then assembled into a dense active layer on track-etched polycarbonate membranes via layer-by-layer deposition. This separating layer was further infiltrated with crosslinking reagents to enhance its integrity (Fig. G). The resulting highly packed artificial channel based composite membranes could be promising energy-efficient separation materials for the future, and showed high water permeability while maintaining selectivity.

Figure. Bioinspired artificial water channels-based membranes are designed for energy-efficient desalination. (A-B) Mangrove lives in the sea and rely on its root system to purify sea water. (C) Biological water channel proteins aquaporins (AQPs) in the cell membranes of the mangrove root system have high water transport rate while rejecting any other solutes and ions. (D) Peptide-appended pillar[5]arene (PAP) artificial water channels mimic the structure of

AQPs. (E) The water conductance of PAP channels are close to that of AQPs and carbon nanotubes (CNTs). (F) PAP channels can form densely packed polymer membranes. (G) Schematic representation of the alternate multilayers artificial channel based 2D materials and polyethylenimine deposited on track-etched membranes by layer-by-layer self-assembly.

References

1 Shannon, M. A. *et al.* Science and technology for water purification in the coming decades. *Nature* **452**, 301-310, (2008).

2 Shen, Y.-x. *et al.* Biomimetic membranes: A review. *J. Membr. Sci.* **454**, 359-381, (2014).

3 Shen, Y.-x. *et al.* Highly permeable artificial water channels that can self-assemble into two-dimensional arrays. *Proc. Natl. Acad. Sci. U.S.A.* **112**, 9810-9815, (2015).

4 Shen, Y.-x. *et al.* Artificial water channels self-assemble into nano-scale domains in block copolymers and retain high permeability. *Submitted to Adv. Mater.*

488879 the Development of Hydrophobic Deep Eutectic Solvents As New Extracting Agents for Furfural and 5-Hydroxymethylfurfural from Aqueous Solutions

Abstract:

The development of hydrophobic deep eutectic solvents as new extracting agents for

furfural and 5-hydroxymethylfurfural

from aqueous solutions

Carin H.J.T. Dietz[a], Maaike C. Kroon[b], Martin van Sint Annaland[a], Fausto Gallucci[a]

[a] Chemical Process Intensification, Dept. Chemical Engineering and Chemistry, Eindhoven University of Technology, PO Box 513, Eindhoven, Netherlands

[b] The Petroleum Institute, Chemical Engineering Department, P.O. Box 2533, Abu Dhabi, United Arab Emirates

A deep eutectic solvent (DES) is a mixture of at least one hydrogen bond donor (HBD) and at least one hydrogen bond acceptor (HBA) with a much lower melting point than its constituents.[1] It is often reported that DESs provide the same advantages as ionic liquids (ILs), with the advantage that the DESs are easier and cheaper to prepare (mixing at moderate temperatures).[2] As DESs are formed because of hydrogen bonding interactions, almost all DESs are hydrophilic (fully miscible with water).[1] The first hydrophobic (water-immiscible) DESs were discovered in 2015, which were a combination of decanoic acid and quaternary ammonium salts.[3]

In this study, new hydrophobic DESs were produced, characterized and screened for the extraction capability of furfural (FF) and hydroxymethylfurfural (HMF) from aqueous solutions.. First, the thermophysical properties (density, viscosity, glass transition temperature, decomposition temperature) of these DESs were measured. Thereafter, the solubilities of various sugar-derived molecules were experimentally assessed. Finally, the extraction performance for FF and HMF from aqueous solutions was determined and compared to commonly used organic extractants. Detailed results will be presented and discussed at the conference.

[1] A. P. Abbot, G. Capper, D. L. Davies, R. K Rasheed, V. Tambyrajah â€œNovel solvent properties of choline

chloride/urea mixturesâ€ Chemical Communication, 2003, 70-71.

[2] M. Francisco, A. van den Bruinhorst, M.C. Kroon, Low-Transition-Temperature Mixtures (LTTMs): A New Generation of Designer Solvents, Angewandte Chemie International Edition, 2012, 52, 3074-3085.

[3] D. J. G. P. van Osch, L. F. Zubeir, A. van den Bruinhorst, M. A. A. Rocha, M. C. Kroon, Hydrophobic deep eutectic solvents as water-immiscible extractants, Green Chemistry, 2015, 17, 4518-4521.

Acknowledgements

This work was performed under the framework of InSciTe-Horizontal project and with contributions from the European Regional Development Fund (ERDF) within the framework of OP-Zuid and with contributions from the province of Brabant and Limburg and the Dutch ministry of economy.

490432 Multiple Morphology Control of Biomacromolecule Crystallization Via Hydrogel Composite Membrane Based Platform

Abstract:

Multiple morphology control of biomacromolecule crystallization via hydrogel composite membrane based platform

Lin Wang, Gaohong He, Xiaobin Jiang*

State Key Laboratory of Fine Chemicals, Research and Development Center of Membrane Science and Technology, School of Chemical Engineering, Dalian University of Technology, Dalian, P. R. China

*Corresponding author: Email: xbjiang@dlut.edu.cn

Abstract

Membrane crystallization (MCr) is an innovative technology in chemical engineering and membrane science, which has been widely applied in seawater desalination, wastewater treatment and pharmaceutical production, due to its advantages on simple equipment, convenient operation, low energy consumption,etc [1,2,3]. Membrane in MCr process not only act as a selective barrier, but also act as the interface to accelerating the heterogeneous nucleation. It is significant for controllable biomacromolecules crystal growth [4,5]. While, the modification and optimization of the membrane interface is still a challenging and promising research. In recent years, some research indicate that hydrogel composite membranes (HCMs) have a great influence on promoting the biomacromolecules crystallization [6,7].

In this study, the HCMs with fingerprint-like interface structure (fabricated by functional monomer and crosslinker through UV initiation) is prepared and applied in lysozyme MCr process. The HCMs prepared by this method possess abundant biomimetic-structure on the surface and the cross-section under different water content and crosslinking degree (shown in Fig. 1), which provide the potential regulable ability to the heterogeneous nucleation acceleration and crystal growth control.

As a results, compared with original PP membrane, the HCMs have a remarkable role on regulating the proportion of the various crystal shapes and promoting the new crystal shapes under different pH value of buffer solution. The hexagon and square shape crystal only observed in HCMs-involved process. The proportion of hexagon and square crystal are 16.0 %, 13.6 % in the pH=3.6 buffer solution (shown in Fig. 1). In addition, in the pH=4.6 buffer solution, no rodlike or long hexagon crystals are observed, the obtained crystals in HCMs platform are all hexagon and square shape, which are account for 22.3 %, 77.7 % (shown in Fig. 2), respectively.

Fig. 1 Comparison of the structure of original PP membrane and HCMs, crystal morphology control results in the pH=3.6 buffer solution.

Fig. 2 Crystal morphology control results of original PP membrane and HCMs in the pH=4.6 buffer solution.

Acknowledgment

We acknowledge financial contribution from from National Natural Science Foundation of China (Grant No. 21527812, 21676043, U1663223), Changjiang Scholars Program (T2012049), the Fundamental Research Funds for the Central Universities (DUT16TD19) and Education Department of the Liaoning Province of China (No. LT2015007).

References

[1] Li, W., Van der Bruggen, B., Luis, P. Integration of reverse osmosis and membrane crystallization for sodium sulphate recovery. *Chem. Eng. Process.: Process Intensification.* 85, 57-68 (2014).

[2] Ye, W., Lin, J., Shen, J., Luis, P., Van der Bruggen, B. Membrane crystallization of sodium carbonate for carbon dioxide recovery: effect of impurities on the crystal morphology. *Cryst. Growth Des.* 13,(6), 2362-2372 (2013).

[3] Meng, S., Ye, Y., Mansouri, J., Chen, V. Crystallization behavior of salts during membrane distillation with hydrophobic and superhydrophobic capillary membranes. *J. Membr. Sci.* 473, 165-176 (2015).

[4] Drioli, E., Di Profio, G., Curcio, E.: Progress in membrane crystallization, *Curr. Opin. Chem. Eng.* 1,(2), 178-182 (2012).

[5] Curcio, E., Di Profio, G., Drioli, E.: Membrane crystallization of macromolecular solutions. *Desalination* 145(1), 173-177 (2002).

[6] Di Profio, G., Polino, M., Nicoletta, F. P., Belviso, B. D., Caliandro, R., Fontananova, E., Filpo, G. D., Curcio, E., Drioli, E.: Tailored hydrogel membranes for efficient protein crystallization. *Adv. Funct. Mater.* 24,(11), 1582-1590 (2014).

[7] Salehi, S. M., Di Profio, G., Fontananova, E., Nicoletta, F. P., Curcio, E., De Filpo, G.: Membrane distillation by novel hydrogel composite membranes. *J. Membr. Sci.* 504, 220-229 (2016).

490957 Bio-Inspired Optimization of Nanochannel Geometry and Surface Chemistry to Improve Water Permeability of Track-Etched Membrane

Abstract:

The high water permeability of aquaporin stems from two main factors: its hourglass-shape channel structure[1], which reduces the resistance to water flow, and the narrow center inside the channel with a hydrophobic surface[2]. The selective permeation of water is further aided by the charge distribution over the channel and the resulting electrostatic interactions. Inspired by aquaporin, we optimize the nanochannel geometry and surface chemistry to improve water permeability of track-etched membranes.

Due to their straight channels, tunable channel size and shape, narrow pore size distribution and modifiable surface, track-etched membranes[3] are not only a common commercial membrane, but also a good model membrane for fundamental membrane studies and analytical separations.

Here, we tune the columnar channels on track-etched membranes to a dumbbell-shape or hourglass-shape by controlling etching time, etching temperature and etchant concentration[4]. We also modify the membrane with a dense hydrophobic layer via chemical grafting or chemical adsorption[5,6]. Both of the two optimizations on geometry and chemistry are investigated in filtration experiments to see to which degree the two factors affect the water permeability. Simulations of water flow through nanochannels are also carried out, to interpret the experimental studies, and for optimization.

REFERENCES

[1] Sui, H.; Han, B.-G.; Lee, J. K.; Walian, P.; Jap, B. K., Nature 2001, 414 (6866), 872-878.

[2] Gravelle, S.; Joly, L.; Detcheverry, F.; Ybert, C.; Cottin-Bizonne, C.; Bocquet, L., Proc. Natl. Acad. Sci. U. S. A. 2013, 110 (41), 16367-16372.

[3] Hou, X.; Zhang, H.; Jiang, L., Angew. Chem. Int. Ed. 2012, 51 (22), 5296-5307.

[4] Orelovich, O. L.; Sartowska, B. A.; Presz, A.; Apel, P. Y., J. Microsc. 2010, 237 (3), 404-406.

[5] Tian, Y.; Hou, X.; Jiang, L., J. Electroanal. Chem. 2011, 656 (1-2), 231-236.

[6] Velleman, L.; Losic, D.; Shapter, J. G., J. Membr. Sci. 2012, 411-412, 211-218.

497408 Adsorption and Diffusion Phenomena in Crystal Size Engineered Zif-8 MOF

Abstract:

ZIF-8 is a flexible zeolitic imidazolate framework (ZIF), member of the metal–organic framework (MOF) porous materials, whose narrow pore structure swings open by reorientation of imidazolate linkers and expands at high loading of guest molecules. This nanoporous material has been identified as a potential candidate for the energy-efficient adsorptive recovery of alcohols from diluted aqueous mixtures, such as obtained with the acetone-butanol-ethanol (ABE) fermentation [1].

In this work, the crystal size dependency of both structural transitions induced by N2 and Ar adsorption was investigated, from which the results, obtained with a volumetric technique, will be presented. On the other hand, the diffusion of n-butanol into well-engineered ZIF-8 crystals with identical surface area and micropore volume was also explored. Results acquired with two gravimetric methods (static and dynamic versions) will be shown. It was found that the crystal downsizing of nanoporous ZIF-8 adsorbent affects its structural flexibility observed from the equilibrium adsorption and desorption data of N2 and Ar. Adsorption kinetics of n-butanol in ZIF-8 showed to be strongly influenced by the crystal size, however, not according to a classical intracrystalline diffusion mechanism. Our results suggest that the structural transitions and uptake rates are dominated by crystal surface effects. Crystal downsizing increases the importance of such surface barriers [2].

References

[1] J. Cousin Saint Remi, T. Remy, V. Van Hunskerken, S. van de Perre, T. Duerinck, M. Maes, D. De Vos, E. Gobechiya, C.E.A. Kirschhock, G.V. Baron, J.F.M. Denayer: *Biobutanol Separation with the Metal–Organic Framework ZIF-8* ChemSusChem **4**, 1074–1077 (2011).

[2] S. Tanaka, K. Fujita, Y. Miyake, M. Miyamoto, Y. Hasegawa, T. Makino, S. Van der Perre, J. Cousin Saint Remi, T. Van Assche, G.V. Baron, J.F.M. Denayer: *Adsorption and Diffusion Phenomena in Crystal Size Engineered ZIF-8 MOF* J. Phys. Chem. C **119**, 28430–28439 (2015).

499581 Development of Concentration-Swing Frequency Response Method for Mass Transfer Studies Â

Abstract:

The design and performance of separation and catalytic processes are highly dependent upon the understanding of the mass transfer inside the adsorbent and catalyst particles. Therefore, the fundamental knowledge of controlling mechanisms and rates is important for process design. Compared to other macroscopic techniques, frequency response (FR) has unique capability to determine dominating resistances by inducing parameter oscillations at various frequencies to help decoupling mass transfer resistances.

A new flow-through FR system, concentration-swing FR (CSFR), has been developed to study mass transfer at an increased range of temperatures and pressures for vapor and gas systems. This CSFR system is housed in an environmental chamber to allow temperature control from -60 °C to 150 °C, and the sample has an additional heat element to allow in-situ regeneration up to 350 °C. The adsorption bed is designed to have flexibility to accommodate various amount samples, and capable of using ~mg samples for zero length column packing. The fast perturbation frequency can be detected up to 1 Hz to enable studies of relatively fast kinetics.

In this work we will present some examples of applying this CSFR technique to study mass transfer of water (vapor) and CO_2 (gas) in zeolite LTA structure.

500626 Measurement of Mass Transfer Parameters for Rapid Pressure Swing Adsorption

Abstract:

Large zeolite particles typically operate under the assumption that the overall mass transfer rate into the particle is controlled by intraparticle diffusion. Since the diffusion path inside the particle is large, it is typically a reasonable estimation for process simulators. While this same assumption is often applied to simulations of rapid pressure swing adsorption (RPSA) processes using much smaller particles (<1mm), it has yet to be determined if it remains a suitable approximation. Since the diffusion path inside the particle has been significantly reduced, assuming intraparticle diffusion controls the mass transfer rate into the particle may no longer be valid. Furthermore, some literature correlations suggest that for small particles, axial dispersion effects significantly increase. The goal of this study is to better understand the significance of axial dispersion for columns of small particles in order to better estimate its effect on RPSA processes for simulation purposes. It is also the aim to evaluate the effectiveness of the mass transfer zone (MTZ) measurement technique for determining mass transfer parmaters relevant to RPSA simulations.

The effect of axial dispersion on MTZ spreading is evaluated through traditional breakthrough experiments. The MTZ length is measured on a column packed with a commercial LiLSX zeolite (~0.5 mm in diameter). Previously we conducted the same experiments using a N_2-He mixture, however, for this study a N_2-O_2 gas mixture was used because of its applicability for air separation. The results from these experiments suggest axial dispersion effects are much more significant than predicted through typical correlations used for larger particles. Mass transfer parameters found from the results are used with an in-house PSA simulator to determine how critical it is to account for the higher level of axial dispersion with small particles. Consequences of not properly estimating the effects of axial dispersion will also be discussed.

495464 Kinetic Measurements on Hierarchical Zeolites Using the Zero Length Column

Abstract:

The Zero Length Column (ZLC) technique, first introduced for diffusion measurements in zeolites [1], has been under continuous development in the past three decades proving to be an essential tool for the investigation of kinetic and equilibrium properties of novel adsorbents. The very small size of the column offers several key advantages compared to other experimental methods. Foremost, the small sample quantity required greatly reduces the synthesis costs of the materials and allows rapid testing of prototype materials. In additionheat resistances and external mass transfer resistances can be neglected and axial dispersion is maximised and as a result, the dynamic response of the ZLC corresponds to that of a perfectly mixed isothermal cell.

Here we present the use of the ZLC for the study of the adsorption kinetics of three hierarchical and the parent Y zeolite. We will show how the novel experimental approach, based on the choice of the most appropriate probe molecule, mesitylene, and the combination of full and partial loading experiments, allows to measure accurately the kinetics of the different zeolites using less than 1 mg of sample. Results show how different synthesis routes lead to different prevailing mass transport mechanisms. The use of the partial loading experiment is used to characterise the combined diffusion and surface resistance mechanisms as well as demonstrate if the system follows a pure diffusion model.

Of the samples tested, one is shown to result in adsorption rates that are four times faster than the parent sample, while one is shown to give slower mass transport kinetics, indicating that simply producing hierarchical materials does not necessarily result in improvements in the mass transport properties and that combining the synthesys process with screening using the ZLC technique can be very effective.

[1] Eic M. and Ruthven D.M. A New Experimental Technique for Measurement of Intracrystalline Diffusivity (1988), 8, 40-45

501735 A Parametric Study of the Adsorption/Desorption Steps for an Adsorptive Reactor (AR) Intensifying the Water Gas Shift (WGS) Reaction

Abstract:

Integration of two operations (reaction and separation) in a single unit is a common intensification step for a variety of processes. Improvement metrics include increased efficiency, lower operating capital costs, energy savings, reduced by-product formation and waste generation. In this work, an Adsorptive Reactor (AR) model is developed. The velocity, temperature and species concentration profiles along the reactorsâ€™ lengths are captured by momentum/species transport models accounting for convection/reaction /diffusion/conduction mechanisms. The modelâ€™s equations are solved using Finite Element Method (FEM).

Â The model is then repeatedly simulated to intensify the water gas shift (WGS) reaction process for hydrogen production with in situ CO_2 capture. AR performance is carried out under dynamic conditions for various operating parameters such as W_{cat}/F_{CO}, H_2O/CO ratio, $W_{cat}/W_{adsorbent}$ and operating pressure and temperature. Moreover, an adsorption/desorption cycling study is carried out to determine process start-up times, the effect of an incompletely desorbed bed on CO_2 and CO breakthrough times on the adsorption/desorption profiles. Complete desorption is CO_2 not required, since there is a trade-off between the actual working capacity of the adsorbent bed, and the adsorbent-agent/CO_2 ratio and operating time

497200 Quantifying Sorption and Diffusion in Polymeric and Non-Polymeric Materials: Experimental Methods and High Fidelity Modeling

Abstract:

Quantifying the sorption and diffusion of vapors and gases in diverse types of materials is important to several industries and applications. Key parameters include the rate of diffusion, the sorption capacity, and the sorption mechanisms. Understanding these phenomena and capturing them in a model allow for predictions of material performance in realistic shapes and dimensions over the life of the system. Here we have developed a modeling and experimental capability to characterize and quantify sorption and diffusion of vapors and gases in materials. Our smallest scale experiments include dynamic vapor sorption and dynamic gas sorption as measured via gravimetric analysis of milligram to gram sized samples. By incrementally stepping up the vapor concentration or gas partial pressure we can monitor the sample mass uptake as a function of time to parametrize our dynamic sorption-diffusion model. Our initial experiments are performed with quasi-1-dimensional samples and the data is used to parameterize the models.Â Our physics based numerical model is a triple-mode sorption model that is coupled to a Fickian diffusion model; the three sorption modes are Henryâ€™s absorption, Langmuir adsorption, and pooling. With this numerical model, one can then predict the sorption and diffusion in more complex, 3-dimensional parts to accurately model the behavior of real articles and systems. Validation experiments are key to building our confidence in our model both in 1D and 3D applications. By measuring 3D shaped samples in our gravimetric analysis instruments or via 3D outgassing experiments, the model can be validated. This model and accompanying experiments were developed to quantify and predict the vapor and gas sorption and outgassing (i.e. sinks and sources) of multiple materials in sealed systems, yet the experimental approach and physical based numerical model support a wide range of applications including adsorption and separation technology, materials designed as barrier materials (e.g. polymeric protection layer on photovoltaic cells), and predictions of material outgassing in sealed systems over long durations (e.g. space and satellites, weapon systems, etc). Here we will discuss our sorption-diffusion numerical model and the accompanying 1D and 3D experiments and simulations on several materials including polymers and non-polymeric samples (e.g. Sylgard-184, Zircar RS-1200 i.e. an alumina ceramic composite, etc).

This work performed under the auspices of the U.S. Department of Energy by Lawrence Livermore National Laboratory under Contract DE-AC52-07NA27344.

499617 Simulation of Lysozyme Concentration Profiles at Contact Lenses and Effect of Homogenous Diffusion Coefficient

Abstract:

Contact lenses have a wide usage in biomedical applications. Lifetime of contact lenses depend on their oxygen permeability and the accumulation of the teardrops components, e.g. proteins, lipids. Especially the proteins adsorbed on the contact lenses may increase the vision defects and risk for serious eye-disorders. Therefore, the information on the amount and conditions of the accumulation of proteins in teardrops is important.

In the present study, the concentration profiles of lysozyme -the protein having the highest concentration in teardrops- in a contact lens during its adsorption and desorption was investigated. The simulation of the protein accumulation was carried out using a new method based on general rate model approach. According to the model, concentration of lysozyme in the bulk liquid phase; liquid and solid concentrations at the interface; and solid phase concentration were individually defined and the equilibrium between these concentrations were described according to the circumstances. The algorithms needed for the concentration profiles on the derived mathematical expressions have been prepared within the study.

In experimental studies, Etafilcon A (Johnson&Johnson) contact lenses were used. Lysozyme was dissolved in phosphate buffered saline (PBS) solution to simulate the human teardrop for the adsorption of the protein on the lenses. In desorption studies, protein-free PBS solution was used as the cleaning solution and the accumulation of lysozyme was observed. Using the proposed model, the information on the period the lenses can be used during the day and that needs to be kept in the cleaning solution were obtained. Additionally, the model enables the calculation of the homogenous diffusion coefficient of lysozyme in the contact lens.

489555 Thermally and Chemically-Treated Polyacrylonitrile Hollow Fiber Membranes for Organic Solvent Nanofiltration

Abstract:

Polyacrylonitrile (PAN) hollow fiber membranes suitable for organic solvent nanofiltration (OSN) were fabricated. To obtain pristine ultrafiltration-level hollow fibers possessing the desired morphology of a sponge-like cross-section with minimal macrovoids, the dope composition was adjusted while hollow fiber spinning parameters such as the take-up speed were changed. Hydrazine monohydrate was subsequently used for the cross-linking process and the effects of cross-linking duration on the hollow fibersâ€™ separation performances were investigated. X-ray photoelectron spectroscopy was used to provide insight into the possible reactions that might have occurred during the modification process and generally indicated the successful cross-linking of the PAN membranes. Using ethanol and dyes as model feed solutions, the hollow fiber membranes were subjected to OSN tests. With this straightforward and effectual modification, the PAN hollow fiber membranes were able to achieve a pure ethanol permeance of 2.32 L m^{-2} h^{-1} bar^{-1} and a remarkable Remazol Brilliant Blue R (M_w 626.54 g mol^{-1}) rejection of more than 99.9%. Considering the relative simplicity of fabrication, these PAN membranes represent a noteworthy advancement in the use of hollow fiber membranes for OSN with potential applications in larger-scale operations.

502710 3D Printing Polyamide Films: an Additive Approach to Making Thin Film Composite Membranes with Tunable Thickness and Roughness

Abstract:

The thin film composite (TFC) membrane is a commonly used platform for making membranes with the capability to desalinate. TFC membranes employ a thin, highly selective aromatic polyamide layer that is formed *in-situ* onto a porous supporting membrane through a poly-condensation reaction between two monomers: a diamine (typically m-phenylene diamine) and an acid chloride (typically trimesoyl chloride). The reaction occurs at the interface between an organic phase containing the acid chloride and the aqueous phase containing the diamine. Remarkably, in 30 years of using this approach, the RO industry has yet to create a polyamide film onto a supporting membrane with controllable thickness, roughness and independence from the substrate. While the film formation using this approach essentially is "self-controlled" as the film formation further monomer reaction, the reaction itself is simply allowed to proceed. Monomers can be changed or additives included to affect selectivity, but generally features of the film, like thickness and roughness, are uncontrolled. Thickness and roughness impact membrane permeance and fouling propensity, respectively, and controlling these parameters would be a great benefit to the membrane community

This work entails demonstration of a new method for making aromatic polyamide TFC membranes using electro-hydrodynamic spray, or, more simply, electrospray. This *additive* approach involves depositing diamine and trifunctional acid chloride monomers sequentially onto a porous supporting surface where they form an ultra-thin, highly selective, and tunable polyamide layer. We demonstrate the use of electrospray to form polyamide selective layers with *adjustable thickness and roughness* while decoupling the film formation from the supporting substrate structural and chemical properties. We demonstrate the ability to control thickness with a resolution of 20 nm and the control of roughness in between 3.5 nm to 35 nm which is comparable to the other novel technique such as m-LBL technique. Additional benefits are garnered from making the polyamide selective layer formation "support independent" by ensuring that support layer properties do not affect film formation. Using this approach, we have fabricated TFC membranes on PAN50 UF membranes (Nanostone Water) and have demonstrated a 66-97 % and 80-98.7 % rejection for sodium chloride, and magnesium sulfate, respectively, with pure water permeance ranging from 0.1 to 0.75 LMH/bar and we believe that such an approach can yield even higher performance. We noted that polyamide films made through our method were more cross-linked compared to those made through the conventional method which may be a reason for their low permeance. Lastly, an additional benefit to this approach is that it is far more environmentally friendly than conventional interfacial polymerization since it requires very little monomer and solution.

495765 Characterization of Composite Pd/Ta Metallic Membranes for Hydrogen Separation

Abstract:

Hydrogen has been considered as an alternative energy carrier as increasing demands of clean energy processes, which could circumvent environmental problems. Gasifier processes are widely used to mass production of hydrogen, but hydrogen produced from catalytic reactions contains massive amount of impurity components such as CO_2. Separation processes using hydrogen selective metallic membranes are considered a promising method for the production of high-purity hydrogen, and especially, Palladium (Pd) based membranes have received great attention because of its high hydrogen permeability and good selectivity at high temperature. In this study, commercial Tantalum (Ta) tube was treated by electroless plating method and fabricated Pd/Ta composite membranes. Prepared membranes were firstly evaluated their hydrogen permeability at different feed composition, reaction temperature and pressure conditions. The diffusion coefficient and surface reaction parameters as a function of membrane thickness and temperature were also calculated by numerical models. The Pd/Ta composite membranes showed high permeability at between 400 and 500 °C and CO-free hydrogen could be obtained from the mixed gas composition. Also, the performance of the Pd/Ta composite membranes could be improved by controlling the thickness of Pd and Ta layers, and the rate-determining step for hydrogen flux through metallic membrane was suggested by the numerical approaches.

497218 Graphene Oxide Membranes: Synthesis, Behavior in Organic Solvent and Application in Environmental Remediation of Toxic Organic Compounds

Abstract:

Graphene oxide (GO) membrane has been synthesized on commercial polysulfone ultrafiltration membranes (Pore size: 17nm) using the drop casting method followed by baking at 90°c for 24 hours. Baking resulted in the reduction of GO and removal of bulk water intercalated in the GO sheets. Deposited GO film showed higher stability under shear stress variation (measured up to 1.7 N/m^2). Microstructural defects in graphene oxide membranes lead to a significantly higher permeation of isopropanol (for initially dry GO membrane). Extensive size exclusion (rejection) studies with various dye and dendrimer molecules showed decrease in pore size from 3.3 nm in the "dry" state to 1.3 nm in the "wet" state of GO membranes. FT-IR characterization of GO membrane suggested adsorption of water on the nanochannels of the active layer. Catalytic property of graphene oxide and reduced graphene oxide to activate persulfate was also investigated in this study. Partially reduced graphene oxide improved the catalytic efficiency of the process and enhanced the degradation rates of Trichloro ethylene (probe toxic organic compound).

Reference:

A. Aher, Y. Cai, M. Majumder, D. Bhattacharyya, Synthesis of graphene oxide membranes and their behavior in water and isopropanol, Carbon 116 (2017) 145-153.

494180 Pilot-Scale Study of Zeolite-Y/Polyethersulfone Substrate for Composite Membrane Fabrication in CO_2 Separation

Abstract:

A vacuum-assisted deposition setup was developed in pilot scale to continuously deposit zeolite-Y (ZY) nanoparticles on a 14-inch wide nanoporous polyethersulfone (PES) substrate. The deposited substrate was used for composite membrane fabrication for carbon capture from flue gas in coal- and/or natural gas-fired power plants. The effects of operating parameters, including ZY dispersion concentration, pressure differential, and web speed, on the ZY layer thickness and uniformity were studied. A ZY dispersion concentration of 0.08 wt.%, a pressure differential of 3.7 inch Hg, and a web speed of 3 feet/min were identified as the optimal operating conditions for depositing a thin and uniform ZY layer with a thickness of approximately 160 nm. A mathematical model with two adjustable parameters was derived from Darcyâ€™s law to correlate the operating parameters to the ZY layer thickness. The model was validified by the parametric study, and the adjustable parameters were determined by fitting the experimental data. After coating an amine-containing polymer selective layer on top of the ZY layer, the composite membrane showed a CO_2 permeance of 800 GPU (1 GPU = 10^{-6} cm^3 (STP)·cm^{-2}·s^{-1}·$cmHg^{-1}$) with a CO_2/N_2 selectivity over 150 at the typical flue gas temperature of 57 °C, which was superior to the performance of the membrane coated on the bare PES substrate. This improvement was due to the reduced selective layer penetration, resulting from the smaller interparticle pore size on the ZY layer. The scale-up vacuum-assisted deposition process developed in this study is also suitable for the scale-up of the deposition of other fine particles onto a porous polymeric substrate. The developed roll-to-roll fabrication process can be applied to the mass production of the nanoparticle layers, which could potentially reduce the fabrication cost of inorganic membranes in a variety of applications.

493240 Preparation of ZIF-8 Membranes Supported on Polymer Hollow Fibers Using Microwave-Assisted Seeding and Secondary Growth

Abstract:

Preparation of ZIF-8 Membranes supported on Polymer Hollow Fibers using Microwave-assisted Seeding and Secondary Growth

Moon Joo Lee[1], Mohamad Rezi Abdul Hamid[1], Jongmyeong Lee[3], Ju Sung Kim[3], Young Moo Lee[3] and Hae-Kwon Jeong*[1,2]

[1]Artie McFerrin Department of Chemical Engineering and [2]Department of Materials Science and Engineering, Texas A&M University, College Station, TX 77843-3122

[3]Department of Energy Engineering, College of Engineering, Hanyang University, Seoul 133-791, Republic of Korea

* Corresponding author: hjeong7@tamu.edu

Separation of olefin/paraffin mixtures, e.g. propylene/propane, is one of the most challenging separations due to the similarity of their physicochemical properties. Currently, an energy intensive cryo-distillation process is employed. Though energy-efficient membrane-based separation processes are drawing significant interest, no membranes are commercially available for olefin/paraffin separations mainly due to the limitations of current membrane materials and processing methods to fabricate their membranes with large membrane areas.[1-3]

Zeolitic-imidazolate frameworks (ZIFs), a sub-class of metal-organic frameworks (MOFs) with zeolite topology, are highly crystalline nanoporous materials with aperture sizes of typically less than 5 Å. ZIFs composed of metal ions (typically Zn^{2+} and Co^{2+}) and imidazole-derived organic linkers,[4] have been extensively studied for gas separation applications. Specifically, ZIF-8 is shown very promising as a new membrane material for propylene/propane separation because the effective pore size is in between the size of propylene and propane. While several ZIF-8 membranes showed high propylene/propane separation factors, however, there are critical issues for their commercial applications: 1) high cost of membranes and 2) low productivities. To increase the productivity and to reduce the cost of ZIF-8 membranes, polymeric hollow fibers have been utilized as substrates, but only a few of the resulting ZIF-8 membranes showed promising propylene/propane separation performances.[6-9]

Herein, we would like to present a microwave-based method to prepare well inter-grown ZIF-8 membranes on polymeric hollow fibers readily. The method is modified based on our previously reported microwave-assisted seeding and secondary growth technique.[10] Densely-packed nano-sized ZIF-8 seed crystals were rapidly prepared on either inner or outer surface of hollow fibers under microwave heating. Subsequently, seed crystals were secondarily grown into continuous ZIF-8 films on polyimide hollow fibers. The thickness of the resulting ZIF-8 membranes was much smaller than any ZIF-8 membranes grown on polymer hollow fibers reported, which are several micrometers in thickness. The effects of synthesis parameters and fiber morphologies on the microstructures of ZIF-8 membranes will be discussed. Finally, the gas separation performances of ZIF-8 membranes will be presented.

References and Notes

1 Koros, William J. and Fleming, G.K. (1993) "Membrane-based gas separation", *Journal of Membrane Science*, 83, pp 1-39

2 Baker RW., (2002) "Future Directions of Membrane Gas Separation Technology" *Industrial & Engineering Chemistry Research.*, 41 (6), pp 1393-1411

3 Tanaka, K., Taguchi, A., Hao, JQ, Kita, H., and Okamoto, K., (1996) "Permeation and separation properties of polyimide membranes to olefins and paraffins", *Journal of Membrane Science*, 121, pp 197-207

4 Park KS, Ni Z, Côté AP, et al., (2006) "Exceptional chemical and thermal stability of zeolitic imidazolate frameworks", *Proceedings of the National Academy of Sciences*, 103 (27), pp 10186-10191

5 Zhang C, Lively RP, Zhang K, Johnson JR, Karvan O, Koros WJ., (2012) "Unexpected Molecular Sieving Properties of Zeolitic Imidazolate Framework-8", *The Journal of Physical Chemistry Letters.*, 3 (16), pp 2130-2134.

6 Andrew J. Brown, Nicholas A. Brunelli, Kiwon Eum, Fereshteh Rashidi, J. R. Johnson, William J. Koros, Christopher W. Jones, Sankar Nair, (2014) "Interfacial microfluidic processing of metal-organic framework hollow fiber membranes", *Science*, 345, 6192, pp 72-75

7 FernandoCacho-Bailo, SilviaCatalán-Aguirre, MirenEtxeberría-Benavides, Oguz Karvan, VictorSebastian, CarlosTéllez, JoaquínCoronas (2015) "Metal-organic framework membranes on the inner-side of a polymeric hollow fiber by microfluidic synthesis", *Journal of membrane science*, 476, pp 277-285

8 Kiwon Eum, Chen Ma, Ali Rownaghi, Christopher W. Jones, and Sankar Nair (2016) "ZIF-8 Membranes via Interfacial Microfluidic Processing in Polymeric Hollow Fibers: Efficient Propylene Separation at Elevated Pressures", *ACS Applied materials and interfaces*, 8, pp 25337 – 25342

9 Anne M. Marti, Wasala Wickramanayake, Ganpat Dahe, Ali Sekizkardes, Tracy L. Bank, David P. Hopkinson, and Surendar R. Venna (2017) "Continuous Flow Processing of ZIF-8 Membranes on Polymeric Porous Hollow Fiber Supports for CO2 Capture", *ACS Applied materials and interfaces*, 9, pp 5678-5682

10 Kwon HT, Jeong H-K., (2013), "Highly propylene-selective supported zeolite-imidazolate framework (ZIF-8) membranes synthesized by rapid microwave-assisted seeding and secondary growth", *Chemical Communications*, 49, pp 3854-3856

A High Free-Volume Pentiptycene-Based Polyimide for CO_2 Removal

Ahmad Arabi Shamsabadi[1], Farzad Seidi[2], Mohammad Nozari[3], and Masoud Soroush[1]

[1]Chemical and Biological Engineering, Drexel University, Philadelphia
[2]Vidyasirimedhi Institute of Science and Technology, Royong, Thailand
[3]Chemistry, Drexel University, Philadelphia

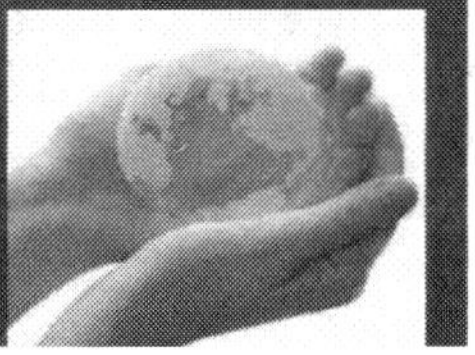

Current Level of CO_2>400 ppm

Average global atmospheric CO_2 level:
February 2016: 402.59 ppm
October 2017: 403.44 ppm

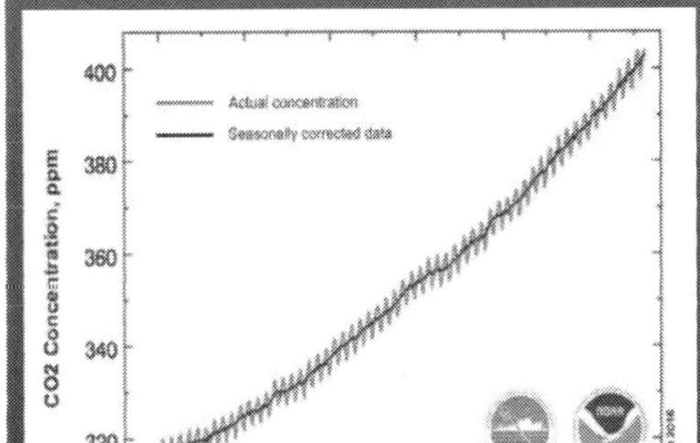

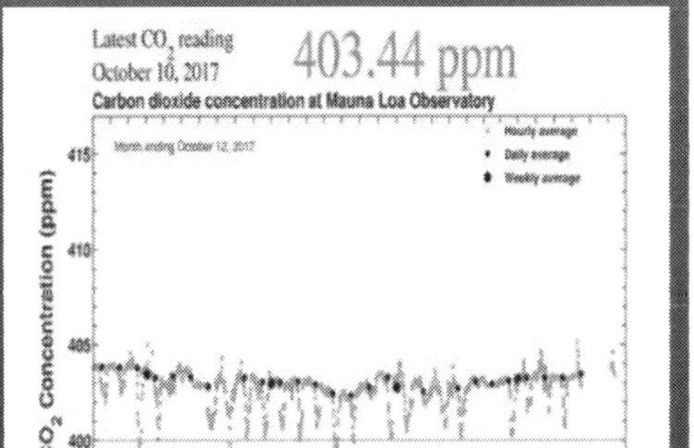

US National Oceanic and Atmospheric Administration's (NOAA)

2

CO_2 Removal Target

Polymers of intrinsic microporosity (PIMs)

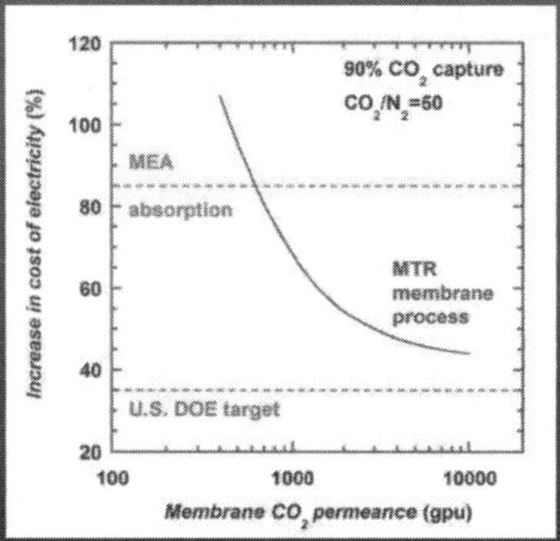

$$N_i = \frac{Pi}{l} A \, \Delta p$$

Thin film composite (TFC) membranes

N_i : *Flux*
A : Surface area
Δp: Pressure difference
l : Membrane thickness
P_i: Gas permeability

Liu et al. Chemistry-A European Journal, 2016

3

PIM-1 and PIM Membranes Moieties

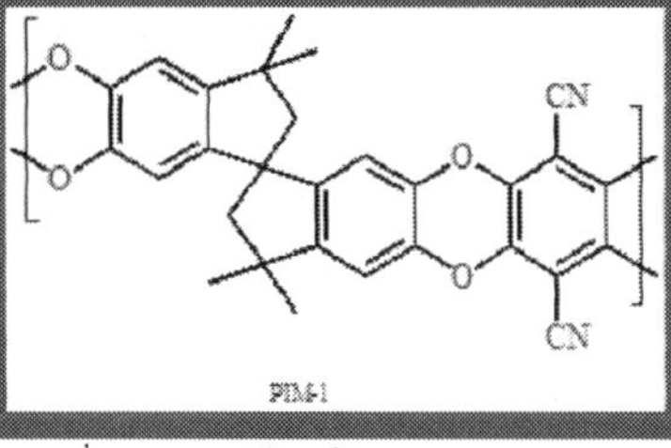

PIM-1 by Budd and McKeown

Budd et al. Advanced Material, 2004

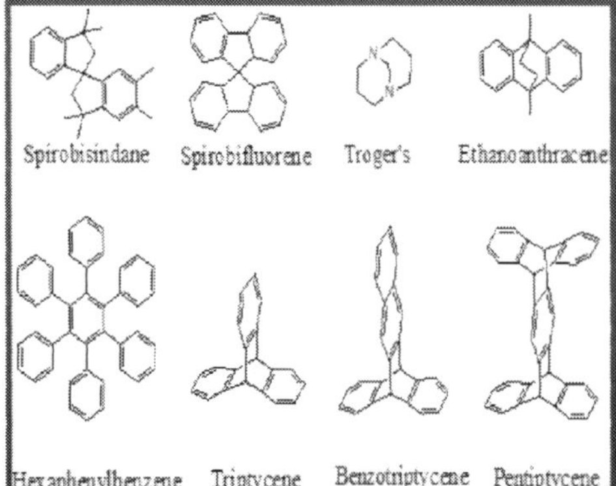

Tetramethyltetrahydronaphthalene (TMN)

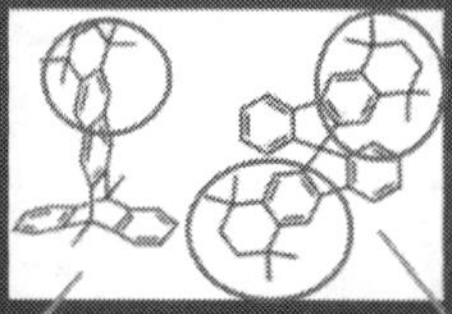

P_{CO2}:33000 Barrer P_{CO2}:17500 Barrer

Rose et al. Nat. Mat. 2017

4

PIM Membranes Challenges

- ❖Low to modest selectivity
- ❖Fast physical aging

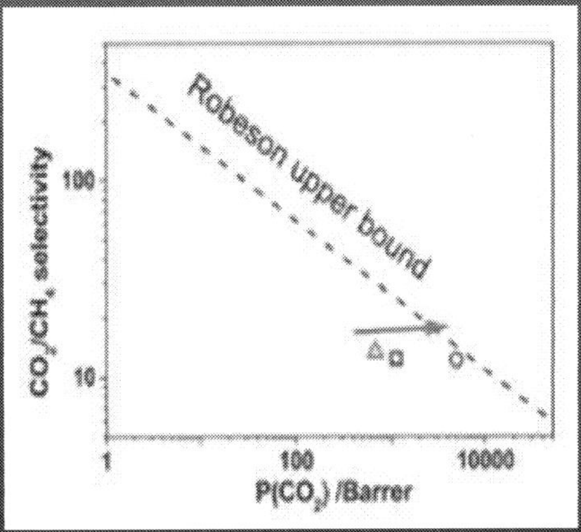

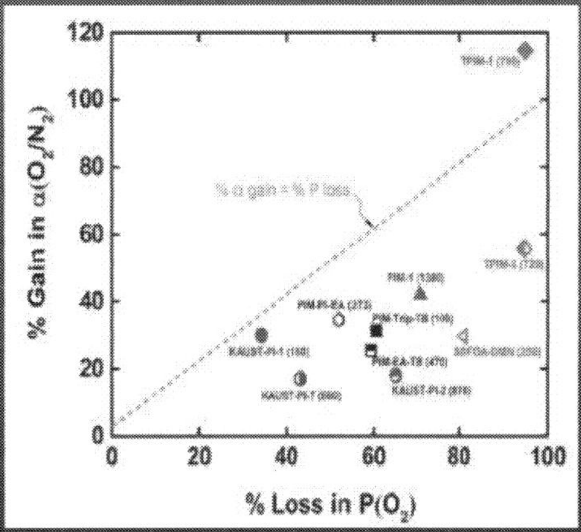

Konnertz et al., *J of Membrane Science, 2017*

Swaidan et al., *Macromolecules*, 2015

5

PIM-Polyimides (PIM-PIs)

High surface area and microporosity of PIMs have motivated researchers to combine the characteristics of PIMs and polyimides to increase selectivity and reduce physical aging

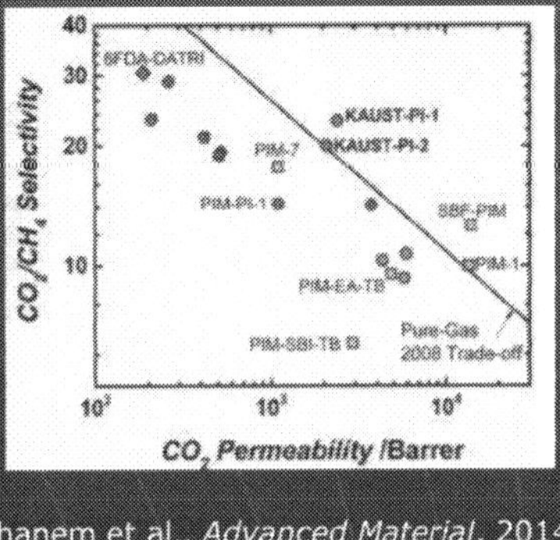

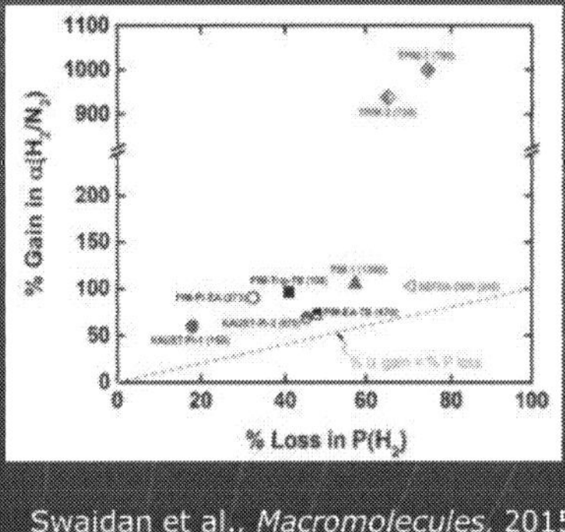

Ghanem et al., *Advanced Material*, 2014

Swaidan et al., *Macromolecules*, 2015

6

Pentiptycene Moiety

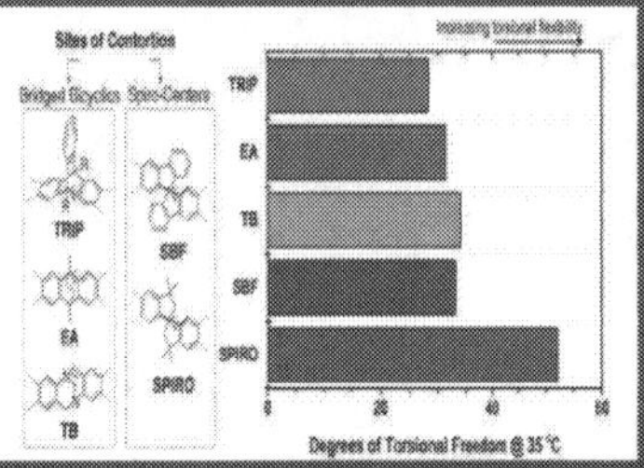

Swaidan et al. , *Macromolecules,* 2014

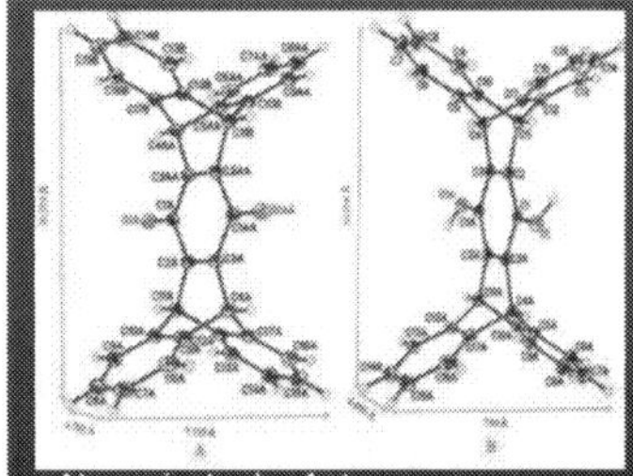

Nozari et al., Acta Crystallographica, 2016
Nozari et al. ,IUCrData , 2016

- ❖ Bulky structure
- ❖ High internal free volume (IFV)
- ❖ permanent IFV and shape-stability

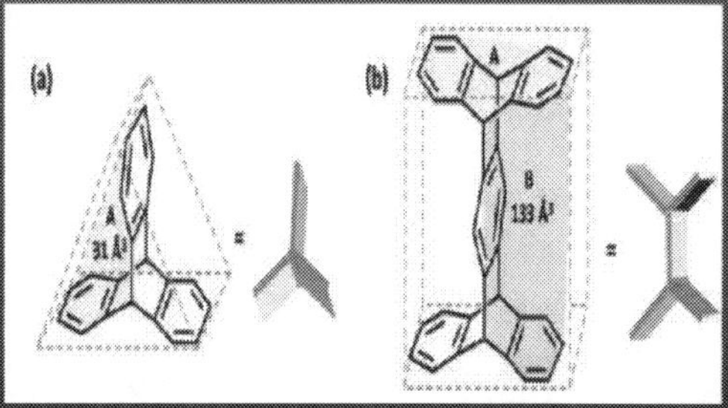

Weidman and Guo, *Industrial & Engineering Chemistry Research,* 2017

7

Existing Pentiptycene-based Membranes

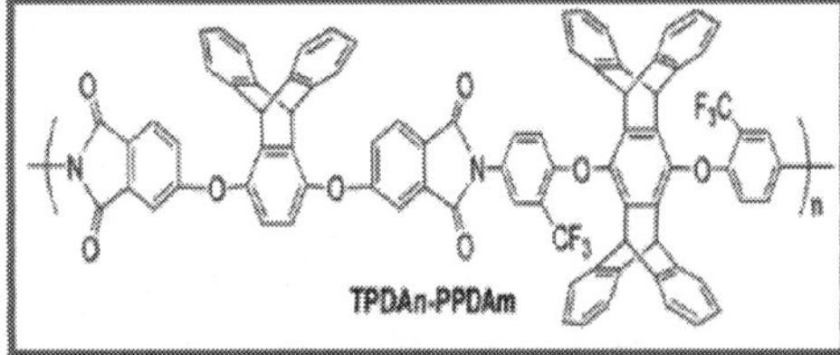

Luo et al., *Macromolecules,* 2016

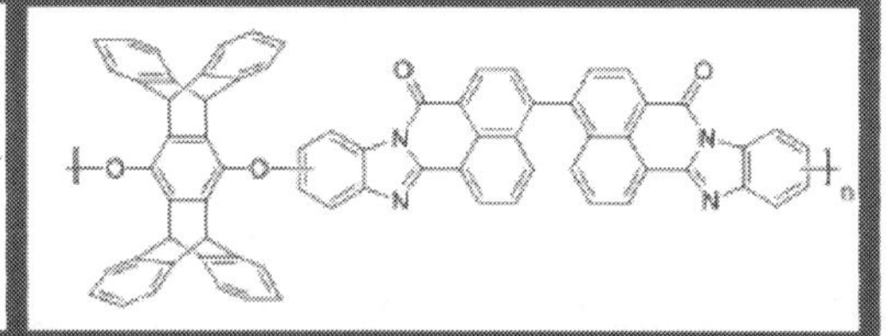

Mao and Zhang ,*Polymer* 2014

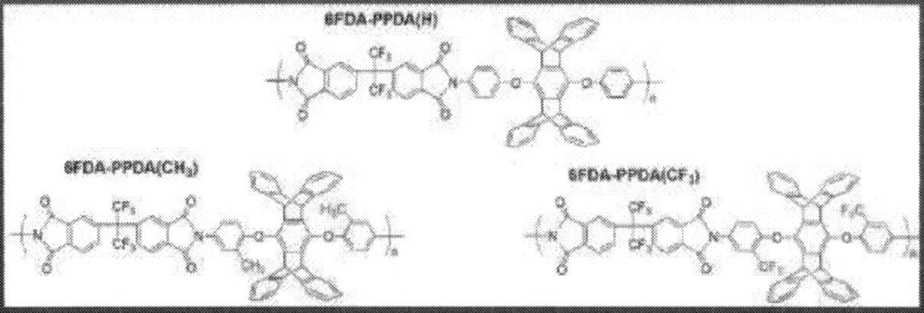

Luo et al., *J. Membr. Sci.,* 2015

FFV=0.171-0.197 CO_2= 39-137.2 Barrer

13

Why Dianhydride?

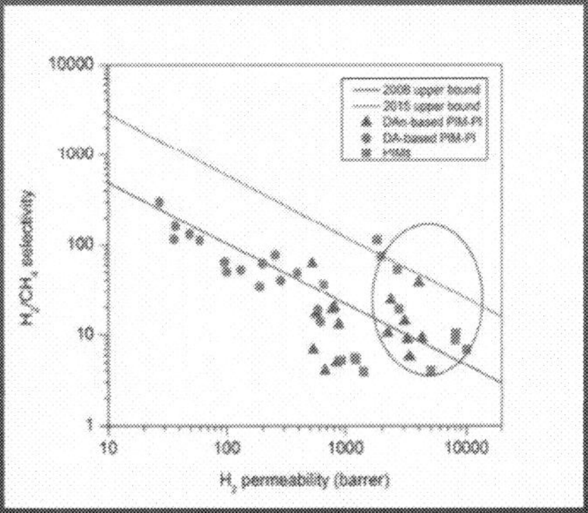

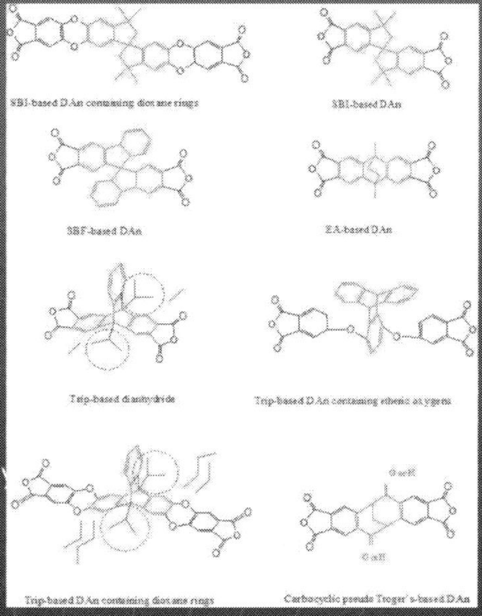

Introducing dianhydrides with rigid moieties into the polymer backbone is a promising strategy to fabricate microporous polyimide membranes with outstanding performance

9

In This Research

How pentiptycene moiety, durene, and imide groups in the polyimide backbone can veil the influence of the flexible etheric oxygen and increase the CO_2 permeability without sacrificing selectivity

- High molecular weight
- Excellent solubility
- High gas permeability

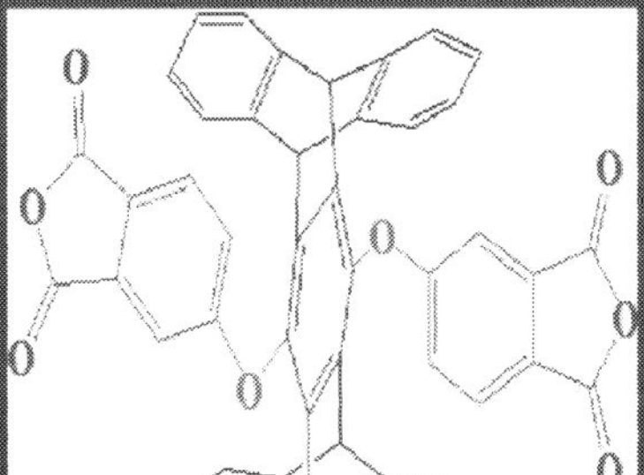

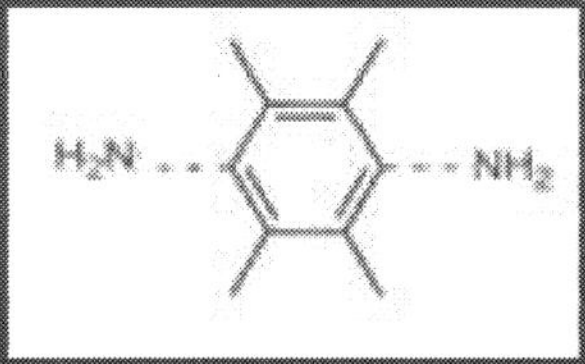

Lin and Chung, *J. Membr. Sci.*, 2001

Swaidan et al. , *Macromolecules*, 2014

Ghanem et al., *Polymer*, 2016

15

Synthesis Reactions

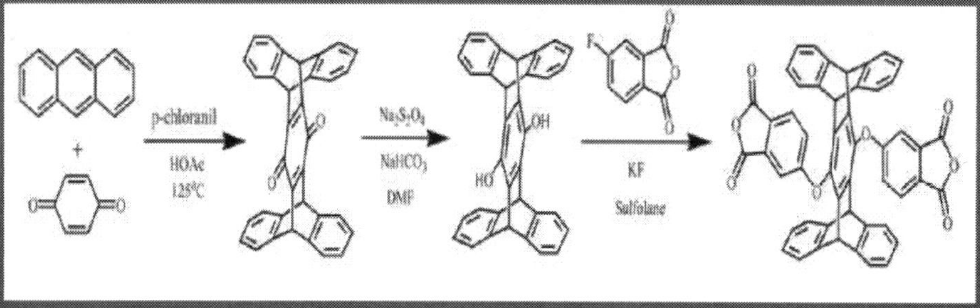

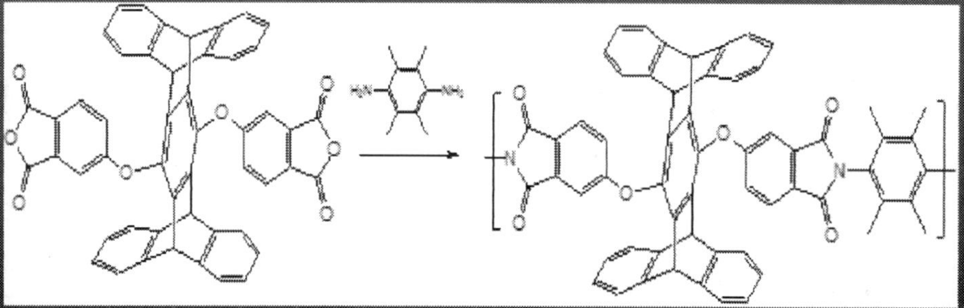

11

Synthesis Reactions

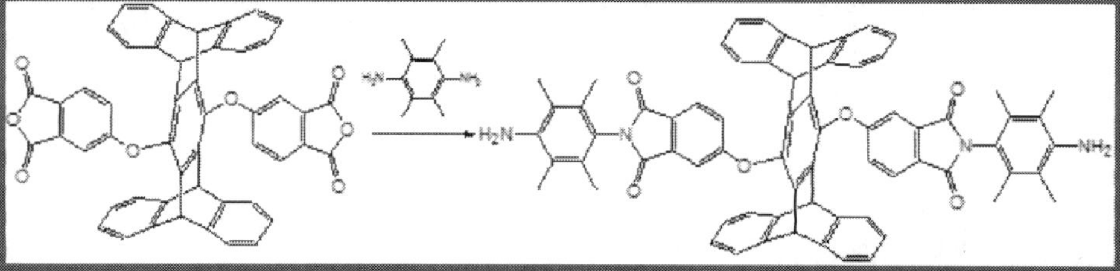

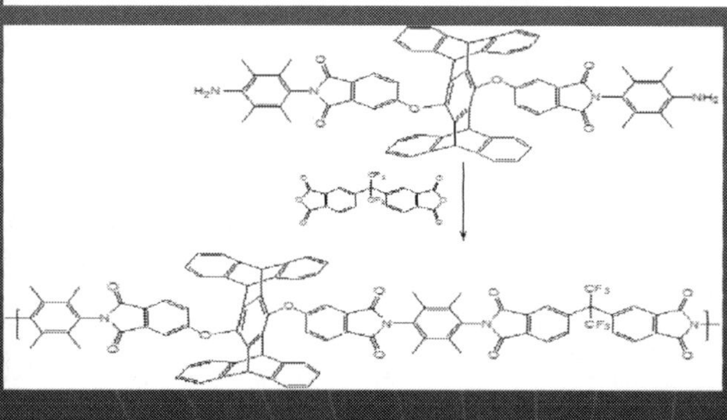

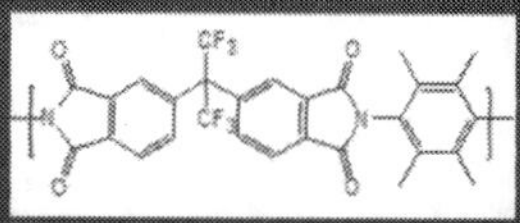

Chung et al., *J. Membr. Sci.* 1994

Mw=85110, PDI=2.3

12

XPS of Membrane

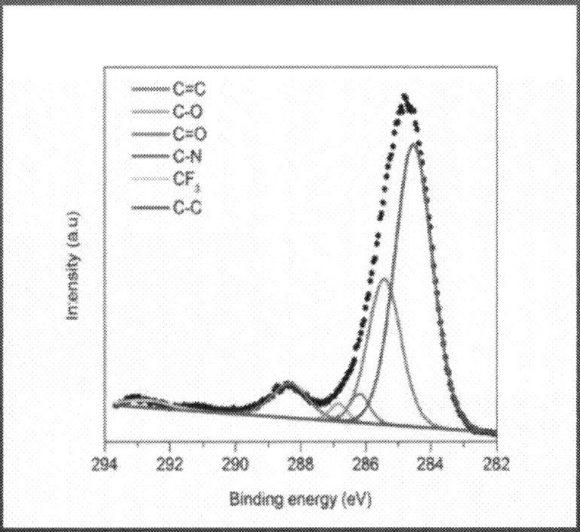

Table 1. Elemental compositions of PPImDA-6FDA.

Sample	C (%)	O (%)	N (%)	F (%)
PPImDA-6FDA (Theory)	81.65	9.17	3.67	5.5
PPImDA-6FDA (Exp)	82.74	8.52	3.28	5.46

Table 2. C-species contents of PPImDA-6FDA based on XPS analysis.

Sample	C-C (%)	C=C (%)	C-O (%)	C-N (%)	C=O (%)	CF_3 (%)
PPImDA-6FDA	59.26	27.12	1.64	3.28	6.88	1.82

13

Membrane Characterization

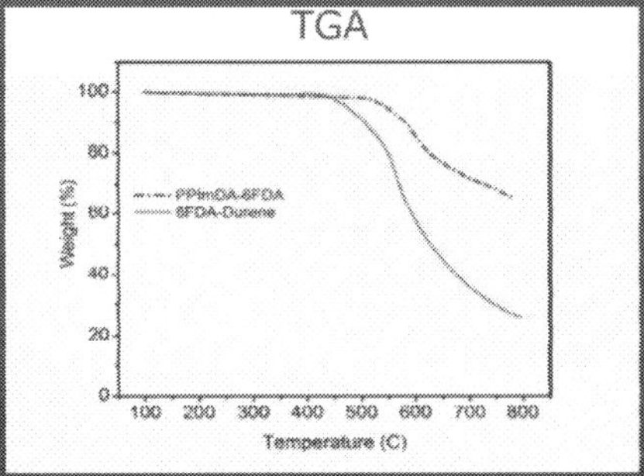

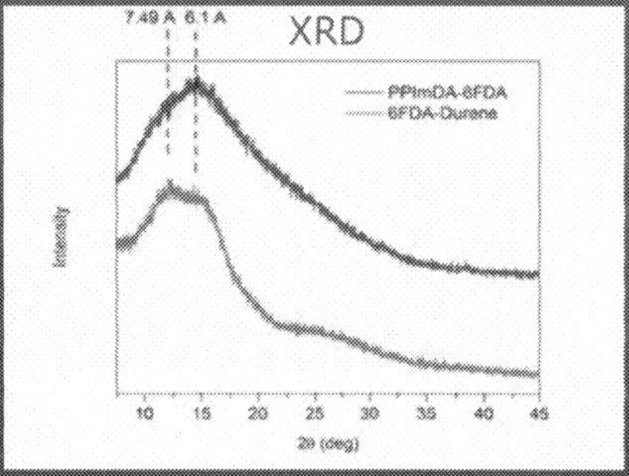

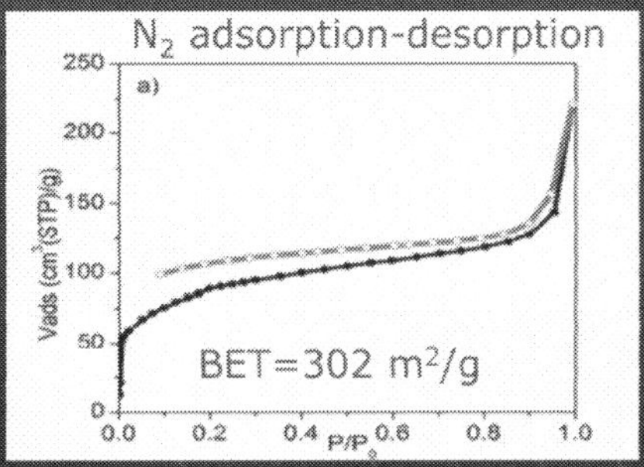

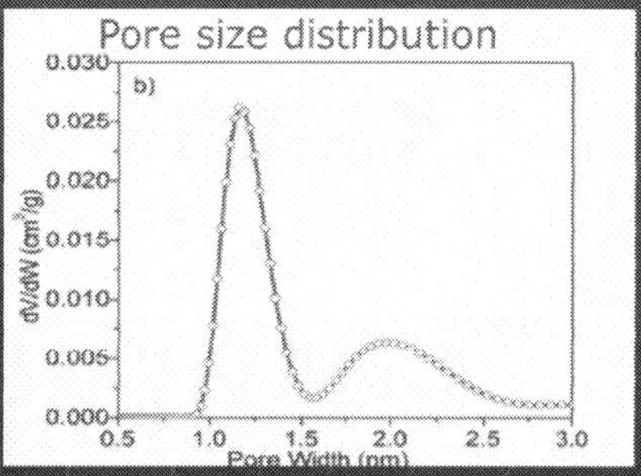

14

Drexel UNIVERSITY

Membrane Performance

17AIChE

Polymer	CO_2	N_2	CH_4	CO_2/N_2	CO_2/CH_4	*FFV*
PPImDA-6FDA						
P	812±9	33.8±0.8	29.6±0.6	24.02±0.83	27.4±0.9	
D	35.34±1.62	15.3±1.05	7.59±0.34	2.31±0.26	4.65±0.42	0.233
S	22.9±0.8	2.2±0.1	3.9±0.1	10.4±0.83	5.87±0.3	
6FDA-durene						
P	477±5	42.5±0.7	38.3±0.5	11.22±0.3	12.45±0.29	
D	19.8±0.94	12.80±0.31	6.7±0.27	1.54±0.11	2.95±0.25	0.224
S	24.1±0.9	3.3±0.1	5.7±0.2	7.3±0.49	4.22±0.3	
PBIBI-PPD						
P	137.2	4.94	3.73	27.77	36.22	
D	6.63	4.62	1.86	1.43	3.56	0.17
S	20.7	1.07	2.03	19.3	10.2	
6FDA-PPDA(CH_3)	55	2.4	2	22.91	28	
P	3.9	1.8	0.48	2	8.1	0.179
D	14.47	1.33	4.16	10.88	3.4	
S						
6FDA-PPDA(H)						
P	73	3.2	2.5	22.81	29	
D	4.9	2.6	0.56	1.88	8.7	0.183
S	14.47	1.23	4.46	11.76	3.4	
6FDA-PPDA(CF_3)						
P	13.2	7	5.5	18.85	24	
D	9.3	5.5	1.4	1.69	6.7	0.197
S	14.47	1.4	3.92	10.33	3.6	
TPDAn-PPDAm						
P	39	1.8	1.5	21.6	26	
D	3.9	2.4	1.1	1.65	3.4	0.187
S	10	0.75	1.36	13.5	7.7	

15

Drexel UNIVERSITY

Membrane Against 2008 Upper Bounds

17AIChE

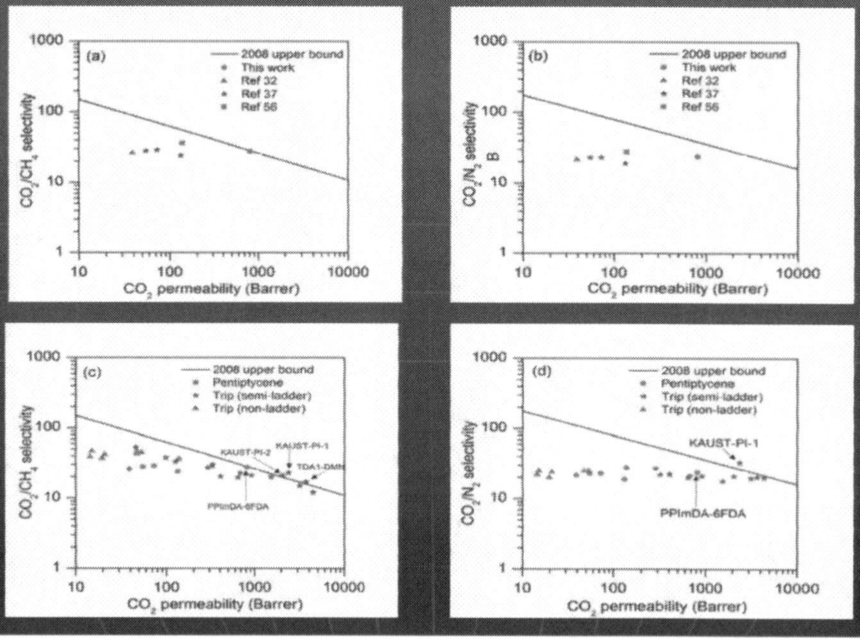

16

Concluding Remarks

Drexel UNIVERSITY — 17 AIChE

- A new pentiptycene-based dianhydride and a pentiptycene imide-containing diamine were synthesized
- A high-molecular weight microporous pentiptycene-based polyimide was prepared and used for membrane fabrication
- The resulting membrane has high fraction volume with bimodal pore size distribution, excellent solubility, and significant thermal resistance

11/17/2017 17

Concluding Remarks (Cont.)

Drexel UNIVERSITY — 17 AIChE

- Pentiptycene moiety, durene, and imide groups in the polyimide backbone veil the influence of the flexible etheric oxygen
- The membrane has the highest CO_2 permeability among all pentiptycene-containing polymers without sacrificing the selectivity touching the 2008 Robeson CO_2/CH_4 upper bound
- Membrane performance is still lower than semi-ladder PIs synthesized by methyl-substituted diamines and triptycene-based dianhydrides with bulky isopropyl bridgeheads

18

Acknowledgment

Prof. Ruilan Guo

19

489506 Design of High Efficiency PVDF-Peg Hollow Fibers for Air Filtration of Ultrafine Particles

Abstract:

The particulate matter (PM) air pollution has become a global issue in the last few decades. Various techniques, such as electron spun air filters have been developed to remove PM from the air. However, limited studies focused on filtration of small particles (< 0.3 mm) which are more toxic to human. Compared to conventional fibers filter media, hollow fibers feature of self-supported properties with tunable pore sizes, making hollow fibers a good candidate for filtration processes. Until now, limited studies have worked on particle-laden gas filtration using hollow fibers. This study reveals as the first attempt to apply hollow fibers for air filtration of ultrafine particles. The asymmetric polyvinylidene fluoride - polyethylene glycol (PVDF-PEG) hollow fibers with high gas permeance have been developed. In order to increase the gas permeance of hollow fibers for air filtration, different molecular weight of PEGs was added into spinning dopes. The addition of PEGs facilitate the formation of loosely connected cross-section and porous outer skin, thus enhances the gas permeance for air filtration. Under the inside-out testing mode, all PVDF-PEG hollow fibers display excellent filtration efficiency of 99.999% against polydispersed NaCl particles with a geometric mean size of ~30 nm (Figure 1). This study may provide useful insights for developing hollow fibers for air filtration with the optimal operation conditions.

500907 Understanding the Effect of Nanoscopic Pore Structure on Transport in Lyotropic Liquid Crystal Membranes

Abstract:

Lyotropic liquid crystals (LLC) are a class of nanostructured materials that can be modified, assembled and crosslinked into stable membranes with uniform straight pores that can perform solute-specific separations. Pore diameters on the order of 1 nm make LLC membranes well-suited for aqueous separations such as desalination and biorefinement. However, without a clear understanding of mechanisms of transport in these complex self-assembled systems, it is challenging to choose momoners to achieve specific separations goals. To advance our knowledge of transport in these systems, we have studied an experimentally characterized model system using molecular dynamics simulations in order to develop a predictive model for membrane design.

We have created a molecular model which is maximally consistent with experimentally measured structural features and material properties, and examined solute transport through the pores. We have developed methods to simulate the crosslinking mechanism, simulate X-ray diffraction (XRD) patterns and to measure ionic conductivity from atomistic simulations. By matching simulated and experimental 2D WAXS patterns, we have discovered an additional metastable state, mainly characterized by its pore structure, which may form under certain experimental conditions. The first state has a condensed, narrow pore with high radial disorder. Although stable out to 100's of nanoseconds, the XRD pattern does not match experimental results. The second, whose XRD pattern gives a good match to experiment, is noticeably more open with aromatic rings stacked in a parallel displaced conformation. Other theorized pore arrangements are not stable in simulation.

The difference in pore structure affects the mechanism by which solutes diffuse across the membrane, which are likely to significantly affect the performance of membranes. We examine transport of a range of solutes with various charges and hydrodynamic radii. With a clear understanding of mechanisms of transport in these complex self-assembled systems, one can choose monomers to achieve specific separation goals. We can use this information to draw correlations between pore structure and selective preferences. These studies will help guide monomer choice for separation-specific objectives.

500983 Understanding the Formation Pathways of Triblock Terpolymer Membranes

Abstract:

Block copolymer-directed self-assembly has enabled the synthesis of porous, nanostructured functional materials with high specific surface areas, tunable morphologies, and periodic nanopores. Increasing pore accessibility, a critical factor in membrane and electrode performance, has been achieved through nonsolvent-induced phase separation (SNIPS), which relies on the partial evaporation of a BCP solution to form a thin, ordered liquid crystalline micelle layer above a low order, asymmetric under-layer. By plunging this graded gel into a nonsolvent bath, the structure is frozen into the solid polymer, an ordered mesoporous surface layer supported by a graded, asymmetric macroporous substructure.

Recently, we have shown that the SNIPS process can be combined with BCP-directed co-assembly of phenol-formaldehyde resols to produce tunable, graded, hierarchically porous carbon material [1]. However, the formation mechanisms of SNIPS derived membranes, even without the complicating inclusions of inorganic additives, are not fully agreed upon. Understanding the solution micelle formation and structure evolution has the potential to significantly advance the ability of the community to target specific pore sizes, expand the variety of functional materials accessible via this approach, and produce better-ordered structures with improved reliability.

Our work in high-resolution ultrafiltration membranes and BCP-derived functional graded materials has focused on the asymmetric triblock terpolymer, polyisoprene-*block*-polystyrene-*block*-poly(4-vinyl pyridine), which can yield cubically packed pores. Via a combination of *ex-situ* and *in-situ* scattering measurements, we have examined the micellar solutions and their transformation from disordered to ordered structures in order to understand the varied morphologies produced by triblock derived membranes and the alterations induced by additive incorporation.

[1] S.A. Hesse, J.G. Werner, U. Wiesner *ACS Macro Lett.* **4**, 477-482 (2015).

493383 Block Polymer Hollow Fiber Membranes Functionalized with Nanoconfined Polyelectrolyte Brushes Achieve Sub-Nanometer Selectivity

Abstract:

Manufacturing membranes with well-defined nanostructures and robust performance in a controllable manner is of critical import to meeting the growing demand for highly selective membranes in the water treatment, pharmaceutical, and electronics industries. Due to their high density of nanoscale pores with a uniform size distribution, self-assembled block polymer membranes are an ideal platform for the development of advanced membrane-based applications. In this study, we demonstrate the development of high-performance self-assembled block polymer nanofiltration membranes in the hollow fiber geometry. Specifically, dual-layer hollow fiber membranes were fabricated by combining the dip-coating archetype and the self-assembly and non-solvent induced phase separation (SNIPS) methodology to coat uniform block polymer films onto the shell side of the hollow fiber membrane support. A set of optimized membrane fabrication conditions (*e.g.*, polymer solution concentration, dip-coating parameters, and solvent evaporation time) that generate high-performance hollow fiber nanofiltration membranes were identified by using a combination of electron microscopy analysis and transport experiments to guide the construction of nanoscale architecture of block polymer thin-film at the interface between the hollow fiber support and the self-assembled membrane. In this manner, ultra-thin (as thin as 200 nm) block polymer thin films with ordered nanostructures were produced. These membranes had a pore radius of 2.5 nm and a hydraulic permeability value of 27 L m^{-2} h^{-1} bar^{-1}.

The block polymer coating was based on a polyisoprene-*b*-polystyrene-*b*-poly(*N,N*-dimethylacrylamide) (PI-PS-PDMA) macromolecular precursor, which allowed the separation selectivity of the membrane to be further tailored using a carbodiimide coupling reaction to covalently attach sulfonic acid moieties to the pore walls of the membrane. The inherently charged, nanoconfined polyelectrolyte brush constrict the pore radius down to a value of 1 nm and result in an extremely high solute selectivity by fully fractionating solutes with only a 4 Å difference in radius. The use of the strong polyelectrolyte also generates a robust membrane that operates reliably in complex environments. Specifically, the transport performance of this membrane was examined over a broad range of solution pH ($1 \le pH \le 13$) and ionic strength solutions ($1mM \le I \le 2.5$ M). The membrane demonstrated a constant hydraulic permeability 4 L m^{-2} h^{-1} bar^{-1}, which is consistent with a polymer brush that lines the pore walls and holds a constant conformation, thus exhibiting a robust separation performance in response to the presence of external stimuli. These dual-layer hollow fiber membranes are promising candidates for fabrication of novel nanofiltration membrane based application devices for a variety of separation needs.

502546 Polymer Nanofilms with Engineered Microporosity By Interfacial Polymerisation for Molecular Separations in Organic Solvent

Abstract:

Molecular design of microporous materials has led to new pathways for improving the molecular differentiation achieved by separation membranes used for gas separation, water desalination, and separations in chemical and pharmaceutical synthesis. Recently developed microporous materials, including polymers of intrinsic microporosity (PIMs), metal organic frameworks, and graphene oxide, all have micropores of less than 1 nm, and so are relevant for developing molecular separating membranes. This paper focuses in membranes for Organic Solvent Nanofiltration (OSN) [1]. IPIMs are a new class of rigid polymers with high free volume, which results from inefficient polymer packing. Several efforts to make dip-coated TFC-PIMs membranes solvent resistant have been reported, including photo-crosslinking, blending with thermally reactive polymers and chemical crosslinking.

In this work we fabricated ultra-thin highly crosslinked PIM-like nanofilms down to 20 nm in thickness by interfacial polymerization. To engineer permeance and selectivity, we sought to design and control the polymer nanofilm structure at a molecular level by incorporating contorted monomers during the interfacial reaction. This results in enhanced microporosity and higher interconnectivity of intermolecular network voids, as rationalised by molecular simulations. Composite membranes comprising these nanofilms with enhanced microporosity fabricated in-situ on crosslinked polyimide UF membranes show outstanding separation performance in organic solvents, with up to two orders of magnitude higher solvent permeance than membranes fabricated with nanofilms made from non-contorted, planar monomers and membranes reported in the literature [2]. A new organic solvent Robinson upper bond has been drawn to show outstanding perm-selective performance in organic solvents.

By tuning cavity width in the range of 5 ~ 10 Å, the molecular separation performances were further improved for driven nanofiltration. Superior selectivities of the micropore nanofilms was further investigated by molecular dynamic simulation, which showed the interplay of pore wall interaction and entrance sieving.

497789 Zwitterionic Copolymer Self-Assembly for Extremely Fouling Resistant, High Flux Membranes with ~1 Nm Pore Size: Understanding Zwitterion Chemistry and Increasing Permeance

Abstract:

Zwitterions, defined as functional groups with equal numbers of positive and negative electrostatic charges, strongly resist biomacromolecular adsorption due to their high degree of hydration. This has led to their incorporation into membranes to prevent fouling by various methods, especially by surface functionalization of existing membranes. Zwitterions also have interesting self-assembly capabilities due to their high dipole moments and strong intra- and inter-molecular interactions. They also interact with small ions and exhibit ionic strength response. Random copolymers of zwitterionic and hydrophobic monomers self-assemble into ~1 nm interconnected domains. This makes them excellent candidates for membrane selective layers with high flux, fouling resistance, size selectivity, and chemical resistance. Composite membranes made by coating these copolymers onto commercial ultrafiltration membrane supports exhibit fluxes comparable with or higher than commercial membranes of similar pore size (~1000 Da MWCO). Based on the rejection of anionic and neutral dyes of varying sizes, they show size-based selectivity with a cut-off around 1 nm, independent of zwitterion chemistry among four zwitterionic monomers surveyed. This pore size closely matches the size of the zwitterionic nanochannels. Zwitterion chemistry, however, has a significant impact on selective layer permeability and fouling resistance. Membranes prepared from copolymers with sulfobetaine methacrylate exhibit excellent fouling resistance, showing little to no flux decline and essentially complete flux recovery with a water rinse upon the filtration of foulants such as protein solutions and oil suspensions. Those prepared with phosphorylcholine methacrylate are even more fesistant to fouling, exhibiting no flux decline even in week-long fouling experiments with oil suspensions. Furthermore, the permeance of these membranes can be boosted 10 times without any changes to the selectivity by using an ionic liquid co-solvent, to about 50 L/m^2.h.bar. This arises from the formation of worm-like supramolecular assemblies in solution that are deposited on the support to form an ultra-thin selective layer. These membranes are highly valuable for bioseparations and wastewater treatment applications.

494135 Modeling the Effects of Mass Transfer on Microstructure Formation in Phase-Inversion Membranes

Abstract:

Self-assembled polymer membranes are typically produced by non-equilibrium processes, making the final microstructure a function of process history. When combined with complex thermodynamics, rheological complexity and a multitude of length and time scales, modeling the formation of these membranes is challenging. One promising path for model development combines continuum fluid dynamics with a field-theoretic description of polymer thermodynamics into a "multi-fluid" model. We have recently developed and validated such a model for a ternary polymer solution undergoing non-solvent induced phase separation. In a first step towards understanding polymer membrane formation, we examine the role of solvent and non-solvent exchange in driving phase inversion. The model predicts a wide variety of qualitative features of the phase inversion as a function of different parameters and initial conditions, including: immediate phase separation, delayed phase separation, single domain (dense) films and multiple domain (porous) films. We will discuss these different qualitative states and potential physical mechanisms of their origin.

483494 MOF Molecular Sieves to Address Challenging Gas/Vapor Separations: Present and Future

Abstract:

The separation of molecules with close physical properties is a challenging task, commonly performed using the conventional low temperature fractional distillation technique which is recognized to be highly energy intensive. After more than 6 decades of revolutionary use of zeolites molecular sieves for separation of physically similar molecules within 1 Å difference in size, researchers from both academia and industry have been dedicating a lot of effort to push the limit of sieving separation to lower than 1 Å. The main purpose of this endeavour is to switch the separation of important isomers and commodities from distillation to more energy efficient adsorption or membranes technologies.

In the last 2 decades, major developments in Metal-Organic Frameworks was dedicated mainly to high surface area materials with large pores rather than molecular sieves with rigid or flexible small pores apertures. In my talk, I will illustrate the progress we have made in the development of tunable platforms with a variety of interesting intrinsic properties to target challenging separation of important isomers in petroleum and petrochemical industries. The optimal structural control at the molecular level of these particular platforms led to the discovery of new generations of MOF molecular sieves, to address challenging separations such as linear paraffin/mono-branched paraffin, mono-branched paraffin /disbranched paraffin and olefin-paraffin.

500253 Understanding CO_2 in MOF-74: Dynamics and Structure of Adsorbed CO_2 with ^{13}C NMR and DFT

Abstract:

Metal-organic frameworks are crystalline organic-inorganic hybrid materials that have been of interest for adsorption applications for their tunability, high crystallinity, and high surface areas. Mg-MOF-74 has long been of interest for its exceptional gravimetric uptake of CO_2 at flue-gas relevant conditions. CO_2 adsorption in Mg-MOF-74 is well studied, but most work has focused on understanding adsorption at the undercoordinated open-metal sites that line the pores. Despite this, loadings of CO_2 in Mg-MOF-74 at even modest partial pressures can exceed one CO_2 per open-metal site, requiring a more nuanced picture to understand the behavior and loading of CO_2. Recent neutron diffraction studies have shown that CO_2 occupies not only these primary adsorption sites, but also "secondary" and even "tertiary" sites that exist within the MOF pore. Here we present a joint experimental NMR and theory study that demonstrates that higher CO_2 loadings qualitatively alter the structure of adsorbed CO_2 within the MOF pore. Constructing special NMR hardware to deliver gases at variable temperature and pressure, we have found that the ^{13}C NMR lineshape evolves in a predictable fashion as these different sites become occupied. We show that the time-averaged value of the second Legendre polynomial, P_2, which maps to the alignment of CO_2 with respect to the MOF pore axis, allows understanding of the so-called "sign" of the NMR lineshape—which shifts in appearance with loading in a highly sensitive and informative manner. The evolution of the NMR lineshape with CO_2 loading, which probes the aggregate dynamics and structure of adsorbed CO_2, agrees exceptionally well with the DFT-predicted loading-dependent structures of CO_2 within the Mg-MOF-74 pore through analysis in terms of P_2. NMR experiments also point to qualitative differences between the behavior of CO_2 in Mg-MOF-74 and Mg—Cd-MOF-74, which we rationalize in terms of theoretical calculations for mixed-metal MOF-74 systems and our calculations on Mg-MOF-74. Importantly, the NMR spectra are difficult to interpret without the detailed picture of adsorption sites provided by DFT, and the combined effect demonstrates how experiments can reinforce the DFT models of collective CO_2 behavior.

489601 Using Metal-Organic Frameworks for CO Purification: Investigation of Key Material Parameters through a Multi-Scale Approach

Abstract:

Carbon Monoxide (CO) production for industrial use is currently accompanied with an energy-intensive separation process. Indeed, its downstream separation from synthesis gas (syngas), a mixture of CO and H_2 as well as impurities like nitrogen, carbon dioxide and water, is a multi-step process involving a cryogenic distillation section aimed at separating CO from H_2. Due to similar sizes, molecular weights and boiling points, it is particularly difficult to separate CO and N_2. Therefore, an expensive cryogenic step is required, and still remains ineffective in purifying CO from syngas streams with high concentrations of N_2. Adsorption separation has the potential to become a more economical alternative.

This study investigates the use of a class of adsorbents, namely metal-organic frameworks (MOFs), for purification and separation of CO from gas streams containing high levels of N_2. Due to the multitude of MOFs that exist, a multi-scale approach - combining molecular simulation and process system modelling - proposed suitable MOFs for the targeted separation. In this presentation, we will show a number of promising MOF structures that were identified, synthesized, characterized and tested for CO adsorption under both equilibrium and dynamic conditions. The effects of the porosity, type of metal sites, group functionalization and incorporation of additional chemisorption sites on the uptake and selectivity were investigated and will be discussed during the talk.

500659 Binary Adsorption of CO_2 and Water on Uio-66 MOF

Abstract:

Metal organic frameworks are frequently examined as potential solutions to complex gas phase separations problems. In many cases, the gas phase adsorption properties of these materials are quantified using single component gas adsorption isotherms and breakthrough experiments. In adsorption separations, however, it is common that the adsorbent participates in a multicomponent adsorption event. In the literature there is a general absence of multicomponent adsorption data with most data predicted via the ideal adsorbed solution theory or molecular simulations. Therefore, in this work, binary adsorption data of CO2 and water on UiO-66 were measured experimentally using a volumetric method at three different temperatures and at three different water loadings. Molecular simulations and ideal adsorbed solution theory calculations were also completed and the results of the simulations and the experiments were compared. The data in this work can be used to examine the applicability of the ideal adsorbed solution theory to metal organic frameworks and can be used to validate molecular simulations of binary adsorption events.

489968 Designing MOF Embedded Electrospun Fibers for Adsorption Applications

Abstract:

Metal-organic-frameworks (MOFs) have demonstrated enormous potential in the adsorption field. Two challenges that prevent these materials from reaching industrial feasibility in a broader range of adsorption-based applications are their stability and typical powder state. Impregnating MOFs into polymeric electrospun fibers has the potential to both attenuate exposure to undesirable chemicals and provide a platform for anchoring MOFs into separation elements with high accessibility. This talk is aimed at demonstrating how design strategies such as polymer selection and developing inter-fiber porosity may be used to develop electrospun fiber/ MOF composites for targeted separation applications.

498038 Butanol Separation from Multicomponent Vapor Mixtures Using Zeolitic Imidazolate Frameworks

Abstract:

Biobutanol is a promising renewable alternative to conventional petroleum-derived liquid fuels, and can be used as either as an additive or as a direct gasoline replacement. It also has many industrial applications, e.g., in the production of paints and polymers. A major challenge in fermentative biobutanol production is to find a cost-effective butanol separation process from the extremely dilute aqueous butanol solution (which is usually in the form of an acetone-butanol-ethanol mixture). Vapor-phase adsorption after gas stripping of the butanol solution can be an effective option to separate butanol from water due to the favorable nature of butanol-water vapor liquid equilibrium. Zeolitic Imidazolate Frameworks (ZIFs), a set of nanoporous metal-organic frameworks (MOFs) with tunable porosity and functionality are possible candidates for this purpose. In previous work by some of the present authors on ZIF-8, ZIF-90 and ZIF-71 adsorbents, it was found from single-component adsorption isotherms and IAST predictions that ZIF-8 and ZIF-71 may be useful for this separation.[1] Previous work from our group has also demonstrated that the effective pore size, hydrophobicity, and organophilicity of ZIFs can be tuned using mixed-linker ZIFs.[2]

However, very little is known about separation of butanol from multicomponent vapor mixtures. In particular, an important issue is the presence of humidified acid gas (CO_2) in differing amounts in the vapor mixture, which may cause degradation of ZIF materials by attacking the Zn-N coordination bonds. To understand these effects and their impact on butanol separation, here we evaluate the separation performance of ZIF-8, ZIF-90, ZIF-71, hybrid ZIF-8_x-90_{100-x} and ZIF-8_x71_{100-x} adsorbents with a 1 mole % butanol-in-water system through vapor phase breakthrough experiments at 35 °C and 1 bar. For simplicity, acetone and ethanol were not included in the mixture. We conducted long-term stability studies of the adsorbents in humid air and CO_2 in addition to vapor breakthrough studies. The separation properties are presented in detail and qualitatively explained based upon the textural properties, hydrophobicity, and organophilicity of the ZIF frameworks. ZIF-8 and ZIF-71 are found to be best materials (in terms of butanol selectivities ~15 and capacities ~4.3 mmol/g) for this separation in the absence of humid CO_2 in the feed stream. However, in the presence of humid CO_2, ZIF-8, ZIF-90 and ZIF-8_x-90_{100-x} hybrids are found to be unstable upon long-term operation. The mechanistic aspects of this degradation are studied by several techniques including FTIR spectroscopy and explained. From a practical viewpoint, ZIF-71 appears to be an excellent candidate for butanol separation owing to its acid gas stability as well as good butanol adsorption capacity and selectivity.

References

1. Zhang, K.; Lively, R. P.; Dose, M. E.; Brown, A. J.; Zhang, C.; Chung, J.; Nair, S.; Koros, W. J.; Chance, R. R., Alcohol and Water Adsorption in Zeolitic Imidazolate Frameworks. *Chem. Commun.* **2013**, *49*, 3245-3247.

2. Eum, K., et al., Highly Tunable Molecular Sieving and Adsorption Properties of Mixed-Linker Zeolitic Imidazolate Frameworks. *J. Am. Chem. Soc.* **2015**, *137*, 4191-4197.

497028 Synthesis of Water-Sensitive Metal-Organic Frameworks within Fiber Sorbent Modules

Abstract:

Metal-Organic Frameworks have garnered significant attention as sorbent materials for a wide variety of industrial gas separations. A potential vehicle for their application is the incorporation of MOFs into fiber sorbent contactors, in which high loadings of sorbent are dispersed within a porous polymer matrix. Due to the structured nature of these contactors, pressure drops across modules are significantly reduced and lend themselves to more rapid pressure and temperature swing cycles. However, fiber sorbents are produced via polymer solution phase inversion techniques and expose the sorbents to water, solvents, and sonication, and therefore is limited to the use of highly stable sorbents. A large portion of MOFs, including those with the highest performance in CO_2 capture, are at least somewhat unstable in humid and aqueous conditions, making them difficult to incorporate without catastrophic loss of porosity. If MOF sorbents were incorporated into the fibers after the spinning process, their degradation under spinning conditions could be circumvented.

In this talk, we present a synthesis route for the production of water-sensitive MOFs in polymer fiber sorbents by the use of metal oxides as sacrificial seeds. Cellulose acetate/ZnO (48 wt %) fibers were spun using a dry-jet wet-quench method and converted into HKUST-1 and ZIF-8 through hydroxy double salt (HDS) intermediates in a post-spinning process. Final MOF loadings within the polymer sorbent reached 85 and 66 wt % respectively as determined by N_2 physisorption. We further demonstrate this process on module-packaged CA/ZnO fibers, in which ready-to-use fiber sorbents are produced in a moisture free environment and with minimal handling. These modules are then employed in proof-of-concept CO_2/N_2 breakthrough experiments.

491107 A Robust Iron Based Metal-Organic Framework PCN-250 for Adsorption Separation of C_2H_6 over C_2H_4

Abstract:

Energy-saving separation C_2H_4/C_2H_6 mixture is of great importance and challenge. In this work, we reported a robust iron based metal-organic framework, PCN-250, as a C_2H_6-selective adsorbent with a high uptake of C_2H_6 and adecent selectivity towards C_2H_6/C_2H_4 separation. Adsorption isotherms of C_2H_4 and C_2H_6 were measured and their adsorption capacities were high up to 4.22 and 5.21 mmol/g at 298 K and 100 kPa, respectively. More importantly, it exhibited preferential adsorption of C_2H_6 over C_2H_4.The adsorption selectivities of C_2H_4/C_2H_6 mixtures (15:1 and 1:1, v/v) were in the range of 1.85-4 on the basis of ideal adsorbed solution theory (IAST). Meanwhile, its separation performance was also verified by the breakthrough experiments, indicating the potential of C_2H_4 purification from C_2H_4/C_2H_6 mixture, particularly for industrial C_2H_4 purification of removing low concentration of C_2H_6 from the cracked gas mixture (C_2H_4/C_2H_6, 15:1). The isosteric heats of C_2H_6 and C_2H_4 calculated by Clausius-Clapeyron equation gradually increased from 23.6 to 27.8 and 21.1 to 25.3 kJ/mol, respectively. The isosteric heats adsorption of C_2H_6 and C_2H_4 on PCN-250 were much lower compared to those Ï€-complexation adsorbents preferentially adsorbing C_2H_4. Configurational-biased grand canonical Monte Carlo (CB-GCMC) method was employed to simulate its adsorption behaviors in PCN-250 and reveal the mechanism underlying the preferential adsorption and separation of C_2H_6 over C_2H_4. Computational simulation results showed that the exceptional C_2H_6 selectivity was attributed to the effect of steric hindrance and revealed that the small pore of PCN-250 was a key factor governing the separation of C_2H_6 and C_2H_4, especially in low pressure. The highethane adsorption capacity, high C_2H_6 /C_2H_4 selectivity and low isosteric heat of adsorption of PCN-250 would make it a promising adsorbent for the effective separation of ethane/ethylene.

496352 Application of a New Zwitterionic Membrane Surface Chemistry for Biofouling Control

Abstract:

Treatment of impaired waters for beneficial use exposes the membranes to feed waters containing biological and abiotic species, which leads to fouling and loss of membrane productivity over time. Since reduction in flux due to fouling is one of the largest costs associated with membrane processes in water treatment, new coatings that limit and reverse fouling would have significant economic and societal impacts. Developing these advanced coatings is the focus of our work.

Prior studies to control biofouling largely have focused either on preventing microorganisms from attaching or killing the microorganisms once they do attach. Based on considerable literature and experience, we know that each of these control strategies alone can be effective but not sufficient for controlling biofouling. Our hypothesis is that membranes coated with dual-mode polymer nanolayers can limit and reverse membrane biofouling by switching reversibly between passive (antifouling) and active (antimicrobial) states of biofouling control. This strategy differs fundamentally from most other surface modification strategies that rely solely on either passive control (i.e., using coatings designed only to weaken foulant adhesion and, thus, limit their accumulation) or active control (e.g., adding biocides).

In this paper, we will describe a method to minimize biofouling on ultrafiltration membranes by coating the membrane surfaces with a new type of responsive zwitterionic polymer. The effectiveness of this new chemistry will be demonstrated with data collected from bacterial deposition and membrane biofouling experiments using appropriate model systems. Bacteria deposition studies showed that the new chemistry performed better than other common anti-fouling chemistries. Biofilm studies showed that PES membranes coated with the new chemistry accumulated half the biomass volume as unmodified membranes. A unique feature of this new polymer coating is that it can switch reversibly between the anti-fouling, zwitterion mode and an anti-microbial, quaternary amine mode. We will explain the chemical switching mechanism and use FTIR to support the reversible switching of the polymer nanolayer between its antifouling and antimicrobial states. We will also explain that by slightly altering the chemistry we can adjust the pH at which the chemistry switches. These new chemistries were challenged on model substrates for biofilm growth.

496664 Facile Grafting of Zwitterions Onto Membrane Surface to Enhance Antifouling Properties for Wastewater Reuse

Abstract:

Facile grafting of zwitterions onto membrane surface to enhance antifouling properties

for wastewater reuse

Nima Shahkaramipour, Chong Cheng, and Haiqing Lin*

[1]Department of Chemical and Biological Engineering, University at Buffalo,

The State University of New York, Buffalo, NY 14260, USA

*Corresponding author. Tel: +1-716-645-1856, Email: haiqingl@buffalo.edu

Abstract

Polymeric membranes for wastewater reuse are often fouled by suspended solids and dissolved organic matters, resulting in dramatic decrease in water flux. Fouling can be mitigated by enhancing hydrophilicity of membrane surface to avoid favorable interactions between the membranes and foulants. However, the modification of membrane surface is often complicate and difficult to integrate into industrial membrane production processes. Herein we demonstrate a facile one-step coating of superhydrophilic zwitterions on the surface of ultrafiltration (UF) and nanofiltration (NF) membranes by coating with dopamine and sulfobetaine methacrylate (SBMA). In the presence of oxygen, dopamine forms polydopamine (PDA) adhering onto the membrane surface and anchors SBMA via Michael addition to form a robust thin superhydrophilic layer, as confirmed by contact angle measurement, thickness measurements, and x-ray photoelectron spectroscopy (XPS). The effect of coating solution composition and coating time on the water flux through the membranes with time was thoroughly studied. The modified UF membranes demonstrate up to 80% higher water flux than the uncoated ones, when tested with water containing bovine serum albumin (BSA) as a model foulant in a crossflow system. This presentation will also discuss the effect of zwitterion chemistry on the antifouling properties of the modified membranes for the wastewater reuse. This facile approach of membrane modification is also adapted for post-modification of a commercial nanofiltration membrane module, which demonstrates superior antifouling properties when tested with real wastewater at a wastewater treatment plant.

498897 Peptoid-Grafted Hollow Fiber Membranes for Improved Biocompatibility

Abstract:

Biofouling is a persistent problem for membranes exposed to blood or other bodily fluids, affecting membrane surface structure and hindering performance. In this work, we demonstrate that polysulfone (PSU) hollow fiber membranes that are grafted with peptoids containing methoxyethyl (NMEG) side chains resist nonspecific protein adsorption. NMEG was attached to surface via polydopamine (PDA). The modified surfaces were characterized to determine grafting density, contact angle, strength, pore size, and smoothness. The long-term stability of the peptoid-modified hollow fibers was confirmed over six months. Peptoid-grafted PSU has increased hydrophilicity compared to unmodified PSU and comparable mechanical stability. The peptoid-grafted hollow fibers had reduced adsorption of single proteins (bovine serum albumin, lysozyme and fibrinogen). Additionally, the effect of side chain length and grafting density on protein adsorption was evaluated. It was determined that there is an optimal grafting density for reduction of protein adsorption, which was dependent on the length of the peptoid. These studies show that peptoid-grafted PSU membraned provide a low-fouling surface with improved biocompatibility for use in biomedical and blood-contacting applications.

496574 Colloidal Foulant Behavior on Membrane Surfaces with Controlled Chemistry and Ordered Roughness

Abstract:

Reduction in water transport (flux) due to fouling is one of the largest costs associated with membrane processes in water treatment. A key to minimizing the energy intensity of membrane-based water treatment systems is designing membranes with high water transport and low fouling propensity. The fouling propensity of a membrane depends greatly on foulant type, operating conditions, and membrane surface properties. Thus, surface modification provides one pathway to improving membrane fouling resistance. Modification can be done by changing surface chemistry or physical properties (e.g., roughness), and these represent orthogonal membrane surface design parameters.

The objective of this project is to develop the basic science needed to design new membranes that are resistant to fouling. In this contribution, we present findings on the relationship between membrane surface properties and fouling that were derived from tests on the hypothesis that combinations of geometric patterns (i.e., ordered roughness) and controlled chemical coatings will reduce membrane fouling compared to the control.

To attain the project objective, we carried out a systematic study to understand the roles of colloidal foulant chemistry and membrane surface properties on membrane performance using crossflow filtration. Nano-patterns were applied by embossing commercial polyamide thin-film composite nanofiltration membranes such as the GE HL series. Silica nanoparticles were functionalized by silane chemistry and applied as a model foulant to study the role of colloidal foulant chemistry. In parallel, interactions between foulants and membrane surfaces were studied by computational and experimental methods to better understand the role of chemistry in membrane fouling. Results of this study support the hypothesis and suggest that combining physical patterning with chemical modification on a membrane surface is an effective strategy for designing new membranes with a low propensity for fouling by colloidal nanoparticles.

498995 Spatial Control of Grafted Polymers on Ultrafiltration Membrane: A New Horizon of Aget-ATRP

Abstract:

Controlled polymerization methods have been shown to be highly desirable when tailoring the three dimensional structure of membrane surfaces. However, selective modification either on the inside pore or outer membrane surface is much more complex. Activator generated electron transfer (AGET) radical polymerization has been demonstrated as a simple and well controlled method for Atom Transfer Radical Polymerization (ATRP) on regenerated cellulose (RC) membrane using 2-hydroxyethyl methacrylate (HEMA) as the monomer unit. The methodology was optimized with respect to initiator concentration, the time of AGET-ATRP and reductant concentration. AGET-ATRP was found to be less sensitive to the presence of air, water or traces of oxygen. The control of the polymer growth inside or outside the pores of RC membrane was investigated using different pore filling solvents having a wide range of viscosity and reactivity. Both of these factors were found to be responsible for surface selective modification of RC membranes. The polymer growth on the outer surface of the RC membranes was evident from FT-IR spectroscopy, while dextran and bovine serum albumin (BSA) rejection tests were used to probe changes in pore dimensions. Dextran rejection was determined by size exclusion chromatography and Total Organic Carbon (TOC) analysis while UV-vis spectroscopy was used for BSA analysis. Results that compare changes in flux and rejection for different pore filling solvents indicate the desirable properties for these solvents. Based on the fact that glycerol has a high viscosity and low miscibility with acetonitrile it was found to be the most promising pore filling solvent. The results obtained here demonstrate the possibility of selective surface modification inside the pores or on the external membrane surface.

VO_x Surface Catalyst for Low Cost, High Performance Hydrogen Permeable Vanadium Membranes

Thomas F. Fuerst, Sean T. B. Lundin, Elias Petsalis, Jennifer Wilcox, J. Douglas Way, and Colin A. Wolden

Department of Chemical and Biological Engineering

Colorado School of Mines, Golden CO 80401 USA

AIChE
11/2/2017
Surface Engineered and Responsive Membranes

COLORADO SCHOOL OF MINES
EARTH • ENERGY • ENVIRONMENT

Overview

- Background
  - Applications
  - Dense Metallic Membranes
  - Catalytic Surface Coatings
  - Vanadium Embrittlement
- V_2O_3 Catalyst Coatings
  - Fabrication / Characterization
  - Membrane Performance
  - Stability
  - V_2O_3 DFT Simulations

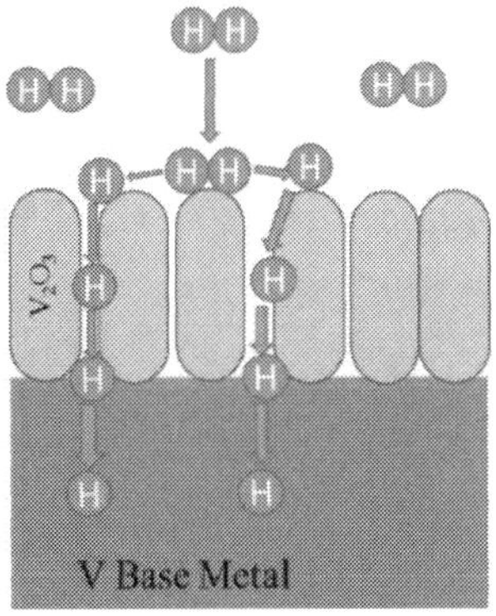

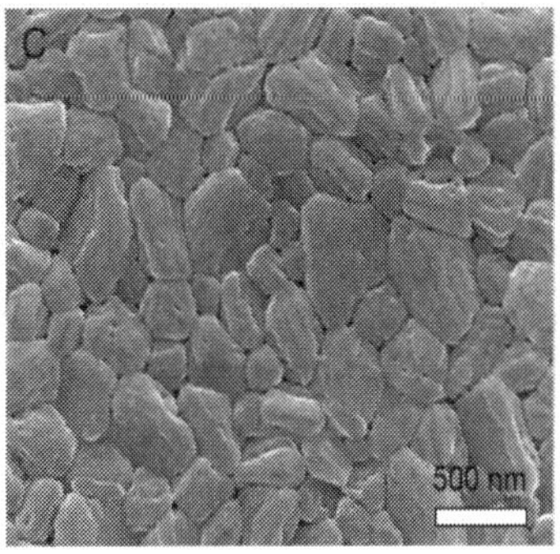

COLORADO SCHOOL OF MINES
EARTH • ENERGY • ENVIRONMENT

H_2 and Applications

3

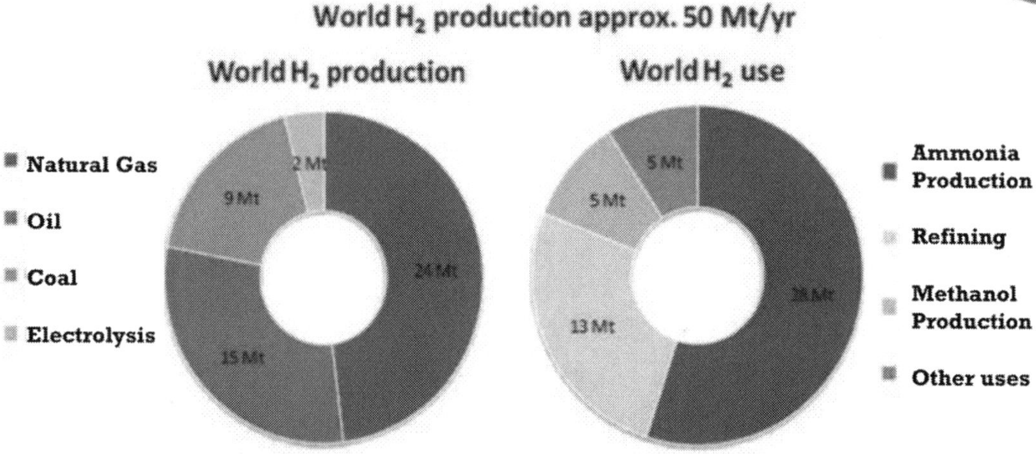

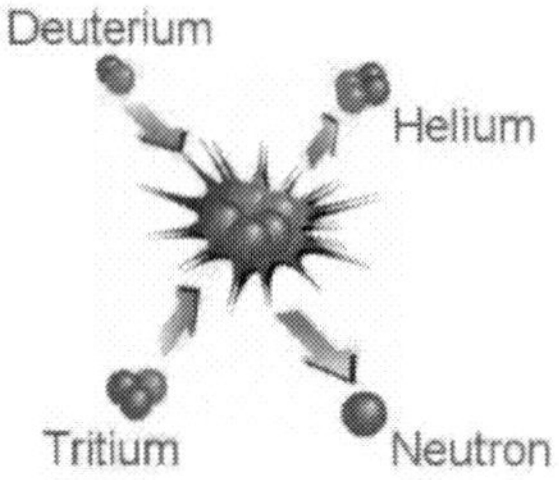

Hydrogen Association; Evers, 2008

COLORADOSCHOOLOFMINES
EARTH • ENERGY • ENVIRONMENT

Hydrogen Transport in Metals

- Several rate processes in series
  - Gas transport of H_2
  - Dissociation, $H_2 \rightarrow 2H$
  - H dissolves in metal
  - Diffusion of H^+
  - H desorbs
  - Recombination, $2H \rightarrow H_2$
  - Desorption of H_2

Governing Equations:

$$\pi_H = S\ x\ D$$

$$J = \frac{\pi_{H(n)}}{L}\left(P^n_{H2,F} - P^n_{H2,P}\right)$$

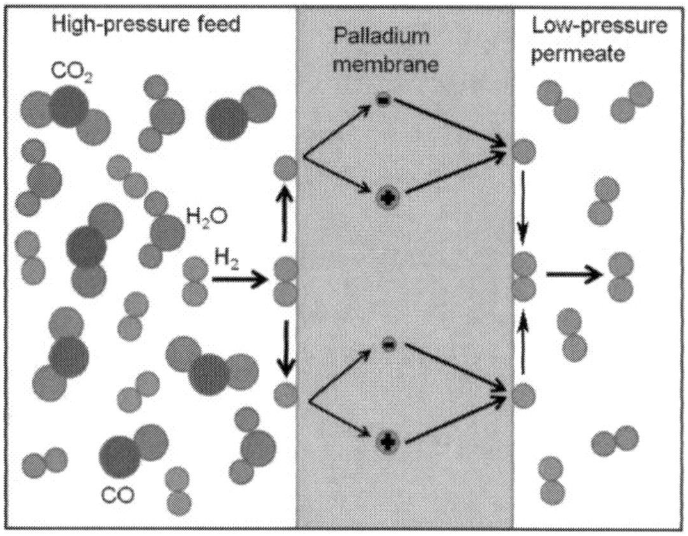

H_2 transport perfectly selective!

H_2 Permeability in Metals

- High H_2 permeability/fluxes
- Very high theoretical selectivity
- Wide temperature range

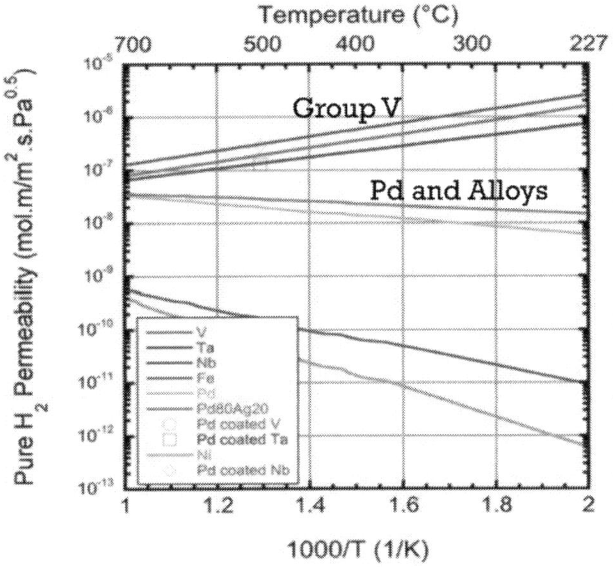

	Comparison of Membrane Cost	
Metal-alloy (wt %)	Alloy Metal Price (US$/ oz)	Membrane Cost (Alloy/ Pure Pd)
Pd	786.5	1.00
Pd-27Ag	17.95	0.74
Pd-20Au	1235.8	1.11
Pd-39Cu	0.19	0.61
V	68*	0.09
Nb	5.6*	0.007
Ta	140*	0.18

*Estimated from ore price March, 2017

COLORADOSCHOOLOFMINES
EARTH • ENERGY • ENVIRONMENT

Pd Catalytic Coatings

- Group V metal surface inactive for H_2 dissociation
- Pd coatings used to impart catalytic properties
- To prevent interdiffusion T $\leq$ 400°C

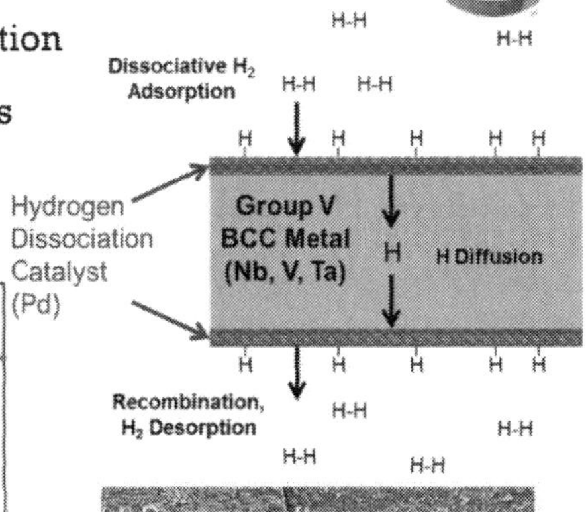

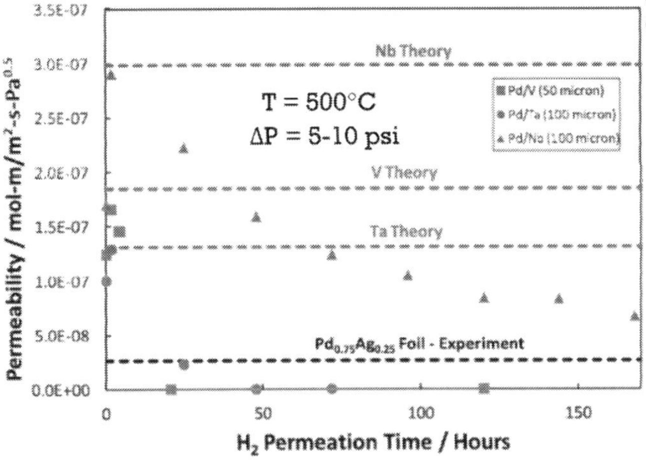

Cooney, D. A.; Way, J. D.; Wolden, C. A. *Int. J. Hydrog. Energy* **2014**, *39* (33), 19009–19017.

COLORADOSCHOOLOFMINES
EARTH • ENERGY • ENVIRONMENT

Embrittlement Issues

Solubility Induced Embrittlement

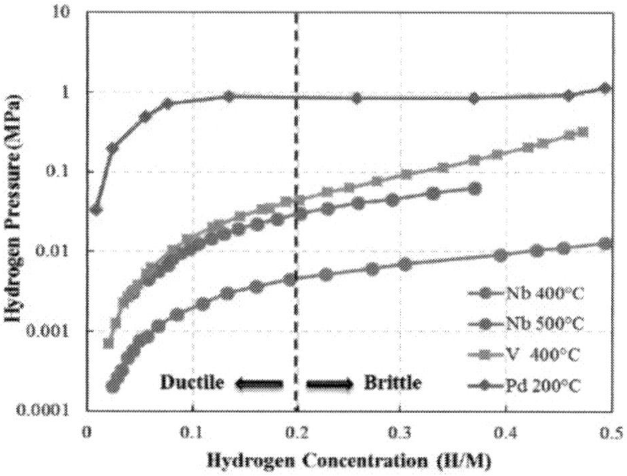

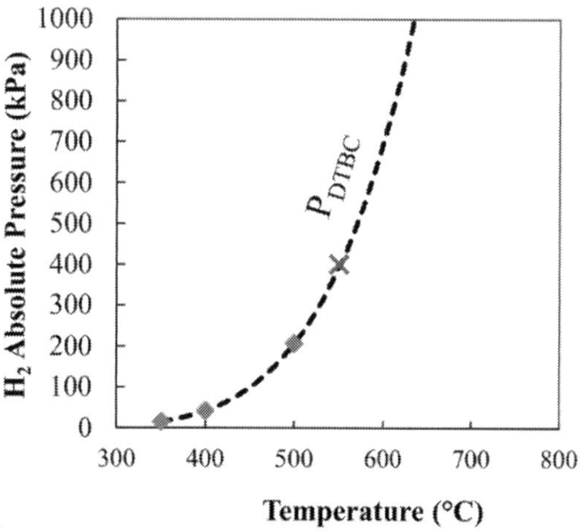

Nambu et al. *J. Alloys Compd.* **2007**, *446-447*, 588-592

Problem

- Forced to operate at low temperature and pressure due to Pd interdiffusion and DBTC

Solutions

- Alloys - DBTC
  - Alloy elements such as: Ni, Al, Mo, W, Ti, Pd, and Ru
  - Suppress solubility
  - Increases DBTC Pressure
  - Allow higher operating pressure at low temperatures < 400°C
- Catalysts – DBTC and Interdiffusion
  - Stable catalysts reduce interdiffusion rate
    - Operate at temperatures > 400°C
    - Reduces solubility to allow higher pressures

V_2O_3 Air Treatment - Background

- Nakamura et al. *J. Alloys Compd.* 2015:
  - Heat V and V-alloy foil to 550°C under vacuum
  - Broke vacuum for 2 min
  - Pump down and expose membrane to 0.4 MPa H_2 for 2 min
  - Repeat once more

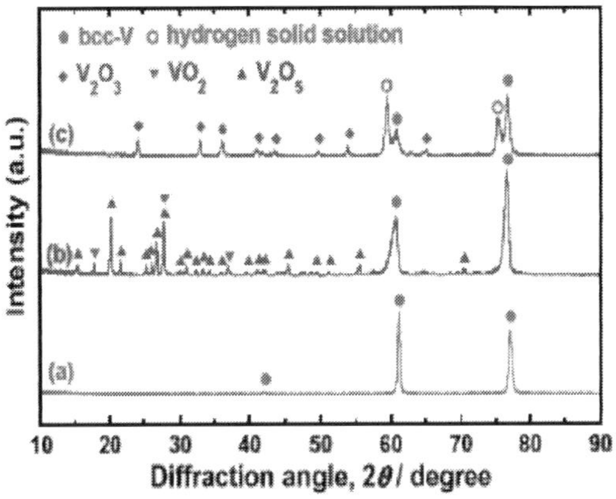

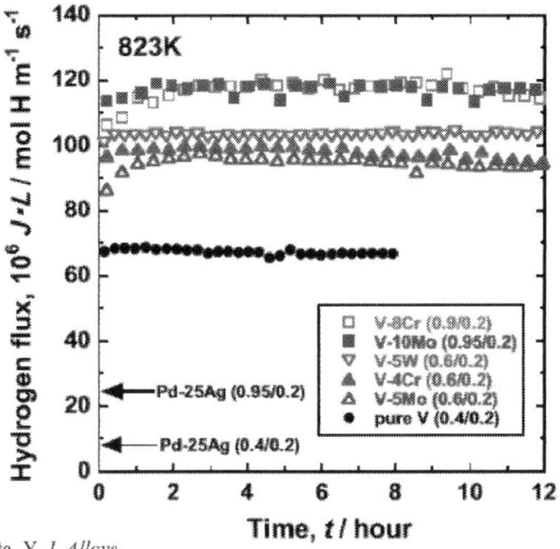

Nakamura, Y.; Yukawa, H.; Suzuki, A.; Nambu, T.; Matsumoto, Y.; Murata, Y. *J. Alloys Compd.* **2015**, *645*

COLORADOSCHOOLOFMINES
EARTH • ENERGY • ENVIRONMENT

V_2O_3 – Spillover Catalyst

Step 1:
5 min Ar sputter clean of 50 μm V foil

Step 2:
Heat up to 550 °C and expose to 0.15 MPa air for 120 s to oxidize surface

Step 3:
Reduce under 0.4 MPa H_2 for 120 s and set up permeation conditions

XPS and XRD: V ⟶ V_2O_5/VO_2 ⟶ V_2O_3

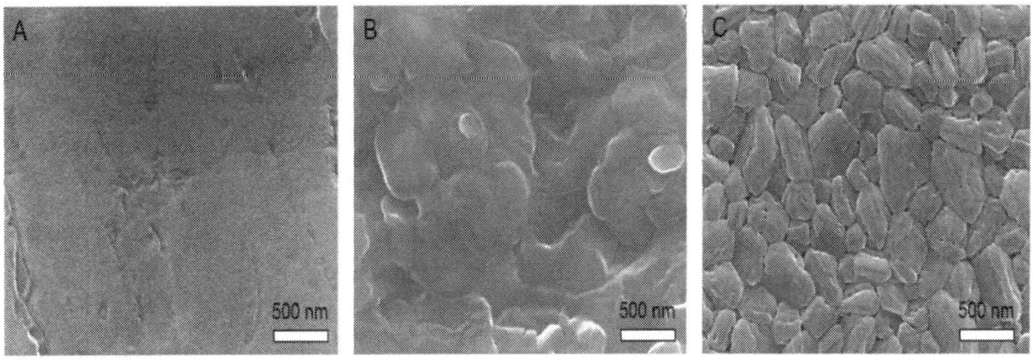

COLORADOSCHOOLOFMINES
EARTH • ENERGY • ENVIRONMENT

Oxidation State

11

- XRD and XPS confirm reduction in oxidation state
- $V_2O_5/VO_2 \rightarrow V_2O_3$

XRD

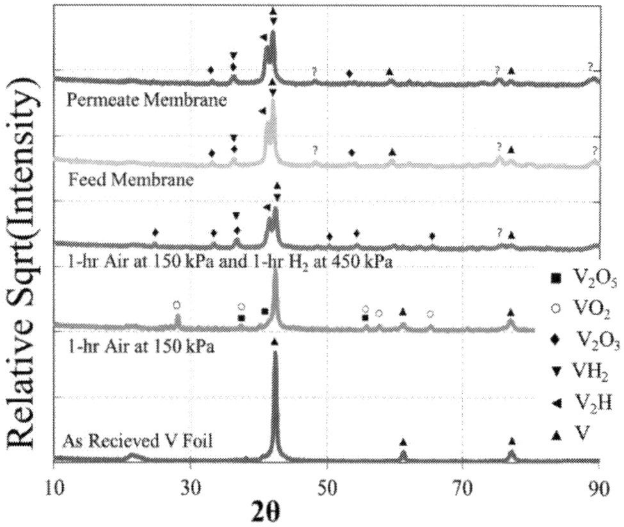

XPS

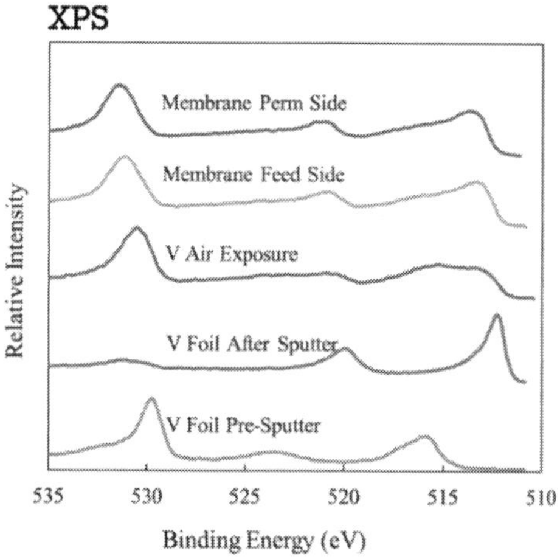

Air Treatment (AT) Analysis

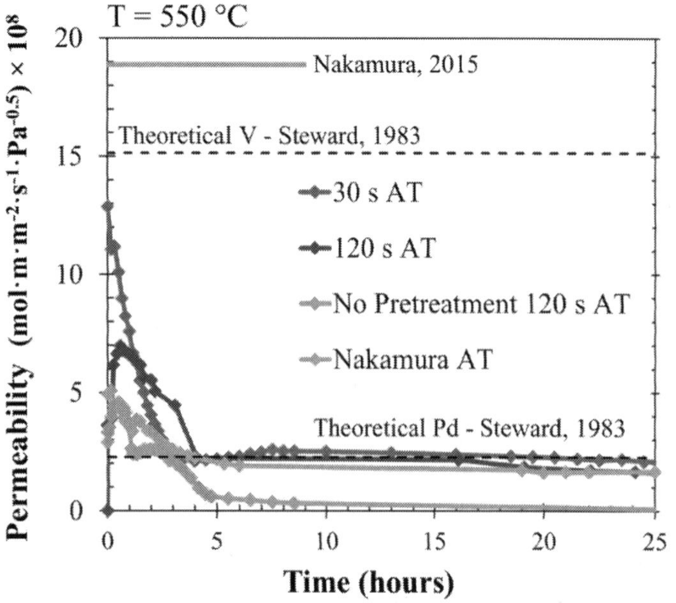

- Nakamura, 2015 higher than theoretical
- Our experiments yield smaller steady-state permeability
- $\sim 2 \times 10^8$ $mol \cdot m \cdot m^{-2} \cdot s^{-1} \cdot Pa^{-0.5}$
- Changing air treatment can improve initial peak in permeability but negligible SS permeability
- Surface clean pretreatment is necessary for stable membrane
- Our permeability on par with Pd

Nakamura, Y.; Yukawa, H.; Suzuki, A.; Nambu, T.; Matsumoto, Y.; Murata, Y. *J. Alloys Compd.* **2015**, 645.

Steward, S.A: *Review of Hydrogen Isotope Permeability Through Materials.* Lawrence Livermore National Lab, **1983**

V_2O_3 Membrane Performance

13

Apparent activation energy

- V_2O_3/V membrane: 36.3 kJ/mol
- H Permeability through V: 24.6 kJ/mol
- Mo_2C/V membrane: 57 kJ/mol

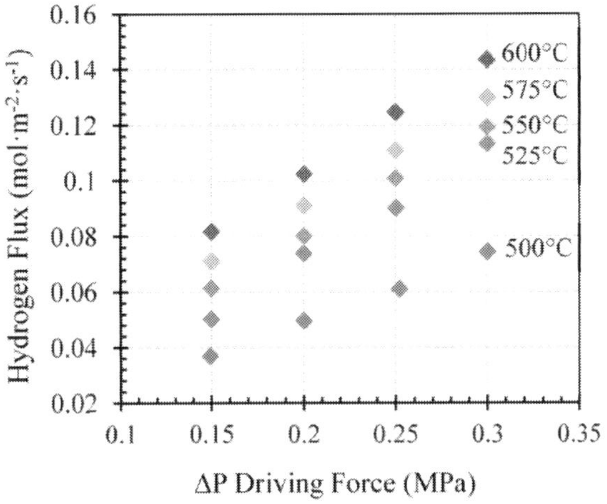

Temp (°C)	Permeability ($mol \cdot m^{-1} \cdot s^{-1} \cdot Pa^{-0.5}$)	Optimized Sievert's Coefficient
500	1.06×10^{-8}	1.0
525	1.57×10^{-8}	1.0
550	1.72×10^{-8}	0.93
575	1.91×10^{-8}	0.75
600	2.14×10^{-8}	0.63

Wolden, C.A. et al. *J. Membr. Sci.* **2013**, 427,150-154.

COLORADOSCHOOLOFMINES
EARTH • ENERGY • ENVIRONMENT

V_2O_3 Thermal Stability/Degradation

14

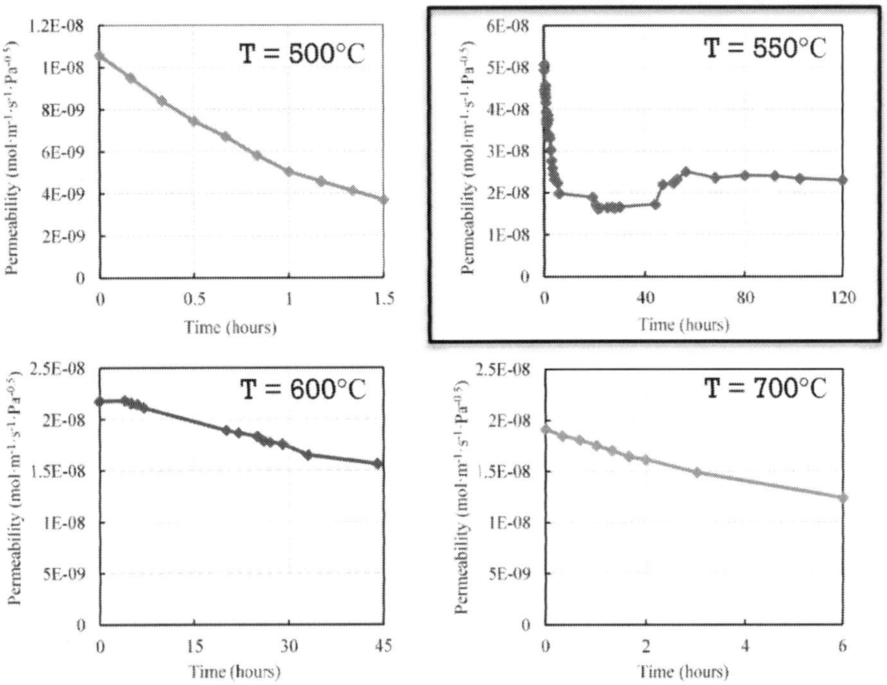

COLORADOSCHOOLOFMINES
EARTH • ENERGY • ENVIRONMENT

DFT Simulations – V_2O_3 [0001] Surface

15

- **Level of Theory**
  - DFT + GGA : PAW – PBE
  - Cutoff Energy = 500 eV
  - K-point mesh: 3 x 3 x 1
- **V_2O_3 Model**
  - 120 atom 2 x 2 x 1 Super Cell
  - Space Group: R -3c
  - a=b: 4.9 Å; c: 3.8 Å

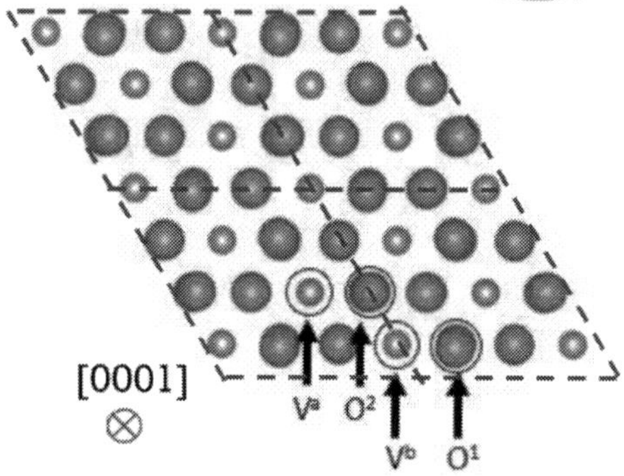

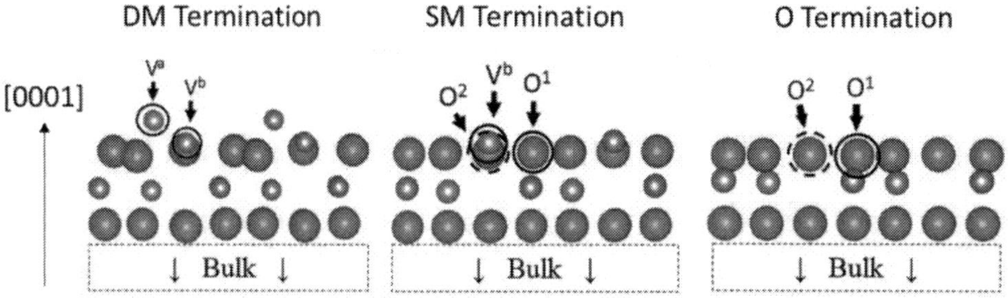

COLORADOSCHOOLOFMINES
EARTH • ENERGY • ENVIRONMENT

H_2 Dissociation on V_2O_3

16

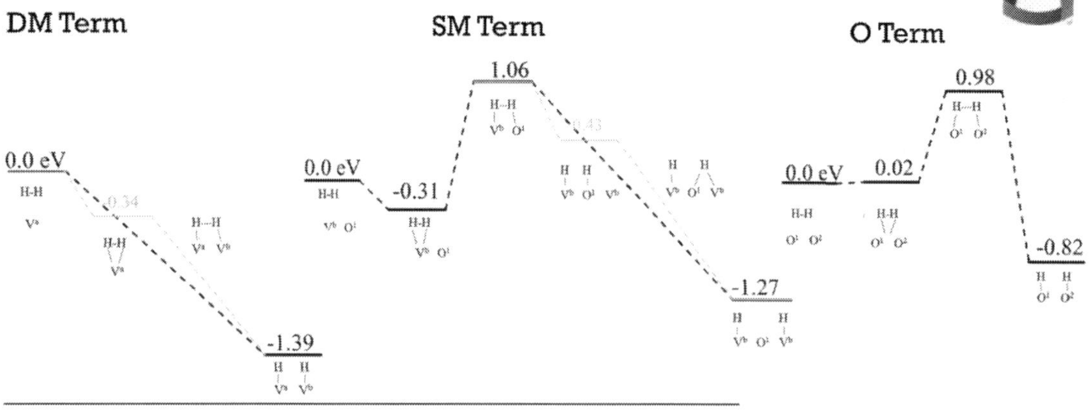

Catalyst	Dissociation Activation Energy (eV)	Dissociated Adsorption Energy (eV)	Reference
V_2O_3	0-1.37	-1.39 - -0.82	This Work
V_2O_5	-	-4.95 - -2.95	Yao, 2015
Pd	-	-1.06	Behm, 1980
Mo2C	0.13-0.18	-1.74 - -0.95	Hao, 2012
TiC	0.51	-0.95	Zhu, 2015

H. *Yao* et al. *Surface Science*. **2012.** 606, 1739-1748

R.J. Behm et al. *Surface Science*. **1980.** 99, 320-340

S. *Hao* et al. *J. Phys. Chem. C.* **2013**. 117, 1217-1223

R. Zhu et al. *J. Phys. Chem C.* **2015**. 119, 7848-7855

COLORADOSCHOOLOFMINES
EARTH • ENERGY • ENVIRONMENT

Summary

- Highly faceted, nanocrystalline V_2O_3 was grown on V surfaces via air exposure and H_2 reduction
- V_2O_3/V membranes have similar permeability to Pd but much smaller than Nakamura et al.
- Small thermal stability window ~550 °C
- DFT simulations show V_2O_3 surfaces as active for H_2 membrane catalysts

Acknowledgements

- Membrane Research Group
  - Advisors: Prof. Colin Wolden, Prof. Doug Way, Prof. Jen Wilcox, and Dr. Simona Liguori
  - Grad Students: Sean Lundin, Neil Patki, and Zhenyu Zhang
  - Undergraduates: Elias Petsalis
- Funding
  - DOE NEUP Fellowship
  - NSF
  - DOE ARPA-E

COLORADOSCHOOLOFMINES
EARTH • ENERGY • ENVIRONMENT

Extra Slides

Extra Slides

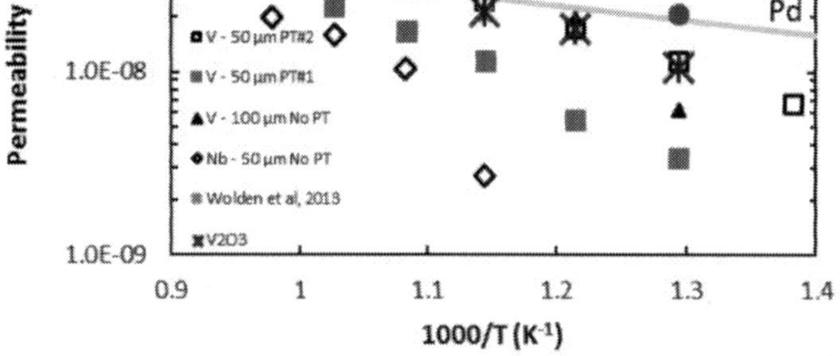

Spillover Catalytic Coatings

- Pt group catalytic activity for H_2 dissociation
- Catalysts are impermeable to H_2
- Thin coatings of V_2O_3 transport H_2 via spillover mechanism

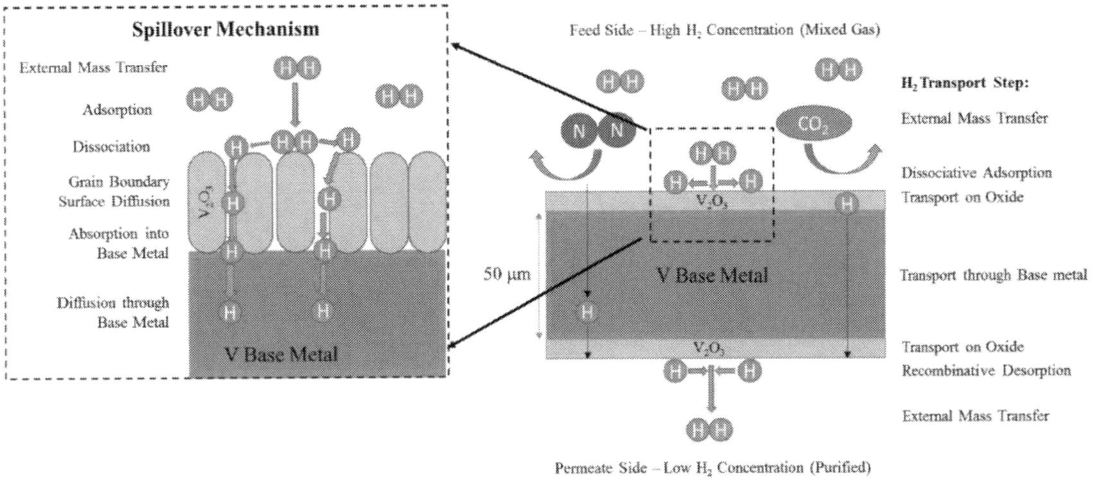

COLORADOSCHOOLOFMINES
EARTH • ENERGY • ENVIRONMENT

H2 Purifiction

Comparison of Hydrogen Purification Techniques

Technique (Ref.)	Principle	Typical feed gas	Hydrogen output per cent		Scale of use	Comments
			Purity	Recovery		
Cryogenic Separation (1)	Partial condensation of gas mixtures at low temperatures	Petrochemical and refinery off-gases	90–98	95	Large scale	Prepurification step necessary to remove CO_2, H_2S and water
Polymer Membrane Diffusion (2)	Differential rate of diffusion of gases through a permeable membrane	Refinery off-gases and ammonia purge gas	92–98	>85	Small to large	He, CO_2 and H_2O may also permeate the membrane
Metal Hydride Separation (3)	Reversible reaction of hydrogen with metals to form hydrides	Ammonia purge gas	99	75–95	Small to medium	Hydrogen absorption poisoned by O_2, N_2, CO and S
Solid Polymer Electrolyte Cell (4)	Electrolytic passage of hydrogen ions across a solid polymer membrane	Purification of hydrogen produced by thermochemical cycles	99.8	95	Small	Sulphur-containing compounds poison the electro-catalysts
Pressure Swing Adsorption (5)	Selective adsorption of impurities from gas stream	Any hydrogen rich gas	99.999	70–85	Large	The recovery is relatively low as hydrogen is lost in the purging step
Catalytic Purification	Removal of oxygen by catalytic reaction with hydrogen	Hydrogen streams with oxygen impurity	99.999	Up to 99	Small to large	Usually used to upgrade electrolytic hydrogen. Organics, Pb-, Hg-, Cd- and S-compounds poison the catalyst. H_2O produced
Palladium Membrane Diffusion	Selective diffusion of hydrogen through a palladium alloy membrane	Any hydrogen containing gas stream	>99.9999	Up to 99	Small to medium	Sulphur-containing compounds and unsaturated hydrocarbon impair permeability

EARTH • ENERGY • ENVIRONMENT

Interdiffusion Pd – Nb

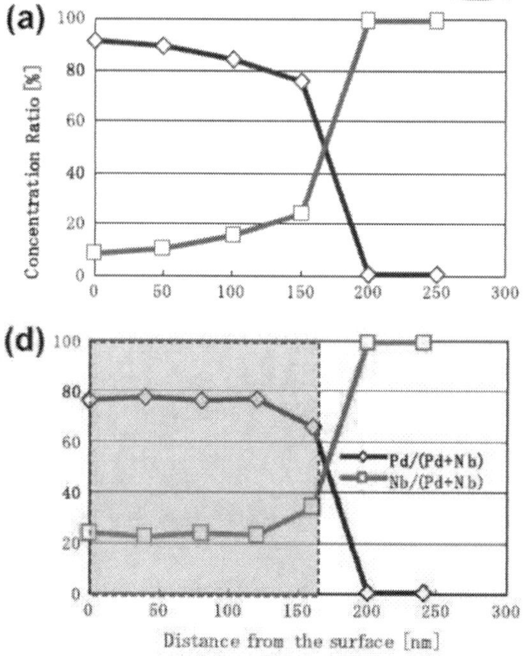

Sievert's Analysis

- *Sievert's Law*
  - $H_{2(gas)} \leftrightarrow 2H_{dissolved\ attoms}$
  - $C_H = \sqrt{KP_{H2}}$
  - Sieverts, Adolf (1929). "The Absorption of Gases by Metals". *Zeitschrift für Metallkunde*. **21**: 37–46
- Fick's First Law:
  - $J = -\frac{kdC}{dx}$
  - $J = \frac{\pi_{H}(n)}{L}(P^n_{H2,F} - P^n_{H2,P})$
  - *n = 0.5 diffusion limited (Sievert's Law)*
  - *n = 1 surface or external mass transfer*

Ward Transport Model

25

- Pd Membranes
  - T.L. Ward and T. Dao (1999). "Model of hydrogen permeation behavior in palladium membranes". *J. Membrane Sci.* **153:** 211-231

1. Transport from bulk gas to gas layer adjacent to surface
2. Dissociative adsorption on surface
3. Transition of atomic H from surface into bulk metal
4. Atomic diffusion through the bul k metal
5. Transition from bulk metal to the surface on the low partial pressure side
6. Recombinative Desorption
7. Gas Transport away from the surface to the bulk gas

COLORADOSCHOOLOFMINES
EARTH • ENERGY • ENVIRONMENT

Ward Model

26

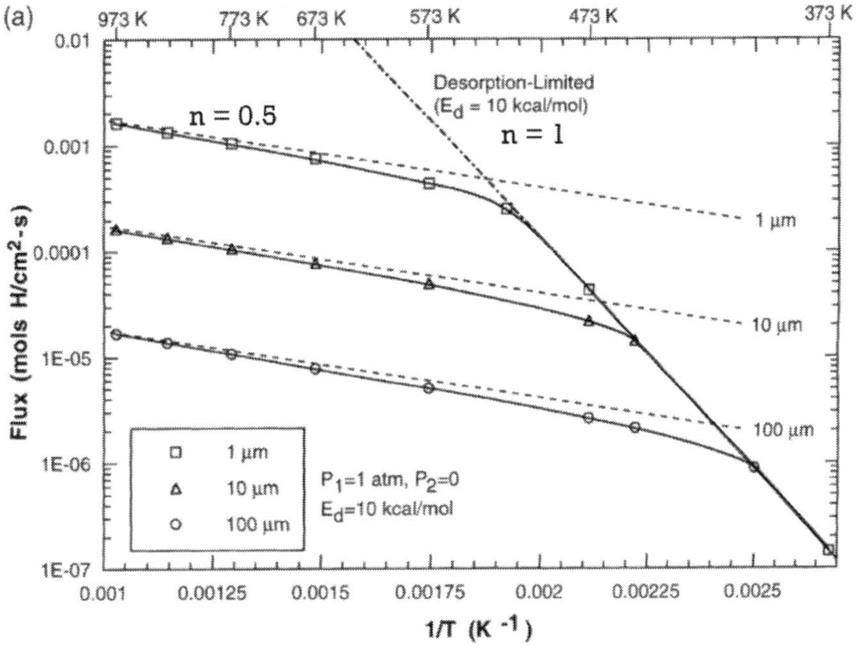

COLORADOSCHOOLOFMINES
EARTH • ENERGY • ENVIRONMENT

- **Step 1 & 7: External Mass Transfer**
  - $J_1 = 2h(C - C_s)$
  - $C[\frac{mol}{vol}] = \frac{P}{RT}$ Assuming Ideal Gas Law
- **Step 2: Dissociated Adsorption on Surface**
  - $J_2 = 2C_s\left(\frac{RT}{2\pi M_{H_2}}\right)^{0.5} S_0\left[1 + K\left(\frac{1}{\theta_{00}} - 1\right)\right]^{-1}$
  - $\theta_{00} = 1 - \theta_1 - \frac{2\theta_1(1-\theta_1)}{\left[1-4\theta_1(1-\theta_1)\left(1-\exp\left(-\frac{w}{kT}\right)\right)\right]^{0.5}+1}$
  - M_{H2}: molecular weight of H2
  - S_0: sticking coefficient at zero coverage (=1 for clean Pd)
  - K: constant that depends on the probabilities of adsorption and desorption over empty and occupied sites. K=0.05 for Pd [Behm Christmann Surf. Sci. 1980]
  - Θ_{00}: probability of two adjacent empty sites based on quasi-chemical equilibrium approximation
  - $\theta 1$: fractional surface coverage (surface H/Pd ratio) on feed surface
  - w: interaction energy (0.5 kcal/mol for Pd)
- **Step 4: Catalyst surface to bulk V transition**
  - $J_4 = N_s N_b v_d \theta_1(1 - X_s)$
  - $v_d = v_0 \exp\left(-\frac{E_A}{RT}\right)$
  - $E_A - E_B = \frac{\Delta E_{ad} - \Delta E_{ab}}{2}$

Nb: bulk V concentration
Xs: H/V ration near feed surface in V
Vo: rate constant
EA: Activation energy surface-bulk

EB: Activation energy bulk-surface
Ead: adsorption activation energy
Eab: absorption activation energy

COLORADOSCHOOLOFMINES
EARTH • ENERGY • ENVIRONMENT

- **Step 5: Atomic diffusion through bulk V**
  - $J_5 = \frac{D_o \exp\left(-\frac{E_{diff}}{RT}\right) N_b (X_1 - X_2)}{\Delta z}$

Do: diffusion rate constant
Ediff: diffusion activation energy
Nb: V atom concentration in bulk

z: membrane thickness
X1: H/V ratio feed surface
X2: H/V ratio permeate surface

- **Step 6: Transition from Bulk V to Mo2C surface**
  - $J_6 = N_s N_b X_{s2}(1 - \theta_2)\beta_d$
  - $\beta_d = \beta_0 \exp(-\frac{E_B}{RT})$
  - $\frac{D_O}{3\alpha a^2} = N_b(1 - \theta_2)\beta_o$

Assumes one third of jumps happen between planes
α: coefficient determined between geometric relationship between interstitial sites
a: lattice parameter
Do: diffusion rate constant

- **Step 7: Recombinative desorption**
  - $J_7 = \frac{k_0 \exp(-\frac{2E_d}{RT}) N_s^2 z \theta_2}{2}\left(1 - \frac{2-2\theta_2}{\left[1-4\theta_2(1-\theta_2)\left(1-\exp\left(-\frac{w}{kT}\right)\right)\right]^{0.5}+1}\right)$

ko: desorption rate constant
Ed: desorption activation energy
Z: number of nearest neighbors on the surface

COLORADOSCHOOLOFMINES
EARTH • ENERGY • ENVIRONMENT

Equipment

Equipment

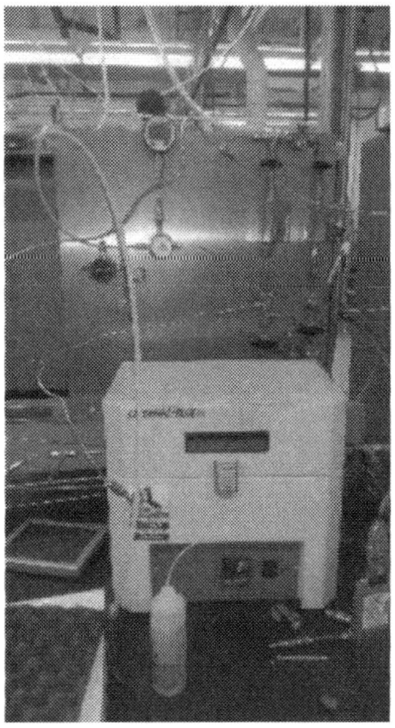

V2O3 Formation

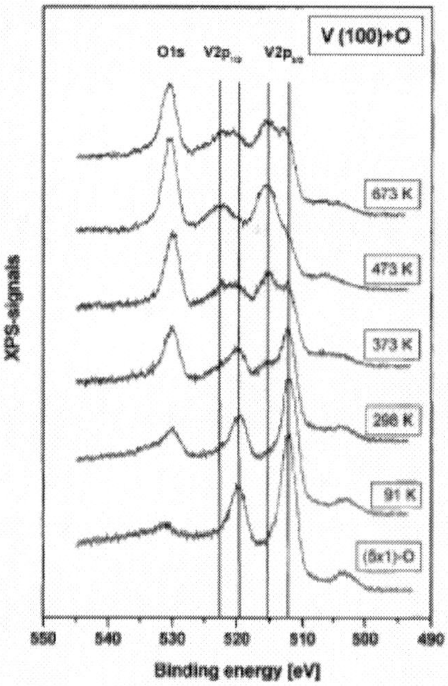

Melting Point:

- V_2O_3 - 1940°C
- VO_2 - 1967°C
- V_2O_5 - 690°C

Core line	Main peak		
	BE (eV)	%L-G	Δ (eV)
O1s	530.0	10	–
$V^{5+}2p_{3/2}$	517.2 ± 0.1	42	12.8 ± 0.1
$V^{5+}2p_{1/2}$	$V^{5+}2p_{3/2}$ + 7.33	42	
$V^{4+}2p_{3/2}$	515.84 ± 0.2	26	14.16 ± 0.2
$V^{4+}2p_{1/2}$	$V^{4+}2p_{3/2}$ + 7.33	26	
$V^{3+}2p_{3/2}$	515.29 ± 0.2	16	14.71 ± 0.2
$V^{3+}2p_{1/2}$	$V^{3+}2p_{3/2}$ + 7.33	16	
$V^{2+/1+}2p_{3/2}$	513.67 ± 0.2	10	16.33 ± 0.2
$V^{2+/1+}2p_{1/2}$	$V^{2+/1+}2p_{3/2}$ + 7.33	10	
$V^{0+}2p_{3/2}$	512.35 ± 0.2	0	17.65 ± 0.2
$V^{0+}2p_{1/2}$	$V^{0+}2p_{3/2}$ + 7.5	0	

Silversmit, G.; Depla, D.; Poelman, H.; Marin, G. B.; De Gryse, R. J. Electron Spectrosc. Relat. Phenom. **2004**, 135 (2–3), 167–175.

H. Schiechl and A. Winkler *J. Anal. Chem.* **2001** 371: 342-347

V_2O_3 Stability

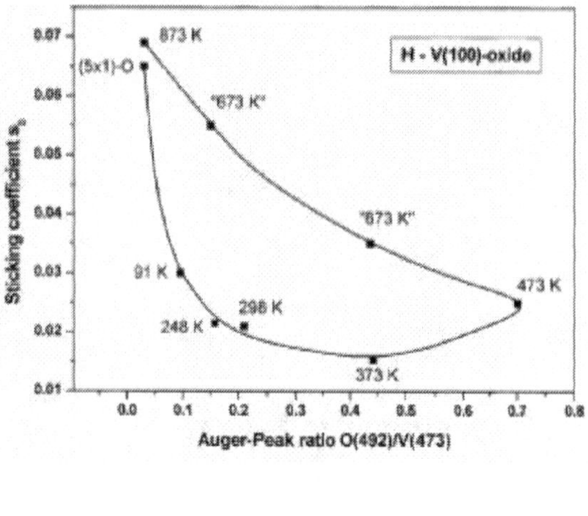

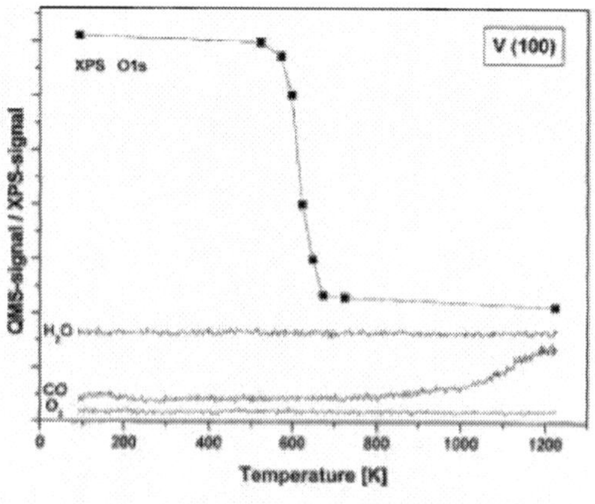

H. Schiechl and A. Winkler *J. Anal. Chem.* **2001** 371: 342-347

Air Treatment Analysis

- Produce functioning membrane! If surface is cleaned.
- Decreasing air exposure time has negligible effect on SS flux

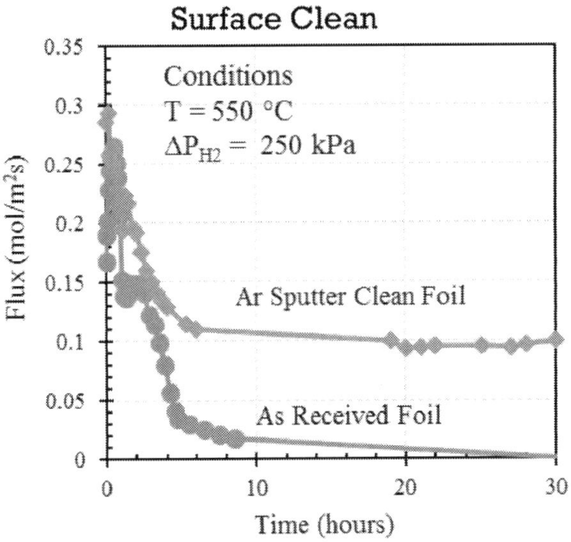

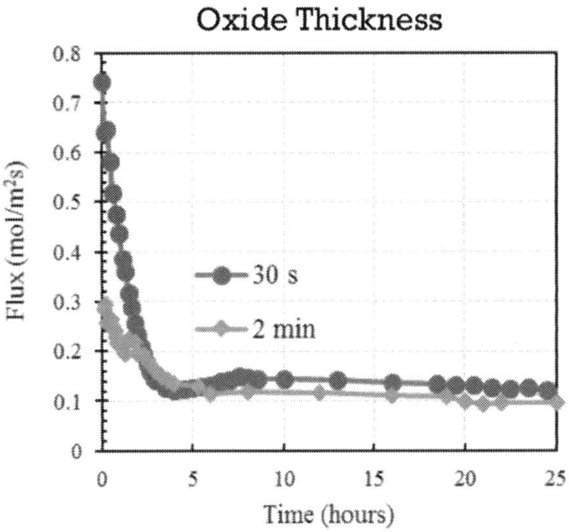

COLORADO SCHOOL OF MINES
EARTH • ENERGY • ENVIRONMENT

AUTHOR INDEX

AUTHOR INDEX

AUTHOR INDEX

AUTHOR INDEX

AUTHOR INDEX

AUTHOR INDEX

AUTHOR INDEX

AUTHOR INDEX

AUTHOR INDEX

AUTHOR INDEX

AUTHOR INDEX

AUTHOR INDEX

AUTHOR INDEX